Wireless Communication Systems

Prentice Hall Communications Engineering and Emerging Technologies Series

Theodore S. Rappaport, *Series Editor*

DOSTERT *Powerline Communications*

DURGIN *Space–Time Wireless Channels*

GANZ, GANZ, & WONGTHAVARAWAT *Multimedia Wireless Networks: Technologies, Standards, and QoS*

GARG *Wireless Network Evolution: 2G to 3G*

GARG *IS-95 CDMA and cdma2000: Cellular/PCS Systems Implementation*

GARG & WILKES *Principles and Applications of GSM*

KIM *Handbook of CDMA System Design, Engineering, and Optimization*

LIBERTI & RAPPAPORT *Smart Antennas for Wireless Communications: IS-95 and Third Generation CDMA Applications*

PAHLAVAN & KRISHNAMURTHY *Principles of Wireless Networks: A Unified Approach*

RAPPAPORT *Wireless Communications: Principles and Practice, Second Edition*

RAZAVI *RF Microelectronics*

REED *Software Radio: A Modern Approach to Radio Engineering*

STARR, CIOFFI, & SILVERMAN *Understanding Digital Subscriber Line Technology*

STARR, SORBARA, CIOFFI, & SILVERMAN *DSL Advances*

TRANTER, SHANMUGAN, RAPPAPORT, & KOSBAR *Principles of Communication Systems Simulation with Wireless Applications*

WANG & POOR *Wireless Communication Systems: Advanced Techniques for Signal Reception*

Wireless Communication Systems

Advanced Techniques for Signal Reception

Xiaodong Wang
H. Vincent Poor

PRENTICE HALL
PROFESSIONAL TECHNICAL REFERENCE
UPPER SADDLE RIVER, NJ 07458
WWW.PHPTR.COM

Library of Congress Cataloging-in-Publication Data

Wang, Xiaodong
Wireless communication systems : advanced techniques for signal reception / Xiaodong Wang, H. Vincent Poor.
p. cm. — (Prentice Hall communications engineering and emerging technologies series)
Includes bibliographical references and index.
ISBN 0-13-021435-3
1. Wireless communication systems. 2. Signal processing—Digital techniques. I. Poor, H. Vincent. II. Title. III. Series.

TK5103.2.W37 2003
621.382–dc21

2003048822

Editorial/production supervision: *Jane Bonnell*
Composition: *Lori Hughes*
Cover design director: *Jerry Votta*
Manufacturing buyer: *Maura Zaldivar*
Publisher: *Bernard M. Goodwin*
Editorial assistant: *Michelle Vincenti*
Marketing manager: *Dan DePasquale*

Publishing as Prentice Hall Professional Technical Reference
Upper Saddle River, NJ 07458

Printed in the United States of America

First Printing

ISBN 0-13-021435-3

Pearson Education LTD.
Pearson Education Australia PTY, Limited
Pearson Education Singapore, Pte. Ltd.
Pearson Education North Asia Ltd.
Pearson Education Canada, Ltd.
Pearson Educación de Mexico, S.A. de C.V.
Pearson Education—Japan
Pearson Education Malaysia, Pte. Ltd.

To our parents

About Prentice Hall Professional Technical Reference

With origins reaching back to the industry's first computer science publishing program in the 1960s, and formally launched as its own imprint in 1986, Prentice Hall Professional Technical Reference (PH PTR) has developed into the leading provider of technical books in the world today. Our editors now publish over 200 books annually, authored by leaders in the fields of computing, engineering, and business.

Our roots are firmly planted in the soil that gave rise to the technical revolution. Our bookshelf contains many of the industry's computing and engineering classics: Kernighan and Ritchie's *C Programming Language*, Nemeth's *UNIX System Adminstration Handbook*, Horstmann's *Core Java*, and Johnson's *High-Speed Digital Design*.

PH PTR acknowledges its auspicious beginnings while it looks to the future for inspiration. We continue to evolve and break new ground in publishing by providing today's professionals with tomorrow's solutions.

PRENTICE HALL PTR

CONTENTS

PREFACE

Wireless communications, together with its applications and underlying technologies, is among today's most active areas of technology development. The very rapid pace of improvements in both custom and programmable integrated circuits for signal processing applications has led to the justifiable view of advanced signal processing as a key enabler of the aggressively escalating capacity demands of emerging wireless systems. Consequently, there has been a tremendous and very widespread effort on the part of the research community to develop novel signal processing techniques that can fulfill this promise. The published literature in this area has grown explosively in recent years, and it has become quite difficult to synthesize the many developments described in this literature. The purpose of this book is to present, in one place and in a unified framework, a number of key recent contributions in this field. Even though these contributions come primarily from the research community, the focus of this presentation is on the development, analysis, and understanding of explicit algorithms for performing advanced processing tasks arising in receiver design for emerging wireless systems.

Although this book is largely self-contained, it is written principally for designers, researchers, and graduate students with some prior exposure to wireless communication systems. Knowledge of the field at the level of Theodore Rappaport's book, *Wireless Communications: Principles and Practice* [405], for example, would be quite useful to the reader of this book, as would some exposure to digital communications at the level of John Proakis's book, *Digital Communications* [396].

Acknowledgments

The authors would like to thank the Army Research Laboratory, the National Science Foundation, the New Jersey Commission on Science and Technology, and the Office of Naval Research for their support of much of the research described in this book.

Chapter 1

INTRODUCTION

1.1 Motivation

Wireless communications is one of the most active areas of technology development of our time. This development is being driven primarily by the transformation of what has been largely a medium for supporting voice telephony into a medium for supporting other services, such as the transmission of video, images, text, and data. Thus, similar to the developments in wireline capacity in the 1990s, the demand for new wireless capacity is growing at a very rapid pace. Although there are, of course, still a great many technical problems to be solved in wireline communications, demands for additional wireline capacity can be fulfilled largely with the addition of new private infrastructure, such as additional optical fiber, routers, switches, and so on. On the other hand, the traditional resources that have been used to add capacity to wireless systems are radio bandwidth and transmitter power. Unfortunately, these two resources are among the most severely limited in the deployment of modern wireless networks: radio bandwidth because of the very tight situation with regard to useful radio spectrum, and transmitter power because mobile and other portable services require the use of battery power, which is limited. These two resources are simply not growing or improving at rates that can support anticipated demands for wireless capacity. On the other hand, one resource that is growing at a very rapid rate is that of processing power. Moore's Law, which asserts a doubling of processor capabilities every 18 months, has been quite accurate over the past 20 years, and its accuracy promises to continue for years to come. Given these circumstances, there has been considerable research effort in recent years aimed at developing new wireless capacity through the deployment of greater intelligence in wireless networks (see, e.g., [145,146,270,376,391] for reviews of some of this work). A key aspect of this movement has been the development of novel signal transmission techniques and advanced receiver signal processing methods that allow for significant increases in wireless capacity without attendant increases in bandwidth or power requirements. The purpose of this book is to present some of the most recent of these receiver signal processing methods in a single place and in a unified framework.

Wireless communications today covers a very wide array of applications. The telecommunications industry is one of the largest industries worldwide, with more than $1 trillion in annual revenues for services and equipment. (To put this in perspective, this number is comparable to the gross domestic product of many of the world's richest countries, including France, Italy, and the United Kingdom.) The largest and most noticeable part of the telecommunications business is telephony. The principal wireless component of telephony is mobile (i.e., cellular) telephony. The worldwide growth rate in cellular telephony is very aggressive, and analysts report that the number of cellular telephony subscriptions worldwide has now surpassed the number of wireline (i.e., fixed) telephony subscriptions. Moreover, at the time of this writing in 2003, the number of cellular telephony subscriptions worldwide is reportedly on the order of 1.2 billion. These numbers make cellular telephony a very important driver of wireless technology development, and in recent years the push to develop new mobile data services, which go collectively under the name *third-generation* (3G) *cellular*, has played a key role in motivating research in new signal processing techniques for wireless. However, cellular telephony is only one of a very wide array of wireless technologies that are being developed very rapidly at the present time. Among other technologies are wireless piconetworking (as exemplified by the Bluetooth radio-on-a-chip) and other personal area network (PAN) systems (e.g., the IEEE 802.15 family of standards), wireless local area network (LAN) systems (exemplified by the IEEE 802.11 and HiperLAN families of standards, called WiFi systems), wireless metropolitan area network (MAN) systems (exemplified by the IEEE 802.16 family of standards, called WiMax systems), other wireless local loop (WLL) systems, and a variety of satellite systems. These additional wireless technologies provide a basis for a very rich array of applications, including local telephony service, broadband Internet access, and distribution of high-rate entertainment content such as high-definition video and high-quality audio to the home, within the home, to automobiles, and so on (see, e.g., [9, 41, 42, 132, 159, 161, 164, 166, 344, 361, 362, 365, 393–395, 429, 437, 449, 457, 508, 558, 559] for further discussion of these and related applications). Like 3G, these technologies have spurred considerable research in signal processing for wireless.

These technologies are supported by a number of transmission and channel-assignment techniques, including time-division multiple access (TDMA), code-division multiple access (CDMA), and other spread-spectrum systems, orthogonal frequency-division multiplexing (OFDM) and other multicarrier systems, and high-rate single-carrier systems. These techniques are chosen primarily to address the physical properties of wireless channels, among the most prominent of which are multipath fading, dispersion, and interference. In addition to these temporal transmission techniques, there are spatial techniques, notably beamforming and space-time coding, that can be applied at the transmitter to exploit the spatial and angular diversity of wireless channels. To obtain maximal benefit from these transmission techniques, to exploit the diversity opportunities of the wireless channel, and to mitigate the impairments of the wireless channel, advanced receiver signal processing techniques are of interest. These include channel equalization to combat dispersion, RAKE combining to exploit resolvable multipath, multiuser detection to mitigate multiple-access interference, suppression methods for co-channel interference, beamforming to exploit spatial diversity, and space-time processing to jointly exploit temporal and spatial properties of the signaling environment. These techniques are all described in the ensuing chapters.

1.2 Wireless Signaling Environment

1.2.1 Single-User Modulation Techniques

To discuss advanced receiver signal processing methods for wireless, it is useful first to specify a general model for the signal received by a wireless receiver. To do so, we can first think of a single transmitter, transmitting a sequence or *frame* $\{b[0], b[1], \ldots, b[M-1]\}$ of channel symbols over a wireless channel. These symbols can be binary (e.g., ± 1), or they may take on more general values from a finite alphabet of complex numbers. In this treatment, we consider only *linear* modulation systems, in which the symbols are transmitted into the channel by being modulated linearly onto a signaling waveform to produce a transmitted signal of this form:

$$x(t) = \sum_{i=0}^{M-1} b[i] w_i(t), \tag{1.1}$$

where $w_i(\cdot)$ is the modulation waveform associated with the ith symbol. In this expression, the waveforms can be quite general. For example, a single-carrier modulation system with carrier frequency ω_c, baseband pulse shape $p(\cdot)$, and symbol rate $1/T$ is obtained by choosing

$$w_i(t) = A\, p(t - iT)\, e^{\jmath(\omega_c t + \phi)}, \tag{1.2}$$

where $A > 0$ and $\phi \in (-\pi, \pi)$ denote carrier amplitude and phase offset, respectively. The baseband pulse shape may, for example, be a simple unit-energy rectangular pulse of duration T:

$$p(t) = p_T(t) \triangleq \begin{cases} \dfrac{1}{\sqrt{T}}, & 0 \le t < T, \\ 0, & \text{otherwise}, \end{cases} \tag{1.3}$$

or it could be a raised-cosine pulse, a bandlimited pulse, and so on. Similarly, a direct-sequence spread-spectrum system is produced by choosing the waveforms as in (1.2) but with the baseband pulse shape chosen to be a spreading waveform:

$$p(t) = \sum_{j=0}^{N-1} c_j \psi(t - j\,T_c), \tag{1.4}$$

where N is the spreading gain, $c_0, c_1, \ldots, c_{N-1}$, is a pseudorandom spreading code (typically, $c_j \in \{+1, -1\}$), $\psi(\cdot)$ is the chip waveform, and $T_c \triangleq T/N$ is the chip interval. The chip waveform may, for example, be a unit-energy rectangular pulse of duration T_c:

$$\psi(t) = p_{T_c}(t). \tag{1.5}$$

Other choices of the chip waveform can also be made to lower the chip bandwidth. The spreading waveform of (1.4) is periodic when used in (1.2), since the same spreading code is repeated in every symbol interval. Some systems (e.g., CDMA systems for cellular telephony) operate with *long spreading codes*, for which the periodicity is much longer than a single symbol interval. This situation can be modeled by (1.1) by replacing $p(t)$ in (1.2) by a variant of (1.4) in which the spreading code varies from symbol to symbol; that is,

$$p_i(t) = \sum_{j=0}^{N-1} c_j^{(i)} \psi(t - j\,T_c). \tag{1.6}$$

Spread-spectrum modulation can also take the form of frequency hopping, in which the carrier frequency in (1.2) is changed over time according to a pseudorandom pattern. Typically, the carrier frequency changes at a rate much slower than the symbol rate, a situation known as *slow frequency hopping*; however, *fast hopping*, in which the carrier changes within a symbol interval, is also possible. Single-carrier systems, including both types of spread spectrum, are widely used in cellular standards, in wireless LANs, Bluetooth, and others (see, e.g., [42, 131, 150, 163, 178, 247, 338, 361, 362, 392, 394, 407, 408, 449, 523, 589]).

Multicarrier systems can also be modeled in the framework of (1.1) by choosing the signaling waveforms $\{w_i(\cdot)\}$ to be sinusoidal signals with different frequencies. In particular, (1.2) can be replaced by

$$w_i(t) = A\,p(t)\,e^{\jmath(\omega_i t + \phi_i)}, \tag{1.7}$$

where now the frequency and phase depend on the symbol number i but all symbols are transmitted simultaneously in time with baseband pulse shape $p(\cdot)$. We can see that (1.2) is the counterpart of this situation with time and frequency reversed: All symbols are transmitted at the same frequency but at different times. (Of course, in practice, multiple symbols are sent in time sequence over each of the multiple carriers in multicarrier systems.) The individual carriers can also be direct-spread, and the baseband pulse shape used can depend on the symbol number i. (For example, the latter situation is used in *multicarrier CDMA*, in which a spreading code is used across the carrier frequencies.) A particular case of (1.7) is OFDM, in which the baseband pulse shape is a unit pulse p_T, the intercarrier spacing is $1/T$ cycles per second, and the phases are chosen so that the carriers are orthogonal at this spacing. (This is the minimal spacing for which such orthogonality can be maintained.) OFDM is widely believed to be among the most effective techniques for wireless broadband applications and is the basis for the IEEE 802.11a high-speed wireless LAN standard (see, e.g., [354] for a discussion of multicarrier systems).

An emerging type of wireless modulation scheme is ultra-wideband (UWB) modulation, in which data are transmitted with no carrier through the modulation of extremely short pulses. Either the timing or amplitude of these pulses can be used to carry the information symbols. Typical UWB systems involve the transmission of many repetitions of the same symbol, possibly with the use of a direct-sequence type of spreading code from transmission to transmission (see, e.g., [569] for a basic description of UWB systems).

Further details on the modulation waveforms above and their properties will be introduced as needed throughout this treatment.

1.2.2 Multiple-Access Techniques

In Section 1.2.1 we discussed ways in which a symbol stream associated with a single user can be transmitted. Many wireless channels, particularly in emerging systems, operate as multiple-access systems, in which multiple users share the same radio resources.

There are several ways in which radio resources can be shared among multiple users. These can be viewed as ways of allocating regions in frequency, space, and time to different users, as shown in Fig. 1.1. For example, a classic multiple-access technique is *frequency-division multiple access* (FDMA), in which the frequency band available for a given service is divided into subbands that are allocated to individual users who wish to use the service. Users are given exclusive use of their subband during their communication session, but they are not allowed to transmit signals within other subbands. FDMA is the principal multiplexing method used in radio and television broadcast and in first-generation (analog voice) cellular telephony systems, such as the Advanced Mobile Phone System (AMPS) and Nordic Mobile Telephone (NMT), developed primarily in the 1970s and 1980s (cf. [458]). FDMA is also used in some form in all other current cellular systems, in tandem with other multiple-access techniques that are used to further allocate the subbands to multiple users.

Similarly, users can share the channel on the basis of *time-division multiple access* (TDMA), in which time is divided into equal-length intervals, which are further divided into equal-length subintervals, or time slots. Each user is allowed to transmit throughout the entire allocated frequency band during a given slot in each interval but is not allowed to transmit during other time slots when other users are transmitting. So, whereas FDMA allows each user to use part of the spectrum all of the time, TDMA allows each user to use all of the spectrum part of the time. This method of channel sharing is widely used in wireless applications, notably in a number of second-generation cellular (i.e., digital voice) sytems, including the widely used Global System for Mobile (GSM) system [178, 407, 408] and in the IEEE 802.16 wireless MAN standards. A form of TDMA is also used in Bluetooth networks, in which one of the Bluetooth devices in the network acts as a network controller to poll the other devices in time sequence.

FDMA and TDMA systems are intended to assign orthogonal channels to all active users by giving each, for their exclusive use, a slice of the available frequency band or transmission time. These channels are said to be *orthogonal* because interference between users does not, in principle, arise in such assignments (although, in practice, there is often such interference, as discussed further below). *Code-division multiple access* (CDMA) assigns channels in a way that allows all users to use all of the available time and frequency resources simultaneously, through the assignment of a pattern or code to each user that specifies the way in which these resources

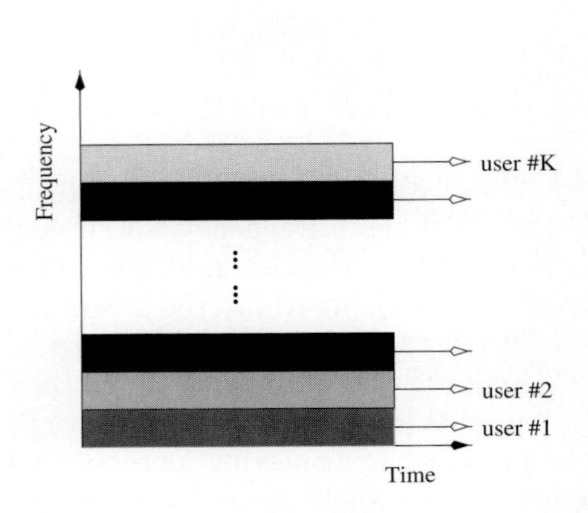

Frequency-Division Multiple-Access (FDMA)

Time-Division Multiple-Access (TDMA)

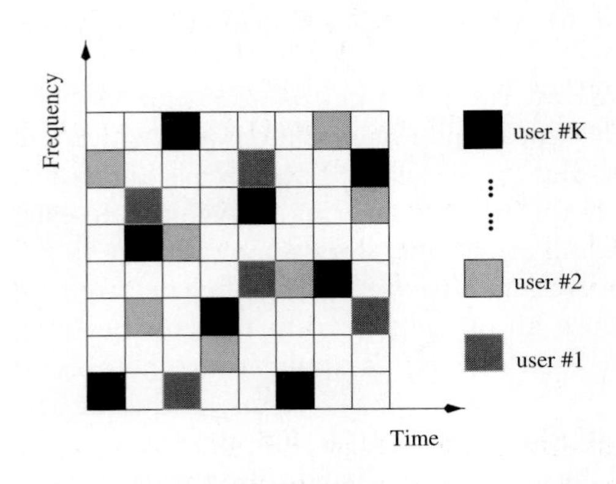

Frequency-Hopping Code-Division Multiple-Access (FH-CDMA)

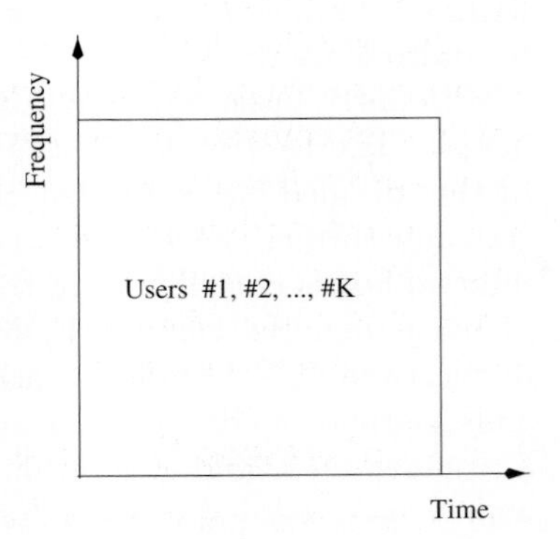

Direct-Sequence Code-Division Multiple-Access (DS-CDMA)

Figure 1.1. Multiple-access schemes.

will be used by that user. Typically, CDMA is implemented via spread-spectrum modulation, in which the pattern is the pseudorandom code that determines the spreading sequence in the case of direct sequence, or the hopping pattern in the case of frequency hopping. In such systems, a channel is defined by a particular pseudorandom code, so each user is assigned a channel by being assigned a pseudorandom code. CDMA is used, notably, in the second-generation cellular standard IS-95 (Interim Standard 95), which makes use of direct-sequence CDMA to allocate subchannels of larger-bandwidth (1.25 MHz) subchannels of the entire cellular band. It is also used, in the form of frequency hopping, in GSM to provide isolation among users in adjacent cells. The spectrum spreading used in wireless LAN systems is also a form of CDMA in that it allows a number of such systems to operate in the same lightly regulated part of the radio spectrum. CDMA is also the basis for the principal standards being developed and deployed for 3G cellular telephony (e.g., [130, 361, 362, 407]).

Any of the multiple-access techniques discussed here can be modeled analytically by considering multiple transmitted signals of the form (1.1). In particular, for a system of K users, we can write a transmitted signal for each user as

$$x_k(t) = \sum_{i=0}^{M-1} b_k[i]\, w_{i,k}(t), \qquad k = 1, 2, \ldots, K, \tag{1.8}$$

where $x_k(\cdot)$, $\{b_k[0], b_k[1], \ldots, b_k[M-1]\}$, and $w_{i,k}(\cdot)$ represent the transmitted signal, symbol stream, and ith modulation waveform, respectively, of user k. That is, each user in a multiple-access system can be modeled in the same way as in a single-user system, but with (usually) differing modulation waveforms (and symbol streams, of course). If the waveforms $\{w_{i,k}(\cdot)\}$ are of the form (1.2) but with different carrier frequencies $\{\omega_k\}$, say, this is FDMA. If they are of the form (1.2) but with time-slotted amplitude pulses $\{p_k(\cdot)\}$, say, this is TDMA. Finally, if they are spread-spectrum signals of this form but with different pseudorandom spreading codes or hopping patterns, this is CDMA. Details of these multiple-access models will be discussed in the sequel as needed.

1.2.3 Wireless Channel

From a technical point of view, the greatest distinction between wireless and wireline communications lies in the physical properties of wireless channels. These physical properties can be described in terms of several distinct phenomena, including ambient noise, propagation losses, multipath, interference, and properties arising from the use of multiple antennas. Here we review these phenomena only briefly. Further discussion and details can be found, for example, in [38,46,148,216,405,450,458,465].

Like all practical communications channels, wireless channels are corrupted by ambient noise. This noise comes from thermal motion of electrons on the antenna and in the receiver electronics and from background radiation sources. This noise is well modeled as having a very wide bandwidth (much wider than the bandwidth of any useful signals in the channel) and no particular deterministic structure (structured noise can be treated separately as interference). A very common and useful model for such noise is additive white Gaussian noise (AWGN), which as the name implies, means that it is additive to the other signals in the receiver, has a flat power spectral density, and induces a Gaussian probability distribution at the output of any linear filter to which it is input. Impulsive noise also occurs in some wireless channels. Such noise is similarly wideband but induces a non-Gaussian amplitude distribution at the output of linear filters. Specific models for such impulsive noise are discussed in Chapter 4.

Propagation losses are also an issue in wireless channels. These are of two basic types: diffusive losses and shadow fading. *Diffusive losses* arise because of the open nature of wireless channels. For example, the energy radiated by a simple point source in free space will spread over an ever-expanding spherical surface as the energy propagates away from the source. This means that an antenna with a given aperture size will collect an amount of energy that decreases with the square of the distance between the antenna and the source. In most terrestrial wireless channels, the diffusion losses are actually greater than this, due to the effects of ground-wave propagation, foliage, and so on. For example, in cellular telephony, the diffusion loss is inverse square with distance within line of sight of the cell tower, and it falls off with a higher power (typically, 3 or 4) at greater distances. As its name implies, *shadow fading* results from the presence of objects (buildings, walls, etc.) between the transmitter and receiver. Shadow fading is typically modeled by an attenuation (i.e., a multiplicative factor) in signal amplitude that follows a log-normal distribution. The variation in this fading is specified by the standard deviation of the logarithm of this attenuation.

Multipath refers to the phenomenon by which multiple copies of a transmitted signal are received at the receiver, due to the presence of multiple radio paths between the transmitter and receiver. These multiple paths arise due to reflections from objects in the radio channel. Multipath is manifested in several ways in communications receivers, depending on the degree of path difference relative to the wavelength of propagation, the degree of path difference relative to the signaling rate, and the relative motion between the transmitter and receiver. Multipath from scatterers that are spaced very close together will cause a random change in the amplitude of the received signal. Due to central-limit effects, the resulting received amplitude is often modeled as being a complex Gaussian random variable. This results in a random amplitude whose envelope has a Rayleigh distribution, and this phenomenon is thus termed *Rayleigh fading.* Other fading distributions also arise, depending on the physical configuration (see, e.g., [396]). When the scatterers are spaced so that the differences in their corresponding path lengths are significant relative to a wavelength of the carrier, the signals arriving at the receiver along different paths can add constructively or destructively. This gives rise to fading that depends on the wavelength (or, equivalently, the frequency) of radiation, which is thus called *frequency-selective fading.* When there is relative motion between the transmitter and receiver, this type of fading also depends on time, since the path length is a function of the radio geometry. This results in *time-selective fading.* (Such motion also causes signal distortion due to Doppler effects.) A related phenomenon arises when the difference in path lengths is such that the time delay of arrival along different paths is significant relative to a symbol interval. This results in dispersion of the transmitted signal, and causes *intersymbol interference* (ISI); that is, contributions from multiple symbols arrive at the receiver at the same time.

Many of the advanced signal transmission and processing methods that have been developed for wireless systems are designed to contravene the effects of multipath. For example, wideband signaling techniques such as spread spectrum are often used as a countermeasure to frequency-selective fading. This both minimizes the effects of deep frequency-localized fades and facilitates the resolvability and subsequent coherent combining of multiple copies of the same signal. Similarly, by dividing a high-rate signal into many parallel lower-rate signals, OFDM mitigates the effects of channel dispersion on high-rate signals. Alternatively, high-data-rate single-carrier systems make use of channel equalization at the receiver to counteract this dispersion. Some of these issues are discussed further in Section 1.3.

Interference, also a significant issue in many wireless channels, is typically one of two types: multiple-access interference and co-channel interference. *Multiple-access interference* (MAI) refers to interference arising from other signals in the same network as the signal of interest. For example, in cellular telephony systems, MAI can arise at the base station when the signals from multiple mobile transmitters are not orthogonal to one another. This happens by design in CDMA systems, and it happens in FDMA or TDMA systems due to channel properties such as multipath or to nonideal system characteristics such as imperfect channelization filters. *Co-channel interference* (CCI) refers to interference from signals from different networks, but operating in the same frequency band as the signal of interest. An example is the interference from adjacent cells in a cellular telephony system. This problem is a chief limitation of using FDMA in cellular systems and was a major factor in moving away from FDMA in second-generation systems. Another example is the interference from other devices operating in the same part of the unregulated spectrum as the signal of interest, such as interference from Bluetooth devices operating in the same 2.4-GHz ISM band as IEEE 802.11 wireless LANs. Interference mitigation is also a major factor in the design of transmission techniques (e.g., the above-noted movement away from FDMA in cellular systems) as well as in the design of advanced signal processing systems for wireless, as we shall see in the sequel.

The phenomena we have discussed above can be incorporated into a general analytical model for a wireless multiple-access channel. In particular, the signal model in a wireless system is illustrated in Fig. 1.2. We can write the signal received at a given receiver in the following form:

$$r(t) = \sum_{k=1}^{K} \sum_{i=0}^{M-1} b_k[i] \int_{-\infty}^{\infty} g_k(t,u) w_{i,k}(u)\, du + i(t) + n(t), \qquad -\infty < t < \infty, \quad (1.9)$$

where $g_k(t,u)$ denotes the impulse response of a linear filter representing the channel between the kth transmitter and the receiver, $i(\cdot)$ represents co-channel interference, and $n(\cdot)$ represents ambient noise. The modeling of the wireless channel as a linear system seems to agree well with the observed behavior of such channels. All of the quantities $g_k(\cdot,\cdot)$, $i(\cdot)$, and $n(\cdot)$ are, in general, random processes. As noted

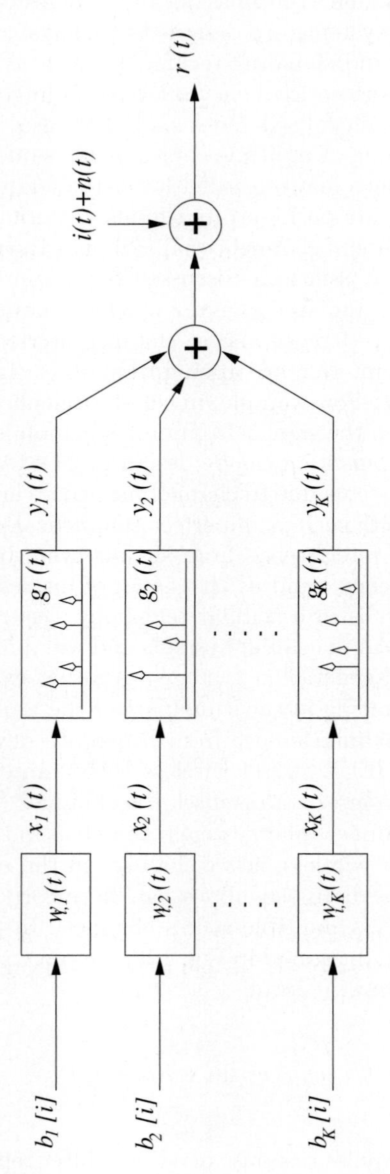

Figure 1.2. Signal model in a wireless system.

above, the ambient noise is typically represented as a white process with very little additional structure. However, the co-channel interference and channel impulse responses are typically structured processes that can be parameterized.

An important special case is that of a pure multipath channel, in which the channel impulse responses can be represented in the form

$$g_k(t,u) = \sum_{\ell=1}^{L_k} \alpha_{\ell,k}\delta(t-u-\tau_{\ell,k}), \tag{1.10}$$

where L_k is the number of paths between user k and the receiver, $\alpha_{\ell,k}$ and $\tau_{\ell,k}$ are the gain and delay, respectively, associated with the ℓth path of the kth user, and $\delta(\cdot)$ denotes the Dirac delta function. Note that this is the situation illustrated in Fig. 1.2, in which we have written the time-invariant impulse response as $g_k(t) \equiv g_k(t,0)$. This model is an idealization of the actual behavior of a multipath channel, which would not have such a sharply defined impulse response. However, it serves as a useful model for signal processor design and analysis. Note that this model gives rise to frequency-selective fading, since the relative delays will cause constructive and destructive interference at the receiver, depending on the wavelength of propagation. Often, the delays $\{\tau_{\ell,k}\}$ are assumed to be known to the receiver or are spaced uniformly at the inverse of the bulk bandwidth of the signaling waveforms. A typical model for the path gains $\{\alpha_{\ell,k}\}$ is that they are independent complex Gaussian random variables, giving rise to Rayleigh fading.

Note that, in general, the receiver will see the following *composite* modulation waveform associated with the symbol $b_k[i]$:

$$f_{i,k}(t) = \int_{-\infty}^{\infty} g_k(t,u)w_{i,k}(u)\,du. \tag{1.11}$$

If these waveforms are not orthogonal for different values of i, ISI will result. Consider, for example, the pure multipath channel of (1.10) with signaling waveforms of the form

$$w_{i,k}(t) = A_k s_k(t-iT), \tag{1.12}$$

where $s_k(\cdot)$ is a normalized signaling waveform $\left[\int |s_k(t)|^2\,dt = 1\right]$, A_k is a complex amplitude, and T is the inverse of the single-user symbol rate. In this case, the composite modulation waveforms are given by

$$f_{i,k}(t) = f_k(t-iT), \tag{1.13}$$

with

$$f_k(t) = A_k \sum_{\ell=1}^{L_k} \alpha_{\ell,k} s_k\left(t-\tau_{\ell,k}\right). \tag{1.14}$$

If the delay spread (i.e., the maximum of the differences of the delays $\{\tau_{\ell,k}\}$ for different values of ℓ) is significant relative to T, ISI may be a factor. Note that for a fixed channel, the delay spread is a function of the physical geometry of the channel, whereas the symbol rate depends on the data rate of the transmitted source. Thus, higher-rate transmissions are more likely to encounter ISI than are lower-rate transmissions. Similarly, if the composite waveforms for different values of k are not orthogonal, MAI will result. This can happen, for example, in CDMA channels when the pseudorandom code sequences used by different users are not orthogonal. It can also happen in CDMA and TDMA channels, due to the effects of multipath or asynchronous transmission. These issues are discussed further in the sequel as the need arises.

This model can be further generalized to account for multiple antennas at the receiver. In particular, we can modify (1.9) as follows:

$$\boldsymbol{r}(t) = \sum_{k=1}^{K} b_k[i] \int_{-\infty}^{\infty} \boldsymbol{g}_k(t,u) w_{i,k}(u)\, du \;+\; \boldsymbol{i}(t) \;+\; \boldsymbol{n}(t), \qquad -\infty < t < \infty, \tag{1.15}$$

where the boldface quantities denote (column) vectors with dimensions equal to the number of antennas at the received array. For example, the pth component of $\boldsymbol{g}_k(t,u)$ is the impulse response of the channel between user k and the pth element of the receiving array. A useful such model is to combine the pure multipath model of (1.10) with a model in which the spatial aspects of the array can be separated from its temporal properties. This yields channel impulse responses of the form

$$\boldsymbol{g}_k(t,u) \;=\; \sum_{\ell=1}^{L_k} \alpha_{\ell,k} \boldsymbol{a}_{\ell,k} \delta(t-u-\tau_{\ell,k}), \tag{1.16}$$

where the complex vector $\boldsymbol{a}_{\ell,k}$ describes the response of the array to the ℓth path of user k. The simplest such situation is the case of a *uniform linear array* (ULA), in which the array elements are uniformly spaced along a line, receiving a single-carrier signal arriving along a planar wavefront and satisfying the *narrowband array assumption.* The essence of this assumption is that the signaling waveforms are sinusoidal carriers carrying narrowband modulation and that all of the variation in the received signal across the array at any given instant in time is due to the carrier (i.e., the modulating waveform is changing slowly enough to be assumed constant across the array). In this case, the array response depends only on the angle $\phi_{\ell,k}$ at which the corresponding path's signal is incident on the array. In particular, the response of a P-element array is given in this case by

$$\boldsymbol{a}_{\ell,k} \;=\; \begin{bmatrix} 1 \\ e^{-\jmath\gamma \sin\phi_{\ell,k}} \\ e^{-\jmath 2\gamma \sin\phi_{\ell,k}} \\ \vdots \\ e^{-\jmath(P-1)\gamma \sin\phi_{\ell,k}} \end{bmatrix}, \tag{1.17}$$

where $\jmath$ denotes the imaginary unit and where $\gamma \triangleq 2\pi d/\lambda$, with λ the carrier wavelength and d the interelement spacing (see [126, 266, 269, 404, 445, 450, 510] for further discussion of systems involving multiple receiver antennas).

It is also of interest to model systems in which there are multiple antennas at both the transmitter and receiver, called *multiple-input/multiple-output* (MIMO) *systems*. In this case the channel transfer functions are matrices, with the number of rows equal to the number of receiving antennas and the number of columns equal to the number of transmitting antennas at each source. There are several ways of handling the signaling in such configurations, depending on the desired effects and the channel conditions. For example, transmitter beamforming can be implemented by transmitting the same symbol simultaneously from multiple antenna elements on appropriately phased versions of the same signaling waveform. Space-time coding can be implemented by transmitting frames of related symbols over multiple antennas. Other configurations are of interest as well. Issues concerning multiple-antenna systems are discussed further in the sequel as they arise.

1.3 Basic Receiver Signal Processing for Wireless

This book is concerned with the design of advanced signal processing methods for wireless receivers, based largely on the models discussed in preceding sections. Before moving to these methods, however, it is of interest to review briefly some basic elements of signal processing for these models. This is not intended to be a comprehensive treatment, and the reader is referred to [145, 146, 270, 376, 381, 385, 391, 396, 510, 520, 523] for further details.

1.3.1 Matched Filter/RAKE Receiver

We consider first the particular case of the model of (1.9), in which there is only a single user (i.e., $K = 1$), the channel impulse $g_1(\cdot, \cdot)$ is known to the receiver, there is no CCI [i.e., $i(\cdot) \equiv 0$], and the ambient noise is AWGN with spectral height σ^2. That is, we have the following model for the received signal:

$$r(t) = \sum_{i=0}^{M-1} b_1[i] f_{i,1}(t) + n(t), \qquad -\infty < t < \infty, \tag{1.18}$$

where $f_{i,1}(\cdot)$ denotes the composite waveform of (1.11), given by

$$f_{i,1}(t) = \int_{-\infty}^{\infty} g_1(t, u) w_{i,1}(u) \, du. \tag{1.19}$$

Let us further restrict attention, for the moment, to the case in which there is only a single symbol to be transmitted (i.e., $M = 1$), in which case we have the received waveform

$$r(t) = b_1[0] f_{0,1}(t) + n(t), \qquad -\infty < t < \infty. \tag{1.20}$$

Optimal inferences about the symbol $b_1[0]$ in (1.20) can be made on the basis of the likelihood function of the observations, conditioned on the symbol $b_1[0]$, which is given in this case by [377]

$$\mathcal{L}\Big(r(\cdot)|b_1[0]\Big) = \exp\left\{\frac{1}{\sigma^2}\left[2\Re\left\{b_1^*[0]\int_{-\infty}^{\infty} f_{0,1}^*(t)r(t)\ \mathrm{d}t\right\} - |b_1[0]|^2\int_{-\infty}^{\infty}|f_{0,1}(t)|^2\ \mathrm{d}t\right]\right\}, \quad (1.21)$$

where the superscript asterisk denotes complex conjugation and $\Re\{\cdot\}$ denotes the real part of its argument.

Optimal inferences about the symbol $b_1[0]$ can be made, for example, by choosing *maximum-likelihood* (ML) or *maximum a posteriori probability* (MAP) values for the symbol. The ML symbol decision is given simply by the argument that maximizes $\mathcal{L}(\ r(\cdot)\mid b_1[0]\ )$ over the symbol alphabet, $\mathcal{A}$:

$$\begin{aligned}\hat{b}_1[0] &= \arg\left\{\max_{b\in\mathcal{A}}\mathcal{L}\Big(\ r(\cdot)\mid b_1[0]=b\ \Big)\right\}\\ &= \arg\left\{\max_{b\in\mathcal{A}}\left[2\Re\left\{b^*\int_{-\infty}^{\infty} f_{0,1}^*(t)r(t)\ \mathrm{d}t\right\} - |b|^2\int_{-\infty}^{\infty}|f_{0,1}(t)|^2\ \mathrm{d}t\right]\right\}. \quad (1.22)\end{aligned}$$

It is easy to see that the corresponding symbol estimate is the solution to the problem

$$\min_{b\in\mathcal{A}} |b-z|^2, \quad (1.23)$$

where

$$z \triangleq \frac{\int_{-\infty}^{\infty} f_{0,1}^*(t)r(t)\ \mathrm{d}t}{\int_{-\infty}^{\infty}|f_{0,1}(t)|^2\ \mathrm{d}t}. \quad (1.24)$$

Thus, the ML symbol estimate is the closest point in the symbol alphabet to the observable z.

Note that the two simplest and most common choices of symbol alphabet are M-ary phase-shift keying (MPSK) and quadrature amplitude modulation (QAM). In MPSK, the symbol alphabet is

$$\mathcal{A} = \Big\{e^{\jmath 2\pi m/M} \mid m \in \{0,1,\ldots,M-1\}\Big\}, \quad (1.25)$$

or some rotation of this set around the unit circle. (M as used in this paragraph should not be confused with the framelength M.) For QAM, a symbol alphabet containing $M \times N$ values is

$$\mathcal{A} = \Big\{b_R + \jmath b_I \mid b_R \in \mathcal{A}_R \text{ and } b_I \in \mathcal{A}_I\Big\}, \quad (1.26)$$

where $\mathcal{A}_R$ and $\mathcal{A}_I$ are discrete sets of amplitudes containing M and N points, respectively; for example, for $M = N$ even, a common choice is

$$\mathcal{A}_R = \mathcal{A}_I = \left\{ \pm\frac{1}{2}, \pm\frac{3}{2}, \ldots, \pm\frac{M}{4} \right\} \quad (1.27)$$

or a scaled version of this choice. A special case of both of these is that of binary phase-shift keying (BPSK), in which $\mathcal{A} = \{-1, +1\}$. The latter case is the one we consider most often in this treatment, primarily for the sake of simplicity. However, most of the results discussed herein extend straightforwardly to these more general signaling alphabets.

ML symbol estimation [i.e., the solution to (1.23)] is very simple for MPSK and QAM. In particular, since the MPSK symbols correspond to phasors at evenly spaced angles around the unit circle, the ML symbol choice is that whose angle is closest to the angle of the complex number z of (1.24). For QAM, the choices of the real and imaginary parts of the ML symbol estimate are decoupled, with $\Re\{b\}$ being chosen to be the closest element of $\mathcal{A}_R$ to $\Re\{z\}$, and similarly for $\Im\{b\}$. For BPSK, the ML symbol estimate is

$$\hat{b}_i[0] = \text{sign}\{\Re\{z\}\} \triangleq \text{sign}\left\{ \Re\left\{ \int_{-\infty}^{\infty} f_{0,1}^*(t) r(t)\, dt \right\} \right\}, \quad (1.28)$$

where sign$\{\cdot\}$ denotes the signum function:

$$\text{sign}\{x\} = \begin{cases} -1 & \text{if } x < 0, \\ 0 & \text{if } x = 0, \\ +1 & \text{if } x > 0. \end{cases} \quad (1.29)$$

MAP symbol detection in (1.20) is also based on the likelihood function of (1.21), after suitable transformation. In particular, if the symbol $b_1[0]$ is a random variable, taking values in $\mathcal{A}$ with known probabilities, the *a posteriori* probability distribution of the symbol conditioned on $r(\cdot)$ is given via Bayes' formula as

$$P\Big(b_1[0] = b \mid r(\cdot)\Big) = \frac{\mathcal{L}(\, r(\cdot) \mid b_1[0] = b\,)\, P(b_1[0] = b)}{\sum_{a\in\mathcal{A}} \mathcal{L}(\, r(\cdot) \mid b_1[0] = a\,)\, P(b_1[0] = a)}, \quad b \in \mathcal{A}. \quad (1.30)$$

The MAP criterion specifies a symbol decision given by

$$\begin{aligned} \hat{b}_1[0] &= \arg\left\{ \max_{b\in\mathcal{A}} P\left(b_1[0] = b \mid r(\cdot)\right) \right\} \\ &= \arg\left\{ \max_{b\in\mathcal{A}} \left[\mathcal{L}(\, r(\cdot) \mid b_1[0] = b\,)\, P(b_1[0] = b)\right] \right\}. \end{aligned} \quad (1.31)$$

Note that in this single-symbol case, if the symbol values are equiprobable, the ML and MAP decisions are the same.

The structure of the ML and MAP decision rules above shows that the main receiver signal processing task in this single-user, single-symbol, known-channel case is the computation of the term

$$y_1[0] \triangleq \int_{-\infty}^{\infty} f_{0,1}^*(t) r(t) \, dt. \tag{1.32}$$

This structure is called a *correlator* because it correlates the received signal $r(\cdot)$ with the known composite signaling waveform $f_{1,0}(\cdot)$. This structure can also be implemented by sampling the output of a time-invariant linear filter:

$$\int_{-\infty}^{\infty} f_{0,1}^*(t) r(t) \, dt = (h \star r)(0), \tag{1.33}$$

where $\star$ denotes convolution and h is the impulse response of the time-invariant linear filter given by

$$h(t) = f_{0,1}^*(-t). \tag{1.34}$$

This structure is called a *matched filter*, since its impulse response is matched to the composite waveform on which the symbol is received. When the composite signaling waveform has a finite duration so that $h(t) = 0$ for $t < -D \leq 0$, the matched-filter receiver can be implemented by sampling at time D the output of the causal filter with the following impulse response:

$$h_D(t) = \begin{cases} f_{0,1}^*(D-t) & \text{if } t \geq 0, \\ 0 & \text{if } t < 0. \end{cases} \tag{1.35}$$

For example, if the signaling waveform $s_{0,1}(t)$ has duration $[0, T]$ and the channel has delay spread τ_d, the composite signaling waveform will have this property with $D = T + \tau_d$.

A special case of the correlator (1.32) arises for a pure multipath channel in which the channel impulse response is given by (1.10). The composite waveform (1.11) in this case is

$$f_{0,1}(t) = \sum_{\ell=1}^{L_1} \alpha_{\ell,1} s_{0,1}(t - \tau_{\ell,1}), \tag{1.36}$$

and the correlator output (1.32) becomes

$$y_1[0] \triangleq \sum_{\ell=1}^{L_1} \alpha_{\ell,1}^* \int_{-\infty}^{\infty} s_{0,1}^*(t - \tau_{\ell,1}) r(t) \, dt, \tag{1.37}$$

a configuration known as a RAKE receiver. Further details on this basic receiver structure can be found, for example, in [396].

1.3.2 Equalization

We now turn to the situation in which there is more than one symbol in the frame of interest (i.e., when $M > 1$). In this case we would like to consider the likelihood function of the observations $r(\cdot)$ conditioned on the entire frame of symbols, $b_1[0], b_1[1], \ldots, b_1[M-1]$, which is given by

$$\mathcal{L}\Big(r(\cdot)|b_1[0], b_1[1], \ldots, b_1[M-1]\Big) = \exp\left\{\frac{1}{\sigma^2}\left[2\Re\left\{\boldsymbol{b}_1^H \boldsymbol{y}_1\right\} - \boldsymbol{b}_1^H \boldsymbol{H}_1 \boldsymbol{b}_1\right]\right\}, \quad (1.38)$$

where the superscript H denotes the conjugate transpose (i.e., the Hermitian transpose), $\boldsymbol{b}_1$ denotes a column vector whose ith component is $b_1[i], i = 0, 1, \ldots, M-1$, $\boldsymbol{y}_1$ denotes a column vector whose ith component is given by

$$y_1[i] \triangleq \int_{-\infty}^{\infty} f_{i,1}^*(t) r(t)\, dt, \qquad i = 0, 1, \ldots, M-1, \quad (1.39)$$

and $\boldsymbol{H}_1$ is an $M \times M$ Hermitian matrix, whose (i, j)th element is the cross-correlation between $f_{i,1}(t)$ and $f_{j,1}(t)$:

$$\boldsymbol{H}_1[i, j] = \int_{-\infty}^{\infty} f_{i,1}^*(t) f_{j,1}(t)\, dt. \quad (1.40)$$

Since the likelihood function depends on $r(\cdot)$ only through the vector $\boldsymbol{y}_1$ of correlator outputs, this vector is a sufficient statistic for making inferences about the vector $\boldsymbol{b}_1$ of symbols [377].

Maximum-likelihood detection in this situation is given by

$$\hat{\boldsymbol{b}}_1 = \arg\left\{\max_{\boldsymbol{b} \in \mathcal{A}^M}\left[2\Re\left\{\boldsymbol{b}^H \boldsymbol{y}_1\right\} - \boldsymbol{b}^H \boldsymbol{H}_1 \boldsymbol{b}\right]\right\}. \quad (1.41)$$

Note that if $\boldsymbol{H}_1$ is a diagonal matrix (i.e., all of its off-diagonal elements are zero), (1.41) decouples into a set of M independent problems of the single-symbol type (1.22). The solution in this case is correspondingly given by

$$\hat{b}_1[i] = \arg\min_{b \in \mathcal{A}} |\, b - z_1[i]|^2, \quad (1.42)$$

where

$$z_1[i] \triangleq \frac{y_i[i]}{\int_{-\infty}^{\infty} |f_{i,1}(t)|^2\, dt}. \quad (1.43)$$

However, in the more general case in which there is intersymbol interference, (1.41) will not decouple and the optimization must take place over the entire frame, a problem known as *sequence detection.*

The problem of (1.41) is an integer quadratic program which is known to be an NP-complete combinatorial optimization problem [380]. This implies that the complexity of (1.41) is potentially quite high: exponential in the frame length M, which is essentially the complexity order of exhausting over the sequence alphabet $\mathcal{A}^M$. This is a prohibitive degree of complexity for most applications, since a typical frame length might be hundreds or even thousands of symbols. Fortunately, this complexity can be mitigated substantially for practical ISI channels. In particular, if the composite signaling waveforms have finite duration D, the matrix $\boldsymbol{H}_1$ is a banded matrix with nonzero elements only on those diagonals that are no more than $\Delta = \lceil D/T \rceil$ diagonals away from the main diagonal (here $\lceil \cdot \rceil$ denotes the smallest integer not less than its argument); that is,

$$|\boldsymbol{H}_1[i,j]| \;=\; 0, \quad \forall\; |i-j| > \Delta. \tag{1.44}$$

This structure of the matrix permits solution of (1.41) with a dynamic program of complexity order $\mathcal{O}\left(|\mathcal{A}|^{\Delta}\right)$, as opposed to the $\mathcal{O}\left(|\mathcal{A}|^{M}\right)$ complexity of direct search. In most situations $\Delta \ll M$, which implies an enormous savings in complexity (see, e.g., [380]). This dynamic programming solution, which can be structured in various ways, is known as a *maximum-likelihood sequence detector* (MLSD).

MAP detection in this model is also potentially of very high complexity. The *a posteriori* probability distribution of a particular symbol, say $b_1[i]$, is given by

$$P\Big(b_1[i] = b|r(\cdot)\Big) = \frac{\sum_{\{\boldsymbol{a}\in\mathcal{A}^M | a_i = b\}} \mathcal{L}\left(\, r(\cdot)|\boldsymbol{b}_1 = \boldsymbol{a}\right) P(\boldsymbol{b}_1 = \boldsymbol{a})}{\sum_{\{\boldsymbol{a}\in\mathcal{A}^M\}} \mathcal{L}\left(\, r(\cdot)|\boldsymbol{b}_1 = \boldsymbol{a}\right) P(\boldsymbol{b}_1 = \boldsymbol{a})}, \qquad b \in \mathcal{A}. \tag{1.45}$$

Note that these summations have $\mathcal{O}\left(|\mathcal{A}|^{M}\right)$ terms and thus are of complexity similar to those of the maximization in (1.41) in general. Fortunately, like (1.41), when $\boldsymbol{H}_1$ is banded these summations can be computed much more efficiently using a generalized dynamic programming technique that results in $\mathcal{O}\left(|\mathcal{A}|^{\Delta}\right)$ complexity (see, e.g., [380]).

The dynamic programs that facilitate (1.41) and (1.45) are of much lower complexity than brute-force computations. However, even this lower complexity is too high for many applications. A number of lower-complexity algorithms have been devised to deal with such situations. These techniques can be discussed easily by examining the sufficient statistic vector $\boldsymbol{y}_1$ of (1.39), which can be written as

$$\boldsymbol{y}_1 \;=\; \boldsymbol{H}_1 \boldsymbol{b}_1 \;+\; \boldsymbol{n}_1, \tag{1.46}$$

where $\boldsymbol{n}_1$ is a complex Gaussian random vector with independent real and imaginary parts having identical $\mathcal{N}(\boldsymbol{0}, \frac{\sigma^2}{2}\boldsymbol{H}_1)$ distributions. Equation (1.46) describes a linear model, and the goal of equalization is thus to fit this model with the data vector $\boldsymbol{b}_1$. The ML and MAP detectors are two ways of doing this fitting, each of which has exponential complexity with exponent equal to the bandwidth of $\boldsymbol{H}_1$. The essential difficulty of this problem arises from the fact that the vector $\boldsymbol{b}_1$ takes on values from a discrete set. One way of easing this difficulty is first to fit the linear model without constraining $\boldsymbol{b}_1$ to be discrete, and then to quantize the resulting (continuous) estimate of $\boldsymbol{b}_1$ into symbol estimates. In particular, we can use a linear fit, $\boldsymbol{M}\boldsymbol{y}_1$, as a continuous estimate of $\boldsymbol{b}_1$, where $\boldsymbol{M}$ is an $M \times M$ matrix. In this way, the ith symbol decision is

$$\hat{b}_1[i] \;=\; q\left([\boldsymbol{M}\boldsymbol{y}_1]_i\right), \tag{1.47}$$

where $[\boldsymbol{M}\boldsymbol{y}_1]_i$ denotes the ith component of $\boldsymbol{M}\boldsymbol{y}_1$ and where $q(\cdot)$ denotes a quantizer mapping the complex numbers to the symbol alphabet $\mathcal{A}$. Various choices of the matrix $\boldsymbol{M}$ lead to different linear equalizers. For example, if we choose $\boldsymbol{M} = \boldsymbol{I}_M$, the $M \times M$ identity matrix, the resulting linear detector is the common matched filter, which is optimal in the absence of ISI. A difficulty with the matched filter is that it ignores the ISI. Alternatively, if $\boldsymbol{H}_1$ is invertible, the choice $\boldsymbol{M} = \boldsymbol{H}_1^{-1}$ forces the ISI to zero,

$$\boldsymbol{H}_1^{-1}\boldsymbol{y}_1 \;=\; \boldsymbol{b}_1 \;+\; \boldsymbol{H}_1^{-1}\boldsymbol{n}_1, \tag{1.48}$$

and is thus known as the *zero-forcing equalizer* (ZFE). Note that this would be optimal (i.e., it would give perfect decisions) in the absence of AWGN. A difficulty with the ZFE is that it can significantly enhance the effects of AWGN by placing high gains on some directions in the set of M-dimensional complex vectors. A trade-off between these extremes is effected by the *minimum-mean-square-error* (MMSE) *linear equalizer*, which chooses $\boldsymbol{M}$ to give an MMSE fit of the model (1.46). Assuming that the symbols are independent of the noise, this results in the choice

$$\boldsymbol{M} \;=\; (\boldsymbol{H}_1 \;+\; \sigma^2\boldsymbol{\Sigma}_b^{-1})^{-1}, \tag{1.49}$$

where $\boldsymbol{\Sigma}_b$ denotes the covariance matrix of the symbol vector $\boldsymbol{b}_1$. (Typically, this covariance matrix will be in the form of a constant times $\boldsymbol{I}_M$.) A number of other techniques for fitting the model (1.46) have been developed, including iterative methods with and without quantization of intermediate results [decision-feedback equalizers (DFEs)], and so on. For a more detailed treatment of equalization methods, see [396].

1.3.3 Multiuser Detection

To finish this section we turn finally to the full multiple-access model of (1.9), within which data detection is referred to as *multiuser detection*. This situation is very similar to the ISI channel described above. In particular, we now consider the likelihood function of the observations $r(\cdot)$ conditioned on all symbols of all users. Sorting these symbols first by symbol number and then by user number, we can collect them in a column vector $\boldsymbol{b}$ given as

$$\boldsymbol{b} \;=\; \begin{bmatrix} b_1[0] \\ b_2[0] \\ \vdots \\ b_K[0] \\ \vdots \\ b_1[M-1] \\ b_2[M-1] \\ \vdots \\ b_K[M-1] \end{bmatrix}, \tag{1.50}$$

so that the nth element of $\boldsymbol{b}$ is given by

$$[\boldsymbol{b}]_n = b_k[i] \text{ with } k \triangleq [n-1]_K \text{ and } i \triangleq \left\lfloor \frac{n-1}{K} \right\rfloor, \qquad n = 1, 2, \ldots, KM, \quad (1.51)$$

where $[\cdot]_K$ denotes reduction of the argument modulo K and $\lfloor\cdot\rfloor$ denotes the integer part of the argument. Analogously with (1.38) we can write the corresponding likelihood function as

$$\mathcal{L}(\, r(\cdot) \mid \boldsymbol{b}\,) \;=\; \exp\left\{\frac{1}{\sigma^2}\left[2\Re\left\{\boldsymbol{b}^H \boldsymbol{y}\right\} \;-\; \boldsymbol{b}^H \boldsymbol{H} \boldsymbol{b}\right]\right\}, \quad (1.52)$$

where $\boldsymbol{y}$ is a column vector that collects the set of observables

$$y_k[i] \;\triangleq\; \int_{-\infty}^{\infty} f_{i,k}^*(t) r(t)\, \mathrm{d}t, \qquad i = 0, 1, \ldots, M-1, \quad k = 1, 2, \ldots, K, \quad (1.53)$$

indexed conformally with $\boldsymbol{b}$, and where $\boldsymbol{H}$ denotes the $KM \times KM$ Hermitian cross-correlation matrix of the composite waveforms associated with the symbols in $\boldsymbol{b}$, again with conformal indexing:

$$\boldsymbol{H}[n, m] \;=\; \int_{-\infty}^{\infty} f_{i,k}^*(t) f_{j,\ell}(t)\, \mathrm{d}t, \quad (1.54)$$

with

$$k \triangleq [n-1]_K, \quad i \triangleq \left\lfloor \frac{n-1}{K} \right\rfloor, \quad \ell \triangleq [m-1]_K, \quad \text{and} \quad j \triangleq \left\lfloor \frac{m-1}{K} \right\rfloor. \quad (1.55)$$

Comparing (1.52), (1.53), and (1.54) with their single-user counterparts (1.38), (1.39), and (1.40), we see that $\boldsymbol{y}$ is a sufficient statistic for making inferences about $\boldsymbol{b}$, and moreover that such inferences can be made in a manner very similar to that for the single-user ISI channel. The principal difference is one of dimensionality: Decisions in the single-user ISI channel involve simultaneous sequence detection with M symbols, whereas decisions in the multiple-access channel involve simultaneous sequence detection with KM symbols. This, of course, can increase the complexity considerably. For example, the complexity of exhaustive search in ML detection, or exhaustive summation in MAP detection, is now on the order of $|\mathcal{A}|^{MK}$. However, as in the single-user case, this complexity can be mitigated considerably if the delay spread of the channel is small. In particular, if the duration of the composite signaling waveforms is D, the matrix $\boldsymbol{H}$ will be a banded matrix with

$$\boldsymbol{H}[m, n] \;=\; 0, \qquad \forall\; |n - m| > K\Delta, \quad (1.56)$$

where, as before, $\Delta = \lceil D/T \rceil$. This bandedness allows the complexity of both ML and MAP detection to be reduced to the order of $|\mathcal{A}|^{K\Delta}$ via dynamic programming.

Although further complexity reduction can be obtained in this problem within additional structural constraints on $\boldsymbol{H}$ (see, e.g., [380]), the $\mathcal{O}\left(|\mathcal{A}|^{K\Delta}\right)$ complexity of ML and MAP multiuser detection is not generally reducible. Consequently, as with the equalization of single-user channels, a number of lower-complexity suboptimal multiuser detectors have been developed. For example, analogously with (1.47), linear multiuser detectors can be written in the form

$$\hat{b}_k[i] = q\left([\boldsymbol{M}\boldsymbol{y}]_n\right), \quad \text{with } k \triangleq [n-1]_K \quad \text{and} \quad i \triangleq \left\lfloor \frac{n-1}{K} \right\rfloor, \tag{1.57}$$

where $\boldsymbol{M}$ is a $KM \times KM$ matrix, $[\boldsymbol{M}\boldsymbol{y}]_n$ denotes the nth component of $\boldsymbol{M}\boldsymbol{y}$, and where, as before, $q(\cdot)$ denotes a quantizer mapping the complex numbers to the symbol alphabet $\mathcal{A}$. The choice $\boldsymbol{M} = \boldsymbol{H}^{-1}$ forces both MAI and ISI to zero and is known as the decorrelating detector, or *decorrelator*. Similarly, the choice

$$\boldsymbol{M} = \left(\boldsymbol{H} + \sigma^2 \boldsymbol{\Sigma}_b^{-1}\right)^{-1}, \tag{1.58}$$

where $\boldsymbol{\Sigma}_b$ denotes the covariance matrix of the symbol vector $\boldsymbol{b}$, is known as the *linear MMSE multiuser detector*. Linear and nonlinear iterative versions of these detectors have also been developed, both to avoid the complexity of inverting $KM \times KM$ matrices and to exploit the finite-alphabet property of the symbols (see, e.g., [520]).

As a final issue here we note that all of the discussion above has involved direct processing of continuous-time observations to obtain a sufficient statistic (in practice, this corresponds to hardware front-end processing), followed by algorithmic processing to obtain symbol decisions. Increasingly, an intermediate step is of interest. In particular, it is often of interest to project continuous-time observations onto a large but finite set of orthonormal functions to obtain a set of observables. These observables can then be processed further using digital signal processing (DSP) to determine symbol decisions (perhaps with intermediate calculation of the sufficient statistic), which is the principal advantage of this approach. A tacit assumption in this process is that the orthonormal set spans all of the composite signaling waveforms of interest, although this will often be only an approximation. A prime example of this kind of processing arises in direct-sequence spread-spectrum systems [see (1.6)], in which the received signal can be passed through a filter matched to the chip waveform and then sampled at the chip rate to produce N samples per symbol interval. These N samples can then be combined in various ways (usually, linearly) for data detection. In this way, for example, the linear equalizer and multiuser detectors discussed above are particularly simple to implement. A significant advantage of this approach is that this combining can often be done adaptively when some aspects of the signaling waveforms are unknown. For example, the channel impulse response may be unknown to the receiver, as may the waveforms of some interfering signals. This kind of processing is a basic element of many of the results discussed in this book and will be revisited in more detail in Chapter 2.

1.4 Outline of the Book

In Section 1.3 we described the basic principles of signal reception for wireless systems. The purpose of this book is to delve into advanced methods for this problem in the contexts of the signaling environments that are of most interest in emerging wireless applications. The scope of the treatment includes advanced receiver techniques for key signaling environments, including multiple-access, MIMO, and OFDM systems, as well as methods that address unique physical issues arising in many wireless channels, including fading, impulsive noise, co-channel interference, and other channel impairments. This material is organized into nine chapters beyond the current chapter. The first five of these deal explicitly with multiuser detection (i.e., with the mitigation of multiple-access interference) combined with other channel features or impairments. The remaining four chapters deal with the treatment of systems involving narrowband co-channel interference, time-selective fading, or multiple carriers, and with a general technique for receiver signal processing based on Monte Carlo Bayesian techniques. These contributions are outlined briefly in the paragraphs below.

Chapter 2 is concerned with the basic problem of adaptive multiuser detection in channels whose principal impairments (aside from multiple-access interference) are additive white Gaussian noise and multipath distortion. Adaptivity is a critical issue in wireless systems because of the dynamic nature of wireless channels. Such dynamism arises from several sources, notably from mobility of the transmitter or receiver and from the fact that the user population of the channel changes due to the entrance and exit of users and interferers from the channels and due to the bursty nature of many information sources. This chapter deals primarily with *blind multiuser detection*, in which the receiver is faced with the problem of demodulating a particular user in a multiple-access system, using knowledge only of the signaling waveform (either the composite receiver waveform or the transmitted waveform) of that user. The "blind" qualifier means that the receiver algorithms to be described are to be adapted without knowledge of the transmitted symbol stream. In this chapter we introduce the basic methods for blind adaptation of the linear multiuser detectors discussed in Section 1.3 via traditional adaptation methods, including least-mean-squares (LMS), recursive least-squares (RLS), and subspace tracking. The combination of multiuser detection with estimation of the channel intervening the desired transmitter and receiver is also treated in this context, as is the issue of correlated noise.

The methods of Chapter 2 are of particular interest in downlink situations (e.g., base to mobile), in which the receiver is interested in the demodulation of only a single user in the system. Another scenario is that the receiver has knowledge of the signaling waveforms used by a group of transmitters and wishes to demodulate this entire group while suppressing the effects of other interfering transmitters. An example of a situation in which this type of problem occurs is the reverse, or mobile-to-base, link in a CDMA cellular telephony system, in which a given base station wishes to demodulate the users in its cell while suppressing interference from users in adjacent cells. Chapter 3 continues with the issue of blind multiuser detection, but in this more general context of group detection. Here, both linear and nonlinear methods are considered, and again the issues of multipath and correlated noise are examined.

Channels in which the ambient noise is assumed to be Gaussian are considered in Chapters 2 and 3. Of course, this assumption of Gaussian noise is a very common one in the design and analysis of communication systems, and there are often good reasons for this assumption, including tractability and a degree of physical reality stemming from phenomena such as thermal noise. However, many practical channels involve noise that is decidedly not Gaussian. This is particularly true in urban and indoor environments, in which there is considerable impulsive noise due to human-made ambient phenomena. Also, in underwater acoustic channels (which are not specifically addressed in this book but which are used for tetherless communications) the ambient noise tends to be non-Gaussian. In systems limited by multiple-access interference, the assumption of Gaussian noise is a reasonable one, since it allows the focus to be placed on the main source of error—multiple-access interference. However, as we shall see in Chapters 2 and 3, the use of multiuser detection can return such channels to channels limited by ambient noise. Thus, the structure of ambient noise is again important, particularly since the performance and design of receiver algorithms can be affected considerably by the shape of the noise distribution even when the noise energy is held constant. In Chapter 4 we consider the problem of adaptive multiuser detection in channels with non-Gaussian ambient noise. This problem is a particularly challenging one because traditional methods for mitigating non-Gaussian noise involve nonlinear front-end processing, whereas methods for mitigating MAI tend to rely on the linear separating properties of the signaling multiplex. Thus, the challenge for non-Gaussian multiple-access channels is to combine these two methodologies without destroying the advantages of either. A powerful approach to this problem based on nonlinear regression is described in Chapter 4. In addition to the design and analysis of basic algorithms for known signaling environments, blind and group-blind methods are also discussed. It is seen that these methods lead to methods for multiuser detection in non-Gaussian environments that perform much better than linear methods in terms of both absolute performance and robustness.

In Chapter 5 we introduce the issue of multiple antennas into the receiver design problem. In particular, we consider the design of optimal and adaptive multiuser detectors for MIMO systems. Here, for known channel and antenna characteristics, the basic sufficient statistic [analogous to (1.53)] is a space-time matched-filter bank, which forms a generic front end for a variety of space-time multiuser detection methods. For adaptive systems, a significant issue that arises beyond those in the single-antenna situation is lack of knowledge of the response of the receiving antenna array. This can be handled through a novel adaptive MMSE multiuser detector described in this chapter. Again, as in the scalar case, the issues of multipath and blind channel identification are considered as well.

In Chapter 6 we treat the problem of signal reception in channel-coded multiple-access systems. In particular, the problem of joint channel decoding and multiuser detection is considered. A turbo-style iterative technique is presented that mitigates the high complexity of optimal processing in this situation. The essential idea of this turbo multiuser detector is to consider the combination of channel coding followed by a multiple-access channel as a concatenated code, which can be decoded by iterating between the constituent decoders—the multiuser detector for the multiple-access channel and a conventional channel decoder for the channel codes—exchanging soft information between each iteration. The constituent algorithms must be soft-input/soft-output (SISO) algorithms, which implies MAP multiuser detection and decoding. In the case of convolutional channel codes, the MAP decoder can be implemented using the well-known Bahl, Cocke, Jelinek, and Raviv (BCJR) algorithm. However, the MAP multiuser detector is quite complex, and thus a SISO MMSE detector is developed to lessen this complexity. A number of issues are treated in this context, including a group-blind implementation to suppress interferers, multipath, and space-time coded systems.

In Chapter 7 we turn to the issue of narrowband interference suppression in spread-spectrum systems. This problem arises for many reasons. For example, in multimedia transmission, signals with different data rates make use of the same radio resources, giving rise to signals of different bandwidths in the same spectrum. Also, some emerging services are being placed in parts of the radio spectrum which are already occupied by existing narrowband legacy systems. Many other systems operate in license-free parts of the spectrum, where signals of all types can share the same spectrum. Similarly, in tactical military systems, jamming gives rise to narrowband interference. The use of spread-spectrum modulation in these types of situations creates a degree of natural immunity to narrowband interference. However, active methods for interference suppression can yield significant performance improvements over systems that rely simply on this natural immunity. This problem is an old one, dating to the 1970s. Here we review the development of this field, which has progressed from methods that exploit only the bandwidth discrepancies between spread and narrowband signals, to more powerful "code-aided" techniques that make use of ideas similar to those used in multiuser detection. We consider several types of narrowband interference, including tonal signals and narrowband digital communication signals, and in all cases it is seen that active methods can offer significant performance gains with relatively small increases in complexity.

Chapter 8 is concerned with the problem of Monte Carlo Bayesian signal processing and its applications in developing adaptive receiver algorithms for tasks such as multiuser detection, equalization, and related tasks. Monte Carlo Bayesian methods have emerged in statistics over the past few years. When adapted to signal processing tasks, they give rise to powerful low-complexity adaptive algorithms whose performance approaches theoretical optima for fast and reliable communications in the dynamic environments in which wireless systems must operate. The chapter begins with a review of the large body of methodology in this area that has been developed over the past decade. It then continues to develop these ideas as signal processing tools, both for batch processing using Markov chain Monte Carlo (MCMC) methods and for online processing using sequential Monte Carlo (SMC) methods. These methods are particularly well suited to problems involving unknown channel conditions, and the power of these techniques is illustrated in the contexts of blind multiuser detection in unknown channels and blind equalization of MIMO channels.

Although most of the methodology discussed in the preceding paragraphs can deal with fading channels, the focus of those methods has been on quasi-static channels in which the fading characteristics of the channel can be assumed to be constant over an entire processing window, such as a data frame. This allows representation of the fading with a set of parameters that can be well estimated by the receiver. An alternative situation arises when the channel fading is fast enough that it can change at a rate comparable to the signaling rate. For such channels, new techniques must be developed in order to mitigate the fast fading, either by tracking it simultaneously with data demodulation or by using modulation techniques that are impervious to fast fading. Chapter 9 is concerned with problems of this type. In particular, after an overview of the physical and mathematical modeling of fading processes, several basic methods for dealing with fast-fading channels are considered. In particular, these methods include application of the expectation-maximization (EM) algorithm and its sequential counterpart, decision-feedback differential detectors for scalar and space-time-coded systems, and sequential Monte Carlo methods for both coded and uncoded systems.

Finally, in Chapter 10, we turn to problems of advanced receiver signal processing for coded OFDM systems. As noted previously, OFDM is becoming the technique of choice for many high-data-rate wireless applications. Recall that OFDM systems are multicarrier systems in which the carriers are spaced as closely as possible while maintaining orthogonality, thereby efficiently using available spectrum. This technique is very useful in frequency-selective channels, since it allows a single high-rate data stream to be converted into a group of many low-rate data streams, each of which can be transmitted without intersymbol interference. The chapter begins with a review of OFDM systems and then considers receiver design for OFDM signaling through unknown frequency-selective channels. In particular, the treatment focuses on turbo receivers in several types of OFDM systems, including systems with frequency offset, a space-time block coded OFDM system, and a space-time coded OFDM system using low-density parity-check (LDPC) codes.

Taken together, the techniques described in these chapters provide a unified methodology for the design of advanced receiver algorithms to deal with the impairments and diversity opportunities associated with wireless channels. Although most of these algorithms represent very recent research contributions, they have generally been developed with an eye toward low complexity and ease of implementation. Thus, it is anticipated that they can be applied readily in the development of practical systems. Moreover, the methodology described herein is sufficiently general that it can be adapted as needed to other problems of receiver signal processing. This is particularly true of the Monte Carlo Bayesian methods described in Chapter 8, which provide a very general toolbox for designing low-complexity yet sophisticated adaptive signal processing algorithms.

Note to the Reader Each chapter of this book describes a number of advanced receiver algorithms. For convenience, the introduction to each chapter contains a list of the algorithms developed in that chapter. Also, the references cited for all chapters are listed near the end of the book. This set of references comprises an extensive, although not exhaustive, bibliography of the literature in this field.

Chapter 2

BLIND MULTIUSER DETECTION

2.1 Introduction

As noted in Chapter 1, code-division multiple access (CDMA) implemented with direct-sequence spread-spectrum (DS-SS) modulation continues to gain popularity as a multiple-access technology for personal, cellular, and satellite communication services. Also as noted in Chapter 1, multiuser detection techniques can substantially increase the capacity of CDMA systems, and a significant body of research has addressed various such schemes. Considerable recent attention has been focused on the problem of adaptive multiuser detection [184, 185]. For example, methods for adapting the linear decorrelating detector that require the transmission of training sequences during adaptation have been proposed in [73, 334, 335]. An alternative linear detector, the linear minimum mean-square-error (MMSE) detector, however, can be adapted either through the use of training sequences [4,307,325,403] or in the blind mode (i.e., with prior knowledge of only the signature waveform and timing of the user of interest) [183, 549]. Timing-free blind multiuser detection has been considered recently in [62], while noncooperative multiuser detection, in which none of the signature waveforms is known, is treated in [596]. Blind adaptation schemes are especially attractive for the downlinks of CDMA systems, since in a dynamic environment it is very difficult for a mobile user to obtain accurate information about other active users in the channel, such as their signature waveforms, and frequent use of training sequences wastes channel bandwidth. There are primarily two approaches to blind multiuser detection: the direct matrix inversion (DMI) approach and the subspace approach. In this chapter we present batch algorithms and adaptive algorithms under both approaches. For the sake of exposition, we first treat simple synchronous single-path CDMA channels and present the principal techniques for blind multiuser detection. We then generalize these methods to the more general asynchronous CDMA channels with multipath effects. The rest of this chapter is organized as follows. In Section 2.2 we introduce the synchronous CDMA signal model and linear multiuser detectors. In Sections 2.3 and 2.4 we discuss the direct approach and the subspace approach to blind multiuser detection, respectively. In Section 2.5 we present analytical performance assessments of direct and subspace multiuser detectors. In Section 2.6 we discuss various subspace tracking algorithms for adaptive implementations of subspace blind multiuser detectors. In Section 2.7 we treat blind multiuser detection in general asynchronous CDMA systems with multipath channels. Finally, in Section 2.8 we present the mathematical derivations and proofs for some results in this chapter.

The following is a list of the algorithms appearing in this chapter.

- *Algorithm 2.1*: DMI blind linear MMSE detector—synchronous CDMA
- *Algorithm 2.2*: LMS blind linear MMSE detector—synchronous CDMA
- *Algorithm 2.3*: RLS blind linear MMSE detector—synchronous CDMA
- *Algorithm 2.4*: QR-RLS blind linear MMSE detector—synchronous CDMA
- *Algorithm 2.5*: Subspace blind linear detector—synchronous CDMA
- *Algorithm 2.6*: Blind adaptive linear MMSE detector based on subspace tracking—synchronous CDMA
- *Algorithm 2.7*: Subspace blind linear multiuser detector—multipath CDMA
- *Algorithm 2.8*: Adaptive blind linear multiuser detector based on subspace tracking—multipath CDMA
- *Algorithm 2.9*: Blind linear MMSE detector in multipath CDMA with correlated noise—SVD-based method
- *Algorithm 2.10*: Blind linear MMSE detector in multipath CDMA with correlated noise—CCD-based method

2.2 Linear Receivers for Synchronous CDMA

2.2.1 Synchronous CDMA Signal Model

We start by considering the most basic multiple-access signal model: a baseband K-user time-invariant synchronous additive white Gaussian noise (AWGN) system, employing periodic (short) spreading sequences and operating with a coherent BPSK modulation format. (An approach to adaptive detection in (long) aperiodic code DS-SS systems is developed in [61].) As noted in Chapter 1, the continuous-time waveform received by a given user in such a system can be modeled as follows:

$$r(t) = \sum_{k=1}^{K} A_k \sum_{i=0}^{M-1} b_k[i] s_k(t - iT) + n(t), \qquad 0 \leq t \leq MT, \tag{2.1}$$

where M is the number of data symbols per user in the data frame of interest; T is the symbol interval; A_k, $\{b_k[i]\}_{i=0}^{M-1}$, and $s_k(t)$ denote, respectively, the received complex amplitude, the transmitted symbol stream, and the normalized signaling waveform of the kth user; and $n(t)$ is the baseband complex Gaussian ambient noise with independent real and imaginary components and with power spectral density σ^2. It is assumed that for each user k, $\{b_k[i]\}_{i=0}^{M-1}$ is a collection of independent equiprobable ± 1 random variables, and the symbol streams of different users are independent. For the direct-sequence spread-spectrum format, each user's signaling waveform is of the form

$$s_k(t) = \frac{1}{\sqrt{N}} \sum_{j=0}^{N-1} c_{j,k} \psi(t - jT_c), \qquad 0 \leq t < T, \tag{2.2}$$

where N is the processing gain, $\{c_{j,k}\}_{j=0}^{N-1}$ is a signature sequence of ± 1's assigned to the kth user, and $\psi(\cdot)$ is a chip waveform of duration $T_c = T/N$ and unit energy [i.e., $\int_0^{T_c} \psi(t)^2 \, dt = 1$].

At the receiver, the received signal $r(t)$ is filtered by a chip-matched filter and then sampled at the chip rate. The sample corresponding to the jth chip of the ith symbol is thus given by

$$r_j[i] \triangleq \int_{iT+jT_c}^{iT+(j+1)T_c} r(t)\psi(t - iT - jT_c) \, dt,$$

$$j = 0, \ldots, N-1; \quad i = 0, \ldots, M-1. \tag{2.3}$$

The resulting discrete-time signal corresponding to the ith symbol is then given by

$$\boldsymbol{r}[i] = \sum_{k=1}^{K} A_k b_k[i] \boldsymbol{s}_k + \boldsymbol{n}[i] \tag{2.4}$$

$$= \boldsymbol{S}\boldsymbol{A}\boldsymbol{b}[i] + \boldsymbol{n}[i], \tag{2.5}$$

with

$$\boldsymbol{r}[i] \triangleq \begin{bmatrix} r_0[i] \\ r_1[i] \\ \vdots \\ r_{N-1}[i] \end{bmatrix}, \qquad \boldsymbol{s}_k \triangleq \frac{1}{\sqrt{N}} \begin{bmatrix} c_{0,k} \\ c_{1,k} \\ \vdots \\ c_{N-1,k} \end{bmatrix}, \qquad \boldsymbol{n}[i] \triangleq \begin{bmatrix} n_0[i] \\ n_1[i] \\ \vdots \\ n_{N-1}[i] \end{bmatrix},$$

where $n_j[i] \triangleq \int_{jT_c}^{(j+1)T_c} n(t)\psi(t - iT - jT_c) \, dt$ is a complex Gaussian random variable with independent real and imaginary components; and $\boldsymbol{n}[i] \sim \mathcal{N}_c(\boldsymbol{0}, \sigma^2 \boldsymbol{I}_N)$. [Here $\mathcal{N}_c(\cdot, \cdot)$ denotes a complex Gaussian distribution and $\boldsymbol{I}_N$ denotes an $N \times N$ identity matrix.] $\boldsymbol{S} \triangleq [\boldsymbol{s}_1 \ \cdots \ \boldsymbol{s}_K]$, $\boldsymbol{A} \triangleq \mathsf{diag}(A_1, \ldots, A_K)$, and $\boldsymbol{b}[i] \triangleq \left[b_1[i] \ \cdots \ b_K[i] \right]^T$.

Suppose that we are interested in demodulating the data bits of a particular user, say user 1, $\{b_1[i]\}_{i=0}^{M-1}$, based on the received waveforms $\{\boldsymbol{r}[i]\}_{i=0}^{M-1}$. A linear receiver for this purpose can be described by a weight vector $\boldsymbol{w}_1 \in \mathbb{C}^N$ such that the desired user's data bits are demodulated according to

$$z_1[i] = \boldsymbol{w}_1^H \boldsymbol{r}[i], \tag{2.6}$$

$$\hat{b}_1[i] = \text{sign}\left\{ \Re\left(A_1^* z_1[i] \right) \right\}. \tag{2.7}$$

Note that the linear equalizers and multiuser detectors discussed in Chapter 1 can all be written in this form, as will be seen below. In case the complex amplitude A_1 of the desired user is unknown, we can resort to differential detection. Define the differential bit as

$$\beta_1[i] \triangleq b_1[i]\, b_1[i-1]. \tag{2.8}$$

Then using the linear detector output[1] (2.6), the following differential detection rule can be used:

$$\hat{\beta}_1[i] = \text{sign}\left\{\Re\left(z_1[i] z_1[i-1]^*\right)\right\}. \tag{2.9}$$

Substituting (2.4) into (2.6), the output of the linear receiver $\boldsymbol{w}_1$ can be written as

$$z_1[i] = A_1\left(\boldsymbol{w}_1^H \boldsymbol{s}_1\right) b_1[i] + \sum_{k=2}^{K} A_k \left(\boldsymbol{w}_1^H \boldsymbol{s}_k\right) b_k[i] + \boldsymbol{w}_1^H \boldsymbol{n}[i]. \tag{2.10}$$

In (2.10), the first term on the right-hand side contains the useful signal of the desired user, the second term contains the signals from other undesired users—the *multiple-access interference* (MAI), and the last term contains the ambient Gaussian noise. The simplest linear receiver is the conventional matched filter, where $\boldsymbol{w}_1 = \boldsymbol{s}_1$. As noted in Chapter 1, such a matched-filter receiver is optimal only in a single-user channel (i.e., $K = 1$). In a multiuser channel (i.e., $K > 1$), this receiver may perform poorly since it makes no attempt to ameliorate the MAI, a limiting source of interference in multiple-access channels. Two popular forms of linear detectors that are capable of suppressing the MAI are the linear decorrelating detector and the linear minimum mean-square-error (MMSE) detector, which are discussed next.

2.2.2 Linear Decorrelating Detector

A linear decorrelating detector for user 1, $\boldsymbol{w}_1 \triangleq \boldsymbol{d}_1 \in \mathbb{C}^N$, is such that when correlated with the received signal $\boldsymbol{r}[i]$, it results in zero MAI [i.e., the second term in (2.10) is zero]. In particular, the linear decorrelating detector $\boldsymbol{d}_1$ for user 1 satisfies

$$\boldsymbol{d}_1^H \boldsymbol{s}_1 = 1, \tag{2.11}$$

$$\boldsymbol{d}_1^H \boldsymbol{s}_k = 0, \qquad k = 2, \ldots, K. \tag{2.12}$$

Denote by $\boldsymbol{e}_k$ a K-vector with all entries zeros except for the kth entry, which is 1. Assume that the user signature sequences are linearly independent [i.e., the matrix $\boldsymbol{S} \triangleq [\boldsymbol{s}_1 \ \cdots \ \boldsymbol{s}_K]$ has full column rank, $\mathsf{rank}(\boldsymbol{S}) = K$]. Let $\boldsymbol{R} \triangleq \boldsymbol{S}^H \boldsymbol{S}$ be the correlation matrix of the user signature sequences. Then $\boldsymbol{R}$ is invertible. The following result gives the expression for the linear decorrelating detector.

[1] For simplicity, we will use the term "detector" to refer to the overall detector (2.6)–(2.7) or (2.6) and (2.9), to the detection statistic (2.6), and to the detector's weight vector.

Proposition 2.1: *The linear decorrelating detector for user 1 is given by*

$$\boldsymbol{d}_1 = \boldsymbol{S}\boldsymbol{R}^{-1}\boldsymbol{e}_1. \tag{2.13}$$

Proof: It is easily verified that

$$\boldsymbol{d}_1^H \boldsymbol{s}_k = \boldsymbol{e}_1^H \boldsymbol{R}^{-1} \underbrace{\boldsymbol{S}^H \boldsymbol{S}}_{\boldsymbol{R}} \boldsymbol{e}_k = \boldsymbol{e}_1^H \boldsymbol{I}_K \boldsymbol{e}_1 = [\boldsymbol{I}_K]_{1,k} = \begin{cases} 1, & k = 1, \\ 0, & k \neq 1. \end{cases} \tag{2.14}$$

Therefore, (2.11) and (2.12) hold. □

The output of the linear decorrelating detector is given by

$$z_1[i] \triangleq \boldsymbol{d}_1^H \boldsymbol{r}[i] \;=\; A_1 b_1[i] + v_1[i], \tag{2.15}$$

with

$$v_1[i] \triangleq \boldsymbol{d}_1^H \boldsymbol{n}[i] \;\sim\; \mathcal{N}_c\left(0,\, \sigma^2 \|\boldsymbol{d}_1\|^2\right), \tag{2.16}$$

where, by (2.13),

$$\|\boldsymbol{d}_1\|^2 = \boldsymbol{e}_1^H \boldsymbol{R}^{-1} \underbrace{\boldsymbol{S}^H \boldsymbol{S}}_{\boldsymbol{R}} \boldsymbol{R}^{-1} \boldsymbol{e}_1 \;=\; \boldsymbol{e}_1^H \boldsymbol{R}^{-1} \boldsymbol{e}_1 \;=\; \left[\boldsymbol{R}^{-1}\right]_{1,1}, \tag{2.17}$$

and where in (2.17), $[\boldsymbol{A}]_{i,j}$ denotes the (i, j)th element of the matrix $\boldsymbol{A}$. Note that by the Cauchy–Schwartz inequality, we have

$$\|\boldsymbol{d}_1\|^2 \cdot \|\boldsymbol{s}_1\|^2 \geq \|\boldsymbol{d}_1^H \boldsymbol{s}_1\|^2. \tag{2.18}$$

Since $\|\boldsymbol{s}_1\| = 1$ and $\boldsymbol{d}_1^H \boldsymbol{s}_1 = 1$, it then follows that $\|\boldsymbol{d}_1\| \geq 1$. Hence, by (2.16), we have $\mathrm{Var}\{v_1[i]\} \geq \sigma^2$ (i.e., the linear decorrelating detector enhances the output noise level).

2.2.3 Linear MMSE Detector

While the linear decorrelating detector is designed to eliminate the MAI completely at the expense of enhancing the ambient noise, the linear MMSE detector, $\boldsymbol{w}_1 \triangleq \boldsymbol{m}_1 \in \mathbb{C}^N$, is designed to minimize the total effect of the MAI and the ambient noise at the detector output. Specifically, the linear MMSE detector for user 1 is given by the solution to the following optimization problem:

$$\boldsymbol{m}_1 = \arg \min_{\boldsymbol{w} \in \mathbb{C}^N} E\left\{\left\|A_1 b_1[i] - \boldsymbol{w}^H \boldsymbol{r}[i]\right\|^2\right\}. \tag{2.19}$$

Denote $|\boldsymbol{A}| \triangleq \mathsf{diag}(|A_1|, \ldots, |A_K|)$. The following result gives the expression for the linear MMSE detector.

Proposition 2.2: *The linear MMSE detector for user 1 is given by*

$$\boldsymbol{m}_1 = \boldsymbol{S}\left(\boldsymbol{R} + \sigma^2 |\boldsymbol{A}|^{-2}\right)^{-1} \boldsymbol{e}_1. \tag{2.20}$$

Proof: First note that any linear detector must lie in the column space of $\boldsymbol{S}$ [i.e., $\boldsymbol{m}_1 \in \mathsf{range}(\boldsymbol{S})$]. This is because any component outside this space does not affect the signal components of the detector output [i.e., the first and second terms of (2.10)], and it merely increases the noise level [i.e., the third term of (2.10)]. Therefore, we can write $\boldsymbol{m}_1 = \boldsymbol{S}\boldsymbol{x}_1$ for some $\boldsymbol{x}_1 \in \mathbb{C}^K$, where

$$\begin{aligned}
\boldsymbol{x}_1 &= \arg\min_{\boldsymbol{x}\in\mathbb{C}^K} E\left\{\left\|A_1 b_1[i] - \boldsymbol{x}^H\boldsymbol{S}^H\boldsymbol{r}[i]\right\|^2\right\} \\
&= \arg\min_{\boldsymbol{x}\in\mathbb{C}^K} \boldsymbol{x}^H\left[\boldsymbol{S}^H E\left\{\boldsymbol{r}[i]\boldsymbol{r}[i]^H\right\}\boldsymbol{S}\right]\boldsymbol{x} - 2\boldsymbol{x}^H\boldsymbol{S}^H\Re\left\{A_1^* E\left(b_1[i]\boldsymbol{r}[i]\right)\right\} \\
&= \arg\min_{\boldsymbol{x}\in\mathbb{C}^K} \boldsymbol{x}^H \underbrace{\left[\boldsymbol{S}^H\left(\boldsymbol{S}|\boldsymbol{A}|^2\boldsymbol{S}^H + \sigma^2\boldsymbol{I}_N\right)\boldsymbol{S}\right]}_{\boldsymbol{R}|\boldsymbol{A}|^2\boldsymbol{R}+\sigma^2\boldsymbol{R}}\boldsymbol{x} - 2\boldsymbol{x}^H\boldsymbol{R}|\boldsymbol{A}|^2\boldsymbol{e}_1 \\
&= \left(\boldsymbol{R} + \sigma^2|\boldsymbol{A}|^{-2}\right)^{-1}\boldsymbol{e}_1. \qquad (2.21)
\end{aligned}$$

Hence (2.20) is obtained. □

The output of the linear MMSE detector is given by

$$z_1[i] \triangleq \boldsymbol{m}_1^H\boldsymbol{r}[i] = A_1\left(\boldsymbol{m}_1^H\boldsymbol{s}_1\right)b_1[i] + \sum_{k=2}^{K} A_k\left(\boldsymbol{m}_1^H\boldsymbol{s}_k\right)b_k[i] + v_1[i], \qquad (2.22)$$

with

$$v_1[i] \triangleq \boldsymbol{m}_1^H\boldsymbol{n}[i] \sim \mathcal{N}_c\left(0, \sigma^2\|\boldsymbol{m}_1\|^2\right), \qquad (2.23)$$

where, using (2.20), we have

$$\boldsymbol{m}_1^H\boldsymbol{s}_k = \left[\left(\boldsymbol{R} + \sigma^2|\boldsymbol{A}|^{-2}\right)^{-1}\boldsymbol{R}\right]_{1,k}, \qquad (2.24)$$

$$\|\boldsymbol{m}_1\|^2 = \left[\left(\boldsymbol{R} + \sigma^2|\boldsymbol{A}|^{-2}\right)^{-1}\boldsymbol{R}\left(\boldsymbol{R} + \sigma^2|\boldsymbol{A}|^{-2}\right)^{-1}\right]_{1,1}. \qquad (2.25)$$

Note that unlike the decorrelator output (2.15), the linear MMSE detector output (2.22) contains some residual MAI. However, we will in general have $\|\boldsymbol{m}_1\| < \|\boldsymbol{d}_1\|$, so that the effects of ambient noise are reduced by the linear MMSE detector.

2.3 Blind Multiuser Detection: Direct Methods

It is seen from (2.13) and (2.20) that these two linear detectors are expressed in terms of a linear combination of the signature sequences of all K users. Recall that for the matched-filter receiver, the only prior knowledge required is the desired user's signature sequence. In the downlink of a CDMA system, the mobile receiver typically has knowledge of only its own signature sequence, but not of those of the other users. Hence it is of interest to consider the problem of *blind* implementation of the linear detectors (i.e., without the requirement of knowing the signature sequences of the interfering users). This problem is relatively easy for the linear

MMSE detector. To see this, consider again the definition (2.19). Directly solving this optimization problem, we obtain

$$\boldsymbol{m}_1 = \arg \min_{\boldsymbol{w} \in \mathbb{C}^N} \Big[\boldsymbol{w}^H \underbrace{E\left\{\boldsymbol{r}[i]\boldsymbol{r}[i]^H\right\}}_{\boldsymbol{C}_r} \boldsymbol{w} - 2\boldsymbol{w}^H \Re\{A_1^* \underbrace{E(\boldsymbol{r}[i]b_1[i])}_{A_1 \boldsymbol{s}_1}\}\Big]$$
$$= |A_1|^2 \, \boldsymbol{C}_r^{-1} \boldsymbol{s}_1, \tag{2.26}$$

where, by (2.5),

$$\boldsymbol{C}_r \triangleq E\left\{\boldsymbol{r}[i]\boldsymbol{r}[i]^H\right\} \;=\; \boldsymbol{S}|\boldsymbol{A}|^2\boldsymbol{S}^H + \sigma^2 \boldsymbol{I}_N \tag{2.27}$$

is the autocorrelation matrix of the received signal. Note that $\boldsymbol{C}_r$ can be estimated from the received signals by the corresponding sample autocorrelation. Note also that the constant $|A_1|^2$ in (2.26) does not affect the linear decision rule (2.7) or (2.9). Hence (2.26) leads straightforwardly to the following blind implementation of the linear MMSE detector—the *direct matrix inversion* (DMI) *blind detector.* Here we do not assume knowledge of the complex amplitude of the desired user, so differential detection will be employed.

Algorithm 2.1: [DMI blind linear MMSE detector—synchronous CDMA]

- *Compute the detector:*

$$\hat{\boldsymbol{C}}_r = \frac{1}{M} \sum_{i=0}^{M-1} \boldsymbol{r}[i]\boldsymbol{r}[i]^H, \tag{2.28}$$

$$\hat{\boldsymbol{m}}_1 = \hat{\boldsymbol{C}}_r^{-1} \boldsymbol{s}_1. \tag{2.29}$$

- *Perform differential detection:*

$$z_1[i] = \hat{\boldsymbol{m}}_1^H \boldsymbol{r}[i], \tag{2.30}$$

$$\hat{\beta}_1[i] = \operatorname{sign}\left\{\Re\left(z_1[i] z_1[i-1]^*\right)\right\}, \qquad i = 1, \ldots, M-1. \tag{2.31}$$

Algorithm 2.1 is a *batch* processing method; it computes the detector only once based on a block of received signals $\{\boldsymbol{r}[i]\}_{i=0}^{M-1}$; and the estimated detector is then used to detect all data bits of the desired user contained in the same signal block, $\{b_1[i]\}_{i=0}^{M-1}$. In what follows we consider *online* implementations of the blind linear MMSE detector. The idea is to perform sequential detector estimation and data detection. That is, suppose that at time $(i-1)$ an estimated detector $\boldsymbol{m}_1[i-1]$ is employed to detect the data bit $b_1[i-1]$. At time i a new signal $\boldsymbol{r}[i]$ is received which is then used to update the detector estimate to obtain $\boldsymbol{m}_1[i]$. The updated detector is used to detect the data bit $b_1[i]$. Hence the blind detector is sequentially updated at the symbol rate. To develop such an adaptive algorithm, we need an alternative characterization of the linear MMSE detector. Consider the following constrained optimization problem:

$$\boldsymbol{m}_1 = \arg \min_{\boldsymbol{w} \in \mathbb{C}^N} E\left\{\left\|\boldsymbol{w}^H \boldsymbol{r}[i]\right\|^2\right\} \qquad \text{subject to } \; \boldsymbol{w}^H \boldsymbol{s}_1 = 1. \tag{2.32}$$

To solve (2.32), define the Lagrangian

$$\begin{aligned}\mathcal{L}(\boldsymbol{w}) &\triangleq E\left\{\left\|\boldsymbol{w}^H\boldsymbol{r}[i]\right\|^2\right\} - 2\lambda\left(\boldsymbol{w}^H\boldsymbol{s}_1 - 1\right)\\ &= \boldsymbol{w}^H\boldsymbol{C}_r\boldsymbol{w} - 2\lambda\boldsymbol{w}^H\boldsymbol{s}_1 + 2\lambda. \end{aligned} \tag{2.33}$$

The solution to (2.32) is then obtained by solving

$$\frac{\mathrm{d}}{\mathrm{d}\boldsymbol{w}}\mathcal{L}(\boldsymbol{w})|_{\boldsymbol{w}=\boldsymbol{m}_1} = \mathbf{0} \Longrightarrow \boldsymbol{m}_1 = \lambda\boldsymbol{C}_r^{-1}\boldsymbol{s}_1, \tag{2.34}$$

where λ is such that $\boldsymbol{m}_1^H\boldsymbol{s}_1 = 1$ [i.e., $\lambda = \left(\boldsymbol{s}_1^H\boldsymbol{C}_r^{-1}\boldsymbol{s}_1\right)^{-1}$]. Comparing the solution above with (2.26), it is seen that they differ only by a positive scaling constant. Since such a scaling constant will not affect the linear decision rule (2.7) or (2.9), (2.32) constitutes an equivalent definition of the linear MMSE detector. The approach to multiuser detection based on (2.32) was proposed in [183] and was termed *minimum-output-energy* (MOE) *detection*. A similar technique was developed for array processing [125,176,511], and in that context is termed the *linearly constrained minimum variance* (LCMV) *array*.

We next consider adaptive algorithms for recursively (online) estimating the linear MMSE detector defined by (2.32).

2.3.1 LMS Algorithm

We first consider the least-mean-squares (LMS) algorithm for recursive estimation of $\boldsymbol{m}_1$ based on (2.32). Define

$$\boldsymbol{P} \triangleq \boldsymbol{I}_N - \boldsymbol{s}_1\left(\boldsymbol{s}_1^H\boldsymbol{s}_1\right)\boldsymbol{s}_1^H = \boldsymbol{I}_N - \boldsymbol{s}_1\boldsymbol{s}_1^H \tag{2.35}$$

as a projection matrix that projects any signal in $\mathbb{C}^N$ onto the orthogonal space of $\boldsymbol{s}_1$. Note that $\boldsymbol{m}_1$ can be decomposed into two orthogonal components:

$$\boldsymbol{m}_1 = \boldsymbol{s}_1 + \boldsymbol{x}_1, \tag{2.36}$$

with

$$\boldsymbol{x}_1 \triangleq \boldsymbol{P}\boldsymbol{m}_1 = \boldsymbol{P}\boldsymbol{x}_1. \tag{2.37}$$

Using the decomposition above, the constrained optimization problem (2.32) can then be converted to the following unconstrained optimization problem:

$$\boldsymbol{x}_1 = \arg\min_{\boldsymbol{x}\in\mathbb{C}^N} E\left\{\left\|\left(\boldsymbol{s}_1 + \boldsymbol{P}\boldsymbol{x}\right)^H\boldsymbol{r}[i]\right\|^2\right\}. \tag{2.38}$$

The LMS algorithm for adapting the vector $\boldsymbol{x}_1$ based on the cost function (2.38) is then given by

$$\boldsymbol{x}_1[i+1] = \boldsymbol{x}_1[i] - \frac{\mu}{2}g\left(\boldsymbol{x}_1[i]\right), \tag{2.39}$$

where μ is the step size and where the stochastic gradient $g(\boldsymbol{x}_1[i])$ is given by

$$
\begin{aligned}
g(\boldsymbol{x}_1[i]) &\triangleq \frac{\mathrm{d}}{\mathrm{d}\boldsymbol{x}} \left\| (\boldsymbol{s}_1 + \boldsymbol{P}\boldsymbol{x})^H \boldsymbol{r}[i] \right\|^2 |_{\boldsymbol{x}=\boldsymbol{x}_1[i]} \\
&= 2\,\boldsymbol{P}\boldsymbol{r}[i] \left[(\boldsymbol{s}_1 + \boldsymbol{P}\boldsymbol{x}_1[i])^H \boldsymbol{r}[i] \right]^* \\
&= 2\,\left(\boldsymbol{I} - \boldsymbol{s}_1\boldsymbol{s}_1^H\right) \boldsymbol{r}[i] \left[(\boldsymbol{s}_1 + \boldsymbol{P}\boldsymbol{x}_1[i])^H \boldsymbol{r}[i] \right]^* \\
&= 2\left[\boldsymbol{r}[i] - \left(\boldsymbol{s}_1^H \boldsymbol{r}[i]\right)\boldsymbol{s}_1\right] \left[(\boldsymbol{s}_1 + \boldsymbol{P}\boldsymbol{x}_1[i])^H \boldsymbol{r}[i] \right]^* . \qquad (2.40)
\end{aligned}
$$

Substituting (2.40) into (2.39), we obtain the following LMS implementation of the blind linear MMSE detector. Suppose that at time i, the estimated blind detector is $\boldsymbol{m}_1[i] = \boldsymbol{s}_1 + \boldsymbol{x}_1[i]$. The algorithm performs the following steps for data detection and detector update.

Algorithm 2.2: [LMS blind linear MMSE detector—synchronous CDMA]

- *Compute the detector output:*

$$
z_1[i] = (\boldsymbol{s}_1 + \boldsymbol{P}\boldsymbol{x}_1[i])^H \boldsymbol{r}[i], \qquad (2.41)
$$

$$
\hat{\beta}_1[i] = \operatorname{sign}\left\{\Re\left(z_1[i] z_1[i-1]^*\right)\right\}. \qquad (2.42)
$$

- *Update the detector:*

$$
\boldsymbol{x}_1[i+1] = \boldsymbol{x}_1[i] - \mu\, z_1[i]^* \left[\boldsymbol{r}[i] - \left(\boldsymbol{s}_1^H \boldsymbol{r}[i]\right)\boldsymbol{s}_1\right]. \qquad (2.43)
$$

The convergence analysis of Algorithm 2.2 is given in [183]. An alternative stochastic gradient algorithm for blind adaptive multiuser detection is developed in [237], which employs the technique of averaging to achieve an accelerated convergence rate (compared with the LMS algorithm). An LMS algorithm for blind adaptive implementation of the linear decorrelating detector is developed in [501]. Moreover, a comparison of the steady-state performance (in terms of output mean-square error) shows that the blind detector incurs a loss compared with a training-based LMS detector [183, 389, 390]. A two-stage adaptive detector is proposed in [63], where symbol-by-symbol predecisions at the output of a first adaptive stage are used to train a second stage, to achieve improved performance.

2.3.2 RLS Algorithm

The LMS algorithm discussed above has a very low computational complexity, on the order of $\mathcal{O}(N)$ operations per update. However, its convergence is usually very slow. We next consider the recursive least-squares (RLS) algorithm for adaptive implementation of the blind linear MMSE detector, which has a much faster convergence rate than the LMS algorithm. Based on the cost function (2.32), at time

i the exponentially windowed RLS algorithm selects the weight vector $\boldsymbol{m}_1[i]$ to minimize the sum of exponentially weighted mean-square output values:

$$\boldsymbol{m}_1[i] = \arg \min_{\boldsymbol{w} \in \mathbb{C}^N} \sum_{n=0}^{i} \lambda^{i-n} \left\| \boldsymbol{w}^H \boldsymbol{r}[n] \right\|^2 \qquad \text{subject to } \boldsymbol{w}^H \boldsymbol{s}_1 = 1, \tag{2.44}$$

where $0 < \lambda < 1$ $(1 - \lambda \ll 1)$ is called the *forgetting factor*. The solution to this optimization problem is given by

$$\boldsymbol{m}_1[i] = \boldsymbol{C}_r[i]^{-1} \boldsymbol{s}_1 \left(\boldsymbol{s}_1^H \boldsymbol{C}_r[i]^{-1} \boldsymbol{s}_1 \right)^{-1}, \tag{2.45}$$

with

$$\boldsymbol{C}_r[i] \triangleq \sum_{n=0}^{i} \lambda^{i-n} \boldsymbol{r}[i] \boldsymbol{r}[i]^H. \tag{2.46}$$

Denote $\boldsymbol{\Phi}[i] \triangleq \boldsymbol{C}_r[i]^{-1}$. Note that since

$$\boldsymbol{C}_r[i] = \lambda \boldsymbol{C}_r[i-1] + \boldsymbol{r}[i] \boldsymbol{r}[i]^H, \tag{2.47}$$

by the matrix inversion lemma we have

$$\boldsymbol{\Phi}[i] = \lambda^{-1} \boldsymbol{\Phi}[i-1] - \frac{\lambda^{-2} \boldsymbol{\Phi}[i-1] \boldsymbol{r}[i] \boldsymbol{r}[i]^H \boldsymbol{\Phi}[i-1]}{1 + \lambda^{-1} \boldsymbol{r}[i]^H \boldsymbol{\Phi}[i-1] \boldsymbol{r}[i]}. \tag{2.48}$$

Hence we obtain the RLS algorithm for adaptive implementation of the blind linear MMSE detector as follows. Suppose that at time $(i-1)$, $\boldsymbol{\Phi}[i-1]$ is available. Then at time i, the following steps are performed to update the detector $\boldsymbol{m}_1[i]$ and to detect the differential bit $\beta_1[i]$.

Algorithm 2.3: [RLS blind linear MMSE detector—synchronous CDMA]

- *Update the detector:*

$$\boldsymbol{k}[i] \triangleq \frac{\lambda^{-1} \boldsymbol{\Phi}[i-1] \boldsymbol{r}[i]}{1 + \lambda^{-1} \boldsymbol{r}[i]^H \boldsymbol{\Phi}[i-1] \boldsymbol{r}[i]}, \tag{2.49}$$

$$\boldsymbol{\Phi}[i] = \lambda^{-1} \left(\boldsymbol{\Phi}[i-1] - \boldsymbol{k}[i] \boldsymbol{r}[i]^H \boldsymbol{\Phi}[i-1] \right), \tag{2.50}$$

$$\boldsymbol{m}_1[i] = \boldsymbol{\Phi}[i] \boldsymbol{s}_1. \tag{2.51}$$

- *Compute the detector output:*

$$z_1[i] = \boldsymbol{m}_1[i]^H \boldsymbol{r}[i], \tag{2.52}$$

$$\hat{\beta}_1[i] = \text{sign} \left\{ \Re \left(z_1[i] z_1[i-1]^* \right) \right\}. \tag{2.53}$$

The convergence properties of Algorithm 2.3 are analyzed in detail in [389].

2.3.3 QR-RLS Algorithm

The RLS approach discussed in Section 2.3.2, which is based on the matrix inversion lemma for recursively updating $\boldsymbol{C}_r[i]^{-1}$, has $\mathcal{O}(N^2)$ complexity per update. Note that although fast RLS algorithms of $\mathcal{O}(N)$ complexity exist [66, 83, 116, 124], all these algorithms exploit a time-index-shifting property of the input data. In this particular application, however, successive input data vectors do not have the shifting relationship; in fact, $\boldsymbol{r}[i]$ and $\boldsymbol{r}[i-1]$ do not overlap at all. Therefore, these standard fast RLS algorithms cannot be applied in this application.

The RLS implementation of the blind linear MMSE detector suffers from two major problems. The first problem is numerical. Recursive estimation of $\boldsymbol{C}_r[i]^{-1}$ is poorly conditioned because it involves inversion of a data correlation matrix. The condition number of a data correlation matrix is the square of the condition number of the corresponding data matrix; hence twice the dynamic range is required in the numerical computation [158]. The second problem is that the form of the recursive update of $\boldsymbol{C}_r[i]^{-1}$ severely limits the parallelism and pipelining that can effectively be applied in implementation.

A well-known approach for overcoming these difficulties associated with the RLS algorithms is the rotation-based QR-RLS algorithm [175, 389, 588]. The QR decomposition transforms the original RLS problem into a problem that uses only transformed data values, by Cholesky factorization of the original least-squares data matrix. This causes the numerical dynamic range of the transformed computational problem to be halved and enables more accurate computation than that with the RLS algorithms that operate directly on $\boldsymbol{C}_r[i]^{-1}$. Another important benefit of the rotation-based QR approaches is that the computation can easily be mapped onto systolic array structures for parallel implementations. We next describe the QR-RLS blind linear MMSE detector, first developed in [389].

QR-RLS Blind Linear MMSE Detector

Assume that $\boldsymbol{C}_r[i]$ is positive definite. Let

$$\boldsymbol{C}_r[i] = \underline{C}[i]^H \underline{C}[i] \tag{2.54}$$

be the Cholesky decomposition (i.e., $\underline{C}[i]$ is the unique upper triangular Cholesky factor with positive diagonal elements). Define the following quantities:

$$\underline{u}[i] \triangleq \underline{C}[i]^{-H} \boldsymbol{s}_1, \tag{2.55}$$

$$\underline{v}[i] \triangleq \underline{C}[i]^{-H} \boldsymbol{r}[i], \tag{2.56}$$

$$\alpha[i] \triangleq \boldsymbol{s}_1^H \boldsymbol{C}_r[i]^{-1} \boldsymbol{s}_1 = \underline{u}[i]^H \underline{u}[i]. \tag{2.57}$$

At time i, the *a posteriori* least-squares (LS) estimate is given by

$$z[i] \triangleq \boldsymbol{m}_1[i]^H \boldsymbol{r}[i] = \frac{\boldsymbol{s}_1^H \boldsymbol{C}_r[i]^{-1} \boldsymbol{r}[i]}{\boldsymbol{s}_1^H \boldsymbol{C}_r[i]^{-1} \boldsymbol{s}_1} \tag{2.58}$$

$$= \underline{u}[i]^H \underline{v}[i] / \alpha[i]. \tag{2.59}$$

The *a priori* LS estimate at time i is given by

$$\xi[i] \triangleq \boldsymbol{m}_1[i-1]^H \boldsymbol{r}[i]. \tag{2.60}$$

It can be shown that $\xi[i]$ and $z[i]$ are related by [389]

$$\xi[i] = \frac{z[i]}{1 - \|\underline{v}[i]\|^2 + \alpha[i]|z[i]|^2}. \tag{2.61}$$

Suppose that $\underline{C}[i-1]$ and $\underline{u}[i-1]$ are available from the previous recursion. At time i the new observation $\boldsymbol{r}[i]$ becomes available. We construct a block matrix consisting of $\underline{C}[i-1]$, $\underline{u}[i-1]$, and $\boldsymbol{r}[i]$ and apply an orthogonal transformation as follows:

$$\underline{Q}[i] \begin{bmatrix} \sqrt{\lambda}\underline{C}[i-1] & \underline{u}[i-1]/\sqrt{\lambda} & \mathbf{0} \\ \boldsymbol{r}[i]^H & 0 & 1 \end{bmatrix} = \begin{bmatrix} \underline{C}[i] & \underline{u}[i] & \underline{v}[i] \\ \mathbf{0}^H & \eta[i] & \gamma[i] \end{bmatrix}. \tag{2.62}$$

In (2.62) the matrix $\underline{Q}[i]$, which zeros the first N elements on the last row of the partitioned matrix appearing on the left-hand side of (2.62), is an orthogonal matrix consisting of N Givens rotations,

$$\underline{Q}[i] = \underline{Q}_N[i] \cdots \underline{Q}_2[i]\underline{Q}_1[i], \tag{2.63}$$

where $\underline{Q}_n[i]$ zeros the nth element in the last row by rotating it with the $(n+1)$th row. An individual rotation is specified by two scalars, c_n and s_n (which can be regarded as the cosine and sine, respectively, of a rotation angle ϕ_n), and affects only the last row and the $(n+1)$th row. The effects on these two rows are

$$\begin{bmatrix} c_n & s_n \\ -s_n^* & c_n \end{bmatrix} \begin{bmatrix} 0 & \cdots & 0 & y_n & y_{n+1} & \cdots \\ 0 & \cdots & 0 & r_n & r_{n+1} & \cdots \end{bmatrix}$$
$$= \begin{bmatrix} 0 & \cdots & 0 & y_n' & y_{n+1}' & \cdots \\ 0 & \cdots & 0 & 0 & r_{n+1}' & \cdots \end{bmatrix}, \begin{matrix} \longleftarrow & (n+1)\text{th row} \\ \longleftarrow & \text{last row} \end{matrix} \tag{2.64}$$

where the rotation factors are defined by

$$c_n = \frac{y_n^*}{\sqrt{|y_n|^2 + |r_n|^2}}, \tag{2.65}$$

$$s_n = \frac{r_n^*}{\sqrt{|y_n|^2 + |r_n|^2}}. \tag{2.66}$$

The correctness of (2.62) is shown in the Appendix (Section 2.8.1). It is seen from (2.62) that the computed quantities appearing on the right-hand side are $\underline{C}[i]$, $\underline{u}[i]$, and $\underline{v}[i]$ at time n. It is also shown in the Appendix (Section 2.8.1) that the quantities $\alpha[i]$, $z[i]$, and $\xi[i]$ can be updated according to the equations

$$\alpha[i] = \alpha[i-1]/\lambda - |\eta[i]|^2, \tag{2.67}$$

$$z[i] = -\eta[i]^* \gamma[i]/\alpha[i], \tag{2.68}$$

$$\xi[i] = \frac{z[i]}{|\gamma[i]|^2 + \alpha[i]|z[i]|^2}. \tag{2.69}$$

Note that $\gamma[i]$ in (2.62) is the last diagonal element of $\underline{Q}[i]$. A direct calculation shows that $\gamma[i] = \prod_{n=1}^{N} c_n$ [175, 316].

The initialization of the QR-RLS blind adaptive algorithm is given by $\underline{C}[-1] = \sqrt{\delta}\, \boldsymbol{I}_N$, $\underline{u}(0) = \boldsymbol{s}_1/\sqrt{\delta}$, and $\alpha[-1] = \delta$, where δ is a small number. This corresponds to the initial condition $\boldsymbol{C}_r[-1] = \delta \boldsymbol{I}_N$ and $\boldsymbol{m}_1[-1] = \boldsymbol{s}_1$ (i.e., the adaptation starts with the matched filter). At each time i, the algorithm proceeds as follows.

Algorithm 2.4: [QR-RLS blind linear MMSE detector—synchronous CDMA]

- *Update the detector: Apply the orthogonal transformation (2.62).*
- *Compute the detector output and perform differential detection:*

$$z_1[i] = \eta[i]^* \gamma[i], \tag{2.70}$$

$$\hat{\beta}_1[i] = \text{sign}\left\{\Re(z_1[i] z_1[i-1]^*)\right\}. \tag{2.71}$$

The orthogonal transformation (2.62) on the block matrix can be mapped onto a triangular systolic array for highly efficient parallel implementation, which is discussed next.

Parallel Implementation on Systolic Arrays

The QR-RLS blind adaptive algorithm derived above has good numerical properties and is well suited for parallel implementation. Figure 2.1 shows systematically a systolic array implementation of this algorithm, using a triangular array first proposed in [316]. It consists of three sections: the basic upper triangular array, which stores and updates $\underline{C}[i]$; the right-hand column of cells, which stores and updates $\underline{u}[i]$; and the final processing cell, which computes the demodulated data bit. The system is initialized as $\underline{C}[-1] = \sqrt{\delta}\, \boldsymbol{I}_N$ and $\underline{u}[-1] = \boldsymbol{s}_1/\sqrt{\delta}$. The received data $\boldsymbol{r}[i]$ are fed from the top and propagate to the bottom of the array. The rotation angles ϕ_n are calculated in left boundary cells and propagate from left to right. The internal cells update their elements by Givens rotations using the angles received from the left. The factor $\gamma[i]$ is calculated along the left boundary cells, where a dot ($\bullet$) represents an extra delay. The final cell extracts the signs of $\eta[i]$ and $\gamma[i]$ and produces the demodulated differential data bit according to (2.71). The computation at each cell is also outlined in Fig. 2.1. The QR-RLS algorithm may also be carried out using the square-root free Givens rotation algorithm to reduce the computational complexity at each cell [158, 316]. For more details on the systolic array implementations, see [175, 316].

The systolic array in Fig. 2.1 operates in a highly pipelined manner. The computational wavefront propagates at the received data symbol rate. The demodulated data bits are also output at the received data symbol rate. Note that the demodulated data bit produced on a given clock corresponds to the received vector entered $2N$ clock cycles earlier.

If multiple synchronous user data streams need to be demodulated, we can simply add more column arrays on the right-hand side and initialize each of them

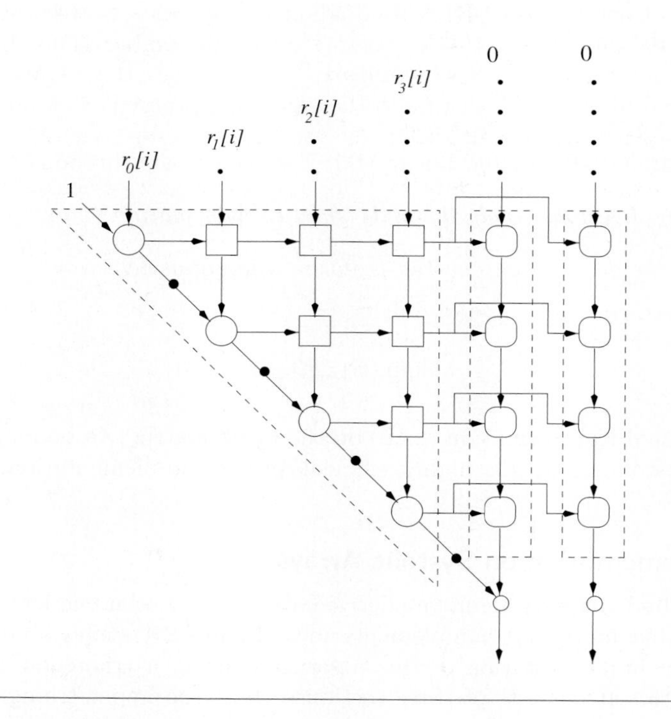

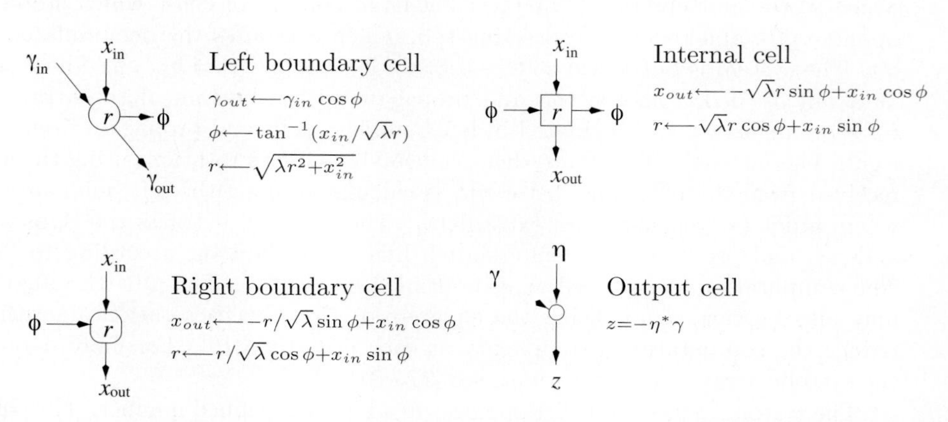

Figure 2.1. Systematic of the systolic array implementation of the QR-RLS blind adaptive multiuser detection algorithm ($N = 4$, $K = 2$) and the operations at each cell.

by the corresponding signature vector of each user. It is clear that by using the same triangular array, multiple users' data can be demodulated simultaneously. This is also illustrated in Fig. 2.1 for the case of two users. (Also, multiple paths of the same signal can be handled by adding appropriate linear arrays to Fig. 2.1.)

2.4 Blind Multiuser Detection: Subspace Methods

In this section we discuss another approach to blind multiuser detection, which was first developed in [549] and is based on estimating the signal subspace spanned by the user signature waveforms. This approach leads to blind implementation of both the linear decorrelating detector and the linear MMSE detector. It also offers a number of advantages over the direct methods discussed in Section 2.3.

Assume that the spreading waveforms $\{\boldsymbol{s}_k\}_{k=1}^{K}$ of K users are linearly independent. Note that $\boldsymbol{C}_r$ of (2.27) is the sum of the rank-K matrix $\boldsymbol{S}|\boldsymbol{A}|^2\boldsymbol{S}^H$ and the identity matrix $\sigma^2\boldsymbol{I}_N$. This matrix then has K eigenvalues that are strictly larger than σ^2 and $(N-K)$ eigenvalues that equal σ^2. Its eigendecomposition can be written as

$$\boldsymbol{C}_r = \boldsymbol{U}_s\boldsymbol{\Lambda}_s\boldsymbol{U}_s^H + \sigma^2\boldsymbol{U}_n\boldsymbol{U}_n^H, \tag{2.72}$$

where $\boldsymbol{\Lambda}_s = \mathsf{diag}(\lambda_1, \ldots, \lambda_K)$ contains the largest K eigenvalues of $\boldsymbol{C}_r$, $\boldsymbol{U}_s = [\boldsymbol{u}_1, \ldots, \boldsymbol{u}_K]$ contains the K orthogonal eigenvectors corresponding to the largest K eigenvalues in $\boldsymbol{\Lambda}_s$; and $\boldsymbol{U}_n = [\boldsymbol{u}_{K+1}, \ldots, \boldsymbol{u}_N]$ contains the $(N-K)$ orthogonal eigenvectors corresponding to the smallest eigenvalue σ^2 of $\boldsymbol{C}_r$. It is easy to see that $\mathsf{range}\,(\boldsymbol{S}) = \mathsf{range}\,(\boldsymbol{U}_s)$. The column space of $\boldsymbol{U}_s$ is called the *signal subspace* and its orthogonal complement, the *noise subspace*, is spanned by the columns of $\boldsymbol{U}_n$. We next derive expressions for the linear decorrelating detector and the linear MMSE detector in terms of the signal subspace parameters $\boldsymbol{U}_s$, $\boldsymbol{\Lambda}_s$, and σ^2.

2.4.1 Linear Decorrelating Detector

The linear decorrelating detector given by (2.13) is characterized by the following results.

Lemma 2.1: *The linear decorrelating detector* $\boldsymbol{d}_1$ *in (2.13) is the unique weight vector* $\boldsymbol{w} \in \mathsf{range}\,(\boldsymbol{U}_s)$, *such that* $\boldsymbol{w}^H\boldsymbol{s}_1 = 1$, *and* $\boldsymbol{w}^H\boldsymbol{s}_k = 0$ *for* $k = 2, \ldots, K$.

Proof: Since $\mathsf{rank}\,(\boldsymbol{U}_s) = K$, the vector $\boldsymbol{w}$ that satisfies the foregoing conditions exists and is unique. Moreover, these conditions have been verified in the proof of Proposition 1 in Section 2.2.2. □

Lemma 2.2: *The decorrelating detector* $\boldsymbol{d}_1$ *in (2.13) is the unique weight vector* $\boldsymbol{w} \in \mathsf{range}\,(\boldsymbol{U}_s)$ *that minimizes* $\varphi(\boldsymbol{w}) \triangleq E\left\{\left\|\boldsymbol{w}^H\left(\boldsymbol{S}\boldsymbol{A}\boldsymbol{b}\right)\right\|^2\right\}$ *subject to* $\boldsymbol{w}^H\boldsymbol{s}_1 = 1$.

Proof: Since

$$\begin{aligned}\varphi(\boldsymbol{w}) &= \boldsymbol{w}^H E\left\{(\boldsymbol{S}\boldsymbol{A}\boldsymbol{b})(\boldsymbol{S}\boldsymbol{A}\boldsymbol{b})^H\right\}\boldsymbol{w} \\ &= \boldsymbol{w}^H\left(\boldsymbol{S}|\boldsymbol{A}|^2\boldsymbol{S}^H\right)\boldsymbol{w} \\ &= |A_1|^2\left|\boldsymbol{w}^H\boldsymbol{s}_1\right|^2 + \sum_{k=2}^{K}|A_k|^2\left|\boldsymbol{w}^H\boldsymbol{s}_k\right|^2 \\ &= |A_1|^2 + \sum_{k=2}^{K}|A_k|^2\left|\boldsymbol{w}^H\boldsymbol{s}_k\right|^2, \end{aligned} \tag{2.73}$$

it then follows that for $\boldsymbol{w} \in \mathsf{range}(\boldsymbol{U}_s) = \mathsf{range}(\boldsymbol{S})$, $\varphi(\boldsymbol{w})$ is minimized if and only if $\boldsymbol{w}^H\boldsymbol{s}_k = 0$ for $k = 2, \dots, K$. By Lemma 2.1 the unique solution is $\boldsymbol{w} = \boldsymbol{d}_1$. □

Proposition 2.3: *The linear decorrelating detector $\boldsymbol{d}_1$ in (2.13) is given in terms of the signal subspace parameters by*

$$\boldsymbol{d}_1 = \alpha_d\boldsymbol{U}_s\left(\boldsymbol{\Lambda}_s - \sigma^2\boldsymbol{I}_K\right)^{-1}\boldsymbol{U}_s^H\boldsymbol{s}_1, \tag{2.74}$$

with

$$\alpha_d \triangleq \left[\boldsymbol{s}_1^H\boldsymbol{U}_s\left(\boldsymbol{\Lambda}_s - \sigma^2\boldsymbol{I}_K\right)^{-1}\boldsymbol{U}_s^H\boldsymbol{s}_1\right]^{-1}. \tag{2.75}$$

Proof: A vector $\boldsymbol{w} \in \mathsf{range}(\boldsymbol{U}_s)$ if and only if it can be written as $\boldsymbol{w} = \boldsymbol{U}_s\boldsymbol{x}$, for some $\boldsymbol{x} \in \mathbb{C}^K$. Then by Lemma 2.2 the linear decorrelating detector $\boldsymbol{d}_1$ has the form $\boldsymbol{d}_1 = \boldsymbol{U}_s\boldsymbol{x}_1$, where

$$\begin{aligned}\boldsymbol{x}_1 &= \arg\min_{\boldsymbol{x}\in\mathbb{C}^K}(\boldsymbol{U}_s\,\boldsymbol{x})^H\left(\boldsymbol{S}|\boldsymbol{A}|^2\boldsymbol{S}^H\right)(\boldsymbol{U}_s\boldsymbol{x}) \quad \text{s.t.} \quad (\boldsymbol{U}_s\boldsymbol{x})^H\boldsymbol{s}_1 = 1, \\ &= \arg\min_{\boldsymbol{x}\in\mathbb{C}^K}\boldsymbol{x}^H\left[\boldsymbol{U}_s^H\left(\boldsymbol{S}|\boldsymbol{A}|^2\boldsymbol{S}^H\right)\boldsymbol{U}_s\right]\boldsymbol{x} \quad \text{s.t.} \quad \boldsymbol{x}^H\left(\boldsymbol{U}_s^H\boldsymbol{s}_1\right) = 1, \\ &= \arg\min_{\boldsymbol{x}\in\mathbb{C}^K}\boldsymbol{x}^H\left(\boldsymbol{\Lambda}_s - \sigma^2\boldsymbol{I}_K\right)\boldsymbol{x} \quad \text{s.t.} \quad \boldsymbol{x}^H\left(\boldsymbol{U}_s^H\boldsymbol{s}_1\right) = 1, \end{aligned} \tag{2.76}$$

where the third equality follows from the fact that

$$\boldsymbol{S}|\boldsymbol{A}|^2\boldsymbol{S}^H = \boldsymbol{U}_s\left(\boldsymbol{\Lambda}_s - \sigma^2\boldsymbol{I}_K\right)\boldsymbol{U}_s^H,$$

which in turn follows directly from (2.27) and (2.72). The optimization problem (2.76) can be solved by the method of Lagrange multipliers. Let

$$\mathcal{L}(\boldsymbol{x}) \triangleq \boldsymbol{x}^H\left(\boldsymbol{\Lambda}_s - \sigma^2\boldsymbol{I}_K\right)\boldsymbol{x} - 2\alpha_d\left[\boldsymbol{x}^H\left(\boldsymbol{U}_s^H\boldsymbol{s}_1\right) - 1\right].$$

Since the matrix $\boldsymbol{\Lambda}_s - \sigma^2\boldsymbol{I}_K$ is positive definite, $\mathcal{L}(\boldsymbol{x})$ is a strictly convex function of $\boldsymbol{x}$. Therefore, the unique global minimum of $\mathcal{L}(\boldsymbol{x})$ is achieved at $\boldsymbol{x}_1$, where $\nabla\mathcal{L}(\boldsymbol{x}_1) = \boldsymbol{0}$, or

$$\left(\boldsymbol{\Lambda}_s - \sigma^2\boldsymbol{I}_K\right)\boldsymbol{x}_1 = \alpha_d\boldsymbol{U}_s^H\boldsymbol{s}_1. \tag{2.77}$$

Therefore, $\boldsymbol{x}_1 = \alpha_d \left(\boldsymbol{\Lambda}_s - \sigma^2 \boldsymbol{I}_K\right)^{-1} \boldsymbol{U}_s^H \boldsymbol{s}_1$, where α_d is determined from the constraint $\left(\boldsymbol{U}_s \boldsymbol{x}_1\right)^H \boldsymbol{s}_1 = 1$; that is, $\alpha_d = \left[\boldsymbol{s}_1^H \boldsymbol{U}_s \left(\boldsymbol{\Lambda}_s - \sigma^2 \boldsymbol{I}_K\right)^{-1} \boldsymbol{U}_s^H \boldsymbol{s}_1\right]^{-1}$. Finally, the weight vector of the linear decorrelating detector is given by $\boldsymbol{d}_1 = \boldsymbol{U}_s \boldsymbol{x}_1 = \alpha_d \boldsymbol{U}_s \left(\boldsymbol{\Lambda}_s - \sigma^2 \boldsymbol{I}_K\right)^{-1} \boldsymbol{U}_s^H \boldsymbol{s}_1$. □

2.4.2 Linear MMSE Detector

The following result gives the subspace form of the linear MMSE detector, defined by (2.32).

Proposition 2.4: *The weight vector $\boldsymbol{m}_1$ of the linear MMSE detector defined by (2.32) is given in terms of the signal subspace parameters by*

$$\boldsymbol{m}_1 = \alpha_m \boldsymbol{U}_s \boldsymbol{\Lambda}_s^{-1} \boldsymbol{U}_s^H \boldsymbol{s}_1, \tag{2.78}$$

with

$$\alpha_m = \left(\boldsymbol{s}_1^H \boldsymbol{U}_s \boldsymbol{\Lambda}_s^{-1} \boldsymbol{U}_s^H \boldsymbol{s}_1\right)^{-1}. \tag{2.79}$$

Proof: From (2.34) the linear MMSE detector defined by (2.32) is given by

$$\boldsymbol{m}_1 = \boldsymbol{C}_r^{-1} \boldsymbol{s}_1 \left(\boldsymbol{s}_1^H \boldsymbol{C}_r^{-1} \boldsymbol{s}_1\right)^{-1}. \tag{2.80}$$

By (2.72),

$$\boldsymbol{C}_r^{-1} = \boldsymbol{U}_s \boldsymbol{\Lambda}_s^{-1} \boldsymbol{U}_s^H + \frac{1}{\sigma^2} \boldsymbol{U}_n \boldsymbol{U}_n^H. \tag{2.81}$$

Substituting (2.81) into (2.80) and using the fact that $\boldsymbol{U}_n^H \boldsymbol{s}_1 = \boldsymbol{0}$, we obtain (2.78).□

Since the decision rules (2.7) and (2.9) are invariant to a positive scaling, the two subspace linear multiuser detectors given by (2.74) and (2.78) can be interpreted as follows. First, the received signal $\boldsymbol{r}[i]$ is projected onto the signal subspace to get $\boldsymbol{y}[i] \triangleq \boldsymbol{U}_s^H \boldsymbol{r}[i] \in \mathbb{C}^K$, which clearly is a sufficient statistic for demodulating the K users' data bits. The spreading waveform $\boldsymbol{s}_1$ of the desired user is also projected onto the signal subspace to obtain $\boldsymbol{p}_1 \triangleq \boldsymbol{U}_s^H \boldsymbol{s}_1 \in \mathbb{C}^K$. The projection of the linear multiuser detector in the signal subspace is then a signal $\boldsymbol{c}_1 \in \mathbb{C}^K$ such that the detector output is $z_1[i] \triangleq \boldsymbol{c}_1^H \boldsymbol{y}[i]$, the data bit is demodulated as $\hat{b}_1[i] = \text{sign}\left\{\Re\left(A_1^* z_1[i]\right)\right\}$ for coherent detection, and $\hat{\beta}_1[i] = \text{sign}\left\{\Re\left(z_1[i] z_1[i-1]^*\right)\right\}$ for differential detection. According to (2.74) and (2.78), the projections of the linear decorrelating detector and that of the linear MMSE detector in the signal subspace are given, respectively, by

$$\boldsymbol{c}_{1,d} = \begin{bmatrix} \frac{1}{\lambda_1 - \sigma^2} & & \\ & \ddots & \\ & & \frac{1}{\lambda_K - \sigma^2} \end{bmatrix} \boldsymbol{p}_1, \tag{2.82}$$

$$\boldsymbol{c}_{1,m} = \begin{bmatrix} \frac{1}{\lambda_1} & & \\ & \ddots & \\ & & \frac{1}{\lambda_K} \end{bmatrix} \boldsymbol{p}_1. \tag{2.83}$$

Therefore, the projection of the linear multiuser detectors in the signal subspace is obtained by projecting the spreading waveform of the desired user onto the signal subspace, followed by scaling the kth component of this projection by a factor of $(\lambda_k - \sigma^2)^{-1}$ (for linear decorrelating detector) or λ_k^{-1} (for linear MMSE detector). Note that as $\sigma^2 \to 0$, the two linear detectors become identical, as we would expect.

Since the autocorrelation matrix $\boldsymbol{C}_r$, and therefore its eigencomponents, can be estimated from the received signals, from the discussion above we see that both the linear decorrelating detector and the linear MMSE detector can be estimated from the received signal with the prior knowledge of only the spreading waveform and the timing of the desired user (i.e., they both can be obtained blindly). We summarize the subspace blind multiuser detection algorithm as follows.

Algorithm 2.5: [Subspace blind linear detector—synchronous CDMA]

- *Compute the detector:*

$$\hat{\boldsymbol{C}}_r \triangleq \frac{1}{M} \sum_{i=0}^{M-1} \boldsymbol{r}[i]\boldsymbol{r}[i]^H, \tag{2.84}$$

$$= \hat{\boldsymbol{U}}_s \hat{\boldsymbol{\Lambda}}_s \hat{\boldsymbol{U}}_s^H + \hat{\boldsymbol{U}}_n \hat{\boldsymbol{\Lambda}}_n \hat{\boldsymbol{U}}_n^H, \tag{2.85}$$

$$\hat{\boldsymbol{d}}_1 = \hat{\boldsymbol{U}}_s \left(\hat{\boldsymbol{\Lambda}}_s - \hat{\sigma^2} \boldsymbol{I}_K \right) \hat{\boldsymbol{U}}_s^H \boldsymbol{s}_1, \quad \text{(linear decorrelating detector)} \tag{2.86}$$

$$\hat{\boldsymbol{m}}_1 = \hat{\boldsymbol{U}}_s \hat{\boldsymbol{\Lambda}}_s \hat{\boldsymbol{U}}_s^H \boldsymbol{s}_1. \quad \text{(linear MMSE detector)} \tag{2.87}$$

- *Perform differential detection:*

$$z_1[i] = \hat{\boldsymbol{w}}_1^H \boldsymbol{r}[i], \qquad \left(\hat{\boldsymbol{w}}_1 = \hat{\boldsymbol{d}}_1 \quad or \quad \hat{\boldsymbol{w}}_1 = \hat{\boldsymbol{m}}_1 \right), \tag{2.88}$$

$$\hat{\beta}_1[i] = \text{sign}\left\{ \Re\left(z_1[i] z_1[i-1]^* \right) \right\}, \qquad i = 1, \ldots, M-1. \tag{2.89}$$

2.4.3 Asymptotics of Detector Estimates

We next examine the consistency and asymptotic variance of the estimates of the two subspace linear detectors. Assuming that the received signal samples are independent and identically distributed (i.i.d.), then by the strong law of large numbers, the sample mean $\hat{\boldsymbol{C}}_r$ converges to $\boldsymbol{C}_r$ almost surely (a.s.) as the number of received signals $M \to \infty$. It then follows [521] that as $M \to \infty$, $\hat{\lambda}_k \to \lambda_k$ a.s., and $\hat{\boldsymbol{u}}_k \to \boldsymbol{u}_k$ a.s., for $k = 1, \ldots, K$. Therefore, we have

$$\hat{\boldsymbol{m}}_1 = \sum_{k=1}^{K} \frac{1}{\hat{\lambda}_k} \hat{\boldsymbol{u}}_k \hat{\boldsymbol{u}}_k^H \boldsymbol{s}_1 \tag{2.90}$$

$$\to \sum_{k=1}^{K} \frac{1}{\lambda_k} \boldsymbol{u}_k \boldsymbol{u}_k^H \boldsymbol{s}_1 \;=\; \alpha_m^{-1}\, \boldsymbol{m}_1 \quad \text{a.s. as } M \to \infty. \tag{2.91}$$

Similarly, $\hat{\boldsymbol{d}}_1 \to \alpha_d^{-1}\, \boldsymbol{d}_1$ a.s. as $M \to \infty$. Hence both the estimated subspace linear multiuser detectors based on the received signals are *strongly consistent*. However, it is in general biased for finite number of samples. We next consider an asymptotic bound on the estimation errors.

First, for all eigenvalues and the K largest eigenvectors of $\hat{\boldsymbol{C}}_r$, the following bounds hold a.s. [521, 609]:

$$\left|\hat{\lambda}_k - \lambda_k\right| = \mathcal{O}(\sqrt{\log\log M / M}), \qquad k = 1, \ldots, N, \tag{2.92}$$

$$\|\hat{\boldsymbol{u}}_k - \boldsymbol{u}_k\| = \mathcal{O}(\sqrt{\log\log M / M}), \qquad k = 1, \ldots, K. \tag{2.93}$$

Using the bounds above, we have

$$\begin{aligned}
\|\alpha_d^{-1} \boldsymbol{m}_1 - \hat{\boldsymbol{m}}_1\| &= \left\| \left(\boldsymbol{U}_s \boldsymbol{\Lambda}_s^{-1} \boldsymbol{U}_s^H - \hat{\boldsymbol{U}}_s \hat{\boldsymbol{\Lambda}}_s^{-1} \hat{\boldsymbol{U}}_s^H \right) \boldsymbol{s}_1 \right\| \\
&\le \left\| \boldsymbol{U}_s \boldsymbol{\Lambda}_s^{-1} \boldsymbol{U}_s^H - \hat{\boldsymbol{U}}_s \hat{\boldsymbol{\Lambda}}_s^{-1} \hat{\boldsymbol{U}}_s^H \right\| \|\boldsymbol{s}_1\| \\
&= \Big\| \left(\boldsymbol{U}_s \boldsymbol{\Lambda}_s^{-1} \boldsymbol{U}_s^H - \hat{\boldsymbol{U}}_s \boldsymbol{\Lambda}_s^{-1} \boldsymbol{U}_s^H \right) \\
&\quad + \left(\hat{\boldsymbol{U}}_s \boldsymbol{\Lambda}_s^{-1} \boldsymbol{U}_s^H - \hat{\boldsymbol{U}}_s \boldsymbol{\Lambda}_s^{-1} \hat{\boldsymbol{U}}_s^H \right) + \hat{\boldsymbol{U}}_s \left(\boldsymbol{\Lambda}_s^{-1} - \hat{\boldsymbol{\Lambda}}_s^{-1} \right) \hat{\boldsymbol{U}}_s^H \Big\| \\
&\le \left\| \boldsymbol{U}_s - \hat{\boldsymbol{U}}_s \right\| \left\| \boldsymbol{\Lambda}_s^{-1} \boldsymbol{U}_s^H \right\| \\
&\quad + \left\| \hat{\boldsymbol{U}}_s \boldsymbol{\Lambda}_s^{-1} \right\| \left\| \boldsymbol{U}_s - \hat{\boldsymbol{U}}_s \right\| + \left\| \hat{\boldsymbol{U}}_s \right\| \left\| \boldsymbol{\Lambda}_s^{-1} - \hat{\boldsymbol{\Lambda}}_s^{-1} \right\| \left\| \hat{\boldsymbol{U}}_s \right\|.
\end{aligned} \tag{2.94}$$

Note that $\left\| \boldsymbol{\Lambda}_s^{-1} \boldsymbol{U}_s^H \right\|$, $\left\| \hat{\boldsymbol{U}}_s \boldsymbol{\Lambda}_s^{-1} \right\|$, and $\left\| \hat{\boldsymbol{U}}_s \right\|$ are all bounded. On the other hand, it is easily seen that

$$\left\| \boldsymbol{U}_s - \hat{\boldsymbol{U}}_s \right\| = \sum_{k=1}^{K} \|\boldsymbol{u}_k - \hat{\boldsymbol{u}}_k\| = \mathcal{O}(\sqrt{\log\log M / M}) \qquad a.s., \tag{2.95}$$

$$\left\| \boldsymbol{\Lambda}_s^{-1} - \hat{\boldsymbol{\Lambda}}_s^{-1} \right\| = \sum_{k=1}^{K} \left| \lambda_k - \hat{\lambda}_k \right| / \left(\lambda_k \hat{\lambda}_k \right) = \mathcal{O}(\sqrt{\log\log M / M}) \qquad a.s. \tag{2.96}$$

Therefore, we obtain the asymptotic estimation error for the linear MMSE detector, and similarly that for the decorrelating detector, given, respectively, by

$$\left\|\hat{\boldsymbol{m}}_1 - \alpha_m^{-1}\boldsymbol{m}_1\right\| = \mathcal{O}\left(\sqrt{\log\log M/M}\right) \qquad a.s.,$$
$$\left\|\hat{\boldsymbol{d}}_1 - \alpha_d^{-1}\boldsymbol{d}_1\right\| = \mathcal{O}\left(\sqrt{\log\log M/M}\right) \qquad a.s.$$

2.4.4 Asymptotic Multiuser Efficiency under Mismatch

We now consider the effect of spreading waveform mismatch on the performance of subspace linear multiuser detectors. Let $\tilde{\boldsymbol{s}}_1$ with $\|\tilde{\boldsymbol{s}}_1\| = 1$ be the assumed spreading waveform of the desired user and $\boldsymbol{s}_1$ be the true spreading waveform of that user. $\tilde{\boldsymbol{s}}_1$ can then be decomposed into components of the signal subspace and the noise subspace; that is,

$$\tilde{\boldsymbol{s}}_1 = \tilde{\boldsymbol{s}}_1^s + \tilde{\boldsymbol{s}}_1^n, \tag{2.97}$$

with

$$\tilde{\boldsymbol{s}}_1^s \triangleq \boldsymbol{U}_s\boldsymbol{U}_s^H\tilde{\boldsymbol{s}}_1 \in \mathsf{range}\left(\boldsymbol{U}_s\right) = \mathsf{range}(\boldsymbol{S}), \tag{2.98}$$
$$\tilde{\boldsymbol{s}}_1^n \triangleq \boldsymbol{U}_n\boldsymbol{U}_n^H\tilde{\boldsymbol{s}}_1 \in \mathsf{range}\left(\boldsymbol{U}_n\right). \tag{2.99}$$

For simplicity, in the following we consider the real-valued signal model [i.e., $A_k > 0$, $k = 1, \ldots, K$, and $\boldsymbol{n}[i] \sim \mathcal{N}(\boldsymbol{0}, \sigma^2\boldsymbol{I}_N)$]. [Here $\mathcal{N}(\cdot,\cdot)$ denotes a real-valued Gaussian distribution.] The signal subspace component $\tilde{\boldsymbol{s}}_1^s$ can then be written as

$$\tilde{\boldsymbol{s}}_1^s = \sum_{k=1}^{K}\psi_k\boldsymbol{s}_k = \boldsymbol{S}\boldsymbol{\psi} \tag{2.100}$$

for some $\boldsymbol{\psi} \in \mathbb{R}^K$ with $\alpha_1 > 0$. A commonly used performance measure for a multiuser detector is the *asymptotic multiuser efficiency* (AME) [520], defined as[2]

$$\eta_1 \triangleq \sup\left\{0 \le r \le 1 : \lim_{\sigma\to 0} P_1(\sigma)/Q\left(\frac{\sqrt{r}A_1}{\sigma}\right) = 0\right\}, \tag{2.101}$$

which measures the exponential decay rate of the error probability as the background noise approaches zero relative to that of a single-user system having the same signal-to-noise ratio. A related performance measure, the *near–far resistance*, is the infimum of AME as the interferers' energies are allowed to vary arbitrarily.

$$\overline{\eta}_1 = \inf_{\substack{A_k \ge 0 \\ k\ne 1}}\{\eta_1\}. \tag{2.102}$$

[2] $P_1(\sigma)$ is the probability of error of the detector for noise level σ; $Q(x) \triangleq (1/\sqrt{2\pi})\int_x^{\infty}\exp\left(-x^2/2\right)$.

Since as $\sigma \to 0$, the linear decorrelating detector and the linear MMSE detector become identical, these two detectors have the same AME and near–far resistance [296, 307]. It is straightforward to compute the AME of the linear decorrelating detector, since its output consists of only the desired user's signal and the ambient Gaussian noise. By (2.15)–(2.17), we conclude that the AME and the near–far resistance of both linear detectors are given by

$$\eta_1 = \bar{\eta}_1 = \frac{1}{\left[\boldsymbol{R}^{-1}\right]_{1,1}}. \tag{2.103}$$

Next we compute the AME and the near-far resistance of the two subspace linear detectors under spreading waveform mismatch. Define the $N \times N$ diagonal matrices

$$\boldsymbol{\Lambda}_0 \triangleq \mathsf{diag}\left(\lambda_1 - \sigma^2, \ldots, \lambda_K - \sigma^2, 0, \ldots, 0\right), \tag{2.104}$$

$$\boldsymbol{\Lambda}_0^\dagger \triangleq \mathsf{diag}\left([\lambda_1 - \sigma^2]^{-1}, \ldots, [\lambda_K - \sigma^2]^{-1}, 0, \ldots, 0\right). \tag{2.105}$$

Denote the singular value decomposition (SVD) of $\boldsymbol{S}$ by

$$\boldsymbol{S} = \boldsymbol{W}\boldsymbol{\Gamma}\boldsymbol{V}^T, \tag{2.106}$$

where the $N \times K$ matrix $\boldsymbol{\Gamma} = [\gamma_{ij}]$ has $\gamma_{ij} = 0$ for all $i \neq j$ and $\gamma_{11} \geq \gamma_{22} \geq \cdots \geq \gamma_{KK}$. The columns of the $N \times N$ matrix $\boldsymbol{W}$ are the orthogonal eigenvectors of $\boldsymbol{S}\boldsymbol{S}^T$, and the columns of the $K \times K$ matrix $\boldsymbol{V}$ are the orthogonal eigenvectors of $\boldsymbol{R} = \boldsymbol{S}^T\boldsymbol{S}$. We have the following result, whose proof is given in the Appendix (Section 2.8.2).

Lemma 2.3: *Let the eigendecomposition of $\boldsymbol{C}_r$ be $\boldsymbol{C}_r = \boldsymbol{U}\boldsymbol{\Lambda}\boldsymbol{U}^T$. Then the $N \times N$ diagonal matrix $\boldsymbol{\Lambda}_0^\dagger$ in (2.105) is given by*

$$\boldsymbol{\Lambda}_0^\dagger = \boldsymbol{U}^T\boldsymbol{W}\boldsymbol{\Gamma}^{\dagger^T}\boldsymbol{V}^T\boldsymbol{A}^{-2}\boldsymbol{V}\boldsymbol{\Gamma}^\dagger\boldsymbol{W}^T\boldsymbol{U}, \tag{2.107}$$

where $\boldsymbol{\Gamma}^\dagger$ is the transpose of $\boldsymbol{\Gamma}$ in which the singular values are replaced by their reciprocals.

Using the result above, we obtain the AME of the subspace linear detectors under spreading waveform mismatch, as follows.

Proposition 2.5: *The AME of the subspace linear decorrelating detector given by (2.74) and that of the subspace linear MMSE detector given by (2.78) under spreading waveform mismatch is given by*

$$\eta_1 = \frac{\max^2\left\{0,\ |\psi_1| - \sum_{k=2}^{K} |\psi_k| A_1/A_k\right\}}{A_1^4 \boldsymbol{\psi}^T \boldsymbol{A}^{-2}\boldsymbol{R}^{-1}\boldsymbol{A}^{-2}\boldsymbol{\psi}}. \tag{2.108}$$

Proof: Since $\boldsymbol{d}_1$ and $\boldsymbol{m}_1$ have the same AME, we need only to compute the AME for $\boldsymbol{d}_1$. Because a positive scaling on the detector does not affect its AME, we consider the AME of the following scaled version of $\boldsymbol{d}_1$ under the signature waveform mismatch:

$$\begin{aligned}\tilde{\boldsymbol{d}}_1 &\triangleq \boldsymbol{U}_s\left(\boldsymbol{\Lambda}_s - \sigma^2 \boldsymbol{I}_K\right)^{-1}\boldsymbol{U}_s^T \tilde{\boldsymbol{s}}_1\\ &= \boldsymbol{U}_s\left(\boldsymbol{\Lambda}_s - \sigma^2 \boldsymbol{I}_K\right)^{-1}\boldsymbol{U}_s^T \tilde{\boldsymbol{s}}_1^s\\ &= \boldsymbol{U}\boldsymbol{\Lambda}_0^{\dagger}\boldsymbol{U}^T\boldsymbol{S}\boldsymbol{\psi},\end{aligned} \tag{2.109}$$

where the second equality follows from the fact that the noise subspace component $\tilde{\boldsymbol{s}}_1^n$ is orthogonal to the signal subspace $\boldsymbol{U}_s$. Substituting (2.106) and (2.107) into (2.109), we have

$$\begin{aligned}\tilde{\boldsymbol{d}}_1^T \boldsymbol{s}_k &= \boldsymbol{\psi}^T\boldsymbol{S}^T\boldsymbol{U}\boldsymbol{\Lambda}_0^{\dagger}\boldsymbol{U}^T\boldsymbol{S}\boldsymbol{e}_k\\ &= \boldsymbol{\psi}^T\left(\boldsymbol{V}\boldsymbol{\Gamma}^T\boldsymbol{W}^T\right)\left(\boldsymbol{W}{\boldsymbol{\Gamma}^{\dagger}}^T\boldsymbol{V}^T\boldsymbol{A}^{-2}\boldsymbol{V}\boldsymbol{\Gamma}^{\dagger}\boldsymbol{W}^T\right)\left(\boldsymbol{W}\boldsymbol{\Gamma}\boldsymbol{V}^T\right)\boldsymbol{e}_k\\ &= \boldsymbol{\psi}^T\boldsymbol{A}^{-2}\boldsymbol{e}_k \;=\; \frac{\psi_k}{A_k^2},\end{aligned} \tag{2.110}$$

$$\begin{aligned}\tilde{\boldsymbol{d}}_1^T\tilde{\boldsymbol{d}}_1 &= \boldsymbol{\psi}^T\left(\boldsymbol{V}\boldsymbol{\Gamma}^T\boldsymbol{W}^T\right)\left(\boldsymbol{W}{\boldsymbol{\Gamma}^{\dagger}}^T\boldsymbol{V}^T\boldsymbol{A}^{-2}\boldsymbol{V}\boldsymbol{\Gamma}^{\dagger}\boldsymbol{W}^T\boldsymbol{U}\right)\\ &\quad\left(\boldsymbol{U}^T\boldsymbol{W}{\boldsymbol{\Gamma}^{\dagger}}^T\boldsymbol{V}^T\boldsymbol{A}^{-2}\boldsymbol{V}\boldsymbol{\Gamma}^{\dagger}\boldsymbol{W}^T\right)\left(\boldsymbol{W}\boldsymbol{\Gamma}\boldsymbol{V}^T\right)\boldsymbol{\psi}\end{aligned} \tag{2.111}$$

$$= \boldsymbol{\psi}^T\boldsymbol{A}^{-2}\underbrace{\boldsymbol{V}\boldsymbol{\Gamma}^{\dagger}{\boldsymbol{\Gamma}^{\dagger}}^T\boldsymbol{V}^T}_{\boldsymbol{R}^{-1}}\boldsymbol{A}^{-2}\boldsymbol{\psi}. \tag{2.112}$$

The output of the detector $\tilde{\boldsymbol{d}}_1$ is given by

$$\begin{aligned}z[i] &\triangleq \tilde{\boldsymbol{d}}_1^T\boldsymbol{r}[i] = \sum_{k=1}^{K} A_k b_k\left(\tilde{\boldsymbol{d}}_1^T\boldsymbol{s}_k\right) + \tilde{\boldsymbol{d}}_1^T\boldsymbol{n}[i]\\ &= \sum_{k=1}^{K}\frac{\psi_k}{A_k}b_k + v[i],\end{aligned} \tag{2.113}$$

where $v[i] \sim \mathcal{N}\left(0,\, \sigma^2\|\tilde{\boldsymbol{d}}_1\|^2\right)$. The probability of error for user 1 is then given by

$$P_1(\sigma) = \frac{1}{2^{K-1}}\sum_{(b_2,\ldots,b_K)\in\{-1,1\}^{K-1}} Q\left(\frac{A_1}{\sigma}\cdot\frac{\psi_1 - \sum_{k=2}^{K}\psi_k b_k A_1/A_k}{\sqrt{A_1^4\,\boldsymbol{\psi}^T\boldsymbol{A}^{-2}\boldsymbol{R}^{-1}\boldsymbol{A}^{-2}\boldsymbol{\psi}}}\right). \tag{2.114}$$

It then follows that the AME is given by (2.108). □

It is seen from (2.114) that spreading waveform mismatch causes MAI leakage at the detector output. Strong interferers ($A_k \gg A_1$) are suppressed at the output, whereas weak interferers ($A_k \ll A_1$) may lead to performance degradation. If the mismatch is not significant, with power control, so that the open-eye condition is satisfied (i.e., $|\psi_1| > \sum_{k=2}^{K}|\psi_k|A_1/A_k$), the performance loss is negligible; otherwise, the effective spreading waveform should be estimated first. Moreover, since the mismatched spreading waveform $\tilde{\boldsymbol{s}}_1$ is first projected onto the signal subspace, its noise subspace component $\tilde{\boldsymbol{s}}_1^n$ is nulled out and does not cause performance degradation; whereas for the blind adaptive MOE detector discussed in Section 2.3, such a noise subspace component may lead to complete cancellation of both the signal and MAI if there is no energy constraint on the detector [183].

2.5 Performance of Blind Multiuser Detectors

2.5.1 Performance Measures

In previous sections we have discussed two approaches to blind multiuser detection: the direct method and the subspace method. These two approaches are based primarily on two equivalent expressions for the linear MMSE detector [i.e., (2.26) and (2.78)]. When the autocorrelation $\boldsymbol{C}_r$ of the received signals is known exactly, the two approaches have the same performance. However, when $\boldsymbol{C}_r$ is replaced by the corresponding sample autocorrelation, quite interestingly, the performance of these two methods is very different. This is due to the fact that these two approaches exhibit different estimation errors on the estimated detector [193, 194, 197]. In this section we present a performance analysis of the two blind multiuser detectors: the DMI blind detector and the subspace blind detector. For simplicity, we consider only real-valued signals [i.e., in (2.4), $A_k > 0, k = 1, \ldots, K$ and $\boldsymbol{n}[i] \sim \mathcal{N}(\boldsymbol{0}, \sigma^2 \boldsymbol{I}_N)$].

Suppose that a linear weight vector $\boldsymbol{w}_1 \in \mathbb{R}^N$ is applied to the received signal $\boldsymbol{r}[i]$ in (2.5). The output is given by (2.10). Since it is assumed that the user bit streams are independent and the noise is independent of the user bits, the signal-to-interference-plus-noise ratio (SINR) at the output of the linear detector is given by

$$\begin{aligned} \text{SINR}(\boldsymbol{w}_1) &= \frac{E\left\{\boldsymbol{w}_1^T \boldsymbol{r}[i] \mid b_1[i]\right\}^2}{E\left\{\text{Var}\left\{\boldsymbol{w}_1^T \boldsymbol{r}[i] \mid b_1[i]\right\}\right\}} \\ &= \frac{A_1^2 \left(\boldsymbol{w}_1^T \boldsymbol{s}_1\right)^2}{\sum_{k=2}^{K} A_k^2 \left(\boldsymbol{w}_1^T \boldsymbol{s}_k\right)^2 + \sigma^2 \|\boldsymbol{w}_1\|^2}. \end{aligned} \tag{2.115}$$

The bit-error probability of the linear detector using weight vector $\boldsymbol{w}_1$ is given by

$$\begin{aligned} P_e(\boldsymbol{w}_1) &= P\left(\hat{b}_1[i] \neq b_1[i]\right) \\ &= \frac{1}{2^{K-1}} \sum_{[b_2 \ldots b_K] \in \{-1,+1\}^{K-1}} Q\left(\frac{A_1 \boldsymbol{w}_1^T \boldsymbol{s}_1 + \sum_{k=2}^{K} A_k b_k \boldsymbol{w}_1^T \boldsymbol{s}_k}{\|\boldsymbol{w}_1\| \sigma}\right). \end{aligned} \tag{2.116}$$

Now suppose that an estimate $\hat{\boldsymbol{w}}_1$ of the weight vector $\boldsymbol{w}_1$ is obtained from the received signals $\{\boldsymbol{r}[i]\}_{i=0}^{M-1}$. Denote

$$\Delta \boldsymbol{w}_1 \triangleq \hat{\boldsymbol{w}}_1 - \boldsymbol{w}_1. \tag{2.117}$$

Obviously, both $\hat{\boldsymbol{w}}_1$ and $\Delta \boldsymbol{w}_1$ are random vectors and are functions of the random quantities $\{\boldsymbol{b}[i], \boldsymbol{n}[i]\}_{i=0}^{M-1}$. In typical *adaptive multiuser detection* scenarios [183,

549], the estimated detector $\hat{\boldsymbol{w}}_1$ is employed to demodulate future received signals, say $\boldsymbol{r}[j], j \geq M$. Then the output is given by

$$\hat{\boldsymbol{w}}_1^T \boldsymbol{r}[j] = \boldsymbol{w}_1^T \boldsymbol{r}[j] + \Delta \boldsymbol{w}_1^T \boldsymbol{r}[j], \qquad j \geq M, \tag{2.118}$$

where the first term in (2.118) represents the output of the true weight vector $\boldsymbol{w}_1$, which has the same form as (2.10). The second term in (2.118) represents an additional noise term caused by the estimation error $\Delta \boldsymbol{w}_1$. Hence from (2.118) the average SINR at the output of any unbiased estimated linear detector $\hat{\boldsymbol{w}}_1$ is given by

$$\overline{\text{SINR}(\hat{\boldsymbol{w}}_1)} = \frac{A_1^2 \left(\boldsymbol{w}_1^T \boldsymbol{s}_1\right)^2}{\sum_{k=2}^{K} A_k^2 \left(\boldsymbol{w}_1^T \boldsymbol{s}_k\right)^2 + \sigma^2 \|\boldsymbol{w}_1\|^2 + E\left\{\left(\Delta \boldsymbol{w}_1^T \boldsymbol{r}[j]\right)^2\right\}}, \tag{2.119}$$

with

$$\begin{aligned} E\left\{\left(\Delta \boldsymbol{w}_1^T \boldsymbol{r}[j]\right)^2\right\} &= \text{tr}\left(E\left\{\Delta \boldsymbol{w}_1^T \boldsymbol{r}[j] \boldsymbol{r}[j]^T \Delta \boldsymbol{w}_1\right\}\right) \\ &= \text{tr}\left(E\left\{\Delta \boldsymbol{w}_1 \Delta \boldsymbol{w}_1^T \boldsymbol{r}[j] \boldsymbol{r}[j]^T\right\}\right) \\ &= \text{tr}\big(\underbrace{E\left\{\Delta \boldsymbol{w}_1 \Delta \boldsymbol{w}_1^T\right\}}_{\boldsymbol{C}_w} \quad \underbrace{E\left\{\boldsymbol{r}[j] \boldsymbol{r}[j]^T\right\}}_{\boldsymbol{C}_r = \boldsymbol{S}\boldsymbol{A}^2\boldsymbol{S}^T + \sigma^2 \boldsymbol{I}_N}\big) \\ &= \frac{1}{M} \text{tr}\left(\boldsymbol{C}_w \boldsymbol{C}_r\right), \end{aligned} \tag{2.120}$$

where $\boldsymbol{C}_w \triangleq M \cdot E\left\{\Delta \boldsymbol{w}_1 \Delta \boldsymbol{w}_1^T\right\}$ and $\boldsymbol{C}_r \triangleq E\left\{\boldsymbol{r}[j] \boldsymbol{r}[j]^T\right\}$. Note that in *batch processing*, on the other hand, the estimated detector is used to demodulate signals $\boldsymbol{r}[i], 0 \leq i \leq M-1$. Since $\Delta \boldsymbol{w}_1$ is a function of $\{\boldsymbol{r}[i]\}_{i=0}^{M-1}$, for fixed i, $\Delta \boldsymbol{w}_1$ and $\boldsymbol{r}[i]$ are in general correlated. For large M, such correlation is small. Therefore, in this case we still use (2.119) and (2.120) as the approximate SINR expression.

If we assume further that $\Delta \boldsymbol{w}_1$ is actually independent of $\boldsymbol{r}[i]$, the average bit-error rate (BER) of this detector is given by

$$\overline{P_e(\hat{\boldsymbol{w}}_1)} = \int P_e(\hat{\boldsymbol{w}}_1) f(\hat{\boldsymbol{w}}_1) \, d\hat{\boldsymbol{w}}_1, \tag{2.121}$$

where $P_e(\hat{\boldsymbol{w}}_1)$ is given by (2.116) and $f(\hat{\boldsymbol{w}}_1)$ denotes the probability density function (pdf) of the estimated weight vector $\hat{\boldsymbol{w}}_1$.

From the discussion above it is seen that to obtain the average SINR at the output of the estimated linear detector $\hat{\boldsymbol{w}}_1$, it suffices to find its covariance matrix $\boldsymbol{C}_w$. On the other hand, the average bit-error rate of the estimated linear detector depends on its distribution through $f(\hat{\boldsymbol{w}}_1)$.

2.5.2 Asymptotic Output SINR

We first present the asymptotic distribution of the two forms of blind linear MMSE detectors for a large number of signal samples, M. Recall that in the direct-matrix-inversion (DMI) method, the blind multiuser detector is estimated according to

$$\hat{\boldsymbol{C}}_r = \frac{1}{M}\sum_{i=0}^{M-1} \boldsymbol{r}[i]\boldsymbol{r}[i]^T, \tag{2.122}$$

$$\hat{\boldsymbol{w}}_1 = \hat{\boldsymbol{C}}_r^{-1}\boldsymbol{s}_1. \qquad \text{(DMI blind linear MMSE detector)} \tag{2.123}$$

In the subspace method, the estimate of the blind detector is given by

$$\begin{aligned}\hat{\boldsymbol{C}}_r &= \frac{1}{M}\sum_{i=0}^{M-1} \boldsymbol{r}[i]\boldsymbol{r}[i]^T \\ &= \hat{\boldsymbol{U}}_s\hat{\boldsymbol{\Lambda}}_s\hat{\boldsymbol{U}}_s^T + \hat{\boldsymbol{U}}_n\hat{\boldsymbol{\Lambda}}_n\hat{\boldsymbol{U}}_n^T, \qquad (2.124)\\ \hat{\boldsymbol{w}}_1 &= \hat{\boldsymbol{U}}_s\hat{\boldsymbol{\Lambda}}_s^{-1}\hat{\boldsymbol{U}}_s^T\boldsymbol{s}_1, \quad \text{(subspace blind linear MMSE detector)} \qquad (2.125)\end{aligned}$$

where $\hat{\boldsymbol{\Lambda}}_s$ and $\hat{\boldsymbol{U}}_s$ contain, respectively, the largest K eigenvalues and the corresponding eigenvectors of $\hat{\boldsymbol{C}}_r$; and where $\hat{\boldsymbol{\Lambda}}_n$ and $\hat{\boldsymbol{U}}_n$ contain, respectively, the remaining eigenvalues and eigenvectors of $\hat{\boldsymbol{C}}_r$. The following result gives the asymptotic distribution of the blind linear MMSE detectors given by (2.123) and (2.125). The proof is given in the Appendix (Section 2.8.3).

Theorem 2.1: *Let $\boldsymbol{w}_1$ be the true weight vector of the linear MMSE detector given by*

$$\boldsymbol{w}_1 = \boldsymbol{C}_r^{-1}\boldsymbol{s}_1 = \boldsymbol{U}_s\boldsymbol{\Lambda}_s^{-1}\boldsymbol{U}_s^T\boldsymbol{s}_1, \tag{2.126}$$

and let $\hat{\boldsymbol{w}}_1$ be the weight vector of the estimated blind linear MMSE detector given by (2.123) or (2.125). Let the eigendecomposition of the autocorrelation matrix $\boldsymbol{C}_r$ of the received signal be

$$\boldsymbol{C}_r = \boldsymbol{U}_s\boldsymbol{\Lambda}_s\boldsymbol{U}_s^T + \sigma^2\boldsymbol{U}_n\boldsymbol{U}_n^T. \tag{2.127}$$

Then

$$\sqrt{M}\,(\hat{\boldsymbol{w}}_1 - \boldsymbol{w}_1) \to \mathcal{N}(\boldsymbol{0}, \boldsymbol{C}_w), \quad \textit{in distribution, as } M \to \infty,$$

with

$$\begin{aligned}\boldsymbol{C}_w &= \left(\boldsymbol{w}_1^T\boldsymbol{s}_1\right)\boldsymbol{U}_s\boldsymbol{\Lambda}_s^{-1}\boldsymbol{U}_s^T \\ &\quad + \boldsymbol{w}_1\boldsymbol{w}_1^T - 2\boldsymbol{U}_s\boldsymbol{\Lambda}_s^{-1}\boldsymbol{U}_s^T\boldsymbol{S}\boldsymbol{D}\boldsymbol{S}^T\boldsymbol{U}_s\boldsymbol{\Lambda}_s^{-1}\boldsymbol{U}_s^T + \tau\boldsymbol{U}_n\boldsymbol{U}_n^T, \qquad (2.128)\end{aligned}$$

where

$$\boldsymbol{D} \triangleq \text{diag}\left\{A_1^4\left(\boldsymbol{w}_1^T\boldsymbol{s}_1\right)^2, A_2^4\left(\boldsymbol{w}_1^T\boldsymbol{s}_2\right)^2, \ldots, A_K^4\left(\boldsymbol{w}_1^T\boldsymbol{s}_K\right)^2\right\}, \tag{2.129}$$

$$\tau \triangleq \begin{cases} \frac{1}{\sigma^2}\boldsymbol{s}_1^T\boldsymbol{U}_s\boldsymbol{\Lambda}_s^{-1}\boldsymbol{U}_s^T\boldsymbol{s}_1, & \text{(DMI blind detector)} \\ \sigma^2\boldsymbol{s}_1^T\boldsymbol{U}_s\boldsymbol{\Lambda}_s^{-1}\left(\boldsymbol{\Lambda}_s-\sigma^2\boldsymbol{I}_K\right)^{-2}\boldsymbol{U}_s^T\boldsymbol{s}_1. & \text{(subspace blind detector)} \end{cases} \tag{2.130}$$

Hence for large M, the covariance of the blind linear detector, $\boldsymbol{C}_w = M \cdot E\left\{\Delta\boldsymbol{w}_1\Delta\boldsymbol{w}_1^T\right\}$, can be approximated by (2.128). Define, as before,

$$\boldsymbol{R} \triangleq \boldsymbol{S}^T\boldsymbol{S}. \tag{2.131}$$

The next result gives an expression for the average output SINR, defined by (2.119), of the blind linear detectors. The proof is given in the Appendix (Section 2.8.3).

Corollary 2.1: *The average output SINR of the estimated blind linear detector is given by*

$$\overline{\text{SINR}(\hat{\boldsymbol{w}}_1)} = \frac{A_1^2\left(\boldsymbol{w}_1^T\boldsymbol{s}_1\right)^2}{\sum_{k=2}^K A_k^2\left(\boldsymbol{w}_1^T\boldsymbol{s}_k\right)^2+\sigma^2\|\boldsymbol{w}_1\|^2+\frac{1}{M}\left[(K+1)\boldsymbol{w}_1^T\boldsymbol{s}_1-2\sum_{k=1}^K A_k^4\left(\boldsymbol{w}_1^T\boldsymbol{s}_k\right)^2\left(\boldsymbol{w}_k^T\boldsymbol{s}_k\right)+(N-K)\tau\sigma^2\right]}, \tag{2.132}$$

where

$$\boldsymbol{w}_l^T\boldsymbol{s}_k = \frac{1}{A_l^2}\left[\boldsymbol{R}\left(\boldsymbol{R}+\sigma^2\boldsymbol{A}^{-2}\right)^{-1}\right]_{k,l}, \qquad k,l = 1,\ldots,K, \tag{2.133}$$

$$\|\boldsymbol{w}_1\|^2 = \frac{1}{A_1^4}\left[\left(\boldsymbol{R}+\sigma^2\boldsymbol{A}^{-2}\right)^{-1}\boldsymbol{R}\left(\boldsymbol{R}+\sigma^2\boldsymbol{A}^{-2}\right)^{-1}\right]_{1,1}, \tag{2.134}$$

$$\tau\sigma^2 = \begin{cases} \boldsymbol{w}_1^T\boldsymbol{s}_1, & \text{(DMI blind detector)} \\ \frac{\sigma^4}{A_1^4}\left[\left(\boldsymbol{R}+\sigma^2\boldsymbol{A}^{-2}\right)^{-1}\boldsymbol{A}^{-2}\boldsymbol{R}^{-1}\right]_{1,1}. & \text{(subspace blind detector)} \end{cases} \tag{2.135}$$

It is seen from (2.132) that the performance difference between the DMI blind detector and the subspace blind detector is caused by the single parameter τ given by (2.130)—the detector with a smaller τ has a higher output SINR. Let $\mu_1, \ldots, \mu_K$ be the eigenvalues of the matrix $\boldsymbol{R}$ given by (2.131). Denote $\mu_{\min} = \min_{1\le k\le K}\{\mu_k\}$ and $\mu_{\max} = \max_{1\le k\le K}\{\mu_k\}$. Denote also $A_{\min} = \min_{1\le k\le K}\{A_k\}$ and $A_{\max} = \max_{1\le k\le K}\{A_k\}$. The next result gives sufficient conditions under which one blind detector outperforms the other in terms of the average output SINR.

Corollary 2.2: *If* $A_{\min}^2/\sigma^2 > \mu_{\max}$, *then* $\overline{\text{SINR}}_{\text{subspace}} > \overline{\text{SINR}}_{\text{DMI}}$; *and if* $A_{\max}^2/\sigma^2 < \mu_{\min}$, *then* $\overline{\text{SINR}}_{\text{subspace}} < \overline{\text{SINR}}_{\text{DMI}}$.

Proof: By rewriting (2.130) as

$$\tau = \begin{cases} \dfrac{1}{\sigma^2}\displaystyle\sum_{k=1}^{K}\frac{1}{\lambda_k}\left(\boldsymbol{s}_1^T\boldsymbol{u}_k\right)^2, & \text{(DMI blind detector)} \\ \sigma^2\displaystyle\sum_{k=1}^{K}\frac{1}{\lambda_k(\lambda_k-\sigma^2)^2}\left(\boldsymbol{s}_1^T\boldsymbol{u}_k\right)^2, & \text{(subspace blind detector)} \end{cases} \tag{2.136}$$

we obtain the following *sufficient* condition under which $\tau_{\text{subspace}} < \tau_{\text{DMI}}$:

$$\lambda_k > 2\sigma^2, \qquad k = 1, \ldots, K. \tag{2.137}$$

On the other hand, note that

$$\boldsymbol{C}_r = \boldsymbol{S}\boldsymbol{A}^2\boldsymbol{S}^T + \sigma^2\boldsymbol{I}_N \succeq A_{\min}^2\boldsymbol{S}\boldsymbol{S}^T + \sigma^2\boldsymbol{I}_N. \tag{2.138}$$

Since the nonzero eigenvalues of $\boldsymbol{S}\boldsymbol{S}^T$ are the same of those of $\boldsymbol{R} = \boldsymbol{S}^T\boldsymbol{S}$, it follows from (2.138) that

$$\lambda_k \geq A_{\min}^2\mu_k + \sigma^2, \qquad k = 1, \ldots, K. \tag{2.139}$$

The first part of the corollary then follows by combining (2.137) and (2.139). The second part of the corollary follows a similar proof. □

The next result gives an upper and a lower bound on the parameter τ in terms of the desired user's amplitude A_1, the noise variance σ^2, and the two extreme eigenvalues of $\boldsymbol{C}_r$.

Corollary 2.3: *The parameter τ defined in (2.130) satisfies*

$$\left(1-\frac{\sigma^2}{\lambda_{\min}}\right)\frac{1}{A_1^2} \leq \tau\sigma^2 \leq \left(1-\frac{\sigma^2}{\lambda_{\max}}\right)\frac{1}{A_1^2}, \quad \text{(DMI blind detector)}$$

$$\frac{1}{\lambda_{\max}\left(\dfrac{\lambda_{\max}}{\sigma^2}-1\right)}\cdot\frac{1}{A_1^2} \leq \tau\sigma^2 \leq \frac{1}{\lambda_{\min}\left(\dfrac{\lambda_{\min}}{\sigma^2}-1\right)}\cdot\frac{1}{A_1^2}. \quad \text{(subspace blind detector)}$$

Proof: The proof follows from (2.136) and the following fact from Chapter 4 [cf. Proposition 4.2]:

$$\frac{1}{A_1^2} = \sum_{k=1}^{K}\frac{\left(\boldsymbol{s}_1^T\boldsymbol{u}_k\right)^2}{\lambda_k-\sigma^2}. \tag{2.140}$$

□

To gain some insight from the result (2.132), we next consider two special cases for which we compare the average output SINRs of the two blind detectors.

Example 1: Orthogonal Signals In this case, we have $\boldsymbol{u}_k = \boldsymbol{s}_k$, $\boldsymbol{R} = \boldsymbol{I}_K$, and $\lambda_k = A_k^2 + \sigma^2$, $k = 1, \ldots, K$. Substituting these into (2.136), we obtain

$$\tau\sigma^2 = \begin{cases} \dfrac{1}{A_1^2+\sigma^2}, & \text{(DMI blind detector)} \\ \left(\dfrac{\sigma^2}{A_1^2}\right)^2\dfrac{1}{A_1^2+\sigma^2}. & \text{(subspace blind detector)} \end{cases} \tag{2.141}$$

Substituting (2.141) into (2.132), and using the fact that in this case $\boldsymbol{w}_k = [1/(A_k^2 + \sigma^2)]\boldsymbol{s}_k$, we obtain the following expressions of the average output SINRs:

$$\overline{\text{SINR}(\hat{\boldsymbol{w}}_1)} = \begin{cases} \dfrac{\phi_1}{1 + \dfrac{1}{M}\left[(\phi_1+1)(N+1) - \dfrac{2\phi_1^2}{1+\phi_1}\right]}, & \text{(DMI blind detector)} \\ \dfrac{\phi_1}{1 + \dfrac{1}{M}\left[(\phi_1+1)\left(K+1+\dfrac{N-K}{\phi_1^2}\right) - \dfrac{2\phi_1^2}{1+\phi_1}\right]}, & \text{(subspace blind detector)} \end{cases} \tag{2.142}$$

where $\phi_1 \triangleq A_1^2/\sigma^2$ is the signal-to-noise ratio (SNR) of the desired user. It is easily seen that in this case, a necessary and sufficient condition for the subspace blind detector to outperform the DMI blind detector is that $\phi_1 > 1$ (i.e., $\text{SNR}_1 > 0$ dB).

Example 2: Equicorrelated Signals with Perfect Power Control In this case it is assumed that $\boldsymbol{s}_k^T \boldsymbol{s}_l = \rho$, for $k \neq l, 1 \leq k, l \leq K$. It is also assumed that $A_1 = \cdots = A_K = A$. It is shown in the Appendix (Section 2.8.3) that the average output SINRs for the two blind detectors are given by

$$\overline{\text{SINR}(\hat{\boldsymbol{w}}_1)} = \frac{1}{(K-1)\alpha + \beta + \frac{1}{M}\left[\dfrac{K+1}{\gamma} - 2\gamma\left[1+(K-1)\alpha\right] + (N-K)\eta\right]}, \tag{2.143}$$

with

$$\alpha \triangleq \left(\frac{\rho(\sigma^2/A^2)}{(\sigma^2/A^2) + (1-\rho)[1+(K-1)\rho]}\right)^2, \tag{2.144}$$

$$\beta \triangleq \frac{\sigma^2}{A^2} \cdot \frac{\left[1+(K-1)\rho + \dfrac{\sigma^2}{A^2}\right]^2 - \rho[1+(K-1)\rho]\left[2+(K-2)\rho + 2\dfrac{\sigma^2}{A^2}\right]}{\left[(1-\rho)[1+(K-1)\rho] + \dfrac{\sigma^2}{A^2}\right]^2}, \tag{2.145}$$

$$\gamma \triangleq \frac{1}{1-\rho+\sigma^2/A^2} + \frac{1+(K-1)\rho}{K}\left[\frac{1}{1+(K-1)\rho+\sigma^2/A^2} - \frac{1}{1-\rho+\sigma^2/A^2}\right], \tag{2.146}$$

and

$$\eta \triangleq \begin{cases} \frac{1}{\gamma}, & \text{(DMI blind detector)} \\ \frac{1}{\gamma^2}\left(\frac{\sigma^2}{A^2}\right)^2 \left(\frac{1}{(1-\rho)^2\left[1-\rho+\frac{\sigma^2}{A^2}\right]} + \frac{1+(K-1)\rho}{K} \left[\frac{1}{[1+(K-1)\rho]^2\left[1+(K-1)\rho+\frac{\sigma^2}{A^2}\right]} - \frac{1}{(1-\rho)^2\left[1-\rho+\frac{\sigma^2}{A^2}\right]} \right] \right) & \text{(subspace blind detector)} \end{cases} \tag{2.147}$$

A necessary and sufficient condition for the subspace blind detector to outperform the DMI blind detector is $\eta_{\text{DMI}} > \eta_{\text{subspace}}$, which after some manipulation reduces to

$$(\mu_1\mu_2)^3\,\phi^3 + (\mu_1\mu_2)^2\,\phi^2 \;>\; \mu_1^3\,\phi + \mu_1\left[\mu_1 + \frac{\mu_2^3 - \mu_1^3}{K}\right], \tag{2.148}$$

where $\phi \triangleq A^2/\sigma^2$ and where $\mu_1 \triangleq 1 + (K-1)\rho$ and $\mu_2 \triangleq 1 - \rho$ are the two distinct eigenvalues of $\boldsymbol{R}$ [cf. the Appendix (Section 2.8.3)]. The region on the SNR–ρ plane where the subspace blind detector outperforms the DMI blind detector is plotted in Fig. 2.2 for different values of K. It is seen that in general the subspace method performs better in the low cross-correlation and high-SNR region.

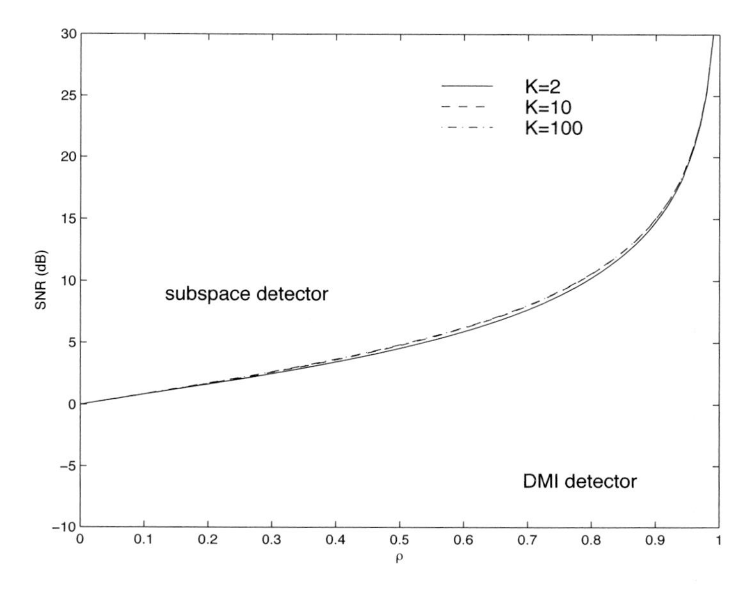

Figure 2.2. Partition of the SNR–ρ plane according to the relative performance of two blind detectors. For each K, in the region above the boundary curve, the subspace blind detector performs better, whereas in the region below the boundary curve, the DMI blind detector performs better.

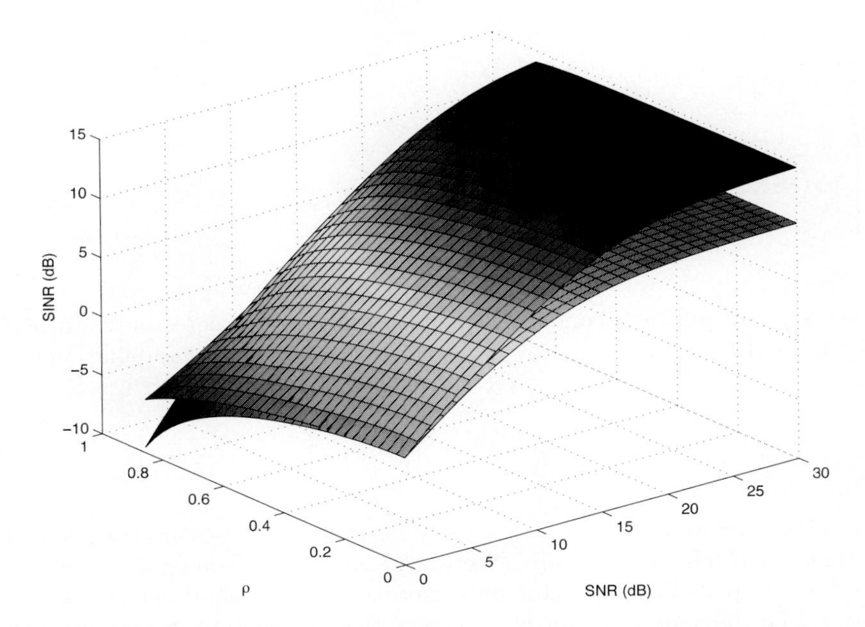

Figure 2.3. Average output SINR versus SNR and ρ for two blind detectors. $N = 16$, $K = 6$, $M = 150$. The upper curve in the high-SNR region represents the performance of the subspace blind detector.

The average output SINR as a function of SNR and ρ for both blind detectors is shown in Fig. 2.3. It is seen that the performance of the subspace blind detector deteriorates in the high-cross-correlation and low-SNR region; the performance of the DMI blind detector is less sensitive to cross-correlation and SNR in this region. This phenomenon is shown more clearly in Figs. 2.4 and 2.5, where the performance of the two blind detectors is compared as a function of ρ and SNR, respectively. The performance of the two blind detectors as a function of the number of signal samples M is plotted in Fig. 2.6, where it is seen that for large M, both detectors converge to the true linear MMSE detector, with the subspace blind detector converging much faster than the DMI blind detector; and the performance gain offered by the subspace detector is quite significant for small values of M. Finally, in Fig. 2.7, the performance of the two blind detectors is plotted as a function of the number of users K. As expected from (2.132), the performance gain offered by the subspace detector is significant for smaller values of K, and the gain diminishes as K increases to N. Moreover, it is seen that the performance of the DMI blind detector is insensitive to K.

Simulation Examples

We consider a system with $K = 11$ users. The users' spreading sequences are randomly generated with processing gain $N = 13$. All users have the same amplitudes.

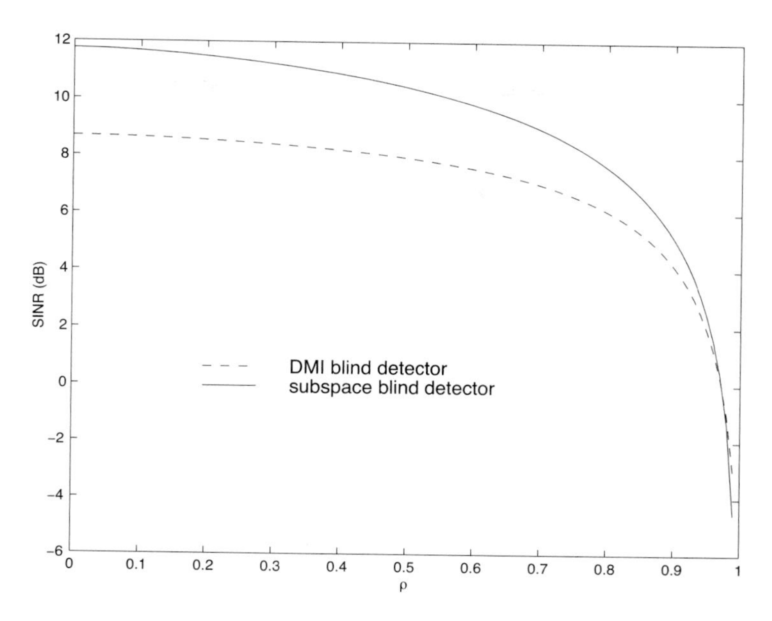

Figure 2.4. Average output SINR versus ρ for two blind detectors. $N = 16$, $K = 6$, $M = 150$, SNR = 15 dB.

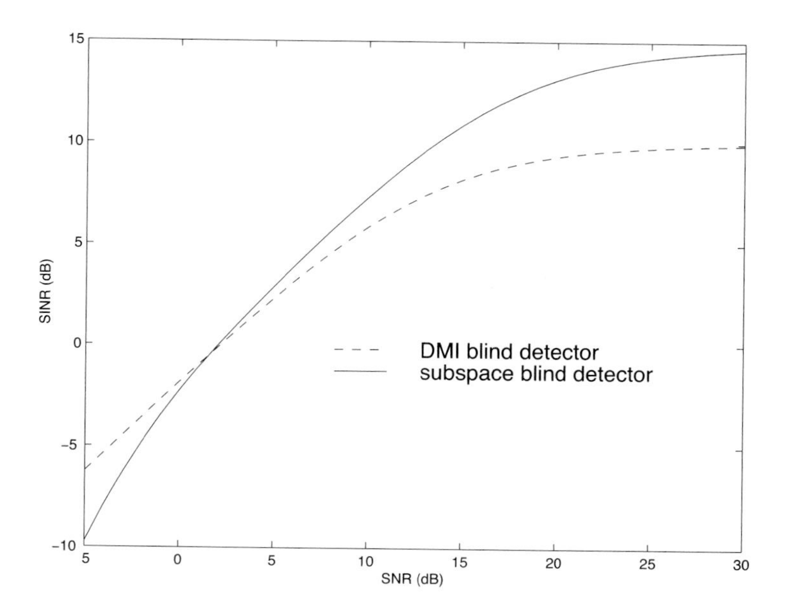

Figure 2.5. Average output SINR versus SNR for two blind detectors. $N = 16$, $K = 6$, $M = 150$, $\rho = 0.4$.

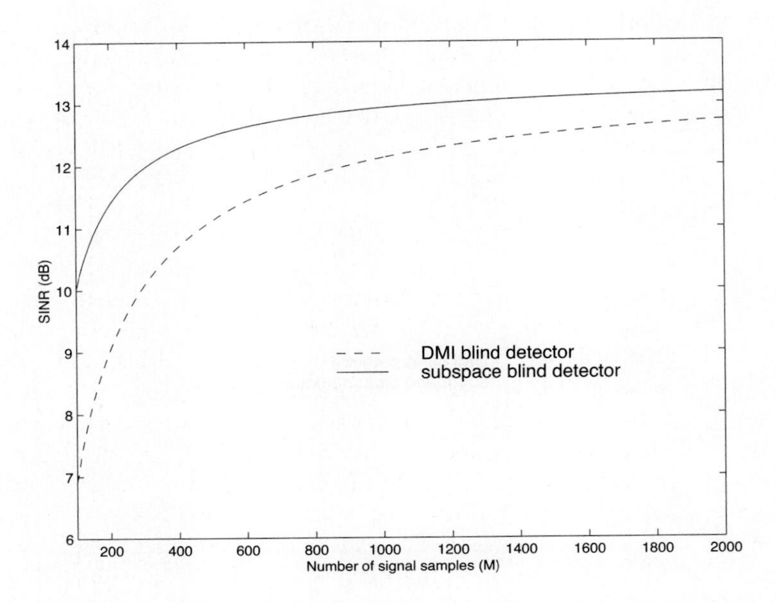

Figure 2.6. Average output SINR versus the number of signal samples M for two blind detectors. $N = 16$, $K = 6$, $\rho = 0.4$, SNR = 15 dB.

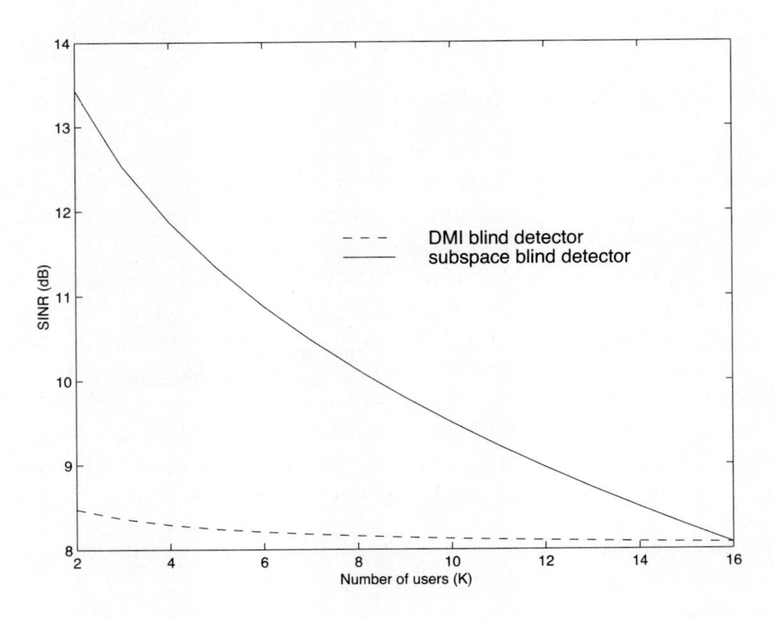

Figure 2.7. Average output SINR versus the number of users K for two blind detectors. $N = 16$, $M = 150$, $\rho = 0.4$, SNR = 15 dB.

Figure 2.8 shows both the analytical and simulated SINR performance for the DMI blind detector and the subspace blind detector. For each detector the SINR is plotted as a function of the number of signal samples (M) used for estimating the detector at some fixed SNR. The simulated and analytical BER performance of these estimated detectors is shown in Fig. 2.9. The analytical BER performance is evaluated using the approximation

$$P_e \cong Q(\sqrt{\text{SINR}}), \tag{2.149}$$

which effectively treats the output interference plus noise of the estimated detector as having a Gaussian distribution. This can be viewed as a generalization of the results in [386], where it is shown that the output of an *exact* linear MMSE detector is well approximated with a Gaussian distribution. From Figs. 2.8 and 2.9 it is seen that the agreement between the analytical performance assessment and the simulation results is excellent for both the SINR and BER. The mismatch between analytical and simulation performance occurs for small values of M, which is not surprising since the analytical performance is based on an asymptotic analysis.

Finally, we note that although in this section we treated only the performance analysis of blind multiuser detection algorithms in simple real-valued synchronous CDMA systems, the analysis for the more realistic complex-valued asynchronous CDMA with multipath channels and blind channel estimation can be found in [196]. Some upper bounds on the achievable performance of various blind multiuser detectors are obtained in [192, 195]. Furthermore, large-system asymptotic performance analysis of blind multiuser detection algorithms is given in [604].

2.6 Subspace Tracking Algorithms

It is seen from Section 2.5 that the linear multiuser detectors are obtained once the signal subspace components are identified. The classic approach to subspace estimation is through batch eigenvalue decomposition (ED) of the sample autocorrelation matrix or batch singular value decomposition (SVD) of the data matrix, both of which are computationally too expensive for adaptive applications. Modern subspace tracking algorithms are recursive in nature and update the subspace in a sample-by-sample fashion. An adaptive blind multiuser detector can be based on subspace tracking by sequentially estimating the signal subspace components and forming the closed-form detector based on these estimates. Specifically, suppose that at time $(i-1)$, the estimated signal subspace rank is $K[i-1]$ and the components are $\boldsymbol{U}_s[i-1]$, $\boldsymbol{\Lambda}_s[i-1]$, and $\sigma^2[i-1]$. Then at time i, the adaptive detector performs the following steps to update the detector and to detect the data.

Algorithm 2.6: [Blind adaptive linear MMSE detector based on subspace tracking—synchronous CDMA]

- *Update the signal subspace: Using a particular signal subspace tracking algorithm, update the signal subspace rank $K[i]$ and the subspace components $\boldsymbol{U}_s[i]$, $\boldsymbol{\Lambda}_s[i]$, and $\sigma^2[i]$.*

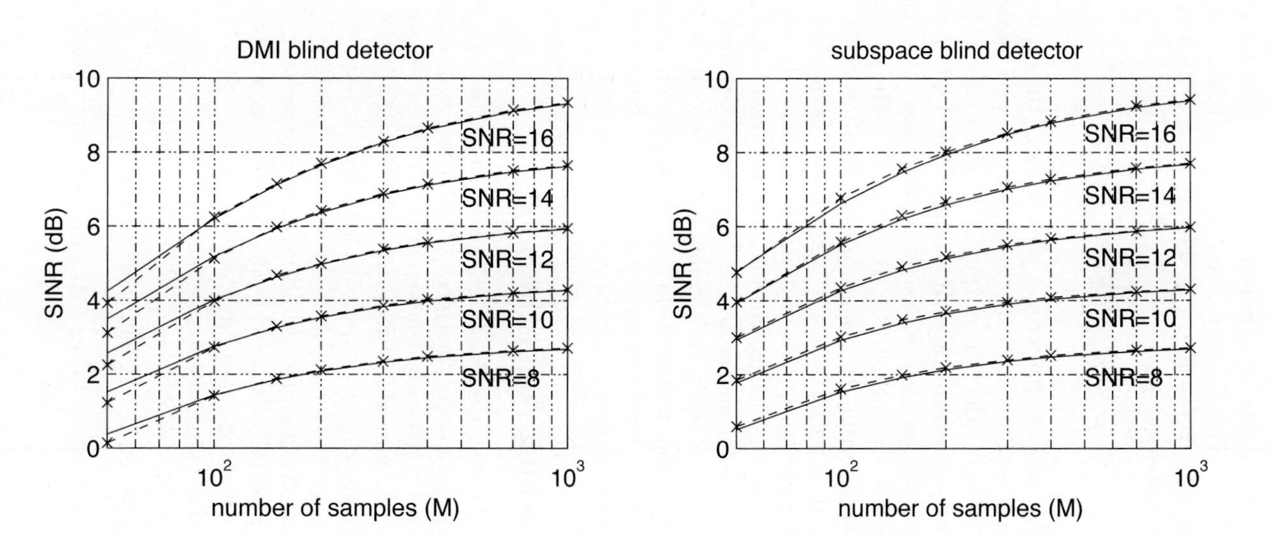

Figure 2.8. Output average SINR versus the number of signal samples M for DMI and subspace detectors. $N = 13, K = 11$. The solid line is the analytical performance, and the dashed line is the simulation performance.

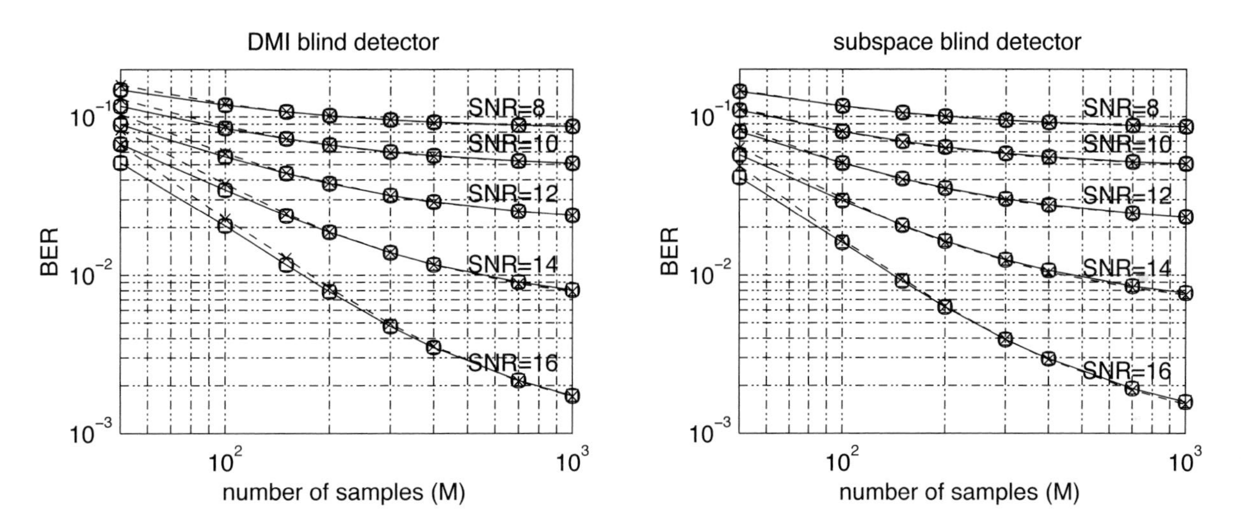

Figure 2.9. BER versus the number of signal samples M for DMI and subspace detectors. $N = 13, K = 11$. The solid line is the analytical performance, and the dashed line is the simulation performance.

- *Form the detector and perform detection:*

$$\boldsymbol{m}_1[i] = \boldsymbol{U}_s[i]\boldsymbol{\Lambda}_s[i]^{-1}\boldsymbol{U}_s[i]^H \boldsymbol{s}_1,$$
$$z_1[i] = \boldsymbol{m}_1[i]^H \boldsymbol{r}[i],$$
$$\hat{\beta}_1[i] = \text{sign}\left\{\Re(z_1[i]z_1[i-1]^*)\right\}.$$

Various subspace tracking algorithms are described in the literature (e.g., [43,87, 97,406,412,461,493,586]). Here we present two low-complexity subspace tracking algorithms: the PASTd algorithm [586] and the more recently developed NAHJ algorithm [412].

2.6.1 PASTd Algorithm

Let $\boldsymbol{r}[i] \in \mathbb{C}^N$ be a random vector with autocorrelation matrix $\boldsymbol{C}_r = E\left\{\boldsymbol{r}[i]\,\boldsymbol{r}[i]^H\right\}$. Consider the scalar function

$$\begin{aligned} \mathcal{J}(\boldsymbol{W}) &= E\left\{\left\|\boldsymbol{r}[i] - \boldsymbol{W}\boldsymbol{W}^H\boldsymbol{r}[i]\right\|^2\right\} \\ &= \text{tr}\left(\boldsymbol{C}_r\right) - 2\,\text{tr}\left(\boldsymbol{W}^H\boldsymbol{C}_r\boldsymbol{W}\right) + \text{tr}\left(\boldsymbol{W}^H\boldsymbol{C}_r\boldsymbol{W}\boldsymbol{W}^H\boldsymbol{W}\right), \end{aligned} \tag{2.150}$$

with a matrix argument $\boldsymbol{W} \in \mathbb{C}^{N\times r}$ $(r < N)$. It can be shown that [586]:

- $\boldsymbol{W}$ is a stationary point of $\mathcal{J}(\boldsymbol{W})$ if and only if $\boldsymbol{W} = \boldsymbol{U}_r\boldsymbol{Q}$, where $\boldsymbol{U}_r \in \mathbb{C}^{N\times r}$ contains any r distinct eigenvectors of $\boldsymbol{C}_r$ and $\boldsymbol{Q} \in \mathbb{C}^{r\times r}$ is any unitary matrix.
- All stationary points of $\mathcal{J}(\boldsymbol{W})$ are saddle points except when $\boldsymbol{U}_r$ contains the r dominant eigenvectors of $\boldsymbol{C}_r$. In that case, $\mathcal{J}(\boldsymbol{W})$ attains the global minimum.

Therefore, for $r = 1$, the solution of minimizing $\mathcal{J}(\boldsymbol{W})$ is given by the most dominant eigenvector of $\boldsymbol{C}_r$. In real applications, only sample vectors $\boldsymbol{r}[i]$ are available. Replacing (2.150) with the exponentially weighted sums yields

$$\mathcal{J}(\boldsymbol{W}[i]) = \sum_{n=0}^{i} \beta^{i-n}\left\|\boldsymbol{r}[n] - \boldsymbol{W}[i]\boldsymbol{W}[i]^H\boldsymbol{r}[n]\right\|^2, \tag{2.151}$$

where $0 < \beta < 1$ is a forgetting factor. The key issue of the PASTd (projection approximation subspace tracking with deflation) approach is to approximate $\boldsymbol{W}[i]^H\boldsymbol{r}[n]$ in (2.151), the unknown projection of $\boldsymbol{r}[n]$ onto the columns of $\boldsymbol{W}[i]$, by $\boldsymbol{y}[n] = \boldsymbol{W}[n-1]^H\boldsymbol{r}[n]$, which can be calculated for $1 \le n \le i$ at time i. This results in a modified cost function

$$\tilde{\mathcal{J}}(\boldsymbol{W}[i]) = \sum_{n=0}^{i} \beta^{i-n}\left\|\boldsymbol{r}[n] - \boldsymbol{W}[i]\boldsymbol{y}[n]\right\|^2. \tag{2.152}$$

The recursive least-squares (RLS) algorithm can then be used to solve for the $\boldsymbol{W}[i]$ that minimizes the exponentially weighted least-squares criterion (2.152).

The PASTd algorithm for tracking the eigenvalues and eigenvectors of the signal subspace is based on the deflation technique and can be described as follows. For $r = 1$, the most dominant eigenvector is updated by minimizing $\tilde{\mathcal{J}}(\boldsymbol{W}[i])$ in (2.152). Then the projection of the current data vector $\boldsymbol{r}[i]$ onto this eigenvector is removed from $\boldsymbol{r}[i]$ itself. Now the second most dominant eigenvector becomes the most dominant one in the updated data vector and it can be extracted similarly. This procedure is applied repeatedly until all the K eigenvectors are estimated sequentially.

Based on the estimated eigenvalues, using information theoretic criteria such as the Akaike information criterion (AIC) or the minimum description length (MDL) criterion [557], the rank of the signal subspace, or equivalently, the number of active users in the channel, can be estimated adaptively as well [585]. The quantities AIC and MDL are defined as follows:

$$\mathrm{AIC}(k) \triangleq (N-k)M \ln \alpha(k) + k(2N-k), \tag{2.153}$$

$$\mathrm{MDL}(k) \triangleq (N-k)M \ln \alpha(k) + \frac{k}{2}(2N-k)\ln M, \qquad k = 1, 2, \ldots, N, \tag{2.154}$$

where M is the number of data samples used in the estimation. When an exponentially weighted window with forgetting factor β is applied to the data, the equivalent number of data samples is $M = 1/(1-\beta)$. $\alpha(k)$ in the definitions above is defined as

$$\alpha(k) = \frac{\left(\sum_{i=k+1}^{N} \hat{\lambda}_i\right) \Big/ (N-k)}{\left(\prod_{i=k+1}^{N} \hat{\lambda}_i\right)^{1/(N-k)}}. \tag{2.155}$$

The AIC (respectively, MDL) estimate of subspace rank is given by the value of k that minimizes the quantity (2.153) [respectively, (2.154)]. Finally, the PASTd algorithm for both rank and signal subspace tracking is summarized in Table 2.1. The computational complexity of this algorithm is $(4K+3)N + \mathcal{O}(K) = \mathcal{O}(NK)$ operations per update. The convergence dynamics of the PASTd algorithm are studied in [587]. It is shown there that with a forgetting factor $\beta = 1$, under mild conditions, this algorithm converges globally and almost surely to the signal eigenvectors and eigenvalues.

Simulation Examples

In what follows we provide two simulation examples to illustrate the performance of the subspace blind adaptive detector employing the PASTd algorithm.

Example 1: Performance Comparison Between Subspace and MOE Blind Detectors This example compares the performance of the subspace-based blind MMSE detector with the performance of the MOE blind adaptive detector. It assumes a real-valued synchronous CDMA system with a processing gain $N = 31$

Table 2.1. PASTd Algorithm [585, 586] for Tracking the Rank and Signal Subspace Components of Received Signal $\boldsymbol{r}[i]$.

Updating the eigenvalues and eigenvectors of signal subspace $\{\lambda_k,\ \boldsymbol{u}_k\}_{k=1}^{K}$			
	$\boldsymbol{x}_1[i]$	$=$	$\boldsymbol{r}[i]$
FOR	$k = 1{:}K_{i-1}$		DO
	$y_k[i]$	$=$	$\boldsymbol{u}_k[i-1]^H \boldsymbol{x}_k[i]$
	$\lambda_k[i]$	$=$	$\beta\lambda_k[i-1] + \lvert y_k[i]\rvert^2$
	$\boldsymbol{u}_k[i]$	$=$	$\boldsymbol{u}_k[i-1] + (\boldsymbol{x}_k[i] - \boldsymbol{u}_k[i-1]y_k[i])\, y_k[i]^*/\lambda_k[i]$
	$\boldsymbol{x}_{k+1}[i]$	$=$	$\boldsymbol{x}_k[i] - \boldsymbol{u}_k[i]y_k[i]$
END			
	$\sigma^2[i]$	$=$	$\beta\sigma^2[i-1] + \lVert\mathbf{x}_{K_{i-1}+1}[i]\rVert^2/\,(N - K_{i-1})$
Updating the rank of signal subspace K_i[a]			
FOR	$k = 1{:}K_{i-1}$		DO
	$\alpha(k)$	$=$	$\left[\sum_{j=k+1}^{N} \lambda_j[i]/N-k\right] / \left(\prod_{j=k+1}^{N} \lambda_j[i]\right)^{1/(N-k)}$
	$\mathrm{AIC}(k)$	$=$	$(N-k)\ln\left[\alpha(k)\right]/(1-\beta) + k(2N-k)$
END			
	K_i	$=$	$\arg\min_{0\leq k\leq N-1} \mathrm{AIC}(k) + 1$
IF	$K_i < K_{i-1}$		THEN
	remove $\{\lambda_k(t), \boldsymbol{u}_k[i]\}_{k=K_i+1}^{K_{i-1}}$		
ELSE IF	$K_i > K_{i-1}$		THEN
	$\mathbf{u}_{K_i}[i]$	$=$	$\boldsymbol{x}_{K_{i-1}+1}(i)/\lVert\boldsymbol{x}_{K_{i-1}+1}[i]\rVert$
	$\lambda_{K_i}[i]$	$=$	$\sigma^2[i]$
END			

[a] The rank estimation is based on the Akaike information criterion.

and six users ($K = 6$). The desired user is user 1. There are four 10-dB multiple-access interferers (MAIs) and one 20-dB MAI (i.e., $A_k^2/A_1^2 = 10$ for $k = 2, \ldots, 5$ and $A_k^2/A_1^2 = 100$ for $k = 6$). The performance measure is the output SINR. For the PASTd subspace tracking algorithm, we found that with a random initialization, the convergence is fairly slow. Therefore, in the simulations, the initial estimates of the eigencomponents of the signal subspace are obtained by applying an SVD to the first 50 data vectors. The PASTd algorithm is then employed thereafter for tracking the signal subspace. The time-averaged output SINR versus number of iterations is plotted in Fig. 2.10.

As a comparison, the simulated performance of the recursive least-squares (RLS) version of the MOE blind adaptive detector is also shown in Fig. 2.10. It has been shown in [389] that the steady-state SINR of this algorithm is given by

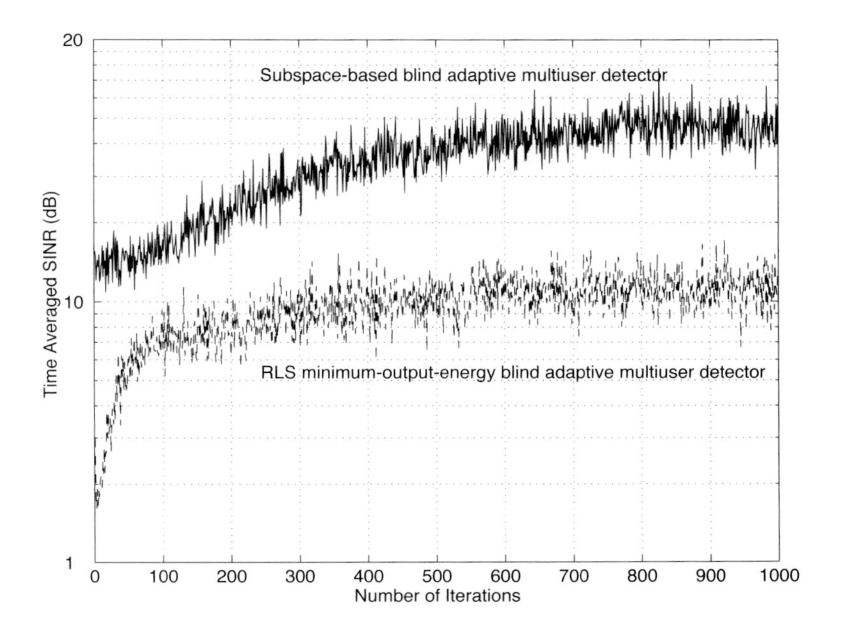

Figure 2.10. Performance comparison between a subspace-based blind linear MMSE multiuser detector and an RLS MOE blind adaptive detector. The processing gain $N = 31$. There are four 10-dB MAIs and one 20-dB MAI in the channel, all relative to the desired user's signal. The signature sequence of the desired user is a m-sequence, whereas the signature sequences of the MAIs are randomly generated. The signal-to-ambient noise ratio after despreading is 20 dB. The forgetting factor used in both algorithms is 0.995. The data plotted are averages over 100 simulations.

$$\mathrm{SINR}^{\infty} = \frac{\mathrm{SINR}^*}{1 + d + d \cdot \mathrm{SINR}^*} = \frac{1}{[(1+d)/\mathrm{SINR}^*] + d} < \frac{1}{d}, \tag{2.156}$$

where SINR^* is the output SINR value of the exact linear MMSE detector, and $d \triangleq [(1-\lambda)/2\lambda]N$ ($0 < \lambda < 1$ is the forgetting factor). Hence the steady-state SINR performance of this algorithm is upper bounded by $1/d$. This behavior can be observed in Fig. 2.10. Although an analytical expression for the steady-state SINR of the subspace-based blind adaptive detector is very difficult to obtain, as the dynamics of the subspace tracking algorithms are fairly complicated, it is seen from Fig. 2.10 that with the same forgetting factor, the subspace blind adaptive detector well outperforms the RLS MOE detector. Moreover, the RLS MOE detector has a computational complexity of $\mathcal{O}(N^2)$ per update, whereas the complexity per update of the subspace detector is $\mathcal{O}(NK)$.

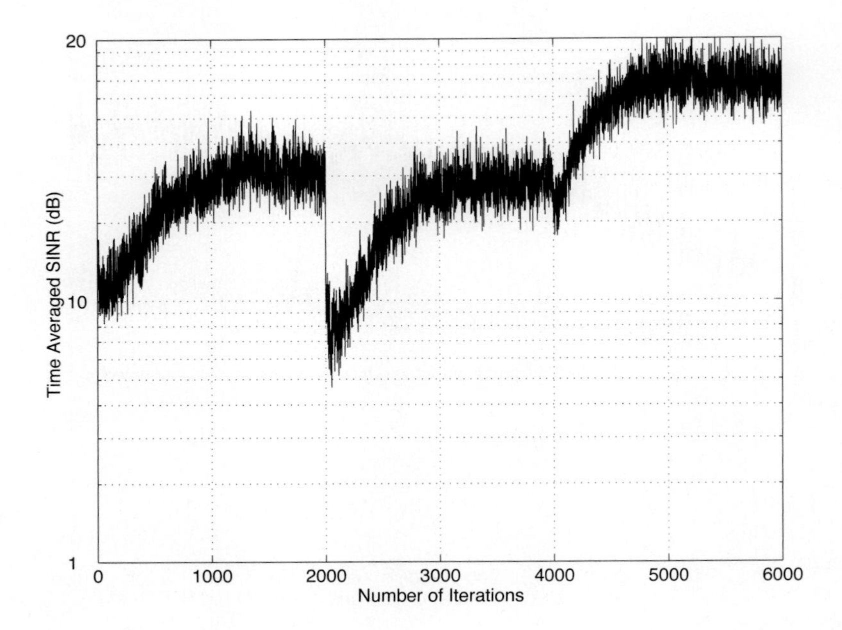

Figure 2.11. Performance of a subspace-based blind linear MMSE multiuser detector in a dynamic multiple-access channel where interferers may enter or exit the channel. At $t = 0$, there are six 10-dB MAIs in the channel; at $t = 2000$, a 20-dB MAI enters the channel; at $t = 4000$, the 20-dB MAI and three of the 10-dB MAIs exit the channel. The processing gain $N = 31$. The signal-to-noise ratio after despreading is 20 dB. The forgetting factor is 0.995. The data plotted are averages over 100 simulations.

Example 2: Tracking Performance in a Dynamic Environment This example illustrates the performance of the subspace blind adaptive detector in a dynamic multiple-access channel, where interferers may enter or exit the channel. The simulation starts with six 10-dB MAIs in the channel; at $t = 2000$, a 20-dB MAI enters the channel; at $t = 4000$, the 20-dB MAI and three of the 10-dB MAIs exit the channel. The performance of the proposed detector is plotted in Fig. 2.11. It is seen that this subspace-based blind adaptive multiuser detector can adapt fairly rapidly to the dynamic channel traffic.

2.6.2 QR-Jacobi Methods

QR-Jacobi methods constitute a family of SVD-based subspace tracking algorithms that rely extensively on Givens rotations during the updating process. This reduces complexity and has the advantage of maintaining the orthogonality of matrices. Members of this family include the algorithms presented in [340, 398, 461].

Let

$$\boldsymbol{Y}[i] = \left[\sqrt{\beta^i}\boldsymbol{r}[0] \ \cdots \ \sqrt{\beta}\,\boldsymbol{r}[i-1] \ \boldsymbol{r}[i]\right] \tag{2.157}$$

denote an $N \times (i+1)$ matrix whose columns contain the exponentially windowed first i snapshots of the received signal. The sample autocorrelation matrix of $\boldsymbol{Y}[i]$ and its eigendecomposition are given by

$$\begin{aligned}\boldsymbol{C}[i] &= \boldsymbol{Y}[i]\boldsymbol{Y}[i]^H \\ &= \boldsymbol{U}_s[i]\boldsymbol{\Lambda}_s[i]\boldsymbol{U}_s[i]^H + \boldsymbol{U}_n[i]\boldsymbol{\Lambda}_n[i]\boldsymbol{U}_n[i]^H.\end{aligned} \tag{2.158}$$

Alternatively, the SVD of the data matrix $\boldsymbol{Y}[i]$ is given by

$$\begin{aligned}\boldsymbol{Y}[i] &= \boldsymbol{U}[i]\boldsymbol{\Sigma}[i]\boldsymbol{V}[i]^H \\ &= \left[\boldsymbol{U}_s[i] \ \boldsymbol{U}_n[i]\right]\begin{bmatrix} \boldsymbol{\Sigma}_s[i] & 0 \\ 0 & \boldsymbol{\Sigma}_n[i] \end{bmatrix}\boldsymbol{V}[i]^H,\end{aligned} \tag{2.159}$$

where in both (2.158) and (2.159) the columns of $\boldsymbol{U}_s[i]$ contain the eigenvectors that span the signal subspace and $\boldsymbol{\Sigma}_s[i] = \sqrt{\boldsymbol{\Lambda}_s[i]}$ contains the square roots of the corresponding eigenvalues. Generally speaking, SVD-based subspace tracking algorithms attempt to track the SVD of a data matrix of growing dimension, defined recursively as

$$\boldsymbol{Y}[i] = \left[\sqrt{\beta}\,\boldsymbol{Y}[i-1] \mid \boldsymbol{r}[i]\right].$$

The matrix $\boldsymbol{V}[i]$ need not be tracked. Furthermore, since the noise subspace does not need to be calculated for the blind multiuser detection algorithm, we do not need to track $\boldsymbol{U}_n[i]$. This allows us to reduce complexity using noise averaging [224]. Since calculating the SVD from scratch at each iteration is time consuming and expensive, the issue then is how to best use the new measurement vector, $\boldsymbol{r}[i+1]$, to update the decomposition in (2.159).

Noise-averaged QR-Jacobi algorithms begin with a Householder transformation that rotates the noise eigenvectors such that the projection of the new measurement vector $\boldsymbol{r}[i+1]$ onto the noise subspace is parallel to the first noise vector, which we denote by $\boldsymbol{u}_n$. Specifically, let

$$\boldsymbol{r}_s = \boldsymbol{U}_s[i]^H\boldsymbol{r}[i+1], \tag{2.160}$$

$$\boldsymbol{u}_n = \frac{\boldsymbol{r}[i+1] - \boldsymbol{U}_s[i]\boldsymbol{r}_s}{\gamma}, \tag{2.161}$$

where $\gamma = \|\boldsymbol{r}[i+1] - \boldsymbol{U}_s[i]\boldsymbol{r}_s\|$. Then we may write the modified factorization

$$\begin{aligned}&\left[\sqrt{\beta}\,\boldsymbol{Y}[i] \mid \boldsymbol{r}[i+1]\right] \\ &\quad = \left[\boldsymbol{U}_s[i] \mid \boldsymbol{u}_n \mid \boldsymbol{U}_n^{\perp}\right]\left[\begin{array}{c|c} \sqrt{\beta}\boldsymbol{\Sigma}[i] & \begin{array}{c}\boldsymbol{r}_s \\ \gamma \\ 0\end{array} \end{array}\right]\left[\begin{array}{c|c} \boldsymbol{V}[i]^H & 0 \\ \hline 0 & 1 \end{array}\right],\end{aligned} \tag{2.162}$$

where $\boldsymbol{U}_n^{\perp}$ represents the subspace of $\boldsymbol{U}_n[i]$ that is orthogonal to $\boldsymbol{u}_n$. The second step in QR-Jacobi methods, sometimes called the *QR step*, involves the use of Givens rotations to zero each entry of the measurement vector's projection onto the signal subspace. We refer the reader to [176] for details concerning the use of Givens matrices for this purpose. The QR step replaces the last row in the middle matrix in the decomposition in (2.162) with zeros. These are row-type transformations involving premultiplication of the middle matrix with a sequence of orthogonal matrices. We do not need to accumulate these transformations in $\boldsymbol{V}[i]$ since it does not need to be tracked.

The next step, the diagonalization step, involves at least one set each of column- and row-type rotations to further concentrate the energy in the middle matrix along its diagonal. Sometimes called the *refinement step*, this is where many of the existing algorithms begin to diverge. The algorithm in [398], for example, performs two fixed sets of rotations in the diagonalization step but leaves the middle matrix in upper triangular form and does not attempt a true diagonalization. This is particularly efficient for applications that do not require a full set of eigenvalues but is not useful here. The algorithm in [398], on the other hand, attempts to optimize the choice of rotations to achieve the best diagonalization possible.

2.6.3 NAHJ Subspace Tracking

The algorithm we present here was developed in [411, 412]. It is a member of the QR-Jacobi family in the sense that it uses Givens rotations during the updating process. However, this algorithm avoids the QR step entirely. Instead of working with the SVD-type decomposition in (2.159), we work with the eigendecomposition of the form

$$\boldsymbol{C}[i] = \boldsymbol{U}[i]\boldsymbol{\Sigma}^2[i]\boldsymbol{U}[i]^H, \tag{2.163}$$

where $\boldsymbol{\Sigma}^2[i]$ is Hermitian and almost diagonal. This is simply the eigendecomposition (2.158) except that we have relaxed the assumption that $\boldsymbol{\Lambda}[i]$ is perfectly diagonal. At each iteration we use a Householder transformation and a vector outer product to update $\boldsymbol{\Sigma}^2[i]$ directly. We then use a single set of two-sided Givens rotations to partially diagonalize the resulting Hermitian matrix. There is no need for a separate QR step. Essentially, the diagonalization process used in this algorithm is a partial implementation of the well-known symmetric Jacobi SVD algorithm [176] (not to be confused with the family of QR-Jacobi update algorithms). This algorithm is used to find the eigenstructure of a general fixed symmetric matrix and is known to generate more accurate eigenvalues and eigenvectors than the symmetric QR SVD algorithm but with a higher computational complexity [313]. However, we do not perform the full sweep of $K(K-1)/2$ rotations required for the symmetric Jacobi algorithm, only a carefully selected set of about K rotations. This is sufficient because the matrix that we wish to diagonalize already has much of its energy concentrated along the diagonal. This is a situation that the Jacobi algorithm can take advantage of but which the QR algorithm cannot. The Jacobi algorithm also has an inherent parallelism, which the QR algorithm does not. Table

Table 2.2. NAHJ Subspace Tracking Algorithm

Given: $\mathbf{\Sigma}_s^2[i-1]$, $\overline{\sigma}^2[i-1]$, and $\boldsymbol{U}_s[i-1]$
1. Calculate $\boldsymbol{r}_s, \boldsymbol{u}_n$, and γ according to (2.160) and (2.161).
2. Dropping the indices, generate the modified factorization: $[\boldsymbol{U}_s \mid \boldsymbol{u}_n \mid \boldsymbol{U}_n^{\perp}]\left(\begin{bmatrix} \beta\mathbf{\Sigma}_s^2 & & \mathbf{0} \\ \mathbf{0} & \beta\overline{\sigma}^2 & \\ \mathbf{0} & & \overline{\sigma}^2\boldsymbol{I}\end{bmatrix} + \begin{bmatrix} \begin{bmatrix} \boldsymbol{r}_s \\ \beta \end{bmatrix} [\boldsymbol{r}_s^H \mid \beta] & \mathbf{0} \\ \mathbf{0} & \mathbf{0}\end{bmatrix}\right)\begin{bmatrix} \boldsymbol{U}_s^H \\ \boldsymbol{u}_n^H \\ \boldsymbol{U}_n^{\perp H}\end{bmatrix}.$
3. Let $\mathbf{\Psi}_s$ be the $K+2$ principal submatrix of the matrix sum in step 2. Apply a sequence of $r+1$ Givens rotations to $\mathbf{\Psi}_s$ to produce $\mathbf{\Psi}_a = \mathbf{\Theta}_{K+1}^T, \ldots, \mathbf{\Theta}_1^T\mathbf{\Psi}_s\mathbf{\Theta}_1, \ldots, \mathbf{\Theta}_{K+1}^T$.
4. Set $\mathbf{\Sigma}_s^2[i]$ equal to the $K+1$ principal submatrix of $\boldsymbol{Y}_a$.
5. Let $\boldsymbol{U}_s[i]$ be composed of the first $K+1$ columns of $[\boldsymbol{U}_s \mid \boldsymbol{u}_n]\,\mathbf{\Theta}_1 \cdots \mathbf{\Theta}_{K+1}$.
6. Reaverage the noise power: $\overline{\sigma}^2[i] = \frac{(N-K-2)(\sqrt{\gamma}\overline{\sigma}^2[i-1]) + \lvert\hat{\sigma}^2\rvert}{N-K-1}$, where $\hat{\sigma}^2 = \boldsymbol{Y}_a(K+2, K+2)$.
7. Let $\mathbf{\Lambda}_s[i]$ be the diagonal matrix whose diagonal is equal to the first K elements of the diagonal of $\mathbf{\Psi}_a$.

2.2 contains a summary of this algorithm, which we term NAHJ (noise-averaged Hermitian–Jacobi) subspace tracking.

Algorithm

The first step in NAHJ subspace tracking is the Householder transformation mentioned previously. The second step involves generating a modified factorization that maintains the equality

$$\boldsymbol{U}[i]\mathbf{\Sigma}^2[i]\boldsymbol{U}^H[i] = \beta\boldsymbol{U}[i-1]\mathbf{\Sigma}^2[i-1]\boldsymbol{U}[i-1]^H + \boldsymbol{r}[i]\boldsymbol{r}[i]^H. \tag{2.164}$$

Step 3 requires that we apply $K+1$ Givens rotations to partially diagonalize $\mathbf{\Psi}_s$. Ideally, we would apply these rotations to those off-diagonal elements having the largest magnitudes. However, since the off-diagonal maxima can be located anywhere in $\mathbf{\Psi}_s$, finding the optimal set of rotations requires an $\mathcal{O}(K^2)$ search for each rotation. This leads to an $\mathcal{O}(K^3)$ complexity algorithm. To maintain low complexity we have implemented a suboptimal alternative that is simple, yet effective. Let $\boldsymbol{z} = [\boldsymbol{r}_s^H \mid \beta]^H$ be the vector whose outer product is used in the modified factorization of step 2. Suppose that $i_0, 1 \leq i_0 \leq K+2$, is the index of the element in $\boldsymbol{z}$ that has the largest magnitude. The set of elements we choose to annihilate with the Givens rotations is given by $\{(\mathbf{\Psi}_s)_{i_0,j}\}_{j=1}^{K+2}, j \neq i_0$. Of course, if $(\mathbf{\Psi}_s)_{i_0,j}$ is annihilated, so is $(\mathbf{\Psi}_s)_{j,i_0}$. This choice of rotations is not optimal; in fact, since we retain the off-diagonal information from the previous iteration we cannot even be sure that we annihilate the off-diagonal element in $\mathbf{\Psi}_s$ with the largest magnitude. Nevertheless, we see that the technique is very simple and is somewhat heuristically pleasing. Ultimately, performance is the measure of merit, and simulations show that it performs very well. The total computational complexity of the NAHJ subspace tracking algorithm is $\mathcal{O}(NK)$ per update.

To adapt to changes in the size of the signal subspace (number of users), the tracking algorithm must be rank-adaptive. As before, both the AIC and the MDL criteria can be used for this purpose. To use this algorithm, we must track at least one extra eigenvalue–eigenvector pair—hence the appearance of $K+1$ in Table 2.2.

Simulation Example

This example compares the performance of the subspace blind adaptive multiuser detector using the NAHJ subspace tracking algorithm with that of the LMS MOE blind adaptive multiuser detector. It assumes a synchronous CDMA system with seven users ($K = 7$), each employing a gold sequence of length 15 ($N = 15$). The desired user is user 1. There are two 0-dB and four 10-dB interferers. The performance measure is the output SINR. The performance is shown in Fig. 2.12. It is seen that the subspace blind detector significantly outperforms the LMS MOE blind detector, in terms of both convergence rate and steady-state SINR. Further applications of the NAHJ subspace tracking algorithm are found in later chapters (cf. Sections 2.7.4, 3.5.2, 5.5.4, and 5.6.3). Simulations show that the NAHJ subspace tracking algorithm substantially outperforms the PASTd algorithm, espe-

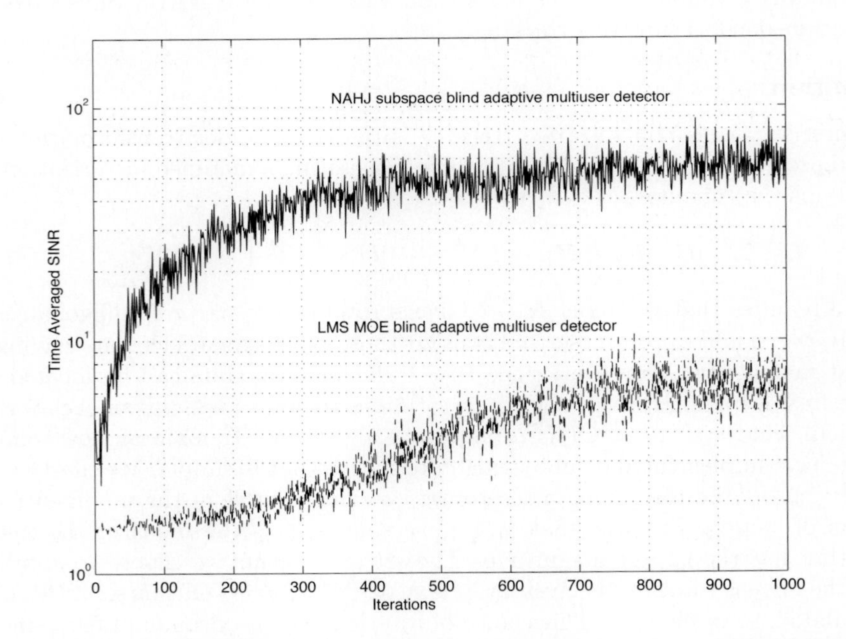

Figure 2.12. Performance comparison between a subspace blind adaptive multiuser detector using the NAHJ subspace tracking algorithm and an LMS MOE blind adaptive multiuser detector.

cially in multipath environments. Note that both algorithms have a computational complexity of $\mathcal{O}(KN)$.

2.7 Blind Multiuser Detection in Multipath Channels

In previous sections we focused on the synchronous CDMA signal model. In a practical wireless CDMA system, however, the users' signals are asynchronous. Moreover, the physical channel exhibits dispersion due to multipath effects that further distort the signals. In this section we address blind multiuser detection in such channels. As will be seen, the principal techniques developed in previous sections can be applied to this more realistic situation as well.

2.7.1 Multipath Signal Model

We now consider a more general multiple-access signal model where the users are asynchronous, and the channel exhibits multipath distortion effects. In particular, the multipath channel impulse response of the kth user is modeled as in (1.10):

$$g_k(t) = \sum_{l=1}^{L} \alpha_{l,k} \delta(t - \tau_{l,k}), \tag{2.165}$$

where L is the total number of paths in the channel, and $\alpha_{l,k}$ and $\tau_{l,k}$ are, respectively, the complex path gain and the delay of the kth user's lth path, $\tau_{1,k} < \tau_{2,k} < \cdots < \tau_{L,k}$. The continuous-time signal received in this case is given by

$$\begin{aligned} r(t) &= \sum_{k=1}^{K} \sum_{i=0}^{M-1} b_k[i] \left\{ s_k(t - iT) \star g_k(t) \right\} + n(t) \\ &= \sum_{k=1}^{K} \sum_{i=0}^{M-1} b_k[i] \sum_{l=1}^{L} \alpha_{l,k} s_k(t - iT - \tau_{l,k}) + n(t), \end{aligned} \tag{2.166}$$

where $\star$ denotes convolution and $s_k(t)$ is the spreading waveform of the kth user given by (2.2).

At the receiver, the received signal $r(t)$ is filtered by a chip-matched filter and sampled at a multiple (p) of the chip rate (i.e., the sampling time interval is $\Delta = T_c/p = T/P$, where $P \triangleq pN$ is the total number of samples per symbol interval). Let

$$\iota \triangleq \max_{1 \le k \le K} \left\{ \left\lceil \frac{\tau_{L,k} + T_c}{T} \right\rceil \right\}$$

be the maximum delay spread in terms of symbol intervals. Substituting (2.2) into (2.166), the qth signal sample during the ith symbol interval is given by

$$\begin{aligned}
r_q[i] &= \int_{iT+q\Delta}^{iT+(q+1)\Delta} r(t)\psi(t-iT-q\Delta)\,\mathrm{d}t \\
&= \int_{iT+q\Delta}^{iT+(q+1)\Delta} \psi(t-iT-q\Delta)\sum_{k=1}^{K}\sum_{m=0}^{M-1} b_k[m]\sum_{l=1}^{L}\alpha_{l,k}\frac{1}{\sqrt{N}} \\
&\qquad \cdot \sum_{j=0}^{N-1} s_{j,k}\psi(t-mT-\tau_{l,k}-jT_c)\,\mathrm{d}t + n_q[i] \\
&= \sum_{k=1}^{K}\sum_{m=i-\iota}^{i} b_k[m]\sum_{l=1}^{L}\alpha_{l,k}\frac{1}{\sqrt{N}}\sum_{j=0}^{N-1} s_{j,k} \\
&\qquad \cdot \int_{iT+q\Delta}^{iT+(q+1)\Delta} \psi(t-iT-q\Delta)\psi(t-mT-\tau_{l,k}-jT_c)\,\mathrm{d}t + n_q[i] \\
&= \sum_{k=1}^{K}\sum_{m=0}^{\iota} b_k[i-m] \\
&\qquad \underbrace{\cdot \sum_{j=0}^{N-1} s_{j,k}\frac{1}{\sqrt{N}}\overbrace{\sum_{l=1}^{L}\alpha_{l,k}\int_0^{\Delta}\psi(t)\psi(t-\tau_{l,k}+mT-jT_c+q\Delta)\,\mathrm{d}t}^{f_k[mP-jp+q]}}_{h_k[mP+q]} + n_q[i], \\
&\qquad q = 0,\ldots,P-1; \quad i = 0,\ldots,M-1,
\end{aligned} \tag{2.167}$$

where $n_q[i] = \int_{iT+q\Delta}^{iT+(q+1)\Delta} n(t)\psi(t-iT-q\Delta)\,\mathrm{d}t$. Denote

$$\underbrace{\underline{r}[i]}_{P\times 1} \triangleq \begin{bmatrix} r_0[i] \\ \vdots \\ r_{P-1}[i] \end{bmatrix}, \quad \underbrace{\underline{b}[i]}_{K\times 1} \triangleq \begin{bmatrix} b_1[i] \\ \vdots \\ b_K[i] \end{bmatrix}, \quad \underbrace{\underline{n}[i]}_{P\times 1} \triangleq \begin{bmatrix} n_0[i] \\ \vdots \\ n_{P-1}[i] \end{bmatrix},$$

$$\underbrace{\underline{H}[j]}_{P\times K} \triangleq \begin{bmatrix} h_1[jP] & \cdots & h_K[jP] \\ \vdots & \vdots & \vdots \\ h_1[jP+P-1] & \cdots & h_K[jP+P-1] \end{bmatrix}, \quad j = 0,\ldots,\iota.$$

Then (2.167) can be written in terms of vector convolution as

$$\underline{r}[i] = \underline{H}[i] \star \underline{b}[i] + \underline{n}[i]. \tag{2.168}$$

By stacking m successive sample vectors, we further define the following quantities:

$$\underbrace{\boldsymbol{r}[i]}_{Pm\times 1} \triangleq \begin{bmatrix} \underline{r}[i] \\ \vdots \\ \underline{r}[i+m-1] \end{bmatrix}, \quad \underbrace{\boldsymbol{n}[i]}_{Pm\times 1} \triangleq \begin{bmatrix} \underline{n}[i] \\ \vdots \\ \underline{n}[i+m-1] \end{bmatrix},$$

$$\underbrace{\boldsymbol{b}[i]}_{K(m+\iota)\times 1} \triangleq \begin{bmatrix} \underline{b}[i-\iota] \\ \vdots \\ \underline{b}[i+m-1] \end{bmatrix},$$

$$\underbrace{\boldsymbol{H}}_{Pm\times K(m+\iota)} \triangleq \begin{bmatrix} \underline{H}[\iota] & \cdots & \underline{H}[0] & \cdots & \mathbf{0} \\ \vdots & \ddots & \ddots & \ddots & \vdots \\ \mathbf{0} & \cdots & \underline{H}[\iota] & \cdots & \underline{H}[0] \end{bmatrix},$$

where the smoothing factor m is chosen according to $m = \lceil (P+K)/(P-K) \rceil \iota$. Note that for such m, the matrix $\boldsymbol{H}$ is a "tall" matrix [i.e., $Pm \geq K(m+\iota)$]. We can then write (2.168) in matrix form as

$$\boldsymbol{r}[i] = \boldsymbol{H}\,\boldsymbol{b}[i] + \boldsymbol{n}[i]. \tag{2.169}$$

2.7.2 Linear Multiuser Detectors

Suppose that we are interested in demodulating the data of user 1. Then (2.168) can be written as

$$\underline{r}[i] = \underline{H}^1[0]b_1[i] + \sum_{j=1}^{\iota} \underline{H}^1[j]b_1[i-j] + \sum_{k=2}^{K}\sum_{j=0}^{\iota} \underline{H}^k[j]b_k[i-j] + \underline{n}[i], \tag{2.170}$$

where $\underline{H}^k[m]$ denotes the kth column of $\underline{H}[m]$. In (2.170), the first term contains the data bit of the desired user at time i; the second term contains the previous data bits of the desired user [i.e., we have intersymbol interference (ISI)]; and the last term contains the signals from other users [i.e., multiple-access interference (MAI)]. Hence, compared with the synchronous model considered in previous sections, the multipath channel introduces ISI, which together with MAI must be contended with at the receiver. Moreover, the augmented signal model (2.169) is very similar to the synchronous signal model (2.5). We proceed to develop linear receivers for this system.

A linear receiver for this purpose can be represented by a Pm-dimensional complex vector $\boldsymbol{w}_1 \in \mathbb{C}^{Pm}$, which is correlated with the received signal $\boldsymbol{r}[i]$ in (2.169) to obtain

$$z_1[i] = \boldsymbol{w}_1^H \boldsymbol{r}[i]. \tag{2.171}$$

The coherent detection rule is then given by

$$\hat{b}_1[i] = \text{sign}\left\{\Re\left(z_1[i]\right)\right\}; \tag{2.172}$$

and the differential detection rule is given by

$$\hat{\beta}_1[i] = \text{sign}\left\{\Re\left(z_1[i]z_1[i-1]^*\right)\right\}. \tag{2.173}$$

As before, two forms of such linear detectors are the linear decorrelating detector and the linear minimum mean-square-error (MMSE) detector, which are described next.

Linear Decorrelating Detector

The linear decorrelating detector for user 1 has the form of (2.171)–(2.173) with the weight vector $\boldsymbol{w}_1 = \boldsymbol{d}_1$, such that both the multiple-access interference (MAI) and the intersymbol interference (ISI) are eliminated completely at the detector output.[3]

Denote by $\boldsymbol{e}_l$ the $K(m+\iota)$-vector with all-zero entries except for the lth entry, which is 1. Recall that the smoothing factor m is chosen such that the matrix $\boldsymbol{H}$ in (2.169) is a tall matrix. Assume that $\boldsymbol{H}$ has full column rank [i.e., $\mathsf{rank}(\boldsymbol{H}) = K(m+\iota) \triangleq r$]. Let $\boldsymbol{H}^\dagger$ be the Moore–Penrose generalized inverse of the matrix $\boldsymbol{H}$:

$$\boldsymbol{H}^\dagger = \left(\boldsymbol{H}^H\boldsymbol{H}\right)^{-1}\boldsymbol{H}^H. \tag{2.174}$$

The linear decorrelating detector for user 1 is then given by

$$\boldsymbol{d}_1 = \boldsymbol{H}^{\dagger H}\boldsymbol{e}_{K\iota+1} = \boldsymbol{H}\left(\boldsymbol{H}^H\boldsymbol{H}\right)^{-1}\boldsymbol{e}_{K\iota+1}. \tag{2.175}$$

Using (2.169) and (2.175), we have

$$\begin{aligned} z_1[i] \triangleq \boldsymbol{d}_1^H\boldsymbol{r}[i] &= \boldsymbol{e}_{K\iota+1}^H\left(\boldsymbol{H}^H\boldsymbol{H}\right)^{-1}\boldsymbol{H}^H\boldsymbol{H}\boldsymbol{b}[i] + \boldsymbol{d}_1^H\boldsymbol{b}[i] \\ &= \boldsymbol{e}_{K\iota+1}^H\boldsymbol{b}[i] + \boldsymbol{d}_1^H\boldsymbol{n}[i] \\ &= b_1[i] + \boldsymbol{d}_1^H\boldsymbol{n}[i]. \end{aligned} \tag{2.176}$$

It is seen from (2.176) that both the MAI and the ISI are eliminated completely at the output of the linear zero-forcing detector. In the absence of noise (i.e., $\boldsymbol{n}[i] = \boldsymbol{0}$), the data bit of the desired user, $b_1[i]$, is recovered perfectly.

Linear MMSE Detector

The linear minimum mean-square-error (MMSE) detector for user 1 has the form of (2.171)–(2.173) with the weight vector $\boldsymbol{w}_1 = \boldsymbol{m}_1$, where $\boldsymbol{m}_1 \in \mathbb{C}^{Pm}$ is chosen to minimize the output mean-square error (MSE):

$$\boldsymbol{m}_1 = \arg\min_{\boldsymbol{w}\in\mathbb{C}^{Pm}} E\left\{\left\|b_1[i] - \boldsymbol{w}^H\boldsymbol{r}[i]\right\|^2\right\} = \boldsymbol{C}_r^{-1}\bar{\boldsymbol{h}}_1, \tag{2.177}$$

[3]In the context of equalization, this detector is known as a *zero-forcing equalizer*, as noted in Chapter 1.

where

$$\boldsymbol{C}_r = E\left\{\boldsymbol{r}[i]\boldsymbol{r}[i]^H\right\} = \boldsymbol{H}\boldsymbol{H}^H + \sigma^2 \boldsymbol{I}_{Pm}, \tag{2.178}$$

$$\begin{aligned} \bar{\boldsymbol{h}}_1 &\triangleq E\left\{\boldsymbol{r}[i]b_1[i]\right\} = \boldsymbol{H}\boldsymbol{e}_{Km+1} \\ &= \Big[\underbrace{h_1[0], \ldots, h_1[P-1], \ldots\ldots, h_1[\iota P], \ldots, h_1[\iota P + P - 1]}_{\boldsymbol{h}_k^T}, \underbrace{0 \cdots 0}_{[P(m-\iota-1)] \text{ 0's}}\Big]^T. \end{aligned} \tag{2.179}$$

Subspace Linear Detectors

Let $\lambda_1 \geq \lambda_2 \geq \cdots \geq \lambda_{Pm}$ be the eigenvalues of $\boldsymbol{C}_r$ in (2.178). Since the matrix $\boldsymbol{H}$ has full column rank $r \triangleq K(m + \iota)$, the signal component of the covariance matrix $\boldsymbol{C}_r$ (i.e., $\boldsymbol{H}\boldsymbol{H}^H$) has rank r. Therefore, we have

$$\begin{aligned} \lambda_i &> \sigma^2 \text{for } i = 1, \ldots, r, \\ \lambda_i &= \sigma^2 \text{for } i = r+1, \ldots, Pm. \end{aligned}$$

By performing an eigendecomposition of the matrix $\boldsymbol{C}_r$, we obtain

$$\boldsymbol{C}_r = \boldsymbol{U}_s \boldsymbol{\Lambda}_s \boldsymbol{U}_s^H + \sigma^2 \boldsymbol{U}_n \boldsymbol{U}_n^H, \tag{2.180}$$

where $\boldsymbol{\Lambda}_s = \mathsf{diag}(\lambda_1, \ldots, \lambda_r)$ contains the r largest eigenvalues of $\boldsymbol{C}_r$ in descending order, $\boldsymbol{U}_s = [\boldsymbol{u}_1 \ \cdots \ \boldsymbol{u}_r]$ contains the corresponding orthogonal eigenvectors, and $\boldsymbol{U}_n = [\boldsymbol{u}_{r+1} \ \cdots \ \boldsymbol{u}_{Pm}]$ contains the $Pm - r$ orthogonal eigenvectors that correspond to the eigenvalue σ^2. It is easy to see that $\mathsf{range}(\boldsymbol{H}) = \mathsf{range}(\boldsymbol{U}_s)$. As before, the column space of $\boldsymbol{U}_s$ is called the *signal subspace* and its orthogonal complement, the *noise subspace*, is spanned by the columns of $\boldsymbol{U}_n$.

Following exactly the same line of development as in the synchronous case, it can be shown that the linear decorrelating detector given by (2.175), and the linear MMSE detector given by (2.177), can be expressed in terms of the signal subspace components above as [548]

$$\boldsymbol{d}_1 = \boldsymbol{U}_s \left(\boldsymbol{\Lambda}_s - \sigma^2 \boldsymbol{I}_r\right)^{-1} \boldsymbol{U}_s^H \bar{\boldsymbol{h}}_1, \tag{2.181}$$

$$\boldsymbol{m}_1 = \boldsymbol{U}_s \boldsymbol{\Lambda}_s^{-1} \boldsymbol{U}_s^H \bar{\boldsymbol{h}}_1. \tag{2.182}$$

Decimation-Combining Linear Detectors

The linear detectors discussed above operate in a Pm-dimensional vector space. As will be seen in the next section, the major computation in channel estimation involves computing the singular value decomposition (SVD) of the autocorrelation matrix $\boldsymbol{C}_r$ of dimensions $Pm \times Pm$, which has computational complexity $\mathcal{O}(P^3 m^3)$. By down-sampling the received signal sample vector $\boldsymbol{r}[i]$ by a factor of p, it is possible to construct the linear detectors in an Nm-dimensional space and to reduce the total computational complexity of channel estimation by a factor of $\mathcal{O}(p^2)$. (Recall that p is the chip oversampling factor.) This technique is described next.

For $q = 0, \ldots, p-1$, denote

$$\underline{r}_q[i] \triangleq \begin{bmatrix} r_q[i] \\ r_{q+p}[i] \\ \vdots \\ r_{q+p(N-1)}[i] \end{bmatrix}_{N\times 1}, \qquad \underline{v}_q[i] \triangleq \begin{bmatrix} n_q[i] \\ n_{q+p}[i] \\ \vdots \\ n_{q+p(N-1)}[i] \end{bmatrix}_{N\times 1},$$

$$\underline{H}_m[j] \triangleq \begin{bmatrix} h_1[mP+q] & \cdots & h_K[mP+q] \\ h_1[mP+q+p] & \cdots & h_K[mP+q+p] \\ \vdots & \vdots & \vdots \\ h_1[mP+q+p(N-1)] & \cdots & h_K[mP+q+p(N-1)] \end{bmatrix}_{N\times K}, \quad m=0,1,\ldots,\iota$$

$$\boldsymbol{r}_q[i] \triangleq \begin{bmatrix} \underline{r}_q[i] \\ \vdots \\ \underline{r}_q[i+m-1] \end{bmatrix}_{Nm\times 1}, \qquad \boldsymbol{n}_q[i] \triangleq \begin{bmatrix} \underline{n}_q[i] \\ \vdots \\ \underline{n}_q[i+m-1] \end{bmatrix}_{Nm\times 1},$$

$$\boldsymbol{H}_q \triangleq \begin{bmatrix} \underline{H}_q[\iota] & \cdots & \underline{H}_q[0] & \cdots & \mathbf{0} \\ \vdots & \ddots & \ddots & \ddots & \vdots \\ \mathbf{0} & \cdots & \underline{H}_q[\iota] & \cdots & \underline{H}_q[0] \end{bmatrix}_{Nm\times K(m+\iota)}.$$

Similar to what we did before, we can write

$$\boldsymbol{r}_q[i] = \boldsymbol{H}_q\,\boldsymbol{b}[i] + \boldsymbol{n}_q[i], \qquad q = 0, \ldots, p-1. \tag{2.183}$$

Assume that $Nm \geq K(m+\iota)$ (i.e., the matrix $\boldsymbol{H}_q$ is a tall matrix), and $\mathsf{rank}(\boldsymbol{H}_q) = K(m+\iota)$ (i.e., $\boldsymbol{H}_q$ has full column rank). For each down-sampled received signal $\boldsymbol{r}_q[i]$, the corresponding weight vectors for user 1's linear decorrelating detector and the linear MMSE detector are given, respectively, by

$$\boldsymbol{d}_{1,q} = \boldsymbol{H}_q \left(\boldsymbol{H}_q^H \boldsymbol{H}_q\right)^{-1} \boldsymbol{e}_{K\iota+1}, \tag{2.184}$$

$$\boldsymbol{m}_{1,q} = \boldsymbol{C}_q^{-1}\bar{\boldsymbol{h}}_{1,q} = \left(\boldsymbol{H}_q\boldsymbol{H}_q^H + \sigma^2\boldsymbol{I}_r\right)^{-1} \boldsymbol{H}_q\boldsymbol{e}_{K\iota+1}, \tag{2.185}$$

where $\boldsymbol{C}_q \triangleq E\left\{\boldsymbol{r}_q[i]\boldsymbol{r}_q[i]^H\right\}$. By computing the subspace components of the autocorrelation matrix $\boldsymbol{C}_q$, subspace versions of the linear detectors above can be constructed in forms similar to (2.181) and (2.182).

To detect user 1's data bits, each down-sampled signal vector $\boldsymbol{r}_q[i]$ is correlated with the corresponding weight vector to obtain

$$z_{1,q}[i] = \boldsymbol{w}_{1,q}^H\boldsymbol{r}_q[i] \;\; (\boldsymbol{w}_{1,q} = \boldsymbol{d}_{1,q} \,\text{or}\, \boldsymbol{w}_{1,q} = \boldsymbol{m}_{1,q}), \qquad q = 0, \ldots, p-1. \tag{2.186}$$

The data bits are then demodulated according to

$$\hat{b}_1[i] = \text{sign}\left\{\Re\left(\sum_{q=0}^{p-1} z_{1,q}[i]\right)\right\}, \qquad \text{(coherent detection)} \tag{2.187}$$

$$\hat{\beta}_1[i] = \text{sign}\left\{\Re\left(\sum_{q=0}^{p-1} z_{1,q}[i]z_{1,q}[i-1]^*\right)\right\}, \qquad \text{(differential detection)} \tag{2.188}$$

In the decimation-combining approach described above, since the signal vectors have dimension Nm, the complexity of estimating each decimated channel response $\boldsymbol{h}_{k,q}, q = 0, \ldots, p-1$, is $\mathcal{O}(N^3 m^3)$. Hence the total complexity of channel estimation is $\mathcal{O}(pN^3 m^3) = \mathcal{O}(P^3 m^3/p^2)$ [i.e., a reduction of $\mathcal{O}(p^2)$ is achieved compared with the Pm-dimensional detectors]. However, the number of users that can be supported by this receiver structure is reduced by a factor of p. That is, for a given smoothing factor m, the number of users that can be accommodated by the Pm-dimensional detector is $\left\lceil \frac{m-\iota}{m+\iota} \cdot P \right\rceil$, whereas by forming p Nm-dimensional detectors and then combining their outputs, the number of users that can be supported is reduced to $\left\lceil \frac{m-\iota}{m+\iota} \cdot N \right\rceil$.

2.7.3 Blind Channel Estimation

It is seen from the discussion above that unlike the synchronous case, where the linear detectors can be written in closed form once the signal subspace components are identified, in multipath channels the composite channel response vector of the desired user, $\bar{\boldsymbol{h}}_1$, is needed to form the blind detector. This vector can be viewed as the channel-distorted original spreading waveform $\boldsymbol{s}_1$. The multipath channel can be estimated by transmitting a training sequence [32, 64, 112, 442, 581, 612]. Alternatively, the channel can be estimated blindly by exploiting the orthogonality between the signal and noise subspaces [31, 272, 484, 548, 551]. We next address the problem of blind channel estimation. From (2.167),

$$h_k[n] = \sum_{j=0}^{N-1} s_{j,k} f_k[n - jp], \qquad n = 0, 1, \ldots, (\iota+1)P - 1, \tag{2.189}$$

with

$$f_k[m] \triangleq \frac{1}{\sqrt{N}} \sum_{l=1}^{L} \alpha_{l,k} \int_0^{\Delta} \psi(t)\psi(t - \tau_{l,k} + m\Delta), \quad m = 0, 1, \ldots, \iota p\mu - 1, \tag{2.190}$$

where $p\mu$ is the length of the channel response $\{f_k[m]\}$, which satisfies

$$p\mu = \left\lceil \frac{\tau_{L,k}}{T_c} \right\rceil = \left\lceil \frac{\tau_{L,k}}{T} \cdot \frac{T}{T_c} \right\rceil \leq \iota N. \tag{2.191}$$

Decimate $h_k[n]$ into p subsequences as

$$h_{k,q}[m] \triangleq h_k[q + mp] = \sum_{j=0}^{N-1} s_{j,k} \underbrace{f_k[q + (m-j)p]}_{f_{k,q}[m-j]},$$
$$q = 0, \ldots, p-1; \; m = 0, \ldots, (\iota+1)N - 1. \tag{2.192}$$

Note that the sequences $f_{k,q}[m]$ are obtained by down-sampling the sequence $\{f_k[m]\}$ by a factor of p:

$$f_{k,q}[m] \triangleq f_k[q+mp], \qquad m = 0, \ldots, \iota N - 1; \;\; q = 0, \ldots, p-1. \tag{2.193}$$

From (2.192), we have

$$\{h_{k,q}[0], \ldots, h_{k,q}[(\iota+1)N-1]\} = \{s_{0,k}, \ldots, s_{N-1,k}\} \star \{f_{k,q}[0], \ldots, f_{k,q}[\mu-1]\}. \tag{2.194}$$

Denote

$$\boldsymbol{h}_{k,q} = \begin{bmatrix} h_{k,q}[0] \\ \vdots \\ h_{k,q}[(\iota+1)N-1] \end{bmatrix}_{(\iota+1)N\times 1}, \qquad \boldsymbol{f}_{k,q} = \begin{bmatrix} f_{k,q}[0] \\ \vdots \\ f_{k,q}[\iota N-1] \end{bmatrix}_{\mu\times 1},$$

$$\boldsymbol{\Xi}_k = \begin{bmatrix} s_{0,k} & & & \\ s_{1,k} & s_{0,k} & & \\ \vdots & s_{1,k} & \ddots & \\ \vdots & \vdots & \ddots & s_{0,k} \\ s_{N-1,k} & \vdots & & s_{1,k} \\ & s_{N-1,k} & & \vdots \\ & & \ddots & \vdots \\ & & & s_{N-1,k} \end{bmatrix}_{(\iota+1)N\times\mu}.$$

Then (2.194) can be written in matrix form as

$$\boldsymbol{h}_{k,q} = \boldsymbol{\Xi}_k \boldsymbol{f}_{k,q}. \tag{2.195}$$

Finally, denote

$$\boldsymbol{h}_k = \begin{bmatrix} h_k[0] \\ \vdots \\ h_k[P-1] \\ \vdots \\ h_k[\iota P] \\ \vdots \\ h_k[(\iota+1)P-1] \end{bmatrix}_{(\iota+1)P\times 1}, \qquad \boldsymbol{f}_k = \begin{bmatrix} f_k[0] \\ \vdots \\ f_k[p\mu-1] \end{bmatrix}_{p\mu\times 1}.$$

Then we have

$$\boldsymbol{h}_k = \widetilde{\boldsymbol{\Xi}}_k \boldsymbol{f}_k, \tag{2.196}$$

where $\tilde{\Xi}_k$ is an $(\iota+1)P \times p\mu$ matrix formed from the signature waveform of the kth user. For instance, when the oversampling factor $p = 2$, we have

$$\boldsymbol{h}_k = \begin{bmatrix} h_{k,0}[0] \\ h_{k,1}[0] \\ \vdots \\ h_{k,0}[N-1] \\ h_{k,1}[N-1] \\ \vdots \\ h_{k,0}[(\iota+1)N-1] \\ h_{k,1}[(\iota+1)N-1] \end{bmatrix}_{2(\iota+1)N\times 1}, \qquad \boldsymbol{f}_k = \begin{bmatrix} f_{k,0}[0] \\ f_{k,1}[0] \\ \vdots \\ f_{k,0}[\mu-1] \\ f_{k,1}[\mu-1] \end{bmatrix}_{2\mu\times 1},$$

$$\tilde{\Xi}_k = \begin{bmatrix} s_{0,k} & & & & & \\ 0 & s_{0,k} & & & & \\ s_{1,k} & 0 & s_{0,k} & & & \\ 0 & s_{1,k} & 0 & s_{0,k} & & \\ \vdots & \ddots & \ddots & \ddots & \ddots & \\ \vdots & & \ddots & \ddots & \ddots & s_{0,k} \\ s_{N-1,k} & & & 0 & s_{1,k} & 0 \\ & s_{N-1,k} & & & 0 & s_{1,k} \\ & & \ddots & & & 0 \\ & & & \ddots & & \vdots \\ & & & & \ddots & \vdots \\ & & & & & s_{N-1,k} \end{bmatrix}_{2(\iota+1)N\times 2\mu}.$$

For other values of p, the matrix $\tilde{\Xi}_k$ is constructed similarly.

Recall that when the ambient channel noise is white, through an eigendecomposition on the autocorrelation matrix of the received signal [cf. (2.180)], the signal subspace and the noise subspace can be identified. The channel response $\boldsymbol{f}_k$ can then be estimated by exploiting the orthogonality between the signal subspace and the noise subspace [31, 272, 484, 548]. Specifically, since $\boldsymbol{U}_n$ is orthogonal to the column space of $\boldsymbol{H}$ and $\bar{\boldsymbol{h}}_k \stackrel{\triangle}{=} \boldsymbol{H}\boldsymbol{e}_{K\iota+k}$ is in the column space of $\boldsymbol{H}$ [cf. (2.179)], we have

$$\boldsymbol{U}_n^H \bar{\boldsymbol{h}}_k = \boldsymbol{U}_n^H \Xi_k \boldsymbol{f}_k = \boldsymbol{0}, \tag{2.197}$$

where

$$\bar{\boldsymbol{h}}_k = \begin{bmatrix} \boldsymbol{h}_k \\ \boldsymbol{0}_{(m-\iota-1)P\times 1} \end{bmatrix} = \underbrace{\begin{bmatrix} \tilde{\Xi}_k \\ \boldsymbol{0}_{(m-\iota-1)P\times p\mu} \end{bmatrix}}_{\Xi_k} \boldsymbol{f}_k. \tag{2.198}$$

Based on the relationship above, we can obtain an estimate of the channel response $\boldsymbol{f}_k$ by computing the minimum eigenvector of the matrix $\overline{\boldsymbol{\Xi}}_k^H \boldsymbol{U}_n \boldsymbol{U}_n^H \overline{\boldsymbol{\Xi}}_k$. The condition for the channel estimate obtained in such a way to be unique is that the matrix $\boldsymbol{U}_n^H \overline{\boldsymbol{\Xi}}_k$ has rank $p\mu - 1$, which necessitates that this matrix be tall (i.e., $[Pm - K(m+\iota)] \geq p\mu_k$). Since $\mu \leq \iota N$ [cf. (2.191)], we therefore choose the smoothing factor m to satisfy

$$Pm - K(m+\iota) \geq \iota P \; = \; \iota_k N p \; \geq \; p\mu. \tag{2.199}$$

That is, $m = \lceil (P-K)/(P+K)\iota \rceil$. On the other hand, the condition (2.199) implies that for fixed m, the total number of users that can be accommodated in the system is $\lceil (m-\iota)(m+\iota)P \rceil$.

Finally, we summarize the batch algorithm for blind linear multiuser detection in multipath CDMA channels as follows.

Algorithm 2.7: [Subspace blind linear multiuser detector—multipath CDMA]

- *Estimate the signal subspace:*

$$\hat{\boldsymbol{C}}_r = \frac{1}{M} \sum_{i=0}^{M-1} \boldsymbol{r}[i] \boldsymbol{r}[i]^H \tag{2.200}$$

$$= \hat{\boldsymbol{U}}_s \hat{\boldsymbol{\Lambda}}_s \hat{\boldsymbol{U}}_s^H + \hat{\boldsymbol{U}}_n \hat{\boldsymbol{\Lambda}}_n \hat{\boldsymbol{U}}_n^H. \tag{2.201}$$

- *Estimate the channel and form the detector:*

$$\hat{\boldsymbol{f}}_1 = \textit{min-eigenvector} \left(\overline{\boldsymbol{\Xi}}_1^H \hat{\boldsymbol{U}}_n \hat{\boldsymbol{U}}_n^H \overline{\boldsymbol{\Xi}}_1 \right), \tag{2.202}$$

$$\hat{\overline{\boldsymbol{h}}}_1 = \overline{\boldsymbol{\Xi}}_1 \hat{\boldsymbol{f}}_1, \tag{2.203}$$

$$\hat{\boldsymbol{d}}_1 = \hat{\boldsymbol{U}}_s \left(\hat{\boldsymbol{\Lambda}}_s - \hat{\sigma}^2 \boldsymbol{I}_r \right) \hat{\boldsymbol{U}}_s^H \hat{\overline{\boldsymbol{h}}}_1, \;\; \textit{(linear decorrelating detector)} \tag{2.204}$$

$$\hat{\boldsymbol{m}}_1 = \hat{\boldsymbol{U}}_s \hat{\boldsymbol{\Lambda}}_s \hat{\boldsymbol{U}}_s^H \hat{\overline{\boldsymbol{h}}}_1. \quad \textit{(linear MMSE detector)} \tag{2.205}$$

- *Perform differential detection:*

$$z_1[i] = \hat{\boldsymbol{w}}_1^H \boldsymbol{r}[i], \qquad \left(\hat{\boldsymbol{w}}_1 = \hat{\boldsymbol{d}}_1 \;\; or \;\; \hat{\boldsymbol{w}}_1 = \hat{\boldsymbol{m}}_1 \right), \tag{2.206}$$

$$\hat{\beta}_1[i] = \text{sign} \left\{ \Re \left(z_1[i] z_1[i-1]^* \right) \right\}, \qquad i = 1, \ldots, M-1. \tag{2.207}$$

Alternatively, if the linear receiver has the decimation-combining structure, the noise subspace $\boldsymbol{U}_{n,q}$ is computed for each $q = 0, \ldots, p-1$. The corresponding channel response $\boldsymbol{f}_{k,q}$ can then be estimated from the orthogonality relationship

$$\boldsymbol{U}_{n,q}^H \boldsymbol{h}_{k,q} \; = \; \boldsymbol{U}_{n,q}^H \boldsymbol{\Xi}_k \boldsymbol{f}_{k,q} \; = \; \boldsymbol{0}, \qquad q = 0, \ldots, p-1. \tag{2.208}$$

Simulation Examples

The simulated system is an asynchronous CDMA system with processing gain $N = 15$. The m-sequences of length 15 and their shifted versions are employed as the user spreading sequences. The chip pulse is a raised cosine pulse with roll-off factor 0.5. Each user's channel has $L = 3$ paths. The delay of each path $\tau_{l,k}$ is uniformly distributed on $[0, 10T_c]$. Hence the maximum delay spread is one symbol interval (i.e., $\iota = 1$). The fading gain of each path in each user's channel is generated from a complex Gaussian distribution and is fixed over the duration of one signal frame. The path gains in each user's channel are normalized so that all users' signals arrive at the receiver with the same power. The oversampling factor is $p = 2$. The smoothing factor is $m = 2$. Hence this system can accommodate up to $\lceil (m - \iota)(m + \iota)P \rceil = 10$ users. If the decimation-combining receiver structure is employed, the maximum number of users is $\lceil (m - \iota)(m + \iota)N \rceil = 5$. The length of each user's signal frame is $M = 200$.

We first consider a five-user system. For the Pm-dimensional implementations, the bit error rates of a particular user incurred by the exact linear MMSE detector, the exact linear zero-forcing detector and the estimated linear MMSE detector are plotted in Fig. 2.13. The bit-error rates of the same user incurred by the three detectors using the decimation-combining structure are plotted in Fig. 2.14. It is seen that for the exact linear zero-forcing and linear MMSE detectors, the performance under the two structures is identical. For the blind linear MMSE receiver, the Pm-

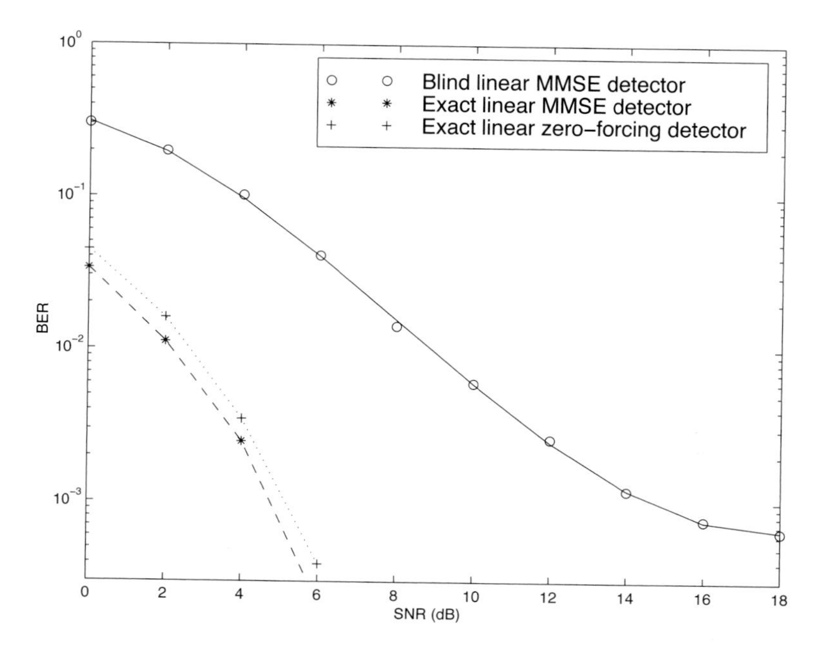

Figure 2.13. Performance of Pm-dimensional linear detectors in a five-user system with white noise.

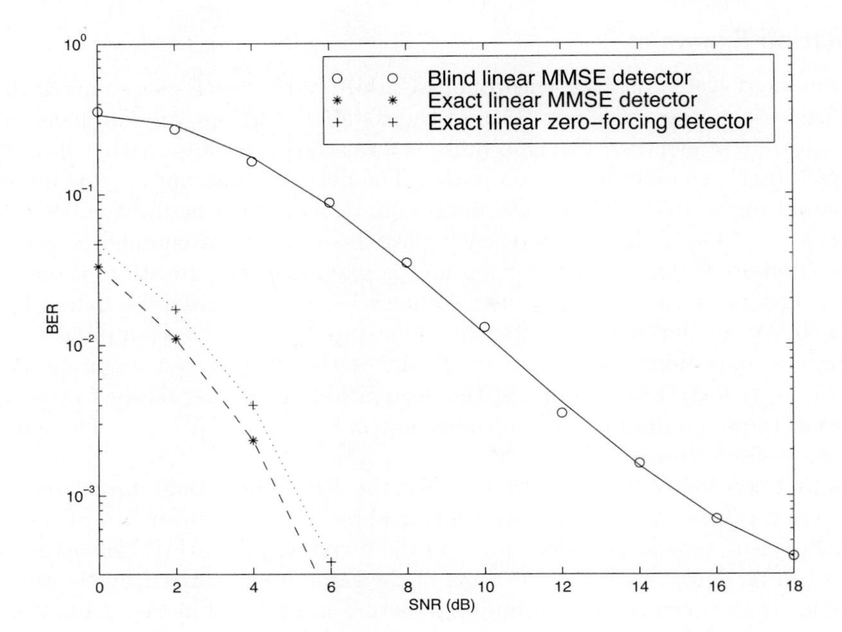

Figure 2.14. Performance of decimation-combining linear detectors in a five-user system with white noise.

dimensional detector achieves an approximately 1-dB performance gain over the decimation-combining detector for SNR in the range 4 to 12 dB. Another observation is that the blind linear MMSE detector tends to exhibit an error floor at high SNR. This is due to the finite length of the signal frame, from which the detector is estimated. Next, a 10-user system is simulated using the Pm-dimensional detectors. The performance of the same user by the three detectors is plotted in Fig. 2.15.

2.7.4 Adaptive Receiver Structures

We next consider adaptive algorithms for sequentially estimating the blind linear detector. First, we address adaptive implementation of the blind channel estimator discussed above. Suppose that the signal subspace $\boldsymbol{U}_s$ is known. Denote by $\boldsymbol{z}[i]$ the projection of the received signal $\boldsymbol{r}[i]$ onto the noise subspace:

$$\boldsymbol{z}[i] \stackrel{\triangle}{=} \boldsymbol{r}[i] - \boldsymbol{U}_s \boldsymbol{U}_s^H \boldsymbol{r}[i] \tag{2.209}$$

$$= \boldsymbol{U}_n \boldsymbol{U}_n^H \boldsymbol{r}[i]. \tag{2.210}$$

Since $\boldsymbol{z}[i]$ lies in the noise subspace, it is orthogonal to any signal in the signal subspace. In particular, it is orthogonal to $\bar{\boldsymbol{h}}_1 = \overline{\boldsymbol{\Xi}}_1 \boldsymbol{f}_1$. Hence $\boldsymbol{f}_1$ is the solution to the following constrained optimization problem:

$$\min_{\boldsymbol{f}_1 \in \mathbb{C}^{p\mu}} E\left\{ \left\| \left(\overline{\boldsymbol{\Xi}}_1 \boldsymbol{f}_1 \right)^H \boldsymbol{z}[i] \right\|^2 \right\} \quad \text{s.t.} \quad \|\boldsymbol{f}_1\| = 1. \tag{2.211}$$

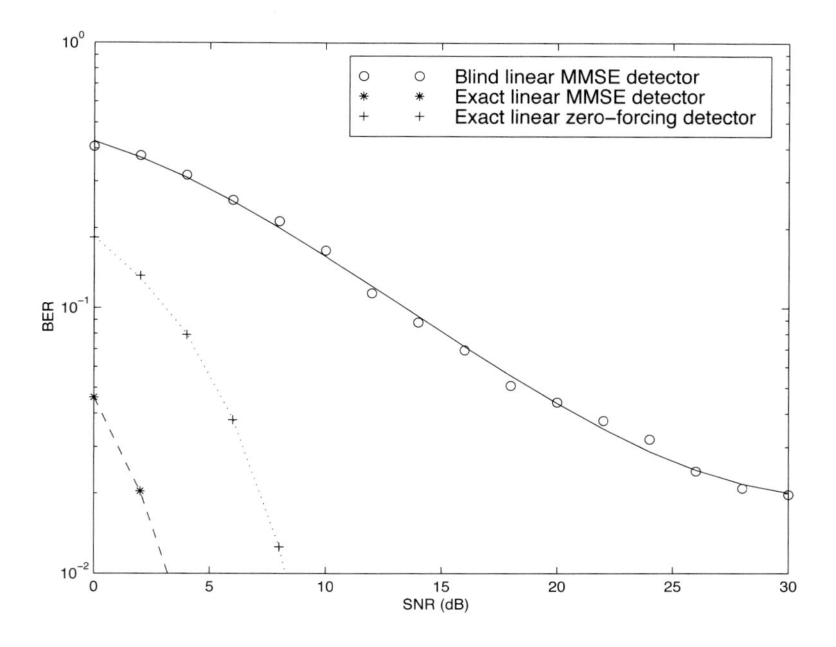

Figure 2.15. Performance of Pm-dimensional linear detectors in a 10-user system with white noise.

In order to obtain a sequential algorithm to solve the optimization problem above, we write it in the following (trivial) state-space form:

$$\boldsymbol{f}_1[i+1] = \boldsymbol{f}_1[i], \qquad \text{(state equation)}$$
$$0 = \left(\boldsymbol{\Xi}_1^H \boldsymbol{z}[i]\right)^H \boldsymbol{f}_1[i]. \qquad \text{(observation equation)}$$

The standard Kalman filter can then be applied to the system above, as follows (we define $\boldsymbol{x}[i] \triangleq \boldsymbol{\Xi}_1^H \boldsymbol{z}[i]$):

$$\boldsymbol{k}[i] = \boldsymbol{\Sigma}[i-1]\boldsymbol{x}[i]\left(\boldsymbol{x}[i]^H \boldsymbol{\Sigma}[i-1]\boldsymbol{x}[i]\right)^{-1}, \tag{2.212}$$
$$\boldsymbol{f}_1[i] = \boldsymbol{f}_1[i-1] - \boldsymbol{k}[i]\left(\boldsymbol{x}[i]^H \boldsymbol{f}_1[i]\right) / \left\|\boldsymbol{f}_1[i-1] - \boldsymbol{k}[i]\left(\boldsymbol{x}[i]^H \boldsymbol{f}_1[i]\right)\right\|, \tag{2.213}$$
$$\boldsymbol{\Sigma}[i] = \boldsymbol{\Sigma}[i-1] - \boldsymbol{k}[i]\boldsymbol{x}[i]^H \boldsymbol{\Sigma}[i-1]. \tag{2.214}$$

Note that (2.213) contains a normalization step to satisfy the constraint $\|\boldsymbol{f}_1[i]\| = 1$.

Since the subspace blind detector may be written in closed form as a function of the signal subspace components, one may use a suitable subspace tracking algorithm in conjuction with this detector and a channel estimator to form an *adaptive* detector that is able to track changes in the number of users and their composite signature waveforms. Figure 2.16 contains a block diagram of such a receiver. The received signal $\boldsymbol{r}[i]$ is fed into a subspace tracker that sequentially estimates the

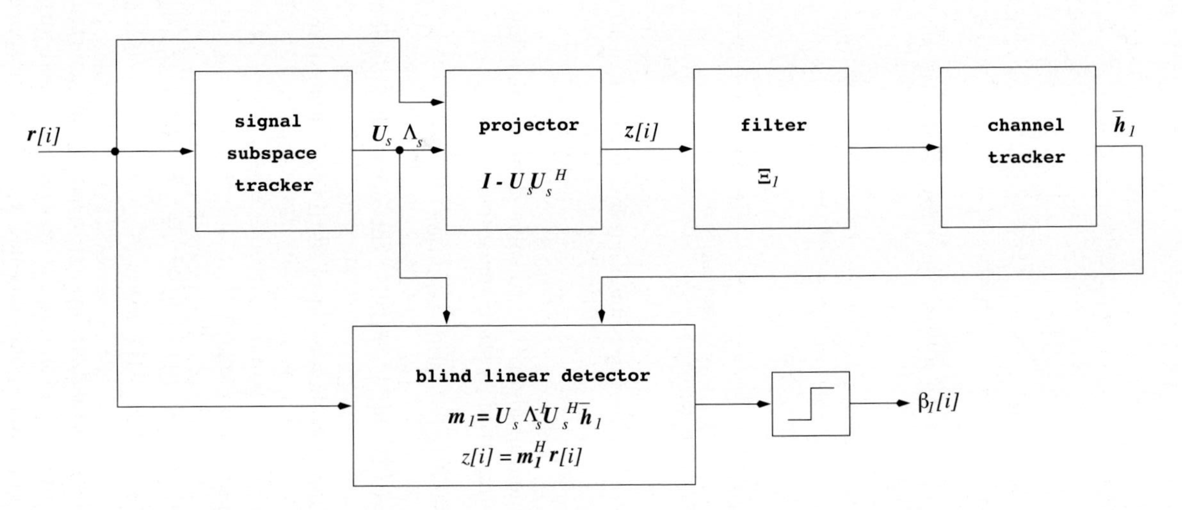

Figure 2.16. Blind adaptive multiuser receiver for multipath CDMA systems.

signal subspace components $(\boldsymbol{U}_s, \boldsymbol{\Lambda}_s)$. The signal $\boldsymbol{r}[i]$ is then projected onto the noise subspace to obtain $\boldsymbol{z}[i]$, which is in turn passed through a linear filter that is determined by the signature sequence of the desired user. The output of this filter is fed into a channel tracker that estimates the channel state of the desired user. Finally, the linear MMSE detector is constructed in closed form based on the estimated signal subspace components and the channel state. The adaptive receiver algorithm is summarized as follows. Suppose that at time $(i-1)$, the estimated signal subspace rank is $r[i-1]$ and the components are $\boldsymbol{U}_s[i-1]$, $\boldsymbol{\Lambda}_s[i-1]$, and $\sigma^2[i-1]$. The estimated channel vector is $\boldsymbol{f}_1[i-1]$. Then at time i, the adaptive detector performs the following steps to update the detector and estimate the data.

Algorithm 2.8: [Adaptive blind linear multiuser detector based on subspace tracking—multipath CDMA]

- *Update the signal subspace: Use a particular signal subspace tracking algorithm to update the signal subspace rank* $r[i]$ *and the subspace components* $\boldsymbol{U}_s[i]$ *and* $\boldsymbol{\Lambda}_s[i]$.
- *Update the channel: Use (2.212)–(2.214) to update the channel estimate* $\boldsymbol{f}_1[i]$.
- *Form the detector and perform differential detection:*

$$\boldsymbol{m}_1[i] = \boldsymbol{U}_s[i]\boldsymbol{\Lambda}_s[i]^{-1}\boldsymbol{U}_s[i]^H\bar{\boldsymbol{h}}_1[i], \tag{2.215}$$

$$z_1[i] = \boldsymbol{m}_1[i]^H\boldsymbol{r}[i], \tag{2.216}$$

$$\hat{\beta}_1[i] = \text{sign}\left\{\Re(z_1[i]z_1[i-1]^*)\right\}. \tag{2.217}$$

Simulation Example

We next give a simulation example illustrating the performance of the blind adaptive receiver in an asynchronous CDMA system with multipath channels. The processing gain $N = 15$ and the spreading codes are Gold codes of length 15. Each user's channel has $L = 3$ paths. The delay of each path $\tau_{l,k}$ is uniformly distributed on $[0, 10T_c]$. Hence, as in the preceding example, the maximum delay spread is one symbol interval (i.e., $\iota = 1$). The fading gain of each path in each user's channel is generated from a complex Gaussian distribution and is fixed for all simulations. The path gains in each user's channel are normalized so that all users' signals arrive at the receiver with the same power. The smoothing factor is $m = 2$. The received signal is sampled at twice the chip rate ($p = 2$). Hence the total number of users that this system can accommodate is 10. Figure 2.17 shows the performance of subspace blind adaptive receiver using the NAHJ subspace tracking algorithm discussed in Section 2.6.3 in terms of output SINR. During the first 1000 iterations there are eight total users. At iteration 1000, four new users are added to the system. At iteration 2000, one additional known user is added and three existing users vanish. We see that this blind adaptive receiver can closely track the dynamics of the channel.

We note that there are many other approaches to blind multiuser detection in multipath CDMA channels, such as constrained optimization methods [59, 60, 80, 187, 300, 305, 306, 427, 485, 487, 490, 498, 583, 584, 605], the auxiliary vector method

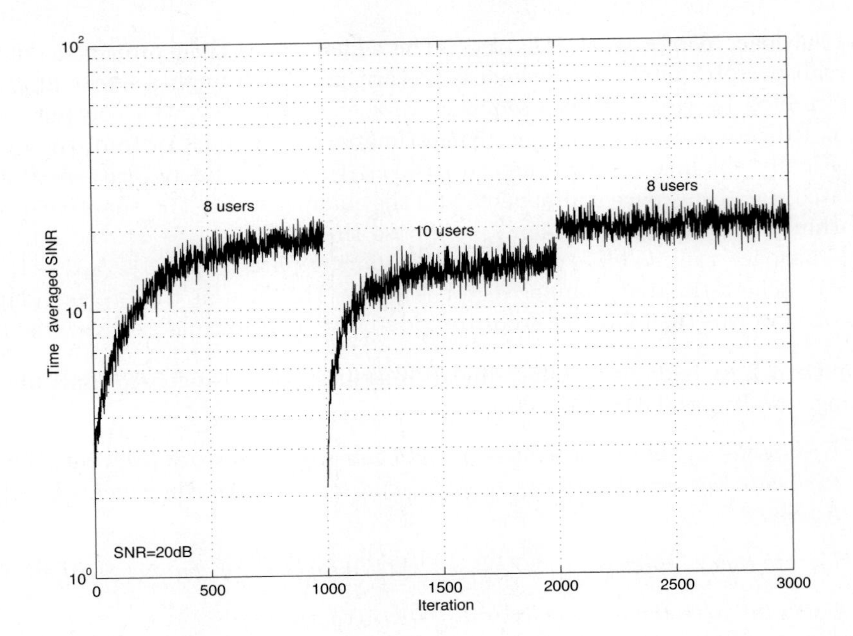

Figure 2.17. Performance of a subspace blind adaptive multiuser detector in an asynchronous CDMA system with multipath.

[364], other subspace methods [10,31,252,258,272,287,446,484,548,551,564], linear prediction methods [69,117,207,606], the multistage Wiener filtering method [156, 186], the constant modulus method [79,218,582], the spreading code design method [435], the maximum-likelihood method [56], the parallel factor method [447], the least-squares smoothing method [483,610], a method based on cyclostationarity [351], and more general methods based on multiple-input/multiple-output (MIMO) blind channel identification [78,214,261,298,462,494–497].

2.7.5 Blind Multiuser Detection in Correlated Noise

So far in developing the subspace-based linear detectors and the channel estimation methods, the ambient channel noise is assumed to be temporally white. In practice, such an assumption may be violated due to, for example, the interference from some narrowband sources. The techniques developed under the white noise assumption are not applicable to channels with correlated ambient noise. In this section we discuss subspace methods for joint suppression of MAI and ISI in multipath CDMA channels with unknown correlated ambient noise, which were first developed in [551]. The key assumption here is that the signal is received by two antennas well separated so that the noise is *spatially uncorrelated.*

We start with the received augmented discrete-time signal model given by (2.169). Assume that the ambient noise vector $\boldsymbol{n}[i]$ has a covariance matrix $\boldsymbol{\Sigma} \triangleq E\left\{\boldsymbol{n}[i]\boldsymbol{n}[i]^H\right\}$.

Then the $Pm \times Pm$ autocorrelation matrix $\boldsymbol{C}_r$ of the received signal $\boldsymbol{r}[i]$ is given by

$$\boldsymbol{C}_r = \boldsymbol{H}\boldsymbol{H}^H + \boldsymbol{\Sigma}. \tag{2.218}$$

The linear MMSE detector $\boldsymbol{m}_1$ for user 1 is given by (2.177) with $\boldsymbol{C}_r$ replaced by (2.218). As before, we must first estimate the desired user's composite signature waveform $\bar{\boldsymbol{h}}_1$ given by (2.198). Notice, however, that when the ambient noise is correlated, it is not possible to separate the signal subspace from the noise subspace based solely on the autocorrelation matrix $\boldsymbol{C}_r$.

To estimate the channel in unknown correlated noise, we make use of two antennas at the receiver. Then the two augmented received signal vectors at the two antennas can be written, respectively, as

$$\boldsymbol{r}_1[i] = \boldsymbol{H}_1\boldsymbol{b}[i] + \boldsymbol{n}_1[i], \tag{2.219}$$
$$\boldsymbol{r}_2[i] = \boldsymbol{H}_2\boldsymbol{b}[i] + \boldsymbol{n}_2[i], \tag{2.220}$$

where $\boldsymbol{H}_1$ and $\boldsymbol{H}_2$ contain the channel information corresponding to the respective antennas. It is assumed that the two antennas are well separated so that the ambient noise is *spatially uncorrelated.* In other words, $\boldsymbol{n}_1[i]$ and $\boldsymbol{n}_2[i]$ are uncorrelated, and their joint covariance is given by

$$E\left\{\begin{bmatrix} \boldsymbol{n}_1[i] \\ \boldsymbol{n}_2[i] \end{bmatrix}\begin{bmatrix} \boldsymbol{n}_1[i]^H & \boldsymbol{n}_2[i]^H \end{bmatrix}\right\} = \begin{bmatrix} \boldsymbol{\Sigma}_1 & \boldsymbol{0} \\ \boldsymbol{0} & \boldsymbol{\Sigma}_2 \end{bmatrix}, \tag{2.221}$$

where $\boldsymbol{\Sigma}_1$ and $\boldsymbol{\Sigma}_2$ are unknown covariance matrices of the noise at the two antennas. The joint autocorrelation matrix of the received signal at the two antennas is then given by

$$\boldsymbol{C} \triangleq E\left\{\begin{bmatrix} \boldsymbol{r}_1[i] \\ \boldsymbol{r}_2[i] \end{bmatrix}\begin{bmatrix} \boldsymbol{r}_1[i]^H & \boldsymbol{r}_2[i]^H \end{bmatrix}\right\} = \begin{bmatrix} \boldsymbol{C}_{11} & \boldsymbol{C}_{12} \\ \boldsymbol{C}_{21} & \boldsymbol{C}_{22} \end{bmatrix}, \tag{2.222}$$

where the submatrices are given by

$$\boldsymbol{C}_{11} \triangleq E\left\{\boldsymbol{r}_1[i]\boldsymbol{r}_1[i]^H\right\} = \boldsymbol{H}_1\boldsymbol{H}_1^H + \boldsymbol{\Sigma}_1, \tag{2.223}$$
$$\boldsymbol{C}_{22} \triangleq E\left\{\boldsymbol{r}_2[i]\boldsymbol{r}_2[i]^H\right\} = \boldsymbol{H}_2\boldsymbol{H}_2^H + \boldsymbol{\Sigma}_2, \tag{2.224}$$
$$\boldsymbol{C}_{12} = \boldsymbol{C}_{21}^H \triangleq E\left\{\boldsymbol{r}_1[i]\boldsymbol{r}_2[i]^H\right\} = \boldsymbol{H}_1\boldsymbol{H}_2^H. \tag{2.225}$$

We next consider two methods for estimating the noise subspaces from the signals received at the two antennas.

Singular Value Decomposition

Assume that both $\boldsymbol{H}_1$ and $\boldsymbol{H}_2$ have full column rank $r \triangleq K(m+\iota)$; then the matrix $\boldsymbol{C}_{12}$ also has rank r. Consider the singular value decomposition of the matrix $\boldsymbol{C}_{12}$,

$$\boldsymbol{C}_{12} = \boldsymbol{H}_1\boldsymbol{H}_2^H = \boldsymbol{U}_1\boldsymbol{\Gamma}\boldsymbol{U}_2^H. \tag{2.226}$$

The $Pm \times Pm$ diagonal matrix $\boldsymbol{\Gamma}$ has the form $\boldsymbol{\Gamma} = \mathsf{diag}(\gamma_1, \ldots, \gamma_r, 0, \ldots, 0)$, with $\gamma_1 \geq \cdots \geq \gamma_r > 0$. Now if we partition the matrix $\boldsymbol{U}_j$ as $\boldsymbol{U}_j = [\boldsymbol{U}_{j,s} \mid \boldsymbol{U}_{j,n}]$ for $j = 1, 2$, where $\boldsymbol{U}_{j,s}$ and $\boldsymbol{U}_{j,n}$ contain the first r columns and the last $Pm - r$ columns of $\boldsymbol{U}_j$, respectively, the column space of $\boldsymbol{U}_{j,n}$ is orthogonal to the column space of $\boldsymbol{H}_j$:

$$\overline{\mathsf{range}(\boldsymbol{H}_j)} = \mathsf{range}(\boldsymbol{U}_{j,n}), \qquad j = 1, 2, \tag{2.227}$$

where $\overline{\mathsf{range}(\boldsymbol{H}_j)}$ denotes the orthogonal complement space of $\mathsf{range}(\boldsymbol{H}_j)$. User 1's channel corresponding to antenna j, $\boldsymbol{f}_{j,1}$, can then be estimated from the orthogonality relationship

$$\boldsymbol{U}_{j,n}^H \bar{\boldsymbol{h}}_{j,1} = \boldsymbol{U}_{j,n}^H \boldsymbol{\Xi}_1 \boldsymbol{f}_{j,1} \; = \; \boldsymbol{0}, \qquad j = 1, 2. \tag{2.228}$$

Canonical Correlation Decomposition

Assume that the matrices $\boldsymbol{C}_{11}$ and $\boldsymbol{C}_{22}$ are both positive definite. The canonical correlation decomposition (CCD) of the matrix $\boldsymbol{C}_{12}$ is given by [18]

$$\boldsymbol{C}_{11}^{-1/2} \boldsymbol{C}_{12} \boldsymbol{C}_{22}^{-1/2} \; = \; \boldsymbol{U}_1 \boldsymbol{\Gamma} \boldsymbol{U}_2^H \tag{2.229}$$

or

$$\boldsymbol{C}_{11}^{-1} \boldsymbol{C}_{12} \boldsymbol{C}_{22}^{-1} \; = \; \boldsymbol{C}_{11}^{-1/2} \boldsymbol{U}_1 \boldsymbol{\Gamma} \boldsymbol{U}_2^H \boldsymbol{C}_{22}^{-1/2}. \tag{2.230}$$

The $Pm \times Pm$ matrix $\boldsymbol{\Gamma}$ has the form $\boldsymbol{\Gamma} = \mathrm{diag}(\gamma_1, \ldots, \gamma_r, 0, \ldots, 0)$, with $\gamma_1 \geq \cdots \geq \gamma_r > 0$. Let

$$\boldsymbol{L}_j = \boldsymbol{C}_{jj}^{-1/2} \boldsymbol{U}_j, \qquad j = 1, 2. \tag{2.231}$$

Partition the matrix $\boldsymbol{L}_j$ such that

$$\boldsymbol{L}_j = [\boldsymbol{L}_{j,s} \mid \boldsymbol{L}_{j,n}] \; = \; \left[\boldsymbol{C}_{jj}^{-1/2} \boldsymbol{U}_{j,s} \mid \boldsymbol{C}_{jj}^{-1/2} \boldsymbol{U}_{j,n} \right], \tag{2.232}$$

where $\boldsymbol{L}_{j,s}$ and $\boldsymbol{L}_{j,n}$ are the first r columns and the last $Pm - r$ columns of $\boldsymbol{L}_j$, respectively. The matrix $\boldsymbol{U}_j$ are similarly partitioned into $\boldsymbol{U}_{j,s}$ and $\boldsymbol{U}_{j,n}$. We have [580]

$$\overline{\mathrm{range}(\boldsymbol{H}_j)} = \mathsf{range}(\boldsymbol{L}_{j,n}), \qquad j = 1, 2. \tag{2.233}$$

Note, however, that $\boldsymbol{L}_{j,s}$ does not necessarily span the signal subspace $\mathsf{range}(\boldsymbol{H}_j)$ [580].

Now suppose that we have estimated the composite signature waveform of the desired user $\bar{\boldsymbol{h}}_{j,1}$, using the identified noise subspace $\boldsymbol{L}_{j,n}$. Since $\bar{\boldsymbol{h}}_{j,1} \in \mathsf{range}(\boldsymbol{H}_j)$, we have

$$\boldsymbol{m}_{j,1} = \boldsymbol{C}_{jj}^{-1} \bar{\boldsymbol{h}}_{j,1} \; = \; \boldsymbol{L}_j \boldsymbol{L}_j^H \bar{\boldsymbol{h}}_{j,1} \; = \; \boldsymbol{L}_{j,s} \boldsymbol{L}_{j,s}^H \bar{\boldsymbol{h}}_{j,1}, \tag{2.234}$$

where the second equality in (2.234) follows from (2.231) and the fact that $\boldsymbol{U}_j$ is a unitary matrix; and the third equality follows from the fact that $\boldsymbol{L}_{j,n}^H \bar{\boldsymbol{h}}_{j,1} = \boldsymbol{0}$.

Let the estimated weight vectors of the linear MMSE detectors at the two antennas be $\hat{\boldsymbol{m}}_{j,1}$, $j = 1, 2$. To make use of the received signal at both antennas, we use the following equal-gain differential combining rule for detecting the differential bit $\beta_1[i]$:

$$z_{j,1}[i] = \hat{\boldsymbol{m}}_{j,1}^H \boldsymbol{r}_j[i], \qquad j = 1, 2, \tag{2.235}$$

$$\hat{\beta}_1[i] = \operatorname{sign}\left\{ \Re\left(\sum_{j=1}^{2} z_{j,1}[i] z_{j,1}[i-1]^* \right) \right\}. \tag{2.236}$$

We next summarize the procedures for computing the linear MMSE detector $\hat{\boldsymbol{m}}_{j,1}$ in unknown correlated noise based on the discussion above. Let

$$\boldsymbol{Y}_j = \Big[\boldsymbol{r}_j[0] \; \boldsymbol{r}_j[1] \; \cdots \; \boldsymbol{r}_j[M-1] \Big], \qquad j = 1, 2, \tag{2.237}$$

be the matrix of M received augmented signal sample vectors at antenna j corresponding to one block of data transmission.

Algorithm 2.9: [Blind linear MMSE detector in multipath CDMA with correlated noise—SVD-based method]

- *Compute the auto- and cross-correlation matrices:*

$$\hat{\boldsymbol{C}}_{ij} = \frac{1}{M} \boldsymbol{Y}_i \boldsymbol{Y}_j^H, \qquad i, j = 1, 2. \tag{2.238}$$

- *Perform an SVD on* $\hat{\boldsymbol{C}}_{12}$ *to get the noise subspace* $\hat{\boldsymbol{U}}_{j,n}$, $j = 1, 2$.

- *Compute the composite signature waveforms* $\hat{\bar{\boldsymbol{h}}}_{j,1}$ *by solving*

$$\hat{\boldsymbol{U}}_{j,n}^H \bar{\boldsymbol{h}}_{j,1} = \boldsymbol{0} \Longrightarrow \hat{\boldsymbol{U}}_{j,n}^H \boldsymbol{\Xi}_1 \boldsymbol{f}_{j,1} = \boldsymbol{0}, \qquad j = 1, 2. \tag{2.239}$$

- *Form the linear MMSE detectors:*

$$\hat{\boldsymbol{m}}_{j,1} = \hat{\boldsymbol{C}}_{jj}^{-1} \hat{\bar{\boldsymbol{h}}}_{j,1}, \qquad j = 1, 2. \tag{2.240}$$

- *Perform differential detection according to (2.235)–(2.236).*

Algorithm 2.10: [Blind linear MMSE detector in multipath CDMA with correlated noise—CCD-based method]

- *Perform QR decomposition:*

$$\frac{1}{\sqrt{M}} \boldsymbol{Y}_j^H = \hat{\boldsymbol{Q}}_j \hat{\boldsymbol{\Upsilon}}_j, \qquad j = 1, 2. \tag{2.241}$$

- *Perform an SVD on* $\left(\hat{\boldsymbol{Q}}_1^H \hat{\boldsymbol{Q}}_2\right)$:

$$\hat{\boldsymbol{Q}}_1^H \hat{\boldsymbol{Q}}_2 = \hat{\boldsymbol{V}}_1 \hat{\boldsymbol{\Gamma}} \hat{\boldsymbol{V}}_2^H. \tag{2.242}$$

- *Compute*

$$\hat{\boldsymbol{L}}_j = \hat{\boldsymbol{\Upsilon}}_j^{-1} \hat{\boldsymbol{V}}_j, \quad j = 1, 2, \tag{2.243}$$

  where $\hat{\boldsymbol{\Upsilon}}_j$ *is an upper triangular matrix.*

- *Partition* $\hat{\boldsymbol{L}}_j = \left[\hat{\boldsymbol{L}}_{j,s} \mid \hat{\boldsymbol{L}}_{j,n}\right]$. *Compute the composite signature waveforms* $\hat{\bar{\boldsymbol{h}}}_{j,1}$, $j = 1, 2$, *by solving*

$$\hat{\boldsymbol{L}}_{j,n}^H \bar{\boldsymbol{h}}_{j,1} = \boldsymbol{0} \Longrightarrow \hat{\boldsymbol{L}}_{j,n}^H \overline{\boldsymbol{\Xi}}_1 \boldsymbol{f}_{j,1} = \boldsymbol{0}. \tag{2.244}$$

- *Form the linear MMSE detectors:*

$$\hat{\boldsymbol{m}}_{j,1} = \hat{\boldsymbol{L}}_{j,s} \hat{\boldsymbol{L}}_{j,s}^H \hat{\bar{\boldsymbol{h}}}_{j,1}, \quad j = 1, 2. \tag{2.245}$$

- *Perform differential detection according to (2.235)–(2.236).*

The procedure above is based on the fast algorithm for computing CCD given in [580].

Note that the two methods above operate on the Pm-dimensional signal vectors $\boldsymbol{r}_j[i], j = 1, 2$. The same procedures can be applied to the decimated received signal vectors to operate on the Nm-dimensional signal vectors $\boldsymbol{r}_{j,q}[i], j = 1, 2, q = 0, \ldots, p-1$. As before, such decimation-combining approach reduces the computational complexity by a factor of $\mathcal{O}(p^2)$. It also reduces the number of users that can be accommodated in the system by a factor of p.

Simulation Examples

We illustrate the performance of the detectors above via simulation examples. The simulated system is the same as that in Section 2.7.3, except that the ambient noise is temporally correlated. The noise at each antenna j is modeled by a second-order autoregressive (AR) model with coefficients $\boldsymbol{a}_j = [a_{j,1}, a_{j,2}]$; that is, the noise field is generated according to

$$n_j[i] = a_{j,1} n_j[i-1] + a_{j,2} n_j[i-2] + w_j[i], \quad j = 1, 2, \tag{2.246}$$

where $n_j[i]$ is the noise at antenna j and sample i, and $w_j[i]$ is a complex white Gaussian noise sample with unit variance. The AR coefficients at the two arrays are chosen as $\boldsymbol{a}_1 = [1, -0.2]$ and $\boldsymbol{a}_2 = [1.2, -0.3]$.

We first consider a five-user system. In Fig. 2.18 the performance of the Pm-dimensional blind linear MMSE detectors is plotted for both SVD- and CCD-based methods. The corresponding performance by the decimation-combining receiver structure is plotted in Fig. 2.19. Next a 10-user system is simulated and the performance of the Pm-dimensional blind linear MMSE detectors is plotted in Fig. 2.20.

It is seen from Figs. 2.18–2.20 that CCD-based detectors are superior in performance to SVD-based detectors. It has been shown that the CCD has the optimality property of maximizing the correlation between the two sets of linearly transformed data [18]. Maximizing the correlation of the two data sets can yield the best estimate of the correlated (i.e., signal) part of the data. CCD makes use of the information of both $\hat{\boldsymbol{C}}_{11}$ and $\hat{\boldsymbol{C}}_{22}$ together with $\hat{\boldsymbol{C}}_{12}$ and creates the maximum correlation between the two data sets. On the other hand, SVD uses only the information $\hat{\boldsymbol{C}}_{12}$ and does not create the maximum correlation between the two data sets, and thus yields inferior performance.

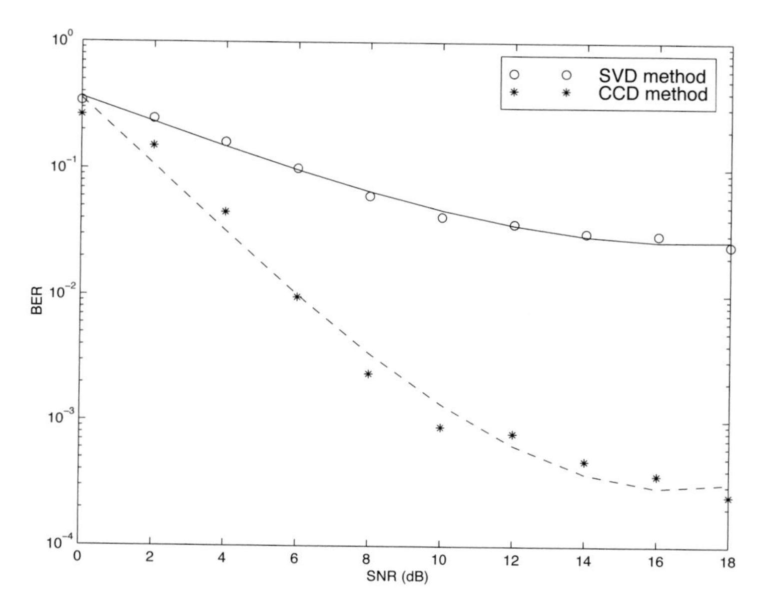

Figure 2.18. Performance of Pm-dimensional blind linear MMSE detectors in a five-user system with correlated noise.

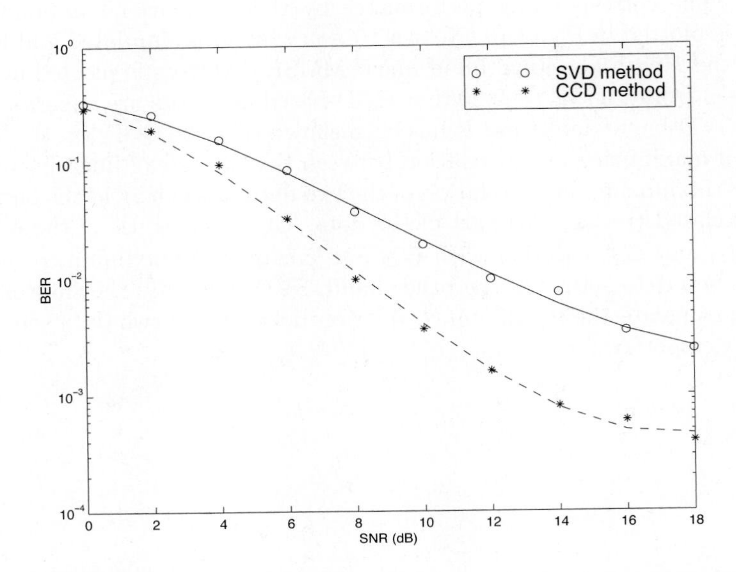

Figure 2.19. Performance of decimation-combining blind linear MMSE detectors in a five-user system with correlated noise.

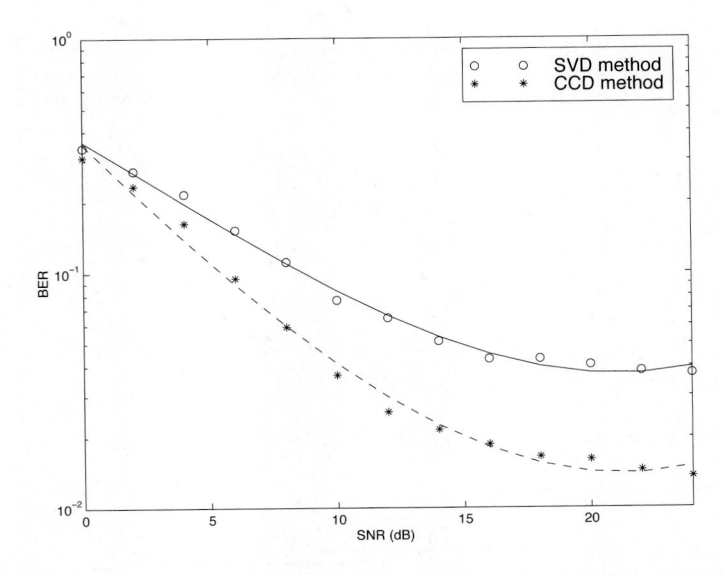

Figure 2.20. Performance of Pm-dimensional blind linear MMSE detectors in a 10-user system with correlated noise.

2.8 Appendix

2.8.1 Derivations for Section 2.3.3

Derivation of Equation (2.61)

Recall that the RLS algorithm for updating the blind linear MMSE algorithm is as follows:

$$\boldsymbol{k}[i] \triangleq \frac{\boldsymbol{C}_r[i-1]^{-1}\boldsymbol{r}[i]}{\lambda + \boldsymbol{r}^H[i]\boldsymbol{C}_r[i-1]^{-1}\boldsymbol{r}[i]}, \tag{2.247}$$

$$\boldsymbol{h}[i] \triangleq \boldsymbol{C}_r[i]^{-1}\boldsymbol{s}_1 \tag{2.248}$$

$$= \frac{1}{\lambda}\left(\boldsymbol{h}[i-1] - \boldsymbol{k}[i]\boldsymbol{r}[i]^H\boldsymbol{h}[i-1]\right), \tag{2.249}$$

$$\boldsymbol{m}_1[i] = \frac{1}{\boldsymbol{s}_1^T\boldsymbol{h}[i]}\boldsymbol{h}[i], \tag{2.250}$$

$$\boldsymbol{C}_r[i]^{-1} = \frac{1}{\lambda}\left(\boldsymbol{C}_r[i-1]^{-1} - \boldsymbol{k}[i]\boldsymbol{r}[i]^H\boldsymbol{C}_r[i-1]^{-1}\right). \tag{2.251}$$

We first derive an explicit recursive relationship between $\boldsymbol{m}_1[i]$ and $\boldsymbol{m}_1[i-1]$. Define

$$\alpha[i] \triangleq \boldsymbol{s}_1^T\boldsymbol{C}_r[i]^{-1}\boldsymbol{s}_1 = \boldsymbol{s}_1^T\boldsymbol{h}[i]. \tag{2.252}$$

Premultiplying both sides of (2.249) by $\boldsymbol{s}_1^T$, we get

$$\alpha[i] = \frac{1}{\lambda}\left(\alpha[i-1] - \boldsymbol{s}_1^T\boldsymbol{k}[i]\boldsymbol{r}[i]^H\boldsymbol{h}[i-1]\right). \tag{2.253}$$

From (2.253), we obtain

$$\begin{aligned}\alpha[i]^{-1} &= \lambda\left(\alpha[i-1]^{-1} + \frac{\alpha[i-1]^{-2}\boldsymbol{s}_1^T\boldsymbol{k}[i]\boldsymbol{r}[i]^H\boldsymbol{h}[i-1]}{1 - \boldsymbol{s}_1^T\boldsymbol{k}[i]\boldsymbol{r}^H[i]\boldsymbol{h}[i-1]\alpha[i-1]^{-1}}\right)\\ &= \lambda\left(\alpha[i-1]^{-1} + \alpha[i-1]^{-1}\beta[i]\boldsymbol{r}[i]^H\boldsymbol{h}[i-1]\right),\end{aligned} \tag{2.254}$$

where

$$\beta[i] \triangleq \frac{\alpha[i-1]^{-1}\boldsymbol{s}_1^T\boldsymbol{k}[i]}{1 - \boldsymbol{s}_1^T\boldsymbol{k}[i]\boldsymbol{r}[i]^H\boldsymbol{h}[i-1]\alpha[i-1]^{-1}}. \tag{2.255}$$

Substituting (2.249) and (2.254) into (2.250), we get

$$\begin{aligned}\boldsymbol{m}_1[i] &= \alpha[i]^{-1}\boldsymbol{h}[i]\\ &= \lambda\alpha[i-1]\boldsymbol{h}[i] + \lambda\beta[i]\underbrace{\left(\alpha[i-1]\boldsymbol{r}[i]^H\boldsymbol{h}[i-1]\right)}_{\xi[i]^*}\boldsymbol{h}[i]\\ &= \alpha[i-1]\left(\boldsymbol{h}[i-1] - \boldsymbol{k}[i]\boldsymbol{r}[i]^H\boldsymbol{h}[i-1]\right) + \lambda\beta[i]\xi[i]^*\boldsymbol{h}[i]\\ &= \boldsymbol{m}_1[i-1] - \xi[i]^*\boldsymbol{k}[i] + \lambda\beta[i]\xi[i]^*\boldsymbol{h}[i],\end{aligned} \tag{2.256}$$

where

$$\xi[i] \triangleq \boldsymbol{m}_1[i-1]^H \boldsymbol{r}[i] \\ = \alpha[i-1]\boldsymbol{h}[i-1]^H \boldsymbol{r}[i] \tag{2.257}$$

is the *a priori* least-squares estimate at time i. It is shown below that

$$\boldsymbol{k}[i] = \boldsymbol{C}_r[i]^{-1}\boldsymbol{r}[i], \tag{2.258}$$

$$\lambda\beta[i] = z[i]. \tag{2.259}$$

Substituting (2.248) and (2.258) into (2.256), we get

$$\boldsymbol{m}_1[i] = \boldsymbol{m}_1[i-1] - \boldsymbol{C}_r[i]^{-1}\boldsymbol{r}[i]\xi[i]^* + \boldsymbol{C}_r[i]^{-1}\boldsymbol{s}_1 z[i]\xi[i]^*. \tag{2.260}$$

Therefore, by (2.260) we have

$$\begin{aligned} z[i] \ &\triangleq \ \boldsymbol{m}_1[i]^H\boldsymbol{r}[i] = \xi[i] - \underbrace{\left(\boldsymbol{r}^H[i]\boldsymbol{C}_r[i]^{-1}\boldsymbol{r}[i]\right)}_{\underline{v}[i]^H\underline{v}[i]}\xi[i] + \underbrace{\left(\boldsymbol{s}_1^T\boldsymbol{C}_r[i]^{-1}\boldsymbol{r}[i]\right)}_{\alpha[i]z[i]} z[i]^*\xi[i] \\ &= \xi[i] - \left(\underline{v}[i]^H\underline{v}[i]\right)\xi[i] + \alpha[i]|z[i]|^2\alpha[i]\xi[i], \end{aligned} \tag{2.261}$$

where $\underline{v}[i]$ is defined in (2.56). Therefore, from (2.261) we get

$$\xi[i] = \frac{z[i]}{1 - (\underline{v}[i]^H\underline{v}[i]) + \alpha[i]|z[i]|^2}. \tag{2.262}$$

Finally, we derive (2.258) and (2.259). Postmultipling both sides of (2.251) by $\boldsymbol{r}[i]$, we get

$$\boldsymbol{C}_r[i]^{-1}\boldsymbol{r}[i] = \frac{1}{\lambda}\left[\boldsymbol{C}_r[i-1]^{-1} - \boldsymbol{k}[i]\boldsymbol{r}[i]^H\boldsymbol{C}[i-1]^{-1}\boldsymbol{r}[i]\right]. \tag{2.263}$$

On the other hand, (2.247) can be rewritten as

$$\boldsymbol{k}[i] = \frac{1}{\lambda}\left[\boldsymbol{C}_r[n-1]^{-1} - \boldsymbol{k}[i]\boldsymbol{r}[i]^H\boldsymbol{C}[i-1]^{-1}\boldsymbol{r}[i]\right]. \tag{2.264}$$

Equation (2.258) is obtained by comparing (2.263) and (2.264). Multiplying both sides of (2.254) by $\boldsymbol{s}_1^T\boldsymbol{k}[i]$, we get

$$\alpha[i]^{-1}\boldsymbol{s}_1^T\boldsymbol{k}[i] = \lambda\left(\alpha[i-1]^{-1}\boldsymbol{s}_1^T\boldsymbol{k}[i] + \alpha[i-1]^{-1}\beta[i]\boldsymbol{r}[i]^H\boldsymbol{h}[i-1]\boldsymbol{s}_1^T\boldsymbol{k}[i]\right). \tag{2.265}$$

Equation (2.255) can be rewritten as

$$\beta[i] = \alpha[i-1]^{-1}\boldsymbol{s}_1^T\boldsymbol{k}[i] + \alpha[i-1]^{-1}\beta[i]\boldsymbol{r}[i]^H\boldsymbol{h}[i-1]\boldsymbol{s}_1^T\boldsymbol{k}[i]. \tag{2.266}$$

Equation (2.259) is obtained comparing (2.265) and (2.266).

Derivation of Equations (2.62)–(2.69)

Suppose that an application of the rotation matrix $\underline{Q}[i]$ yields the following form:

$$\underline{Q}[i][\boldsymbol{A}_1 \ \boldsymbol{A}_2] = [\boldsymbol{B}_1 \ \boldsymbol{B}_2]. \tag{2.267}$$

Then because of the orthogonality property of $\underline{Q}[i]$ (i.e., $\underline{Q}^H[i]\underline{Q}[i] = \boldsymbol{I}$), taking the outer products of each side of (2.267) with their respective Hermitians, we get the following identities:

$$\boldsymbol{A}_1^H \boldsymbol{A}_1 = \boldsymbol{B}_1^H \boldsymbol{B}_1, \tag{2.268}$$
$$\boldsymbol{A}_1^H \boldsymbol{A}_2 = \boldsymbol{B}_1^H \boldsymbol{B}_2, \tag{2.269}$$
$$\boldsymbol{A}_2^H \boldsymbol{A}_2 = \boldsymbol{B}_2^H \boldsymbol{B}_2. \tag{2.270}$$

Associating $\boldsymbol{A}_1$ with the first N columns of the partitioned matrix on the left-hand side of (2.62), and $\boldsymbol{B}_1$ with the first N columns of the partitioned matrix on the right-hand side of (2.62), then (2.268), (2.269), and (2.270) yield

$$\underline{C}[i]^H \underline{C}[i] = \lambda \underline{C}[i-1]^H \underline{C}[i-1] + \boldsymbol{r}[i]\boldsymbol{r}[i]^H, \tag{2.271}$$
$$\underline{C}[i]^H \underline{u}[i] = \underline{C}[i-1]^H \underline{u}[i-1], \tag{2.272}$$
$$\underline{C}[i]^H \underline{v}[i] = \boldsymbol{r}[i], \tag{2.273}$$
$$\lambda \underline{u}[i]^H \underline{u}[i] + \lambda |\eta[i]|^2 = \underline{u}[i-1]^H \underline{u}[i-1], \tag{2.274}$$
$$\underline{u}[i]^H \underline{v}[i] + \eta[i]^* \gamma[i] = 0, \tag{2.275}$$
$$\underline{v}[i]^H \underline{v}[i] + |\gamma[i]|^2 = 1. \tag{2.276}$$

A comparison of (2.271)–(2.273) with (2.54)–(2.56) shows that $\underline{C}[i]$, $\underline{u}[i]$, and $\underline{v}[i]$ in (2.62) are the correct updated quantities at time n. Moreover, (2.67) follows from (2.274) and (2.57), (2.68) follows from (2.275) and (2.59), and (2.69) follows from (2.276) and (2.262).

2.8.2 Proofs for Section 2.4.4

Proof of Lemma 2.3

Denote

$$\boldsymbol{H} \triangleq \boldsymbol{S}\boldsymbol{A}\boldsymbol{S}^T,$$
$$\boldsymbol{G} \triangleq \boldsymbol{W}\boldsymbol{\Sigma}^{\dagger^T}\boldsymbol{V}^T|\boldsymbol{A}|^{-2}\boldsymbol{V}\boldsymbol{\Sigma}^{\dagger}\boldsymbol{W}^T.$$

Note that the eigendecomposition of $\boldsymbol{H}$ is given by

$$\boldsymbol{H} \triangleq \boldsymbol{S}\boldsymbol{A}\boldsymbol{S}^T = \boldsymbol{U}\boldsymbol{\Lambda}_0\boldsymbol{U}^T. \tag{2.277}$$

Then the Moore–Penrose generalized inverse [189] of matrix $\boldsymbol{H}$ is given by

$$\boldsymbol{H}^{\dagger} = \left(\boldsymbol{S}\boldsymbol{A}\boldsymbol{S}^T\right)^{\dagger} = \boldsymbol{U}\boldsymbol{\Lambda}_0^{\dagger}\boldsymbol{U}^T. \tag{2.278}$$

On the other hand, the Moore–Penrose generalized inverse $\boldsymbol{H}^{\dagger}$ of a matrix $\boldsymbol{H}$ is the unique matrix that satisfies [189] (a) $\boldsymbol{H}\boldsymbol{H}^{\dagger}$ and $\boldsymbol{H}^{\dagger}\boldsymbol{H}$ are symmetric; (b) $\boldsymbol{H}\boldsymbol{H}^{\dagger}\boldsymbol{H} = \boldsymbol{H}$; and (c) $\boldsymbol{H}^{\dagger}\boldsymbol{H}\boldsymbol{H}^{\dagger} = \boldsymbol{H}^{\dagger}$. Next we show that $\boldsymbol{G} = \boldsymbol{H}^{\dagger}$ by verifying these three conditions. We first verify condition (a). Using (2.106), we have

$$\begin{aligned} \boldsymbol{H}\boldsymbol{G} &= \left(\boldsymbol{W}\boldsymbol{\Sigma}\boldsymbol{V}^T\boldsymbol{A}\boldsymbol{V}\boldsymbol{\Sigma}^T\boldsymbol{W}^T\right)\left(\boldsymbol{W}{\boldsymbol{\Sigma}^{\dagger}}^T\boldsymbol{V}^T|\boldsymbol{A}|^{-2}\boldsymbol{V}\boldsymbol{\Sigma}^{\dagger}\boldsymbol{W}^T\right) \\ &= \boldsymbol{W}\boldsymbol{\Sigma}\boldsymbol{\Sigma}^{\dagger}\boldsymbol{W}^T, \end{aligned} \tag{2.279}$$

where the second equality follows from the facts that $\boldsymbol{W}^T\boldsymbol{W} = \boldsymbol{I}_N$ and $\boldsymbol{\Sigma}^T{\boldsymbol{\Sigma}^{\dagger}}^T = \boldsymbol{V}^T\boldsymbol{V} = \boldsymbol{V}\boldsymbol{V}^T = \boldsymbol{I}_K$. Since the $N \times N$ diagonal matrix $\boldsymbol{\Sigma}\boldsymbol{\Sigma}^{\dagger} = \mathsf{diag}\,(\boldsymbol{I}_K, \boldsymbol{0})$, it follows from (2.279) that $\boldsymbol{H}\boldsymbol{G}$ is symmetric. Similarly, $\boldsymbol{G}\boldsymbol{H}$ is also symmetric. Next we verify condition (b).

$$\begin{aligned} \boldsymbol{H}\boldsymbol{G}\boldsymbol{H} &= \left(\boldsymbol{W}\boldsymbol{\Sigma}\boldsymbol{V}^T\boldsymbol{A}\boldsymbol{V}\boldsymbol{\Sigma}^T\boldsymbol{W}^T\right)\left(\boldsymbol{W}{\boldsymbol{\Sigma}^{\dagger}}^T\boldsymbol{V}^T|\boldsymbol{A}|^{-2}\boldsymbol{V}\boldsymbol{\Sigma}^{\dagger}\boldsymbol{W}^T\right)\left(\boldsymbol{W}\boldsymbol{\Sigma}\boldsymbol{V}^T\boldsymbol{A}\boldsymbol{V}\boldsymbol{\Sigma}^T\boldsymbol{W}^T\right) \\ &= \boldsymbol{W}\boldsymbol{\Sigma}\boldsymbol{\Sigma}^{\dagger}\boldsymbol{\Sigma}\boldsymbol{V}^T\boldsymbol{A}\boldsymbol{V}\boldsymbol{\Sigma}^T\boldsymbol{W}^T \\ &= \boldsymbol{W}\boldsymbol{\Sigma}\boldsymbol{V}^T\boldsymbol{A}\boldsymbol{V}\boldsymbol{\Sigma}^T\boldsymbol{W}^T \\ &= \boldsymbol{S}\boldsymbol{A}\boldsymbol{S}^T \;=\; \boldsymbol{H}, \end{aligned} \tag{2.280}$$

where in the second equality, the following facts are used: $\boldsymbol{W}^T\boldsymbol{W} = \boldsymbol{I}_N$, $\boldsymbol{\Sigma}^T{\boldsymbol{\Sigma}^{\dagger}}^T = \boldsymbol{I}_K$, and $\boldsymbol{V}^T\boldsymbol{V} = \boldsymbol{V}\boldsymbol{V}^T = \boldsymbol{I}_K$; the third equality follows from the fact that $\boldsymbol{\Sigma}\boldsymbol{\Sigma}^{\dagger}\boldsymbol{\Sigma} = \boldsymbol{\Sigma}$. Condition (c) can be similarly verified (i.e., $\boldsymbol{G}\boldsymbol{H}\boldsymbol{G} = \boldsymbol{G}$). Therefore, we have

$$\boldsymbol{U}\boldsymbol{\Lambda}_0^{\dagger}\boldsymbol{U}^T \;=\; \boldsymbol{H}^{\dagger} \;=\; \boldsymbol{G} \;=\; \boldsymbol{W}{\boldsymbol{\Sigma}^{\dagger}}^T\boldsymbol{V}^T|\boldsymbol{A}|^{-2}\boldsymbol{V}\boldsymbol{\Sigma}^{\dagger}\boldsymbol{W}^T. \tag{2.281}$$

Now (2.107) follows immediately from (2.281) and the fact that $\boldsymbol{U}^T\boldsymbol{U} = \boldsymbol{U}\boldsymbol{U}^T = \boldsymbol{I}_N$.

2.8.3 Proofs for Section 2.5.2

Some Useful Lemmas

We first list some lemmas that will be used in proving the results in Section 2.5.2. A random matrix is said to be Gaussian distributed if the joint distribution of all its elements is Gaussian. First we have the following vector form of the central limit theorem.

Lemma 2.4: (Theorem 1.9.1B in [443]) *Let $\{\boldsymbol{x}_i\}$ be i.i.d. random vectors with mean $\boldsymbol{\mu}$ and covariance matrix $\boldsymbol{\Sigma}$. Then*

$$\sqrt{M}\left(\frac{1}{M}\sum_{i=1}^{M}\boldsymbol{x}_i - \boldsymbol{\mu}\right) \to \mathcal{N}(\boldsymbol{0}, \boldsymbol{\Sigma}), \quad \textit{in distribution}, \quad \textit{as } M \to \infty.$$

Next we establish that the sample autocorrelation matrix $\hat{\boldsymbol{C}}_r$ given by (2.122) is asymptotically Gaussian distributed as the sample size $M \to \infty$.

Lemma 2.5: *Denote*

$$C_r = E\left\{r[i]r[i]^T\right\}, \tag{2.282}$$

$$\hat{C}_r = \frac{1}{M}\sum_{i=0}^{M-1} r[i]r[i]^T, \tag{2.283}$$

$$\Delta C_r \triangleq \hat{C}_r - C_r. \tag{2.284}$$

Then $\sqrt{M}\Delta C_r$ converges in probability toward a Gaussian matrix with mean $\mathbf{0}$ and an $N^2 \times N^2$ covariance matrix whose elements are specified by

$$M \cdot \mathrm{Cov}\left\{[\Delta C_r]_{i,j}, [\Delta C_r]_{m,n}\right\}$$
$$= [C_r]_{i,m}[C_r]_{j,n} + [C_r]_{i,n}[C_r]_{j,m} - 2\sum_{\alpha=1}^{K} A_\alpha^4 [s_\alpha]_i [s_\alpha]_j [s_\alpha]_m [s_\alpha]_n. \tag{2.285}$$

Proof: Since $\hat{C}_r$ given by (2.285) has $E\{\hat{C}_r\} = C_r$, and it is a sum of i.i.d. terms $\left(r[i]r[i]^T\right)$, by Lemma 2.4, it is asymptotically Gaussian, with an $N^2 \times N^2$ covariance matrix whose elements are given by the covariance of the zero-mean random matrix $\left(r[i]r[i]^T\right)$. To calculate this covariance, note that (for notational convenience, in what follows we drop the time index i)

$$\left[rr^T\right]_{i,j} = \sum_{\alpha=1}^{K}\sum_{\beta=1}^{K} A_\alpha A_\beta [s_\alpha]_i [s_\beta]_j b_\alpha b_\beta$$
$$+ \sum_{\alpha=1}^{K} A_\alpha [s_\alpha]_i b_\alpha n_j + \sum_{\alpha=1}^{K} A_\alpha [s_\alpha]_j b_\alpha n_i + n_i n_j. \tag{2.286}$$

We have

$$\mathrm{Cov}\left\{\left[rr^T\right]_{i,j}, \left[rr^T\right]_{m,n}\right\}$$
$$= \sum_{\alpha=1}^{K}\sum_{\beta=1}^{K}\sum_{\gamma=1}^{K}\sum_{\lambda=1}^{K} A_\alpha A_\beta A_\gamma A_\lambda [s_\alpha]_i [s_\beta]_j [s_\gamma]_m [s_\lambda]_n \quad \underbrace{\mathrm{Cov}\{b_\alpha b_\beta, b_\gamma b_\lambda\}}_{\delta_{\alpha=\gamma}\delta_{\beta=\lambda} + \delta_{\alpha=\lambda}\delta_{\beta=\gamma} - 2\,\delta_{\alpha=\beta=\gamma=\lambda}}$$
$$+ \sum_{\alpha=1}^{K}\sum_{\beta=1}^{K} A_\alpha A_\beta [s_\alpha]_i [s_\beta]_m \underbrace{\mathrm{Cov}\{b_\alpha n_i, b_\beta n_m\}}_{\sigma^2\,\delta_{\alpha=\beta}\delta_{i=m}}$$
$$+ \sum_{\alpha=1}^{K}\sum_{\beta=1}^{K} A_\alpha A_\beta [s_\alpha]_i [s_\beta]_n \underbrace{\mathrm{Cov}\{b_\alpha n_i, b_\beta n_n\}}_{\sigma^2\,\delta_{\alpha=\beta}\delta_{i=n}}$$

$$+\sum_{\alpha=1}^{K}\sum_{\beta=1}^{K} A_\alpha A_\beta [\boldsymbol{s}_\alpha]_j [\boldsymbol{s}_\beta]_m \underbrace{\mathrm{Cov}\{b_\alpha n_j, b_\beta n_m\}}_{\sigma^2\,\delta_{\alpha=\beta}\delta_{j=m}}$$

$$+\sum_{\alpha=1}^{K}\sum_{\beta=1}^{K} A_\alpha A_\beta [\boldsymbol{s}_\alpha]_j [\boldsymbol{s}_\beta]_n \underbrace{\mathrm{Cov}\{b_\alpha n_j, b_\beta n_n\}}_{\sigma^2\,\delta_{\alpha=\beta}\delta_{j=n}}$$

$$+ \underbrace{\mathrm{Cov}\{n_i n_j, n_m n_n\}}_{\sigma^4\,(\delta_{i=m}\delta_{j=n}+\delta_{i=n}\delta_{j=m})}$$

$$= \sum_{\alpha=1}^{K}\sum_{\beta=1}^{K} A_\alpha^2 A_\beta^2 \left\{[\boldsymbol{s}_\alpha]_i[\boldsymbol{s}_\alpha]_m[\boldsymbol{s}_\beta]_j[\boldsymbol{s}_\beta]_n + [\boldsymbol{s}_\alpha]_i[\boldsymbol{s}_\alpha]_n[\boldsymbol{s}_\beta]_j[\boldsymbol{s}_\beta]_m\right\}$$

$$+ \sigma^2 \sum_{\alpha=1}^{K} A_\alpha^2 \{[\boldsymbol{s}_\alpha]_i[\boldsymbol{s}_\alpha]_m\delta_{j=n} + [\boldsymbol{s}_\alpha]_i[\boldsymbol{s}_\alpha]_n\delta_{j=m} + [\boldsymbol{s}_\alpha]_j[\boldsymbol{s}_\alpha]_m\delta_{i=n} + [\boldsymbol{s}_\alpha]_j[\boldsymbol{s}_\alpha]_n\delta_{i=m}\}$$

$$- 2\sum_{\alpha=1}^{K} A_\alpha^4 [\boldsymbol{s}_\alpha]_i[\boldsymbol{s}_\alpha]_j[\boldsymbol{s}_\alpha]_m[\boldsymbol{s}_\alpha]_n + \sigma^4\,(\delta_{i=m}\delta_{j=n} + \delta_{i=n}\delta_{j=m})$$

$$= [\boldsymbol{C}_r]_{i,m}[\boldsymbol{C}_r]_{j,n} + [\boldsymbol{C}_r]_{i,n}[\boldsymbol{C}_r]_{j,m} - 2\sum_{\alpha=1}^{K} A_\alpha^4 [\boldsymbol{s}_\alpha]_i[\boldsymbol{s}_\alpha]_j[\boldsymbol{s}_\alpha]_m[\boldsymbol{s}_\alpha]_n, \tag{2.287}$$

where the last equality follows from the fact that

$$[\boldsymbol{C}_r]_{i,j} = \left[\sum_{k=1}^{K} A_k^2 \boldsymbol{s}_k \boldsymbol{s}_k^T + \sigma^2 \boldsymbol{I}_N\right]_{i,j} = \sum_{\alpha=1}^{K} A_\alpha^2 [\boldsymbol{s}_\alpha]_i[\boldsymbol{s}_\alpha]_j + \sigma^2\delta_{i=j}. \tag{2.288}$$

□

Note that the last term of (2.285) is due to the nonnormality of the received signal $\boldsymbol{r}[i]$. If the signal had been Gaussian, the result would have been the first two terms of (2.285) only (compare this result with Theorem 3.4.4 in [18]). Using a different modulation scheme (other than BPSK) will result in a different form for the last term in (2.285).

In what follows we make frequent use of the differential of a matrix function (cf. [421], Chap. 14). Consider a function $\boldsymbol{f} : \mathbb{R}^n \to \mathbb{R}^m$. Recall that the differential of $\boldsymbol{f}$ at a point $\boldsymbol{x}_0$ is a linear function $\boldsymbol{L}_f(\cdot;\boldsymbol{x}_0) : \mathbb{R}^n \to \mathbb{R}^m$ such that

$$\forall\epsilon > 0,\ \exists\delta > 0 :\ \|\boldsymbol{x} - \boldsymbol{x}_0\| < \delta \Rightarrow \|\boldsymbol{f}(\boldsymbol{x}) - \boldsymbol{f}(\boldsymbol{x}_0) - \boldsymbol{L}_f(\boldsymbol{x} - \boldsymbol{x}_0;\boldsymbol{x}_0)\| < \epsilon. \tag{2.289}$$

If the differential exists, it is given by $\boldsymbol{L}_f(\boldsymbol{x};\boldsymbol{x}_0) = \boldsymbol{T}(\boldsymbol{x}_0)\boldsymbol{x}$, where $\boldsymbol{T}(\boldsymbol{x}_0) \triangleq (\partial\boldsymbol{f}/\partial\boldsymbol{x})|_{\boldsymbol{x}=\boldsymbol{x}_0}$. Let $\boldsymbol{y} = \boldsymbol{f}(\boldsymbol{x})$ and consider its differential at $\boldsymbol{x}_0$. Denote $\Delta\boldsymbol{x} \triangleq \boldsymbol{x} - \boldsymbol{x}_0$ and $\Delta\boldsymbol{y} \triangleq \boldsymbol{L}_f(\Delta\boldsymbol{x};\boldsymbol{x}_0)$. Hence for fixed $\boldsymbol{x}_0$, $\Delta\boldsymbol{y}$ is a function of $\Delta\boldsymbol{x}$; and for fixed $\boldsymbol{x}_0$, if $\boldsymbol{x}$ is random, so is $\Delta\boldsymbol{y}$. We have the following lemma regarding the asymptotic distribution of a function of a sequence of asymptotically Gaussian vectors.

Lemma 2.6: (Theorem 3.3A in [443]) *Suppose that* $\boldsymbol{x}(M) \in \mathbb{R}^n$ *is asymptotically Gaussian; that is,*

$$\sqrt{M}\,[\boldsymbol{x}(M) - \boldsymbol{x}_0] \rightarrow \mathcal{N}(\mathbf{0}, \boldsymbol{C}_x), \quad \textit{in distribution,} \quad \textit{as } M \rightarrow \infty.$$

Let $\boldsymbol{f} : \mathbb{R}^n \rightarrow \mathbb{R}^m$ *be a function. Denote* $\boldsymbol{y}(M) = \boldsymbol{f}[\boldsymbol{x}(M)]$. *Suppose that* $\boldsymbol{f}$ *has a nonzero differential* $\boldsymbol{L}_f(\boldsymbol{x}; \boldsymbol{x}_0) = \boldsymbol{T}(\boldsymbol{x}_0)\boldsymbol{x}$ *at* $\boldsymbol{x}_0$. *Denote* $\Delta\boldsymbol{x}(M) \triangleq \boldsymbol{x}(M) - \boldsymbol{x}_0$ *and* $\Delta\boldsymbol{y}(M) = \boldsymbol{T}(\boldsymbol{x}_0)\Delta\boldsymbol{x}(M)$. *Then*

$$\sqrt{M}\,[\boldsymbol{y}(M) - \boldsymbol{f}(\boldsymbol{x}_0)] \rightarrow \mathcal{N}(\mathbf{0}, \boldsymbol{C}_y), \quad \textit{in distribution,} \quad \textit{as } M \rightarrow \infty, \tag{2.290}$$

where

$$\boldsymbol{C}_y = \boldsymbol{T}(\boldsymbol{x}_0)\boldsymbol{C}_x\boldsymbol{T}(\boldsymbol{x}_0)^T \triangleq \lim_{M\rightarrow\infty} \boldsymbol{T}(\boldsymbol{x}_0)E\left\{\Delta\boldsymbol{x}(M)\Delta\boldsymbol{x}(M)\right\}\boldsymbol{T}(\boldsymbol{x}_0)^T \tag{2.291}$$

$$= \lim_{M\rightarrow\infty} E\left\{\Delta\boldsymbol{y}(M)\Delta\boldsymbol{y}(M)^T\right\}. \tag{2.292}$$

To calculate $\boldsymbol{C}_y$ we can use either (2.291) or (2.292). When dealing with functions of matrices, however, it is usually easier to use (2.292). In what follows we make use of the following identities of matrix differentials:

$$\boldsymbol{C} = \boldsymbol{f}(\boldsymbol{X}) \triangleq \boldsymbol{M}\boldsymbol{X} \Longrightarrow \Delta\boldsymbol{C} = \boldsymbol{M}\Delta\boldsymbol{X}, \tag{2.293}$$

$$\boldsymbol{C} = \boldsymbol{f}(\boldsymbol{X}, \boldsymbol{Y}) \triangleq \boldsymbol{X}\boldsymbol{Y} \Longrightarrow \Delta\boldsymbol{C} = \boldsymbol{X}\Delta\boldsymbol{Y} + \Delta\boldsymbol{X}\boldsymbol{Y}, \tag{2.294}$$

$$\boldsymbol{C} = \boldsymbol{f}(\boldsymbol{X}) \triangleq \boldsymbol{X}^{-1} \Longrightarrow \Delta\boldsymbol{C} = -\boldsymbol{X}^{-1}\Delta\boldsymbol{X}\boldsymbol{X}^{-1}. \tag{2.295}$$

Finally, we have the following lemma regarding the differentials of the eigencomponents of a symmetric matrix. It is a generalization of Theorem 13.5.1 in [18]. Its proof can be found in [197].

Lemma 2.7: *Let the* $N \times N$ *symmetric matrix* $\boldsymbol{C}_0$ *have an eigendecomposition* $\boldsymbol{C}_0 = \boldsymbol{U}_0\boldsymbol{\Lambda}_0\boldsymbol{U}_0^T$, *where the eigenvalues satisfy* $\lambda_1^0 > \lambda_2^0 > \cdots > \lambda_K^0 > \lambda_{K+1}^0 = \lambda_{K+2}^0 = \cdots = \lambda_N^0$. *Let* $\Delta\boldsymbol{C}$ *be a symmetric variation of* $\boldsymbol{C}_0$ *and denote* $\boldsymbol{C} \triangleq \boldsymbol{C}_0 + \Delta\boldsymbol{C}$. *Let* $\boldsymbol{T}$ *be a unitary transformation of* $\boldsymbol{C}$ *as*

$$\boldsymbol{T}(\boldsymbol{C}) \triangleq \boldsymbol{U}_0^T\boldsymbol{C}\boldsymbol{U}_0. \tag{2.296}$$

Denote the eigendecomposition of $\boldsymbol{T}$ *as*

$$\boldsymbol{T} = \boldsymbol{W}\boldsymbol{\Lambda}\boldsymbol{W}^T. \tag{2.297}$$

(Note that if $\boldsymbol{C} = \boldsymbol{C}_0$, *then* $\boldsymbol{W} = \boldsymbol{I}_N$ *and* $\boldsymbol{\Lambda} = \boldsymbol{\Lambda}_0$.*) The differential of* $\boldsymbol{\Lambda}$ *at* $\boldsymbol{\Lambda}_0$, *and the differential of* $\boldsymbol{W}$ *at* $\boldsymbol{I}_N$, *as a function of* $\Delta\boldsymbol{T} = \boldsymbol{U}_0\Delta\boldsymbol{C}\boldsymbol{U}_0^T$, *are given, respectively, by*

$$\Delta\lambda_k = [\Delta\boldsymbol{T}]_{k,k}, \qquad 1 \le k \le K, \tag{2.298}$$

$$[\Delta\boldsymbol{W}]_{i,k} = \begin{cases} 0, & i = k, \\ \dfrac{1}{\lambda_k^0 - \lambda_i^0}[\Delta\boldsymbol{T}]_{i,k}, & i \neq k, \end{cases} \qquad 1 \le i \le N,\ 1 \le k \le K. \tag{2.299}$$

Proof of Theorem 2.1

DMI Blind Detector Consider the function $\hat{\boldsymbol{C}}_r \rightarrow \hat{\boldsymbol{w}}_1 = \hat{\boldsymbol{C}}_r^{-1}\boldsymbol{s}_1$. The differential of $\hat{\boldsymbol{w}}_1$ at $\boldsymbol{C}_r$ is given by

$$\Delta \boldsymbol{w}_1 = -\boldsymbol{C}_r^{-1}\Delta\boldsymbol{C}_r\boldsymbol{C}_r^{-1}\boldsymbol{s}_1, \tag{2.300}$$

where $\Delta\boldsymbol{C}_r \triangleq \hat{\boldsymbol{C}}_r - \boldsymbol{C}_r$. Then according to Lemma 2.6, $\sqrt{M}\,(\hat{\boldsymbol{w}}_1 - \boldsymbol{w}_1)$ is asymptotically Gaussian as $M \rightarrow \infty$, with zero mean and covariance matrix given by (2.292)[4]

$$\begin{aligned}\boldsymbol{C}_w \triangleq M \cdot E\left\{\Delta\boldsymbol{w}_1\Delta\boldsymbol{w}_1^T\right\} &= M \cdot E\left\{\boldsymbol{C}_r^{-1}\Delta\boldsymbol{C}_r\boldsymbol{C}_r^{-1}\boldsymbol{s}_1\boldsymbol{s}_1^T\boldsymbol{C}_r^{-1}\Delta\boldsymbol{C}_r\boldsymbol{C}_r^{-1}\right\}\\ &= M \cdot \boldsymbol{C}_r^{-1}E\left\{\Delta\boldsymbol{C}_r\boldsymbol{w}_1\boldsymbol{w}_1^T\Delta\boldsymbol{C}_r\right\}\boldsymbol{C}_r^{-1}.\end{aligned} \tag{2.301}$$

Now, by Lemma 2.5, we have

$$\begin{aligned}&M \cdot E\left\{\Delta\boldsymbol{C}_r\boldsymbol{w}_1\boldsymbol{w}_1^T\Delta\boldsymbol{C}_r\right\}_{i,j}\\ &= M \cdot E\left\{\sum_{m=1}^{N}\sum_{n=1}^{N}[\Delta\boldsymbol{C}_r]_{i,m}[\boldsymbol{w}_1]_m[\Delta\boldsymbol{C}_r]_{j,n}[\boldsymbol{w}_1]_n\right\}\\ &= \sum_{m=1}^{N}\sum_{n=1}^{N}\\ &\quad\left([\boldsymbol{C}_r]_{i,j}[\boldsymbol{C}_r]_{m,n} + [\boldsymbol{C}_r]_{i,n}[\boldsymbol{C}_r]_{m,j} - 2\sum_{\alpha=1}^{K}A_\alpha^4[\boldsymbol{s}_\alpha]_i[\boldsymbol{s}_\alpha]_j[\boldsymbol{s}_\alpha]_l[\boldsymbol{s}_\alpha]_m\right)[\boldsymbol{w}_1]_m[\boldsymbol{w}_1]_n\\ &= [\boldsymbol{C}_r]_{i,j}\underbrace{\left(\sum_{m=1}^{N}\sum_{n=1}^{N}[\boldsymbol{C}_r]_{m,n}[\boldsymbol{w}_1]_m[\boldsymbol{w}_1]_n\right)}_{\boldsymbol{w}_1^T\boldsymbol{C}_r\boldsymbol{w}_1} + \underbrace{\left(\sum_{m=1}^{N}[\boldsymbol{C}_r]_{m,j}[\boldsymbol{w}_j]_k\right)}_{[\boldsymbol{C}_r\boldsymbol{w}_1]_j}\underbrace{\left(\sum_{n=1}^{N}[\boldsymbol{C}_r]_{i,n}[\boldsymbol{w}_1]_i\right)}_{[\boldsymbol{C}_r\boldsymbol{w}_1]_i}\\ &\quad - 2\sum_{\alpha=1}^{K}[\boldsymbol{s}_\alpha]_i[\boldsymbol{s}_\alpha]_jA_\alpha^4\underbrace{\left(\sum_{m=1}^{N}[\boldsymbol{s}_\alpha]_m[\boldsymbol{w}_1]_m\right)^2}_{\boldsymbol{s}_\alpha^T\boldsymbol{w}_1}.\end{aligned} \tag{2.302}$$

Writing (2.302) in a matrix form, we have

$$M \cdot E\left\{\Delta\boldsymbol{C}_r\boldsymbol{w}_1\boldsymbol{w}_1^T\Delta\boldsymbol{C}_r\right\} = \boldsymbol{C}_r\left(\boldsymbol{w}_1^T\boldsymbol{C}_r\boldsymbol{w}_1\right) + \boldsymbol{C}_r\boldsymbol{w}_1\boldsymbol{w}_1^T\boldsymbol{C}_r - 2\boldsymbol{S}\boldsymbol{D}\boldsymbol{S}^T, \tag{2.303}$$

with

$$\boldsymbol{D} \triangleq \mathsf{diag}\left\{A_1^4(\boldsymbol{s}_1^T\boldsymbol{w}_1)^2, A_2^4(\boldsymbol{s}_2^T\boldsymbol{w}_1)^2, \ldots, A_K^4(\boldsymbol{s}_K^T\boldsymbol{w}_1)^2\right\}.$$

The eigendecomposition of $\boldsymbol{C}_r$ is

$$\boldsymbol{C}_r = \boldsymbol{U}_s\boldsymbol{\Lambda}_s\boldsymbol{U}_s^T + \sigma^2\boldsymbol{U}_n\boldsymbol{U}_n^T. \tag{2.304}$$

[4] We do not need the limit here, since the covariance matrix of $\sqrt{M}\Delta\boldsymbol{C}_r$ is independent of M.

Substituting (2.303) and (2.304) into (2.301), we get

$$
\begin{aligned}
&M \cdot \boldsymbol{C}_w \\
&= (\boldsymbol{w}_1^T \boldsymbol{C}_r \boldsymbol{w}_1)\boldsymbol{C}_r^{-1} + \boldsymbol{w}_1\boldsymbol{w}_1^T - 2\boldsymbol{C}_r^{-1}\boldsymbol{S}\boldsymbol{D}\boldsymbol{S}^T\boldsymbol{C}_r^{-1} \\
&= (\boldsymbol{w}_1^T\boldsymbol{s}_1)\left(\boldsymbol{U}_s\boldsymbol{\Lambda}_s^{-1}\boldsymbol{U}_s^T + \frac{1}{\sigma^2}\boldsymbol{U}_n\boldsymbol{U}_n^T\right) + \boldsymbol{w}_1\boldsymbol{w}_1^T \\
&\quad - 2\left(\boldsymbol{U}_s\boldsymbol{\Lambda}_s^{-1}\boldsymbol{U}_s^T + \frac{1}{\sigma^2}\boldsymbol{U}_n\boldsymbol{U}_n^T\right)\boldsymbol{S}\boldsymbol{D}\boldsymbol{S}^T\left(\boldsymbol{U}_s\boldsymbol{\Lambda}_s^{-1}\boldsymbol{U}_s^T + \frac{1}{\sigma^2}\boldsymbol{U}_n\boldsymbol{U}_n^T\right) \\
&= (\boldsymbol{w}_1^T\boldsymbol{s}_1)\,\boldsymbol{U}_s\boldsymbol{\Lambda}_s^{-1}\boldsymbol{U}_s^T + \boldsymbol{w}_1\boldsymbol{w}_1^T - 2\boldsymbol{U}_s\boldsymbol{\Lambda}_s^{-1}\boldsymbol{U}_s^T\boldsymbol{S}\boldsymbol{D}\boldsymbol{S}^T\boldsymbol{U}_s\boldsymbol{\Lambda}_s^{-1}\boldsymbol{U}_s^T \\
&\quad + \underbrace{\left[\frac{\boldsymbol{w}_1^T\boldsymbol{s}_1}{\sigma^2}\right]}_{\tau}\boldsymbol{U}_n\boldsymbol{U}_n^T,
\end{aligned}
$$

where the last equality follows from the fact that $\boldsymbol{U}_n^T\boldsymbol{S} = \boldsymbol{0}$.

Subspace Blind Detector We will prove the following more general proposition, which will be used in later proofs. The part of Theorem 2.1 for the subspace blind detector follows with $\boldsymbol{v} = \boldsymbol{s}_1$.

Proposition 2.6: *Let* $\boldsymbol{w}_1 = \boldsymbol{U}_s\boldsymbol{\Lambda}_s^{-1}\boldsymbol{U}_s^T\boldsymbol{v}$ *be the weight vector of a detector,* $\boldsymbol{v} \in \mathsf{range}(\boldsymbol{S})$*, and let* $\hat{\boldsymbol{w}}_1 = \hat{\boldsymbol{U}}_s\hat{\boldsymbol{\Lambda}}_s^{-1}\hat{\boldsymbol{U}}_s^T\boldsymbol{v}$ *be the weight vector of the corresponding estimated detector. Then*

$$
\sqrt{M}\,(\hat{\boldsymbol{w}}_1 - \boldsymbol{w}_1) \to \mathcal{N}(\boldsymbol{0}, \boldsymbol{C}_w), \quad \textit{in distribution,} \qquad \textit{as } M \to \infty,
$$

with

$$
\begin{aligned}
\boldsymbol{C}_w = &\ (\boldsymbol{w}_1^T\boldsymbol{v}_1)\,\boldsymbol{U}_s\boldsymbol{\Lambda}_s^{-1}\boldsymbol{U}_s^T \\
&+ \boldsymbol{w}_1\boldsymbol{w}_1^T - 2\boldsymbol{U}_s\boldsymbol{\Lambda}_s^{-1}\boldsymbol{U}_s^T\boldsymbol{S}\boldsymbol{D}\boldsymbol{S}^T\boldsymbol{U}_s\boldsymbol{\Lambda}_s^{-1}\boldsymbol{U}_s^T + \tau\boldsymbol{U}_n\boldsymbol{U}_n^T,
\end{aligned}
\tag{2.305}
$$

where

$$
\boldsymbol{D} \triangleq \mathsf{diag}\left\{A_1^4\left(\boldsymbol{s}_1^T\boldsymbol{w}_1\right)^2, A_2^4\left(\boldsymbol{s}_2^T\boldsymbol{w}_1\right)^2, \ldots, A_K^4\left(\boldsymbol{s}_K^T\boldsymbol{w}_1\right)^2\right\}, \tag{2.306}
$$

$$
\tau \triangleq \sigma^2\boldsymbol{v}^T\boldsymbol{U}_s\boldsymbol{\Lambda}_s^{-1}\left(\boldsymbol{\Lambda}_s - \sigma^2\boldsymbol{I}_K\right)^{-2}\boldsymbol{U}_s^T\boldsymbol{v}. \tag{2.307}
$$

Proof: Consider the function $(\hat{\boldsymbol{U}}_s, \hat{\boldsymbol{\Lambda}}_s) \to \hat{\boldsymbol{w}}_1 = \hat{\boldsymbol{U}}_s\hat{\boldsymbol{\Lambda}}_s^{-1}\hat{\boldsymbol{U}}_s^T\boldsymbol{v}$. By Lemma 2.6, $\sqrt{M}\,(\hat{\boldsymbol{w}}_1 - \boldsymbol{w}_1)$ is asymptotically Gaussian as $M \to \infty$, with zero mean and covariance matrix given by $\boldsymbol{C}_w \triangleq M \cdot E\{\Delta\boldsymbol{w}_1\Delta\boldsymbol{w}_1^T\}$, where $\Delta\boldsymbol{w}_1$ is the differential of $\hat{\boldsymbol{w}}_1$ at $(\boldsymbol{U}_s, \boldsymbol{\Lambda}_s)$. Denote $\boldsymbol{U} = [\boldsymbol{U}_s\ \boldsymbol{U}_n]$. Define

$$
\boldsymbol{T} \triangleq \boldsymbol{U}^T\hat{\boldsymbol{C}}_r\boldsymbol{U} = \boldsymbol{U}^T\left(\hat{\boldsymbol{U}}_s\hat{\boldsymbol{\Lambda}}_s\hat{\boldsymbol{U}}_s^T + \hat{\boldsymbol{U}}_n\hat{\boldsymbol{\Lambda}}_n\hat{\boldsymbol{U}}_n^T\right)\boldsymbol{U}. \tag{2.308}
$$

Since $\boldsymbol{T}$ is a unitary transformation of $\hat{\boldsymbol{C}}_r$, its eigenvalues are the same as those of $\hat{\boldsymbol{C}}_r$. Hence its eigendecomposition can be written as

$$
\boldsymbol{T} = \boldsymbol{W}_s\hat{\boldsymbol{\Lambda}}_s\boldsymbol{W}_s^T + \boldsymbol{W}_n\hat{\boldsymbol{\Lambda}}_n\boldsymbol{W}_n^T, \tag{2.309}
$$

where $\boldsymbol{W} = [\boldsymbol{W}_s \ \boldsymbol{W}_n] \triangleq \boldsymbol{U}^T[\hat{\boldsymbol{U}}_s\hat{\boldsymbol{U}}_n]\boldsymbol{U}$ are eigenvectors of $\boldsymbol{T}$. From (2.308) and (2.309), we have

$$\hat{\boldsymbol{U}}_s\hat{\boldsymbol{\Lambda}}_s^{-1}\hat{\boldsymbol{U}}_s^T = \boldsymbol{U}\boldsymbol{W}_s\hat{\boldsymbol{\Lambda}}_s^{-1}\boldsymbol{W}_s^T\boldsymbol{U}^T. \tag{2.310}$$

Thus we have

$$\begin{aligned}\Delta\boldsymbol{w}_1 &= \Delta\left(\hat{\boldsymbol{U}}_s\hat{\boldsymbol{\Lambda}}_s^{-1}\hat{\boldsymbol{U}}_s^T\right)\boldsymbol{v} \\ &= \boldsymbol{U}\underbrace{\Delta\left(\boldsymbol{W}_s\hat{\boldsymbol{\Lambda}}_s^{-1}\boldsymbol{W}_s^T\right)\boldsymbol{U}^T\boldsymbol{v}}_{\boldsymbol{z}}.\end{aligned} \tag{2.311}$$

The differential in (2.311) at $(\boldsymbol{I}_N, \boldsymbol{\Lambda})$ is given by

$$\Delta\left(\boldsymbol{W}_s\hat{\boldsymbol{\Lambda}}_s^{-1}\boldsymbol{W}_s^T\right) = \Delta\boldsymbol{W}_s\boldsymbol{\Lambda}_s^{-1}\boldsymbol{E}_s^T + \boldsymbol{E}_s\boldsymbol{\Lambda}_s^{-1}\Delta\boldsymbol{W}_s^T - \boldsymbol{E}_s\boldsymbol{\Lambda}_s^{-2}\Delta\boldsymbol{\Lambda}_s\boldsymbol{E}_s^T, \tag{2.312}$$

where $\boldsymbol{E}_s$ is composed of the first K columns of $\boldsymbol{I}_N$. Using Lemma 2.7, after some manipulations, we have

$$\begin{aligned}&[\boldsymbol{z}]_i \\ &= \sum_{k=1,k\neq i}^{K} \frac{1}{\lambda_k(\lambda_k-\lambda_i)}[\Delta\boldsymbol{T}]_{i,k}[\boldsymbol{U}^T\boldsymbol{v}]_k \\ &\quad + \left(\frac{1}{\lambda_i}\sum_{k=1,k\neq i}^{K}\frac{1}{\lambda_i-\lambda_k}[\Delta\boldsymbol{T}]_{k,i}[\boldsymbol{U}^T\boldsymbol{v}]_k - \frac{1}{\lambda_i^2}[\Delta\boldsymbol{T}]_{i,i}[\boldsymbol{U}^T\boldsymbol{v}]_i\right)\delta_{i\le K} \\ &= \left[-\frac{1}{\lambda_i}\sum_{k=1}^{K}\frac{1}{\lambda_k}[\Delta\boldsymbol{T}]_{i,k}[\boldsymbol{U}^T\boldsymbol{v}]_k\right]\delta_{i\le K} \\ &\quad + \left[\sum_{k=1}^{K}\frac{1}{\lambda_k(\lambda_k-\lambda_i)}[\Delta\boldsymbol{T}]_{i,k}[\boldsymbol{U}^T\boldsymbol{v}]_k\right]\delta_{i>K} \\ &= \sum_{k=1}^{K}\frac{1}{\lambda_k\gamma_{k,i}}[\Delta\boldsymbol{T}]_{i,k}[\boldsymbol{U}^T\boldsymbol{v}]_k,\end{aligned} \tag{2.313}$$

with

$$\gamma_{k,i} \triangleq -\lambda_i\,\delta_{i\le K} + \left(\lambda_k - \sigma^2\right)\delta_{i>K}, \tag{2.314}$$

where we have used the fact that $\Delta\boldsymbol{T}$ is symmetric (i.e., $[\Delta\boldsymbol{T}]_{i,j} = [\Delta\boldsymbol{T}]_{j,i}$). Denote

$$\boldsymbol{y} \triangleq \boldsymbol{U}^T\boldsymbol{r}. \tag{2.315}$$

Then $\boldsymbol{C}_y = \boldsymbol{U}^T \boldsymbol{C}_r \boldsymbol{U} = \boldsymbol{\Lambda}$. Moreover, we have $\Delta \boldsymbol{T} = \Delta \boldsymbol{C}_y$. Since $E\{\Delta \boldsymbol{T}\} = \boldsymbol{0}$, by Lemma 2.5 for $1 \le i, j \le N$,

$$
\begin{aligned}
M \cdot E\{[\Delta \boldsymbol{T}]_{i,k}, [\Delta \boldsymbol{T}]_{j,l}\}
&= M \cdot \operatorname{Cov}\{[\Delta \boldsymbol{C}_y]_{i,k}, [\Delta \boldsymbol{C}_y]_{j,l}\} \\
&= [\boldsymbol{C}_y]_{i,j}[\boldsymbol{C}_y]_{k,l} + [\boldsymbol{C}_y]_{i,l}[\boldsymbol{C}_y]_{k,j} \\
&\quad - 2\sum_{\alpha=1}^{K} A_\alpha^4 [\boldsymbol{U}^T \boldsymbol{s}_\alpha]_i [\boldsymbol{U}^T \boldsymbol{s}_\alpha]_k [\boldsymbol{U}^T \boldsymbol{s}_\alpha]_j [\boldsymbol{U}^T \boldsymbol{s}_\alpha]_l \\
&= \lambda_i \lambda_k (\delta_{i=j}\delta_{k=l} + \delta_{i=l}\delta_{k=j}) \\
&\quad - 2\sum_{\alpha=1}^{K} A_\alpha^4 [\boldsymbol{U}^T \boldsymbol{s}_\alpha]_i [\boldsymbol{U}^T \boldsymbol{s}_\alpha]_k [\boldsymbol{U}^T \boldsymbol{s}_\alpha]_j [\boldsymbol{U}^T \boldsymbol{s}_\alpha]_l .
\end{aligned}
\tag{2.316}
$$

Using (2.313) and (2.316), we have

$$
\begin{aligned}
M \cdot E&\left\{ \left[\boldsymbol{z}\boldsymbol{z}^T\right]_{i,j} \right\} \\
&= \sum_{k=1}^{K}\sum_{l=1}^{K} \frac{M \cdot E\{[\Delta \boldsymbol{T}]_{i,k}[\Delta \boldsymbol{T}]_{j,l}\}}{\lambda_k \gamma_{k,i} \lambda_l \gamma_{l,j}} [\boldsymbol{U}^T \boldsymbol{v}]_k [\boldsymbol{U}^T \boldsymbol{v}]_l \\
&= \lambda_i\, \delta_{i=j} \sum_{k=1}^{K} \frac{[\boldsymbol{U}^T \boldsymbol{v}]_k^2}{\lambda_k \gamma_{k,i}\gamma_{k,j}} + \frac{[\boldsymbol{U}^T \boldsymbol{v}]_i [\boldsymbol{U}^T \boldsymbol{v}]_j}{\gamma_{i,j}\gamma_{j,i}} \\
&\quad - 2\sum_{\alpha=1}^{K} A_\alpha^4 [\boldsymbol{U}^T \boldsymbol{s}_\alpha]_i [\boldsymbol{U}^T \boldsymbol{s}_\alpha]_j \left(\sum_{k=1}^{K} \frac{[\boldsymbol{U}^T \boldsymbol{s}_\alpha]_k [\boldsymbol{U}^T \boldsymbol{v}]_k}{\lambda_k \gamma_{k,i}} \right) \left(\sum_{k=1}^{K} \frac{[\boldsymbol{U}^T \boldsymbol{s}_\alpha]_k [\boldsymbol{U}^T \boldsymbol{v}]_k}{\lambda_k \gamma_{k,j}} \right) \\
&= \left[\frac{\delta_{i=j}}{\lambda_i} \sum_{k=1}^{K} \frac{[\boldsymbol{U}_s^T \boldsymbol{v}]_k^2}{\lambda_k} \right] \delta_{i\le K} + \left[\delta_{i=j}\, \sigma^2 \sum_{k=1}^{K} \frac{[\boldsymbol{U}_s^T \boldsymbol{v}]_k^2}{\lambda_k \left(\lambda_k - \sigma^2\right)^2} \right] \delta_{i>K} \\
&\quad + \left[\frac{[\boldsymbol{U}_s^T \boldsymbol{v}]_i [\boldsymbol{U}_s^T \boldsymbol{v}]_j}{\lambda_i \lambda_j} \right] \delta_{i\le K}\delta_{j\le K} \\
&\quad - 2\left[\sum_{\alpha=1}^{K} \frac{A_\alpha^4 [\boldsymbol{U}_s^T \boldsymbol{s}_\alpha]_i [\boldsymbol{U}_s^T \boldsymbol{s}_\alpha]_j}{\lambda_i \lambda_j} \left(\sum_{k=1}^{K} \frac{1}{\lambda_k} [\boldsymbol{U}_s^T \boldsymbol{s}_\alpha]_k [\boldsymbol{U}_s^T \boldsymbol{v}]_k \right)^2 \right] \delta_{i\le K}\delta_{j\le K},
\end{aligned}
\tag{2.317}
$$

where (2.317) follows from the fact that

$$
\boldsymbol{U}^T \boldsymbol{v} = \begin{bmatrix} \boldsymbol{U}_s^T \boldsymbol{v} \\ \boldsymbol{U}_n^T \boldsymbol{v} \end{bmatrix} = \begin{bmatrix} \boldsymbol{U}_s^T \boldsymbol{v} \\ \boldsymbol{0} \end{bmatrix},
\tag{2.318}
$$

since it is assumed that $\boldsymbol{v} \in \mathsf{range}(\boldsymbol{S})$; a similar relationship holds for $\boldsymbol{U}^T\boldsymbol{s}_\alpha$. Writing (2.317) in matrix form, we obtain

$$
\begin{aligned}
M \cdot E\left\{\boldsymbol{z}\boldsymbol{z}^T\right\} \\
&= \mathsf{diag}(\mu\boldsymbol{\Lambda}_s, \tau\boldsymbol{I}_{N-K}) + \left[\boldsymbol{\Lambda}_s^{-1}\boldsymbol{U}_s^T\boldsymbol{v}\right]\left[\boldsymbol{\Lambda}_s^{-1}\boldsymbol{U}_s^T\boldsymbol{v}\right]^T \\
&\quad - 2\,\boldsymbol{\Lambda}_s^{-1}\boldsymbol{U}_s^T\boldsymbol{S}\boldsymbol{D}\boldsymbol{S}^T\boldsymbol{U}_s\boldsymbol{\Lambda}_s^{-1},
\end{aligned}
\tag{2.319}
$$

where

$$
\mu \triangleq \sum_{k=1}^{K} \frac{[\boldsymbol{U}_s^T\boldsymbol{v}]_k^2}{\lambda_k} = \boldsymbol{v}^T\boldsymbol{U}_s\boldsymbol{\Lambda}_s^{-1}\boldsymbol{U}_s^T\boldsymbol{v} = \boldsymbol{v}^T\boldsymbol{w}_1, \tag{2.320}
$$

$$
\tau \triangleq \sigma^2 \sum_{k=1}^{K} \frac{[\boldsymbol{U}_s^T\boldsymbol{v}]_k^2}{\lambda_k\left(\lambda_k - \sigma^2\right)^2} = \sigma^2\boldsymbol{v}^T\boldsymbol{U}_s\boldsymbol{\Lambda}_s^{-1}\left(\boldsymbol{\Lambda}_s - \sigma^2\boldsymbol{I}_K\right)^{-2}\boldsymbol{U}_s^T\boldsymbol{v}, \tag{2.321}
$$

$$
\begin{aligned}
\boldsymbol{D} &\triangleq \mathsf{diag}\left\{A_\alpha^4\left(\sum_{k=1}^{K}\frac{1}{\lambda_k}[\boldsymbol{U}_s^T\boldsymbol{s}_\alpha]_k[\boldsymbol{U}_s^T\boldsymbol{v}]_k\right)^2\right\}_{\alpha=1}^{K} \\
&= \mathsf{diag}\left\{A_\alpha^4\left(\boldsymbol{s}_\alpha^T\boldsymbol{U}_s\boldsymbol{\Lambda}_s^{-1}\boldsymbol{U}_s^{-1}\boldsymbol{v}\right)^2\right\}_{\alpha=1}^{K} = \mathsf{diag}\left\{A_\alpha^4\left(\boldsymbol{s}_\alpha^T\boldsymbol{w}_1\right)^2\right\}_{\alpha=1}^{K}.
\end{aligned}
\tag{2.322}
$$

Finally, by (2.311), $M \cdot E\left\{\Delta\boldsymbol{w}_1\Delta\boldsymbol{w}_1^T\right\} = M \cdot \boldsymbol{U}E\left\{\boldsymbol{z}\boldsymbol{z}^T\right\}\boldsymbol{U}^T$. Substituting (2.319) into this expansion, we obtain (2.305).

Proof of Corollary 2.1

First we compute the term given by (2.120). Using (2.304) and (2.128) and the fact that $\boldsymbol{U}_s^T\boldsymbol{U}_n = \boldsymbol{0}$, we have

$$
\begin{aligned}
\mathrm{tr}(\boldsymbol{C}_w\boldsymbol{C}_r) &= \boldsymbol{s}_1^T\boldsymbol{w}_1\,\mathrm{tr}(\underbrace{\boldsymbol{U}_s\boldsymbol{\Lambda}_s^{-1}\boldsymbol{U}_s^T\boldsymbol{U}_s\boldsymbol{\Lambda}_s\boldsymbol{U}_s^T}_{\boldsymbol{U}_s\boldsymbol{U}_s^T}) + \mathrm{tr}(\boldsymbol{w}_1\boldsymbol{s}_1^T\,\underbrace{\boldsymbol{U}_s\boldsymbol{\Lambda}_s^{-1}\boldsymbol{U}_s^T\boldsymbol{U}_s\boldsymbol{\Lambda}_s\boldsymbol{U}_s^T}_{\boldsymbol{U}_s\boldsymbol{U}_s^T}) \\
&\quad - 2\,\mathrm{tr}(\boldsymbol{U}_s\boldsymbol{\Lambda}_s^{-1}\boldsymbol{U}_s^T\boldsymbol{S}\boldsymbol{D}\boldsymbol{S}^T\,\underbrace{\boldsymbol{U}_s\boldsymbol{\Lambda}_s^{-1}\boldsymbol{U}_s^T\boldsymbol{U}_s\boldsymbol{\Lambda}_s\boldsymbol{U}_s^T}_{\boldsymbol{U}_s\boldsymbol{U}_s^T}) \\
&\quad + \tau\sigma^2\,\mathrm{tr}(\underbrace{\boldsymbol{U}_n\boldsymbol{U}_n^T\boldsymbol{U}_n\boldsymbol{U}_n^T}_{\boldsymbol{U}_n\boldsymbol{U}_n^T}) \\
&= A + B - 2C + D,
\end{aligned}
\tag{2.323}
$$

with

$$A = \boldsymbol{s}_1^T \boldsymbol{w}_1 \operatorname{tr}(\boldsymbol{U}_s^T \boldsymbol{U}_s) = K \boldsymbol{s}_1^T \boldsymbol{w}_1, \tag{2.324}$$

$$B = \operatorname{tr}(\boldsymbol{s}_1^T \boldsymbol{U}_s \boldsymbol{U}_s^T \boldsymbol{U}_s \boldsymbol{\Lambda}_s^{-1} \boldsymbol{U}_s^T \boldsymbol{s}_1) = \operatorname{tr}(\boldsymbol{s}_1^T \boldsymbol{U}_s \boldsymbol{\Lambda}_s^{-1} \boldsymbol{U}_s^T \boldsymbol{s}_1) = \boldsymbol{s}_1^T \boldsymbol{w}_1, \tag{2.325}$$

$$C = \operatorname{tr}(\boldsymbol{S}^T \boldsymbol{U}_s \boldsymbol{U}_s^T \boldsymbol{U}_s \boldsymbol{\Lambda}_s^{-1} \boldsymbol{U}_s^T \boldsymbol{S} \boldsymbol{D}) = \operatorname{tr}(\boldsymbol{S}^T \underbrace{\boldsymbol{U}_s \boldsymbol{\Lambda}_s^{-1} \boldsymbol{U}_s^T \boldsymbol{S}}_{\boldsymbol{W} = [\boldsymbol{w}_1 \ldots \boldsymbol{w}_K]} \boldsymbol{D})$$

$$= \sum_{k=1}^{K} A_k^4 (\boldsymbol{s}_k^T \boldsymbol{w}_1)^2 (\boldsymbol{s}_k^T \boldsymbol{w}_k), \tag{2.326}$$

$$D = \tau \sigma^2 \operatorname{tr}(\boldsymbol{U}_n^T \boldsymbol{U}_n) = (N - K) \tau \sigma^2. \tag{2.327}$$

Hence we have

$$\operatorname{tr}(\boldsymbol{C}_w \boldsymbol{C}_r) = (K+1) \boldsymbol{s}_1^T \boldsymbol{w}_1 - 2 \sum_{k=1}^{K} A_k^4 \left(\boldsymbol{s}_k^T \boldsymbol{w}_1\right)^2 \left(\boldsymbol{s}_k^T \boldsymbol{w}_k\right) + (N-K) \tau \sigma^2. \tag{2.328}$$

Next note that the linear MMSE detector can also be written in terms of $\boldsymbol{R}$ as [520]

$$\boldsymbol{W} = [\boldsymbol{w}_1 \cdots \boldsymbol{w}_K] \triangleq \boldsymbol{C}^{-1} \boldsymbol{S} = \boldsymbol{S} \left(\boldsymbol{R} + \sigma^2 \boldsymbol{A}^{-2}\right)^{-1} \boldsymbol{A}^{-2}. \tag{2.329}$$

Therefore, we have

$$\boldsymbol{s}_k^T \boldsymbol{w}_l = A_l^{-2} \left[\boldsymbol{S}^T \boldsymbol{S} \left(\boldsymbol{R} + \sigma^2 \boldsymbol{A}^{-2}\right)^{-1}\right]_{k,l} = A_l^{-2} \left[\boldsymbol{R} \left(\boldsymbol{R} + \sigma^2 \boldsymbol{A}^{-2}\right)^{-1}\right]_{k,l}, \tag{2.330}$$

$$\|\boldsymbol{w}_1\|^2 = A_1^{-4} \left[\left(\boldsymbol{R} + \sigma^2 \boldsymbol{A}^{-2}\right)^{-1} \boldsymbol{R} \left(\boldsymbol{R} + \sigma^2 \boldsymbol{A}^{-2}\right)^{-1}\right]_{1,1}. \tag{2.331}$$

By (2.130), for the DMI blind detector, we have $\tau \sigma^2 = \boldsymbol{s}_1^T \boldsymbol{w}_1$, and for the subspace blind detector,

$$\tau \sigma^2 = \sigma^4 \Big[\underbrace{\boldsymbol{S}^T \boldsymbol{U}_s \boldsymbol{\Lambda}_s^{-1} \boldsymbol{U}_s^T}_{\boldsymbol{W}^T} \boldsymbol{U}_s \left(\boldsymbol{\Lambda}_s - \sigma^2 \boldsymbol{I}_K\right)^{-1} \boldsymbol{U}_s^T \underbrace{\boldsymbol{U}_s \left(\boldsymbol{\Lambda}_s - \sigma^2 \boldsymbol{I}_K\right)^{-1} \boldsymbol{U}_s^T \boldsymbol{S}}_{\boldsymbol{S}\boldsymbol{R}^{-1}\boldsymbol{A}^{-2}} \Big]_{1,1}$$

$$= \sigma^4 \Big[\boldsymbol{A}^{-2} \left(\boldsymbol{R} + \sigma^2 \boldsymbol{A}^{-2}\right)^{-1} \boldsymbol{S}^T \underbrace{\boldsymbol{U}_s \left(\boldsymbol{\Lambda}_s - \sigma^2 \boldsymbol{I}_K\right)^{-1} \boldsymbol{U}_s^T \boldsymbol{S}}_{\boldsymbol{D} = \boldsymbol{S}\boldsymbol{R}^{-1}\boldsymbol{A}^{-2}} \boldsymbol{R}^{-1} \boldsymbol{A}^{-2} \Big]_{1,1}$$

$$= \frac{\sigma^4}{A_1^4} \left[\left(\boldsymbol{R} + \sigma^2 \boldsymbol{A}^{-2}\right)^{-1} \boldsymbol{A}^{-2} \boldsymbol{R}^{-1}\right]_{1,1}, \tag{2.332}$$

where we have used the fact that the decorrelating detector can be written as [549]

$$\boldsymbol{D} = \boldsymbol{U}_s \left(\boldsymbol{\Lambda}_s - \sigma^2 \boldsymbol{I}_K\right)^{-1} \boldsymbol{U}_s^T \boldsymbol{S} = \boldsymbol{S} \boldsymbol{R}^{-1} \boldsymbol{A}^{-2}. \tag{2.333}$$

Finally, substituting (2.328)–(2.332) into (2.119), we obtain (2.132).

SINR for Equicorrelated Signals

In this case, $\boldsymbol{R}$ is given by

$$\boldsymbol{R} \triangleq \boldsymbol{S}^T \boldsymbol{S} = \rho \boldsymbol{1}\boldsymbol{1}^T + (1-\rho)\boldsymbol{I}_K, \tag{2.334}$$

where $\boldsymbol{1}$ is an all-1 K-vector. It is straightforward to verify the following eigenstructure of $\boldsymbol{R}$:

$$\boldsymbol{R} = \sum_{k=1}^{K} \mu_k \boldsymbol{v}_k \boldsymbol{v}_k^T, \tag{2.335}$$

with

$$\mu_1 = 1 + (K-1)\rho, \qquad \boldsymbol{v}_1 = \frac{1}{\sqrt{K}}\boldsymbol{1}, \tag{2.336}$$

$$\mu_k = 1 - \rho, \quad k = 2, \ldots, K. \tag{2.337}$$

Since $\boldsymbol{A}^2 = A^2 \boldsymbol{I}_K$, we have

$$\begin{aligned}
\boldsymbol{R}\left(\boldsymbol{R} + \sigma^2 \boldsymbol{A}^{-2}\right)^{-1} &= \left(\sum_{i=1}^{K} \mu_i \boldsymbol{v}_i \boldsymbol{v}_i^T\right)\left(\sum_{j=1}^{K} \frac{1}{\mu_j + \sigma^2/A^2} \boldsymbol{v}_j \boldsymbol{v}_j^T\right) \\
&= \sum_{i=1}^{K} \frac{\mu_i}{\mu_i + \sigma^2/A^2} \boldsymbol{v}_i \boldsymbol{v}_i^T \\
&= \frac{1}{\mu_2 + \sigma^2/A^2} \underbrace{\sum_{i=1}^{K} \mu_i \boldsymbol{v}_i \boldsymbol{v}_i^T}_{\boldsymbol{R}} + \mu_1 \left(\frac{1}{\mu_1 + \sigma^2/A^2} - \frac{1}{\mu_2 + \sigma^2/A^2}\right) \boldsymbol{v}_1 \boldsymbol{v}_1^T \\
&= \underbrace{\frac{1}{\mu_2 + \sigma^2/A^2}}_{a} \boldsymbol{R} \\
&\quad + \underbrace{\frac{\mu_1}{K}\left(\frac{1}{\mu_1 + \sigma^2/A^2} - \frac{1}{\mu_2 + \sigma^2/A^2}\right)}_{b} \boldsymbol{1}\boldsymbol{1}^T.
\end{aligned} \tag{2.338}$$

Similarly, we obtain

$$\left(\boldsymbol{R}+\sigma^2\boldsymbol{A}^{-2}\right)^{-1}\boldsymbol{R}\left(\boldsymbol{R}+\sigma^2\boldsymbol{A}^{-2}\right)^{-1}$$
$$=\underbrace{\frac{1}{\left(\mu_2+\sigma^2/A^2\right)^2}}_{a'}\boldsymbol{R}+\underbrace{\frac{\mu_1}{K}\left[\frac{1}{\left(\mu_1+\sigma^2/A^2\right)^2}-\frac{1}{\left(\mu_2+\sigma^2/A^2\right)^2}\right]}_{b'}\boldsymbol{1}\boldsymbol{1}^T, \tag{2.339}$$

$$A^2\cdot\left(\boldsymbol{R}+\sigma^2\boldsymbol{A}^{-2}\right)^{-1}\boldsymbol{A}^{-2}\boldsymbol{R}^{-1}$$
$$=\underbrace{\frac{1}{\left(\mu_2+\sigma^2/A^2\right)\mu_2^2}}_{a''}\boldsymbol{R}+\underbrace{\frac{\mu_1}{K}\left[\frac{1}{\left(\mu_1+\sigma^2/A^2\right)\mu_1^2}-\frac{1}{\left(\mu_2+\sigma^2/A^2\right)\mu_2^2}\right]}_{b''}\boldsymbol{1}\boldsymbol{1}^T. \tag{2.340}$$

Substituting (2.338)–(2.340) into (2.132)–(2.135), and defining

$$\alpha \triangleq \left(\frac{\boldsymbol{w}_1^T\boldsymbol{s}_2}{\boldsymbol{w}_1^T\boldsymbol{s}_1}\right)^2, \tag{2.341}$$

$$\beta \triangleq \frac{\sigma^2}{A^2}\frac{\|\boldsymbol{w}_1\|^2}{\left(\boldsymbol{w}_1^T\boldsymbol{s}_1\right)^2}, \tag{2.342}$$

$$\gamma \triangleq A^2\cdot\boldsymbol{w}_1^T\boldsymbol{s}_1, \tag{2.343}$$

$$\eta \triangleq \frac{\tau\sigma^2}{A^2\left(\boldsymbol{w}_1^T\boldsymbol{s}_1\right)^2}, \tag{2.344}$$

we obtain expression (2.143) for the average output SINRs of the DMI blind detector and the subspace blind detector.

Chapter 3

GROUP-BLIND MULTIUSER DETECTION

3.1 Introduction

The blind multiuser detection techniques discussed in Chapter 2 are especially useful for interference suppression in CDMA downlinks, where a mobile receiver knows only its own spreading sequence. In CDMA uplinks, however, typically the base station receiver has knowledge of the spreading sequences of a group of users (e.g., the users within its own cell) but not those of the users from other cells. It is natural to expect that some performance gains can be achieved over the blind methods (which exploit only the spreading sequence of a single user) in detecting each user's data if information about the spreading sequences of the other known users are also exploited [190, 191, 198, 545]. In this chapter we discuss group-blind multiuser detection techniques that suppress intracell interference using knowledge of the spreading sequences and the estimated multipath channels of a group of known users while suppressing intercell interference blindly. Several forms of linear and nonlinear group-blind detectors are developed based on different criteria. These group-blind techniques offer significant performance improvement over the blind methods in a CDMA uplink environment.

The rest of this chapter is organized as follows. In Section 3.2 we introduce various linear group-blind multiuser detectors for synchronous CDMA systems. In Section 3.3 we present analytical performance assessment for linear group-blind multiuser detectors. In Section 3.4 we discuss nonlinear group-blind multiuser detection based on local likelihood search. In Section 3.5 we treat group-blind multiuser detection in general asynchronous CDMA systems with multipath channels. Finally, Section 3.6 contains the mathematical derivations and proofs for some results in this chapter.

The following is a list of the algorithms appearing in this chapter.

- *Algorithm 3.1*: Group-blind linear hybrid detector (form I)—synchronous CDMA

- *Algorithm 3.2*: Group-blind linear hybrid detector (form II)—synchronous CDMA

- *Algorithm 3.3*: Slowest-descent-search multiuser detector
- *Algorithm 3.4*: Nonlinear group-blind multiuser detector—synchronous CDMA
- *Algorithm 3.5*: Group-blind linear hybrid detector (form I)—multipath CDMA
- *Algorithm 3.6*: Group-blind linear hybrid detector (form II)—multipath CDMA
- *Algorithm 3.7*: Adaptive group-blind linear hybrid multiuser detector—multipath CDMA
- *Algorithm 3.8*: Group-blind linear hybrid detector—multipath CDMA and correlated noise
- *Algorithm 3.9*: Nonlinear group-blind detector—multipath CDMA

3.2 Linear Group-Blind Multiuser Detection for Synchronous CDMA

We start by considering the following discrete-time signal model for a synchronous CDMA system:

$$\boldsymbol{r}[i] = \sum_{k=1}^{K} A_k b_k[i] \boldsymbol{s}_k + \boldsymbol{n}[i] \tag{3.1}$$

$$= \boldsymbol{S}\boldsymbol{A}\boldsymbol{b}[i] + \boldsymbol{n}[i], \qquad i = 0, 1, \ldots, M-1, \tag{3.2}$$

where, as before, K is the total number of users; A_k, $b_k[i]$, and $\boldsymbol{s}_k$ are, respectively, the complex amplitude, ith transmitted bit, and signature waveform of the kth user; $\boldsymbol{n}[i] \sim \mathcal{N}_c(\boldsymbol{0}, \sigma^2 \boldsymbol{I}_N)$ is a complex Gaussian noise vector; $\boldsymbol{S} \triangleq [\boldsymbol{s}_1 \ \cdots \ \boldsymbol{s}_K]$; $\boldsymbol{A} \triangleq \mathsf{diag}(A_1, \ldots, A_K)$; and $\boldsymbol{b}[i] \triangleq \left[b_1[i] \ \cdots \ b_K[i]\right]^T$. In this chapter it is assumed that the receiver has knowledge of the signature waveforms of the first $\tilde{K}$ users ($\tilde{K} \leq K$), whose data bits are to be demodulated, whereas the signature waveforms of the remaining $K - \tilde{K}$ users are unknown to the receiver. Denote

$$\tilde{\boldsymbol{S}} \triangleq [\boldsymbol{s}_1 \ \cdots \ \boldsymbol{s}_{\tilde{K}}],$$
$$\bar{\boldsymbol{S}} \triangleq [\boldsymbol{s}_{\tilde{K}+1} \ \cdots \ \boldsymbol{s}_K],$$
$$\tilde{\boldsymbol{A}} \triangleq \mathsf{diag}(A_1, \ldots, A_{\tilde{K}}),$$
$$\bar{\boldsymbol{A}} \triangleq \mathsf{diag}(A_{\tilde{K}+1}, \ldots, A_K),$$
$$\tilde{\boldsymbol{b}}[i] \triangleq \left[b_1[i] \ \cdots \ b_{\tilde{K}}[i]\right],$$
$$\bar{\boldsymbol{b}}[i] \triangleq \left[b_{\tilde{K}+1}[i] \ \cdots \ b_K[i]\right].$$

It is assumed that the users' signature waveforms are linearly independent (i.e., $\boldsymbol{S}$ has full column rank). Hence both $\tilde{\boldsymbol{S}}$ and $\bar{\boldsymbol{S}}$ also have full column ranks. Then (3.2) can be written as

$$\boldsymbol{r}[i] = \tilde{\boldsymbol{S}}\tilde{\boldsymbol{A}}\tilde{\boldsymbol{b}}[i] + \bar{\boldsymbol{S}}\bar{\boldsymbol{A}}\bar{\boldsymbol{b}}[i] + \boldsymbol{n}[i]. \tag{3.3}$$

The problem of linear group-blind multiuser detection can be stated as follows. Given prior knowledge of the signature waveforms $\tilde{\boldsymbol{S}}$ of the $\tilde{K}$ desired users, find a weight vector $\boldsymbol{w}_k \in \mathbb{C}^N$ for each desired user k, $1 \leq k \leq \tilde{K}$, such that the data bits of these users can be demodulated according to

$$z_k[i] = \boldsymbol{w}_k^H \boldsymbol{r}[i], \tag{3.4}$$

and

$$\hat{b}_k[i] = \text{sign}\left\{\Re\left(A_k^* z_k[i]\right)\right\}, \quad \text{(coherent detection)} \tag{3.5}$$

or

$$\hat{\beta}_k[i] = \text{sign}\left\{\Re\left(z_k[i] z_k[i-1]^*\right)\right\}, \quad \text{(differential detection)} \qquad k = 1, \ldots, \tilde{K}. \tag{3.6}$$

The basic idea behind the solution to the problem above is to suppress the interference from known users based on the signature waveforms of these users and to suppress the interference from other unknown users using subspace-based blind methods. We first consider the linear decorrelating detector, which eliminates the multiple-access interference (MAI) completely, at the expense of enhancing the noise level. To facilitate the derivation of its group-blind form, we need the following alternative definition of this detector. In this section we denote $\tilde{\boldsymbol{e}}_k$ as a $\tilde{K}$-vector with all elements zeros, except for the kth element, which is 1.

Definition 3.1: [Group-blind linear decorrelating detector—synchronous CDMA] *The weight vector $\boldsymbol{d}_k$ of the linear decorrelating detector for user k is given by the solution to the following constrained optimization problem:*

$$\min_{\boldsymbol{w} \in \text{range}(\boldsymbol{S})} \left\|\boldsymbol{w}^H \boldsymbol{S} \boldsymbol{A}\right\|^2 \quad s.t. \quad \boldsymbol{w}^H \tilde{\boldsymbol{S}} = \tilde{\boldsymbol{e}}_k^H, \qquad k = 1, \ldots, \tilde{K}. \tag{3.7}$$

This definition is equivalent to the one given in Section 2.2.2. To see this, it suffices to show that $\boldsymbol{d}_k^H \boldsymbol{s}_k = 1$, and $\boldsymbol{d}_k^H \boldsymbol{s}_l = 0$ for $l \neq k$. Since $\tilde{\boldsymbol{S}}$ contains the first $\tilde{K}$ columns of $\boldsymbol{S}$, then for any $\boldsymbol{w}$ we have

$$\left\|\boldsymbol{w}^H \boldsymbol{S} \boldsymbol{A}\right\|^2 = \left\|\boldsymbol{w}^H \tilde{\boldsymbol{S}} \tilde{\boldsymbol{A}}\right\|^2 + \sum_{l=\tilde{K}+1}^{K} |A_l|^2 \left|\boldsymbol{w}^H \boldsymbol{s}_l\right|^2. \tag{3.8}$$

Under the constraint $\boldsymbol{w}^H \tilde{\boldsymbol{S}} = \tilde{\boldsymbol{e}}_k^H$, we have $\left\|\boldsymbol{w}^H \tilde{\boldsymbol{S}} \tilde{\boldsymbol{A}}\right\|^2 = A_k^2$. It then follows that for $\boldsymbol{w} \in \text{range}(\boldsymbol{S})$, $\left\|\boldsymbol{w}^H \boldsymbol{S} \boldsymbol{A}\right\|^2$ is minimized subject to $\boldsymbol{w}^H \tilde{\boldsymbol{S}} = \tilde{\boldsymbol{e}}_k^H$ if and only if

$\boldsymbol{w}^H \boldsymbol{s}_l = 0$ for $l = \tilde{K}+1, \ldots, K$. Since $\mathsf{rank}(\boldsymbol{S}) = K$, such a $\boldsymbol{w} \in \mathsf{range}(\boldsymbol{S})$ is unique and is indeed the linear decorrelating detector.

The second linear group-blind detector considered here is a hybrid detector that zero-forces the interference caused by the $\tilde{K}$ known users and suppresses the interference from unknown users according to the MMSE criterion.

Definition 3.2: [Group-blind linear hybrid detector—synchronous CDMA] *The weight vector $\boldsymbol{w}_k$ of the group-blind linear hybrid detector for user k is given by the solution to the following constrained optimization problem:*

$$\min_{\boldsymbol{w} \in \mathsf{range}(\boldsymbol{S})} E\left\{\left|A_k b_k[i] - \boldsymbol{w}^H \boldsymbol{r}[i]\right|^2\right\} \quad s.t. \quad \boldsymbol{w}^H \tilde{\boldsymbol{S}} = \tilde{\boldsymbol{e}}_k^H, \qquad k = 1, \ldots, \tilde{K}. \tag{3.9}$$

Another form of linear group-blind detector is analogous to the linear MMSE detector introduced in Section 2.2.3. It suppresses the interference from the known users and that from the unknown users separately, both in the MMSE sense. First define the following projection matrix:

$$\bar{\boldsymbol{P}} \triangleq \boldsymbol{I}_N - \tilde{\boldsymbol{S}} \left(\tilde{\boldsymbol{S}}^H \tilde{\boldsymbol{S}}\right)^{-1} \tilde{\boldsymbol{S}}^H, \tag{3.10}$$

which projects any signal onto the subspace $\mathsf{null}(\tilde{\boldsymbol{S}}^H)$. Recall that the autocorrelation matrix of the received signal in (3.1) is given by

$$\boldsymbol{C}_r \triangleq E\left\{\boldsymbol{r}[i]\boldsymbol{r}[i]^H\right\} = \boldsymbol{S}|\boldsymbol{A}|^2\boldsymbol{S}^H + \sigma^2 \boldsymbol{I}_N, \tag{3.11}$$

where $|\boldsymbol{A}|^2 \triangleq \mathsf{diag}\left(|A_1|^2, \ldots, |A_K|^2\right)$. It is then easily seen that the matrix $\bar{\boldsymbol{P}}\boldsymbol{C}_r\bar{\boldsymbol{P}}$ has an eigenstructure of the form

$$\bar{\boldsymbol{P}}\boldsymbol{C}_r\bar{\boldsymbol{P}} = \left[\bar{\boldsymbol{U}}_s \, \bar{\boldsymbol{U}}_n \, \bar{\boldsymbol{U}}_o\right] \begin{bmatrix} \bar{\boldsymbol{\Lambda}}_s & \mathbf{0} & \mathbf{0} \\ \mathbf{0} & \sigma^2 \boldsymbol{I}_{N-K} & \mathbf{0} \\ \mathbf{0} & \mathbf{0} & \mathbf{0} \end{bmatrix} \begin{bmatrix} \bar{\boldsymbol{U}}_s^H \\ \bar{\boldsymbol{U}}_n^H \\ \bar{\boldsymbol{U}}_o^H \end{bmatrix}, \tag{3.12}$$

where $\bar{\boldsymbol{\Lambda}}_s = \mathsf{diag}\left(\bar{\lambda}_1, \ldots, \bar{\lambda}_{K-\tilde{K}}\right)$, with $\bar{\lambda}_i > \sigma^2$, $i = 1, \ldots, K - \tilde{K}$; and the columns of $\bar{\boldsymbol{U}}_s$ form an orthogonal basis of the subspace $\mathsf{range}(\boldsymbol{S}) \bigcap \mathsf{null}(\tilde{\boldsymbol{S}}^H)$. We next define the linear group-blind MMSE detector. As noted in Chapter 2, any linear detector must lie in the space $\mathsf{range}(\boldsymbol{S}) = \mathsf{range}(\tilde{\boldsymbol{S}}) + \mathsf{range}(\bar{\boldsymbol{U}}_s)$. The group-blind linear MMSE detector for the kth user has the form $\boldsymbol{m}_k = \tilde{\boldsymbol{m}}_k + \bar{\boldsymbol{m}}_k$, where $\tilde{\boldsymbol{m}}_k \in \mathsf{range}(\tilde{\boldsymbol{S}})$ and $\bar{\boldsymbol{m}}_k \in \mathsf{range}(\bar{\boldsymbol{U}}_s)$, such that $\tilde{\boldsymbol{m}}_k$ suppresses interference from known users in the MMSE sense, and $\bar{\boldsymbol{m}}_k$ suppresses interference from unknown users in the MMSE sense. Formally, we have the following definition.

Definition 3.3: [Group-blind linear MMSE detector—synchronous CDMA] *Let $\tilde{\boldsymbol{r}}[i] = \tilde{\boldsymbol{S}}\tilde{\boldsymbol{A}}\tilde{\boldsymbol{b}}[i] + \boldsymbol{n}[i]$ be the components of the received signal $\boldsymbol{r}[i]$ in (3.3) consisting of the signals from known users plus the noise. The weight vector of the group-blind* *linear MMSE detector for user k is given by* $\boldsymbol{m}_k = \tilde{\boldsymbol{m}}_k + \bar{\boldsymbol{m}}_k$, *where* $\tilde{\boldsymbol{m}}_k \in \mathsf{range}(\tilde{\boldsymbol{S}})$ *and* $\bar{\boldsymbol{m}}_k \in \mathsf{range}(\bar{\boldsymbol{U}}_s)$, *such that*

$$\tilde{\boldsymbol{m}}_k = \arg \min_{\boldsymbol{w} \in \mathsf{range}(\tilde{\boldsymbol{S}})} E\left\{ \left| A_k b_k[i] - \boldsymbol{w}^H \tilde{\boldsymbol{r}}[i] \right|^2 \right\}, \tag{3.13}$$

$$\bar{\boldsymbol{m}}_k = \arg \min_{\boldsymbol{w} \in \mathsf{range}(\bar{\boldsymbol{U}}_s)} E\left\{ \left| A_k b_k[i] - (\boldsymbol{w} + \tilde{\boldsymbol{m}}_k)^H \boldsymbol{r}[i] \right|^2 \right\}, \quad k = 1, \ldots, \tilde{K}. \tag{3.14}$$

Note that in general the linear group-blind MMSE detector $\boldsymbol{m}_k$ defined above is different from the linear MMSE detector defined in Section 2.2.3, due to the specific structure that the former imposes.

We next give expressions for the three linear group-blind detectors defined above in terms of the known users' signature waveforms $\tilde{\boldsymbol{S}}$ and the unknown users' signal subspace components $\bar{\boldsymbol{\Lambda}}_s$ and $\bar{\boldsymbol{U}}_s$ defined in (3.12).

Proposition 3.1: [Group-blind linear decorrelating detector (form I)—synchronous CDMA] *The weight vector of the group-blind linear decorrelating detector for user k is given by*

$$\boldsymbol{d}_k = \left[\boldsymbol{I}_N - \bar{\boldsymbol{U}}_s \left(\bar{\boldsymbol{\Lambda}}_s - \sigma^2 \boldsymbol{I}_{K-\tilde{K}} \right)^{-1} \bar{\boldsymbol{U}}_s^H \boldsymbol{C}_r \right] \tilde{\boldsymbol{S}} \left(\tilde{\boldsymbol{S}}^H \tilde{\boldsymbol{S}} \right)^{-1} \tilde{\boldsymbol{e}}_k, \quad k = 1, \ldots, \tilde{K}. \tag{3.15}$$

Proof: Decompose $\boldsymbol{d}_k$ as $\boldsymbol{d}_k = \tilde{\boldsymbol{d}}_k + \bar{\boldsymbol{d}}_k$, where $\tilde{\boldsymbol{d}}_k \in \mathsf{range}(\tilde{\boldsymbol{S}})$ and $\bar{\boldsymbol{d}}_k \in \mathsf{range}(\bar{\boldsymbol{U}}_s)$. Substituting this into the constraint $\boldsymbol{w}^H \tilde{\boldsymbol{S}} = \tilde{\boldsymbol{e}}_k^H$ in (3.7), we have

$$\tilde{\boldsymbol{d}}_k = \tilde{\boldsymbol{S}} \left(\tilde{\boldsymbol{S}}^H \tilde{\boldsymbol{S}} \right)^{-1} \tilde{\boldsymbol{e}}_k. \tag{3.16}$$

Hence $\boldsymbol{d}_k$ has the form $\boldsymbol{d}_k = \bar{\boldsymbol{U}}_s \boldsymbol{c}_k + \tilde{\boldsymbol{d}}_k$ for some $\boldsymbol{c}_k \in \mathbb{C}^{K-\tilde{K}}$. Substituting this into the minimization problem in (3.7), we get

$$\begin{aligned}
\boldsymbol{c}_k &= \arg \min_{\boldsymbol{c} \in \mathbb{C}^{K-\tilde{K}}} \left\| \left(\bar{\boldsymbol{U}}_s \boldsymbol{c} + \tilde{\boldsymbol{d}}_k \right)^H \boldsymbol{S} \boldsymbol{A} \right\|^2 \\
&= \arg \min_{\boldsymbol{c} \in \mathbb{C}^{K-\tilde{K}}} \left(\bar{\boldsymbol{U}}_s \boldsymbol{c} + \tilde{\boldsymbol{d}}_k \right)^H \left(\boldsymbol{C}_r - \sigma^2 \boldsymbol{I}_N \right) \left(\bar{\boldsymbol{U}}_s \boldsymbol{c} + \tilde{\boldsymbol{d}}_k \right) && (3.17) \\
&= - \left[\bar{\boldsymbol{U}}_s^H \left(\boldsymbol{C}_r - \sigma^2 \boldsymbol{I}_N \right) \bar{\boldsymbol{U}}_s \right]^{-1} \bar{\boldsymbol{U}}_s^H \left(\boldsymbol{C}_r - \sigma^2 \boldsymbol{I}_N \right) \tilde{\boldsymbol{d}}_k \\
&= - \left[\bar{\boldsymbol{U}}_s^H \bar{\boldsymbol{P}} \left(\boldsymbol{C}_r - \sigma^2 \boldsymbol{I}_N \right) \bar{\boldsymbol{P}} \bar{\boldsymbol{U}}_s \right]^{-1} \bar{\boldsymbol{U}}_s^H \left(\boldsymbol{C}_r - \sigma^2 \boldsymbol{I}_N \right) \tilde{\boldsymbol{d}}_k && (3.18) \\
&= - \left(\bar{\boldsymbol{\Lambda}}_s - \sigma^2 \boldsymbol{I}_{K-\tilde{K}} \right)^{-1} \bar{\boldsymbol{U}}_s^H \left(\boldsymbol{C}_r - \sigma^2 \boldsymbol{I}_N \right) \tilde{\boldsymbol{d}}_k && (3.19) \\
&= - \left(\bar{\boldsymbol{\Lambda}}_s - \sigma^2 \boldsymbol{I}_{K-\tilde{K}} \right)^{-1} \bar{\boldsymbol{U}}_s^H \boldsymbol{C}_r \tilde{\boldsymbol{d}}_k, && (3.20)
\end{aligned}$$

where (3.17) follows from (3.11); (3.18) follows from the fact that $\bar{\boldsymbol{P}} \bar{\boldsymbol{U}}_s = \bar{\boldsymbol{U}}_s$; (3.19) follows from (3.12); and (3.20) follows from the fact that $\bar{\boldsymbol{U}}_s^H \tilde{\boldsymbol{d}}_k = \boldsymbol{0}$. Hence

$$\boldsymbol{d}_k = \bar{\boldsymbol{U}}_s \boldsymbol{c}_k + \tilde{\boldsymbol{d}}_k = \left[\boldsymbol{I}_N - \left(\bar{\boldsymbol{\Lambda}}_s - \sigma^2 \boldsymbol{I}_{K-\tilde{K}}\right)^{-1} \bar{\boldsymbol{U}}_s^H \boldsymbol{C}_r\right] \tilde{\boldsymbol{S}} \left(\tilde{\boldsymbol{S}}^H \tilde{\boldsymbol{S}}\right)^{-1} \tilde{\boldsymbol{e}}_k. \tag{3.21}$$

□

Proposition 3.2: [Group-blind linear hybrid detector (form I)—synchronous CDMA] *The weight vector of the group-blind linear hybrid detector for user k is given by*

$$\boldsymbol{w}_k = \left(\boldsymbol{I}_N - \bar{\boldsymbol{U}}_s \bar{\boldsymbol{\Lambda}}_s^{-1} \bar{\boldsymbol{U}}_s^H \boldsymbol{C}_r\right) \tilde{\boldsymbol{S}} \left(\tilde{\boldsymbol{S}}^H \tilde{\boldsymbol{S}}\right)^{-1} \tilde{\boldsymbol{e}}_k, \qquad k = 1, \ldots, \tilde{K}. \tag{3.22}$$

Proof: Decompose $\boldsymbol{w}_k$ as $\boldsymbol{w}_k = \tilde{\boldsymbol{w}}_k + \bar{\boldsymbol{w}}_k$, where $\tilde{\boldsymbol{w}}_k \in \mathsf{range}(\tilde{\boldsymbol{S}})$ and $\bar{\boldsymbol{w}}_k \in \mathsf{range}(\bar{\boldsymbol{U}}_s)$. Substituting this into the constraint $\boldsymbol{w}^H \tilde{\boldsymbol{S}} = \tilde{\boldsymbol{e}}_k^H$ in (3.9), we have

$$\tilde{\boldsymbol{w}}_k = \tilde{\boldsymbol{S}} \left(\tilde{\boldsymbol{S}}^H \tilde{\boldsymbol{S}}\right)^{-1} \tilde{\boldsymbol{e}}_k. \tag{3.23}$$

Hence $\boldsymbol{w}_k = \bar{\boldsymbol{U}}_s \boldsymbol{c}_k + \tilde{\boldsymbol{w}}_k$ for some $\boldsymbol{c}_k \in \mathbb{C}^{K-\tilde{K}}$. Substituting this into the minimization problem in (3.9), we get

$$\begin{aligned}
\boldsymbol{c}_k &= \arg \min_{\boldsymbol{c} \in \mathbb{C}^{K-\tilde{K}}} E\left\{\left|A_k b_k[i] - \left(\bar{\boldsymbol{U}}_s \boldsymbol{c} + \tilde{\boldsymbol{w}}_k\right)^H \boldsymbol{r}[i]\right|^2\right\} \\
&= \arg \min_{\boldsymbol{c} \in \mathbb{C}^{K-\tilde{K}}} \left\{\left(\bar{\boldsymbol{U}}_s \boldsymbol{c} + \tilde{\boldsymbol{w}}_k\right)^H \boldsymbol{C}_r \left(\bar{\boldsymbol{U}}_s \boldsymbol{c} + \tilde{\boldsymbol{w}}_k\right) - 2|A_k|^2 \boldsymbol{c}^H \bar{\boldsymbol{U}}_s^H \boldsymbol{s}_k\right\} \\
&= -\left(\bar{\boldsymbol{U}}_s^H \boldsymbol{C}_r \bar{\boldsymbol{U}}_s\right)^{-1} \bar{\boldsymbol{U}}_s^H \boldsymbol{C}_r \tilde{\boldsymbol{w}}_k \qquad (3.24) \\
&= -\bar{\boldsymbol{\Lambda}}_s^{-1} \bar{\boldsymbol{U}}_s^H \boldsymbol{C}_r \tilde{\boldsymbol{w}}_k, \qquad (3.25)
\end{aligned}$$

where (3.24) follows from the fact that $\bar{\boldsymbol{U}}_s^H \boldsymbol{s}_k = \boldsymbol{0}$, and (3.25) follows from (3.12). Hence

$$\boldsymbol{w}_k = \bar{\boldsymbol{U}}_s \boldsymbol{c}_k + \tilde{\boldsymbol{w}}_k = \left(\boldsymbol{I}_N - \bar{\boldsymbol{U}}_s \bar{\boldsymbol{\Lambda}}_s^{-1} \bar{\boldsymbol{U}}_s^H \boldsymbol{C}_r\right) \tilde{\boldsymbol{S}} \left(\tilde{\boldsymbol{S}}^H \tilde{\boldsymbol{S}}\right)^{-1} \tilde{\boldsymbol{e}}_k. \tag{3.26}$$

□

Proposition 3.3: [Group-blind linear MMSE detector (form I)—synchronous CDMA] *The weight vector of the group-blind linear MMSE detector for user k is given by*

$$\boldsymbol{m}_k = \left(\boldsymbol{I}_N - \bar{\boldsymbol{U}}_s \bar{\boldsymbol{\Lambda}}_s^{-1} \bar{\boldsymbol{U}}_s^H \boldsymbol{C}_r\right) \tilde{\boldsymbol{S}} \left(\tilde{\boldsymbol{S}}^H \tilde{\boldsymbol{S}} + \sigma^2 |\tilde{\boldsymbol{A}}|^{-2}\right)^{-1} \tilde{\boldsymbol{e}}_k, \quad k = 1, \ldots, \tilde{K}. \tag{3.27}$$

Proof: We first solve for $\tilde{\boldsymbol{m}}_k$ in (3.13). Since $\tilde{\boldsymbol{m}}_k \in \mathsf{range}(\tilde{\boldsymbol{S}})$, and $\tilde{\boldsymbol{S}}$ has full column rank $\tilde{K}$, we can write $\tilde{\boldsymbol{m}}_k = \tilde{\boldsymbol{S}} \tilde{\boldsymbol{c}}_k$ for some $\tilde{\boldsymbol{c}}_k \in \mathbb{C}^{\tilde{K}}$. Substituting this into (3.13), we have

$$\begin{aligned}
\tilde{\boldsymbol{c}}_k &= \arg \min_{\boldsymbol{c} \in \mathbb{C}^{\tilde{K}}} E\left\{\left|A_k b_k[i] - \boldsymbol{c}^H \tilde{\boldsymbol{S}}^H \tilde{\boldsymbol{r}}[i]\right|^2\right\} \\
&= \arg \min_{\boldsymbol{c} \in \mathbb{C}^{\tilde{K}}} \left\{\boldsymbol{c}^H \left[\tilde{\boldsymbol{S}}^H \left(\tilde{\boldsymbol{S}} |\tilde{\boldsymbol{A}}|^2 \tilde{\boldsymbol{S}}^H + \sigma^2 \boldsymbol{I}_N\right) \tilde{\boldsymbol{S}}\right] \boldsymbol{c} - 2|A_k|^2 \boldsymbol{s}_k^H \tilde{\boldsymbol{S}} \boldsymbol{c}\right\} \\
&= \arg \min_{\boldsymbol{c} \in \mathbb{C}^{\tilde{K}}} \left\{\boldsymbol{c}^H \left[\left(\tilde{\boldsymbol{S}}^H \tilde{\boldsymbol{S}}\right) |\tilde{\boldsymbol{A}}|^2 \left(\tilde{\boldsymbol{S}}^H \tilde{\boldsymbol{S}}\right) + \sigma^2 \left(\tilde{\boldsymbol{S}}^H \tilde{\boldsymbol{S}}\right)\right] \boldsymbol{c} - 2\tilde{\boldsymbol{e}}_k^H |\tilde{\boldsymbol{A}}|^2 \left(\tilde{\boldsymbol{S}}^H \tilde{\boldsymbol{S}}\right) \boldsymbol{c}\right\} \\
&= \left(\tilde{\boldsymbol{S}}^H \tilde{\boldsymbol{S}} + \sigma^2 |\tilde{\boldsymbol{A}}|^{-2}\right)^{-1} \tilde{\boldsymbol{e}}_k. \qquad (3.28)
\end{aligned}$$

Next we solve $\bar{\boldsymbol{m}}_k = \bar{\boldsymbol{U}}_s \bar{\boldsymbol{c}}_k$ in (3.14) for some $\bar{\boldsymbol{c}}_k \in \mathbb{C}^{K-\tilde{K}}$. Following the same derivation as that of (3.25), we obtain

$$\bar{\boldsymbol{c}}_k = -\bar{\boldsymbol{\Lambda}}_s^{-1} \bar{\boldsymbol{U}}_s^H \boldsymbol{C}_r \tilde{\boldsymbol{m}}_k. \tag{3.29}$$

Therefore, we have

$$\boldsymbol{m}_k = \bar{\boldsymbol{U}}_s \bar{\boldsymbol{c}}_k + \tilde{\boldsymbol{S}} \tilde{\boldsymbol{c}}_k = \left(\boldsymbol{I}_N - \bar{\boldsymbol{U}}_s \bar{\boldsymbol{\Lambda}}_s^{-1} \bar{\boldsymbol{U}}_s^H \boldsymbol{C}_r \right) \tilde{\boldsymbol{S}} \left(\tilde{\boldsymbol{S}}^H \tilde{\boldsymbol{S}} + \sigma^2 |\tilde{\boldsymbol{A}}|^{-2} \right)^{-1} \tilde{\boldsymbol{e}}_k. \tag{3.30}$$

□

Based on the results above, we can implement the linear group-blind multiuser detection algorithms based on the received signals $\{\boldsymbol{r}[i]\}_{i=0}^{M-1}$ and the signature waveforms $\tilde{\boldsymbol{S}}$ of the desired users. For example, the batch algorithm for the group-blind linear hybrid detector (form I) is summarized as follows.

Algorithm 3.1: [Group-blind linear hybrid detector (form I)—synchronous CDMA]

- *Compute the unknown users' signal subspace:*

$$\hat{\boldsymbol{C}}_r = \sum_{i=0}^{M-1} \boldsymbol{r}[i]\boldsymbol{r}[i]^H, \tag{3.31}$$

$$\bar{\boldsymbol{P}} \hat{\boldsymbol{C}}_r \bar{\boldsymbol{P}} = \hat{\bar{\boldsymbol{U}}}_s \hat{\bar{\boldsymbol{\Lambda}}}_s \hat{\bar{\boldsymbol{U}}}_s^H + \hat{\bar{\boldsymbol{U}}}_n \hat{\bar{\boldsymbol{\Lambda}}}_n \hat{\bar{\boldsymbol{U}}}_n^H + \hat{\bar{\boldsymbol{U}}}_o \hat{\bar{\boldsymbol{\Lambda}}}_o \hat{\bar{\boldsymbol{U}}}_o^H, \tag{3.32}$$

where $\bar{\boldsymbol{P}}$ is given by (3.10).

- *Form the detectors:*

$$\hat{\boldsymbol{w}}_k = \left(\boldsymbol{I}_N - \hat{\bar{\boldsymbol{U}}}_s \hat{\bar{\boldsymbol{\Lambda}}}_s^{-1} \hat{\bar{\boldsymbol{U}}}_s^H \boldsymbol{C}_r \right) \tilde{\boldsymbol{S}} \left(\tilde{\boldsymbol{S}}^H \tilde{\boldsymbol{S}} \right)^{-1} \tilde{\boldsymbol{e}}_k, \qquad k = 1, \ldots, \tilde{K}. \tag{3.33}$$

- *Perform differential detection:*

$$z_k[i] = \hat{\boldsymbol{w}}_k^H \boldsymbol{r}[i], \tag{3.34}$$

$$\hat{\beta}_k[i] = \text{sign}\left\{ \Re\left(z_k[i] z_k[i-1]^* \right) \right\}, \quad i = 1, \ldots, M-1; \; k = 1, \ldots, \tilde{K}. \tag{3.35}$$

The group-blind linear decorrelating detector and the group-blind linear MMSE detector can be implemented similarly. Note that both of them require an estimate of the noise variance σ^2. A simple estimator of σ^2 is the average of the $N - K$ eigenvalues in $\hat{\bar{\boldsymbol{\Lambda}}}_n$. Note also that the group-blind linear MMSE detector requires an estimate of the inverse of the energy of the desired users, $|\tilde{\boldsymbol{A}}|^{-2}$, as well. The following result can be found in Section 4.5 (cf. Proposition 4.2):

$$|\boldsymbol{A}|^{-2} = \tilde{\boldsymbol{S}}^H \boldsymbol{U}_s \left(\boldsymbol{\Lambda}_s - \sigma^2 \boldsymbol{I}_K \right)^{-1} \boldsymbol{U}_s^H \tilde{\boldsymbol{S}}. \tag{3.36}$$

Hence $|\tilde{\boldsymbol{A}}|^{-2} \triangleq \mathsf{diag}(|A_1|^{-2}, \ldots, |A_{\tilde{K}}|^{-2})$ can be estimated by using (3.36) with the signal subspace parameters replaced by their respective sample estimates.

In the results above, the linear group-blind detectors are expressed in terms of the known users' signature waveforms $\tilde{\boldsymbol{S}}$ and the unknown users' signal subspace components $\bar{\boldsymbol{\Lambda}}_s$ and $\bar{\boldsymbol{U}}_s$ defined in (3.12). Let the eigendecomposition of the autocorrelation matrix $\boldsymbol{C}_r$ in (3.11) be

$$\boldsymbol{C}_r = \boldsymbol{U}_s \boldsymbol{\Lambda}_s \boldsymbol{U}_s^H + \sigma^2 \boldsymbol{U}_n \boldsymbol{U}_n^H. \tag{3.37}$$

The linear group-blind detectors can also be expressed in terms of the signal subspace components $\boldsymbol{\Lambda}_s$ and $\boldsymbol{U}_s$ of all users' signals defined in (3.37), as given by the following three results.

Proposition 3.4: [Group-blind linear decorrelating detector (form II)—synchronous CDMA] *The weight vector of the group-blind linear decorrelating detector for user k is given by*

$$\boldsymbol{d}_k = \boldsymbol{U}_s \left(\boldsymbol{\Lambda}_s - \sigma^2 \boldsymbol{I}_K\right)^{-1} \boldsymbol{U}_s^H \tilde{\boldsymbol{S}} \left[\tilde{\boldsymbol{S}}^H \boldsymbol{U}_s \left(\boldsymbol{\Lambda}_s - \sigma^2 \boldsymbol{I}_K\right) \boldsymbol{U}_s^H \tilde{\boldsymbol{S}}\right]^{-1} \tilde{\boldsymbol{e}}_k, \qquad k = 1, \ldots, \tilde{K}. \tag{3.38}$$

Proof: Using the method of Lagrange multipliers to solve the constrained optimization problem (3.7), we obtain

$$\begin{aligned} \boldsymbol{d}_k &= \arg \min_{\boldsymbol{w} \in \mathrm{range}(\boldsymbol{S})} \boldsymbol{w}^H \boldsymbol{S} |\boldsymbol{A}|^2 \boldsymbol{S}^H \boldsymbol{w} + \boldsymbol{\lambda}^H \left(\tilde{\boldsymbol{S}}^H \boldsymbol{w} - \tilde{\boldsymbol{e}}_k\right) \\ &= \left(\boldsymbol{S} |\boldsymbol{A}|^2 \boldsymbol{S}^H\right)^{\dagger} \tilde{\boldsymbol{S}} \boldsymbol{\lambda}, \end{aligned} \tag{3.39}$$

where $\boldsymbol{\lambda} \in \mathbb{C}^{\tilde{K}}$. Substituting (3.39) into the constraint that $\boldsymbol{d}_k^H \tilde{\boldsymbol{S}} = \tilde{\boldsymbol{e}}_k^H$, we obtain

$$\boldsymbol{\lambda} = \left[\tilde{\boldsymbol{S}}^H \left(\boldsymbol{S} |\boldsymbol{A}|^2 \boldsymbol{S}^H\right)^{\dagger} \tilde{\boldsymbol{S}}\right]^{-1} \tilde{\boldsymbol{e}}_k. \tag{3.40}$$

Hence

$$\begin{aligned} \boldsymbol{d}_k &= \left(\boldsymbol{S} |\boldsymbol{A}|^2 \boldsymbol{S}^H\right)^{\dagger} \tilde{\boldsymbol{S}} \left[\tilde{\boldsymbol{S}}^H \left(\boldsymbol{S} |\boldsymbol{A}|^2 \boldsymbol{S}^H\right)^{\dagger} \tilde{\boldsymbol{S}}\right]^{-1} \tilde{\boldsymbol{e}}_k \\ &= \boldsymbol{U}_s \left(\boldsymbol{\Lambda}_s - \sigma^2 \boldsymbol{I}_K\right)^{-1} \boldsymbol{U}_s^H \tilde{\boldsymbol{S}} \left[\tilde{\boldsymbol{S}} \boldsymbol{U}_s \left(\boldsymbol{\Lambda}_s - \sigma^2 \boldsymbol{I}_K\right)^{-1} \boldsymbol{U}_s^H \tilde{\boldsymbol{S}}\right]^{-1} \tilde{\boldsymbol{e}}_k, \end{aligned} \tag{3.41}$$

where (3.41) follows from (3.11), (3.37), and the fact that $\boldsymbol{U}_n^H \tilde{\boldsymbol{S}} = \boldsymbol{0}$. □

Proposition 3.5: [Group-blind linear hybrid detector (form II)—synchronous CDMA] *The weight vector of the group-blind linear hybrid detector for user k is given by*

$$\boldsymbol{w}_k = \boldsymbol{U}_s \boldsymbol{\Lambda}_s^{-1} \boldsymbol{U}_s^H \tilde{\boldsymbol{S}} \left(\tilde{\boldsymbol{S}}^H \boldsymbol{U}_s \boldsymbol{\Lambda}_s^{-1} \boldsymbol{U}_s^H \tilde{\boldsymbol{S}}\right)^{-1} \tilde{\boldsymbol{e}}_k, \qquad k = 1, \ldots, \tilde{K}. \tag{3.42}$$

Proof: Using the method of Lagrange multipliers to solve the relaxed optimization problem (3.9) over $\boldsymbol{w} \in \mathbb{C}^N$, we obtain

$$\begin{aligned}
\boldsymbol{w}_k &= \arg \min_{\boldsymbol{w} \in \mathbb{C}^N} \left[E\left\{ \left| A_k b_k[i] - \boldsymbol{w}^H \boldsymbol{r}[i] \right|^2 \right\} + \boldsymbol{\lambda}^H \left(\tilde{\boldsymbol{S}}^H \boldsymbol{w} - \boldsymbol{e}_k \right) \right] \\
&= \arg \min_{\boldsymbol{w} \in \mathbb{C}^N} \left[\boldsymbol{w}^H \boldsymbol{C}_r \boldsymbol{w} - 2|A_1|^2 \boldsymbol{s}_k^H \boldsymbol{w} + \boldsymbol{\lambda}^H \tilde{\boldsymbol{S}}^H \boldsymbol{w} \right] \\
&= \arg \min_{\boldsymbol{w} \in \mathbb{C}^N} \left[\boldsymbol{w}^H \boldsymbol{C}_r \boldsymbol{w} + \left(\boldsymbol{\lambda} - 2|A_k|^2 \tilde{\boldsymbol{e}}_k \right)^H \tilde{\boldsymbol{S}}^H \boldsymbol{w} \right] \\
&= \boldsymbol{C}_r^{-1} \tilde{\boldsymbol{S}} \boldsymbol{\mu}, \qquad (3.43)
\end{aligned}$$

where $\boldsymbol{\lambda} \in \mathbb{C}^{\tilde{K}}$ is the Lagrange multiplier and $\boldsymbol{\mu} \triangleq \boldsymbol{\lambda} - 2|A_k|^2 \tilde{\boldsymbol{e}}_k$. Substituting (3.43) into the constraint that $\boldsymbol{d}_k^H \tilde{\boldsymbol{S}} = \tilde{\boldsymbol{e}}_k$ we obtain

$$\boldsymbol{\mu} = \left(\tilde{\boldsymbol{S}}^H \boldsymbol{C}_r^{-1} \tilde{\boldsymbol{S}} \right)^{-1} \tilde{\boldsymbol{e}}_k. \qquad (3.44)$$

Hence

$$\begin{aligned}
\boldsymbol{w}_k &= \boldsymbol{C}_r^{-1} \tilde{\boldsymbol{S}} \left(\tilde{\boldsymbol{S}}^H \boldsymbol{C}_r^{-1} \tilde{\boldsymbol{S}} \right)^{-1} \tilde{\boldsymbol{e}}_k \\
&= \boldsymbol{U}_s \boldsymbol{\Lambda}_s^{-1} \boldsymbol{U}_s^H \tilde{\boldsymbol{S}} \left(\tilde{\boldsymbol{S}}^H \boldsymbol{U}_s \boldsymbol{\Lambda}_s^{-1} \boldsymbol{U}_s^H \tilde{\boldsymbol{S}} \right)^{-1} \tilde{\boldsymbol{e}}_k, \qquad (3.45)
\end{aligned}$$

where (3.45) follows from (3.11), (3.37), and the fact that $\boldsymbol{U}_n^H \tilde{\boldsymbol{S}} = \boldsymbol{0}$. It is seen from (3.45) that $\boldsymbol{w}_k \in \mathsf{range}(\boldsymbol{U}_s) = \mathsf{range}(\boldsymbol{S})$; therefore, it is the solution to the constrained optimization problem (3.9). □

To form the group-blind linear MMSE detector in terms of the signal subspace $\boldsymbol{U}_s$, we first need to find a basis for the subspace $\mathsf{range}\left(\bar{\boldsymbol{U}}_s\right)$. Clearly, $\mathsf{range}\left(\bar{\boldsymbol{P}} \boldsymbol{U}_s\right) = \mathsf{range}\left(\bar{\boldsymbol{U}}_s\right)$. Consider the (rank-deficient) QR factorization of the $N \times K$ matrix $\bar{\boldsymbol{P}} \boldsymbol{U}_s$:

$$\bar{\boldsymbol{P}} \boldsymbol{U}_s = \begin{bmatrix} \boldsymbol{Q}_s & \boldsymbol{Q}_o \end{bmatrix} \begin{bmatrix} \boldsymbol{R}_s & \boldsymbol{R}_o \\ \boldsymbol{0} & \boldsymbol{0} \end{bmatrix} \boldsymbol{\Pi}, \qquad (3.46)$$

where $\boldsymbol{Q}_s$ is an $N \times \tilde{K}$ matrix, $\boldsymbol{R}_s$ is a $\tilde{K} \times \tilde{K}$ nonsingular upper triangular matrix, and $\boldsymbol{\Pi}$ is a permutation matrix. Then the columns of $\boldsymbol{Q}_s$ form an orthogonal basis of $\mathsf{range}\left(\bar{\boldsymbol{U}}_s\right)$.

Proposition 3.6: [Group-blind linear MMSE detector (form II)—synchronous CDMA] *The weight vector of the group-blind linear MMSE detector for user k is given by*

$$\boldsymbol{m}_k = \left[\boldsymbol{I}_N - \left(\boldsymbol{Q}_s \boldsymbol{R}_s^{-H} \right) \left(\boldsymbol{\Pi} \boldsymbol{\Lambda}_s \boldsymbol{\Pi}^H \right)^{-1} \left(\boldsymbol{Q}_s \boldsymbol{R}_s^{-H} \right)^H \boldsymbol{C}_r \right] \tilde{\boldsymbol{S}} \left(\tilde{\boldsymbol{S}}^H \tilde{\boldsymbol{S}} + \sigma^2 |\tilde{\boldsymbol{A}}|^{-2} \right)^{-1} \tilde{\boldsymbol{e}}_k,$$
$$k = 1, \ldots, \tilde{K}. \qquad (3.47)$$

Proof: Since the columns of $\boldsymbol{Q}_s$ form an orthogonal basis of $\mathsf{range}\left(\bar{\boldsymbol{U}}_s\right)$, following the same derivation as (3.30), we have

$$\boldsymbol{m}_k = \left[\boldsymbol{I}_N - \boldsymbol{Q}_s \left(\boldsymbol{Q}_s^H \boldsymbol{C}_r \boldsymbol{Q}_s \right)^{-1} \boldsymbol{Q}_s^H \boldsymbol{C}_r \right] \tilde{\boldsymbol{S}} \left(\tilde{\boldsymbol{S}}^H \tilde{\boldsymbol{S}} + \sigma^2 |\tilde{\boldsymbol{A}}|^{-2} \right)^{-1} \tilde{\boldsymbol{e}}_k. \qquad (3.48)$$

Furthermore, we have

$$
\begin{aligned}
\boldsymbol{Q}_s^H \boldsymbol{C}_r \boldsymbol{Q}_s &= \boldsymbol{Q}_s^H \left(\boldsymbol{U}_s \boldsymbol{\Lambda}_s \boldsymbol{U}_s^H + \sigma^2 \boldsymbol{U}_n \boldsymbol{U}_n^H \right) \boldsymbol{Q}_s \\
&= \boldsymbol{Q}_s^H \left(\boldsymbol{U}_s \boldsymbol{\Lambda}_s \boldsymbol{U}_s^H \right) \boldsymbol{Q}_s && (3.49) \\
&= \boldsymbol{Q}_s^H \left(\bar{\boldsymbol{P}} \boldsymbol{U}_s \boldsymbol{\Lambda}_s \boldsymbol{U}_s^H \bar{\boldsymbol{P}} \right) \boldsymbol{Q}_s && (3.50) \\
&= \boldsymbol{Q}_s^H \left(\bar{\boldsymbol{P}} \boldsymbol{U}_s \right) \boldsymbol{\Lambda}_s \left(\bar{\boldsymbol{P}} \boldsymbol{U}_s \right)^H \boldsymbol{Q}_s \\
&= \boldsymbol{Q}_s^H \left[\ \boldsymbol{Q}_s \ \ \boldsymbol{Q}_o \ \right] \begin{bmatrix} \boldsymbol{R}_s & \boldsymbol{R}_o \\ \boldsymbol{0} & \boldsymbol{0} \end{bmatrix} \boldsymbol{\Pi} \boldsymbol{\Lambda}_s \boldsymbol{\Pi}^H \\
&\qquad \begin{bmatrix} \boldsymbol{R}_s & \boldsymbol{R}_o \\ \boldsymbol{0} & \boldsymbol{0} \end{bmatrix}^H \left[\ \boldsymbol{Q}_s \ \ \boldsymbol{Q}_o \ \right]^H \boldsymbol{Q}_s && (3.51) \\
&= \boldsymbol{R}_s \boldsymbol{\Pi} \boldsymbol{\Lambda}_s \boldsymbol{\Pi}^H \boldsymbol{R}_s^H, && (3.52)
\end{aligned}
$$

where (3.49) follows from $\boldsymbol{U}_n^H \boldsymbol{Q}_s = \boldsymbol{0}$, (3.50) follows from $\bar{\boldsymbol{P}} \boldsymbol{Q}_s = \boldsymbol{0}$, and (3.51) follows from (3.46). Substituting (3.52) into (3.48), we obtain (3.47). □

Based on the results above, we can implement the form II linear group-blind multiuser detection algorithms based on the received signals $\{\boldsymbol{r}[i]\}_{i=0}^{M-1}$ and the signature waveforms $\tilde{\boldsymbol{S}}$ of the desired users. For example, the batch algorithm for the linear hybrid group-blind detector (form II) is as follows. (The group-blind linear decorrelating detector and the group-blind linear MMSE detector can be implemented similarly.)

Algorithm 3.2: [Group-blind linear hybrid detector (form II)—synchronous CDMA]

- *Compute the signal subspace:*

$$
\begin{aligned}
\hat{\boldsymbol{C}}_r &= \sum_{i=0}^{M-1} \boldsymbol{r}[i] \boldsymbol{r}[i]^H && (3.53) \\
&= \hat{\boldsymbol{U}}_s \hat{\boldsymbol{\Lambda}}_s \hat{\boldsymbol{U}}_s^H + \hat{\boldsymbol{U}}_n \hat{\boldsymbol{\Lambda}}_n \hat{\boldsymbol{U}}_n^H . && (3.54)
\end{aligned}
$$

- *Form the detectors:*

$$
\hat{\boldsymbol{w}}_k = \hat{\boldsymbol{U}}_s \hat{\boldsymbol{\Lambda}}_s^{-1} \hat{\boldsymbol{U}}_s^T \tilde{\boldsymbol{S}} \left(\tilde{\boldsymbol{S}}^H \hat{\boldsymbol{U}}_s \hat{\boldsymbol{\Lambda}}_s^{-1} \hat{\boldsymbol{U}}_s^H \tilde{\boldsymbol{S}} \right)^{-1} \tilde{\boldsymbol{e}}_k, \qquad k = 1, \ldots, \tilde{K}. \quad (3.55)
$$

- *Perform differential detection:*

$$
\begin{aligned}
z_k[i] &= \hat{\boldsymbol{w}}_k^H \boldsymbol{r}[i], && (3.56) \\
\hat{\beta}_k[i] &= \operatorname{sign}\left\{ \Re\left(z_k[i] z_k[i-1]^* \right) \right\}, \ i = 1, \ldots, M-1; \ k = 1, \ldots, \tilde{K}. && (3.57)
\end{aligned}
$$

In summary, for both the group-blind zero-forcing detector and the group-blind hybrid detector, the interfering signals from known users are nulled out by a projection of the received signal onto the orthogonal subspace of these users' signal subspace. The unknown interfering users' signals are then suppressed by identifying the subspace spanned by these users, followed by a linear transformation in this subspace based on the zero-forcing or MMSE criterion. In the group-blind MMSE detector, the interfering users from the known and unknown users are suppressed separately under the MMSE criterion. The suppression of the unknown users again relies on identification of the signal subspace spanned by these users.

3.3 Performance of Group-Blind Multiuser Detectors

In this section we consider the performance of group-blind linear multiuser detection. Specifically, we focus on the performance of the group-blind linear hybrid detector defined by (3.9). As in Section 2.5, for simplicity, we consider only real-valued signals. The analytical framework presented in this section was developed in [197].

3.3.1 Form II Group-Blind Hybrid Detector

The following result gives the asymptotic distribution of the estimated weight vector of the form II group-blind hybrid detector. The proof is given in the Appendix (Section 3.6.1).

Theorem 3.1: *Let the sample autocorrelation of the received signals and its eigendecomposition be*

$$\hat{\boldsymbol{C}}_r = \frac{1}{M}\sum_{i=0}^{M-1} \boldsymbol{r}[i]\boldsymbol{r}[i]^T \tag{3.58}$$

$$= \hat{\boldsymbol{U}}_s\hat{\boldsymbol{\Lambda}}_s\hat{\boldsymbol{U}}_s^T + \hat{\boldsymbol{U}}_n\hat{\boldsymbol{\Lambda}}_n\hat{\boldsymbol{U}}_n^T. \tag{3.59}$$

Let $\hat{\boldsymbol{w}}_1$ be the estimated weight vector of the form II group-blind linear hybrid detector, given by

$$\hat{\boldsymbol{w}}_1 = \hat{\boldsymbol{U}}_s\hat{\boldsymbol{\Lambda}}_s^{-1}\hat{\boldsymbol{U}}_s^T\tilde{\boldsymbol{S}}\left(\tilde{\boldsymbol{S}}^T\hat{\boldsymbol{U}}_s\hat{\boldsymbol{\Lambda}}_s^{-1}\hat{\boldsymbol{U}}_s^T\tilde{\boldsymbol{S}}\right)^{-1}\tilde{\boldsymbol{e}}_1. \tag{3.60}$$

Then

$$\sqrt{M}\left(\hat{\boldsymbol{w}}_1 - \boldsymbol{w}_1\right) \to \mathcal{N}(\boldsymbol{0}, \boldsymbol{C}_w), \quad \textit{in distribution}, \qquad \textit{as } M \to \infty,$$

with

$$\boldsymbol{C}_w =$$

$$\boldsymbol{Q}\left[\left(\boldsymbol{w}_1^T\boldsymbol{v}_1\right)\boldsymbol{U}_s\boldsymbol{\Lambda}_s^{-1}\boldsymbol{U}_s^T - 2\boldsymbol{U}_s\boldsymbol{\Lambda}_s^{-1}\boldsymbol{U}_s^T\boldsymbol{S}\boldsymbol{D}\boldsymbol{S}^T\boldsymbol{U}_s\boldsymbol{\Lambda}_s^{-1}\boldsymbol{U}_s^T\right]\boldsymbol{Q}^T + \tau\boldsymbol{U}_n\boldsymbol{U}_n^T, \tag{3.61}$$

where

$$\boldsymbol{v}_1 \triangleq \tilde{\boldsymbol{S}}\left(\tilde{\boldsymbol{S}}^T \boldsymbol{U}_s \boldsymbol{\Lambda}_s^{-1} \boldsymbol{U}_s^T \tilde{\boldsymbol{S}}\right)^{-1} \tilde{\boldsymbol{e}}_1, \tag{3.62}$$

$$\boldsymbol{D} \triangleq \mathsf{diag}\left\{A_1^4 \left(\boldsymbol{w}_1^T \boldsymbol{s}_1\right)^2, \ldots, A_K^4 \left(\boldsymbol{w}_1^T \boldsymbol{s}_K\right)^2\right\}, \tag{3.63}$$

$$\tau = \sigma^2 \boldsymbol{v}_1^T \boldsymbol{U}_s \boldsymbol{\Lambda}_s^{-1} \left(\boldsymbol{\Lambda}_s - \sigma^2 \boldsymbol{I}_K\right)^{-2} \boldsymbol{U}_s^T \boldsymbol{v}_1, \tag{3.64}$$

$$\boldsymbol{Q} \triangleq \boldsymbol{I}_N - \boldsymbol{U}_s \boldsymbol{\Lambda}_s^{-1} \boldsymbol{U}_s^T \tilde{\boldsymbol{S}}\left(\tilde{\boldsymbol{S}}^T \boldsymbol{U}_s \boldsymbol{\Lambda}_s^{-1} \boldsymbol{U}_s^T \tilde{\boldsymbol{S}}\right)^{-1} \tilde{\boldsymbol{S}}^T. \tag{3.65}$$

Define the partition of the following matrix:

$$\boldsymbol{R}\left(\boldsymbol{R} + \sigma^2 \boldsymbol{A}^{-2}\right)^{-1} \boldsymbol{A}^{-2} = \begin{bmatrix} \boldsymbol{\Psi}_{11} & \boldsymbol{\Psi}_{12} \\ \boldsymbol{\Psi}_{12}^T & \boldsymbol{\Psi}_{22} \end{bmatrix}, \tag{3.66}$$

where the dimension of $\boldsymbol{\Psi}_{11}$ is $\tilde{K} \times \tilde{K}$. Note that the left-hand side of (3.66) is equal to $\left(\boldsymbol{S}^T \boldsymbol{U}_s \boldsymbol{\Lambda}_s^{-1} \boldsymbol{U}_s^T \boldsymbol{S}\right)$ [cf. the Appendix (Section 3.6.1)], and therefore it is indeed symmetric. Define further

$$\boldsymbol{\Pi} \triangleq \left[\boldsymbol{A}^{-2}\left(\boldsymbol{R} + \sigma^2 \boldsymbol{A}^{-2}\right)^{-1} \boldsymbol{R}\left(\boldsymbol{R} + \sigma^2 \boldsymbol{A}^{-2}\right)^{-1} \boldsymbol{A}^{-2}\right]_{1:\tilde{K},1:\tilde{K}}, \tag{3.67}$$

$$\boldsymbol{\Xi} \triangleq \left[\boldsymbol{A}^{-2}\left(\boldsymbol{R} + \sigma^2 \boldsymbol{A}^{-2}\right)^{-1} \boldsymbol{A}^{-2} \boldsymbol{R}^{-1} \boldsymbol{A}^{-2}\right]_{1:\tilde{K},1:\tilde{K}}. \tag{3.68}$$

The next result gives an expression for the average output SINR of the form II group-blind hybrid detector. The proof is given in the Appendix (Section 3.6.1).

Corollary 3.1: *The average output SINR of the estimated form II group-blind linear hybrid detector is given by*

$$\overline{\mathrm{SINR}(\hat{\boldsymbol{w}}_1)} = \frac{A_1^2}{\sum_{k=1}^{K-\tilde{K}} A_{\tilde{K}+k}^2 \left(\boldsymbol{w}_1^T \boldsymbol{s}_{\tilde{K}+k}\right)^2 + \sigma^2 \|\boldsymbol{w}_1\|^2 + \frac{1}{M}\mathrm{tr}(\boldsymbol{C}_w \boldsymbol{C}_r)}, \tag{3.69}$$

where

$$\boldsymbol{w}_1^T \boldsymbol{s}_{\tilde{K}+k} = \left[\boldsymbol{\Psi}_{12}^T \boldsymbol{\Psi}_{11}^{-1}\right]_{k,1}, \tag{3.70}$$

$$\|\boldsymbol{w}_1\|^2 = \left[\boldsymbol{\Psi}_{11}^{-1} \boldsymbol{\Pi} \boldsymbol{\Psi}_{11}^{-1}\right]_{1,1}, \tag{3.71}$$

$$\begin{aligned} \mathrm{tr}(\boldsymbol{C}_w \boldsymbol{C}_r) = {} & (K - \tilde{K})\left[\boldsymbol{\Psi}_{11}^{-1}\right]_{1,1} \\ & - 2\sum_{k=1}^{K-\tilde{K}} A_{\tilde{K}+k}^4 \left[\boldsymbol{\Psi}_{12}^T \boldsymbol{\Psi}_{11}^{-1}\right]_{k,1}^2 \left[\boldsymbol{\Psi}_{22} - \boldsymbol{\Psi}_{12}^T \boldsymbol{\Psi}_{11}^{-1} \boldsymbol{\Psi}_{12}\right]_{k,k} \\ & + (N - K)\sigma^4 \left[\boldsymbol{\Psi}_{11}^{-1} \boldsymbol{\Xi} \boldsymbol{\Psi}_{11}^{-1}\right]_{1,1}. \end{aligned} \tag{3.72}$$

As in Section 2.5, in order to gain insights from the result (3.69), we next compute the average output SINR of the form II group-blind hybrid detector for two special cases: orthogonal signals and equicorrelated signals.

Example 1: Orthogonal Signals In this case $\boldsymbol{w}_1 = \boldsymbol{s}_1$ and $\boldsymbol{R} = \boldsymbol{I}_K$. After some manipulations, the average output SINR in this case is

$$\overline{\text{SINR}\left(\hat{\boldsymbol{w}}_1\right)} = \frac{\phi_1}{1 + \frac{1}{M}\left(\phi_1 + 1\right)\left(K - \tilde{K} + \frac{N-K}{\phi_1^2}\right)}, \tag{3.73}$$

where $\phi_1 \triangleq A_1^2/\sigma^2$ is the SNR of the desired user. On comparing (3.73) with (2.142), we obtain the following necessary and sufficient condition for the group-blind hybrid detector to outperform the subspace blind detector:

$$\tilde{K} + 1 > \frac{2\phi_1^2}{(1+\phi_1)^2}. \tag{3.74}$$

Since $\tilde{K} \geq 1$, the condition above is always satisfied. Hence we conclude that in this case the group-blind hybrid detector always outperforms the subspace blind detector. On the other hand, based on (3.73) and (2.142), we can also obtain the following necessary and sufficient condition under which the group-blind hybrid detector outperforms the DMI blind detector:

$$\left(1 - \frac{1}{\phi_1^2}\right)(N - K) + \tilde{K} + 1 > \frac{2\phi_1^2}{(1+\phi_1)^2}. \tag{3.75}$$

It is seen from (3.75) that at very low SNR (e.g., $\phi_1 \ll 1$), the DMI detector will outperform the group-blind hybrid detector. Moreover, a sufficient condition for the group-blind hybrid detector to outperform the DMI detector is $\phi_1 \geq 1\,(= 0$ dB).

Example 2: Equicorrelated Signals with Perfect Power Control Recall that in this case, it is assumed that $\boldsymbol{s}_k^T \boldsymbol{s}_l = \rho$ for $k \neq l$; and $A_1 = \cdots = A_K = A$. Denote

$$a \triangleq \frac{1}{1 - \rho + \frac{\sigma^2}{A^2}}, \tag{3.76}$$

$$b \triangleq \frac{1 + (K-1)\rho}{K}\left[\frac{1}{1 + (K-1)\rho + \frac{\sigma^2}{A^2}} - \frac{1}{1 - \rho + \frac{\sigma^2}{A^2}}\right], \tag{3.77}$$

$$a' \triangleq \frac{1}{\left(1-\rho+\frac{\sigma^2}{A^2}\right)^2}, \tag{3.78}$$

$$b' \triangleq \frac{1+(K-1)\rho}{K}\left[\frac{1}{\left(1+(K-1)\rho+\frac{\sigma^2}{A^2}\right)^2} - \frac{1}{\left(1-\rho+\frac{\sigma^2}{A^2}\right)^2}\right], \tag{3.79}$$

$$a'' \triangleq \frac{1}{\left(1-\rho+\frac{\sigma^2}{A^2}\right)(1-\rho)^2}, \tag{3.80}$$

$$b'' \triangleq \frac{1+(K-1)\rho}{K} \left[\frac{1}{\left(1+(K-1)\rho+\frac{\sigma^2}{A^2}\right)[1+(K-1)\rho]^2} - \frac{1}{\left(1-\rho+\frac{\sigma^2}{A^2}\right)(1-\rho)^2}\right]. \tag{3.81}$$

It is shown in the Appendix (Section 3.6.1) that the average output SINR of the form II group-blind hybrid detector in this case is given by

$$\overline{\text{SINR}(\hat{\boldsymbol{w}}_1)} = \frac{1}{(K-\tilde{K})\alpha^2+\beta+\frac{1}{M}\left[(K-\tilde{K})\left(\gamma-2\eta\alpha^2\right)+(N-K)\mu\right]}, \tag{3.82}$$

where

$$\alpha \triangleq \frac{a\rho+b}{a(1-\rho)+\tilde{K}(a\rho+b)}, \tag{3.83}$$

$$\beta \triangleq \left(\frac{\sigma^2}{A^2}\right)\frac{1}{a^2(1-\rho)^2}\left\{a'+b'-\alpha\left[2(1-\rho)a'+(\tilde{K}+1)(a'\rho+b')\right]\right.$$
$$\left.+\alpha^2\tilde{K}\left[a'(1-\rho)+\tilde{K}(a'\rho+b')\right]\right\}, \tag{3.84}$$

$$\gamma \triangleq \frac{1-\alpha}{a(1-\rho)}, \tag{3.85}$$

$$\eta \triangleq a+b-\alpha\tilde{K}(a\rho+b), \tag{3.86}$$

$$\mu \triangleq \left(\frac{\sigma^2}{A^2}\right)^2\frac{1}{a^2(1-\rho)^2}\left\{a''+b''-\alpha\left[2(1-\rho)a''+(\tilde{K}+1)(a''\rho+b'')\right]\right.$$
$$\left.+\alpha^2\tilde{K}\left[a''(1-\rho)+\tilde{K}(a''\rho+b'')\right]\right\}. \tag{3.87}$$

The average output SINR as a function of SNR and ρ for the form II group-blind hybrid detector and the subspace blind detector is shown in Fig. 3.1. It is seen that the group-blind hybrid detector outperforms the subspace blind detector. The

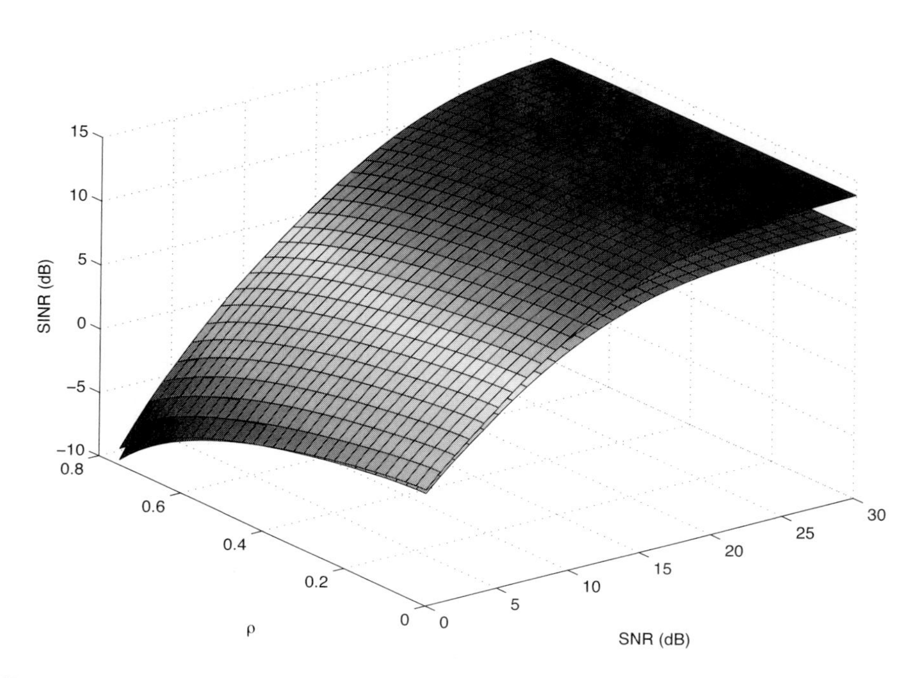

Figure 3.1. Average output SINR versus SNR and ρ for a subspace blind detector and form II group-blind hybrid detector. $N = 32$, $K = 16$, $\tilde{K} = 8$, $M = 200$. The upper curve represents the performance of the form II group-blind detector.

performance of this group-blind detector in the high cross-correlation and low SNR region is more clearly seen in Figs. 3.2 and 3.3, where its performance under different numbers of known users, as well as the performance of the two blind detectors, is compared as a function of ρ and SNR, respectively. Interestingly, it is seen from Fig. 3.2 that like the DMI blind detector, the group-blind detector is insensitive to the signal cross-correlation. Moreover, for the SNR values considered here, the group-blind detector outperforms both blind detectors for all ranges of ρ, even for the case that the numbers of known users $\tilde{K} = 1$. Note that when $\tilde{K} = 1$, the form II group-blind hybrid detector (3.60) becomes

$$\hat{\boldsymbol{w}}_1 = \hat{\boldsymbol{U}}_s \hat{\boldsymbol{\Lambda}}_s^{-1} \hat{\boldsymbol{U}}_s^T \boldsymbol{s}_1 \left(\boldsymbol{s}_1^T \hat{\boldsymbol{U}}_s \hat{\boldsymbol{\Lambda}}_s^{-1} \hat{\boldsymbol{U}}_s^T \boldsymbol{s}_1 \right)^{-1}. \tag{3.88}$$

This is essentially the *constrained subspace blind detector*, with the constraint being $\hat{\boldsymbol{w}}_1^T \boldsymbol{s}_1 = 1$. It is seen that by imposing such a constraint on the subspace blind detector, the detector becomes more resistant to high signal cross-correlation. However, from Fig. 3.3, in the low-SNR region, the group-blind detector behaves similarly to the subspace blind detector (e.g., the performance of both detectors deteriorates quickly as SNR drops below 0 dB), whereas the performance degradation of the DMI blind detector in this region is more graceful.

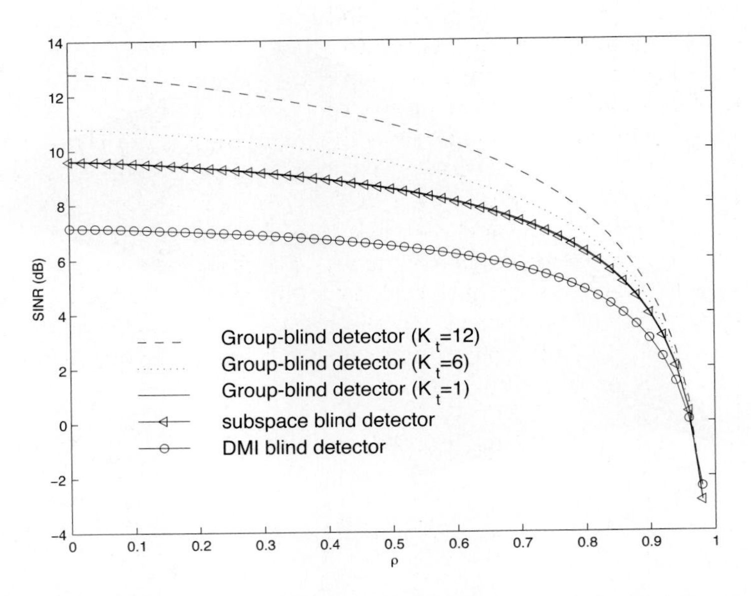

Figure 3.2. Average output SINR versus ρ for a form II group-blind hybrid detector and two blind detectors. $N = 32$, $K = 16$, $M = 200$, SNR $= 15\,$dB. (In the figure $K_t \triangleq \tilde{K}$.)

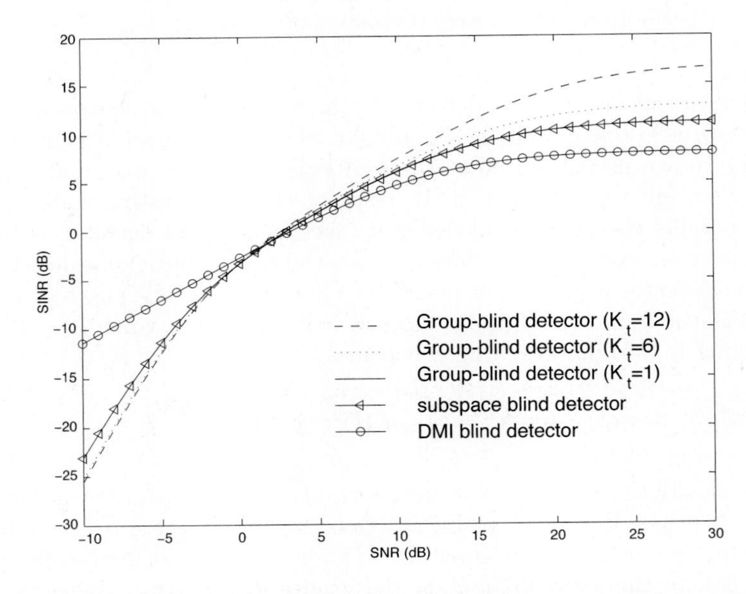

Figure 3.3. Average output SINR versus SNR for a form II group-blind hybrid detector and two blind detectors. $N = 32$, $K = 16$, $M = 200$, $\rho = 0.4$. (In the figure $K_t \triangleq \tilde{K}$.)

Next, the performance of the group-blind and blind detectors as a function of the number of signal samples, M, is plotted in Fig. 3.4, where it is seen that as the number of known users $\tilde{K}$ increases, both the asymptotic SINR (as $M \to \infty$) of the group-blind hybrid detector and its convergence rate increase. Finally, the performance of blind and group-blind detectors as a function of the number of users K is plotted in Fig. 3.5, where it is seen that for the values of SNR and ρ considered here, when the number of known users $\tilde{K} > 1$, the group-blind hybrid detector outperforms both blind detectors, even in a fully loaded system (i.e., $K = N$). In summary, we have seen that except for the very low SNR region (e.g., below 0 dB), where the DMI blind detector performs the best (however, such a region is *not* of practical interest), in general, by incorporating the knowledge of the spreading sequences of other users, the group-blind detector offers performance improvement over both DMI and subspace blind detectors.

3.3.2 Form I Group-Blind Detectors

Define

$$\tilde{\boldsymbol{d}}_1 \triangleq \tilde{\boldsymbol{S}} \left(\tilde{\boldsymbol{S}}^T \tilde{\boldsymbol{S}} \right)^{-1} \tilde{\boldsymbol{e}}_1, \tag{3.89}$$

$$\tilde{\boldsymbol{m}}_1 \triangleq \tilde{\boldsymbol{S}} \left(\tilde{\boldsymbol{S}}^T \tilde{\boldsymbol{S}} + \sigma^2 \boldsymbol{I}_{\tilde{K}} \right)^{-1} \tilde{\boldsymbol{e}}_1, \tag{3.90}$$

and let $\bar{\boldsymbol{S}} \triangleq [\boldsymbol{s}_{\tilde{K}+1}, \ldots, \boldsymbol{s}_K]$. The following result gives the asymptotic distribution of the estimated weight vector of the form I linear group-blind hybrid detector and that of the form I linear group-blind MMSE detector. The proof is given in the Appendix (Section 3.6.2).

Theorem 3.2: *Let*

$$\hat{\boldsymbol{C}}_r = \frac{1}{M} \sum_{i=0}^{M-1} \boldsymbol{r}[i] \boldsymbol{r}[i]^T \tag{3.91}$$

be the sample autocorrelation matrix of the received signals based on M samples. Define

$$\bar{\boldsymbol{P}} \triangleq \boldsymbol{I}_N - \tilde{\boldsymbol{S}} \left(\tilde{\boldsymbol{S}}^T \tilde{\boldsymbol{S}} \right)^{-1} \tilde{\boldsymbol{S}}^T. \tag{3.92}$$

Let $\hat{\bar{\boldsymbol{\Lambda}}}_s$ and $\hat{\bar{\boldsymbol{U}}}_s$ contain, respectively, the largest $K - \tilde{K}$ eigenvalues of $\bar{\boldsymbol{P}} \hat{\boldsymbol{C}}_r \bar{\boldsymbol{P}}$ and the corresponding eigenvectors. Let $\hat{\boldsymbol{w}}_1$ be the estimated weight vector of the form I group-blind linear detector, given by

$$\hat{\boldsymbol{w}}_1 = \left(\boldsymbol{I}_N - \hat{\bar{\boldsymbol{U}}}_s \hat{\bar{\boldsymbol{\Lambda}}}_s^{-1} \hat{\bar{\boldsymbol{U}}}_s^T \hat{\boldsymbol{C}}_r \right) \boldsymbol{v}, \tag{3.93}$$

where $\boldsymbol{v} \triangleq \tilde{\boldsymbol{d}}_1$ for the group-blind linear hybrid detector and $\boldsymbol{v} \triangleq \tilde{\boldsymbol{m}}_1$ for the group-blind linear MMSE detector. Then

$$\sqrt{M} \left(\hat{\boldsymbol{w}}_1 - \boldsymbol{w}_1 \right) \to \mathcal{N}(\boldsymbol{0}, \boldsymbol{C}_w), \quad \textit{in distribution}, \quad \textit{as } M \to \infty,$$

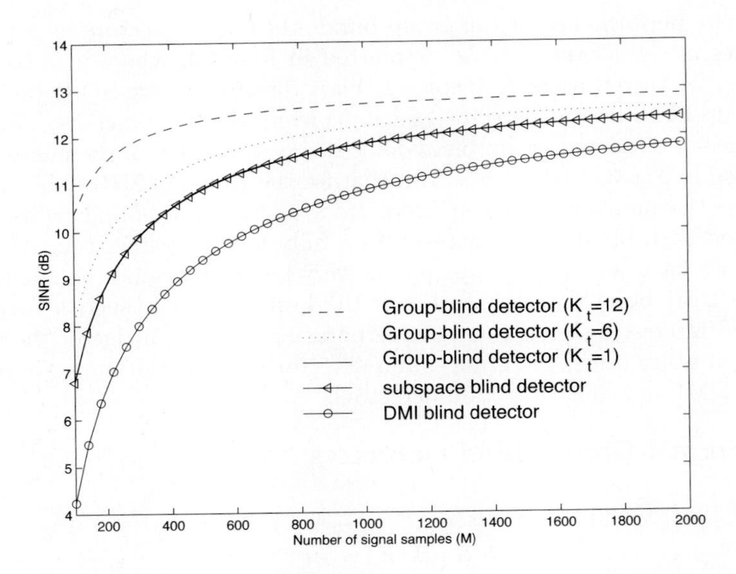

Figure 3.4. Average output SINR versus the number of signal samples M for a form II group-blind hybrid detector and two blind detectors. $N = 32$, $K = 16$, $\rho = 0.4$, $\mathrm{SNR} = 15\,\mathrm{dB}$. (In the figure $K_t \triangleq \tilde{K}$.)

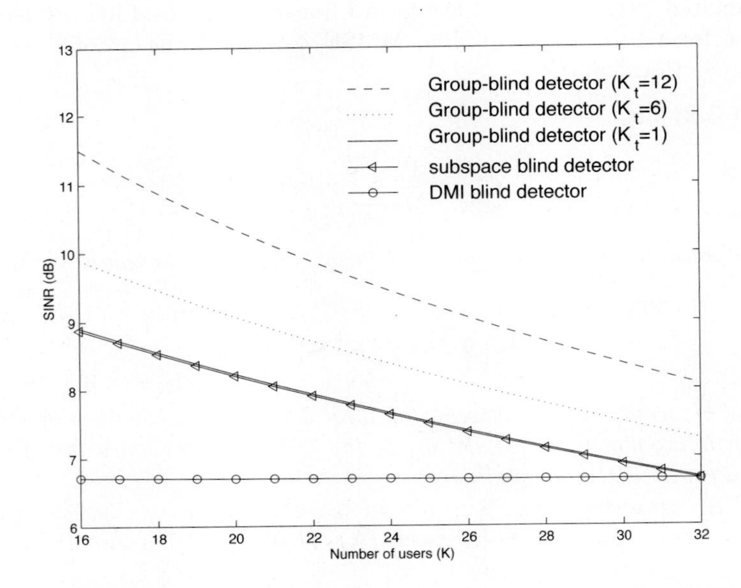

Figure 3.5. Average output SINR versus the number of users K for a form II group-blind hybrid detector and two blind detectors. $N = 32$, $M = 200$, $\rho = 0.4$, $\mathrm{SNR} = 15\,\mathrm{dB}$. (In the figure $K_t \triangleq \tilde{K}$.)

with

$$\boldsymbol{C}_w = \left(\boldsymbol{w}_1^T \boldsymbol{C}_r \boldsymbol{v}\right) \bar{\boldsymbol{U}}_s \bar{\boldsymbol{\Lambda}}_s^{-1} \bar{\boldsymbol{U}}_s^T - 2\bar{\boldsymbol{U}}_s \bar{\boldsymbol{\Lambda}}_s^{-1} \bar{\boldsymbol{U}}_s^T \bar{\boldsymbol{S}} \bar{\boldsymbol{D}}^2 \bar{\boldsymbol{S}} \bar{\boldsymbol{U}}_s \bar{\boldsymbol{\Lambda}}_s^{-1} \bar{\boldsymbol{U}}_s^T + \tau \bar{\boldsymbol{U}}_n \bar{\boldsymbol{U}}_n^T, \quad (3.94)$$

where

$$\tau = \sigma^2 \left(\boldsymbol{C}_r \boldsymbol{v}\right)^T \bar{\boldsymbol{U}}_s \bar{\boldsymbol{\Lambda}}_s^{-1} \left(\bar{\boldsymbol{\Lambda}}_s - \sigma^2 \boldsymbol{I}_{K-\tilde{K}}\right)^{-2} \bar{\boldsymbol{U}}_s^T \left(\boldsymbol{C}_r \boldsymbol{v}\right), \quad (3.95)$$

$$\bar{\boldsymbol{D}} = \mathsf{diag}\left\{A_{\tilde{K}+1}^2 \left(\boldsymbol{w}_1^T \boldsymbol{s}_{\tilde{K}+1}\right), \ldots, A_K^2 \left(\boldsymbol{w}_1^T \boldsymbol{s}_K\right)\right\}. \quad (3.96)$$

As before, the SINRs for the form I group-blind detectors can be expressed in terms of $\boldsymbol{R}$, σ^2, and $\boldsymbol{A}$. However, the closed-form SINR expressions are too complicated for this case and we therefore do not present them here. Nevertheless, the SINR of the group-blind linear hybrid detector for orthogonal signals can be obtained explicitly, as in the following example.

Example 1: Orthogonal Signals We consider the form I linear hybrid detector. In this case $\boldsymbol{w}_1 = \boldsymbol{v} = \boldsymbol{s}_1$, $\bar{\boldsymbol{U}}_s = [\boldsymbol{s}_{\tilde{K}+1} \; \cdots \; \boldsymbol{s}_K]$, and $\bar{\boldsymbol{\Lambda}}_s = \mathsf{diag}\left\{A_{\tilde{K}+1}^2, \ldots, A_K^2\right\} + \sigma^2 \boldsymbol{I}_{K-\tilde{K}}$. Moreover, $\bar{\boldsymbol{U}}_s^T \boldsymbol{C}_r \boldsymbol{v} = \boldsymbol{0}$, so that $\tau = 0$, $\boldsymbol{w}_1^T \boldsymbol{C}_r \boldsymbol{v} = A_1^2 + \sigma^2$, and $\bar{\boldsymbol{D}} = \boldsymbol{0}$. Hence $\boldsymbol{C}_w = \dfrac{A_1^2 + \sigma^2}{M} \bar{\boldsymbol{U}}_s \bar{\boldsymbol{\Lambda}}_s^{-1} \bar{\boldsymbol{U}}_s^T$. Substituting these into (2.119), and after some manipulation, we get

$$\overline{\mathrm{SINR}\left(\hat{\boldsymbol{w}}_1\right)} = \frac{\phi_1}{1 + \frac{1}{M}\left(\phi_1 + 1\right)\left(K - \tilde{K}\right)}. \quad (3.97)$$

Comparing (3.73) and (3.97), we see that for the orthogonal-signal case, the form I group-blind hybrid detector always outperforms the form II group-blind hybrid detector.

In Fig. 3.6 the output SINR of the two blind detectors and that of the two forms of group-blind hybrid detectors [given, respectively, by (2.142), (3.73), and (3.97)] are plotted as functions of the desired user's SNR, ϕ_1. It is seen that in the high-SNR region, the DMI blind detector has the worst performance among these detectors. In the low-SNR region, however, both the form II group-blind detector and the subspace blind detector perform worse than the DMI blind detector. The form I group-blind detector performs the best in this case.

Example 2: Equicorrelated Signals with Perfect Power Control Although we do not present a closed-form expression for the output SINR for the form I group-blind detector, we can still evaluate the SINR for this case as follows. As noted above, the SINR is a function of the user spreading sequences $\boldsymbol{S}$ only through the correlation matrix $\boldsymbol{R} \triangleq \boldsymbol{S}^T \boldsymbol{S}$. In other words, with the same $\boldsymbol{A}$ and σ^2, systems employing different set of spreading sequences $\boldsymbol{S}$ and $\boldsymbol{S}'$ will have the same SINR as long as $\boldsymbol{S}^T \boldsymbol{S} = \boldsymbol{S}'^T \boldsymbol{S}'$ [even if the spreading sequences take real values rather than

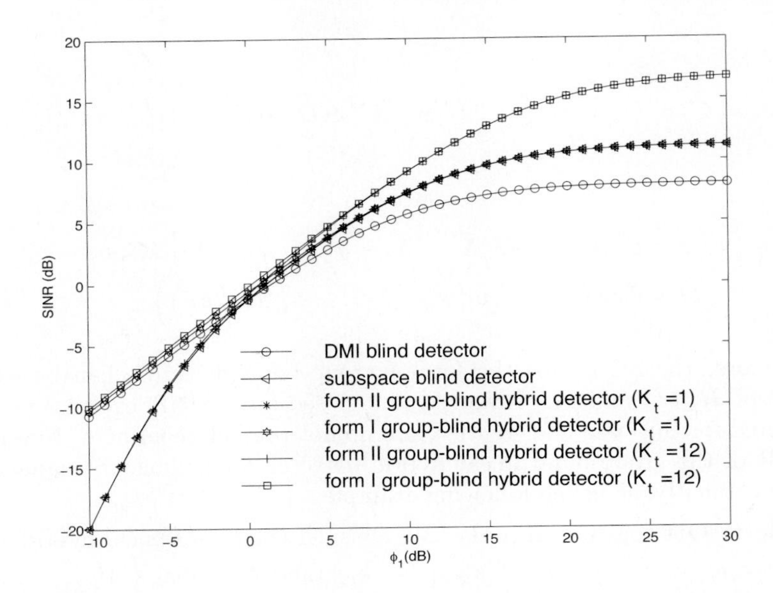

Figure 3.6. Average output SINR versus ϕ_1 of blind and group-blind detectors for the orthogonal signal case. $N = 32$, $K = 16$, $M = 200$. (In the figure $K_t \triangleq \tilde{K}$.)

the form $(1/\sqrt{N})[s_{0,k}, \ldots, s_{N-1,k}]^T$, $s_{j,k} \in \{+1, -1\}]$. Hence given $\boldsymbol{R}$, $\boldsymbol{A}$, and σ^2, we can, for example, designate $\boldsymbol{S}$ to be of the form

$$\boldsymbol{S} = \begin{bmatrix} \mathbf{0}_{(N-K)\times K} \\ \sqrt{\boldsymbol{R}} \end{bmatrix} \tag{3.98}$$

(where $\sqrt{\boldsymbol{R}}$ denotes the Cholesky factor of $\boldsymbol{R}$) and then use (2.119) and (3.94) to compute the SINR. Note that each column of $\boldsymbol{S}$ in (3.98) has unit norm since the diagonal elements of $\boldsymbol{R}$ are all 1s. Our computation shows that the performance of the form I group-blind hybrid detector is similar to that of the form II group-blind hybrid detector, with the exception that the form I detector behaves similarly to the DMI blind detector in the very low SNR region—namely, it does not deteriorate as much as do the form I group-blind and subspace blind detectors. This is shown in Fig. 3.7. (The performance of the form I group-blind MMSE detector is indistinguishable from that of the form I group-blind hybrid detector in this case.)

In summary, we have seen that the performance of the subspace blind detector deteriorates in the low-SNR and high-cross-correlation region, the form II group-blind detector is resistant to high cross-correlation but not to low SNR, and the form I group-blind detector is resistant to both high cross-correlation and low SNR. Although the DMI blind detector is also insensitive to both high cross-correlation and low SNR, its performance in other regions is inferior to all the subspace-based blind and group-blind detectors. Hence we conclude that the form I group-blind detector achieves the best overall performance among all the detectors considered here.

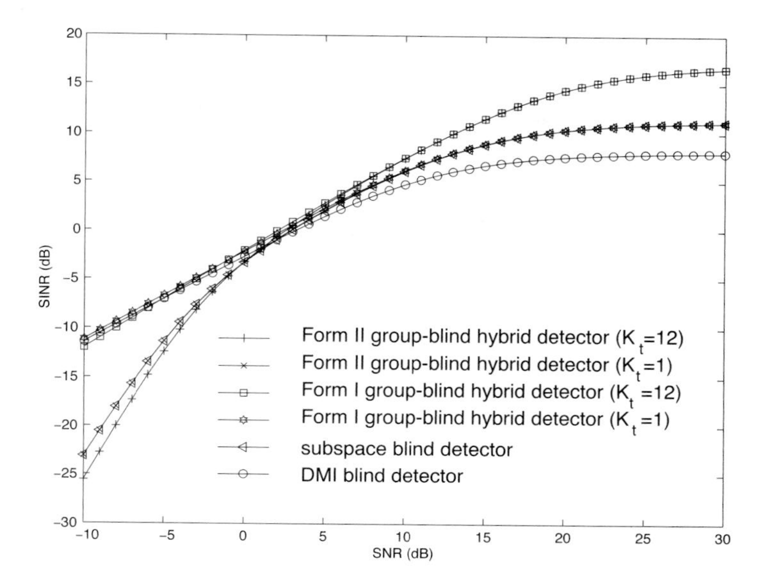

Figure 3.7. Average output SINR versus SNR for form I and form II group-blind detectors and blind detectors. $N = 32$, $K = 16$, $M = 200$, $\rho = 0.4$. (In the figure $K_t \triangleq \tilde{K}$.)

Finally, we compare the analytical performance expressions given in this section with the simulation results. The simulated system is the same as that in Section 2.5 ($N = 13$, $K = 11$). The analytical and simulated SINR performance of the form I and form II group-blind detectors is shown in Fig. 3.8. For each detector, the SINR is plotted as a function of the number of signal samples (M) used for estimating the detector at some fixed SNR. The simulated and analytical BER performance of these estimated detectors is shown in Fig. 3.9. As before, the analytical BER performance is based on a Gaussian approximation of the output of the estimated linear detector. It is seen that as is true for the DMI blind detector and the subspace detector treated in Section 2.5, the analytical performance expressions discussed in this section for group-blind detectors match the simulation results very well. Performance analysis for the group-blind detectors in the more realistic complex-valued asynchronous CDMA with multipath channels and blind channel estimation can be found in [196]. Some upper bounds on the achievable performance of various group blind multiuser detectors are obtained in [192, 195]. Moreover, large-system asymptotic performance of the group-blind multiuser detectors is given in [604].

3.4 Nonlinear Group-Blind Multiuser Detection for Synchronous CDMA

In Section 3.2 we developed linear receivers for detecting a group of $\tilde{K}$ users' data bits in the presence of unknown interfering users. In this section we develop non-

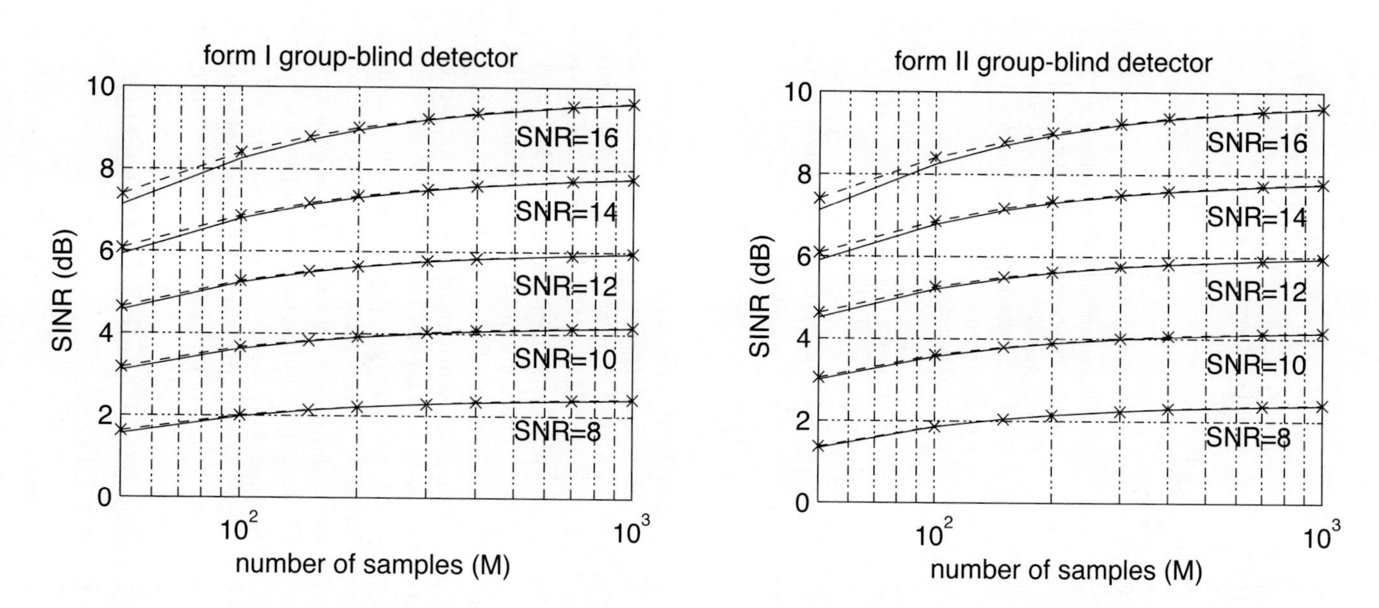

Figure 3.8. Output average SINR versus the number of signal samples M for form I and form II group-blind hybrid detectors. $N = 13, K = 11$. The solid line is the analytical performance, and the dashed line is the simulation performance.

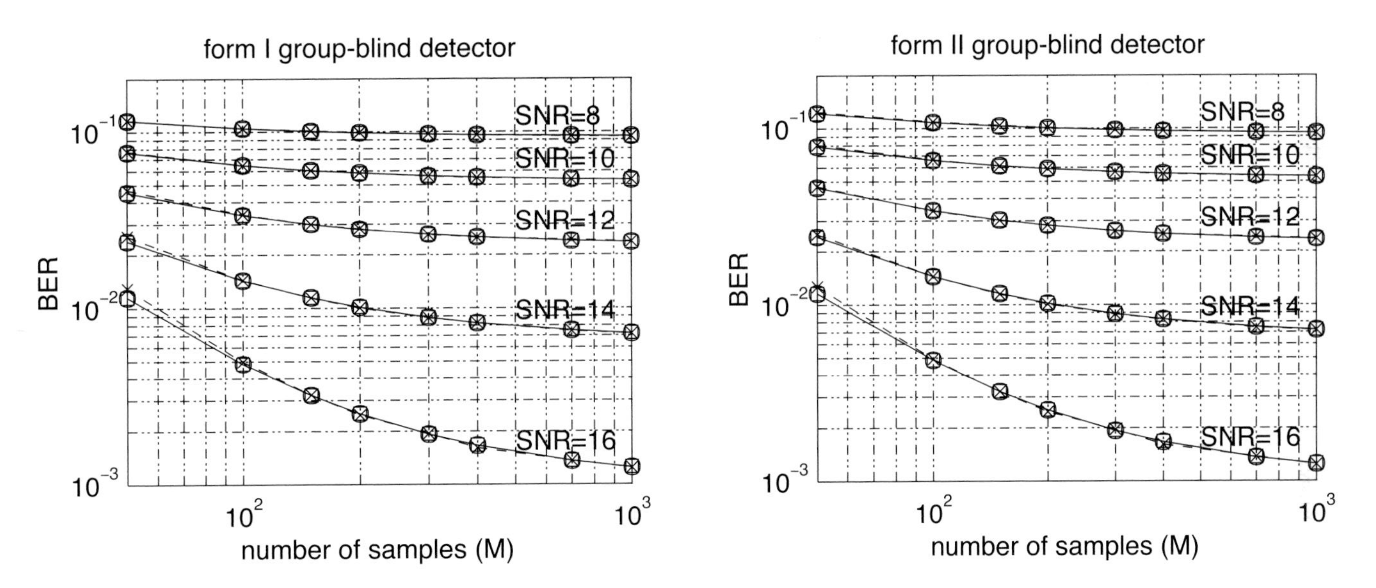

Figure 3.9. BER versus the number of signal samples M for form I and form II group-blind hybrid detectors. $N = 13, K = 11$. The solid line is the analytical performance, and the dashed line is the simulation performance.

linear methods for joint detection of the desired users' data. The basic idea is to construct a likelihood function for these users' data and then to perform a local search over such a likelihood surface, starting from the estimate closest to the unconstrained maximizer of the likelihood function, and along mutually orthogonal directions where the likelihood function drops at the slowest rate. The techniques described in this section were developed in [456].

Consider the signal model (3.2). Since the transmitted symbols $\boldsymbol{b} \in \{+1, -1\}^K \subset \mathbb{R}^K$, for the convenience of the development in this section, we write (3.2) in terms of real-valued signals. Specifically, denote (recall that $\boldsymbol{S}$ is real valued)

$$\boldsymbol{y}[i] \triangleq \begin{bmatrix} \Re\{\boldsymbol{r}[i]\} \\ \Im\{\boldsymbol{r}[i]\} \end{bmatrix}, \boldsymbol{\Psi} \triangleq \begin{bmatrix} \boldsymbol{S}\Re\{\boldsymbol{A}\} \\ \boldsymbol{S}\Im\{\boldsymbol{A}\} \end{bmatrix}, \boldsymbol{v}[i] \triangleq \begin{bmatrix} \Re\{\boldsymbol{n}[i]\} \\ \Im\{\boldsymbol{n}[i]\} \end{bmatrix}, \tag{3.99}$$

where $\boldsymbol{v}[i] \sim \mathcal{N}\left(\boldsymbol{0}, (\sigma^2/2)\boldsymbol{I}_{2N}\right)$ is a real-valued noise vector. Then (3.2) can be written as

$$\boldsymbol{y}[i] = \boldsymbol{\Psi}\boldsymbol{b}[i] + \boldsymbol{v}[i]. \tag{3.100}$$

For notational simplicity in what follows, we drop the symbol index i. In this case the maximum-likelihood estimate of the transmitted symbols (of all users) is given by

$$\begin{aligned} \widehat{\boldsymbol{b}}_{ML} &= \arg \max_{\boldsymbol{b} \in \{+1,-1\}^K} p(\boldsymbol{y} \mid \boldsymbol{b}) \\ &= \arg \min_{\boldsymbol{b} \in \{+1,-1\}^K} \underbrace{\|\boldsymbol{y} - \boldsymbol{\Psi}\boldsymbol{b}\|^2}_{\ell(\boldsymbol{b})} \\ &= \arg \min_{\boldsymbol{b} \in \{+1,-1\}^K} \ell(\boldsymbol{\theta}) + (\boldsymbol{b} - \boldsymbol{\theta})^T \nabla_\ell^2 (\boldsymbol{b} - \boldsymbol{\theta}), \end{aligned} \tag{3.101}$$

where $\boldsymbol{\theta}$ is the stationary point of $\ell(\boldsymbol{b})$:

$$\nabla_\ell(\boldsymbol{\theta}) = \boldsymbol{0} \Longrightarrow \boldsymbol{\theta} = \left(\boldsymbol{\Psi}^T\boldsymbol{\Psi}\right)^{-1} \boldsymbol{\Psi}^T \boldsymbol{y}. \tag{3.102}$$

In (3.101) the Hessian matrix of the log-likelihood function is given by

$$\begin{aligned} \nabla_\ell^2 &= \boldsymbol{\Psi}^T\boldsymbol{\Psi} \\ &= \Re\{\boldsymbol{A}\} \left(\boldsymbol{S}^T\boldsymbol{S}\right) \Re\{\boldsymbol{A}\} + \Im\{\boldsymbol{A}\} \left(\boldsymbol{S}^T\boldsymbol{S}\right) \Im\{\boldsymbol{A}\}. \end{aligned} \tag{3.103}$$

It is well known that the combinatorial optimization problem (3.101) is computationally hard (i.e., it is NP-complete) [519]. We next consider a local-search approach to approximating its solution. The basic idea is to search the optimal solution in a subset Ω of the discrete parameter set $\{-1, +1\}^K$ that is close to the stationary point $\boldsymbol{\theta}$. More precisely, we approximate the solution to (3.101) by

$$\begin{aligned} \widehat{\boldsymbol{b}} &\cong \arg \max_{\boldsymbol{b} \in \Omega \subset \{+1,-1\}^K} p(\boldsymbol{y} \mid \boldsymbol{b}) \\ &= \arg \min_{\boldsymbol{b} \in \Omega} (\boldsymbol{b} - \boldsymbol{\theta})^H \nabla_\ell^2 (\boldsymbol{b} - \boldsymbol{\theta}). \end{aligned} \tag{3.104}$$

In the *slowest-descent method* [453, 454], the candidate set Ω consists of the discrete parameters chosen such that they are in the neighborhood of Q ($Q \leq K$) lines in $\mathbb{R}^K$, which are defined by the stationary point $\boldsymbol{\theta}$ and the Q eigenvectors of ∇_ℓ^2 corresponding to the Q smallest eigenvalues. The basic idea of this method is explained next.

3.4.1 Slowest-Descent Search

The basic idea of the slowest-descent search method is to choose the candidate points in Ω such that they are closest to a line $(\boldsymbol{\theta} + \mu \boldsymbol{g})$ in $\mathbb{R}^K$, originating from $\boldsymbol{\theta}$ and along a direction $\boldsymbol{g}$, where the likelihood function $p(\boldsymbol{y}|\boldsymbol{b})$ drops at the slowest rate. Given any line in $\mathbb{R}^K$, there are at most K points where the line intersects the coordinate hyperplanes (e.g., $\boldsymbol{\theta}^1$ and $\boldsymbol{\theta}^2$ in Fig. 3.10 for $K = 2$). The set of intersection points corresponding to a line defined by $\boldsymbol{\theta}$ and $\boldsymbol{g}$ can be expressed as

$$\left\{\boldsymbol{\theta}^i = \boldsymbol{\theta} - \mu^i \boldsymbol{g} : \ \mu^i = \theta_i / g_i \right\}_{i=1}^{K}, \tag{3.105}$$

where θ_i and g_i denote the ith elements of the respective vectors $\boldsymbol{\theta}$ and $\boldsymbol{g}$. Each intersection point $\boldsymbol{\theta}^i$ has only its ith component equal to zero (i.e., $\theta_i^i = 0$). For simplicity we do not consider lines that simultaneously intersect more than one coordinate hyperplane since this event occurs with probability zero.

Any point on the line except for an intersection point has a unique closest candidate point in $\{+1, -1\}^K$. An intersection point is of equal distance from its two neighboring candidate points [e.g., $\boldsymbol{\theta}^1$ is equidistant to $\boldsymbol{b}^1$ and $\boldsymbol{b}^2$ in Fig. 3.10(a)].

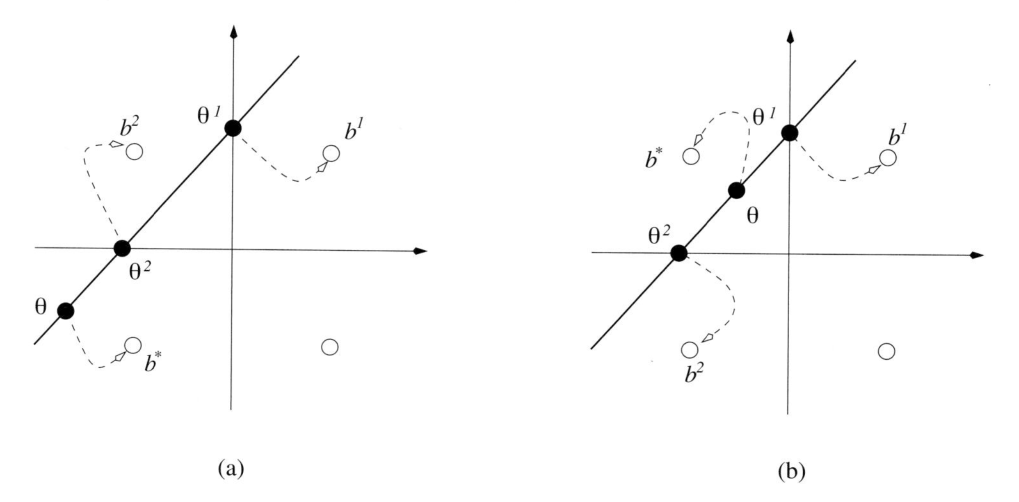

Figure 3.10. One-to-one mapping from $\{\boldsymbol{\theta}, \boldsymbol{\theta}^1, \ldots, \boldsymbol{\theta}^K\}$ to $\Omega \triangleq \{\boldsymbol{b}^*, \boldsymbol{b}^1, \ldots, \boldsymbol{b}^K\}$ for $K = 2$. Each intersection point $\boldsymbol{\theta}^i$ is of equal distance from its two neighboring candidate points. $\boldsymbol{b}^i$ is chosen to be one of these two candidate points that is on the opposite side of the ith coordinate hyperplane with respect to $\boldsymbol{b}^*$.

Two neighboring intersection points share a unique closest candidate point [e.g., $\boldsymbol{\theta}^1$ and $\boldsymbol{\theta}^2$ share the nearest candidate point $\boldsymbol{b}^2$ in Fig. 3.10(a)]. Define

$$\boldsymbol{b}^* \triangleq \text{sign}(\boldsymbol{\theta}) \tag{3.106}$$

as the candidate point closest to $\boldsymbol{\theta}$, which is also the decision given by the decorrelating multiuser detector in a coherent channel. By carefully selecting one of the two candidate points closest to each intersection point to avoid choosing the same point twice, one can specify K distinct candidate points in $\{+1, -1\}^K$ that are closest to the line $(\boldsymbol{\theta} + \mu \boldsymbol{g})$. To that end, consider the following set:

$$\left\{ \boldsymbol{b}^i \in \{-1, +1\}^K : b_k^i = \begin{cases} \text{sign}\left(\theta_k^i\right), & k \neq i \\ -b_i^*, & k = i \end{cases} \right\}_{i=1}^K . \tag{3.107}$$

It is seen that (3.107) assigns to each intersection point $\boldsymbol{\theta}^i$ a closest candidate point $\boldsymbol{b}^i$ that is on the opposite side of the ith coordinate hyperplane from $\boldsymbol{b}^*$ [cf. Fig. 3.10(a) and (b)]. We next show that the K points in (3.107) are distinct. To see this, we use proof by contradiction. Suppose that otherwise the set in (3.107) contains two identical candidate points, say $\boldsymbol{b}^1 = \boldsymbol{b}^2$. It then follows from the definitions in (3.105) and (3.107) that

$$\text{sign}\left(\theta_1 - \frac{g_1}{g_2}\theta_2\right) \triangleq b_1^2 = b_1^1 \triangleq -b_1^* \triangleq -\text{sign}(\theta_1), \tag{3.108}$$

$$\text{sign}\left(\theta_2 - \frac{g_2}{g_1}\theta_1\right) \triangleq b_2^1 = b_2^2 \triangleq -b_2^* \triangleq -\text{sign}(\theta_2). \tag{3.109}$$

Consider the case $\theta_1 > 0$ and $\theta_2 > 0$. By (3.108) and (3.109) we have

$$\theta_1 - \frac{g_1}{g_2}\theta_2 < 0 \Longrightarrow \frac{\theta_1}{\theta_2} < \frac{g_1}{g_2}, \tag{3.110}$$

$$\theta_2 - \frac{g_2}{g_1}\theta_1 < 0 \Longrightarrow \frac{\theta_1}{\theta_2} > \frac{g_1}{g_2}. \tag{3.111}$$

Hence, (3.110) and (3.111) contradict each other. The same contradiction arises for the other three choices of polarities for θ_1 and θ_2.

In general, the slowest-descent search method chooses the candidate set Ω in (3.104) as follows:

$$\Omega = \{\boldsymbol{b}^*\} \cup \bigcup_{q=1}^{Q} \left\{ \boldsymbol{b}^{q,\mu} \in \{-1, +1\}^K : b_k^{q,\mu} = \begin{cases} \text{sign}\left(\theta_k - \mu g_k^q\right) & \text{if } \theta_k - \mu g_k^q \neq 0, \\ -b_k^* & \text{if } \theta_k - \mu g_k^q = 0, \end{cases} \right.$$

$$\left. \boldsymbol{g}^q \text{ is the } q\text{th smallest eigenvector of } \nabla_\ell^2,\ \mu \in \left\{ \frac{\theta_1}{g_1^q}, \ldots, \frac{\theta_K}{g_K^q} \right\} \right\}, \tag{3.112}$$

where θ_k and g_k^q denote the kth elements of the respective vectors $\boldsymbol{\theta}$ and $\boldsymbol{g}^q$. Hence, $\{\boldsymbol{b}^{q,\mu}\}_\mu$ contains the K closest neighbors of $\boldsymbol{\theta}$ in $\{-1, +1\}^K$ along the direction of

$\boldsymbol{g}^q$. Note that $\{\boldsymbol{g}^q\}_{q=1}^Q$ represents the Q mutually orthogonal directions where the likelihood function $p(\boldsymbol{y}|\boldsymbol{b})$ drops most slowly from the peak point $\boldsymbol{\theta}$. Intuitively, the maximum-likelihood solution $\widehat{\boldsymbol{b}}_{ML}$ in (3.101) is probably in this neighborhood. The multiuser detection algorithm based on the slowest-descent-search method is summarized as follows (assuming that the signature waveforms $\boldsymbol{S}$ and the complex amplitudes $\boldsymbol{A}$ of all users are known).

Algorithm 3.3: [Slowest-descent-search multiuser detector]

- *Compute the Hessian matrix ∇_ℓ^2 given by (3.103) and its Q smallest eigenvectors $\boldsymbol{g}^1, \dots, \boldsymbol{g}^Q$.*
- *Compute the stationary point $\boldsymbol{\theta}$ given by (3.102).*
- *Solve the discrete optimization problem defined by (3.104) and (3.112) by an exhaustive search [over $(KQ+1)$ points].*

The first step involves calculating the eigenvectors of a $K \times K$ symmetric matrix, the second step involves inverting a $K \times K$ matrix, and the third step involves evaluating the likelihood values at $KQ + 1$ points. Note that the first two steps need to be performed only once if a block of M data bits need to be demodulated. Hence the dominant computational complexity of the algorithm above is $\mathcal{O}(KQ)$ per bit for K users.

Simulation Examples

For simulations, we assume a processing gain $N = 15$, the number of users $K = 8$, and equal amplitudes of user signals (i.e., $|A_k| = 1$, $k = 1, \dots, K$). The signature matrix $\boldsymbol{S}$ and the user phase offsets $\{\angle A_k\}_{k=1}^K$ are chosen at random and kept fixed throughout the simulations. Figure 3.11 demonstrates that the slowest-descent method with only one search direction ($Q = 1$) offers a significant performance gain over the linear decorrelator. Searching one more direction ($Q = 2$) results in some additional performance improvement. Further increase in the number of search directions only results in a diminishing improvement in performance.

3.4.2 Nonlinear Group-Blind Multiuser Detection

In group-blind multiuser detection, only the first $\tilde{K}$ users' signals need to be demodulated. As before, denote as $\tilde{\boldsymbol{S}}$ and $\bar{\boldsymbol{S}}$ matrices containing, respectively, the first $\tilde{K}$ and the last $K - \tilde{K}$ columns of $\boldsymbol{S}$. Similarly define the quantities $\tilde{\boldsymbol{A}}$, $\tilde{\boldsymbol{b}}$, $\bar{\boldsymbol{A}}$, and $\bar{\boldsymbol{b}}$. Then (3.2) can be rewritten as (again, we drop the symbol index i for convenience)

$$\boldsymbol{r} = \boldsymbol{S}\boldsymbol{A}\boldsymbol{b} + \boldsymbol{n} \tag{3.113}$$

$$= \tilde{\boldsymbol{S}}\tilde{\boldsymbol{A}}\tilde{\boldsymbol{b}} + \bar{\boldsymbol{S}}\bar{\boldsymbol{A}}\bar{\boldsymbol{b}} + \boldsymbol{n}. \tag{3.114}$$

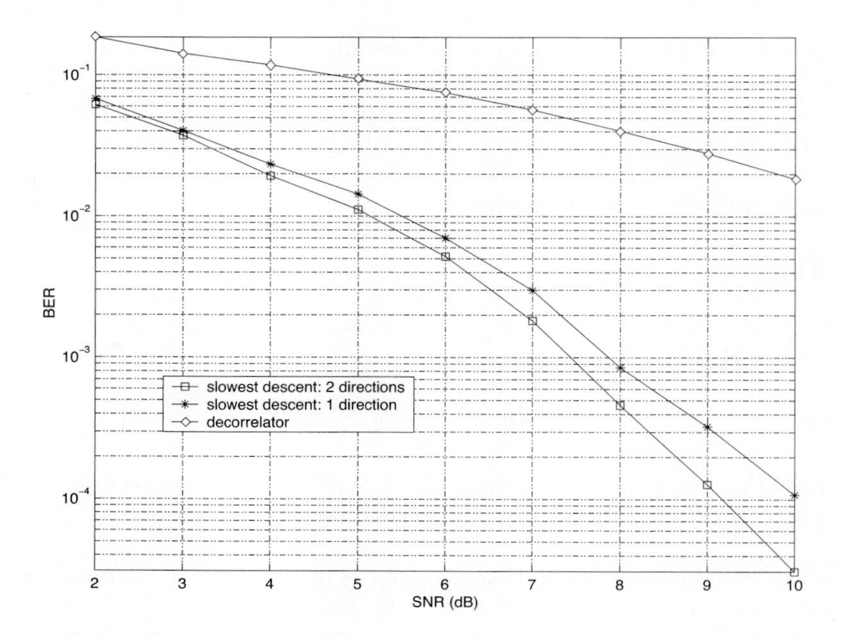

Figure 3.11. Performance of a slowest-descent-based multiuser detector in a synchronous CDMA system. $N = 15, K = 8$. The spreading waveforms $\boldsymbol{S}$ and complex amplitudes $\boldsymbol{A}$ of all users are assumed known to the receiver. The bit-error-rate (BER) curves of a linear decorrelator and slowest-descent detector with $Q = 1$ and $Q = 2$ are shown.

Let $\tilde{\boldsymbol{D}}$ denote the decorrelating detectors of the desired users, given by

$$\begin{aligned} \tilde{\boldsymbol{D}} &= [\boldsymbol{d}_1 \; \cdots \; \boldsymbol{d}_{\tilde{K}}] \\ &= \left[\boldsymbol{S} \left(\boldsymbol{S}^T \boldsymbol{S} \right)^{-1} \right]_{:,1:\tilde{K}}, \end{aligned} \tag{3.115}$$

where $[\boldsymbol{X}]_{:,n_1:n_2}$ denotes columns n_1 to n_2 of the matrix $\boldsymbol{X}$. It is easily seen that $\tilde{\boldsymbol{D}}$ satisfies

$$\tilde{\boldsymbol{D}}^T \bar{\boldsymbol{S}} = \boldsymbol{0} \quad \text{and} \quad \tilde{\boldsymbol{D}}^T \tilde{\boldsymbol{S}} = \boldsymbol{I}_{\tilde{K}}. \tag{3.116}$$

In group-blind multiuser detection, the undesired users' signals are first nulled out from the received signal by the following projection operation:

$$\boldsymbol{z} \triangleq \tilde{\boldsymbol{D}}^T \boldsymbol{r} \;=\; \tilde{\boldsymbol{A}} \tilde{\boldsymbol{b}} + \tilde{\boldsymbol{D}}^T \boldsymbol{n}, \tag{3.117}$$

where the second equality in (3.117) follows from (3.114) and (3.116). Denote

$$\boldsymbol{y} \triangleq \begin{bmatrix} \Re\{\boldsymbol{z}\} \\ \Im\{\boldsymbol{z}\} \end{bmatrix}, \boldsymbol{\Phi} \triangleq \begin{bmatrix} \Re\{\tilde{\boldsymbol{A}}\} \\ \Im\{\tilde{\boldsymbol{A}}\} \end{bmatrix}, \boldsymbol{v} \triangleq \begin{bmatrix} \tilde{\boldsymbol{D}}^T \Re\{\boldsymbol{n}\} \\ \tilde{\boldsymbol{D}}^T \Im\{\boldsymbol{n}\} \end{bmatrix}.$$

Then (3.117) can be written as

$$\boldsymbol{y} = \boldsymbol{\Phi}\tilde{\boldsymbol{b}} + \boldsymbol{v}. \tag{3.118}$$

Note that the covariance matrix of $\boldsymbol{v}$ is given by

$$\text{Cov}\{\boldsymbol{v}\} = \frac{\sigma^2}{2} \underbrace{\begin{bmatrix} \boldsymbol{Q} & \boldsymbol{0} \\ \boldsymbol{0} & \boldsymbol{Q} \end{bmatrix}}_{\boldsymbol{\Sigma}}, \tag{3.119}$$

with

$$\boldsymbol{Q} \triangleq \tilde{\boldsymbol{D}}^T \tilde{\boldsymbol{D}} = \left[\left(\boldsymbol{S}^T \boldsymbol{S} \right)^{-1} \right]_{1:\tilde{K},1:\tilde{K}}. \tag{3.120}$$

In what follows we consider nonlinear estimation of $\tilde{\boldsymbol{b}}$ from (3.118) based on the slowest-descent search. We also discuss the problem of estimating $\tilde{\boldsymbol{D}}$ and $\tilde{\boldsymbol{A}}$ from the received signals.

The maximum-likelihood estimate of $\tilde{\boldsymbol{b}}$ based on $\boldsymbol{y}$ in (3.118) is given by

$$\begin{aligned}
\widehat{\tilde{\boldsymbol{b}}}_{ML} &= \arg \max_{\tilde{\boldsymbol{b}} \in \{+1,-1\}^{\tilde{K}}} p(\boldsymbol{y} \mid \tilde{\boldsymbol{b}}) \\
&= \arg \min_{\tilde{\boldsymbol{b}} \in \{+1,-1\}^{\tilde{K}}} \underbrace{\left(\boldsymbol{y} - \boldsymbol{\Phi}\tilde{\boldsymbol{b}}\right)^T \boldsymbol{\Sigma}^{-1} \left(\boldsymbol{y} - \boldsymbol{\Phi}\tilde{\boldsymbol{b}}\right)}_{\tilde{\ell}(\tilde{\boldsymbol{b}})} \\
&= \arg \min_{\tilde{\boldsymbol{b}} \in \{+1,-1\}^{\tilde{K}}} \tilde{\ell}(\tilde{\boldsymbol{\theta}}) + \left(\tilde{\boldsymbol{b}} - \tilde{\boldsymbol{\theta}}\right)^T \nabla^2_{\tilde{\ell}} \left(\tilde{\boldsymbol{b}} - \tilde{\boldsymbol{\theta}}\right),
\end{aligned} \tag{3.121}$$

where the Hessian matrix is given by

$$\begin{aligned}
\nabla^2_{\tilde{\ell}} &= \boldsymbol{\Phi}^T \boldsymbol{\Sigma}^{-1} \boldsymbol{\Phi} \\
&= \Re\{\tilde{\boldsymbol{A}}\} \boldsymbol{Q}^{-1} \Re\{\tilde{\boldsymbol{A}}\} + \Im\{\tilde{\boldsymbol{A}}\} \boldsymbol{Q}^{-1} \Im\{\tilde{\boldsymbol{A}}\},
\end{aligned} \tag{3.122}$$

and $\tilde{\boldsymbol{\theta}}$ is the stationary point of $\tilde{\ell}(\tilde{\boldsymbol{b}})$:

$$\begin{aligned}
\nabla_{\tilde{\ell}}(\tilde{\boldsymbol{\theta}}) = \boldsymbol{0} \Longrightarrow \tilde{\boldsymbol{\theta}} &= \left(\boldsymbol{\Phi}^T \boldsymbol{\Sigma}^{-1} \boldsymbol{\Phi}\right)^{-1} \boldsymbol{\Phi}^T \boldsymbol{\Sigma}^{-1} \boldsymbol{y} \\
&= \left(\nabla^2_{\tilde{\ell}}\right)^{-1} \left[\Re\{\tilde{\boldsymbol{A}}\} \boldsymbol{Q}^{-1} \Re\{\boldsymbol{z}\} + \Im\{\tilde{\boldsymbol{A}}\} \boldsymbol{Q}^{-1} \Im\{\boldsymbol{z}\} \right].
\end{aligned} \tag{3.123}$$

As before, we approximate the solution to (3.121) by

$$\begin{aligned}
\widehat{\tilde{\boldsymbol{b}}} &\cong \arg \max_{\tilde{\boldsymbol{b}} \in \tilde{\Omega} \subset \{+1,-1\}^{\tilde{K}}} p(\boldsymbol{y} \mid \tilde{\boldsymbol{b}}) \\
&= \arg \min_{\tilde{\boldsymbol{b}} \in \tilde{\Omega}} \left(\tilde{\boldsymbol{b}} - \tilde{\boldsymbol{\theta}}\right)^T \nabla^2_{\tilde{\ell}} \left(\tilde{\boldsymbol{b}} - \tilde{\boldsymbol{\theta}}\right),
\end{aligned} \tag{3.124}$$

where $\tilde{\Omega}$ contains $\tilde{K}Q+1$ closest neighbors of $\tilde{\boldsymbol{\theta}}$ in $\{-1,+1\}^{\tilde{K}}$ along the slowest-descent directions of the likelihood function $p(\boldsymbol{y}|\tilde{\boldsymbol{b}})$, given by

$$\tilde{\boldsymbol{b}}^* \triangleq \text{sign}(\tilde{\boldsymbol{\theta}}), \tag{3.125}$$

$$\tilde{\Omega} = \{\tilde{\boldsymbol{b}}^*\} \cup \bigcup_{q=1}^{Q} \left\{ \tilde{\boldsymbol{b}}^{q,\mu} \in \{+1,-1\}^{\tilde{K}} : \tilde{b}_k^{q,\mu} = \begin{cases} \text{sign}\left(\tilde{\theta}_k - \mu g_k^q\right) & \text{if } \tilde{\theta}_k - \mu g_k^q \neq 0, \\ -\tilde{b}_k^* & \text{if } \tilde{\theta}_k - \mu g_k^q = 0, \end{cases} \right.$$

$$\left. \boldsymbol{g}^q \text{ is the } q\text{th smallest eigenvector of } \nabla^2_{\tilde{\ell}}\ , \mu \in \left\{ \frac{\tilde{\theta}_1}{g_1^q}, \ldots, \frac{\tilde{\theta}_{\tilde{K}}}{g_{\tilde{K}}^q} \right\} \right\}. \tag{3.126}$$

Estimation of $\tilde{D}$ and $\tilde{A}$

To implement the local-search-based group-blind multiuser detection algorithm above, we must first estimate the decorrelator matrix $\tilde{\boldsymbol{D}}$ and the complex amplitudes $\tilde{\boldsymbol{A}}$. Note that the decorrelating detectors $\tilde{\boldsymbol{D}}$ for the desired users are simply the group-blind linear decorrelating detectors discussed in Section 3.2. For example, based on the eigendecomposition (3.37) of the autocorrelation matrix of the received signal, $\tilde{\boldsymbol{D}}$ is given in terms of the signal subspace parameters by (3.38).

We next consider estimation of the complex amplitudes $\tilde{\boldsymbol{A}}$ of the desired users. Consider the decorrelator output (3.117); we now have [recall that $\tilde{\boldsymbol{A}} \triangleq (A_1, \ldots, A_{\tilde{K}})$]

$$z_k = A_k\, b_k + \tilde{n}_k, \qquad k = 1, \ldots, \tilde{K}, \tag{3.127}$$

where $\tilde{n}_k \triangleq \left[\tilde{\boldsymbol{D}}^T \boldsymbol{n}\right]_k$. Since $b_k \in \{+1,-1\}$, it follows from (3.127) that the decorrelator outputs corresponding to the kth user form two clusters centered, respectively, at A_k and $-A_k$. Let $A_k = \rho_k e^{\jmath\phi_k}$; a simple estimator of A_k is given by $\hat{A}_k = \hat{\rho}_k e^{\jmath\hat{\phi}_k}$ with

$$\hat{\rho}_k = \hat{E}\{|z_k|\}, \tag{3.128}$$

$$\hat{\phi}_k = \begin{cases} \hat{E}\{\angle\left[z_k \text{sign}\left(\Re\{z_k\}\right)\right]\} & \text{if } \hat{E}\{|\Re\{z_k\}|\} > \hat{E}\{|\Im\{z_k\}|\}, \\ \hat{E}\{\angle\left[z_k \text{sign}\left(\Im\{z_k\}\right)\right]\} & \text{if } \hat{E}\{|\Re\{z_k\}|\} < \hat{E}\{|\Im\{z_k\}|\}, \end{cases} \tag{3.129}$$

where $\hat{E}\{\cdot\}$ denotes the sample average operation. Note that the foregoing estimate of the phase ϕ_k has an ambiguity of π, which necessitates differential encoding and decoding of data. Finally, we summarize the nonlinear group-blind multiuser detection algorithm for synchronous CDMA as follows.

Algorithm 3.4: [Nonlinear group-blind multiuser detector—synchronous CDMA]

- *Compute the signal subspace:*

$$\hat{\boldsymbol{C}}_r = \frac{1}{M} \sum_{i=0}^{M-1} \boldsymbol{r}[i]\boldsymbol{r}[i]^H \tag{3.130}$$

$$= \hat{\boldsymbol{U}}_s \hat{\boldsymbol{\Lambda}}_s \hat{\boldsymbol{U}}_s^H + \hat{\boldsymbol{U}}_n \hat{\boldsymbol{\Lambda}}_n \hat{\boldsymbol{U}}_n^H. \tag{3.131}$$

- *Form the linear group-blind decorrelating detectors:*

$$\hat{\tilde{\boldsymbol{D}}} = \hat{\boldsymbol{U}}_s \left(\hat{\boldsymbol{\Lambda}}_s - \hat{\sigma^2} \boldsymbol{I}_{\tilde{K}} \right)^{-1} \hat{\boldsymbol{U}}_s^H \tilde{\boldsymbol{S}} \left[\tilde{\boldsymbol{S}}^T \hat{\boldsymbol{U}}_s \left(\hat{\boldsymbol{\Lambda}}_s - \hat{\sigma^2} \boldsymbol{I}_{\tilde{K}} \right)^{-1} \hat{\boldsymbol{U}}_s^H \tilde{\boldsymbol{S}} \right]^{-1}, \quad (3.132)$$

$$\hat{\boldsymbol{Q}} = \Re\left\{\tilde{\boldsymbol{D}}\right\}^T \Re\left\{\tilde{\boldsymbol{D}}\right\} + \Im\left\{\tilde{\boldsymbol{D}}\right\}^T \Im\left\{\tilde{\boldsymbol{D}}\right\}, \quad (3.133)$$

where $\hat{\sigma^2}$ *is given by the mean of the* $N-K$ *eigenvalues in* $\hat{\boldsymbol{\Lambda}}_n$.

- *Estimate the complex amplitudes* $\tilde{\boldsymbol{A}}$:

$$\boldsymbol{z}[i] = \hat{\tilde{\boldsymbol{D}}}^H \boldsymbol{r}[i], \qquad i = 0, \ldots, M-1, \quad (3.134)$$

$$\hat{\rho}_k = \frac{1}{M} \sum_{i=0}^{M-1} |z_k[i]|, \quad (3.135)$$

$$R_k = \frac{1}{M} \sum_{i=0}^{M-1} \Re\{z_k[i]\}, \quad (3.136)$$

$$I_k = \frac{1}{M} \sum_{i=0}^{M-1} \Im\{z_k[i]\}, \quad (3.137)$$

$$\hat{\phi}_k = \begin{cases} \frac{1}{M} \sum_{i=0}^{M-1} \angle \left[z_k[i] \text{sign} \left(\Re\left\{z_k[i]\right\} \right) \right] & \textit{if } R_k \geq I_k, \\ \frac{1}{M} \sum_{i=0}^{M-1} \angle \left[z_k[i] \text{sign} \left(\Im\left\{z_k[i]\right\} \right) \right] & \textit{if } R_k < I_k, \end{cases} \quad (3.138)$$

$$\hat{A}_k = \hat{\rho}_k e^{\jmath \hat{\phi}_k}, \qquad k = 1, \ldots, \tilde{K}. \quad (3.139)$$

Let

$$\hat{\tilde{\boldsymbol{A}}} \triangleq \mathsf{diag}(\hat{A}_1, \ldots, \hat{A}_{\tilde{K}}). \quad (3.140)$$

- *Compute the Hessian*

$$\hat{\nabla}^2 = \Re\{\hat{\tilde{\boldsymbol{A}}}\} \hat{\boldsymbol{Q}}^{-1} \Re\{\hat{\tilde{\boldsymbol{A}}}\} + \Im\{\hat{\tilde{\boldsymbol{A}}}\} \hat{\boldsymbol{Q}}^{-1} \Im\{\hat{\tilde{\boldsymbol{A}}}\} \quad (3.141)$$

and the Q *smallest eigenvectors* $\boldsymbol{g}^1, \ldots, \boldsymbol{g}^Q$ *of* $\hat{\nabla}^2$.

- *Detect each symbol by solving the following discrete optimization problem using an exhaustive search (over* $\tilde{K}Q+1$ *points):*

$$\boldsymbol{\theta}[i] = \left(\hat{\nabla}^2\right)^{-1} \left[\Re\{\hat{\tilde{\boldsymbol{A}}}\} \hat{\boldsymbol{Q}}^{-1} \Re\{\boldsymbol{z}[i]\} + \Im\{\hat{\tilde{\boldsymbol{A}}}\} \hat{\boldsymbol{Q}}^{-1} \Im\{\boldsymbol{z}[i]\} \right], \quad (3.142)$$

$$\hat{\tilde{\boldsymbol{b}}}^*[i] = \text{sign}(\boldsymbol{\theta}[i]), \quad (3.143)$$

$$\widehat{\tilde{\boldsymbol{b}}}[i] = \arg\min_{\tilde{\boldsymbol{b}} \in \tilde{\Omega}[i]} \left(\tilde{\boldsymbol{b}} - \boldsymbol{\theta}[i]\right)^T \hat{\nabla}^2 \left(\tilde{\boldsymbol{b}} - \boldsymbol{\theta}[i]\right), \tag{3.144}$$

$$\tilde{\Omega}[i] = \{\hat{\tilde{\boldsymbol{b}}}^*[i]\} \cup \bigcup_{q=1}^{Q} \left\{\tilde{\boldsymbol{b}}^{q,\mu} \in \{-1,+1\}^{\tilde{K}} : \tilde{b}_k^{q,\mu}\right.$$

$$= \begin{cases} \text{sign}\,(\theta_k[i] - \mu g_k^q) & \textit{if } \theta_k[i] - \mu g_k^q \neq 0, \\ -\hat{\tilde{b}}_k^*[i] & \textit{if } \theta_k[i] - \mu g_k^q = 0, \end{cases}$$

$$\left.\mu \in \left\{\frac{\theta_1[i]}{g_1^q}, \ldots, \frac{\theta_{\tilde{K}}[i]}{g_{\tilde{K}}^q}\right\}\right\}, \qquad i = 0, \ldots, M-1. \tag{3.145}$$

- *Perform differential decoding:*

$$\hat{\beta}_k[i] = \hat{b}_k[i]\hat{b}_k[i-1], \qquad k = 1, \ldots, \tilde{K};\ i = 1, \ldots, M-1. \tag{3.146}$$

It is seen that compared with linear group-blind detectors discussed in Section 3.2, the additional computation incurred by the nonlinear detector involves a complex amplitude estimation step for all desired users and a likelihood search step, both of which have computational complexities linear in $\tilde{K}$.

Simulation Examples

In this simulation we assume a processing gain $N = 15$, number of users $K = 8$, number of desired users $\tilde{K} = 4$, and equal amplitudes of user signals. The signature matrix $\boldsymbol{S}$ and the user phase offsets are chosen at random and kept fixed throughout simulations. Figure 3.12 demonstrates the considerable performance gain offered by the nonlinear group-blind multiuser detector developed above over the group-blind linear detector. Again, it is seen that it suffices for the nonlinear group-blind detector to search along only one (i.e., the slowest-descent) direction ($Q = 1$).

3.5 Group-Blind Multiuser Detection in Multipath Channels

In this section we extend the linear and nonlinear group-blind multiuser detection methods developed in previous sections to general asynchronous CDMA systems with multipath channel distortion. The signal model for multipath CDMA systems is developed in Section 2.7.1. At the receiver, the received signal is filtered by a chip-matched filter and sampled at a multiple (p) of the chip rate. Denote $r_q[i]$ as the qth signal sample during the ith symbol [cf. (2.167)]. Recall that by denoting

$$\underbrace{\underline{r}[i]}_{P\times 1} \triangleq \begin{bmatrix} r_0[i] \\ \vdots \\ r_{P-1}[i] \end{bmatrix}, \quad \underbrace{\underline{b}[i]}_{K\times 1} \triangleq \begin{bmatrix} b_1[i] \\ \vdots \\ b_K[i] \end{bmatrix}, \quad \underbrace{\underline{n}[i]}_{P\times 1} \triangleq \begin{bmatrix} n_0[i] \\ \vdots \\ n_{P-1}[i] \end{bmatrix},$$

$$\underbrace{\underline{H}[j]}_{P\times K} \triangleq \begin{bmatrix} h_1[jP] & \ldots & h_K[jP] \\ \vdots & \vdots & \vdots \\ h_1[jP+P-1] & \ldots & h_K[jP+P-1] \end{bmatrix}, \quad j = 0, \ldots, \iota,$$

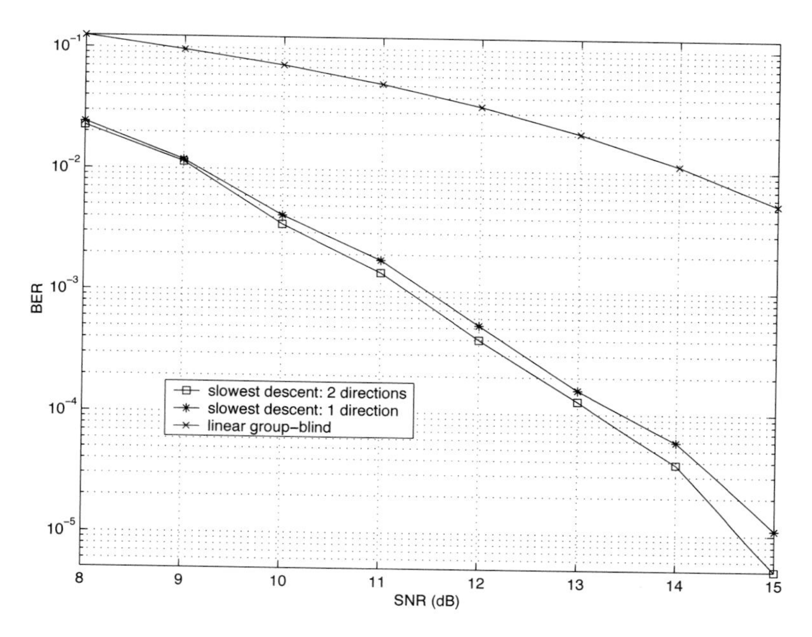

Figure 3.12. Performance of a slowest-descent-based group-blind multiuser detector in a synchronous CDMA system. $N = 15, K = 8, \tilde{K} = 4$. Only the spreading waveforms $\tilde{\boldsymbol{S}}$ of the desired users are assumed known to the receiver. The BER curves of a linear group-blind detector and slowest-descent (nonlinear) group-blind detector with $Q = 1$ and $Q = 2$ are shown.

we have the following discrete-time signal model:

$$\underline{r}[i] = \underline{H}[i] \star \underline{b}[i] + \underline{n}[i]. \tag{3.147}$$

By stacking m successive sample vectors, we further define the following quantities:

$$\underbrace{\boldsymbol{r}[i]}_{Pm\times 1} \triangleq \begin{bmatrix} \underline{r}[i] \\ \vdots \\ \underline{r}[i+m-1] \end{bmatrix}, \quad \underbrace{\boldsymbol{n}[i]}_{Pm\times 1} \triangleq \begin{bmatrix} \underline{n}[i] \\ \vdots \\ \underline{n}[i+m-1] \end{bmatrix},$$

$$\underbrace{\boldsymbol{b}[i]}_{K(m+\iota)\times 1} \triangleq \begin{bmatrix} \underline{b}[i-\iota] \\ \vdots \\ \underline{b}[i+m-1] \end{bmatrix}, \quad \underbrace{\boldsymbol{H}}_{Pm\times K(m+\iota)} \triangleq \begin{bmatrix} \underline{H}[\iota] & \cdots & \underline{H}[0] & \cdots & \mathbf{0} \\ \vdots & \ddots & \ddots & \ddots & \vdots \\ \mathbf{0} & \cdots & \underline{H}[\iota] & \cdots & \underline{H}[0] \end{bmatrix},$$

where the smoothing factor m is chosen according to

$$m = \left\lceil \frac{P+K}{P-K} \right\rceil \iota,$$

such that the matrix $\boldsymbol{H}$ is a "tall" matrix [i.e., $Pm \geq K(m+\iota)$]. We can then write (3.147) in matrix form as

$$\boldsymbol{r}[i] = \boldsymbol{H}\,\boldsymbol{b}[i] + \boldsymbol{n}[i]. \tag{3.148}$$

Assume that $\boldsymbol{H}$ has full column rank; that is,

$$r \triangleq K(m+\iota).$$

The autocorrelation matrix of the signal $\boldsymbol{r}[i]$ and its eigendecomposition are given, respectively, by

$$\boldsymbol{C}_r \triangleq E\left\{\boldsymbol{r}[i]\boldsymbol{r}[i]^H\right\} = \boldsymbol{H}\boldsymbol{H}^H + \sigma^2 \boldsymbol{I}_{Pm} \tag{3.149}$$

$$= \boldsymbol{U}_s \boldsymbol{\Lambda}_s \boldsymbol{U}_s + \sigma^2 \boldsymbol{U}_n \boldsymbol{U}_n^H, \tag{3.150}$$

where $\boldsymbol{\Lambda}_s = \mathsf{diag}(\lambda_1, \ldots, \lambda_r)$ contains the r largest eigenvalues of $\boldsymbol{C}_r$.

In what follows it is assumed that the receiver has knowledge of the first $\tilde{K}$ ($\tilde{K} \leq K$) users' signature waveforms, $\tilde{\boldsymbol{S}}$, whereas the signature waveforms of the remaining $(K - \tilde{K})$ users are unknown to the receiver. Denote by $\underline{\tilde{H}}[m]$ and $\tilde{\boldsymbol{H}}$ the submatrics of $\underline{H}[m]$ and $\boldsymbol{H}$, respectively, corresponding to the desired users:

$$\underbrace{\underline{\tilde{H}}[m]}_{P\times\tilde{K}} \triangleq \begin{bmatrix} h_1[mP] & \cdots & h_{\tilde{K}}[mP] \\ \vdots & \vdots & \vdots \\ h_1[mP+P-1] & \cdots & h_{\tilde{K}}[mP+P-1] \end{bmatrix}, \quad m = 0, \ldots, \iota,$$

$$\underbrace{\tilde{\boldsymbol{H}}}_{Pm\times\tilde{K}(m+\iota)} \triangleq \begin{bmatrix} \underline{\tilde{H}}[\iota] & \cdots & \underline{\tilde{H}}[0] & \cdots & \boldsymbol{0} \\ \vdots & \ddots & \ddots & \ddots & \vdots \\ \boldsymbol{0} & \cdots & \underline{\tilde{H}}[\iota] & \cdots & \underline{\tilde{H}}[0] \end{bmatrix}.$$

It is assumed that $\tilde{\boldsymbol{H}}$ has full column rank:

$$\tilde{r} \triangleq \tilde{K}(m+\iota).$$

As in the synchronous case, the following projection matrix is needed in the definition of the form I group-blind linear MMSE detector:

$$\bar{\boldsymbol{P}} \triangleq \boldsymbol{I}_{Pm} - \tilde{\boldsymbol{H}}\left(\tilde{\boldsymbol{H}}^H \tilde{\boldsymbol{H}}\right)^{-1} \tilde{\boldsymbol{H}}^H. \tag{3.151}$$

Note that $\bar{\boldsymbol{P}}$ projects any signal onto the subspace $\mathsf{null}(\tilde{\boldsymbol{H}}^H)$. It is then easily seen that the matrix $\bar{\boldsymbol{P}}\boldsymbol{C}_r\bar{\boldsymbol{P}}$ has an eigenstructure of the form

$$\bar{\boldsymbol{P}}\boldsymbol{C}_r\bar{\boldsymbol{P}} = \left[\bar{\boldsymbol{U}}_s\, \bar{\boldsymbol{U}}_n\, \bar{\boldsymbol{U}}_o\right] \begin{bmatrix} \bar{\boldsymbol{\Lambda}}_s & \boldsymbol{0} & \boldsymbol{0} \\ \boldsymbol{0} & \sigma^2 \boldsymbol{I}_{Pm-r} & \boldsymbol{0} \\ \boldsymbol{0} & \boldsymbol{0} & \boldsymbol{0} \end{bmatrix} \begin{bmatrix} \bar{\boldsymbol{U}}_s^H \\ \bar{\boldsymbol{U}}_n^H \\ \bar{\boldsymbol{U}}_o^H \end{bmatrix}, \tag{3.152}$$

where $\bar{\boldsymbol{\Lambda}}_s = \mathsf{diag}(\bar{\lambda}_1, \ldots, \bar{\lambda}_{r-\tilde{r}})$, with $\bar{\lambda}_i > \sigma^2$, $i = 1, \ldots, r - \tilde{r}$, and the columns of $\bar{\boldsymbol{U}}_s$ form an orthogonal basis of the subspace $\mathsf{range}(\boldsymbol{H}) \bigcap \mathsf{null}(\tilde{\boldsymbol{H}}^H)$.

3.5.1 Linear Group-Blind Detectors

As before, the basic idea behind group-blind linear detectors is to suppress the interference from known users based on the spreading sequences of these users and to suppress the interference from other, unknown users using subspace-based blind methods. Analogous to the synchronous case, we have the following three types of linear group-blind detectors. (In this section, $\boldsymbol{e}_k$ denotes an $\tilde{r}$-vector with all elements zeros except for the kth, which is 1.)

Definition 3.4: [Group-blind linear decorrelating detector—multipath CDMA] *The weight vector of the group-blind linear decorrelating detector for user k is given by the solution to the following constrained optimization problem:*

$$\boldsymbol{d}_k = \arg \min_{\boldsymbol{w} \in \text{range}(\boldsymbol{H})} \left\| \boldsymbol{d}^H \boldsymbol{H} \right\|^2 \quad s.t. \; \boldsymbol{w}^H \tilde{\boldsymbol{H}} = \boldsymbol{e}^T_{\tilde{K}\iota+k}, \qquad k = 1, \ldots, \tilde{K}. \tag{3.153}$$

Definition 3.5: [Group-blind linear hybrid detector—multipath CDMA] *The weight vector of the group-blind linear hybrid detector for user k is given by the solution to the following constrained optimization problem:*

$$\boldsymbol{w}_k = \arg \min_{\boldsymbol{w} \in \text{range}(\boldsymbol{H})} E\left\{ \left| b_k[i] - \boldsymbol{w}^H \boldsymbol{r}[i] \right|^2 \right\} \; s.t. \; \boldsymbol{w}^H \tilde{\boldsymbol{H}} = \boldsymbol{e}^T_{\tilde{K}\iota+k},$$
$$k = 1, \ldots, \tilde{K}. \tag{3.154}$$

Definition 3.6: [Group-blind linear MMSE detector—multipath CDMA] *Let $\tilde{\boldsymbol{r}}[i] = \tilde{\boldsymbol{H}}\tilde{\boldsymbol{b}}[i] + \boldsymbol{n}[i]$ be the components of the received signal $\boldsymbol{r}[i]$ in (3.148) consisting of signals from known users plus noise. ($\tilde{\boldsymbol{b}}[i]$ is the subvector of $\boldsymbol{b}[i]$ containing bits of the desired users.) The weight vector of the group-blind linear MMSE detector for user k is given by $\boldsymbol{m}_k = \tilde{\boldsymbol{m}}_k + \bar{\boldsymbol{m}}_k$, where $\tilde{\boldsymbol{m}}_k \in \text{range}(\tilde{\boldsymbol{H}})$ and $\bar{\boldsymbol{m}}_k \in \text{range}(\bar{\boldsymbol{U}}_s)$ [note that $\bar{\boldsymbol{U}}_s$ is given in (3.152)], such that*

$$\tilde{\boldsymbol{m}}_k = \arg \min_{\boldsymbol{w} \in \text{range}(\tilde{\boldsymbol{H}})} E\left\{ \left| b_k[i] - \boldsymbol{w}^H \tilde{\boldsymbol{r}}[i] \right|^2 \right\}, \tag{3.155}$$

$$\bar{\boldsymbol{m}}_k = \arg \min_{\boldsymbol{w} \in \text{range}(\bar{\boldsymbol{U}}_s)} E\left\{ \left| b_k[i] - (\boldsymbol{w} + \tilde{\boldsymbol{m}}_k)^H \boldsymbol{r}[i] \right|^2 \right\}, \qquad k = 1, \ldots, \tilde{K}. \tag{3.156}$$

The following results give expressions for the three group-blind linear detectors defined above in terms of the known users' channel matrix $\tilde{\boldsymbol{H}}$ and the unknown users' signal subspace components $\bar{\boldsymbol{\Lambda}}_s$ and $\bar{\boldsymbol{U}}_s$ defined in (3.152). The proofs of these results are similar to those corresponding to the synchronous case.

Proposition 3.7: [Group-blind linear decorrelating detector (form I)—multipath CDMA] *The weight vector of the group-blind linear decorrelating detector for the kth user is given by*

$$\boldsymbol{d}_k = \left[\boldsymbol{I}_{Pm} - \bar{\boldsymbol{U}}_s \left(\bar{\boldsymbol{\Lambda}}_s - \sigma^2 \boldsymbol{I}_{r-\tilde{r}} \right)^{-1} \bar{\boldsymbol{U}}_s^H \boldsymbol{C}_r \right] \tilde{\boldsymbol{H}} \left(\tilde{\boldsymbol{H}}^H \tilde{\boldsymbol{H}} \right)^{-1} \boldsymbol{e}_{\tilde{K}\iota+k},$$
$$k = 1, \ldots, \tilde{K}. \tag{3.157}$$

Proposition 3.8: [Group-blind linear hybrid detector (form I)—multipath CDMA] *The weight vector of the group-blind linear hybrid detector for the kth user is given by*

$$\boldsymbol{w}_k = \left(\boldsymbol{I}_{Pm} - \bar{\boldsymbol{U}}_s \bar{\boldsymbol{\Lambda}}_s^{-1} \bar{\boldsymbol{U}}_s^H \boldsymbol{C}_r\right) \tilde{\boldsymbol{H}} \left(\tilde{\boldsymbol{H}}^H \tilde{\boldsymbol{H}}\right)^{-1} \boldsymbol{e}_{\tilde{K}\iota+k}, \qquad k = 1, \ldots, \tilde{K}. \quad (3.158)$$

Proposition 3.9: [Group-blind linear MMSE detector (form I)—multipath CDMA] *The weight vector of the group-blind linear MMSE detector for the kth user is given by*

$$\boldsymbol{m}_k = \left(\boldsymbol{I}_{Pm} - \bar{\boldsymbol{U}}_s \bar{\boldsymbol{\Lambda}}_s^{-1} \bar{\boldsymbol{U}}_s^H \boldsymbol{C}_r\right) \tilde{\boldsymbol{H}} \left(\tilde{\boldsymbol{H}}^H \tilde{\boldsymbol{H}} + \sigma^2 \boldsymbol{I}_{\tilde{r}}\right)^{-1} \boldsymbol{e}_{\tilde{K}\iota+k},$$
$$k = 1, \ldots, \tilde{K}. \quad (3.159)$$

Note that to implement these group-blind linear detectors, the matrix $\tilde{\boldsymbol{H}}$ must be estimated first. The blind channel estimation procedure is discussed in Section 2.7.3. The channel estimator discussed there can be used to estimate the channel for each desired user. Once the desired users' channels are estimated, the matrix $\tilde{\boldsymbol{H}}$ can be formed. As before, the blind channel estimator has an arbitrary phase ambiguity, which necessitates the use of differential encoding and decoding of the data bits. We next summarize the group-blind linear hybrid multiuser detection algorithm in multipath channels.

Algorithm 3.5: [Group-blind linear hybrid detector (form I)—multipath CDMA]

- *Compute the signal subspace:*

$$\hat{\boldsymbol{C}}_r = \frac{1}{M} \sum_{i=0}^{M-1} \boldsymbol{r}[i] \boldsymbol{r}[i]^H \quad (3.160)$$

$$= \hat{\boldsymbol{U}}_s \hat{\boldsymbol{\Lambda}}_s \hat{\boldsymbol{U}}_s^H + \hat{\boldsymbol{U}}_n \hat{\boldsymbol{\Lambda}}_n \hat{\boldsymbol{U}}_s^H. \quad (3.161)$$

- *Estimate the desired users' channels (cf. Section 2.7.3):*

$$\hat{\boldsymbol{f}}_k = \textit{min-eigenvector}\left(\boldsymbol{\Xi}_k^H \hat{\boldsymbol{U}}_n \hat{\boldsymbol{U}}_n^H \boldsymbol{\Xi}_k\right), \quad (3.162)$$

$$\hat{\boldsymbol{h}}_k = \boldsymbol{\Xi}_k \hat{\boldsymbol{f}}_k, \qquad k = 1, \ldots, \tilde{K}. \quad (3.163)$$

Form $\hat{\tilde{\boldsymbol{H}}}$ *using* $\hat{\boldsymbol{h}}_1, \ldots, \hat{\boldsymbol{h}}_{\tilde{K}}$.

- *Compute the unknown users' subspace:*

$$\hat{\tilde{\boldsymbol{P}}} = \boldsymbol{I}_{Pm} - \hat{\tilde{\boldsymbol{H}}} \left(\hat{\tilde{\boldsymbol{H}}}^H \hat{\tilde{\boldsymbol{H}}}\right)^{-1} \hat{\tilde{\boldsymbol{H}}}^H, \quad (3.164)$$

$$\hat{\tilde{\boldsymbol{P}}} \hat{\boldsymbol{C}}_r \hat{\tilde{\boldsymbol{P}}} = \hat{\bar{\boldsymbol{U}}}_s \hat{\bar{\boldsymbol{\Lambda}}}_s \hat{\bar{\boldsymbol{U}}}_s^H + \hat{\bar{\boldsymbol{U}}}_n \hat{\bar{\boldsymbol{\Lambda}}}_n \hat{\bar{\boldsymbol{U}}}_n^H + \hat{\bar{\boldsymbol{U}}}_o \hat{\bar{\boldsymbol{\Lambda}}}_o \hat{\bar{\boldsymbol{U}}}_o^H. \quad (3.165)$$

- *Form the detectors:*

$$\hat{\boldsymbol{w}}_k = \left(\boldsymbol{I}_{Pm} - \hat{\boldsymbol{U}}_s \hat{\boldsymbol{\Lambda}}_s^{-1} \hat{\boldsymbol{U}}_s^H \hat{\boldsymbol{C}}_r\right) \hat{\tilde{\boldsymbol{H}}} \left(\hat{\tilde{\boldsymbol{H}}}^H \hat{\tilde{\boldsymbol{H}}}\right)^{-1} \boldsymbol{e}_{\tilde{K}\iota+k}, \quad k = 1, \ldots, \tilde{K}. \tag{3.166}$$

- *Perform differential detection:*

$$z_k[i] = \hat{\boldsymbol{w}}_k^H \boldsymbol{r}[i], \tag{3.167}$$

$$\hat{\beta}_k[i] = \text{sign}\left\{\Re\left(z_k[i] z_k[i-1]^*\right)\right\}, \quad i = 1, \ldots, M-1; \; k = 1, \ldots, \tilde{K}. \tag{3.168}$$

Note that the group-blind linear decorrelating detector and the group-blind linear MMSE detector can be implemented similarly. Both require an estimate of σ^2, which can be obtained simply as the mean of the noise subspace eigenvalues $\hat{\boldsymbol{\Lambda}}_n$.

Alternatively, the group-blind linear detectors can be expressed in terms of the signal subspace components $\boldsymbol{\Lambda}_s$ and $\boldsymbol{U}_s$ of all users defined in (3.150), as given by the following three results. The proofs are again similar to their counterparts in the synchronous case.

Proposition 3.10: [Group-blind linear decorrelating detector (form II)—multipath CDMA] *The group-blind linear decorrelating detector for the kth user is given by*

$$\boldsymbol{d}_k = \boldsymbol{U}_s \left(\boldsymbol{\Lambda}_s - \sigma^2 \boldsymbol{I}_K\right)^{-1} \boldsymbol{U}_s^H \tilde{\boldsymbol{H}} \left[\tilde{\boldsymbol{H}}^H \boldsymbol{U}_s \left(\boldsymbol{\Lambda}_s - \sigma^2 \boldsymbol{I}_K\right) \boldsymbol{U}_s \tilde{\boldsymbol{H}}\right]^{-1} \boldsymbol{e}_{\tilde{K}\iota+k}, \quad k = 1, \ldots, \tilde{K}. \tag{3.169}$$

Proposition 3.11: [Group-blind linear hybrid detector (form II)—multipath CDMA] *The group-blind linear hybrid detector for the kth user is given by*

$$\boldsymbol{w}_k = \boldsymbol{U}_s \boldsymbol{\Lambda}_s^{-1} \boldsymbol{U}_s^H \tilde{\boldsymbol{H}} \left(\tilde{\boldsymbol{H}}^H \boldsymbol{U}_s \boldsymbol{\Lambda}_s^{-1} \boldsymbol{U}_s^H \tilde{\boldsymbol{H}}\right)^{-1} \boldsymbol{e}_{\tilde{K}\iota+k}, \qquad k = 1, \ldots, \tilde{K}. \tag{3.170}$$

Proposition 3.12: [Group-blind linear MMSE detector (form II)—multipath CDMA] *Let the (rank-deficient) QR factorization of the $Pm \times r$ matrix $\bar{\boldsymbol{P}}\boldsymbol{U}_s$ be*

$$\bar{\boldsymbol{P}}\boldsymbol{U}_s = \left[\begin{array}{cc} \boldsymbol{Q}_s & \boldsymbol{Q}_o \end{array}\right] \left[\begin{array}{cc} \boldsymbol{R}_s & \boldsymbol{R}_o \\ \boldsymbol{0} & \boldsymbol{0} \end{array}\right] \boldsymbol{\Pi}, \tag{3.171}$$

where $\boldsymbol{Q}_s$ is a $Pm \times \tilde{r}$ matrix, $\boldsymbol{R}_s$ is a $\tilde{r} \times \tilde{r}$ nonsingular upper triangular matrix, and $\boldsymbol{\Pi}$ is a permutation matrix. The group-blind linear MMSE detector for the kth user is given by

$$\boldsymbol{m}_k = \left[\boldsymbol{I}_{Pm} - \left(\boldsymbol{Q}_s \boldsymbol{R}_s^{-H}\right) \left(\boldsymbol{\Pi} \boldsymbol{\Lambda}_s \boldsymbol{\Pi}^T\right)^{-1} \left(\boldsymbol{Q}_s \boldsymbol{R}_s^{-H}\right)^H \boldsymbol{C}_r\right] \tilde{\boldsymbol{H}} \left(\tilde{\boldsymbol{H}}^H \tilde{\boldsymbol{H}} + \sigma^2 \boldsymbol{I}_{\tilde{r}}\right)^{-1} \boldsymbol{e}_{\tilde{K}\iota+k}, \qquad k = 1, \ldots, \tilde{K}. \tag{3.172}$$

Finally, we summarize the form II group-blind linear hybrid multiuser detection algorithm in multipath channels as follows.

Algorithm 3.6: [Group-blind linear hybrid detector (form II)—multipath CDMA]

- *Compute the signal subspace:*

$$\hat{\boldsymbol{C}}_r = \frac{1}{M}\sum_{i=0}^{M-1} \boldsymbol{r}[i]\boldsymbol{r}[i]^H \tag{3.173}$$

$$= \hat{\boldsymbol{U}}_s\hat{\boldsymbol{\Lambda}}_s\hat{\boldsymbol{U}}_s^H + \hat{\boldsymbol{U}}_n\hat{\boldsymbol{\Lambda}}_n\hat{\boldsymbol{U}}_s^H. \tag{3.174}$$

- *Estimate the desired users' channels (cf. Section 2.7.3):*

$$\hat{\boldsymbol{f}}_k = \textit{min-eigenvector}\left(\boldsymbol{\Xi}_k^H\hat{\boldsymbol{U}}_n\hat{\boldsymbol{U}}_n^H\boldsymbol{\Xi}_k\right), \tag{3.175}$$

$$\hat{\boldsymbol{h}}_k = \boldsymbol{\Xi}_k\hat{\boldsymbol{f}}_k, \qquad k = 1,\ldots,\tilde{K}. \tag{3.176}$$

Form $\hat{\tilde{\boldsymbol{H}}}$ *using* $\hat{\boldsymbol{h}}_1,\ldots,\hat{\boldsymbol{h}}_{\tilde{K}}$.

- *Form the detectors:*

$$\hat{\boldsymbol{w}}_k = \hat{\boldsymbol{U}}_s\hat{\boldsymbol{\Lambda}}_s^{-1}\hat{\boldsymbol{U}}_s^H\hat{\tilde{\boldsymbol{H}}}\left(\hat{\tilde{\boldsymbol{H}}}^H\hat{\boldsymbol{U}}_s\hat{\boldsymbol{\Lambda}}_s^{-1}\hat{\boldsymbol{U}}_s^H\hat{\tilde{\boldsymbol{H}}}\right)^{-1}\boldsymbol{e}_{\tilde{K}\iota+k}, \quad k = 1,\ldots,\tilde{K}. \tag{3.177}$$

- *Perform differential detection:*

$$z_k[i] = \hat{\boldsymbol{w}}_k^H\boldsymbol{r}[i], \tag{3.178}$$

$$\hat{\beta}_k[i] = \operatorname{sign}\left\{\Re\left(z_k[i]z_k[i-1]^*\right)\right\}, \quad i = 1,\ldots,M-1;\ k = 1,\ldots,\tilde{K}. \tag{3.179}$$

It is seen that form I group-blind detectors are based on an estimate of the signal subspace of the matrix $\bar{\boldsymbol{P}}\boldsymbol{C}_r\bar{\boldsymbol{P}}$, whereas form II group-blind detectors are based on an estimate of the signal subspace of the matrix $\boldsymbol{C}_r$. If the signal subspace dimension $K-\tilde{K}$ of $\bar{\boldsymbol{P}}\boldsymbol{C}_r\bar{\boldsymbol{P}}$ is less than that of $\boldsymbol{C}_r$, which is K, form I implementation in general gives a more accurate estimation of group-blind detectors. On the other hand, for multipath channels, estimation of the given users' channels is based on eigendecomposition of $\boldsymbol{C}_r$. Hence form II group-blind detectors are more efficient in terms of implementation since they do not require the eigendecomposition (3.152), which is required by form I group-blind detectors. If however, the channels are estimated by some other means not involving the eigendecomposition of $\boldsymbol{C}_r$, form I detectors can be computationally less complex than form II detectors, since the dimension of the estimated signal subspace of the former is less than that of the latter. (That is, of course, if computationally efficient subspace tracking algorithms [98] are used instead of the conventional eigendecomposition.)

Simulation Examples

Next, we provide computer simulation results to demonstrate the performance of the proposed blind and group-blind linear multiuser detectors under a number of channel conditions. The simulated system is an asynchronous CDMA system with processing gain $N = 15$. Employed as the user spreading sequences are m-sequences of length 15 and their shifted versions. The chip pulse is a raised cosine pulse with roll-off factor 0.5. Each user's channel has $L = 3$ paths. The delay of each path is uniform on $[0, 10T_c]$. Hence the maximum delay spread is one symbol interval (i.e., $\iota = 1$). The fading gain of each path in each user's channel is generated from a complex Gaussian distribution and fixed for all simulations. The path gains in each user's channel are normalized so that all users' signals arrive at the receiver with the same power. The oversampling factor is $p = 2$. The smoothing factor is $m = 2$. Hence this system can accommodate up to $\lceil \frac{m+\iota}{m-\iota} \cdot P \rceil = 10$ users. The number of users in the simulation is 10, with seven known users (i.e., $K = 10$ and $\tilde{K} = 7$). The length of each user's signal frame is $M = 200$.

In each simulation, an eigendecomposition is performed on the sample autocorrelation matrix of the received signals. The signal subspace consists of the eigenvectors corresponding to the largest r eigenvalues. [Recall that $r \triangleq K(m + \iota)$ is the dimension of the signal subspace.] The remaining eigenvectors constitute the noise subspace. An estimate of the noise variance σ^2 is given by the average of the $Pm - r$ smallest eigenvalues.

We first compare the performance of four *exact* detectors (i.e., assuming that $\boldsymbol{H}$ and σ^2 are known):

1. The linear MMSE detector
2. The linear zero-forcing detector
3. The group-blind linear hybrid detector
4. The group-blind linear MMSE detector

For each of these detectors, and for each SNR value, the minimum and maximum bit-error rate (BER) among the seven known users is plotted in Fig. 3.13. It is seen from this figure that, as expected, the closer the detector is to the true linear MMSE detector, the better its performance is.

Next, the performance of the various estimated group-blind detectors (i.e., the detectors are estimated based on the M received signal vectors) is shown in Fig. 3.14. It is seen that at low SNR, the group-blind MMSE detectors perform best, whereas at high SNR, the group-blind hybrid detectors perform best. This is because the hybrid detector zero-forces the known users' signals and enhances the noise level, whereas the group-blind linear MMSE detector suppresses both the interference and the noise. At high SNR, the group-blind hybrid and group-blind MMSE detectors tend to become the same. However, implementation of the latter requires an estimate of the noise level. When the noise level is low, this estimate is noisy, which causes the performance of the group-blind MMSE detector to deteriorate. It

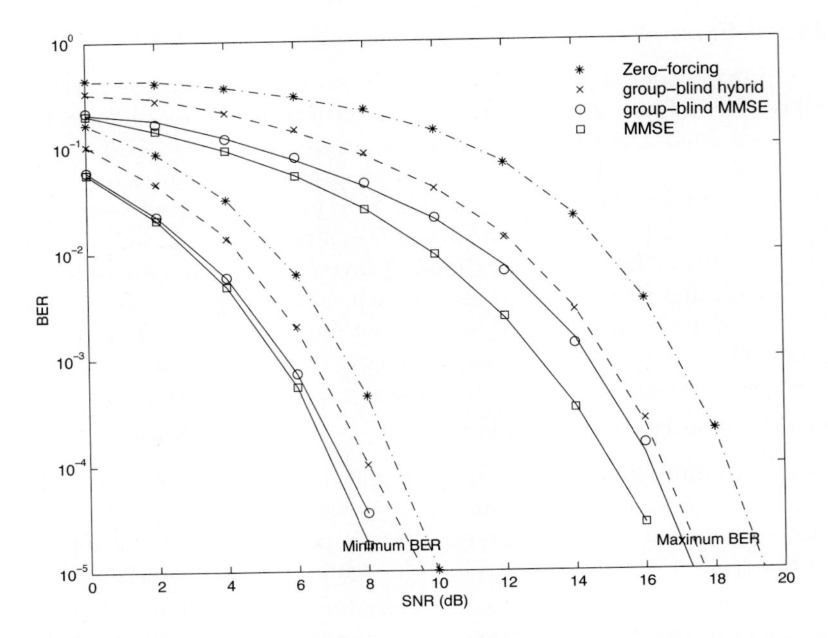

Figure 3.13. Comparison of the performance of four *exact* linear detectors in white noise. $K = 10, \tilde{K} = 7$.

is also seen that the performance of form I detectors is only slightly better than that of the corresponding form II detectors, at the expense of higher computational complexity.

Comparing Fig. 3.13 with Fig. 3.14, it is seen that the performance of the estimated detectors is substantially different from that of the corresponding exact detectors for the block size considered here (i.e., $M = 200$). It is known that the subspace detectors converge to the exact detectors at a rate of $\mathcal{O}\left(\sqrt{\log\log M/M}\right)$. It is also seen from Fig. 3.14 that the form II hybrid detector performs very well compared with other forms of group-blind detectors, even though it has the lowest computational complexity. Hence in subsequent simulation studies, we compare the performance of the form II hybrid detector with that of some multiuser detectors proposed previously.

We next compare the performance of the group-blind hybrid detector with that of the blind detector for the same system. The result is shown in Fig. 3.15, where the BER curves for the blind linear MMSE detector, the form II group-blind linear hybrid detector, and a partial MMSE detector are plotted. The partial MMSE detector ignores the unknown users and forms the linear MMSE detector for the $\tilde{K}$ known users using the estimated matrix $\tilde{\boldsymbol{H}}$. It is seen that the group-blind detector significantly outperforms the blind MMSE detector and the partial MMSE detector. Indeed, the blind MMSE detector exhibits an error floor at high SNR values. This is due to the finite length of the received signal frame, from which the detector is

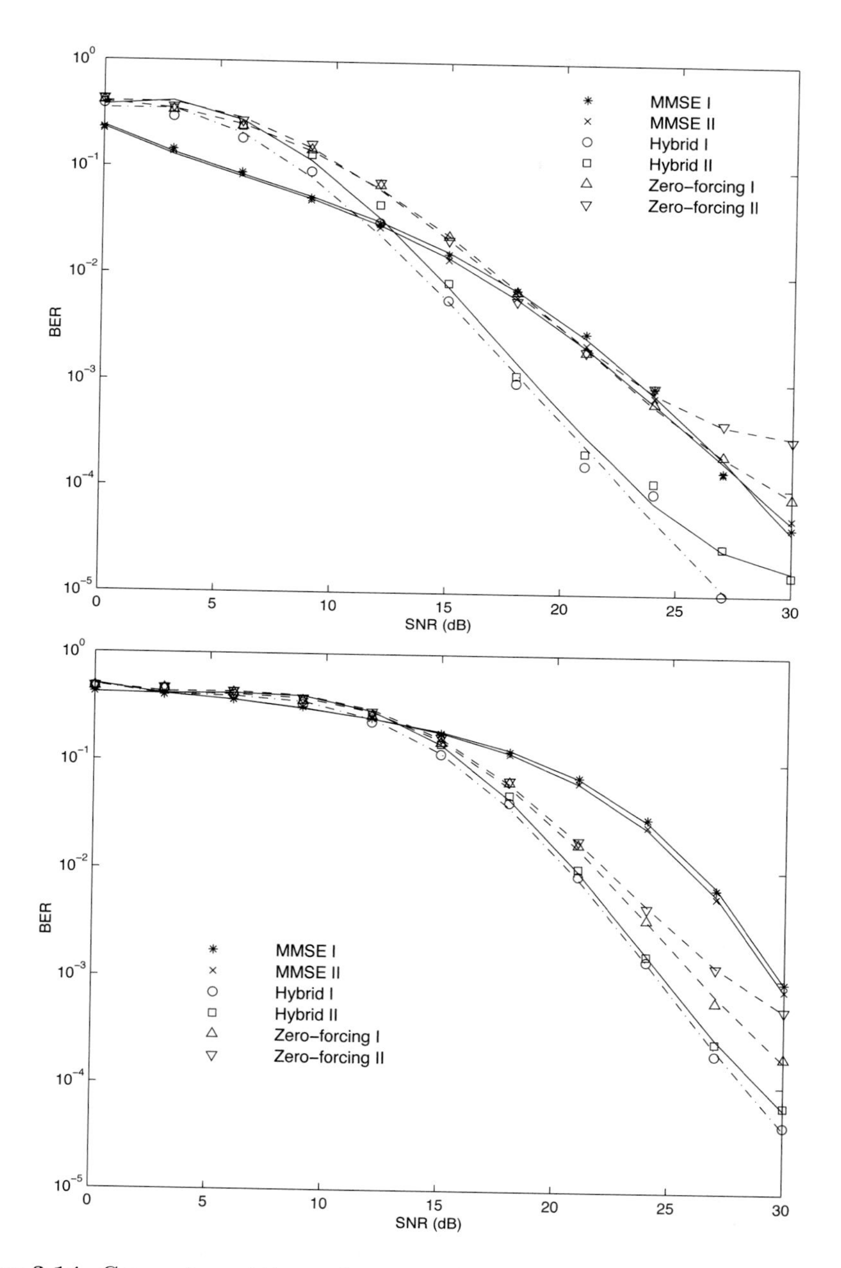

Figure 3.14. Comparison of the performance of various estimated group-blind detectors in white noise. $K = 10, \tilde{K} = 7$. (Top: minimum BER; bottom: maximum BER.)

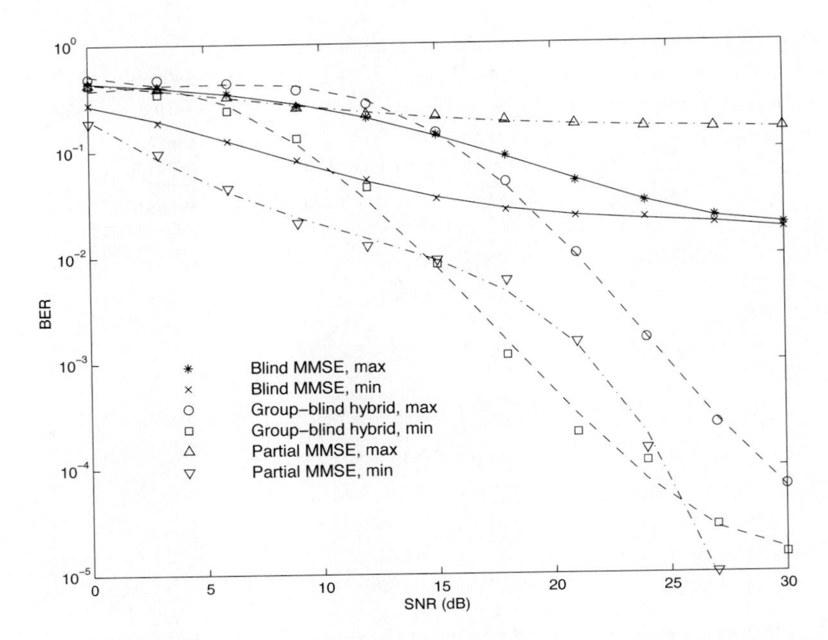

Figure 3.15. Comparison of the performance of blind and group-blind linear detectors in white noise. $K = 10, \tilde{K} = 7$.

estimated. The group-blind hybrid MMSE detector does not show an error floor in the BER range considered here. Of course, due to the finite frame length, the group-blind detector also has an error floor. But such a floor is much lower than that of the blind linear MMSE detector.

Theoretically, both the blind and group-blind detectors converge to the true linear MMSE detector (at a high signal-to-noise ratio) as the signal frame size $M \to \infty$. Hence the asymptotic performance of the two detectors is the same at high signal-to-noise ratio. However, for a finite frame length M, the group-blind detector performs significantly better than the blind detector, as seen from the simulation results above. An intuitive explanation for such performance improvement is that more information about the multiuser environment is incorporated in forming the group-blind detector. For example, the computations for subspace decomposition and channel estimation involved in the two detectors are exactly the same. However, the blind detector is formed based solely on the composite channel of the desired user, whereas the group-blind detector is formed based on the composite channels of all known users. By incorporating more information about the multiuser channel, the estimated group-blind detector is more accurate than the estimated blind detector (i.e., the former is "closer" than the latter to the exact detector).

It is seen from Fig. 3.13 that when the spreading waveforms and the channels of all users are known, all three forms of the exact group-blind detectors perform worse than the linear MMSE detector, which is the exact blind detector. This is because the zero-forcing and hybrid group-blind detectors zero-force all or some users' signals and enhance the noise level, whereas the group-blind MMSE detector is defined in terms of a specific constrained form which in general is different from the true MMSE detector. However, with imperfect channel information, the roles are reversed and the group-blind detectors outperform the blind detector. Of course, both the blind and group-blind detectors are developed based on the assumption that the multiuser channel is not perfectly known, and a study of the performance of the exact detectors is only of theoretical interest. Nevertheless, it is interesting to observe that by changing the assumption on prior knowledge about the channel, the relative performance of two detectors can be different.

3.5.2 Adaptive Group-Blind Linear Multiuser Detection

As for the blind linear multiuser detector discussed in Chapter 2, the group-blind linear multiuser detectors can also be implemented adaptively. Specifically, for example, since the form II linear hybrid detector can be written in closed form as a function of the signal subspace components, we can use a suitable subspace tracking algorithm in conjunction with this detector and a channel estimator to form an adaptive detector that is able to track changes in the number of users and their composite signature waveforms [412]. Figure 3.16 contains a block diagram of such a receiver. The received signal $\boldsymbol{r}[i]$ is fed into a subspace tracker which sequentially estimates the signal subspace components $(\boldsymbol{U}_s, \boldsymbol{\Lambda}_s)$. The received signal $\boldsymbol{r}[i]$ is then projected onto the noise subspace to obtain $\boldsymbol{z}[i]$, which is in turn passed through a bank of parallel linear filters, each determined by the signature waveform of a desired user. The output of each filter is fed into a channel tracker which estimates the channel state of that particular user. Finally, the linear hybrid group-blind detector is constructed in closed form based on the estimated signal subspace components and the channel states of the desired users. This adaptive algorithm is summarized as follows. Suppose that at time $i-1$, the estimated signal subspace rank is $r[i-1]$ and the signal subspace components are $\boldsymbol{U}_s[i-1]$, $\boldsymbol{\Lambda}_s[i-1]$, and $\sigma^2[i-1]$. The estimated channel states for the desired users are $\boldsymbol{f}_k[i-1]$, $1 \leq k \leq \tilde{K}$. Then at time i, the adaptive detector performs the following steps to update the detector and detect the data.

Algorithm 3.7: [Adaptive group-blind linear hybrid multiuser detector—multipath CDMA]

- *Update the signal subspace: Using a particular signal subspace tracking algorithm, update the signal subspace rank* $r[i]$ *and the signal subspace components* $\boldsymbol{U}_s[i]$, $\boldsymbol{\Lambda}_s[i]$, *and* $\sigma^2[i]$.

- *Estimate the desired users' channels (cf. Section 2.7.4):*

$$\boldsymbol{x}_k[i] = \overline{\boldsymbol{\Xi}}_k^H \boldsymbol{z}[i], \tag{3.180}$$

$$\boldsymbol{k}_k[i] = \boldsymbol{\Sigma}_k[i-1]\boldsymbol{x}_k[i] \left(\boldsymbol{x}_k[i]^H \boldsymbol{\Sigma}_k[i-1]\boldsymbol{x}_k[i]\right)^{-1}, \tag{3.181}$$

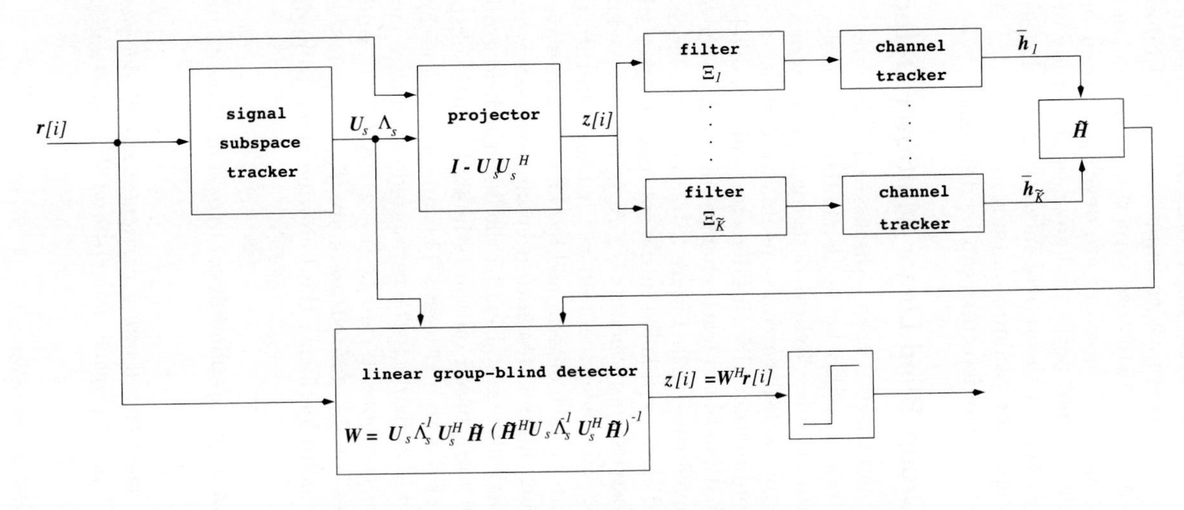

Figure 3.16. Adaptive group-blind receiver structure.

$$\boldsymbol{f}_k[i] = \boldsymbol{f}_k[i-1] - \boldsymbol{k}_k[i] \left(\boldsymbol{x}_k[i]^H \boldsymbol{f}_k[i]\right) \left\| \boldsymbol{f}_k[i-1] - \boldsymbol{k}_k[i] \left(\boldsymbol{x}_k[i]^H \boldsymbol{f}_k[i]\right) \right\|, \quad (3.182)$$

$$\boldsymbol{\Sigma}_k[i] = \boldsymbol{\Sigma}_k[i-1] - \boldsymbol{k}_k[i] \boldsymbol{x}_k[i]^H \boldsymbol{\Sigma}_k[i-1], \qquad k = 1, \ldots, \tilde{K}. \quad (3.183)$$

Form $\hat{\tilde{\boldsymbol{H}}}[i]$ *using* $\boldsymbol{f}_1[i], \ldots, \boldsymbol{f}_{\tilde{K}}[i]$.

- *Form the detectors:*

$$\hat{\boldsymbol{w}}_k[i] = \hat{\boldsymbol{U}}_s[i] \hat{\boldsymbol{\Lambda}}_s[i]^{-1} \hat{\boldsymbol{U}}_s[i]^H \hat{\tilde{\boldsymbol{H}}}[i] \left(\hat{\tilde{\boldsymbol{H}}}[i]^H \hat{\boldsymbol{U}}_s[i] \hat{\boldsymbol{\Lambda}}_s[i]^{-1} \hat{\boldsymbol{U}}_s[i]^H \hat{\tilde{\boldsymbol{H}}}[i] \right)^{-1} \boldsymbol{e}_{\tilde{K}\iota + k}, \qquad k = 1, \ldots, \tilde{K}. \quad (3.184)$$

- *Perform differential detection:*

$$z_k[i] = \hat{\boldsymbol{w}}_k[i]^H \boldsymbol{r}[i], \quad (3.185)$$

$$\hat{\beta}_k[i] = \operatorname{sign}\left\{\Re\left(z_k[i] z_k[i-1]^*\right)\right\}, \qquad k = 1, \ldots, \tilde{K}. \quad (3.186)$$

Simulation Examples

We next illustrate the performance of the adaptive receiver in an asynchronous CDMA system. The processing gain $N = 15$ and the spreading codes are Gold codes of length 15. The chip pulse waveform is a raised cosine pulse with a roll-off factor of 0.5. Each user's channel has $L = 3$ paths. The delay of each path is distributed uniformly on $[0, 10T_c]$. Hence, the maximum delay spread is one symbol interval (i.e., $\iota = 1$). The channel gain of each path in each user's channel is generated from a complex Gaussian distribution and is fixed for all simulations. The path gains in each user's channel are normalized so that all users' signals arrive at the receiver with the same power. The oversampling factor is $p = 2$ and the smoothing factor is $m = 2$. The performance measures are the SINR and the BER.

Figure 3.17 is a comparison of the adaptive performance of the MMSE and hybrid group-blind detectors using the NAHJ subspace tracking algorithm discussed in Section 2.6.3. During the first 1000 iterations there are eight total users, six of which are known by the group-blind detector. At iteration 1000, four new users are added to the system. At iteration 2000, one additional known user is added and three unknown users vanish. We see that there is a substantial performance gain using the group-blind detector at each stage and that convergence occurs in less than 500 iterations.

Figure 3.18 is created with parameters identical to Fig. 3.17 except that the tracking algorithm used is an exact rank-one SVD update. Again we see a significant improvement in performance using the group-blind detector. More important, when we compare Figs. 3.17 and 3.18 we see very little difference between the performance we obtain using the NAHJ subspace tracking and that we obtain using an exact SVD update.

Figure 3.19 shows the steady-state BER performance of our receiver using NAHJ subspace tracking and the exact SVD update for both blind and group-blind multiuser detection. The number of users is eight and the number of known users is

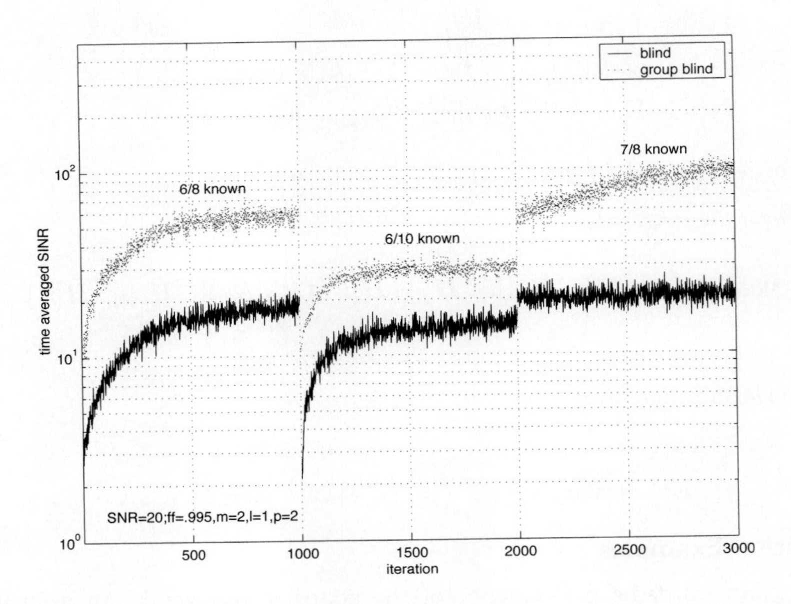

Figure 3.17. Performance of an adaptive group-blind receiver employing the NAHJ subspace tracking algorithm.

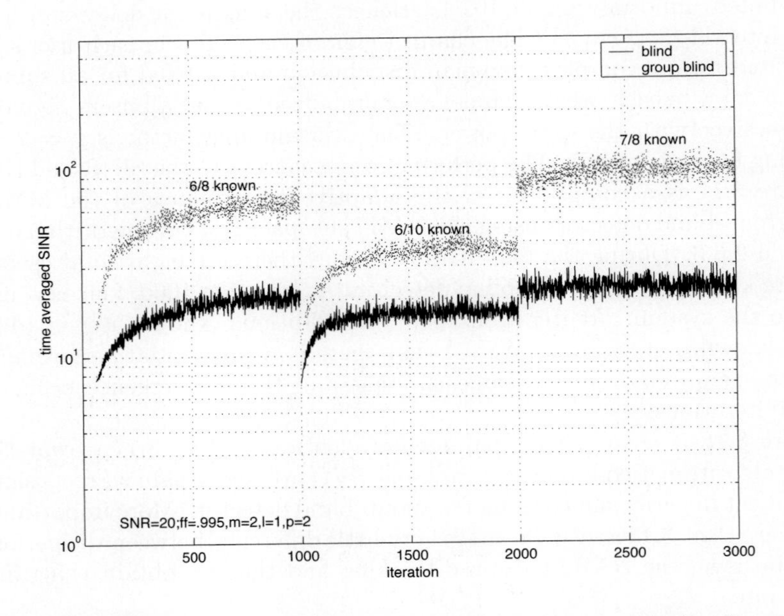

Figure 3.18. Performance of an adaptive group-blind receiver employing the exact SVD update.

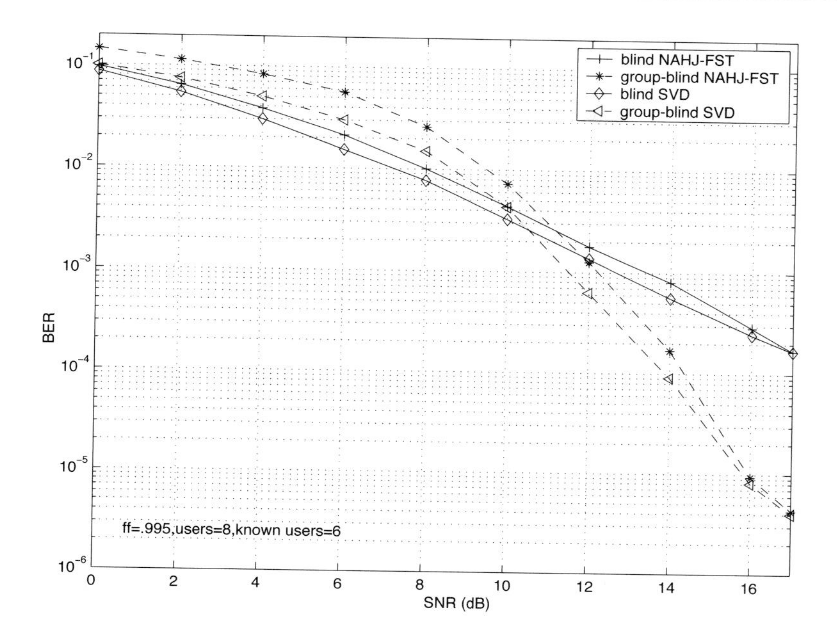

Figure 3.19. Steady-state performance of adaptive group-blind receivers.

six. At SNR above about 11 dB we see that the group-blind detectors provide a substantial improvement in BER. At lower SNR, the group-blind detectors seem to suffer from the noise enhancement problems that often accompany zero-forcing detectors. Recall that the hybrid group-blind detector zero-forces interference from known users and suppresses interference from unknown users via the MMSE criterion. Once again, note the relatively small difference between the performance of NAHJ and that of exact SVD, especially at high SNR.

3.5.3 Linear Group-Blind Detection in Correlated Noise

The problem of blind linear multiuser detection in unknown correlated noise is discussed in Section 2.7.5. In this section we consider the problem of group-blind linear multiuser detection in the same environment, which was first treated in [551]. Recall that in this case it is assumed that the signal is received by two well-separated antennas, so that the noise is spatially uncorrelated. The two augmented received signal vectors at the two antennas are given, respectively, by

$$\boldsymbol{r}_1[i] = \boldsymbol{H}_1 \boldsymbol{b}[i] + \boldsymbol{n}_1[i], \tag{3.187}$$

$$\boldsymbol{r}_2[i] = \boldsymbol{H}_2 \boldsymbol{b}[i] + \boldsymbol{n}_2[i], \tag{3.188}$$

where $\boldsymbol{H}_1$ and $\boldsymbol{H}_2$ contain the channel information corresponding to the respective antennas; $\boldsymbol{n}_1[i]$ and $\boldsymbol{n}_2[i]$ are the Gaussian noise vectors at the two antennas with the following correlations:

$$E\{\boldsymbol{n}_1[i]\boldsymbol{n}_1[i]_H\} = \boldsymbol{\Sigma}_1, \quad (3.189)$$

$$E\{\boldsymbol{n}_2[i]\boldsymbol{n}_2[i]^H\} = \boldsymbol{\Sigma}_2, \quad (3.190)$$

$$E\{\boldsymbol{n}_1[i]\boldsymbol{n}_2[i]^H\} = \boldsymbol{0}. \quad (3.191)$$

Define

$$\boldsymbol{C}_{11} = E\{\boldsymbol{r}_1[i]\boldsymbol{r}_1[i]^H\} = \boldsymbol{H}_1\boldsymbol{H}_1^H + \boldsymbol{\Sigma}_1, \quad (3.192)$$

$$\boldsymbol{C}_{22} = E\{\boldsymbol{r}_2[i]\boldsymbol{r}_2[i]^H\} = \boldsymbol{H}_2\boldsymbol{H}_2^H + \boldsymbol{\Sigma}_2, \quad (3.193)$$

$$\boldsymbol{C}_{12} = E\{\boldsymbol{r}_1[i]\boldsymbol{r}_2[i]^H\} = \boldsymbol{H}_1\boldsymbol{H}_2^H. \quad (3.194)$$

The canonical correlation decomposition (CCD) of the matrix $\boldsymbol{C}_{12}$ is given by

$$\boldsymbol{C}_{11}^{-1/2}\boldsymbol{C}_{12}\boldsymbol{C}_{22}^{-1/2} = \boldsymbol{U}_1\boldsymbol{\Gamma}\boldsymbol{U}_2^H, \quad (3.195)$$

$$\Longrightarrow \quad \boldsymbol{C}_{11}^{-1}\boldsymbol{C}_{12}\boldsymbol{C}_{22}^{-1} = \underbrace{\left(\boldsymbol{C}_{11}^{-1/2}\boldsymbol{U}_1\right)}_{\boldsymbol{L}_1}\boldsymbol{\Gamma}\underbrace{\left(\boldsymbol{C}_{22}^{-1/2}\boldsymbol{U}_2\right)^H}_{\boldsymbol{L}_2}. \quad (3.196)$$

The $Pm \times Pm$ matrix $\boldsymbol{\Gamma}$ has the form $\boldsymbol{\Gamma} = \mathsf{diag}(\gamma_1, \ldots, \gamma_r, 0, \ldots, 0)$, with $\gamma_1 \geq \cdots \geq \gamma_r > 0$. Define $\boldsymbol{L}_{j,s}$ and $\boldsymbol{L}_{j,n}$ as, respectively, the first r columns and the last $Pm - r$ columns of $\boldsymbol{L}_j$, $j = 1, 2$. It is known then that

$$\mathsf{range}\left(\boldsymbol{L}_{j,n}\right) = \mathsf{null}\left(\boldsymbol{H}_j^H\right), \qquad j = 1, 2. \quad (3.197)$$

As discussed in Section 2.7.5, the composite signature waveform $\bar{\boldsymbol{h}}_{j,k}$ of the desired user k, $1 \leq k \leq \tilde{K}$, can be estimated based on the orthogonality relationship $\boldsymbol{L}_{j,n}^H\bar{\boldsymbol{h}}_{j,k} = \boldsymbol{0}$.

We next consider the group-blind linear detector in correlated ambient noise based on the CCD method. Since the signal subspace cannot be identified directly in the CCD, we will not consider the group-blind linear zero-forcing or MMSE detectors, which require the identification of some signal subspace. Nevertheless, the form II group-blind linear hybrid detector can easily be constructed for correlated noise, as given by the following result.

Proposition 3.13: [Group-blind linear hybrid detector in correlated noise (form II)] *The weight vector of the group-blind linear hybrid detector for the kth user at the jth antenna in correlated noise is given by*

$$\boldsymbol{w}_{j,k} = \boldsymbol{L}_{j,s}\boldsymbol{L}_{j,s}^H\tilde{\boldsymbol{H}}_j\left(\tilde{\boldsymbol{H}}_j^H\boldsymbol{L}_{j,s}\boldsymbol{L}_{j,s}^H\tilde{\boldsymbol{H}}_j\right)^{-1}\boldsymbol{e}_{\tilde{K}\iota+k}, \quad j = 1, 2;\ k = 1, \ldots, \tilde{K}. \quad (3.198)$$

Proof: By definition, the group-blind linear hybrid detector is given by the following constrained optimization problem:

$$\boldsymbol{w}_{j,k} = \arg\min_{\boldsymbol{w} \in \mathbb{C}^{Pm}} E\left\{\left|b_k[i] - \boldsymbol{w}^H\boldsymbol{r}_j[i]\right|^2\right\} \text{ s.t. } \boldsymbol{w}^H\tilde{\boldsymbol{H}}_j = \boldsymbol{e}_{\tilde{K}\iota+k}^T. \quad (3.199)$$

Using the method of Lagrange multipliers to solve (3.199), we obtain

$$\begin{aligned}
\boldsymbol{w}_{j,k} &= \arg\min_{\boldsymbol{w}\in\mathbb{C}^{Pm}} E\left\{\left|b_k[i] - \boldsymbol{w}^H \boldsymbol{r}_j[i]\right|^2\right\} + \boldsymbol{\lambda}^H \left(\tilde{\boldsymbol{H}}_j^H \boldsymbol{w} - \boldsymbol{e}_{\tilde{K}\iota+k}\right) \\
&= \arg\min_{\boldsymbol{w}\in\mathbb{C}^{Pm}} \boldsymbol{w}^H \boldsymbol{C}_{jj} \boldsymbol{w} - 2\left(\tilde{\boldsymbol{H}}_j \boldsymbol{e}_{\tilde{K}\iota+k}\right)^H \boldsymbol{w} + \boldsymbol{\lambda}^H \tilde{\boldsymbol{H}}_j^H \boldsymbol{w} \\
&= \arg\min_{\boldsymbol{w}\in\mathbb{C}^{Pm}} \boldsymbol{w}^H \boldsymbol{C}_{jj} \boldsymbol{w} + \left(\boldsymbol{\lambda} - 2\boldsymbol{e}_{\tilde{K}\iota+k}\right)^H \tilde{\boldsymbol{H}}_j^H \boldsymbol{w} \\
&= \boldsymbol{C}_{jj}^{-1} \tilde{\boldsymbol{H}}_j \boldsymbol{\mu},
\end{aligned} \tag{3.200}$$

where $\boldsymbol{\lambda} \in \mathbb{C}^{\tilde{r}}$ is the Lagrange multiplier, and $\boldsymbol{\mu} \triangleq \boldsymbol{\lambda} - 2\boldsymbol{e}_{\tilde{K}\iota+k}$. Substituting (3.200) into the constraint that $\boldsymbol{w}_k^H \tilde{\boldsymbol{H}}_j = \boldsymbol{e}_{\tilde{K}\iota+k}$, we obtain

$$\boldsymbol{\mu} = \left(\tilde{\boldsymbol{H}}_j^H \boldsymbol{C}_{jj}^{-1} \tilde{\boldsymbol{H}}_j\right)^{-1} \boldsymbol{e}_{\tilde{K}\iota+k}.$$

Hence

$$\boldsymbol{w}_{j,k} = \boldsymbol{C}_{jj}^{-1} \tilde{\boldsymbol{H}}_j \left(\tilde{\boldsymbol{H}}_j^H \boldsymbol{C}_{jj}^{-1} \tilde{\boldsymbol{H}}_j\right)^{-1} \boldsymbol{e}_{\tilde{K}\iota+k}. \tag{3.201}$$

Moreover, by definition,

$$\begin{aligned}
\boldsymbol{L}_j = [\boldsymbol{L}_{j,s} \mid \boldsymbol{L}_{j,n}] &= \boldsymbol{C}_{jj}^{-1/2} \boldsymbol{U}_j \\
\Longrightarrow \quad \boldsymbol{C}_{jj}^{-1} = \boldsymbol{L}_j \boldsymbol{L}_j^H &= \boldsymbol{L}_{j,s} \boldsymbol{L}_{j,s}^H + \boldsymbol{L}_{j,n} \boldsymbol{L}_{j,n}^H.
\end{aligned} \tag{3.202}$$

Substituting (3.202) into (3.201), and using the fact that $\boldsymbol{L}_{j,n}^H \tilde{\boldsymbol{H}}_j = \boldsymbol{0}$, we obtain (3.198). □

The group-blind linear multiuser detection algorithm in multipath channels with correlated noise is summarized as follows.

Algorithm 3.8: [Group-blind linear hybrid detector—multipath CDMA and correlated noise]

- *Compute the CCD: Let* $\boldsymbol{Y}_j \triangleq \left[\boldsymbol{r}_j[0], \ldots, \boldsymbol{r}_j[M-1]\right]$, $j = 1, 2$:

$$\frac{1}{\sqrt{M}} \boldsymbol{X}_j^H = \hat{\boldsymbol{Q}}_j \hat{\boldsymbol{\Upsilon}}_j, \qquad j = 1, 2, \qquad \textit{(QR decomposition of } \boldsymbol{X}_j\textit{)} \tag{3.203}$$

$$\hat{\boldsymbol{Q}}_1^H \hat{\boldsymbol{Q}}_2 = \hat{\boldsymbol{V}}_1 \hat{\boldsymbol{\Gamma}} \hat{\boldsymbol{V}}_2^H. \quad \textit{(SVD of } \hat{\boldsymbol{Q}}_1^H \hat{\boldsymbol{Q}}_2\textit{)} \tag{3.204}$$

$$\hat{\boldsymbol{L}}_j = \hat{\boldsymbol{\Upsilon}}_j^{-1} \hat{\boldsymbol{V}}_j = \left[\hat{\boldsymbol{L}}_{j,s} \mid \hat{\boldsymbol{L}}_{j,n}\right], \qquad j = 1, 2. \tag{3.205}$$

- *Estimate the desired users' channels (cf. Section 2.7.3):*

$$\hat{\boldsymbol{f}}_{j,k} = \textit{min-eigenvector}\left(\boldsymbol{\Xi}_k^H \hat{\boldsymbol{L}}_{j,n} \hat{\boldsymbol{L}}_{j,n}^H \boldsymbol{\Xi}_k\right), \tag{3.206}$$

$$\hat{\boldsymbol{h}}_{j,k} = \boldsymbol{\Xi}_k \hat{\boldsymbol{f}}_{j,k}, \qquad j = 1, 2;\ k = 1, \ldots, \tilde{K}. \tag{3.207}$$

Form $\hat{\tilde{\boldsymbol{H}}}_j$ *using* $\hat{\boldsymbol{h}}_{j,1}, \ldots, \hat{\boldsymbol{h}}_{j,\tilde{K}}$.

- *Form the detector:*

$$\hat{\boldsymbol{w}}_{j,k} = \hat{\mathbf{L}}_{j,s}\hat{\mathbf{L}}_{j,s}^H \hat{\tilde{\boldsymbol{H}}}_j \left(\hat{\tilde{\boldsymbol{H}}}_j^H \hat{\mathbf{L}}_{j,s}\hat{\mathbf{L}}_{j,s}^H \hat{\tilde{\boldsymbol{H}}}_j \right)^{-1} \boldsymbol{e}_{\tilde{K}\iota+k},$$
$$j = 1, 2; \; k = 1, \ldots, \tilde{K}. \quad (3.208)$$

- *Perform differential detection:*

$$z_{j,k}[i] = \hat{\boldsymbol{w}}_{j,k}^H \boldsymbol{r}_j[i], \qquad j = 1, 2.$$
$$\hat{\beta}_k[i] = \text{sign}\left\{ \Re\left(\sum_{j=1}^{2} z_{j,k}[i] z_{j,k}[i-1]^* \right) \right\}, \quad i = 1, \ldots, M-1; \; k = 1, \ldots, \tilde{K}.$$

Simulation Example

The simulated system is the same as that described in Section 3.5.1. The noise at each antenna j is modeled by a second-order autoregressive (AR) model with coefficients $[a_{j,1}, a_{j,2}]$; that is, the noise field is generated according to

$$v_j[n] = a_{j,1}\, v_j[n-1] + a_{j,2}\, v_j[n-2] + w_j[n], \quad j = 1, 2, \qquad (3.209)$$

where $v_j[n]$ is the noise at antenna j and sample n, and $w_j[n]$ is a complex white Gaussian noise sample. The AR coefficients at the two antennas are chosen as $[a_{1,1}, a_{1,2}] = [1, -0.2]$ and $[a_{2,1}, a_{2,2}] = [1.2, -0.3]$. The performance of the group-blind linear hybrid detector is compared with that of the blind linear MMSE detector. The result is shown in Fig. 3.20. It is seen that similar to the white noise case, the proposed group-blind linear detector offers substantial performance gain over the blind linear detector.

3.5.4 Nonlinear Group-Blind Detection

In this section we extend the nonlinear multiuser detection methods discussed in Section 3.4 to asynchronous CDMA systems with multipath [456]. The idea is essentially the same as in the synchronous case. We first estimate the decorrelating detectors of the desired users, given by

$$\hat{\tilde{\boldsymbol{D}}} = \hat{\boldsymbol{U}}_s \left(\hat{\mathbf{\Lambda}}_s - \hat{\sigma^2} \boldsymbol{I}_r \right)^{-1} \hat{\boldsymbol{U}}_s^H \hat{\tilde{\boldsymbol{H}}}$$
$$\left[\hat{\tilde{\boldsymbol{H}}}^H \hat{\boldsymbol{U}}_s \left(\hat{\mathbf{\Lambda}}_s - \hat{\sigma^2} \boldsymbol{I}_r \right)^{-1} \hat{\boldsymbol{U}}_s^H \hat{\tilde{\boldsymbol{H}}} \right]^{-1} [\boldsymbol{e}_{\tilde{K}\iota+1} \; \cdots \; \boldsymbol{e}_{\tilde{K}\iota+\tilde{K}}]. \quad (3.210)$$

Note that $\hat{\tilde{\boldsymbol{H}}}$ can be estimated only up to a phase ambiguity. Denote the output of the decorrelating detector as

$$\boldsymbol{z}[i] \triangleq \hat{\tilde{\boldsymbol{D}}}^H \boldsymbol{r}[i] = \tilde{\boldsymbol{A}}\tilde{\boldsymbol{b}}[i] + \hat{\tilde{\boldsymbol{D}}}^H \boldsymbol{n}[i]$$
$$\Longrightarrow z_k[i] = \alpha_k b_k[i] + \tilde{n}_k[i], \qquad k = 1, \ldots, \tilde{K}, \quad (3.211)$$

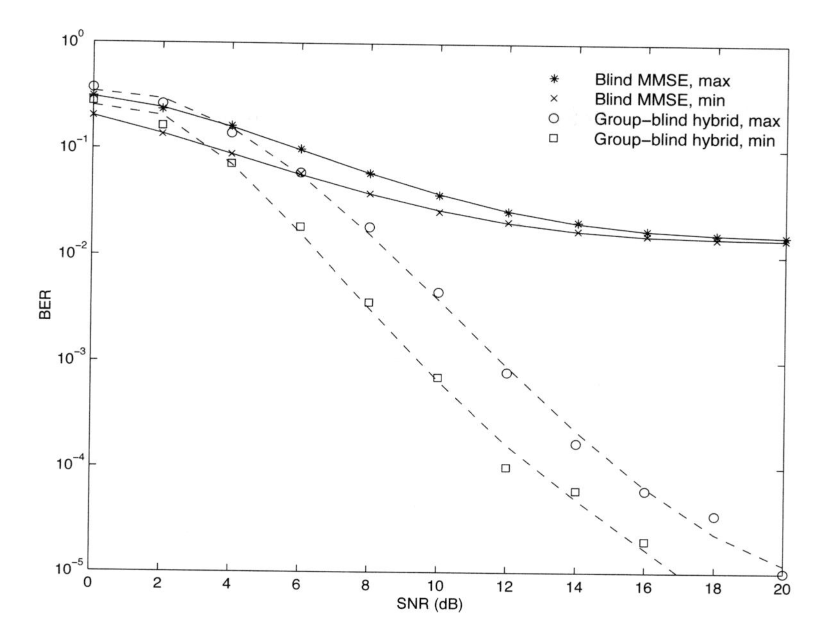

Figure 3.20. Comparison of the performance of blind and group-blind linear detectors in correlated noise. $K = 10, \tilde{K} = 7$.

where $\tilde{\boldsymbol{A}} \triangleq \text{diag}(\alpha_1, \ldots, \alpha_{\tilde{K}})$ is the phase ambiguity induced by the channel estimator which can be estimated using (3.129).

Denote

$$\boldsymbol{y}[i] \triangleq \begin{bmatrix} \Re\{\boldsymbol{\theta}[i]\} \\ \Im\{\boldsymbol{\theta}[i]\} \end{bmatrix}, \boldsymbol{\Phi} \triangleq \begin{bmatrix} \Re\left\{\tilde{\boldsymbol{A}}\right\} \\ \Im\left\{\tilde{\boldsymbol{A}}\right\} \end{bmatrix}, \boldsymbol{v}[i] \triangleq \begin{bmatrix} \Re\left\{\tilde{\boldsymbol{D}}^H \boldsymbol{n}[i]\right\} \\ \Im\left\{\tilde{\boldsymbol{D}}^H \boldsymbol{n}[i]\right\} \end{bmatrix}.$$

Then (3.211) can be written as

$$\boldsymbol{y}[i] = \boldsymbol{\Phi}\tilde{\boldsymbol{b}}[i] + \boldsymbol{v}[i]. \tag{3.212}$$

Note that the covariance of $\boldsymbol{v}$ is given by

$$\text{Cov}\{\boldsymbol{v}\} = \frac{\sigma^2}{2} \begin{bmatrix} \boldsymbol{Q} & \boldsymbol{0} \\ \boldsymbol{0} & \boldsymbol{Q} \end{bmatrix}, \tag{3.213}$$

with

$$\boldsymbol{Q} \triangleq \Re\left\{\hat{\tilde{\boldsymbol{D}}}\right\}^T \Re\left\{\hat{\tilde{\boldsymbol{D}}}\right\} + \Im\left\{\hat{\tilde{\boldsymbol{D}}}\right\}^T \Im\left\{\hat{\tilde{\boldsymbol{D}}}\right\}. \tag{3.214}$$

Based on (3.212), the slowest-descent-search method for estimating $\tilde{\boldsymbol{b}}[i]$ is given by the same procedure as (3.121)–(3.126), with the covariance matrix given by (3.214). The algorithm is summarized as follows.

Algorithm 3.9: [Nonlinear group-blind detector—multipath CDMA]

- *Compute the signal subspace:*

$$\hat{\boldsymbol{C}}_r = \frac{1}{M}\sum_{i=0}^{M-1} \boldsymbol{r}[i]\boldsymbol{r}[i]^H \tag{3.215}$$

$$= \hat{\boldsymbol{U}}_s\hat{\boldsymbol{\Lambda}}_s\hat{\boldsymbol{U}}_s^H + \hat{\boldsymbol{U}}_n\hat{\boldsymbol{\Lambda}}_n\hat{\boldsymbol{U}}_s^H. \tag{3.216}$$

- *Estimate the desired users' channels (cf. Section 2.7.3):*

$$\hat{\boldsymbol{f}}_k = \textit{min-eigenvector}\left(\boldsymbol{\Xi}_k^H\hat{\boldsymbol{U}}_n\hat{\boldsymbol{U}}_n^H\boldsymbol{\Xi}_k\right), \tag{3.217}$$

$$\hat{\boldsymbol{h}}_k = \boldsymbol{\Xi}_k\hat{\boldsymbol{f}}_k, \qquad k = 1, \ldots, \tilde{K}. \tag{3.218}$$

Form $\hat{\tilde{\boldsymbol{H}}}$ *using* $\hat{\boldsymbol{h}}_1, \ldots, \hat{\boldsymbol{h}}_{\tilde{K}}$.

- *Form the decorrelating detectors using (3.210).*
- *Estimate the complex amplitudes* $\tilde{\boldsymbol{A}}$*:*

$$\boldsymbol{z}[i] = \hat{\tilde{\boldsymbol{D}}}^H\boldsymbol{r}[i], \qquad i = 0, \ldots, M-1. \tag{3.219}$$

$$\hat{\rho}_k = \frac{1}{M}\sum_{i=0}^{M-1} |z_k[i]|, \tag{3.220}$$

$$R_k = \frac{1}{M}\sum_{i=0}^{M-1} \Re\{z_k[i]\}, \tag{3.221}$$

$$I_k = \frac{1}{M}\sum_{i=0}^{M-1} \Im\{z_k[i]\}, \tag{3.222}$$

$$\hat{\phi}_k = \begin{cases} \frac{1}{M}\sum_{i=0}^{M-1} \angle\left[z_k[i]\text{sign}\left(\Re\left\{z_k[i]\right\}\right)\right] & \textit{if } R_k \geq I_k, \\ \frac{1}{M}\sum_{i=0}^{M-1} \angle\left[z_k[i]\text{sign}\left(\Im\left\{z_k[i]\right\}\right)\right] & \textit{if } R_k < I_k, \end{cases} \tag{3.223}$$

$$\hat{A}_k = \hat{\rho}_k e^{\jmath\hat{\phi}_k}, \qquad k = 1, \ldots, \tilde{K}. \tag{3.224}$$

- *Compute the Hessian:*

$$\hat{\nabla}^2 = \Re\left\{\hat{\tilde{\boldsymbol{A}}}\right\}\hat{\boldsymbol{Q}}^{-1}\Re\left\{\hat{\tilde{\boldsymbol{A}}}\right\} + \Im\left\{\hat{\tilde{\boldsymbol{A}}}\right\}\hat{\boldsymbol{Q}}^{-1}\Im\left\{\hat{\tilde{\boldsymbol{A}}}\right\}, \tag{3.225}$$

and the Q *smallest eigenvectors* $\boldsymbol{g}^1, \ldots, \boldsymbol{g}^Q$ *of* $\hat{\nabla}^2$.

- *Detect each symbol by solving the following discrete optimization problem using an exhaustive search [over $(\tilde{K}Q+1)$ points]:*

$$\hat{\tilde{\boldsymbol{\theta}}}[i] = \left(\hat{\nabla}^2\right)^{-1}\left[\Re\{\hat{\tilde{\boldsymbol{A}}}\}\hat{\boldsymbol{Q}}^{-1}\Re\{\boldsymbol{z}[i]\} + \Im\{\hat{\tilde{\boldsymbol{A}}}\}\hat{\boldsymbol{Q}}^{-1}\Im\{\boldsymbol{z}[i]\}\right], \tag{3.226}$$

$$\hat{\tilde{\boldsymbol{b}}}^*[i] = \text{sign}(\hat{\tilde{\boldsymbol{\theta}}}[i]), \tag{3.227}$$

$$\widehat{\tilde{\boldsymbol{b}}}[i] = \arg\min_{\tilde{\boldsymbol{b}}\in\tilde{\Omega}[i]} \left(\tilde{\boldsymbol{b}} - \hat{\tilde{\boldsymbol{\theta}}}[i]\right)^T \hat{\nabla}^2 \left(\tilde{\boldsymbol{b}} - \hat{\tilde{\boldsymbol{\theta}}}[i]\right), \tag{3.228}$$

$$\tilde{\Omega}[i] = \{\hat{\tilde{\boldsymbol{b}}}^*[i]\} \cup \bigcup_{q=1}^{Q} \left\{\tilde{\boldsymbol{b}}^{q,\mu} \in \{-1,+1\}^{\tilde{K}} : \tilde{b}_k^{q,\mu}\right.$$
$$= \begin{cases} \text{sign}\left(\hat{\tilde{\beta}}_k[i] - \mu g_k^q\right) & \textit{if } \hat{\tilde{\beta}}_k[i] - \mu g_k^q \neq 0, \\ -\hat{\tilde{b}}_k^*[i] & \textit{if } \hat{\tilde{\beta}}_k[i] - \mu g_k^q = 0, \end{cases}$$
$$\left.\mu \in \left\{\frac{\hat{\tilde{\beta}}_1[i]}{g_1^q}, \ldots, \frac{\hat{\tilde{\beta}}_{\tilde{K}}[i]}{g_{\tilde{K}}^q}\right\}\right\}, \qquad i = 0, \ldots, M-1. \tag{3.229}$$

- *Perform differential decoding:*

$$\hat{\beta}_k[i] = \hat{b}_k[i]\hat{b}_k[i-1], \qquad k = 1, \ldots, \tilde{K};\ i = 1, \ldots, M-1. \tag{3.230}$$

Simulation Examples

The simulation set is the same as that in Section 3.5.2. Figure 3.21 shows that similar to the synchronous case, in multipath channels, the nonlinear group-blind multiuser detector outperforms the linear group-blind detector by a significant margin. Furthermore, most of the performance gain offered by the slowest-descent method is obtained by searching along only one direction.

3.6 Appendix

3.6.1 Proofs for Section 3.3.1

Proof of Theorem 3.1

Denote $\boldsymbol{Y} \triangleq \boldsymbol{U}_s\boldsymbol{\Lambda}_s^{-1}\boldsymbol{U}_s^T$ and $\hat{\boldsymbol{Y}} \triangleq \hat{\boldsymbol{U}}_s\hat{\boldsymbol{\Lambda}}_s^{-1}\hat{\boldsymbol{U}}_s^T$. We have the following differential (at $\boldsymbol{Y}$):

$$\Delta\left(\tilde{\boldsymbol{S}}^T\hat{\boldsymbol{Y}}\tilde{\boldsymbol{S}}\right)^{-1} = -\left(\tilde{\boldsymbol{S}}^T\boldsymbol{Y}\tilde{\boldsymbol{S}}\right)^{-1}\tilde{\boldsymbol{S}}^T\Delta\boldsymbol{Y}\tilde{\boldsymbol{S}}\left(\tilde{\boldsymbol{S}}^T\boldsymbol{Y}\tilde{\boldsymbol{S}}\right)^{-1}. \tag{3.231}$$

The differential of the form II group-blind detector,

$$\hat{\boldsymbol{w}}_1 \triangleq \hat{\boldsymbol{Y}}\tilde{\boldsymbol{S}}\left(\tilde{\boldsymbol{S}}^T\hat{\boldsymbol{Y}}\tilde{\boldsymbol{S}}\right)^{-1}\tilde{\boldsymbol{e}}_1, \tag{3.232}$$

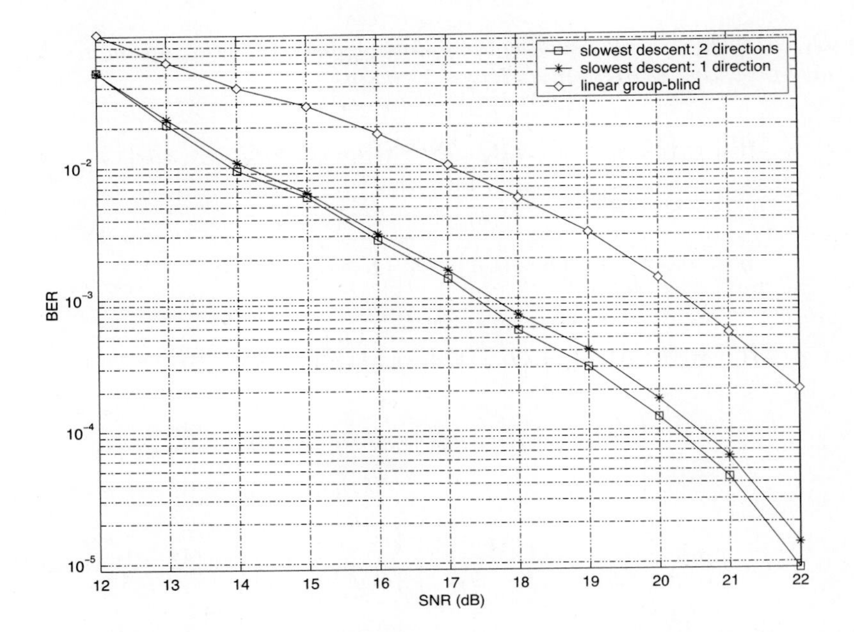

Figure 3.21. Performance of a slowest-descent-based group-blind multiuser detector in a multipath channel. $N = 15, K = 8, \tilde{K} = 4$. Each user's channel consists of three paths with randomly generated complex gains and delays. Only the spreading waveforms $\tilde{\boldsymbol{S}}$ of the desired users are assumed known to the receiver. The BER curves of a linear group-blind detector and slowest-descent (nonlinear) group-blind detector with $Q = 1$ and $Q = 2$ are shown.

is then given by

$$\begin{aligned} \Delta \boldsymbol{w}_1 &= \Delta \boldsymbol{Y} \tilde{\boldsymbol{S}} \left(\tilde{\boldsymbol{S}}^T \boldsymbol{Y} \tilde{\boldsymbol{S}} \right)^{-1} \boldsymbol{e}_1 - \boldsymbol{Y} \tilde{\boldsymbol{S}} \left(\tilde{\boldsymbol{S}}^T \boldsymbol{Y} \tilde{\boldsymbol{S}} \right)^{-1} \tilde{\boldsymbol{S}}^T \Delta \boldsymbol{Y} \tilde{\boldsymbol{S}} \left(\tilde{\boldsymbol{S}}^T \boldsymbol{Y} \tilde{\boldsymbol{S}} \right)^{-1} \boldsymbol{e}_1 \\ &= \underbrace{\left[\boldsymbol{I}_N - \boldsymbol{Y} \tilde{\boldsymbol{S}} \left(\tilde{\boldsymbol{S}}^T \boldsymbol{Y} \tilde{\boldsymbol{S}} \right)^{-1} \tilde{\boldsymbol{S}}^T \right]}_{\boldsymbol{Q}} \Delta \boldsymbol{Y} \underbrace{\tilde{\boldsymbol{S}} \left(\tilde{\boldsymbol{S}}^T \boldsymbol{Y} \tilde{\boldsymbol{S}} \right)^{-1} \boldsymbol{e}_1}_{\boldsymbol{v}_1} . \end{aligned} \tag{3.233}$$

It then follows from Lemma 2.6 that $\hat{\boldsymbol{w}}_1$ is asymptotically Gaussian. To find $\boldsymbol{C}_w$, notice that $\boldsymbol{v}_1 \in \mathsf{range}(\boldsymbol{U}_s)$. Hence, by Proposition 2.6, we have

$$M \cdot \text{cov}\left\{ \Delta \boldsymbol{Y} \boldsymbol{v}_1 \right\} = \left(\boldsymbol{v}_1^T \boldsymbol{Y} \boldsymbol{v}_1 \right) \boldsymbol{Y} + \left(\boldsymbol{Y} \boldsymbol{v}_1 \right) \left(\boldsymbol{Y} \boldsymbol{v}_1 \right)^T - 2 \boldsymbol{Y} \boldsymbol{S} \boldsymbol{D} \boldsymbol{S}^T \boldsymbol{Y} + \tau \boldsymbol{U}_n \boldsymbol{U}_n^T, \tag{3.234}$$

with

$$\boldsymbol{D} \triangleq \mathsf{diag}\left\{ A_1^4 \left(\boldsymbol{s}_1^T \boldsymbol{Y} \boldsymbol{v}_1 \right)^2, \ldots, A_K^4 \left(\boldsymbol{s}_K^T \boldsymbol{Y} \boldsymbol{v}_1 \right)^2 \right\}, \tag{3.235}$$

$$\tau = \sigma^2 \boldsymbol{v}_1^T \boldsymbol{U}_s \boldsymbol{\Lambda}_s^{-1} \left(\boldsymbol{\Lambda}_s - \sigma^2 \boldsymbol{I}_K \right)^{-2} \boldsymbol{U}_s^T \boldsymbol{v}_1. \tag{3.236}$$

Therefore, the asymptotic covariance of $\hat{\boldsymbol{w}}_1$ is given by

$$M \cdot \boldsymbol{C}_w = \boldsymbol{Q} \left(M \cdot \mathrm{Cov} \left\{ \Delta \boldsymbol{Y} \boldsymbol{v}_1 \right\} \right) \boldsymbol{Q}^T. \tag{3.237}$$

It is easily verified that $\boldsymbol{Q}\boldsymbol{Y}\boldsymbol{v}_1 = \boldsymbol{0}$. Using this and the facts that $\boldsymbol{w}_1 = \boldsymbol{Y}\boldsymbol{v}_1$ and $\boldsymbol{Q}\boldsymbol{U}_n = \boldsymbol{U}_n$, by substituting (3.234) into (3.237), we obtain (3.61).

Proof of Corollary 3.1

In this appendix $\boldsymbol{e}_k$ denotes the kth unit vector in $\mathbb{R}^K$ and $\tilde{\boldsymbol{e}}_k$ denotes the kth unit vector in $\mathbb{R}^{\tilde{K}}$. Denote $\tilde{\boldsymbol{S}} \triangleq [\boldsymbol{s}_1 \cdots \boldsymbol{s}_{\tilde{K}}]$ and $\bar{\boldsymbol{S}} \triangleq [\boldsymbol{s}_{\tilde{K}+1} \cdots \boldsymbol{s}_K]$. Denote further that $\boldsymbol{Y} \triangleq \boldsymbol{U}_s \boldsymbol{\Lambda}_s^{-1} \boldsymbol{U}_s^T$ and $\boldsymbol{X} \triangleq \boldsymbol{U}_s \boldsymbol{\Lambda}_s \boldsymbol{U}_s^T$. Note that $(\boldsymbol{Y}\boldsymbol{s}_k)$ is the linear MMSE detector for the kth user, given by

$$\boldsymbol{Y}\boldsymbol{s}_k = \boldsymbol{S} \left(\boldsymbol{R} + \sigma^2 \boldsymbol{A}^{-2} \right)^{-1} \boldsymbol{A}^{-2} \boldsymbol{e}_k. \tag{3.238}$$

Denote

$$\tilde{\boldsymbol{S}}^T \boldsymbol{Y} \tilde{\boldsymbol{S}} = \left[\boldsymbol{R} \left(\boldsymbol{R} + \sigma^2 \boldsymbol{A}^{-2} \right)^{-1} \boldsymbol{A}^{-2} \right]_{1:\tilde{K},1:\tilde{K}} = \boldsymbol{\Psi}_{11}, \tag{3.239}$$

$$\tilde{\boldsymbol{S}}^T \boldsymbol{Y} \bar{\boldsymbol{S}} = \left[\boldsymbol{R} \left(\boldsymbol{R} + \sigma^2 \boldsymbol{A}^{-2} \right)^{-1} \boldsymbol{A}^{-2} \right]_{1:\tilde{K},\tilde{K}+1:K} = \boldsymbol{\Psi}_{12}, \tag{3.240}$$

$$\bar{\boldsymbol{S}}^T \boldsymbol{Y} \tilde{\boldsymbol{S}} = \left[\boldsymbol{R} \left(\boldsymbol{R} + \sigma^2 \boldsymbol{A}^{-2} \right)^{-1} \boldsymbol{A}^{-2} \right]_{\tilde{K}+1:K,1:\tilde{K}} = \boldsymbol{\Psi}_{12}^T, \tag{3.241}$$

$$\bar{\boldsymbol{S}}^T \boldsymbol{Y} \bar{\boldsymbol{S}} = \left[\boldsymbol{R} \left(\boldsymbol{R} + \sigma^2 \boldsymbol{A}^{-2} \right)^{-1} \boldsymbol{A}^{-2} \right]_{\tilde{K}+1:K,\tilde{K}+1:K} = \boldsymbol{\Psi}_{22}. \tag{3.242}$$

First we compute the term $\mathrm{tr}(\boldsymbol{C}_w \boldsymbol{C}_r)$. Using (3.61) and the facts that $\boldsymbol{C}_r = \boldsymbol{U}_s \boldsymbol{\Lambda}_s \boldsymbol{U}_s + \sigma^2 \boldsymbol{U}_n \boldsymbol{U}_n$ and $\boldsymbol{U}_s^T \boldsymbol{U}_n = \boldsymbol{0}$, we have

$$\mathrm{tr}(\boldsymbol{C}_w \boldsymbol{C}_r) = \boldsymbol{v}_1^T \boldsymbol{w}_1 \,\mathrm{tr} \left(\boldsymbol{Q}\boldsymbol{Y}\boldsymbol{Q}^T \boldsymbol{X} \right) - 2\,\mathrm{tr} \left(\boldsymbol{Q}\boldsymbol{Y}\boldsymbol{S}\boldsymbol{D}\boldsymbol{S}^T \boldsymbol{Y}\boldsymbol{Q}^T \boldsymbol{X} \right) + \tau\sigma^2 \underbrace{\mathrm{tr}(\boldsymbol{U}_n \boldsymbol{U}_n^T)}_{N-K}. \tag{3.243}$$

The term $(\boldsymbol{v}_1^T \boldsymbol{w}_1)$ in (3.243) can be computed as

$$\begin{aligned} \boldsymbol{v}_1^T \boldsymbol{w}_1 &= \mathrm{tr} \left(\boldsymbol{w}_1 \boldsymbol{v}_1^T \right) = \mathrm{tr} \left(\boldsymbol{Y} \boldsymbol{v}_1 \boldsymbol{v}_1^T \right) \\ &= \mathrm{tr} \left(\boldsymbol{Y} \tilde{\boldsymbol{S}} \boldsymbol{\Psi}_{11}^{-1} \tilde{e}_1 \tilde{e}_1^T \boldsymbol{\Psi}_{11}^{-1} \tilde{\boldsymbol{S}}^T \right) = \left[\boldsymbol{\Psi}_{11}^{-1} \right]_{1,1}. \end{aligned} \tag{3.244}$$

Using the fact that $\boldsymbol{X}\boldsymbol{Y}\tilde{\boldsymbol{S}} = \boldsymbol{U}_s \boldsymbol{U}_s^T \tilde{\boldsymbol{S}} = \tilde{\boldsymbol{S}}$, the term $\mathrm{tr} \left(\boldsymbol{Q}\boldsymbol{Y}\boldsymbol{Q}^T \boldsymbol{X} \right)$ in (3.243) is given by

$$\begin{aligned} \mathrm{tr} \left(\boldsymbol{Q}\boldsymbol{Y}\boldsymbol{Q}^T \boldsymbol{X} \right) &= \mathrm{tr} \left\{ \left[\boldsymbol{Y} - \boldsymbol{Y} \tilde{\boldsymbol{S}} \boldsymbol{\Psi}_{11}^{-1} \tilde{\boldsymbol{S}}^T \boldsymbol{Y} \right] \left[\boldsymbol{X} - \tilde{\boldsymbol{S}} \boldsymbol{\Psi}_{11}^{-1} \tilde{\boldsymbol{S}}^T \boldsymbol{Y} \boldsymbol{X} \right] \right\} \\ &= \mathrm{tr} \left(\boldsymbol{U}_s^T \boldsymbol{U}_s \right) - \mathrm{tr} \left(\tilde{\boldsymbol{S}}^T \boldsymbol{Y} \tilde{\boldsymbol{S}} \boldsymbol{\Psi}_{11}^{-1} \right) \\ &= \mathrm{tr}(\boldsymbol{I}_K) - \mathrm{tr}(\boldsymbol{I}_{\tilde{K}}) = K - \tilde{K}. \end{aligned} \tag{3.245}$$

To compute the second term in (3.243), first note that $\boldsymbol{S}\boldsymbol{D}\boldsymbol{S}^T = \tilde{\boldsymbol{S}}\tilde{\boldsymbol{D}}\tilde{\boldsymbol{S}} + \bar{\boldsymbol{S}}\bar{\boldsymbol{D}}\bar{\boldsymbol{S}}$. We have

$$\begin{aligned}\operatorname{tr}\left(\boldsymbol{Q}\boldsymbol{Y}\boldsymbol{S}\boldsymbol{D}\boldsymbol{S}^T\boldsymbol{Y}\boldsymbol{Q}^T\boldsymbol{X}\right) &= \operatorname{tr}\left(\boldsymbol{Q}\boldsymbol{Y}\bar{\boldsymbol{S}}\bar{\boldsymbol{D}}\bar{\boldsymbol{S}}^T\boldsymbol{Y}\boldsymbol{Q}^T\boldsymbol{X}\right) \\ &= \operatorname{tr}\left(\bar{\boldsymbol{S}}^T\boldsymbol{Y}\boldsymbol{Q}^T\boldsymbol{X}\boldsymbol{Q}\boldsymbol{Y}\bar{\boldsymbol{S}}\bar{\boldsymbol{D}}\right), \end{aligned} \tag{3.246}$$

where the first equality follows from the fact that $\boldsymbol{Q}\boldsymbol{Y}\tilde{\boldsymbol{S}} = \boldsymbol{0}$. Moreover,

$$\begin{aligned}\bar{\boldsymbol{S}}^T\boldsymbol{Y}\boldsymbol{Q}^T\boldsymbol{X}\boldsymbol{Q}\boldsymbol{Y}\bar{\boldsymbol{S}} &= \left(\bar{\boldsymbol{S}}^T\boldsymbol{Y} - \boldsymbol{\Psi}_{12}^T\boldsymbol{\Psi}_{11}^{-1}\tilde{\boldsymbol{S}}^T\boldsymbol{Y}\right)\boldsymbol{X}\left(\boldsymbol{Y}\bar{\boldsymbol{S}} - \boldsymbol{Y}\tilde{\boldsymbol{S}}\boldsymbol{\Psi}_{11}^{-1}\boldsymbol{\Psi}_{12}\right) \\ &= \boldsymbol{\Psi}_{22} - \boldsymbol{\Psi}_{12}^T\boldsymbol{\Psi}_{11}^{-1}\boldsymbol{\Psi}_{12}, \end{aligned} \tag{3.247}$$

where we used the fact that $\boldsymbol{Y}\boldsymbol{X}\boldsymbol{Y} = \boldsymbol{Y}$. In (3.246) $\bar{\boldsymbol{D}}$ is given by

$$\bar{\boldsymbol{D}} \triangleq \mathsf{diag}\left\{A_{\tilde{K}+1}^4\left(\boldsymbol{s}_{\tilde{K}+1}^T\boldsymbol{w}_1\right)^2, \ldots, A_K^4\left(\boldsymbol{s}_K^T\boldsymbol{w}_1\right)^2\right\}, \tag{3.248}$$

$$\boldsymbol{s}_{\tilde{K}+k}^T\boldsymbol{w}_1 = \left[\bar{\boldsymbol{S}}^T\boldsymbol{Y}\tilde{\boldsymbol{S}}\boldsymbol{\Psi}_{11}^{-1}\right]_{k,1} = \left[\boldsymbol{\Psi}_{12}^T\boldsymbol{\Psi}_{11}^{-1}\right]_{k,1}. \tag{3.249}$$

Substituting (3.247) and (3.249) into (3.246) we obtain the second term in (3.243):

$$\begin{aligned}\operatorname{tr}\left(\boldsymbol{Q}\boldsymbol{Y}\boldsymbol{S}\boldsymbol{D}\boldsymbol{S}^T\boldsymbol{Y}\boldsymbol{Q}^T\boldsymbol{X}\right) &= \operatorname{tr}\left[\left(\boldsymbol{\Psi}_{22} - \boldsymbol{\Psi}_{12}^T\boldsymbol{\Psi}_{11}^{-1}\boldsymbol{\Psi}_{12}\right)\bar{\boldsymbol{D}}\right], \\ &= \sum_{k=1}^{K-\tilde{K}} A_{\tilde{K}+k}^4\left[\boldsymbol{\Psi}_{12}^T\boldsymbol{\Psi}_{11}^{-1}\right]_{k,1}^2 \\ &\quad \cdot \left[\boldsymbol{\Psi}_{22} - \boldsymbol{\Psi}_{12}^T\boldsymbol{\Psi}_{11}^{-1}\boldsymbol{\Psi}_{12}\right]_{k,k}. \end{aligned} \tag{3.250}$$

Finally, we compute τ in the last term in (3.243). By definition,

$$\begin{aligned}\tau &= \sigma^2\boldsymbol{v}_1^T\boldsymbol{U}_s\boldsymbol{\Lambda}_s^{-1}\left(\boldsymbol{\Lambda}_s - \sigma^2\boldsymbol{I}_K\right)^{-2}\boldsymbol{U}_s^T\boldsymbol{v}_1 \\ &= \sigma^2\,\tilde{\boldsymbol{e}}_1^T\boldsymbol{\Psi}_{11}^{-1}\underbrace{\tilde{\boldsymbol{S}}^T\boldsymbol{U}_s\boldsymbol{\Lambda}_s^{-1}\boldsymbol{U}_s^T}_{\tilde{\boldsymbol{M}}^T}\boldsymbol{U}_s \\ &\quad \left(\boldsymbol{\Lambda}_s - \sigma^2\boldsymbol{I}_K\right)^{-1}\boldsymbol{U}_s^T\underbrace{\boldsymbol{U}_s\left(\boldsymbol{\Lambda}_s - \sigma^2\boldsymbol{I}_K\right)^{-1}\boldsymbol{U}_s^T\tilde{\boldsymbol{S}}}_{\tilde{\boldsymbol{D}}}\boldsymbol{\Psi}_{11}^{-1}\tilde{\boldsymbol{e}}_1 \\ &= \sigma^2\,\tilde{\boldsymbol{e}}_1^T\boldsymbol{\Psi}_{11}^{-1}\tilde{\boldsymbol{E}}^T\boldsymbol{A}^{-2} \\ &\quad \left(\boldsymbol{R} + \sigma^2\boldsymbol{A}^{-2}\right)^{-1}\boldsymbol{S}^T\underbrace{\boldsymbol{U}_s\left(\boldsymbol{\Lambda}_s - \sigma^2\boldsymbol{I}_K\right)^{-1}\boldsymbol{U}_s^T\boldsymbol{S}}_{\boldsymbol{D}=\boldsymbol{S}\boldsymbol{R}^{-1}\boldsymbol{A}^{-2}}\boldsymbol{R}^{-1}\boldsymbol{A}^{-2}\tilde{\boldsymbol{E}}\boldsymbol{\Psi}_{11}^{-1}\tilde{\boldsymbol{e}}_1 \\ &= \sigma^2\left[\boldsymbol{\Psi}_{11}^{-1}\boldsymbol{\Xi}\boldsymbol{\Psi}_{11}^{-1}\right]_{1,1}, \end{aligned} \tag{3.251}$$

where

$$\tilde{\boldsymbol{M}} \triangleq \boldsymbol{U}_s \boldsymbol{\Lambda}_s^{-1} \boldsymbol{U}_s^T \tilde{\boldsymbol{S}} = \boldsymbol{S} \left(\boldsymbol{R} + \sigma^2 \boldsymbol{A}^{-2}\right)^{-1} \boldsymbol{A}^{-2} \tilde{\boldsymbol{E}}, \tag{3.252}$$

$$\tilde{\boldsymbol{D}} \triangleq \boldsymbol{U}_s \left(\boldsymbol{\Lambda}_s - \sigma^2 \boldsymbol{I}_K\right)^{-1} \boldsymbol{U}_s^T \tilde{\boldsymbol{S}} = \boldsymbol{S} \boldsymbol{R}^{-1} \boldsymbol{A}^{-2} \tilde{\boldsymbol{E}}, \tag{3.253}$$

$$\boldsymbol{\Xi} \triangleq \left[\boldsymbol{A}^{-2} \left(\boldsymbol{R} + \sigma^2 \boldsymbol{A}^{-2}\right)^{-1} \boldsymbol{A}^{-2} \boldsymbol{R}^{-1} \boldsymbol{A}^{-2}\right]_{1:\tilde{K},1:\tilde{K}}, \tag{3.254}$$

with $\tilde{\boldsymbol{E}} \triangleq [\boldsymbol{e}_1 \cdots \boldsymbol{e}_{\tilde{K}}]$. Substituting (3.244), (3.245), (3.250), and (3.251) into (3.243), we have

$$\begin{aligned} \operatorname{tr}(\boldsymbol{C}_w \boldsymbol{C}_r) = (K - \tilde{K}) & \left[\boldsymbol{\Psi}_{11}^{-1}\right]_{1,1} \\ & - 2 \sum_{k=1}^{K-\tilde{K}} A_{\tilde{K}+k}^4 \left[\boldsymbol{\Psi}_{12}^T \boldsymbol{\Psi}_{11}^{-1}\right]_{k,1}^2 \left[\boldsymbol{\Psi}_{22} - \boldsymbol{\Psi}_{12}^T \boldsymbol{\Psi}_{11}^{-1} \boldsymbol{\Psi}_{12}\right]_{k,k} \\ & + (N - K) \sigma^4 \left[\boldsymbol{\Psi}_{11}^{-1} \boldsymbol{\Xi} \boldsymbol{\Psi}_{11}^{-1}\right]_{1,1}. \end{aligned} \tag{3.255}$$

Moreover, we have

$$\boldsymbol{s}_k^T \boldsymbol{w}_1 = \begin{cases} 1, & k = 1, \\ 0, & 1 < k \le \tilde{K}, \\ \left[\boldsymbol{\Psi}_{12}^T \boldsymbol{\Psi}_{11}^{-1}\right]_{k-\tilde{K},1}, & \tilde{K} < k \le K. \end{cases} \tag{3.256}$$

Next we compute $\|\boldsymbol{w}_1\|^2$. Since

$$\boldsymbol{w}_1 = \boldsymbol{Y} \tilde{\boldsymbol{S}} \left(\tilde{\boldsymbol{S}}^T \boldsymbol{Y} \tilde{\boldsymbol{S}}\right)^{-1} \tilde{\boldsymbol{e}}_1 = \tilde{\boldsymbol{M}} \boldsymbol{\Psi}_{11}^{-1} \tilde{\boldsymbol{e}}_1, \tag{3.257}$$

$$\tilde{\boldsymbol{M}}^T \tilde{\boldsymbol{M}} = \left[\boldsymbol{A}^{-2} \left(\boldsymbol{R} + \sigma^2 \boldsymbol{A}^{-2}\right)^{-1} \boldsymbol{R} \left(\boldsymbol{R} + \sigma^2 \boldsymbol{A}^{-2}\right)^{-1} \boldsymbol{A}^{-2}\right]_{1:\tilde{K},1:\tilde{K}} \triangleq \boldsymbol{\Pi}, \tag{3.258}$$

we have

$$\|\boldsymbol{w}_1\|^2 = \left[\boldsymbol{\Psi}_{11}^{-1} \boldsymbol{\Pi} \boldsymbol{\Psi}_{11}^{-1}\right]_{1,1}. \tag{3.259}$$

By (3.255)–(3.259) we obtain the corollary.

SINR Calculation for Example 2

Substituting (2.338)–(2.340) and $\boldsymbol{A}^2 = A^2 \boldsymbol{I}_K$ into (3.66)–(3.68), we have

$$A^2 \cdot \boldsymbol{\Psi}_{11} = a(1 - \rho) \tilde{\boldsymbol{I}} + (a\rho + b) \tilde{\mathbf{1}} \tilde{\mathbf{1}}^T, \tag{3.260}$$

$$A^2 \cdot \boldsymbol{\Psi}_{22} = a(1 - \rho) \bar{\boldsymbol{I}} + (a\rho + b) \bar{\mathbf{1}} \bar{\mathbf{1}}^T, \tag{3.261}$$

$$A^2 \cdot \boldsymbol{\Psi}_{12} = (a\rho + b) \tilde{\mathbf{1}} \bar{\mathbf{1}}^T, \tag{3.262}$$

$$A^4 \cdot \boldsymbol{\Pi} = a'(1 - \rho) \tilde{\boldsymbol{I}} + (a'\rho + b') \tilde{\mathbf{1}} \tilde{\mathbf{1}}^T, \tag{3.263}$$

$$A^6 \cdot \boldsymbol{\Xi} = a''(1 - \rho) \tilde{\boldsymbol{I}} + (a''\rho + b'') \tilde{\mathbf{1}} \tilde{\mathbf{1}}^T, \tag{3.264}$$

where $\tilde{\boldsymbol{I}} \triangleq \boldsymbol{I}_{\tilde{K}}$, $\bar{\boldsymbol{I}} \triangleq \boldsymbol{I}_{K-\tilde{K}}$, $\tilde{\mathbf{1}}$ denotes an all-1 $\tilde{K}$-vector, and $\bar{\mathbf{1}}$ denotes an all-1 $(K-\tilde{K})$-vector. After some manipulations, we obtain the following expressions:

$$A^{-2} \cdot \boldsymbol{\Psi}_{11}^{-1} = \frac{1}{a(1-\rho)} \left[\tilde{\boldsymbol{I}} - \frac{a\rho + b}{a(1-\rho) + \tilde{K}(a\rho+b)} \tilde{\mathbf{1}}\tilde{\mathbf{1}}^T \right], \tag{3.265}$$

$$A^{-2} \cdot \boldsymbol{\Psi}_{22}^{-1} = \frac{1}{a(1-\rho)} \left[\bar{\boldsymbol{I}} - \frac{a\rho + b}{a(1-\rho) + (K-\tilde{K})(a\rho+b)} \bar{\mathbf{1}}\bar{\mathbf{1}}^T \right], \tag{3.266}$$

$$\boldsymbol{\Psi}_{12}^{T}\boldsymbol{\Psi}_{11}^{-1} = \frac{a\rho + b}{a(1-\rho) + \tilde{K}(a\rho+b)} \bar{\mathbf{1}}\tilde{\mathbf{1}}^T, \tag{3.267}$$

$$A^{2} \cdot \boldsymbol{\Psi}_{12}^{T}\boldsymbol{\Psi}_{11}^{-1}\boldsymbol{\Psi}_{12} = \frac{\tilde{K}(a\rho + b)^2}{a(1-\rho) + \tilde{K}(a\rho+b)} \bar{\mathbf{1}}\bar{\mathbf{1}}^T, \tag{3.268}$$

$$\begin{aligned}\boldsymbol{\Psi}_{11}^{-1}\boldsymbol{\Pi}\boldsymbol{\Psi}_{11}^{-1} = \frac{1}{a^2(1-\rho)^2} \Bigg\{ & a'\rho + b' - \frac{a\rho+b}{a(1-\rho)+\tilde{K}(a\rho+b)} \Big[2(1-\rho)a' \\ & + (\tilde{K}+1)(a'\rho+b')\Big] + \left[\frac{a\rho+b}{a(1-\rho)+\tilde{K}(a\rho+b)}\right]^2 \tilde{K}\Big[a'(1-\rho) \\ & + \tilde{K}(a'\rho+b')\Big]\Bigg\} \tilde{\mathbf{1}}\tilde{\mathbf{1}}^T + \frac{a'}{a^2(1-\rho)}\tilde{\boldsymbol{I}},\end{aligned} \tag{3.269}$$

$$\begin{aligned}A^2 \cdot \boldsymbol{\Psi}_{11}^{-1}\boldsymbol{\Xi}\boldsymbol{\Psi}_{11}^{-1} = \frac{1}{a^2(1-\rho)^2} \Bigg\{ & a''\rho + b'' - \frac{a\rho+b}{a(1-\rho)+\tilde{K}(a\rho+b)} \Big[2(1-\rho)a'' \\ & + (\tilde{K}+1)(a''\rho+b'')\Big] + \left[\frac{a\rho+b}{a(1-\rho)+\tilde{K}(a\rho+b)}\right]^2 \tilde{K}\Big[a''(1-\rho) \\ & + \tilde{K}(a''\rho+b'')\Big]\Bigg\} \tilde{\mathbf{1}}\tilde{\mathbf{1}}^T + \frac{a''}{a^2(1-\rho)}\tilde{\boldsymbol{I}}.\end{aligned} \tag{3.270}$$

Substituting (3.265)–(3.270) into (3.69)–(3.72), and letting

$$\alpha \triangleq \left[\boldsymbol{\Psi}_{12}^{T}\boldsymbol{\Psi}_{11}^{-1}\right]_{k,1}, \tag{3.271}$$

$$\beta \triangleq \frac{\sigma^2}{A^2}\left[\boldsymbol{\Psi}_{11}^{-1}\boldsymbol{\Pi}\boldsymbol{\Psi}_{11}^{-1}\right]_{1,1}, \tag{3.272}$$

$$\gamma \triangleq A^{-2}\left[\boldsymbol{\Psi}_{11}^{-1}\right]_{1,1}, \tag{3.273}$$

$$\eta \triangleq A^{2}\left[\boldsymbol{\Psi}_{22} - \boldsymbol{\Psi}_{12}^{T}\boldsymbol{\Psi}_{11}^{-1}\boldsymbol{\Psi}_{12}\right]_{1,1}, \tag{3.274}$$

$$\mu \triangleq A^{2}\left[\boldsymbol{\Psi}_{11}^{-1}\boldsymbol{\Xi}\boldsymbol{\Psi}_{11}^{-1}\right]_{1,1}, \tag{3.275}$$

we obtain (3.82).

3.6.2 Proofs for Section 3.3.2

Proof of Theorem 3.2

We prove this theorem for the case of a linear group-blind hybrid detector (i.e., $\boldsymbol{v} = \tilde{\boldsymbol{d}}_1$). The proof for a linear group-blind MMSE detector is essentially the same.

Denote $\boldsymbol{e}_k$ as the kth unit vector in $\mathbb{R}^N$. Let $\boldsymbol{Q}_1$ be an orthogonal transformation such that

$$\boldsymbol{Q}_1^T \boldsymbol{s}_k = \boldsymbol{e}_{N-\tilde{K}+k}, \qquad k = 1, \ldots, \tilde{K}. \tag{3.276}$$

For any $\boldsymbol{v} \in \mathbb{R}^N$, denote $\boldsymbol{v}^{q_1} \triangleq \boldsymbol{Q}_1^T \boldsymbol{v}$. The corresponding projection matrix in the $\boldsymbol{Q}_1$-rotated coordinate system is

$$\bar{\boldsymbol{P}}^{q_1} \triangleq \boldsymbol{I}_N - \tilde{\boldsymbol{S}}^{q_1} \left[(\tilde{\boldsymbol{S}}^{q_1})^T \tilde{\boldsymbol{S}}^{q_1}\right]^{-1} (\tilde{\boldsymbol{S}}^{q_1})^T \tag{3.277}$$

$$= \boldsymbol{I}_N - \mathsf{diag}(\boldsymbol{0}, \boldsymbol{I}_{\tilde{K}}) \tag{3.278}$$

$$= \mathsf{diag}(\boldsymbol{I}_{N-\tilde{K}}, \boldsymbol{0}). \tag{3.279}$$

Denote

$$\boldsymbol{C}_r^{q_1} \triangleq E\left\{\boldsymbol{r}^{q_1}(\boldsymbol{r}^{q_1})^T\right\} = \boldsymbol{Q}_1^T \boldsymbol{C}_r \boldsymbol{Q}_1 \tag{3.280}$$

$$= \begin{bmatrix} \boldsymbol{C}_{11}^{q_1} & \boldsymbol{C}_{12}^{q_1} \\ \boldsymbol{C}_{12}^{q_1 T} & \boldsymbol{C}_{22}^{q_1} \end{bmatrix}, \tag{3.281}$$

where the dimension of $\boldsymbol{C}_{11}^{q_1}$ is $(N - \tilde{K}) \times (N - \tilde{K})$. Hence

$$\bar{\boldsymbol{P}}^{q_1} \boldsymbol{C}_r^{q_1} \bar{\boldsymbol{P}}^{q_1} = \begin{bmatrix} \boldsymbol{C}_{11}^{q_1} & \boldsymbol{0} \\ \boldsymbol{0} & \boldsymbol{0} \end{bmatrix}. \tag{3.282}$$

Let the eigendecomposition of $\boldsymbol{C}_{11}^{q_1}$ be[1]

$$\boldsymbol{C}_{11} = \boldsymbol{U}_1 \bar{\boldsymbol{\Lambda}} \boldsymbol{U}_1^T = \boldsymbol{U}_{s,1} \bar{\boldsymbol{\Lambda}}_s \boldsymbol{U}_{s,1}^T + \sigma^2 \boldsymbol{U}_{n,1} \boldsymbol{U}_{n,1}^T. \tag{3.283}$$

Define another orthogonal transformation,

$$\boldsymbol{Q}_2 = \begin{bmatrix} \boldsymbol{U}_1^T & \boldsymbol{0} \\ \boldsymbol{0} & \boldsymbol{I}_{\tilde{K}} \end{bmatrix}. \tag{3.284}$$

For any $\boldsymbol{v} \in \mathbb{R}^N$, denote $\boldsymbol{v}^q \triangleq \boldsymbol{Q}_2^T \boldsymbol{Q}_1^T \boldsymbol{v}$. In what follows, we compute the asymptotic covariance matrix of the detector in the $\boldsymbol{Q}_1\boldsymbol{Q}_2$-rotated coordinate system. In this new coordinate system, we have

$$\boldsymbol{C}_r^q \triangleq E\left\{\boldsymbol{r}^q(\boldsymbol{r}^q)^T\right\} = \boldsymbol{Q}_2^T \boldsymbol{Q}_1^T \boldsymbol{C}_r \boldsymbol{Q}_1 \boldsymbol{Q}_2 \tag{3.285}$$

$$= \begin{bmatrix} \boldsymbol{C}_{11}^q & \boldsymbol{C}_{12}^q \\ \boldsymbol{C}_{12}^{qT} & \boldsymbol{C}_{22}^q \end{bmatrix} = \begin{bmatrix} \bar{\boldsymbol{\Lambda}} & \boldsymbol{U}_1^T \boldsymbol{C}_{12}^{q_1} \\ \left(\boldsymbol{U}_1^T \boldsymbol{C}_{12}^{q_1}\right)^T & \boldsymbol{C}_{22}^{q_1} \end{bmatrix}, \tag{3.286}$$

$$\bar{\boldsymbol{P}}^q \triangleq \boldsymbol{I}_N - \tilde{\boldsymbol{S}}^q \left[(\tilde{\boldsymbol{S}}^q)^T \tilde{\boldsymbol{S}}^q\right]^{-1} \tilde{\boldsymbol{S}}^{qT} = \mathsf{diag}(\boldsymbol{I}_{N-\tilde{K}}, \boldsymbol{0}), \tag{3.287}$$

$$\bar{\boldsymbol{P}}^q \boldsymbol{C}_r^q \bar{\boldsymbol{P}}^q = \begin{bmatrix} \bar{\boldsymbol{\Lambda}} & \boldsymbol{0} \\ \boldsymbol{0} & \boldsymbol{0} \end{bmatrix}. \tag{3.288}$$

[1] The eigenvalues are unchanged by similarity transformations.

Furthermore, after rotation, $\tilde{\boldsymbol{d}}_1^q \in \mathbb{R}^N$ has the form

$$\tilde{\boldsymbol{d}}_1^q \triangleq \tilde{\boldsymbol{S}}^q \left[(\tilde{\boldsymbol{S}}^q)^T \tilde{\boldsymbol{S}}\right]^{-1} \boldsymbol{e}_1 = \begin{bmatrix} \boldsymbol{0} \\ \boldsymbol{p} \end{bmatrix}, \tag{3.289}$$

for some $\boldsymbol{p} \in \mathbb{R}^{\tilde{K}}$. After some manipulations, the form I group-blind hybrid detector in the new coordinate system has the form

$$\boldsymbol{w}_1^q \triangleq \left(\boldsymbol{I}_N - \bar{\boldsymbol{U}}_s^q \, [\bar{\boldsymbol{\Lambda}}_s^q]^{-1} \, [\bar{\boldsymbol{U}}_s^q]^T \boldsymbol{C}_r^q\right) \tilde{\boldsymbol{d}}_1^q = \begin{bmatrix} -\boldsymbol{E}_s \bar{\boldsymbol{\Lambda}}_s^{-1} \boldsymbol{E}_s^T \boldsymbol{C}_{12}^q \boldsymbol{p} \\ \boldsymbol{p} \end{bmatrix}, \tag{3.290}$$

where $\boldsymbol{E}_s$ consists of the first $K - \tilde{K}$ columns of $\boldsymbol{I}_{N-\tilde{K}}$ (i.e., $\boldsymbol{I}_{N-\tilde{K}} = [\boldsymbol{E}_s \boldsymbol{E}_n]$). Let the estimated autocorrelation matrix in the rotated coordinate system be

$$\hat{\boldsymbol{C}}_r^q \triangleq \boldsymbol{Q}_2^T \boldsymbol{Q}_1^T \hat{\boldsymbol{C}}_r \boldsymbol{Q}_1 \boldsymbol{Q}_2 = \begin{bmatrix} \hat{\boldsymbol{C}}_{11}^q & \hat{\boldsymbol{C}}_{12}^q \\ \hat{\boldsymbol{C}}_{12}^{qT} & \hat{\boldsymbol{C}}_{22}^q \end{bmatrix}. \tag{3.291}$$

Let the corresponding eigendecomposition of $\hat{\boldsymbol{C}}_{11}^q$ be

$$\hat{\boldsymbol{C}}_{11} = \hat{\boldsymbol{U}}_{s,1} \hat{\bar{\boldsymbol{\Lambda}}}_s \hat{\boldsymbol{U}}_{s,1}^T + \hat{\boldsymbol{U}}_{n,1} \hat{\bar{\boldsymbol{\Lambda}}}_s \hat{\boldsymbol{U}}_{n,1}^T. \tag{3.292}$$

Then the estimated detector in the same coordinate system is given by

$$\hat{\boldsymbol{w}}_1^q = \begin{bmatrix} -\hat{\boldsymbol{U}}_{s,1} \hat{\bar{\boldsymbol{\Lambda}}}_s^{-1} \hat{\boldsymbol{U}}_{s,1}^T \hat{\boldsymbol{C}}_{12}^q \boldsymbol{p} \\ \boldsymbol{p} \end{bmatrix}.$$

Note that in such a rotated coordinate system, estimation error occurs only in the first $K - \tilde{K}$ elements of $\hat{\boldsymbol{w}}_1^q$. Denote

$$\boldsymbol{m} \triangleq \underbrace{\boldsymbol{E}_s \bar{\boldsymbol{\Lambda}}_s^{-1} \boldsymbol{E}_s^T}_{\boldsymbol{Y}} \boldsymbol{C}_{12}^q \boldsymbol{p}, \tag{3.293}$$

$$\hat{\boldsymbol{m}} \triangleq \underbrace{\hat{\boldsymbol{U}}_{s,1} \hat{\bar{\boldsymbol{\Lambda}}}_s^{-1} \hat{\boldsymbol{U}}_{s,1}^T}_{\hat{\boldsymbol{Y}}} \hat{\boldsymbol{C}}_{12}^q \boldsymbol{p}. \tag{3.294}$$

Hence $\hat{\boldsymbol{m}}$ is a function of $\hat{\boldsymbol{C}}_r^q$, and its differential at $\boldsymbol{C}_r^q$ (i.e., at $\hat{\boldsymbol{U}}_{s,1} = \boldsymbol{E}_s$ and $\hat{\bar{\boldsymbol{\Lambda}}}_s = \bar{\boldsymbol{\Lambda}}_s$) is given by

$$\Delta \hat{\boldsymbol{m}} = \boldsymbol{Y} \Delta \boldsymbol{C}_{12}^q \boldsymbol{p} + \Delta \boldsymbol{Y} \boldsymbol{C}_{12}^q \boldsymbol{p}. \tag{3.295}$$

By Lemma 2.6, $\hat{\boldsymbol{m}}$ is then asymptotically Gaussian with a covariance matrix given by

$$\begin{aligned} \boldsymbol{C}_m \triangleq\ & \underbrace{E\left\{\Delta \boldsymbol{Y} \boldsymbol{C}_{12}^q \boldsymbol{p}\boldsymbol{p}^T [\boldsymbol{C}_{12}^q]^T \Delta \boldsymbol{Y}^T\right\}}_{\boldsymbol{T}_1} + \boldsymbol{Y} \underbrace{E\left\{\Delta \boldsymbol{C}_{12}^q \boldsymbol{p}\boldsymbol{p}^T \Delta [\boldsymbol{C}_{12}^q]^T\right\}}_{\boldsymbol{T}_2} \boldsymbol{Y}^T \\ & + \underbrace{E\left\{\Delta \boldsymbol{Y} \boldsymbol{C}_{12}^q \boldsymbol{p}\boldsymbol{p}^T \Delta [\boldsymbol{C}_{12}^q]^T\right\} \boldsymbol{Y}^T}_{\boldsymbol{T}_3} + \boldsymbol{Y} \underbrace{E\left\{\Delta \boldsymbol{C}_{12}^q \boldsymbol{p}\boldsymbol{p}^T [\boldsymbol{C}_{12}^q]^T \Delta \boldsymbol{Y}^T\right\}}_{\boldsymbol{T}_3^T}. \end{aligned} \tag{3.296}$$

We next compute the three terms $\boldsymbol{T}_1$, $\boldsymbol{T}_2$, and $\boldsymbol{T}_3$ in (3.296).

We first compute $\boldsymbol{T}_1$. Denote $\boldsymbol{z}_k$ and $\boldsymbol{x}_k$ as the subvectors of $\boldsymbol{s}_k^q$ containing, respectively, the first $N-\tilde{K}$ and the last $\tilde{K}$ elements of $\boldsymbol{s}_k^q$ (i.e., $\boldsymbol{s}_k^q = [\boldsymbol{z}_k^T \ \boldsymbol{x}_k^T]^T$) for $k = \tilde{K}+1, \ldots, K$. Let $\boldsymbol{Z} \triangleq [\boldsymbol{z}_{\tilde{K}+1} \cdots \boldsymbol{z}_K]$. It is clear that $\boldsymbol{C}_r \tilde{\boldsymbol{d}}_1 \in \mathsf{range}(\boldsymbol{U}_s)$, and therefore $\bar{\boldsymbol{P}} \boldsymbol{C}_r \tilde{\boldsymbol{d}}_1 \in \mathsf{range}(\bar{\boldsymbol{U}}_s)$. Expressed in the rotated coordinate system, we have $\boldsymbol{C}_{12}^q \boldsymbol{p} \in \mathsf{range}(\boldsymbol{E}_s)$. We can therefore apply Proposition 2.6 to $\boldsymbol{T}_1$ to obtain

$$
\begin{aligned}
\boldsymbol{T}_1 &= \left(\boldsymbol{m}^T \boldsymbol{C}_{12}^q \boldsymbol{p}\right) \boldsymbol{E}_s \bar{\boldsymbol{\Lambda}}_s^{-1} \boldsymbol{E}_s^T + \boldsymbol{m}\boldsymbol{m}^T \\
&\quad - 2\boldsymbol{E}_s \bar{\boldsymbol{\Lambda}}_s^{-1} \boldsymbol{E}_s^T \boldsymbol{Z} \boldsymbol{D}_1 \boldsymbol{Z}^T \boldsymbol{E}_s \bar{\boldsymbol{\Lambda}}_s^{-1} \boldsymbol{E}_s^T + \tau \boldsymbol{E}_n \boldsymbol{E}_n^T,
\end{aligned} \tag{3.297}
$$

with

$$
\tau \triangleq \sigma^2 \boldsymbol{p}^T \boldsymbol{C}_{12}^{qT} \boldsymbol{E}_s \bar{\boldsymbol{\Lambda}}_s^{-1} \left(\bar{\boldsymbol{\Lambda}}_s - \sigma^2 \boldsymbol{I}_{K-\tilde{K}}\right)^{-2} \boldsymbol{E}_s^T \boldsymbol{C}_{12}^q \boldsymbol{p}, \tag{3.298}
$$

$$
\boldsymbol{D}_1 \triangleq \mathsf{diag}\left\{ A_{\tilde{K}+1}^4 \left(\boldsymbol{z}_{\tilde{K}+1}^T \boldsymbol{m}\right)^2, \ldots, A_K^4 \left(\boldsymbol{z}_K^T \boldsymbol{m}\right)^2 \right\}. \tag{3.299}
$$

The term $\boldsymbol{T}_2$ can be computed following a similar derivation as in the proof of Theorem 1 for the DMI blind detector. Specifically, we have, similar to (2.302),

$$
\begin{aligned}
[\boldsymbol{T}_2]_{i,j} &= E\left\{\Delta \boldsymbol{C}_{12}^q \boldsymbol{p}\boldsymbol{p}^T \Delta [\boldsymbol{C}_{12}^q]^T\right\}_{i,j} \\
&= E\left\{ \sum_{m=1}^{\tilde{K}} \sum_{n=1}^{\tilde{K}} [\Delta \boldsymbol{C}_r^q]_{i,m+N-\tilde{K}} [\boldsymbol{p}]_m [\Delta \boldsymbol{C}_r^q]_{j,n+N-\tilde{K}} [\boldsymbol{p}]_n \right\} \\
&= \sum_{m=1}^{\tilde{K}} \sum_{n=1}^{\tilde{K}} \Bigg([\boldsymbol{C}_r^q]_{i,j} [\boldsymbol{C}_r^q]_{m+N-\tilde{K},n+N-\tilde{K}} + [\boldsymbol{C}_r^q]_{i,n+N-\tilde{K}} [\boldsymbol{C}_r^q]_{m+N-\tilde{K},j} \\
&\quad - 2 \sum_{\alpha=\tilde{K}+1}^{K} \sum_{m=1}^{\tilde{K}} \sum_{n=1}^{\tilde{K}} A_\alpha^4 [\boldsymbol{s}_\alpha^q]_i [\boldsymbol{s}_\alpha^q]_j [\boldsymbol{s}_\alpha^q]_{m+N-\tilde{K}} [\boldsymbol{s}_\alpha^q]_{n+N-\tilde{K}} \Bigg) [\boldsymbol{p}]_m [\boldsymbol{p}]_n \\
&= [\boldsymbol{C}_r^q]_{i,j} \underbrace{\left(\sum_{m=1}^{\tilde{K}} \sum_{n=1}^{\tilde{K}} [\boldsymbol{C}_r^q]_{m+N-\tilde{K},n+N-\tilde{K}} [\boldsymbol{p}]_m [\boldsymbol{p}]_n \right)}_{\boldsymbol{p}^T \boldsymbol{C}_{22}^q \boldsymbol{p}} \\
&\quad + \underbrace{\left(\sum_{m=1}^{\tilde{K}} [\boldsymbol{C}_r^q]_{m+N-\tilde{K},j} [\boldsymbol{p}]_m \right)}_{[\boldsymbol{C}_{12}^q \boldsymbol{p}]_j} \underbrace{\left(\sum_{n=1}^{\tilde{K}} [\boldsymbol{C}_r^q]_{i,n+N-\tilde{K}} [\boldsymbol{p}]_n \right)}_{[\boldsymbol{C}_{12}^q \boldsymbol{p}]_i} \\
&\quad - 2 \sum_{\alpha=\tilde{K}+1}^{K} [\boldsymbol{z}_\alpha]_i [\boldsymbol{z}_\alpha]_j A_\alpha^4 \underbrace{\left(\sum_{m=1}^{\tilde{K}} [\boldsymbol{x}_\alpha]_m [\boldsymbol{p}]_m \right)}_{\boldsymbol{x}_\alpha^T \boldsymbol{p}}^2 .
\end{aligned} \tag{3.300}
$$

Writing (3.300) in matrix form, we have

$$\boldsymbol{T}_2 = \left(\boldsymbol{p}^T \boldsymbol{C}_{22}^q \boldsymbol{p}\right) \bar{\boldsymbol{\Lambda}}_s + \left(\boldsymbol{C}_{12}^q \boldsymbol{p}\right)\left(\boldsymbol{C}_{12}^q \boldsymbol{p}\right)^T + \boldsymbol{Z}\boldsymbol{D}_2\boldsymbol{Z}^T, \tag{3.301}$$

with

$$\boldsymbol{D}_2 = \mathsf{diag}\left\{A_{\tilde{K}+1}^4 \left(\boldsymbol{x}_{\tilde{K}+1}^T \boldsymbol{p}\right)^2, \ldots, A_K^4 \left(\boldsymbol{x}_K^T \boldsymbol{p}\right)^2\right\}. \tag{3.302}$$

Hence the second term in (3.296) is

$$\begin{aligned}\boldsymbol{Y}\boldsymbol{T}_2\boldsymbol{Y}^T &= \left(\boldsymbol{E}_s \bar{\boldsymbol{\Lambda}}_s^{-1} \boldsymbol{E}_s^T\right) \boldsymbol{T}_2 \left(\boldsymbol{E}_s \bar{\boldsymbol{\Lambda}}_s^{-1} \boldsymbol{E}_s^T\right) \\ &= \left(\boldsymbol{p}^T \boldsymbol{C}_{22}^q \boldsymbol{p}\right) \boldsymbol{E}_{s,1} \bar{\boldsymbol{\Lambda}}_s^{-1} \boldsymbol{E}_{s,1}^T \\ &\quad + \boldsymbol{m}\boldsymbol{m}^T - 2\boldsymbol{E}_s \bar{\boldsymbol{\Lambda}}_s^{-1} \boldsymbol{E}_s^T \boldsymbol{Z}\boldsymbol{D}_2\boldsymbol{Z}^T \boldsymbol{E}_s \bar{\boldsymbol{\Lambda}}_s^{-1} \boldsymbol{E}_s^T,\end{aligned} \tag{3.303}$$

where we have used the fact that $\boldsymbol{E}_s^T \boldsymbol{E}_s = \boldsymbol{I}_{K-\tilde{K}}$, and the definition in (3.293).

Finally, we calculate $\boldsymbol{T}_3$. Denote $\boldsymbol{q} \triangleq \boldsymbol{C}_{12}^q \boldsymbol{p}$. By following the same derivation leading to (2.313), we get for $i \leq K - \tilde{K}$,

$$\left[\Delta \boldsymbol{Y} \boldsymbol{C}_{12}^q \boldsymbol{p}\right]_i = -\frac{1}{\bar{\lambda}_i} \sum_{k=1}^{K-\tilde{K}} \frac{1}{\bar{\lambda}_k} [\Delta \boldsymbol{C}_r^q]_{i,k} [\boldsymbol{q}]_k. \tag{3.304}$$

As before, we only have to consider $[\boldsymbol{T}_3]_{i,j}$ for $i, j \leq K - \tilde{K}$ or $i, j > K - \tilde{K}$. However, all terms corresponding to $i, j > K - \tilde{K}$ will be nulled out because of the multiplication of $\boldsymbol{Y}$ on $\boldsymbol{T}_3$. Using Lemma 2.5, we then get (for $i, j \leq K - \tilde{K}$)

$$\begin{aligned}[\boldsymbol{T}_3]_{i,j} &= E\left\{\Delta \boldsymbol{Y} \boldsymbol{C}_{12}^q \boldsymbol{p}\boldsymbol{p}^T [\Delta \boldsymbol{C}_{12}^q]^T\right\}_{i,j} \\ &= -\frac{1}{\bar{\lambda}_i} \sum_{k=1}^{K-\tilde{K}} \sum_{l=1}^{\tilde{K}} \frac{1}{\bar{\lambda}_k} E\{[\Delta \boldsymbol{C}_r^q]_{i,k} [\Delta \boldsymbol{C}_r^q]_{j,l+N-\tilde{K}}\} [\boldsymbol{q}]_k [\boldsymbol{p}]_l \\ &= -\frac{1}{\bar{\lambda}_i} \sum_{k=1}^{K-\tilde{K}} \sum_{l=1}^{\tilde{K}} \frac{1}{\bar{\lambda}_k} \Bigg(\bar{\lambda}_i \delta_{i=j} [\boldsymbol{C}_r^q]_{k,l+N-\tilde{K}} + \bar{\lambda}_j \delta_{k=j} [\boldsymbol{C}_r^q]_{i,l+N-\tilde{K}} \\ &\quad - 2 \sum_{\alpha=\tilde{K}+1}^{K} A_\alpha^4 [\boldsymbol{s}_\alpha^q]_i [\boldsymbol{s}_\alpha^q]_j [\boldsymbol{s}_\alpha^q]_k [\boldsymbol{s}_\alpha^q]_{l+N-\tilde{K}}\Bigg) [\boldsymbol{q}]_k [\boldsymbol{p}]_l \\ &= -\delta_{i=j} \sum_{k=1}^{K-\tilde{K}} \sum_{l=1}^{\tilde{K}} \frac{1}{\bar{\lambda}_k} [\boldsymbol{C}_r^q]_{k,l+N-\tilde{K}} [\boldsymbol{q}]_k [\boldsymbol{p}]_l - \frac{1}{\bar{\lambda}_i} [\boldsymbol{q}]_j \sum_{l=1}^{\tilde{K}} [\boldsymbol{C}_r^q]_{i,l+N-\tilde{K}} [\boldsymbol{p}]_l \\ &\quad + \frac{2}{\bar{\lambda}_i} \sum_{\alpha=\tilde{K}+1}^{K} A_\alpha^4 [\boldsymbol{s}_\alpha^q]_i [\boldsymbol{s}_\alpha^q]_j \sum_{k=1}^{K-\tilde{K}} \sum_{l=1}^{\tilde{K}} \frac{1}{\bar{\lambda}_k} [\boldsymbol{s}_\alpha^q]_k [\boldsymbol{s}_\alpha^q]_{l+N-\tilde{K}} [\boldsymbol{q}]_k [\boldsymbol{p}]_l\end{aligned}$$

$$= -\delta_{i=j} \sum_{k=1}^{K-\tilde{K}} \frac{1}{\bar{\lambda}_k} [\boldsymbol{q}]_k^2 - \frac{1}{\bar{\lambda}_i} [\boldsymbol{q}]_i [\boldsymbol{q}]_j$$

$$+ \frac{2}{\bar{\lambda}_i} \sum_{\alpha=\tilde{K}+1}^{K} A_\alpha^4 [\boldsymbol{z}_\alpha]_i [\boldsymbol{z}_\alpha]_j \underbrace{\left(\boldsymbol{z}_\alpha^T \boldsymbol{E}_s \bar{\boldsymbol{\Lambda}}_s^{-1} \boldsymbol{E}_s^T \boldsymbol{C}_{12}^q \boldsymbol{p}\right)}_{\boldsymbol{z}_\alpha^T \boldsymbol{m}} \left(\boldsymbol{x}_\alpha^T \boldsymbol{p}\right). \tag{3.305}$$

Writing this in matrix form, we have

$$\boldsymbol{T}_3 = -\left(\boldsymbol{q}^T \bar{\boldsymbol{\Lambda}}_s^{-1} \boldsymbol{q}\right) \boldsymbol{E}_s \boldsymbol{E}_s^T - \boldsymbol{E}_s \bar{\boldsymbol{\Lambda}}_s^{-1} \boldsymbol{E}_s^T \boldsymbol{q}\boldsymbol{q}^T + 2\boldsymbol{E}_s \bar{\boldsymbol{\Lambda}}_s^{-1} \boldsymbol{E}_s^T \boldsymbol{Z} \boldsymbol{D}_3 \boldsymbol{Z}^T, \tag{3.306}$$

$$\boldsymbol{D}_3 \triangleq \mathsf{diag}\left\{A_{\tilde{K}+1}^4 \left(\boldsymbol{z}_{\tilde{K}+1}^T \boldsymbol{m}\right)\left(\boldsymbol{x}_{\tilde{K}+1}^T \boldsymbol{p}\right), \ldots, A_K^4 \left(\boldsymbol{z}_K^T \boldsymbol{m}\right)\left(\boldsymbol{x}_K^T \boldsymbol{p}\right)\right\}. \tag{3.307}$$

Hence the third term in (3.296) is given by

$$\boldsymbol{T}_3 \boldsymbol{Y}^T = -\left(\boldsymbol{m}^T \boldsymbol{C}_{12}^q \boldsymbol{p}\right) \boldsymbol{E}_s \bar{\boldsymbol{\Lambda}}_s^{-1} \boldsymbol{E}_s^T - \boldsymbol{m}\boldsymbol{m}^T + 2\boldsymbol{E}_s \bar{\boldsymbol{\Lambda}}_s^{-1} \boldsymbol{E}_s^T \boldsymbol{Z} \boldsymbol{D}_3 \boldsymbol{Z}^T \boldsymbol{E}_s \bar{\boldsymbol{\Lambda}}_s^{-1} \boldsymbol{E}_s^T. \tag{3.308}$$

Substituting (3.297), (3.303), and (3.308) into (3.296), we obtain

$$\boldsymbol{C}_m = \left(\boldsymbol{p}^T \boldsymbol{C}_{22}^q \boldsymbol{p} - \boldsymbol{m}^T \boldsymbol{C}_{12}^q \boldsymbol{p}\right) \boldsymbol{E}_s \bar{\boldsymbol{\Lambda}}_s^{-1} \boldsymbol{E}_s^T + 2\boldsymbol{E}_s \bar{\boldsymbol{\Lambda}}_s^{-1} \boldsymbol{E}_s^T \boldsymbol{Z} \left(\sqrt{\boldsymbol{D}_1} - \sqrt{\boldsymbol{D}_2}\right)^2 \boldsymbol{Z}^T \boldsymbol{E}_s \bar{\boldsymbol{\Lambda}}_s^{-1} \boldsymbol{E}_s^T + \tau \boldsymbol{E}_n \boldsymbol{E}_n^T, \tag{3.309}$$

where $\boldsymbol{D}_1$ and $\boldsymbol{D}_2$ are given, respectively, by (3.299) and (3.302), and τ is given by (3.298). Theorem 3 is now easily obtained by transforming (3.309) back to the original coordinate system according to the following mappings: $\boldsymbol{E}_s \rightarrow \bar{\boldsymbol{U}}_s$, $\boldsymbol{E}_n \rightarrow \bar{\boldsymbol{U}}_n$, $\boldsymbol{Z} \rightarrow \bar{\boldsymbol{S}}$, $\boldsymbol{p}^T \boldsymbol{C}_{22}^q \boldsymbol{p} \rightarrow \tilde{\boldsymbol{d}}_1^T \boldsymbol{C}_r \tilde{\boldsymbol{d}}_1$, $\boldsymbol{m}^T \boldsymbol{C}_{12}^q \boldsymbol{p} \rightarrow \left[\bar{\boldsymbol{U}}_s \bar{\boldsymbol{\Lambda}}_s^{-1} \bar{\boldsymbol{U}}_s^T \boldsymbol{C}_r \tilde{\boldsymbol{d}}_1\right]^T \boldsymbol{C}_r \tilde{\boldsymbol{d}}_1$, and $\left(\boldsymbol{z}_k^T \boldsymbol{m} - \boldsymbol{x}_k^T \boldsymbol{p}\right) \rightarrow \boldsymbol{s}_k^T \boldsymbol{w}_1$.

Chapter 4

ROBUST MULTIUSER DETECTION IN NON-GAUSSIAN CHANNELS

4.1 Introduction

As we have seen in preceding chapters, the use of multiuser detection (or derivative signal processing techniques) can return performance in multiuser channels to that of corresponding single-user channels, or at least to a situation in which performance is no longer limited by the multiple-access interference (MAI). Thus far, our discussion of these problems has focused on the situation in which the ambient noise is additive white Gaussian noise (AWGN). This was an appropriate model in previous chapters, since the focus there was on the mitigation of the most severe noise source, the MAI. However, as increasingly practical techniques for multiuser detection become available, such as the methods discussed in Chapters 2 and 3, the situation in which practical multiple-access channels will be ambient-noise limited can be realistically envisioned.

In many physical channels, such as urban and indoor radio channels [44, 45, 319, 320, 323] and underwater acoustic channels [52, 321], the ambient noise is known through experimental measurements to be decidedly non-Gaussian, due to the impulsive nature of the human-made electromagnetic interference and of a great deal of natural noise as well. (For measurement results of impulsive noise in outdoor/indoor mobile and portable radio communications, see [44, 45] and the references therein.) It is widely known in the single-user context that non-Gaussian noise can be quite detrimental to the performance of conventional systems designed on the basis of a Gaussian noise assumption, whereas it can actually be beneficial to performance if appropriately modeled and ameliorated. Neither of these properties is surprising. The first is a result of the lack of robustness of linear and quadratic signal processing procedures to many types of non-Gaussian statistical behavior [226]. The second is a manifestation of the well-known least favorability of Gaussian channels [128].

In view of the lack of realism of an AWGN model for ambient noise arising in many practical channels in which multiuser detection techniques may be applied, natural questions arise concerning the applicability, robustness, and performance of multiuser detection techniques for non-Gaussian multiple-access channels. Although performance indices such as mean-square error (MSE) and signal-to-interference-plus-noise ratio (SINR) for linear multiuser detectors are not affected by the amplitude distribution of the noise (only the spectrum matters), the more crucial bit-error rate can depend heavily on the shape of the noise distribution. Results of an early study of error rates in non-Gaussian direct-sequence code-division multiple-access (DS-CDMA) channels are given in [1–3], in which the performance of the conventional and modified conventional (linear matched filter) detectors is shown to depend significantly on the shape of the ambient noise distribution. In particular, impulsive noise can severely degrade the error probability for a given level of ambient noise variance. In the context of multiple-access capability, this implies that fewer users can be supported with conventional detection in an impulsive channel than in a Gaussian channel. However, since non-Gaussian noise can, in fact, be beneficial to system performance if treated properly, the problem of joint mitigation of structured interference and non-Gaussian ambient noise is of interest [378]. An approach to this problem for narrowband interference (NBI) suppression in spread-spectrum systems is described in [133]. A further study [383] has shown that the performance gains afforded by maximum-likelihood (ML) multiuser detection in impulsive noise can be substantial compared to optimum multiuser detection based on a Gaussian noise assumption. However, the computational complexity of ML detection is quite high (even more so with non-Gaussian ambient noise), and therefore effective near-optimal multiuser detection techniques in non-Gaussian noise are needed. In this chapter we address the MAI mitigation problem in DS-CDMA channels with non-Gaussian ambient noise.

In the past, considerable research has been conducted to model the non-Gaussian phenomena encountered in practice which are characterized by sharp spikes, occasional bursts, and heavy outliers, resulting in a variety of statistical models, the most common of which include the statistically and physically derived Middleton mixture models [319–323], the empirical Gaussian mixtures, and other heavy-tailed distributions, such as the Weibull, the K, and the log-normal, as well as the stable models [357]. Particularly accurate are the Middleton models, which are based on a filtered-impulse mechanism and can be classified into three classes, A, B, and C. Interference in class A is coherent in narrowband receivers, causing a negligible number of transients. Interference in class B is impulsive, consisting of a large number of overlapping transients. Interference in class C is the sum of the other two types of interference. The Middleton model has been shown to describe actual impulsive interference phenomena with high fidelity; however, it is mathematically involved for signal processing applications. In this chapter we use the widely adopted two-term Gaussian mixture distribution (which gives a good approximation to the Middleton models) to model the non-Gaussian noise, and discuss various robust multiuser detection techniques based on such a model. In the end, we will show that these robust signal processing techniques are also very effective in ameliorating other types of non-Gaussian noise, such as symmetric stable noise.

This chapter is organized as follows. In Section 4.2 we discuss robust multiuser detection techniques based on M-regression. In Section 4.3 we present asymptotic performance analyses for the robust multiuser detectors. In Section 4.4 we discuss implementation issues arising in robust multiuser detection. In Section 4.5 we treat the topic of robust blind multiuser detection. In Section 4.6 we present improved versions of robust multiuser detectors based on local likelihood search. In Section 4.7 we discuss robust group-blind multiuser detection. In Section 4.8 we consider robust multiuser detection in multipath channels. Finally, in Section 4.9 we briefly introduce α-stable noise and illustrate the performance of various robust multiuser detectors in such noise. The proofs of some results in this chapter are appended in Section 4.10.

The following is a list of the algorithms appearing in this chapter.

- *Algorithm 4.1*: Robust multiuser detector—synchronous CDMA
- *Algorithm 4.2*: Robust blind multiuser detector—synchronous CDMA
- *Algorithm 4.3*: Adaptive robust blind multiuser detector—synchronous CDMA
- *Algorithm 4.4*: Robust multiuser detector based on slowest-descent-search—synchronous CDMA
- *Algorithm 4.5*: Robust group-blind multiuser detector—synchronous CDMA
- *Algorithm 4.6*: Robust blind multiuser detector—multipath CDMA
- *Algorithm 4.7*: Robust group-blind multiuser detector—multipath CDMA

4.2 Multiuser Detection via Robust Regression

4.2.1 System Model

For the sake of simplicity, we start the discussion in this chapter by focusing on a real-valued discrete-time synchronous CDMA signal model. At any time instant (until needed in Section 4.5, we will suppress the symbol index i), the received signal is the superposition of K users' signals, plus the ambient noise, given by (see Section 2.2.1)

$$\boldsymbol{r} = \sum_{k=1}^{K} A_k\, b_k \boldsymbol{s}_k + \boldsymbol{n} \tag{4.1}$$

$$= \boldsymbol{S}\boldsymbol{A}\boldsymbol{b} + \boldsymbol{n}, \tag{4.2}$$

where, as before, $\boldsymbol{s}_k = (1/\sqrt{N})[c_{0,k} \ \cdots \ c_{N-1,k}]^T$, $c_{j,k} \in \{+1, -1\}$, is the normalized signature waveform of the kth user; N is the processing gain; $b_k \in \{+1, -1\}$ and A_k are, respectively, the data bit and the amplitude of the kth user; $\boldsymbol{S} \triangleq [\boldsymbol{s}_1 \ \cdots \ \boldsymbol{s}_K]$; $\boldsymbol{A} \triangleq \mathsf{diag}(A_1, \ldots, A_K)$; $\boldsymbol{b} \triangleq [b_1 \ \cdots \ b_K]^T$; and $\boldsymbol{n} = [n_1 \cdots n_N]^T$ is a vector of independent and identically distributed (i.i.d.) ambient noise samples. As noted above, we adopt the commonly used two-term Gaussian mixture model for the additive noise samples $\{n_j\}$. The marginal probability density function (pdf) of this noise model has the form

$$f = (1-\epsilon)\mathcal{N}\left(0, \nu^2\right) + \epsilon\mathcal{N}\left(0, \kappa\nu^2\right), \tag{4.3}$$

with $\nu > 0$, $0 \le \epsilon < 1$, and $\kappa > 1$. Here the $\mathcal{N}\left(0, \nu^2\right)$ term represents the nominal background noise, and the $\mathcal{N}\left(0, \kappa\nu^2\right)$ term represents the impulsive component, with ϵ representing the probability that impulses occur. It is usually of interest to study the effects of variation in the shape of a distribution on the performance of the system, by varying the parameters ϵ and κ with fixed total noise variance

$$\sigma^2 \triangleq (1-\epsilon)\,\nu^2 + \epsilon\,\kappa\,\nu^2. \tag{4.4}$$

This model serves as an approximation to the more fundamental Middleton class A noise model [321, 600] and has been used extensively to model physical noise arising in radar, acoustic, and radio channels. In what follows we discuss some robust techniques for multiuser detection in non-Gaussian ambient noise CDMA channels, which are essentially robustified versions of the linear decorrelating multiuser detector.

4.2.2 Least-Squares Regression and Linear Decorrelator

Consider the synchronous signal model (4.2). Denote $\theta_k \triangleq A_k b_k$. Then (4.2) can be rewritten as

$$r_j = \sum_{k=1}^{K} c_{j,k}\theta_k + n_j, \qquad j = 1, \ldots, N, \tag{4.5}$$

or in matrix notation,

$$\boldsymbol{r} = \boldsymbol{S}\,\boldsymbol{\theta} + \boldsymbol{n}, \tag{4.6}$$

where $\boldsymbol{\theta} \triangleq [\theta_1\, \theta_2\, \cdots\, \theta_K]^T$. Consider the linear regression problem of estimating the K unknown parameters $\theta_1, \theta_2, \ldots, \theta_K$ from the N observations $r_1, r_2, \ldots, r_N$ in (4.5). Classically, this problem can be solved by minimizing the sum of squared errors (or squared residuals) [i.e., through the least-squares (LS) method]:

$$\begin{aligned} \hat{\boldsymbol{\theta}}_{LS} &= \arg\min_{\boldsymbol{\theta}} \sum_{j=1}^{N} \left(r_j - \sum_{k=1}^{K} c_{j,k}\theta_k \right)^2 \\ &= \arg\min_{\boldsymbol{\theta}} \left\| \boldsymbol{r} - \boldsymbol{S}\,\boldsymbol{\theta} \right\|^2. \end{aligned} \tag{4.7}$$

If $n_j \sim \mathcal{N}(0,\ \sigma^2)$, the pdf of the received signal $\boldsymbol{r}$ under the true parameters $\boldsymbol{\theta}$ is given by

$$f_{\boldsymbol{\theta}}(\boldsymbol{r}) = \left(2\pi\sigma^2\right)^{-(N/2)} \exp\left(-\frac{\left\|\boldsymbol{r} - \boldsymbol{S}\,\boldsymbol{\theta}\right\|^2}{2\sigma^2} \right). \tag{4.8}$$

It is easily seen from (4.8) that the maximum-likelihood estimate of $\boldsymbol{\theta}$ under the i.i.d. Gaussian noise assumption is given by the LS solution $\hat{\boldsymbol{\theta}}_{LS}$ in (4.7). Upon differentiating (4.7), $\hat{\boldsymbol{\theta}}_{LS}$ is then the solution to the following linear system equations

$$\sum_{j=1}^{N}\left(r_j - \sum_{l=1}^{K} c_{j,l}\,\theta_l\right) c_{j,k} = 0, \qquad k = 1, \ldots, K, \tag{4.9}$$

or in matrix form,

$$\boldsymbol{S}^T \boldsymbol{S}\, \boldsymbol{\theta} = \boldsymbol{S}^T \boldsymbol{r}. \tag{4.10}$$

Define the cross-correlation matrix of the signature waveforms of all users as $\boldsymbol{R} \triangleq \boldsymbol{S}^T \boldsymbol{S}$. Assuming that the user signature waveforms are linearly independent (i.e., $\boldsymbol{S}$ has a full column rank K), $\boldsymbol{R}$ is invertible, and the LS solution to (4.9) or (4.10) is given by

$$\begin{aligned}\hat{\boldsymbol{\theta}}_{LS} &= \left(\boldsymbol{S}^T \boldsymbol{S}\right)^{-1} \boldsymbol{S}^T \boldsymbol{r} \\ &= \boldsymbol{R}^{-1} \boldsymbol{S}^T \boldsymbol{r}.\end{aligned} \tag{4.11}$$

We observe from (4.11) that the LS estimate $\hat{\boldsymbol{\theta}}_{LS}$ is exactly the output of the linear decorrelating multiuser detector for the K users (cf. Proposition 2.1). This is not surprising, since the linear decorrelating detector gives the maximum likelihood estimate of the product of the amplitude and the data bit $\theta_k = A_k b_k$ in Gaussian noise [296]. Given the estimate $\hat{\theta}_k$, the estimated amplitude and the data bit are then determined by

$$\hat{A}_k = \left|\hat{\theta}_k\right|, \tag{4.12}$$

$$\hat{b}_k = \text{sign}\left(\hat{\theta}_k\right). \tag{4.13}$$

4.2.3 Robust Multiuser Detection via *M*-Regression

It is well known that the LS estimate is very sensitive to the tail behavior of the noise density [499]. Its performance depends significantly on the Gaussian assumption and even a slight deviation of the noise density from the Gaussian distribution can, in principle, cause a substantial degradation of the LS estimate. Since the linear decorrelating multiuser detector is in the form of the LS solution to a linear regression problem, it can be concluded that its performance is also sensitive to the tail behavior of the noise distribution. As will be demonstrated later, the performance of the linear decorrelating detector degrades substantially if the ambient noise deviates even slightly from a Gaussian distribution. In this section we consider some robust versions of the decorrelating multiuser detector, first developed in [553], which are nonlinear in nature. Robustness of an estimator refers to its performance insensitivity to small deviations in actual statistical behavior from the assumed underlying statistical model.

The LS estimate corresponding to (4.7) and (4.9) can be robustified by using the class of M-estimators proposed by Huber [203]. Instead of minimizing a sum of squared residuals as in (4.7), Huber proposed to minimize a sum of a less rapidly increasing function, ρ, of the residuals:

$$\hat{\boldsymbol{\theta}} = \arg \min_{\boldsymbol{\theta} \in \mathbb{R}^K} \sum_{j=1}^{N} \rho \left(r_j - \sum_{k=1}^{K} c_{j,k} \, \theta_k \right). \tag{4.14}$$

Suppose that ρ has a derivative $\psi \triangleq \rho'$, then the solution to (4.14) satisfies the implicit equation

$$\sum_{j=1}^{N} \psi \left(r_j - \sum_{l=1}^{K} c_{j,l} \, \theta_l \right) c_{j,k} = 0, \qquad k = 1, \ldots, K, \tag{4.15}$$

or in vector form,

$$\boldsymbol{S}^T \, \psi \, (\boldsymbol{r} - \boldsymbol{S} \, \boldsymbol{\theta}) = \boldsymbol{0}, \tag{4.16}$$

where $\psi(\boldsymbol{x}) \triangleq [\psi(x_1), \ldots, \psi(x_K)]^T$ for any $\boldsymbol{x} \in \mathbb{R}^K$ and $\boldsymbol{0}$ denotes an all-zero vector. An estimator defined by (4.14) or (4.15) is called an *M-estimator*, from "maximum-likelihood-type estimator" [203], since the choice of $\rho(x) = -\log f(x)$ gives the ordinary maximum-likelihood estimator. If ρ is convex, then (4.14) and (4.15) are equivalent; otherwise, (4.15) is still very useful in searching for the solution to (4.14). To achieve robustness, it is necessary that ψ be bounded and continuous. Usually, to achieve high efficiency when the noise is actually Gaussian, we require that $\psi(x) \approx x$ for x small. Consistency of the estimate requires that $E\{\psi(n_j)\} = 0$. Hence for symmetric noise densities, ψ is usually odd-symmetric. We next consider some specific choices of the penalty function ρ and the corresponding derivative ψ.

Linear Decorrelating Detector

The linear decorrelating detector, which is simply the LS estimator, corresponds to choosing the penalty function and its derivative, respectively, as

$$\rho_{LS}(x) = \frac{x^2}{2\alpha}, \tag{4.17}$$

$$\psi_{LS}(x) = \frac{x}{\alpha}, \tag{4.18}$$

where α is any positive constant. Notice that the linear decorrelating detector is scale invariant.

Maximum-Likelihood Decorrelating Detector

Assume that the i.i.d. noise samples have a pdf f. Then the likelihood function of the received signal $\boldsymbol{r}$ under the true parameters $\boldsymbol{\theta}$ is given by

$$\begin{aligned}\mathcal{L}_{\boldsymbol{\theta}}(\boldsymbol{r};\, f) &\triangleq -\log \prod_{j=1}^{N} f\left(r_j - \sum_{k=1}^{K} c_{j,k}\theta_k\right)\\ &= -\sum_{j=1}^{N} \log f\left(r_j - \sum_{k=1}^{K} c_{j,k}\theta_k\right). \end{aligned} \tag{4.19}$$

Therefore, the maximum-likelihood decorrelating detector in non-Gaussian noise with pdf f (in the sense that it gives the maximum-likelihood estimate of the product of the amplitude and data bit $\theta_k \triangleq A_k b_k$) is given by the M-estimator with the penalty function and its derivative, respectively, chosen as

$$\rho_{ML}(x) = -\log f(x), \tag{4.20}$$

$$\psi_{ML}(x) = -\frac{f'(x)}{f(x)}. \tag{4.21}$$

Minimax Decorrelating Detector

We next consider a robust decorrelating detector in a minimax sense based on Huber's minimax M-estimator [203]. Huber considered the robust location estimation problem. Suppose that we have one-dimensional i.i.d. observations $X_1, \ldots, X_n$. The observations belong to some subset $\mathcal{X}$ of the real line $\mathbb{R}$. A parametric model consists of a family of probability distributions F_θ on $\mathcal{X}$, where the unknown parameter θ belongs to some parameter space Θ. When estimating location in the model $\mathcal{X} = \mathbb{R}$, $\Theta = \mathbb{R}$, the parametric model is $F_\theta(x) = F(x - \theta)$, and the M-estimator is determined by a ψ-function of the type $\psi(x, \theta) = \psi(x - \theta)$; that is, the M-estimate of the location parameter θ is given by the solution to the equation

$$\sum_{i=1}^{n} \psi(x_i - \theta) = 0, \tag{4.22}$$

assuming that the noise distribution function belongs to the set of *ϵ-contaminated Gaussian models* given by

$$\mathcal{P}_\epsilon \triangleq \left\{(1-\epsilon)\mathcal{N}(0, \nu^2) + \epsilon H;\, H \text{ is a symmetric distribution}\right\}, \tag{4.23}$$

where $0 < \epsilon < 1$ is fixed and ν^2 is the variance of the nominal Gaussian distribution. It can be shown that within mild regularity, the asymptotic variance of an M-estimator of the location parameter θ defined by (4.22) at a distribution function $F \in \mathcal{P}_\epsilon$ is given by [203]

$$V(\psi;\, F) = \frac{\displaystyle\int \psi^2 \, \mathrm{d}F}{\left(\displaystyle\int \psi' \, \mathrm{d}F\right)^2}. \tag{4.24}$$

Huber's idea was to minimize the maximal asymptotic variance over $\mathcal{P}_\epsilon$, that is, to find the M-estimator ψ_0 that satisfies

$$\sup_{F \in \mathcal{P}_\epsilon} V(\psi_0; F) = \inf_{\psi} \sup_{F \in \mathcal{P}_\epsilon} V(\psi; F). \tag{4.25}$$

This is achieved by finding the least favorable distribution F_0, that is, the distribution function that minimizes the Fisher information

$$I(F) = \int \left(\frac{F''}{F'} \right)^2 \mathrm{d}F, \tag{4.26}$$

over all $F \in \mathcal{P}_\epsilon$. Then $\psi_0 = -F_0''/F_0'$ yields the maximum-likelihood estimator for this least favorable distribution. Huber showed that the Fisher information (4.26) is minimized over $\mathcal{P}_\epsilon$ the distribution with pdf

$$f_0(x) = \begin{cases} \dfrac{1-\epsilon}{\sqrt{2\pi}\nu} \exp\left(\dfrac{x^2}{2\nu^2}\right) & \text{for } |x| \leq \gamma\nu^2, \\ \dfrac{1-\epsilon}{\sqrt{2\pi}\nu} \exp\left(\dfrac{\gamma^2\nu^2}{2} - \gamma|x|\right) & \text{for } |x| > \gamma\nu^2, \end{cases} \tag{4.27}$$

where k, ϵ, and ν are connected through

$$\frac{\phi(\gamma\nu)}{\gamma\nu} - Q(\gamma\nu) = \frac{\epsilon}{2(1-\epsilon)}, \tag{4.28}$$

with $\phi(x) \triangleq (1/\sqrt{2\pi}) \exp\left(-x^2/2\right)$, and $Q(x) \triangleq (1/\sqrt{2\pi}) \int_x^\infty \exp\left(-t^2/2\right) \mathrm{d}t$. The corresponding minimax M-estimator is then determined by the Huber penalty function and its derivative, given, respectively, by

$$\rho_H(x) = \begin{cases} \dfrac{x^2}{2\nu^2} & \text{for } |x| \leq \gamma\nu^2, \\ \dfrac{\gamma^2\nu^2}{2} - \gamma|x| & \text{for } |x| > \gamma\nu^2, \end{cases} \tag{4.29}$$

$$\psi_H(x) = \begin{cases} \dfrac{x}{\nu^2} & \text{for } |x| \leq \gamma\nu^2, \\ \gamma\, \mathrm{sign}(x) & \text{for } |x| > \gamma\nu^2. \end{cases} \tag{4.30}$$

The minimax robust decorrelating detector is obtained by substituting ρ_H and ψ_H into (4.14) and (4.15).

Assuming that the noise distribution has the ϵ-mixture density (4.3), in Fig. 4.1 we plot the ψ functions for the three types of decorrelating detectors discussed above for the cases $\epsilon = 0.1$ and $\epsilon = 0.01$, respectively. Note that for small measurement x, both $\psi_{ML}(x)$ and $\psi_H(x)$ are essentially linear, and they coincide with $\psi_{LS}(x)$; for large measurement x, $\psi_{ML}(x)$ approximates a blanker, whereas $\psi_H(x)$ acts as a clipper. Thus the action of the nonlinear function ψ in the nonlinear decorrelators defined by (4.15) relative to the linear decorrelator defined by (4.9) is clear in this case. Namely, the linear decorrelator incorporates the residuals linearly into the signal estimate, whereas the nonlinear decorrelators incorporate small residuals linearly but blank or clip larger residuals that are likely to be the result of noise impulses.

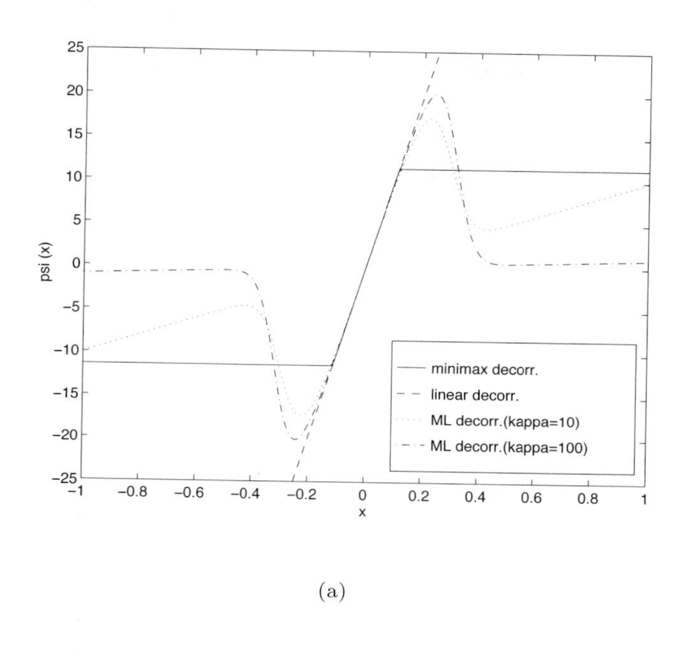

(a)

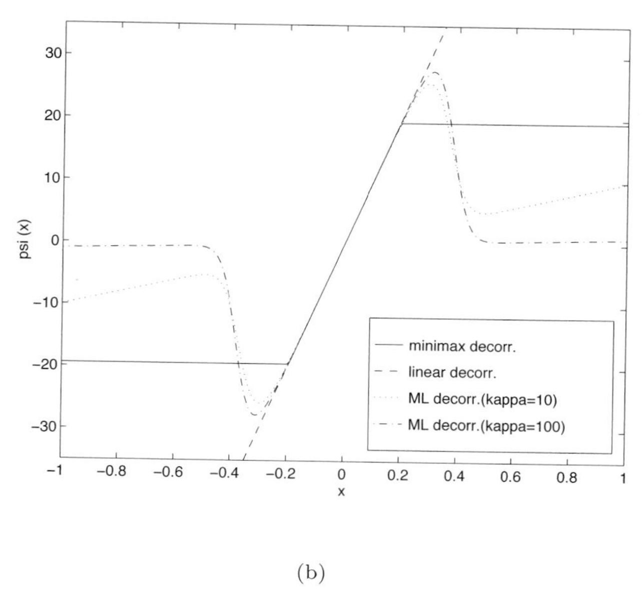

(b)

Figure 4.1. The ψ functions for a linear decorrelator, maximum-likelihood decorrelator, and minimax decorrelator under the Gaussian mixture noise model. The variance of the nominal Gaussian distribution is $\nu^2 = 0.01$. (a) $\epsilon = 0.1$. The cutoff point for the Huber estimator is obtained by solving (4.28), resulting in $\gamma = 11.40$. (b) $\epsilon = 0.01$, $\gamma = 19.45$.

4.3 Asymptotic Performance of Robust Multiuser Detection

4.3.1 Influence Function

The influence function (IF) introduced by Hampel [170, 203], is an important tool used to study robust estimators. It measures the influence of a vanishingly small contamination of the underlying distribution on the estimator. It is assumed that the estimator can be defined as a functional T, operating on the empirical distribution function of the observation F_n, [i.e., $T = T(F_n)$] and that the estimator is consistent as $n \to \infty$ [i.e., $T(F) = \lim_{n\to\infty} T(F_n)$, where F is the underlying distribution]. The IF is defined as

$$\mathrm{IF}(x; T, F) \triangleq \lim_{t\to 0} \frac{T[(1-t)F + t\Delta_x] - T(F)}{t}, \tag{4.31}$$

where Δ_x is the distribution that puts a unit mass at x. Roughly speaking, the influence function $\mathrm{IF}(x; T, F)$ is the first derivative of the statistic T at an underlying distribution F and at the coordinate x. We next compute the influence function of the nonlinear decorrelating multiuser detectors defined by (4.15).

Denote the jth row of the matrix $\boldsymbol{S}$ by $\boldsymbol{\xi}_j^T$ (i.e., $\boldsymbol{\xi}_j \triangleq [c_{j,1} \cdots c_{j,K}]^T$). Assume that the signature waveforms of all users are chosen randomly and let $q(\boldsymbol{\xi})$ be the probability density function of $\boldsymbol{\xi}_j$. Assume further that the noise distribution has marginal density f. Denote the joint distribution of the received signal r_j and the chip samples of the K users $\boldsymbol{\xi}_j$ under the true parameter $\boldsymbol{\theta}$ by $G_{\boldsymbol{\theta}}(r, \boldsymbol{\xi})$, with density

$$g_{\boldsymbol{\theta}}(r, \boldsymbol{\xi}) = f\left(r - \boldsymbol{\xi}^T\boldsymbol{\theta}\right) q(\boldsymbol{\xi}). \tag{4.32}$$

If G_n is the empirical distribution function generated by the signal samples $\{r_j, \boldsymbol{\xi}_j\}_{j=1}^n$, the solution $\hat{\boldsymbol{\theta}}_n$ to (4.15) can also be written as $\hat{\boldsymbol{\theta}}(G_n)$, where $\hat{\boldsymbol{\theta}}$ is the K-dimensional functional determined by

$$\int \psi\left(r - \boldsymbol{\xi}^T\hat{\boldsymbol{\theta}}(G)\right) \boldsymbol{\xi}\, \mathrm{d}G(r, \boldsymbol{\xi}) = \mathbf{0}, \tag{4.33}$$

for all distributions G for which the integral is defined. Let the distribution be

$$G = (1-t)G_{\boldsymbol{\theta}} + t\Delta_{r,\xi}. \tag{4.34}$$

Substituting this distribution into (4.33), differentiating with respect to t, and evaluating the derivative at $t = 0$, we get

$$\begin{aligned}
\mathbf{0} &= \int \psi\left[r - \boldsymbol{\xi}^T\hat{\boldsymbol{\theta}}(G_{\boldsymbol{\theta}})\right] \boldsymbol{\xi}\, \mathrm{d}(\Delta_{r,\xi} - G_{\boldsymbol{\theta}}) \\
&\quad - \int \psi'\left[r - \boldsymbol{\xi}^T\hat{\boldsymbol{\theta}}(G_{\boldsymbol{\theta}})\right] \boldsymbol{\xi}\boldsymbol{\xi}^T f(r - \boldsymbol{\xi}^T\boldsymbol{\theta})\, q(\boldsymbol{\xi})\, \mathrm{d}\boldsymbol{\xi}\, \mathrm{d}r \cdot \frac{\partial}{\partial t}\left[\hat{\boldsymbol{\theta}}(G)\right]|_{t=0} \\
&= \psi\left[r - \boldsymbol{\xi}^T\hat{\boldsymbol{\theta}}(G_{\boldsymbol{\theta}})\right] \boldsymbol{\xi} - \int \psi\left[r - \boldsymbol{\xi}^T\hat{\boldsymbol{\theta}}(G_{\boldsymbol{\theta}})\right] \boldsymbol{\xi}\, \mathrm{d}G_{\boldsymbol{\theta}} \\
&\quad - \int \psi'\left[r - \boldsymbol{\xi}^T\hat{\boldsymbol{\theta}}(G_{\boldsymbol{\theta}})\right] f(r - \boldsymbol{\xi}^T\boldsymbol{\theta})\, \boldsymbol{\xi}\boldsymbol{\xi}^T q(\boldsymbol{\xi})\, \mathrm{d}\boldsymbol{\xi}\, \mathrm{d}r \cdot \mathrm{IF}(r, \boldsymbol{\xi}; \psi, G_{\boldsymbol{\theta}}),
\end{aligned} \tag{4.35}$$

where by definition,

$$\frac{\partial}{\partial t}\left[\hat{\boldsymbol{\theta}}(G)\right]\Big|_{t=0} \triangleq \lim_{t\to 0} \frac{\hat{\boldsymbol{\theta}}[(1-t)G_{\boldsymbol{\theta}} + t\Delta_{r,\xi}] - \hat{\boldsymbol{\theta}}(G_{\boldsymbol{\theta}})}{t}$$
$$\triangleq \mathrm{IF}(r, \boldsymbol{\xi}; \hat{\boldsymbol{\theta}}, G_{\boldsymbol{\theta}}). \tag{4.36}$$

Note that by (4.33) the second term on the right-hand side of (4.35) equals zero:

$$\int \psi\left(r - \boldsymbol{\xi}^T \hat{\boldsymbol{\theta}}(G_{\boldsymbol{\theta}})\right) \boldsymbol{\xi} \, \mathrm{d}G_{\boldsymbol{\theta}} = \mathbf{0}. \tag{4.37}$$

Now assume that the functional $\hat{\boldsymbol{\theta}}$ is Fisher consistent [170] [i.e., $\hat{\boldsymbol{\theta}}(G_{\boldsymbol{\theta}}) = \boldsymbol{\theta}$], which means that at the model the estimator $\hat{\boldsymbol{\theta}}_n : n \geq 1$ asymptotically measures the right quantity when applied to the model distribution. We proceed with (4.35) to obtain

$$\mathbf{0} = \psi\left(r - \boldsymbol{\xi}^T\boldsymbol{\theta}\right)\boldsymbol{\xi} - \int \psi'\left(r - \boldsymbol{\xi}^T\boldsymbol{\theta}\right) f\left(r - \boldsymbol{\xi}^T\boldsymbol{\theta}\right) \boldsymbol{\xi}\boldsymbol{\xi}^T q(\boldsymbol{\xi})\, \mathrm{d}\boldsymbol{\xi}\, \mathrm{d}r \cdot \mathrm{IF}(r, \boldsymbol{\xi}; \psi, G_{\boldsymbol{\theta}})$$
$$= \psi\left(r - \boldsymbol{\xi}^T\boldsymbol{\theta}\right)\boldsymbol{\xi} - \int \psi'(u) f(u)\, \mathrm{d}u \cdot \boldsymbol{R}^* \cdot \mathrm{IF}(r, \boldsymbol{\xi}; \hat{\boldsymbol{\theta}}, G_{\boldsymbol{\theta}}), \tag{4.38}$$

where

$$\boldsymbol{R}^* \triangleq \int \boldsymbol{\xi}\boldsymbol{\xi}^T q(\boldsymbol{\xi})\, \mathrm{d}\boldsymbol{\xi} \tag{4.39}$$

is the cross-correlation matrix of the random infinite-length signature waveforms of the K users. From (4.38) we obtain the influence function of the nonlinear decorrelating multiuser detectors determined by (4.15) as

$$\mathrm{IF}(r, \boldsymbol{\xi}; \hat{\boldsymbol{\theta}}, G_{\boldsymbol{\theta}}) = \frac{\psi\left(r - \boldsymbol{\xi}^T\boldsymbol{\theta}\right)}{\int \psi'(u) f(u)\, \mathrm{d}u} \boldsymbol{R}^{*-1} \boldsymbol{\xi}. \tag{4.40}$$

The influence function above is instrumental to deriving the asymptotic performance of the robust multiuser detectors, as explained below.

4.3.2 Asymptotic Probability of Error

Under certain regularity conditions, the M-estimators defined by (4.14) or (4.15) are consistent and asymptotically Gaussian [170]; that is (here we denote $\hat{\boldsymbol{\theta}}_N$ as the estimate of $\boldsymbol{\theta}$ based on N chip samples),

$$\sqrt{N}\left(\hat{\boldsymbol{\theta}}_N - \boldsymbol{\theta}\right) \sim \mathcal{N}\left(\mathbf{0}, \boldsymbol{V}\left(\hat{\boldsymbol{\theta}}, G_{\boldsymbol{\theta}}\right)\right), \qquad \text{as } n \to \infty, \tag{4.41}$$

where the asymptotic covariance matrix is given by

$$\begin{aligned} \boldsymbol{V}\left(\hat{\boldsymbol{\theta}}, G_{\boldsymbol{\theta}}\right) &= \int \mathrm{IF}(r, \boldsymbol{\xi}; \hat{\boldsymbol{\theta}}, G_{\boldsymbol{\theta}}) \cdot \mathrm{IF}(r, \boldsymbol{\xi}; \hat{\boldsymbol{\theta}}, G_{\boldsymbol{\theta}})^T \, \mathrm{d}G_{\boldsymbol{\theta}}(r, \boldsymbol{\xi}) \\ &= \frac{\int \psi^2(u) f(u) \, \mathrm{d}u}{\left[\int \psi'(u) f(u) \, \mathrm{d}u\right]^2} \boldsymbol{R}^{*-1}, \end{aligned} \tag{4.42}$$

and where (4.42) follows from (4.32) and (4.40).

We can also compute the Fisher information matrix for the parameters $\boldsymbol{\theta}$ at the underlying noise distribution. Define the likelihood score vector as

$$\begin{aligned} \boldsymbol{s}\left(r, \boldsymbol{\xi}; \boldsymbol{\theta}\right) &\triangleq \frac{\partial}{\partial \boldsymbol{\theta}} \ln g_{\boldsymbol{\theta}}\left(r, \boldsymbol{\xi}\right) \\ &= \frac{\partial}{\partial \boldsymbol{\theta}} \ln f\left(r - \boldsymbol{\xi}^T \boldsymbol{\theta}\right) \\ &= \frac{f'\left(r - \boldsymbol{\xi}^T \boldsymbol{\theta}\right)}{f\left(r - \boldsymbol{\xi}^T \boldsymbol{\theta}\right)} \boldsymbol{\xi}. \end{aligned} \tag{4.43}$$

The Fisher information matrix is then given by

$$\begin{aligned} \boldsymbol{J}(\boldsymbol{\theta}) &\triangleq \int \boldsymbol{s}\left(r, \boldsymbol{\xi}; \boldsymbol{\theta}\right) \boldsymbol{s}\left(r, \boldsymbol{\xi}; \boldsymbol{\theta}\right)^T g_{\boldsymbol{\theta}}\left(r, \boldsymbol{\xi}\right) \, \mathrm{d}r \, \mathrm{d}\boldsymbol{\xi} \\ &= \boldsymbol{R}^* \int \frac{f'(u)^2}{f(u)} \, \mathrm{d}u. \end{aligned} \tag{4.44}$$

It is known that the maximum likelihood estimate based on i.i.d. samples is asymptotically unbiased and the asymptotic covariance matrix is $\boldsymbol{J}(\boldsymbol{\theta})^{-1}$ [377]. As discussed earlier, the maximum likelihood estimate of $\boldsymbol{\theta}$ corresponds to having $\psi(x) = -f'(x)/f(x)$. Hence we can deduce that the asymptotic covariance matrix $\boldsymbol{V}(\hat{\boldsymbol{\theta}}, G_{\boldsymbol{\theta}}) = \boldsymbol{J}(\boldsymbol{\theta})^{-1}$, when $\psi(x) = -f'(x)/f(x)$. To verify this, substitute $\psi(x) = -f'(x)/f(x)$ into (4.42); we obtain

$$\begin{aligned} \boldsymbol{V}(\hat{\boldsymbol{\theta}}, G_{\boldsymbol{\theta}}) &= \boldsymbol{R}^{*-1} \frac{\int \frac{f'(u)^2}{f(u)} \, \mathrm{d}u}{\left[\int \frac{f'(u)^2}{f(u)} \, \mathrm{d}u - \int f''(u) \, \mathrm{d}u\right]^2} \\ &= \boldsymbol{R}^{*-1} \left[\int \frac{f'(u)^2}{f(u)} \, \mathrm{d}u\right]^{-1} \\ &= \boldsymbol{J}(\boldsymbol{\theta})^{-1}, \end{aligned} \tag{4.45}$$

where we have assumed that $f'(-\infty) = f'(\infty) = 0$.

Next we consider the asymptotic probability of error for the class of decorrelating detectors defined by (4.15) as the processing gain $N \to \infty$. Using the asymptotic normality condition (4.41), $\hat{\boldsymbol{\theta}}_N \sim \mathcal{N}(\boldsymbol{\theta}, \boldsymbol{V})$. The asymptotic probability of error for the kth user is then given by

$$P_e^k \triangleq P\left(\hat{\theta}_k < 0 \mid \theta_k > 0\right) = Q\left(\frac{A_k}{\upsilon\sqrt{[\boldsymbol{R}^{*-1}]_{k,k}}}\right), \tag{4.46}$$

where υ is the asymptotic variance given by

$$\upsilon^2 \triangleq \frac{\int \psi^2(u) f(u)\, \mathrm{d}u}{\left[\int \psi'(u) f(u)\, \mathrm{d}u\right]^2}. \tag{4.47}$$

Hence for the class of M-decorrelators defined by (4.15), their asymptotic probabilities of detection error are determined by the parameter υ. We next compute υ for the three decorrelating detectors discussed in Section 4.2.3, under the Gaussian mixture noise model (4.3).

Linear Decorrelating Detector

The asymptotic variance for the linear decorrelator is given by

$$\begin{aligned} \upsilon_{LS}^2 &= \int u^2 f(u)\, \mathrm{d}u \;=\; \mathrm{Var}(n_j) \\ &= (1-\epsilon)\,\nu^2 + \epsilon\,\kappa\,\nu^2. \end{aligned} \tag{4.48}$$

That is, asymptotically, the performance of the linear decorrelating detector is determined completely by the noise variance, independent of the noise distribution. However, as will be seen later, the noise distribution does affect substantially the finite sample performance of the linear decorrelating detector.

Maximum-Likelihood Decorrelating Detector

The variance of the estimate used in the maximum-likelihood decorrelating detector achieves the Fisher information covariance matrix, and we have

$$\upsilon_{ML}^2 = \left[\int \frac{f'(u)^2}{f(u)}\, \mathrm{d}u\right]^{-1}. \tag{4.49}$$

In fact, (4.49) gives the minimum achievable v^2. To see this, we use the Cauchy–Schwarz inequality, to yield

$$\begin{aligned}
\int \psi(u)^2 f(u)\,\mathrm{d}u \cdot \int \frac{f'(u)^2}{f(u)}\,\mathrm{d}u &\geq \left(\int |\psi(u)\, f'(u)|\,\mathrm{d}u\right)^2 \\
&\geq \left(\int \psi(u)\, f'(u)\,\mathrm{d}u\right)^2 \\
&= \left(\psi(u)f(u)|_{-\infty}^{+\infty} - \int \psi'(u)\, f(u)\,\mathrm{d}u\right)^2 \\
&= \left(\int \psi'(u)\, f(u)\,\mathrm{d}u\right)^2, \qquad (4.50)
\end{aligned}$$

where the last equality follows from the fact that $\psi(u)f(u) \to 0$, as $|u| \to \infty$. To see this, we use (4.3) and (4.21) to obtain

$$\begin{aligned}
f(u)\,\psi(u) \triangleq -f(u)\,\frac{f'(u)}{f(u)} &= -f'(u) \\
&= \frac{u}{\nu^2\sqrt{2\pi\nu^2}}\left[(1-\epsilon)\exp\left(-\frac{u^2}{2\nu^2}\right) + \frac{\epsilon}{\kappa\sqrt{\kappa}}\exp\left(-\frac{u^2}{2\kappa\nu^2}\right)\right] \\
&\to 0, \qquad \text{as } |u| \to \infty. \qquad (4.51)
\end{aligned}$$

Hence it follows from (4.50) that

$$\frac{\int \psi^2(u)f(u)\,\mathrm{d}u}{\left[\int \psi'(u)f(u)\,\mathrm{d}u\right]^2} \geq \left[\int \frac{f'(u)^2}{f(u)}\,\mathrm{d}u\right]^{-1}. \qquad (4.52)$$

Minimax Decorrelating Detector

For the minimax decorrelating detector, we have

$$\psi(x) = \frac{x}{\nu^2}\cdot 1_{\{|x|\leq\gamma\nu^2\}} + \gamma\,\mathrm{sign}(x)\cdot 1_{\{|x|>\gamma\nu^2\}}, \qquad (4.53)$$

$$\psi'(x) = \frac{1}{\nu^2}\cdot 1_{\{|x|\leq\gamma\nu^2\}} + \gamma\,\delta_{-\gamma\nu^2} - \gamma\,\delta_{\gamma\nu^2}, \qquad (4.54)$$

where $1_\Omega(x)$ denotes the indicator function of the set Ω, and δ_x denotes the Dirac delta function at x. After some algebra, we obtain

$$\int \psi^2(u)f(u)\,du = \frac{2}{\nu^2}\left[\frac{1+(\kappa-1)\epsilon}{2} + (1-\epsilon)\left(\gamma^2\nu^2 - 1\right)Q(\gamma\nu)\right.$$
$$+\epsilon\left(\gamma^2\nu^2-\kappa\right)Q\left(\frac{\gamma\nu}{\sqrt{\kappa}}\right)$$
$$\left.-\frac{(1-\epsilon)\gamma\nu}{\sqrt{2\pi}}\exp\left(-\frac{\gamma^2\nu^2}{2}\right) - \sqrt{\frac{\kappa}{2\pi}}\epsilon\gamma\nu\exp\left(-\frac{\gamma^2\nu^2}{2\kappa}\right)\right], \quad (4.55)$$
$$\int \psi'(u)f(u)\,du = \frac{2}{\nu^2}\left[\frac{1}{2} - (1-\epsilon)\,Q(\gamma\nu) - \epsilon\,Q\left(\frac{\gamma\nu}{\sqrt{\kappa}}\right)\right]. \quad (4.56)$$

The asymptotic variance v_H^2 of the minimax decorrelating detector is obtained by substituting (4.55) and (4.56) into (4.47).

In Fig. 4.2 we plot the asymptotic variance v^2 of the maximum-likelihood decorrelator and the minimax robust decorrelator as a function of ϵ and κ under the Gaussian mixture noise model (4.3). The total noise variance is kept constant as ϵ and κ vary [i.e., $\sigma^2 \triangleq (1-\epsilon)\nu^2 + \epsilon\kappa\nu^2 = (0.1)^2$]. From the two plots we see that the two nonlinear detectors have very similar asymptotic performance. Moreover, in this case the asymptotic variance v^2 is a decreasing function of either ϵ or κ when one of them is fixed. The asymptotic variances of both nonlinear decorrelators are strictly less than that of the linear decorrelator, which corresponds to a plane at $v^2 = \sigma^2 = (0.1)^2$. In Fig. 4.3 we plot the asymptotic variance v^2 of the three decorrelating detectors as a function of κ with fixed ϵ; and in Fig. 4.4 we plot the asymptotic variance v^2 of the three decorrelating detectors as a function of ϵ with fixed κ. As before, the total variance of the noise for both figures is fixed at $\sigma^2 = (0.1)^2$. From these figures we see that the asymptotic variance of the minimax decorrelator is very close to that of the maximum-likelihood decorrelator for the cases of small contamination (e.g., $\epsilon \leq 0.1$), while both of the detectors can outperform the linear detector by a substantial margin.

4.4 Implementation of Robust Multiuser Detectors

In this section we discuss computational procedures for obtaining the output of the nonlinear decorrelating multiuser detectors [i.e., the solution to (4.15)]. Assume that the penalty function $\rho(x)$ in (4.14) has a bounded second-order derivative:

$$|\rho''(x)| \;=\; |\psi'(x)| \leq \alpha \quad (4.57)$$

for some $\alpha > 0$. Then (4.15) can be solved iteratively by the following modified residual method [203]. Let $\boldsymbol{\theta}^l$ be the estimate at the lth step of the iteration; then it is updated according to

$$\boldsymbol{z}^l \triangleq \psi\left(\boldsymbol{r} - \boldsymbol{S}\boldsymbol{\theta}^l\right), \quad (4.58)$$
$$\boldsymbol{\theta}^{l+1} = \boldsymbol{\theta}^l + \frac{1}{\mu}\left(\boldsymbol{S}^T\boldsymbol{S}\right)^{-1}\boldsymbol{S}^T\boldsymbol{z}^l, \qquad l = 0, 1, 2, \ldots \quad (4.59)$$

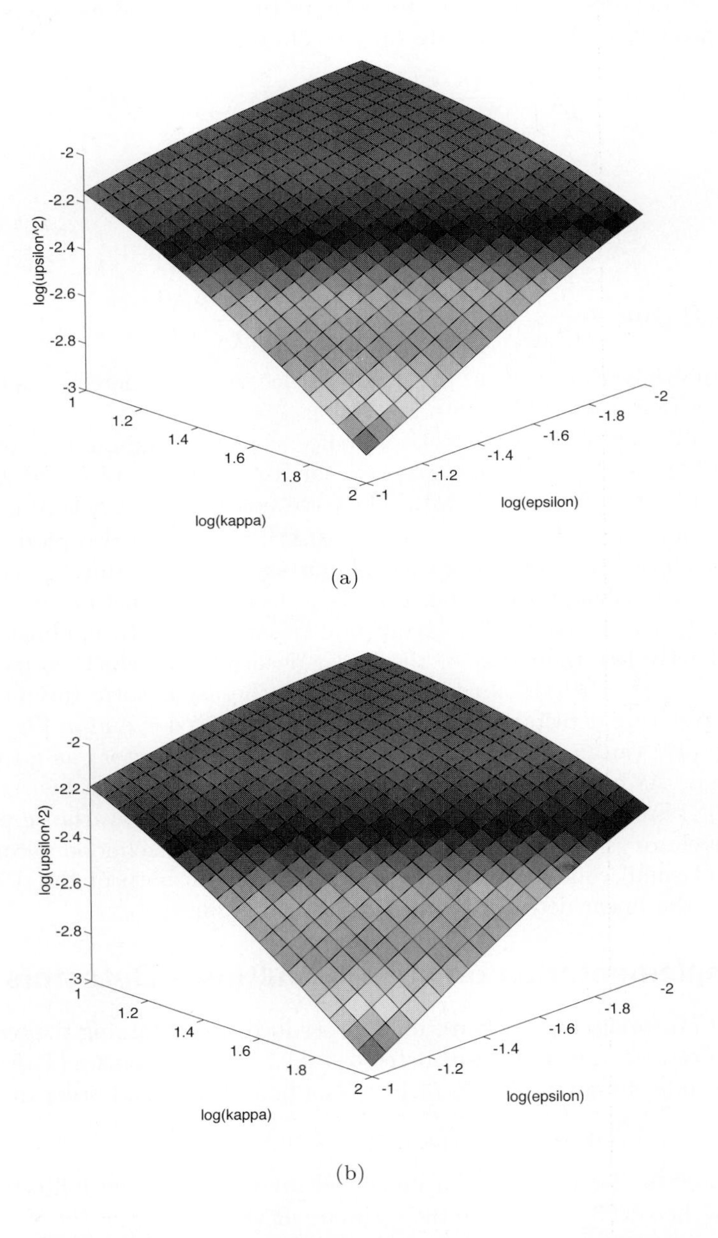

Figure 4.2. Asymptotic variance v^2 of (a) a minimax robust decorrelating detector, and (b) a maximum-likelihood decorrelating detector, as a function of ϵ and κ, under the Gaussian mixture noise model, with variance of the noise fixed at $\sigma^2 \triangleq (1-\epsilon)\nu^2 + \epsilon\kappa\nu^2 = (0.1)^2$.

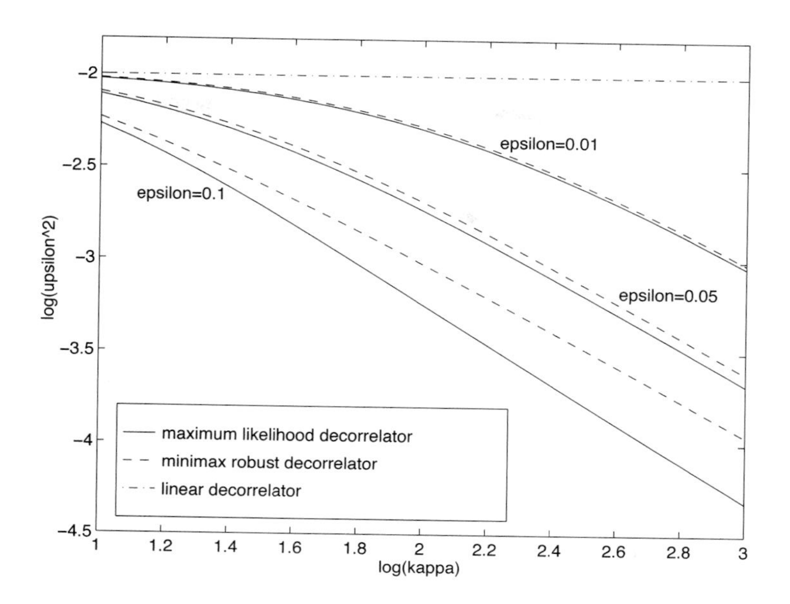

Figure 4.3. Asymptotic variance υ^2 of three decorrelating detectors as a function of κ with fixed parameter ϵ. The variance of the noise is fixed at $\sigma^2 \triangleq (1-\epsilon)\nu^2 + \epsilon\kappa\nu^2 = (0.1)^2$.

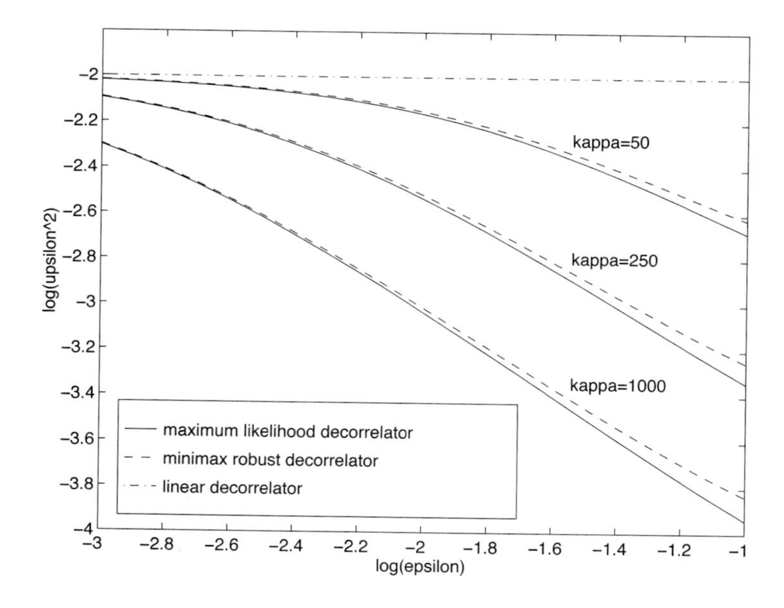

Figure 4.4. Asymptotic variance υ^2 of three decorrelating detectors as a function of ϵ with fixed parameter κ. The variance of the noise is fixed at $\sigma^2 \triangleq (1-\epsilon)\nu^2 + \epsilon\kappa\nu^2 = (0.1)^2$.

where $\mu \geq \alpha$ is a step-size parameter. Denote the cost function in (4.14) by

$$\mathcal{C}(\boldsymbol{\theta}) \triangleq \sum_{j=1}^{N} \rho\left(r_j - \boldsymbol{\xi}_j^T \boldsymbol{\theta}\right). \tag{4.60}$$

We have the following result regarding the convergence behavior of the iterative procedure above. The proof is given in the Appendix (Section 4.10.1).

Proposition 4.1: *If $|\psi'(x)| \leq \alpha \leq \mu$, the iterative procedure defined by (4.58) and (4.59) satisfies*

$$\begin{aligned} \mathcal{C}\left(\boldsymbol{\theta}^l\right) - \mathcal{C}\left(\boldsymbol{\theta}^{l+1}\right) &\geq \frac{\mu}{2}\left(\boldsymbol{\theta}^l - \boldsymbol{\theta}^{l+1}\right)^T \boldsymbol{R}\left(\boldsymbol{\theta}^l - \boldsymbol{\theta}^{l+1}\right) \\ &= \frac{1}{2\mu}\boldsymbol{z}\left(\boldsymbol{\theta}^l\right)^T \boldsymbol{S}\boldsymbol{R}^{-1}\boldsymbol{S}^T \boldsymbol{z}\left(\boldsymbol{\theta}^l\right), \end{aligned} \tag{4.61}$$

where $\boldsymbol{R} \triangleq \boldsymbol{S}^T\boldsymbol{S}$ is assumed to be positive definite and $\boldsymbol{z}(\boldsymbol{\theta}) \triangleq \psi\left(\boldsymbol{r} - \boldsymbol{S}\boldsymbol{\theta}\right)$. Furthermore, if $\rho(x)$ is convex and bounded from below, then with probability 1, $\boldsymbol{\theta}^l \to \boldsymbol{\theta}^*$ as $l \to \infty$, where $\boldsymbol{\theta}^*$ is the unique minimum point of the cost function $\mathcal{C}(\boldsymbol{\theta})$ [i.e., the unique solution to (4.15)].

Notice that for the minimax robust decorrelating detector, the Huber penalty function $\rho_H(x)$ does not have second-order derivatives at the two "corner" points (i.e., $x = \pm\gamma\nu^2$). In principle, this can be resolved by defining a smoothed version of the Huber penalty function, for example, as follows:

$$\tilde{\rho}_H(x) = \begin{cases} \dfrac{x^2}{2\nu^2} & \text{if } |x| \leq (\gamma-\eta)\nu^2, \\ (\gamma-\eta)x + \eta^2\nu^2 \ln\cosh\left[\dfrac{x-(\gamma-\eta)\nu^2}{\eta\nu^2}\right] & \text{if } x > (\gamma-\eta)\nu^2, \\ -(\gamma-\eta)x + \eta^2\nu^2 \ln\cosh\left[\dfrac{x+(\gamma-\eta)\nu^2}{\eta\nu^2}\right] & \text{if } x < -(\gamma-\eta)\nu^2, \end{cases} \tag{4.62}$$

where η is a small number. The first- and second-order derivatives of this smoothed Huber penalty function are given, respectively, by

$$\tilde{\psi}_H(x) \triangleq \tilde{\rho}'_H(x) = \begin{cases} \frac{x}{\nu^2} & \text{if } |x| \leq (\gamma - \eta)\nu^2, \\ \gamma - \eta + \eta \tanh\left[\frac{x - (\gamma-\eta)\nu^2}{\eta\nu^2}\right] & \text{if } x > (\gamma - \eta)\nu^2, \\ -(\gamma - \eta) + \eta \tanh\left[\frac{x + (\gamma-\eta)\nu^2}{\eta\nu^2}\right] & \text{if } x < -(\gamma - \eta)\nu^2, \end{cases} \tag{4.63}$$

$$\begin{aligned} \tilde{\psi}'_H(x) = \tilde{\rho}''_H(x) &= \begin{cases} \frac{1}{\nu^2} & \text{if } |x| \leq (\gamma - \eta)\nu^2, \\ \frac{1}{\nu^2}\left\{1 - \tanh^2\left[\frac{x - (\gamma-\eta)\nu^2}{\eta\nu^2}\right]\right\} & \text{if } x > (\gamma - \eta)\nu^2, \\ \frac{1}{\nu^2}\left\{1 - \tanh^2\left[\frac{x + (\gamma-\eta)\nu^2}{\eta\nu^2}\right]\right\} & \text{if } x < -(\gamma - \eta)\nu^2, \end{cases} \\ &\leq \frac{1}{\nu^2}. \end{aligned} \tag{4.64}$$

We can then apply the iterative procedure (4.58)–(4.59) using this smoothed penalty function and the step size $1/\mu = \nu^2$. In practice, however, convergence is always observed even if the nonsmooth nonlinearity $\psi_H(x)$ is used.

Notice that the matrix $(1/\mu)\left(\boldsymbol{S}^T\boldsymbol{S}\right)^{-1}\boldsymbol{S}^T$ in (4.59) can be computed off-line, and the major computation involved at each iteration is the product of this $K \times K$ matrix with a K-vector $\boldsymbol{z}^l$. For the initial estimate $\boldsymbol{\theta}^0$ we can take the least-squares solution:

$$\boldsymbol{\theta}^0 = \left(\boldsymbol{S}^T\boldsymbol{S}\right)^{-1}\boldsymbol{S}^T\boldsymbol{r}. \tag{4.65}$$

The iteration is stopped if $\|\boldsymbol{\theta}^l - \boldsymbol{\theta}^{l-1}\| \leq \delta$ for some small number δ. Simulations show that on average it takes fewer than 10 iterations for the algorithm to converge. Finally, we summarize the robust multiuser detection algorithm as follows.

Algorithm 4.1: [Robust multiuser detector—synchronous CDMA]

- *Compute the decorrelating detector output (as before, $\boldsymbol{R} \triangleq \boldsymbol{S}^T\boldsymbol{S}$):*

$$\boldsymbol{\theta}^0 = \boldsymbol{R}^{-1}\boldsymbol{S}^T\boldsymbol{r}. \tag{4.66}$$

- *Compute the robust detector output:*

Do

$$\boldsymbol{z}^l = \psi\left(\boldsymbol{r} - \boldsymbol{S}\boldsymbol{\theta}^l\right), \tag{4.67}$$

$$\boldsymbol{\theta}^{l+1} = \boldsymbol{\theta}^l + \frac{1}{\mu}\boldsymbol{R}^{-1}\boldsymbol{S}^T\boldsymbol{z}^l, \tag{4.68}$$

While $\quad \left\|\boldsymbol{\theta}^{l+1} - \boldsymbol{\theta}^l\right\| > \delta$

Let $\hat{\boldsymbol{\theta}} = \boldsymbol{\theta}^l$.

- *Perform detection:*

$$\hat{\boldsymbol{b}} = \operatorname{sign}\left(\hat{\boldsymbol{\theta}}\right). \tag{4.69}$$

The operations of the M-decorrelating multiuser detector are depicted in Fig. 4.5. It is evident that it is essentially a robust version of the linear decorrelating detector. At each iteration, the residual signal, which is the difference between the received signal $\boldsymbol{r}$ and the remodulated signal $\boldsymbol{S}\boldsymbol{\theta}^l$, is passed through the nonlinearity $\psi(\cdot)$. Then the modified residual $\boldsymbol{z}^l$ is passed through the linear decorrelating filter to get the modification on the previous estimate.

Simulation Examples

In this section we provide some simulation examples to demonstrate the performance of the nonlinear robust multiuser detectors against multiple-access interference and non-Gaussian additive noise. We consider a synchronous system with $K = 6$ users. The spreading sequence of each user is a shifted version of an m-sequence of length $N = 31$.

We first demonstrate the performance degradation of the linear multiuser detectors in non-Gaussian ambient noise. Two popular linear multiuser detectors are the linear decorrelating and linear MMSE detectors. The performance of the linear decorrelating detector in several different ϵ-mixture channels is depicted in Fig. 4.6. In this figure we plot the BER versus the SNR (defined as A_1^2/σ^2) corresponding to user 1, assuming that all users have the same amplitudes. The performance of the linear MMSE multiuser detector is indistinguishable in this case from that of the linear decorrelating detector. It is seen that the impulsive character of the ambient noise can substantially degrade the performance of both linear multiuser detectors. Similar situations have been observed for the conventional matched filter receiver in [1]. In [383] it is observed that non-Gaussian-based optimal detection can achieve significant performance gain (more than 10 dB in some cases) over Gaussian-based optimal detection in multiple-access channels when the ambient noise is impulsive. However, this gain is obtained with a significant penalty on complexity. The robust techniques discussed in this chapter constitute some low-complexity multiuser detectors that account for non-Gaussian ambient noise. We next demonstrate the performance gain afforded by this non-Gaussian-based suboptimal detection technique over its Gaussian-based counterpart (i.e., the linear decorrelator).

The next example demonstrates the performance gains achieved by the minimax robust decorrelating detector over the linear decorrelator in impulsive noise. The noise distribution parameters are $\epsilon = 0.01$ and $\kappa = 100$. The BER performance of the two detectors is plotted in Fig. 4.7. Also shown in this figure is the performance of an "approximate" minimax decorrelating detector, in which the nonlinearity $\psi(\cdot)$ is taken as

$$\psi(x) = \begin{cases} \dfrac{x}{\sigma^2} & \text{for } |x| \le \gamma\sigma^2, \\ \gamma \operatorname{sign}(x) & \text{for } |x| > \gamma\sigma^2, \end{cases} \tag{4.70}$$

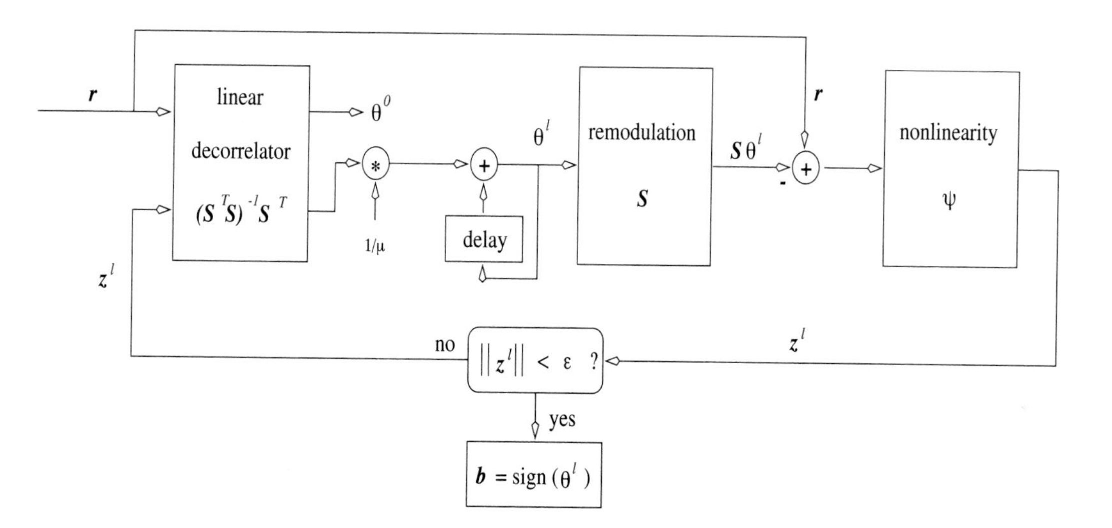

Figure 4.5. M-decorrelating multiuser detector, a robust version of the linear decorrelating multiuser detector.

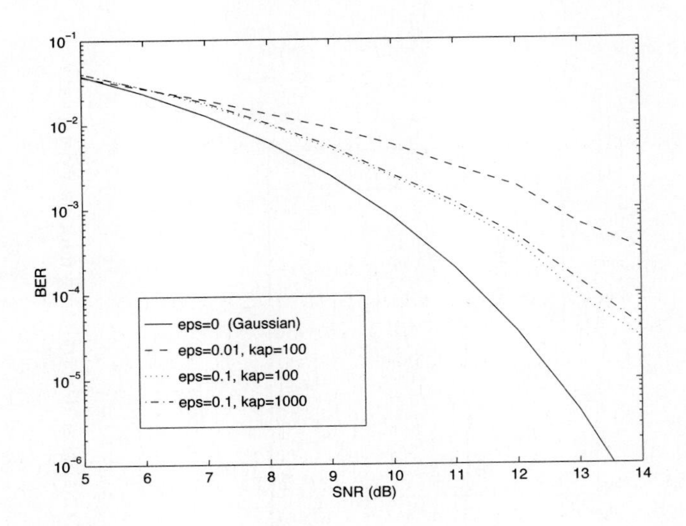

Figure 4.6. BER performance of a linear decorrelating detector for user 1 in a synchronous CDMA channel with Gaussian and ϵ-mixture ambient noise. $N = 31, K = 6$. All users have the same amplitudes.

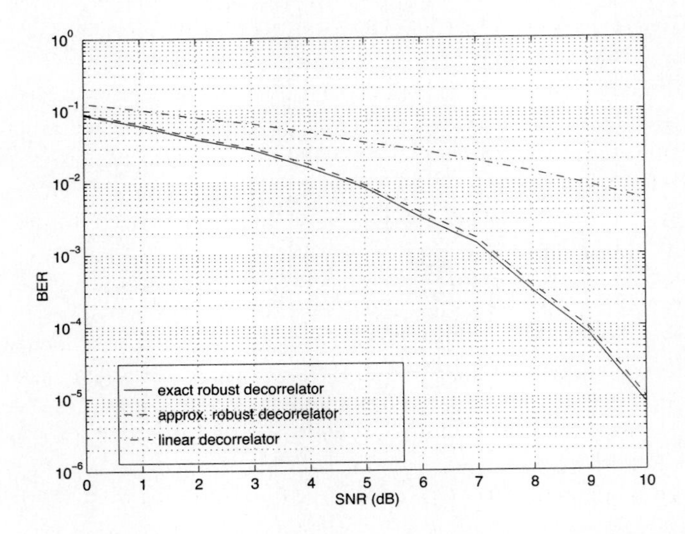

Figure 4.7. BER performance of user 1 for an exact minimax decorrelating detector, approximate minimax decorrelating detector, and linear decorrelating detector in a synchronous CDMA channel with impulsive noise. $N = 31, K = 6, \epsilon = 0.01, \kappa = 100$. The powers of the interferers are 10 dB above the power of user 1 (i.e., $A_k^2/A_1^2 = 10$ for $k \neq 1$).

where the parameter γ is taken as

$$\gamma = \frac{3}{2\sigma}, \tag{4.71}$$

and the step-size parameter μ in the modified residual method (4.59) is set as

$$\mu = \sigma^2. \tag{4.72}$$

The reason for studying such an approximate robust detector is that in practice, it is unlikely that the exact parameters ϵ and ν in the noise model (4.3) are known to the receiver. However, the total noise variance σ^2 can be estimated from the received signal (as discussed in the next section). Hence if we could set some simple rule for choosing the nonlinearity $\psi(\cdot)$ and μ, this approximate robust detector is much easier to implement than the exact one. It is seen from Fig. 4.7 that the robust decorrelating multiuser detector offers significant performance gains over the linear decorrelating detector. Moreover, this performance gain increases as the SNR increases. Another important observation is that the performance of the robust multiuser detector is insensitive to the parameters ϵ and κ in the noise model, which is evidenced by the fact that the performance of the approximate robust detector is very close to that of the exact robust detector. We next consider a synchronous system with 20 users ($K = 20$). The spreading sequence of each user is still a shifted version of an m-sequence of length $N = 31$. The performance of the approximate robust decorrelator and that of the linear decorrelator is shown in Fig. 4.8. Again it is seen that the robust detector offers a substantial performance gain over the linear detector.

In the third example we consider the performance of the approximate robust decorrelator in Gaussian noise. Shown in Fig. 4.9 are the BER curves for the robust decorrelator and the linear decorrelator in a six-user system ($K = 6$). It is seen that there is only a very slight performance degradation by the robust decorrelator in Gaussian channels, relative to the linear decorrelator, which is the optimal decorrelating detector in Gaussian noise. By comparing the BER curves of the robust decorrelator in Figs. 4.7 and 4.9, it is seen that the robust detector performs better in impulsive noise than in Gaussian noise with the same noise variance. This is because in an impulsive environment, a portion of the total noise variance is due to impulses, which have large amplitudes. Such impulses are clipped by the nonlinearity in the detector. Therefore, the effective noise variance at the output of the robust detector is smaller than the input total noise variance. In fact, the asymptotic performance gain by the robust detector in impulsive noise over Gaussian noise is quantified by the asymptotic variance v^2 in (4.47) [cf. Figs. 4.2, 4.3, and 4.4].

In summary, we have seen that the performance of the linear decorrelating detector degrades substantially when the distribution of the ambient channel noise deviates even slightly from Gaussian. By using the robust decorrelating detector, such performance loss is prevented and this detector thus offers significant performance gains over the linear detectors, which translates into an increase in capacity

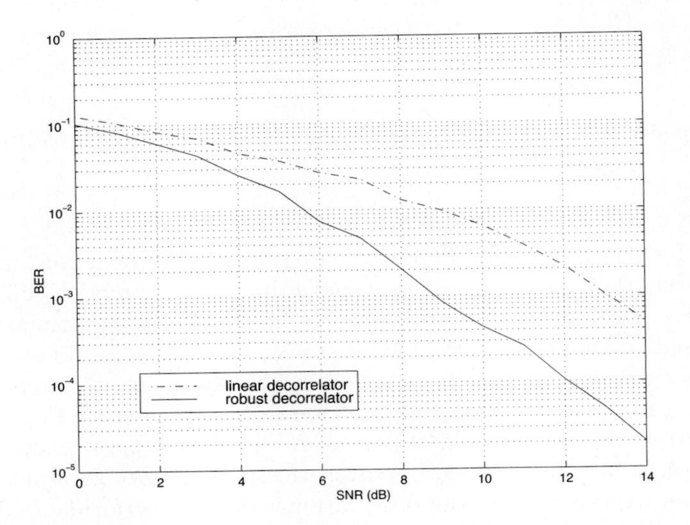

Figure 4.8. BER performance of user 1 for an approximate minimax decorrelating detector and linear decorrelating detector in a synchronous CDMA channel with impulsive noise. $N = 31, K = 20, \epsilon = 0.01, \kappa = 100$. All users have the same amplitudes.

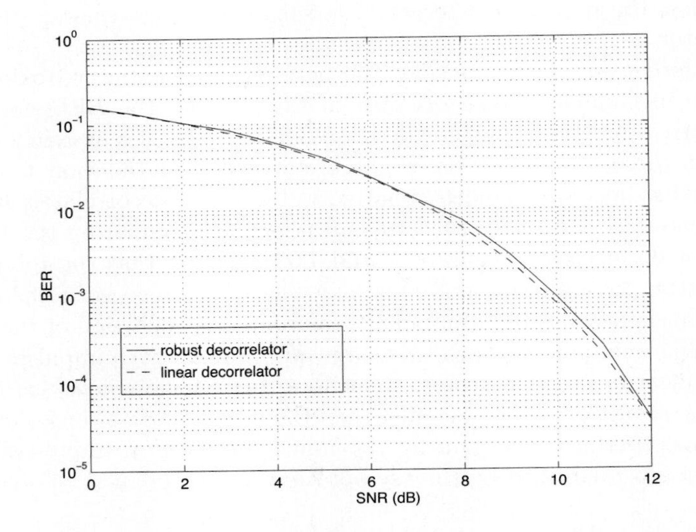

Figure 4.9. BER performance of user 1 for a robust decorrelating detector and linear decorrelating detector in a synchronous CDMA channel with Gaussian noise. $N = 31, K = 6$. The powers of the interferers are 10 dB above the power of user 1.

in multiple-access channels. On the other hand, even when the ambient noise distribution is indeed Gaussian, the robust detector incurs only negligible performance loss relative to the linear detectors.

A number of other techniques have been proposed in the literature to combat impulsive ambient noise in multiple-access channels. These include adaptive receivers with certain nonlinearities [27, 28], a neural network approach [81], maximum-likelihood methods based on the expectation-maximization (EM) algorithm [47, 236, 607], a Bayesian approach based on the Markov chain Monte Carlo technique [540], and extensions to fading channels [382, 384].

4.5 Robust Blind Multiuser Detection

The robust multiuser detection procedure developed in previous sections offers substantial performance gain over linear multiuser detectors when the ambient noise becomes impulsive. So far in this chapter, we have assumed that the signature waveforms of all users, as well as the distribution of the ambient noise, are known to the receiver in order to implement the robust multiuser detectors. The requirement of knowledge of the exact noise distribution can be alleviated, since as demonstrated in Section 4.4, little performance degradation is incurred if we simply adopt in the robust multiuser detector some nonlinearity ψ, which depends only on the total noise variance, not on the shape of the distribution. In this section we develop a technique to alleviate the requirement of knowledge of all users' signatures. As discussed in Chapter 2, one remarkable feature of linear multiuser detectors is that there exist blind techniques that can be used to adapt these detectors, which allow one to use a linear multiuser detector for a given user with no knowledge beyond that required for implementation of the conventional matched-filter detector for that user. In this section we show that the robust multiuser detector can also be implemented blindly (i.e., with the prior knowledge of the signature waveform of only one user of interest).

As discussed in Chapter 2, there are two major approaches to blind adaptive multiuser detection. In the first approach the received signal is passed through a linear filter, which is chosen to minimize, within a constraint, the mean-square value of its output [183]. Adaptation algorithms such as least mean squares (LMS) or recursive least squares (RLS) can be applied to update the filter weights. Ideally, the adaptation will lead the filter to converge to the linear MMSE multiuser detector, irrespective of the noise distribution. (In practice, the impulsiveness of the noise will slow down the convergence.) Therefore, this approach cannot be used to adapt the robust multiuser detector.

Another approach to blind multiuser detection is the subspace-based method proposed in [549], through which both the linear decorrelating detector and the linear MMSE detector can be obtained blindly. As discussed in this section, this approach is more fruitful in leading to a blind adaptive robust multiuser detection method. The blind robust multiuser detection method discussed in this section was first proposed in [553] in a CDMA context, and was subsequently generalized to develop robust adaptive antenna array in a TDMA context in [550].

The autocorrelation matrix of the received signal $\boldsymbol{r}$ in (4.2) is given by

$$\boldsymbol{C}_r \triangleq E\left\{\boldsymbol{r}\,\boldsymbol{r}^T\right\} \;=\; \boldsymbol{S}\boldsymbol{A}^2\boldsymbol{S}^T + \sigma^2\boldsymbol{I}_N. \tag{4.73}$$

By performing an eigendecomposition of the matrix $\boldsymbol{C}_r$, we can write

$$\boldsymbol{C}_r = \boldsymbol{U}_s\boldsymbol{\Lambda}_s\boldsymbol{U}_s^T + \sigma^2\boldsymbol{U}_n\boldsymbol{U}_n^T, \tag{4.74}$$

where $\boldsymbol{\Lambda}_s = \mathsf{diag}(\lambda_1, \ldots, \lambda_K)$ contains the K largest eigenvalues of $\boldsymbol{C}_r$ in descending order, $\boldsymbol{U}_s = [\boldsymbol{u}_1 \;\cdots\; \boldsymbol{u}_K]$ contains the corresponding orthogonal eigenvectors, and $\boldsymbol{U}_n = [\boldsymbol{u}_{K+1} \;\cdots\; \boldsymbol{u}_N]$ contains the $N-K$ orthogonal eigenvectors that correspond to the smallest eigenvalue σ^2. The following result is instrumental to developing the subspace-based blind robust multiuser detector. The proof is given in the Appendix (Section 4.10.2).

Proposition 4.2: *Given the eigendecomposition (4.74) of the autocorrelation matrix $\boldsymbol{C}_r$, suppose that*

$$\sum_{k=1}^{K} \theta_k\, \boldsymbol{s}_k = \sum_{j=1}^{K} \zeta_j\, \boldsymbol{u}_j, \qquad \theta_k \in \mathbb{R}, \qquad \zeta_j \in \mathbb{R}. \tag{4.75}$$

Then we have

$$\theta_k = \alpha_k \sum_{j=1}^{K} \frac{\boldsymbol{u}_j^T\boldsymbol{s}_k}{\lambda_j - \sigma^2}\,\zeta_j, \qquad k = 1, \ldots, K, \tag{4.76}$$

where α_k is a positive constant, given by

$$\alpha_k \;=\; \left[\sum_{j=1}^{K} \frac{\left(\boldsymbol{u}_j^T\boldsymbol{s}_k\right)^2}{\lambda_j - \sigma^2}\right]^{-1} \;=\; A_k^2 \tag{4.77}$$

or in matrix form,

$$\boldsymbol{\theta} = \underbrace{\left[\boldsymbol{S}^T\boldsymbol{U}_s\left(\boldsymbol{\Lambda}_s - \sigma^2\boldsymbol{I}_K\right)^{-1}\boldsymbol{U}_s^T\boldsymbol{S}\right]^{-1}}_{\boldsymbol{A}^2} \boldsymbol{S}^T\boldsymbol{U}_s\left(\boldsymbol{\Lambda}_s - \sigma^2\boldsymbol{I}_K\right)\boldsymbol{\zeta}. \tag{4.78}$$

The result above leads to a subspace-based blind robust multiuser detection technique as follows. From the received signals, we can estimate the signal subspace components (i.e., $\hat{\boldsymbol{\Lambda}}_s = \mathsf{diag}(\hat{\lambda}_1, \ldots, \hat{\lambda}_K)$, and $\hat{\boldsymbol{U}}_s = [\hat{\boldsymbol{u}}_1 \;\cdots\; \hat{\boldsymbol{u}}_K]$). The received signal $\boldsymbol{r}$ in (4.6) can be expressed as

$$\boldsymbol{r} = \boldsymbol{S}\boldsymbol{\theta} + \boldsymbol{n} \tag{4.79}$$

$$= \boldsymbol{U}_s\,\boldsymbol{\zeta} + \boldsymbol{n}, \tag{4.80}$$

for some $\boldsymbol{\zeta} \triangleq [\zeta_1, \ldots, \zeta_K]^T \in \mathbb{R}^K$. Now instead of robustly estimating the parameters $\boldsymbol{\theta}$ using the signature waveforms $\boldsymbol{S}$ of all users, as is done in Section 4.4, we can robustly estimate the parameters $\boldsymbol{\zeta}$ using the estimated signal subspace coordinates $\hat{\boldsymbol{U}}_s$. Denote such a robust estimate as $\hat{\boldsymbol{\zeta}} \triangleq [\hat{\zeta}_1, \ldots, \hat{\zeta}_K]^T$. Suppose that user 1 is the user of interest. Finally, we compute the estimate of the parameter θ_1 of this user (up to a positive scaling factor) according to

$$\hat{\theta}_1 = \sum_{j=1}^{K} \frac{\hat{\boldsymbol{u}}_j^T \boldsymbol{s}_1}{\hat{\lambda}_j - \hat{\sigma^2}} \hat{\zeta}_j. \tag{4.81}$$

Note that using this method to demodulate the desired user's data bit b_1, the only prior knowledge required at the receiver is the signature waveform $\boldsymbol{s}_1$ of this user, and thus the term *blind robust multiuser detector*. Note also that since the columns of $\hat{\boldsymbol{U}}_s$ are orthogonal, the modified residual method for updating the robust estimate of $\boldsymbol{\zeta}$ is given by

$$\boldsymbol{z}^l = \psi\left(\boldsymbol{r} - \hat{\boldsymbol{U}}_s \boldsymbol{\zeta}^l\right), \tag{4.82}$$

$$\boldsymbol{\zeta}^{l+1} = \boldsymbol{\zeta}^l + \frac{1}{\mu} \hat{\boldsymbol{U}}_s^T \boldsymbol{z}^l, \qquad l = 0, 1, 2, \ldots. \tag{4.83}$$

The following is a summary of the robust blind multiuser detection algorithm. [We re-introduce the symbol index i into the model (4.2).]

Algorithm 4.2: [Robust blind multiuser detector—synchronous CDMA]

- *Estimate the signal subspace:*

$$\hat{\boldsymbol{C}}_r \triangleq \frac{1}{M} \sum_{i=0}^{M-1} \boldsymbol{r}[i]\boldsymbol{r}[i]^T \tag{4.84}$$

$$= \hat{\boldsymbol{U}}_s \hat{\boldsymbol{\Lambda}}_s \hat{\boldsymbol{U}}_s^T + \hat{\boldsymbol{U}}_n \hat{\boldsymbol{\Lambda}}_n \hat{\boldsymbol{U}}_n^T. \tag{4.85}$$

Set $\hat{\sigma^2}$ to be the mean of the diagonal elements of $\hat{\boldsymbol{\Lambda}}_n$.

- *Compute the robust estimate of $\boldsymbol{\zeta}[i]$:*

$$\boldsymbol{\zeta}^0[i] = \boldsymbol{S}^T \hat{\boldsymbol{U}}_s \left(\hat{\boldsymbol{\Lambda}}_s - \hat{\sigma^2} \boldsymbol{I}_K\right)^{-1} \hat{\boldsymbol{U}}_s^T \boldsymbol{r}[i], \tag{4.86}$$

Do

$$\boldsymbol{z}^l[i] \triangleq \psi\left(\boldsymbol{r}[i] - \hat{\boldsymbol{U}}_s \boldsymbol{\zeta}^l[i]\right), \tag{4.87}$$

$$\boldsymbol{\zeta}^{l+1}[i] = \boldsymbol{\zeta}^l[i] + \frac{1}{\mu} \hat{\boldsymbol{U}}_s^T \boldsymbol{z}^l[i], \tag{4.88}$$

While $\quad \left\|\boldsymbol{\zeta}^{l+1}[i] - \boldsymbol{\zeta}^l[i]\right\| > \delta, \qquad i = 0, \ldots, M-1.$

Set $\hat{\boldsymbol{\zeta}}[i] = \boldsymbol{\zeta}^l[i]$.

- *Compute the robust estimate of* $\theta_1[i]$*:*

$$\hat{\theta}_1[i] = \sum_{j=1}^{K} \frac{\hat{\boldsymbol{u}}_j^T \boldsymbol{s}_1}{\hat{\lambda}_j - \hat{\sigma^2}} \hat{\zeta}_j[i], \tag{4.89}$$

$$\hat{b}_1[i] = \text{sign}\left(\hat{\theta}_1[i]\right), \qquad i = 0, \ldots, M-1. \tag{4.90}$$

Alternatively, robust blind multiuser detection can also be implemented adaptively based on sequential signal subspace tracking. For instance, suppose that at time $i-1$, the estimated signal subspace rank is $K[i-1]$ and the components are $\boldsymbol{U}_s[i-1]$, $\boldsymbol{\Lambda}_s[i-1]$, and $\sigma^2[i-1]$. Then at time i, the adaptive detector performs the following steps to update the detector and to estimate the data bit of the desired user.

Algorithm 4.3: [Adaptive robust blind multiuser detector—synchronous CDMA]

- *Update the signal subspace: Using a particular signal subspace tracking algorithm, update the signal subspace rank* $K[i]$ *and the subspace components* $\boldsymbol{U}_s[i]$, $\boldsymbol{\Lambda}_s[i]$, *and* $\sigma^2[i]$.
- *Compute the robust estimate of* $\boldsymbol{\zeta}[i]$*:*

$$\boldsymbol{\zeta}^0[i] = \boldsymbol{U}_s[i]\left(\boldsymbol{\Lambda}_s[i] - \sigma^2[i]\boldsymbol{I}_K\right)^{-1}\boldsymbol{U}_s[i]^T\boldsymbol{r}[i], \tag{4.91}$$

Do

$$\boldsymbol{z}^l[i] \triangleq \psi\left(\boldsymbol{r}[i] - \boldsymbol{U}_s[i]\,\boldsymbol{\zeta}^l[i]\right), \tag{4.92}$$

$$\boldsymbol{\zeta}^{l+1}[i] = \boldsymbol{\zeta}^l[i] + \frac{1}{\mu}\boldsymbol{U}_s[i]^T\boldsymbol{z}^l[i], \tag{4.93}$$

While $\quad \left\|\boldsymbol{\zeta}^{l+1}[i] - \boldsymbol{\zeta}^l[i]\right\| > \delta.$

Set $\hat{\boldsymbol{\zeta}}[i] = \boldsymbol{\zeta}^l[i]$.

- *Compute the robust estimate of* $\theta_1[i]$ *and perform detection:*

$$\hat{\theta}_1[i] = \sum_{j=1}^{K} \frac{\boldsymbol{u}_j[i]^T \boldsymbol{s}_1}{\lambda_j[i] - \sigma^2[i]} \hat{\zeta}_j[i], \tag{4.94}$$

$$\hat{b}_1[i] = \text{sign}\left(\hat{\theta}_1[i]\right). \tag{4.95}$$

Simulation Examples

As before, we consider a synchronous system with $K = 6$ users and spreading gain $N = 31$. First we illustrate the performance of the blind robust multiuser detector based on batch eigendecomposition. The size of the data block is $M = 200$. The noise distribution parameters are $\epsilon = 0.01$ and $\kappa = 100$. The BER performance of user 1 is plotted in Fig. 4.10 for both the blind linear MMSE detector and the blind robust detector. The powers of all interferers are 10 dB above user 1. The performance of the blind adaptive robust multiuser detector based on subspace tracking is shown in Fig. 4.11, where the PASTd algorithm from [586] is used for tracking the signal subspace parameters (see Section 2.6.1). The forgetting factor used in this algorithm is 0.999. It is seen from these two figures that as in the nonadaptive case, the robust multiuser detector offers significant performance gain over the linear multiuser detector in impulsive noise. Furthermore, in this example, the adaptive version of the blind robust detector based on subspace tracking outperforms the batch SVD-based approach [this is because with a forgetting factor 0.999, the effective data window size is $1/(1-0.999) = 1000$, whereas the window size in the batch method is $M = 200$], while it has a practical computational complexity and incurs no delay in data demodulation.

4.6 Robust Multiuser Detection Based on Local Likelihood Search

Recall that in Section 3.4 we introduced a nonlinear multiuser detection method based on local likelihood search, which offers significant performance improvement over linear multiuser detection methods with comparable computational complexity. When combined with the subspace technique, this method also leads to a nonlinear group-blind multiuser detector. In this section we discuss the application of such a local likelihood search method in robust multiuser detection and group-blind robust multiuser detection. The materials in this section were developed in [455, 456].

4.6.1 Exhaustive-Search and Decorrelative Detection

Consider the following *complex-valued discrete-time synchronous CDMA signal model.* At any time instant, the received signal is the superposition of K users' signals, plus the ambient noise, given by

$$\boldsymbol{r} = \sum_{k=1}^{K} A_k b_k \boldsymbol{s}_k + \boldsymbol{n} \tag{4.96}$$

$$= \boldsymbol{S}\boldsymbol{A}\boldsymbol{b} + \boldsymbol{n}, \tag{4.97}$$

where, as before, $\boldsymbol{s}_k = (1/\sqrt{N})[c_{0,k} \cdots c_{N-1,k}]^T$, $c_{j,k} \in \{+1, -1\}$, is the normalized signature sequence of the kth user; N is the processing gain; $b_k \in \{+1, -1\}$ and A_k are, respectively, the data symbol and the complex amplitude of the kth user; $\boldsymbol{S} \triangleq [\boldsymbol{s}_1 \ \cdots \ \boldsymbol{s}_K]$; $\boldsymbol{A} \triangleq \mathsf{diag}(A_1, \ldots, A_K)$; $\boldsymbol{b} \triangleq [b_1 \ \cdots \ b_K]^T$; and $\boldsymbol{n} \triangleq [n_1 \cdots n_N]^T$ is a complex vector of i.i.d. ambient noise samples with independent real and imaginary components. Denote

$$\boldsymbol{y} \triangleq \begin{bmatrix} \Re\{\boldsymbol{r}\} \\ \Im\{\boldsymbol{r}\} \end{bmatrix}, \boldsymbol{\Psi} \triangleq \begin{bmatrix} \boldsymbol{S}\Re\{\boldsymbol{A}\} \\ \boldsymbol{S}\Im\{\boldsymbol{A}\} \end{bmatrix}, \boldsymbol{v} \triangleq \begin{bmatrix} \Re\{\boldsymbol{n}\} \\ \Im\{\boldsymbol{n}\} \end{bmatrix},$$

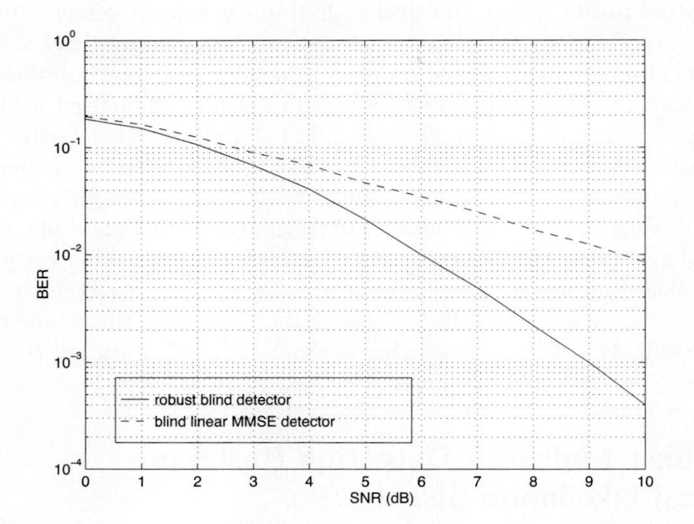

Figure 4.10. BER performance of user 1 for a blind robust detector and blind linear detector, using batch eigendecomposition, in a synchronous CDMA channel with non-Gaussian noise. $N = 31, K = 6, \epsilon = 0.01, \kappa = 100$. The powers of all interferers are 10 dB above the power of user 1.

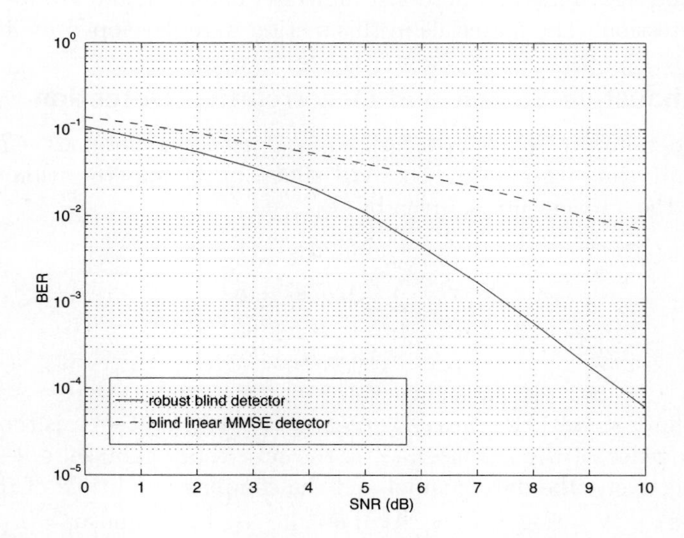

Figure 4.11. BER performance of user 1 for a blind robust detector and blind linear detector, using subspace tracking, in a synchronous CDMA channel with non-Gaussian noise. $N = 31, K = 6, \epsilon = 0.01, \kappa = 100$. The powers of the interferers are 10 dB above the power of user 1.

where $\boldsymbol{v}$ is a real noise vector consisting of $2N$ i.i.d. samples. Then (4.97) can be written as

$$\boldsymbol{y} = \boldsymbol{\Psi b} + \boldsymbol{v}. \tag{4.98}$$

It is assumed that each element v_j of $\boldsymbol{v}$ follows a two-term Gaussian mixture distribution:

$$v_j \sim (1-\epsilon)\mathcal{N}\left(0, \nu^2\right) + \epsilon\mathcal{N}\left(0, \kappa\nu^2\right), \tag{4.99}$$

with $0 \leq \epsilon < 1$ and $\kappa > 1$. Note that the overall variance of the noise sample v_j is

$$\frac{\sigma^2}{2} \triangleq (1-\epsilon)\nu^2 + \epsilon\kappa\nu^2. \tag{4.100}$$

We have $\text{Cov}(\boldsymbol{v}) = (\sigma^2/2)\boldsymbol{I}_{2N}$ and $\text{Cov}(\boldsymbol{n}) = \sigma^2\boldsymbol{I}_N$.

Recall from the preceding sections that there are primarily two categories of multiuser detectors for estimating $\boldsymbol{b}$ from $\boldsymbol{y}$ in (4.98), all based on minimizing the sum of a certain function $\rho(\cdot)$ of the chip residuals

$$\mathcal{C}(\boldsymbol{b}; \boldsymbol{y}) \triangleq \sum_{j=1}^{2N} \rho\left(y_j - \boldsymbol{\xi}_j^T \boldsymbol{b}\right), \tag{4.101}$$

where $\boldsymbol{\xi}_j^T$ denotes the jth row of the matrix $\boldsymbol{\Psi}$. These are as follows.

- *Exhaustive-search detectors:*

$$\boldsymbol{b}^e = \arg \min_{\boldsymbol{b} \in \{+1,-1\}^K} \mathcal{C}(\boldsymbol{b}; \boldsymbol{y}). \tag{4.102}$$

- *Decorrelative detectors:*

$$\boldsymbol{\beta} = \arg \min_{\boldsymbol{b} \in \mathbb{R}^K} \mathcal{C}(\boldsymbol{b}; \boldsymbol{y}), \tag{4.103}$$

$$\boldsymbol{b}^* = \text{sign}(\boldsymbol{\beta}). \tag{4.104}$$

Note that exhaustive-search detection is based on the discrete minimization of the cost function $\mathcal{C}(\boldsymbol{b}; \boldsymbol{y})$, over 2^K candidate points, whereas decorrelative detection is based on the continuous minimization of the same cost function. As before, let $\psi = \rho'$ be the derivative of the penalty function ρ. In general, the optimization problem (4.103) can be solved iteratively according to the following steps [553]:

$$\boldsymbol{z}^l = \psi\left(\boldsymbol{y} - \boldsymbol{\Psi}\boldsymbol{\beta}^l\right), \tag{4.105}$$

$$\boldsymbol{\beta}^{l+1} = \boldsymbol{\beta}^l + \frac{1}{\mu}\left(\boldsymbol{\Psi}^T\boldsymbol{\Psi}\right)^{-1}\boldsymbol{\Psi}^T\boldsymbol{z}^l, \qquad l = 0, 1, \ldots. \tag{4.106}$$

Recall further from Section 4.2 the following three choices of the penalty function $\rho(\cdot)$ in (4.101), corresponding to different forms of detectors:

- *Log-likelihood penalty function:*

$$\rho_{ML}(x) \triangleq -\log f(x), \tag{4.107}$$

$$\psi_{ML}(x) = -\frac{f'(x)}{f(x)}, \tag{4.108}$$

where $f(\cdot)$ denotes the pdf of the noise sample v_j. In this case, the exhaustive-search detector (4.102) corresponds to the ML detector, and the decorrelative detector (4.104) corresponds to the ML decorrelator.

- *Least-squares penalty function:*

$$\rho_{LS}(x) \triangleq \frac{1}{2}x^2, \tag{4.109}$$

$$\psi_{LS}(x) = x. \tag{4.110}$$

In this case, the exhaustive-search detector (4.102) corresponds to the ML detector based on a Gaussian noise assumption, and the decorrelative detector (4.104) corresponds to the linear decorrelator.

- *Huber penalty function:*

$$\rho_H(x) = \begin{cases} \dfrac{x^2}{\sigma^2} & \text{if } |x| \le \dfrac{c\sigma^2}{2}, \\ c|x| - \dfrac{c^2\sigma^2}{4} & \text{if } |x| > \dfrac{c\sigma^2}{2}, \end{cases} \tag{4.111}$$

$$\psi_H(x) = \begin{cases} \dfrac{x}{\sigma^2} & \text{if } |x| \le \dfrac{c\sigma^2}{2}, \\ c\ \text{sign}(x) & \text{if } |x| > \dfrac{c\sigma^2}{2}. \end{cases} \tag{4.112}$$

where $\sigma^2/2$ is the noise variance given by (4.100) and

$$c = \frac{3}{\sqrt{2}\sigma} \tag{4.113}$$

is a constant. In this case, the exhaustive-search detector (4.102) corresponds to the discrete minimizer of the Huber cost function, and the decorrelative detector (4.104) corresponds to the robust decorrelator.

4.6.2 Local-Search Detection

Clearly, the optimal performance is achieved by the exhaustive-search detector with the log-likelihood penalty function (i.e., the ML detector). As will be seen later, the performance of the exhaustive-search detector with the Huber penalty function is close to that of the ML detector, while this detector does not require knowledge of the exact noise pdf. However, the computational complexity of the exhaustive-search detector (4.102) is on the order of $\mathcal{O}(2^K)$. We next discuss a local-search approach to approximating the solution to (4.102), based on the slowest-descent search method discussed in Section 3.4. The basic idea is to minimize the cost function $\mathcal{C}(\boldsymbol{b};\boldsymbol{y})$ over a subset Ω of the discrete parameter set $\{+1,-1\}^K$ that is close to the continuous stationary point $\boldsymbol{\beta}$ given by (4.103). More precisely, we approximate the solution to (4.102) by a local one:

$$\boldsymbol{b}^l \triangleq \arg\min_{\boldsymbol{b}\in\Omega} \mathcal{C}(\boldsymbol{b};\boldsymbol{y}). \tag{4.114}$$

In the slowest-descent-search method, the candidate set Ω consists of the discrete parameters chosen such that they are in the neighborhood of Q ($Q \le K$) lines in $\mathbb{R}^K$, which are defined by the stationary point $\boldsymbol{\beta}$ and the Q eigenvectors of the Hessian matrix $\nabla^2_{\mathcal{C}}(\boldsymbol{\beta})$ of $\mathcal{C}(\boldsymbol{b};\boldsymbol{y})$ at $\boldsymbol{\beta}$ corresponding to the Q smallest eigenvalues.

For the three types of penalty functions, the Hessian matrix at the stationary points are given, respectively, by

$$\rho_{ML}: \nabla^2_{\mathcal{C}}(\boldsymbol{\beta}) = \boldsymbol{\Psi}^T \text{diag}\left\{\rho''_{ML}\left(y_j - \boldsymbol{\xi}_j^T\boldsymbol{\beta}\right), j=1,\ldots,2N.\right\}\boldsymbol{\Psi}, \tag{4.115}$$

$$\rho_{LS}: \nabla^2_{\mathcal{C}}(\boldsymbol{\beta}) = \boldsymbol{\Psi}^T\boldsymbol{\Psi}, \tag{4.116}$$

$$\rho_{H}: \nabla^2_{\mathcal{C}}(\boldsymbol{\beta}) = \boldsymbol{\Psi}^T \text{diag}\left\{I\left(\left|y_j - \boldsymbol{\xi}_j^T\boldsymbol{\beta}\right| \le \frac{c\sigma^2}{2}\right), j=1,\ldots,2N.\right\}\boldsymbol{\Psi}, \tag{4.117}$$

where, in (4.115),

$$\rho''_{ML}(x) = \psi^2_{ML}(x) - f''(x)/f(x), \tag{4.118}$$

and in (4.117) the indicator function $I(y \le a) = 1$ if $y \le a$ and 0 otherwise; hence in this case those rows of $\boldsymbol{\Psi}$ with large residual signals as a possible result of impulsive noise are nullified, whereas other rows of $\boldsymbol{\Psi}$ are not affected.

Denote $\boldsymbol{b}^* \triangleq \text{sign}(\boldsymbol{\beta})$. In general, the slowest-descent-search method chooses the candidate set Ω in (4.114) as follows:

$$\Omega = \{\boldsymbol{b}^*\} \cup \bigcup_{q=1}^{Q} \left\{\boldsymbol{b}^{q,\mu} \in \{-1,+1\}^K : b_k^{q,\mu} = \begin{cases} \text{sign}\left(\beta_k - \mu g_k^q\right) & \text{if } \beta_k - \mu g_k^q \neq 0, \\ -b_k^* & \text{if } \beta_k - \mu g_k^q = 0, \end{cases}\right.$$

$$\left. \boldsymbol{g}^q \text{ is the } q\text{th smallest eigenvector of } \nabla^2_{\mathcal{C}},\ \mu \in \left\{\frac{\beta_1}{g_1^q}, \ldots, \frac{\beta_K}{g_K^q}\right\}\right\}. \tag{4.119}$$

Hence, $\{\boldsymbol{b}^{q,\mu}\}_\mu$ contains the K closest neighbors of $\boldsymbol{\beta}$ in $\{-1,+1\}^K$ along the direction of $\boldsymbol{g}^q$. Note that $\{\boldsymbol{g}^q\}_{q=1}^Q$ represent the Q mutually orthogonal directions where the cost function $\mathcal{C}(\boldsymbol{b};\boldsymbol{y})$ grows the slowest from the minimum point $\boldsymbol{\beta}$.

Finally, we summarize the slowest-descent-search algorithm for robust multiuser detection in non-Gaussian noise. Given a penalty function $\rho(\cdot)$, this algorithm solves the discrete optimization problem (4.114) according to the following steps.

Algorithm 4.4: [Robust multiuser detector based on slowest-descent-search—synchronous CDMA]

- *Compute the continuous stationary point* $\boldsymbol{\beta}$ *in (4.103):*

$$\boldsymbol{\beta}^0 = \left(\boldsymbol{\Psi}^T \boldsymbol{\Psi}\right)^{-1} \boldsymbol{\Psi}^T \boldsymbol{y}, \tag{4.120}$$

Do

$$\boldsymbol{z}^l = \psi\left(\boldsymbol{y} - \boldsymbol{\Psi}\,\boldsymbol{\beta}^l\right), \tag{4.121}$$

$$\boldsymbol{\beta}^{l+1} = \boldsymbol{\beta}^l + \frac{1}{\mu}\left(\boldsymbol{\Psi}^T \boldsymbol{\Psi}\right)^{-1} \boldsymbol{\Psi}^T \boldsymbol{z}^l, \tag{4.122}$$

While $\left\|\boldsymbol{\beta}^{l+1} - \boldsymbol{\beta}^l\right\| > \delta.$

Set $\hat{\boldsymbol{\beta}} = \boldsymbol{\beta}^l$ *and* $\boldsymbol{b}^* = \text{sign}\left(\hat{\boldsymbol{\beta}}\right)$.

- *Compute the Hessian matrix* $\nabla^2_{\mathcal{C}}(\hat{\boldsymbol{\beta}})$ *given by (4.115) or (4.116) or (4.117), and its* Q *smallest eigenvectors* $\boldsymbol{g}^1, \ldots, \boldsymbol{g}^Q$.

- *Solve the discrete optimization problem defined by (4.114) and (4.119) by an exhaustive search [over* $(KQ+1)$ *points].*

Simulation Results

For simulations, we consider a synchronous CDMA system with a processing gain $N = 15$, number of users $K = 6$, and no phase offset and equal amplitudes of user signals (i.e., $\alpha_k = 1$, $k = 1, \ldots, K$). The signature sequence $\boldsymbol{s}_1$ of user 1 is generated randomly and kept fixed throughout simulations. The signature sequences of other users are generated by circularly shifting the sequence of user 1.

For each of the three penalty functions, Fig. 4.12 presents the BER performance of the decorrelative detector, slowest-descent-search detector with two search directions, and exhaustive-search detector. Searching further slowest-descent directions does not improve the performance in this case. We observe that for all three criteria, the performance of the slowest-descent-search detector is close to that of its respective exhaustive-search version. Moreover, the LS-based detectors have the worst performance. Note that the detector based on the Huber penalty function and the slowest-descent search offers significant performance gain over the robust decorrelator developed in Section 4.4 (Algorithm 4.1). For example, at the BER of 10^{-3}, it is only less than 1 dB from the ML detector, whereas the robust decorrelator is about 3 dB from the ML detector.

4.7 Robust Group-Blind Multiuser Detection

Consider the received signal of (4.97). As noted in Chapter 3, in group-blind multiuser detection, only a subset of the K users' signals need to be demodulated.

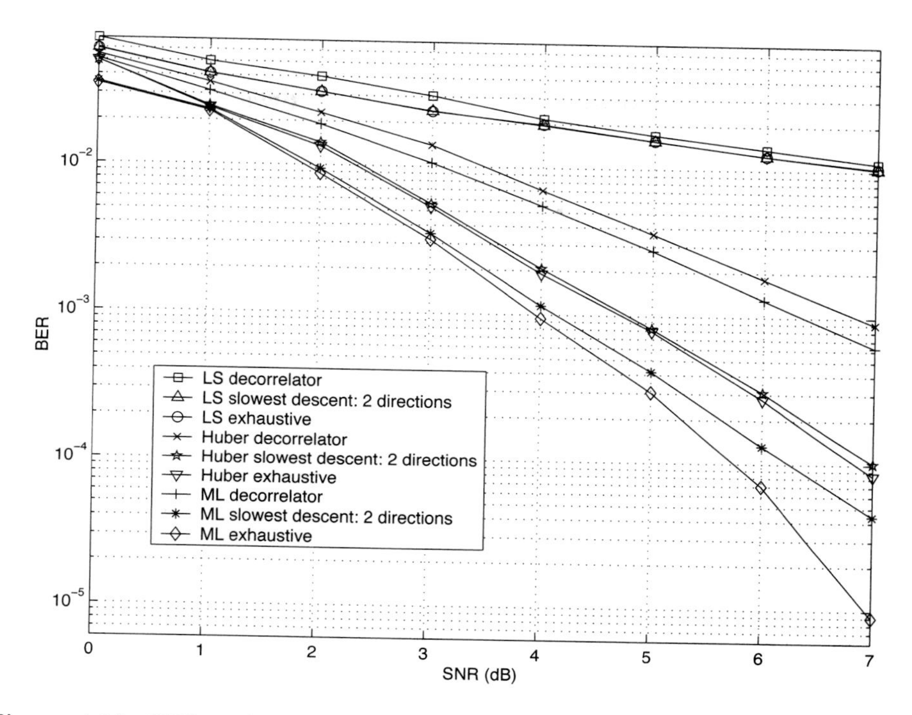

Figure 4.12. BER performance of various detectors in a DS-CDMA system with non-Gaussian ambient noise. $N = 15, K = 8, \epsilon = 0.01, \kappa = 100$.

Specifically, suppose that the first $\tilde{K}$ ($\tilde{K} \leq K$) users are the users of interest. Denote $\tilde{\boldsymbol{S}}$ and $\bar{\boldsymbol{S}}$ as matrices containing, respectively, the first $\tilde{K}$ and the last $(K - \tilde{K})$ columns of $\boldsymbol{S}$. Similarly define the quantities $\tilde{\boldsymbol{A}}$, $\tilde{\boldsymbol{b}}$, $\bar{\boldsymbol{A}}$, and $\bar{\boldsymbol{b}}$. Then (4.97) can be rewritten as

$$\begin{aligned} \boldsymbol{r} &= \boldsymbol{S}\boldsymbol{A}\boldsymbol{b} + \boldsymbol{n} && (4.123) \\ &= \tilde{\boldsymbol{S}}\tilde{\boldsymbol{A}}\tilde{\boldsymbol{b}} + \bar{\boldsymbol{S}}\bar{\boldsymbol{A}}\bar{\boldsymbol{b}} + \boldsymbol{n}. && (4.124) \end{aligned}$$

Let the autocorrelation matrix of the received signal and its eigendecomposition be

$$\boldsymbol{C}_r \triangleq E\left\{\boldsymbol{r}\boldsymbol{r}^H\right\} = \boldsymbol{U}_s \boldsymbol{\Lambda}_s \boldsymbol{U}_s^H + \sigma^2 \boldsymbol{U}_n \boldsymbol{U}_n^H. \quad (4.125)$$

We next consider the problem of nonlinear group-blind multiuser detection in non-Gaussian noise. Denote $\tilde{\boldsymbol{\theta}} \triangleq \tilde{\boldsymbol{A}}\tilde{\boldsymbol{b}}$ and $\bar{\boldsymbol{\theta}} \triangleq \bar{\boldsymbol{A}}\bar{\boldsymbol{b}}$. Then (4.124) can be written as

$$\begin{aligned} \boldsymbol{r} &= \tilde{\boldsymbol{S}}\tilde{\boldsymbol{\theta}} + \bar{\boldsymbol{S}}\bar{\boldsymbol{\theta}} + \boldsymbol{n} && (4.126) \\ &= \boldsymbol{U}_s \boldsymbol{\zeta} + \boldsymbol{n}, && (4.127) \end{aligned}$$

for some $\boldsymbol{\zeta} \in \mathbb{C}^K$. The basic idea here is to get an estimate of the sum of the undesired users' signals, $\bar{\boldsymbol{S}}\bar{\boldsymbol{\theta}}$, and to subtract it from $\boldsymbol{r}$. This effectively reduces the problem to the form treated in Section 4.6. To that end, denote

$$\underbrace{\begin{bmatrix} \Re\{\boldsymbol{r}\} \\ \Im\{\boldsymbol{r}\} \end{bmatrix}}_{\boldsymbol{y}} = \underbrace{\begin{bmatrix} \tilde{\boldsymbol{S}}\Re\{\tilde{\boldsymbol{A}}\} \\ \tilde{\boldsymbol{S}}\Im\{\tilde{\boldsymbol{A}}\} \end{bmatrix}}_{\tilde{\boldsymbol{\Psi}}} \tilde{\boldsymbol{b}} + \underbrace{\begin{bmatrix} \bar{\boldsymbol{S}}\Re\{\bar{\boldsymbol{A}}\} \\ \bar{\boldsymbol{S}}\Im\{\bar{\boldsymbol{A}}\} \end{bmatrix}}_{\bar{\boldsymbol{\Psi}}} \bar{\boldsymbol{b}} + \underbrace{\begin{bmatrix} \Re\{\boldsymbol{n}\} \\ \Im\{\boldsymbol{n}\} \end{bmatrix}}_{\boldsymbol{v}} \tag{4.128}$$

$$= \underbrace{\begin{bmatrix} \Re\{\boldsymbol{U}_s\} & -\Im\{\boldsymbol{U}_s\} \\ \Im\{\boldsymbol{U}_s\} & \Re\{\boldsymbol{U}_s\} \end{bmatrix}}_{\boldsymbol{\Xi}} \underbrace{\begin{bmatrix} \Re\{\boldsymbol{\zeta}\} \\ \Im\{\boldsymbol{\zeta}\} \end{bmatrix}}_{\boldsymbol{\phi}} + \underbrace{\begin{bmatrix} \Re\{\boldsymbol{n}\} \\ \Im\{\boldsymbol{n}\} \end{bmatrix}}_{\boldsymbol{v}}. \tag{4.129}$$

Next, we outline the method for estimating the signal $\bar{\boldsymbol{S}}\bar{\boldsymbol{\theta}}$ in (4.127). Denote

$$\boldsymbol{\phi}^0 \triangleq \left(\boldsymbol{\Xi}^T\boldsymbol{\Xi}\right)^{-1} \boldsymbol{\Xi}^T \boldsymbol{y}. \tag{4.130}$$

In what follows we assume that the Huber penalty function is used. We first obtain a robust estimate of $\boldsymbol{\phi}$ by the following iterative procedure:

$$\boldsymbol{z}^l = \psi\left(\boldsymbol{y} - \boldsymbol{\Xi}\boldsymbol{\phi}^l\right), \tag{4.131}$$

$$\boldsymbol{\phi}^{l+1} = \boldsymbol{\phi}^l + \frac{1}{\mu}\left(\boldsymbol{\Xi}^T\boldsymbol{\Xi}\right)^{-1} \boldsymbol{\Xi}^T \boldsymbol{z}^l, \quad l = 0, 1, 2, \ldots. \tag{4.132}$$

The robust estimate of $\boldsymbol{\phi}$ translates into a robust estimate of $\boldsymbol{\zeta}$, which by Proposition 4.2, in turn translates into a robust estimate of $\tilde{\boldsymbol{\theta}}$, as

$$\tilde{\boldsymbol{\theta}} = \left[\tilde{\boldsymbol{S}}^T \boldsymbol{U}_s \left(\boldsymbol{\Lambda}_s - \sigma^2 \boldsymbol{I}_K\right)^{-1} \boldsymbol{U}_s^H \tilde{\boldsymbol{S}}\right]^{-1} \tilde{\boldsymbol{S}}^T \boldsymbol{U}_s \left(\boldsymbol{\Lambda}_s - \sigma^2 \boldsymbol{I}_K\right)^{-1} \boldsymbol{\zeta}. \tag{4.133}$$

Using the above-estimated $\tilde{\boldsymbol{\theta}}$, the desired users' signals are then subtracted from the received signal to obtain

$$\bar{\boldsymbol{r}} \triangleq \boldsymbol{r} - \tilde{\boldsymbol{S}}\tilde{\boldsymbol{\theta}}. \tag{4.134}$$

Next, the subspace components of the undesired users' signals are identified as follows. Let

$$\bar{\boldsymbol{C}} \triangleq E\left\{\bar{\boldsymbol{r}}\bar{\boldsymbol{r}}^H\right\} = \bar{\boldsymbol{U}}_s \bar{\boldsymbol{\Lambda}}_s \bar{\boldsymbol{U}}_s^H + \sigma^2 \bar{\boldsymbol{U}}_n \bar{\boldsymbol{U}}_n^H, \tag{4.135}$$

where the dimension of the signal subspace in (4.135) is $K - \tilde{K}$. We then have

$$\bar{\boldsymbol{r}} = \bar{\boldsymbol{S}}\bar{\boldsymbol{\theta}} + \boldsymbol{n} \tag{4.136}$$

$$= \bar{\boldsymbol{U}}_s \bar{\boldsymbol{\zeta}} + \boldsymbol{n}, \tag{4.137}$$

for some $\bar{\zeta} \in \mathbb{C}^{K-\tilde{K}}$, or in its real-valued form,

$$\underbrace{\begin{bmatrix} \Re\{\bar{r}\} \\ \Im\{\bar{r}\} \end{bmatrix}}_{y} = \underbrace{\begin{bmatrix} \bar{S}\Re\{\bar{A}\} \\ \bar{S}\Im\{\bar{A}\} \end{bmatrix}}_{\bar{\Psi}} \bar{b} + \underbrace{\begin{bmatrix} \Re\{n\} \\ \Im\{n\} \end{bmatrix}}_{v} \tag{4.138}$$

$$= \underbrace{\begin{bmatrix} \Re\{\bar{U}_s\} & -\Im\{\bar{U}_s\} \\ \Im\{\bar{U}_s\} & \Re\{\bar{U}_s\} \end{bmatrix}}_{\bar{\Xi}} \underbrace{\begin{bmatrix} \Re\{\bar{\zeta}\} \\ \Im\{\bar{\zeta}\} \end{bmatrix}}_{\bar{\phi}} + \underbrace{\begin{bmatrix} \Re\{n\} \\ \Im\{n\} \end{bmatrix}}_{v}. \tag{4.139}$$

A robust estimate of $\bar{\phi}$ is then obtained from (4.139) using an iterative procedure similar to (4.131)–(4.132). Finally, the estimated undesired users' signals are subtracted from the received signal to obtain

$$\tilde{y} = y - \bar{\Xi}\bar{\phi} \tag{4.140}$$

$$= \tilde{\Psi}\tilde{b} + v. \tag{4.141}$$

Note that in order to form $\tilde{\Psi}$, the complex amplitudes of the desired users, $\tilde{A}$, must be estimated, which can be done based on $\tilde{\theta}$, as discussed in Section 3.4. Note also that such an estimate has a phase ambiguity of π, which necessitates differential encoding and decoding of data. The signal model (4.141) is the same as the one treated in Section 4.6. Accordingly, define the following cost function based on the Huber penalty function:

$$\tilde{\ell}\left(\tilde{b}\right) \triangleq \sum_{j=1}^{2N} \rho_H \left(\tilde{y}_j - \eta_j^T \tilde{b}\right), \tag{4.142}$$

where η_j^T denotes the jth row of the matrix $\tilde{\Psi}$. Let $\tilde{\beta}$ be the stationary point of $\tilde{\ell}(\cdot)$, which can also be found using an iterative method similar to (4.105)–(4.106). The Hessian of $\tilde{\ell}(\cdot)$ at the stationary point is given by

$$\nabla_{\tilde{\ell}}^2 \left(\tilde{\beta}\right) = \tilde{\Psi}^T P\left(\tilde{\beta}\right)\tilde{\Psi}, \tag{4.143}$$

with

$$P\left(\tilde{\beta}\right) = \operatorname{diag}\left\{ I\left(\left|\tilde{y}_j - \eta_j^T\tilde{\beta}\right| \le c\sigma^2/2\right), j = 1, \ldots, 2N\right\}. \tag{4.144}$$

The estimate of the desired users' bits $\tilde{b}$ based on the slowest-descent search is now given by [let $\tilde{b}^* \triangleq \operatorname{sign}\left(\tilde{\beta}\right)$]

$$\hat{\tilde{b}} \cong \arg \min_{\tilde{b} \in \tilde{\Omega} \subset \{+1,-1\}^{\tilde{K}}} \tilde{\ell}\left(\tilde{b}\right), \tag{4.145}$$

with

$$\tilde{\Omega} = \left\{ \tilde{\boldsymbol{b}}^* \right\} \cup \bigcup_{q=1}^{Q} \left\{ \tilde{\boldsymbol{b}}^{q,\mu} \in \{+1,-1\}^{\tilde{K}} : \tilde{b}_k^{q,\mu} = \begin{cases} \text{sign}\left(\tilde{\beta}_k - \mu g_k^q\right) & \text{if } \tilde{\beta}_k - \mu g_k^q \neq 0, \\ -\tilde{b}_k^* & \text{if } \tilde{\beta}_k - \mu g_k^q = 0, \end{cases}\right.$$

$$\boldsymbol{g}^q \text{ is the } q\text{th smallest eigenvector of } \nabla_{\tilde{\ell}}^2 \left(\tilde{\boldsymbol{\beta}}\right), \ \mu \in \left\{ \frac{\tilde{\beta}_1}{g_1^q}, \ldots, \frac{\tilde{\beta}_K}{g_K^q} \right\} \Bigg\}. \quad (4.146)$$

The robust group-blind multiuser detection algorithm for synchronous CDMA with non-Gaussian noise is summarized below.

Algorithm 4.5: [Robust group-blind multiuser detector—synchronous CDMA]

- *Compute the sample autocorrelation matrix of the received signal and its eigendecomposition.*

- *Compute the robust estimate of* $\boldsymbol{\phi}$ *using (4.130)–(4.132); compute the robust estimate of* $\tilde{\boldsymbol{\theta}}$ *using (4.133).*

- *Compute the estimate of the complex amplitudes* $\tilde{\boldsymbol{A}}$ *based on the robust estimate of* $\tilde{\boldsymbol{\theta}}$ *using (3.127)–(3.129) [cf. (3.134)–(3.140)].*

- *Obtain the robust estimate of the undesired users' signals according to (4.134)–(4.139), by applying an iterative procedure similar to (4.130)–(4.132); subtract the undesired users' signals from the received signal to obtain* $\tilde{\boldsymbol{y}}$ *in (4.141).*

- *Compute the stationary point* $\tilde{\boldsymbol{\beta}}$ *from* $\tilde{\boldsymbol{y}}$ *using an iterative procedure similar to (4.105)–(4.106); compute the Hessian* $\nabla_{\tilde{\ell}}^2(\tilde{\boldsymbol{\beta}})$ *using (4.143) and (4.144).*

- *Solve the discrete optimization problem defined by (4.145) and (4.146) using an exhaustive search [over* $(\tilde{K}Q+1)$ *points]; perform differential decoding.*

Simulation Examples

We consider a synchronous CDMA system with a processing gain $N = 15$, the number of users $K = 8$, and random phase offset and equal amplitudes of user signals. The number of desired users is $\tilde{K} = 4$. Only the spreading waveforms $\tilde{\boldsymbol{S}}$ of the desired users are assumed to be known to the receiver. The noise parameters are $\epsilon = 0.01, \kappa = 100$. The BER curves of the robust blind detector of Section 4.5 (Algorithm 4.2) and the slowest-descent-search robust group-blind detector with $Q = 1$ and $Q = 2$ are shown in Fig. 4.13. It is seen that significant performance improvement is offered by the robust group-blind local-search-based multiuser detector in non-Gaussian noise channels over the (nonlinear) blind robust detector discussed in Section 4.5.

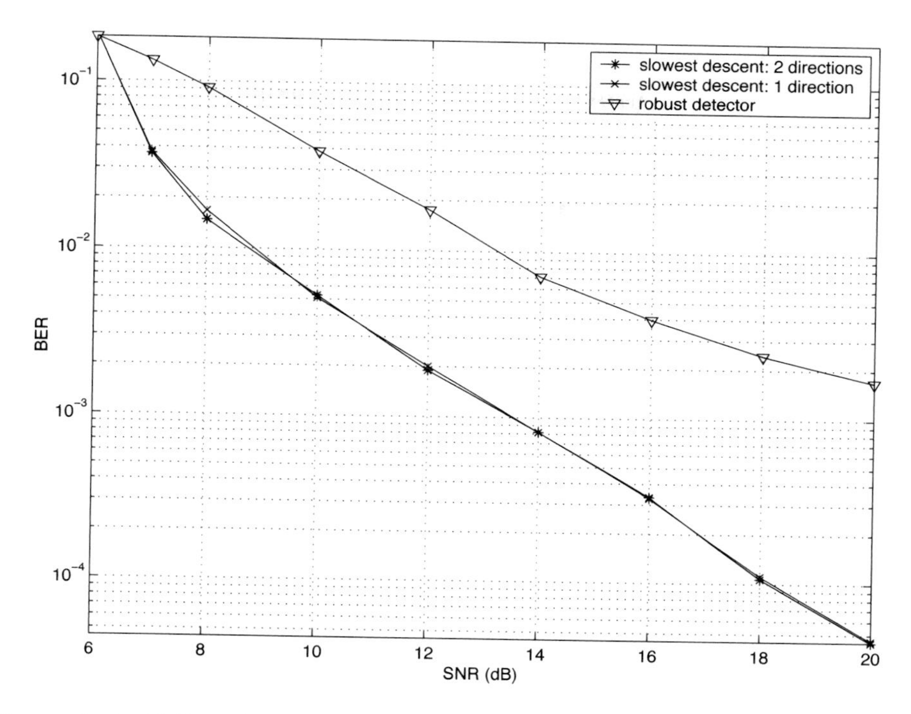

Figure 4.13. BER performance of a slowest-descent-based group-blind multiuser detector in non-Gaussian noise: synchronous case. $N = 15, K = 8, \tilde{K} = 4, \epsilon = 0.01, \kappa = 100$.

4.8 Extension to Multipath Channels

In this section we extend the robust group-blind multiuser detection techniques developed in previous sections to general asynchronous CDMA channels with multipath distortion. Let the impulse response of the kth user's multipath channel be given by

$$g_k(t) = \sum_{\ell=1}^{L} \alpha_{\ell,k} \delta(t - \tau_{\ell,k}), \tag{4.147}$$

where L is the total number of paths in the channel, and where $\alpha_{\ell,k}$ and $\tau_{\ell,k}$ are, respectively, the complex path gain and the delay of the kth user's ℓth path. It is assumed that $\tau_{1,k} < \tau_{2,k} < \cdots < \tau_{L,k}$. The received continuous-time signal is then given by

$$r(t) = \sum_{k=1}^{K} \sum_{i=0}^{M-1} b_k[i] \underbrace{[s_k(t - iT) \star g_k(t)]}_{h_k(t-iT)} + n(t), \tag{4.148}$$

where $\star$ denotes convolution.

As discussed in Section 2.7.1, at the receiver, the received signal is filtered by a chip-matched filter and sampled at a multiple (p) of the chip rate. Denote $r_q[i]$ as the qth signal sample during the ith symbol [cf. (2.167)]. Recall that by denoting

$$\underbrace{\underline{r}[i]}_{P\times 1} \triangleq \begin{bmatrix} r_0[i] \\ \vdots \\ r_{P-1}[i] \end{bmatrix}, \quad \underbrace{\underline{b}[i]}_{K\times 1} \triangleq \begin{bmatrix} b_1[i] \\ \vdots \\ b_K[i] \end{bmatrix}, \quad \underbrace{\underline{n}[i]}_{P\times 1} \triangleq \begin{bmatrix} n_0[i] \\ \vdots \\ n_{P-1}[i] \end{bmatrix},$$

$$\underbrace{\underline{H}[j]}_{P\times K} \triangleq \begin{bmatrix} h_1[jP] & \cdots & h_K[jP] \\ \vdots & \vdots & \vdots \\ h_1[jP+P-1] & \cdots & h_K[jP+P-1] \end{bmatrix}, \quad j = 0, \ldots, \iota,$$

we have the following discrete-time signal model:

$$\underline{r}[i] = \underline{H}[i] \star \underline{b}[i] + \underline{n}[i]. \tag{4.149}$$

By stacking m successive sample vectors, we further define the following quantities:

$$\underbrace{\boldsymbol{r}[i]}_{Pm\times 1} \triangleq \begin{bmatrix} \underline{r}[i] \\ \vdots \\ \underline{r}[i+m-1] \end{bmatrix}, \quad \underbrace{\boldsymbol{n}[i]}_{Pm\times 1} \triangleq \begin{bmatrix} \underline{n}[i] \\ \vdots \\ \underline{n}[i+m-1] \end{bmatrix},$$

$$\underbrace{\boldsymbol{b}[i]}_{K(m+\iota)\times 1} \triangleq \begin{bmatrix} \underline{b}[i-\iota] \\ \vdots \\ \underline{b}[i+m-1] \end{bmatrix}, \quad \underbrace{\boldsymbol{H}}_{Pm\times K(m+\iota)} \triangleq \begin{bmatrix} \underline{H}[\iota] & \cdots & \underline{H}[0] & \cdots & \boldsymbol{0} \\ \vdots & \ddots & \ddots & \ddots & \vdots \\ \boldsymbol{0} & \cdots & \underline{H}[\iota] & \cdots & \underline{H}[0] \end{bmatrix},$$

where $m = \left\lceil \frac{P+K}{P-K} \right\rceil \iota$; $r \triangleq K(m+\iota)$, and where ι is the maximum delay spread in terms of symbol intervals. We can then have the following matrix form of the discrete-time signal model:

$$\boldsymbol{r}[i] = \boldsymbol{H}\,\boldsymbol{b}[i] + \boldsymbol{n}[i]; \tag{4.150}$$

and as before, we write the eigendecomposition of the autocorrelation matrix of the received signal as

$$\boldsymbol{C}_r \triangleq E\left\{\boldsymbol{r}[i]\boldsymbol{r}[i]^H\right\} = \boldsymbol{H}\boldsymbol{H}^H + \sigma^2 \boldsymbol{I}_{Pm} \tag{4.151}$$

$$= \boldsymbol{U}_s \boldsymbol{\Lambda}_s \boldsymbol{U}_s^H + \sigma^2 \boldsymbol{U}_n \boldsymbol{U}_n^H, \tag{4.152}$$

where the signal subspace $\boldsymbol{U}_s$ has r columns.

We next discuss robust blind multiuser detection and robust group-blind multiuser detection in multipath channels.

4.8.1 Robust Blind Multiuser Detection in Multipath Channels

Suppose that user 1 is the user of interest. Then we can rewrite (4.150) as

$$\boldsymbol{r}[i] = \bar{\boldsymbol{h}}_1 b_1[i] + \boldsymbol{H}_0 \boldsymbol{b}_0[i] + \boldsymbol{n}[i] \tag{4.153}$$

$$= \boldsymbol{U}_s \boldsymbol{\zeta}[i] + \boldsymbol{n}[i], \tag{4.154}$$

where $\bar{\boldsymbol{h}}_1$ denotes the $(K\iota + 1)$th column of $\boldsymbol{H}$ (corresponding to the bit $b_1[i]$), $\boldsymbol{H}_0$ denotes the submatrix of $\boldsymbol{H}$ obtained by striking out the $(K\iota + 1)$th column, and $\boldsymbol{b}_0[i]$ denotes the subvector of $\boldsymbol{b}[i]$ obtained by striking out the $(K\iota + 1)$th element. As before, the basic idea behind robust blind multiuser detection is first to obtain a robust estimate of $\boldsymbol{\zeta}[i]$ using the identified signal subspace $\boldsymbol{U}_s$. On the other hand, as discussed in Section 2.7.3, given the spreading waveform $\boldsymbol{s}_1$ of the desired user, by exploiting the orthogonality between the signal subspace and noise subspace, the composite signature waveform $\bar{\boldsymbol{h}}_1$ of this user can be estimated (up to a complex scaling factor). Once an estimate of $\bar{\boldsymbol{h}}_1$ is available, the robust estimate of $\boldsymbol{\zeta}[i]$ can then be translated into a robust estimate of $b_1[i]$ (up to a complex scaling factor) by Proposition 4.2, as

$$\theta_1[i] \triangleq \bar{\boldsymbol{h}}_1^H \boldsymbol{U}_s \left(\boldsymbol{\Lambda}_s - \sigma^2 \boldsymbol{I}_r\right)^{-1} \boldsymbol{\zeta}[i]. \tag{4.155}$$

Finally, differential detection is performed according to

$$\hat{\beta}_1[i] = \text{sign}\left(\Re\left\{\theta_1[i]\theta_1[i-1]^*\right\}\right). \tag{4.156}$$

The algorithm is summarized as follows.

Algorithm 4.6: [Robust blind multiuser detector—multipath CDMA]

- *Compute the sample autocorrelation matrix of the received augmented signal $\boldsymbol{r}[i]$ and its eigendecomposition.*
- *Compute the robust estimate of $\boldsymbol{\zeta}[i]$ following a procedure similar to (4.128)–(4.132).*
- *Compute an blind estimate of $\bar{\boldsymbol{h}}_1$ according to (2.202)–(2.203).*
- *Compute the output of the robust blind detector according to (4.155).*
- *Perform differential detection according to (4.156).*

4.8.2 Robust Group-Blind Multiuser Detection in Multipath Channels

We now turn to the group-blind version of the robust multiuser detector for the multipath channel. As before, we can rewrite (4.150) as

$$\begin{aligned} \boldsymbol{r}[i] &= \tilde{\boldsymbol{H}}\tilde{\boldsymbol{b}}[i] + \bar{\boldsymbol{H}}\bar{\boldsymbol{b}}[i] + \boldsymbol{n}[i] & (4.157) \\ &= \boldsymbol{U}_s\boldsymbol{\zeta}[i] + \boldsymbol{n}[i], & (4.158) \end{aligned}$$

where $\tilde{\boldsymbol{b}}[i]$ and $\bar{\boldsymbol{b}}[i]$ contain the data bits in $\boldsymbol{b}[i]$ corresponding to sets of desired users and undesired users, respectively; $\tilde{\boldsymbol{H}}$ and $\bar{\boldsymbol{H}}$ contain columns of $\boldsymbol{H}$ corresponding to desired users and undesired users, respectively. As discussed in Section 2.7.3, based on the knowledge of the spreading waveforms $\tilde{\boldsymbol{S}}$ of the desired users, by exploiting the orthogonality between the signal subspace and the noise subspace, we can blindly estimate $\tilde{\boldsymbol{H}}$ up to a scale and phase ambiguity for each user. With such an estimate, we can write

$$\tilde{\boldsymbol{H}}\boldsymbol{b}[i] = \tilde{\boldsymbol{H}}_0\tilde{\boldsymbol{A}}\,\underline{\tilde{b}}[i] + \tilde{\boldsymbol{H}}_I\boldsymbol{\theta}_I[i], \tag{4.159}$$

where the term $\tilde{\boldsymbol{H}}_0\tilde{\boldsymbol{A}}\underline{\tilde{b}}[i]$ contains the signal carrying the current bits $\underline{\tilde{b}}[i] \triangleq [b_1[i]\cdots b_{\tilde{K}}[i]]^T$ of the desired users; and the term $\tilde{\boldsymbol{H}}_I\boldsymbol{\theta}_I[i]$ contains the signal carrying the previous and future bits $\{\underline{\tilde{b}}[l]\}_{l\neq i}$ (i.e., the intersymbol interference). Note that in (4.159) the term $\tilde{\boldsymbol{H}}_0$ represents the estimated channel for the desired users' current bits, and $\tilde{\boldsymbol{A}}$ is a diagonal matrix containing the complex scalars of ambiguities; the term $\tilde{\boldsymbol{H}}_I$ represents the estimated channel for the desired users' past and future bits, and $\boldsymbol{\theta}_I[i]$ contains the products of those bits and the complex ambiguities of the corresponding channels. Following the method outlined in Section 4.7, we first obtain a robust estimate of $\boldsymbol{\zeta}[i]$ and then translate it into the estimate of $\tilde{\boldsymbol{\theta}}[i]$ by again applying Proposition 4.2:

$$\tilde{\boldsymbol{\theta}}[i] \triangleq \tilde{\boldsymbol{H}}^H\boldsymbol{U}_s\left(\boldsymbol{\Lambda}_s - \sigma^2\boldsymbol{I}_r\right)^{-1}\boldsymbol{\zeta}[i]. \tag{4.160}$$

Next, we obtain a robust estimate of the sum of the undesired users' signals based on the relationship

$$\bar{\boldsymbol{r}}[i] = \boldsymbol{r}[i] - \tilde{\boldsymbol{H}}\tilde{\boldsymbol{\theta}}[i] \tag{4.161}$$

$$= \bar{\boldsymbol{U}}_s\bar{\boldsymbol{\zeta}}[i] + \boldsymbol{n}[i], \tag{4.162}$$

where $\bar{\boldsymbol{U}}_s$ represents the signal subspace obtained from the eigendecomposition of the autocorrelation matrix of $\bar{\boldsymbol{r}}[i]$. Finally, we subtract the estimated undesired users' signals and the intersymbol interference from $\boldsymbol{r}[i]$ to obtain

$$\tilde{\boldsymbol{r}}[i] \triangleq \boldsymbol{r}[i] - \bar{\boldsymbol{U}}_s\bar{\boldsymbol{\zeta}}[i] - \tilde{\boldsymbol{H}}_I\tilde{\boldsymbol{\theta}}_I[i] \tag{4.163}$$

$$= \tilde{\boldsymbol{H}}_0\tilde{\boldsymbol{A}}\,\underline{\tilde{b}}[i] + \boldsymbol{n}[i]. \tag{4.164}$$

Note that the complex ambiguities in $\tilde{\boldsymbol{A}}$ can be estimated based on the estimate of $\tilde{\boldsymbol{\theta}}[i]$, as discussed in Section 4.7. Note also that (4.164) has the same form as (4.141), and hence similarly to (4.143)–(4.146), the slowest-descent-search method can then be employed to obtain a robust estimate of $\underline{\tilde{b}}[i]$ from (4.164). The algorithm is summarized below.

Algorithm 4.7: [Robust group-blind multiuser detector—multipath CDMA]

- *Compute the sample autocorrelation matrix of the received augmented signal* $\boldsymbol{r}[i]$ *and its eigendecomposition.*

- *Compute the robust estimate of* $\boldsymbol{\zeta}[i]$ *following a procedure similar to (4.128)–(4.132).*

- *Compute a blind estimate of* $\tilde{\boldsymbol{H}}$ *according to (3.162)–(3.163).*

- *Compute the output of the robust blind detector according to (4.160).*
- *Compute the sum of the undesired users' signals* $\bar{\mathbf{r}}[i]$ *according to (4.161); compute the sample autocorrelation matrix of the signal* $\bar{\mathbf{r}}[i]$ *and its eigendecomposition.*
- *Compute the robust estimate of* $\bar{\boldsymbol{\zeta}}[i]$ *in (4.162) following a procedure similar to (4.128)–(4.132).*
- *Compute the sum of the desired users' signals* $\tilde{\mathbf{r}}[i]$ *according to (4.163).*
- *Estimate the complex amplitudes of ambiguities* $\tilde{\boldsymbol{A}}$ *introduced by the blind estimator based on the robust estimate of* $\tilde{\boldsymbol{\theta}}[i]$ *using (3.127)–(3.129) [cf. (3.134)–(3.140)].*
- *Form the Huber penalty function and apply the slowest-descent search of* $\underline{\tilde{b}}[i]$*, similar to (4.143)–(4.146).*
- *Perform differential decoding.*

Simulation Examples

In the following simulation, the number of users is $K = 8$ and the spreading gain is $N = 15$. Each user's channel is assumed to have $L = 3$ paths and a delay spread of up to one symbol. The complex gains and the delays of each user's channel are generated randomly. The chip pulse is a raised cosine pulse with roll-off factor 0.5. The path gains are normalized so that each user's signal arrives at the receiver with unit power. The channel is normalized in such a way that the composite of the multipath channel and the spreading waveform has unit power. The noise parameters are $\epsilon = 0.01$ and $\kappa = 100$. The smoothing factor is $m = 2$ and the oversampling factor is $p = 2$. Shown in Fig. 4.14 is the BER performance of the robust blind multiuser detector and that of the robust group-blind multiuser detector ($\tilde{K} = 4$). It is seen that in the presence of both non-Gaussian noise and multipath channel distortion, the group-blind robust detector substantially improves the performance of the blind robust detector. Furthermore, most of the performance gain offered by the slowest-descent search is obtained by searching along only one direction.

4.9 Robust Multiuser Detection in Stable Noise

So far in this chapter we have modeled non-Gaussian ambient noise using a mixture Gaussian distribution. Recently, the stable noise model has been proposed as a statistical model for the impulsive noise in several applications, including wireless communications [357,491]. In this section we first give a brief description of the stable distribution. We then demonstrate that the various robust multiuser detection techniques discussed in this chapter are also very effective in combating impulsive noise modeled by a stable distribution.

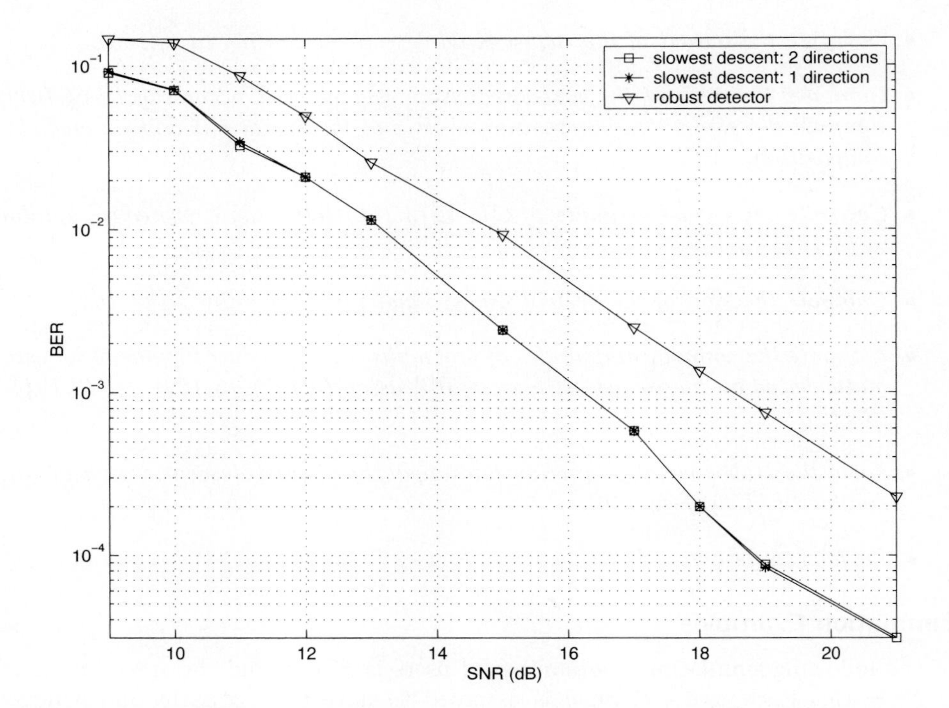

Figure 4.14. BER performance of a group-blind robust multiuser detector in non-Gaussian noise: multipath channel. $N = 15, K = 8, \tilde{K} = 4, \epsilon = 0.01, \kappa = 100$. Each user's channel consists of three paths with randomly generated complex gains and delays. Only the spreading waveforms $\tilde{\boldsymbol{S}}$ of the desired users are assumed known to the receiver. The BER curves of the robust blind detector (Algorithm 4.2) and robust group-blind detector (Algorithm 4.7) with one and two search directions are shown.

4.9.1 Symmetric Stable Distribution

A symmetric stable distribution is defined through its characteristic function as follows.

Definition 4.1: [Symmetric stable distribution] *A random variable X has a symmetric stable distribution if and only if its characteristic function has the form*

$$\begin{aligned}\phi(t; \alpha, \gamma, \theta) &= E\left\{\exp(\jmath t X)\right\} \\ &= \exp\left(\jmath \theta t - \gamma |t|^{\alpha}\right),\end{aligned} \tag{4.165}$$

where

$$\gamma > 0, \quad 0 < \alpha \leq 2, \quad -\infty < \theta < \infty.$$

Thus, a symmetric stable random variable is completely characterized by three parameters, α, γ, and θ, where:

- α is called the *characteristic exponent*, which indicates the "heaviness" of the tails of the distribution—a small value of α implies a heavier tail. The case $\alpha = 2$ corresponds to a Gaussian distribution, whereas $\alpha = 1$ corresponds to a Cauchy distribution.
- γ is called the *dispersion*. For the Gaussian case (i.e., $\alpha = 2$), $\gamma = \frac{1}{2}\text{Var}(X)$.
- θ is a *location parameter*, which is the mean when $1 < \alpha \leq 2$ and the median when $0 < \alpha < 1$.

By taking the Fourier transform of the characteristic function, we can obtain the probability density function (pdf) of the symmetric stable random variable X:

$$f(x; \alpha, \gamma, \theta) = \frac{1}{2\pi} \int_{-\infty}^{\infty} \phi(t; \alpha, \gamma, \theta) \exp(-\jmath x t)\, dt. \tag{4.166}$$

No closed-form expressions exist for general stable pdf's except for the Gaussian ($\alpha = 2$) and Cauchy ($\alpha = 1$) pdf's. For these two pdf's, closed-form expression exist:

$$f(x; \alpha = 2, \gamma, \theta) = \frac{1}{\sqrt{4\pi\gamma}} \exp\left[-\frac{(x-\theta)^2}{4\gamma}\right], \tag{4.167}$$

$$f(x; \alpha = 1, \gamma, \theta) = \frac{1}{\pi} \frac{\gamma}{\gamma^2 + (x-\theta)^2}. \tag{4.168}$$

It is known that for a non-Gaussian ($\alpha < 2$) symmetric stable random variable X with location parameter $\theta = 0$ and dispersion γ, we have the asymptote

$$\lim_{x \to \infty} x^{\alpha} P(|X| > x) = \gamma C(\alpha), \tag{4.169}$$

where $C(\alpha)$ is a positive constant depending on α. Thus, stable distributions with $\alpha < 2$ have inverse power tails, whereas Gaussian distributions have exponential tails. Hence the tails of the stable distributions are significantly heavier than those of the Gaussian distributions. In fact, the smaller is α, the slower does its tail drop to zero, as shown in Figs. 4.15 and 4.16.

As a consequence of (4.169), stable distributions do not have second-order moments except for the limiting case of $\alpha = 2$. More specifically, let X be a symmetric stable random variable with characteristic exponent α. If $0 < \alpha < 2$, then

$$E\{|X|^m\} \begin{cases} = \infty, & m \geq \alpha \\ < \infty, & 0 \leq m < \alpha. \end{cases} \tag{4.170}$$

If $\alpha = 2$, then

$$E\{|X|^m\} < \infty \tag{4.171}$$

for all $m \geq 0$. Hence for $0 < \alpha \leq 1$, stable distributions have no finite first- or higher-order moments; for $1 < \alpha < 2$, they have the first moments; and for $\alpha = 2$, all moments exist. In particular, all non-Gaussian stable distributions have infinite variance. The reader is referred to [357] for further details of these properties of α-stable distribution.

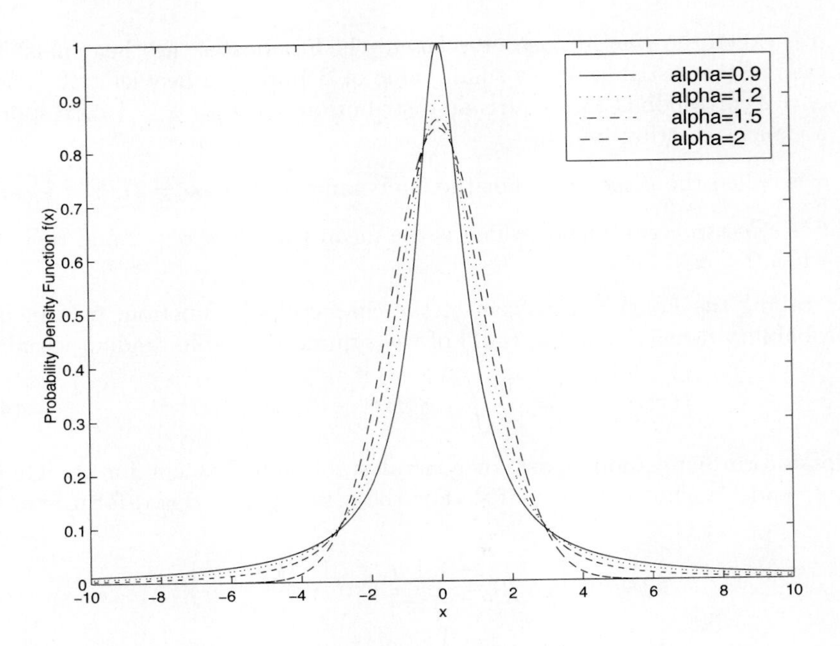

Figure 4.15. Symmetric stable pdf's for different values of α. $\gamma = 1$.

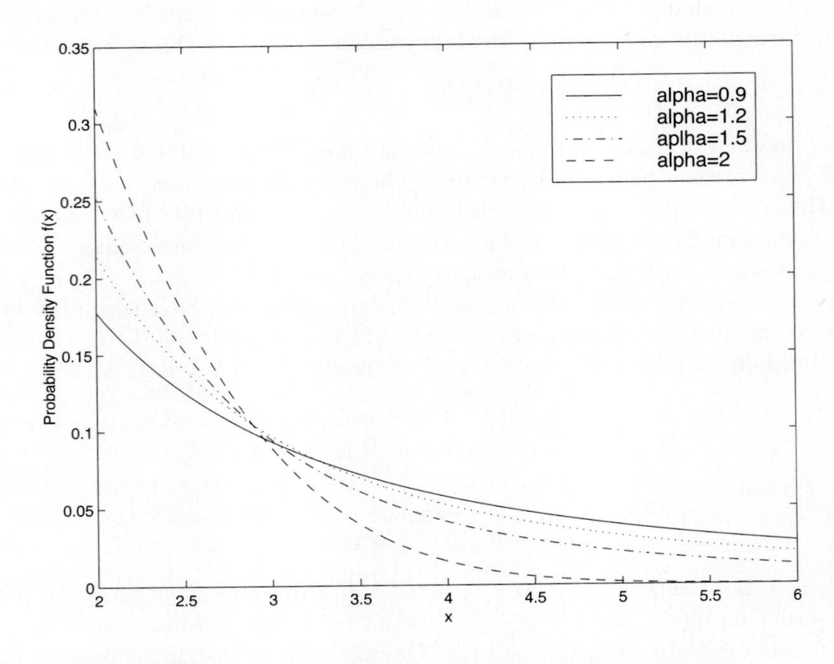

Figure 4.16. Tails of the symmetric stable pdf's for different values of α. $\gamma = 1$.

Generation of Symmetric Stable Random Variables

The following procedure generates a standard symmetric stable random variable X with characteristic exponent α, dispersion $\gamma = 1$ and location parameter $\theta = 0$ (see [357]):

$$\Phi \sim \text{uniform}\left(-\frac{\pi}{2}, \frac{\pi}{2}\right), \tag{4.172}$$

$$W \sim \exp(1), \tag{4.173}$$

$$\epsilon = 1 - \alpha, \tag{4.174}$$

$$a = \tan\left(\frac{\Phi}{2}\right), \tag{4.175}$$

$$b = \tan\left(\frac{\epsilon\Phi}{2}\right), \tag{4.176}$$

$$B = \frac{2b}{\epsilon\Phi}, \tag{4.177}$$

$$z = \frac{\cos(\epsilon\Phi)}{W\cos(\Phi)}, \tag{4.178}$$

$$X = \frac{2(a-b)(1+ab)}{(1-a^2)(1+b^2)} z^{\epsilon/\alpha}. \tag{4.179}$$

Now in order to generate a symmetric stable random variable Y with parameters (α, γ, θ), we first generate a standard symmetric stable random variable X with parameters $(\alpha, 1, 0)$, using the procedure above. Then Y can be generated from X according to the following transformation:

$$Y = \gamma^{1/\alpha} X + \theta. \tag{4.180}$$

4.9.2 Performance of Robust Multiuser Detectors in Stable Noise

We consider the performance of the robust multiuser detection techniques discussed in previous sections in symmetric stable noise. In particular, we consider the performance of the linear decorrelator, the maximum-likelihood decorrelator, and the Huber decorrelator, as well as their improved versions based on local likelihood search. First, the ψ functions for these three decorrelative detectors are plotted in Fig. 4.17. For the Huber decorrelator, the variance σ^2, the original definition of $\psi_H(\cdot)$ in (4.112), is replaced by the dispersion parameter γ. Note that since the pdf of the symmetric stable distribution does not have a closed form, we have to resort to a numerical method to compute $\psi_{\text{ML}}(x)$ given by (4.108). In particular, we can use discrete Fourier transform (DFT) to calculate samples of $f(x)$ and $f'(x)$, as follows. Recall that the characteristic function is given by

$$\phi(t) = \exp\left(-\gamma|t|^\alpha\right). \tag{4.181}$$

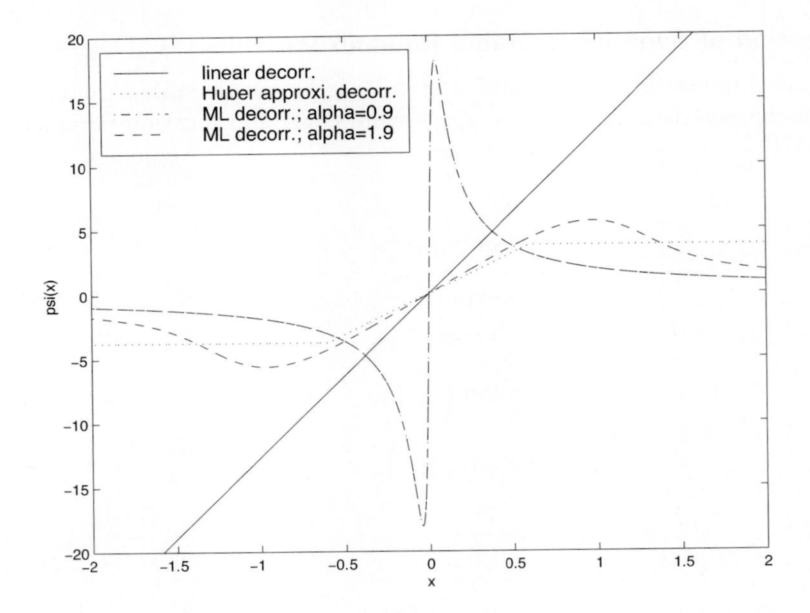

Figure 4.17. The ψ functions for a linear decorrelator, Huber decorrelator, and maximum-likelihood decorrelator under symmetric stable noise. $\gamma = 0.0792$.

The pdf and its derivative are related to the characteristic function through

$$f(x) = \frac{1}{2\pi}\int_{-\infty}^{+\infty} \phi(t) e^{-\jmath x t}\, \mathrm{d}t$$

$$= \mathcal{F}^{-1}\{\phi(-t)\} \;=\; \mathcal{F}^{-1}\{\phi(t)\}, \tag{4.182}$$

$$f'(x) = \jmath\, x\, \mathcal{F}^{-1}\{\phi(t)\}. \tag{4.183}$$

Hence by sampling the characteristic function $\phi(t)$ and then perform (inverse) DFT, we can get samples of $f(t)$ and $f'(t)$, which in turns give $\psi_{\mathrm{ML}}(x)$.

First we demonstrate the performance degradation of the linear decorrelator in symmetric stable noise. The BER performance of the linear decorrelator in several symmetric stable noise channels is depicted in Fig. 4.18. Here the SNR is defined as A_1^2/γ. It is seen that the smaller is α (i.e., the more impulsive is the noise), the more severe is the performance degradation incurred by the linear decorrelator. We next demonstrate the performance gain achieved by the Huber decorrelator. Figure 4.19 shows the BER performance of the Huber decorrelator. It is seen that as the noise becomes more impulsive (i.e., α becomes smaller), the Huber deccorrelator offers more performance improvement over the linear decorrelator. Finally, we depict the BER performance of the linear decorrelator, the Huber decorrelator, and the ML decorrelator, as well as their their improved versions based on the slowest-descent search, in Fig. 4.20. It is seen that the performance of the improved/unimproved linear decorrelator is substantially worse than that of the Huber decorrelator and the

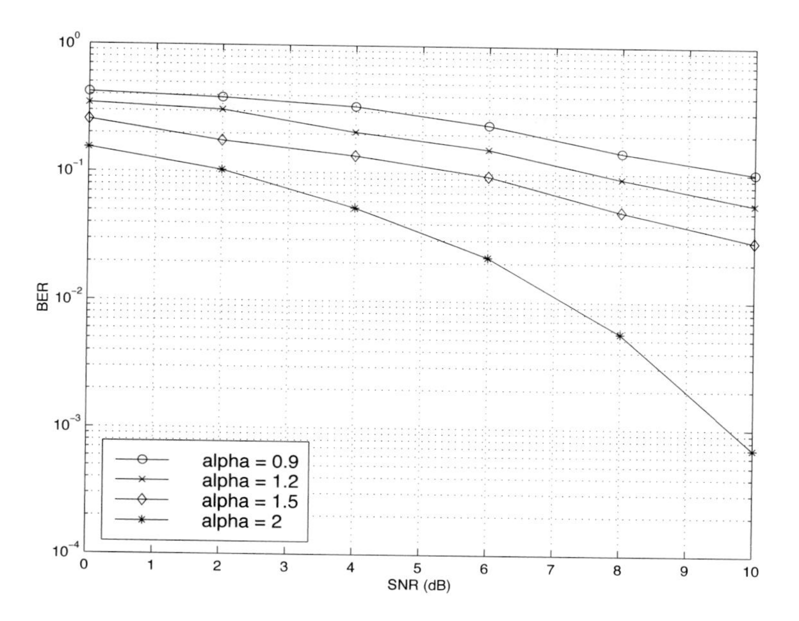

Figure 4.18. BER performance of a linear decorrelator in α-stable noise. $N = 31, K = 6$. The powers of the interferers are 10 dB above the power of user 1.

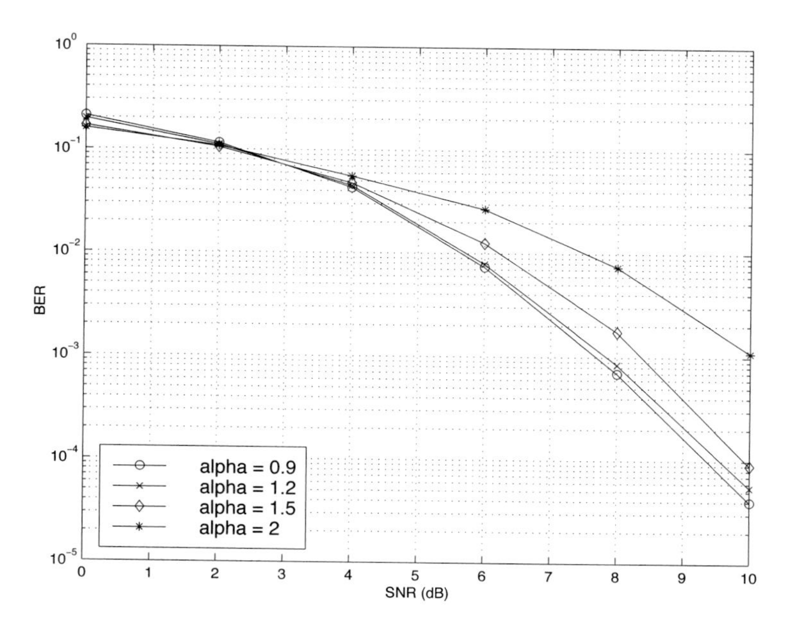

Figure 4.19. BER performance of a Huber decorrelator in α-stable noise. $N = 31, K = 6$. The powers of the interferers are 10 dB above the power of user 1.

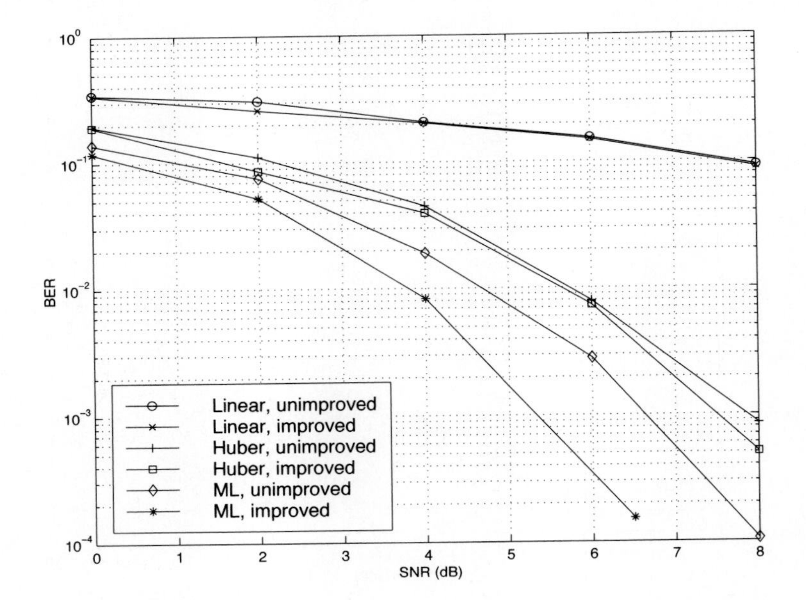

Figure 4.20. BER performance of three decorrelative detectors and their local-likelihood-search versions. $N = 31, K = 6, \alpha = 1.2$. The powers of the interferers are 10 dB above the power of user 1.

ML decorrelator. The improved Huber decorrelator performs more closely to the ML decorrelator.

4.10 Appendix

4.10.1 Proof of Proposition 4.1 in Section 4.4

We follow the technique used in [203] by defining the function

$$d(\boldsymbol{\tau}) = \frac{1}{\mu}\left[C\left(\boldsymbol{\theta}^l\right) - C\left(\boldsymbol{\theta}^l + \boldsymbol{\tau}\right)\right] + \frac{1}{2}\left[\boldsymbol{\tau}^T\left(\boldsymbol{S}^T\boldsymbol{S}\right)\boldsymbol{\tau} - \frac{2}{\mu}\boldsymbol{\tau}^T\boldsymbol{S}^T\boldsymbol{z}\left(\boldsymbol{\theta}^l\right)\right], \quad \boldsymbol{\tau} \in \mathbb{R}^K. \tag{4.184}$$

Notice that

$$d(\mathbf{0}) = 0, \tag{4.185}$$

$$\frac{\partial}{\partial \boldsymbol{\tau}} d(\boldsymbol{\tau})\bigg|_{\boldsymbol{\tau}=0} = \frac{1}{\mu}\sum_{j=1}^{N}\psi\left[r_j - \boldsymbol{\xi}_j^T\left(\boldsymbol{\theta}^l + \boldsymbol{\tau}\right)\right]\boldsymbol{\xi}_j + \boldsymbol{S}^T\boldsymbol{S}\,\boldsymbol{\tau} - \frac{1}{\mu}\boldsymbol{S}^T\boldsymbol{z}\left(\boldsymbol{\theta}^l\right)|_{\boldsymbol{\tau}=\mathbf{0}} \tag{4.186}$$

$$= \frac{1}{\mu} \boldsymbol{S}^T \boldsymbol{z} \left(\boldsymbol{\theta}^l + \boldsymbol{\tau} \right) + \boldsymbol{S}^T \boldsymbol{S} \boldsymbol{\tau} - \frac{1}{\mu} \boldsymbol{S}^T \boldsymbol{z} \left(\boldsymbol{\theta}^l \right) |_{\boldsymbol{\tau}=\mathbf{0}} = \mathbf{0}, \quad (4.187)$$

$$\frac{\partial^2}{\partial^2 \boldsymbol{\tau}} d(\boldsymbol{\tau}) = -\frac{1}{\mu} \sum_{j=1}^{N} \psi' \left[r_j - \boldsymbol{\xi}_j^T \left(\boldsymbol{\theta}^l + \boldsymbol{\tau} \right) \right] \boldsymbol{\xi}_j \boldsymbol{\xi}_j^T + \boldsymbol{S}^T \boldsymbol{S}$$

$$\succeq - \sum_{j=1}^{N} \boldsymbol{\xi}_j \boldsymbol{\xi}_j^T + \boldsymbol{S}^T \boldsymbol{S} = \mathbf{0}, \quad (4.188)$$

where (4.188) follows from the assumption that $\psi'(x) \leq \mu$. In (4.188), $\boldsymbol{A} \succeq \boldsymbol{B}$ denotes that the matrix $(\boldsymbol{A} - \boldsymbol{B})$ is positive semidefinite. It then follows from (4.185), (4.187), and (4.188) that $d(\boldsymbol{\tau}) \geq 0$, for any $\boldsymbol{\tau} \in \mathbb{R}^K$. Now on setting

$$\boldsymbol{\tau} = \boldsymbol{\theta}^{l+1} - \boldsymbol{\theta}^l$$
$$= \frac{1}{\mu} \left(\boldsymbol{S}^T \boldsymbol{S} \right)^{-1} \boldsymbol{S}^T \boldsymbol{z}(\boldsymbol{\theta}^l), \quad (4.189)$$

we obtain

$$0 \leq d(\boldsymbol{\tau}) = \frac{1}{\mu} \left[\mathcal{C} \left(\boldsymbol{\theta}^l \right) - \mathcal{C} \left(\boldsymbol{\theta}^{l+1} \right) \right] - \frac{1}{2\mu^2} \boldsymbol{z}(\boldsymbol{\theta}^l)^T \boldsymbol{S} \left(\boldsymbol{S}^T \boldsymbol{S} \right)^{-1} \boldsymbol{S}^T \boldsymbol{z}(\boldsymbol{\theta}^l)$$
$$= \frac{1}{\mu} \left[\mathcal{C} \left(\boldsymbol{\theta}^l \right) - \mathcal{C} \left(\boldsymbol{\theta}^{l+1} \right) \right] - \frac{1}{2} \boldsymbol{\tau}^T \left(\boldsymbol{S}^T \boldsymbol{S} \right) \boldsymbol{\tau}. \quad (4.190)$$

Assume that the penalty function $\rho(x)$ is convex and bounded from below; then the cost function $\mathcal{C}(\boldsymbol{\theta})$ is convex and has a unique minimum $\mathcal{C}(\boldsymbol{\theta}^*)$. Therefore, $\boldsymbol{\theta}^*$ is the unique solution to (4.15) such that $\boldsymbol{z}(\boldsymbol{\theta}^*) = \mathbf{0}$. Since the sequence $\mathcal{C}\left(\boldsymbol{\theta}^l\right)$ is decreasing and bounded from below, it converges. Therefore, from (4.190) we have

$$\boldsymbol{z} \left(\boldsymbol{\theta}^l \right)^T \left[\boldsymbol{S} \boldsymbol{R}^{-1} \boldsymbol{S}^T \right] \boldsymbol{z} \left(\boldsymbol{\theta}^l \right) \leq 2\mu \left[\mathcal{C} \left(\boldsymbol{\theta}^l \right) - \mathcal{C} \left(\boldsymbol{\theta}^{l+1} \right) \right] \to 0, \quad \text{as } l \to \infty. \quad (4.191)$$

Since for any realization of $\boldsymbol{r}$, the probability that $\boldsymbol{z}\left(\boldsymbol{\theta}^l\right)$ falls in the null space of the matrix $\left(\boldsymbol{S}\boldsymbol{R}^{-1}\boldsymbol{S}^T\right)$ is zero, then (4.191) implies that $\boldsymbol{z}\left(\boldsymbol{\theta}^l\right) \to \mathbf{0}$ with probability 1. Since $\boldsymbol{z}(\boldsymbol{\theta})$ is a continuous function of $\boldsymbol{\theta}$ and has a unique minimum point $\boldsymbol{\theta}^*$, we then have $\boldsymbol{\theta}^l \to \boldsymbol{\theta}^*$ with probability 1, as $l \to \infty$.

4.10.2 Proof of Proposition 4.2 in Section 4.5

Denote $\boldsymbol{\zeta} \triangleq [\zeta_1 \ \cdots \ \zeta_K]^T$. Then (4.75) can be written in matrix form as

$$\boldsymbol{S}\boldsymbol{\theta} = \boldsymbol{U}_s \boldsymbol{\zeta}. \quad (4.192)$$

Denote $\boldsymbol{\Lambda}_0 \triangleq \boldsymbol{\Lambda}_s - \sigma^2 \boldsymbol{I}_K$. Then from (4.73) and (4.74) we obtain

$$\boldsymbol{S}\boldsymbol{A}^2\boldsymbol{S}^T = \boldsymbol{U}_s \boldsymbol{\Lambda}_0 \boldsymbol{U}_s^T. \quad (4.193)$$

Using (4.192) and (4.193), we obtain

$$\boldsymbol{S}^T \left(\boldsymbol{S}\boldsymbol{A}\boldsymbol{S}^T\right)^\dagger \boldsymbol{S}\boldsymbol{\theta} = \boldsymbol{S}^T \left(\boldsymbol{U}_s \boldsymbol{\Lambda}_0^{-1} \boldsymbol{U}_s^T\right) \boldsymbol{U}_s \boldsymbol{\zeta} \tag{4.194}$$

$$\Longrightarrow \left(\boldsymbol{S}^T \boldsymbol{S}^{T^\dagger}\right) \boldsymbol{A}^{-2} \left(\boldsymbol{S}^\dagger \boldsymbol{S}\right) \boldsymbol{\theta} = \boldsymbol{S}^T \boldsymbol{U}_s \boldsymbol{\Lambda}_0^{-1} \left(\boldsymbol{U}_s^T \boldsymbol{U}_s\right) \boldsymbol{\zeta} \tag{4.195}$$

$$\Longrightarrow \boldsymbol{\theta} = \boldsymbol{A}^2 \boldsymbol{S}^T \boldsymbol{U}_s \boldsymbol{\Lambda}_0^{-1} \boldsymbol{\zeta}, \tag{4.196}$$

where in (4.194) $\dagger$ denotes the Moore–Penrose generalized matrix inverse [189]; in (4.195) we have used the fact that $(\boldsymbol{S}\boldsymbol{A}^2\boldsymbol{S}^T)^\dagger = \boldsymbol{S}^{T^\dagger} \boldsymbol{A}^{-2} \boldsymbol{S}^\dagger$, which can easily be verified by using the definition of the Moore–Penrose generalized matrix inverse [189]; in (4.196) we have used the facts that $(\boldsymbol{S}^T \boldsymbol{S}^{T^\dagger}) = (\boldsymbol{S}^\dagger \boldsymbol{S}) = (\boldsymbol{U}_s^T \boldsymbol{U}_s) = \boldsymbol{I}_K$; and (4.196) is the matrix form of (4.76). Finally, we notice that

$$\begin{aligned} \boldsymbol{A}^{-2} &= \boldsymbol{S}^T \left(\boldsymbol{S}\boldsymbol{A}^2\boldsymbol{S}^T\right)^\dagger \boldsymbol{S} \\ &= \boldsymbol{S}^T \left(\boldsymbol{U}_s \boldsymbol{\Lambda}_0^{-1} \boldsymbol{U}_s^T\right) \boldsymbol{S}. \end{aligned} \tag{4.197}$$

It follows from (4.197) that the kth diagonal element A_k^{-2} of the diagonal matrix $\boldsymbol{A}^{-2}$ satisfies

$$A_k^{-2} = \sum_{j=1}^{K} \frac{\left(\boldsymbol{u}_j^T \boldsymbol{s}_k\right)^2}{\lambda_j - \sigma^2}. \tag{4.198}$$

Chapter 5

SPACE-TIME MULTIUSER DETECTION

5.1 Introduction

It is anticipated that transmit and receive antenna arrays together with adaptive space-time processing techniques will be used in future high-capacity cellular communication systems, to combat interference, time dispersion and multipath fading. There has been a significant amount of recent interest in developing adaptive array techniques for wireless communications (e.g., [48, 154, 356, 570, 571]). These studies have shown that substantial performance gains and capacity increases can be achieved by employing antenna arrays and space-time signal processing to suppress multiple-access interference, co-channel interference, and intersymbol interference, and at the same time to provide spatial diversity to combat multipath fading. In this chapter we discuss a number of signal processing techniques for space-time processing in wireless communication systems. We first discuss adaptive antenna array processing technques for TDMA systems. We then discuss space-time multiuser detection for CDMA systems.

Due to multipath propagation effects and the movement of mobile units, the array steering vector in a multiple-antenna system changes with time, and it is of interest to estimate and track it during communication sessions. One attractive approach to steering vector estimation is to exploit a known portion of the data stream (e.g., the synchronization data stream). For instance, the TDMA mobile radio systems IS-54/136 use 14 known synchronization symbols in each time slot of 162 symbols. These known symbols are very useful for estimating the steering vector and computing the optimal array combining weights. We discuss a number of approaches to adaptive array processing in such systems.

Many advanced signal processing techniques have been proposed for combating interference and multipath channel distortion in CDMA systems, and these techniques fall largely into two categories: multiuser detection (cf. Chapters 1–4) and space-time processing [370]. Recall that multiuser detection techniques exploit the underlying structure induced by the spreading waveforms of the DS-CDMA user signals for interference suppression. In antenna array processing, on the other hand, the signal structure induced by multiple receiving antennas (i.e., the spatial signatures) is exploited for interference suppression [34, 231, 273, 348, 467, 505, 574]. Combined multiuser detection and array processing has also been addressed in a number of works [90, 121, 142, 202, 347, 504, 505]. In this chapter we provide a comprehensive treatment of space-time multiuser detection in multipath CDMA channels with both transmitter and receiver antenna arrays. We derive several space-time multiuser detection structures, including the maximum likelihood multiuser sequence detector, linear space-time multiuser detectors, and adaptive space-time multiuser detectors.

The rest of this chapter is organized as follows. In Section 5.2 we discuss adaptive antenna array techniques for interference suppression in TDMA systems. In Section 5.3 we treat the problem of optimal space-time processing in CDMA systems employing multiple receive antennas. In Section 5.4 we discuss linear space-time receiver techniques for CDMA systems with multiple receive antennas. In Section 5.5 we discuss space-time processing methods in synchronous CDMA systems that employ multiple transmit and receive antennas, and their adaptive implementations. Finally, in Section 5.6 we present adaptive space-time receiver structures in multipath CDMA channels with multiple transmit and receive antennas.

The following is a list of the algorithms appearing in this chapter.

- *Algorithm 5.1:* LMS adaptive array
- *Algorithm 5.2:* DMI adaptive array
- *Algorithm 5.3:* Subspace-based adaptive array for TDMA
- *Algorithm 5.4:* Batch blind linear space-time multiuser detector—synchronous CDMA, two transmit antennas, and two receive antennas
- *Algorithm 5.5:* Blind adaptive linear space-time multiuser detector—synchronous CDMA, two transmit antennas, and two receive antennas
- *Algorithm 5.6:* Blind adaptive linear space-time multiuser detector—asynchronous multipath CDMA, two transmit antennas, and two receive antennas

5.2 Adaptive Array Processing in TDMA Systems

5.2.1 Signal Model

A wireless cellular communication system employing adaptive antenna arrays at the base station is shown in Fig. 5.1, where a base with P antenna elements receives signals from K users. The K users operate in the same bandwidth at the same time. One of the signals is destined to the base. The other signals are destined to other bases, and they interfere with the desired signal; that is, they constitute co-channel interference. Note that although here we consider the uplink scenario (mobile to base), where antenna arrays are most likely to be employed, the adaptive array techniques discussed in this section apply to the downlink (base to mobile) as well, provided that a mobile receiver is equipped with multiple antennas. The general structure can be applied to other systems as well.

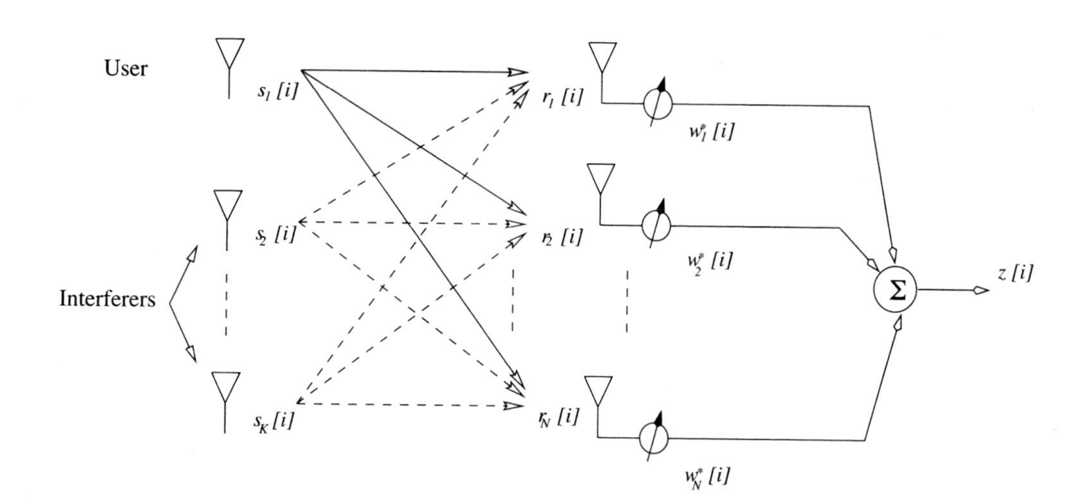

Figure 5.1. Wireless communication system employing adaptive arrays at the base station. An array of P antenna elements at the base receives signals from K co-channel users, one of which is the desired user's signal, and the rest are interfering signals.

The received signal at the antenna array is the superposition of K co-channel signals from the desired user and the interferers, plus the ambient channel noise. Assume that the signal bandwidth of the desired user and the interferers is smaller than the channel coherence bandwidth, so that the signals are subject to flat fading. Assume also that the fading is slow, such that the channel remains constant during one time slot containing M data symbol intervals. To focus on the spatial processing, we assume for the time being that all users employ the same modulation waveform,[1] so that after matched filtering with this waveform, the P-vector of received complex signal at the antenna array during the ith symbol interval within a time slot can be expressed as

$$\boldsymbol{r}[i] = \sum_{k=1}^{K} \boldsymbol{g}_k \, b_k[i] + \boldsymbol{n}[i], \quad i = 0, \ldots, M-1, \tag{5.1}$$

where $b_k[i]$ is the ith symbol transmitted by the kth user, $\boldsymbol{g}_k = [g_{1,k} \cdots g_{P,k}]^T$ is a complex vector (the steering vector) representing the response of the channel and array to the kth user's signal, and $\boldsymbol{n}[i] \sim \mathcal{N}_c(\boldsymbol{0}, \sigma^2 \boldsymbol{I}_P)$ is a vector of complex Gaussian noise samples. It is assumed that all users employ phase-shift-keying (PSK) modulation with all symbol values being equiprobable. Thus, we have

$$E\{b_k[i]\} = 0 \text{ and } E\left\{|b_k[i]|^2\right\} = 1.$$

The nth element of the steering vector $\boldsymbol{g}_k$ can be expressed as

$$g_{n,k} = A_k \, \alpha_{n,k} \, a_{n,k}, \quad n = 1, \ldots, P, \tag{5.2}$$

where A_k is the transmitted complex amplitude of the kth user's signal, $g_{n,k}$ is the complex fading gain between the kth user's transmitter and the nth antenna at the receiver, and $a_{n,k}$ is the response of the nth antenna to the kth user's signal. It is also assumed that the data symbols of all users $\{b_k[i]\}$ are mutually independent and that they are independent of the ambient noise $\boldsymbol{n}[i]$. The noise vectors $\{\boldsymbol{n}[i]\}$ are assumed to be i.i.d. with independent real and imaginary components. Note that, mathematically, the model (5.1) is identical to the synchronous CDMA model of (2.1). However, the different physical interpretation of the various quantities in (5.1) leads to somewhat different algorithms than those discussed previously. Nevertheless, this mathematical equivalence will be exploited in the sequel.

5.2.2 Linear MMSE Combining

Throughout this section we assume that user 1 is the desired user. In adaptive array processing, the received signal $\boldsymbol{r}[i]$ is combined linearly through a complex weight vector $\boldsymbol{w} \in \mathbb{C}^P$, to yield the array output signal $z[i]$:

$$z[i] = \boldsymbol{w}^H \boldsymbol{r}[i].$$

[1] In Section 5.3, where we consider both spatial and temporal processing, we drop the assumption.

In linear MMSE combining [570], the weight vector $\boldsymbol{w}$ is chosen such that the mean-square error between the transmitted symbol $b_1[i]$ and the array output $z[i]$ is minimized:

$$\begin{aligned}\boldsymbol{w} &= \arg \min_{\boldsymbol{w} \in \mathbb{C}^P} E\left\{\left|b_1[i] - z[i]\right|^2\right\} \\ &= \underbrace{E\left\{\boldsymbol{r}[i]\boldsymbol{r}[i]^H\right\}^{-1}}_{\boldsymbol{C}^{-1}} \underbrace{E\left\{\boldsymbol{r}[i]b_1[i]^*\right\}}_{\boldsymbol{g}_1}, \end{aligned} \tag{5.3}$$

where the expectation is taken with respect to the symbols of interfering users $\{b_k[i] : k \neq 1\}$ and the ambient noise $\boldsymbol{n}[i]$.

In practice, the autocorrelation matrix $\boldsymbol{C}$ and the steering vector of the desired user $\boldsymbol{g}_1$ are not known *a priori* to the receiver, and therefore they must be estimated in order to compute the optimal combining weight $\boldsymbol{w}$ in (5.3). In several TDMA-based wireless communication systems (e.g., GSM, IS-54 and IS-136), the information symbols in each slot are preceded by a preamble of known synchronization symbols, which can be used for training the optimal weight vector. The trained weight vector is then used for combining during the demodulation of the information symbols in the same slot.

Assume that in each time slot there are m_t training symbols and $M - m_t$ information symbols. Two popular methods for training the combining weights are the least-mean-squares (LMS) algorithm and the direct matrix inversion (DMI) algorithm [570]. The LMS training algorithm is as follows.

Algorithm 5.1: [LMS adaptive array]

- *Compute the combining weight:*

$$\epsilon[i] = b_1[i] - \hat{\boldsymbol{w}}[i-1]^H \boldsymbol{r}[i], \tag{5.4}$$

$$\hat{\boldsymbol{w}}[i+1] = \hat{\boldsymbol{w}}[i] + \mu\, \epsilon[i]^* \,\boldsymbol{r}[i], \qquad i = 0, 1, \ldots, m_t - 1, \tag{5.5}$$

where μ is a step-size parameter. Set $\hat{\boldsymbol{w}} \triangleq \hat{\boldsymbol{w}}[m_t]$.

- *Perform data detection: Obtain $\hat{b}_1[i]$ by quantizing $\hat{\boldsymbol{w}}^H \boldsymbol{r}[i]$ for $i = m_t, \ldots, M-1$.*

Although the LMS algorithm has a very low computational complexity, it also has a slow convergence rate. Given that the number of training symbols in each time slot is usually small, it is unlikely that the LMS algorithm will converge to the optimum weight vector within the training period.

The DMI algorithm for training the optimum weight vector essentially forms the sample estimates of the autocorrelation matrix $\boldsymbol{C}$ and the steering vector $\boldsymbol{g}_1$ using the signal received during the training period and the known training symbols, and then computes the combining weight vector according to (5.3) using these estimates. Specifically, it proceeds as follows.

Algorithm 5.2: [DMI adaptive array]

- *Compute the combining weight:*

$$\hat{\boldsymbol{C}} \triangleq \frac{1}{m_t} \sum_{i=0}^{m_t-1} \boldsymbol{r}[i]\, \boldsymbol{r}[i]^H, \tag{5.6}$$

$$\hat{\boldsymbol{g}}_1 \triangleq \frac{1}{m_t} \sum_{i=0}^{m_t-1} \boldsymbol{r}[i]\, b_1[i]^*, \tag{5.7}$$

$$\hat{\boldsymbol{w}} = \hat{\boldsymbol{C}}^{-1} \hat{\boldsymbol{g}}_1. \tag{5.8}$$

- *Perform data detection: Obtain* $\hat{b}_1[i]$ *by quantizing* $\hat{\boldsymbol{w}}^H \boldsymbol{r}[i]$ *for* $i = m_t, \ldots, M-1$.

It is easily seen that the sample estimates $\hat{\boldsymbol{C}}$ and $\hat{\boldsymbol{g}}_1$ are unbiased [i.e., $E\{\hat{\boldsymbol{C}}\} = \boldsymbol{C}$ and $E\{\hat{\boldsymbol{g}}_1\} = \boldsymbol{g}_1$]. They are also strongly consistent; that is, they converge, respectively, to the true autocorrelation matrix $\boldsymbol{C}$ and the true steering vector $\boldsymbol{g}_1$ almost surely as $m_t \to \infty$. Notice that both the LMS algorithm and the DMI algorithm compute the combining weights based only on the signal received during the training period. Since in practice the training period is short compared with the slot length (i.e., $m_t \ll M$), the weight vector $\hat{\boldsymbol{w}}$ obtained by such an algorithm can be very noisy. In what follows, we consider a more powerful technique for computing the steering vector and the combining weights that exploits the received signal corresponding to the unknown $(M - m_t)$ information symbols as well.

5.2.3 Subspace-Based Training Algorithm

Notice that the sample correlation matrix $\hat{\boldsymbol{C}}$ in (5.6) does *not* depend on the training symbols of the desired user $\{b_1[i] : i = 0, \ldots, m_t - 1\}$, and therefore we can use the received signals during the entire user time slot to get a better sample estimate of $\boldsymbol{C}$:

$$\tilde{\boldsymbol{C}} \triangleq \frac{1}{M} \sum_{i=0}^{M-1} \boldsymbol{r}[i]\, \boldsymbol{r}[i]^H. \tag{5.9}$$

However, the sample estimate $\hat{\boldsymbol{g}}_1$ of the steering vector given by (5.7) *does* depend on the training symbols, and therefore this estimator cannot make use of the received signals corresponding to the unknown information symbols. In this section we present a more powerful subspace-based technique for computing the steering vector and the array combining weight vector. This method first appeared in [550].

Steering Vector Estimation

In what follows it is assumed that the number of antennas is greater than the number of interferers (i.e., $P \geq K$). A typical way to treat the case of $P < K$ is to oversample the received signal to increase the dimensionality of the signal for processing [342]. For convenience and without loss of generality, we assume that the steering vectors $\{\boldsymbol{g}_k, k = 1, \ldots, K\}$ are linearly independent. The autocorrelation matrix $\boldsymbol{C}$ of the receive signal in (5.1) is given by

$$\begin{aligned}\boldsymbol{C} &\triangleq E\left\{\boldsymbol{r}[i]\boldsymbol{r}[i]^H\right\} \\ &= \sum_{k=1}^{K} \boldsymbol{g}_k \boldsymbol{g}_k^H + \sigma^2 \boldsymbol{I}_P.\end{aligned} \tag{5.10}$$

The eigendecomposition of $\boldsymbol{C}$ is given by

$$\begin{aligned}\boldsymbol{C} &= \boldsymbol{U}\boldsymbol{\Lambda}\boldsymbol{U}^H \\ &= \boldsymbol{U}_s \boldsymbol{\Lambda}_s \boldsymbol{U}_s^H + \sigma^2 \boldsymbol{U}_n \boldsymbol{U}_n^H,\end{aligned} \tag{5.11}$$

where as in previous chapters, $\boldsymbol{U} = [\boldsymbol{U}_s\ \boldsymbol{U}_n]$, $\boldsymbol{\Lambda} = \mathsf{diag}\left\{\boldsymbol{\Lambda}_s,\ \sigma^2 \boldsymbol{I}_{P-K}\right\}$; $\boldsymbol{\Lambda}_s = \mathsf{diag}\{\lambda_1, \ldots, \lambda_K\}$ contains the K largest eigenvalues of $\boldsymbol{C}$ in descending order, and $\boldsymbol{U}_s = [\boldsymbol{u}_1\ \cdots\ \boldsymbol{u}_K]$ contains the corresponding orthogonal eigenvectors; and $\boldsymbol{U}_n = [\boldsymbol{u}_{K+1}\ \ldots\ \boldsymbol{u}_P]$ contains the $(N-K)$ orthogonal eigenvectors that correspond to the smallest eigenvalue σ^2. Denote $\boldsymbol{G} \triangleq [\boldsymbol{g}_1\ \cdots\ \boldsymbol{g}_K]$. It is easy to see that $\mathsf{range}(\boldsymbol{G}) = \mathsf{range}(\boldsymbol{U}_s)$. Thus the range space of $\boldsymbol{U}_s$ is a signal subspace and its orthogonal complement, the noise subspace, is spanned by $\boldsymbol{U}_n$. Note that in contrast to the signal and noise subspaces discussed in preceding chapters, which are based on temporal structure, here the subspaces describe the spatial structure of the received signals. The following result is instrumental to developing the alternative steering vector estimator for the desired user.

Proposition 5.1: *Given the eigendecomposition (5.11) of the autocorrelation matrix $\boldsymbol{C}$, suppose that a received noise-free signal is given by*

$$\boldsymbol{y}[i] \triangleq \sum_{k=1}^{K} \boldsymbol{g}_k\, b_k[i] = \sum_{j=1}^{K} \boldsymbol{u}_j\, q_j[i]. \tag{5.12}$$

Then the kth user's transmitted symbol can be expressed as

$$b_k[i] = \sum_{j=1}^{K} \frac{\boldsymbol{u}_j^H \boldsymbol{g}_k}{\lambda_j - \sigma^2} q_j[i], \qquad k = 1, \ldots, K. \tag{5.13}$$

Proof: Denote $\boldsymbol{b}[i] \triangleq \left[b_1[i]\ \cdots\ b_K[i]\right]^T$ and $\boldsymbol{q}[i] \triangleq \left[q_1[i]\ \cdots\ q_K[i]\right]^T$. Then (5.12) can be written in matrix form as

$$\boldsymbol{y}[i] = \boldsymbol{G}\boldsymbol{s}[i] = \boldsymbol{U}_s \boldsymbol{q}[i]. \tag{5.14}$$

Denote further

$$\boldsymbol{\Lambda}_0 \triangleq \boldsymbol{\Lambda}_s - \sigma^2 \boldsymbol{I}_K = \mathsf{diag}\{\lambda_1 - \sigma^2, \ldots, \lambda_K - \sigma^2\}. \tag{5.15}$$

Then from (5.10) and (5.11) we have

$$\boldsymbol{G}\boldsymbol{G}^H = \boldsymbol{U}_s \boldsymbol{\Lambda}_0 \boldsymbol{U}_s^H. \tag{5.16}$$

Taking the Moore–Penrose generalized matrix inverse [189] on both sides of (5.16), we obtain

$$\begin{aligned} \left(\boldsymbol{G}\boldsymbol{G}^H\right)^\dagger &= \left(\boldsymbol{U}_s \boldsymbol{\Lambda}_0 \boldsymbol{U}_s^H\right)^\dagger \\ \Longrightarrow \boldsymbol{G}^{H\dagger}\boldsymbol{G}^\dagger &= \boldsymbol{U}_s \boldsymbol{\Lambda}_0^{-1} \boldsymbol{U}_s^H. \end{aligned} \tag{5.17}$$

From (5.14) and (5.17) we then have

$$\begin{aligned} \boldsymbol{G}^H \left(\boldsymbol{G}^{H\dagger}\boldsymbol{G}^\dagger\right) \boldsymbol{G}\,\boldsymbol{b}[i] &= \boldsymbol{G}^H \left(\boldsymbol{U}_s \boldsymbol{\Lambda}_0^{-1} \boldsymbol{U}_s^H\right) \boldsymbol{U}_s \boldsymbol{q}[i] \\ \Longrightarrow \left(\boldsymbol{G}^H\boldsymbol{G}^{H\dagger}\right)\left(\boldsymbol{G}^\dagger\boldsymbol{G}\right)\boldsymbol{b}[i] &= \boldsymbol{G}^H \boldsymbol{U}_s \boldsymbol{\Lambda}_0^{-1} \left(\boldsymbol{U}_s^H \boldsymbol{U}_s\right) \boldsymbol{q}[i] \\ \Longrightarrow \boldsymbol{b}[i] &= \boldsymbol{G}^H \boldsymbol{U}_s \boldsymbol{\Lambda}_0^{-1} \boldsymbol{q}[i], \end{aligned} \tag{5.18}$$

where the last equality follows from the fact that $\boldsymbol{G}^H\boldsymbol{G}^{H\dagger} = \boldsymbol{G}^\dagger\boldsymbol{G} = \boldsymbol{U}_s^H\boldsymbol{U}_s = \boldsymbol{I}_K$. Note that (5.18) is the matrix form of (5.13). □.

Suppose now that the signal subspace parameters $\boldsymbol{U}_s$, $\boldsymbol{\Lambda}_s$, and σ^2 are known. We next consider the problem of estimating the steering vector $\boldsymbol{g}_1$ of the desired user, given m_t training symbols $\{b_1[i], i = 0, \ldots, m_t - 1\}$, where $m_t \geq K$. The next result shows that in the absence of ambient noise, K linearly independent received signals suffice to determine the steering vector *exactly*.

Proposition 5.2: *Let* $\boldsymbol{\beta}_1 \triangleq \left[b_1[0] \;\cdots\; b_1[m_t - 1]\right]^T$ *be the vector of training symbols of the desired user, and* $\boldsymbol{Y} \triangleq \left[\boldsymbol{y}[0] \;\cdots\; \boldsymbol{y}[m_t - 1]\right]$ *be the matrix of* m_t *noise-free received signals during the training stage. Assume that* $\mathsf{rank}(\boldsymbol{Y}) = K$. *Then the steering vector of the desired user can be expressed as*

$$\boldsymbol{g}_1 = \boldsymbol{U}_s \boldsymbol{\Lambda}_0 \left(\boldsymbol{Y}^H \boldsymbol{U}_s\right)^\dagger \boldsymbol{\beta}_1, \tag{5.19}$$

where $\boldsymbol{U}_s$ *and* $\boldsymbol{\Lambda}_0$ *are defined in (5.11) and (5.15), respectively.*

Proof: The ith received noise-free array output vector can be expressed as

$$\boldsymbol{y}[i] = \sum_{k=1}^{K} \boldsymbol{g}_k\, b_k[i] = \sum_{j=1}^{K} \boldsymbol{u}_j\, q_j[i], \qquad i = 0, \ldots, m_t - 1. \tag{5.20}$$

Denote $\boldsymbol{q}[i] \triangleq \left[q_1[i] \;\cdots\; q_K[i]\right]^T$. It then follows from (5.20) that

$$\begin{aligned} \boldsymbol{y}[i] &= \boldsymbol{U}_s\, \boldsymbol{q}[i] \\ \Longrightarrow \boldsymbol{q}[i] &= \boldsymbol{U}_s^H\, \boldsymbol{y}[i], \qquad i = 0, \ldots, m_t - 1, \end{aligned} \tag{5.21}$$

since $\boldsymbol{U}_s^H \boldsymbol{U}_s = \boldsymbol{I}_K$. On substituting (5.21) into (5.13), we obtain

$$b_1[i] = \left(\boldsymbol{U}_s^H \boldsymbol{g}_1\right)^H \boldsymbol{\Lambda}_0^{-1} \boldsymbol{q}[i] \tag{5.22}$$

$$= \left(\boldsymbol{U}_s{}^H \boldsymbol{g}_1\right)^H \boldsymbol{\Lambda}_0^{-1} \boldsymbol{U}_s^H \boldsymbol{y}[i], \qquad i = 0, \ldots, m_t - 1. \tag{5.23}$$

Equation (5.23) can be written in matrix form as

$$\boldsymbol{Y}^H \boldsymbol{U}_s \boldsymbol{\Lambda}_0^{-1} \left(\boldsymbol{U}_s^H \boldsymbol{g}_1\right) = \boldsymbol{\beta}_1. \tag{5.24}$$

Since $\mathsf{rank}(\boldsymbol{Y}) = K$ and $\mathsf{rank}(\boldsymbol{U}_s) = K$, we have $\mathsf{rank}(\boldsymbol{Y}^H \boldsymbol{U}_s) = K$. Therefore, $\boldsymbol{g}_1$ can be obtained uniquely from (5.24) by

$$\begin{aligned} \boldsymbol{U}_s^H \boldsymbol{g}_1 &= \boldsymbol{\Lambda}_0 \left(\boldsymbol{Y}^H \boldsymbol{U}_s\right)^\dagger \boldsymbol{\beta}_1 \\ \Longrightarrow \boldsymbol{g}_1 &= \boldsymbol{U}_s \boldsymbol{\Lambda}_0 \left(\boldsymbol{Y}^H \boldsymbol{U}_s\right)^\dagger \boldsymbol{\beta}_1, \end{aligned}$$

where the last equality follows from the fact that $\boldsymbol{g}_1 = \boldsymbol{U}_s \boldsymbol{U}_s^H \boldsymbol{g}_1$, since $\boldsymbol{g}_1 \in \mathsf{range}(\boldsymbol{U}_s)$. □

We can interpret the result above as follows. If the length of the data frame tends to infinity (i.e., $M \to \infty$), the sample estimate $\tilde{\boldsymbol{C}}$ in (5.9) converges to the true autocorrelation matrix $\boldsymbol{C}$ almost surely, and an eigendecomposition of the corresponding $\tilde{\boldsymbol{C}}$ will give the true signal subspace parameters $\boldsymbol{U}_s$ and $\boldsymbol{\Lambda}_0$. The result above then indicates that in the absence of background noise, a perfect estimate of the steering vector of the desired user $\boldsymbol{g}_1$ can be obtained by using K linearly independent received signals and the corresponding training symbols for the desired user. The steering vector estimator $\hat{\boldsymbol{g}}_1$ in the DMI method, given by (5.7), however, cannot achieve perfect steering vector estimation even in the absence of noise (i.e., $\sigma = 0$), unless the number of training symbols tends to infinity (i.e., $m_t \to \infty$). In fact, it is easily seen that the covariance matrix of that estimator is given by

$$\begin{aligned} E\left\{(\hat{\boldsymbol{g}}_1 - \boldsymbol{g}_1)(\hat{\boldsymbol{g}}_1 - \boldsymbol{g}_1)^H\right\} &= \frac{1}{m_t}\left(\sum_{k=2}^{K} \boldsymbol{g}_k \boldsymbol{g}_k^H + \sigma^2 \boldsymbol{I}_P\right) \\ &\overset{\sigma \to 0}{\longrightarrow} \frac{1}{m_t} \sum_{k=2}^{K} \boldsymbol{g}_k \boldsymbol{g}_k^H. \end{aligned}$$

In practice, the received signals are corrupted by ambient noise:

$$\begin{aligned} \boldsymbol{r}[i] &= \boldsymbol{y}[i] + \boldsymbol{n}[i] = \sum_{k=1}^{K} \boldsymbol{g}_k b_k[i] + \boldsymbol{n}[i] \\ &= \sum_{j=1}^{K} \boldsymbol{u}_j q_j[i] + \boldsymbol{n}[i], \qquad i = 0, \ldots, m_t - 1. \end{aligned} \tag{5.25}$$

Since $\boldsymbol{n}[i] \sim \mathcal{N}_c(\mathbf{0}, \sigma^2 \boldsymbol{I}_P)$, the log-likelihood function of the received signal $\boldsymbol{r}[i]$ conditioned on $\boldsymbol{q}[i]$, is given by

$$\mathcal{L}\left(\boldsymbol{r}[i] \mid \boldsymbol{q}[i]\right) = -\frac{1}{\sigma^2}\left\|\boldsymbol{r}[i] - \boldsymbol{U}_s\, \boldsymbol{q}[i]\right\|^2.$$

Hence the maximum-likelihood estimate of $\boldsymbol{q}[i]$ from $\boldsymbol{r}[i]$ is given by

$$\begin{aligned} \tilde{\boldsymbol{q}}[i] &= \left(\boldsymbol{U}_s^H \boldsymbol{U}_s\right)^{-1} \boldsymbol{U}_s^H \boldsymbol{r}[i] \\ &= \boldsymbol{U}_s^H \boldsymbol{r}[i], \end{aligned} \tag{5.26}$$

where the last equality follows from the fact that $\boldsymbol{U}_s^H \boldsymbol{U}_s = \boldsymbol{I}_K$. Similar to (5.23), we can set up the following equations for estimating the steering vector $\boldsymbol{g}_1$ from the noisy signal:

$$\begin{aligned} b_1[i] &\cong \left(\boldsymbol{U}_s^H \boldsymbol{g}_1\right)^H \boldsymbol{\Lambda}_0^{-1} \tilde{\boldsymbol{q}}[i] \\ &= \left(\boldsymbol{U}_s{}^H \boldsymbol{g}_1\right)^H \boldsymbol{\Lambda}_0^{-1} \boldsymbol{U}_s^H\, \boldsymbol{r}[i], \qquad i = 0, \ldots, m_t - 1. \end{aligned} \tag{5.27}$$

Denote $\boldsymbol{\Gamma} \triangleq \left[\boldsymbol{r}[0] \ \cdots \ \boldsymbol{r}[m_t - 1]\right]$; then (5.27) can be written in matrix form as

$$\boldsymbol{\beta}_1 \cong \boldsymbol{\Gamma}^H \boldsymbol{U}_s \boldsymbol{\Lambda}_0^{-1} \left(\boldsymbol{U}_s^H \boldsymbol{g}_1\right). \tag{5.28}$$

Solving $\boldsymbol{g}_1$ from (5.28), we obtain

$$\begin{aligned} \boldsymbol{g}_1 &\cong \boldsymbol{U}_s \boldsymbol{\Lambda}_0 \left(\boldsymbol{\Gamma}^H \boldsymbol{U}_s\right)^{\dagger} \boldsymbol{\beta}_1 \\ &= \boldsymbol{U}_s\, \boldsymbol{\Lambda}_0 \left(\boldsymbol{U}_s^H \boldsymbol{\Gamma}\boldsymbol{\Gamma}^H \boldsymbol{U}_s\right)^{-1} \boldsymbol{U}_s^H \boldsymbol{\Gamma} \boldsymbol{\beta}_1. \end{aligned} \tag{5.29}$$

To implement an estimator of $\boldsymbol{g}_1$ based on (5.29), we first compute the sample autocorrelation matrix $\tilde{\boldsymbol{C}}$ of the received signal according to (5.9). An eigendecomposition on $\tilde{\boldsymbol{C}}$ is then performed to get

$$\tilde{\boldsymbol{C}} = \tilde{\boldsymbol{U}}_s \tilde{\boldsymbol{\Lambda}}_s \tilde{\boldsymbol{U}}_s^H + \tilde{\boldsymbol{U}}_n \tilde{\boldsymbol{\Lambda}}_n \tilde{\boldsymbol{U}}_n^H. \tag{5.30}$$

The steering vector estimator for the desired user is then given by

$$\tilde{\boldsymbol{g}}_1 = \tilde{\boldsymbol{U}}_s \tilde{\boldsymbol{\Lambda}}_s \left(\tilde{\boldsymbol{U}}_s^H \boldsymbol{\Gamma}\boldsymbol{\Gamma}^H \tilde{\boldsymbol{U}}_s\right)^{-1} \tilde{\boldsymbol{U}}_s^H \boldsymbol{\Gamma} \boldsymbol{\beta}_1. \tag{5.31}$$

Note that $\tilde{\boldsymbol{\Lambda}}_s$ is used in (5.31) instead of $\tilde{\boldsymbol{\Lambda}}_0$ as in (5.29). The reason for this is to make this estimator strongly consistent.That is, if we let $m_t = M \to \infty$, we have $\hat{\boldsymbol{U}}_s \overset{a.s.}{\rightarrow} \boldsymbol{U}_s$, $\hat{\boldsymbol{\Lambda}}_s \overset{a.s.}{\rightarrow} \boldsymbol{\Lambda}_s$, $(1/m_t)\boldsymbol{\Gamma\Gamma}^H \overset{a.s}{\rightarrow} \boldsymbol{C}$, and $(1/m_t)\boldsymbol{\Gamma\beta}_1 \overset{a.s.}{\rightarrow} \boldsymbol{g}_1$. Hence from (5.31) we have

$$\begin{aligned}\tilde{\boldsymbol{g}}_1 &\overset{a.s.}{\rightarrow} \boldsymbol{U}_s\boldsymbol{\Lambda}_s \left(\boldsymbol{U}_s^H\boldsymbol{C}\boldsymbol{U}_s\right)^{-1} \boldsymbol{U}_s^H\boldsymbol{g}_1 \\ &= \boldsymbol{U}_s\boldsymbol{\Lambda}_s\boldsymbol{\Lambda}_s^{-1}\boldsymbol{U}_s^H\boldsymbol{g}_1 \\ &= \boldsymbol{U}_s\boldsymbol{U}_s^H\boldsymbol{g}_1 \\ &= \boldsymbol{g}_1.\end{aligned}$$

Interestingly, if on the other hand, we replace $\tilde{\boldsymbol{U}}_s$ and $\tilde{\boldsymbol{\Lambda}}_s$ in (5.31) by the corresponding sample estimates obtained from an eigendecomposition of $\hat{\boldsymbol{C}}$ in (5.6), then in the absence of noise, we obtain the same steering vector estimate $\hat{\boldsymbol{g}}_1$ as in (5.7); while with noise, we obtain a less noisy estimate of $\boldsymbol{g}_1$ than (5.7). Formally, we have the following result.

Proposition 5.3: *Let the eigendecomposition of the sample autocorrelation matrix* $\hat{\boldsymbol{C}}$ *in (5.6) of the received training signals be*

$$\hat{\boldsymbol{C}} = \hat{\boldsymbol{U}}_s\hat{\boldsymbol{\Lambda}}_s\hat{\boldsymbol{U}}_s^H + \hat{\boldsymbol{U}}_n\hat{\boldsymbol{\Lambda}}_n\hat{\boldsymbol{U}}_n^H. \tag{5.32}$$

If we form the following estimator for the steering vector $\mathbf{p}_1$,

$$\tilde{\hat{\boldsymbol{g}}}_1 = \hat{\boldsymbol{U}}_s\hat{\boldsymbol{\Lambda}}_s \left(\hat{\boldsymbol{U}}_s^H\boldsymbol{\Gamma\Gamma}^H\hat{\boldsymbol{U}}_s\right)^{-1} \hat{\boldsymbol{U}}_s^H\boldsymbol{\Gamma\beta}_1, \tag{5.33}$$

then $\tilde{\hat{\boldsymbol{g}}}_1$ *is related to* $\hat{\boldsymbol{g}}_1$ *in (5.8) by*

$$\tilde{\hat{\boldsymbol{g}}}_1 = \hat{\boldsymbol{U}}_s\hat{\boldsymbol{U}}_s^H\hat{\boldsymbol{g}}_1. \tag{5.34}$$

Proof: Using (5.33), we have

$$\begin{aligned}\tilde{\hat{\boldsymbol{g}}}_1 &= \hat{\boldsymbol{U}}_s\hat{\boldsymbol{\Lambda}}_s \left(\hat{\boldsymbol{U}}_s^H\boldsymbol{\Gamma}\,\boldsymbol{\Gamma}^H\hat{\boldsymbol{U}}_s\right)^{-1} \hat{\boldsymbol{U}}_s^H\boldsymbol{\Gamma\beta}_1 \\ &= \hat{\boldsymbol{U}}_s\hat{\boldsymbol{\Lambda}}_s \left[\hat{\boldsymbol{U}}_s^H\left(\frac{1}{m_t}\boldsymbol{\Gamma\Gamma}^H\right)\hat{\boldsymbol{U}}_s\right]^{-1} \hat{\boldsymbol{U}}_s^H\left(\frac{1}{m_t}\boldsymbol{\Gamma\beta}_1\right) \\ &= \hat{\boldsymbol{U}}_s\hat{\boldsymbol{\Lambda}}_s\left[\hat{\boldsymbol{U}}_s^H\hat{\boldsymbol{C}}\hat{\boldsymbol{U}}_s\right]^{-1}\hat{\boldsymbol{U}}_s^H\hat{\boldsymbol{g}}_1 \\ &= \hat{\boldsymbol{U}}_s\hat{\boldsymbol{\Lambda}}_s\hat{\boldsymbol{\Lambda}}_s^{-1}\hat{\boldsymbol{U}}_s^H\hat{\boldsymbol{g}}_1 \\ &= \hat{\boldsymbol{U}}_s\hat{\boldsymbol{U}}_s^H\hat{\boldsymbol{g}}_1,\end{aligned}$$

where in the third equality we have used (5.6) and (5.7), and where the fourth equality follows from (5.32). Therefore, in the absence of noise, $\tilde{\hat{\boldsymbol{g}}}_1 = \hat{\boldsymbol{g}}_1$; whereas with noise, $\tilde{\hat{\boldsymbol{g}}}_1$ is the projection of $\hat{\boldsymbol{g}}_1$ onto the estimated signal subspace and therefore is a less noisy estimate of $\boldsymbol{g}_1$. □

Weight Vector Calculation

The linear MMSE array combining weight vector in (5.3) can be expressed in terms of the signal subspace components, as stated by the following result.

Proposition 5.4: *Let $\boldsymbol{U}_s$ and $\boldsymbol{\Lambda}_s$ be the signal subspace parameters defined in (5.11); then the linear MMSE combining weight vector for the desired user 1 is given by*

$$\boldsymbol{w} = \boldsymbol{U}_s \boldsymbol{\Lambda}_s^{-1} \boldsymbol{U}_s^H \boldsymbol{g}_1. \tag{5.35}$$

Proof: The linear MMSE weight vector is given in (5.3). Substituting (5.11) into (5.3), we have

$$\begin{aligned}\boldsymbol{w} &= \boldsymbol{C}^{-1}\boldsymbol{g}_1\\ &= \left(\boldsymbol{U}_s \boldsymbol{\Lambda}_s^{-1} \boldsymbol{U}_s^H + \frac{1}{\sigma^2}\boldsymbol{U}_n\boldsymbol{U}_n^H\right)\boldsymbol{g}_1\\ &= \boldsymbol{U}_s \boldsymbol{\Lambda}_s^{-1} \boldsymbol{U}_s^H \boldsymbol{g}_1,\end{aligned}$$

where the last equality follows from the fact that the steering vector is orthogonal to the noise subspace (i.e., $\boldsymbol{U}_n^H \boldsymbol{g}_1 = \boldsymbol{0}$). □

By replacing $\boldsymbol{U}_s$, $\boldsymbol{\Lambda}_s$, and $\boldsymbol{g}_1$ in (5.35) by the corresponding estimates [i.e., $\tilde{\boldsymbol{U}}_s$ and $\tilde{\boldsymbol{\Lambda}}_s$ in (5.30) and $\tilde{\boldsymbol{g}}_1$ in (5.31)], we can compute the linear MMSE combining weight vector as follows:

$$\begin{aligned}\tilde{\boldsymbol{w}} &= \tilde{\boldsymbol{U}}_s \tilde{\boldsymbol{\Lambda}}_s^{-1} \tilde{\boldsymbol{U}}_s^H \tilde{\boldsymbol{g}}_1\\ &= \tilde{\boldsymbol{U}}_s \tilde{\boldsymbol{\Lambda}}_s^{-1} \tilde{\boldsymbol{U}}_s^H \tilde{\boldsymbol{U}}_s \tilde{\boldsymbol{\Lambda}}_s \left(\tilde{\boldsymbol{U}}_s^H \boldsymbol{\Gamma}\boldsymbol{\Gamma}^H \tilde{\boldsymbol{U}}_s\right)^{-1} \tilde{\boldsymbol{U}}_s^H \boldsymbol{\Gamma}\boldsymbol{\beta}_1\\ &= \tilde{\boldsymbol{U}}_s \left(\tilde{\boldsymbol{U}}_s^H \boldsymbol{\Gamma}\boldsymbol{\Gamma}^H \tilde{\boldsymbol{U}}_s\right)^{-1} \tilde{\boldsymbol{U}}_s^H \boldsymbol{\Gamma}\boldsymbol{\beta}_1.\end{aligned} \tag{5.36}$$

Finally, we summarize the subspace-based adaptive array algorithm discussed in this section as follows.

Algorithm 5.3: [Subspace-based adaptive array for TDMA] *Denote* $\boldsymbol{\beta}_1 \triangleq \left[b_1[0] \cdots b_1[m_t-1]\right]^T$ *as the training symbols and* $\boldsymbol{\Gamma} \triangleq \left[\boldsymbol{r}[0] \; \cdots \; \boldsymbol{r}[m_t-1]\right]$ *as the corresponding received signals during the training period.*

- *Compute the signal subspace:*

$$\tilde{\boldsymbol{C}} = \frac{1}{M}\sum_{i=0}^{M-1} \boldsymbol{r}[i]\boldsymbol{r}[i]^H \tag{5.37}$$

$$= \tilde{\boldsymbol{U}}_s \tilde{\boldsymbol{\Lambda}}_s \tilde{\boldsymbol{U}}_s^H + \tilde{\boldsymbol{U}}_n \tilde{\boldsymbol{\Lambda}}_n \tilde{\boldsymbol{U}}_n^H. \tag{5.38}$$

- *Compute the combining weight vector:*

$$\tilde{\boldsymbol{w}} = \tilde{\boldsymbol{U}}_s \left(\tilde{\boldsymbol{U}}_s^H \boldsymbol{\Gamma}\boldsymbol{\Gamma}^H \tilde{\boldsymbol{U}}_s\right)^{-1} \tilde{\boldsymbol{U}}_s^H \boldsymbol{\Gamma}\boldsymbol{\beta}_1. \tag{5.39}$$

- *Perform data detection: Obtain* $\hat{b}_1[i]$ *by quantizing* $\hat{\boldsymbol{w}}^H \boldsymbol{r}[i]$ *for* $i = m_t, \ldots,$ $M-1$.

5.2.4 Extension to Dispersive Channels

So far we have assumed that the channels are nondispersive [i.e., there is no intersymbol interference (ISI)]. We next extend the techniques considered in previous subsections to dispersive channels and develop space-time processing techniques for suppressing both co-channel interference and intersymbol interference.

Let Δ be the delay spread of the channel (in units of symbol intervals). Then the received signal at the antenna array during the ith symbol interval can be expressed as

$$\underline{r}[i] = \sum_{\ell=0}^{\Delta-1} \sum_{k=1}^{K} \underline{g}_{\ell,k}\, b_k[i-\ell] + \underline{n}[i], \tag{5.40}$$

where $\underline{g}_{\ell,k}$ is the array steering vector for the kth user's ℓth symbol delay and $\underline{n}[i] \sim \mathcal{N}_c(\mathbf{0}, \sigma^2 \boldsymbol{I}_P)$. Denote $\underline{b}[i] \triangleq \begin{bmatrix} b_1[i] & \cdots & b_K[i] \end{bmatrix}^T$. By stacking m successive data samples, we define the following quantities:

$$\boldsymbol{r}[i] \triangleq \begin{bmatrix} \underline{r}[i] \\ \vdots \\ \underline{r}[i+m-1] \end{bmatrix}_{Pm \times 1}, \quad \boldsymbol{n}[i] \triangleq \begin{bmatrix} \underline{n}[i] \\ \vdots \\ \underline{n}[i+m-1] \end{bmatrix}_{Pm \times 1},$$

$$\boldsymbol{b}[i] \triangleq \begin{bmatrix} \underline{b}[i-\Delta+1] \\ \vdots \\ \underline{b}[i+m-1] \end{bmatrix}_{K(m+\Delta-1) \times 1}.$$

Then from (5.40) we can write

$$\boldsymbol{r}[i] = \boldsymbol{G}\, \boldsymbol{b}[i] + \boldsymbol{n}[i], \tag{5.41}$$

where $\boldsymbol{P}$ is a matrix of the form

$$\boldsymbol{G} \triangleq \begin{bmatrix} \underline{G}_{\Delta-1} & \cdots & \underline{G}_0 & \mathbf{0} & \mathbf{0} \\ \mathbf{0} & \underline{G}_{\Delta-1} & \cdots & \underline{G}_0 & \mathbf{0} \\ \vdots & \ddots & \ddots & \ddots & \vdots \\ \mathbf{0} & \mathbf{0} & \underline{G}_{\Delta-1} & \cdots & \underline{G}_0 \end{bmatrix}_{Pm \times K(m+\Delta-1)},$$

with

$$\underline{G}_\ell \triangleq \begin{bmatrix} \underline{g}_{\ell,1} & \cdots & \underline{g}_{\ell,K} \end{bmatrix}_{P \times K}.$$

Here, as before, m is the smoothing factor and is chosen such that the matrix $\boldsymbol{G}$ is a "tall" matrix [i.e., $Pm \geq K(m+\Delta-1)$]. Hence $m \triangleq \left\lceil \dfrac{K(\Delta-1)}{P-K} \right\rceil$. We assume that $\boldsymbol{G}$ has full column rank. From the signal model (5.41), it is evident that the techniques discussed in previous subsections can be applied straightforwardly to dispersive channels, with signal processing carried out on signal vectors of higher dimension. For example, the linear MMSE combining method for estimating the transmitted symbol $b_1[i]$ is based on quantizing the correlator output $\boldsymbol{w}^H \boldsymbol{r}[i]$, where $\boldsymbol{w} = \boldsymbol{C}^{-1} \boldsymbol{g}_1$, with

$$\boldsymbol{C} \triangleq E\left\{\boldsymbol{r}[i]\,\boldsymbol{r}[i]^H\right\} = \boldsymbol{G}\boldsymbol{G}^H + \sigma^2 \boldsymbol{I}_{Pm},$$

with

$$\boldsymbol{g}_1 \triangleq \left[\underline{g}_{0,1}^T \;\cdots\; \underline{g}_{\Delta-1,1}^T \; 0 \cdots 0\right]^T .$$

To apply the subspace-based adaptive array algorithm, we first estimate the signal subspace $(\boldsymbol{U}_s, \boldsymbol{\Lambda}_s)$ of $\boldsymbol{C}$ by forming the sample autocorrelation matrix of $\boldsymbol{r}[i]$ and then performing an eigendecomposition. Notice that the rank of the signal subspace is $K \times (m + \Delta - 1)$. Once the signal subspace is estimated, it is straightforward to apply the algorithms listed in Section 5.2.3 to estimate the data symbols.

Simulation Examples

In what follows we provide some simulation examples to demonstrate the performance of the subspace-based adaptive array algorithm discussed above. In the following simulations, it is assumed that an array of $P = 10$ antenna elements is employed at the base station. The number of symbols in each time slot is $M = 162$ with $m_t = 14$ training symbols, as in IS-54/136 systems. The modulation scheme is binary PSK (BPSK). The channel is subject to Rayleigh fading, so that the steering vectors $\{\boldsymbol{g}_k, k = 1, \ldots, K\}$ are i.i.d. complex Gaussian vectors, $\boldsymbol{g}_k \sim \mathcal{N}_c(\boldsymbol{0}, A_k^2\, \boldsymbol{I}_P)$, where A_k^2 is the received power of the kth user. The desired user is user 1. The interfering signal powers are assumed to be 6 dB below the desired signal power (i.e., $A_k = A_1/2$, for $k = 2, \ldots, K$). The ambient noise process $\{\boldsymbol{n}[i]\}$ is a sequence of i.i.d. complex Gaussian vectors, $\boldsymbol{n}[i] \sim \mathcal{N}_c(\boldsymbol{0}, \sigma^2\, \boldsymbol{I}_P)$.

In the first example we compare the performance of the two steering vector estimators $\tilde{\boldsymbol{g}}_1$ in (5.31) and $\hat{\boldsymbol{g}}_1$ in (5.7). The number of users is six (i.e., $K = 6$) and the channels have no dispersion. For each SNR value, the normalized root-mean-square error (MSE) is computed for each estimator. For the subspace estimator, we consider its performance under both the exact signal subspace parameters $(\boldsymbol{U}_s, \boldsymbol{\Lambda}_s)$ and the estimated signal subspace parameters $(\tilde{\boldsymbol{U}}_s, \tilde{\boldsymbol{\Lambda}}_s)$. The results are plotted in Fig. 5.2. It is seen that the subspace-based steering vector estimator offers significant performance improvement over the conventional correlation estimator, especially in the high-SNR region. Notice that although both estimators tend to exhibit error floors at high SNR values, their causes are different. The floor of the sample correlation estimator is due to the finite length of the training preamble m_t, whereas the floor of the subspace estimator is due to the finite length of the time slot M. It is also seen that the performance loss due to inexact signal subspace parameters is not significant in this case.

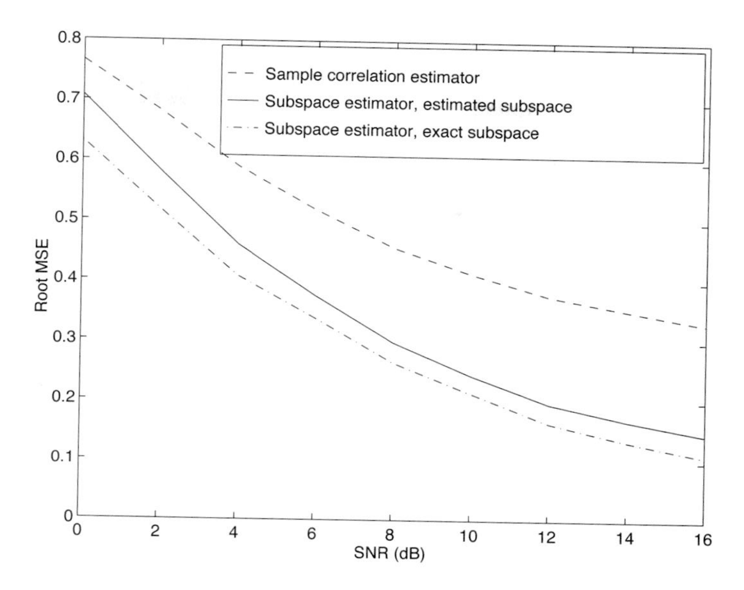

Figure 5.2. Comparison of normalized root MSEs of the subspace steering vector estimator and sample correlation steering vector estimator.

In the next example we compare the BER performance of the subspace training method and that of the DMI training method. The simulated system is the same as in the previous example. The BER curves of the three array combining methods, namely, the exact MMSE combining (5.3), the subspace algorithm, and the DMI method (5.6)–(5.8), are plotted in Fig. 5.3. It is evident from this figure that the subspace training method offers substantial performance gain over the DMI method.

Finally, we illustrate the performance of the subspace-based spatial-temporal technique for jointly suppressing co-channel interference (CCI) and intersymbol interference (ISI). The simulated system is the same as above, except now the channel is dispersive with $\Delta = 1$. It is assumed that $\underline{g}_{0,k} \sim \mathcal{N}_c(\mathbf{0}, A_k^2 \boldsymbol{I}_P)$ and $\underline{g}_{1,k} \sim \mathcal{N}_c(\mathbf{0}, (A_k^2/4)\boldsymbol{I}_P)$ for $k = 1, \ldots, K$, where A_k^2 is the received power of the kth user. As before, it is assumed that $A_k = A_1/2$ for $k = 2, \ldots, K$. The smoothing factor is taken to be $m = 2$. In Fig. 5.4 the BER performance is plotted for the DMI algorithm, subspace algorithm, and exact linear MMSE algorithm. (Note here that for the DMI method, the number of training symbols must satisfy $m_t \geq Km$ in order to get an invertible autocorrelation matrix $\boldsymbol{C}$.) It is seen again that the subspace method achieves considerable performance gain over the DMI method.

5.3 Optimal Space-Time Multiuser Detection

In Section 5.2, we considered (linear) spatial processing as a mechanism for separating multiple users sharing identical temporal signatures. In the remainder of this chapter we examine the situation in which both temporal and spatial signatures of

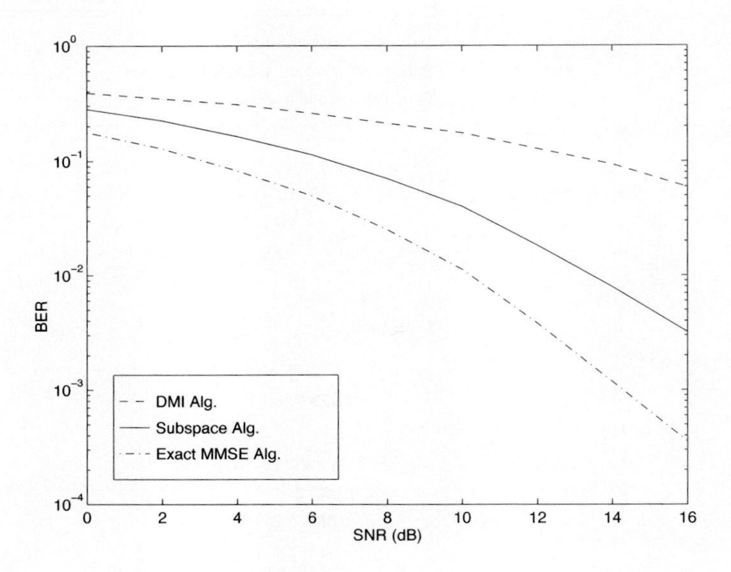

Figure 5.3. BER performance of the subspace training algorithm, DMI algorithm, and exact MMSE algorithm in a nondispersive channel.

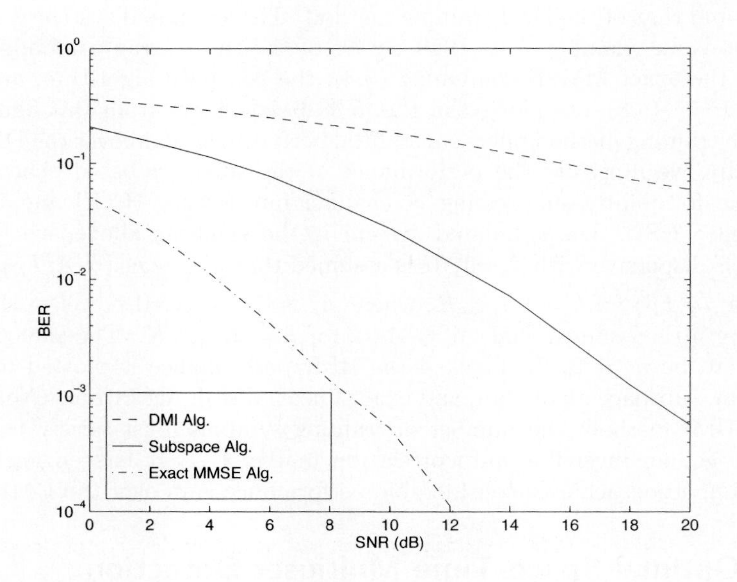

Figure 5.4. BER performance of the subspace training algorithm, DMI algorithm, and exact MMSE algorithm in a dispersive channel.

the users differ and consider the joint exploitation of these differences to separate users. Such joint processing is known as *space-time processing*. In this and the following sections, we discuss such processing in the context of multiuser detection in a CDMA system with multipath channel distortion and multiple receive antennas. We begin in this section with consideration of optimal (nonlinear) processing, turning in subsequent sections to linear and adaptive linear methods. The materials in this and the next sections first appeared in [554].

5.3.1 Signal Model

Consider a DS-CDMA mobile radio network with K users, employing normalized spreading waveforms $s_1, s_2, \ldots, s_K$, and transmitting sequences of BPSK symbols through their respective multipath channels. The transmitted baseband signal due to the kth user is given by

$$x_k(t) = A_k \sum_{i=0}^{M-1} b_k[i]\, s_k(t - iT), \qquad k = 1, \ldots, K, \tag{5.42}$$

where M is the number of data symbols per user per frame, T is the symbol interval, $b_k[i] \in \{+1, -1\}$ is the ith symbol transmitted by the kth user, and A_k and $s_k(t)$ denote, respectively, the amplitude and normalized signaling waveform of the kth user. It is assumed that $s_k(t)$ is supported only on the interval $[0, T]$ and has unit energy. (Here, for simplicity, we assume that periodic spreading sequences are employed in the system. The generalization to the aperiodic spreading case is straightforward.) It is also assumed that each user transmits independent equiprobable symbols and that the symbol sequences from different users are independent. Recall that in the direct-sequence spread-spectrum multiple-access format, the user signaling waveforms are of the form

$$s_k(t) = \sum_{j=0}^{N-1} c_{j,k}\, \psi(t - jT_c), \quad 0 \le t \le T, \tag{5.43}$$

where N is the processing gain, $\{c_{j,k} : j = 0, \ldots, N-1\}$ is a signature sequence of ± 1's assigned to the kth user, and ψ is a normalized chip waveform of duration $T_c = T/N$.

At the receiver an antenna array of P elements is employed. Assuming that each transmitter is equipped with a single antenna, the baseband multipath channel between the kth user's transmitter and the base station receiver can be modeled as a single-input/multiple-output channel with the following vector impulse response:

$$\underline{g}_k(t) = \sum_{\ell=1}^{L} \underline{a}_{\ell,k}\, \alpha_{\ell,k}\, \delta(t - \tau_{\ell,k}), \tag{5.44}$$

where L is the number of paths in each user's channel, $\alpha_{\ell,k}$ and $\tau_{\ell,k}$ are, respectively, the complex gain and delay of the ℓth path of the kth user's signal, and $\underline{a}_{\ell,k} =$

$[a_{\ell,k,1} \cdots a_{\ell,k,P}]^T$ is the array response vector corresponding to the ℓth path of the kth user's signal. The total received signal at the receiver is then the superposition of the signals from the K users plus the additive ambient noise, given by

$$\begin{aligned} \underline{r}(t) &= \sum_{k=1}^{K} x_k(t) \star \underline{g}_k(t) + \underline{n}(t) \\ &= \sum_{i=0}^{M-1} \sum_{k=1}^{K} A_k \, b_k[i] \sum_{\ell=1}^{L} \underline{a}_{\ell,k} \, \alpha_{\ell,k} \, s_k(t - iT - \tau_{\ell,k}) + \underline{n}(t), \end{aligned} \tag{5.45}$$

where $\star$ denotes convolution; $\underline{n}(t) = [n_1(t) \;\cdots\; n_P(t)]^T$ is a vector of independent zero-mean complex white Gaussian noise processes, each with power spectral density σ^2.

5.3.2 Sufficient Statistic

We next derive a sufficient statistic for demodulating the multiuser symbols from the space-time signal model (5.45). To do so, we first denote the useful signal in (5.45) by

$$\underline{S}(t; \boldsymbol{b}) \triangleq \sum_{i=0}^{M-1} \sum_{k=1}^{K} A_k \, b_k[i] \sum_{\ell=1}^{L} \underline{a}_{\ell,k} \, \alpha_{\ell,k} \, s_k(t - iT - \tau_{\ell,k}), \tag{5.46}$$

where $\boldsymbol{b} \triangleq \left[\underline{b}[0]^T \;\cdots\; \underline{b}[M-1]^T\right]^T$ and $\underline{b}[i] \triangleq \left[b_1[i] \;\cdots\; b_K[i]\right]^T$. Using the Cameron–Martin formula [377], the likelihood function of the received waveform $\underline{r}(t)$ in (5.45) conditioned on all the transmitted symbols $\boldsymbol{b}$ of all users can be written as

$$\mathcal{L}\Big(\{\underline{r}(t) : -\infty < t < \infty\} \mid \boldsymbol{b}\Big) \propto \exp\left[\Omega(\boldsymbol{b})/\sigma^2\right], \tag{5.47}$$

where

$$\Omega(\boldsymbol{b}) \triangleq 2\,\Re\left\{\int_{-\infty}^{\infty} \underline{S}(t; \boldsymbol{b})^H \, \underline{r}(t) \, \mathrm{d}t\right\} - \int_{-\infty}^{\infty} \mid \underline{S}(t; \boldsymbol{b}) \mid^2 \, \mathrm{d}t. \tag{5.48}$$

The first integral in (5.48) can be expressed as

$$\begin{aligned} &\int_{-\infty}^{\infty} \underline{S}(t; \boldsymbol{b})^H \, \underline{r}(t) \, \mathrm{d}t \\ &\triangleq \sum_{i=0}^{M-1} \sum_{k=1}^{K} A_k \, b_k[i] \overbrace{\sum_{\ell=1}^{L} \alpha_{\ell,k}^* \, \underline{a}_{\ell,k}^H \underbrace{\int_{-\infty}^{\infty} \underline{r}(t) \, s_k(t - iT - \tau_{\ell,k}) \, \mathrm{d}t}_{\underline{z}_{\ell,k}[i]}}^{y_k[i]}. \end{aligned} \tag{5.49}$$

Since the second integral in (5.48) does not depend on the received signal $\underline{r}(t)$, by (5.49) we see that $\{y_k[i]\}$ is a sufficient statistic for detecting the multiuser symbols $\boldsymbol{b}$. From (5.49) it is seen that this sufficient statistic is obtained by passing the received signal vector $\underline{r}(t)$ through KL beamformers directed at each path of each user's signal, followed by a bank of K maximum-ratio multipath combiners (i.e., RAKE receivers). Since this beamformer is a spatial matched filter for the array antenna receiver, and a RAKE receiver is a temporal matched filter for multipath channels, the sufficient statistic $\{y_k[i]\}_{i;k}$ is simply the output of a *space-time matched filter*. Next we derive an explicit expression for this sufficient statistic in terms of the multiuser channel parameters and transmitted symbols, which is instrumental to developing various space-time multiuser receivers in subsequent sections.

Assume that the multipath delay spread of any user signal is limited to at most Δ symbol intervals, where Δ is a positive integer. That is,

$$\tau_{\ell,k} \leq \Delta T, \qquad 1 \leq k \leq K, \quad 1 \leq \ell \leq L. \tag{5.50}$$

Define the following cross-correlations of the delayed user signaling waveforms:

$$\rho^{[j]}_{(k,\ell)(k',\ell')} \triangleq \int_{-\infty}^{\infty} s_k(t-\tau_{\ell,k})\, s_{k'}(t-jT-\tau_{\ell',k'})\,\mathrm{d}t,$$
$$-\Delta \leq j \leq \Delta, \quad 1 \leq k,k' \leq K, \quad 1 \leq \ell,\ell' \leq L. \tag{5.51}$$

Since $\tau_{\ell,k} \leq \Delta T$ and $s_k(t)$ is nonzero only for $t \in [0,T]$, it then follows that $\rho^{[j]}_{(k,\ell)(k',\ell')} = 0$ for $|j| > \Delta$. Now substituting (5.45) into (5.49), we have

$$\begin{aligned}
&\underline{a}_{\ell,k}^H \underline{z}_{\ell,k}[i] \\
&= \sum_{i'=0}^{M-1} \sum_{k'=1}^{K} A_{k'} b_{k'}[i'] \sum_{\ell'=1}^{L} \underline{a}_{\ell,k}^H \underline{a}_{\ell',k'}\, \alpha_{\ell',k'} \int_{-\infty}^{\infty} s_{k'}(t-i'T-\tau_{\ell',k'}) s_k(t-iT-\tau_{\ell,k})\,\mathrm{d}t \\
&\quad \underbrace{+\, \underline{a}_{\ell,k}^H \int_{-\infty}^{\infty} \underline{n}(t) s_k(t-iT-\tau_{\ell,k})\,\mathrm{d}t}_{u_{\ell,k}[i]} \\
&= \sum_{j=-\Delta}^{\Delta} \sum_{k'=1}^{K} A_{k'} b_{k'}[i+j] \sum_{\ell'=1}^{L} \underline{a}_{\ell,k}^H \underline{a}_{\ell',k'}\, \alpha_{\ell',k'}\, \rho^{[j]}_{(k,\ell)(k',l')} + u_{\ell,k}[i],
\end{aligned} \tag{5.52}$$

where $\{u_{\ell,k}[i]\}$ are zero-mean complex Gaussian random sequences with the following covariance:

$$\begin{aligned}
&E\{u_{\ell,k}[i]\, u_{\ell',k'}[i']^*\} \\
&= E\left\{\left[\underline{a}_{\ell,k}^H \int_{-\infty}^{\infty} \underline{n}(t) s_k(t-iT-\tau_{\ell,k})\,\mathrm{d}t\right]\left[\underline{a}_{\ell',k'}^T \int_{-\infty}^{\infty} \underline{n}^*(t') s_{k'}(t'-i'T-\tau_{\ell',k'})\,\mathrm{d}t'\right]\right\} \\
&= \underline{a}_{\ell,k}^H \left[\int_{-\infty}^{\infty}\int_{-\infty}^{\infty} E\{\underline{n}(t)\underline{n}^H(t')\}\, s_k(t-iT-\tau_{\ell,k})\, s_{k'}(t'-i'T-\tau_{\ell',k'})\,\mathrm{d}t\,\mathrm{d}t'\right] \underline{a}_{\ell',k'}
\end{aligned}$$

$$= \underline{a}_{\ell,k}^H \left[\int_{-\infty}^{\infty} \int_{-\infty}^{\infty} \underline{I}_p \, \delta(t-t') \, s_k(t - iT - \tau_{\ell,k}) \, s_{k'}(t' - i'T - \tau_{\ell',k'}) \, \mathrm{d}t \, \mathrm{d}t' \right] \underline{a}_{\ell',k'}$$

$$= \underline{a}_{\ell,k}^H \, \underline{a}_{\ell',k'} \int_{-\infty}^{\infty} s_k(t - iT - \tau_{\ell,k}) \, s_{k'}(t - i'T - \tau_{\ell',k'}) \, \mathrm{d}t$$

$$= \rho_{(k,l)(k',\ell')}^{[i'-i]} \, \underline{a}_{\ell,k}^H \, \underline{a}_{\ell',k'}, \tag{5.53}$$

where $\underline{I}_p$ denotes a $p \times p$ identity matrix and $\delta(t)$ is the Dirac delta function. Define the following quantities:

$$\underline{R}^{[j]} \triangleq \begin{bmatrix} \rho_{(1,1)(1,1)}^{[j]} & \cdots & \rho_{(1,1)(1,L)}^{[j]} & \cdots & \rho_{(1,1)(K,1)}^{[j]} & \cdots & \rho_{(1,1)(K,L)}^{[j]} \\ \rho_{(2,1)(1,1)}^{[j]} & \cdots & \rho_{(2,1)(1,L)}^{[j]} & \cdots & \rho_{(2,1)(K,1)}^{[j]} & \cdots & \rho_{(2,1)(K,L)}^{[j]} \\ \vdots & \vdots & \vdots & \vdots & \vdots & \vdots & \vdots \\ \rho_{(K,L)(1,1)}^{[j]} & \cdots & \rho_{(K,L)(1,L)}^{[j]} & \cdots & \rho_{(K,L)(K,1)}^{[j]} & \cdots & \rho_{(K,L)(K,L)}^{[j]} \end{bmatrix}, \quad (KL \times KL \text{ matrix})$$

$$\underline{\Phi} \triangleq \left[\underline{a}_{1,1} \cdots \underline{a}_{L,1} \cdots \underline{a}_{1,K} \cdots \underline{a}_{L,K} \right], \quad (P \times KL \text{ matrix})$$

$$\underline{\zeta}[i] \triangleq \left[\underline{a}_{1,1}^H \underline{z}_{1,1}[i] \cdots \underline{a}_{L,1}^H \underline{z}_{L,1}[i] \cdots \underline{a}_{1,K}^H \underline{z}_{1,K}[i] \cdots \underline{a}_{L,K}^H \underline{z}_{L,K}[i] \right]^T, \quad (KL\text{-vector})$$

$$\underline{u}[i] \triangleq \left[u_{1,1}[i] \cdots u_{L,1}[i] \cdots u_{1,K}[i] \cdots u_{L,K}[i] \right]^T, \quad (KL\text{-vector})$$

$$\underline{g}_k \triangleq \left[\alpha_{1,k} \cdots \alpha_{L,k} \right]^T, \quad (L\text{-vector})$$

$$\underline{G} \triangleq \mathrm{diag}\left(\underline{g}_1, \ldots, \underline{g}_K \right), \quad (KL \times K \text{ matrix})$$

$$\underline{A} \triangleq \mathrm{diag}\left\{ A_1, \ldots, A_K \right\}, \quad (K \times K \text{ matrix})$$

$$\underline{y}[i] \triangleq \left[y_1[i] \cdots y_K[i] \right]^T. \quad (K\text{-vector})$$

We can then write (5.52) in the following vector form:

$$\underline{\zeta}[i] = \sum_{j=-\Delta}^{\Delta} \left[\underline{R}^{[j]} \circ \left(\underline{\Phi}^H \underline{\Phi} \right) \right] \underline{G} \, \underline{A} \, \underline{b}[i+j] + \underline{u}[i], \tag{5.54}$$

where $\circ$ denotes the Schur matrix product (i.e., elementwise product), and from (5.53) the covariance matrix of the complex Gaussian vector sequence $\{\underline{u}[i]\}$ is

$$E\left\{ \underline{u}[i] \, \underline{u}[i+j]^H \right\} = \underline{R}^{[j]} \circ \left(\underline{\Phi}^H \underline{\Phi} \right). \tag{5.55}$$

Substituting (5.54) into (5.49), we obtain a useful expression for the sufficient statistic $\underline{y}[i]$, given by

$$\underline{y}[i] \triangleq \underline{G}^H \, \underline{\zeta}[i] = \sum_{j=-\Delta}^{\Delta} \underbrace{\underline{G}^H \left[\underline{R}^{[j]} \circ \left(\underline{\Phi}^H \underline{\Phi} \right) \right] \underline{G}}_{\underline{H}^{[j]}} \, \underline{A} \, \underline{b}[i+j] + \underbrace{\underline{G}^H \underline{u}[i]}_{\underline{v}[i]}, \tag{5.56}$$

where $\{\underline{v}[i]\}$ is a sequence of zero-mean complex Gaussian vectors with covariance matrix

$$E\left\{\underline{v}[i]\,\underline{v}[i+j]^H\right\} = \underline{G}^H \left[\underline{R}^{[j]} \circ \left(\underline{\Phi}^H \underline{\Phi}\right)\right] \underline{G} \triangleq \underline{H}^{[j]}. \quad (5.57)$$

Note that by definition (5.51) we have $\rho^{[j]}_{(k,l)(k',l')} = \rho^{[-j]}_{(k',l')(k,l)}$. From this it follows that $\underline{R}^{[-j]} = \underline{R}^{[j]T}$, and therefore $\underline{H}^{[-j]} = \underline{H}^{[j]H}$.

5.3.3 Maximum-Likelihood Multiuser Sequence Detector

We now use the sufficient statistic above to derive the maximum-likelihood detector for symbols in $\boldsymbol{b}$. The maximum-likelihood sequence decision rule chooses $\boldsymbol{b}$ that maximizes the log-likelihood function (5.48). Using (5.46), the second integral in (5.48) can be computed as

$$\begin{aligned}\int_{-\infty}^{\infty} \left|\underline{S}(t;\boldsymbol{b})\right|^2 \mathrm{d}t &= \sum_{i=0}^{M-1}\sum_{i'=0}^{M-1}\sum_{k=1}^{K}\sum_{k'=1}^{K} A_k A_{k'} b_k[i] b_{k'}[i'] \\ &\quad \sum_{\ell=1}^{L}\sum_{\ell'=1}^{L} \underline{a}^H_{\ell,k}\,\underline{a}_{\ell',k'}\,\alpha^*_{\ell,k}\alpha_{\ell',k'}\,\rho^{[i'-i]}_{(k,\ell)(k',\ell')} \\ &= \boldsymbol{b}^T \boldsymbol{A}\boldsymbol{H}\boldsymbol{A}\boldsymbol{b},\end{aligned} \quad (5.58)$$

where $\boldsymbol{H}$ denotes the following $MK \times MK$ block Jacobi matrix:

$$\boldsymbol{H} \triangleq \begin{bmatrix} \underline{H}^{[0]} & \underline{H}^{[1]} & \cdots & \underline{H}^{[\Delta]} & & & \\ \underline{H}^{[-1]} & \underline{H}^{[0]} & \underline{H}^{[1]} & \cdots & \underline{H}^{[\Delta]} & & \\ & & & & & & \\ & \underline{H}^{[-\Delta]} & \cdots & \underline{H}^{[0]} & \cdots & \underline{H}^{[\Delta]} & \\ & & & & & & \\ & & \underline{H}^{[-\Delta]} & \cdots & \underline{H}^{[-1]} & \underline{H}^{[0]} & \underline{H}^{[1]} \\ & & & \underline{H}^{[-\Delta]} & \cdots & \underline{H}^{[-1]} & \underline{H}^{[0]} \end{bmatrix}, \quad (5.59)$$

$\boldsymbol{A} \triangleq \underline{I}_M \otimes \underline{A}$ ($\otimes$ denotes the Kronecker matrix product), $\boldsymbol{y} \triangleq \left[\underline{y}[0]^T \; \cdots \; \underline{y}[M-1]^T\right]^T$, and recall that $\boldsymbol{b} \triangleq \left[\underline{b}[0]^T \; \cdots \; \underline{b}[M-1]^T\right]^T$.

Substituting (5.49) and (5.58) into (5.48), the log-likelihood function $\Omega(\boldsymbol{b})$ can then be written as

$$\Omega(\boldsymbol{b}) = 2\Re\left\{\boldsymbol{b}^T \boldsymbol{A}\boldsymbol{y}\right\} - \boldsymbol{b}^T \boldsymbol{A}\boldsymbol{H}\boldsymbol{A}\boldsymbol{b}. \quad (5.60)$$

For any integer n satisfying $1 \leq n \leq MK$, denote its modulo-K decomposition with remainder $\kappa(n) = 1, \ldots, K$, by $n = \eta(n)K + \kappa(n)$ [518]. Then we can write[2]

$$\begin{aligned} &\boldsymbol{b}^T \boldsymbol{A} \boldsymbol{y} \\ &\quad = \sum_{n=1}^{MK} \boldsymbol{A}[n,n]\boldsymbol{b}[n]\boldsymbol{y}[n] \\ &\quad = \sum_{n=1}^{MK} A_{\kappa(n)} b_{\kappa(n)}[\eta(n)] y_{\kappa(n)}[\eta(n)], \end{aligned} \tag{5.61}$$

$$\begin{aligned} &\boldsymbol{b}^T \boldsymbol{A}\boldsymbol{H}\boldsymbol{A}\boldsymbol{b} \\ &\quad = \sum_{n=1}^{MK} \boldsymbol{A}[n,n]\boldsymbol{b}[n] \left(\boldsymbol{H}[n,n]\boldsymbol{A}[n,n]\boldsymbol{b}[n] + 2\Re\left\{ \sum_{j=1}^{n-1} \boldsymbol{H}[n,j]\boldsymbol{A}[j,j]\boldsymbol{b}[j] \right\} \right) \\ &\quad = \sum_{n=1}^{MK} A_{\kappa(n)} b_{\kappa(n)}[\eta(n)] \left[A_{\kappa(n)} h^{[0]}_{\kappa(n),\kappa(n)} b_{\kappa(n)}[\eta(n)] + 2\Re\left\{ \boldsymbol{f}_n^H \boldsymbol{x}_n \right\} \right], \end{aligned} \tag{5.62}$$

where in (5.62) the vectors $\boldsymbol{x}_n$ and $\boldsymbol{f}_n$ have dimensions $\Delta K + K - 1$, given, respectively, by

$$\boldsymbol{x}_n \triangleq \Big[\overbrace{0 \ldots 0}^{K-\kappa(n)} \; \underline{b}[\eta(n)-\Delta]^T \cdots \underline{b}[\eta(n)-1]^T b_1[\eta(n)] \cdots b_{\kappa(n)-1}[\eta(n)] \Big]^T,$$

$$\boldsymbol{f}_n \triangleq \Big[\overbrace{0 \ldots 0}^{K-\kappa(n)} \left(\underline{Ah}^{[-\Delta]}_{\kappa(n)} \right)^T \cdots \left(\underline{Ah}^{[-1]}_{\kappa(n)} \right)^T \left(A_1 h^{[0]}_{\kappa(n),1} \right) \cdots \left(A_{\kappa(n)-1} h^{[0]}_{\kappa(n),\kappa(n)-1} \right) \Big]^T,$$

where $\underline{h}_k^{[j]}$ denotes the kth column of matrix $\underline{H}^{[j]}$, and $h^{[j]}_{k,k'}$ denotes the (k,k')th entry of $\underline{H}^{[j]}$. Substituting (5.61) and (5.62) into (5.60), the log-likelihood function can then be decomposed as follows:

$$\Omega(\boldsymbol{b}) = \sum_{n=1}^{MK} \lambda_n \left(\xi_n, \boldsymbol{x}_n \right), \tag{5.63}$$

where $\xi_n \triangleq b_{\kappa(n)}[\eta(n)]$ and

$$\lambda_n\left(\xi, \boldsymbol{x}\right) = \xi A_{\kappa(n)} \Re\left\{ 2\, y_{\kappa(n)}[\eta(n)] - 2\boldsymbol{f}_n^H \boldsymbol{x} - A_{\kappa(n)}\, h^{[0]}_{\kappa(n),\kappa(n)}\, \xi \right\}, \tag{5.64}$$

with the state vector recursively defined according to $\boldsymbol{x}_{n+1} = \Big[\boldsymbol{x}_n[2], \ldots, \boldsymbol{x}_n[\Delta K + K - 1], \; \xi_n \Big]^T$ and $\boldsymbol{x}_1 = \underline{0}_{(\Delta K + K - 1)}$, where $\underline{0}_m$ denotes a zero vector of dimension m.

Given the additive decomposition (5.63) of the log-likelihood function, it is straightforward to apply the dynamic programming to compute the sequence $\hat{\boldsymbol{b}}$ that

[2]Notation: $\boldsymbol{A}[i,j]$ denotes the (i,j)th element of matrix $\boldsymbol{A}$; $\boldsymbol{b}[i]$ denotes the ith element of vector $\boldsymbol{b}$.

maximizes $\Omega(\boldsymbol{b})$ (i.e., the maximum-likelihood estimate of the transmitted multiuser symbol sequences). Since the dimensionality of the state vector is $\Delta K + K - 1$, the computational complexity of the maximum-likelihood sequence detector is on the order of $\mathcal{O}\left(2^{(\Delta+1)K}\right)$. Note that in the absence of multipath (i.e., $L = 1$ and $\Delta = 1$), if the users are numbered according to their relative delays in ascending order (i.e., $0 \leq \tau_{1,1} \leq \cdots \leq \tau_{1,K} < T$), the matrix $\underline{H}^{[-1]}$ becomes strictly upper triangular. In this case, the dimension of the state vector is reduced to $K - 1$ and the computational complexity of the corresponding maximum likelihood sequence detection algorithm is $\mathcal{O}\left(2^{K}\right)$ [324, 518]. However, in the presence of multipath, even if the multipath delays are still within one symbol interval (i.e., $\Delta = 1$), the matrix $\underline{H}^{[-1]}$ no longer has an upper triangular structure in general. Hence the dimension of the state vector in this case is $2K - 1$ and the complexity of the dynamic programming is $\mathcal{O}\left(2^{2K}\right)$. Even though the $\mathcal{O}\left(2^{(\Delta+1)K}\right)$ complexity is generally much lower than the $\mathcal{O}(2^{KM})$ complexity of brute-force maximization of (5.60) (Δ is typically only a few symbols, while the frame length M can be hundreds or even thousands of symbols), this complexity is still prohibitively high if the number of users is even moderate (say a dozen). Thus, it is of interest to find low-complexity alternatives.

5.4 Linear Space-Time Multiuser Detection

As seen in Section 5.3, the optimal space-time multiuser detection algorithm typically has a prohibitive computational complexity. In this section we discuss linear space-time multiuser detection techniques that mitigate this complexity significantly. To consider such detectors we will assume for now that the receiver has knowledge of the spreading waveforms and the channel parameters of all users. The method discussed here is based on iterative interference cancellation and has a low computational complexity. For comparison, we also discuss single-user-based linear space-time processing methods.

5.4.1 Linear Multiuser Detection via Iterative Interference Cancellation

From (5.56) we can write the expression for the sufficient statistic vector $\boldsymbol{y}$ in matrix form as

$$\boldsymbol{y} = \boldsymbol{H}\boldsymbol{A}\boldsymbol{b} + \boldsymbol{v}, \tag{5.65}$$

where by (5.57), $\boldsymbol{v} \sim \mathcal{N}_c\left(\boldsymbol{0}, \sigma^2\boldsymbol{H}\right)$. Recall that in linear multiuser detection, a linear transformation is applied to the sufficient statistic vector $\boldsymbol{y}$, followed by local decisions for each user. That is, the multiuser data bits are demodulated according to

$$\hat{\boldsymbol{b}} = \text{sign}\left[\Re\left\{\boldsymbol{W}\,\boldsymbol{y}\right\}\right], \tag{5.66}$$

where $\boldsymbol{W}$ is an $MK \times MK$ complex matrix. As discussed in Chapter 2, two popular linear multiuser detectors are the linear decorrelating (i.e., zero-forcing) detector, which chooses the weight matrix $\boldsymbol{W}$ to eliminate the interference completely (at the expense of enhancing the noise), and the linear MMSE detector, which chooses the weight matrix $\boldsymbol{W}$ to minimize the mean-square error between the transmitted symbols and the outputs of the linear transformation (i.e., $E\left\{\|\boldsymbol{A}\,\boldsymbol{b} - \boldsymbol{W}\,\boldsymbol{y}\|^2\right\}$). The corresponding weight matrices for these two linear multiuser detectors are given, respectively, by

$$\boldsymbol{W}_d = \boldsymbol{H}^{-1}, \qquad \text{(linear decorrelating detector)} \tag{5.67}$$

$$\boldsymbol{W}_m = \left(\boldsymbol{H} + \sigma^2 \boldsymbol{A}^{-2}\right)^{-1}. \qquad \text{(linear MMSE detector)} \tag{5.68}$$

Since the frame length M is typically large, direct inversion of the $MK \times MK$ matrices above is too costly for practical purposes. Moreover, the complexity cannot generally be mitigated over multiple frames, since the matrices $\boldsymbol{H}$ and $\boldsymbol{A}$ may vary from frame to frame due to mobility, aperiodic spreading codes, and so on. This complexity can be mitigated, however, by using iterative methods of equation solving, which we now discuss. We first consider Gauss–Seidel iteration to obtain the linear multiuser detector output. This method effectively performs serial interference cancellation on the sufficient statistic vector $\boldsymbol{y}$ and recursively refines the estimates of the multiuser signals $\{x_k[i] \triangleq A_k b_k[i]\}$. Denote such an estimate at the mth iteration as $x_k^m[i]$. Also denote $\underline{x}^m[i] \triangleq \left[x_1^m[i] \ \cdots \ x_K^m[i]\right]^T$ and $\boldsymbol{x}^m \triangleq \left[\underline{x}^m[0]^T \ \cdots \ \underline{x}^m[M-1]^T\right]^T$. The algorithm is listed in Table 5.1.

The convergence properties of this serial interference cancellation algorithm are characterized by the following result.

Proposition 5.5: *(1) If $\gamma_k = h_{kk}^{[0]}$ and if $\mathbf{H}$ is positive definite, then $\boldsymbol{x}^m \to \boldsymbol{W}_d\,\mathbf{y}$, as $m \to \infty$; (2) if $\gamma_k = h_{kk}^{[0]} + \sigma^2/A_k^2$, then $\boldsymbol{x}^m \to \boldsymbol{W}_m\,\boldsymbol{y}$, as $m \to \infty$.*

Proof: Consider the following system of linear equations:

$$\boldsymbol{H}\boldsymbol{x} = \boldsymbol{y}. \tag{5.69}$$

The Gauss–Seidel procedure [598] for iteratively solving $\boldsymbol{x}$ from (5.69) is given by

$$\boldsymbol{x}^m[i'] = \frac{1}{\boldsymbol{H}[i',i']}\left(\boldsymbol{y}[i'] - \sum_{j'<i'} \boldsymbol{H}[i',j']\,\boldsymbol{x}^m[j'] - \sum_{j'>i'} \boldsymbol{H}[i',j']\,\boldsymbol{x}^{m-1}[j']\right),$$

$$i' = 1,\ldots,MK, \quad m = 1,2,\ldots \tag{5.70}$$

Substituting in (5.70) the notations $\boldsymbol{x}^m[Ki+k] = x_k^m[i]$, $\boldsymbol{y}[Ki+k] = y_k[i]$ for $k = 1,\ldots,K$, $i = 0,\ldots,M-1$, and the elements of the matrix $\boldsymbol{H}$ given in (5.59), it then follows that the serial interference cancellation procedure in Table 5.1 is the same as the Gauss–Seidel iteration (5.70) if we choose $\gamma_k = h_{kk}^{[0]}$. Then by the Ostrowski–Reich theorem [598], a sufficient condition for the Gauss–Seidel iteration (5.70) to converge to the solution of (5.69) (i.e., the output of the linear decorrelating detector, $\boldsymbol{x}^m \to \boldsymbol{H}^{-1}\boldsymbol{y} \triangleq \boldsymbol{W}_d\,\boldsymbol{y}$) is that $\boldsymbol{H}$ be positive definite.

Table 5.1. Iterative Implementation of Linear Space-Time Multiuser Detection Using Serial Interference Cancellation

$x_k[i] \;=\; y_k[i], \quad k = 1, \ldots, K; \;\; i = 0, \ldots, M-1$

for $\quad m = 1, 2, \ldots$

 for $\quad i = 0, \ldots, M-1$

  for $\quad k = 1, \ldots, K$

$$
\begin{aligned}
x_k^m[i] \;&=\; \tfrac{1}{\gamma_k}\Big(y_k[i] - \textstyle\sum_{j=-\Delta}^{-1}\sum_{k'=1}^{K} h_{k,k'}^{[j]}\, x_{k'}^{m}[i+j] - \sum_{k'=1}^{k-1} h_{k,k'}^{[0]}\, x_{k'}^{m}[i] \\
&\quad - \textstyle\sum_{k'=k+1}^{K} h_{k,k'}^{[0]}\, x_{k'}^{m-1}[i] - \sum_{j=1}^{\Delta}\sum_{k'=1}^{K} h_{k,k'}^{[j]}\, x_{k'}^{m-1}[i+j]\Big)
\end{aligned}
$$

  end

 end

end

linear decorrelating detector: $\gamma_k \;=\; h_{kk}^{[0]}$

linear MMSE detector: $\gamma_k \;=\; h_{kk}^{[0]} + \sigma^2/A_k^2$

Similarly, consider the system of linear equations

$$\left(\boldsymbol{H} + \sigma^2 \boldsymbol{A}^{-2}\right) \boldsymbol{x} = \boldsymbol{y}. \tag{5.71}$$

The corresponding Gauss–Seidel iteration is given by

$$\boldsymbol{x}^m[i'] = \frac{1}{\boldsymbol{H}[i',i'] + \sigma^2/\boldsymbol{A}[i',i']^2} \left(\boldsymbol{y}[i'] - \sum_{j'<i'} \boldsymbol{H}[i',j']\, \boldsymbol{x}^m[j'] - \sum_{j'>i'} \boldsymbol{H}[i',j']\, \boldsymbol{x}^{m-1}[j'] \right),$$

$$i' = 1, \ldots, MK, \quad m = 1, 2, \ldots, \tag{5.72}$$

which is the same as the serial interference cancellation procedure in Table 5.1 with $\gamma_k = h_{kk}^{[0]} + \sigma^2/A_k^2$. It is seen from (5.58) that the matrix $\boldsymbol{H}$ is positive semidefinite, as $\int_{-\infty}^{\infty} |\underline{S}(t;\boldsymbol{b})|^2 \, \mathrm{d}t = \boldsymbol{x}^H \boldsymbol{H} \boldsymbol{x} \geq 0$, where $\boldsymbol{x} \triangleq \boldsymbol{A}\boldsymbol{b}$. Therefore, $\left(\boldsymbol{H} + \sigma^2 \boldsymbol{A}^{-2}\right)$ is positive definite, and by the Ostrowski–Reich theorem, iteration (5.72) converges to the solution to (5.71) [i.e., the output of the linear MMSE detector, $\boldsymbol{x}^m \to \left(\boldsymbol{H} + \sigma^2 \boldsymbol{A}^{-2}\right)^{-1} \boldsymbol{y} \triangleq \boldsymbol{W}_m \boldsymbol{y}$]. □

The computational complexity of the iterative serial interference cancellation algorithm above per user per bit is $\left[\bar{m}M(2\Delta+1)K^2\right]/KM = \mathcal{O}\left(\bar{m}\Delta K\right)$, where $\bar{m}$ is the total number of iterations. The complexity per user per bit of direct inversion of the matrices in (5.67) or (5.68) is $\mathcal{O}\left(K^3M^3/KM\right) = \mathcal{O}\left(K^2M^2\right)$. By exploiting the Hermitian $(2\Delta+1)$-block Toeplitz structure of the matrix $\boldsymbol{H}$, this complexity can be reduced to $\mathcal{O}\left(K^2 M \Delta\right)$ [324]. Since in practice the number of iterations is a small number (e.g., $\bar{m} \leq 5$), the iterative method for linear multiuser detection above achieves significant complexity reduction over the direct matrix inversion method.

A natural alternative to the serial interference cancellation method is the following *parallel interference cancellation method*,

$$x_k^m[i] = \frac{1}{\gamma_k} \left(y_k[i] - \sum_{j=-\Delta}^{\Delta} \sum_{\substack{k'=1 \\ (k',j)\neq(k,0)}}^{K} h_{k,k'}^{[j]}\, x_{k'}^{m-1}[i+j] \right),$$

$$k = 1, \ldots, K, \quad i = 0, \ldots, M-1, \quad m = 1, 2, \ldots \tag{5.73}$$

Unlike the serial method, in which the new estimate $x_k^m[i]$ is used to update the subsequent estimates as soon as it is available, in the parallel method, at the mth iteration, $x_k^m[i]$ is updated using the estimates only from the previous iteration. Parallel interference cancellation corresponds to Jacobi's method [598] for solving the linear system (5.69) or (5.71):

$$\boldsymbol{x}^m[i'] = \frac{1}{\gamma[i']} \left(\boldsymbol{y}[i'] - \sum_{j'\neq i'} \boldsymbol{H}[i',j']\, \boldsymbol{x}^{m-1}[j'] \right),$$

$$i' = 1, \ldots, MK, \quad m = 1, 2, \ldots \tag{5.74}$$

with $\gamma[i'] = \boldsymbol{H}[i', i']$ or $\gamma[i'] = \boldsymbol{H}[i', i'] + \sigma^2/\boldsymbol{A}[i', i']^2$. However, the convergence of Jacobi's method (5.74) and hence that of the parallel interference cancellation method (5.73) is *not* guaranteed in general. To see this, for example, let $\boldsymbol{D}$ be the diagonal matrix containing the diagonal elements of $\boldsymbol{H}$ and let $\boldsymbol{H} = \boldsymbol{D} + \boldsymbol{B}$ be the splitting of $\boldsymbol{H}$ into its diagonal and off-diagonal elements. Suppose that $\boldsymbol{H} = \boldsymbol{D}+\boldsymbol{B}$ is positive definite; then a necessary and sufficient condition for the convergence of Jacobi's iteration is that $\boldsymbol{D} - \boldsymbol{B}$ be positive definite [598]. In general, this condition may not be satisfied, and hence the parallel interference cancellation method (5.74) is not guaranteed to produce the linear multiuser detector output. An analysis of parallel interference cancellation is found in [53]. Other iterative methods for solving (5.71), such as conjugate-gradient techniques, are described in [92].

5.4.2 Single-User Linear Space-Time Detection

In what follows we consider several single-user-based linear space-time processing methods. These methods have been advocated in the recent literature, as they lead to several space-time adaptive receiver structures [34, 273, 348, 574]. We derive the exact forms of these single-user detectors in terms of multiuser channel parameters. We then compare the performance of these single-user space-time receivers with that of the multiuser space-time receivers discussed in Section 5.4.1.

Denote $r_p(t)$ as the received signal at the pth antenna element [i.e., the pth element of the received vector signal $\underline{r}(t)$ in (5.45)]:

$$r_p(t) = \sum_{i=0}^{M-1} \sum_{k=1}^{K} A_k b_k[i] \sum_{\ell=1}^{L} a_{\ell,k,p} \alpha_{\ell,k} s_k(t - iT - \tau_{\ell,k}) + n_p(t), \qquad p = 1, \ldots, P. \quad (5.75)$$

Suppose that the user of interest is the kth user. In the single-user approach, in order to demodulate the ith symbol of the kth user, that user's matched-filter output corresponding to each path at each antenna element is first computed:

$$\begin{aligned} z_{\ell,k,p}[i] &\triangleq \int_{-\infty}^{\infty} r_p(t) s_k(t - iT - \tau_{\ell,k})\, dt \\ &= \sum_{j=-\Delta}^{\Delta} \sum_{k'=1}^{K} A_{k'}\, b_{k'}[i+j] \sum_{\ell'=1}^{L} a_{\ell',k',p}\, \alpha_{\ell',k'}\, \rho_{(k,\ell)(k',\ell')}^{[j]} + n_{\ell,k,p}[i], \\ &\qquad \ell = 1, 2, \ldots, L, \quad p = 1, 2, \ldots, P, \end{aligned} \quad (5.76)$$

where $\{n_{\ell,k,p}[i]\}$ are zero-mean complex Gaussian random sequences with covariance

$$E\Big\{ n_{\ell,k,p}[i]\, n_{\ell',k,p'}[i']^* \Big\} = \begin{cases} 0 & \text{if } p \neq p' \text{ or } |i - i'| > \Delta, \\ \rho_{(k,l),(k,l')}^{[i'-i]} & \text{otherwise.} \end{cases} \quad (5.77)$$

Note that $z_{\ell,k,p}[i]$ is the pth element of the vector $\underline{z}_{\ell,k}[i]$ defined in (5.49). Denote[3]

$$\begin{aligned}
\tilde{\underline{z}}_{kp}[i] &\triangleq \left[z_{1,1,p}[i] \ \cdots \ z_{L,1,p}[i]\right]^T, && (L\text{-vector})\\
\tilde{\underline{n}}_{kp}[i] &\triangleq \left[n_{1,1,p}[i] \ \cdots \ n_{L,1,p}[i]\right]^T, && (L\text{-vector})\\
\underline{\theta}_{kp} &\triangleq [a_{1,k,p} \ \cdots \ a_{L,k,p}]^T, && (L\text{-vector})\\
\underline{\boldsymbol{\Theta}}_p &\triangleq \text{diag}\{\underline{\theta}_{1p}, \ \cdots \underline{\theta}_{Kp}\}, && (KL \times K \text{ matrix})\\
\underline{R}_k^{[j]} &\triangleq \underline{R}^{[j]}[kL-L:kL, \ 1:KL], && (L \times KL \text{ matrix})\\
\tilde{\underline{R}}_k^{[j]} &\triangleq \underline{R}^{[j]}[kL-L:kL, \ kL-L:kL]. && (L \times L \text{ matrix})
\end{aligned}$$

Then we can write (5.76) in the following matrix form:

$$\tilde{\underline{z}}_{kp}[i] = \sum_{j=-\Delta}^{\Delta} \underline{R}_k^{[j]} \left(\underline{\Theta}_p \circ \underline{G}\right) \underline{A}\, \underline{b}[i+j] + \tilde{\underline{n}}_{kp}[i], \quad p = 1, \ldots, P, \tag{5.78}$$

where by (5.77) the complex Gaussian vector sequence $\{\tilde{\underline{n}}_{kp}[i]\}$ has the following covariance matrix:

$$E\left\{\tilde{\underline{n}}_{kp}[i]\, \tilde{\underline{n}}_{kp'}[i']^H\right\} = \begin{cases} \mathbf{0} & \text{if } p \neq p' \text{ or } |i-i'| > \Delta, \\ \tilde{\underline{R}}_k^{[i'-i]} & \text{otherwise.} \end{cases} \tag{5.79}$$

From (5.78) we then have

$$\underbrace{\begin{bmatrix} \tilde{\underline{z}}_{k1}[i] \\ \vdots \\ \tilde{\underline{z}}_{kP}[i] \end{bmatrix}}_{\boldsymbol{z}_k[i]} = \sum_{j=-\Delta}^{\Delta} \underbrace{\begin{bmatrix} \underline{R}_k^{[j]} \left(\underline{\Theta}_1 \circ \underline{G}\right) \\ \vdots \\ \underline{R}_k^{[j]} \left(\underline{\Theta}_P \circ \underline{G}\right) \end{bmatrix}}_{\boldsymbol{\Xi}_k^{[j]}} \underline{A}\, \underline{b}[i+j] + \underbrace{\begin{bmatrix} \underline{n}_{k1}[i] \\ \vdots \\ \underline{n}_{kP}[i] \end{bmatrix}}_{\boldsymbol{n}_k[i]}, \tag{5.80}$$

where, by (5.79), $\boldsymbol{n}_k[i] \sim \mathcal{N}_c\left(\underline{0}_P, \underline{I}_P \otimes \tilde{\underline{R}}_k^{[0]}\right)$.

In the single-user-based linear space-time processing methods, the kth user's ith bit is demodulated according to the following rule:

$$\hat{b}_k[i] = \text{sign}\left[\Re\left\{\boldsymbol{w}_k^H\, \boldsymbol{z}_k[i]\right\}\right], \tag{5.81}$$

where $\boldsymbol{w}_k \in \mathbb{C}^{LP}$. We next consider three choices of the weight vector $\boldsymbol{w}_k$ according to different criteria.

---

[3]Notation: $\underline{R}[i_0 : i_1, j_0 : j_1]$ denotes the submatrix of $\underline{R}$ consisting of rows i_0 to i_1 and columns j_0 to j_1.

Space-Time Matched Filter

The simplest linear combining strategy is the space-time matched filter, which chooses the weight vector as

$$\boldsymbol{w}_k = \boldsymbol{h}_k \triangleq \left[(\underline{\theta}_{k1} \circ \underline{g}_k)^T \; \cdots \; (\underline{\theta}_{kP} \circ \underline{g}_k)^T \right]^T. \tag{5.82}$$

Note that the output of this space-time matched filter is $y_k[i] = \boldsymbol{h}_k^H \, \boldsymbol{z}_k[i]$, a quantity that first appeared in (5.49).

Linear Minimum Mean-Square Error Combiner

In linear MMSE combining, the weight vector is chosen to minimize the mean-square error between the kth user's transmitted signal and the output of the linear combiner:

$$\boldsymbol{w}_k = \arg \min_{\boldsymbol{w} \in \mathbb{C}^{LP}} E\left\{ \left| b_k[i] - \boldsymbol{w}^H \, \boldsymbol{z}_k[i] \right|^2 \right\} \; = \; \boldsymbol{\Sigma}_k^{-1} \, \boldsymbol{p}_k, \tag{5.83}$$

where using (5.80), we have

$$\boldsymbol{\Sigma}_k \triangleq E\left\{ \boldsymbol{z}_k[i] \, \boldsymbol{z}_k[i]^H \right\} \; = \; \sum_{j=-\Delta}^{\Delta} \boldsymbol{\Xi}_k^{[j]} \, \underline{A}^2 \, \boldsymbol{\Xi}_k^{[j]H} + \sigma^2 \, \underline{I}_p \otimes \underline{\tilde{R}}_k^{[0]}, \tag{5.84}$$

$$\boldsymbol{p}_k \triangleq E\left\{ \boldsymbol{z}_k[i] \, b_k[i] \right\} \; = \; \boldsymbol{\Xi}_k^{[0]} \, \boldsymbol{e}_k, \tag{5.85}$$

with $\boldsymbol{e}_k$ a K-vector of all zeros except for the kth entry, which is 1.

Maximum Signal-to-Interference Ratio Combiner

In MSIR combining, the weight vector $\boldsymbol{w}_k$ is chosen to maximize the signal-to-interference ratio,

$$\begin{aligned} \boldsymbol{w}_k &= \arg \max_{\boldsymbol{w} \in \mathbb{C}^{LP}} \frac{\left\| E\left\{ \boldsymbol{w}^H \boldsymbol{z}_k \right\} \right\|^2}{E\left\{ \left\| \boldsymbol{w}^H \boldsymbol{z}_k \right\|^2 \right\} - \left\| E\left\{ \boldsymbol{w}^H \boldsymbol{z}_k \right\} \right\|^2} \\ &= \arg \max_{\boldsymbol{w} \in \mathbb{C}^{LP}} \frac{\boldsymbol{w}^H \, \boldsymbol{p}_k \, \boldsymbol{p}_k^H \, \boldsymbol{w}}{\boldsymbol{w}^H \left(\boldsymbol{\Sigma}_k - \boldsymbol{p}_k \, \boldsymbol{p}_k^H \right) \boldsymbol{w}} \\ &= \arg \max_{\boldsymbol{w} \in \mathbb{C}^{LP}} \frac{\boldsymbol{w}^H \, \boldsymbol{p}_k \, \boldsymbol{p}_k^H \, \boldsymbol{w}}{\boldsymbol{w}^H \, \boldsymbol{\Sigma}_k \, \boldsymbol{w}}. \end{aligned} \tag{5.86}$$

The solution to (5.86) is then given by the generalized eigenvector associated with the largest generalized eigenvalue of the matrix pencil $\left(\boldsymbol{p}_k \, \boldsymbol{p}_k^H, \, \boldsymbol{\Sigma}_k \right)$:

$$\boldsymbol{p}_k \boldsymbol{p}_k^H \boldsymbol{w} = \lambda \boldsymbol{\Sigma}_k \boldsymbol{w}. \tag{5.87}$$

From (5.87) it is immediate that the largest generalized eigenvalue is $\boldsymbol{p}_k^H \boldsymbol{\Sigma}_k^{-1} \boldsymbol{p}_k$, and the corresponding generalized eigenvector is $\boldsymbol{w}_k = \alpha \, \boldsymbol{\Sigma}_k^{-1} \, \boldsymbol{p}_k$, with some scalar constant α. Since scaling the combining weight by a positive constant does not affect the decision (5.81), the MSIR weight vector is the same as the MMSE weight vector (5.83).

The space-time matched filter is data independent (assuming that the array responses and the multipath gains are known) and the single-user MMSE (MSIR) method is data dependent. Hence, in general, the latter outperforms the former. In essence the single-user MMSE method exploits the *spatial signatures* introduced into the different user signals by the array responses and the multipath gains to suppress the interference. For example, such a spatial signature for the kth user is given by (5.82). The kth user's MMSE receiver then correlates the signal vector $\boldsymbol{z}_k[i]$ along a direction in the space spanned by such spatial signatures of all users, such that the SIR of the kth user is maximized. Moreover, this approach admits several interesting blind adaptive implementations, even for systems that employ aperiodic spreading sequences [273, 574].

However, the interference suppression capability of such a single-user approach is limited, since it does not exploit the inherent signal structure induced by the multiuser spreading waveforms. This method can still suffer from the near–far problem, as in matched-filter detection, because the degree of freedom provided by the spatial signature is limited. Furthermore, since users' signals are originally designed to separate from each other by their spreading waveforms, the multiuser space-time approach, which exploits the structure of users' signals in terms of both spreading waveforms and spatial signatures, can significantly outperform the single-user approach. This is illustrated later by simulation examples.

5.4.3 Combined Single-User/Multiuser Linear Detection

The linear space-time multiuser detection methods discussed in Section 5.4.1 are based on the assumption that the receiver has knowledge of the spreading signatures and channel parameters (multipath delays and gains, array responses) of all users. In a practical cellular system, however, there might be a few external interfering signals (e.g., signals from other cells), whose spreading waveforms and channel parameters are not known to the receiver. In this section we consider space-time processing in such a scenario by combining the single-user and multiuser approaches. The basic strategy is first to suppress the known interferers' signals through the iterative interference cancellation technique discussed in Section 5.4.1, and then to apply the single-user MMSE method discussed in Section 5.4.2 to the residual signal to further suppress the unknown interfering signals. Thus these procedures are group space-time multiuser detectors.

Consider the received signal model (5.45). Assume that the users of interest are users $k = 1, \ldots, \tilde{K} < K$ and that the spreading waveforms as well as the channel parameters of these users are known to the receiver. Users $k = \tilde{K}+1, \ldots, K$ are unknown external interferers whose data are not to be demodulated. For each user of interest, the receiver first computes the LP-vectors of matched-filter outputs, $\boldsymbol{z}_k[i]$, $1 \leq k \leq \tilde{K}$, $i = 0, \ldots, M-1$, [cf. (5.80)]. The space-time matched-filter outputs $y_k[i]$ [cf. (5.49)] are then computed by correlating $\boldsymbol{z}_k[i]$ with the space-time matched filter given in (5.82).

Next, the iterative serial interference cancellation algorithm discussed in Section 4.1 (here the total number of users K is replaced by the total number of users of interest $\tilde{K}$) is applied to the data $\{y_k[i] : 1 \le k \le \tilde{K}, i = 0, \dots, M-1\}$ to suppress interference from known users. This is equivalent to implementing a linear multiuser detector assuming only $\tilde{K}$ (instead of K) users present. As a result, only the known interferers' signals are suppressed at the detector output. Denote by $\hat{x}_k[i]$ the converged estimate of the kth user's signal (i.e., $\hat{x}_k[i] \triangleq \lim_{m\to\infty} x_k^m[i]$, $1 \le k \le \tilde{K}$, $i = 0, \dots, M-1$). Note that $\hat{x}_k[i]$ contains the desired user's signal, the unknown interferers' signals, and the ambient noise. Using these estimates and based on the signal model (5.80), we next cancel the known interferers' signals from the vector $\boldsymbol{z}_k[i]$ to obtain

$$\hat{\boldsymbol{z}}_k[i] \triangleq \boldsymbol{z}_k[i] - \sum_{j=-\Delta}^{\Delta} \sum_{\substack{k'=1 \\ (j,k')\neq(0,k)}}^{K_0} \boldsymbol{\Xi}_k^{[j]}\, \boldsymbol{e}_{k'}\, \hat{x}_{k'}[i+j],$$

$$k = 1, \dots, K_0, \quad i = 0, \dots, M-1. \tag{5.88}$$

Finally, a single-user combining weight $\boldsymbol{w}_k$ is applied to the vector $\hat{\boldsymbol{z}}_k[i]$ and the decision rule is given by

$$\hat{b}_k[i] = \text{sign}\left[\Re\left\{\boldsymbol{w}_k^H \hat{\boldsymbol{z}}_k[i]\right\}\right]. \tag{5.89}$$

If the weight vector $\boldsymbol{w}_k$ is chosen to be a scaled version of the matched filter (5.82) [i.e., $\boldsymbol{w}_k = (1/\gamma_k)\boldsymbol{h}_k$], the output of this matched filter is simply $\hat{x}_k[i]$ (i.e., $(1/\gamma_k)\boldsymbol{h}_k^H\, \hat{\boldsymbol{z}}_k[i] = \hat{x}_k[i]$). To see this, first using (5.80) and (5.82), we have the following identity:

$$\begin{aligned}
\boldsymbol{w}_k^H\, \boldsymbol{\Xi}_k^{[j]}\, \boldsymbol{e}_{k'} &= \frac{1}{\gamma_k} \sum_{p=1}^{P} \sum_{\ell=1}^{L} \left(a_{\ell,k,p}^*\, \alpha_{\ell,k}^*\right) \left(\sum_{\ell'=1}^{L} \rho_{(k,\ell)(k',\ell')}^{[j]}\, \alpha_{\ell',k'}\, a_{\ell',k',p} \right) \\
&= \frac{1}{\gamma_k} \sum_{\ell=1}^{L} \sum_{\ell'=1}^{L} \left(\sum_{p=1}^{P} a_{\ell,k,p}^*\, a_{\ell',k',p} \right) \alpha_{\ell,k}^*\, \alpha_{\ell',k'}\, \rho_{(k,\ell)(k',\ell')}^{[j]} \\
&= \frac{1}{\gamma_k} \sum_{\ell=1}^{L} \sum_{\ell'=1}^{L} \underline{a}_{\ell,k}^H\, \underline{a}_{\ell',k'}\, \alpha_{\ell,k}^*\, \alpha_{\ell',k'}\, \rho_{(k,\ell)(k',\ell')}^{[j]} \;=\; \frac{1}{\gamma_k}\, h_{k,k'}^{[j]}.
\end{aligned} \tag{5.90}$$

Now apply the matched filter (5.82) to both sides of (5.88); we have

$$\boldsymbol{w}_k^H\, \hat{\boldsymbol{z}}_k[i] \triangleq \frac{1}{\gamma_k}\left(y_k[i] - \sum_{j=-\Delta}^{\Delta} \sum_{\substack{k'=1 \\ (j,k')\neq(0,k)}}^{K_0} h_{k,k'}^{[j]}\, \hat{x}_{k'}[i+j] \right) \tag{5.91}$$

$$= \hat{x}_k[i], \quad i = 0, \dots, M-1, \quad k = 1, \dots, \tilde{K}, \tag{5.92}$$

where (5.91) follows from (5.90) and $y_k[i] = \gamma_k\, \boldsymbol{h}_k^H\, \hat{\boldsymbol{z}}_k[i]$, and (5.92) follows from the fact that $\{\hat{x}_k[i]\}$ are the converged outputs of the iterative serial interference cancellation algorithm.

On the other hand, if the combining weight is chosen according to the MMSE criterion, it is given by

$$\begin{aligned} \boldsymbol{w}_k &= \arg \min_{\boldsymbol{w} \in \mathbb{C}^{LP}} E\left\{ \left| b_k[i] - \boldsymbol{w}_k^H \, \boldsymbol{z}_k[i] \right|^2 \right\} \\ &= E\left\{ \hat{\boldsymbol{z}}_k[i] \, \hat{\boldsymbol{z}}_k[i]^H \right\}^{-1} \cdot E\left\{ \hat{\boldsymbol{z}}_k[i] \, b_k[i] \right\} \\ &\cong \left(\sum_{i=0}^{M-1} \hat{\boldsymbol{z}}_k[i] \, \hat{\boldsymbol{z}}_k[i]^H \right)^{-1} \boldsymbol{p}_k. \end{aligned} \tag{5.93}$$

It is clear from the discussion above that in this combined approach, interference due to the known users is suppressed by serial multiuser interference cancellation, whereas the residual interference due to the unknown users is suppressed by the single-user MMSE combiner.

Simulation Examples

In what follows we use computer simulations to assess the performance of the various multiuser and single-user space-time processing methods discussed in this section. We first outline the simulated system in Examples 1, 2, and 3. It consists of eight users ($K = 8$) with a spreading gain 16 ($N = 16$). Each user's propagation channel consists of three paths ($L = 3$). The receiver employs a linear antenna array with three elements ($P = 3$) and half-wavelength spacing. Let the direction of arrival (DOA) of the kth user's signal along the ℓth path with respect to the antenna array be $\phi_{\ell,k}$; then, assuming a uniform linear array with half-wavelength space [i.e., (1.17) with $d = \lambda/2$], the array response is given by

$$a_{\ell,k,p} = \exp\left\{ \jmath(p-1)\pi \sin\left(\phi_{\ell,k}\right) \right\}. \tag{5.94}$$

The spreading sequences, multipath delays, complex gains, and DOAs of all user signals in the simulated system are tabulated in Table 5.2. These parameters are randomly generated and kept fixed for all the simulations. All users have equal transmitted power (i.e., $A_1 = \cdots = A_K$). However, the received signal powers are unequal, due to the unequal strength of the multipath gain for each user. The total strength of each user's multipath channel, measured by the norm of the channel gain vector $\|\underline{g}_k\|$, is also listed in Table 5.2. Note that this system has a near–far situation (i.e., user 3 is the weakest user and user 6 is the strongest).

Example 1: Performance Comparison of Multiuser versus Single-User Space-Time Processing We first compare the BER performance of the multiuser linear space-time detector and that of the single-user linear space-time detector. Three receivers are considered: the single-user space-time matched filter given by (5.82), the single-user space-time MMSE receiver given by (5.83), and the multiuser MMSE receiver implemented by the iterative interference cancellation algorithm (5.70) (five iterations are used). Figure 5.5 shows the performance of the weak users (users 1, 3, 4, and 8). It is seen that, in general, the single-user MMSE receiver outperforms the matched-filter receiver. (Interestingly though, for user

Table 5.2. Simulated Multipath CDMA System for Examples 1, 2, 3, 6, and 8

k	Signature $\{c_k(j)\}$	Delay (T_c)			DOA (°)			Multipath Gain			
		τ_{k1}	τ_{k2}	τ_{k3}	α_{k1}	α_{k2}	α_{k3}	g_{k1}	g_{k2}	g_{k3}	$\|\underline{g}_k\|$
1	0011011100000011	0	2	3	34	−16	−14	$0.193 - j\,0.714$	$0.131 - j\,0.189$	$0.353 - j\,0.079$	0.85
2	1101011100011101	1	4	5	2	42	−9	$0.508 - j\,0.113$	$-0.103 + j\,0.807$	$0.143 + j\,0.013$	0.98
3	1110110000110001	2	3	6	−33	−13	35	$0.125 - j\,0.064$	$0.187 - j\,0.249$	$-0.196 + j\,0.092$	0.41
4	0100001101111100	2	4	5	58	13	61	$0.354 - j\,0.121$	$0.141 - j\,0.455$	$-0.618 + j\,0.004$	0.87
5	0011101101100000	4	6	7	−72	69	1	$0.597 + j\,0.395$	$0.470 + j\,0.115$	$-0.069 + j\,0.255$	0.90
6	1111111001100001	5	7	8	3	18	−55	$0.084 + j\,1.205$	$0.106 - j\,0.181$	$0.167 + j\,0.007$	1.24
7	1110010000010010	6	8	9	−79	−53	70	$-0.428 + j\,0.188$	$-0.711 + j\,0.064$	$0.562 - j\,0.111$	1.03
8	1101101001011000	8	9	12	53	25	−20	$-0.575 + j\,0.018$	$-0.320 + j\,0.081$	$-0.139 + j\,0.199$	0.70

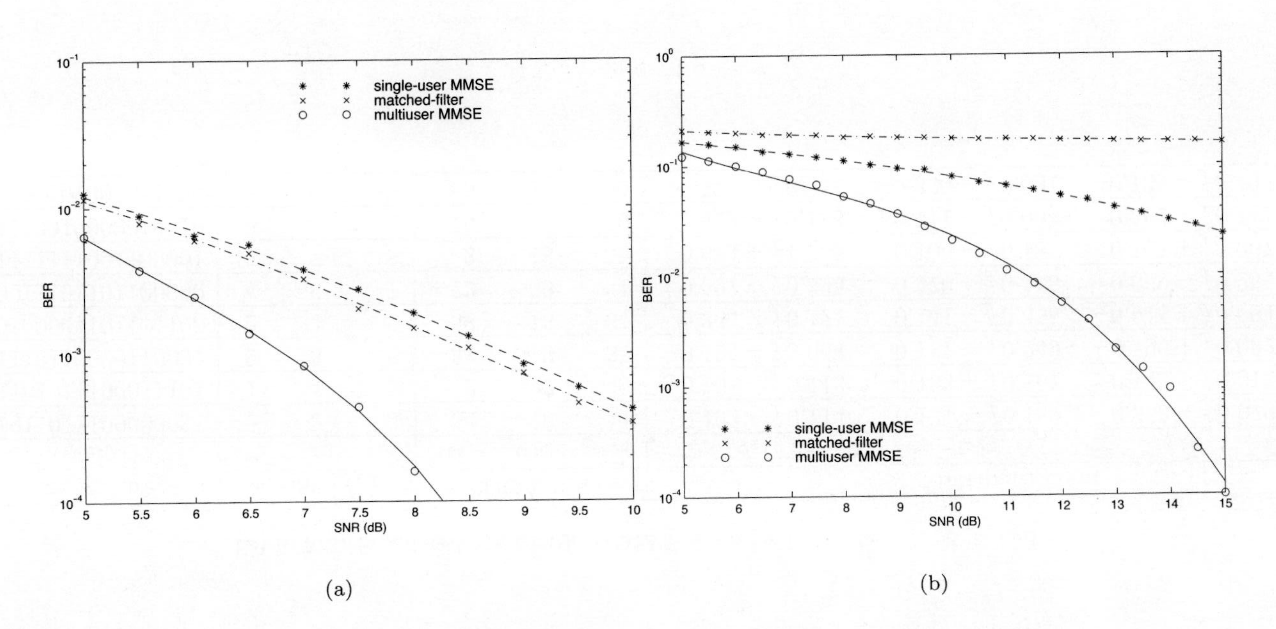

Figure 5.5. Comparisons of the BER performance of three receivers: single-user space-time matched-filter, single-user space-time MMSE receiver, and multiuser space-time MMSE receiver. (a) User 1; (b) user 3; (c) user 4; (d) user 8.

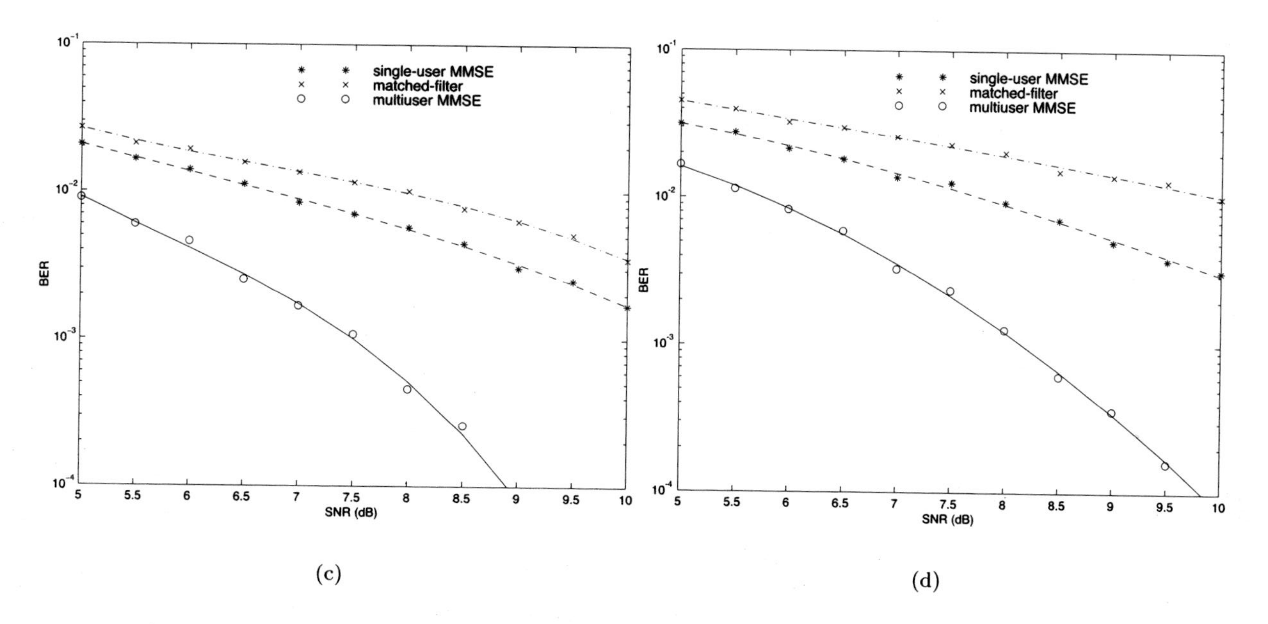

Figure 5.5. *Continued*

1 the matched filter actually slightly outperforms the single-user MMSE receiver. This is not surprising, since due to the interference, the detector output distribution is not Gaussian, and minimizing the mean-square error does not necessarily lead to minimum bit-error probability.) It is also evident that the multiuser approach offers substantial performance improvement over single-user methods.

Example 2: Convergence of the Iterative Interference Cancellation Method This example serves to illustrate the convergence behavior of the iterative interference cancellation method (5.70). The BER performance corresponding to the first four iterations for users 4 and 8 is shown in Fig. 5.6. It is seen that the algorithm converges within four or five iterations. It is also seen that the most significant performance improvement occurs at the second iteration.

Example 3: Performance of the Combined Multiuser/Single-User Space-Time Processing In this example it is assumed that users 7 and 8 are external interferers and that their signature waveforms and channel parameters are not known to the receiver. Therefore, the combined multiuser/single-user space-time processing method discussed in Section 5.4.3 is employed at the receiver. Figure 5.7 illustrates the BER performance of users 3 and 4. Four methods are considered here: the single-user matched filter (5.82), single-user MMSE receiver (5.83), and partial interference cancellation followed by a matched-filter or single-user MMSE receiver. It is seen that the combined multiuser/single-user space-time processing achieves the best performance among the four methods.

Example 4: Performance versus Number of Antennas/Number of Users Next, we illustrate how performance varies with the number of users and receive antennas for both the multiuser space-time detector and single-user space-time detector. The simulated system has $K = 16$ users and the processing gain $N = 16$. The number of paths for each user is $L = 3$. The performance of user 5 in this system using the single-user MMSE receiver and that using the multiuser MMSE receiver are plotted in Fig. 5.8, with the number of antennas $P = 2, 4$, and 6. It is seen that whereas in the single-user approach, the performance improvement due to the increasing number of receive antennas is only marginal, such performance improvement in the multiuser approach is of orders of magnitude. Next, we fix the number of antennas as $P = 4$ and vary the number of users in the system. The processing gain is still $N = 16$ and the number of paths for each user is $L = 3$. The performance of the single-user MMSE receiver and the multiuser receiver for user 2 is plotted in Fig. 5.9, with the number of users $N = 10, 20$, and 30. Again we see that in all these cases the multiuser method significantly outperforms the single-user method.

Example 5: Performance versus Spreading Gain/Number of Antennas In this example we consider the performance of the single-user and multiuser methods by varying the processing gain N and number of receive antennas P while keeping their product (NP) fixed. The simulated system has $K = 16$ users, and the number of paths for each user is $L = 3$. Three cases are simulated: $N = 64, P = 1$;

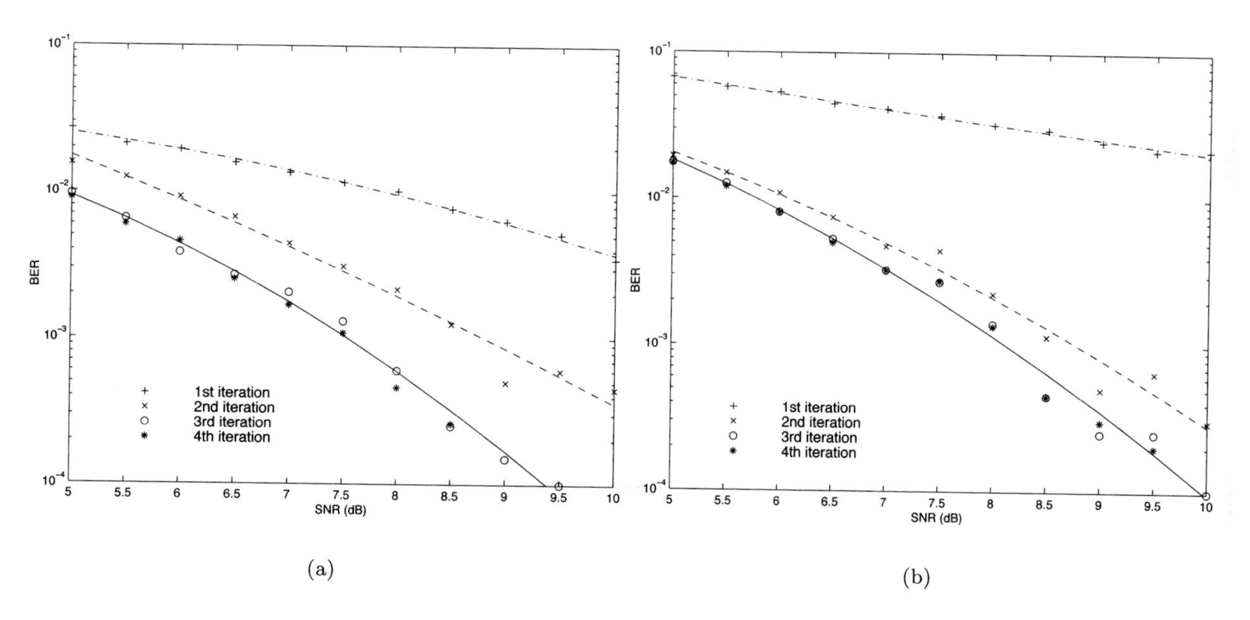

Figure 5.6. BER performance of the iterative interference cancellation method (first four iterations). (a) User 4; (b) user 8.

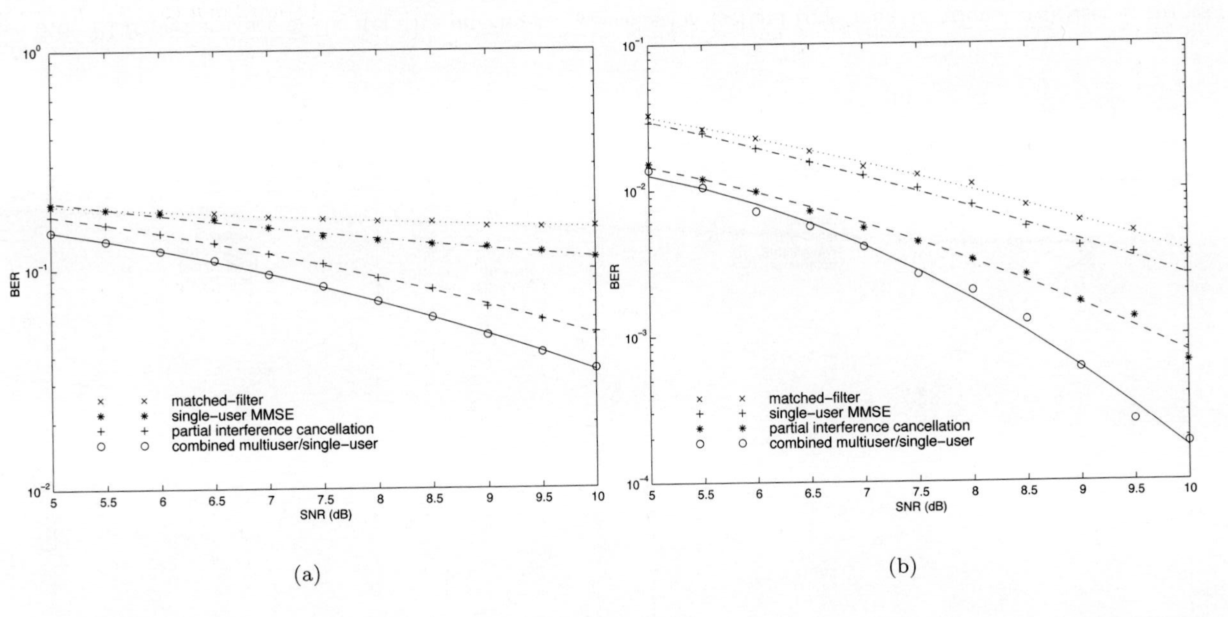

Figure 5.7. BER performance of four receivers in the presence of unknown interferers. (a) User 3; (b) user 4.

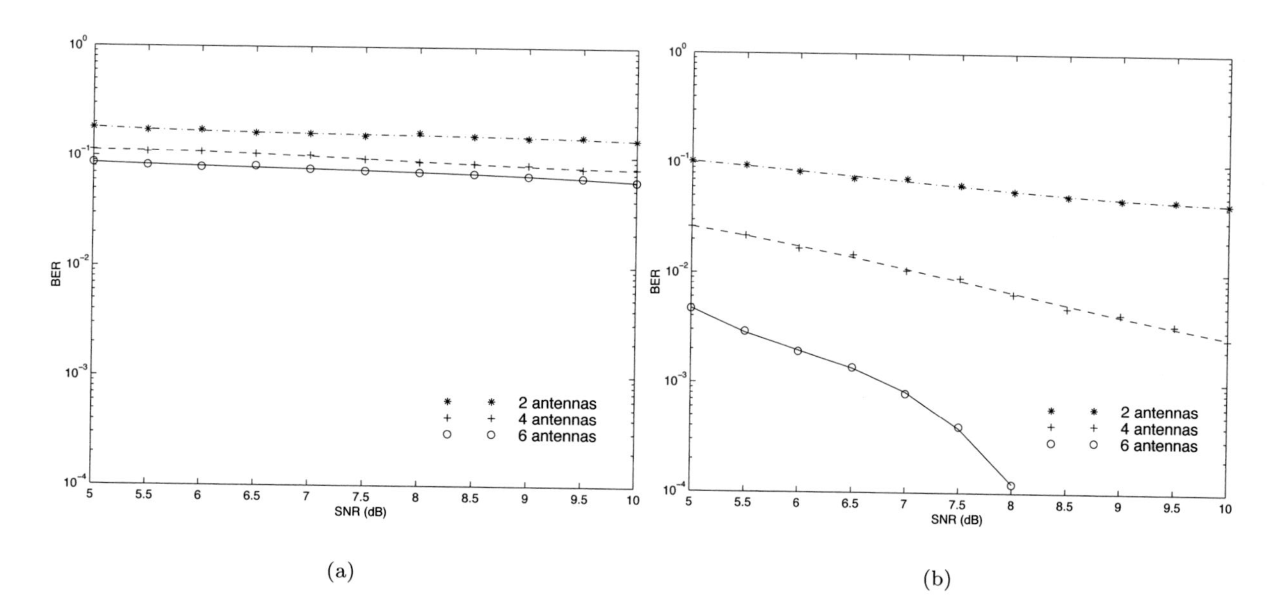

(a)

(b)

Figure 5.8. Single-user and multiuser receiver performance under a different number of antennas $K = 16, N = 16$. (a) Single-user MMSE receiver, user 5; (b) multiuser MMSE receiver, user 5.

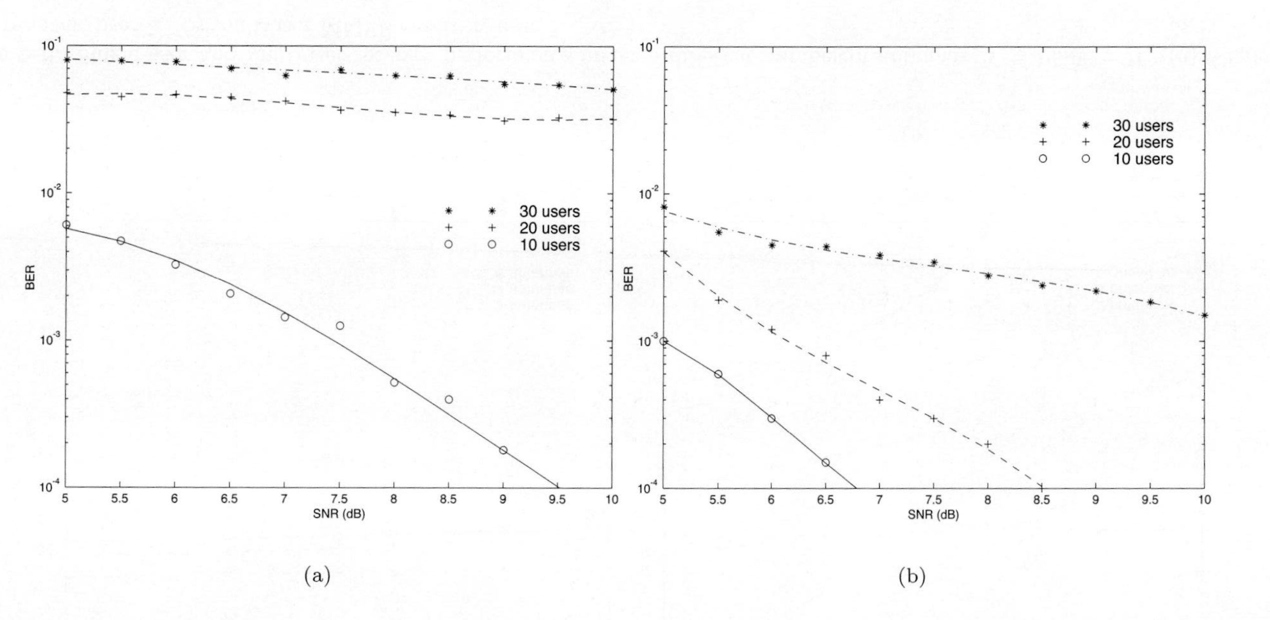

Figure 5.9. Single-user and multiuser receiver performance under different number of users. $N = 16, P = 4$. (a) Single-user MMSE receiver, user 2; (b) multiuser MMSE receiver, user 2.

$N = 32, P = 2$; and $N = 16, P = 4$. The performance for user 2 is shown in Fig. 5.10. It is seen that in this case, this user's signal is best separated from others when $N = 16, P = 4$ for both the single-user and multiuser methods. Moreover, the multiuser approach offers orders-of-magnitude performance improvement over the single-user method.

In summary, in this and the preceding section we have discussed multiuser space-time receiver structures based on the sufficient statistic, which is illustrated in Fig. 5.11. It is seen that the front end of the receiver consists of a bank of matched filters, followed by a bank of array combiners, followed by a bank of multipath combiners, which produces the sufficient statistic. The maximum-likelihood multiuser sequence detector and linear multiuser detectors based on serial iterative interference cancellation have been derived. Note that since the detection algorithms in Fig. 5.11 operate on the sufficient statistic, their complexities are functions of only the number of users (K) and the length of the data block (M), not of the number of antennas (P).

5.5 Adaptive Space-Time Multiuser Detection in Synchronous CDMA

Generally speaking, space-time processing involves the exploitation of spatial diversity using multiple transmit and/or receive antennas and, perhaps, some form of coding. In previous sections we have focused on systems that employ one transmit antenna and multiple receive antennas. Recently, however, much of the work in this area has focused on transmit diversity schemes that use multiple transmit antennas. They include delay schemes [444, 572, 573] in which copies of the same symbol are transmitted through multiple antennas at different times, the space-time trellis coding algorithm in [477], and the simple space-time block coding (STBC) scheme developed in [12], which has been adopted in third-generation (3G) wideband CDMA (WCDMA) standards [294, 479]. A generalization of this simple space-time block coding concept is developed in [475, 476]. It has been shown that these techniques can significantly increase capacity [122, 478].

In this section we discuss adaptive receiver structures for synchronous CDMA systems with multiple transmit antennas and multiple receive antennas. Specifically, we focus on three configurations: (1) one transmit antenna, two receive antennas; (2) two transmit antennas, one receive antenna; and (3) two transmit antennas, two receive antennas. It is assumed that the orthogonal space-time block code [12] is employed in systems with two transmit antennas. For each of these configurations, we discuss two possible linear receiver structures and compare their performance in terms of diversity gain and signal separation capability. We also describe blind adaptive receiver structures for such multiple-antenna CDMA systems. The methods discussed in this section are generalized in the next section to mutipath CDMA systems. The materials discussed in this and the following sections first appeared in [415].

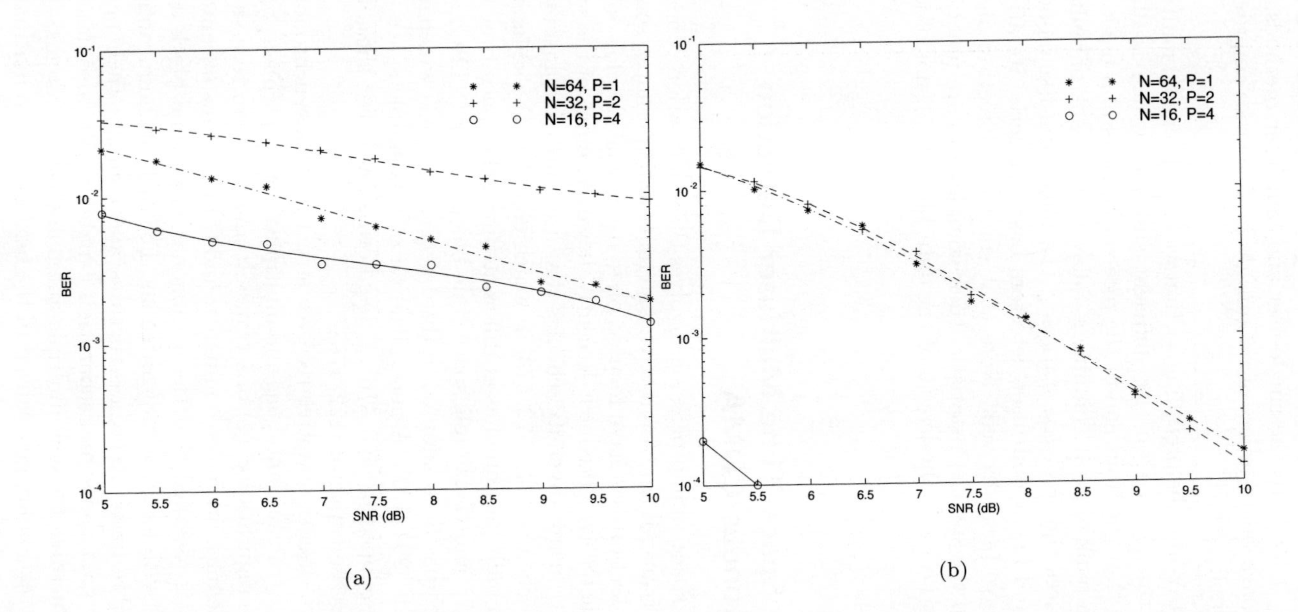

Figure 5.10. Single-user and multiuser receiver performance under different space-time gains. $K = 16$. (a) Single-user MMSE receiver, user 2; (b) multiuser MMSE receiver, user 2.

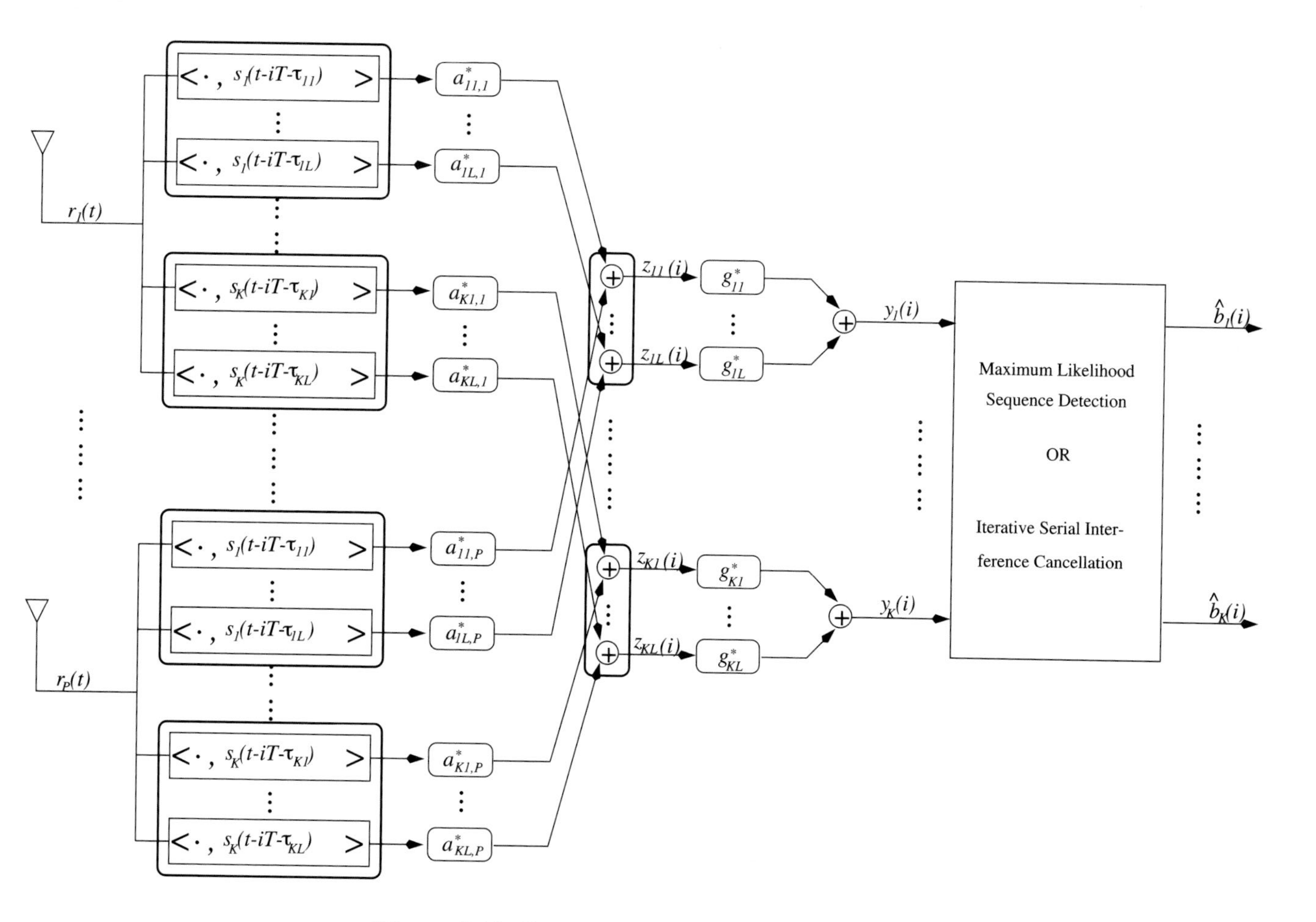

Figure 5.11. Space-time multiuser receiver structure.

5.5.1 One Transmit Antenna, Two Receive Antennas

Consider the following discrete-time K-user synchronous CDMA channel with one transmit antenna and two receive antennas. The received baseband signal at the pth antenna can be modeled as

$$\boldsymbol{r}_p = \sum_{k=1}^{K} g_{p,k}\, b_k\, \boldsymbol{s}_k + \boldsymbol{n}_p, \quad p = 1, 2, \tag{5.95}$$

where $\boldsymbol{s}_k$ is the N-vector of the discrete-time signature waveform of the kth user with unit norm (i.e., $\|\boldsymbol{s}_k\| = 1$), $b_k \in \{+1, -1\}$ is the data bit of the kth user, $g_{p,k}$ is the complex channel response of the pth receive antenna element to the kth user's signal, and $\boldsymbol{n}_p \sim \mathcal{N}_c(\boldsymbol{0}, \sigma^2 \boldsymbol{I}_N)$ is the ambient noise vector at antenna p. It is assumed that $\boldsymbol{n}_1$ and $\boldsymbol{n}_2$ are independent.

Linear Diversity Multiuser Detector

Denote

$$\begin{aligned} \boldsymbol{g}_k &\triangleq [g_{1,k}\ g_{2,k}]^T, \\ \boldsymbol{S} &\triangleq [\boldsymbol{s}_1\ \cdots\ \boldsymbol{s}_K], \\ \boldsymbol{R} &\triangleq \boldsymbol{S}^T \boldsymbol{S}. \end{aligned}$$

Suppose that user 1 is the user of interest. We first consider the linear diversity multiuser detection scheme, which first applies a linear multiuser detector to the received signal $\boldsymbol{r}_p$ in (5.95) at each antenna $p = 1, 2$, and then combines the outputs of these linear detectors to make a decision. For example, a linear decorrelating detector for user 1 based on the signal in (5.95) is simply

$$\boldsymbol{w}_1 = \boldsymbol{S}\boldsymbol{R}^{-1}\boldsymbol{e}_1, \tag{5.96}$$

where $\boldsymbol{e}_1$ denotes the first unit vector in $\mathbb{R}^K$. This detector is applied to the received signal at each antenna $p = 1, 2$, to obtain $\boldsymbol{z} = [z_1\ z_2]^T$, where

$$z_p \ \triangleq\ \boldsymbol{w}_1^T \boldsymbol{r}_p = g_{p,1}\, b_1 + u_p, \tag{5.97}$$

with

$$u_p \ \triangleq\ \boldsymbol{w}_1^T \boldsymbol{n}_p \sim \mathcal{N}_c\left(\boldsymbol{0}, \sigma^2 \|\boldsymbol{w}_1\|^2\right), \qquad p = 1, 2, \tag{5.98}$$

where $\|\boldsymbol{w}_1\|^2 = \left[\boldsymbol{R}^{-1}\right]_{1,1}$. Denote

$$\eta_1 \triangleq \frac{1}{\sqrt{\left[\boldsymbol{R}^{-1}\right]_{1,1}}} \tag{5.99}$$

and $\boldsymbol{g}_k \triangleq [g_{1,k}\ g_{2,k}]^T$. Since the noise vectors from different antennas are independent, we can write

$$\boldsymbol{z} = b_1 \boldsymbol{g}_1 + \boldsymbol{u}, \tag{5.100}$$

with

$$\boldsymbol{u} \sim \mathcal{N}_c\left(\mathbf{0}, \frac{\sigma^2}{\eta_1^2} \cdot \boldsymbol{I}_2\right). \tag{5.101}$$

The maximum-likelihood (ML) decision rule for b_1 based on $\boldsymbol{z}$ in (5.100) is then

$$\hat{b}_1 = \text{sign}\left(\Re\left\{\boldsymbol{g}_1^H \boldsymbol{z}\right\}\right). \tag{5.102}$$

Let $E_1 \triangleq \|\boldsymbol{g}_1\|^2$ be the total received desired user's signal energy. The decision statistic in (5.102) can be expressed as

$$\xi \triangleq \boldsymbol{g}_1^H \boldsymbol{z} = E_1 b_1 + v, \tag{5.103}$$

with

$$v \triangleq \boldsymbol{g}_1^H \boldsymbol{u} \sim \mathcal{N}_c\left(0, E_1 \sigma^2 / \eta_1^2\right). \tag{5.104}$$

The probability of detection error is computed as

$$\begin{aligned} P_1^{\text{DC}}(e) &= P\Big(\Re\left\{\xi\right\} < 0 \mid b_1 = 1\Big) \\ &= P\Big(\Re\{v\} < -E_1\Big) \\ &= Q\left(\frac{\sqrt{2E_1}}{\sigma} \cdot \eta_1\right). \end{aligned} \tag{5.105}$$

Linear Space-Time Multiuser Detector

Denote

$$\begin{aligned} \boldsymbol{b} &\triangleq [b_1 \cdots b_K]^T, \\ \boldsymbol{G} &\triangleq [\boldsymbol{g}_1 \cdots \boldsymbol{g}_K], \\ \tilde{\boldsymbol{s}}_k &\triangleq \boldsymbol{g}_k \otimes \boldsymbol{s}_k, \\ \tilde{\boldsymbol{S}} &\triangleq [\tilde{\boldsymbol{s}}_1 \cdots \tilde{\boldsymbol{s}}_K], \\ \tilde{\boldsymbol{R}} &\triangleq \tilde{\boldsymbol{S}}^T \tilde{\boldsymbol{S}}, \\ \tilde{\boldsymbol{r}} &\triangleq \left[\boldsymbol{r}_1^T\ \boldsymbol{r}_2^T\right]^T, \\ \tilde{\boldsymbol{n}} &\triangleq \left[\boldsymbol{n}_1^T\ \boldsymbol{n}_2^T\right]^T. \end{aligned}$$

Then, by augmenting the received signals at two antennas, (5.95) can be written as

$$\begin{aligned}\tilde{\boldsymbol{r}} &= \sum_{k=1}^{K} b_k \tilde{\boldsymbol{s}} + \tilde{\boldsymbol{n}} \\ &= \tilde{\boldsymbol{S}}\boldsymbol{b} + \tilde{\boldsymbol{n}},\end{aligned} \tag{5.106}$$

with $\tilde{\boldsymbol{n}} \sim \mathcal{N}_c(\boldsymbol{0}, \sigma^2 \boldsymbol{I}_{2N})$. A linear space-time multiuser detector operates on the augmented received signal $\tilde{\boldsymbol{r}}$ directly. For example, the linear decorrelating detector for user 1 in this case is given by

$$\tilde{\boldsymbol{w}}_1 = \tilde{\boldsymbol{S}}\tilde{\boldsymbol{R}}^{-1}\boldsymbol{e}_1. \tag{5.107}$$

This detector is applied to the augmented received signal $\tilde{\boldsymbol{r}}$ to obtain

$$\tilde{z} \triangleq \tilde{\boldsymbol{w}}_1^H \tilde{\boldsymbol{r}} = b_1 + \tilde{u}, \tag{5.108}$$

with

$$\tilde{u} \triangleq \tilde{\boldsymbol{w}}_1^H \tilde{\boldsymbol{n}} \sim \mathcal{N}_c\left(\boldsymbol{0}, \sigma^2 \|\tilde{\boldsymbol{w}}_1\|^2\right), \tag{5.109}$$

where $\|\tilde{\boldsymbol{w}}_1\|^2 = \left[\tilde{\boldsymbol{R}}^{-1}\right]_{1,1}$. Denote

$$\tilde{\eta}_1 \triangleq \frac{1}{\sqrt{E_1}} \cdot \frac{1}{\sqrt{\left[\tilde{\boldsymbol{R}}^{-1}\right]_{1,1}}}. \tag{5.110}$$

An expression for $\tilde{\boldsymbol{R}}$ can be found as follows. Note that

$$\begin{aligned}[\tilde{\boldsymbol{R}}]_{i,j} \triangleq \left[\tilde{\boldsymbol{S}}^H \tilde{\boldsymbol{S}}\right]_{i,j} &= \tilde{\boldsymbol{s}}_i^H \tilde{\boldsymbol{s}}_j \\ &= (\boldsymbol{g}_i \otimes \boldsymbol{s}_i)^H (\boldsymbol{g}_j \otimes \boldsymbol{s}_j) & (5.111) \\ &= (\boldsymbol{g}_i^H \otimes \boldsymbol{s}_i^H)(\boldsymbol{g}_j \otimes \boldsymbol{s}_j) & (5.112) \\ &= (\boldsymbol{g}_i^H \boldsymbol{g}_j) \otimes (\boldsymbol{s}_i^H \boldsymbol{s}_j) \\ &= \left[\boldsymbol{G}^H \boldsymbol{G}\right]_{i,j} \cdot [\boldsymbol{R}]_{i,j},\end{aligned}$$

where (5.111) and (5.112) follow, respectively, from the following two matrix identities:

$$(\boldsymbol{A} \otimes \boldsymbol{B})^H = \boldsymbol{A}^H \otimes \boldsymbol{B}^H, \tag{5.113}$$

$$(\boldsymbol{A} \otimes \boldsymbol{B})(\boldsymbol{C} \otimes \boldsymbol{D}) = (\boldsymbol{A}\boldsymbol{C}) \otimes (\boldsymbol{B}\boldsymbol{D}). \tag{5.114}$$

Hence

$$\tilde{\boldsymbol{R}} \triangleq \tilde{\boldsymbol{S}}^H \tilde{\boldsymbol{S}} = \boldsymbol{R} \circ \left(\boldsymbol{G}^H \boldsymbol{G}\right), \tag{5.115}$$

where $\circ$ denotes the Schur matrix product (i.e., elementwise product).

The ML decision rule for b_1 based on $\tilde{z}$ in (5.108) is then

$$\hat{b}_1 = \text{sign}\left(\Re\{\tilde{z}\}\right). \tag{5.116}$$

The probability of detection error is computed as

$$\begin{aligned} P_1^{\text{ST}}(e) &= P\Big(\Re\{\tilde{z}\} < 0 \mid b_1 = 1\Big) \\ &= P\Big(\Re\{\tilde{u}\} < -1\Big) \\ &= Q\left(\frac{\sqrt{2E_1}}{\sigma} \cdot \tilde{\eta}_1\right). \end{aligned} \tag{5.117}$$

Performance Comparison

From the discussion above it is seen that the linear space-time multiuser detector exploits the signal structure in both the time domain (i.e., induced by the signature waveform $\boldsymbol{s}_k$) and the spatial domain (i.e., induced by the channel response $\boldsymbol{g}_k$) for interference rejection; whereas for the linear diversity multiuser detector, interference rejection is performed only in the time domain, and the spatial domain is used only for diversity combining. The next result, which first appeared in [324], shows that the linear space-time multiuser detector always outperforms the linear diversity multiuser detector.

Proposition 5.6: *Let $P_1^{\text{DC}}(e)$ given by (5.105) and $P_k^{\text{ST}}(e)$ given by (5.117) be, respectively, the probability of detection error of the linear diversity detector and the linear space-time detector. Then*

$$P_1^{\text{ST}}(e) \leq P_1^{\text{DC}}(e).$$

Proof: By (5.105) and (5.117) it suffices to show that

$$\left[\tilde{\boldsymbol{R}}^{-1}\right]_{1,1} \leq \frac{1}{E_1}\left[\boldsymbol{R}^{-1}\right]_{1,1}.$$

We make use of the following facts. Denote by $\boldsymbol{A}_{i,j}$ the submatrix of $\boldsymbol{A}$ obtained by striking out the ith row and the jth column. Then it is known that

$$\boldsymbol{A} \succeq \boldsymbol{0} \Longrightarrow \boldsymbol{A} - \frac{\boldsymbol{1}}{\left[\boldsymbol{A}^{-1}\right]_{\mathbf{k},\mathbf{k}}} \boldsymbol{e}_{\mathbf{k}} \boldsymbol{e}_{\mathbf{k}}^{\mathbf{T}} \succeq \boldsymbol{0}. \tag{5.118}$$

It is also known that

$$\boldsymbol{A} \succeq \boldsymbol{0},\ \boldsymbol{B} \succeq \boldsymbol{0} \Longrightarrow \boldsymbol{A} \circ \boldsymbol{B} \succeq \boldsymbol{0}. \tag{5.119}$$

Assuming that $\boldsymbol{R} \succeq \mathbf{0}$ and $\boldsymbol{Q} \triangleq \boldsymbol{H}^H \boldsymbol{H} \succeq \mathbf{0}$, and using the two results above, we have

$$\begin{aligned}
\mathbf{0} \le \det & \left[\left(\boldsymbol{R} - \frac{1}{[\boldsymbol{R}^{-1}]_{1,1}} \boldsymbol{e}_1 \boldsymbol{e}_1^T \right) \circ \boldsymbol{Q} \right] \\
&= \det \left[\tilde{\boldsymbol{R}} - \frac{E_1}{[\boldsymbol{R}^{-1}]_{1,1}} \boldsymbol{e}_1 \boldsymbol{e}_1^T \right] && (5.120) \\
&= \det \tilde{\boldsymbol{R}} \left(1 - \frac{E_1}{[\boldsymbol{R}^{-1}]_{k,k}} \boldsymbol{e}_k^T \tilde{\boldsymbol{R}}^{-1} \boldsymbol{e}_1 \right) && (5.121) \\
&= \det \tilde{\boldsymbol{R}} - \frac{E_1}{[\boldsymbol{R}^{-1}]_{1,1}} \det \tilde{\boldsymbol{R}}_{1,1}, && (5.122)
\end{aligned}$$

where (5.120) follows from the fact that $\tilde{\boldsymbol{R}} \triangleq \boldsymbol{R} \circ \boldsymbol{Q}$ and $\left(\boldsymbol{e}_1 \boldsymbol{e}_1^T\right) \circ \boldsymbol{Q} = E_1 \boldsymbol{e}_1 \boldsymbol{e}_1^T$; (5.121) follows from the matrix identity

$$\det\left(\boldsymbol{A} + \boldsymbol{B}\boldsymbol{C}\boldsymbol{D}\right) = \det \boldsymbol{A} \det \boldsymbol{C} \det\left(\boldsymbol{C}^{-1} + \boldsymbol{D}\boldsymbol{A}^{-1}\boldsymbol{B}\right); \qquad (5.123)$$

and (5.122) follows from

$$\boldsymbol{e}_1^T \tilde{\boldsymbol{R}}^{-1} \boldsymbol{e}_1 = \left[\tilde{\boldsymbol{R}}^{-1}\right]_{1,1} = \frac{\det \tilde{\boldsymbol{R}}_{1,1}}{\det \tilde{\boldsymbol{R}}}. \qquad (5.124)$$

Hence we have

$$\frac{1}{\left[\tilde{\boldsymbol{R}}^{-1}\right]_{1,1}} = \frac{\det \tilde{\boldsymbol{R}}}{\det \tilde{\boldsymbol{R}}_{1,1}} \ge \frac{E_1}{\left[\boldsymbol{R}^{-1}\right]_{1,1}}. \qquad (5.125)$$

□

We next consider a simple example to demonstrate the performance difference between the two receivers discussed above. Consider a two-user system with

$$\boldsymbol{R} = \begin{bmatrix} 1 & \rho \\ \rho & 1 \end{bmatrix}, \quad \boldsymbol{G} = \begin{bmatrix} 1 & 1 \\ e^{\jmath\theta_1} & e^{\jmath\theta_2} \end{bmatrix},$$

where ρ is the correlation of the signature waveforms of the two users and θ_1 and θ_2 are the directions of arrival of the two users' signals. Define $\alpha \triangleq \theta_2 - \theta_1$. Then we have $E_1 = E_2 = 1$ and

$$\boldsymbol{Q} \triangleq \boldsymbol{G}^H \boldsymbol{G} = \begin{bmatrix} 2 & 1 + e^{\jmath\alpha} \\ 1 + e^{-\jmath\alpha} & 2 \end{bmatrix}, \qquad (5.126)$$

$$\tilde{\boldsymbol{R}} \triangleq \boldsymbol{R} \circ \boldsymbol{Q} = \begin{bmatrix} 2 & \rho(1 + e^{\jmath\alpha}) \\ \rho(1 + e^{-\jmath\alpha}) & 2 \end{bmatrix}, \qquad (5.127)$$

$$\eta_1 \triangleq \frac{1}{\sqrt{\left[\boldsymbol{R}^{-1}\right]_{1,1}}} = \sqrt{1-\rho^2}, \tag{5.128}$$

$$\tilde{\eta}_1 \triangleq \frac{1}{\sqrt{2}} \cdot \frac{1}{\sqrt{\left[\tilde{\boldsymbol{R}}^{-1}\right]_{1,1}}} = \sqrt{1-\rho^2 \cos^2 \frac{\alpha}{2}}. \tag{5.129}$$

These expressions are plotted in Fig. 5.12. It is seen that while the multiuser space-time receiver can exploit both the temporal signal separation (along the ρ-axis) and the spatial signal separation (along the α-axis), the multiuser diversity receiver can exploit only the temporal signal separation. For example, for large ρ, the performance of the multiuser diversity receiver is poor, no matter what value α takes; but the performance of the multiuser space-time receiver can be quite good as long as α is large.

5.5.2 Two Transmit Antennas, One Receive Antenna

When two antennas are employed at the transmitter, we must first specify how the information bits are transmitted across the two antennas. Here we adopt the well-known orthogonal space-time block coding scheme [12, 475]. Specifically, for user k, two information symbols, $b_{k,1}$ and $b_{k,2}$, are transmitted over two symbol intervals. At the first time interval, the symbol pair $(b_{k,1}, b_{k,2})$ is transmitted across the two transmit antennas; and at the second time interval, the symbol pair $(-b_{k,2}, b_{k,1})$ is transmitted. The received signals corresponding to these two time intervals are given by

$$\boldsymbol{r}_1 = \sum_{k=1}^{K} \Big(g_{1,k}\, b_{1,k} + g_{2,k}\, b_{2,k} \Big) \boldsymbol{s}_k + \boldsymbol{n}_1, \tag{5.130}$$

$$\boldsymbol{r}_2 = \sum_{k=1}^{K} \Big(- g_{1,k}\, b_{2,k} + g_{2,k}\, b_{1,k} \Big) \boldsymbol{s}_k + \boldsymbol{n}_2, \tag{5.131}$$

where $g_{1,k}$ ($g_{2,k}$) is the complex channel response between the first (second) transmit antenna and the receive antenna; $\boldsymbol{n}_1$ and $\boldsymbol{n}_2$ are independent received $\mathcal{N}_c(\boldsymbol{0}, \boldsymbol{I}_N)$ noise vectors at the two time intervals.

Linear Diversity Multiuser Detector

We first consider the linear diversity multiuser detection scheme, which first applies the linear multiuser detector $\boldsymbol{w}_1$ in (5.96) to the received signals $\boldsymbol{r}_1$ and $\boldsymbol{r}_2$ during the two time intervals, and then performs a space-time decoding. Specifically, denote

$$z_1 \triangleq \boldsymbol{w}_1^H \boldsymbol{r}_1 = g_{1,k}\, b_{1,k} + g_{2,k}\, b_{2,k} + u_1, \tag{5.132}$$

$$z_2 \triangleq \left(\boldsymbol{w}_1^H \boldsymbol{r}_2\right)^* = -g_{1,k}^*\, b_{2,k} + g_{2,k}^*\, b_{1,k} + u_2^*, \tag{5.133}$$

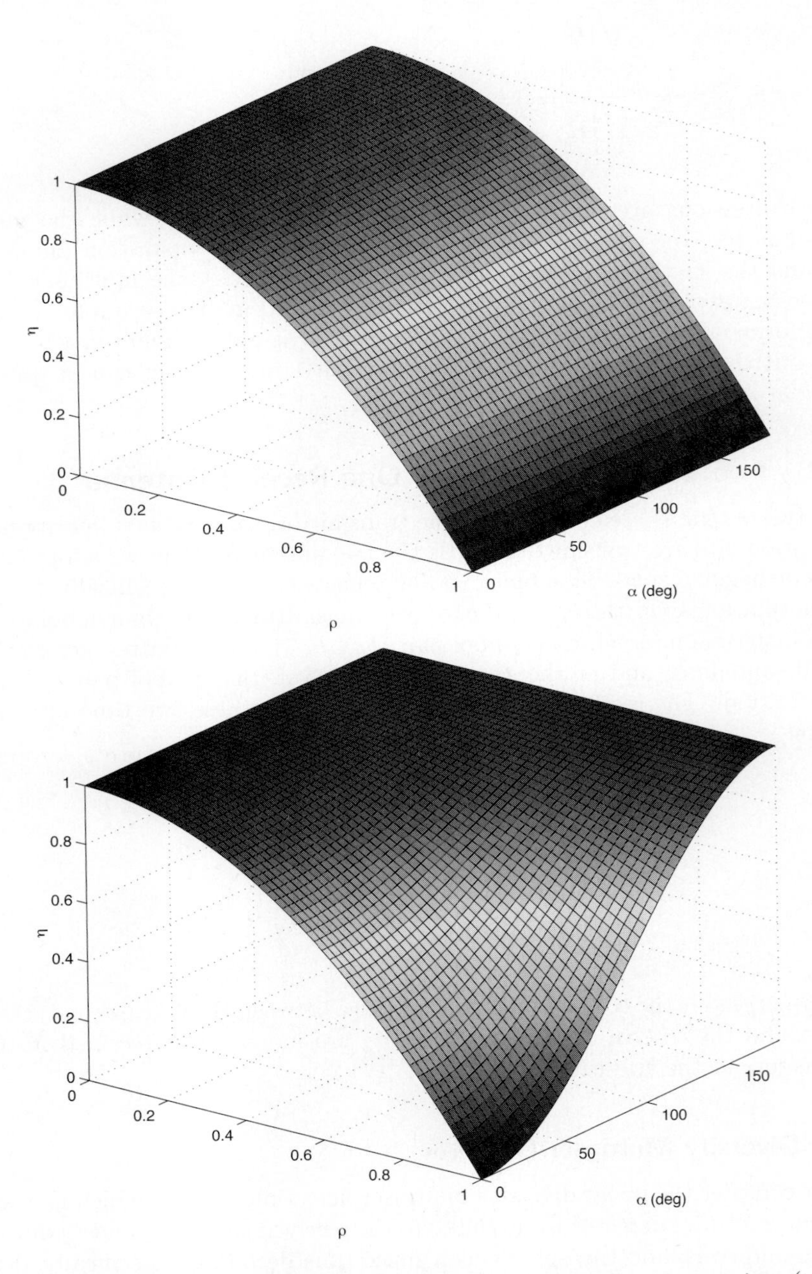

Figure 5.12. Performance comparison between a multiuser diversity receiver (top) and multiuser space-time receiver (bottom).

with

$$u_p \triangleq \boldsymbol{w}_1^H \boldsymbol{n}_p \sim \mathcal{N}_c\left(\mathbf{0}, \sigma^2 \|\boldsymbol{w}_1\|^2\right), \qquad p = 1, 2, \tag{5.134}$$

where $\|\boldsymbol{w}_1\|^2 = \left[\boldsymbol{R}^{-1}\right]_{1,1}$.

Denote $\boldsymbol{z} \triangleq [z_1\ z_2]^T$, $\boldsymbol{u} \triangleq [u_1\ u_2^*]^T$,

$$\boldsymbol{g}_k \triangleq [g_{1,k}\ \ g_{2,k}^*]^T,$$
$$\bar{\boldsymbol{g}}_k \triangleq [g_{2,k}\ \ -g_{1,k}^*]^T.$$

It is easily seen that $\boldsymbol{g}_k^H \bar{\boldsymbol{g}}_k = 0$. Then (5.132)–(5.134) can be written as

$$\boldsymbol{z} = [\boldsymbol{g}_1\ \bar{\boldsymbol{g}}_1] \begin{bmatrix} b_{1,1} \\ b_{2,1} \end{bmatrix} + \boldsymbol{u}, \tag{5.135}$$

with

$$\boldsymbol{u} \sim \mathcal{N}_c\left(\mathbf{0}, \frac{\sigma^2}{\eta_1^2} \boldsymbol{I}_2\right). \tag{5.136}$$

As before, denote $E_1 \triangleq \|\boldsymbol{g}_1\|^2 = \|\bar{\boldsymbol{g}}_1\|^2$. Note that

$$[\boldsymbol{g}_1\ \bar{\boldsymbol{g}}_1]^H [\boldsymbol{g}_1\ \bar{\boldsymbol{g}}_1] = \begin{bmatrix} E_1 & 0 \\ 0 & E_1 \end{bmatrix}. \tag{5.137}$$

The ML decision rule for $b_{1,1}$ and $b_{2,1}$ based on $\boldsymbol{z}$ in (5.135) is then given by

$$\begin{aligned} \begin{bmatrix} \hat{b}_{1,1} \\ \hat{b}_{2,1} \end{bmatrix} &= \operatorname{sign}\left(\Re\left\{[\boldsymbol{g}_1\ \bar{\boldsymbol{g}}_1]^H \boldsymbol{z}\right\}\right) \\ &= \operatorname{sign}\left(\Re\left\{\begin{bmatrix} \boldsymbol{g}_1^H \boldsymbol{z} \\ \bar{\boldsymbol{g}}_1^H \boldsymbol{z} \end{bmatrix}\right\}\right). \end{aligned} \tag{5.138}$$

Using (5.135), it is easily seen that the decision statistic in (5.138) is distributed according to

$$\frac{1}{\sqrt{E_1}} \boldsymbol{g}_1^H \boldsymbol{z} \sim \mathcal{N}_c\left(\sqrt{E_1} b_{1,1}, \frac{\sigma^2}{\eta_1^2}\right), \tag{5.139}$$

$$\frac{1}{\sqrt{E_1}} \bar{\boldsymbol{g}}_1^H \boldsymbol{z} \sim \mathcal{N}_c\left(\sqrt{E_1} b_{2,1}, \frac{\sigma^2}{\eta_1^2}\right). \tag{5.140}$$

Hence the probability of error is given by

$$P_1^{\text{DC}}(e) = Q\left(\frac{\sqrt{2E_1}}{\sigma} \eta_1\right). \tag{5.141}$$

This is the same expression as (5.117) for the linear diversity receiver with one transmit antenna and two receive antennas.

Linear Space-Time Multiuser Detector

Denote $\tilde{\boldsymbol{r}} \triangleq \left[\boldsymbol{r}_1^T \ \boldsymbol{r}_2^H\right]^T$ and $\tilde{\boldsymbol{n}} \triangleq \left[\boldsymbol{n}_1^T \ \boldsymbol{n}_2^H\right]^T$. Then (5.130) and (5.131) can be written as

$$\tilde{\boldsymbol{r}} = \sum_{k=1}^{K} \left(b_{1,k}\boldsymbol{g}_k \otimes \boldsymbol{s}_k + b_{2,k}\bar{\boldsymbol{g}}_k \otimes \boldsymbol{s}_k\right) + \tilde{\boldsymbol{n}}. \tag{5.142}$$

On denoting

$$\begin{aligned}\tilde{\boldsymbol{S}} &= \left[\boldsymbol{g}_1 \otimes \boldsymbol{s}_1,\ \bar{\boldsymbol{g}}_1 \otimes \boldsymbol{s}_1,\ \ldots,\ \boldsymbol{g}_K \otimes \boldsymbol{s}_K,\ \bar{\boldsymbol{g}}_K \otimes \boldsymbol{s}_K\right]_{N\times 2K},\\ \tilde{\boldsymbol{R}} &= \tilde{\boldsymbol{S}}^H \tilde{\boldsymbol{S}},\end{aligned}$$

the decorrelating detector for detecting the bit $b_{1,1}$ based on $\tilde{\boldsymbol{r}}$ in (5.142) is given by

$$\tilde{\boldsymbol{w}}_{1,1} = \tilde{\boldsymbol{S}}\tilde{\boldsymbol{R}}^{-1}\tilde{\boldsymbol{e}}_1, \tag{5.143}$$

where $\tilde{\boldsymbol{e}}_1$ is the first unit vector in $\mathbb{R}^{2K}$. We have the following result.

Proposition 5.7: *The decorrelating detector in (5.143) is given by*

$$\tilde{\boldsymbol{w}}_{1,1} = \frac{\boldsymbol{g}_1 \otimes \boldsymbol{w}_1}{\|\boldsymbol{g}_1\|^2}, \tag{5.144}$$

where $\boldsymbol{w}_1$ is given by (5.96).

Proof: We need to verify that

$$\left(\frac{\boldsymbol{g}_1 \otimes \boldsymbol{w}_1}{\|\boldsymbol{g}_1\|^2}\right)^H \tilde{\boldsymbol{S}} = \tilde{\boldsymbol{e}}_1. \tag{5.145}$$

We have

$$\frac{1}{\|\boldsymbol{g}_1\|^2}\left(\boldsymbol{g}_1 \otimes \boldsymbol{w}_1\right)^H\left(\boldsymbol{g}_1 \otimes \boldsymbol{s}_1\right) = \frac{1}{\|\boldsymbol{g}_1\|^2}\left(\boldsymbol{g}_1^H\boldsymbol{g}_1\right)\underbrace{\left(\boldsymbol{w}_1^H\boldsymbol{s}_1\right)}_{1} = 1, \tag{5.146}$$

$$\frac{1}{\|\boldsymbol{g}_1\|^2}\left(\boldsymbol{g}_1 \otimes \boldsymbol{w}_1\right)^H\left(\bar{\boldsymbol{g}}_1 \otimes \boldsymbol{s}_1\right) = \frac{1}{\|\boldsymbol{h}_1\|^2}\underbrace{\left(\boldsymbol{g}_1^H\bar{\boldsymbol{g}}_1\right)}_{0}\underbrace{\left(\boldsymbol{w}_1^H\boldsymbol{s}_1\right)}_{1} = 0, \tag{5.147}$$

$$\frac{1}{\|\boldsymbol{g}_1\|^2}\left(\boldsymbol{g}_1 \otimes \boldsymbol{w}_1\right)^H\left(\boldsymbol{g}_k \otimes \boldsymbol{s}_k\right) = \frac{1}{\|\boldsymbol{g}_1\|^2}\left(\boldsymbol{g}_1^H\boldsymbol{g}_k\right)\underbrace{\left(\boldsymbol{w}_1^H\boldsymbol{s}_k\right)}_{0} = 0, \quad k = 2, \ldots, K, \tag{5.148}$$

$$\frac{1}{\|\boldsymbol{g}_1\|^2}\left(\boldsymbol{g}_1 \otimes \boldsymbol{w}_1\right)^H\left(\bar{\boldsymbol{g}}_k \otimes \boldsymbol{s}_k\right) = \frac{1}{\|\boldsymbol{g}_1\|^2}\left(\boldsymbol{g}_1^H\bar{\boldsymbol{g}}_k\right)\underbrace{\left(\boldsymbol{w}_1^H\boldsymbol{s}_k\right)}_{0} = 0, \quad k = 2, \ldots, K. \tag{5.149}$$

This verifies (5.145), so that (5.144) is indeed the decorrelating detector given by (5.143). □

Thus the output of the linear space-time detector in this case is given by

$$\tilde{z}_1 = \tilde{\boldsymbol{w}}_{1,1}^H \tilde{\boldsymbol{r}} \;=\; b_{1,1} + u_1, \tag{5.150}$$

with

$$u_1 \triangleq \tilde{\boldsymbol{w}}_{1,1}^H \tilde{\boldsymbol{n}} \;\sim\; \mathcal{N}\left(0,\;\sigma^2 \|\tilde{\boldsymbol{w}}_{1,1}\|^2\right), \tag{5.151}$$

where using (5.99) and (5.144), we have

$$\begin{aligned} \|\tilde{\boldsymbol{w}}_{1,1}\|^2 &= \frac{\|\boldsymbol{g}_1 \otimes \boldsymbol{w}_1\|^2}{\|\boldsymbol{g}_1\|^4} \\ &= \frac{\|\boldsymbol{w}_1\|^2}{\|\boldsymbol{g}_1\|^2} \;=\; \frac{1}{E_1 \eta_1^2}. \end{aligned} \tag{5.152}$$

Therefore, the probability of detection error is given by

$$\begin{aligned} P_1^{\mathrm{ST}}(e) &= P\Big(\Re\left\{\tilde{z}_1\right\} < 0 \mid b_{1,1} = 1\Big) \\ &= P\Big(\Re\{u_1\} < -1\Big) \\ &= Q\left(\frac{\sqrt{2E_1}}{\sigma}\eta_1\right). \end{aligned} \tag{5.153}$$

On comparing (5.141) with (5.153) we see that for the case of two transmit antennas and one receive antenna, the linear diversity receiver and the linear space-time receiver have the same performance. Hence the multiple transmit antennas with space-time block coding provide only diversity gain and no signal separation capability.

5.5.3 Two Transmit and Two Receive Antennas

We combine the results from the two preceding sections to investigate an environment in which we use two transmit antennas and two receive antennas. We adopt the space-time block coding scheme used in the preceding section. The received signals at antenna 1 during the two symbol intervals are

$$\boldsymbol{r}_1^{(1)} = \sum_{k=1}^{K} \left[g_k^{(1,1)} b_{1,k} + g_k^{(2,1)} b_{2,k}\right] \boldsymbol{s}_k + \boldsymbol{n}_1^{(1)}, \tag{5.154}$$

$$\boldsymbol{r}_2^{(1)} = \sum_{k=1}^{K} \left[-g_k^{(1,1)} b_{2,k} + g_k^{(2,1)} b_{1,k}\right] \boldsymbol{s}_k + \boldsymbol{n}_2^{(1)}, \tag{5.155}$$

and the corresponding signals received at antenna 2 are

$$\boldsymbol{r}_1^{(2)} = \sum_{k=1}^{K} \left[g_k^{(1,2)} b_{1,k} + g_k^{(2,2)} b_{2,k}\right] \boldsymbol{s}_k + \boldsymbol{n}_1^{(2)}, \tag{5.156}$$

$$\boldsymbol{r}_2^{(2)} = \sum_{k=1}^{K} \left[-g_k^{(1,2)} b_{2,k} + g_k^{(2,2)} b_{1,k} \right] \boldsymbol{s}_k + \boldsymbol{n}_2^{(2)}, \tag{5.157}$$

where $g_k^{(i,j)}, i, j \in \{1, 2\}$ is the complex channel response between transmit antenna i and receive antenna j for user k. The noise vectors $\boldsymbol{n}_1^{(1)}, \boldsymbol{n}_1^{(2)}, \boldsymbol{n}_2^{(1)}$, and $\boldsymbol{n}_2^{(2)}$ are independent and identically distributed with distribution $\mathcal{N}_c(\boldsymbol{0}, \sigma^2 \boldsymbol{I}_N)$.

Linear Diversity Multiuser Detector

As before, we first consider the linear diversity multiuser detection scheme for user 1, which applies the linear multiuser detector $\boldsymbol{w}_1$ in (5.96) to each of the four received signals $\boldsymbol{r}_1^{(1)}, \boldsymbol{r}_1^{(2)}, \boldsymbol{r}_2^{(1)}$, and $\boldsymbol{r}_2^{(2)}$ and then performs a space-time decoding. Specifically, denote the filter outputs as

$$\begin{aligned} z_1^{(1)} &\triangleq \boldsymbol{w}_1^T \boldsymbol{r}_1^{(1)} \\ &= g_1^{(1,1)} b_{1,1} + g_1^{(2,1)} b_{2,1} + u_1^{(1)}, \end{aligned} \tag{5.158}$$

$$\begin{aligned} z_2^{(1)} &\triangleq \left(\boldsymbol{w}_1^T \boldsymbol{r}_2^{(1)} \right)^* \\ &= -\left(g_1^{(1,1)} \right)^* b_{2,1} + \left(g_1^{(2,1)} \right)^* b_{1,1} + \left(u_2^{(1)} \right)^*, \end{aligned} \tag{5.159}$$

$$\begin{aligned} z_1^{(2)} &\triangleq \boldsymbol{w}_1^T \boldsymbol{r}_1^{(2)} \\ &= g_1^{(1,2)} b_{1,1} + g_1^{(2,2)} b_{2,1} + u_1^{(2)}, \end{aligned} \tag{5.160}$$

$$\begin{aligned} z_2^{(2)} &\triangleq \left(\boldsymbol{w}_1^T \boldsymbol{r}_2^{(2)} \right)^* \\ &= -\left(g_1^{(1,2)} \right)^* b_{2,1} + \left(g_1^{(2,2)} \right)^* b_{1,1} + \left(u_2^{(2)} \right)^*, \end{aligned} \tag{5.161}$$

with

$$\begin{aligned} u_i^{(j)} &\triangleq \boldsymbol{w}_1^T \boldsymbol{n}_i^{(j)} \\ &\sim \mathcal{N}_c \left(\boldsymbol{0}, \frac{\sigma^2}{\eta_1^2} \right), \qquad i, j = 1, 2, \end{aligned} \tag{5.162}$$

where, as before, $\eta_1^2 \triangleq 1 / \left[\boldsymbol{R}^{-1} \right]_{1,1}$.

We define the following quantities:

$$\begin{aligned} \boldsymbol{z} &\triangleq \left[z_1^{(1)} \ z_2^{(1)} \ z_1^{(2)} \ z_2^{(2)} \right]^T, \\ \boldsymbol{u} &\triangleq \left[u_1^{(1)} \ \left(u_2^{(1)} \right)^* \ u_1^{(2)} \ \left(u_2^{(2)} \right)^* \right]^T, \\ \boldsymbol{g}_1^{(1)} &\triangleq \left[g_1^{(1,1)} \ g_1^{(2,1)} \right]^H, \\ \bar{\boldsymbol{g}}_1^{(1)} &\triangleq \left[g_1^{(2,1)} \ - g_1^{(1,1)} \right]^T, \end{aligned}$$

$$\boldsymbol{g}_1^{(2)} \triangleq \left[g_1^{(1,2)} \; g_1^{(2,2)} \right]^H,$$

$$\bar{\boldsymbol{g}}_1^{(2)} \triangleq \left[g_1^{(2,2)} \; -g_1^{(1,2)} \right]^T.$$

Then (5.158)–(5.162) can be written as

$$\boldsymbol{z} = \underbrace{\left[\boldsymbol{g}_1^{(1)} \; \bar{\boldsymbol{g}}_1^{(1)} \; \boldsymbol{g}_1^{(2)} \; \bar{\boldsymbol{g}}_1^{(2)} \right]^H}_{\boldsymbol{G}_1^H} \begin{bmatrix} b_{1,1} \\ b_{2,1} \end{bmatrix} + \boldsymbol{u}, \tag{5.163}$$

with

$$\boldsymbol{u} \sim \mathcal{N}_c \left(\boldsymbol{0}, \frac{\sigma^2}{\eta_1^2} \cdot \boldsymbol{I}_4 \right). \tag{5.164}$$

It is readily verified that

$$\boldsymbol{G}_1 \boldsymbol{G}_1^H = \begin{bmatrix} E_1 & 0 \\ 0 & E_1 \end{bmatrix}, \tag{5.165}$$

with

$$E_1 \triangleq \left| g_1^{(1,1)} \right|^2 + \left| g_1^{(1,2)} \right|^2 + \left| g_1^{(2,1)} \right|^2 + \left| g_1^{(2,2)} \right|^2. \tag{5.166}$$

To form the ML decision statistic, we premultiply $\boldsymbol{z}$ by $\boldsymbol{G}_1$ and obtain

$$\begin{bmatrix} d_{1,1} \\ d_{2,1} \end{bmatrix} \triangleq \boldsymbol{G}_1 \boldsymbol{z} = E_1 \begin{bmatrix} b_{1,1} \\ b_{2,1} \end{bmatrix} + \boldsymbol{v}, \tag{5.167}$$

with

$$\boldsymbol{v} \sim \mathcal{N}_c \left(\boldsymbol{0}, \frac{E_1 \, \sigma^2}{\eta_1^2} \cdot \boldsymbol{I}_2 \right). \tag{5.168}$$

The corresponding bit estimates are given by

$$\begin{bmatrix} \hat{b}_{1,1} \\ \hat{b}_{2,1} \end{bmatrix} = \text{sign} \left(\Re \left\{ \begin{bmatrix} d_{1,1} \\ d_{2,1} \end{bmatrix} \right\} \right). \tag{5.169}$$

The bit error probability is then given by

$$\begin{aligned} P_1^{\mathrm{DC}}(e) &= P\Big(\Re\{d_{1,1}\} < 0 \mid b_{1,1} = +1 \Big) \\ &= P\left[E_1 + \mathcal{N}\left(0, \; \frac{E_1 \sigma^2}{2\eta_1^2} \right) < 0 \right] \\ &= Q\left(\frac{\sqrt{2E_1}}{\sigma} \eta_1 \right). \end{aligned} \tag{5.170}$$

Linear Space-Time Multiuser Detector

We denote

$$\tilde{\boldsymbol{r}} \triangleq \begin{bmatrix} \boldsymbol{r}_1^{(1)} \\ \left(\boldsymbol{r}_2^{(1)}\right)^* \\ \boldsymbol{r}_1^{(2)} \\ \left(\boldsymbol{r}_2^{(2)}\right)^* \end{bmatrix}, \quad \tilde{\boldsymbol{n}} \triangleq \begin{bmatrix} \boldsymbol{n}_1^{(1)} \\ \left(\boldsymbol{n}_2^{(1)}\right)^* \\ \boldsymbol{n}_1^{(2)} \\ \left(\boldsymbol{n}_2^{(2)}\right)^* \end{bmatrix},$$

$$\boldsymbol{g}_k \triangleq \begin{bmatrix} g_k^{(1,1)} \\ \left(g_k^{(2,1)}\right)^* \\ g_k^{(1,2)} \\ \left(g_k^{(2,2)}\right)^* \end{bmatrix}, \quad \bar{\boldsymbol{g}}_k \triangleq \begin{bmatrix} g_k^{(2,1)} \\ \left(-g_k^{(1,1)}\right)^* \\ g_k^{(2,2)} \\ \left(-g_k^{(1,2)}\right)^* \end{bmatrix}.$$

Then (5.154)–(5.157) may be written as

$$\tilde{\boldsymbol{r}} = \sum_{k=1}^{K} \left(b_{1,k}\boldsymbol{g}_k \otimes \boldsymbol{s}_k + b_{2,k}\bar{\boldsymbol{g}}_k \otimes \boldsymbol{s}_k\right) + \tilde{\boldsymbol{n}} \tag{5.171}$$

$$= \tilde{\boldsymbol{S}}\boldsymbol{b} + \tilde{\boldsymbol{n}}, \tag{5.172}$$

where

$$\tilde{\boldsymbol{S}} \triangleq [\boldsymbol{g}_1 \otimes \boldsymbol{s}_1, \;\; \bar{\boldsymbol{g}}_1 \otimes \boldsymbol{s}_1, \ldots, \;\; \boldsymbol{g}_K \otimes \boldsymbol{s}_K, \;\; \bar{\boldsymbol{g}}_K \otimes \boldsymbol{s}_K]_{4N\times 2K}$$

$$\boldsymbol{b} \triangleq [b_{1,1} \;\; b_{2,1} \;\; b_{1,2} \;\; b_{2,2} \cdots b_{1,K} \;\; b_{2,K}]^T.$$

Since $\boldsymbol{g}_k^H \bar{\boldsymbol{g}}_k = 0$ and (5.171) has the same form as (5.142), it is easy to show that the decorrelating detector for detecting the bit $b_{1,1}$ based on $\tilde{\boldsymbol{r}}$ is given by

$$\tilde{\boldsymbol{w}}_{1,1} = \frac{\boldsymbol{g}_1 \otimes \boldsymbol{w}_1}{\|\boldsymbol{g}_1\|^2}. \tag{5.173}$$

Hence the output of the linear space-time detector in this case is given by

$$\tilde{z}_1 = \tilde{\boldsymbol{w}}_{1,1}^H \tilde{\boldsymbol{r}} \;=\; b_{1,1} + u_1, \tag{5.174}$$

with

$$u_1 \triangleq \tilde{\boldsymbol{w}}_{1,1}^H \tilde{\boldsymbol{n}} \;\sim\; \mathcal{N}_c\left(0,\, \sigma^2 \|\tilde{\boldsymbol{w}}_{1,1}\|^2\right), \tag{5.175}$$

where

$$\|\tilde{\boldsymbol{w}}_{1,1}\|^2 = \frac{\|\boldsymbol{w}_1\|^2}{\|\boldsymbol{g}_1\|^2} \;=\; \frac{1}{E_1 \eta_1^2}. \tag{5.176}$$

Therefore, the probability of detection error is given by

$$\begin{aligned} P_1^{\mathrm{ST}}(e) &= P\Big(\Re\{\tilde{z}_1\} < 0 \mid b_{1,1} = +1\Big) \\ &= P\left[1 + \mathcal{N}\left(0, \frac{1}{2E_1\eta_1^2}\right) < 0\right] \\ &= Q\left(\frac{\sqrt{2E_1}}{\sigma}\eta_1\right). \end{aligned} \quad (5.177)$$

Comparing (5.177) with (5.170), it is seen that when two transmit antennas and two receive antennas are employed and the signals are transmitted in the form of a space-time block code, the linear diversity receiver and the linear space-time receiver have identical performance.

Remarks

We have seen that the performance of space-time multiuser detection (STMUD) and linear diversity multiuser detection (LDMUD) are similar for two transmit/one receive and two transmit/two receive antenna configurations. What, then, are the benefits of the space-time detection technique? They are as follows:

1. Although LDMUD and STMUD perform similarly for the 2×1 and 2×2 cases, the performance of STMUD is superior for configurations with one transmit antenna and $P \geq 2$ receive antennas.

2. User capacity for CDMA systems is limited by correlations among composite signature waveforms. This multiple-access interference will tend to decrease as the dimension of the vector space in which the signature waveforms reside increases. The signature waveforms for linear diversity detection are of length N (i.e., they reside in $\mathbb{C}^N$). Since the received signals are stacked for space-time detection, these signature waveforms reside in $\mathbb{C}^{2N}$ for two transmit and one receive antennas or $\mathbb{C}^{4N}$ for two transmit and two receive antennas. As a result, the space-time structure can support more users than linear diversity detection for a given performance threshold.

3. For adaptive configurations (Section 5.5.4 and Section 5.6.2), LDMUD requires four independent subspace trackers operating simultaneously since the receiver performs detection on each of the four received signals, and each has a different signal subspace. The space-time structure requires only one subspace tracker.

5.5.4 Blind Adaptive Implementations

We next develop both batch and sequential *blind adaptive* implementations of the linear space-time receiver. These implementations are blind in the sense that they require only knowledge of the signature waveform of the user of interest. Instead of the decorrelating detector used in previous sections, we will use a linear MMSE detector for the adaptive implementations because the MMSE detector is more suitable for adaptation and its performance is comparable to that of the decorrelating detector. We consider only the environment in which we have two transmit antennas and two receive antennas. The other cases can be derived in a similar manner. Note that inherent to any *blind* receiver in multiple transmit antenna systems is an ambiguity issue. That is, if the same spreading waveform is used for a user at both transmit antennas, the blind receiver cannot distinguish which bit is from which antenna. To resolve such an ambiguity, here we use two different spreading waveforms for each user (i.e., $\boldsymbol{s}_{j,k}, j \in \{1,2\}$ is the spreading code for user k for the transmission of bit $b_{j,k}$).

There are two bits, $b_{1,k}[i]$ and $b_{2,k}[i]$, associated with each user at each time slot i, and the difference in time between slots is $2T$, where T is the symbol interval. The received signal at antenna 1 during the two symbol periods for time slot i is

$$\boldsymbol{r}_1^{(1)}[i] = \sum_{k=1}^{K} \left(g_k^{(1,1)} b_{1,k}[i] \boldsymbol{s}_{1,k} + g_k^{(2,1)} b_{2,k}[i] \boldsymbol{s}_{2,k} \right) + \boldsymbol{n}_1^{(1)}[i], \tag{5.178}$$

$$\boldsymbol{r}_2^{(1)}[i] = \sum_{k=1}^{K} \left(-g_k^{(1,1)} b_{2,k}[i] \boldsymbol{s}_{2,k} + g_k^{(2,1)} b_{1,k}[i] \boldsymbol{s}_{1,k} \right) + \boldsymbol{n}_2^{(1)}[i]; \tag{5.179}$$

and the corresponding signals received at antenna 2 are

$$\boldsymbol{r}_1^{(2)}[i] = \sum_{k=1}^{K} \left(g_k^{(1,2)} b_{1,k}[i] \boldsymbol{s}_{1,k} + g_k^{(2,2)} b_{2,k}[i] \boldsymbol{s}_{2,k} \right) + \boldsymbol{n}_1^{(2)}[i], \tag{5.180}$$

$$\boldsymbol{r}_2^{(2)}[i] = \sum_{k=1}^{K} \left(-g_k^{(1,2)} b_{2,k}[i] \boldsymbol{s}_{2,k} + g_k^{(2,2)} b_{1,k}[i] \boldsymbol{s}_{1,k} \right) + \boldsymbol{n}_2^{(2)}[i]. \tag{5.181}$$

We stack these received signal vectors and denote

$$\tilde{\boldsymbol{r}}[i] \triangleq \begin{bmatrix} \boldsymbol{r}_1^{(1)}[i] \\ \left(\boldsymbol{r}_2^{(1)}[i]\right)^* \\ \boldsymbol{r}_1^{(2)}[i] \\ \left(\boldsymbol{r}_2^{(2)}[i]\right)^* \end{bmatrix}, \quad \tilde{\boldsymbol{n}}[i] \triangleq \begin{bmatrix} \boldsymbol{n}_1^{(1)}[i] \\ \left(\boldsymbol{n}_2^{(1)}[i]\right)^* \\ \boldsymbol{n}_1^{(2)}[i] \\ \left(\boldsymbol{n}_2^{(2)}[i]\right)^* \end{bmatrix},$$

$$\boldsymbol{g}_k \triangleq \begin{bmatrix} g_k^{(1,1)} \\ \left(g_k^{(2,1)}\right)^* \\ g_k^{(1,2)} \\ \left(g_k^{(2,2)}\right)^* \end{bmatrix}, \quad \bar{\boldsymbol{g}}_k \triangleq \begin{bmatrix} g_k^{(2,1)} \\ \left(-g_k^{(1,1)}\right)^* \\ g_k^{(2,2)} \\ \left(-g_k^{(1,2)}\right)^* \end{bmatrix}.$$

Then we may write

$$\begin{aligned} \tilde{\boldsymbol{r}}[i] &= \sum_{k=1}^{K} \left(b_{1,k}[i] \boldsymbol{g}_k \otimes \boldsymbol{s}_{1,k} + b_{2,k}[i] \bar{\boldsymbol{g}}_k \otimes \boldsymbol{s}_{2,k} \right) + \tilde{\boldsymbol{n}}[i] \\ &= \tilde{\boldsymbol{S}} \boldsymbol{b}[i] + \tilde{\boldsymbol{n}}[i], \end{aligned} \tag{5.182}$$

where

$$\tilde{\boldsymbol{S}} \triangleq \Big[\boldsymbol{g}_1 \otimes \boldsymbol{s}_{1,1},\ \bar{\boldsymbol{g}}_1 \otimes \boldsymbol{s}_{2,1}, \ldots, \boldsymbol{g}_K \otimes \boldsymbol{s}_{1,K},\ \bar{\boldsymbol{g}}_K \otimes \boldsymbol{s}_{2,K}\Big]_{4N\times 2K},$$

$$\boldsymbol{b}[i] \triangleq \Big[b_{1,1}[i]\ b_{2,1}[i]\ b_{1,2}[i]\ b_{2,2}[i] \cdots b_{1,K}[i]\ b_{2,K}[i]\Big]^T_{2K\times 1}.$$

The autocorrelation matrix of the stacked signal $\tilde{\boldsymbol{r}}[i]$, $\boldsymbol{C}$, and its eigendecomposition are given by

$$\boldsymbol{C} = E\left\{\tilde{\boldsymbol{r}}[i]\tilde{\boldsymbol{r}}[i]^H\right\} = \tilde{\boldsymbol{S}}\tilde{\boldsymbol{S}}^H + \sigma^2 \boldsymbol{I}_{4N} \tag{5.183}$$

$$= \boldsymbol{U}_s \boldsymbol{\Lambda}_s \boldsymbol{U}_s^H + \sigma^2 \boldsymbol{U}_n \boldsymbol{U}_n^H, \tag{5.184}$$

where $\boldsymbol{\Lambda}_s = \mathrm{diag}\{\lambda_1, \lambda_2, \ldots, \lambda_{2K}\}$ contains the largest $(2K)$ eigenvalues of $\boldsymbol{C}$, the columns of $\boldsymbol{U}_s$ are the corresponding eigenvectors, and the columns of $\boldsymbol{U}_n$ are the $4N - 2K$ eigenvectors corresponding to the smallest eigenvalue σ^2.

The blind linear MMSE detector for detecting $\Big[\boldsymbol{b}[i]\Big]_1 = b_{1,1}[i]$ is given by the solution to the optimization problem

$$\boldsymbol{w}_{1,1} \triangleq \arg \min_{\boldsymbol{w}\in\mathbb{C}^{4N}} E\left\{\left|b_{1,1}[i] - \boldsymbol{w}^H \tilde{\boldsymbol{r}}[i]\right|^2\right\}. \tag{5.185}$$

From Chapter 2, a scaled version of the solution can be written in terms of the signal subspace components as

$$\boldsymbol{w}_{1,1} = \boldsymbol{U}_s \boldsymbol{\Lambda}_s^{-1} \boldsymbol{U}_s^H \left(\boldsymbol{g}_1 \otimes \boldsymbol{s}_{1,1}\right), \tag{5.186}$$

and the decision is made according to

$$z_{1,1}[i] = \boldsymbol{w}_{1,1}^H \tilde{\boldsymbol{r}}[i], \tag{5.187}$$

and

$$\hat{b}_{1,1}[i] = \mathrm{sign}\Big[\Re\Big(z_{1,1}[i]\Big)\Big], \quad \text{(coherent detection)} \tag{5.188}$$

or

$$\hat{\beta}_{1,1}[i] = \mathrm{sign}\Big[\Re\Big(z_{1,1}[i-1]^* z_{1,1}[i]\Big)\Big]. \quad \text{(differential detection)} \tag{5.189}$$

Before we address specific batch and sequential adaptive algorithms, we note that these algorithms can also be implemented using linear group-blind multiuser detectors instead of blind MMSE detectors. This would be appropriate, for example, in uplink environments in which the base station has knowledge of the signature waveforms of all of the users in the cell, but not those of users outside the cell. Specifically, we may rewrite (5.182) as

$$\tilde{\boldsymbol{r}}[i] = \breve{\boldsymbol{S}}\breve{\boldsymbol{b}}[i] + \bar{\boldsymbol{S}}\bar{\boldsymbol{b}}[i] + \tilde{\boldsymbol{n}}[i], \tag{5.190}$$

where we have separated the users into two groups. The composite signature sequences of the known users are the columns of $\breve{\boldsymbol{S}}$. The unknown users' composite sequences are the columns of $\bar{\boldsymbol{S}}$. Then, from Chapter 3, the group-blind linear hybrid detector for bit $b_{1,1}[i]$ is given by

$$\boldsymbol{w}_{1,1}^{\text{GB}} = \boldsymbol{U}_s \boldsymbol{\Lambda}_s^{-1} \boldsymbol{U}_s^H \breve{\boldsymbol{S}} \left[\breve{\boldsymbol{S}}^H \boldsymbol{U}_s \boldsymbol{\Lambda}_s^{-1} \boldsymbol{U}_s^H \breve{\boldsymbol{S}} \right]^{-1} (\boldsymbol{g}_1 \otimes \boldsymbol{s}_{1,1}) . \tag{5.191}$$

This detector offers a significant performance improvement over (5.186) for environments in which the signature sequences of some of the interfering users are known.

Batch Blind Linear Space-Time Multiuser Detection

To obtain an estimate of $\boldsymbol{g}_1$, we make use of the orthogonality between the signal and noise subspaces [i.e., the fact that $\boldsymbol{U}_n^H (\boldsymbol{g}_1 \otimes \boldsymbol{s}_{1,1}) = \boldsymbol{0}$]. In particular, we have

$$\begin{aligned}
\hat{\boldsymbol{g}}_1 &= \arg \min_{\boldsymbol{g} \in \mathbb{C}^4} \left\| \boldsymbol{U}_n^H (\boldsymbol{g} \otimes \boldsymbol{s}_{1,1}) \right\|^2 \qquad \text{s.t. } \|\boldsymbol{g}\| = 1 \\
&= \arg \max_{\boldsymbol{g} \in \mathbb{C}^4} \left\| \boldsymbol{U}_s^H (\boldsymbol{g} \otimes \boldsymbol{s}_{1,1}) \right\|^2 \\
&= \arg \max_{\boldsymbol{g} \in \mathbb{C}^4} \left(\boldsymbol{g}^H \otimes \boldsymbol{s}_{1,1}^H \right) \boldsymbol{U}_s \boldsymbol{U}_s^H \left(\boldsymbol{g} \otimes \boldsymbol{s}_{11} \right) \\
&= \arg \max_{\boldsymbol{g} \in \mathbb{C}^4} \boldsymbol{g}^H \underbrace{\left[(\boldsymbol{I}_4 \otimes \boldsymbol{s}_{1,1}^H) \boldsymbol{U}_s \boldsymbol{U}_s^H (\boldsymbol{I}_4 \otimes \boldsymbol{s}_{1,1}) \right]}_{\boldsymbol{Q}} \boldsymbol{h} \qquad (5.192) \\
&= \text{principal eigenvector of } \boldsymbol{Q}. \qquad (5.193)
\end{aligned}$$

In (5.193), $\hat{\boldsymbol{g}}_1$ specifies $\boldsymbol{g}_1$ up to an arbitrary complex scale factor α (i.e., $\hat{\boldsymbol{g}}_1 = \alpha \, \boldsymbol{g}_1$). The following is a summary of a batch blind space-time multiuser detection algorithm for the two transmit antenna/two receive antenna configuration.

Algorithm 5.4: [Batch blind linear space-time multiuser detector—synchronous CDMA, two transmit antennas, and two receive antennas]

- *Estimate the signal subspace:*

$$\begin{aligned}
\hat{\boldsymbol{C}} &= \frac{1}{M} \sum_{i=0}^{M-1} \tilde{\boldsymbol{r}}[i] \tilde{\boldsymbol{r}}[i]^H \qquad (5.194) \\
&= \hat{\boldsymbol{U}}_s \hat{\boldsymbol{\Lambda}}_s \hat{\boldsymbol{U}}_s^H + \hat{\boldsymbol{U}}_n \hat{\boldsymbol{\Lambda}}_n \hat{\boldsymbol{U}}_n^H . \qquad (5.195)
\end{aligned}$$

- *Estimate the channels:*

$$\begin{aligned}
\hat{\boldsymbol{Q}}_1 &= (\boldsymbol{I}_4 \otimes \boldsymbol{s}_{1,1}^H) \hat{\boldsymbol{U}}_s \hat{\boldsymbol{U}}_s^H (\boldsymbol{I}_4 \otimes \boldsymbol{s}_{1,1}) , \qquad (5.196) \\
\hat{\boldsymbol{Q}}_2 &= (\boldsymbol{I}_4 \otimes \boldsymbol{s}_{2,1}^H) \hat{\boldsymbol{U}}_s \hat{\boldsymbol{U}}_s^H (\boldsymbol{I}_4 \otimes \boldsymbol{s}_{2,1}) , \qquad (5.197) \\
\hat{\boldsymbol{g}}_1 &= \text{principal eigenvector of } \hat{\boldsymbol{Q}}_1, \qquad (5.198) \\
\hat{\hat{\boldsymbol{g}}}_1 &= \text{principal eigenvector of } \hat{\boldsymbol{Q}}_2. \qquad (5.199)
\end{aligned}$$

- *Form the detectors:*

$$\hat{\boldsymbol{w}}_{1,1} = \hat{\boldsymbol{U}}_s \hat{\boldsymbol{\Lambda}}_s^{-1} \hat{\boldsymbol{U}}_s^H \left(\hat{\boldsymbol{g}}_1 \otimes \boldsymbol{s}_{1,1} \right), \tag{5.200}$$

$$\hat{\boldsymbol{w}}_{2,1} = \hat{\boldsymbol{U}}_s \hat{\boldsymbol{\Lambda}}_s^{-1} \hat{\boldsymbol{U}}_s^H \left(\hat{\hat{\boldsymbol{g}}}_1 \otimes \boldsymbol{s}_{2,1} \right). \tag{5.201}$$

- *Perform differential detection:*

$$z_{1,1}[i] = \hat{\boldsymbol{w}}_{1,1}^H \tilde{\boldsymbol{r}}[i], \tag{5.202}$$

$$z_{2,1}[i] = \hat{\boldsymbol{w}}_{2,1}^H \tilde{\boldsymbol{r}}[i], \tag{5.203}$$

$$\hat{\beta}_{1,1}[i] = \text{sign}\left(\Re\left\{ z_{1,1}[i] z_{1,1}[i-1]^* \right\} \right), \tag{5.204}$$

$$\hat{\beta}_{2,1}[i] = \text{sign}\left(\Re\left\{ z_{2,1}[i] z_{2,1}[i-1]^* \right\} \right), \qquad i = 0, \ldots, M-1. \tag{5.205}$$

A batch group-blind space-time multiuser detector algorithm can be implemented with simple modifications to (5.200) and (5.201).

Adaptive Blind Linear Space-Time Multiuser Detection

To form a sequential blind adaptive receiver, we need adaptive algorithms for sequentially estimating the channel and the signal subspace components $\boldsymbol{U}_s$ and $\boldsymbol{\Lambda}_s$. First, we address sequential adaptive channel estimation. Denote by $\boldsymbol{z}[i]$ the projection of the stacked signal $\tilde{\boldsymbol{r}}[i]$ onto the noise subspace:

$$\boldsymbol{z}[i] = \tilde{\boldsymbol{r}}[i] - \boldsymbol{U}_s \boldsymbol{U}_s^H \tilde{\boldsymbol{r}}[i] \tag{5.206}$$

$$= \boldsymbol{U}_n \boldsymbol{U}_n^H \tilde{\boldsymbol{r}}[i]. \tag{5.207}$$

Since $\boldsymbol{z}[i]$ lies in the noise subspace, it is orthogonal to any signal in the signal subspace, and in particular, it is orthogonal to $(\boldsymbol{g}_1 \otimes \boldsymbol{s}_{1,1})$. Hence $\boldsymbol{g}_1$ is the solution to the following constrained optimization problem:

$$\begin{aligned}
&\min_{\boldsymbol{g}_1 \in \mathbb{C}^4} E\left\{ \left\| \boldsymbol{z}[i]^H (\boldsymbol{g}_1 \otimes \boldsymbol{s}_{1,1}) \right\|^2 \right\} \\
&\quad = \min_{\boldsymbol{g}_1 \in \mathbb{C}^4} E\left\{ \left\| \boldsymbol{z}[i]^H (\boldsymbol{I}_4 \otimes \boldsymbol{s}_{1,1}) \, \boldsymbol{g}_1 \right\|^2 \right\} \\
&\quad = \min_{\boldsymbol{g}_1 \in \mathbb{C}^4} E\left\{ \left\| \left[\left(\boldsymbol{I}_4 \otimes \boldsymbol{s}_{1,1}^H \right) \boldsymbol{z}[i] \right]^H \boldsymbol{g}_1 \right\|^2 \right\} \qquad \text{s.t.} \quad \|\boldsymbol{g}_1\| = 1.
\end{aligned} \tag{5.208}$$

To obtain a sequential algorithm to solve the optimization problem above, we write it in the following (trivial) state space form:

$$\boldsymbol{g}_1[i+1] = \boldsymbol{g}_1[i], \qquad \text{(state equation)}$$

$$0 = \left[\left(\boldsymbol{I}_4 \otimes \boldsymbol{s}_{1,1}^H \right) \boldsymbol{z}[i] \right]^H \boldsymbol{g}_1[i], \qquad \text{(observation equation)}$$

The standard Kalman filter can then be applied to the system above as follows. Denote $\boldsymbol{x}[i] \triangleq \left(\boldsymbol{I}_4 \otimes \boldsymbol{s}_{1,1}^T\right) \boldsymbol{z}[i]$. We have

$$\boldsymbol{k}[i] = \boldsymbol{\Sigma}[i-1]\boldsymbol{x}[i] \left(\boldsymbol{x}[i]^H \boldsymbol{\Sigma}[i-1]\boldsymbol{x}[i]\right)^{-1}, \tag{5.209}$$

$$\boldsymbol{g}_1[i] = \boldsymbol{g}_1[i-1] - \boldsymbol{k}[i] \left(\boldsymbol{x}[i]^H \boldsymbol{g}_1[i-1]\right) / \left\|\boldsymbol{g}_1[i-1] - \boldsymbol{k}[i]\left(\boldsymbol{x}[i]^H \boldsymbol{g}_1[i-1]\right)\right\|, \tag{5.210}$$

$$\boldsymbol{\Sigma}[i] = \boldsymbol{\Sigma}[i-1] - \boldsymbol{k}[i]\, \boldsymbol{x}[i]^H \boldsymbol{\Sigma}[i-1]. \tag{5.211}$$

Once we have obtained channel estimates at time slot i, we can combine them with estimates of the signal subspace components to form the detector in (5.186). Since we are stacking received signal vectors, and subspace tracking complexity increases at least linearly with signal subspace dimension, it is imperative that we choose an algorithm with minimal complexity. The best existing low-complexity algorithm for this purpose appears to be the NAHJ subspace tracking algorithm discussed in Section 2.6.3. This algorithm has the lowest complexity of any algorithm used for similar purposes and has performed well when used for signal subspace tracking in multipath fading environments. Since the size of $\boldsymbol{U}_s$ is $4N \times 2K$, the complexity is $40 \cdot 4N \cdot 2K + 3 \cdot 4N + 7.5(2K)^2 + 7 \cdot 2K$ floating point operations per iteration.

Algorithm 5.5: [Blind adaptive linear space-time multiuser detector—synchronous CDMA, two transmit antennas, and two receive antennas]

- *Using a suitable signal subspace tracking algorithm (e.g., NAHJ), update the signal subspace components* $\boldsymbol{U}_s[i]$ *and* $\boldsymbol{\Lambda}_s[i]$ *at each time slot* i.

- *Track the channel* $\boldsymbol{g}_1[i]$ *and* $\bar{\boldsymbol{g}}_1[i]$ *according to the following:*

$$\boldsymbol{z}[i] = \tilde{\boldsymbol{r}}[i] - \boldsymbol{U}_s[i]\boldsymbol{U}_s[i]^H \tilde{\boldsymbol{r}}[i], \tag{5.212}$$

$$\boldsymbol{x}[i] = \left(\boldsymbol{I}_4 \otimes \boldsymbol{s}_{1,1}^H\right) \boldsymbol{z}[i], \tag{5.213}$$

$$\bar{\boldsymbol{x}}[i] = \left(\boldsymbol{I}_4 \otimes \boldsymbol{s}_{2,1}^H\right) \boldsymbol{z}[i], \tag{5.214}$$

$$\boldsymbol{k}[i] = \boldsymbol{\Sigma}[i-1]\, \boldsymbol{x}[i] \left(\boldsymbol{x}[i]^H \boldsymbol{\Sigma}[i-1]\boldsymbol{x}[i]\right)^{-1}, \tag{5.215}$$

$$\bar{\boldsymbol{k}}[i] = \bar{\boldsymbol{\Sigma}}[i-1]\, \bar{\boldsymbol{x}}[i] \left(\bar{\boldsymbol{x}}[i]^H \bar{\boldsymbol{\Sigma}}[i-1]\bar{\boldsymbol{x}}[i]\right)^{-1}, \tag{5.216}$$

$$\boldsymbol{g}_1[i] = \boldsymbol{g}_1[i-1] - \boldsymbol{k}[i] \left(\boldsymbol{x}[i]^H \boldsymbol{g}_1[i-1]\right) / \left\|\boldsymbol{g}_1[i-1] - \boldsymbol{k}[i]\left(\boldsymbol{x}[i]^H \boldsymbol{g}_1[i-1]\right)\right\|, \tag{5.217}$$

$$\bar{\boldsymbol{g}}_1[i] = \bar{\boldsymbol{h}}_1[i-1] - \bar{\boldsymbol{k}}[i] \left(\bar{\boldsymbol{x}}[i]^H \bar{\boldsymbol{g}}_1[i-1]\right) / \left\|\bar{\boldsymbol{g}}_1[i-1] - \bar{\boldsymbol{k}}[i]\left(\bar{\boldsymbol{x}}[i]^H \bar{\boldsymbol{g}}_1[i-1]\right)\right\|, \tag{5.218}$$

$$\boldsymbol{\Sigma}[i] = \boldsymbol{\Sigma}[i-1] - \boldsymbol{k}[i]\, \boldsymbol{x}[i]^H \boldsymbol{\Sigma}[i-1], \tag{5.219}$$

$$\bar{\boldsymbol{\Sigma}}[i] = \bar{\boldsymbol{\Sigma}}[i-1] - \bar{\boldsymbol{k}}[i]\, \bar{\boldsymbol{x}}[i]^H \bar{\boldsymbol{\Sigma}}[i-1]. \tag{5.220}$$

- *Form the detectors:*

$$\hat{\boldsymbol{w}}_{1,1}[i] = \boldsymbol{U}_s[i]\boldsymbol{\Lambda}_s^{-1}[i]\boldsymbol{U}_s[i]^H\Big(\boldsymbol{g}_1[i] \otimes \boldsymbol{s}_{1,1}\Big), \tag{5.221}$$

$$\hat{\boldsymbol{w}}_{2,1}[i] = \boldsymbol{U}_s[i]\boldsymbol{\Lambda}_s^{-1}[i]\boldsymbol{U}_s[i]^H\Big(\bar{\boldsymbol{g}}_1[i] \otimes \boldsymbol{s}_{2,1}\Big). \tag{5.222}$$

- *Perform differential detection:*

$$z_{1,1}[i] = \hat{\boldsymbol{w}}_{1,1}[i]^H \tilde{\boldsymbol{r}}[i], \tag{5.223}$$

$$z_{2,1}[i] = \hat{\boldsymbol{w}}_{2,1}[i]^H \tilde{\boldsymbol{r}}[i], \tag{5.224}$$

$$\hat{\beta}_{1,1}[i] = \text{sign}\left(\Re\Big\{z_{1,1}[i]\, z_{1,1}[i-1]^*\Big\}\right), \tag{5.225}$$

$$\hat{\beta}_{2,1}[i] = \text{sign}\left(\Re\Big\{z_{2,1}[i]\, z_{2,1}[i-1]^*\Big\}\right). \tag{5.226}$$

A group-blind sequential adaptive space-time multiuser detector can be implemented similarly. The adaptive receiver structure is illustrated in Fig. 5.13.

5.6 Adaptive Space-Time Multiuser Detection in Multipath CDMA

5.6.1 Signal Model

In this section we develop adaptive space-time multiuser detectors for asynchronous CDMA systems with two transmit and two receive antennas. The continuous-time signal transmitted from antennas 1 and 2 due to the kth user for time interval $i \in \{0, 1, \ldots\}$ is given by

$$x_k^{(1)}(t) = \sum_{i=0}^{M-1} \Big[b_{1,k}[i]s_{1,k}(t - 2iT) - b_{2,k}[i]s_{2,k}(t - (2i+1)T)\Big], \tag{5.227}$$

$$x_k^{(2)}(t) = \sum_{i=0}^{M-1} \Big[b_{2,k}[i]s_{2,k}(t - 2iT) + b_{1,k}[i]s_{1,k}(t - (2i+1)T)\Big], \tag{5.228}$$

where M denotes the length of the data frame, T denotes the information symbol interval, and $\{b_k[i]\}_i$ is the symbol stream of user k. Although this is an asynchronous system, we have, for notational simplicity, suppressed the delay associated with each user's signal and incorporated it into the path delays in (5.230). We assume that for each k the symbol stream, $\{b_k[i]\}_i$, is a collection of independent random variables that take on values of $+1$ and -1 with equal probability. Furthermore, we assume that the symbol streams of different users are independent. As discussed in Chapter 2, for the direct-sequence spread-spectrum (DS-SS) format, the user signaling waveforms have the form

$$s_{q,k}(t) = \sum_{j=0}^{N-1} c_{q,k}[j]\psi(t - jT_c), \qquad 0 \le t \le T, \tag{5.229}$$

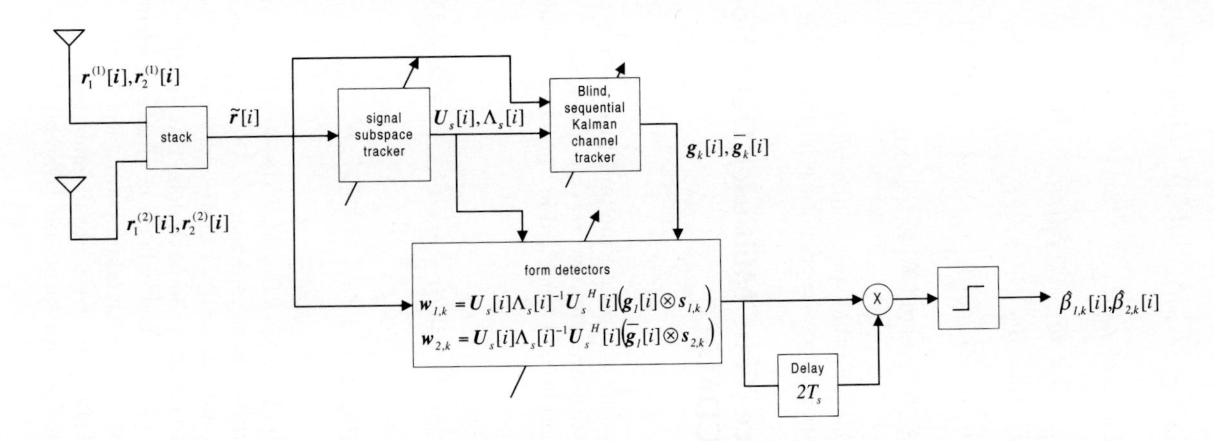

Figure 5.13. Adaptive receiver structure for linear space-time multiuser detectors.

where N is the processing gain, $\{c_{q,k}[j]\}, q \in \{1,2\}$ is a signature sequence of ± 1's assigned to the kth user for bit $b_{q,k}[i]$, and $\psi(t)$ is a normalized chip waveform of duration $T_c = T/N$. The kth user's signals, $x_k^{(1)}(t)$ and $x_k^{(2)}(t)$, propagate from transmit antenna a to receive antenna b through a multipath fading channel whose impulse response is given by

$$g_k^{(a,b)}(t) = \sum_{\ell=1}^{L} \alpha_{\ell,k}^{(a,b)} \delta\left(t - \tau_{\ell,k}^{(a,b)}\right), \tag{5.230}$$

where $\alpha_{\ell,k}^{(a,b)}$ is the complex path gain from antenna a to antenna b associated with the ℓth path for the kth user, and $\tau_{\ell,k}^{(a,b)}, \tau_{1,k}^{(a,b)} < \tau_{2,k}^{(a,b)} < \cdots < \tau_{L,k}^{(a,b)}$ is the sum of the corresponding path delay and initial transmission delay of user k. It is assumed that the channel is slowly varying, so that the path gains and delays remain constant over the duration of one signal frame (MT).

The received signal component due to the transmission of $x_k^{(1)}(t)$ and $x_k^{(2)}(t)$ through the channel at receive antennas 1 and 2 is given by

$$y_k^{(1)}(t) = x_k^{(1)}(t) \star g_k^{(1,1)}(t) + x_k^{(2)}(t) \star g_k^{(2,1)}(t), \tag{5.231}$$
$$y_k^{(2)}(t) = x_k^{(1)}(t) \star g_k^{(1,2)}(t) + x_k^{(2)}(t) \star g_k^{(2,2)}(t). \tag{5.232}$$

Substituting (5.228) and (5.230) into (5.232), we have for receive antenna $b \in \{1,2\}$,

$$\begin{aligned} y_k^{(b)}(t) = &\sum_{i=0}^{M-1} \left[b_{1,k}[i] s_{1,k}(t - 2iT) \star g_k^{(1,b)}(t) - b_{2,k}[i] s_{2,k}(t-(2i+1)T) \star g_k^{(1,b)}(t)\right] \\ &+ \sum_{i=0}^{M-1} \left[b_{2,k}[i] s_{2,k}(t-2iT) \star g_k^{(2,b)}(t)\right. \\ &\left. + b_{1,k}[i] s_{1,k}(t-(2i+1)T) \star g_k^{(2,b)}(t)\right]. \end{aligned} \tag{5.233}$$

For $a, b, q \in \{1,2\}$, we define

$$\begin{aligned} h_{q,k}^{(a,b)}(t) &\triangleq s_{q,k}(t) \star g_k^{(a,b)}(t) \\ &= \sum_{j=0}^{N-1} c_{q,k}[j] \underbrace{\left[\sum_{\ell=1}^{L} \alpha_{\ell,k}^{(a,b)} \psi\left(t - jT_c - \tau_{\ell,k}^{(a,b)}\right)\right]}_{\bar{g}_k^{(a,b)}(t-jT_c)}. \end{aligned} \tag{5.234}$$

In (5.234), $\overline{g}_k^{(a,b)}(t)$ is the composite channel response for the channel between transmit antenna a and receive antenna b, taking into account the effects of the chip pulse waveform and the multipath channel. Then we have

$$\begin{aligned} y_k^{(b)}(t) = &\sum_{i=0}^{M-1} \left[b_{1,k}[i] h_{1,k}^{(1,b)}(t-2iT) - b_{2,k}[i] h_{2,k}^{(1,b)}(t-(2i+1)T) \right] \\ &+ \sum_{i=0}^{M-1} \left[b_{2,k}[i] h_{2,k}^{(2,b)}(t-2iT) + b_{1,k}[i] h_{1,k}^{(2,b)}(t-(2i+1)T) \right]. \end{aligned} \tag{5.235}$$

The total received signal at receive antenna $b \in \{1,2\}$ is given by

$$r^{(b)}(t) = \sum_{k=1}^{K} y_k^{(b)}(t) + v^{(b)}(t). \tag{5.236}$$

At the receiver, the received signal is match-filtered to the chip waveform and sampled at the chip rate (i.e., the sampling interval is T_c, N is the total number of samples per symbol interval, and $2N$ is the total number of samples per time slot). The nth matched-filter output during the ith time slot is given by

$$\begin{aligned} r^{(b)}[i,n] &\triangleq \int_{2iT+nT_c}^{2iT+(n+1)T_c} r^{(b)}(t)\psi(t-2iT-nT_c)\,\mathrm{d}t \\ &= \sum_{k=1}^{K} \underbrace{\left\{ \int_{2iT+nT_c}^{2iT+(n+1)T_c} \psi(t-2iT-nT_c) y_k^{(b)}(t)\,\mathrm{d}t \right\}}_{y_k^{(b)}[i,n]} \\ &+ \underbrace{\int_{2iT+nT_c}^{2iT+(n+1)T_c} v^{(b)}(t)\psi(t-2iT-nT_c)\,\mathrm{d}t}_{v^{(b)}[i,n]}. \end{aligned} \tag{5.237}$$

Denote the maximum delay (in symbol intervals) as

$$\iota_k^{(a,b)} \triangleq \left\lceil \frac{\tau_{L,k}^{(a,b)} + T_c}{T} \right\rceil \quad \text{and} \quad \iota \triangleq \max_{k,a,b} \iota_k^{(a,b)}. \tag{5.238}$$

Substituting (5.235) into (5.237), we obtain

$$\begin{aligned} &y_k^{(b)}[i,n] \\ &= \sum_{p=0}^{M-1} \left\{ b_{1,k}[p] \int_{2iT+nT_c}^{2iT+(n+1)T_c} h_{1,k}^{(1,b)}(t-2pT)\psi(t-2iT-nT_c)\,\mathrm{d}t \right. \\ &- b_{2,k}[p] \int_{2iT+nT_c}^{2iT+(n+1)T_c} h_{2,k}^{(1,b)}(t-(2p+1)T)\psi(t-2iT-nT_c)\,\mathrm{d}t \end{aligned}$$

$$+ b_{2,k}[p] \int_{2iT+nT_c}^{2iT+(n+1)T_c} h_{2,k}^{(2,b)}(t - 2pT)\psi(t - 2iT - nT_c)\, dt$$
$$+ b_{1,k}[p] \int_{2iT+nT_c}^{2iT+(n+1)T_c} h_{1,k}^{(2,b)}(t - (2p-1)T)\psi(t - 2iT - nT_c)\, dt\Big\}. \quad (5.239)$$

Further substitution of (5.234) into (5.239) shows that

$$
\begin{aligned}
& y_k^{(b)}[i,n] \\
&= \sum_{p=0}^{M-1} \Big\{ b_{1,k}[p] \sum_{j=0}^{N-1} c_{1,k}[j] \sum_{\ell=1}^{L} \alpha_{\ell,k}^{(1,b)} \\
&\quad \cdot \int_{2iT+nT_c}^{2iT+(n+1)T_c} \psi(t - 2iT_s - nT_c)\psi(t - 2pT - jT_c - \tau_{\ell,k}^{(1,b)})\, dt \\
&\quad - b_{2,k}[p] \sum_{j=0}^{N-1} c_{2,k}[j] \sum_{\ell=1}^{L} \alpha_{\ell,k}^{(1,b)} \\
&\quad \cdot \int_{2iT+nT_c}^{2iT+(n+1)T_c} \psi(t - 2iT - nT_c)\psi(t - (2p+1)T - jT_c - \tau_{\ell,k}^{(1,b)})\, dt \\
&\quad + b_{2,k}[p] \sum_{j=0}^{N-1} c_{2,k}[j] \sum_{\ell=1}^{L} \alpha_{\ell,k}^{(2,b)} \\
&\quad \cdot \int_{2iT+nT_c}^{2iT+(n+1)T_c} \psi(t - 2iT - nT_c)\psi(t - 2pT - jT_c - \tau_{\ell,k}^{(2,b)})\, dt \\
&\quad + b_{1,k}[p] \sum_{j=0}^{N-1} c_{1,k}[j] \sum_{\ell=1}^{L} \alpha_{\ell,k}^{(2,b)} \\
&\quad \cdot \int_{2iT+nT_c}^{2iT+(n+1)T_c} \psi(t - 2iT - nT_c)\psi(t - (2p+1)T - jT_c - \tau_{\ell,k}^{(2,b)})\, dt \Big\} \\
&= \sum_{p=0}^{\lceil \iota/2 \rceil} \Big\{ b_{1,k}[i-p] \\
&\underbrace{\sum_{j=0}^{N-1} c_{1,k}[j] \overbrace{\sum_{\ell=1}^{L} \alpha_{\ell,k}^{(1,b)} \int_0^{T_c} \psi(t)\psi(t - jT_c - \tau_{\ell,k}^{(1,b)} + 2pT + nT_c)\, dt}^{f_k^{(1,b)}[n+2pN-j]}]}_{h_{1,k}^{(1,b)}[p,n]}
\end{aligned}
$$

$$
- b_{2,k}[i-p] \underbrace{\sum_{j=0}^{N-1} c_{2,k}[j] \overbrace{\sum_{\ell=1}^{L} \alpha_{\ell,k}^{(1,b)} \int_0^{T_c} \psi(t)\psi(t - jT_c - \tau_{\ell,k}^{(1,b)} + 2pT - T + nT_c)\,\mathrm{d}t}^{f_k^{(1,b)}[n+2pN-N-j]}}_{h_{2,k}^{(1,b)}[p,n]}
$$

$$
+ b_{2,k}[i-p] \underbrace{\sum_{j=0}^{N-1} c_{2,k}[j] \overbrace{\sum_{\ell=1}^{L} \alpha_{\ell,k}^{(2,b)} \int_0^{T_c} \psi(t)\psi(t - jT_c - \tau_{\ell,k}^{(2,b)} + 2pT + nT_c)\,\mathrm{d}t}^{f_k^{(2,b)}[n+2pN-j]}}_{h_{2,k}^{(2,b)}[p,n]}
$$

$$
+ b_{1,k}[i-p] \underbrace{\sum_{j=0}^{N-1} c_{1,k}[j] \overbrace{\sum_{\ell=1}^{L} \alpha_{\ell,k}^{(2,b)} \int_0^{T_c} \psi(t)\psi(t - jT_c - \tau_{\ell,k}^{(2,b)} + 2pT - T + nT_c)\,\mathrm{d}t}^{f_k^{(2,b)}[n+2pN-N-j]}}_{h_{1,k}^{(2,b)}[p,n]} \Bigg\}. \tag{5.240}
$$

We may write $y_k^{(b)}[i,n]$ more compactly as

$$
\begin{aligned}
y_k^{(b)}[i,n] &= \sum_{j=0}^{\lceil \iota/2 \rceil} \Big(h_{1,k}^{(1,b)}[j,n] b_{1,k}[i-j] - h_{2,k}^{(1,b)}[j,n] b_{2,k}[i-j] \\
&\quad + h_{2,k}^{(2,b)}[j,n] b_{2,k}[i-j] + h_{1,k}^{(2,b)}[j,n] b_{1,k}[i-j] \Big) \\
&= \sum_{j=0}^{\lceil \iota/2 \rceil} b_{1,k}[i-j] g_{1,k}^{(b)}[j,n] + \sum_{j=0}^{\lceil \iota/2 \rceil} b_{2,k}[i-j] g_{2,k}^{(b)}[j,n],
\end{aligned} \tag{5.241}
$$

where

$$
g_{1,k}^{(b)}[j,n] \triangleq h_{1,k}^{(1,b)}[j,n] + h_{1,k}^{(2,b)}[j,n], \tag{5.242}
$$

$$
g_{2,k}^{(b)}[j,n] \triangleq h_{2,k}^{(2,b)}[j,n] - h_{2,k}^{(1,b)}[j,n]. \tag{5.243}
$$

For $j = 0, 1, \ldots, \lceil \iota/2 \rceil$, define the following notation:

$$\underline{H}^{(b)}[j] \triangleq \begin{bmatrix} g_{1,1}^{(b)}[j,0] & \cdots & g_{1,K}^{(b)}[j,0] & g_{2,1}^{(b)}[j,0] & \cdots & g_{2,K}^{(b)}[j,0] \\ \vdots & \vdots & \vdots & \vdots & \vdots & \vdots \\ g_{1,1}^{(b)}[j,2N-1] & \cdots & g_{1,K}^{(b)}[j,2N-1] & g_{2,1}^{(b)}[j,2N-1] & \cdots & g_{2,K}^{(b)}[j,2N-1] \end{bmatrix}_{2N\times 2K}. \tag{5.244}$$

$$\underline{r}^{(b)}[i] \triangleq \begin{bmatrix} r^{(b)}[i,0] \\ \vdots \\ r^{(b)}[i,2N-1] \end{bmatrix}_{2N\times 1}, \qquad \underline{v}^{(b)}[i] \triangleq \begin{bmatrix} v^{(b)}[i,0] \\ \vdots \\ v^{(b)}[i,2N-1] \end{bmatrix}_{2N\times 1}, \tag{5.245}$$

$$\underline{b}[i] \triangleq \begin{bmatrix} b_{1,1}[i] \\ \vdots \\ b_{1,K}[i] \\ b_{2,1}[i] \\ \vdots \\ b_{2,K}[i] \end{bmatrix}_{2K\times 1}. \tag{5.246}$$

Then we have

$$\underline{r}^{(b)}[i] = \underbrace{\sum_{j=0}^{\lceil \iota/2 \rceil} \underline{H}^{(b)}[j]\underline{b}[i-j]}_{\underline{H}^{(b)}[i]\star\underline{b}[i]} + \underline{v}^{(b)}[i].$$

To exploit both temporal and spatial diversity, we stack the vectors received from both receive antennas,

$$\underline{r}[i] \triangleq \begin{bmatrix} \underline{r}^{(1)}[i] \\ \underline{r}^{(2)}[i] \end{bmatrix}_{4N\times 1}$$

and observe that

$$\underline{r}[i] = \underline{H}[i] \star \underline{b}[i] + \underline{v}[i], \tag{5.247}$$

where

$$\underline{H}[j] \triangleq \begin{bmatrix} \underline{H}^{(1)}[j] \\ \underline{H}^{(2)}[j] \end{bmatrix}_{4N\times 2K}, \quad j = 0, 1, \ldots, \lceil \iota/2 \rceil \quad \text{and} \quad \underline{v}[i] \triangleq \begin{bmatrix} \underline{v}^{(1)}[i] \\ \underline{v}^{(2)}[i] \end{bmatrix}_{4N\times 1}.$$

By stacking m successive received sample vectors, we create the following quantities:

$$\boldsymbol{r}[i] \triangleq \begin{bmatrix} \underline{r}[i] \\ \vdots \\ \underline{r}[i+m-1] \end{bmatrix}_{4Nm\times 1}, \qquad \boldsymbol{v}[i] \triangleq \begin{bmatrix} \underline{v}[i] \\ \vdots \\ \underline{v}[i+m-1] \end{bmatrix}_{4Nm\times 1},$$

$$\boldsymbol{b}[i] \triangleq \begin{bmatrix} \underline{b}[i - \lceil \iota/2 \rceil] \\ \vdots \\ \underline{b}[i + m - 1] \end{bmatrix}_{r \times 1},$$

$$\boldsymbol{H} \triangleq \begin{bmatrix} \underline{H}[\lceil \iota/2 \rceil] & \cdots & \underline{H}[0] & \cdots & \mathbf{0} \\ \vdots & \ddots & \ddots & \ddots & \vdots \\ \mathbf{0} & \cdots & \underline{H}[\lceil \iota/2 \rceil] & \cdots & \underline{H}[0] \end{bmatrix}_{4Nm \times r},$$

where $r \triangleq 2K(m + \lceil \iota/2 \rceil)$. We can write (5.247) in matrix form as

$$\boldsymbol{r}[i] = \boldsymbol{H}\boldsymbol{b}[i] + \boldsymbol{v}[i]. \tag{5.248}$$

We will see in Section 5.6.3 that the smoothing factor, m, is chosen such that

$$m \geq \left\lceil \frac{N(\iota + 1) + K\lceil \iota/2 \rceil + 1}{2N - K} \right\rceil \tag{5.249}$$

for channel identifiability. Note that the columns of $\boldsymbol{H}$ (the composite signature vectors) contain information about both the timings and the complex path gains of the multipath channel of each user. Hence an estimate of these waveforms eliminates the need for separate estimates of the timing information $\left\{\tau_{\ell,k}^{(a,b)}\right\}_{\ell=1}^{L}$.

5.6.2 Blind MMSE Space-Time Multiuser Detection

Since the ambient noise is white (i.e., $E\{\boldsymbol{v}[i]\boldsymbol{v}[i]^H\} = \sigma^2 \boldsymbol{I}_{4Nm}$), the autocorrelation matrix of the received signal in (5.248) is

$$\boldsymbol{R} \triangleq E\{\boldsymbol{r}[i]\boldsymbol{r}[i]^H\} = \boldsymbol{H}\boldsymbol{H}^H + \sigma^2 \boldsymbol{I}_{4Nm} \tag{5.250}$$

$$= \boldsymbol{U}_s \boldsymbol{\Lambda}_s \boldsymbol{U}_s^H + \sigma^2 \boldsymbol{U}_n \boldsymbol{U}_n^H, \tag{5.251}$$

where (5.251) is the eigendecomposition of $\boldsymbol{R}$. The matrix $\boldsymbol{U}_s$ has dimensions $4Nm \times r$ and $\boldsymbol{U}_n$ has dimensions $4Nm \times (4Nm - r)$.

The linear MMSE space-time multiuser detector and corresponding bit estimate for $b_{a,k}[i], a \in \{1, 2\}$ are given, respectively, by

$$\boldsymbol{w}_{a,k} \triangleq \arg \min_{\boldsymbol{w} \in \mathbb{C}^{4Pm}} E\left\{\left|b_{a,k}[i] - \boldsymbol{w}^H \boldsymbol{r}[i]\right|^2\right\}, \tag{5.252}$$

$$\hat{b}_{a,k}[i] = \text{sign}\left(\Re\left\{\boldsymbol{w}_{a,k}^H \boldsymbol{r}[i]\right\}\right). \tag{5.253}$$

The solution to (5.252) can be written in terms of the signal subspace components as

$$\boldsymbol{w}_{a,k} = \boldsymbol{U}_s \boldsymbol{\Lambda}_s^{-1} \boldsymbol{U}_s^H \boldsymbol{h}_{a,k}, \tag{5.254}$$

where $\boldsymbol{h}_{a,k} \triangleq \boldsymbol{H}\boldsymbol{e}_{K(2\lceil \iota/2 \rceil + a - 1) + k}$ is the composite signature waveform of user k for bit $a \in \{1, 2\}$. This detector is termed *blind* since it requires knowledge only of the signature sequence of the user of interest. Of course, we also require estimates of the signal subspace components and of the channel. We address the issue of channel estimation next.

5.6.3 Blind Adaptive Channel Estimation

In this section we extend the blind adaptive channel estimation technique described in Section 5.6.2 to the asynchronous multipath case. First, however, we describe the discrete-time channel model in order to formulate an analog to the optimization problem in (5.208).

Discrete-Time Channel Model

Using (5.244), it is easy to see that

$$\boldsymbol{h}_{a,k} = \begin{bmatrix} g_{a,k}^{(1)}[0,0] \\ \vdots \\ g_{a,k}^{(1)}[0,2N-1] \\ g_{a,k}^{(2)}[0,0] \\ \vdots \\ g_{a,k}^{(2)}[0,2N-1] \\ \vdots \\ g_{a,k}^{(1)}[\lceil \iota/2 \rceil,0] \\ \vdots \\ g_{a,k}^{(1)}[\lceil \iota/2 \rceil,2N-1] \\ g_{a,k}^{(2)}[\lceil \iota/2 \rceil,0] \\ \vdots \\ g_{a,k}^{(2)}[\lceil \iota/2 \rceil,2N-1] \end{bmatrix}_{4N(\lceil \iota/2 \rceil+1)\times 1} . \tag{5.255}$$

From (5.241) we have for $j = 0, \ldots, \lceil \iota/2 \rceil; n = 0, \ldots, 2N-1; b = 1, 2,$

$$g_{1,k}^{(b)}[j,n] = h_{1,k}^{(1,b)}[j,n] + h_{1,k}^{(2,b)}[j,n], \tag{5.256}$$
$$g_{2,k}^{(b)}[j,n] = h_{2,k}^{(2,b)}[j,n] - h_{2,k}^{(1,b)}[j,n]. \tag{5.257}$$

We will develop the discrete-time channel model for $g_{1,k}^{b}[j,n]$. The development for $g_{2,k}^{b}[j,n]$ follows similarly. From (5.240) we see that

$$g_{1,k}^{b}[j,n] \\ = \sum_{q=0}^{N-1} c_{1,k}[q] f_k^{(1,b)}[n+2jN-q] + \sum_{q=0}^{N-1} c_{1,k}[q] f_k^{(2,b)}[n+2jN-N-q]. \tag{5.258}$$

From (5.240) we can also see that the sequences $f_k^{1,b}[i]$ and $f_k^{2,b}[i]$ are zero whenever $i < 0$ or $i > \iota N$. With this in mind we define the following vectors:

$$\boldsymbol{g}_{1,k}^{(b)} \triangleq \left[g_{1,k}^{(b)}[0,0] \cdots g_{1,k}^{(b)}[0,2N-1] \cdots g_{1,k}^{(b)}[\lceil \iota/2 \rceil, 0] \cdots g_{1,k}^{(b)}[\lceil \iota/2 \rceil, 2N-1]\right]^T,$$

$$\boldsymbol{f}_{1,k}^{(1,b)} \triangleq \left[f_k^{(1,b)}[0] \cdots f_k^{(1,b)}[\iota N] \underbrace{0 \cdots 0}_{N\text{zeros}}\right]^T,$$

$$\boldsymbol{f}_{1,k}^{(2,b)} \triangleq \left[\underbrace{0 \ldots 0}_{N\text{zeros}} \; f_k^{(2,b)}[0] \cdots f_k^{(2,b)}[\iota N]\right]^T.$$

Then we have

$$\boldsymbol{g}_{1,k}^{(b)} = \boldsymbol{C}_{1,k} \cdot \underbrace{\left[\boldsymbol{f}_{1,k}^{(1,b)} + \boldsymbol{f}_{1,k}^{(2,b)}\right]}_{\boldsymbol{f}_{1,k}^{(b)}}, \tag{5.259}$$

where

$$\boldsymbol{C}_{1,k} \triangleq \begin{bmatrix} c_{1,k}[0] & & \\ c_{1,k}[0] & \ddots & \\ \vdots & \ddots & c_{1,k}[0] \\ \vdots & & c_{1,k}[1] \\ c_{1,k}[N-1] & & \vdots \\ & \ddots & \vdots \\ & & c_{1,k}[N-1] \end{bmatrix}_{(2N(\lceil \iota/2 \rceil+1)) \times (N(\iota+1)+1)}.$$

A similar development for $g_{2,k}^b[j,n]$ produces the result

$$\boldsymbol{g}_{2,k}^{(b)} = \boldsymbol{C}_{2,k} \cdot \underbrace{\left[\boldsymbol{f}_{2,k}^{(2,b)} - \boldsymbol{f}_{2,k}^{(1,b)}\right]}_{\boldsymbol{f}_{2,k}^{(b)}}, \tag{5.260}$$

where

$$\boldsymbol{f}_{2,k}^{(2,b)} \triangleq \left[f_k^{(2,b)}[0] \cdots f_k^{(2,b)}[\iota N] \underbrace{0 \cdots 0}_{N\text{zeros}}\right]^T,$$

$$\boldsymbol{f}_{2,k}^{(1,b)} \triangleq \left[\underbrace{0 \cdots 0}_{N\text{zeros}} \; f_k^{(1,b)}[0] \cdots f_k^{(1,b)}[\iota N]\right]^T.$$

The final task in the section is to form expressions for the composite signature waveforms $\boldsymbol{h}_{1,k}$ and $\boldsymbol{h}_{2,k}$ in terms of the signature matrices $\boldsymbol{C}_{1,k}, \boldsymbol{C}_{2,k}$ and the channel response vectors $\boldsymbol{f}_{1,k}^{(b)}$ and $\boldsymbol{f}_{2,k}^{(b)}$. Denote by $\boldsymbol{C}_{a,k}[j], j = 0, 1, \ldots, \lceil \iota/2 \rceil, a \in$

$\{1, 2\}$ the submatrix of $\boldsymbol{C}_{a,k}$ consisting of rows $2Nj + 1$ through $2(j + 1)N$. Then it is easy to show that

$$\boldsymbol{h}_{a,k} = \overline{\boldsymbol{C}}_{a,k} \boldsymbol{f}_{a,k}, \tag{5.261}$$

where

$$\overline{\boldsymbol{C}}_{a,k} \triangleq \begin{bmatrix} \boldsymbol{C}_{a,k}[0] & \mathbf{0} \\ \mathbf{0} & \boldsymbol{C}_{a,k}[0] \\ \boldsymbol{C}_{a,k}[1] & \mathbf{0} \\ \mathbf{0} & \boldsymbol{C}_{a,k}[1] \\ \vdots & \vdots \\ \boldsymbol{C}_{a,k}[\lceil \iota/2 \rceil] & \mathbf{0} \\ \mathbf{0} & \boldsymbol{C}_{a,k}[\lceil \iota/2 \rceil] \end{bmatrix}_{4N(\lceil \iota/2 \rceil + 1) \times (2N(\iota+1)+2)},$$

$$\boldsymbol{f}_{a,k} \triangleq \begin{bmatrix} \boldsymbol{f}_{a,k}^{(1)} \\ \boldsymbol{f}_{a,k}^{(2)} \end{bmatrix}_{(2N(\iota+1)+2) \times 1}.$$

Blind Adaptive Channel Estimation

The blind channel estimation problem for the asynchronous multipath case involves the estimation of $\boldsymbol{f}_{a,k}$ from the received signal $\boldsymbol{r}[i]$. As we did for the synchronous case, we exploit the orthogonality between the signal subspace and noise subspace. Specifically, since $\boldsymbol{U}_n$ is orthogonal to the column space of $\boldsymbol{H}$, we have

$$\boldsymbol{U}_n^H \boldsymbol{h}_{a,k} = \boldsymbol{U}_n^H \overline{\boldsymbol{C}}_{a,k} \boldsymbol{f}_{a,k} = \mathbf{0}. \tag{5.262}$$

Denote by $\boldsymbol{z}[i]$ the projection of the received signal $\boldsymbol{r}[i]$ onto the noise subspace:

$$\boldsymbol{z}[i] = \boldsymbol{r}[i] - \boldsymbol{U}_s \boldsymbol{U}_s^H \boldsymbol{r}[i] \tag{5.263}$$

$$= \boldsymbol{U}_n \boldsymbol{U}_n^H \boldsymbol{r}[i]. \tag{5.264}$$

Using (5.262), we have

$$\boldsymbol{f}_{a,k}^H \overline{\boldsymbol{C}}_{a,k}^H \boldsymbol{z}[i] = 0. \tag{5.265}$$

Our channel estimation problem, then, involves solution of the optimization problem

$$\hat{\boldsymbol{f}}_{a,k} = \arg\min_{\boldsymbol{f}} E\left\{ \left| \boldsymbol{f}^H \overline{\boldsymbol{C}}_{a,k}^H \boldsymbol{z}[i] \right|^2 \right\} \tag{5.266}$$

subject to the constraint $\|\boldsymbol{f}\| = 1$. If we denote $\boldsymbol{x}[i] \triangleq \overline{\boldsymbol{C}}_{a,k}^H \boldsymbol{z}[i]$, we can use the Kalman-type algorithm described in (5.209)–(5.211), where $\boldsymbol{h}_1[i]$ is replaced with $\boldsymbol{f}_{a,k}[i]$.

Note that a necessary condition for the channel estimate to be unique is that the matrix $\boldsymbol{U}_n^H \overline{\boldsymbol{C}}_{a,k}$ is tall [i.e., $4Nm - 2K(m + \lceil \iota/2 \rceil) \geq 2N(\iota + 1) + 2$]. Therefore, we choose the smoothing factor m such that

$$m \geq \left\lceil \frac{N(\iota + 1) + K\lceil \iota/2 \rceil + 1}{2N - K} \right\rceil. \tag{5.267}$$

Using the same constraint, we find that for a fixed m, the maximum number of users that can be supported is

$$\min\left\{\left\lfloor\frac{N(2m-\iota-1)-1}{m+\lceil\iota/2\rceil}\right\rfloor, \frac{N}{2}\right\}. \tag{5.268}$$

Equation (5.268) illustrates one of the benefits of the proposed space-time receiver structure over the linear diversity structure. Notice that for reasonable choices of m and ι, (5.268) is larger than the maximum number of users for the linear diversity receiver structure, given by

$$\left\lfloor\frac{N(m-\iota)}{2(m+\iota)}\right\rfloor. \tag{5.269}$$

Another significant benefit of the space-time receiver is reduced complexity. The diversity structure requires four independent subspace trackers operating simultaneously since the receiver performs detection on each of the four received signals, and each has a different signal subspace. Since we stack received signal vectors before detection, the space-time structure requires only one subspace tracker.

Once an estimate of the channel state, $\hat{\boldsymbol{f}}_{a,k}$, is obtained, the composite signature vector of the kth user for bit a is given by (5.261). Note that there is an arbitrary phase ambiguity in the estimated channel state, which necessitates differential encoding and decoding of the transmitted data. Finally, we summarize the blind adaptive linear space-time multiuser detection algorithm in multipath CDMA as follows.

Algorithm 5.6: [Blind adaptive linear space-time multiuser detector—asynchronous multipath CDMA, two transmit antennas, and two receive antennas]

- *Stack matched-filter outputs in (5.237) to form* $\mathbf{r}[i]$ *in (5.248).*
- *Form* $\overline{\boldsymbol{C}}_{a,k}$ *as discussed in Section 5.6.3.*
- *Using a suitable signal subspace tracking algorithm (e.g., NAHJ), update the signal subspace components* $\boldsymbol{U}_s[i]$ *and* $\boldsymbol{\Lambda}_s[i]$ *at each time slot* i.
- *Track the channel* $\boldsymbol{f}_{a,k}$ *according to the following*

$$\boldsymbol{z}[i] = \boldsymbol{r}[i] - \boldsymbol{U}_s[i]\boldsymbol{U}_s[i]^H\boldsymbol{r}[i], \tag{5.270}$$

$$\boldsymbol{x}[i] = \overline{\boldsymbol{C}}_{a,k}^H\boldsymbol{z}[i], \tag{5.271}$$

$$\boldsymbol{k}[i] = \boldsymbol{\Sigma}[i-1]\boldsymbol{x}[i]\left(\boldsymbol{x}[i]^H\boldsymbol{\Sigma}[i-1]\boldsymbol{x}[i]\right)^{-1}, \tag{5.272}$$

$$\boldsymbol{f}_{a,k}[i] = \boldsymbol{f}_{a,k}[i-1] - \boldsymbol{k}[i] \left(\boldsymbol{x}[i]^H\boldsymbol{f}_{a,k}[i-1]\right) / \left\|\boldsymbol{f}_{a,k}[i-1] - \boldsymbol{k}[i]\left(\boldsymbol{x}[i]^H\boldsymbol{f}_{a,k}[i-1]\right)\right\|, \tag{5.273}$$

$$\boldsymbol{\Sigma}[i] = \boldsymbol{\Sigma}[i-1] - \boldsymbol{k}[i]\,\boldsymbol{x}[i]^H\boldsymbol{\Sigma}[i-1]. \tag{5.274}$$

- *Form the detectors:*

$$\boldsymbol{w}_{a,k}[i] = \boldsymbol{U}_s[i]\boldsymbol{\Lambda}_s^{-1}[i]\boldsymbol{U}_s[i]^H\overline{\boldsymbol{C}}_{a,k}\boldsymbol{f}_{a,k}[i]. \tag{5.275}$$

- *Perform differential detection:*

$$z_{a,k}[i] = \boldsymbol{w}_{a,k}[i]^H \boldsymbol{r}[i], \tag{5.276}$$

$$\hat{\beta}_{a,k}[i] = \text{sign}\left(\Re\left\{z_{a,k}[i]\, z_{a,k}[i-1]^*\right\}\right). \tag{5.277}$$

Simulation Results

In what follows we present simulation results to illustrate the performance of blind adaptive space-time multiuser detection. We first look at the synchronous flat-fading case; then we consider the asynchronous multipath-fading scenario. For all simulations we use the two transmit/two receive antenna configuration. Gold codes of length 15 are used for each user. The chip pulse is a raised cosine with roll-off factor 0.5. For the multipath case, each user has $L = 3$ paths. The delay of each path is uniform on $[0, T]$. Hence, the maximum delay spread is one symbol interval (i.e., $\iota = 1$). The fading gain for each user's channel is generated from a complex Gaussian distribution and is fixed for all simulations. The path gains in each user's channel are normalized so that each user's signal arrives at the receiver with the same power. The smoothing factor is $m = 2$. For SINR plots, the number of users for the first 1500 iterations is four. At iteration 1501, three users are added so that the system is fully loaded. At iteration 3001, five users are removed. For the steady-state bit-error-probability plots, the frame size is 200 bits and the system is allowed 1000 bits to reach steady state before errors are checked. The forgetting factor for the subspace tracking algorithm for all simulations is 0.995.

The performance measures are bit-error probability and the signal-to-interference-plus-noise ratio, defined by SINR $\triangleq E^2\{\boldsymbol{w}^H \boldsymbol{r}\}/\text{Var}\{\boldsymbol{w}^H \boldsymbol{r}\}$, where the expectation is with respect to the data bits of interfering users, ISI bits, and ambient noise. In the simulations, the expectation operation is replaced by time averaging. SINR is a particularly appropriate figure of merit for MMSE detectors since it has been shown [386] that the output of an MMSE detector is approximately Gaussian distributed. Hence, the SINR values translate directly and simply to bit-error probabilities [i.e., $P_e(\text{SINR}) \approx Q\left(\sqrt{\text{SINR}}\right)$]. The labeled horizontal lines on the SINR plots (Figs. 5.14 and 5.16) represent BER thresholds.

Figure 5.14 illustrates the adaptation performance for the synchronous case. The SNR is 8 dB. Notice that the BER does not drop below 10^{-3} even when users enter or leave the system.

Figure 5.15 shows the steady-state performance for the synchronous case for different system loads. We see that the performance changes little as the system load changes.

Figure 5.16 shows the adaptation performance for the asynchronous multipath case. The SNR for this simulation is 11 dB. Again, notice that the BER does not drop significantly as users enter and leave the system.

Figure 5.17 shows the steady-state performance for the asynchronous multipath case for different system loads. It is seen that system loading has a more significant effect on performance for the asynchronous multipath case than it does for the synchronous case.

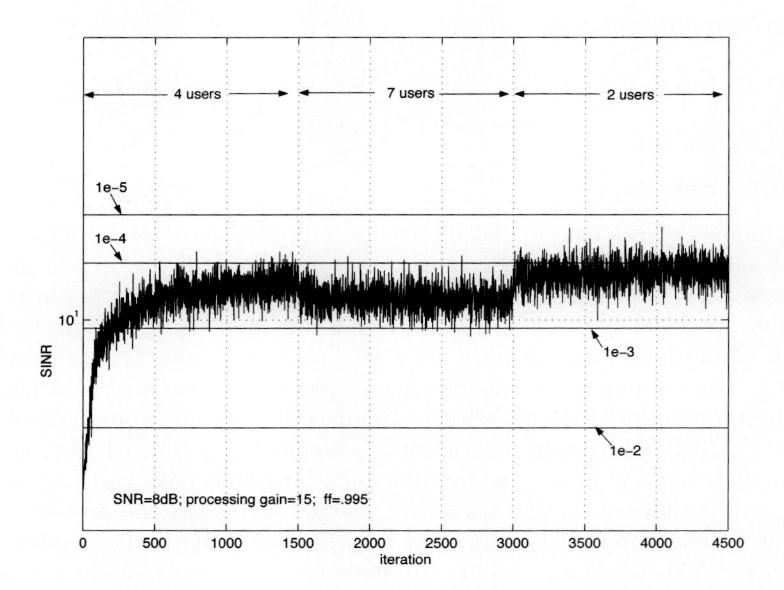

Figure 5.14. Adaptation performance of space-time multiuser detection for synchronous CDMA. The labeled horizontal lines represent bit-error-probability thresholds.

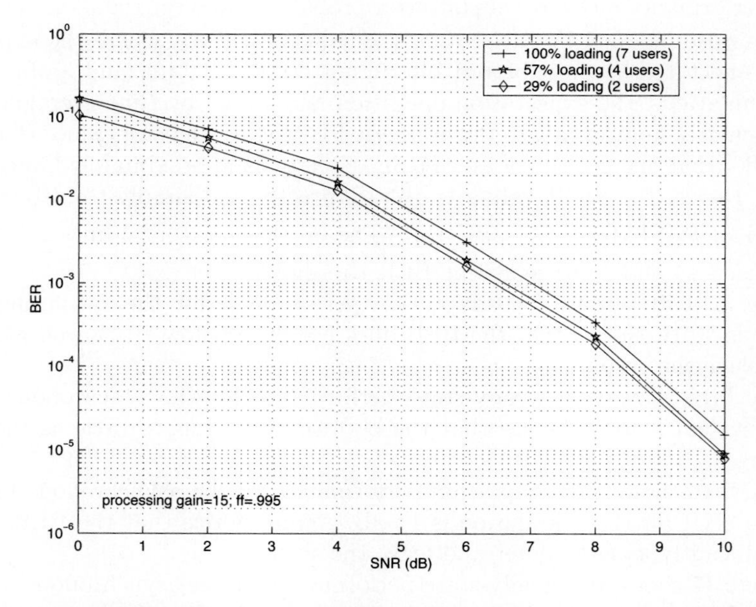

Figure 5.15. Steady-state performance of space-time multiuser detection for synchronous CDMA.

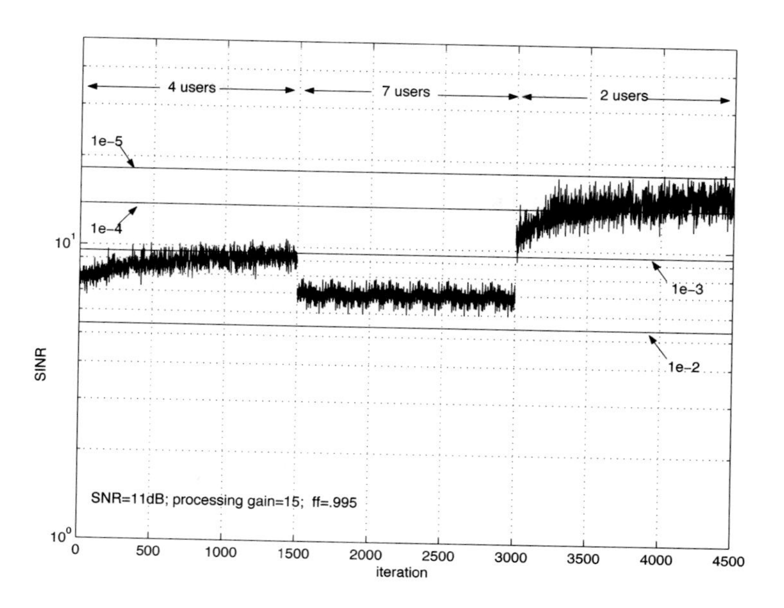

Figure 5.16. Adaptation performance of space-time multiuser detection for asynchronous multipath CDMA. The labeled horizontal lines represent BER thresholds.

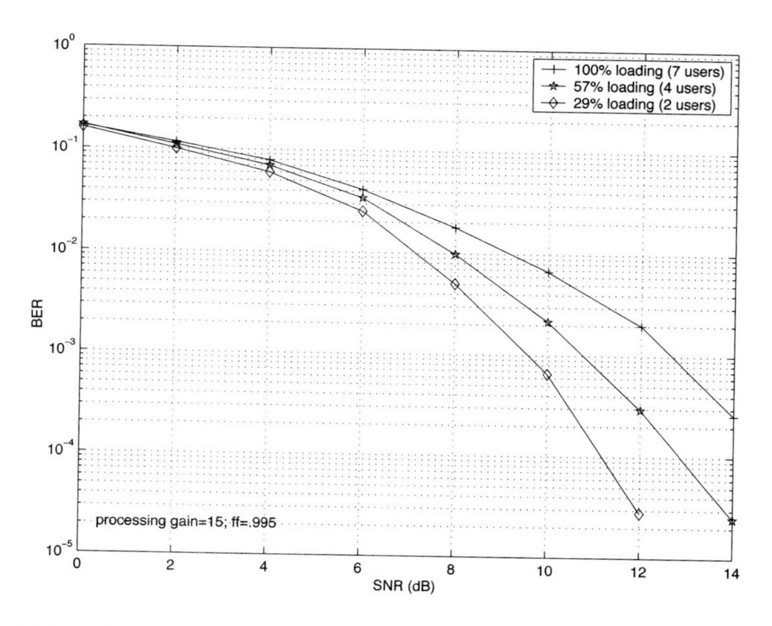

Figure 5.17. Steady-state performance of space-time multiuser detection for asynchronous multipath CDMA.

Chapter 6

TURBO MULTIUSER DETECTION

6.1 Introduction to Turbo Processing

In recent years, iterative (turbo) processing techniques have received considerable attention, stimulated by the discovery of powerful turbo codes [35, 36]. The *turbo principle* can be applied successfully to many detection/decoding problems, such as serial concatenated decoding, equalization, coded modulation, multiuser detection, and joint source and channel decoding [169, 379, 448]. We start the discussion in this chapter by illustrating the general concept of turbo processing for concatenated systems using a simple example.

A typical communications system in general consists of a cascade of subsystems with different signal processing functionalities. Consider, for example, the simple communication system employing channel coding and signaling over an intersymbol interference (ISI) channel, as shown in Fig. 6.1. In a conventional receiver, the demodulator makes hard decisions about the transmitted bits $\{b[i]\}$ based on the received signal $r(t)$, which are then passed to the channel decoder to decode the transmitted information. The problem with this approach is that by making hard decisions of the bits, information loss is incurred in each subsystem (i.e., demodulator and decoder). This is because while the subsystem only indicates whether it believes that a given bit is a 0 or a 1, it usually has sufficient information to estimate the degree of confidence in its decisions. One straightforward way to reduce the loss of information, and the resulting loss in performance, is to pass the confidence level along with the decision (i.e., to render soft decisions). This is often done when passing information from a demodulator to a channel decoder, which is known to result in approximately a 2-dB performance gain in the additive white Gaussian noise (AWGN) channel [396]. However, even if optimal bit-by-bit soft decisions are passed between all the subsystems in the receiver, the overall performance can still be far from optimal. This is due to the fact that while later stages (e.g., the channel decoder) can use the information gleaned from previous stages (e.g., the demodulator), the reverse is not generally true. While the optimal performance can be achieved by performing a joint detection, taking all receiver processing into account simultaneously (e.g., maximum likelihood detection based

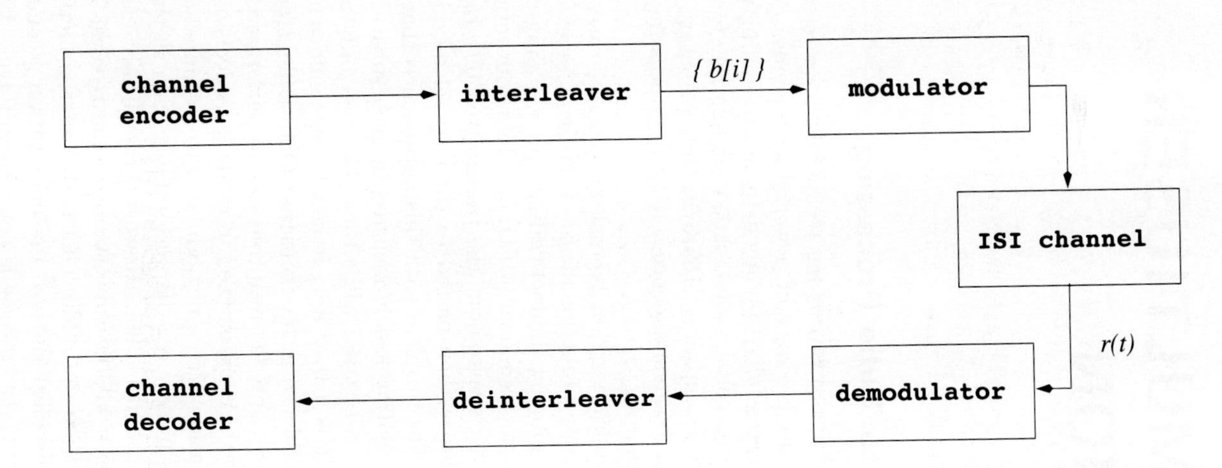

Figure 6.1. Coded communication system signaling through an intersymbol interference (ISI) channel.

on the supertrellis of both the channel code and the ISI channel), the complexity of such a joint approach is usually prohibitive. This motivates an iterative (turbo) processing approach which allows earlier stages (e.g., the demodulator) to refine their processing based on information obtained from later stages (e.g., the channel decoder).

To employ turbo processing in the system shown in Fig. 6.1, both the demodulator and channel decoder are of the maximum *a posteriori* probability (MAP) type. The function of a MAP demodulator is to produce soft decisions which reflect the probability that a given bit is a 0 or a 1. At the lth iteration, the information available to the MAP demodulator consists of the received signal $r(t)$ and the *a priori* probabilities of the input bits, the latter of which are obtained by the MAP channel decoder based on its output from the $(l-1)$th iteration. The MAP demodulator uses this information, combined with knowledge of the chosen modulation and of the channel structure, to produce the *a posteriori* probabilities (APPs) of the channel bits:

$$P^l(b[i] = 1 \mid r(t)) = \frac{p(r(t) \mid b[i] = 1)\, P^{l-1}(b[i] = 1)}{p(r(t))}, \tag{6.1}$$

$$P^l(b[i] = 0 \mid r(t)) = \frac{p(r(t) \mid b[i] = 0)\, P^{l-1}(b[i] = 0)}{p(r(t))}, \tag{6.2}$$

for all $\{b[i]\}_i$. Consider the log-likelihood ratio (LLR) formed from the *a posteriori* probabilities of (6.1) and (6.2):

$$\begin{aligned} \Lambda_1^l(b[i]) &\triangleq \log \frac{P^l(b[i] = 1 \mid r(t))}{P^l(b[i] = 0 \mid r(t))} \\ &= \underbrace{\log \frac{p(r(t) \mid b[i] = 1)}{p(r(t) \mid b[i] = 0)}}_{\lambda_1^l(b[i])} + \underbrace{\log \frac{P^{l-1}(b[i] = 1)}{P^{l-1}(b[i] = 0)}}_{\lambda_2^{l-1}(b[i])}. \end{aligned} \tag{6.3}$$

It is seen from (6.3) that the LLR is the sum of two distinct quantities. The first term, $\lambda_1^l(b[i])$, is the *extrinsic information* produced by the first-stage subsystem in the receiver (i.e., the MAP demodulator), which is information that the MAP demodulator gleans about $b[i]$ from the received signal $r(t)$ and the *a priori* probabilities of the other transmitted bits, without using the *a priori* probability of $b[i]$. The second term, $\lambda_2^{l-1}(b[i])$, contains the *a priori* probability of $b[i]$. Note that typically, for the first iteration ($l = 1$), we set $P^0(b[i] = 1) = P^0(b[i] = 0) = \frac{1}{2}$ [i.e., $\lambda_2^0(b[i]) = 0$ for all i]. The extrinsic information $\{\lambda_1^l(b[i])\}$ produced by the MAP demodulator is sent to the second-stage subsystem (i.e., the MAP channel decoder) as the *a priori* information for channel decoding.

Based on the *a priori* information provided by the MAP demodulator, and the channel code constraints, the MAP channel decoder computes the *a posteriori* LLR of each code bit:

$$\Lambda_2^l(b[i]) \triangleq \log \frac{P(b[i]=1 \mid \{\lambda_1^l(b[i])\}; \text{code structure})}{P(b[i]=0 \mid \{\lambda_2^l(b[i])\}; \text{code structure})}$$
$$= \lambda_2^l(b[i]) + \lambda_1^l(b[i]). \tag{6.4}$$

The factorization (6.4) will be shown in Section 6.2. Here again we see that the output of the channel decoder is the sum of the extrinsic information $\lambda_2^l(b[i])$ obtained by the second-stage subsystem (i.e., the MAP channel decoder) and the prior information $\lambda_1^l(b[i])$ delivered by the preceding stage (i.e., the MAP demodulator). The extrinsic information $\lambda_2^l(b[i])$ is then fed back to the MAP demodulator as the *a priori* information in the next [i.e., $(l+1)$th] iteration. It is important to note that (6.3) and (6.4) hold only if the inputs to the demodulator or the decoder are independent. Since both the ISI channel and the channel encoder have memory, this independence assumption will not be valid; therefore, interleaving (i.e., permutation of time order) must be present between the demodulator and the decoder in order to provide approximate independence. Finally, the turbo receiver structure for the coded ISI system is illustrated in Fig. 6.2. This scheme was introduced in [105] and is termed a *turbo equalizer*. The name *turbo* is justified because both the demodulator and the decoder use their processed output values as *a priori* input for the next iteration, similar to the operation of a turbo engine. Application of the turbo processing principle for joint demodulation and decoding in fading channels can be found in [139, 181].

The turbo principle can similarly be applied in coded multiple-access channels, resulting in procedures known as *turbo multiuser detectors*. In this chapter we discuss applications of such techniques in a variety of multiple-access communication systems with different coding schemes (convolutional codes, turbo codes, space-time codes), signaling structures [CDMA, TDMA, space-division multiple-access (SDMA)] and channel conditions (AWGN, fading, multipath).

The rest of this chapter is organized as follows. In Section 6.2 we present a maximum *a posteriori* (MAP) decoding algorithm for convolutional codes. In Section 6.3 we discuss turbo multiuser detectors in synchronous CDMA systems. In Section 6.4 we treat the problem of turbo multiuser detection in the presence of unknown interferers. In Section 6.5 we discuss turbo multiuser detection in general asynchronous CDMA systems with multipath fading channels. In Section 6.6 we discuss turbo multiuser detection for turbo-coded CDMA systems. In Sections 6.7 and 6.8 we discuss applications of turbo multiuser detection in space-time block-coded systems and space-time trellis-coded systems, respectively. Some mathematical proofs and derivations are given in Section 6.9.

The following is a list of the algorithms appearing in this chapter.

- *Algorithm 6.1:* MAP decoding algorithm for convolutional codes
- *Algorithm 6.2:* Low-complexity SISO multiuser detector—synchronous CDMA

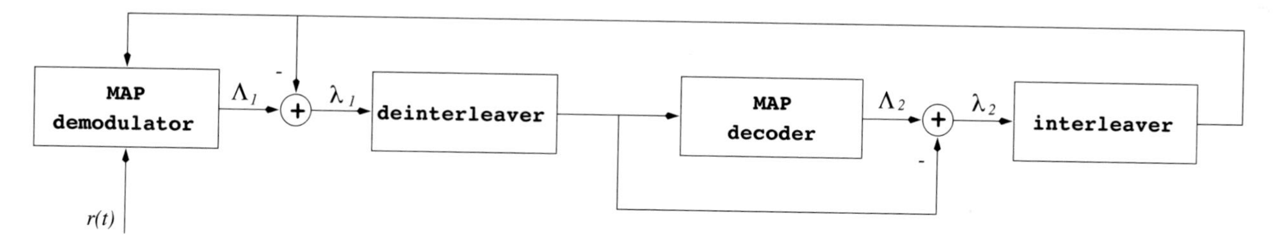

Figure 6.2. Turbo receiver for coded communication over ISI channel.

- *Algorithm 6.3:* Group-blind SISO multiuser detector—synchronous CDMA
- *Algorithm 6.4:* SISO multiuser detector—multipath fading channel

6.2 MAP Decoding Algorithm for Convolutional Codes

The input–output relationship of the MAP channel decoder is illustrated in Fig. 6.3. The MAP decoder takes as input the *a priori* LLRs (or equivalently, probability distributions) of the code (i.e., channel) bits. It delivers as outputs updates of the LLRs of the code bits as well as the LLRs of the information bits, based on the code constraints. In this section we outline a recursive algorithm for computing the LLRs of the code bits and the information bits, which is essentially a slight modification of the celebrated BCJR algorithm [25].

Consider a binary rate-k_0/n_0 convolutional encoder of overall constraint length $k_0\nu$. The input to the encoder at time t is a block of k_0 information bits,

$$\underline{d}_t = \left(d_t^1, \ldots, d_t^{k_0}\right),$$

and the corresponding output is a block of n_0 code bits,

$$\underline{b}_t = \left(b_t^1, \ldots, b_t^{n_0}\right).$$

The state of the trellis at time t can be represented by a $[k_0(\nu-1)]$-tuple, as

$$S_t = \left(s_t^1, \ldots, s_t^{k_0(\nu-1)}\right) = \left(\underline{d}_{t-1}, \ldots, \underline{d}_{t-\nu+1}\right).$$

The dynamics of a convolutional code is completely specified by its trellis representation, which describes the transitions between the states at time instants t and $t+1$. A path segment in the trellis from $t = a$ to $t = b > a$ is determined by the states it traverses at each time $a \leq t \leq b$ and is denoted by

$$\mathcal{L}_a^b \triangleq (S_a, S_{a+1}, \ldots, S_b).$$

Denote the input information bits that cause the state transition from $S_{t-1} = s'$ to $S_t = s$ by $\underline{d}(s', s)$ and the corresponding output code bits by $\underline{b}(s', s)$. Assuming that a systematic code is used, the pair $(s', \underline{b})$ uniquely determines the state transition (s', s). We next consider computing the probability $P\left(S_{t-1} = s', S_t = s\right)$ based on

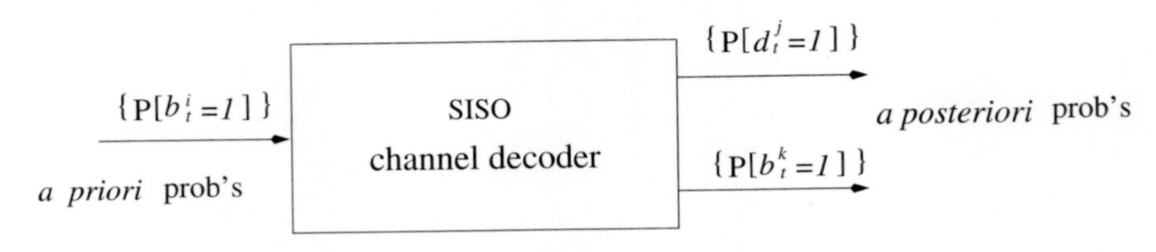

Figure 6.3. Input–output relationship of a MAP channel decoder.

the *a priori* probabilities of the code bits $\{P(\underline{b}_t)\}$ and the constraints imposed by the trellis structure of the code. Suppose that the encoder starts in state $S_0 = \underline{0}$. An information bit stream $\{\underline{d}_t\}_{t=1}^{T}$ is the input to the encoder, followed by ν blocks of all zero inputs, causing the encoder to end in state $S_\tau = \underline{0}$, where $\tau = T + \nu$. Let $\underline{b}_t$ denote the output of the channel encoder at time t. We use the notation

$$P\left[\underline{b}_t(s', s)\right] \triangleq P\left[\underline{b}_t = \underline{b}(s', s)\right]. \tag{6.5}$$

Then we have

$$\begin{aligned}
P\left(S_{t-1} = s', S_t = s\right) &= \sum_{\mathcal{L}_0^\tau :\, S_{t-1}=s', S_t=s} P\left(\mathcal{L}_0^\tau\right) \\
&= \sum_{\mathcal{L}_0^t :\, S_{t-1}=s'} \; \sum_{\mathcal{L}_{t+1}^\tau :\, S_t=s} P\left(\mathcal{L}_0^t\right) P\left[\underline{b}_t(s', s)\right] P\left(\mathcal{L}_{t+1}^\tau\right) \\
&= \underbrace{\left[\sum_{\mathcal{L}_0^t :\, S_{t-1}=s'} P\left(\mathcal{L}_0^t\right)\right]}_{\alpha_{t-1}(s')} P\left[\underline{b}_t(s', s)\right] \underbrace{\left[\sum_{\mathcal{L}_{t+1}^\tau :\, S_t=s} P\left(\mathcal{L}_{t+1}^\tau\right)\right]}_{\beta_t(s)} \\
&= \alpha_{t-1}(s')\, \beta_t(s) \prod_{i=1}^{n_0} P\left[b_t^i(s', s)\right],
\end{aligned} \tag{6.6}$$

where $\alpha_t(s)$ denotes the total probability of all path segments starting from the origin of the trellis that terminate in state s at time t; and where $\beta_t(s)$ denotes the total probability of all path segments terminating at the end of the trellis that originate from state s at time t. In (6.6) we have assumed that the interleaving is ideal and therefore the joint distribution of $\underline{b}_t$ factors into the product of its marginals:

$$P\left[\underline{b}_t(s', s)\right] = \prod_{i=1}^{n_0} P\left[b_t^i(s', s)\right]. \tag{6.7}$$

The quantities $\alpha_t(s)$ and $\beta_t(s)$ in (6.6) can be computed through the following forward and backward recursions [25]:

$$\alpha_t(s) = \sum_{s'} \alpha_{t-1}(s') P\left[\underline{b}_t(s', s)\right], \qquad t = 1, 2, \ldots, \tau, \tag{6.8}$$

$$\beta_t(s) = \sum_{s'} \beta_{t+1}(s') P\left[\underline{b}_{t+1}(s, s')\right], \qquad t = \tau - 1, \tau - 2, \ldots, 0, \tag{6.9}$$

with boundary conditions $\alpha_0(\underline{0}) = 1$, $\alpha_0(s) = 0$ for $s \neq \underline{0}$; and $\beta_\tau(\underline{0}) = 1$, $\beta_\tau(s) = 0$ for $s \neq \underline{0}$. In (6.8) the summation is taken over all the states s' where the transition (s', s) is possible, and similarly for the summation in (6.9).

Let $\mathcal{S}_j^+$ be the set of state pairs (s', s) such that the jth bit of the code symbol $\underline{b}(s', s)$ is $+1$. Define $\mathcal{S}_j^-$ similarly. Using (6.6), the *a posteriori* LLR of the code bit b_t^j at the output of the channel decoder is given by

$$\begin{aligned}
\Lambda_2\left(b_t^j\right) &\triangleq \log \frac{P\left(b_t^j = +1 \mid \{P(\underline{b}_t)\}_t ;\ \text{code structure}\right)}{P\left(b_t^j = -1 \mid \{P(\underline{b}_t)\}_t ;\ \text{code structure}\right)} \\
&= \log \frac{\sum_{(s',s)\in\mathcal{S}_j^+} \alpha_{t-1}(s') \cdot \beta_t(s) \cdot \prod_{i=1}^{n_0} P\left[b_t^i(s', s)\right]}{\sum_{(s',s)\in\mathcal{S}_j^-} \alpha_{t-1}(s') \cdot \beta_t(s) \cdot \prod_{i=1}^{n_0} P\left[b_t^i(s', s)\right]} \\
&= \underbrace{\log \frac{\sum_{(s',s)\in\mathcal{S}_j^+} \alpha_{t-1}(s') \cdot \beta_t(s) \cdot \prod_{i\neq j} P\left[b_t^i(s', s)\right]}{\sum_{(s',s)\in\mathcal{S}_j^-} \alpha_{t-1}(s') \cdot \beta_t(s) \cdot \prod_{i\neq j} P\left[b_t^i(s', s)\right]}}_{\lambda_2\left(b_t^j\right)} \\
&\quad + \underbrace{\log \frac{P\left(b_t^j = +1\right)}{P\left(b_t^j = -1\right)}}_{\lambda_1\left(b_t^j\right)} .
\end{aligned} \tag{6.10}$$

It is seen from (6.10) that the output of the MAP channel decoder is the sum of the prior information $\lambda_1\left(b_t^j\right)$ of the code bit b_t^j and the extrinsic information $\lambda_2\left(b_t^j\right)$. The extrinsic information is the information about the code bit b_t^j gleaned from the prior information about the other code bits based on the trellis structure of the code.

A direct implementation of the recursions (6.8) and (6.9) is numerically unstable, since both $\alpha_t(s)$ and $\beta_t(s)$ drop toward zero exponentially. For sufficiently large τ, the dynamic range of these quantities will exceed the range of any machine. To obtain a numerically stable algorithm, these quantities must be scaled as the computation proceeds. Let $\tilde{\alpha}_t(s)$ denote the scaled version of $\alpha_t(s)$. Initially, $\alpha_1(s)$ is computed according to (6.8), and we set

$$\hat{\alpha}_1(s) = \alpha_1(s), \tag{6.11}$$

$$\tilde{\alpha}_1 = c_1\, \hat{\alpha}_1(s), \tag{6.12}$$

with

$$c_1 \triangleq \frac{1}{\sum_s \hat{\alpha}_1(s)}. \tag{6.13}$$

For each $t \geq 2$, we compute $\tilde{\alpha}_t(s)$ according to

$$\hat{\alpha}_t(s) = \sum_{s'} \tilde{\alpha}_{t-1}(s') P\left[\underline{b}_t(s', s)\right], \tag{6.14}$$

$$\tilde{\alpha}_t(s) = c_t \, \hat{\alpha}_t(s), \tag{6.15}$$

with

$$c_t = \frac{1}{\sum_s \hat{\alpha}_t(s)}, \qquad t = 2, \ldots, \tau. \tag{6.16}$$

Now by a simple induction, we obtain

$$\tilde{\alpha}_{t-1}(s) = \underbrace{\left(\prod_{i=1}^{t-1} c_i\right)}_{C_{t-1}} \alpha_{t-1}(s). \tag{6.17}$$

Thus we can write $\tilde{\alpha}_t(s)$ as

$$\begin{aligned} \tilde{\alpha}_t(s) &= \frac{\sum_{s'} C_{t-1} \alpha_{t-1}(s') \, P[\underline{b}_t(s', s)]}{\sum_s \sum_{s'} C_{t-1} \alpha_{t-1}(s') \, P[\underline{b}_t(s', s)]} \\ &= \frac{\alpha_t(s)}{\sum_s \alpha_t(s)}. \end{aligned} \tag{6.18}$$

That is, each $\alpha_t(s)$ is effectively scaled by the sum over all states of $\alpha_t(s)$.

Let $\tilde{\beta}_t(s)$ denote the scaled version of $\beta_t(s)$. Initially, $\beta_{\tau-1}(s)$ is computed according to (6.9), and we set $\tilde{\beta}_{\tau-1}(s) = \beta_{\tau-1}(s)$. For each $t < \tau - 1$, we compute $\tilde{\beta}_t(s)$ according to

$$\hat{\beta}_t(s) = \sum_{s'} \tilde{\beta}_{t+1}(s') P\left[\underline{b}_{t+1}(s, s')\right], \tag{6.19}$$

$$\tilde{\beta}_t(s) = c_t \, \hat{\beta}_t(s), \qquad t = \tau - 2, \ldots, 0. \tag{6.20}$$

Because the scale factor c_t effectively restores the sum of $\alpha_t(s)$ over all states to 1, and because the magnitudes of $\alpha_t(s)$ and $\beta_t(s)$ are comparable, using the same scaling factor is an effective way to keep the computation within a reasonable range. Furthermore, by induction, we can write

$$\tilde{\beta}_t(s) = \underbrace{\left(\prod_{i=t}^{\tau} c_i\right)}_{D_t} \beta_t(s). \tag{6.21}$$

Using the fact that

$$C_{t-1}D_t = \prod_{i=1}^{t-1} c_i \cdot \prod_{i=t}^{\tau} c_i = \prod_{i=1}^{\tau} c_i \tag{6.22}$$

is a constant that is independent of t, we can then rewrite (6.10) in terms of the scaled variables as

$$\begin{aligned}\Lambda_2\left(b_t^j\right) &= \log \frac{\sum_{(s',s)\in\mathcal{S}_j^+} C_{t-1}\,\alpha_{t-1}(s')\,D_t\,\beta_t(s)\cdot\prod_{i\neq j} P\left[b_t^i(s',s)\right]}{\sum_{(s',s)\in\mathcal{S}_j^-} C_{t-1}\,\alpha_{t-1}(s')\,D_t\,\beta_t(s)\cdot\prod_{i\neq j} P\left[b_t^i(s',s)\right]} + \lambda_1\left(b_t^j\right) \\ &= \underbrace{\log \frac{\sum_{(s',s)\in\mathcal{S}_j^+} \tilde{\alpha}_{t-1}(s')\cdot\tilde{\beta}_t(s)\cdot\prod_{i\neq j} P\left[b_t^i(s',s)\right]}{\sum_{(s',s)\in\mathcal{S}_j^-} \tilde{\alpha}_{t-1}(s')\cdot\tilde{\beta}_t(s)\cdot\prod_{i\neq j} P\left[b_t^i(s',s)\right]}}_{\lambda_2\left(b_t^j\right)} + \lambda_1\left(b_t^j\right).\end{aligned} \tag{6.23}$$

We can also compute the *a posteriori* LLR of the information symbol bit. Let $\mathcal{U}_j^+$ be the set of state pairs (s', s) such that the jth bit of the information symbol $\underline{d}(s', s)$ is $+1$. Define $\mathcal{U}_j^-$ similarly. Then we have

$$\Lambda_2\left(d_t^j\right) = \log \frac{\sum_{(s',s)\in\mathcal{U}_j^+} \tilde{\alpha}_{t-1}(s')\tilde{\beta}_t(s)\prod_{i=1}^{n_0} P\left[b_t^i(s',s)\right]}{\sum_{(s',s)\in\mathcal{U}_j^-} \tilde{\alpha}_{t-1}(s')\tilde{\beta}_t(s)\prod_{i=1}^{n_0} P\left[b_t^i(s',s)\right]}. \tag{6.24}$$

Note that the LLRs of the information bits are computed only at the last iteration. The information bit d_t^j is then decoded according to

$$\hat{d}_t^j = \text{sign}\left[\Lambda_2\left(d_t^j\right)\right]. \tag{6.25}$$

Finally, since the input to the MAP channel decoder is the LLR of the code bits, $\{\lambda_1(b_t^i)\}$, as will be shown in the next section, the code bit distribution $P\left[b_t^i(s',s)\right]$ can be expressed in terms of its LLR as [cf. (6.39)]

$$P\left[b_t^i(s',s)\right] = \frac{1}{2}\left\{1 + b^i(s',s)\tanh\left[\frac{1}{2}\lambda_1\left(b_t^i\right)\right]\right\}. \tag{6.26}$$

The following is a summary of the MAP decoding algorithm for convolutional codes.

Algorithm 6.1: [MAP decoding algorithm for convolutional codes]

- *Compute the code bit probabilities from the corresponding LLRs using (6.26).*
- *Initialize the forward and backward recursions:*

$$\begin{aligned} \alpha_0(\underline{0}) &= 1, \qquad \alpha_0(s) = 0, \ \textit{for } s \neq 0; \\ \beta_\tau(\underline{0}) &= 1, \qquad \beta_\tau(s) = 0, \ \textit{for } s \neq 0. \end{aligned}$$

- *Compute the forward recursion using (6.8), (6.11)–(6.16).*
- *Compute the backward recursion using (6.9), (6.19)–(6.20).*
- *Compute the LLRs of the code bits and the information bits using (6.23) and (6.24).*

6.3 Turbo Multiuser Detection for Synchronous CDMA

6.3.1 Turbo Multiuser Receiver

We consider a convolutionally coded synchronous real-valued CDMA system with K users, employing normalized modulation waveforms $s_1, s_2, \ldots, s_K$, and signaling through an additive white Gaussian noise channel. A block diagram of the transmitter end of such a system is shown in Fig. 6.4. The binary information symbols $\{d_k[m]\}_m$ for user k, $k = 1, \ldots, K$, are convolutionally encoded with code rate R_k. A code-bit interleaver is used to reduce the influence of the error bursts at the input of each channel decoder. The interleaved code bits of the kth user are BPSK modulated, yielding data symbols of duration T. Each data symbol $b_k[i]$ is then spread by a signature waveform $s_k(t)$ and transmitted through the channel.

As seen from preceding chapters, the received continuous-time signal can be written as

$$r(t) = \sum_{k=1}^{K} A_k \sum_{i=0}^{M-1} b_k[i] s_k(t - iT) + n(t), \tag{6.27}$$

where $n(t)$ is a zero-mean white Gaussian noise process with power spectral density σ^2, and A_k is the amplitude of the kth user.

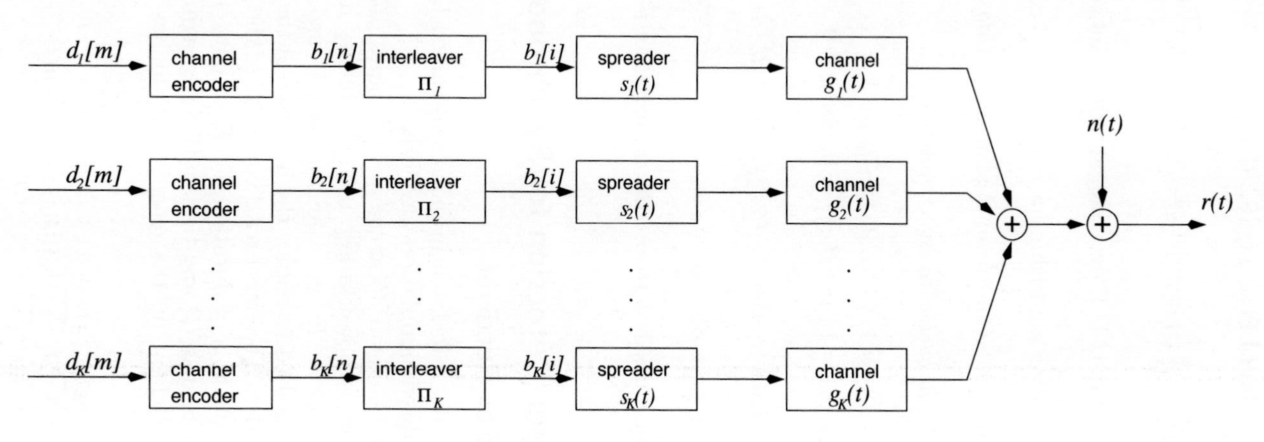

Figure 6.4. Coded CDMA system.

The turbo receiver structure is shown in Fig. 6.5. It consists of two stages: a soft-input/soft-output (SISO) multiuser detector, followed by K parallel single-user MAP channel decoders. The two stages are separated by deinterleavers and interleavers. The SISO multiuser detector delivers the *a posteriori* log-likelihood ratio (LLR) of a transmitted "+1" and a transmitted "−1" for every code bit of each user,

$$\Lambda_1(b_k[i]) \triangleq \log \frac{P(b_k[i] = +1 \mid r(t))}{P(b_k[i] = -1 \mid r(t))}, \qquad k = 1, \ldots, K, \quad i = 0, \ldots, M-1. \tag{6.28}$$

As before, using Bayes' formula, (6.28) can be written as

$$\Lambda_1(b_k[i]) = \underbrace{\log \frac{p[r(t) \mid b_k[i] = +1]}{p[r(t) \mid b_k[i] = -1]}}_{\lambda_1(b_k[i])} + \underbrace{\log \frac{P(b_k[i] = +1)}{P(b_k[i] = -1)}}_{\lambda_2(b_k[i])}, \tag{6.29}$$

where the second term in (6.29), denoted by $\lambda_2[b_k(i)]$, represents the *a priori* LLR of the code bit $b_k[i]$, which is computed by the MAP channel decoder of the kth user in the previous iteration, interleaved and then fed back to the SISO multiuser detector. For the first iteration, assuming equally likely code bits (i.e., no prior information available), we then have $\lambda_2(b_k[i]) = 0$, for $1 \leq k \leq K$ and $0 \leq i < M$. The first term in (6.29), denoted by $\lambda_1(b_k[i])$, represents the *extrinsic* information delivered by the SISO multiuser detector, based on the received signal $r(t)$, the structure of the multiuser signal given by (6.27), the prior information about the code bits of all other users, $\{\lambda_2(b_l[i])\}_{i;\, l \neq k}$, and the prior information about the code bits of the kth user other than the ith bit, $\{\lambda_2(b_k[j])\}_{j \neq i}$. The extrinsic information $\{\lambda_1(b_k[i])\}_i$ of the kth user, which is not influenced by the *a priori* information $\{\lambda_2(b_k[i])\}_i$ provided by the MAP channel decoder, is then reverse interleaved and fed into the kth user's channel decoder as the *a priori* information in the next iteration. Denote the code-bit sequence of the kth user after deinterleaving as $\{b_k^{\pi}[i]\}_i$.

Based on the prior information $\{\lambda_1(b_k^{\pi}[i])\}_i$ and the trellis structure of the channel code (i.e., the constraints imposed by the code), the kth user's MAP channel decoder computes the *a posteriori* LLR of each code bit,

$$\begin{aligned}\Lambda_2\left(b_k^{\pi}[i]\right) &\triangleq \log \frac{P\left(b_k^{\pi}[i] = +1 \mid \{\lambda_1(b_k^{\pi}[i])\}_i; \text{ code structure}\right)}{P\left(b_k^{\pi}[i] = -1 \mid \{\lambda_1(b_k^{\pi}[i])\}_i; \text{ code structure}\right)} \\ &= \lambda_2(b_k^{\pi}[i]) + \lambda_1(b_k^{\pi}[i]), \qquad i = 0, \ldots, M-1; \quad k = 1, \ldots, K,\end{aligned} \tag{6.30}$$

where the second equality was shown in Section 6.2 [cf. (6.10)]. It is seen from (6.30) that the output of the MAP channel decoder is the sum of the prior information $\lambda_1(b_k^{\pi}[i])$ and the extrinsic information $\lambda_2(b_k^{\pi}[i])$ delivered by the MAP channel decoder. As discussed in Section 6.2, this extrinsic information is the information about the code bit $b_k^{\pi}[i]$ gleaned from prior information about the other code bits, $\{\lambda_1(b_k^{\pi}[j])\}_{j \neq i}$, based on the trellis constraint of the code. The MAP channel decoder also computes the *a posteriori* LLR of every information bit, which is used to make a decision on the decoded bit at the last iteration. After interleaving,

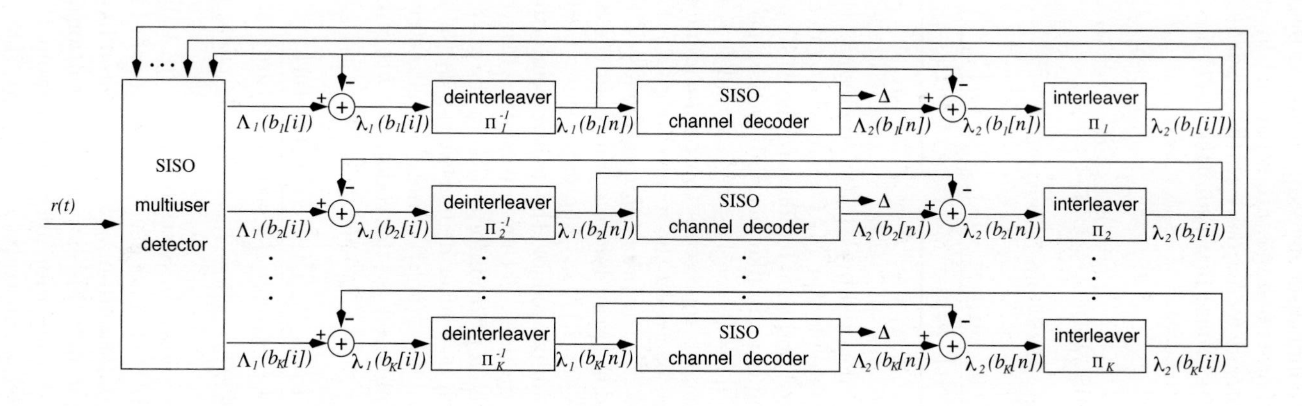

Δ : decoded information bits

Figure 6.5. Turbo multiuser receiver.

the extrinsic information delivered by the K MAP channel decoders $\{\lambda_2(b_k[i])\}_{i;k}$ is then fed back to the SISO multiuser detector, as the prior information about the code bits of all users, in the next iteration. Note that at the first iteration, the extrinsic information quantities $\{\lambda_1(b_k[i])\}_{i;k}$ and $\{\lambda_2(b_k[i])\}_{i;k}$ are statistically independent. But subsequently, since they use the same information indirectly, they will become more and more correlated, and finally, the improvement through iteration will diminish.

6.3.2 Optimal SISO Multiuser Detector

For the synchronous CDMA system (6.27), it is easily seen that a sufficient statistic for demodulating the ith code bits of the K users is given by the K-vector $\boldsymbol{y}[i] = \left[y_1[i] \cdots y_K[i]\right]^T$, whose kth component is the output of a filter matched to $s_k(\cdot)$ in the ith code-bit interval:

$$y_k[i] \triangleq \int_{iT}^{(i+1)T} r(t)\, s_k(t - iT)\, \mathrm{d}t. \tag{6.31}$$

This sufficient statistic vector $\boldsymbol{y}[i]$ can be written as [520]

$$\boldsymbol{y}[i] = \boldsymbol{R}\boldsymbol{A}\boldsymbol{b}[i] + \boldsymbol{n}[i], \tag{6.32}$$

where $\boldsymbol{R}$ denotes the normalized cross-correlation matrix of the signal set $s_1, \ldots, s_K$:

$$[\boldsymbol{R}]_{k,l} = \rho_{kl} \triangleq \int_0^T s_k(t) s_l(t)\, \mathrm{d}t; \tag{6.33}$$

$\boldsymbol{A} \triangleq \mathrm{diag}(A_1, \ldots, A_K)$; $\boldsymbol{b}[i] = \left[b_1[i] \cdots b_K[i]\right]^T$; and $\boldsymbol{n}[i] \sim \mathcal{N}(\boldsymbol{0}, \sigma^2\boldsymbol{R})$ is a Gaussian noise vector, independent of $\boldsymbol{b}[i]$.

In what follows, for notational simplicity, we drop the symbol index i whenever possible. Denote

$$\mathcal{B}_k^+ \triangleq \Big\{(b_1, \ldots, b_{k-1}, +1, b_{k+1}, \ldots, b_K) : b_j \in \{+1, -1\}\Big\},$$
$$\mathcal{B}_k^- \triangleq \Big\{(b_1, \ldots, b_{k-1}, -1, b_{k+1}, \ldots, b_K) : b_j \in \{+1, -1\}\Big\}.$$

From (6.32), the extrinsic information $\lambda_1(b_k)$ delivered by the SISO multiuser detector [cf. (6.29)] is then given by

$$\lambda_1(b_k) \triangleq \log \frac{p(\boldsymbol{y} \mid b_k = +1)}{p(\boldsymbol{y} \mid b_k = -1)}$$
$$= \log \frac{\displaystyle\sum_{\boldsymbol{b} \in \mathcal{B}_k^+} \exp\left[-\frac{1}{2\sigma^2}(\boldsymbol{y} - \boldsymbol{R}\boldsymbol{A}\boldsymbol{b})^T \boldsymbol{R}^{-1} (\boldsymbol{y} - \boldsymbol{R}\boldsymbol{A}\boldsymbol{b})\right] \prod_{j \neq k} P(b_j)}{\displaystyle\sum_{\boldsymbol{b} \in \mathcal{B}_k^-} \exp\left[-\frac{1}{2\sigma^2}(\boldsymbol{y} - \boldsymbol{R}\boldsymbol{A}\boldsymbol{b})^T \boldsymbol{R}^{-1} (\boldsymbol{y} - \boldsymbol{R}\boldsymbol{A}\boldsymbol{b})\right] \prod_{j \neq k} P(b_j)}, \tag{6.34}$$

where we have used the notation $P(b_j) \triangleq P(b_j[i] = b_j)$ for $b_j \in \{+1, -1\}$. The summations in the numerator (respectively, denominator) in (6.34) are over all the 2^{K-1} possible vectors $\underline{b}$ in $\mathcal{B}_k^+$ (respectively, $\mathcal{B}_k^-$). We have

$$\exp\left[-\frac{1}{2\sigma^2}(\boldsymbol{y} - \boldsymbol{RAb})^T \boldsymbol{R}^{-1}(\boldsymbol{y} - \boldsymbol{RAb})\right]$$
$$= \exp\left(-\frac{1}{2\sigma^2}\boldsymbol{y}^T\boldsymbol{R}^{-1}\boldsymbol{y}\right)\exp\left(-\frac{1}{2\sigma^2}\boldsymbol{b}^T\boldsymbol{ARAb}\right)\exp\left(\frac{1}{\sigma^2}\boldsymbol{b}^T\boldsymbol{Ay}\right). \tag{6.35}$$

Note that the first term in (6.35) is independent of $\boldsymbol{b}$ and therefore will be canceled in (6.34). The third term in (6.35) can be written as

$$\exp\left(\frac{1}{\sigma^2}\boldsymbol{b}^T\boldsymbol{Ay}\right) = \exp\left(\frac{1}{\sigma^2}\sum_{j=1}^{K} A_j y_j b_j\right) = \prod_{j=1}^{K}\exp\left(\frac{A_j y_j b_j}{\sigma^2}\right)$$
$$= \prod_{j=1}^{K}\left[\frac{1+b_j}{2}\exp\left(\frac{A_j y_j}{\sigma^2}\right) + \frac{1-b_j}{2}\exp\left(-\frac{A_j y_j}{\sigma^2}\right)\right] \tag{6.36}$$
$$= \prod_{j=1}^{K}\left\{\frac{1}{2}\left[\exp\left(\frac{A_j y_j}{\sigma^2}\right) + \exp\left(-\frac{A_j y_j}{\sigma^2}\right)\right] + \frac{b_j}{2}\left[\exp\left(\frac{A_j y_j}{\sigma^2}\right) - \exp\left(-\frac{A_j y_j}{\sigma^2}\right)\right]\right\}$$
$$= \prod_{j=1}^{K}\left[\cosh\left(\frac{A_j y_j}{\sigma^2}\right)\right]\left[1 + b_j \tanh\left(\frac{A_j y_j}{\sigma^2}\right)\right], \tag{6.37}$$

where (6.36) follows from the fact that $b_j \in \{+1, -1\}$. The first term in (6.37) is also independent of $\boldsymbol{b}$ and will be canceled in (6.34). In (6.34) the *a priori* probabilities of the code bits can be expressed in terms of their LLRs $\lambda_2(b_j[i])$, as follows. Since

$$\lambda_2(b_j[i]) \triangleq \log\frac{P(b_j[i] = +1)}{P(b_j[i] = -1)},$$

after some manipulations, we have for $b_j \in \{+1, -1\}$,

$$P(b_j) \triangleq P(b_j[i] = b_j)$$
$$= \frac{\exp\left[b_j\,\lambda_2(b_j[i])\right]}{1 + \exp\left[b_j\,\lambda_2(b_j[i])\right]}$$
$$= \frac{\exp\left[\frac{1}{2}b_j\,\lambda_2(b_j[i])\right]}{\exp\left[-\frac{1}{2}b_j\,\lambda_2(b_j[i])\right] + \exp\left[\frac{1}{2}b_j\,\lambda_2(b_j[i])\right]}$$
$$= \frac{\cosh\left[\frac{1}{2}\lambda_2(b_j[i])\right]\left\{1 + b_j\,\tanh\left[\frac{1}{2}\lambda_2(b_j[i])\right]\right\}}{2\cosh\left[\frac{1}{2}\lambda_2(b_j[i])\right]} \tag{6.38}$$
$$= \frac{1}{2}\left\{1 + b_j\,\tanh\left[\frac{1}{2}\lambda_2(b_j[i])\right]\right\}, \tag{6.39}$$

where (6.38) follows from a derivation similar to that of (6.37). Substituting (6.35), (6.37), and (6.39) into (6.34), we obtain

$$\lambda_1(b_k[i]) = \frac{2A_k\, y_k[i]}{\sigma^2} + \log \frac{\sum_{\boldsymbol{b} \in \mathcal{B}_k^+} \left\{ \exp\left(-\frac{1}{2\sigma^2}\boldsymbol{b}^T \boldsymbol{ARAb}\right) \prod_{j \neq k} \left[1 + b_j \tanh\left(\frac{A_j\, y_j[i]}{\sigma^2}\right)\right] \left[1 + b_j \tanh\left(\frac{1}{2}\lambda_2(b_j[i])\right)\right] \right\}}{\sum_{\boldsymbol{b} \in \mathcal{B}_k^-} \left\{ \exp\left(-\frac{1}{2\sigma^2}\boldsymbol{b}^T \boldsymbol{ARAb}\right) \prod_{j \neq k} \left[1 + b_j \tanh\left(\frac{A_j\, y_j[i]}{\sigma^2}\right)\right] \left[1 + b_j \tanh\left(\frac{1}{2}\lambda_2(b_j[i])\right)\right] \right\}}. \tag{6.40}$$

It is seen from (6.40) that the extrinsic information $\lambda_1(b_k[i])$ at the output of the SISO multiuser detector consists of two parts; the first term contains the channel value of the desired user $y_k[i]$, and the second term is the information extracted from the other users' channel values $\{y_j[i]\}_{j \neq k}$ as well as their prior information $\{\lambda_2(b_j[i])\}_{j \neq k}$.

6.3.3 Low-Complexity SISO Multiuser Detector

It is clear from (6.40) that the computational complexity of the optimal SISO multiuser detector is exponential in terms of the number of users K, which is certainly prohibitive for channels with moderate to large numbers of users. In what follows we describe a low-complexity SISO multiuser detector based on a novel technique of combined soft interference cancellation and linear MMSE filtering, which was first developed in [552].

Soft Interference Cancellation and Instantaneous Linear MMSE Filtering

Based on the *a priori* LLR of the code bits of all users, $\{\lambda_2(b_k[i])\}_k$, provided by the MAP channel decoder from the previous iteration, we first form soft estimates of the code bits of all users as

$$\begin{aligned} \tilde{b}_k[i] \triangleq E\{b_k[i]\} &= \sum_{b \in \{+1,-1\}} b\, P(b_k[i] = b) \\ &= \sum_{b \in \{+1,-1\}} \frac{b}{2} \left\{ 1 + b \tanh\left[\frac{1}{2}\lambda_2(b_j[i])\right] \right\} && (6.41) \\ &= \tanh\left[\frac{1}{2}\lambda_2(b_j[i])\right], \qquad k = 1, \ldots, K, && (6.42) \end{aligned}$$

where (6.41) follows from (6.39). Define

$$\tilde{\boldsymbol{b}}[i] \triangleq \left[\tilde{b}_1[i] \;\cdots\; \tilde{b}_K[i]\right]^T, \tag{6.43}$$

$$\begin{aligned} \tilde{\boldsymbol{b}}_k[i] &\triangleq \tilde{\boldsymbol{b}}[i] - \tilde{b}_k[i]\boldsymbol{e}_k \\ &= \left[\tilde{b}_1[i] \;\cdots\; \tilde{b}_{k-1}[i]\; 0\; \tilde{b}_{k+1}[i] \;\cdots\; \tilde{b}_K[i]\right]^T, && (6.44) \end{aligned}$$

where $\boldsymbol{e}_k$ denotes a K-vector of all zeros, except for the kth element, which is 1. Therefore, $\tilde{\boldsymbol{b}}_k[i]$ is obtained from $\tilde{\boldsymbol{b}}[i]$ by setting the kth element to zero. For each user, a soft interference cancellation is performed on the matched-filter output $\boldsymbol{y}[i]$ in (6.32), to obtain

$$\begin{aligned}\boldsymbol{y}_k[i] &\triangleq \boldsymbol{y}[i] - \boldsymbol{R}\boldsymbol{A}\tilde{\boldsymbol{b}}_k[i] \\ &= \boldsymbol{R}\boldsymbol{A}\left(\boldsymbol{b}[i] - \tilde{\boldsymbol{b}}_k[i]\right) + \boldsymbol{n}[i], \qquad k = 1, \ldots, K. \end{aligned} \tag{6.45}$$

Such a soft interference cancellation scheme was first proposed in [168]. Next, in order to further suppress the residual interference in $\boldsymbol{y}_k[i]$, an instantaneous linear minimum mean-square error (MMSE) filter $\boldsymbol{w}_k[i]$ is applied to $\boldsymbol{y}_k[i]$, to obtain

$$z_k[i] = \boldsymbol{w}_k[i]^T \, \boldsymbol{y}_k[i], \tag{6.46}$$

where the filter $\boldsymbol{w}_k[i] \in \mathbb{R}^K$ is chosen to minimize the mean-square error between the code bit $b_k[i]$ and the filter output $z_k[i]$:

$$\begin{aligned}\boldsymbol{w}_k[i] &= \arg\min_{\boldsymbol{w}\in\mathbb{R}^K} E\left\{\left(b_k[i] - \boldsymbol{w}^T \boldsymbol{y}_k[i]\right)^2\right\} \\ &= \arg\min_{\boldsymbol{w}\in\mathbb{R}^K} \boldsymbol{w}^T E\left\{\boldsymbol{y}_k[i]\,\boldsymbol{y}_k[i]^T\right\} \boldsymbol{w} - 2\,\boldsymbol{w}^T E\left\{b_k[i]\,\boldsymbol{y}_k[i]\right\}, \end{aligned} \tag{6.47}$$

where using (6.45), we have

$$E\left\{\boldsymbol{y}_k[i]\,\boldsymbol{y}_k[i]^T\right\} = \boldsymbol{R}\boldsymbol{A}\,\mathrm{Cov}\left\{\boldsymbol{b}[i] - \tilde{\boldsymbol{b}}_k[i]\right\}\boldsymbol{A}\boldsymbol{R} + \sigma^2\boldsymbol{R}, \tag{6.48}$$

$$\begin{aligned} E\left\{b_k[i]\,\boldsymbol{y}_k[i]\right\} &= \boldsymbol{R}\boldsymbol{A}E\left\{b_k[i]\left(\boldsymbol{b}[i] - \tilde{\boldsymbol{b}}_k[i]\right)\right\} \\ &= \boldsymbol{R}\boldsymbol{A}\boldsymbol{e}_k, \end{aligned} \tag{6.49}$$

and in (6.48),

$$\begin{aligned} &\mathrm{Cov}\left\{\boldsymbol{b}[i] - \tilde{\boldsymbol{b}}_k[i]\right\} \\ &= \mathrm{diag}\Big\{\mathrm{Var}\{b_1[i]\}, \ldots, \mathrm{Var}\{b_{k-1}[i]\}, 1, \mathrm{Var}\{b_{k+1}[i]\}, \ldots, \mathrm{Var}\{b_K[i]\}\Big\} \\ &= \mathrm{diag}\left\{1 - \tilde{b}_1[i]^2, \ldots, 1 - \tilde{b}_{k-1}[i]^2, 1, 1 - \tilde{b}_{k+1}[i]^2, \ldots, 1 - \tilde{b}_K[i]^2\right\}, \end{aligned} \tag{6.50}$$

because

$$\mathrm{Var}\{b_j[i]\} = E\left\{b_j[i]^2\right\} - \left(E\{b_j[i]\}\right)^2 = 1 - \tilde{b}_j[i]^2. \tag{6.51}$$

Denote

$$\begin{aligned} \boldsymbol{V}_k[i] &\triangleq \boldsymbol{A}\,\mathrm{Cov}\left\{\boldsymbol{b}[i] - \tilde{\boldsymbol{b}}_k[i]\right\}\boldsymbol{A} \\ &= \sum_{j\neq k} A_j^2\left(1 - \tilde{b}_j[i]^2\right)\boldsymbol{e}_j\boldsymbol{e}_j^T + A_k^2\boldsymbol{e}_k\boldsymbol{e}_k^T. \end{aligned} \tag{6.52}$$

Substituting (6.48) and (6.49) into (6.47), we have

$$\begin{aligned} \boldsymbol{w}_k[i] &= \left(\boldsymbol{R}\boldsymbol{V}_k[i]\boldsymbol{R} + \sigma^2\boldsymbol{R}\right)^{-1}\boldsymbol{R}\boldsymbol{A}\boldsymbol{e}_k \\ &= A_k\,\boldsymbol{R}^{-1}\left(\boldsymbol{V}_k[i] + \sigma^2\boldsymbol{R}^{-1}\right)^{-1}\boldsymbol{e}_k. \end{aligned} \tag{6.53}$$

Substituting (6.45) and (6.53) into (6.46), we obtain

$$z_k[i] = A_k\boldsymbol{e}_k^T\left(\boldsymbol{V}_k[i] + \sigma^2\boldsymbol{R}^{-1}\right)^{-1}\left(\boldsymbol{R}^{-1}\boldsymbol{y}[i] - \boldsymbol{A}\tilde{\boldsymbol{b}}_k[i]\right). \tag{6.54}$$

Notice that the term $\boldsymbol{R}^{-1}\boldsymbol{y}[i]$ in (6.54) is the output of a linear decorrelating multiuser detector. Next, we consider some special cases of the output $z_k[i]$.

1. No prior information on the code bits of the interfering users; that is, $\lambda_2(b_k[i]) = 0$ for $1 \le k \le K$. In this case, $\tilde{\boldsymbol{b}}_k[i] = \boldsymbol{0}$, and $\boldsymbol{V}_k[i] = \boldsymbol{A}^2$. Then (6.54) becomes

$$z_k[i] = A_k\boldsymbol{e}_k^T\left(\boldsymbol{R} + \sigma^2\boldsymbol{A}^{-2}\right)^{-1}\boldsymbol{y}[i], \tag{6.55}$$

   which is simply the output of the linear MMSE multiuser detector for user k.

2. Perfect prior information on the code bits of the interfering users; that is, $\lambda_2(b_k[i]) = \pm\infty$ for $1 \le k \le K$. In this case,

$$\begin{aligned} \tilde{\boldsymbol{b}}_k[i] &= \left[b_1[i] \;\cdots\; b_{k-1}[i]\; 0\; b_{k+1}[i] \;\cdots\; b_K[i]\right]^T, \\ \boldsymbol{V}_k[i] &= A_k^2\boldsymbol{e}_k\boldsymbol{e}_k^T. \end{aligned}$$

   Substituting these into (6.53), we obtain

$$\begin{aligned} \boldsymbol{w}_k[i] &= A_k\boldsymbol{R}^{-1}\left(A_k^2\boldsymbol{e}_k\boldsymbol{e}_k^T + \sigma^2\boldsymbol{R}^{-1}\right)^{-1}\boldsymbol{e}_k \\ &= \frac{A_k}{\sigma^2}\boldsymbol{R}^{-1}\left(\boldsymbol{R} - \frac{A_k^2}{A_k^2+\sigma^2}\boldsymbol{R}\boldsymbol{e}_k\boldsymbol{e}_k^T\boldsymbol{R}\right)\boldsymbol{e}_k \qquad (6.56) \\ &= \frac{A_k}{A_k^2+\sigma^2}\boldsymbol{e}_k, \qquad (6.57) \end{aligned}$$

   where (6.56) follows from the matrix inversion lemma[1] and (6.57) follows from the fact that $\boldsymbol{e}_k^T\boldsymbol{R}\boldsymbol{e}_k = [\boldsymbol{R}]_{kk} = 1$. The output of the soft instantaneous MMSE filter is then given by

$$\begin{aligned} z_k[i] &= \boldsymbol{w}_k[i]^T\boldsymbol{y}_k[i] \\ &= \frac{A_k}{A_k^2+\sigma^2}\boldsymbol{e}_k^T\boldsymbol{y}_k[i] \\ &= \frac{A_k}{A_k^2+\sigma^2}\left(y_k[i] - \sum_{j\ne k} A_j\rho_{kj}b_j[i]\right). \end{aligned} \tag{6.58}$$

---

[1] $\left(\boldsymbol{A} + \alpha\boldsymbol{b}\boldsymbol{c}^T\right)^{-1} = \boldsymbol{A}^{-1} - \frac{1}{\alpha^{-1}+\boldsymbol{c}^T\boldsymbol{A}^{-1}\boldsymbol{b}}\boldsymbol{A}^{-1}\boldsymbol{b}\boldsymbol{c}^T\boldsymbol{A}^{-1}$.

That is, in this case, the output of the soft instantaneous MMSE filter is a scaled version of the kth user's matched filter output after ideal interference cancellation.

3. In general, the prior information provided by the MAP channel decoder satisfies $0 < |\lambda_2(b_k[i])| < \infty$, $1 \le k \le K$. The signal-to-interference-plus-noise ratio (SINR) at the soft instantaneous MMSE filter output is defined as

$$\text{SINR}(z_k[i]) \triangleq \frac{E^2\{z_k[i]\}}{\text{Var}\{z_k[i]\}}. \tag{6.59}$$

Denote by $\underline{\text{SINR}}(z_k[i])$ the output SINR when there is no prior information on the code bits of interfering users (i.e., the SINR of the linear MMSE detector). Also denote by $\overline{\text{SINR}}(z_k[i])$ the output SINR when there is perfect prior information on the code bits of interfering users [i.e., the input signal-to-noise ratio (SNR) for the kth user]. Then we have the following result, whose proof is given in the Appendix (Section 6.9.1).

Proposition 6.1: *If* $0 < |\lambda_2(b_k[i])| < \infty$ *for* $1 \le k \le K$, *we have*

$$\overline{\text{SINR}}(z_k[i]) > \text{SINR}(z_k[i]) > \underline{\text{SINR}}(z_k[i]). \tag{6.60}$$

Gaussian Approximation of Linear MMSE Filter Output

It is shown in [386] that the distribution of the residual interference plus noise at the output of a linear MMSE multiuser detector is well approximated by a Gaussian distribution. In what follows, we assume that the output $z_k[i]$ of the instantaneous linear MMSE filter in (6.46) represents the output of an equivalent additive white Gaussian noise channel having $b_k[i]$ as its input symbol. This equivalent channel can be represented as

$$z_k[i] = \mu_k[i]\, b_k[i] + \eta_k[i], \tag{6.61}$$

where $\mu_k[i]$ is the equivalent amplitude of the kth user's signal at the output and $\eta_k[i] \sim \mathcal{N}\left(0, \nu_k^2[i]\right)$ is a Gaussian noise sample. Using (6.45) and (6.46), the parameters $\mu_k[i]$ and $\nu_k^2[i]$ can be computed as follows, where the expectation is taken with respect to the code bits of interfering users $\{b_j[i]\}_{j \ne k}$ and the channel noise vector $\boldsymbol{n}[i]$:

$$\begin{aligned}
\mu_k[i] &\triangleq E\{z_k[i]\, b_k[i]\} \\
&= A_k \boldsymbol{e}_k^T \left(\boldsymbol{V}_k[i] + \sigma^2 \boldsymbol{R}^{-1}\right)^{-1} E\left\{b_k[i]\boldsymbol{A}\left(\boldsymbol{b}[i] - \tilde{\boldsymbol{b}}_k[i]\right) + b_k[i]\boldsymbol{n}[i]\right\} \\
&= A_k^2 \boldsymbol{e}_k^T \left(\boldsymbol{V}_k[i] + \sigma^2 \boldsymbol{R}^{-1}\right)^{-1} \boldsymbol{e}_k \\
&= A_k^2 \left[\left(\boldsymbol{V}_k[i] + \sigma^2 \boldsymbol{R}^{-1}\right)^{-1}\right]_{kk},
\end{aligned} \tag{6.62}$$

$$\begin{aligned}
\nu_k^2[i] &\triangleq \mathrm{Var}\{z_k[i]\} = E\{z_k[i]^2\} - \mu_k[i]^2 \\
&= \boldsymbol{w}_k[i]^T E\left\{\boldsymbol{y}_k[i]\boldsymbol{y}_k[i]^T\right\}\boldsymbol{w}_k[i] - \mu_k[i]^2 \\
&= A_k^2 \boldsymbol{e}_k^T \left(\boldsymbol{V}_k[i] + \sigma^2 \boldsymbol{R}^{-1}\right)^{-1} \boldsymbol{R}^{-1}\left(\boldsymbol{R}\boldsymbol{V}_k[i]\boldsymbol{R} + \sigma^2\boldsymbol{R}\right) \\
&\quad \boldsymbol{R}^{-1}\left(\boldsymbol{V}_k[i] + \sigma^2 \boldsymbol{R}^{-1}\right)^{-1}\boldsymbol{e}_k - \mu_k[i]^2 \\
&= A_k^2 \boldsymbol{e}_k^T \left(\boldsymbol{V}_k[i] + \sigma^2 \boldsymbol{R}^{-1}\right)^{-1}\boldsymbol{e}_k - \mu_k[i]^2 \\
&= \mu_k[i] - \mu_k[i]^2.
\end{aligned} \tag{6.63}$$

Using (6.61) and (6.63) the extrinsic information delivered by the instantaneous linear MMSE filter is then

$$\begin{aligned}
\lambda_1(b_k[i]) &\triangleq \log\frac{p(z_k[i] \mid b_k[i] = +1)}{p(z_k[i] \mid b_k[i] = -1)} \\
&= -\frac{(z_k[i] - \mu_k[i])^2}{2\nu_k^2[i]} + \frac{(z_k[i] + \mu_k[i])^2}{2\nu_k^2[i]} \\
&= \frac{2\,\mu_k[i]\,z_k[i]}{\nu_k^2[i]} \\
&= \frac{2\,z_k[i]}{1 - \mu_k[i]}.
\end{aligned} \tag{6.64}$$

Recursive Procedure for Computing Soft Output

It is seen from (6.64) that to form the extrinsic LLR $\lambda_1(b_k[i])$ at the instantaneous linear MMSE filter, we must first compute $z_k[i]$ and $\mu_k[i]$. From (6.54) and (6.62) the computation of $z_k[i]$ and $\mu_k[i]$ involves inverting a $K \times K$ matrix:

$$\boldsymbol{\Phi}_k[i] \triangleq \left(\boldsymbol{V}_k[i] + \sigma^2 \boldsymbol{R}^{-1}\right)^{-1}. \tag{6.65}$$

Next we outline a recursive procedure for computing $\boldsymbol{\Phi}_k[i]$. Define $\boldsymbol{\Psi}^{(0)} \triangleq (1/\sigma^2)\boldsymbol{R}$, and

$$\boldsymbol{\Psi}^{(k)} \triangleq \left[\sigma^2\boldsymbol{R}^{-1} + \sum_{j=1}^{k} A_j^2\left(1 - \tilde{b}_j[i]^2\right)\boldsymbol{e}_j\,\boldsymbol{e}_j^T\right]^{-1}, \qquad k = 1, \ldots, K. \tag{6.66}$$

Using the matrix inversion lemma, $\boldsymbol{\Psi}^{(k)}$ can be computed recursively as

$$\boldsymbol{\Psi}^{(k)} = \boldsymbol{\Psi}^{(k-1)} - \frac{1}{A_k^{-2}\left(1 - \tilde{b}_j[i]^2\right)^{-1} + \left[\boldsymbol{\Psi}^{(k-1)}\right]_{kk}}\left[\boldsymbol{\Psi}^{(k-1)}\boldsymbol{e}_k\right]\left[\boldsymbol{\Psi}^{(k-1)}\boldsymbol{e}_k\right]^T,$$

$$k = 1, \ldots, K. \tag{6.67}$$

Denote $\boldsymbol{\Psi} \triangleq \boldsymbol{\Psi}^{(K)}$. Using the definition of $\boldsymbol{V}_k[i]$ given by (6.52), we can then compute $\boldsymbol{\Phi}_k[i]$ from $\boldsymbol{\Psi}$ as follows:

$$\begin{aligned}\boldsymbol{\Phi}_k[i] &= \left(\boldsymbol{\Psi}^{-1} + A_k^2 \tilde{b}_k[i]^2 \boldsymbol{e}_k \boldsymbol{e}_k^T\right)^{-1} \\ &= \boldsymbol{\Psi} - \frac{1}{\left(A_k \tilde{b}_k[i]\right)^{-2} + [\boldsymbol{\Psi}]_{kk}} [\boldsymbol{\Psi}\boldsymbol{e}_k][\boldsymbol{\Psi}\boldsymbol{e}_k]^T, \qquad k = 1, \ldots, K. \end{aligned} \tag{6.68}$$

Finally, we summarize the low-complexity SISO multiuser detection algorithm as follows.

Algorithm 6.2: [Low-complexity SISO multiuser detector—synchronous CDMA]

- *Given* $\{\lambda_2(b_k[i])\}_{k=1}^K$, *form the soft bit vectors* $\tilde{\boldsymbol{b}}[i]$ *and* $\{\tilde{\boldsymbol{b}}_k[i]\}_{k=1}^K$ *according to (6.42)–(6.44).*

- *Compute the matrix inversion* $\boldsymbol{\Phi}_k[i] \triangleq \left(\boldsymbol{V}_k[i] + \sigma^2 \boldsymbol{R}^{-1}\right)^{-1}$, $k = 1, \ldots, K$, *according to (6.66)–(6.68).*

- *Compute the extrinsic information* $\{\lambda_1(b_k[i])\}_{k=1}^K$ *according to (6.54), (6.62), and (6.64):*

$$z_k[i] = A_k \boldsymbol{e}_k^T \boldsymbol{\Phi}_k[i] \left(\boldsymbol{R}^{-1}\boldsymbol{y}[i] - \boldsymbol{A}\tilde{\boldsymbol{b}}_k[i]\right), \tag{6.69}$$

$$\mu_k[i] = A_k^2 \{\boldsymbol{\Phi}_k[i]\}_{kk}, \tag{6.70}$$

$$\lambda_1(b_k[i]) = \frac{2\, z_k[i]}{1 - \mu_k[i]}, \qquad k = 1, \ldots, K. \tag{6.71}$$

Next we examine the computational complexity of the low-complexity SISO multiuser detector discussed in this section. From the discussion above, it is seen that at each symbol time i, the dominant computation involved in computing the matrix $\boldsymbol{\Phi}_k[i]$ for $k = 1, \ldots, K$ consists of $2K$ K-vector outer products [i.e., K outer products in computing $\boldsymbol{\Psi}^{(k)}$ as in (6.67), and K outer products in computing $\boldsymbol{\Phi}_k[i]$ as in (6.68)]. From (6.62) and (6.64), in order to obtain the soft output $\lambda_1(b_k[i])$, we also need to compute the soft instantaneous MMSE filter output $z_k[i]$, which by (6.54), is dominated by two K-vector inner products, one in computing the kth user's decorrelating filter output and another in computing the final $z_k[i]$. Therefore, in computing the soft output of the SISO multiuser detector, the dominant computation per user per symbol involves two K-vector outer products and two K-vector inner products.

Note that a number of studies have addressed different aspects of turbo multiuser detection in CDMA systems. In particular, in [336], an optimal turbo multiuser detector is derived based on iterative techniques for cross-entropy minimization. Turbo multiuser detection methods based on different interference cancellation schemes are proposed in [13, 14, 88, 129, 157, 200, 265, 397, 409, 410, 471, 552, 576, 608].

An interesting framework that unifies these approaches to iterative multiuser detection is given in [50]. Moreover, techniques for turbo multiuser detection in unknown channels are developed in [540, 593], which are based on the Markov chain Monte Carlo (MCMC) method for Bayesian computation. Application of the low-complexity SISO detection scheme discussed in this section to equalization of ISI channels with long memory is given in [413].

Simulation Examples

In this section we present some simulation examples to illustrate the performance of the turbo multiuser receiver in synchronous CDMA systems. Of particular interest is the receiver that employs the low-complexity SISO multiuser detector of the preceding section. All users employ the same rate-$\frac{1}{2}$ constraint-length $\nu = 5$ convolutional code (with generators 23 and 35 in octal notation). Each user uses a different interleaver generated randomly. The same set of interleavers is used for all simulations. The block size of the information bits for each user is 128.

First we consider a four-user system with equal cross-correlations $\rho_{ij} = 0.7$, for $1 \leq i, j \leq 4$. All the users have the same power. In Fig. 6.6 the BER performance of the turbo receiver that employs the optimal SISO multiuser detector (6.40) is shown for the first five iterations. In Fig. 6.7, the BER performance of the turbo receiver that employs the low-complexity SISO multiuser detector is shown for the same channel. In each of the these figures, the single-user BER performance (i.e., $\rho_{ij} = 0$) is also shown. It is seen that the performance of both receivers converges toward single-user performance at high SNR. Moreover, the performance loss due to using the low-complexity SISO multiuser detector is small. Next, we consider a near–far situation, where there are two equal-power strong users and two equal-power weak users. The strong users' powers are 3 dB above the powers of the weak users. The user cross-correlations remain the same. Figures 6.8 and 6.9 show, respectively, the BER performance of strong and weak users under the turbo receiver that employs the low-complexity SISO multiuser detectors. It is seen that in such a near–far situation, the weak users actually benefit from the strong interference, whereas the strong users suffer performance loss from the weak interference, a phenomenon previously also observed in the optimal multiuser detector [518] and the multistage multiuser detector [512]. Note that with $\mathcal{O}(2^K)$ computational complexity the optimal SISO multiuser detector (6.40) is not feasible for practical implementation in channels with moderate to large numbers of users K, whereas the low-complexity SISO multiuser detector has a reasonable complexity that can be implemented easily even for very large K. Figure 6.10 illustrates the BER performance of the turbo receiver that employs a low-complexity SISO multiuser detector in an eight-user system. The user cross-correlations are still $\rho_{ij} = 0.7$. All users have the same power. Note that the performance of such a receiver after the first iteration corresponds to the performance of a traditional noniterative receiver structure consisting of a linear MMSE multiuser detector followed by K parallel (soft) channel decoders. It is seen from these figures that at a reasonably high SNR, the turbo receiver offers significant performance gain over traditional noniterative receivers.

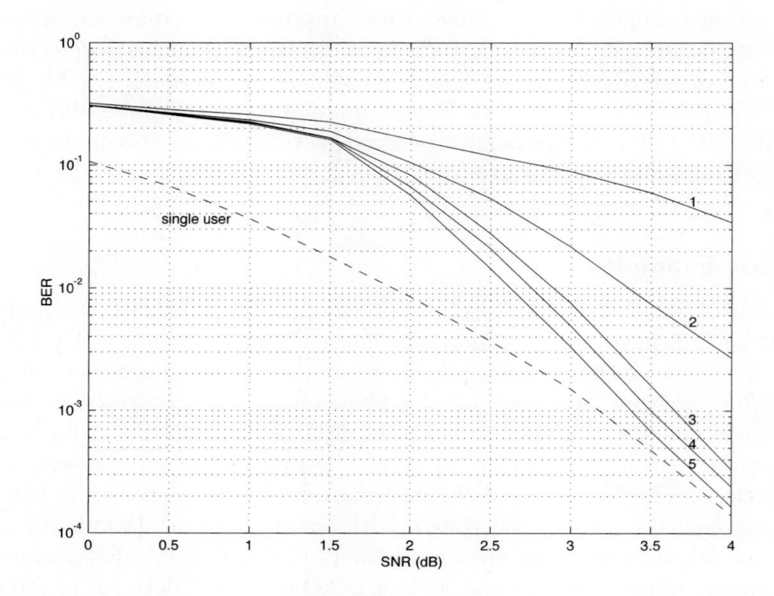

Figure 6.6. Performance of a turbo multiuser receiver that employs an optimal SISO multiuser detector. $K = 4$, $\rho_{ij} = 0.7$. All users have equal power.

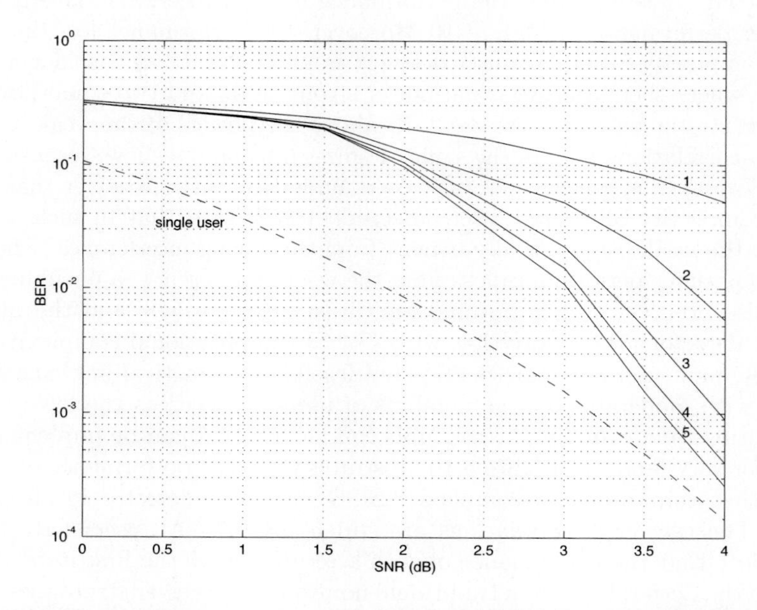

Figure 6.7. Performance of a turbo multiuser receiver that employs a low-complexity SISO multiuser detector. $K = 4$, $\rho_{ij} = 0.7$. All users have equal power.

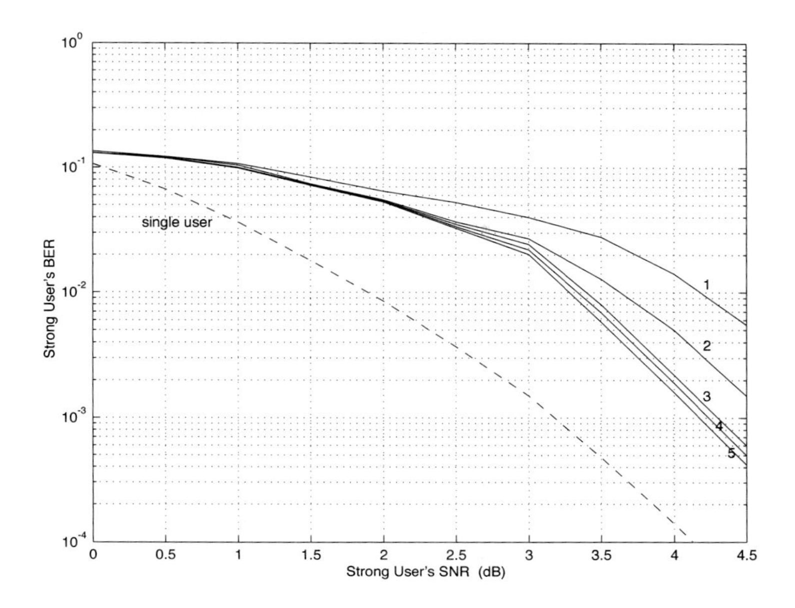

Figure 6.8. Strong user performance under a turbo multiuser receiver that employs a low-complexity SISO multiuser detector. $K = 4$, $\rho_{ij} = 0.7$. Two users are 3 dB stronger than the other two.

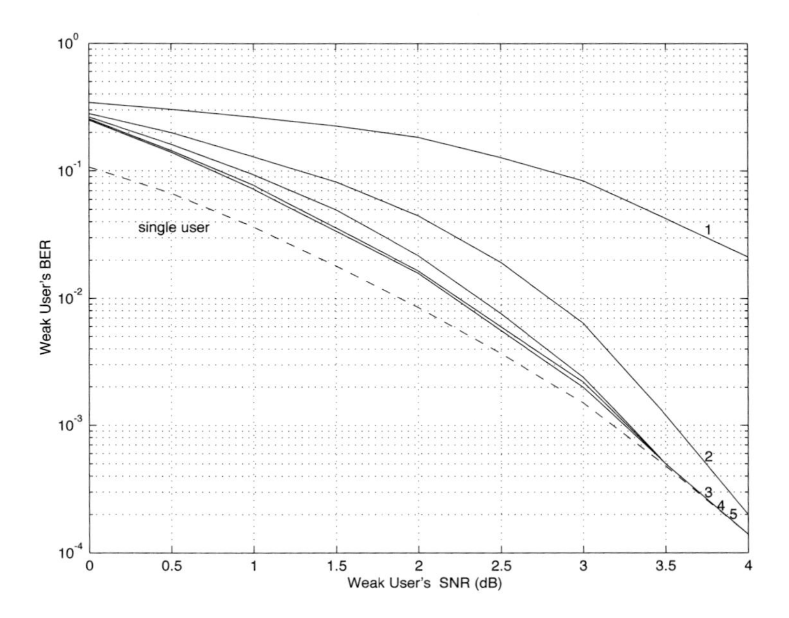

Figure 6.9. Weak user performance under a turbo multiuser receiver that employs a low-complexity SISO multiuser detector. $K = 4$, $\rho_{ij} = 0.7$. Two users are 3 dB stronger than the other two.

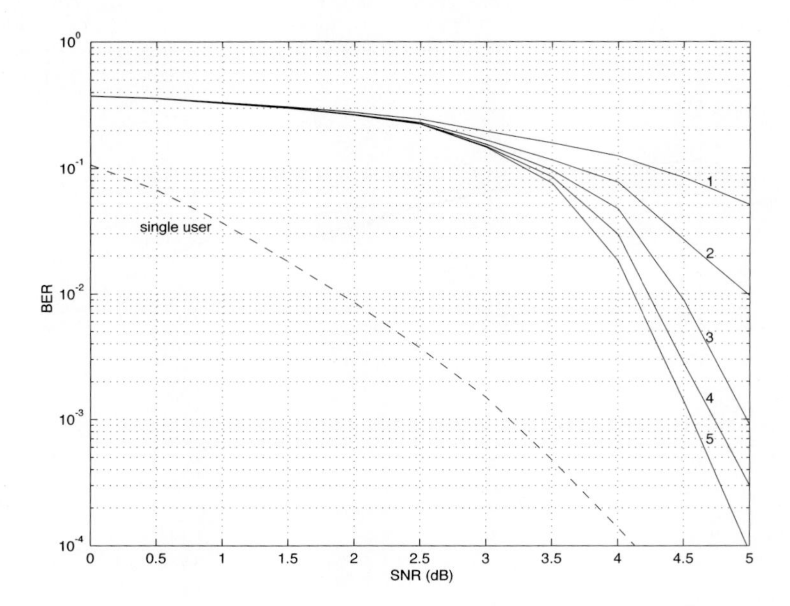

Figure 6.10. Performance of a turbo multiuser receiver that employs a low-complexity SISO multiuser detector. $K = 8$, $\rho_{ij} = 0.7$. All users have equal power.

6.4 Turbo Multiuser Detection with Unknown Interferers

The turbo multiuser detection techniques developed so far assume that the spreading waveforms of all users are known to the receiver. Another important scenario, discussed in Chapter 3, is that the receiver has knowledge of the spreading waveforms of some but not all of the users in a system. Such a situation arises, for example, in a cellular system where the base station receiver knows the spreading waveforms of the in-cell users but not those of the out-of-cell users. In this section we discuss a turbo multiuser detection method that can be applied in the presence of unknown interference, which was first developed in [414].

6.4.1 Signal Model

Consider again the synchronous CDMA signal model (6.27). Here we assume that the spreading waveforms and the received amplitudes of the first $\tilde{K}$ ($\tilde{K} < K$) users are known to the receiver, whereas the rest of the users are unknown to the receiver. Since some of the spreading waveforms are unknown, we cannot form the sufficient statistic (6.32). Instead, as done in Chapters 2 and 3, we sample the received continuous-time signal $r(t)$ at the chip rate to convert it to discrete-time signal. The sample that corresponds to the jth chip of the ith symbol is given by

$$r_j[i] \triangleq \int_{iT+jT_c}^{iT+(j+1)T_c} r(t)\psi(t - iT - jT_c)\,\mathrm{d}t,$$
$$j = 0, \ldots, N-1; \quad i = 0, \ldots, M-1. \quad (6.72)$$

The resulting discrete-time signal corresponding to the ith symbol is then given by

$$\boldsymbol{r}[i] = \sum_{k=1}^{K} A_k b_k[i] \boldsymbol{s}_k + \boldsymbol{n}[i] \quad (6.73)$$
$$= \boldsymbol{S}\boldsymbol{A}\boldsymbol{b}[i] + \boldsymbol{n}[i], \quad (6.74)$$

with

$$\boldsymbol{r}[i] \triangleq \begin{bmatrix} r_0[i] \\ r_1[i] \\ \vdots \\ r_{N-1}[i] \end{bmatrix}, \quad \boldsymbol{s}_k \triangleq \frac{1}{\sqrt{N}} \begin{bmatrix} c_{0,k} \\ c_{1,k} \\ \vdots \\ c_{N-1,k} \end{bmatrix}, \quad \boldsymbol{n}[i] \triangleq \begin{bmatrix} n_0[i] \\ n_1[i] \\ \vdots \\ n_{N-1}[i] \end{bmatrix},$$

where

$$n_j[i] = \int_{jT_c}^{(j+1)T_c} n(t)\psi(t - iT - jT_c)\,\mathrm{d}t, \quad (6.75)$$

is a Gaussian random variable; $\boldsymbol{n}[i] \sim \mathcal{N}(\boldsymbol{0}, \sigma^2 \boldsymbol{I}_N)$; $\boldsymbol{s}_k$ is the normalized discrete-time spreading waveform of the kth user, with $c_{n,k} \in \{+1, -1\}$; $\boldsymbol{S} \triangleq [\boldsymbol{s}_1 \ \cdots \ \boldsymbol{s}_K]$; $\boldsymbol{A} \triangleq \mathsf{diag}(A_1, \ldots, A_K)$; and $\boldsymbol{b}[i] \triangleq \left[b_1[i] \ \cdots \ b_K[i]\right]^T$.

Denote by $\tilde{\boldsymbol{S}}$ the matrix consisting of the first $\tilde{K}$ columns of $\boldsymbol{S}$. Denote the remaining $\bar{K} = K - \tilde{K}$ columns of $\boldsymbol{S}$ by $\bar{\boldsymbol{S}}$. These first $\tilde{K}$ signature sequences are unknown to the receiver. Let $\tilde{\boldsymbol{b}}[i]$ be the $\tilde{K}$-vector containing the first $\tilde{K}$ bits of $\boldsymbol{b}[i]$, and let $\bar{\boldsymbol{b}}[i]$ contain the remaining $\bar{K}$ bits. Then we may write (6.74) as

$$\boldsymbol{r}[i] = \tilde{\boldsymbol{S}}\tilde{\boldsymbol{A}}\tilde{\boldsymbol{b}}[i] + \bar{\boldsymbol{S}}\bar{\boldsymbol{A}}\bar{\boldsymbol{b}}[i] + \boldsymbol{n}[i]. \quad (6.76)$$

Since we do not have knowledge of $\bar{\boldsymbol{S}}$, we cannot hope to demodulate $\bar{\boldsymbol{b}}[i]$. We therefore write (6.76) as

$$\boldsymbol{r}[i] = \tilde{\boldsymbol{S}}\tilde{\boldsymbol{A}}\tilde{\boldsymbol{b}}[i] + \boldsymbol{I}[i] + \boldsymbol{n}[i], \quad (6.77)$$

where $\boldsymbol{I}[i] \triangleq \bar{\boldsymbol{S}}\bar{\boldsymbol{A}}\bar{\boldsymbol{b}}[i]$ is regarded as an interference term that is to be estimated and removed by the multiuser detector before it computes the *a posteriori* log-likelihood ratios (LLRs) for the bits in $\tilde{\boldsymbol{b}}[i]$.

6.4.2 Group-Blind SISO Multiuser Detector

The heart of the turbo group-blind receiver is the soft-input/soft-output (SISO) group-blind multiuser detector. The detector accepts, as inputs, the *a priori* LLRs for the code bits of the known users delivered by the SISO MAP channel decoders of these users, and produces, as outputs, updated LLRs for these code bits. This is accomplished by soft interference cancellation and MMSE filtering. Specifically, using the *a priori* LLRs and knowledge of the signature sequences and received amplitudes of the known users, the detector performs a soft-interference cancellation for each user, in which estimates of the multiuser interference from the other known users and an estimate for the interference caused by the unknown users are subtracted from the received signal. Residual interference is suppressed by passing the resulting signal through an instantaneous MMSE filter. The *a posteriori* LLR can then be computed from the MMSE filter output.

The detector first forms soft estimates of the user code bits as

$$\tilde{b}_k[i] \triangleq E\{b_k[i]\} = \tanh\left[\frac{1}{2}\lambda_2(b_k[i])\right],$$
$$i = 0, \ldots, M-1, \quad k = 1, \ldots, \tilde{K}, \quad (6.78)$$

where $\lambda_2(b_k[i])$ is the *a priori* LLR of the kth user's ith bit delivered by the MAP channel decoder. We denote hard estimates of the code bits as

$$\hat{b}_k[i] \triangleq \operatorname{sign}\left(\tilde{b}_k[i]\right), \qquad k = 1, \ldots, \tilde{K}, \quad (6.79)$$

and denote $\hat{\tilde{\boldsymbol{b}}}[i] \triangleq \left[\hat{b}_1[i]\ \hat{b}_2[i] \cdots \hat{b}_{\tilde{K}}[i]\right]^T$.

In the next step we form an estimate of interference of the unknown users, $\boldsymbol{I}[i]$, which we denote by $\hat{\boldsymbol{I}}[i]$. We begin by forming the preliminary estimate

$$\begin{aligned}\boldsymbol{\gamma}[i] &\triangleq \boldsymbol{r}[i] - \tilde{\boldsymbol{S}}\tilde{\boldsymbol{A}}\hat{\tilde{\boldsymbol{b}}}[i] \\ &= \tilde{\boldsymbol{S}}\tilde{\boldsymbol{A}}\underbrace{\left(\tilde{\boldsymbol{b}}[i] - \hat{\tilde{\boldsymbol{b}}}[i]\right)}_{\boldsymbol{d}[i]} + \bar{\boldsymbol{S}}\bar{\boldsymbol{A}}\bar{\boldsymbol{b}}[i] + \boldsymbol{n}[i],\end{aligned} \quad (6.80)$$

where $\boldsymbol{d}[i] \triangleq \left[d_1[i]\ d_2[i] \cdots d_{\tilde{K}}[i]\right]^T$ and $d_k[i]$ is a random variable defined by

$$d_k[i] \triangleq b_k[i] - \hat{b}_k[i], \qquad k = 1, \ldots, \tilde{K}. \quad (6.81)$$

It will be seen that our ability to form a soft estimate for $d_k[i]$ will allow us to perform the soft interference cancellation mentioned above. Clearly, $d_k[i]$ can take on one of two values, 0 or $2b_k[i]$. The probability that $d_k[i]$ is equal to zero is the probability that the hard estimate is correct and is given by

$$P(d_k[i] = 0) = P\left(b_k[i] = \operatorname{sign}\left\{\tanh\left[\frac{1}{2}\lambda_2(b_k[i])\right]\right\}\right). \quad (6.82)$$

Recall that for $b \in \{+1, -1\}$, the probability that $b_k[i] = b$ is related to the corresponding LLR by [cf. (6.42)]

$$P(b_k[i] = b) = \frac{1}{2} + \frac{b}{2}\tanh\left[\frac{1}{2}\lambda_2(b_k[i])\right]. \quad (6.83)$$

On substituting $b = \text{sign}\left\{\tanh\left[\frac{1}{2}\lambda_2(b_k[i])\right]\right\}$ in (6.83), we find that

$$P(d_k[i] = 0) = \frac{1}{2}\left\{1 + \text{sign}\left(\tanh\left[\frac{1}{2}\lambda_2(b_k[i])\right]\right)\tanh\left[\frac{1}{2}\lambda_2(b_k[i])\right]\right\}$$
$$= \frac{1}{2}\left\{1 + \tanh\left(\frac{1}{2}|\lambda_2(b_k[i])|\right)\right\}. \tag{6.84}$$

Therefore, $d_k[i]$ is a random variable that can be described as

$$d_k[i] = \begin{cases} 0 & \text{with probability} \quad \frac{1}{2} + \frac{1}{2}\tanh\left[\frac{1}{2}|\lambda_2(b_k[i])|\right], \\ 2b_k[i] & \text{with probability} \quad \frac{1}{2} - \frac{1}{2}\tanh\left[\frac{1}{2}|\lambda_2(b_k[i])|\right]. \end{cases} \tag{6.85}$$

We now perform an eigendecomposition on $\boldsymbol{\Gamma}\boldsymbol{\Gamma}^T/M$ where $\boldsymbol{\Gamma} \triangleq [\boldsymbol{\gamma}[0] \ldots \boldsymbol{\gamma}[M-1]]$. We denote by $\boldsymbol{U}_u$ the matrix of eigenvectors corresponding to the $\bar{K}$ largest eigenvalues. The span of the columns of $\boldsymbol{U}_u$ represents an estimate of the subspace of the unknown users (i.e., the interference subspace). Ideally, that is, when $\boldsymbol{d}[i] = \boldsymbol{0}$ in (6.80), $\boldsymbol{U}_u$ contains the signal subspace spanned by the unknown interference $\bar{\boldsymbol{S}}$. To refine our estimate of $\boldsymbol{I}[i]$, we project $\boldsymbol{\gamma}[i]$ onto $\boldsymbol{U}_u$. The result is

$$\hat{\boldsymbol{I}}[i] = \boldsymbol{U}_u\boldsymbol{U}_u^T\left\{\tilde{\boldsymbol{S}}\tilde{\boldsymbol{A}}\boldsymbol{d}[i] + \bar{\boldsymbol{S}}\bar{\boldsymbol{A}}\bar{\boldsymbol{b}}[i] + \boldsymbol{n}[i]\right\}. \tag{6.86}$$

Denote $\tilde{\boldsymbol{S}}_u \triangleq \boldsymbol{U}_u\boldsymbol{U}_u^T\tilde{\boldsymbol{S}}$ and $\boldsymbol{n}_u[i] \triangleq \boldsymbol{U}_u\boldsymbol{U}_u^T\boldsymbol{n}[i]$. Since, ideally, $\boldsymbol{U}_u\boldsymbol{U}_u^T\bar{\boldsymbol{S}} = \bar{\boldsymbol{S}}$, we have

$$\hat{\boldsymbol{I}}[i] = \tilde{\boldsymbol{S}}_u\tilde{\boldsymbol{A}}\boldsymbol{d}[i] + \bar{\boldsymbol{S}}\bar{\boldsymbol{A}}\bar{\boldsymbol{b}}[i] + \boldsymbol{n}_u[i]. \tag{6.87}$$

Now we subtract the interference estimate from the received signal and form a new signal

$$\boldsymbol{\zeta}[i] \triangleq \boldsymbol{r}[i] - \hat{\boldsymbol{I}}[i]$$
$$= \tilde{\boldsymbol{S}}\tilde{\boldsymbol{A}}\tilde{\boldsymbol{b}}[i] - \tilde{\boldsymbol{S}}_u\tilde{\boldsymbol{A}}\boldsymbol{d}[i] + \boldsymbol{v}[i], \tag{6.88}$$

where

$$\boldsymbol{v}[i] \triangleq \boldsymbol{n}[i] - \boldsymbol{n}_u[i] \sim \mathcal{N}(\boldsymbol{0}, \boldsymbol{\Sigma}_v), \tag{6.89}$$

with

$$\boldsymbol{\Sigma}_v = \sigma^2\left(\boldsymbol{I}_N - \boldsymbol{U}_u\boldsymbol{U}_u^T\right). \tag{6.90}$$

For each known user we perform a soft interference cancellation on $\boldsymbol{\zeta}[i]$ to obtain

$$\boldsymbol{r}_k[i] \triangleq \boldsymbol{\zeta}[i] - \tilde{\boldsymbol{S}}\tilde{\boldsymbol{A}}\tilde{\boldsymbol{b}}_k[i] + \tilde{\boldsymbol{S}}_u\tilde{\boldsymbol{A}}\tilde{\boldsymbol{d}}[i], \qquad k = 1, \ldots, \tilde{K}, \tag{6.91}$$

where

$$\tilde{\boldsymbol{b}}_k[i] \triangleq \left[\tilde{b}_1[i] \cdots \tilde{b}_{k-1}[i]\ 0\ \tilde{b}_{k+1}[i] \cdots \tilde{b}_{\tilde{K}}[i]\right]^T,$$
$$\tilde{\boldsymbol{d}}[i] \triangleq \left[\tilde{d}_1[i]\ \tilde{d}_2[i] \cdots \tilde{d}_{\tilde{K}}[i]\right]^T,$$

with $\tilde{d}_k[i]$ a soft estimate for $d_k[i]$, given via (6.85) by

$$\begin{aligned}\tilde{d}_k[i] &\triangleq E\{d_k[i]\} \\ &= E\{E\{d_k[i] \mid b_k[i]\}\} \\ &= \tilde{b}_k[i]\left\{1 - \tanh\left[\frac{1}{2}|\lambda_2(b_k[i])|\right]\right\}, \qquad k = 1, \dots, \tilde{K}.\end{aligned} \tag{6.92}$$

Substituting (6.88) into (6.91), we obtain

$$\boldsymbol{r}_k[i] = \underbrace{\tilde{\boldsymbol{S}}\tilde{\boldsymbol{A}}}_{\tilde{\boldsymbol{H}}}\left(\tilde{\boldsymbol{b}}[i] - \tilde{\boldsymbol{b}}_k[i]\right) - \underbrace{\tilde{\boldsymbol{S}}_u\tilde{\boldsymbol{A}}}_{\tilde{\boldsymbol{H}}_u}\left(\boldsymbol{d}[i] - \tilde{\boldsymbol{d}}[i]\right) + \boldsymbol{v}[i]. \tag{6.93}$$

An instantaneous linear MMSE filter is then applied to $\boldsymbol{r}_k[i]$ to obtain

$$z_k[i] \triangleq \boldsymbol{w}_k[i]^T \boldsymbol{r}_k[i]. \tag{6.94}$$

The filter $\boldsymbol{w}_k[i] \in \mathbb{R}^N$ is chosen to minimize the mean-square error between the code bit $b_k[i]$ and the filter output $z_k[i]$:

$$\boldsymbol{w}_k[i] \triangleq \arg\min_{\boldsymbol{w}\in\mathbb{R}^N} E\left\{\left(b_k[i] - \boldsymbol{w}^T\boldsymbol{r}_k[i]\right)^2\right\}, \tag{6.95}$$

where the expectation is with respect to the ambient noise and the interfering users. The solution to (6.95) is given by

$$\boldsymbol{w}_k[i] = E\left\{\boldsymbol{r}_k[i]\boldsymbol{r}_k[i]^T\right\}^{-1} E\left\{b_k[i]\boldsymbol{r}_k[i]\right\}. \tag{6.96}$$

It is easy to show that

$$\begin{aligned}&E\left\{\boldsymbol{r}_k[i]\boldsymbol{r}_k[i]^T\right\} \\ &= E\left\{\left[\tilde{\boldsymbol{H}}\ \tilde{\boldsymbol{H}}_u\right]\begin{bmatrix} -\left(\tilde{\boldsymbol{b}}[i] - \tilde{\boldsymbol{b}}_k[i]\right) \\ \boldsymbol{d}[i] - \tilde{\boldsymbol{d}}[i]\end{bmatrix}\begin{bmatrix} -\left(\tilde{\boldsymbol{b}}[i] - \tilde{\boldsymbol{b}}_k[i]\right) \\ \boldsymbol{d}[i] - \tilde{\boldsymbol{d}}[i]\end{bmatrix}^T\begin{bmatrix}\tilde{\boldsymbol{H}}^T \\ \tilde{\boldsymbol{H}}_u^T\end{bmatrix}\right\} + \boldsymbol{\Sigma}_v \\ &= \boldsymbol{\mathcal{H}}\underbrace{\mathrm{Cov}\left\{\begin{bmatrix}\tilde{\boldsymbol{b}}_k[i] - \tilde{\boldsymbol{b}}[i] \\ \boldsymbol{d}[i] - \tilde{\boldsymbol{d}}[i]\end{bmatrix}\right\}}_{\boldsymbol{\Delta}[i]}\boldsymbol{\mathcal{H}}^T + \boldsymbol{\Sigma}_v,\end{aligned} \tag{6.97}$$

where $\boldsymbol{\mathcal{H}} \triangleq \left[\tilde{\boldsymbol{H}}\ \tilde{\boldsymbol{H}}_u\right]$. The covariance matrix $\boldsymbol{\Delta}[i]$ has the dimensions $2\tilde{K} \times 2\tilde{K}$ and may be partitioned into four diagonal $\tilde{K} \times \tilde{K}$ blocks in the following manner:

$$\boldsymbol{\Delta}[i] = \begin{bmatrix}\boldsymbol{\Delta}_{11}[i] & \boldsymbol{\Delta}_{12}[i] \\ \boldsymbol{\Delta}_{21}[i] & \boldsymbol{\Delta}_{22}[i]\end{bmatrix}. \tag{6.98}$$

The diagonal elements of $\boldsymbol{\Delta}_{11}[i]$ are given by

$$\begin{aligned}\left[\boldsymbol{\Delta}_{11}[i]\right]_{kk} &= \mathrm{Var}\{b_k[i]\} \\ &= 1 - \tilde{b}_k[i]^2, \qquad k = 1, \dots, \tilde{K}.\end{aligned} \tag{6.99}$$

Using (6.85), the diagonal elements of $\boldsymbol{\Delta}_{22}[i]$ are given by

$$\begin{aligned}\left[\boldsymbol{\Delta}_{22}[i]\right]_{kk} &= \text{Var}\{d_k[i]\} \\ &= 2\alpha_k[i] - \tilde{b}_k[i]^2\alpha_k[i]^2, \qquad k = 1, \ldots, \tilde{K},\end{aligned} \tag{6.100}$$

where

$$\alpha_k[i] \triangleq 1 - \tanh\left[\frac{1}{2}|\lambda_2(b_k[i])|\right]. \tag{6.101}$$

The diagonal elements of $\boldsymbol{\Delta}_{12}[i]$ and $\boldsymbol{\Delta}_{21}[i]$ are identical and are given by

$$\begin{aligned}\left[\boldsymbol{\Delta}_{12}[i]\right]_{kk} &= \text{Cov}\{b_k[i], d_k[i]\} \\ &= \alpha_k[i]\left(1 - \tilde{b}_k[i]^2\right), \qquad k = 1, \ldots, \tilde{K}.\end{aligned} \tag{6.102}$$

It is also easy to see that

$$E\{b_k[i]\boldsymbol{r}_k[i]\} = \tilde{\boldsymbol{H}}\boldsymbol{e}_k - \alpha_k[i]\tilde{\boldsymbol{H}}_u\boldsymbol{e}_k \;\triangleq\; \boldsymbol{p}_k, \tag{6.103}$$

where $\boldsymbol{e}_k$ is a $\tilde{K}$-vector whose elements are all zero except for the kth element, which is 1. Substituting (6.97) and (6.103) into (6.96), we may write the instantaneous MMSE filter for user k as

$$\boldsymbol{w}_k[i] = \left(\boldsymbol{\mathcal{H}}\boldsymbol{\Delta}[i]\boldsymbol{\mathcal{H}}^T + \boldsymbol{\Sigma}_v\right)^{-1}\boldsymbol{p}_k. \tag{6.104}$$

As before, we make the assumption that the MMSE filter output is Gaussian; we may write

$$\begin{aligned}z_k[i] &\triangleq \boldsymbol{w}_k^T[i]\boldsymbol{r}_k[i] \\ &= \mu_k[i]b_k[i] + \eta_k[i],\end{aligned} \tag{6.105}$$

where $\mu_k[i]$ is the equivalent amplitude of the kth user's signal at the filter output, and $\eta_k[i] \sim \mathcal{N}\left(0, \nu_k^2[i]\right)$ is a Gaussian noise sample. Using (6.97) and (6.104), the parameter $\mu_k[i]$ is computed as

$$\begin{aligned}\mu_k[i] &= E\{z_k[i]b_k[i]\} = \boldsymbol{w}_k^T E\{b_k[i]\boldsymbol{r}_k[i]\} \\ &= \boldsymbol{p}_k^T\left(\boldsymbol{\mathcal{H}}\boldsymbol{\Delta}[i]\boldsymbol{\mathcal{H}}^T + \boldsymbol{\Sigma}_v\right)^{-1}\boldsymbol{p}_k,\end{aligned} \tag{6.106}$$

$$\begin{aligned}\nu_k^2[i] &\triangleq \text{Var}\{z_k[i]\} = E\left\{z_k[i]^2\right\} - \mu_k[i]^2 \\ &= \boldsymbol{w}_k[i]^T E\left\{\boldsymbol{r}_k[i]\boldsymbol{r}_k[i]^T\right\}\boldsymbol{w}_k[i] - \mu_k[i]^2 \\ &= \mu_k[i] - \mu_k[i]^2,\end{aligned} \tag{6.107}$$

where (6.107) follows from (6.97), (6.104), and (6.106).

Finally, exactly the same as (6.64), the extrinsic information, $\lambda_1(b_k[i])$, delivered by the SISO multiuser detector is given by

$$\lambda_1(b_k[i]) \triangleq \log \frac{p(z_k[i] \mid b_k[i] = +1)}{p(z_k[i] \mid b_k[i] = -1)} = \frac{2z_k[i]}{1 - \mu_k[i]}, \qquad k = 1, \ldots, \tilde{K}. \tag{6.108}$$

This group-blind SISO multiuser detection algorithm is summarized as follows.

Algorithm 6.3: [Group-blind SISO multiuser detector—synchronous CDMA]

- *Given* $\{\lambda_2(b_k[i])\}$, *form soft and hard estimates of the code bits:*

$$\tilde{b}_k[i] = \tanh\left[\frac{1}{2}\lambda_2(b_k[i])\right], \tag{6.109}$$

$$\hat{b}_k[i] = \text{sign}\left(\tilde{b}_k[i]\right), \qquad k = 1, \ldots, \tilde{K}; \quad i = 0, \ldots, M-1. \tag{6.110}$$

Denote

$$\hat{\boldsymbol{b}}[i] \triangleq \left[\hat{b}_1[i]\ \hat{b}_2[i] \cdots \hat{b}_{\tilde{K}}[i]\right]^T,$$

$$\tilde{\boldsymbol{b}}_k[i] \triangleq \left[\tilde{b}_1[i] \cdots \tilde{b}_{k-1}[i]\ 0\ \tilde{b}_{k+1}[i] \ldots \tilde{b}_{\tilde{K}}[i]\right]^T.$$

- *Let*

$$\boldsymbol{\gamma}[i] \triangleq \boldsymbol{r}[i] - \tilde{\boldsymbol{S}}\tilde{\boldsymbol{A}}\hat{\tilde{\boldsymbol{b}}}[i], \qquad i = 0, \ldots, M-1, \tag{6.111}$$

$$\boldsymbol{\Gamma} \triangleq \left[\boldsymbol{\gamma}[0] \cdots \boldsymbol{\gamma}[M-1]\right]. \tag{6.112}$$

Perform an eigendecomposition on $\boldsymbol{\Gamma}\boldsymbol{\Gamma}^T/M$,

$$\frac{1}{M}\boldsymbol{\Gamma}\boldsymbol{\Gamma}^T = \boldsymbol{U}\boldsymbol{\Lambda}\boldsymbol{U}^T. \tag{6.113}$$

Set $\boldsymbol{U}_u$ *equal to the first* $\bar{K}$ *columns of* $\boldsymbol{U}$.

- *For* $i = 0, 1, \ldots, M-1$:
  - *Refine the estimate of the unknown interference by projection:*

$$\hat{\boldsymbol{I}}[i] = \boldsymbol{U}_u\boldsymbol{U}_u^T\boldsymbol{\gamma}[i]. \tag{6.114}$$

  - *Compute* $\tilde{d}_k[i]$ *according to*

$$\tilde{d}_k[i] = \tilde{b}_k[i]\alpha_k[i], \qquad k = 1, \ldots, \tilde{K}, \tag{6.115}$$

where $\alpha_k[i]$ *is defined in (6.101). Define*

$$\tilde{\boldsymbol{d}}[i] \triangleq \left[\tilde{d}_1[i]\ \tilde{d}_2[i] \cdots \tilde{d}_{\tilde{K}}[i]\right]^T.$$

– *Subtract* $\hat{\boldsymbol{I}}[i]$ *from* $\boldsymbol{r}[i]$ *and perform soft interference cancellation:*

$$\boldsymbol{r}_k[i] = \boldsymbol{r}[i] - \hat{\boldsymbol{I}}[i] - \tilde{\boldsymbol{S}}\tilde{\boldsymbol{A}}\tilde{\boldsymbol{b}}_k[i] + \tilde{\boldsymbol{S}}_u\tilde{\boldsymbol{A}}\tilde{\boldsymbol{d}}[i], \qquad k = 1, \ldots, \tilde{K}, \tag{6.116}$$

where $\tilde{\boldsymbol{S}}_u \triangleq \boldsymbol{U}_u\boldsymbol{U}_u^T\tilde{\boldsymbol{S}}$.

– *Calculate* $\boldsymbol{\Delta}[i]$ *according to (6.99)–(6.102).*

– *Calculate and apply the MMSE filters:*

$$\boldsymbol{w}_k[i] = \left(\boldsymbol{\mathcal{H}}\boldsymbol{\Delta}[i]\boldsymbol{\mathcal{H}}^T + \boldsymbol{\Sigma}_v\right)^{-1}\left(\tilde{\boldsymbol{H}}\boldsymbol{e}_k - \alpha_k[i]\tilde{\boldsymbol{H}}_u\boldsymbol{e}_k\right), \tag{6.117}$$

$$z_k[i] = \boldsymbol{w}_k[i]^T\boldsymbol{r}_k[i]. \tag{6.118}$$

where $\boldsymbol{\Sigma}_v \triangleq \sigma^2\left(\boldsymbol{I}_N - \boldsymbol{U}_u\boldsymbol{U}_u^T\right)$, $\boldsymbol{\mathcal{H}} \triangleq \left[\tilde{\boldsymbol{H}}\ \tilde{\boldsymbol{H}}_u\right]$, *and where* $\tilde{\boldsymbol{H}} \triangleq \tilde{\boldsymbol{S}}\tilde{\boldsymbol{A}}$ *and* $\tilde{\boldsymbol{H}}_u \triangleq \boldsymbol{U}_u\boldsymbol{U}_u^H\tilde{\boldsymbol{H}}$.

– *Compute* $\mu_k[i]$ *according to (6.106).*

– *Compute the a posteriori LLRs for code bit* $b_k[i]$ *according to (6.108).*

6.4.3 Sliding Window Group-Blind Detector for Asynchronous CDMA

It is not difficult to extend the results of Section 6.4.2 to asynchronous CDMA. The received signal due to user $k(1 \leq k \leq K)$ is given by

$$y_k(t) = A_k \sum_{i=0}^{M-1} b_k[i] \sum_{j=0}^{N-1} c_{j,k}\psi(t - jT_c - iT - \tau_k), \tag{6.119}$$

where τ_k is the delay of the kth user's signal, $\{c_{j,k}\}_{j=0}^{N-1}$ is a signature sequence of ±1's assigned to the kth user, and $\psi(t)$ is a normalized chip waveform of duration $T_c = T/N$. The total received signal, given by

$$r(t) = \sum_{k=1}^{K} y_k(t) + v(t) \tag{6.120}$$

is match-filtered to the chip waveform and sampled at the chip rate. The nth matched-filter output during the ith symbol interval is

$$\begin{aligned} r[i,n] &\triangleq \int_{iT+nT_c}^{iT+(n+1)T_c} r(t)\psi(t - iT - nT_c)\,\mathrm{d}t \\ &= \sum_{k=1}^{K} \underbrace{\left\{\int_{iT+nT_c}^{iT+(n+1)T_c} \psi(t - iT - nT_c)y_k(t)\,\mathrm{d}t\right\}}_{y_k[i,n]} \\ &\quad + \underbrace{\int_{iT+nT_c}^{iT_s+(n+1)T_c} v(t)\psi(t - iT - nT_c)\,\mathrm{d}t}_{v[i,n]}. \end{aligned} \tag{6.121}$$

Substituting (6.119) into (6.121), we obtain

$$
\begin{aligned}
&y_k[i,n] \\
&= A_k \sum_{p=0}^{M-1} b_k[p] \sum_{j=0}^{N-1} c_k[j] \int_{iT+nT_c}^{iT+(n+1)T_c} \psi(t-iT-nT_c)\psi(t-jT_c-pT-\tau_k)\,\mathrm{d}t \\
&= \sum_{p=0}^{\iota_k-1} b_k[i-p] \underbrace{A_k \sum_{j=0}^{N-1} c_k[j] \int_0^{T_c} \psi(t)\psi(t-jT_c+nT_c+pT-\tau_k)\,\mathrm{d}t}_{h_k[p,n]}, \qquad (6.122)
\end{aligned}
$$

where $\iota_k \triangleq 1 + \lceil (\tau_k + T_c)/T \rceil$. Then

$$
r[i,n] = h_k[0,n]b_k[i] + \underbrace{\sum_{j=1}^{\iota_k-1} h_k[j,n]b_k[i-j]}_{\text{ISI}} + \underbrace{\sum_{k'\neq k} y_{k'}[i,n]}_{\text{MAI}} + v[i,n]. \qquad (6.123)
$$

Denote

$$
\underline{r}[i] \triangleq \begin{bmatrix} r[i,0] \\ \vdots \\ r[i,N-1] \end{bmatrix}, \quad \underline{v}[i] \triangleq \begin{bmatrix} v[i,0] \\ \vdots \\ v[i,N-1] \end{bmatrix}, \quad \underline{b}[i] \triangleq \begin{bmatrix} b_1[i] \\ \vdots \\ b_K[i] \end{bmatrix}, \qquad (6.124)
$$

and for $j = 0, 1, \ldots, \iota_k - 1$,

$$
\underline{H}[j] \triangleq \begin{bmatrix} h_1[j,0] & \cdots & h_{\tilde{K}}[j,0] & \cdots & h_K[j,0] \\ \vdots & \vdots & \vdots & \vdots & \vdots \\ h_1[j,N-1] & \cdots & h_{\tilde{K}}[j,N-1] & \cdots & h_K[j,N-1] \end{bmatrix}. \qquad (6.125)
$$

Then

$$
\underline{r}[i] = \underline{H}[i] \star \underline{b}[i] + \underline{v}[i]. \qquad (6.126)
$$

By stacking $\iota \triangleq \max_k \iota_k$ successive received sample vectors, we define

$$
\underbrace{\underline{\boldsymbol{r}}[i]}_{N\iota\times 1} \triangleq \begin{bmatrix} \underline{r}[i] \\ \vdots \\ \underline{r}[i+\iota-1] \end{bmatrix}, \quad \underbrace{\underline{\boldsymbol{v}}[i]}_{N\iota\times 1} \triangleq \begin{bmatrix} \underline{v}[i] \\ \vdots \\ \underline{v}[i+\iota-1] \end{bmatrix}, \quad \underbrace{\underline{\boldsymbol{b}}[i]}_{r\times 1} \triangleq \begin{bmatrix} \underline{b}[i-\iota+1] \\ \vdots \\ \underline{b}[i+\iota-1] \end{bmatrix}, \qquad (6.127)
$$

$$
\underbrace{\underline{\boldsymbol{H}}}_{N\iota\times r} \triangleq \begin{bmatrix} \underline{H}[\iota-1] & \cdots & \underline{H}[0] & \cdots & \mathbf{0} \\ \vdots & \ddots & \ddots & \ddots & \vdots \\ \mathbf{0} & \cdots & \underline{H}[\iota-1] & \cdots & \underline{H}[0] \end{bmatrix}, \qquad (6.128)
$$

where $r \triangleq K(2\iota - 1)$. Then we can write the received signal in matrix form as

$$\underline{\boldsymbol{r}}[i] = \underline{\boldsymbol{H}}\underline{\boldsymbol{b}}[i] + \underline{\boldsymbol{v}}[i]. \tag{6.129}$$

Define the set of matrices $\{\underline{\tilde{\boldsymbol{H}}}_j\}_{j=0}^{2\iota-2}$ such that $\underline{\tilde{\boldsymbol{H}}}_j$ is the $N\iota \times \tilde{K}$ matrix composed of columns $jK+1$ through $jK+\tilde{K}$ of the matrix $\underline{\boldsymbol{H}}$. We define the matrix $\underline{\tilde{\boldsymbol{H}}} \triangleq [\underline{\tilde{\boldsymbol{H}}}_0 \; \underline{\tilde{\boldsymbol{H}}}_1 \cdots \underline{\tilde{\boldsymbol{H}}}_{2\iota-2}]$. The size of $\underline{\tilde{\boldsymbol{H}}}$ is $N\iota \times \tilde{K}(2\iota-1)$. We denote by $\underline{\bar{\boldsymbol{H}}}$ the matrix that contains the remaining $\bar{K}(2\iota-1)$ columns of $\underline{\boldsymbol{H}}$. We define $\underline{\tilde{\boldsymbol{b}}}[i]$ and $\underline{\bar{\boldsymbol{b}}}[i]$ by performing a similar separation of the elements of $\underline{\boldsymbol{b}}[i]$. Then we may write (6.129) as

$$\underline{\boldsymbol{r}}[i] = \underline{\tilde{\boldsymbol{H}}}\underline{\tilde{\boldsymbol{b}}}[i] + \underline{\bar{\boldsymbol{H}}}\underline{\bar{\boldsymbol{b}}}[i] + \underline{\boldsymbol{v}}[i]. \tag{6.130}$$

This equation is the asynchronous analog to (6.76). We can obtain estimates of $b_1[i], b_2[i], \ldots, b_{\tilde{K}}[i]$ with straightforward modifications to Algorithm 6.3.

Simulation Examples

We next present simulation results to demonstrate the performance of the proposed turbo group-blind multiuser receiver for asynchronous CDMA. The processing gain of the system is seven and the total number of users is seven. The number of known users is either two or five, as noted on the figures. The spreading sequences are randomly generated and the same sequences are used for all simulations. All users employ the same rate-$\frac{1}{2}$, constraint-length-3 convolutional code (with generators $g_1 = [110]$ and $g_2 = [111]$). Each user uses a different random interleaver, and the same interleavers are used in all simulations. The block size of information bits for each user is 128. The maximum delay in symbol intervals is 1. All users use the same transmitted power and the chip pulse waveform is a raised cosine with roll-off factor 0.5.

Figure 6.11 illustrates the average bit-error-rate performance of the known users for the group-blind turbo receiver and the conventional turbo receiver discussed in Section 6.3 for the first four iterations. The number of known users is five. For the sake of comparison, we have included plots for the conventional turbo receiver when *all* of the users are known. The three sets of plots in this figure are denoted in the legend by "GBMUD," "TMUD," and "ALL KNOWN," respectively. Note that the curves for the first iteration are identical for GBMUD and TMUD. Hence we have suppressed the plot of the first iteration for TMUD, to improve clarity. Notice that iteration does not significantly improve the performance of the conventional turbo receiver, whereas the group-blind receiver provides significant gains through iteration at moderate and high signal-to-noise ratios. We can also see that the use of more than three iterations does not provide significant benefits.

In Fig. 6.12, the number of known users has been changed to two. As we would expect, there is performance degradation for both conventional and group-blind turbo receivers. In fact, the conventional receiver gains nothing through iteration for this scenario because there are now five users whose interference is simply ignored. It is also apparent that the group-blind turbo receiver will not be able to mitigate

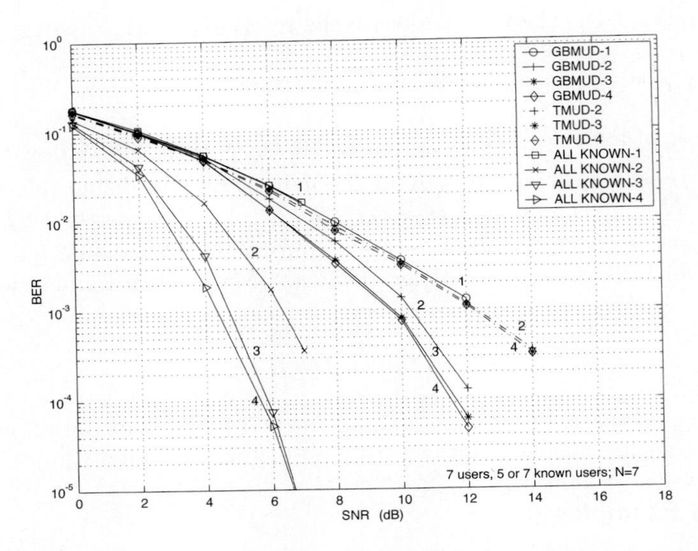

Figure 6.11. Performance of a group-blind iterative multiuser receiver with five known users. Curves denoted GBMUD are produced using a turbo group-blind multiuser receiver and those denoted TMUD are produced using a standard turbo multiuser receiver. Also included are plots for TMUD when all users are known.

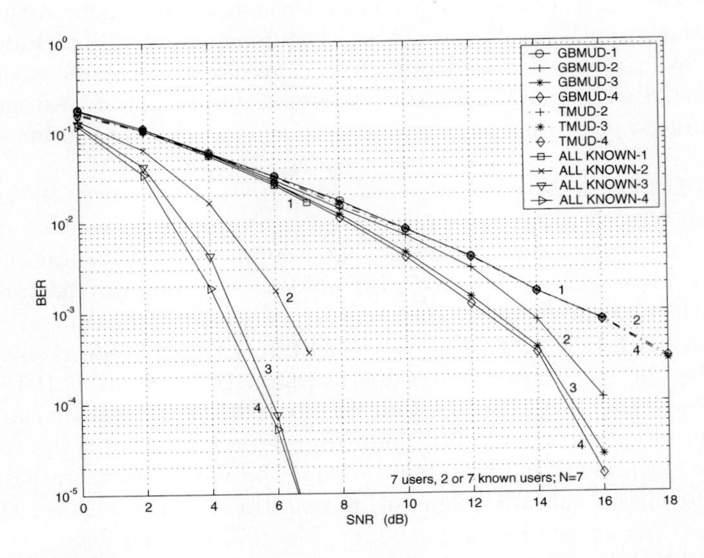

Figure 6.12. Performance of a group-blind iterative multiuser receiver with two known users. Curves denoted GBMUD are produced using a turbo group-blind multiuser receiver and those denoted TMUD are produced using a standard turbo multiuser receiver. Also included are plots for TMUD when all users are known.

all of the interference of unknown users, even for a large number of iterations. This is due, in part, to the use of an imperfect interference subspace estimate in the SISO group-blind multiuser detector.

6.5 Turbo Multiuser Detection in CDMA with Multipath Fading

In this section we generalize the low-complexity SISO multiuser detector developed in Section 6.3.3 for synchronous CDMA systems to general asynchronous CDMA systems with multipath fading channels. The method discussed in this section was first developed in [257].

6.5.1 Signal Model and Sufficient Statistics

We consider a K-user asynchronous CDMA system employing aperiodic spreading waveforms and signaling over multipath fading channels. The transmitted signal due to the kth user is given by

$$x_k(t) = A_k \sum_{i=0}^{M-1} b_k[i]\, s_{i,k}(t - iT), \tag{6.131}$$

where M is the number of data symbols per user per frame, T is the symbol interval, and A_k, $b_k[i]$, and $\{s_{i,k}(t);\ 0 \leq t \leq T\}$ denote, respectively, the amplitude, ith transmitted bit, and normalized signature waveform during the ith symbol interval of the kth user. It is assumed that $s_{i,k}(t)$ is supported only on the interval $[0, T]$ and has unit energy. Note here that we allow the possibility that aperiodic spreading waveforms are employed in the system, and hence the spreading waveforms vary with symbol index i.

The kth user's signal $x_k(t)$ propagates through a multipath channel with impulse response

$$g_k(t) = \sum_{\ell=1}^{L} \alpha_{\ell,k}(t)\delta(t - \tau_{\ell,k}), \tag{6.132}$$

where L is the number of paths in the kth user's channel, and where $\alpha_{\ell,k}(t)$ and $\tau_{\ell,k}$ are, respectively, the complex fading process and the delay of the ℓth path of the kth user's signal. It is assumed that the fading is slow:

$$\alpha_{\ell,k}(t) = \alpha_{\ell,k}[i] \quad \text{for } iT \leq t < (i+1)T,$$

which is a reasonable assumption in many practical situations. At the receiver, the received signal due to the kth user is then given by

$$\begin{aligned} y_k(t) &= x_k(t) \star g_k(t) \\ &= A_k \sum_{i=0}^{M-1} b_k[i] \sum_{\ell=1}^{L} \alpha_{\ell,k}[i]\, s_{i,k}(t - iT - \tau_{\ell,k}), \end{aligned} \tag{6.133}$$

where $\star$ denotes convolution. The received signal at the receiver is the superposition of the K users' signals plus the additive white Gaussian noise, given by

$$r(t) = \sum_{k=1}^{K} y_k(t) + n(t), \tag{6.134}$$

where $n(t)$ is a zero-mean complex white Gaussian noise process with power spectral density σ^2.

Denote $\underline{b}[i] \triangleq \left[b_1[i] \; \cdots \; b_K[i]\right]^T$ and $\boldsymbol{b} \triangleq \left[\underline{b}[0]^T \; \cdots \; \underline{b}[M-1]^T\right]^T$. Define

$$S(t;\boldsymbol{b}) \triangleq \sum_{k=1}^{K} A_k \sum_{i=0}^{M-1} b_k[i] \sum_{\ell=1}^{L} \alpha_{\ell,k}[i]\, s_{i,k}(t - iT - \tau_{\ell,k}). \tag{6.135}$$

Similar to the situations in earlier chapters, using the Cameron–Martin formula [377], the likelihood function of the received waveform $r(t)$ in (6.134) conditioned on all the transmitted symbols $\boldsymbol{b}$ of all users can be written as

$$\mathcal{L}\left(\{r(t) : -\infty < t < \infty\} \mid \boldsymbol{b}\right) = C \exp\left[\Omega(\boldsymbol{b})/\sigma^2\right], \tag{6.136}$$

where C is some positive scalar constant, and

$$\Omega(\boldsymbol{b}) \triangleq 2\,\Re\left\{\int_{-\infty}^{\infty} S(t;\boldsymbol{b})^* r(t)\, dt\right\} - \int_{-\infty}^{\infty} |\, S(t;\boldsymbol{b})\,|^2 \, dt. \tag{6.137}$$

The first integral in (6.137) can be expressed as

$$\int_{-\infty}^{\infty} S(t;\boldsymbol{b})^*\, r(t)\, dt$$

$$\triangleq \sum_{k=1}^{K} A_k \sum_{i=0}^{M-1} b_k[i] \overbrace{\sum_{\ell=1}^{L} \alpha_{\ell,k}[i]^* \underbrace{\int_{-\infty}^{\infty} r(t) s_{i,k}(t - iT - \tau_{\ell,k})\, dt}_{z_{\ell,k}[i]}}^{y_k[i]}. \tag{6.138}$$

Since the second integral in (6.137) does not depend on the received signal $r(t)$, by (6.138) a sufficient statistic for detecting the multiuser symbols $\boldsymbol{b}$ is $\{y_k[i]\}_{i;k}$. From (6.138) it is seen that this sufficient statistic is obtained by passing the received signal $r(t)$ through a bank of K maximal-ratio multipath combiners (i.e., RAKE receivers). Next, we derive an explicit expression for this sufficient statistic in terms of the multiuser channel parameters and transmitted symbols, which is instrumental to developing the SISO multiuser detector. Note that the derivations below are similar to those in Section 5.3.1 for space-time CDMA systems.

Assume that the multipath spread of any user's channel is limited to at most Δ symbol intervals, where Δ is a positive integer. That is,

$$\tau_{\ell,k} \leq \Delta T, \qquad 1 \leq k \leq K, \quad 1 \leq \ell \leq L. \tag{6.139}$$

Define the following correlation of the delayed signaling waveforms:

$$\rho^{[j]}_{(k,\ell)(k',\ell')}[i] \triangleq \int_{-\infty}^{\infty} s_{i,k}(t-\tau_{\ell,k})\, s_{i-j,k'}(t+jT-\tau_{\ell',k'})\,\mathrm{d}t,$$
$$-\Delta \le j \le \Delta, \quad 1 \le k,k' \le K, \quad 1 \le \ell,\ell' \le L. \tag{6.140}$$

Since $\tau_{\ell,k} \le \Delta T$ and $s_{i,k}(t)$ is nonzero only for $t \in [0,T]$, it then follows that $\rho^{[j]}_{(k,\ell)(k',\ell')}[i] = 0$ for $|j| > \Delta$. Now substituting (6.134) into (6.138), we have

$$z_{\ell,k}[i] = \sum_{i'=0}^{M-1}\sum_{k'=1}^{K} A_{k'} b_{k'}[i'] \sum_{\ell'=1}^{L} \alpha_{\ell',k'}[i] \int_{-\infty}^{\infty} s_{i',k'}(t-i'T-\tau_{\ell',k'}) s_{i,k}(t-iT-\tau_{\ell,k})\,\mathrm{d}t$$
$$+ \underbrace{\int_{-\infty}^{\infty} n(t) s_{i,k}(t-iT-\tau_{\ell,k})\,\mathrm{d}t}_{u_{\ell,k}[i]}$$
$$= \sum_{j=-\Delta}^{\Delta}\sum_{k'=1}^{K} A_{k'} b_{k'}[i+j] \sum_{\ell'=1}^{L} \alpha_{\ell',k'}[i]\, \rho^{[-j]}_{(k,\ell)(k',\ell')}[i] + u_{\ell,k}[i], \tag{6.141}$$

where $\{u_{k,\ell}[i]\}$ are zero-mean complex Gaussian random sequences with the following covariance:

$$E\left\{u_{\ell,k}[i]\, u_{\ell',k'}[i']^*\right\}$$
$$= E\left\{\left[\int_{-\infty}^{\infty} n(t) s_{i,k}(t-iT-\tau_{\ell,k})\,\mathrm{d}t\right]\left[\int_{-\infty}^{\infty} n^*(t') s_{i',k'}(t'-i'T-\tau_{\ell',k'})\,\mathrm{d}t'\right]\right\}$$
$$= \left[\int_{-\infty}^{\infty}\int_{-\infty}^{\infty} E\left\{n(t)n^*(t')\right\} s_{i,k}(t-iT-\tau_{\ell,k})\, s_{i',k'}(t'-i'T-\tau_{\ell',k'})\,\mathrm{d}t\,\mathrm{d}t'\right]$$
$$= \left[\int_{-\infty}^{\infty}\int_{-\infty}^{\infty} \underline{I}_p\, \delta(t-t')\, s_{i,k}(t-iT-\tau_{\ell,k})\, s_{i',k'}(t'-i'T-\tau_{\ell',k'})\,\mathrm{d}t\,\mathrm{d}t'\right]$$
$$= \int_{-\infty}^{\infty} s_{i,k}(t-iT-\tau_{\ell,k})\, s_{i',k'}(t-i'T-\tau_{\ell',k'})\,\mathrm{d}t$$
$$= \rho^{[i-i']}_{(k,\ell)(k',\ell')}[i], \tag{6.142}$$

where $\underline{I}_p$ denotes a $p \times p$ identity matrix and $\delta(t)$ is the Dirac delta function. Define the following quantities:

$$\underline{R}^{[j]}[i] \triangleq \begin{bmatrix} \rho^{[j]}_{(1,1)(1,1)}[i] & \cdots & \rho^{[j]}_{(1,1)(1,L)}[i] & \cdots & \rho^{[j]}_{(1,1)(K,1)}[i] & \cdots & \rho^{[j]}_{(1,1)(K,L)}[i] \\ \rho^{[j]}_{(2,1)(1,1)}[i] & \cdots & \rho^{[j]}_{(2,1)(1,L)}[i] & \cdots & \rho^{[j]}_{(2,1)(K,1)}[i] & \cdots & \rho^{[j]}_{(2,1)(K,L)}[i] \\ \vdots & \vdots & \vdots & \vdots & \vdots & \vdots & \vdots \\ \rho^{[j]}_{(K,L)(1,1)}[i] & \cdots & \rho^{[j]}_{(K,L)(1,L)}[i] & \cdots & \rho^{[j]}_{(K,L)(K,1)}[i] & \cdots & \rho^{[j]}_{(K,L)(K,L)}[i] \end{bmatrix},$$

$(KL \times KL$ matrix$)$

$$\underline{\zeta}[i] \triangleq \left[z_{1,1}[i] \ldots z_{L,1}[i] \ldots z_{1,K}[i] \ldots z_{L,K}[i]\right]^T, \qquad (KL\text{-vector})$$

$$\underline{u}[i] \triangleq \left[u_{1,1}[i] \ldots u_{L,1}[i] \ldots u_{1,K}[i] \ldots u_{L,K}[i]\right]^T, \qquad (KL\text{-vector})$$

$$\underline{g}_k[i] \triangleq \left[\alpha_{1,k}[i] \ldots \alpha_{L,k}[i]\right]^T, \qquad (L\text{-vector})$$

$$\underline{G}[i] \triangleq \text{diag}\left\{\underline{g}_1[i], \ldots, \underline{g}_K[i]\right\}, \qquad (KL \times K \text{ matrix})$$

$$\underline{A} \triangleq \text{diag}\left\{A_1, \ldots, A_K\right\}, \qquad (K \times K \text{ matrix})$$

$$\underline{y}[i] \triangleq \left[y_1[i] \ldots y_K[i]\right]^T, \qquad (K\text{-vector})$$

$$\underline{b}[i] \triangleq \left[b_1[i] \ldots b_K[i]\right]^T. \qquad (K\text{-vector})$$

We can then write (6.141) in the following vector form:

$$\underline{\zeta}[i] = \sum_{j=-\Delta}^{\Delta} \underline{R}^{[-j]}[i]\, \underline{G}[i]\, \underline{A}\, \underline{b}[i+j] + \underline{u}[i], \tag{6.143}$$

and from (6.142), the covariance matrix of the complex Gaussian vector sequence $\{\underline{u}[i]\}$ is

$$E\left\{\underline{u}[i]\, \underline{u}[i+j]^H\right\} = \sigma^2\, \underline{R}^{[-j]}[i]. \tag{6.144}$$

Substituting (6.143) into (6.138), we obtain an expression for the sufficient statistic $\underline{y}[i]$, given by

$$\underline{y}[i] \triangleq \underline{G}[i]^H \underline{\zeta}[i] = \sum_{j=-\Delta}^{\Delta} \underbrace{\underline{G}[i]^H \underline{R}^{[-j]}[i] \underline{G}[i]}_{\underline{H}^{[-j]}[i]} \underline{A}\underline{b}[i+j] + \underbrace{\underline{G}[i]^H \underline{u}[i]}_{\underline{v}[i]}, \tag{6.145}$$

where $\underline{v}[i]$ is a sequence of zero-mean complex Gaussian vectors with covariance matrix

$$E\left\{\underline{v}[i]\underline{v}[i+j]^H\right\} = \sigma^2 \underline{G}[i]^H \underline{R}^{[-j]}[i] \underline{G}[i] \triangleq \sigma^2 \underline{H}^{[-j]}[i]. \tag{6.146}$$

Note that by definition (6.140), we have $\rho^{[j]}_{(k,\ell)(k',\ell')}[i] = \rho^{[-j]}_{(k',\ell')(k,\ell)}[i]$. It then follows that $\underline{R}^{[-j]}[i] = \underline{R}^{[j]}[i]^T$, and therefore, $\underline{H}^{[-j]}[i] = \underline{H}^{[j]}[i]^H$.

6.5.2 SISO Multiuser Detector in Multipath Fading Channels

In what follows we assume that the multipath spread is within one symbol interval (i.e., $\Delta = 1$). Define the following quantities:

$$\boldsymbol{H}[i] \triangleq \left[\ \underline{H}^{[1]}[i] \quad \underline{H}^{[0]}[i] \quad \underline{H}^{[-1]}[i]\ \right], \qquad (K \times 3K \text{ matrix})$$

$$\boldsymbol{A} \triangleq \text{diag}\left\{\underline{A}, \underline{A}, \underline{A}\right\}, \qquad (3K \times 3K \text{ matrix})$$

$$\boldsymbol{b}[i] \triangleq \left[\underline{b}[i-1]^T \ \ \underline{b}[i]^T \ \ \underline{b}[i+1]^T\right]^T. \qquad (3K\text{-vector})$$

We can then write (6.145) in matrix form as

$$\underline{y}[i] = \boldsymbol{H}[i]\boldsymbol{A}\boldsymbol{b}[i] + \underline{v}[i], \tag{6.147}$$

where by (6.146), $\underline{v}(i) \sim \mathcal{N}_c\left(\underline{0}, \sigma^2\, \underline{H}^0[i]\right)$.

Based on the *a priori* LLR of the code bits of all users, $\{\lambda_2(b_k[i])\}_{i;k}$, provided by the MAP channel decoder, we first form soft estimates of the user code bits:

$$\tilde{b}_k[i] \triangleq \tanh\left[\frac{1}{2}\lambda_2(b_k[i])\right], \qquad i = 0, \ldots, M-1; \quad k = 1, \ldots, K. \tag{6.148}$$

Denote

$$\tilde{\underline{b}}[i] \triangleq \left[\tilde{b}_1[i] \;\cdots\; \tilde{b}_K[i]\right]^T, \tag{6.149}$$

$$\tilde{\boldsymbol{b}}[i] = \left[\tilde{\underline{b}}[i-1]^T \; \tilde{\underline{b}}[i]^T \; \tilde{\underline{b}}[i+1]^T \;\right]^T, \tag{6.150}$$

$$\tilde{\boldsymbol{b}}_k[i] \triangleq \tilde{\boldsymbol{b}}[i] - \tilde{b}_k[i]\boldsymbol{e}_k, \tag{6.151}$$

where $\boldsymbol{e}_k$ denotes a $3K$-vector of all zeros, except for the $(K+k)$th element, which is 1.

At symbol time i, for each user, a soft interference cancellation is performed on the received discrete-time signal $\underline{y}[i]$ in (6.147), to obtain

$$\underline{y}_k[i] \triangleq \underline{y}[i] - \boldsymbol{H}[i]\,\boldsymbol{A}\,\tilde{\boldsymbol{b}}_k[i] \tag{6.152}$$

$$= \boldsymbol{H}[i]\,\boldsymbol{A}\left(\boldsymbol{b}[i] - \tilde{\boldsymbol{b}}_k[i]\right) + \underline{v}[i], \qquad k = 1, \ldots, K. \tag{6.153}$$

An instantaneous linear MMSE filter is then applied to $\underline{y}_k[i]$, to obtain

$$z_k[i] = \underline{w}_k[i]^H\, \underline{y}_k[i], \tag{6.154}$$

where the filter $\underline{w}_k[i] \in \mathbb{C}^K$ is chosen to minimize the mean-square error between the code bit $b_k[i]$ and the filter output $z_k[i]$:

$$\begin{aligned}\underline{w}_k[i] &= \arg\min_{\underline{w}\in\mathbb{C}^K} E\left\{\left|b_k[i] - \underline{w}^H\, \underline{y}_k[i]\right|^2\right\} \\ &= \arg\min_{\underline{w}\in\mathbb{C}^K} \underline{w}^H\, E\left\{\underline{y}_k[i]\,\underline{y}_k[i]^H\right\}\,\underline{w} - 2\Re\left[\underline{w}^H\, E\left\{b_k[i]\,\underline{y}_k[i]\right\}\right],\end{aligned} \tag{6.155}$$

where

$$E\left\{\underline{y}_k[i]\,\underline{y}_k[i]^H\right\} = \boldsymbol{H}[i]\,\boldsymbol{A}\,\boldsymbol{\Delta}_k[i]\,\boldsymbol{A}\,\boldsymbol{H}[i]^H + \sigma^2\,\underline{H}^0[i], \tag{6.156}$$

$$E\left\{b_k[i]\,\underline{y}_k[i]\right\} = \boldsymbol{H}[i]\,\boldsymbol{A}\,\boldsymbol{e}_k \;= A_k\,\boldsymbol{H}[i]\,\boldsymbol{e}_k, \tag{6.157}$$

with

$$\boldsymbol{\Delta}_k[i] \triangleq \text{Cov}\left\{\boldsymbol{b}[i] - \tilde{\boldsymbol{b}}_k[i]\right\} = \text{diag}\left\{\Delta_k[i-1],\, \Delta_k[i],\, \Delta_k[i+1]\right\},$$

$$\Delta_k[i-l] \triangleq \text{diag}\left\{1 - \tilde{b}_1[i-l]^2, \ldots, 1 - \tilde{b}_K[i-l]^2\right\}, \qquad l = \pm 1,$$

$$\Delta_k[i] \triangleq \text{diag}\left\{1 - \tilde{b}_1[i]^2, \ldots, 1 - \tilde{b}_{k-1}[i]^2,\, 1,\, 1 - \tilde{b}_{k+1}[i]^2,\, 1 - \tilde{b}_K[i]^2\right\}.$$

The solution to (6.155) is given by

$$\underline{w}_k[i] = A_k \left(\boldsymbol{H}[i]\boldsymbol{A}\boldsymbol{\Delta}_k[i]\boldsymbol{A}\boldsymbol{H}[i]^H + \sigma^2 \underline{H}^0[i]\right)^{-1} \boldsymbol{H}[i]\boldsymbol{e}_k. \tag{6.158}$$

As before, in order to form the LLR of the code bit $b_k[i]$, we approximate the instantaneous linear MMSE filter output $z_k[i]$ in (6.154) as being Gaussian [i.e., $z_k[i] \sim \mathcal{N}_c\left(\mu_k[i]b_k[i], \nu_k^2[i]\right)$]. Conditioned on the code bit $b_k[i]$, the mean and variance of $z_k[i]$ are given, respectively, by

$$\begin{aligned}
\mu_k[i] &\triangleq E\{z_k(i)b_k(i)\} \\
&= \boldsymbol{e}_k^H \boldsymbol{H}[i]^H \left(\boldsymbol{H}[i]\boldsymbol{\Delta}_k[i]\boldsymbol{H}[i]^H + \sigma^2 \underline{H}^0[i]\right)^{-1} \boldsymbol{H}[i] E\left\{\boldsymbol{b}[i] - \tilde{\boldsymbol{b}}_k[i]\right\} \\
&= \boldsymbol{e}_k^T \boldsymbol{H}[i]^H \left(\boldsymbol{H}[i]\boldsymbol{\Delta}_k[i]\boldsymbol{H}^H + \sigma^2 \underline{H}^0[i]\right)^{-1} \boldsymbol{H}[i]\boldsymbol{e}_k,
\end{aligned} \tag{6.159}$$

$$\begin{aligned}
\nu_k^2[i] &\triangleq \text{Var}\{z_k[i]\} = E\left\{\left|z_k[i]\right|^2\right\} - \mu_k[i]^2 \\
&= \boldsymbol{w}_k^H E\left\{\boldsymbol{y}_k[i]\boldsymbol{y}_k[i]^H\right\} \boldsymbol{w}_k - \mu_k[i]^2 \\
&= \boldsymbol{e}_k^T \boldsymbol{H}[i]^H \left(\boldsymbol{H}[i]\boldsymbol{\Delta}_k[i]\boldsymbol{H}[i]^H + \sigma^2 \underline{H}^0[i]\right)^{-1} \boldsymbol{H}[i]\boldsymbol{e}_k - \mu_k[i]^2 \\
&= \mu_k[i] - \mu_k[i]^2.
\end{aligned} \tag{6.160}$$

Therefore, the extrinsic information $\lambda_1(b_k[i])$ delivered by the instantaneous linear MMSE filter is given by

$$\begin{aligned}
\lambda_1[b_k(i)] &= -\frac{\left|z_k[i] - \mu_k[i]\right|^2}{\nu_k^2[i]} + \frac{\left|z_k[i] - \mu_k[i]\right|^2}{\nu_k^2[i]} \\
&= \frac{4\Re\{\mu_k[i]\, z_k[i]\}}{\nu_k^2[i]} \\
&= \frac{4\Re\{z_k[i]\}}{1 - \mu_k[i]}.
\end{aligned} \tag{6.161}$$

The SINR at the instantaneous linear MMSE filter output is given by

$$\begin{aligned}
\text{SINR}(z_k[i]) &\triangleq \frac{E^2\left\{\Re(z_k[i])\right\}}{\text{Var}\left\{\Re(z_k[i])\right\}} \\
&= \frac{\mu_k[i]^2}{\frac{1}{2}\nu_k^2[i]} = \frac{2}{1/\mu_k[i] - 1}.
\end{aligned} \tag{6.162}$$

Recursive Algorithm for Computing Soft Output

Similar to our earlier discussion, computation of the extrinsic information can be implemented efficiently. In particular, the major computation involved is the following $K \times K$ matrix inversion:

$$\boldsymbol{\Psi}_k[i] \triangleq \left(\boldsymbol{H}[i]\boldsymbol{\Delta}_k[i]\boldsymbol{H}[i]^H + \sigma^2 \underline{H}^0[i]\right)^{-1}. \tag{6.163}$$

Note that $\boldsymbol{\Delta}_k[i]$ can be written as

$$\boldsymbol{\Delta}_k[i] = \boldsymbol{\Delta}[i] + \tilde{b}_k[i]^2 \boldsymbol{e}_k \boldsymbol{e}_k^T, \tag{6.164}$$

where

$$\boldsymbol{\Delta}[i] \triangleq \text{diag}\left\{1 - \tilde{b}_1[i-1]^2, \ \ldots, \ 1 - \tilde{b}_K[i-1]^2, \ 1 - \tilde{b}_1[i]^2, \ \ldots, \right.$$
$$\left. 1 - \tilde{b}_K[i]^2, \ 1 - \tilde{b}_1[i+1]^2, \ \ldots, \ 1 - \tilde{b}_K[i+1]^2 \right\}. \tag{6.165}$$

Substituting (6.164) into (6.163), we have

$$\boldsymbol{\Psi}_k[i] = \left(\boldsymbol{H}[i]\boldsymbol{\Delta}[i]\boldsymbol{H}[i]^H + \sigma^2 \underline{H}^0[i] + \tilde{b}_k[i]^2 \boldsymbol{H}_{(:,K+k)}[i] \boldsymbol{H}_{(:,K+k)}[i]^H\right)^{-1}, \tag{6.166}$$

where $\boldsymbol{H}_{(:,K+k)}[i]$ denotes the $(K+k)$th column of $\boldsymbol{H}[i]$. Define

$$\boldsymbol{\Psi}[i] \triangleq \left(\boldsymbol{H}[i]\boldsymbol{\Delta}[i]\boldsymbol{H}[i]^H + \sigma^2 \underline{H}^0[i]\right)^{-1}. \tag{6.167}$$

Then by the matrix inversion lemma, (6.166) can be written as

$$\boldsymbol{\Psi}_k[i] = \boldsymbol{\Psi}[i] - \frac{1}{\tilde{b}_k[i]^{-2} + \boldsymbol{H}_{(:,K+k)}[i]^H \boldsymbol{\Psi}[i] \boldsymbol{H}_{(:,K+k)}[i]}$$
$$\left(\boldsymbol{\Psi}[i]\boldsymbol{H}_{(:,K+k)}[i]\right)\left(\boldsymbol{\Psi}[i]\boldsymbol{H}_{(:,K+k)}[i]\right)^H, \qquad k = 1, \ldots, K. \tag{6.168}$$

Equations (6.167) and (6.168) constitute a recursive procedure for computing $\boldsymbol{\Psi}_k[i]$ in (6.163).

Next, we summarize the SISO multiuser detection algorithm in multipath fading channels ($\Delta = 1$) as follows.

Algorithm 6.4: [SISO multiuser detector—multipath fading channel]

- *Form the soft bit estimates using (6.148)–(6.151).*
- *Compute the matrix inversions using (6.167) and (6.168).*
- *For $i = 0, \ldots, M-1$ and for $k = 1, \ldots, K$, compute $z_k[i]$ using (6.152), (6.154), and (6.158); compute $\mu_k[i]$ using (6.159); and compute $\lambda_1(b_k[i])$ using (6.161).*

Finally, we examine the computational complexity of the SISO multiuser detector in multipath fading channels. By (6.167), it takes $\mathcal{O}(K^3)$ multiplications to obtain $\boldsymbol{\Psi}[i]$ using direct matrix inversion. After $\boldsymbol{\Psi}[i]$ is obtained, by (6.168), it takes $\mathcal{O}(K^2)$ more multiplications to get $\boldsymbol{\Psi}_k[i]$ for each k. Since at each time i, $\boldsymbol{\Psi}[i]$ is computed only once for all K users, it takes $\mathcal{O}(K^2)$ multiplications per user per code bit to obtain $\boldsymbol{\Psi}_k[i]$. After $\boldsymbol{\Psi}_k[i]$ is computed, by (6.154), (6.159), and (6.161), it takes $\mathcal{O}(K^2)$ multiplications to obtain $\lambda_1(b_k[i])$. Therefore, the total time complexity of the SISO multiuser detector is $\mathcal{O}(K^2)$ per user per code bit.

6.6 Turbo Multiuser Detection in CDMA with Turbo Coding

In this section we discuss turbo multiuser detection for CDMA systems employing turbo codes. Parallel concatenated codes, called *turbo codes*, constitute the most important breakthrough in coding of the 1990s [177]. Since these powerful codes can achieve near-Shannon-limit error correction performance with relatively low complexity, they have been adopted as an optional coding technique standardized in third-generation (3G) CDMA systems [201]. We first give a brief introduction to turbo codes and describe the turbo decoding algorithm for computing the extrinsic information. We then compare the performance of a turbo multiuser receiver with that of a conventional RAKE receiver followed by turbo decoding, in a turbo-coded CDMA system with mulitpath fading channels. The material discussed in this section was developed in [256, 257].

6.6.1 Turbo Code and Soft Decoding Algorithm

Turbo Encoder

A typical parallel concatenated convolutional (PCC) turbo encoder consists of two (or more) simple constituent recursive convolutional encoders linked by an interleaver (or different interleavers). A block diagram of such an encoder is shown in Fig. 6.13. The interleavers can be random, nonrandom, or semirandom.

The turbo encoder works as follows. Suppose that all constituent encoders start from the zero state and the first constituent encoder terminates in the zero state. For user k, the frame of input binary information bits, denoted by $\boldsymbol{d}_k = \left[d_k[0], \ldots, d_k[I-1]\right]$, is encoded by the constituent encoders, where I is the size of the information bit frame. Let $\underline{x}_k[i] = \left[x_k^0[i], \ldots, x_k^J[i]\right]$ denote the systematic bit and output parity bits of the constituent encoders, corresponding to $d_k[i]$, where $x_k^0[i]$ is the systematic bit; $x_k^j[i]$, $j \neq 0$, is the parity bit generated by the jth constituent encoder; and J is the number of constituent encoders. To generate a desired code rate k_0/n_0, $\{\underline{x}_k[i]\}_{0=1}^{I-1}$ is punctured. The punctured bits are BPSK mapped and then transmitted serially. The output bit frame is denoted by $\boldsymbol{b}_k = \left[b_k[0], \ldots, b_k[M-1]\right]$, where M is the size of the code-bit frame. To terminate the first constituent encoder in the zero state, the last ν bits of $\boldsymbol{d}_k$ are termination bits, where ν is the number of shift registers in the first encoder.

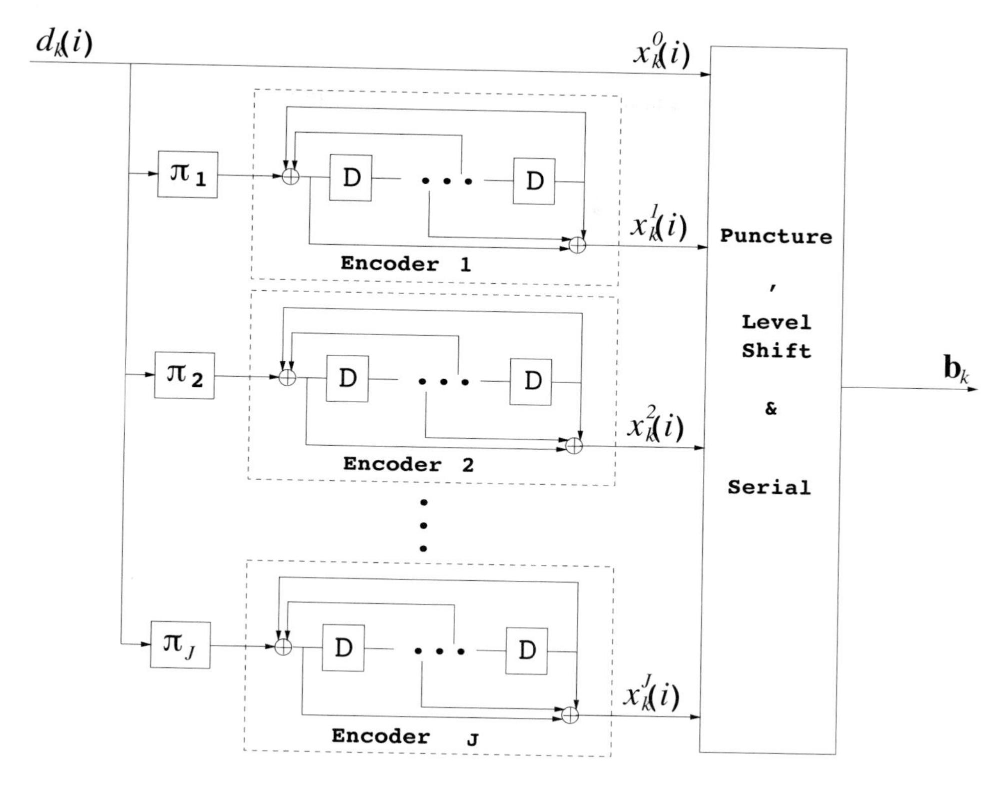

Figure 6.13. Typical turbo encoder.

Soft Turbo Decoder

Corresponding to the turbo encoder in Fig. 6.13, the block diagram of an iterative soft turbo decoder is shown in Fig. 6.14. The turbo decoder consists of J MAP decoders. Each MAP decoder is a slight modification of the MAP decoding algorithm for multiple turbo codes given in [29,100]. The signal flow is shown in Fig. 6.14. The deinterleaved LLRs $\{\lambda_1(b_k[i])\}_i$ of the kth user's code bits delivered by the SISO multiuser detector are distributed to the J MAP decoders as follows. The LLRs of the systematic bits, $\{\lambda_1(x_k^0[i])\}_i$, are sent to all MAP decoders after going through different interleavers. The LLRs of the jth parity bits, $\left\{\lambda_1\left(x_k^j[i]\right)\right\}_i$, are sent to the jth MAP decoder. Note that for a punctured bit $x_k^j[i]$, we let $\lambda_1\left(x_k^j[i]\right) = 0$, since no information is obtained by the soft multiuser detector for this bit.

The soft turbo decoder is itself an iterative algorithm. The jth MAP decoder in the turbo decoder computes the partial extrinsic information for the systematic bit and the jth parity bit, $\lambda_2^j(x_k^0[i])$ and $\lambda_2^j\left(x_k^j[i]\right)$, based on the code constraints, the input LLRs given by the SISO multiuser detector, and the partial extrinsic information given by other modified MAP decoders. This partial extrinsic information

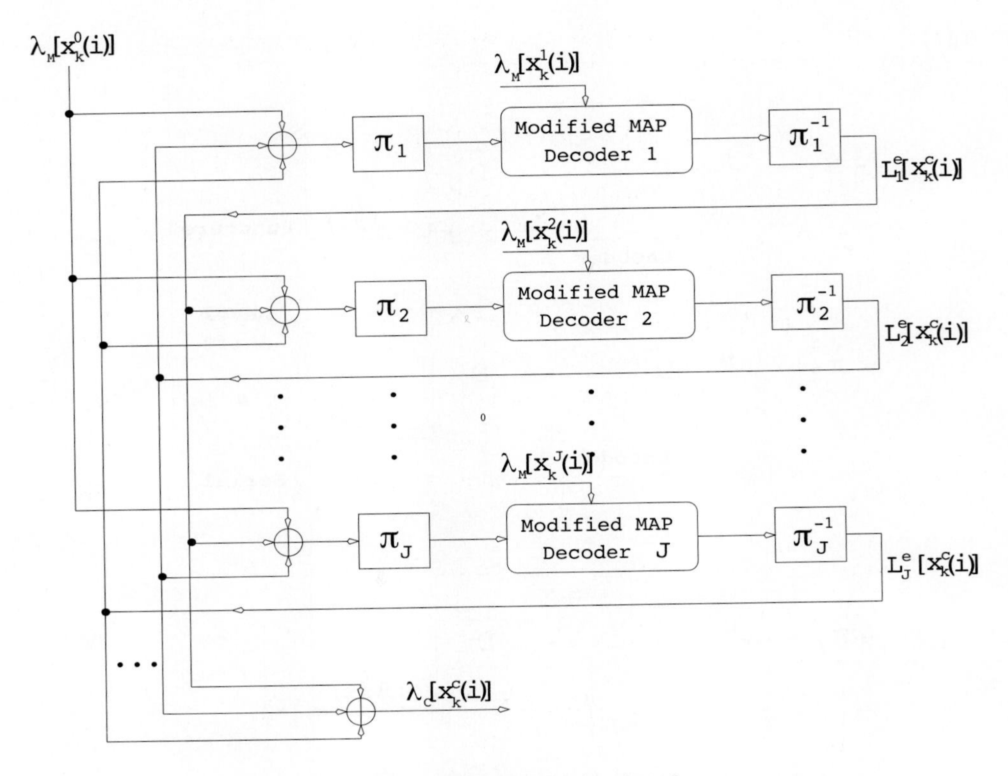

Figure 6.14. Soft turbo decoder.

is then sent to the other modified MAP decoders for the next iteration within a soft turbo decoding stage. After some iterations, the combined partial extrinsic information, which is the sum of all J modified MAP decoders' partial extrinsic information, is sent to the SISO multiuser detector as the *a priori* information for the next iteration of turbo multiuser detection. A more detailed description of the soft turbo decoder is as follows.

Denote the LLR of a code bit at the jth MAP decoder as

$$\begin{aligned}
&\Lambda_2^j\left(x_k^c[i]\right) \\
&\triangleq \log \frac{P\Big(x_k^c[i]=+1|\left\{\lambda_1(x_k^0[i])\right\}_i,\left\{\lambda_1(x_k^j[i])\right\}_i,\left\{\lambda_2^l(x_k^0[i])\right\}_{i;l\neq j},\ j\text{th encoder structure}\Big)}{P\Big(x_k^c[i]=-1|\left\{\lambda_1(x_k^0[i])\right\}_i,\left\{\lambda_1(x_k^j[i])\right\}_i,\left\{\lambda_2^l(x_k^0[i])\right\}_{i;l\neq j},\ j\text{th encoder structure}\Big)} \\
&= \log \frac{\displaystyle\sum_{(s',s)\in\mathcal{S}_{i,c}^+} \alpha_{i-1}(s')\gamma_i(s',s)\,\beta_i(s)}{\displaystyle\sum_{(s',s)\in\mathcal{S}_{i,c}^-} \alpha_{i-1}(s')\,\gamma_i(s',s)\beta_i(s)}, \qquad c=0 \text{ or } j; \quad j=1,\ldots,J, \qquad (6.169)
\end{aligned}$$

where $\mathcal{S}_{i,c}^{+}$ and $\mathcal{S}_{i,c}^{-}$ denote the sets of state transition pairs (s', s) such that the code bit $x_k^c[i]$ is $+1$ and -1, respectively. Define

$$\lambda_2^j\left(x_k^c[i]\right) \triangleq \Lambda_2^j\left(x_k^c[i]\right) - \lambda_1\left(x_k^c[i]\right), \qquad c = 0 \text{ or } j \tag{6.170}$$

as the partial extrinsic information of bit $x_k^c[i]$ delivered by the jth MAP decoder.

As before, $\alpha_i(s)$ and $\beta_{i-1}(s)$ can be computed by the following forward and backward recursions, respectively:

$$\alpha_i(s) = \sum_{s' \in \mathcal{S}} \alpha_{i-1}(s')\gamma_i(s', s), \qquad i = 1, \dots, I - 1 + \nu; \tag{6.171}$$

$$\beta_{i-1}(s) = \sum_{s' \in \mathcal{S}} \beta_i(s')\gamma_i(s, s'), \qquad i = I - 2 + \nu, \dots, 0, \tag{6.172}$$

where $\mathcal{S}$ is the set of all 2^ν constituent encoder states. The quantity γ_i is defined as

$$\begin{aligned}
\gamma_i(s', s) &= P\left[d_k[i] = d(s', s) \mid \lambda_1(x_k^0[i]),\ \lambda_1(x_k^j[i]),\ \left\{\lambda_2^l(x_k^0[i])\right\}_{l \neq j}\right] \\
&= P\left[x_k^0[i] = d(s', s),\ x_k^j[i] = x_k^j(s', s) \mid \lambda_1(x_k^0[i]),\ \lambda_1(x_k^j[i]),\ \left\{\lambda_2^l(x_k^0[i])\right\}_{l \neq j}\right] \\
&= P\left[x_k^0[i] = d(s', s) \mid \lambda_1(x_k^0[i]) + \sum_{l \neq j} \lambda_2^l(x_k^0[i])\right] \\
&\quad P\left[x_k^j[i] = x_k^j(s', s) \mid \lambda_1(x_k^j[i])\right].
\end{aligned} \tag{6.173}$$

Note that, by definition,

$$\lambda(x) \triangleq \log \frac{P(x = +1)}{P(x = -1)}. \tag{6.174}$$

Then for $b \in \{+1, -1\}$, we have

$$\begin{aligned}
P(x = b) &= \frac{\exp\left[b\,\lambda(x)\right]}{1 + \exp\left[b\,\lambda(x)\right]} \\
&= \frac{\exp\left[(b/2)\lambda(x)\right]}{\exp\left[-(b/2)\lambda(x)\right] + \exp\left[(b/2)\lambda(x)\right]} &(6.175)\\
&= \frac{1}{\exp\left[-\frac{1}{2}\lambda(x)\right] + \exp\left[\frac{1}{2}\lambda(x)\right]}\ \exp\left[\frac{b}{2}\lambda(x)\right] &(6.176)\\
&\propto \exp\left[\frac{b}{2}\lambda(x)\right], &(6.177)
\end{aligned}$$

where (6.176) follows from the fact that $b \in \{+1, -1\}$. Using (6.177) in (6.173), we obtain

$$\gamma_i(s', s) \propto \underbrace{\exp\left\{\frac{1}{2}\, d(s', s)\left[\lambda_1(x_k^0[i]) + \sum_{l \neq j} \lambda_2^l(x_k^0[i])\right]\right\}}_{\gamma_i^0(s', s)} \cdot \underbrace{\exp\left\{\frac{1}{2}\, x_k^j(s', s)\, \lambda_1(x_k^j[i])\right\}}_{\gamma_i^1(s', s)}. \quad (6.178)$$

Substituting (6.178) into (6.169), we have

$$\Lambda_2^j(x_k^0[i]) = \lambda_1(x_k^0[i]) + \underbrace{\sum_{l \neq j} \lambda_2^l(x_k^0[i]) + \overbrace{\log \frac{\displaystyle\sum_{(s',s) \in \mathcal{S}_{i,0}^+} \alpha_{i-1}(s')\, \gamma_i^1(s', s)\, \beta_i(s)}{\displaystyle\sum_{(s',s) \in \mathcal{S}_{i,0}^-} \alpha_{i-1}(s')\, \gamma_i^1(s', s)\, \beta_i(s)}}^{\lambda_2^j(x_k^0[i])}}_{\lambda_2(x_k^0[i])}, \quad (6.179)$$

$$\Lambda_2^j(x_k^j[i]) = \lambda_1(x_k^j[i]) + \underbrace{\log \frac{\displaystyle\sum_{(s',s) \in \mathcal{S}_{i,j}^+} \alpha_{i-1}(s')\, \gamma_i^0(s', s)\, \beta_i(s)}{\displaystyle\sum_{(s',s) \in \mathcal{S}_{i,j}^-} \alpha_{i-1}(s')\, \gamma_i^0(s', s)\, \beta_i(s)}}_{\lambda_2(x_k^j[i])}, \quad (6.180)$$

where the term $\lambda_2^j(x_k^c[i])$ is the partial extrinsic information obtained by the jth MAP decoder which will be sent to the other MAP decoders, as shown in Fig. 6.14; after some iterations within the turbo decoder, the total extrinsic information, $\lambda_2(x_k^c[i])$, is sent to the soft multiuser detector as the *a priori* information about $x_k^c[i]$, if $x_k^c(i)$ is not unpunctured. At the end of the turbo multiuser receiver iteration, a hard decision is made on each information bit $d_k[i] = x_k^0[i]$, according to

$$\hat{d}_k[i] = \text{sign}\left[\Lambda_2(x_k^0[i])\right]. \quad (6.181)$$

For numerical stability, (6.179) and (6.180) should be scaled as computation proceeds, in a manner similar to that discussed in Section 6.3.3.

6.6.2 Turbo Multiuser Receiver in Turbo-Coded CDMA with Multipath Fading

In this section we demonstrate the performance of the turbo multiuser receiver in a turbo-coded CDMA system with multipath fading. We consider a K-user CDMA system employing random aperiodic spreading waveforms and signaling through multipath fading channels. Each user's information data bits are encoded by a turbo encoder and then randomly interleaved. The interleaved code bits are then BPSK mapped and spread by a random signature waveform before being sent to the multipath fading channel. A block diagram of the system is illustrated in Fig. 6.15. The turbo multiuser receiver for this system iterates between the SISO multiuser detection stage (as discussed in Section 6.5.2) and the soft turbo decoding stage

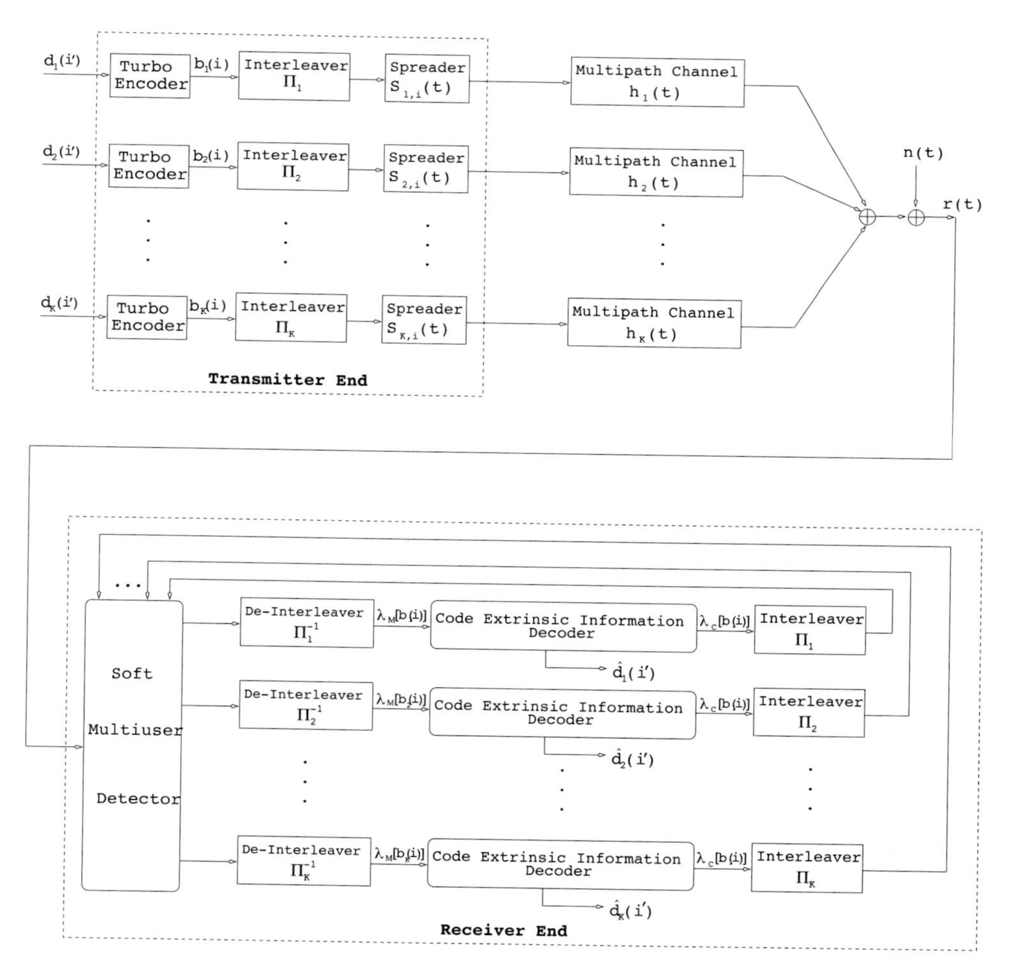

Figure 6.15. Turbo-coded CDMA system with a turbo multiuser receiver.

(as discussed in Section 6.6.1) by passing the extrinsic information of the code bits between the two stages.

Single-User RAKE Receiver

To compare the performance of a turbo multiuser receiver with that of a conventional technique used in practical systems, a single-user RAKE receiver employing maximal-ratio combining followed by a turbo decoder for the turbo-coded CDMA system is described next. The received signal in this system is given by (6.133) and (6.134). In a single-user RAKE receiver, the decision statistic for the kth user's ith code bit, $b_k[i]$, is given by $y_k[i]$ defined in (6.138):

$$y_k[i] \triangleq \sum_{\ell=1}^{L} \alpha_{\ell,k}[i]^* \int r(t) s_{i,k}(t - iT - \tau_{\ell,k})\, dt. \tag{6.182}$$

To obtain the LLR of the code bit $b_k[i]$ based on $y_k[i]$, a Gaussian assumption is made on the distribution of $y_k[i]$. Moreover, assume that the user spreading waveforms contain i.i.d. random chips and that the time delay $\tau_{\ell,k}$ is distributed uniformly over a symbol interval. Assume also that the multipath fading gains are independent between different users and are normalized such that $\sum_{\ell=1}^{L} E\left\{\left|\alpha_{\ell,k}[i]\right|^2\right\} = 1$. It is shown in the Appendix (Section 6.9.2) that the LLR of $b_k[i]$ based on the assumption above is given by

$$\lambda_1(b_k[i]) = \frac{4\, A_k\, \Re\{y_k[i]\}}{\sigma^2 + \dfrac{1}{N} \displaystyle\sum_{j \neq k}^{K} A_j^2}. \tag{6.183}$$

The LLRs $\{\lambda_1(b_k[i])\}_i$ of the kth user's code bits are then sent to the corresponding turbo decoder to obtain the estimated information bits.

Note that the SISO multiuser detector discussed in Section 6.5.2 operates on the same decision statistic as the conventional RAKE receiver (i.e., the outputs of the maximum ratio combiners $\{y_k[i]\}_{k;i}$). The RAKE receiver demodulates the kth user's data bits based only on $\{y_k[i]\}_i$, whereas the SISO multiuser detector demodulates all users' data bits *jointly* using all decision statistics $\{y_k[i]\}_{i;k}$.

Simulation Examples

Next we demonstrate the performance of the proposed turbo multiuser receiver in multipath fading CDMA channels by some simulation examples. The multipath channel model is given by (6.132). The number of paths for each user is three ($L = 3$). The delays of all users' paths are randomly generated. The time-varying fading coefficients are randomly generated to simulate channels with different data rates and vehicle speeds. The parameters are chosen based on the prospective services of wideband CDMA systems [431].

We consider a reverse link of an asynchronous CDMA system with six users ($K = 6$). The spreading sequence of each different user's different coded bit is

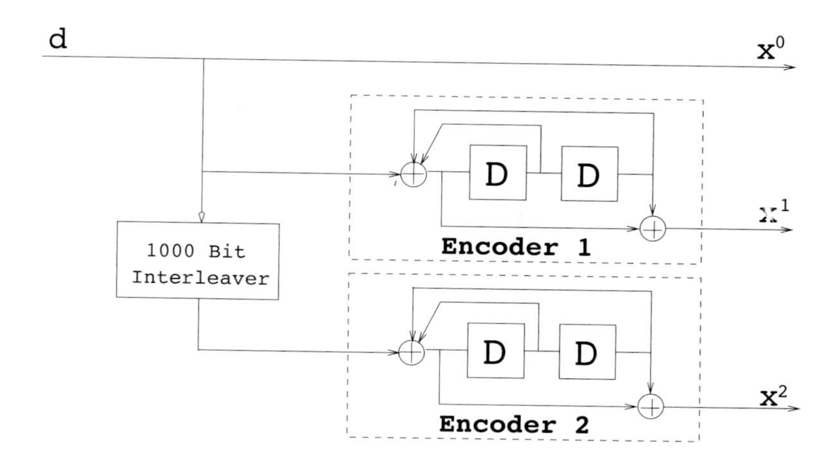

Figure 6.16. Rate-$\frac{1}{3}$ turbo encoder.

independently and randomly generated. The processing gain is $N = 16$. Each user uses a different random interleaver to permute its code bits. In all simulations, the same set of interleavers is used, and all users have equal signal amplitudes. The number of iterations within each soft turbo decoder is five.

The code we choose is a rate-$\frac{1}{3}$ binary turbo code, whose encoder is shown in Fig. 6.16. The two recursive convolutional constituent encoders have a generator polynomial,

$$G = \frac{n(D)}{d(D)} = \frac{1 + D^2}{1 + D + D^2}$$

with effective free distance 10 [30]. An S-random interleaver, π_j, shown in Fig. 6.14, is used and explained below. The interleaver size is $I = 1000$ and $S = 22$. (Hence the symbol frame length $M = 3000$.)

The *S-random interleaver* [103] is one type of semirandom interleaver. It is constructed as follows. To obtain a new interleaver index, a number is randomly selected from the numbers that have not previously been selected as interleaver indices. The number selected is accepted if and only if the absolute values of the differences between the number currently selected and the S numbers accepted previously are greater than S. If the number selected is rejected, a new number is selected randomly. This process is repeated until all I (interleaver size) indices are obtained. The searching time increases with S. Choosing $S < \sqrt{I/2}$ usually produces a solution in reasonable time. Note that the minimum weight of the code words increases as S increases. This equivalently increases the effective free distance [30] of parallel concatenated codes, which improves the weight distribution and thus the performance of the code. In Example 1 we will see that S-random interleavers offer significant interleaver gains over random interleavers.

Example 1: Effect of the S-Interleaver The BER performance of the turbo code used in this study with random interleavers and an S-random interleaver in

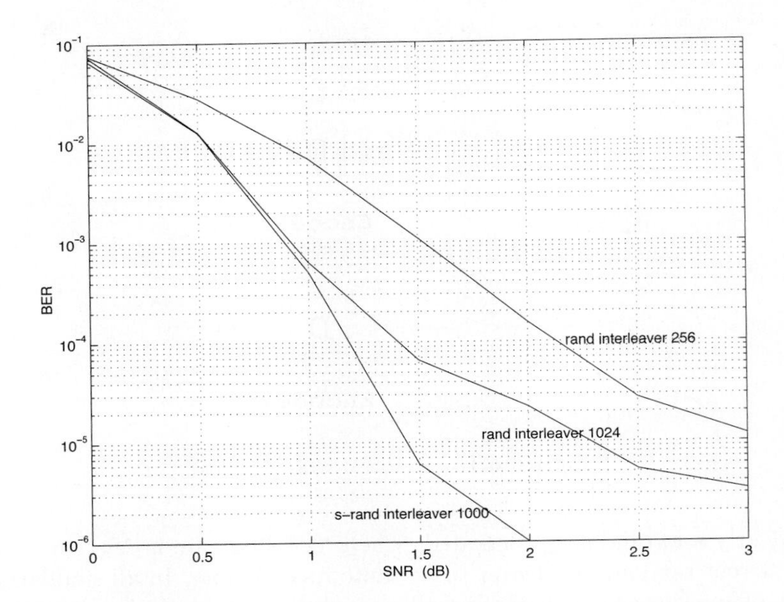

Figure 6.17. BER performance of the turbo code with different interleavers. (Random interleavers with size 256 and 1024, S-random interleaver with size 1000.)

a single-user AWGN channel is plotted in Fig. 6.17. It is seen that the S-random interleaver offers a significant interleaver gain over random interleavers.

In the following three examples, the performance of a turbo multiuser receiver is compared with that of a conventional single-user RAKE receiver. The single-user RAKE receiver computes the code-bit LLRs of the kth user using (6.183); these are then fed to a turbo decoder to decode the information bits. The BER averaged over all six users is plotted.

Example 2: Fast Vehicle Speed and Low Data Rate In this example we consider a Rayleigh fading channel with vehicle speed of 120 km/h, data rate of 9.6 kb/s, and carrier frequency of 2.0 GHz (the effective bandwidth–time product is $BT = 0.0231$). The results are plotted in Fig. 6.18.

Example 3: Medium Vehicle Speed and Medium Data Rate Next, we consider a multipath Rayleigh fading channel with vehicle speed 60 km/h, data rate 38.4 kb/s, and carrier frequency 2.0 GHz ($BT = 0.00289$). The results are plotted in Fig. 6.19.

Example 4: Very Slow Fading Finally, we consider a very slow fading channel (a time-invariant channel). The fading coefficients $\{\alpha_{\ell,k}\}$ of paths are randomly generated and kept fixed, and every user has equal received signal energy. The results are plotted in Fig. 6.20.

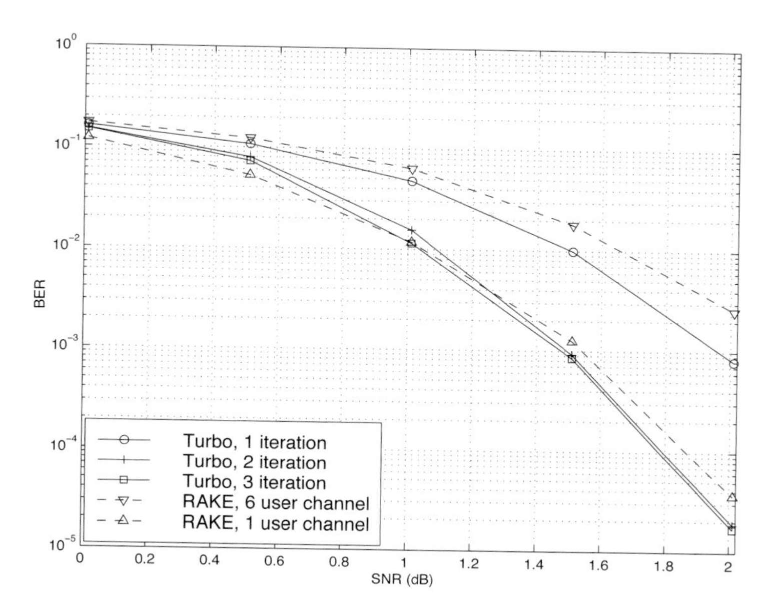

Figure 6.18. BER performance comparison between a turbo multiuser receiver and a RAKE receiver in a multipath fading channel with $K = 6$, processing gain $N = 16$, vehicle speed 120 km/h, data rate 9.6 kb/s, and carrier frequency 2.0 GHz.

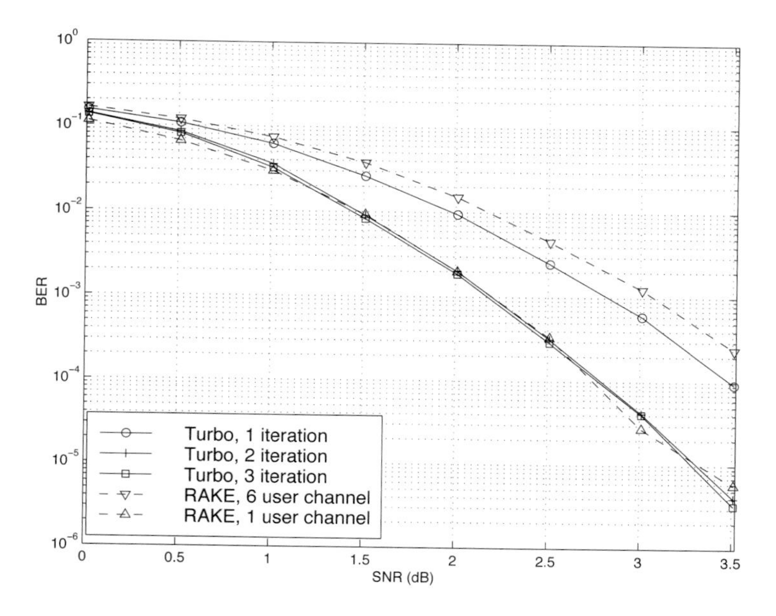

Figure 6.19. BER performance comparison between a turbo multiuser receiver and a RAKE receiver in a multipath fading channel with $K = 6$, processing gain $N = 16$, vehicle speed 60 km/h, data rate 38.4 kb/s, and carrier frequency 2.0 GHz.

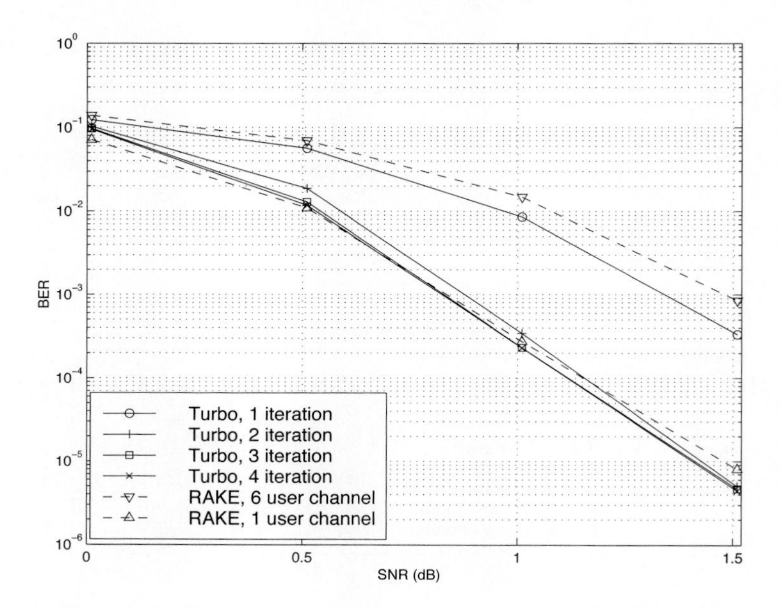

Figure 6.20. BER performance comparison between a turbo multiuser receiver and a RAKE receiver in a time-invariant multipath channel with $K = 6$ and processing gain $N = 16$.

From Examples 2, 3, and 4 it is seen that significant performance gain is achieved by a turbo multiuser receiver compared with a conventional noniterative receiver (i.e., the RAKE receiver followed by a turbo decoder). The performance of a turbo multiuser receiver with two iterations is very close to that of a RAKE receiver in a single-user channel. Moreover, at high SNR, the detrimental effects of multiple-access and intersymbol interference in the channel can be eliminated almost completely. Furthermore, it is seen from the simulation results that a turbo multiuser receiver in a multiuser channel even outperforms a RAKE receiver in a single-user channel. This is because the RAKE receiver makes the assumption that the delayed signals from different paths for each user are orthogonal, which effectively neglects the intersymbol interference.

6.7 Turbo Multiuser Detection in Space-Time Block-Coded Systems

The recently developed space-time coding (STC) techniques [350] integrate the methods of transmitter diversity and channel coding and provide significant capacity gains over traditional communication systems in fading wireless channels. STC has been developed along two major directions: *space-time block coding* (STBC) and *space-time trellis coding* (STTC). The common features of STBC and STTC lie in their realizations of spatial diversity (i.e., both methods transmit a vector of complex code symbols simultaneously from multiple transmitter antennas). Their differences, on the other hand, lie in their realizations of temporal diversity: In STBC, the temporal constraint is represented in matrix form; whereas in STTC, the temporal constraint is represented in the form of a trellis tree, which is akin to the trellis-coded modulation (TCM) code.

From the coding perspective, the single-user performance of STBC and STTC has been studied in [476, 477], and some code design criteria have been developed. However, in wireless communication systems, sharing the limited radio resources among multiple users is inevitable. Indeed, the emerging wireless systems with multiple transmitter and receiver antennas enable a new dimension for multiple accessing: space-division multiple access (SDMA) [492], which when employed with the more conventional TDMA or CDMA techniques, can substantially increase system capacity. However, if not properly ameliorated, the presence of multiuser interference can significantly degrade receiver performance as well as system capacity. Therefore, the development of efficient detection and decoding techniques for multiuser STC systems (illustrated in Fig. 6.21) is the key to bringing STC techniques into the practical arena of wireless communications. Research results along this direction first appeared in [349, 476], where some techniques for combined array processing, interference cancellation, and space-time decoding for multiuser STC systems were proposed.

In this and the following sections we discuss turbo receiver structures for joint detection and decoding in multiuser STC systems, based on the techniques developed in previous sections. Such iterative receivers and their variants, which were first developed in [288], are described for both STBC and STTC systems. During iterations, extrinsic information is computed and exchanged between a soft multiuser demodulator and a bank of MAP decoders to achieve successively refined estimates of the users' signals. Further developments of low-complexity turbo structures for multiuser STTC systems are given in [217].

6.7.1 Multiuser STBC System

The transmitter end of a multiuser STBC system is shown in Fig. 6.22. The information bit stream for the kth user, $\{d_k[n]\}_n$, is encoded by a convolutional encoder; the resulting convolutional code-bit stream $\{b_k[i]\}_i$ is then interleaved by a code-bit interleaver. After interleaving, the interleaved code-bit stream is then fed to an M-PSK modulator, which maps the binary bits into complex symbols $\{c_k[l]\}_l$, where $c_k[l] \in \Omega_C \triangleq \{C_1, C_2, \ldots, C_{|\Omega_C|}\}$, and Ω_C is the M-PSK symbol constellation set ($M = |\Omega_C|$). The symbol stream $\{c_k[l]\}_l$ is partitioned into blocks, with each block consisting of N symbols. Due to the existence of the interleaver, we can ignore the temporal constraint induced by the outer convolutional encoder and assume that the set $\{c_k[l]\}_l$ contains independent symbols. Hence, from the STBC decoder's perspective, we need only consider one block of symbols in the code symbol stream: the code vector $\underline{c}_k \triangleq \Big[c_k[1], c_k[2], \ldots, c_k[N]\Big]^T$.

STBC was first proposed in [12] and was later generalized systematically in [476]. Following [476], the kth user's STBC is defined by a $P \times N$ code matrix $\mathcal{G}_k$, where N denotes the number of transmitter antennas or the *spatial* transmitter diversity

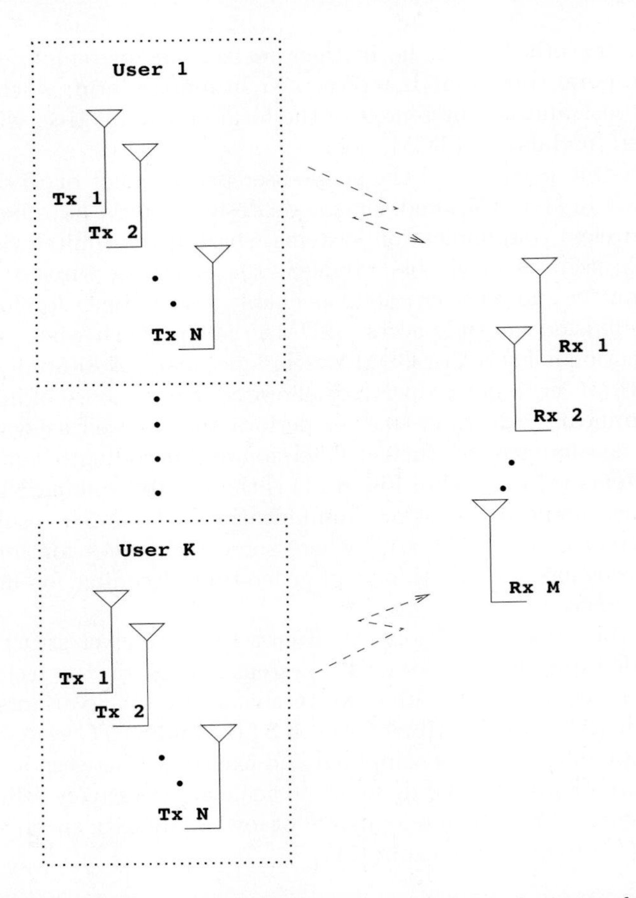

Figure 6.21. Multiuser wireless communication system employing multiple transmitter and receiver antennas. There are K users in the system, each user employing N transmitter antennas. At the receiver side, there are M receiver antennas.

order, and P denotes the number of time slots for transmitting an STBC code word or the *temporal* transmitter diversity order. Each row of $\mathcal{G}_k$ is a permuted and transformed (i.e., negated and/or conjugated) form of the code vector $\underline{c}_k$. An STBC encoder takes as input the code vector $\underline{c}_k$ and transmits each row of symbols in $\mathcal{G}_k$ at P consecutive time slots. At each time slot, the symbols contained in an N-dimensional row vector of $\mathcal{G}_k$ are transmitted through N transmitter antennas simultaneously.

As a simple example, we consider a particular user employing a 2×2 STBC (i.e., $P = 2, N = 2$). Its code matrix $\mathcal{G}_1$ is defined by

$$\mathcal{G}_1 = \begin{bmatrix} c[1] & c[2] \\ -c[2]^* & c[1]^* \end{bmatrix}. \tag{6.184}$$

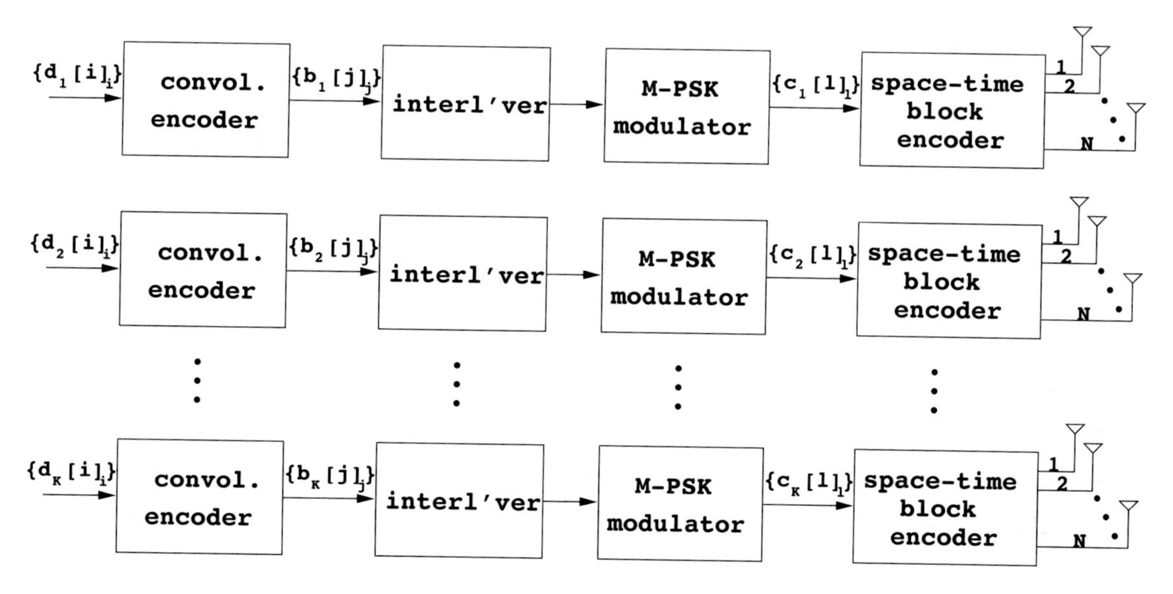

Figure 6.22. Transmitter structure for a multiuser STBC system.

The input to this STBC is the code vector $\underline{c} = \left[c[1]\ c[2]\right]^T$. During the first time slot, the two symbols in the first row of $\mathcal{G}_1$ (i.e., $c[1]$ and $c[2]$) are transmitted simultaneously at the two transmitter antennas; during the second time slot, the symbols in the second row of $\mathcal{G}_1$ (i.e., $-c[2]^*$ and $c[1]^*$) are transmitted.

We assume a flat-fading channel between each transmitter–receiver pair. Specifically, denote $\alpha_{m,n}$ as the complex fading gain from the nth transmitter antenna to the mth receiver antenna, where $\alpha_{m,n} \sim \mathcal{N}_c(0,1)$ is assumed to be a zero-mean circularly symmetric complex Gaussian random variable with unit variance. It is also assumed that the fading gains remain constant over an entire signal frame, but they may vary from one frame to another.

In general, we consider an STBC system with K users, each employing N transmitter antennas. At the receiver side, there are M receiver antennas. In this case, the received signal can be written as

$$\underbrace{\begin{bmatrix} \underline{r}^1 \\ \underline{r}^2 \\ \vdots \\ \underline{r}^M \end{bmatrix}}_{\boldsymbol{r}}{}_{MP\times 1} = \underbrace{\begin{bmatrix} \boldsymbol{H}_1 & \boldsymbol{H}_2 & \cdots \boldsymbol{H}_K \end{bmatrix}}_{\boldsymbol{H}}{}_{MP\times NK} \underbrace{\begin{bmatrix} \underline{c}_1 \\ \underline{c}_2 \\ \vdots \\ \underline{c}_K \end{bmatrix}}_{\boldsymbol{c}}{}_{NK\times 1} + \underbrace{\begin{bmatrix} \underline{n}^1 \\ \underline{n}^2 \\ \vdots \\ \underline{n}^M \end{bmatrix}}_{\boldsymbol{n}}{}_{MP\times 1}. \tag{6.185}$$

In (6.185), $\underline{r}^m \triangleq \left[r^m[1], r^m[2], \ldots, r^m[P]\right]^T$, $m = 1, 2, \ldots, M$, consists of the received signal from time slots 1 to P, at the mth receiver antenna; $\boldsymbol{H}_k$, $k = 1, 2, \ldots, K$, is the channel response matrix for the kth user, as explained below; $\underline{c}_k \triangleq \left[c_k[1], c_k[2], \ldots, c_k[N]\right]^T$ is the code vector for the kth user; and $\underline{n}^m \triangleq \left[n^m[1], n^m[2], \ldots, n^m[P]\right]^T$ contains the additive Gaussian noise samples from time slots 1 to P at the mth receiver antenna.

As a simple example, consider a single-user ($K = 1$) STBC system with two ($N = 2$) transmitter antennas and M receiver antennas, employing the code matrix $\mathcal{G}_1$ in (6.184), the received signal at the mth receiver antenna for this single user can be written as

$$\begin{bmatrix} r^m[1] \\ r^m[2] \end{bmatrix} = \begin{bmatrix} c[1] & c[2] \\ -c[2]^* & c[1]^* \end{bmatrix} \begin{bmatrix} \alpha_{m,1} \\ \alpha_{m,2} \end{bmatrix} + \begin{bmatrix} n^m[1] \\ n^m[2] \end{bmatrix}, \qquad m = 1, 2, \ldots, M. \tag{6.186}$$

For notational convenience, we write (6.186) in an alternative form by conjugating $r^m[1]$:

$$\underbrace{\begin{bmatrix} r^m[1] \\ r^m[2]^* \end{bmatrix}}_{\underline{r}^m} = \underbrace{\begin{bmatrix} \alpha_{m,1} & \alpha_{m,2} \\ \alpha_{m,2}^* & -\alpha_{m,1}^* \end{bmatrix}}_{\underline{H}_1^m} \underbrace{\begin{bmatrix} c[1] \\ c[2] \end{bmatrix}}_{\underline{c}_1} + \underbrace{\begin{bmatrix} n^m[1] \\ n^m[2]^* \end{bmatrix}}_{\underline{n}^m}, \quad m = 1, 2, \ldots, M. \tag{6.187}$$

We can see that $\underline{H}_1^m$ contains information not only of the channel response related to the mth receiver antenna, but also the code constraint of the STBC $\mathcal{G}_1$. Finally, by stacking all the $\underline{r}^m$ in (6.187), we obtain

$$\underbrace{\begin{bmatrix} \underline{r}^1 \\ \underline{r}^2 \\ \vdots \\ \underline{r}^M \end{bmatrix}}_{\boldsymbol{r} \quad 2M\times 1} = \underbrace{\begin{bmatrix} \underline{H}_1^1 \\ \underline{H}_1^2 \\ \vdots \\ \underline{H}_1^M \end{bmatrix}}_{\boldsymbol{H}_1 \quad 2M\times 2} \underbrace{\begin{bmatrix} c[1] \\ c[2] \end{bmatrix}}_{\underline{c}_1 \quad 2\times 1} + \underbrace{\begin{bmatrix} \underline{n}^1 \\ \underline{n}^2 \\ \vdots \\ \underline{n}^M \end{bmatrix}}_{\boldsymbol{n} \quad 2M\times 1}. \tag{6.188}$$

The signal model in (6.188) can easily be extended to the general model (6.185) of a K-user $P \times N$ STBC system, in which each user employs the $\mathcal{G}$ code defined in [476]. The analogy between this multiuser STBC signal model and the synchronous CDMA signal model (6.74) is evident. Note that to effectively suppress the interfering signals in model (6.185), the size of the receiver signal $\boldsymbol{r}$ should be larger than the number of symbol to be decoded (i.e., $MP \geq NK$).

6.7.2 Turbo Multiuser Receiver for STBC System

The iterative receiver structure for a multiuser STBC system is illustrated in Fig. 6.23. It consists of a soft multiuser demodulator, followed by K parallel MAP convolutional decoders. The two stages are separated by interleavers and deinterleavers. The soft multiuser demodulator takes as input the signals received from the M receiver antennas and the interleaved extrinsic log-likelihood ratios (LLRs) of the code bits of all users $\{\lambda_2(b_k[i])\}_{i;k}$ (which are fed back by the K single-user MAP decoders), and computes as output the *a posteriori* LLRs of the code bits of all users, $\{\Lambda_1(b_k[i])\}_{i;k}$. The MAP decoder of the kth user takes as input the deinterleaved extrinsic LLRs of the code bits $\{\lambda_1(d_k^\pi[i])\}_i$ from the soft multiuser demodulator and computes as output the *a posteriori* LLRs of the code bits $\{\Lambda_2(b_k^\pi[i])\}_i$, as well as the LLRs of the information bits $\{\Lambda_2(d_k^\pi[n])\}_n$. We next describe each component of the receiver in Fig. 6.23.

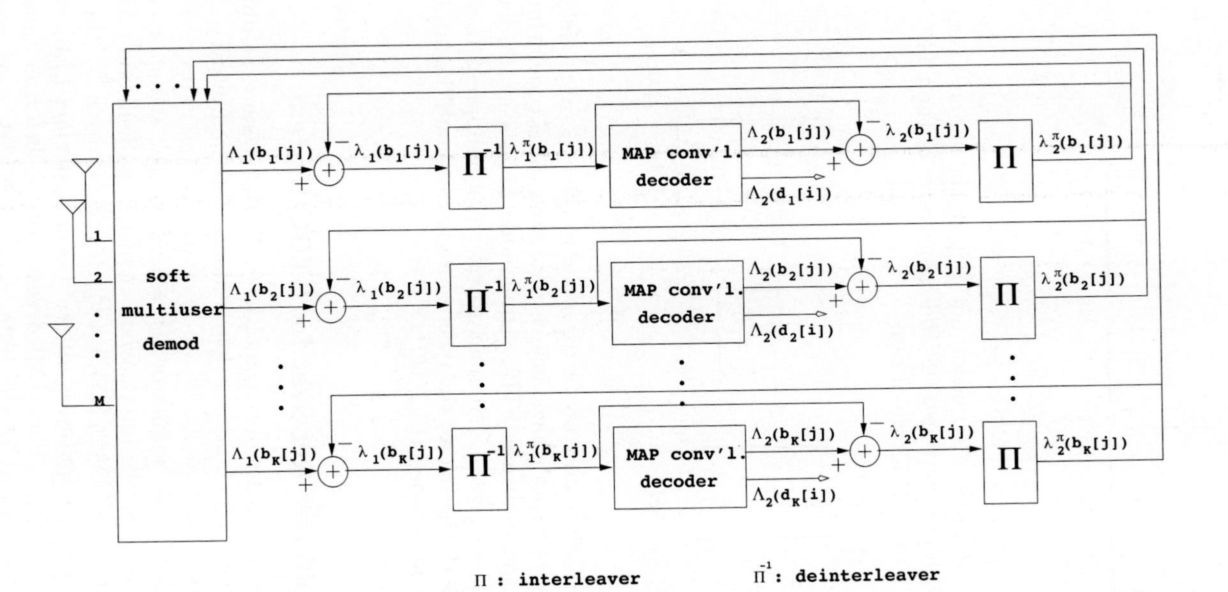

Figure 6.23. Iterative receiver structure for a multiuser STBC system.

Soft Multiuser Demodulator

The soft multiuser demodulator is based on the technique described in Section 6.3.3. First, a *soft* estimate $\tilde{c}_k[l]$ of the kth user's lth code symbol $c_k(l)$ is formed by

$$\tilde{c}_k[l] \triangleq E\{c_k[l]\} = \sum_{C_i \in \Omega_C} C_i \, P(c_k[l] = C_i), \tag{6.189}$$

where Ω_C is the set of all code symbols. At the first iteration, no prior information about code symbols is available; thus code symbols are assumed to be equiprobable [i.e., $P(c_k[l] = C_i) = 1/|\Omega_C|$]. In subsequent iterations, the probability $P(c_k[i] = C_i)$ is computed from the extrinsic information delivered by the MAP decoder, as will be explained later [cf. (6.204)].

For the K-user STBC system (6.185), define an NK-dimensional soft code vector

$$\begin{aligned}\tilde{\boldsymbol{c}} &\triangleq \left[\underline{\tilde{c}}_1^T, \underline{\tilde{c}}_2^T, \ldots, \underline{\tilde{c}}_K^T\right]^T \\ &= \left[\tilde{c}_1[1], \ldots, \tilde{c}_1[N], \tilde{c}_2[1], \ldots, \tilde{c}_2[N], \ldots\ldots, \tilde{c}_K[1], \ldots, \tilde{c}_K[N]\right]^T.\end{aligned}$$

The basic idea is to treat every element in $\tilde{\boldsymbol{c}}$ as a *virtual user*, and therefore there are a total of NK virtual users in the system (6.185). Viewing it this way, the model (6.185) is similar to the synchronous CDMA signal model treated in Section 6.3.3. Henceforth in this section, the notation ${}_k[l]$ is used to index a virtual user. Define

$$\tilde{\boldsymbol{c}}_k[l] \triangleq \tilde{\boldsymbol{c}} - \tilde{c}_k[l]\boldsymbol{e}_k[l]. \tag{6.190}$$

In (6.190), $\boldsymbol{e}_k[l]$ is an NK-vector of all zeros, except for the "1" element in the corresponding entry of the ${}_k[l]$th virtual user. That is, $\tilde{\boldsymbol{c}}_k[l]$ is obtained from $\tilde{\boldsymbol{c}}$ by setting the ${}_k[l]$th element to zero.

Subtracting the soft estimate of the interfering signals of other virtual users from the received signal $\boldsymbol{r}$ in (6.185), we get

$$\tilde{\boldsymbol{r}}_k[l] \triangleq \boldsymbol{r} - \boldsymbol{H}\tilde{\boldsymbol{c}}_k[l] \tag{6.191}$$

$$= \boldsymbol{H}\left(\boldsymbol{c} - \tilde{\boldsymbol{c}}_k[l]\right) + \boldsymbol{n}. \tag{6.192}$$

As before, to further suppress the residual interference in $\tilde{\boldsymbol{r}}_k[l]$, we apply an instantaneous linear minimum mean-square error (MMSE) filter to $\tilde{\boldsymbol{r}}_k[l]$. The linear MMSE weight vector $\boldsymbol{w}_k[l]$ is chosen to minimize the MSE between the transmitted symbol $c_k[l]$ and the filter output:

$$\begin{aligned}\boldsymbol{w}_k[l] &= \arg\min_{\boldsymbol{w} \in \mathbb{C}^{NK}} E\left\{\left|c_k[l] - \boldsymbol{w}^H \tilde{\boldsymbol{r}}_k[l]\right|^2\right\} \\ &= E\left\{\tilde{\boldsymbol{r}}_k[l]\tilde{\boldsymbol{r}}_k^H[l]\right\}^{-1} E\left\{c_k[l]^* \tilde{\boldsymbol{r}}_k[l]\right\}.\end{aligned} \tag{6.193}$$

Using (6.191) and assuming that the M-PSK symbol $c_k[l]$ is of unit energy (i.e., $\left|c_k[l]\right|^2 = 1$) and $E\{\boldsymbol{n}\,\boldsymbol{n}^H\} = \sigma^2 \boldsymbol{I}_{MP}$, we have

$$E\left\{c_k[l]^* \tilde{\boldsymbol{r}}_k[l]\right\} = \boldsymbol{H} E\left\{c_k[l]^* \left(\boldsymbol{c} - \tilde{\boldsymbol{c}}_k[l]\right)\right\} = \boldsymbol{H}\boldsymbol{e}_k[l]\,, \tag{6.194}$$

$$E\left\{\tilde{\boldsymbol{r}}_k[l]\tilde{\boldsymbol{r}}_k[l]^H\right\} = \boldsymbol{H}\boldsymbol{V}_k[l]\boldsymbol{H}^H + \sigma^2 \boldsymbol{I}_{MP}\,, \tag{6.195}$$

with

$$\boldsymbol{V}_k[l] \triangleq \mathrm{Cov}\Big\{\boldsymbol{c} - \tilde{\boldsymbol{c}}_k[l]\Big\} = \mathrm{diag}\Big\{1 - \Big|\tilde{c}_1[1]\Big|^2, \ldots, 1 - \Big|\tilde{c}_1[N]\Big|^2, \ldots, 1 - \Big|\tilde{c}_k[l-1]\Big|^2,$$

$$1, 1 - \Big|\tilde{c}_k[l+1]\Big|^2, \ldots, 1 - \Big|\tilde{c}_K[N]\Big|^2\Big\}. \tag{6.196}$$

Using (6.193)–(6.196), the instantaneous MMSE estimate for $c_k[l]$ is then given by

$$\hat{c}_k[l] \triangleq \boldsymbol{w}_k[l]^H \tilde{\boldsymbol{r}}_k[l] = \boldsymbol{e}_k[l]^T \boldsymbol{H}^H \left(\boldsymbol{H}\boldsymbol{V}_k[l]\boldsymbol{H}^H + \sigma^2 \boldsymbol{I}_{MP}\right)^{-1} \tilde{\boldsymbol{r}}_k[l]. \tag{6.197}$$

The instantaneous MMSE filter output is modeled by an equivalent additive white Gaussian noise channel having $c_k[l]$ as its input symbol. The output of this filter can then be written as

$$\hat{c}_k[l] = \mu_k[l] c_k[l] + \eta_k[l], \tag{6.198}$$

with

$$\mu_k[l] \triangleq E\left\{\hat{c}_k[l] c_k[l]^*\right\} = \left[\boldsymbol{H}^H \left(\boldsymbol{H}\boldsymbol{V}_k[l]\boldsymbol{H}^H + \sigma^2 \boldsymbol{I}_{MP}\right)^{-1} \boldsymbol{H}\right]_{kk}, \tag{6.199}$$

$$\nu_k^2[l] \triangleq \mathrm{Var}\left\{\hat{c}_k[l]\right\} = \mu_k[l] - \mu_k[l]^2. \tag{6.200}$$

Note that $\mu_k[l]$ and $\nu_k^2[l]$ are real numbers. Equations (6.198)–(6.200) give the probability distribution of the code symbol $\hat{c}_k[l]$, based on which the *a posteriori* probability of the code bits are computed, as discussed below.

From the discussion above, the major computation involved in the soft multiuser demodulator is the $MP \times MP$ matrix inversion, $\left(\boldsymbol{H}\boldsymbol{V}_k[l]\boldsymbol{H}^H + \sigma^2 \boldsymbol{I}_{MP}\right)^{-1}$. As before, this can be done recursively by making use of the matrix inversion lemma. As a result, the computational complexity of the proposed soft multiuser demodulator per user per symbol is $\mathcal{O}\left[(MP)^3/NK\right]$. (Recall that M is the number of receiver antennas, N is the number of transmitter antennas, P is the number of time slots in an STBC code word, and K is the number of users.)

Computing a posteriori Code-Bit LLRs

The convolutional code is chosen as the outer channel code in the system considered here (Fig. 6.22). First we need to compute the *a posteriori* LLRs of the code bits based on the estimated code symbols given by the soft multiuser demodulator. Since each user decodes its convolutional code independently, henceforth we drop the subscript k, the user index, to simplify the notation.

Every complex symbol $c[l]$ can be represented by a J-dimensional binary bit vector, $\Big[b[l,1], \ \ldots \ b[l,J]\Big]$, where $J \triangleq \log_2 |\Omega_C|$ and $b[l,j] \in \{+1, -1\}$ denotes the jth binary bit of the lth complex code symbol. By (6.177),

$$P(b[l,j] = B) \propto \exp\left\{\frac{B}{2}\lambda(b[l,j])\right\}, \tag{6.201}$$

with

$$\lambda(b[l,j]) \triangleq \log \frac{P(b[l,j]=+1)}{P(b[l,j]=-1)} \quad \text{and} \quad B \in \{+1,-1\}\,.$$

Due to the existence of the interleaver, we can assume that $b[l,j]$ is independent of $c[l']$, $l' \neq l$; and $b[l,j]$ is independent of $b[l,j']$, $j \neq j'$. Based on the Gaussian model (6.198), we have

$$p\Big(\hat{c}[l] \mid c[l] = C_i\Big) = \frac{1}{\pi \nu^2[l]} \exp\left(-\frac{\left|\hat{c}[l] - \mu[l] C_i\right|^2}{\nu^2[l]} \right), \qquad i = 1, 2, \ldots, |\Omega_{\mathcal{C}}|, \quad (6.202)$$

where $C_i \in \Omega_{\mathcal{C}}$ and with a binary representation, $C_i \equiv \Big[B[i,1] \ \cdots \ B[i,J]\Big]$. Then the *a posteriori* LLR of $b[l,j]$ at the output of soft multiuser demodulator can be computed as

$$\begin{aligned}
\Lambda_1(b[l,j]) &\triangleq \log \frac{P[b(l,j] = +1 \mid \hat{c}[l])}{P(b[l,j] = -1 \mid \hat{c}[l])} \\
&= \log \frac{\sum_{C_i \in \mathcal{C}_j^+} p\left(\hat{c}[l] \mid c[l] = C_i\right) P(C_i)}{\sum_{C_i \in \mathcal{C}_j^-} p\left(\hat{c}[l] \mid c[l] = C_i\right) P(C_i)} \\
&= \log \frac{\sum_{C_i \in \mathcal{C}_j^+} p\left(\hat{c}[l] \mid c[l] = C_i\right) \prod_{j'=1, j' \neq j}^{J} P(b[l,j'] = B[i,j'])}{\sum_{C_i \in \mathcal{C}_j^-} p\left(\hat{c}[l] \mid c[l] = C_i\right) \prod_{j'=1, j' \neq j}^{J} P(b[l,j'] = B[i,j'])} \\
&\quad + \log \frac{P(b[l,j] = +1)}{P(b[l,j] = -1)} \\
&= \underbrace{\log \frac{\sum_{C_i \in \mathcal{C}_j^+} \exp\left(-\frac{\left|\hat{c}[l] - \mu[l] C_i\right|^2}{\nu^2[l]} + \sum_{j' \neq j} \frac{B[i,j']}{2} \lambda_2(b[l,j']) \right)}{\sum_{C_i \in \mathcal{C}_j^-} \exp\left(-\frac{\left|\hat{c}[l] - \mu[l] C_i\right|^2}{\nu^2[l]} + \sum_{j' \neq j} \frac{B[i,j']}{2} \lambda_2(b[l,j']) \right)}}_{\lambda_1[b(l,j)]} + \lambda_2(b[l,j]), \\
&\qquad\qquad j = 1, 2, \ldots, J, \qquad (6.203)
\end{aligned}$$

where $\mathcal{C}_j^+$ is the set of the complex code symbols whose jth binary bit equals $+1$; and $\mathcal{C}_j^-$ is defined similarly. The last equality in (6.203) follows from (6.201) and

(6.202). The term $\lambda_2(b[l,j])$ in (6.203) is the interleaved extrinsic LLR of the jth code bit for the lth complex code symbol in the previous iteration, and is computed by bitwise-subtracting the code-bit LLR at the input of the decoder from the corresponding code-bit LLR at the output (cf. Fig. 6.23). At the first iteration, no prior information about code bits is available; thus $\lambda_2(b[l,j]) = 0$. It is seen from (6.203) that the output of the soft multiuser demodulator is the sum of the *a priori* information $\lambda_2(b[l,j])$ provided by the MAP convolutional decoder in the previous iteration and the extrinsic information $\lambda_1(b[l,j])$. Finally, the extrinsic LLRs of the code bits calculated in (6.203) are deinterleaved and then fed to the MAP convolutional decoder.

MAP Decoding Algorithm for Convolutional Code

Consider a rate-k_0/n_0 binary convolutional code of overall constraint length $k_0\nu_0$. At each time t, the input to the encoder is the k_0-dimensional binary vector $\underline{d}_t = (d_t^1, \ldots, d_t^{k_0})$ and the corresponding output is the n_0-dimensional binary vector $\underline{b}_t = (b_t^1, \ldots, b_t^{n_0})$.

As shown in Fig. 6.23, the deinterleaved extrinsic LLRs of the code bits $\{\lambda_1(b^\pi[i])\}_i$ are fed as input to the MAP convolutional decoder. We partition this stream into n_0-sized blocks, each block consisting of n_0 code bit LLRs, corresponding to the n_0 output code bits at one time instant. Denote each block by a vector $\lambda_1(\underline{b}_t) \triangleq \left[\lambda_1\left(b_t^{\pi,1}\right), \ldots, \lambda_1\left(b_t^{\pi,n_0}\right)\right]$, $t = 1, 2, \ldots, \tau_0$, where τ_0 blocks of code bits are transmitted in each signal frame. The partitioned code-bit LLR stream is then denoted as $\{\lambda_1(\underline{b}_t^\pi)\}_t$.

The extrinsic LLRs of the code bits $\{\lambda_2(b^\pi[i])\}_i$ are computed based on $\{\lambda_1(\underline{b}_t^\pi)\}_t$ and the convolutional code structure by the MAP decoding algorithm described in Section 6.2. Based on the interleaved extrinsic LLRs of the code bits in the previous iteration, $\lambda_2(b[l,j]), j = 1, 2, \ldots, J$, the soft estimate $\tilde{c}[l]$ [cf. (6.189)] of the code symbol $c[l]$ can then be computed as

$$
\begin{aligned}
\tilde{c}[l] = E\{c[l]\} &= \sum_{C_i \in \Omega_C} C_i \, P(c[l] = C_i) \\
&= \sum_{C_i \in \Omega_C} C_i \prod_{j=1}^{J} P(b[l,j] = B[i,j]) \\
&= \sum_{C_i \in \Omega_C} C_i \prod_{j=1}^{J} \frac{\exp\left\{B[i,j]\lambda_2(b[l,j])\right\}}{1 + \exp\left\{B[i,j]\lambda_2(b[l,j])\right\}},
\end{aligned} \tag{6.204}
$$

where (6.204) follows from (6.175). At the first iteration, no prior information is available, and thus $\lambda_2(b[l,j]) = 0$. At the last iteration, the information bits are recovered by the MAP decoding algorithm.

6.7.3 Projection-Based Turbo Multiuser Detection

In Section 6.7.2 we considered the problem of decoding the information of *all* users in the system. In some cases, however, we are interested in decoding the information of specific users only and are not willing to pay extra receiver complexity for decoding the information of undesired users. One approach to addressing this problem is to null out the signals of the K_u undesired users at the front end and then to apply the iterative soft multiuser demodulation algorithm on the rest of $K_d = K - K_u$ users' signals [439]. In [476], a projection-based technique was proposed for interference cancellation in STC multiuser systems. Here, we discuss applying soft multiuser demodulation and iterative processing on the projected signal to further enhance the receiver performance.

Consider again the signal model (6.185). Divide the users into two groups, desired users and undesired users. Rewrite (6.185) as

$$\boldsymbol{r} = [\boldsymbol{H}_d \boldsymbol{H}_u] \begin{bmatrix} \boldsymbol{c}_d \\ \boldsymbol{c}_u \end{bmatrix} + \boldsymbol{n}, \tag{6.205}$$

where the subscript d denotes desired users and u denotes undesired users. Define $\boldsymbol{H}_u^{\perp} \triangleq \boldsymbol{I}_{MP} - \boldsymbol{H}_u \boldsymbol{H}_u^{\dagger}$, $\tilde{\boldsymbol{r}} \triangleq \boldsymbol{H}_u^{\perp} \boldsymbol{r}$, and $\widetilde{\boldsymbol{H}} \triangleq \boldsymbol{H}_u^{\perp} \boldsymbol{H}_d$ (where $\boldsymbol{H}_u^{\dagger}$ is the Moore–Penrose generalized inverse of $\boldsymbol{H}_u$ [189]). It is assumed that the matrix $\boldsymbol{H}_u$ is "tall" (i.e., $MP > NK_u$) and has full column rank. It is easily seen that

$$\tilde{\boldsymbol{r}} = \widetilde{\boldsymbol{H}} \boldsymbol{c}_d + \tilde{\boldsymbol{n}}, \qquad \tilde{\boldsymbol{n}} \sim \mathcal{N}_c(0, \sigma^2 \boldsymbol{H}_u^{\perp} \boldsymbol{H}_u^{\perp H}) \tag{6.206}$$

(i.e., the undesired users' signals are nulled out by this projection operation). Moreover, before projection, the number of linearly independent rows of $\boldsymbol{H}$ is MP, whereas after projection, the number of linearly independent rows of $\widetilde{\boldsymbol{H}}$ becomes $MP - NK_u$, which implies that in the projected system the effective number of receiver antennas (the effective receiver diversity order) is reduced to $M' \triangleq (MP - NK_u)/P$. Hence, the projection operation incurs a diversity loss. Following the same derivations as in Section 6.7.2, we can apply the soft multiuser demodulator and MAP decoder on the projected signal $\tilde{\boldsymbol{r}}$ in (6.206) to iteratively detect and decode the information bits of the desired users.

Since we assume that the fading channel remains static within an entire signal frame, which normally contains hundreds of code blocks (of N symbols), we need only compute the projection matrix $\boldsymbol{H}_u^{\perp}$ once per signal frame. Therefore, the dominant computation of the projection-based soft multiuser demodulator is the same as before. However, the overall computational complexity of the multiuser receiver is reduced, since now we need only decode K_d users' information [i.e., only K_d (instead of K) MAP decoders are needed].

Simulation Examples

Next we provide computer simulation results to illustrate the performance of the turbo receivers in multiuser STBC systems. It is assumed that the fading processes are uncorrelated among all transmitter–receiver antenna pairs of all users; and for each user, the fading processes are uncorrelated from frame to frame but remain static within each frame. It is also assumed that the channel response matrix $\boldsymbol{H}$ in (6.185) is known perfectly. All users employ the same STBC code, but each user uses a different random interleaver. Furthermore, all users transmit M-PSK symbols with equal powers, a scenario in space-division multiple-access (SDMA) systems. Such an equal-power setup is also the worst-case scenario from the interference mitigation point of view.

We consider a four-user ($K = 4$) STBC system, as shown in Fig. 6.22. Each user employs the STBC $\mathcal{G}_1$ defined in (6.184) and two transmitter antennas ($N = 2$). An 8-PSK signal constellation is used in the modulator. The outer convolutional code, which is the same for all users, is a four-state, rate-$\frac{1}{2}$ code with generator $(5, 7)$ in octal notation. The encoder is forced to the all-zero state at the end of every signal frame. Each signal frame contains 128 8-PSK symbols. At the receiver side, four antennas ($M = 4$) are used.

Assume that all K users' signals are to be decoded ($K = 4$). We first demonstrate the performance of the iterative receiver discussed in Section 6.7.2. The frame-error rate (FER) and the bit-error rate (BER) are shown in Fig. 6.24. For the purpose of comparison, we also include the performance of the single-user STBC system with iterative decoding. The dotted lines (denoted as SU1-1) in Fig. 6.24 represent the performance of the single-user system with two transmitter antennas ($N = 2$) and four receiver antennas ($M = 4$) after its first iteration [i.e., the conventional (noniterative) single-user performance]. The dash-dotted lines (denoted as SU1-6) in Figs. 6.24 and 6.25 represent the single-user performance after six iterations using the same iterative structure discussed in Section 6.7.2.

Since a single user transmits different STC code symbols from its N transmitter antennas, it could be viewed as a virtual N-user system as discussed in Section 6.7.2. Then the iterative receiver structure for the multiuser STC systems can also be applied to single-user STC systems. Note that the optimal receiver of the proposed multiuser STBC system involves joint decoding of the multiuser STBC and outer convolutional codes, which has a prohibitive complexity $\mathcal{O}\left[|\Omega_{\mathcal{C}}|^{NK}2^{\nu}\right]$. However, the STBC signal model in (6.188), either single-user or multiuser, is analogous to the synchronous CDMA multiuser system model. As seen from previous sections, at high SNR, the iterative technique for interference suppression and decoding in a multiuser system can approach the performance of a single-user system (which is a lower bound for optimal performance). Hence it is reasonable to view the performance of the iterative single-user STBC system as an approximate lower bound on optimal joint decoding performance. It is seen from Fig. 6.24 that after six iterations the performance, in terms of both FER and BER, of both the single-user and multiuser STBC systems is significantly improved compared with that of the noniterative receivers (i.e., the performance after the first iteration). More impressively, the performance of the iterative receiver in a multiuser system approaches that of the iterative single-user receiver at high SNR.

We next demonstrate the performance of the projection-based turbo receiver. Assume that K_d out of all K users' signals are to be decoded ($K = 4, K_d = 2$). In this scenario, two users are first nulled out by a projection operation, and

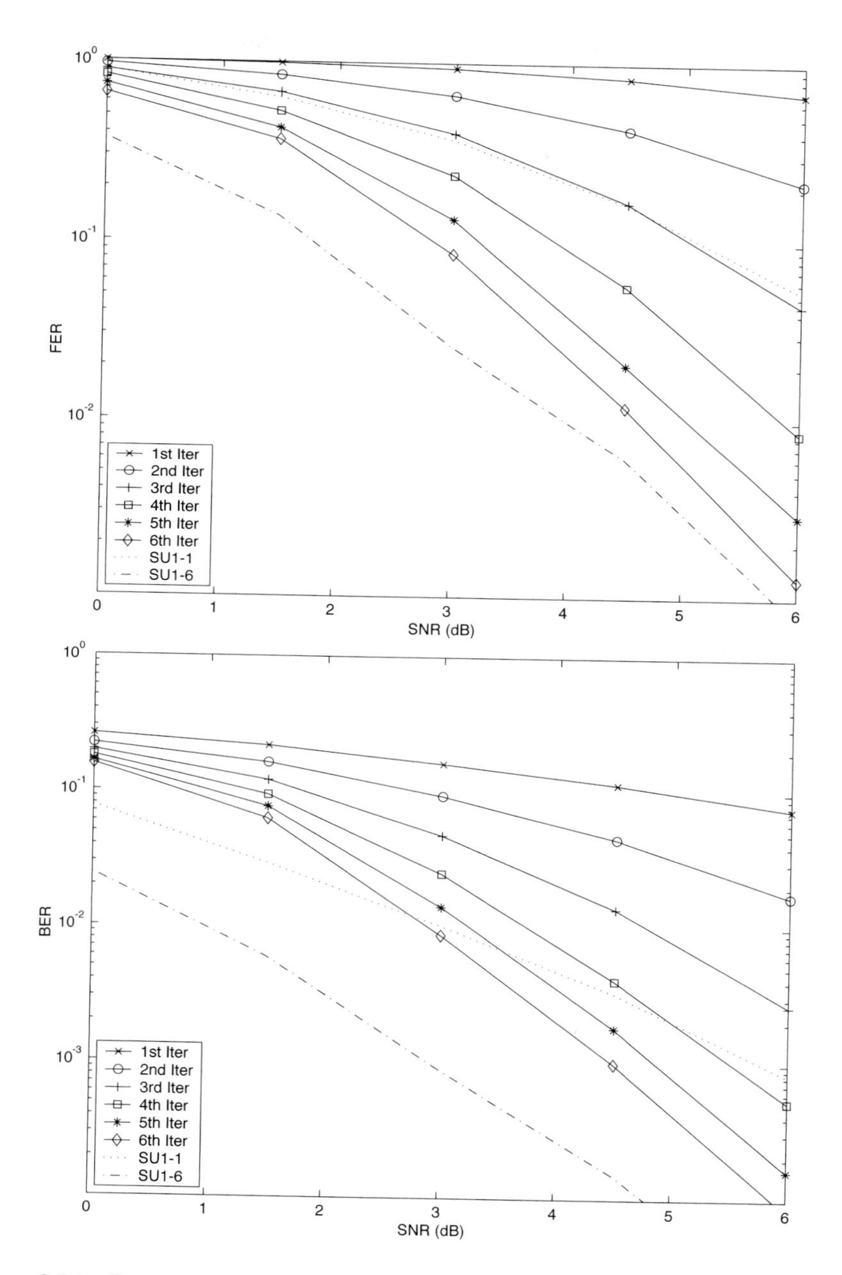

Figure 6.24. Frame-error rate (FER) and bit-error rate (BER) for a four-user STBC system. $K = 4, N = 2, M = 4$. All four users are iteratively detected and decoded. SU1-1 and SU1-6 denote the iterative decoding performance of the single-user system with $K = 1, N = 2, M = 4$.

the other two users are iteratively detected and decoded, as discussed in Section 6.7.3. The performance is shown in Fig. 6.25. It is shown in [476] that due to the projection operation, the equivalent receiver antenna number (the receiver diversity) is reduced from MP to $MP - NK_u$. So, for a fair comparison, in Fig. 6.25 we also present the iterative decoding performance after the first iteration (denoted by SU2-1) and after the sixth iteration (denoted by SU2-6) of the single-user system with two transmitter antennas ($N = 2$) and two receiver antennas ($M' = 2$), where M' denotes the effective number of receiver antennas for the projected system, with $M' \triangleq (MP - NK_u)/P$. It is seen that the projection-based turbo receiver still significantly outperforms the projection-based noniterative receiver. However, compared with the turbo receiver discussed above, the projection operation incurs a substantial performance loss. The reason for such a performance loss is twofold. First, the projection operation causes a diversity loss by suppressing the interference from other K_u users; second, the projection operation enhances the background ambient noise. It is therefore preferable to use turbo receiver operating on all users' signals in STBC systems.

6.8 Turbo Multiuser Detection in Space-Time Trellis-Coded Systems

In Section 6.7 we developed an iterative receiver structure for multiuser STBC systems based on the turbo multiuser detection technique. In this section we apply the same ideas to multiuser space-time trellis-coded (STTC) systems. Moreover, by exploiting the intrinsic structures of STTC, we consider a more compact system without introducing any outer channel code.

6.8.1 Multiuser STTC System

An STTC is basically a TCM code, which can be defined in terms of a trellis tree. The input to the encoder at time t is the k_0-dimensional binary vector $\underline{d}_t = \left(d_t^1, \dots, d_t^{k_0}\right)$ and the corresponding output is an n_0-dimensional complex symbol vector $\underline{c}_t = \left(c_t^1, \dots, c_t^{n_0}\right)$, where the symbol $c_t^l \in \Omega_\mathcal{C}$, $l = 1, 2, \dots, n_0$. At each time t, every k_0 binary input bits determine a state transition, and with each state transition there are n_0 output M-PSK symbols ($M = |\Omega_\mathcal{C}|$). Rather than transmitting the output code symbols serially from a single transmitter antenna as in the traditional TCM scheme, in STTC all the output code symbols at each time t are transmitted simultaneously from the $N = n_0$ transmitter antennas. Hence, the rate of the STTC at each transmitter antenna is $k_0 / \log_2 |\Omega_\mathcal{C}|$, which is equal to 1 for all the STTCs designed so far. The first single-user STTC communication system was proposed in [477]. Some design criteria and performance analysis for STTC in flat-fading channels were also given there.

In what follows we discuss a multiuser STTC communication system that employs a turbo receiver structure. The transmitter end of the proposed multiuser STTC system is depicted in Fig. 6.26. Note that there is a complex symbol inter-

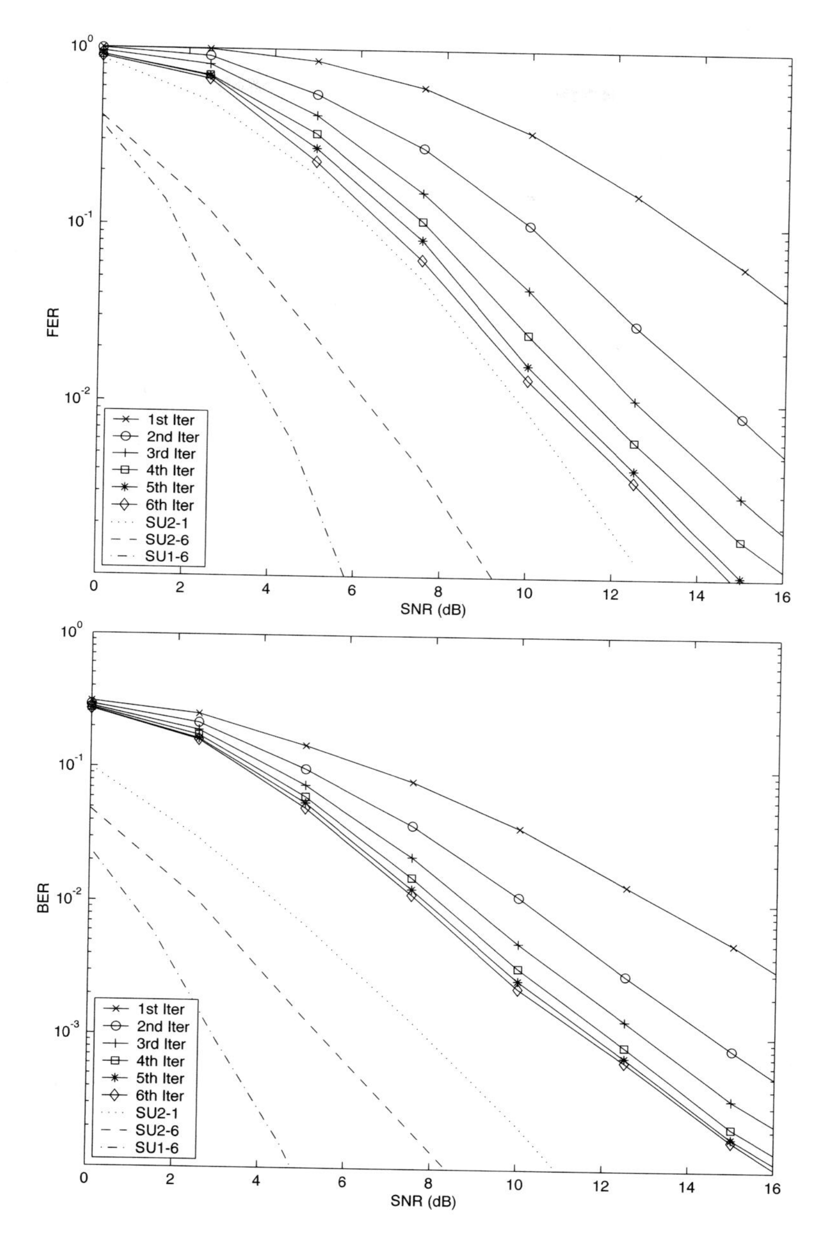

Figure 6.25. Frame-error rate (FER) and bit-error rate (BER) for a four-user STBC system. $K = 4, N = 2, M = 4, M' = 2$. Two users are first nulled out, the other two users are iteratively detected and decoded. SU2-1 and SU2-6 denote the iterative decoding performance of the single-user system with $K = 1, N = 2, M' = 2$. The gap between SU1-6 and SU2-6 constitutes the diversity loss caused by the projection operation.

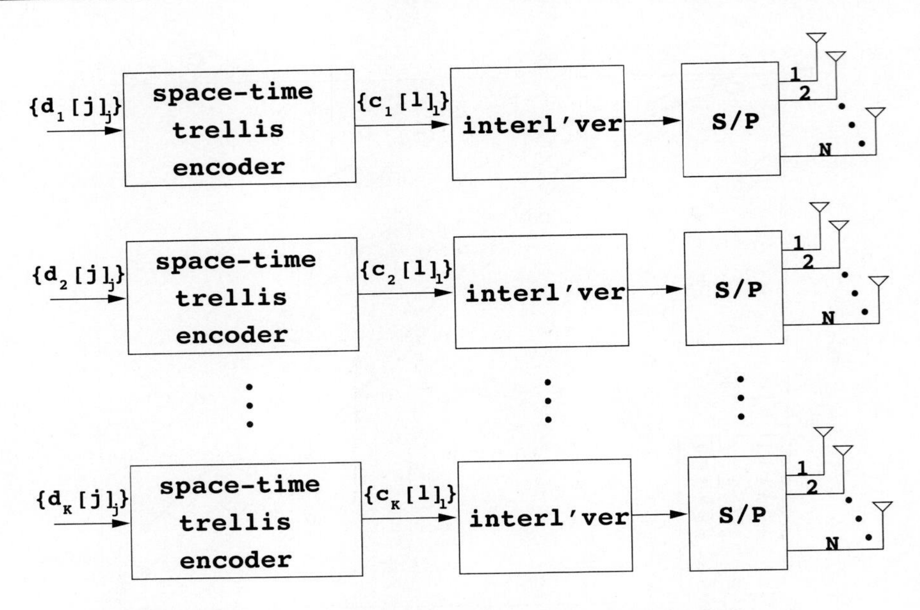

Figure 6.26. Transmitter structure for a multiuser STTC system.

leaver between the STTC encoder and the multiple transmitter antennas for each user. Such an interleaver is the key to reducing the influence of error bursts at the input of each user's MAP STTC decoder.

For the multiuser STTC system, the channel model is similar to (6.185), except that matrix-based temporal constraints no longer exist. We have the following system model for the multiuser STTC system:

$$\underbrace{\begin{bmatrix} r^1 \\ r^2 \\ \vdots \\ r^M \end{bmatrix}}_{\boldsymbol{r} \quad M\times 1} = \underbrace{\begin{bmatrix} \boldsymbol{H}_1 & \boldsymbol{H}_2 & \cdots & \boldsymbol{H}_K \end{bmatrix}}_{\boldsymbol{H} \quad M\times NK} \underbrace{\begin{bmatrix} \underline{c}_1 \\ \underline{c}_2 \\ \vdots \\ \underline{c}_K \end{bmatrix}}_{\boldsymbol{c} \quad NK\times 1} + \underbrace{\begin{bmatrix} n^1 \\ n^2 \\ \vdots \\ n^M \end{bmatrix}}_{\boldsymbol{n} \quad M\times 1} . \tag{6.207}$$

In (6.207), r^m, $m = 1, 2, \ldots, M$, is the received signal at the mth receiver antenna; $\boldsymbol{H}_k$, $k = 1, 2, \ldots, K$, is the channel response matrix for the kth user; the (m, n)th element of $\boldsymbol{H}_k$ is the fading gain from the nth, $n = 1, 2, \ldots, N$, transmitter antenna to the mth, $m = 1, 2, \ldots, M$, receiver antenna of the kth user;

$\underline{c}_k \triangleq \left[c_k[1] \;\; c_k[2] \;\; \cdots \;\; c_k[N]\right]^T$ is the code vector of the kth user; and n^m is the additive Gaussian noise sample at the mth receiver antenna.

Comparing (6.185) and (6.207), the key differences between the STBC and STTC systems are as follows:

- The matrix $\boldsymbol{H}$ in (6.185) contains both the channel fading gains and the STBC temporal constraints. By contrast, the matrix $\boldsymbol{H}$ in (6.207) contains only channel fading gains.

- The vector $\boldsymbol{r}$ in (6.185) is the stack of received signals over the consecutive P time slots, whereas in (6.207) the vector $\boldsymbol{r}$ contains only the signals received at one time slot. Therefore, (6.185) defines the received signals of an entire STBC code word and contains all the information for decoding an STBC code word. By contrast, (6.207) contains only the spatial information of the multiuser STTC system, whereas the STTC temporal information of each user is contained in the trellis tree that defines the STTC.

- In both signal models (6.185) and (6.207), in order to have a sufficient degree of freedom to suppress interference, the matrix $\boldsymbol{H}$ should be "tall," which implies that for STBC, $MP \geq NK$, whereas for STTC, $M \geq NK$.

6.8.2 Turbo Multiuser Receiver for STTC System

The turbo receiver structure for a multiuser STTC system is illustrated in Fig. 6.27. It consists of a soft multiuser demodulator, followed by K parallel MAP STTC decoders. The two stages are separated by interleavers and deinterleavers. The soft multiuser demodulator takes as input the received signals from the M receiver antennas and the interleaved extrinsic log probabilities (LPs) of the complex code symbols (i.e., $\{\underline{\Lambda}_2(c_k[l])\}_{l;k}$, which are fed back by the K single-user MAP STTC decoders), and computes as output the *a posteriori* LPs of the complex code symbols $\{\underline{\Lambda}_1(c_k[l])\}_{l;k}$. The MAP STTC decoder of the kth user takes as input the deinterleaved extrinsic LPs of complex code symbols $\{\underline{\Lambda}_1\left(c_k^{\pi}[l]\right)\}_l$ from the soft multiuser demodulator, and computes as output the *a posteriori* LPs of the complex code symbols $\{\underline{\Lambda}_2\left(c_k^{\pi}[l]\right)\}_l$, as well as the LLRs of the information bits $\{\Lambda_2\left(d_k^{\pi}[j]\right)\}_j$. Here, a scaling factor Q is introduced in the multiuser STTC iterative receiver (in Fig. 6.27) to enhance the iterative processing performance. Such an idea first appeared in [105]. Although the receiver in Fig. 6.27 appears similar to that in Fig. 6.23, there are some important differences between the STTC multiuser receiver and the STBC multiuser receiver. In what follows, we elaborate on those differences.

Soft Multiuser Demodulator

In terms of algorithmic operations, the soft multiuser demodulator for the STTC system (6.207) is exactly the same as that discussed in Section 6.7.2. However, since the system (6.185) contains both the information about the channel and the STBC temporal constraints, the soft multiuser demodulator for the multiuser STBC

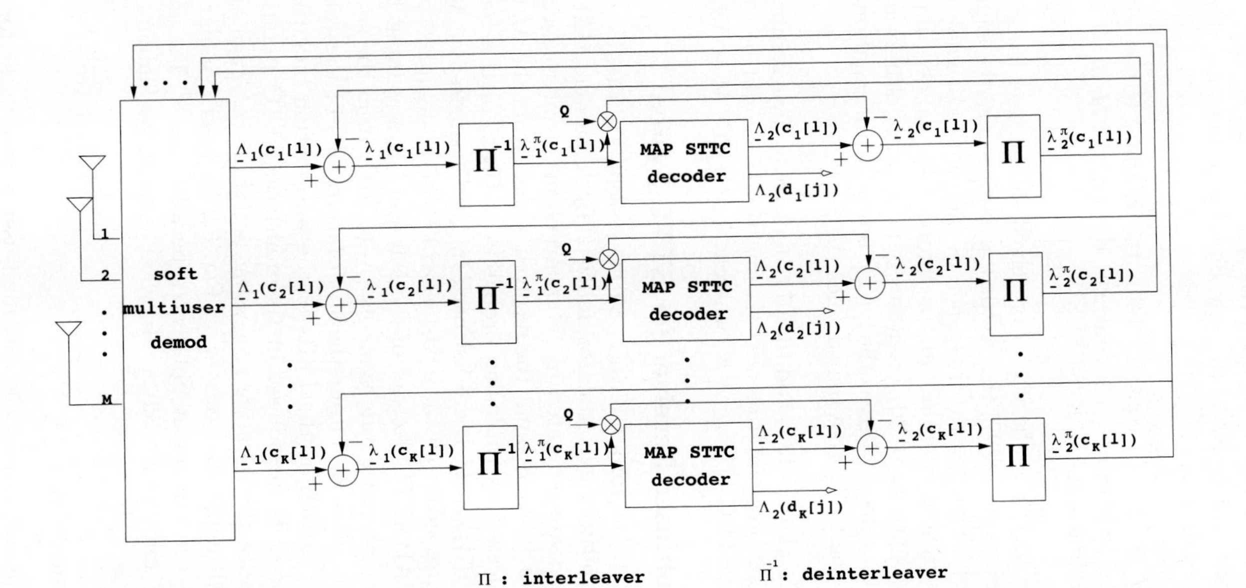

Figure 6.27. Iterative receiver structure for a multiuser STTC system.

system decodes the STBC code words of all users by estimating the code symbols of all users (which is realized implicitly through the instantaneous MMSE filtering of virtual users' signals in Section 6.7.2). By contrast, the system (6.207) contains only the information about the channel, and therefore the soft multiuser demodulator for the multiuser STTC system estimates only the code symbols of all users, and each user needs to apply the MAP decoder further to decode its STTC.

Computing a posteriori Code Symbol LPs

The output of the soft multiuser demodulator is a soft estimate $\hat{c}[l]$ of the code symbol $c[l]$, given by (6.198). (For simplicity we drop the subscript k, the user index.) In this section, the concept of log probability is further elaborated upon for nonbinary code symbols, which is equivalent to the concept of LLR for binary code bits. Due to the presence of the interleaver, all estimated symbols obtained from the soft multiuser demodulator are assumed to be mutually independent. For each code symbol $c[l]$, define an $|\Omega_C|$-dimensional *a posteriori* LP vector as

$$\underline{\Lambda}(c[l]) = \Big[\Lambda(c[l,1]), \ldots, \Lambda(c[l, |\Omega_C|])\Big]^T,$$

with

$$\Lambda(c[l,i]) \triangleq \log P\Big(c[l] = C_i \mid \hat{c}[l]\Big), \qquad i = 1, \ldots, |\Omega_C|, \tag{6.208}$$

where the index $[l,i]$ denotes the ith element in the $|\Omega_C|$-dimensional LP vector for the lth code symbol. Note that

$$\sum_{i=1}^{|\Omega_C|} \exp\Big\{\Lambda(c[l,i])\Big\} \equiv 1. \tag{6.209}$$

At the output of the soft multiuser demodulator, the ith element of the *a posteriori* LP vector for code symbol $c[l]$ is computed as

$$\begin{aligned}
\Lambda_1(c[l,i]) &\triangleq \log P\Big(c[l] = C_i \mid \hat{c}[l]\Big) \\
&= \log p\Big(\hat{c}[l] \mid c[l] = C_i\Big) - \log p\left(\hat{c}[l]\right) + \log P(C_i) \\
&= \underbrace{\left(-\log \pi\nu^2[l] - \frac{\Big|\hat{c}[l] - \mu[l] C_i\Big|^2}{\nu^2[l]}\right) - \log p[\hat{c}[l]]}_{\lambda_1(c[l,i])} + \lambda_2^{\pi}(c[l,i]),
\end{aligned}$$

$$i = 1, 2, \ldots, |\Omega_C|, \tag{6.210}$$

where $\log p(\hat{c}[l])$ is a constant term for all the elements in the LP vector $\underline{\Lambda}_1(c[l])$, and need not be evaluated explicitly. [Its value can be computed from (6.209).] The term $\lambda_2(c[l,i])$, which substitutes the *a priori* log probability $\log P(C_i)$ in (6.210), is the ith element of the interleaved extrinsic LP vector for code symbol $c[l]$ (which is provided by MAP STTC decoders) from the previous iteration, and is given by (cf. Fig. 6.27)

$$\lambda_2(c[l,i]) = \Lambda_2(c[l,i]) - Q \cdot \lambda_1(c[l,i]). \tag{6.211}$$

In (6.211), Q is a coefficient and will be discussed later. At the first iteration, no prior information about code symbols is available; thus $\lambda_2(c[l,i]) = \log 1/|\Omega_C|$, $i = 1, 2, \ldots, |\Omega_C|$. It is seen from (6.210) that the output of the soft multiuser demodulator is the sum of the *a priori* information $\lambda_2(c[l,i])$ provided by the MAP STTC decoder in the previous iteration and the extrinsic information $\lambda_1(c[l,i])$. Finally, the extrinsic LPs, calculated in (6.210), are deinterleaved and then fed to the MAP STTC decoder. Note that at the STTC multiuser receiver side, the interleavers/deinterleavers perform interleaving/deinterleaving on the LP vectors rather than on the elements of the LP vector.

MAP Decoding Algorithm for STTC

The MAP STTC decoding algorithm is very similar to the MAP decoding algorithm for the convolutional code except for some minor modifications. For the sake of conciseness, we omit the derivation of the MAP STTC decoding algorithm here. Similar to our work in Section 6.7.2, we partition the deinterleaved extrinsic LP's stream $\{\underline{\lambda}_1(c^\pi[l])\}_l$ into blocks, with each block consisting of n_O code symbol LP vectors, corresponding to the n_O output code symbols at one time instant. This reorganized code symbol LP stream is denoted as $\{\underline{\lambda}_1(\underline{c}_t^\pi)\}_t \triangleq \left\{\underline{\lambda}_1\left(c_t^{\pi,1}\right), \ldots, \underline{\lambda}_1\left(c_t^{\pi,n_O}\right)\right\}_t$. Then the extrinsic LPs for the code symbols are computed by the MAP decoding algorithm for STTC.

Finally, based on the interleaved extrinsic LP vector $\underline{\lambda}_2(c[l])$ of the lth code symbol, the soft estimated $\tilde{c}[l]$ [cf. (6.189)] is calculated as

$$\begin{aligned}\tilde{c}[l] &\triangleq E\{c[l]\} = \sum_{C_i \in \Omega_C} C_i \, P(c[l] = C_i) \\ &= \sum_{C_i \in \Omega_C} C_i \exp\Big\{\lambda_2(c[l,i])\Big\}.\end{aligned} \tag{6.212}$$

At the first iteration, no prior information is available; thus $\lambda_2(c[l,i]) = \log 1/|\Omega_C|$. At the last iteration, the information bits are recovered by the MAP algorithm.

Projection-Based Soft Multiuser Demodulator

The projection operation involves processing only the spatial information that is contained in the channel response matrix $\boldsymbol{H}$ in (6.207). Therefore, in terms of algorithmic operations, the projection-based soft multiuser demodulator for the STTC system is exactly the same as that discussed for STBC in Section 6.7.2.

Simulation Examples

We consider a four-user ($K = 4$) STTC system, as shown in Fig. 6.26. Each user employs the same 8-PSK eight-state STTC, as defined in Fig. 7 in [477], and two transmitter antennas ($N = 2$). The STTC encoder is forced to the all-zero state at the end of every signal frame, where each frame contains 128 8-PSK code symbols. At the receiver side, eight antennas ($M = 8$) are used.

Assume that all K users' signals are to be decoded ($K = 4$). The performance of the turbo receiver discussed in Section 6.8.2 is shown in Fig. 6.28. In computing the extrinsic code symbol information obtained from the MAP STTC decoder, we have introduced a scaling factor Q (cf. Fig. 6.27), which takes the same value between 0 and 1 for all users. In our simulations we find that in STTC systems, introducing such a Q factor significantly improves the performance of the iterative receiver. (The best performance is achieved for $Q \approx 0.85$.) Interestingly, we have found that for STBC systems, such a scaling factor does not offer performance improvement.

As before, the dotted lines (denoted as SU1-1) and the dash-dotted lines (denoted as SU1-6) in Figs. 6.28 and 6.29 represent, respectively, the iterative decoding performance after the first and sixth iterations in a single-user system ($N = 2, M = 8$). In contrast to the STBC system, the STTC system proposed does not include the outer channel code; therefore, the optimal single-user STTC receiver is straightforward to implement [477]. We also include its performance (in circle-dotted lines and denoted as MLLB) in Figs. 6.28 and 6.29. Similarly as in STBC systems, it is seen that the performance of the iterative receiver is significantly improved compared with that of the noniterative receiver, and it approaches the optimal single-user performance as well as the single-user iterative decoding performance.

We next consider the performance of the projection-based turbo receiver. Assume that K_d out of all K users' signals are to be decoded ($K = 4, K_d = 2$). In this scenario, two users are first nulled out by a projection operation, and the other two users are iteratively detected and decoded. The performance is shown in Fig. 6.29. Again the dotted lines (denoted as SU2-1) and the dash-dashed lines (denoted as SU2-6) in this figure represent, respectively, the single-user iterative decoding performance ($N = 2, M' = 4$) after the first and sixth iterations. As in STBC systems, it is seen that the projection operation incurs a substantial performance loss compared with the turbo receiver operating on all users in the STTC system. Hence it is the best to avoid the use of such a projection operation whenever possible in order to achieve optimal performance.

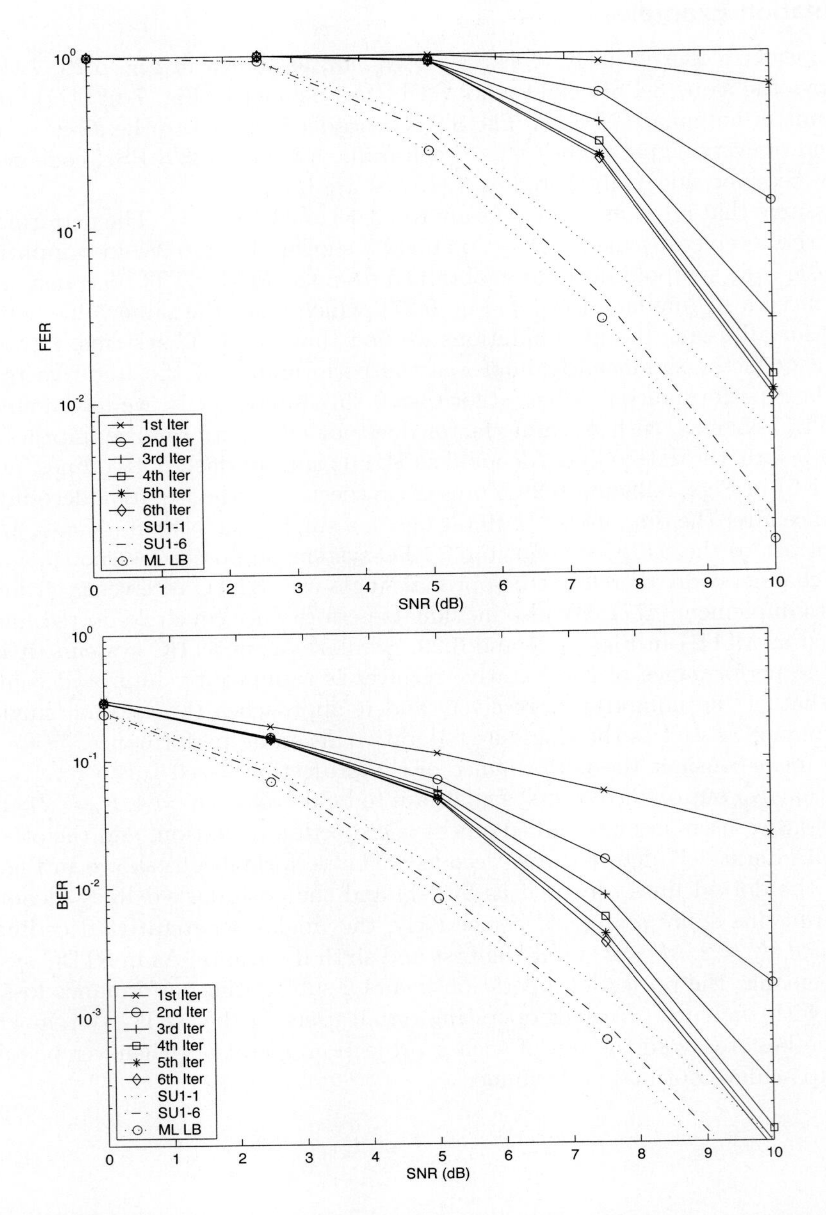

Figure 6.28. Frame-error rate (FER) and bit-error rate (BER) for a four-user STTC system. $K = 4, N = 2, M = 8, Q = 0.85$. All four users are iteratively detected and decoded. SU1-1 and SU1-6 denote the iterative decoding performance of the single-user system with $K = 1, N = 2, M = 8, Q = 0.85$. MLLB denotes the optimal single-user performance with $K = 1, N = 2, M = 8$.

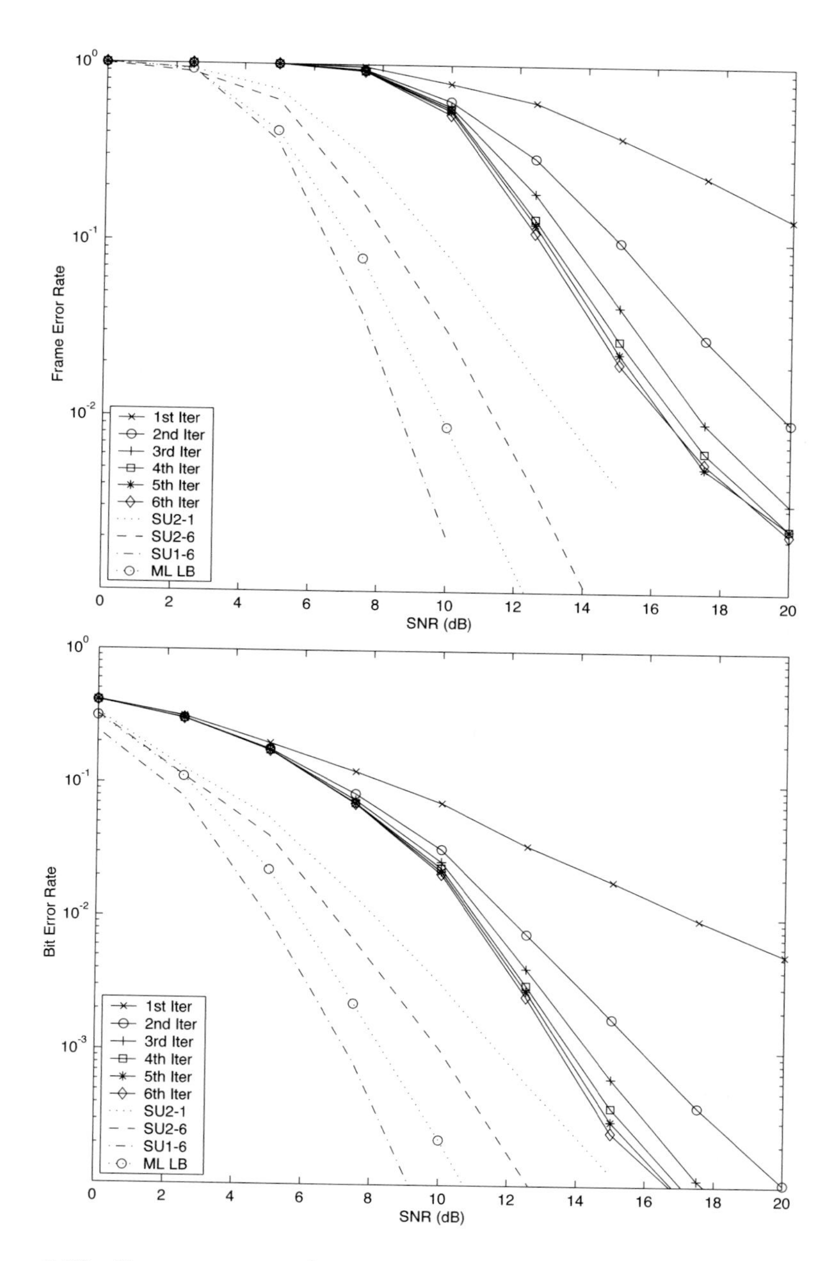

Figure 6.29. Frame-error rate (FER) and bit-error rate (BER) for a four-user STTC system. $K = 4, N = 2, M = 8, M' = 4, Q = 0.85$. Two users are first nulled out; the other two users are iteratively detected and decoded. SU2-1 and SU2-6 denote the iterative decoding performance of the single-user system with $K = 1, N = 2, M' = 4, Q = 0.85$. MLLB denotes the optimal single-user performance with $K = 1, N = 2, M' = 4$. The gap between SU1-6 and SU2-6 constitutes the diversity loss caused by the projection operation.

6.9 Appendix

6.9.1 Proofs for Section 6.3.3

Proof of Proposition 6.1

The proof of this result is based on the following lemma.

Lemma 6.1: *Let $\boldsymbol{X}$ be a $K \times K$ positive-definite matrix. Denote by $\boldsymbol{X}_k$ the $(K-1) \times (K-1)$ submatrix obtained from $\boldsymbol{X}$ by deleting the kth row and kth column. Also, denote by $\boldsymbol{x}_k$ the kth column of $\boldsymbol{X}$ with the kth entry x_{kk} removed. Then we have*

$$\left[\boldsymbol{X}^{-1}\right]_{kk} = \frac{1}{1 - \boldsymbol{x}_k^T \boldsymbol{X}_k^{-1} \boldsymbol{x}_k}. \tag{6.213}$$

Proof: Since $\boldsymbol{X}_k$ is a principal submatrix of $\boldsymbol{X}$, and $\boldsymbol{X}$ is positive definite, $\boldsymbol{X}_k$ is also positive definite. Hence $\boldsymbol{X}_k^{-1}$ exists.

Denote the above-mentioned partitioning of the symmetric matrix $\boldsymbol{X}$ with respect to the kth column and row by

$$\boldsymbol{X} = (\boldsymbol{X}_k, \boldsymbol{x}_k, x_{kk}).$$

In the same way, we partition its inverse:

$$\boldsymbol{Y} \triangleq \boldsymbol{X}^{-1} = (\boldsymbol{Y}_k, \boldsymbol{y}_k, y_{kk}).$$

Now from the fact that $\boldsymbol{X}\boldsymbol{Y} = \boldsymbol{I}_K$, it follows that

$$\boldsymbol{X}_k \boldsymbol{y}_k + y_{kk} \boldsymbol{x}_k = \boldsymbol{0}, \tag{6.214}$$

$$\boldsymbol{y}_k^T \boldsymbol{x}_k + y_{kk} = 1. \tag{6.215}$$

Solving for y_{kk} from (6.214) and (6.215), we obtain (6.213).

Proof of (6.60)

Using (6.62) and (6.63), by definition we have

$$\text{SINR}(z_k[i]) = \frac{\mu_k[i]^2}{\nu_k^2[i]} = \frac{1}{1/\mu_k[i] - 1}. \tag{6.216}$$

From (6.62) and (6.216) it is immediate that (6.60) is equivalent to

$$\left[\left(A_k^2 \boldsymbol{e}_k \boldsymbol{e}_k^T + \sigma^2 \boldsymbol{R}^{-1}\right)^{-1}\right]_{kk} > \left[\left(\boldsymbol{V}_k[i] + \sigma^2 \boldsymbol{R}^{-1}\right)^{-1}\right]_{kk} \tag{6.217}$$

$$> \left[\left(\boldsymbol{A}^2 + \sigma^2 \boldsymbol{R}^{-1}\right)^{-1}\right]_{kk}. \tag{6.218}$$

Partition the three matrices above with respect to the kth column and the kth row to get

$$A_k^2 \boldsymbol{e}_k \boldsymbol{e}_k^T + \sigma^2 \boldsymbol{R}^{-1} = (\boldsymbol{O}_k, \boldsymbol{o}_k, \alpha),$$

$$\boldsymbol{V}_k[i] + \sigma^2 \boldsymbol{R}^{-1} = (\boldsymbol{P}_k, \boldsymbol{p}_k, \beta),$$

$$\boldsymbol{A}^2 + \sigma^2 \boldsymbol{R}^{-1} = (\boldsymbol{Q}_k, \boldsymbol{q}_k, \gamma).$$

By (6.213), (6.218) is then equivalent to

$$\boldsymbol{o}_k^T \boldsymbol{O}_k^{-1} \boldsymbol{o}_k \;>\; \boldsymbol{p}_k^T \boldsymbol{P}_k^{-1} \boldsymbol{p}_k \;>\; \boldsymbol{q}_k^T \boldsymbol{Q}_k^{-1} \boldsymbol{q}_k. \tag{6.219}$$

Since

$$\boldsymbol{A}^2 = \sum_{j=1}^{K} A_j^2 \boldsymbol{e}_j \boldsymbol{e}_j^T, \tag{6.220}$$

$$\boldsymbol{V}_k[i] = A_k^2 \boldsymbol{e}_k \boldsymbol{e}_k^T + \sum_{j \neq k} A_j^2 \left(1 - \tilde{b}_j[i]^2\right) \boldsymbol{e}_j \boldsymbol{e}_j^T, \tag{6.221}$$

we then have

$$\boldsymbol{o}_k \;=\; \boldsymbol{p}_k \;=\; \boldsymbol{q}_k. \tag{6.222}$$

Therefore, to show (6.219), it suffices to show that

$$\boldsymbol{O}_k^{-1} \;\succ\; \boldsymbol{P}_k^{-1} \;\succ\; \boldsymbol{Q}_k^{-1}, \tag{6.223}$$

which is in turn equivalent to [189]

$$\boldsymbol{Q}_k \;\succ\; \boldsymbol{P}_k \;\succ\; \boldsymbol{O}_k, \tag{6.224}$$

where $\boldsymbol{X} \succ \boldsymbol{Y}$ means that the matrix $\boldsymbol{X} - \boldsymbol{Y}$ is positive definite. Since by assumption, $0 < |\lambda_2(b_j[i])| < \infty$, we have $0 < \tilde{b}_j[i] < 1$, $j = 1, \ldots, K$. It is easy to check that

$$\begin{aligned}\boldsymbol{Q}_k - \boldsymbol{P}_k &= \text{diag}\left\{A_1^2 \tilde{b}_1[i]^2, \ldots, A_{k-1}^2 \tilde{b}_{k-1}[i]^2, A_{k+1}^2 \tilde{b}_{k+1}[i]^2, \ldots, A_K^2 \tilde{b}_K[i]^2\right\} \\ &\succ \mathbf{0},\end{aligned} \tag{6.225}$$

$$\begin{aligned}\boldsymbol{P}_k - \boldsymbol{O}_k &= \text{diag}\left\{A_1^2[1 - \tilde{b}_1[i]^2], \ldots, A_{k-1}^2[1 - \tilde{b}_{k-1}[i]^2], A_{k+1}^2[1 - \tilde{b}_{k+1}[i]^2], \right. \\ &\quad \left. \ldots, A_K^2[1 - \tilde{b}_K[i]^2]\right\} \;\succ\; \mathbf{0}.\end{aligned} \tag{6.226}$$

Hence (6.224) holds and so does (6.60).

6.9.2 Derivation of the LLR for the RAKE Receiver in Section 6.6.2

To obtain the code-bit LLR for the RAKE receiver, a Gaussian assumption is made on the distribution of $y_k[i]$ in (6.182); that is, we assume that

$$y_k[i] = \mu_k[i] b_k[i] + \eta_k[i], \qquad \eta_k[i] \;\sim\; \mathcal{N}_c\left(0, \nu_k^2[i]\right), \tag{6.227}$$

where $\mu_k[i]$ is the equivalent signal amplitude and $\nu_k^2[i]$ is the equivalent noise variance.

As in typical RAKE receivers [396], we assume that the signals from different paths are orthogonal for a particular user. Conditioned on $b_k[i]$, using (6.133) and (6.134), the mean $\mu_k[i]$ in (6.227) is given by

$$\begin{aligned}\mu_k[i] &\triangleq E\left\{b_k[i]\, y_k[i]\right\}\\ &= E\left\{\sum_{\ell=1}^{L}\int b_k[i] r(t)\alpha_{\ell,k}[i]^* s_{i,k}(t-iT-\tau_{\ell,k})\,\mathrm{d}t\right\}\\ &= A_k \underbrace{\sum_{\ell=1}^{L} E\left\{\left|\alpha_{\ell,k}[i]\right|^2\right\}}_{1} = A_k,\end{aligned} \tag{6.228}$$

where the expectation is taken with respect to channel noise and all the code bits other than $b_k[i]$. The variance $\nu_k^2[i]$ in (6.227) can be computed as

$$\begin{aligned}\nu_k^2[i] &= E\left\{\left|y_k[i]-\mu_k[i]b_k[i]\right|^2\right\}\\ &= E\{|\underbrace{b_k[i]\textstyle\sum_{\ell=1}^{L}\int \alpha_{\ell,k}[i]^* s_{i,k}(t-iT-\tau_{\ell,k})\sum_{\ell'=1}^{L} A_k\alpha_{\ell',k}[i]s_{i,k}(t-iT-\tau_{\ell',k})\,\mathrm{d}t-\mu_k[i]b_k[i]}_{A}\\ &+\underbrace{\textstyle\sum_{\ell=1}^{L}\int\sum_{k'\neq k}\sum_{\ell'=1}^{L}\sum_{i'=0}^{M-1} b_{k'}[i']A_{k'}\alpha_{\ell',k'}[i']s_{i',k'}(t-i'T-\tau_{\ell',k'})\alpha_{\ell,k}[i]^* s_{i,k}(t-iT-\tau_{\ell,k})\,\mathrm{d}t}_{B}\\ &+\underbrace{\sum_{\ell=1}^{L}\int n(t)\alpha_{\ell,k}[i]^* s_{i,k}(t-iT-\tau_{\ell,k})\,\mathrm{d}t}_{C}\Big|^2\Bigg\}.\end{aligned} \tag{6.229}$$

Using the orthogonality assumption, it is easy to check that $A=0$ in (6.229). Since term B and term C in (6.229) are two zero-mean independent random variables, the variance is then given by

$$\nu_k^2[i] = E\left\{|B|^2\right\} + E\left\{|C|^2\right\}. \tag{6.230}$$

Due to the orthogonality assumption, the second term in (6.230) is given by

$$E\left\{|C|^2\right\} = \sigma^2\sum_{\ell=1}^{L} E\left\{\left|\alpha_{\ell,k}[i]\right|^2\right\} = \sigma^2. \tag{6.231}$$

For simplicity, we assume that the time delay $\tau_{l,k}$ is an integer multiple of the chip duration. Then the first term in (6.230) can be written as

$$E\left\{|B|^2\right\} = E\left\{\left|\sum_{\ell=1}^{L}\alpha_{\ell,k}[i]^*\sum_{k'\neq k}A_{k'}\sum_{\ell'=1}^{L}\alpha_{\ell',k'}[i']\sum_{n=0}^{N-1}\rho_{(k,\ell),(k',\ell')}[n]\right|^2\right\}, \tag{6.232}$$

where

$$\rho_{(k,\ell),(k',\ell')}[n] \triangleq \int_{T_c} b_{k'}[i'] s_{i',k'}(t - i'T - nT_c - \tau_{\ell',k'}) s_{i,k}(t - iT - nT_c - \tau_{\ell,k})\, \mathrm{d}t. \quad (6.233)$$

Assuming that the signature waveforms contain i.i.d. antipodal chips, $\rho_{(k,\ell),(k',\ell')}[n]$ is an i.i.d. binary random variable taking values of $\pm(1/N)$ with equal probability. Since $\alpha_{\ell,k}[i]$, $\alpha_{\ell',k'}[i']$, and $\rho_{(k,\ell),(k',\ell')}[n]$ are independent, (6.232) can be written as

$$\begin{aligned} E\left\{|B|^2\right\} &= \sum_{k' \neq k} A_{k'}^2 \underbrace{\sum_{\ell=1}^{L} E\left\{\left|\alpha_{\ell,k}[i]\right|^2\right\}}_{1} \underbrace{\sum_{\ell'=1}^{L} E\left\{\left|\alpha_{\ell',k'}[i']\right|^2\right\}}_{1} \\ &\quad \sum_{n=0}^{N-1} \underbrace{E\left\{\left|\rho_{(k,\ell),(k',\ell')}[n]\right|^2\right\}}_{1/N^2} \\ &= \frac{1}{N} \sum_{k' \neq k} A_{k'}^2. \end{aligned} \quad (6.234)$$

Substituting (6.231) and (6.234) into (6.230), we have

$$\nu_k^2[i] = \sigma^2 + \frac{1}{N} \sum_{j \neq k} A_j^2. \quad (6.235)$$

Hence the LLR of $b_k[i]$ can be written as

$$\begin{aligned} \lambda_1(b_k[i]) &= -\frac{\left|y_k[i] - \mu_k[i]\right|^2}{\nu_k^2[i]} + \frac{\left|y_k[i] + \mu_k[i]\right|^2}{\nu_k^2[i]} \\ &= \frac{4 A_k \Re\{y_k[i]\}}{\sigma^2 + \frac{1}{N} \sum_{j \neq k} {A_j}^2}. \end{aligned} \quad (6.236)$$

Chapter 7

NARROWBAND INTERFERENCE SUPPRESSION

7.1 Introduction

As we noted in Chapter 1, spread spectrum, in the form of direct sequence (DS) or frequency hopping (FH), is one of the most common signaling schemes in current and emerging wireless services. Such services include both second-generation (IS-95) and third-generation (WCDMA, cdma2000) cellular telephony [130, 329], piconets (Bluetooth) [41], and wireless LANs (IEEE802.11 and HiperLAN) [41]. Among the reasons that spread spectrum is so useful in wireless channels are its use as a countermeasure to frequency-selective fading caused by multipath and its favorable performance in shared channels. In this chapter we are concerned with the latter aspect of spread spectrum, which includes several particular advantages, including flexibility in the allocation of channels and the ability to operate asynchronously in multiuser systems, frequency reuse in cellular systems, increased capacity in bursty or fading channels, and the ability to share bandwidth with narrowband communication systems without undue degradation of either system's performance. More particularly, this chapter is concerned with this last aspect of spread spectrum, specifically with the suppression of narrowband interference (NBI) from the spread-spectrum partner of such shared-access systems.

NBI arises in a number of types of practical spread-spectrum systems. A classical and still important instance of this is narrowband jamming in tactical spread-spectrum communications; and of course, the antijamming capabilities of spread spectrum was an early motivator for its development as a military communications technique. A further situation in which NBI can be a significant factor in spread-spectrum systems is in systems deployed in unregulated bands, such as the industrial, scientific, and medical (ISM) bands, in which wireless LANs, cordless phones, and Bluetooth piconets operate as spread-spectrum systems. Similarly, shared access gives rise to NBI in military very high frequency (VHF) systems that must contend with civilian VHF traffic. An example of this arises in littoral sonobuoy networks that must contend with onshore commercial VHF systems, such as dispatch systems. Yet another situation of interest is that in which traffic with multiple signaling rates is generated by heterogeneous users sharing the same CDMA network. Finally, in some parts of the world the spectrum for third-generation (3G) systems is being allocated in bands not yet vacated by existing narrowband services, creating NBI with which 3G spread-spectrum systems must contend. In all but the first of these examples, the spread-spectrum systems are operating as *overlay systems*, in which the combination of wideband signaling, low spectral energy density, and natural immunity to NBI of spread-spectrum systems, are being exploited to make more efficient use of a slice of the radio spectrum. These advantages are so compelling that we can expect the use of such systems to continue to rise in the future. Thus, the issue of NBI in spread-spectrum overlay systems is one that is of increasing importance in the development of future advanced wideband telecommunications systems [58, 229, 230, 234, 331, 332, 358, 368, 400, 401, 466, 528–534, 562, 566].

The ability of spread-spectrum systems to coexist with narrowband systems can easily be explained with the help of Fig. 7.1. We first note that spreading the spread-spectrum data signal over a wide bandwidth gives it a low spectral density that assures that it will cause little damage to the narrowband signal beyond that already caused by the ambient wideband noise in the channel. On the other hand, although the narrowband signal has very high spectral density, this energy is concentrated near one frequency and is of very narrow bandwidth. The despreading operation of the spread-spectrum receiver has the effect of spreading this narrowband energy over a wide bandwidth while it collapses the energy of the originally spread data signal down to its original data bandwidth. So, after despreading, the situation is reversed between the original narrowband interferer (now wideband) and the originally spread data signal (now narrowband). A bandpass filter can be employed so that only the interferer power that falls within the bandwidth of the despread signal causes any interference. This will be only a fraction (the inverse of the spreading gain) of the original NBI that could have occupied this same bandwidth before despreading.

Although spread-spectrum systems are naturally resistant to narrowband interference, it has been known for decades that active methods of NBI suppression can significantly improve the performance of such systems. Not only does active suppression of NBI improve error-rate performance [37], it also leads to increased CDMA cellular system capacity [373], improved acquisition capability [328], and so on. Existing active NBI suppression techniques can be grouped into three basic types: frequency-domain techniques, predictive techniques, and code-aided techniques. To illustrate these three types, let us consider a basic received waveform

$$r(t) = S(t) + I(t) + N(t), \tag{7.1}$$

consisting of the useful (wideband) data signal $\{S(t)\}$, NBI signal $\{I(t)\}$, and wideband ambient noise $\{N(t)\}$. As the name implies, *frequency-domain techniques* operate by transforming the received signal $\{r(t)\}$ into the frequency domain, masking frequency bands in which the NBI $\{I(t)\}$ is dominant and then passing the signal off for subsequent despreading and demodulation. This process is illustrated in Fig. 7.2. Alternatively, *predictive systems* operate in the time domain. The basic

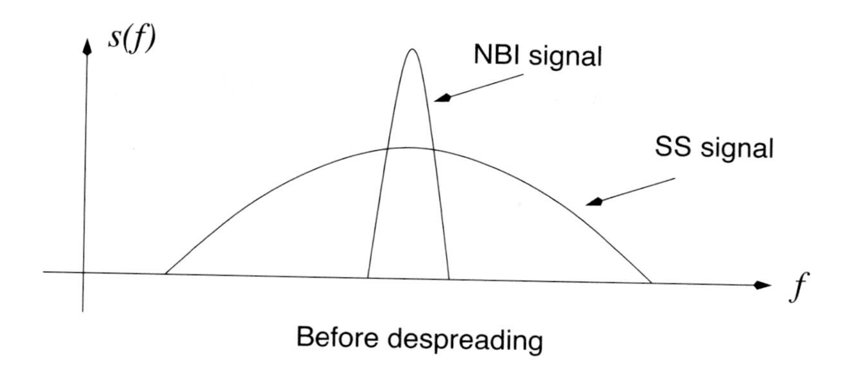

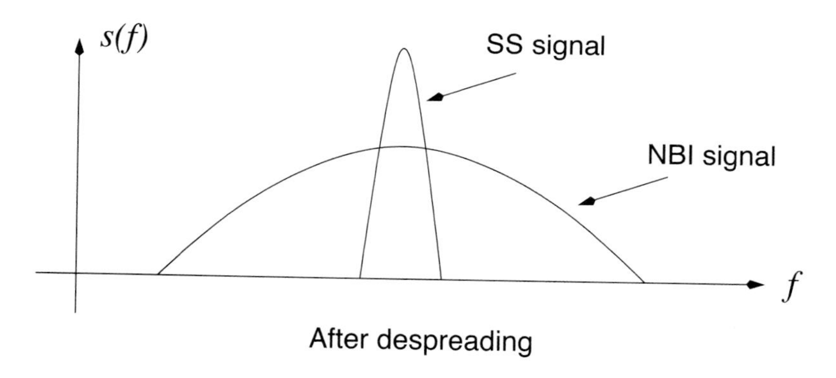

Figure 7.1. Spectral characteristics of a narrowband interference (NBI) signal and spread-spectrum (SS) signal before and after despreading.

idea of such systems is to exploit the discrepancy in predictability of narrowband signals and wideband signals to form an accurate replica of the NBI that can be subtracted from the received signal to suppress the NBI. In particular, the received signal $\{r(t)\}$ consists of the wideband component $\{S(t)+N(t)\}$ and the narrowband component $\{I(t)\}$. If one generates a prediction (e.g., a linear prediction) of $\{r(t)\}$, the values predicted will consist primarily of a prediction of $\{I(t)\}$ since the wideband parts of the signal are largely unpredictable (without making explicit use of the structure of $\{S(t)\}$). Thus, such a prediction forms a replica of the NBI, which can then be suppressed from the received signal. That is, if we form a residual signal

$$r(t) - \hat{r}(t)$$

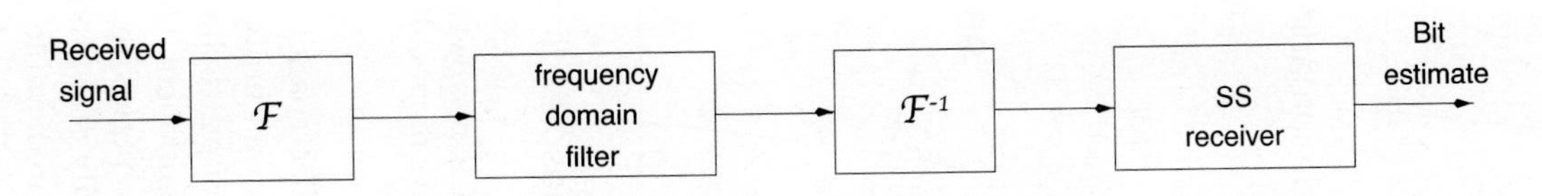

Figure 7.2. Transform domain NBI suppression.

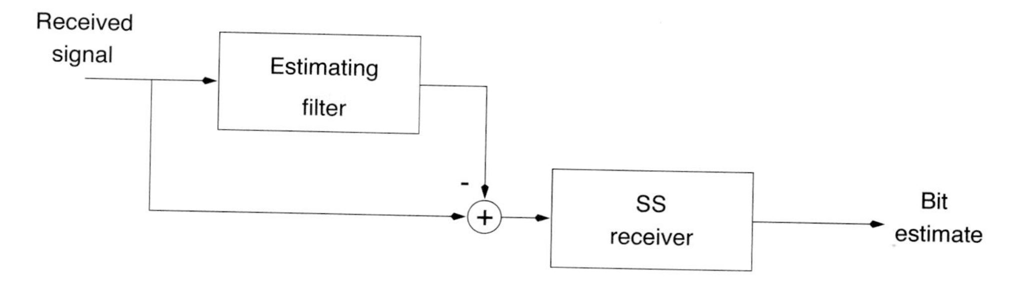

Figure 7.3. Predictive method of NBI suppression.

where $\hat{r}(t)$ is a prediction of $r(t)$ from past observations, the effect of the subtraction is to significantly reduce the narrowband component of $\{r(t)\}$. The prediction residual is then passed on for despreading and demodulation. Interpolators can also be used to produce the replica in such a scheme, with somewhat better performance and with less distortion of the useful data signal. Prediction-based methods take advantage of the difference in bandwidths of the spread-spectrum signal and the NBI without making use of any knowledge of the specific structure of the spread-spectrum data signal. Figure 7.3 illustrates this process, which is described in more detail in Sections 7.2 and 7.3. *Code-aided techniques* get further performance improvement by making explicit use of the structure of the useful data signal and, where possible, of the NBI. To date, these methods have made use primarily of techniques from linear multiuser detection, such as those described in Chapter 2.

Progress in the area of NBI suppression for spread-spectrum systems up until the late 1980s is reviewed in [327]. The principal techniques of that era were frequency-domain techniques and predictive or interpolative techniques based on linear predictors or interpolators. In the past decade or so there have been a number of developments in this field, the main thrust of which has been to take further advantage of the signaling structure. This has led to techniques that improve the performance of predictive and interpolative methods, and more recently to the more powerful code-aided techniques mentioned above. In this chapter we discuss these latter developments. Since these results have been concerned primarily with direct-sequence spread-spectrum systems (exceptions are found in [211], which considers frequency-hopping systems, and [423], which considers multicarrier systems), we restrict our attention to such systems throughout most of this chapter. We also focus here on predictive and code-aided techniques; for discussions of frequency-domain and other transform-domain techniques (including time-frequency methods), the reader is referred to [95,140,152,225,251,317,327,330,402,418,430,432,433,590,611]. We also refer the reader to a recent survey paper [57], which discusses a number of aspects of code-aided NBI suppression.

The remainder of this chapter is organized as follows. In Sections 7.2, 7.3, and 7.4 we discuss, respectively, linear predictive techniques, nonlinear predictive techniques, and code-aided techniques for NBI suppression. In Section 7.5 we present performance comparisons of the foregoing three families of NBI suppression techniques. In Section 7.6 we discuss the near–far resistance of the linear MMSE detector to both NBI and MAI. In Section 7.7 we present the adaptive linear MMSE NBI suppression algorithm. In Section 7.8 we discuss briefly a maximum-likelihood code-aided NBI suppression method. Finally, some mathematical derivations are collected in Section 7.9.

The following is a list of the algorithms appearing in this chapter.

- *Algorithm 7.1:* Kalman–Bucy prediction-based NBI suppression
- *Algorithm 7.2:* LMS linear prediction-based NBI suppression
- *Algorithm 7.3:* ACM-filter-based NBI suppression
- *Algorithm 7.4:* LMS nonlinear prediction-based NBI suppression

7.2 Linear Predictive Techniques

7.2.1 Signal Models

We next refine the model of (7.1) to account more completely for the structure of the useful data signal $\{S(t)\}$ and of the narrowband interference $\{I(t)\}$. It is the exploitation of such structure that has led to many of the improvements in NBI suppression that have been developed in the past decade.

Let us, then, reconsider the model (7.1) and examine its components in more detail. (These components are assumed throughout to be independent of one another.) We first consider the useful data signal $\{S(t)\}$. In this chapter we treat primarly the case in which this signal is a multiuser, linearly modulated, digital communications signal in the real baseband, which can be written more explicitly as (see also Chapter 2)

$$S(t) = \sum_{k=1}^{K} A_k \sum_{i=0}^{M-1} b_k[i] s_k(t - \tau_k - iT), \tag{7.2}$$

where K is the number of active (wideband) users in the channel, M is the number of symbols per user in a data frame of interest, $b_k[i]$ is the ith binary (± 1) symbol transmitted by user k, $A_k > 0$ and τ_k are the respective amplitude and delay with which user k's signal is received, $s_k(t)$ is user k's normalized ($\int |s_k(t)|^2 \, dt = 1$) transmitted waveform, and $1/T$ is the per-user symbol rate. It is also assumed that the support of $s_k(t)$ is completely within the interval $[0, T]$. The signaling waveforms are assumed to be direct-sequence spread-spectrum signals of the form

$$s_k(t) = \frac{1}{\sqrt{N}} \sum_{n=0}^{N-1} c_{n,k} \psi(t - nT_c), \qquad 0 \le t \le T, \tag{7.3}$$

where N, $\{c_{0,k}, c_{1,k}, \ldots, c_{N-1,k}\}$, and $1/T_c$ are the respective spreading ratio, binary (± 1) spreading code, and chip rate of the spread-spectrum signal $\{s_k(t)\}$; and where $\psi(t)$ is a unit-energy pulse of duration T_c.

It should be noted that this model accounts for asynchrony and slow fading but not for other possible channel features and impairments, such as multipath, dispersion, carrier offsets, multiple antennas, aperiodic spreading codes, fast fading, higher-order signaling, and so on. All of these phenomena can be incorporated into a more general model for a linearly modulated signal in the complex baseband:

$$S(t) = \sum_{k=1}^{K} \sum_{i=0}^{M-1} b_k[i] f_{i,k}(t), \tag{7.4}$$

in which the symbols $\{b_k[i]\}$ are complex and $f_{i,k}(t)$ is the (possibly vector-valued) waveform received from user k in the ith symbol interval. Here, the collection of waveforms $\{f_{i,k}(t) : i = 0, 1, \ldots, M-1; k = 1, 2, \ldots, K\}$ contains all information about the signaling waveforms transmitted by the users and all information about the channels intervening the users and the receiver (see Chapter 1). Many of the results discussed in this chapter can be transferred directly to this more general model, although we will not always explicitly mention such generalizations.

It is also of interest to model more explicitly the narrowband interference signal $\{I(t)\}$ appearing in (7.1). Here, we can consider three basic types of NBI: tonal signals, narrowband digital communication signals, and entropic narrowband stochastic processes. *Tonal signals* are those that consist of the sum of pure sinusoidal signals. These signals are useful for modeling tone jammers and other harmonic interference phenomena. *Narrowband digital communication signals* generalize tonal signals to include digitally modulated carriers. This leads to signals with nonzero-bandwidth components, and as we will see in the sequel, the digital signaling structure can be exploited to improve the NBI suppression capability. Less structure can be assumed by modeling the NBI as *entropic narrowband stochastic processes* such as narrowband autoregressions. Such processes do not have specific deterministic structure. Typical models that can be used in this framework are ideal narrowband processes (with brick-wall spectra) or processes generated by linear stochastic models. Further discussion of the details of these models is deferred until they arise in the following sections. Finally, for convenience, we will assume almost exclusively that the ambient noise $\{N(t)\}$ is a white Gaussian process, although in the following section we mention briefly the situation in which this noise may have impulsive components.

As noted in Section 7.1, narrowband signals can be suppressed from wideband signals by exploiting the difference in predictability in these two types of signals. In this section we develop this idea in more detail. To focus on this issue, we consider the specific situation of (7.1)–(7.2), in which there is only a single spread-spectrum signal in the channel (i.e., $K = 1$). It is also useful to convert the continuous-time signal of (7.1) to discrete time by passing it through an arrangement of a filter matched to the *chip* waveform $\psi(t)$, followed by a chip-rate sampler. That is, we convert the signal (7.1) to a discrete-time signal

$$\begin{aligned} r_n &= \int_{nT_c+\tau_1}^{(n+1)T_c+\tau_1} r(t)\psi(t-nT_c-\tau_1)\,\mathrm{d}t \\ &= c_n + i_n + u_n, \qquad n = 0, 1, \ldots, NM-1, \end{aligned} \tag{7.5}$$

where $\{c_n\}$, $\{i_n\}$, and $\{u_n\}$ represent the converted spread-spectrum data signal, narrowband interferer, and white Gaussian noise, respectively. Note that for the single-user channel ($K = 1$) and in the absence of NBI, a sufficient statistic for detecting the data bit $b_1[i]$ is the signaling-waveform matched-filter output

$$y_1[i] = \int_{iT+\tau_1}^{(i+1)T+\tau_1} r(t)s_1(t-iT-\tau_1)\,\mathrm{d}t, \tag{7.6}$$

which can be written in terms of this sampled signal as

$$y_1[i] = \sum_{n=0}^{N-1} c_{n,1} r_{n+iN}. \tag{7.7}$$

Thus, this conversion to discrete time can be thought of as an intermediate step in the calculation of the sufficient statistic vector and is therefore lossless in the absence of NBI.

Narrowband interference suppression in this type of signal can be based on the following idea. Since the spread-spectrum signal has a nearly flat spectrum, it cannot be predicted accurately from its past values (unless, of course, we were to make use of our knowledge of the spreading code, as discussed in Section 7.4). On the other hand, the interfering signal, being narrowband, can be predicted accurately. Hence, a prediction of the received signal based on previously received values will, in effect, be an estimate of the narrowband interfering signal. Thus, by subtracting a prediction of the received signal obtained at each sampling instant from the signal received during the subsequent instant and using the resulting prediction error as the input to the matched filter (7.7), the effect of the interfering signal can be reduced. Thus, in such a scheme the signal $\{r_n\}$ is replaced in the matched filter (7.7) by the *prediction residual* $\{r_n - \hat{r}_n\}$, where $\hat{r}_n$ denotes the prediction of the received signal at time n, and the data detection scheme becomes

$$\hat{b}_1[i] = \operatorname{sign}\left\{ \sum_{n=0}^{N-1} c_{n,1}[r_{n+iN} - \hat{r}_{n+iN}] \right\}. \tag{7.8}$$

7.2.2 Linear Predictive Methods

This technique for narrowband interference suppression has been explored in detail through the use of fixed and adaptive linear predictors (e.g., [17, 19, 20, 37, 199, 209, 210, 221, 228, 235, 253, 255, 310, 312, 327, 346, 366, 451, 480]; see [6, 244, 327, 332] for reviews). Two basic architectures for fixed linear predictors are Kalman–Bucy predictors, based on a state-space model for the interference, and finite-impulse-response (FIR) linear predictors, based on a tapped-delay-line structure.

Kalman–Bucy Predictors

To use Kalman–Bucy prediction (cf. [235]) in this application, it is useful to model the narrowband interference as a pth order Gaussian autoregressive [AR(p)] process:

$$i_n = \sum_{i=1}^{p} \phi_i i_{n-i} + e_n, \tag{7.9}$$

where $\{e_n\}$ is a white Gaussian sequence, $e_n \sim \mathcal{N}(0, \nu^2)$, and where the AR parameters $\phi_1, \phi_2, \ldots, \phi_p$ are assumed to be constant or slowly varying.

Under this model, the received discrete-time signal (7.5) has a state-space representation as follows (assuming one spread-spectrum user, i.e., $K = 1$):

$$\boldsymbol{x}_n = \boldsymbol{\Phi} \boldsymbol{x}_{n-1} + \boldsymbol{z}_n, \tag{7.10}$$

$$r_n = \boldsymbol{c}^T \boldsymbol{x}_n + v_n, \tag{7.11}$$

where

$$\begin{aligned}
\boldsymbol{x}_n &\triangleq \left[i_n i_{n-1} \cdots i_{n-p+1} \right]^T, \\
\boldsymbol{\Phi} &\triangleq \begin{pmatrix} \phi_1 & \phi_2 & \cdots & \phi_{p-1} & \phi_p \\ 1 & 0 & \cdots & 0 & 0 \\ 0 & 1 & \cdots & 0 & 0 \\ \vdots & \vdots & \ddots & \vdots & \vdots \\ 0 & 0 & \cdots & 0 & 1 \end{pmatrix}, \\
\boldsymbol{z}_n &\triangleq [e_n \ \ 0 \ \ \cdots \ \ 0\ ]^T, \\
\boldsymbol{c} &\triangleq [\ 1 \ \ 0 \ \ \cdots \ \ 0\ ]^T, \\
v_n &\triangleq c_n + u_n,
\end{aligned} \tag{7.12}$$

with

$$c_n \in \left\{ \frac{A_1}{\sqrt{N}}, -\frac{A_1}{\sqrt{N}} \right\}, \qquad u_n \sim \mathcal{N}(0, \sigma^2).$$

Given this state-space formalism, the linear minimum mean-square-error (MMSE) prediction of the received signal (and hence of the interference) can be computed recursively via the Kalman–Bucy filtering equations (e.g., [377]), which predicts the nth observation r_n as $\hat{r}_n = \boldsymbol{c}^T \hat{\boldsymbol{x}}_n$, where $\hat{\boldsymbol{x}}_n$ denotes the state prediction in (7.10)–(7.11), given recursively through the update equations

$$\hat{\boldsymbol{x}}_{n+1} = \boldsymbol{\Phi} \bar{\boldsymbol{x}}_n, \tag{7.13}$$

$$\bar{\boldsymbol{x}}_n = \hat{\boldsymbol{x}}_n + \frac{r_n - \hat{r}_n}{\sigma_n^2} \boldsymbol{M}_n \boldsymbol{c}, \tag{7.14}$$

with $\sigma_n^2 = \boldsymbol{c}^T \boldsymbol{M}_n \boldsymbol{c} + A_1^2/N + \sigma^2$, denoting the variance of the prediction residual, and where the matrix $\boldsymbol{M}_n$ (which is the covariance of the state prediction error $\boldsymbol{x}_n - \hat{\boldsymbol{x}}_n$) is computed via the recursion

$$\boldsymbol{M}_{n+1} = \boldsymbol{\Phi} \boldsymbol{P}_n \boldsymbol{\Phi}^T + \boldsymbol{Q}, \tag{7.15}$$

$$\boldsymbol{P}_n = \boldsymbol{M}_n - \frac{1}{\sigma_n^2} \boldsymbol{M}_n \boldsymbol{c}\boldsymbol{c}^T \boldsymbol{M}_n, \tag{7.16}$$

with

$$\boldsymbol{Q} = E\left\{\boldsymbol{z}_n \boldsymbol{z}_n^T\right\} = \nu^2 \boldsymbol{e}_1 \boldsymbol{e}_1^T, \tag{7.17}$$

where $\boldsymbol{e}_1$ denotes a p-vector with all entries being zeros, except for the first entry, which is 1. The Kalman–Bucy prediction-based NBI suppression algorithm based on the state-space model (7.10)–(7.11) is summarized as follows. (Note that it is assumed that the model parameters are known.)

Algorithm 7.1: [Kalman–Bucy prediction-based NBI suppression] *At time i, N received samples $\{r_{iN}, r_{iN+1}, \ldots, r_{iN+N-1}\}$ are obtained at the chip-matched filter output (7.5).*

- *For $n = iN, iN+1, \ldots, iN+N-1$ perform the following steps:*

$$\hat{r}_n = \boldsymbol{c}^T \hat{\boldsymbol{x}}_n, \tag{7.18}$$

$$\sigma_n^2 = \boldsymbol{c}^T \boldsymbol{M}_n \boldsymbol{c} + \frac{A_1^2}{N} + \sigma^2, \tag{7.19}$$

$$\boldsymbol{P}_n = \boldsymbol{M}_n - \frac{1}{\sigma_n^2} \boldsymbol{M}_n \boldsymbol{c}\boldsymbol{c}^T \boldsymbol{M}_n, \tag{7.20}$$

$$\bar{\boldsymbol{x}}_n = \hat{\boldsymbol{x}}_n + \frac{r_n - \hat{r}_n}{\sigma_n^2} \boldsymbol{M}_n \boldsymbol{c}, \tag{7.21}$$

$$\hat{\boldsymbol{x}}_{n+1} = \boldsymbol{\Phi} \bar{\boldsymbol{x}}_n, \tag{7.22}$$

$$\boldsymbol{M}_{n+1} = \boldsymbol{\Phi} \boldsymbol{P}_n \boldsymbol{\Phi}^T + \boldsymbol{Q}. \tag{7.23}$$

- *Detect the ith bit $b_1[i]$ according to*

$$\hat{b}_1[i] = \operatorname{sign}\left\{\sum_{n=0}^{N-1} c_{n,1}\left[r_{n+iN} - \hat{r}_{n+iN}\right]\right\}. \tag{7.24}$$

Linear FIR Predictor

The Kalman–Bucy filter is, of course, an infinite-impulse-response (IIR) filter. A simpler linear structure is a tapped-delay-line (TDL) configuration, which makes one-step predictions via the FIR filter

$$\hat{r}_n = \sum_{\ell=1}^{L} \alpha_\ell \, r_{n-\ell}, \tag{7.25}$$

where L is the data length used by the predictor, and $\alpha_1, \alpha_2, \ldots, \alpha_L$, are tap weights. In the stationary case, the tap weights can be chosen optimally via the Levinson algorithm (see, e.g., [377]). More important, though, the FIR structure (7.25) can easily be adapted using, for example, the least-mean-squares (LMS) algorithm (e.g., [463]). Denote $\boldsymbol{\alpha} \triangleq [\alpha_1 \ \cdots \ \alpha_L]^T$. Let $\boldsymbol{\alpha}[n]$ denote the tap-weight vector to be applied at the nth chip sample (i.e., to predict r_{n+1}). Also denote $\boldsymbol{r}_n \triangleq [r_{n-1} \ r_{n-2} \ \cdots \ r_{n-L}]^T$. Then the predictor coefficients can be updated according to

$$\boldsymbol{\alpha}[n] = \boldsymbol{\alpha}[n-1] + \mu\Big(r_n - \hat{r}_n\Big)\boldsymbol{r}_n, \tag{7.26}$$

where μ is a tuning constant. Although the Kalman–Bucy filter can also be adapted, the ease and stability with which the FIR structure can be adapted makes it a useful choice for this application. To make the choice of tuning constant invariant to changes in the input signal levels, the LMS algorithm (7.26) can be normalized as follows:

$$\boldsymbol{\alpha}[n] = \boldsymbol{\alpha}[n-1] + \frac{\mu_0}{p_n}\Big(r_n - \hat{r}_n\Big)\boldsymbol{r}_n, \tag{7.27}$$

where p_n is an estimate of the input power obtained by

$$p_n = p_{n-1} + \mu_0\Big(\|\boldsymbol{r}_n\|^2 - p_{n-1}\Big). \tag{7.28}$$

The estimate of the signal power p_n is an exponentially weighted estimate. The constant μ_0 is chosen small enough to ensure convergence, and the initial condition p_0 should be large enough so that the denominator never shrinks so small as to make the step size large enough for the adaptation to become unstable.

A block diagram of a TDL-based linear predictor is shown in Fig. 7.4. The LMS linear prediction-based NBI suppression algorithm is summarized as follows.

Algorithm 7.2: [LMS linear prediction-based NBI suppression] *At time i, N received samples $\{r_{iN}, r_{iN+1}, \ldots, r_{iN+N-1}\}$ are obtained at the chip-matched filter output (7.5).*

- *For $n = iN, iN+1, \ldots, iN+N-1$ perform the following steps:*

$$\hat{r}_n = \boldsymbol{\alpha}[n-1]^T \boldsymbol{r}_{n-1}, \tag{7.29}$$

$$p_n = p_{n-1} + \mu_0\Big(\|\boldsymbol{r}_n\|^2 - p_{n-1}\Big), \tag{7.30}$$

$$\boldsymbol{\alpha}[n] = \boldsymbol{\alpha}[n-1] + \frac{\mu_0}{p_n}\Big(r_n - \hat{r}_n\Big)\boldsymbol{r}_n. \tag{7.31}$$

- *Detect the ith bit $b_1[i]$ according to*

$$\hat{b}_1[i] = \text{sign}\left\{\sum_{n=0}^{N-1} c_{n,1}[r_{n+iN} - \hat{r}_{n+iN}]\right\}. \tag{7.32}$$

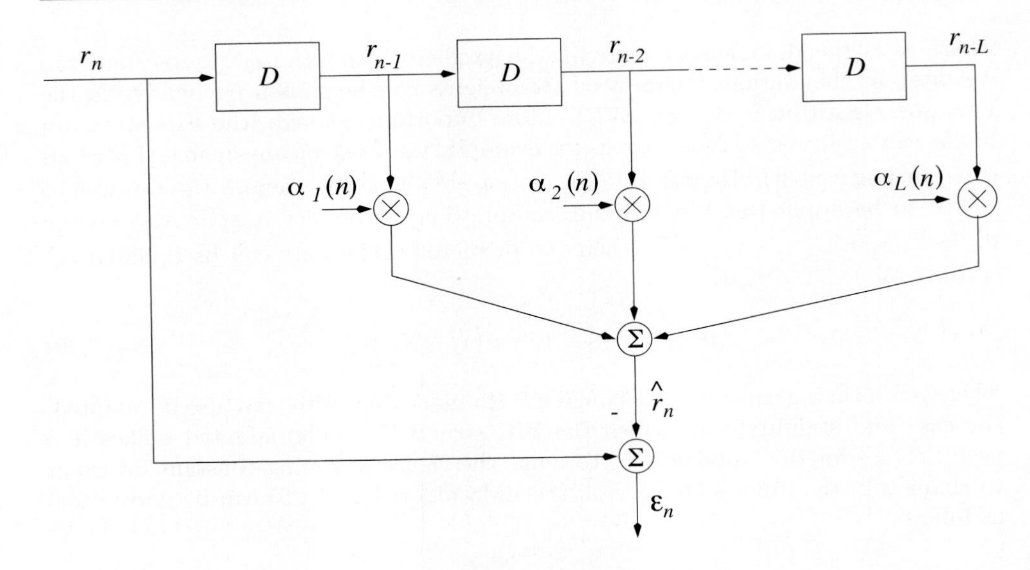

Figure 7.4. Tapped-delay-line linear predictor.

Performance and convergence analyses of these types of linear predictor–subtractor systems have shown that considerable signal-to-interference-plus-noise ratio (SINR) improvement can be obtained by these methods. (See the above-cited references and the results in Section 7.4.) Linear interpolation filters can also be used in this context, leading to further improvements in SINR and to better phase characteristics compared with linear prediction filters (e.g., [311]). For example, a simple linear interpolator of order $L_1 + L_2$ for estimating r_n is given by

$$\hat{r}_n = \sum_{\ell=-L_1,\ell\neq n}^{L_2} \alpha_\ell \, r_{n-\ell}, \tag{7.33}$$

where $\alpha_{-L_1}, \ldots, \alpha_{L_2}$, are tap weights. Such an interpolator can be adapted similarly via the LMS algorithm.

7.3 Nonlinear Predictive Techniques

Linear predictive methods exploit the wideband nature of the useful data signal to suppress the interference. In doing so, they are exploiting only the spectral structure of the spread data signal, not its further structure. These techniques can be improved upon in this application by exploiting such further structure of the useful data signal as it manifests itself in the sampled observations (7.5). In particular, on examining (7.1), (7.2), (7.3), and (7.5), we see that for the single-user case (i.e., $K = 1$), the discrete-time data signal $\{c_n\}$ takes on values of only $\pm A_1/\sqrt{N}$. Although linear prediction would be optimal in the model of (7.5) in the case in which all signals are Gaussian, this binary-valued direct-sequence data signal $\{c_n\}$ is highly non-Gaussian. So, even if the NBI and background noise are assumed to be Gaussian, the optimal filter for performing the required prediction will, in general, be nonlinear (e.g., [377]). This non-Gaussian structure of direct-sequence signals can be exploited to obtain nonlinear filters that exhibit significantly better suppression of narrowband interference than do linear filters under conditions where this non-Gaussian-ness is of sufficient import. In the following paragraphs we elaborate on this idea, which was introduced in [522] and explored further in [133, 376, 387, 425, 535–538].

Consider again the state-space model of (7.10)–(7.11). The Kalman–Bucy estimator discussed above is the best linear predictor of r_n from its past values. If the observation noise $\{v_n\}$ of (7.12) were a Gaussian process, this filter would also give the global MMSE (or conditional mean) prediction of the received signal (and hence of the interference). However, since $\{v_n\}$ is not Gaussian but rather is the sum of two independent random variables, one of which is Gaussian and the other of which is binary ($\pm A_1/\sqrt{N}$), its probability density is the weighted sum of two Gaussian densities. In this case, the exact conditional mean estimator can be shown to have a complexity that increases exponentially in time [452], which renders it unsuitable for practical implementation.

7.3.1 ACM Filter

In [309], Masreliez proposed an *approximate conditional mean* (ACM) *filter* for estimating the state of a linear system with Gaussian state noise and non-Gaussian measurement noise. In particular, Masreliez proposed that some, but not all, of the Gaussian assumptions used in derivation of the Kalman filter be retained in defining a nonlinearly recursively updated filter. He retained a Gaussian distribution for the conditional mean, although it is not a consequence of the probability densities of the system (as is the case for Gaussian observation noise), hence the name *approximate conditional mean* that is applied to this filter. In [133, 387, 522] this ACM filter was developed for the model (7.10)–(7.11). To describe this filter, first denote the prediction residual by

$$\epsilon_n = r_n - \hat{r}_n. \tag{7.34}$$

This filter operates just as that of (7.13)–(7.17), except that the measurement update equation (7.14) is replaced with

$$\bar{\boldsymbol{x}}_n = \hat{\boldsymbol{x}}_n + g_n(\epsilon_n)\boldsymbol{M}_n\boldsymbol{c}, \tag{7.35}$$

and the update equation (7.16) is replaced with

$$\boldsymbol{P}_n = \boldsymbol{M}_n - G_n(\epsilon_n)\boldsymbol{M}_n\boldsymbol{c}\boldsymbol{c}^T\boldsymbol{M}_n. \tag{7.36}$$

The terms g_n and G_n are nonlinearities arising from the non-Gaussian distribution of the observation noise and are given by

$$g_n(z) \triangleq -\frac{\partial p\left(z \mid \underline{r}_0^{n-1}\right)/\partial z}{p\left(z \mid \underline{r}_0^{n-1}\right)}, \tag{7.37}$$

$$G_n(z) \triangleq \frac{\partial g_n(z)}{\partial z}, \tag{7.38}$$

where we have used the notation $\underline{r}_a^b \triangleq \{r_a, r_{a+1}, \ldots, r_b\}$, and $p\left(z|\underline{r}_0^{n-1}\right)$ denotes the measurement prediction density. The measurement updates reduce to the standard equations for the Kalman–Bucy filter when the observation noise is Gaussian.

For the single-user system, the density of the observation noise in (7.12) is given by the following Gaussian mixture:

$$v_n \sim \frac{1}{2}\mathcal{N}\left(\frac{A_1}{\sqrt{N}}, \sigma^2\right) + \frac{1}{2}\mathcal{N}\left(-\frac{A_1}{\sqrt{N}}, \sigma^2\right). \tag{7.39}$$

Let $\tilde{\sigma}_n^2$ be the variance of the innovation (or residual) signal in (7.34):

$$\tilde{\sigma}_n^2 = \boldsymbol{c}^T \boldsymbol{M}_n \boldsymbol{c} + \sigma^2, \tag{7.40}$$

we can then write the functions g_n and G_n in this case as

$$g_n(z) = z - \frac{A_1}{\sqrt{N}}\tanh\left(\frac{A_1}{\sqrt{N}}\frac{z}{\tilde{\sigma}_n^2}\right), \tag{7.41}$$

$$G_n(z) = 1 - \frac{A_1^2}{N\tilde{\sigma}_n^2}\operatorname{sech}^2\left(\frac{A_1}{\sqrt{N}}\frac{z}{\tilde{\sigma}_n^2}\right). \tag{7.42}$$

The ACM filter is thus seen to have a structure similar to that of the standard Kalman–Bucy filter. The time updates (7.13) and (7.15) are identical to those in the Kalman–Bucy filter. The measurement updates (7.35) and (7.36) involve correcting the predicted value by a nonlinear function of the prediction residual ϵ_n. This correction essentially acts like a soft-decision feedback to suppress the spread-spectrum signal from the measurements. That is, it corrects the measurement by a factor in the range $\left[-A_1/\sqrt{N}, A_1/\sqrt{N}\right]$ that estimates the spread-spectrum signal. When the filter is performing well, the variance term in the denominator of $\tanh(\cdot)$ is low. This means that the argument of $\tanh(\cdot)$ is larger, driving $\tanh(\cdot)$ into a region where it behaves like the $\text{sign}(\cdot)$ function, and thus estimates the spread-spectrum signal to be $A_1/\sqrt{N}$ if the residual signal ϵ_n is positive and $-A_1/\sqrt{N}$ if the residual is negative. On the other hand, when the filter is not making good estimates, the variance is high and $\tanh(\cdot)$ is in a linear region of operation. In this region, the filter hedges its bet on the accuracy of $\text{sign}(\epsilon_n)$ as an estimate of the spread-spectrum signal. Here the filter behaves essentially like the (linear) Kalman filter. The ACM-filter-based NBI suppression algorithm based on the state-space model (7.10)–(7.11) is summarized as follows.

Algorithm 7.3: [ACM-filter-based NBI suppression] *At time i, N received samples $\{r_{iN}, r_{iN+1}, \ldots, r_{iN+N-1}\}$ are obtained at the chip-matched filter output (7.5).*

- *For $n = iN, iN+1, \ldots, iN+N-1$ perform the following steps:*

$$\hat{r}_n = \boldsymbol{c}^T \hat{\boldsymbol{x}}_n, \tag{7.43}$$

$$\sigma_n^2 = \boldsymbol{c}^T \boldsymbol{M}_n \boldsymbol{c} + \sigma^2, \tag{7.44}$$

$$\boldsymbol{P}_n = \boldsymbol{M}_n - G_n(\epsilon_n) \boldsymbol{M}_n \boldsymbol{c}\boldsymbol{c}^T \boldsymbol{M}_n, \tag{7.45}$$

$$\bar{\boldsymbol{x}}_n = \hat{\boldsymbol{x}}_n + g_n(\epsilon_n) \boldsymbol{M}_n \boldsymbol{c}, \tag{7.46}$$

$$\hat{\boldsymbol{x}}_{n+1} = \boldsymbol{\Phi} \bar{\boldsymbol{x}}_n, \tag{7.47}$$

$$\boldsymbol{M}_{n+1} = \boldsymbol{\Phi} \boldsymbol{P}_n \boldsymbol{\Phi}^T + \boldsymbol{Q}, \tag{7.48}$$

where g_n and G_n are defined in (7.41) and (7.42), respectively.

- *Detect the ith bit $b_1[i]$ according to*

$$\hat{b}_1[i] = \operatorname{sign}\left\{ \sum_{n=0}^{N-1} c_{n,1} [r_{n+iN} - \hat{r}_{n+iN}] \right\}. \tag{7.49}$$

Simulation Examples

When the interference is modeled as a first-order autoregressive process, which does not have a very sharply peaked spectrum, the performance of the ACM filter does not seem to be appreciably better than that of the Kalman–Bucy filter. However, when the spectrum of the interference is made to be more sharply peaked by increasing the order of the autoregression, the ACM filter is found to give significant performance gains over the Kalman filter. Simulations were run for a second-order AR interferer with both poles at 0.99:

$$i_n = 1.98 i_{n-1} - 0.9801 i_{n-2} + e_n,$$

where $\{e_n\}$ is an i.i.d. Gaussian sequence. The ambient noise power is held constant at $\sigma^2 = 0.01$, while the total of noise plus interference power varies from 5 to 20 dB (all relative to a unity power spread-spectrum signal). In comparing filtering methods, the figure of merit is the ratio of SINR at the output of filtering to the SINR at the input, which reduces to

$$\text{SINR improvement} \triangleq \frac{E\left\{|r_n - s_n|^2\right\}}{E\left\{|\epsilon_n - s_n|^2\right\}},$$

where ϵ_n is defined as in (7.34). The results from the Kalman and ACM predictors are shown in Fig. 7.5. The filters were run for 1500 points. The results reflect the last 500 points, and the values given represent averages over 4000 independent simulations.

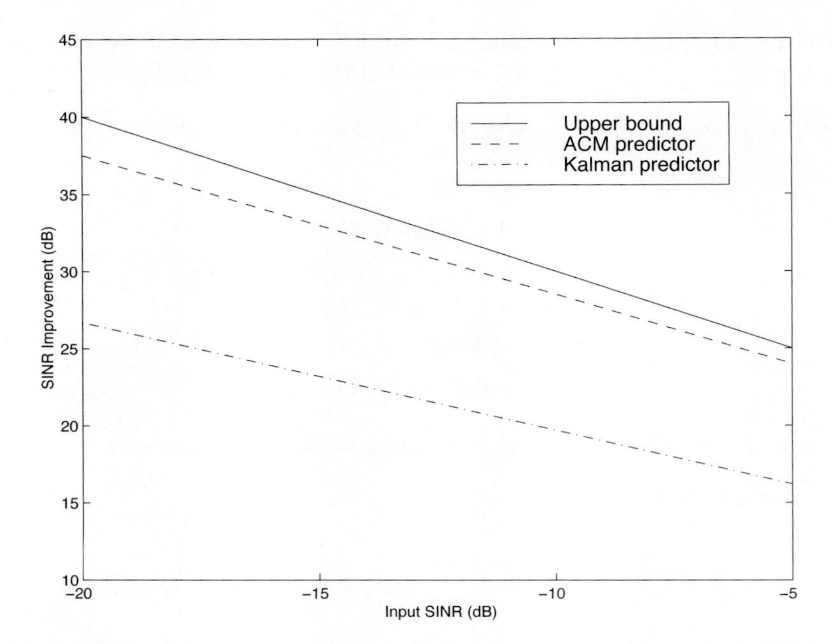

Figure 7.5. Performance of the Kalman filter– and ACM filter–based NBI suppression methods.

To stress the effectiveness against the narrowband interferer (versus the background noise), the solid line in Fig. 7.5 gives an upper bound on SNR improvement, assuming that the narrowband interference is predicted with noiseless accuracy. This is calculated by setting $E\Big\{|\epsilon_n - s_n|^2\Big\}$ equal to the power of the AWGN driving the AR process (i.e., the unpredictable portion of the interference). Note that the SINR improvement due to using the ACM filter is quite substantial and is very near the theoretical bound.

7.3.2 Adaptive Nonlinear Predictor

It is seen that in the ACM filter, the predicted value of the state is obtained as a linear function of the previous estimate modified by a nonlinear function of the prediction error. We now use the same approach to modify the adaptive linear predictive filter described in Section 7.2.2. This technique was first developed in [387, 522]. To show the influence of the prediction error explicitly, using (7.34) we rewrite (7.25) as

$$\hat{r}_n = \sum_{\ell=1}^{L} \alpha_\ell \Big(\hat{r}_{n-\ell} + \epsilon_{n-\ell}\Big). \tag{7.50}$$

We make the assumption, similar to that made in the derivation of the ACM filter, that the prediction residual ϵ_n is the sum of a Gaussian random variable and a

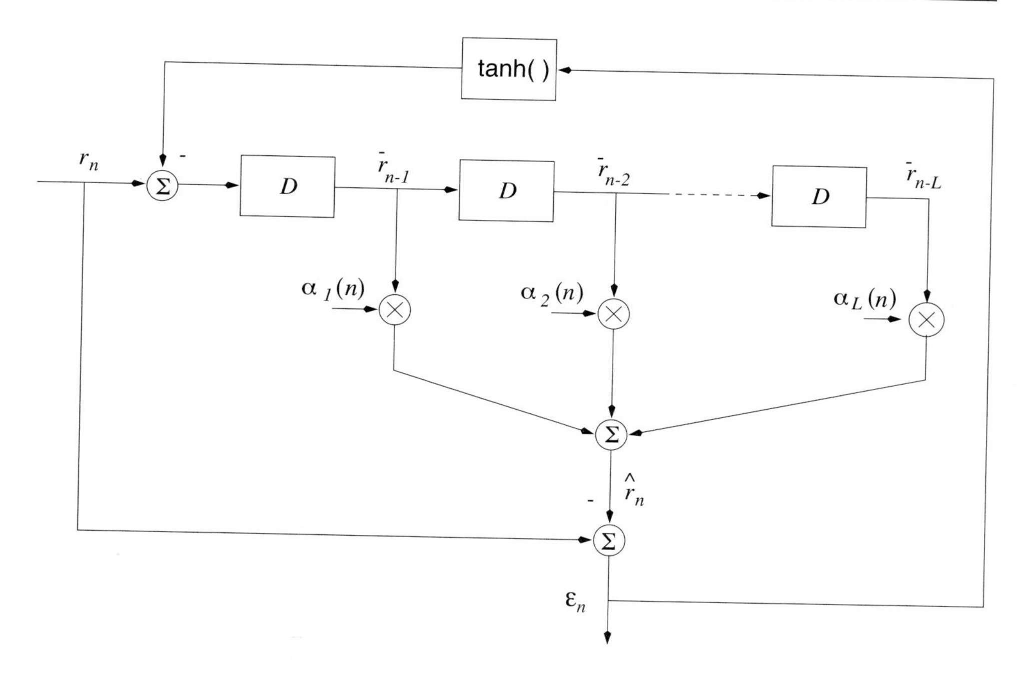

Figure 7.6. Tapped-delay-line nonlinear predictor.

binary random variable. If the variance of the Gaussian random variable is $\tilde{\sigma}_n^2$, the nonlinear transformation appearing in the ACM filter can be written as

$$g_n(\epsilon_n) = \epsilon_n - \frac{A_1}{\sqrt{N}} \tanh\left(\frac{A_1}{\sqrt{N}} \frac{\epsilon_n}{\tilde{\sigma}_n^2}\right). \tag{7.51}$$

By transforming the prediction error in (7.50) using the nonlinearity above, we get a nonlinear transversal filter for the prediction of r_n, namely,

$$\hat{r}_n = \sum_{\ell=1}^{L} \alpha_\ell \underbrace{\left[\hat{r}_{n-\ell} + g_{n-\ell}(\epsilon_{n-\ell})\right]}_{\bar{r}_{n-\ell}}, \tag{7.52}$$

where $\bar{r}_n$ is given by

$$\begin{aligned} \bar{r}_n &\triangleq \hat{r}_n + g_n(\epsilon_n), \\ &= \underbrace{\hat{r}_n + \epsilon_n}_{r_n} - \frac{A_1}{\sqrt{N}} \tanh\left(\frac{A_1}{\sqrt{N}} \frac{\epsilon_n}{\tilde{\sigma}_n^2}\right). \end{aligned} \tag{7.53}$$

The structure of this filter is shown in Fig. 7.6. To implement the filter of (7.52), an estimate of the parameter $\tilde{\sigma}_n^2$ and an algorithm for updating the tap weights must be obtained. A useful estimate for $\tilde{\sigma}_n^2$ is $\widehat{\tilde{\sigma}^2} = \Delta_n - A_n^2/N$, where Δ_n is a sample estimate of the prediction error variance [e.g., $\Delta_n = (1/L)\sum_{\ell=1}^{L} \epsilon_{n-\ell}^2$]. On the other hand, the tap-weight vector can be updated according to the modified LMS algorithm

$$\boldsymbol{\alpha}[n] = \boldsymbol{\alpha}[n-1] + \frac{\mu_0}{p_n}\Big(\bar{r}_n - \hat{r}_n\Big)\bar{\boldsymbol{r}}_n, \tag{7.54}$$

where $\bar{\boldsymbol{r}}_n = \begin{bmatrix} \bar{r}_{n-1} & \bar{r}_{n-1} & \cdots & \bar{r}_{n-L} \end{bmatrix}^T$ and p_n is given by (7.28). Note that the nonlinear prediction given by (7.52) is recursive in the sense that the prediction depends explicitly on the previous predicted values as well as on the previous input to the filter. This is in contrast to the linear prediction of (7.25), which depends explicitly only on the previous inputs to the filter, although it depends on the previous outputs implicitly through their influence on the tap-weight updates. The nonlinear prediction-based NBI suppression algorithm is summarized as follows.

Algorithm 7.4: [LMS nonlinear prediction-based NBI suppression] *At time i, N received samples $\{r_{iN}, r_{iN+1}, \ldots, r_{iN+N-1}\}$ are obtained at the chip-matched filter output (7.5).*

- *For $n = iN, iN+1, \ldots, iN+N-1$ perform the following steps:*

$$\hat{r}_n = \boldsymbol{\alpha}[n-1]^T \bar{\boldsymbol{r}}_{n-1}, \tag{7.55}$$

$$\bar{r}_n = r_n - \frac{A_1}{\sqrt{N}} \tanh\left(\frac{A_1}{\sqrt{N}}\frac{\epsilon_n}{\tilde{\sigma}_n^2}\right), \tag{7.56}$$

$$p_n = p_{n-1} + \mu_0\Big(\|\boldsymbol{r}_n\|^2 - p_{n-1}\Big), \tag{7.57}$$

$$\boldsymbol{\alpha}[n] = \boldsymbol{\alpha}[n-1] + \frac{\mu_0}{p_n}\Big(\bar{r}_n - \hat{r}_n\Big)\bar{\boldsymbol{r}}_n. \tag{7.58}$$

- *Detect the ith bit $b_1[i]$ according to*

$$\hat{b}_1[i] = \text{sign}\left\{\sum_{n=0}^{N-1} c_{n,1}[r_{n+iN} - \hat{r}_{n+iN}]\right\}. \tag{7.59}$$

It is interesting to note that the predictor (7.52) can be viewed as a generalization of both linear and hard-decision-feedback (see, e.g., [107, 108, 254]) adaptive predictors, in which we use our knowledge of the prediction error statistics to make a soft decision about the binary signal, which is then fed back to the predictor. As noted above, introduction of this nonlinearity improves the prediction performance over the linear version. As discussed in [387], softening of this feedback nonlinearity improves the convergence properties of the adaptation over the use of hard-decision feedback.

Simulation Examples

To assess the nonlinear adaptive NBI suppression algorithm above, simulations were performed on the AR model for interference given in Section 7.2. The results are

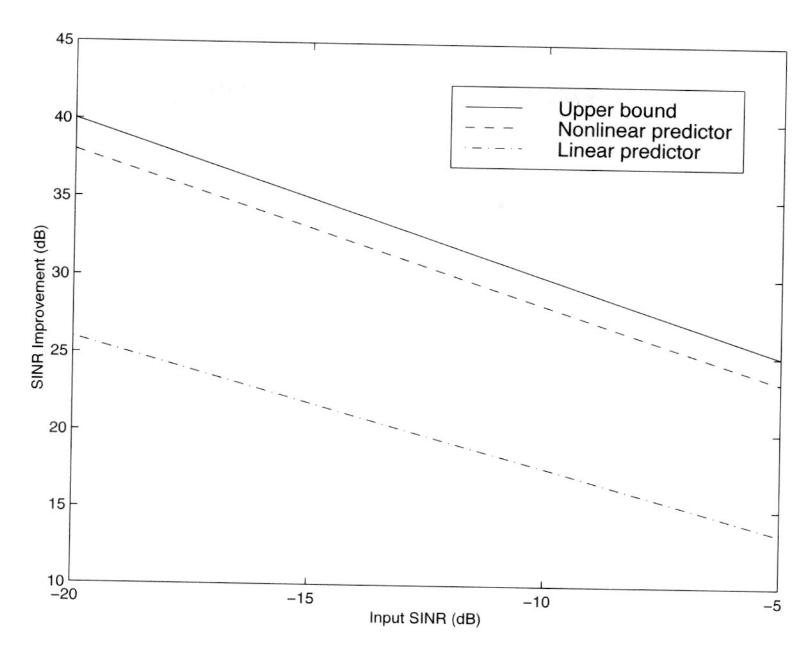

Figure 7.7. Performance of adaptive linear predictor– and adaptive nonlinear predictor–based NBI suppression methods.

shown in Fig. 7.7. It is seen that, as in the case where the interference statistics are known, the nonlinear adaptive NBI suppression method significantly outperforms its linear counterpart.

7.3.3 Nonlinear Interpolating Filters

ACM Interpolator

Nonlinear interpolative interference suppression filters have been developed in [425]. We next derive the interpolating ACM filter. We consider the density of the current state conditioned on previous and following states. We have

$$
\begin{aligned}
p\left(i_n \mid \underline{r}_0^{n-1}, \underline{r}_{n+1}^{M-1}\right) &= \frac{p\left(\underline{r}_0^{n-1}, \underline{r}_{n+1}^{M-1} \mid i_n\right) p(i_n)}{p\left(\underline{r}_0^{n-1}, \underline{r}_{n+1}^{M-1}\right)} \\
&\cong \frac{p\left(\underline{r}_0^{n-1} \mid i_n\right) p\left(\underline{r}_{n+1}^{M-1} \mid i_n\right) p(i_n)}{p\left(\underline{r}_0^{n-1}, \underline{r}_{n+1}^{M-1}\right)} \qquad (7.60) \\
&= \frac{p\left(i_n \mid \underline{r}_0^{n-1}\right) p\left(i_n \mid \underline{r}_{n+1}^{M-1}\right)}{p(i_n)} \, \frac{p\left(\underline{r}_{n+1}^{M-1}\right)}{p\left(\underline{r}_{n+1}^{M-1} \mid \underline{r}_0^{n-1}\right)}, \qquad (7.61)
\end{aligned}
$$

where in (7.60) we made the approximation that, conditioned on i_n, $\underline{r}_0^{n-1}$ and $\underline{r}_{n+1}^{M-1}$ are independent. The second term in (7.61) is independent of i_n. If it is assumed (analogously to what is done in the ACM filter) that the two densities in the numerator of the first term in (7.61) are Gaussian, the interpolated estimate is also Gaussian. Therefore, if we assume that the densities are (where f indicates the forward prediction and b indicates the backward prediction)

$$p\left(i_n \mid \underline{r}_0^{n-1}\right) \sim \mathcal{N}(\mu_{f,n},\ \Sigma_{f,n}),$$
$$p\left(i_n \mid \underline{r}_{n+1}^{M-1}\right) \sim \mathcal{N}(\mu_{b,n},\ \Sigma_{b,n}),$$
$$p(i_n) \sim \mathcal{N}(\mu_0,\ \Sigma_0),$$

the interpolated estimate is still Gaussian:

$$p\left(i_n \mid \underline{r}_0^{n-1},\ \underline{r}_{n+1}^{M-1}\right) \sim \mathcal{N}(\mu_n,\ \Sigma_n), \tag{7.62}$$

with

$$\Sigma_n \triangleq \left(\Sigma_{f,n}^{-1} + \Sigma_{b,n}^{-1} - \Sigma_0^{-1}\right)^{-1}, \tag{7.63}$$
$$\mu_n \triangleq \left(\mu_{f,n}\Sigma_{f,n}^{-1} + \mu_{b,n}\Sigma_{b,n}^{-1}\right)\Sigma_n. \tag{7.64}$$

While the mean and variance of the interpolated estimate at each sample n can be computed via the equations above, recall that the forward and backward means and variances are determined by the nonlinear ACM filter recursions.

Simulation Examples

The equations above can be used for both the linear Kalman filter and the ACM filter to generate interpolative predictions from the forward and backward predicted estimates. As in the ACM prediction filter, we have approximated the conditional densities as being Gaussian, although the observation noise is not Gaussian. The filters are run forward on a block of data and then backward on the same data. The two results are combined to form the interpolated prediction via (7.63)–(7.64).

Simulations were run on the same AR model for interference as that given in Section 7.2. Figure 7.8 gives results for interpolative filtering over predictive filtering for the known statistics case. The filters were run forward and backward for all 1500 points in the block. Interpolator SINR gain was calculated over the middle 500 points (when both forward and backward predictors were in steady state).

Adaptive Nonlinear Block Interpolator

Recall that the ACM predictor uses the interference prediction at time n, $\hat{r}_n$, to generate a prediction of the observation less the spread spectrum signal $\bar{r}_n$. This estimate $\bar{r}_n$ is used in subsequent samples to generate new interference predictions. Since the estimates of $\bar{r}_{n+\ell}$ are not available for $\ell > 0$ at time n (i.e., samples

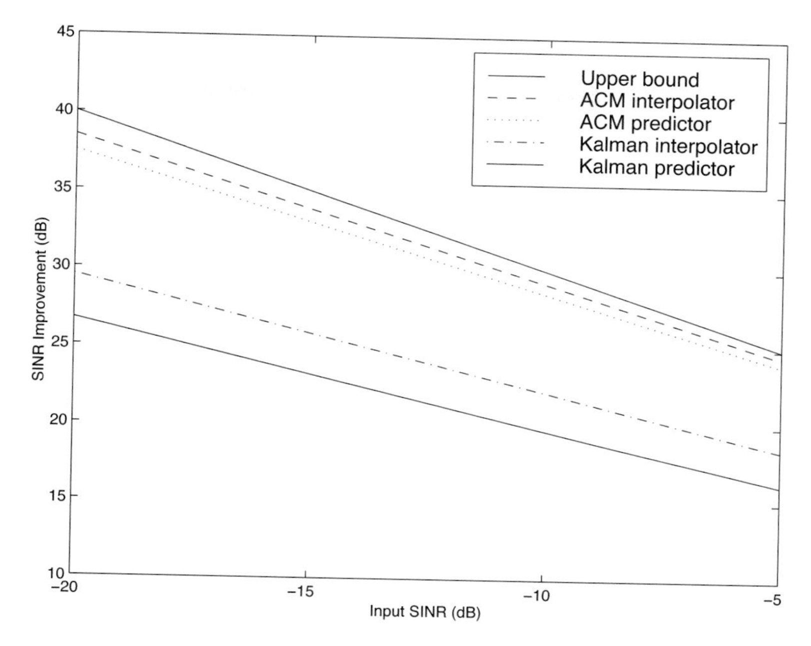

Figure 7.8. Performance of Kalman interpolator– and ACM interpolator–based NBI suppression methods.

that occur after the current one), the ACM filter cannot be cast directly in the interpolator structure. However, an approach similar to the one for the known-statistics ACM interpolator can be used. In this approach the data are segmented into blocks and run through a forward filter of length L to give predictions $\hat{r}_n^f$ and $\bar{r}_n^f$. The same data are run through a backward adaptive ACM filter with a separate tap-weight vector, also of length L, to generate estimates $\hat{r}_n^b$ and $\bar{r}_n^b$. After these calculations are made for the entire block, the data are combined to form an interpolated prediction according to

$$\hat{r}_n = \frac{1}{2}\left(\hat{r}_n^f + \hat{r}_n^b\right), \tag{7.65}$$

$$\bar{r}_n = r_n - \frac{A_1}{\sqrt{N}}\tanh\left(\frac{A_1}{\sqrt{N}}\frac{r_n - \bar{r}_n}{\tilde{\sigma}_n^2}\right). \tag{7.66}$$

The next block of data follows the same procedure. However, when the next block is initialized, the previous tap weights are used to start the forward predictor, and the interpolated predictions $\{\bar{r}_n\}$ are used to initialize the forward prediction. This "head start" on the adaptation can only take place in the forward direction. We do not have any information on the following block of data to give us insight into the backward prediction. Therefore, the backward prediction is less reliable than the forward prediction. To compensate for this effect, consecutive blocks are overlapped,

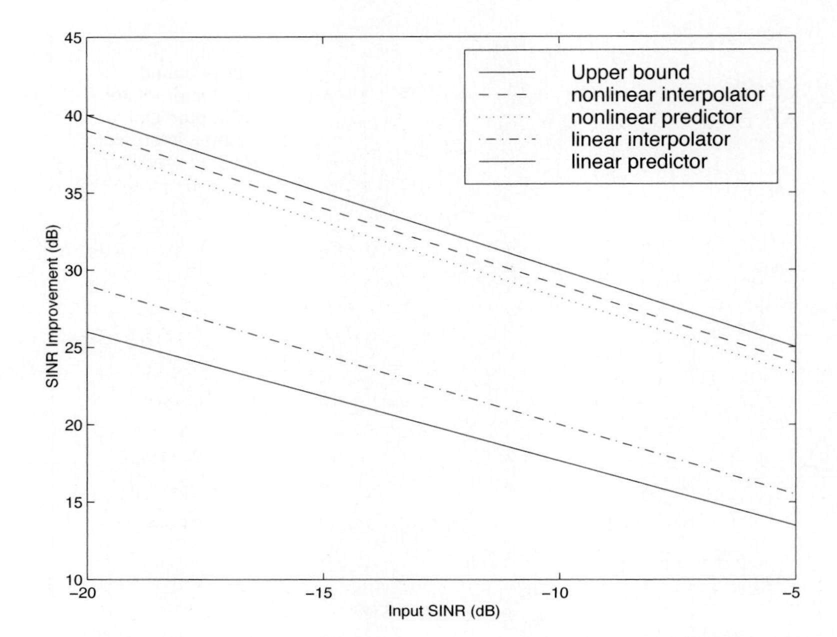

Figure 7.9. Performance of linear interpolator– and nonlinear interpolator–based NBI suppression methods.

with the overlap being used to allow the backward predictor some startup time to begin good predictions of the spread-spectrum signal [381, 425].

Simulation Examples

Results for the same simulation when the statistics are unknown are given in Fig. 7.9. The adaptive interpolator had a block length of 250 samples, with 100 samples being overlapped. That is, for each block of 250 samples, 150 interpolated estimates were made. For the case of known statistics, the ACM predictor already performs well, and there is little margin for improvement via use of an interpolator. The adaptive filter shows greater margin for improvement, on which the interpolator capitalizes. However, in either case, the interpolator does offer improved phase characteristics and some performance gain at the cost of additional complexity and a delay in processing.

A number of further results have been developed using and expanding the ideas discussed above. For example, performance analysis methods have been developed for both predictive [537] and interpolative [538] nonlinear suppression filters. Predictive filters for the further situation in which the ambient noise $\{N(t)\}$ has impulsive components have been developed in [133]. The multiuser case, in which $K > 1$, has been considered in [425]. Further results can be found in [11, 15, 238, 535, 536].

7.3.4 HMM-Based Methods

In the prediction-based methods discussed above, the narrowband interference environment is assumed to be stationary or, at worst, slowly varying. In some applications, however, the interference environment is dynamic in that narrowband interferers enter and leave the channel at random and at arbitrary frequencies within the spread bandwidth. An example of such an application arises in the littoral sonobuoy arrays mentioned in Section 7.1, in which shore-based commercial VHF traffic, such as dispatch traffic, appears throughout the spread bandwidth in a very bursty fashion. A similar phenomenon arises when the direct-sequence system coexists with a frequency-hopping system, which happens, for example, when wireless LANs and Bluetooth systems operate in the same location. A difficulty with use of adaptive prediction filters of the type noted above is that when an interferer suddenly drops out of the channel, the "notch" that the adaptation algorithm created to suppress it will persist for some time after the signal leaves the channel. This is because, while the energy of the narrowband source drives the adaptation algorithms to suppress an interferer when it enters the channel, there is no counterbalancing energy to drive the adaptation algorithm back to normalcy when an interferer exits the channel. That is, there is an asymmetry between what happens when an interferer enters the channel and what happens when an interferer exits the channel. If interferers enter and exit randomly across a wide band, this asymmetry will cause the appearance of notches across a large fraction of the spread bandwidth, which will result in a significantly degraded signal of interest. Thus, a more sophisticated approach is needed for such cases. One such approach, described in [67], is based on a hidden-Markov model (HMM) for the process controlling the exit and entry of NBIs in the channel. An HMM filter is then used to detect the subchannels that are hit by interferers, and a suppression filter is placed in each such subchannel as it is hit. When an exit is detected in a subchannel, the suppression filter is removed from that subchannel. Related ideas for interference suppression based on HMMs and other "hidden data" models have been explored in [219, 238, 355, 375].

7.4 Code-Aided Techniques

In preceding sections we discussed the use of linear predictive interference suppression methods, which make use of the spectral properties of the spread data signal. We also discussed improvement in these predictive methods by making use of a more accurate model for the spread-spectrum signal. In the latter situation, we considered in particular the first-order probability distribution of the data signal (i.e., binary valued) which led to the ACM filter and its adaptive transversal form. Further improvements in NBI suppression can be made by going beyond random modeling at the chip level and taking advantage of the fact that we must know the spreading code of at least one user of interest in order to begin data demodulation. Techniques for taking advantage of this are termed *code-aided techniques*, a term coined in [388].

This approach was first proposed in [381, 426] and has been explored further in several works, including [388–390]. These works have been based primarily on detectors originally designed for linear multiuser detection. As noted in Section 2.2, in the context of multiuser detection, linear detectors operate by estimating the data sequence via linear model-fitting techniques and then quantizing the resulting estimates to get estimates of the data symbols themselves. Two of the principal such multiuser detection techniques are the zero-forcing detector, or decorrelator, and the linear MMSE detector. Recall that the decorrelator completely eliminates the multiple-access interference (MAI), with the attendant disadvantage of possibly enhancing the ambient noise, while the linear MMSE detector reduces the latter effect by minimizing the mean-square error between the linear estimate and the transmitted symbols. The linear MMSE detector has the further advantage of being more easily adapted than the decorrelator and it results in a lower bit-error rate under most practical circumstances [343, 386].

Although developed originally for the suppression of intersymbol interference and (later) MAI, these two methods can also be applied to the problem of suppressing NBI. This idea was first proposed in [381, 426] for the case in which the NBI signal is also a digital communications signal, but with a data rate much lower than the spread-spectrum chip rate. This digital NBI model finds applications, for example, in modeling the interference in multirate CDMA systems in which multiple spreading gains and multiple chip rates may be employed (e.g., [74, 333]). In [425], the decorrelator was employed to suppress the NBI in such cases, and comparison with even ideal predictive techniques showed signifcant performance gains from this method. The linear MMSE detector, in both fixed and adaptive forms, was proposed for suppression of digital NBI in [390], again resulting in significant performance gains over predictive techniques. The linear MMSE detector was explored further in [388, 389] for suppression of tonal and entropic narrowband interferers and for the joint suppression of NBI and MAI.

7.4.1 NBI Suppression via the Linear MMSE Detector

As before, we begin by considering the case of (7.1)–(7.2) in which there is only a single spread-spectrum signal in the channel (i.e., $K = 1$) in addition to the NBI signal and white Gaussian noise. We again adopt the discrete-time model (7.5), and (without loss of generality) restrict attention to the observations in a single symbol interval, say the zeroth one: $[0, T]$. It is convenient here to represent the corresponding samples in vector form:

$$\boldsymbol{r} = \left[r_0 \; r_1 \; \cdots \; r_{N-1}\right]^T = A b \boldsymbol{s} + \boldsymbol{i} + \boldsymbol{n}, \tag{7.67}$$

where for convenience we denote $A \equiv A_1$; $b \equiv b_1[0]$; $\boldsymbol{s}$ contains the (normalized) spreading code of user 1:

$$\boldsymbol{s} = \frac{1}{\sqrt{N}}\left[c_{0,1} \; c_{1,1} \; \cdots \; c_{N-1,1}\right]^T; \tag{7.68}$$

$\boldsymbol{i} = [i_0 \; i_1 \; \cdots \; i_{N-1}]^T$ is a vector containing the NBI samples; and $\boldsymbol{n} = [u_0 \; u_1 \; \cdots \; u_{N-1}]^T \sim \mathcal{N}(\mathbf{0}, \sigma^2 \boldsymbol{I}_N)$ is a vector containing the corresponding ambient noise samples. Denote by $\boldsymbol{R}_i$ the covariance matrix of $\boldsymbol{i}$ (i.e., $\boldsymbol{R}_i \triangleq E\left\{\boldsymbol{i}\,\boldsymbol{i}^T\right\}$). For simplicity, we assume for the remainder of this section that the sampled interference signal is wide-sense stationary with zero mean, although some of the results given do not require this.

In this framework the linear MMSE detector has the form

$$\hat{b} = \text{sign}\left(\boldsymbol{w}^T \boldsymbol{r}\right), \tag{7.69}$$

where $\boldsymbol{w} \in \mathbb{R}^N$ is a weight vector chosen to minimize the mean-square error

$$\text{MSE} \triangleq E\left\{\left(\boldsymbol{w}^T \boldsymbol{r} - b\right)^2\right\}. \tag{7.70}$$

As we have noted before, the motivation for this criterion is that we would like for the continuous estimator $\boldsymbol{w}^T \boldsymbol{r}$ of the symbol b to be as close to the symbol as possible in some sense before quantizing it. The MSE is a convenient and tractable measure of closeness for this purpose. Using (7.67), and the assumption that b, $\boldsymbol{i}$, and $\boldsymbol{n}$ are mutually independent, (7.70) can be written as

$$\text{MSE} = \boldsymbol{w}^T \left(\boldsymbol{R}_i + \sigma^2 \boldsymbol{I}_N + A^2 \boldsymbol{s}\boldsymbol{s}^T\right) \boldsymbol{w} - 2A\boldsymbol{w}^T \boldsymbol{s} + 1. \tag{7.71}$$

Taking the gradient of the MSE with respect to $\boldsymbol{w}$ and setting it to zero, we get

$$\left(\boldsymbol{R}_i + \sigma^2 \boldsymbol{I}_N + A^2 \boldsymbol{s}\boldsymbol{s}^T\right) \boldsymbol{w} - A\boldsymbol{s} = \boldsymbol{0}. \tag{7.72}$$

Solving for $\boldsymbol{w}$ in (7.72), and using the matrix inversion lemma, we obtain the minimizing weights as

$$\boldsymbol{w} = \underbrace{\frac{A}{1 + A^2 \boldsymbol{s}^T \left(\boldsymbol{R}_i + \sigma^2 \boldsymbol{I}_N\right)^{-1} \boldsymbol{s}}}_{\alpha} \left(\boldsymbol{R}_i + \sigma^2 \boldsymbol{I}_N\right)^{-1} \boldsymbol{s}. \tag{7.73}$$

A useful figure of merit for assessing the NBI suppression capability of the linear detector with weights $\boldsymbol{w}$ is the output signal-to-interference-plus-noise ratio, which is defined in this situation as

$$\text{SINR} \triangleq \frac{E\left\{E\left\{\boldsymbol{w}^T \boldsymbol{r} \mid b\right\}^2\right\}}{E\left\{\text{Var}\left\{\boldsymbol{w}^T \boldsymbol{r} \mid b\right\}\right\}}. \tag{7.74}$$

Using (7.67), (7.73), and the assumption that b, $\boldsymbol{i}$, and $\boldsymbol{n}$ are independent and zero mean, we have

$$\begin{aligned} E\left\{\boldsymbol{w}^T \boldsymbol{r} \mid b\right\} &= Ab\boldsymbol{w}^T \boldsymbol{s}, \\ &= \alpha A b \boldsymbol{s}^T \left(\boldsymbol{R}_i + \sigma^2 \boldsymbol{I}_N\right)^{-1} \boldsymbol{s}, \end{aligned} \tag{7.75}$$

$$\begin{aligned} \text{Var}\left\{\boldsymbol{w}^T \boldsymbol{r} \mid b\right\} &= \text{Var}\left\{\boldsymbol{w}^T \boldsymbol{i}\right\} + \text{Var}\left\{\boldsymbol{w}^T \boldsymbol{n}\right\} \\ &= \boldsymbol{w}^T \left(\boldsymbol{R}_i + \sigma^2 \boldsymbol{I}_N\right) \boldsymbol{w} \\ &= \alpha^2 \boldsymbol{s}^T \left(\boldsymbol{R}_i + \sigma^2 \boldsymbol{I}_N\right)^{-1} \boldsymbol{s}. \end{aligned} \tag{7.76}$$

Substituting (7.75) and (7.76) into (7.74), the output SINR for the linear MMSE detector is then given by

$$\text{SINR} = A^2 \left[\boldsymbol{s}^T \left(\boldsymbol{R}_i + \sigma^2 \boldsymbol{I}_N \right)^{-1} \boldsymbol{s} \right]. \tag{7.77}$$

As noted previously, the NBI signal can be modeled in one of three basic ways: a tonal signal, an entropic narrowband stochastic process, or a digital data signal with data rate much lower than the spread-spectrum chip rate. We next analyze the performance of the linear MMSE detector against each of these three types of narrowband interference.

7.4.2 Tonal Interference

For mathematical convenience, we assume that the narrowband interference signal consists of m complex sinusoids of the form

$$i_n = \sum_{l=1}^{m} \sqrt{P_l}\, e^{\jmath(2\pi f_l n + \phi_l)}, \tag{7.78}$$

where P_l and f_l are the power and normalized frequency of the lth sinusoid, and the $\{\phi_l\}$ are independent random phases distributed uniformly on $(0, 2\pi)$. The covariance matrix $\boldsymbol{R}_i$ of the multitone interference signal $\boldsymbol{i}$ can be represented as

$$\boldsymbol{R}_i = \sum_{l=1}^{m} P_l \boldsymbol{g}_l\, \boldsymbol{g}_l^H, \tag{7.79}$$

where

$$\boldsymbol{g}_l \triangleq \left[1, e^{\jmath 2\pi f_l}, e^{\jmath 4\pi f_l}, \ldots, e^{\jmath 2\pi (N-1) f_l} \right]^T. \tag{7.80}$$

Denote $\boldsymbol{R}_m \triangleq \sum_{l=1}^{m} P_l \boldsymbol{g}_l \boldsymbol{g}_l^H + \sigma^2 \boldsymbol{I}_N$, and $\boldsymbol{K}_m \triangleq \boldsymbol{R}_m^{-1}$. Then $\boldsymbol{R}_m = \boldsymbol{R}_{m-1} + P_m \boldsymbol{g}_m \boldsymbol{g}_m^H$, and hence we have

$$\boldsymbol{K}_m = \boldsymbol{K}_{m-1} - \frac{\boldsymbol{K}_{m-1} \boldsymbol{g}_m \boldsymbol{g}_m^H \boldsymbol{K}_{m-1}}{P_m^{-1} + \boldsymbol{g}_m^H \boldsymbol{K}_{m-1} \boldsymbol{g}_m}, \tag{7.81}$$

where $\boldsymbol{K}_{m-1} \triangleq \boldsymbol{R}_{m-1}^{-1}$. According to (7.77), let $\text{SINR}_m \triangleq A^2 \left(\boldsymbol{s}^T \boldsymbol{K}_m \boldsymbol{s} \right)$. Then from (7.81) we can write

$$\text{SINR}_m = \text{SINR}_{m-1} - \frac{A^2 \left| \boldsymbol{s}^T \boldsymbol{K}_{m-1} \boldsymbol{g}_m \right|^2}{P_m^{-1} + \boldsymbol{g}_m^H \boldsymbol{K}_{m-1} \boldsymbol{g}_m}. \tag{7.82}$$

Assuming that the spread-spectrum user has a random signature sequence, we next derive expressions for the expected values of the output SINR with respect to the random signature vector $\boldsymbol{s}$, for several special cases.

Case 1: $m = 1$. We have $\text{SINR}_0 = A^2\sigma^{-2}$, $\boldsymbol{K}_0 = \sigma^{-2}\boldsymbol{I}_N$, $\boldsymbol{g}_1^H \boldsymbol{K}_0 \boldsymbol{g}_1 = N\sigma^{-2}$, and

$$E\left\{\left|\boldsymbol{s}\boldsymbol{K}_0\boldsymbol{g}_1\right|^2\right\} = \sigma_n^{-4} E\left\{\left|\boldsymbol{s}^T\boldsymbol{g}_1\right|^2\right\} = \sigma_n^{-4}\,\text{tr}\left(E\left\{\boldsymbol{s}\boldsymbol{s}^T\right\}\boldsymbol{g}_1\boldsymbol{g}_1^H\right)$$
$$= \frac{\sigma_n^{-4}}{N}\boldsymbol{g}_1^H\boldsymbol{g}_1 = \sigma_n^{-4}, \tag{7.83}$$

where we have used $E\left\{\boldsymbol{s}\boldsymbol{s}^T\right\} = \frac{1}{N}\boldsymbol{I}_N$ and $\boldsymbol{g}_1^H\boldsymbol{g}_1 = N$. Substituting these into (7.82), we obtain

$$E\left\{\text{SINR}_1\right\} = \left[1 - \frac{(P_1/\sigma^2)}{1 + N\left(P_1/\sigma^2\right)}\right]\frac{A^2}{\sigma^2}$$
$$\to \left(1 - \frac{1}{N}\right)\frac{A^2}{\sigma^2}, \qquad \text{as } P_1 \to \infty. \tag{7.84}$$

Therefore, when N is large, the energy of a strong interferer is almost completely suppressed by the linear MMSE detector.

Case 2: $m = 2$. From (7.81) we have

$$\boldsymbol{K}_1 = \sigma^{-2}\left[\boldsymbol{I}_N - \frac{\gamma_1}{1 + N\gamma_1}\boldsymbol{g}_1\boldsymbol{g}_1^H\right], \tag{7.85}$$

where $\gamma_1 \triangleq A^2\sigma^{-2}$. Now using (7.82), we obtain

$$E\left\{\text{SINR}_2\right\} = E\left\{\text{SINR}_1\right\} - \frac{P_2 E\left\{\left|\boldsymbol{s}^T\boldsymbol{K}_1\boldsymbol{g}_2\right|^2\right\}}{1 + P_2\left(\boldsymbol{g}_2^H\boldsymbol{K}_1\boldsymbol{g}_2\right)}A^2. \tag{7.86}$$

Using (7.85), we have

$$A^2\left(\boldsymbol{g}_2^H\boldsymbol{K}_1\boldsymbol{g}_2\right) = \gamma_2\left(N - \frac{\gamma_1}{1 + N\gamma_1}\beta_{12}\right), \tag{7.87}$$

where $\gamma_2 \triangleq P_2\sigma^{-2}$, and

$$\beta_{12} \triangleq \left|\boldsymbol{g}_2^H\boldsymbol{g}_1\right|^2 = \left(\sum_{k=0}^{N-1} e^{\jmath 2\pi\Delta f k}\right)\left(\sum_{k=0}^{N-1} e^{-\jmath 2\pi\Delta f k}\right)$$
$$= \frac{\left(1 - e^{\jmath 2\pi\Delta f N}\right)\left(1 - e^{-\jmath 2\pi\Delta f N}\right)}{\left(1 - e^{\jmath 2\pi\Delta f}\right)\left(1 - e^{-\jmath 2\pi\Delta f}\right)} = \left(\frac{\sin\pi\Delta f N}{\sin\pi\Delta f}\right)^2, \tag{7.88}$$

where $\Delta f \triangleq f_1 - f_2$. On the other hand, using (7.85), we can write

$$E\left\{\left|\boldsymbol{s}^T\boldsymbol{K}_1\boldsymbol{g}_2\right|^2\right\} = \text{tr}\left(E\left\{\boldsymbol{s}\boldsymbol{s}^T\right\}\boldsymbol{K}_1\boldsymbol{g}_2\boldsymbol{g}_2^H\boldsymbol{K}_1\right) = \frac{1}{N}\boldsymbol{g}_2^H\boldsymbol{K}_1\boldsymbol{K}_1\boldsymbol{g}_2$$
$$= \sigma^{-4}\left[1 - \frac{2}{N}\left(\frac{\gamma_1}{1 + N\gamma_1}\right)\beta_{12} + \left(\frac{\gamma_1}{1 + N\gamma_1}\right)^2\beta_{12}\right]. \tag{7.89}$$

Substituting (4.55) and (7.89) into (7.86), we then have

$$E\{\text{SINR}_2\} = \left\{ 1 - \frac{\gamma_1}{1+N\gamma_1} - \frac{\gamma_2\left[1 - \frac{2}{N}\left(\frac{\gamma_1}{1+N\gamma_1}\right)\beta_{12} + \left(\frac{\gamma_1}{1+N\gamma_1}\right)^2\beta_{12}\right]}{1+\gamma_2\left[N - \left(\frac{\gamma_1}{1+N\gamma_1}\right)\beta_{12}\right]} \right\}\frac{A^2}{\sigma^2}$$

$$\rightarrow \left(1 - \frac{2}{N}\right)\frac{A^2}{\sigma^2}, \qquad \text{as } \gamma_1 \rightarrow \infty \text{ and } \gamma_2 \rightarrow \infty. \tag{7.90}$$

Again we see that for large N, the interfering energy is almost completely suppressed. In general, it is difficult to obtain an explicit expression for $E\{\text{SINR}_m\}$ for $m > 2$. However, for the special case when the $\{\boldsymbol{g}_l\}$ are mutually orthogonal, a closed-form expression for $E\{\text{SINR}_m\}$ can easily be found.

Case 3: Orthogonal $\{\boldsymbol{g}_l\}$. Assume that

$$\boldsymbol{g}_l^H \boldsymbol{g}_k = \begin{cases} N, & l = k, \\ 0, & l \neq k. \end{cases} \tag{7.91}$$

This condition is met, for example, when $f_l - f_k$ is a multiple of $1/N$ for all $l \neq k$. Under this condition of orthogonality, it follows straightforwardly that

$$\boldsymbol{K}_m = \sigma^{-2}\left(\boldsymbol{I}_N - \sum_{l=1}^{m} \frac{P_l}{\sigma^2 + NP_l}\boldsymbol{g}_l\boldsymbol{g}_l^H\right). \tag{7.92}$$

The expected value of the output SINR with respect to the random signature vector $\boldsymbol{s}$ is

$$\begin{aligned} E\{\text{SINR}_m\} &= A^2 E\left\{\boldsymbol{s}^T\boldsymbol{K}_m\boldsymbol{s}\right\} \\ &= \left[1 - \sum_{l=1}^{m}\frac{\gamma_l}{1+N\gamma_l}E\left\{\left|\boldsymbol{s}^T\boldsymbol{g}_l\right|^2\right\}\right]\frac{A^2}{\sigma^2} \\ &= \left[1 - \sum_{l=1}^{m}\frac{\gamma_l}{1+N\gamma_l}\right]\frac{A^2}{\sigma^2} \\ &\geq \frac{1+(N-m)\gamma_{\max}}{1+N\gamma_{\max}}\frac{A^2}{\sigma^2} \\ &\rightarrow \left(\frac{N-m}{N}\right)\frac{A^2}{\sigma^2}, \qquad \text{as } \gamma_{\max} \rightarrow \infty, \end{aligned} \tag{7.93}$$

where $\gamma_l \triangleq P_l\sigma^{-2}$, and $\gamma_{\max} \triangleq \max_{1\leq l\leq m}\{\gamma_l\}$. Figure 7.10 shows some numerical examples of tone suppression by the linear MMSE detector. In both plots the SINRs without interference are 10dB (after despreading). Each curve in the plots corresponds to one set of three-tone (or seven-tone) frequencies $\{f_l\}$ randomly chosen. (The vectors $\{\boldsymbol{g}_l\}$ are not necessarily orthogonal.) Interestingly, it is seen from Fig. 7.10 that the output SINRs are centered at the value given by (7.93).

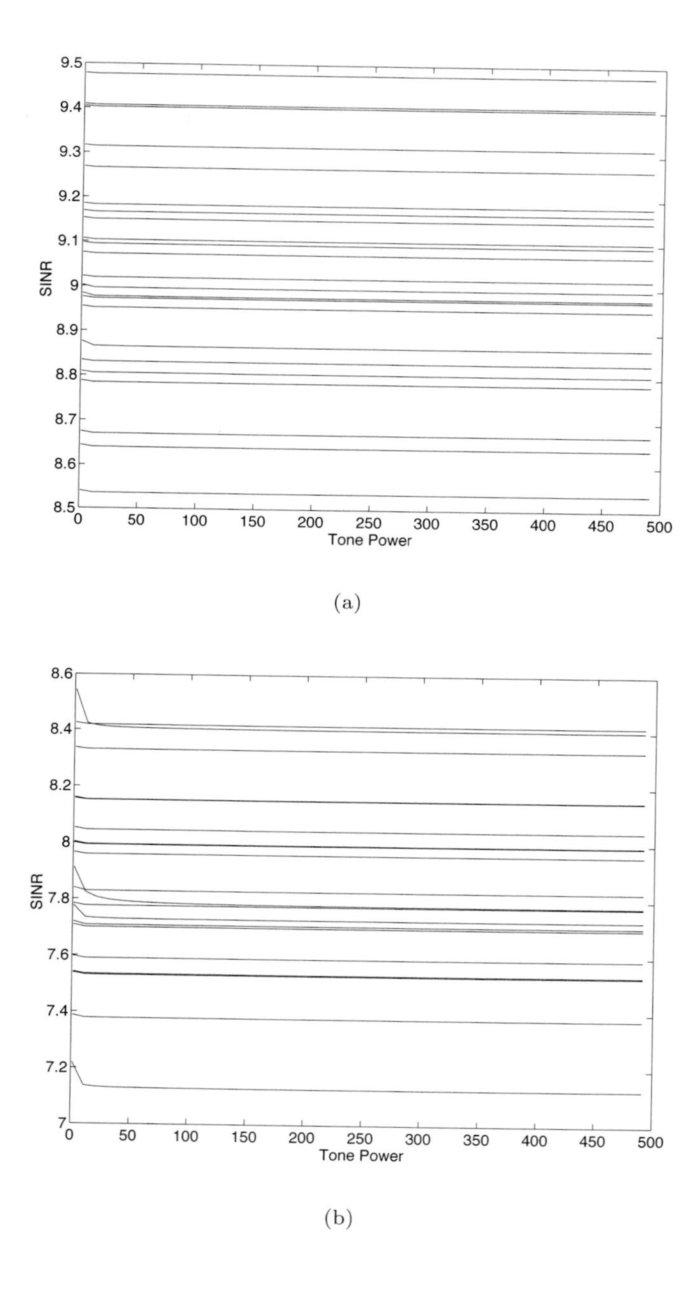

Figure 7.10. Numerical examples of multitone interference suppression in a CDMA system by the linear MMSE detector. The parameters are $N = 31$, $A^2 = 1$, and $\sigma^2 = 0.1$. The interfering tones have the same power P_l ranging from 1 to 500. Each curve in the plots corresponds to one set of (a) three-tone or (b) seven-tone frequencies $\{f_l\}$ randomly chosen. The SINRs are calculated by using (7.77).

7.4.3 Autoregressive Interference

Let us assume that the NBI signal is modeled as a pth order AR process, where $p \ll N$:

$$i_n = -\sum_{j=1}^{p} \phi_j i_{n-j} + e_n, \tag{7.94}$$

where $\{e_n\}$ is an i.i.d. Gaussian sequence with variance ν^2. Supposing that $\boldsymbol{R}_i$ is positive definite, we first derive a closed-form expression for $\boldsymbol{R}_i^{-1}$. Using (7.94), we can write the following:

$$\begin{bmatrix} 1 & \phi_1 & \phi_2 & \cdots & \phi_p & & & & \\ & 1 & \phi_1 & \phi_2 & \cdots & \phi_p & & & \\ & & \ddots & \ddots & & \ddots & & & \\ & & & 1 & \phi_1 & \phi_2 & \cdots & \phi_p \\ & & & & 1 & & & \\ & & & & & 1 & & \\ & & & & & & \ddots & \\ & & & & & & & 1 \end{bmatrix} \begin{bmatrix} i_n \\ i_{n-1} \\ \vdots \\ i_{n-N+p+1} \\ i_{n-N+p} \\ i_{n-N+p-1} \\ \vdots \\ i_{n-N+1} \end{bmatrix} = \begin{bmatrix} e_n \\ e_{n-1} \\ \vdots \\ e_{n-N+p+1} \\ i_{n-N+p} \\ i_{n-N+p-1} \\ \vdots \\ i_{n-N+1} \end{bmatrix}, \tag{7.95}$$

or, in compact form,

$$\boldsymbol{A} \begin{bmatrix} \boldsymbol{i}_{N-p} \\ \boldsymbol{i}_p \end{bmatrix} = \begin{bmatrix} \boldsymbol{e}_{N-p} \\ \boldsymbol{i}_p \end{bmatrix}, \tag{7.96}$$

where $\boldsymbol{A}$ is the matrix appearing on the left-hand side of (7.95), $\boldsymbol{i}_{N-p} = [i_n, i_{n-1}, \ldots, i_{n-N+p+1}]^T$, $\boldsymbol{i}_p = [i_{n-N+p}, i_{n-N+p-1}, \ldots, i_{n-N+1}]^T$, and $\boldsymbol{e}_{N-p} = [e_n, e_{n-1}, \ldots, e_{n-N+p+1}]^T$. Multiplying both sides of (7.96) by their transposes and taking expectations, we obtain

$$\boldsymbol{A}\, E\left\{ \begin{bmatrix} \boldsymbol{i}_{N-p} \\ \boldsymbol{i}_p \end{bmatrix} \begin{bmatrix} \boldsymbol{i}_{N-p}^T & \boldsymbol{i}_p^T \end{bmatrix} \right\} \boldsymbol{A}^T = E\left\{ \begin{bmatrix} \boldsymbol{e}_{N-p} \\ \boldsymbol{i}_p \end{bmatrix} \begin{bmatrix} \boldsymbol{e}_{N-p}^T & \boldsymbol{i}_p^T \end{bmatrix} \right\}; \tag{7.97}$$

that is,

$$\boldsymbol{A}\boldsymbol{R}_i^{(N)}\boldsymbol{A}^T = \begin{bmatrix} \nu^2 \boldsymbol{I}_{N-p} & \boldsymbol{0} \\ \boldsymbol{0} & \boldsymbol{R}_i^{(p)} \end{bmatrix}, \tag{7.98}$$

where $\boldsymbol{R}_i^{(N)}$ and $\boldsymbol{R}_i^{(p)}$ are, respectively, the $N \times N$ and $p \times p$ autocorrelation matrices of the interference signal. Since $\boldsymbol{A}$ is nonsingular, then

$$\boldsymbol{R}_i^{(N)} = \boldsymbol{A}^{-1} \begin{bmatrix} \nu^2 \boldsymbol{I}_{N-p} & \boldsymbol{0} \\ \boldsymbol{0} & \boldsymbol{R}_i^{(p)} \end{bmatrix} \boldsymbol{A}^{-T}, \tag{7.99}$$

$$\boldsymbol{R}_i^{(N)^{-1}} = \boldsymbol{A}^T \begin{bmatrix} \frac{1}{\nu^2} \boldsymbol{I}_{N-p} & \boldsymbol{0} \\ \boldsymbol{0} & \boldsymbol{R}_i^{(p)^{-1}} \end{bmatrix} \boldsymbol{A}. \tag{7.100}$$

Partition the $N \times N$ matrix $\boldsymbol{A}$ into the following four blocks:

$$\boldsymbol{A} = \begin{bmatrix} \boldsymbol{A}_{11} & \boldsymbol{A}_{12} \\ \boldsymbol{0} & \boldsymbol{I}_p \end{bmatrix}, \tag{7.101}$$

where $\boldsymbol{A}_{11}$ is of dimension $(N-p) \times (N-p)$, and $\boldsymbol{A}_{12}$ is of dimension $(N-p) \times p$. Substituting (7.101) into (7.100), we can write

$$\boldsymbol{R}_i^{(N)^{-1}} = \frac{1}{\nu^2} \begin{bmatrix} \boldsymbol{A}_{11}^T \boldsymbol{A}_{11} & \boldsymbol{A}_{11}^T \boldsymbol{A}_{12} \\ \boldsymbol{A}_{12}^T \boldsymbol{A}_{11} & \boldsymbol{A}_{12}^T \boldsymbol{A}_{12} + \nu^2 \boldsymbol{R}_p^{-1} \end{bmatrix}. \tag{7.102}$$

Now most of the elements of $\boldsymbol{R}_i^{(N)^{-1}}$ are explicitly given by (7.102), except for the southeast $p \times p$ block. But notice that $\boldsymbol{R}_i^{(N)}$ is a Toeplitz matrix, and the inverse of a nonsingular Toeplitz matrix is *persymmetric* (i.e., it is symmetric about its northeast-southwest diagonal) [158]. Therefore, the elements of the southeast $p \times p$ block of $\boldsymbol{R}_i^{(N)^{-1}}$ can be found in the northwest $p \times p$ block, which have already been determined. Hence, with the aid of persymmetry, $\boldsymbol{R}_i^{(N)^{-1}}$ is completely specified by (7.102). Straightforward calculation of (7.102) then shows that $\boldsymbol{R}_i^{(N)^{-1}}$ is a bandlimited matrix, with bandwidth $2p+1$. Since it is symmetric, we need only to specify the upper $p+1$ nonzero diagonals, as follows:

$$\begin{aligned}
D_0 &= \frac{1}{\nu^2} \mathrm{diag}\Big\{1,\ 1+\phi_1^2,\ \ldots,\ 1+\phi_1^2+\cdots+\phi_p^2,\ \ldots,\ 1+\phi_1^2+\cdots+\phi_p^2, \\
&\qquad \ldots,\ 1+\phi_1^2,\ 1\Big\}, \\
D_1 &= \frac{1}{\nu^2} \mathrm{diag}\Big\{\phi_1,\ \phi_1+\phi_1\phi_2,\ \ldots,\ \phi_1+\phi_1\phi_2+\cdots+\phi_{p-1}\phi_p, \\
&\qquad \ldots,\ \phi_1+\phi_1\phi_2+\cdots+\phi_{p-1}\phi_p,\ \ldots,\ \phi_1+\phi_1\phi_2,\ \phi_1\Big\}, \\
D_2 &= \frac{1}{\nu^2} \mathrm{diag}\Big\{\phi_2,\ \phi_2+\phi_1\phi_3,\ \ldots,\ \phi_2+\phi_1\phi_3+\cdots+\phi_{p-2}\phi_p, \\
&\qquad \ldots,\ \phi_2+\phi_1\phi_3+\cdots+\phi_{p-2}\phi_p,\ \ldots,\ \phi_2+\phi_1\phi_3,\ \phi_2\Big\}, \\
&\ \vdots \quad \vdots \\
D_p &= \frac{1}{\nu^2} \mathrm{diag}\Big\{\phi_p,\ \ldots,\ \phi_p\Big\},
\end{aligned} \tag{7.103}$$

where D_k contains the $(N-k)$ elements on the kth upper (lower) diagonal of $\boldsymbol{R}_i^{(N)^{-1}}$, $k = 0, 1, \ldots, p$.

Next we consider the output SINR of the linear MMSE detector when the interferer is an AR signal. For the sake of analytical tractability, and to stress the effectiveness of the MMSE detector against the narrowband AR interference (versus the background noise), we consider the output SINR when there is no background noise (i.e., $\sigma^2 \to 0$). Using (7.103), we have

$$\begin{aligned} \boldsymbol{s}^T \boldsymbol{R}_i^{(N)^{-1}} \boldsymbol{s} &= \frac{1}{N} \sum_{i=0}^{N-1} D_0[i] + 2 \sum_{k=1}^{p} \sum_{i=0}^{N-1-k} D_k[i] s[i] s[i+k] \\ &\cong D_0[\lceil N/2 \rceil] + 2 \sum_{k=1}^{p} D_k[\lceil N/2 \rceil] \sum_{i=0}^{N-1-k} s[i] s[i+k] \qquad (7.104) \\ &\cong \frac{1 + \phi_1^2 + \cdots + \phi_p^2}{\nu^2}, \qquad (7.105) \end{aligned}$$

where in (7.104), we have made the approximation that $D_k[i] = D_k[\lceil N/2 \rceil]$, $0 \le k \le p$, $0 \le i \le N-k-1$, since when $N \gg p$, it is seen from (7.103) that on each nonzero diagonal most of the elements are the same; and in (7.105) we used the approximation $\sum_{n=0}^{N-1-k} c_{n,1} c_{n+k,1} \cong \pm 1/N$ and thus dropped the second term in (7.104). The output SINR is then

$$\begin{aligned} \text{SINR} &= A^2 \left(\boldsymbol{s}^T \boldsymbol{R}_i^{(N)^{-1}} \boldsymbol{s} \right) \\ &\cong \left(1 + \phi_1^2 + \cdots + \phi_p^2 \right) \frac{A^2}{\nu^2}. \qquad (7.106) \end{aligned}$$

As will be seen in Section 7.5, this SINR value is the same as an SINR *upper bound* given by the nonlinear interpolator NBI suppression method in the absence of background noise.

7.4.4 Digital Interference

Now let us consider a system with one spread-spectrum (SS) signal and one narrowband binary signal in an otherwise additive white Gaussian noise (AWGN) channel. We assume for now that the narrowband signal is synchronized with the SS signal. Furthermore, we assume a relationship between the data rates of the two users (i.e., m bits of the narrowband user occur for each bit of the SS user). (Given the typical data rates employed in many wireless systems, it is often reasonable to assume an integer relationship between the bit rates. Situations in which this is not true are considered in [57].) As shown in Fig. 7.11, the narrowband digital signal can be regarded as m virtual users, each with its virtual signature sequence. The first virtual user's signature sequence equals 1 during the first narrowband user's bit interval (i.e., a virtual chip interval) and zero everywhere else. Similarly, each other narrowband user's bit can be thought of as a signal arising from a virtual user with a signature sequence with only one nonzero entry. It is obvious from this construction that the signature waveforms of the virtual users are orthogonal to each other. However, in general, the kth virtual user has some cross-correlation with the spread-spectrum user. If we use $\boldsymbol{\rho}$ to denote the vectors formed by the cross-correlations, defined explicitly in (7.108), the cross-correlation matrix $\boldsymbol{R}$ of

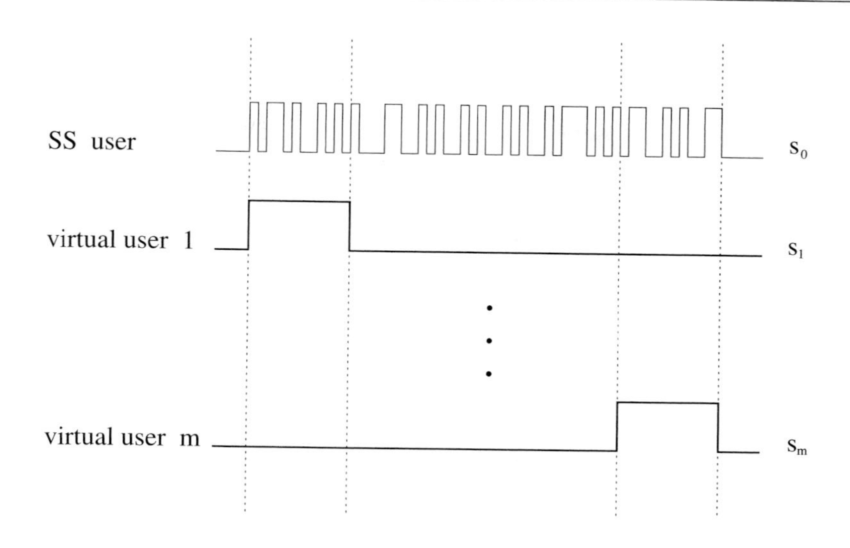

Figure 7.11. Virtual CDMA system: synchronous case.

this virtual multiuser system has the following simple structure (note that the SS user is numbered 0, and the m virtual users are numbered from 1 to m):

$$\boldsymbol{R} = \begin{bmatrix} 1 & \boldsymbol{\rho}^T \\ \boldsymbol{\rho} & \boldsymbol{I}_m \end{bmatrix}. \tag{7.107}$$

We have assumed that the narrowband user had a faster data rate than the SS user (but this rate is still much slower than the chip rate). The opposite case can also hold, and our analysis applies to it as well, although we do not discuss that case explicitly. The covariance matrix of the system in that case has the same structure as (7.107).

Let T be the bit duration of the SS user, so that T/m is the bit duration of the narrowband user. Similarly, let N be the processing gain of the SS signal, so that the chip interval has length T/N. By our assumption that the interferer is narrowband, we have $N \gg m$. Let $s(t)$ be the normalized signature waveform of the SS user [i.e., $s(t)$ is zero outside the interval $[0, T]$ and has unity energy]. Similarly, let $p(t)$ be the normalized bit waveform of the narrowband user [i.e., $p(t)$ is zero outside the interval $[0, T/m]$ and has unity energy]. Then the normalized signature waveform of the kth virtual user is $p_k(t) = p(t-(k-1)T/m)$. The cross-correlation vector mentioned earlier is $\boldsymbol{\rho} = [\rho_1 \; \rho_2 \; \cdots \; \rho_m]^T$, where ρ_k is the cross-correlation between the kth virtual user and the SS user, defined as

$$\rho_k = \langle s, p_k \rangle. \tag{7.108}$$

where the inner product notation denotes $\langle x, y \rangle = \int_0^T x(t)y(t)\,\mathrm{d}t$.

We assume that the SS user and the narrowband user are sending digital data through the same channel characterized by AWGN with power spectral density σ^2. Let A_I be the received amplitude of the narrowband signal and A be the received amplitude of the SS signal. We use the notation that the narrowband user data bits during the interval $(0, T)$ are $d_1, d_2, \ldots, d_m$, and the SS bit is b. When the users are synchronous, it is sufficient to consider the one-shot version of the received signal:

$$r(t) = A\, b\, s(t) + A_I \sum_{k=1}^{m} d_k p_k(t) + n(t), \quad t \in [0, T]. \tag{7.109}$$

where $n(t)$ is the white Gaussian noise with power spectral density σ^2.

The linear MMSE detector for user 0 (i.e., the SS user) is characterized by the impulse response $w \in L_2[0, T]$, such that the decision on b_0 is

$$\hat{b} = \mathrm{sign}(\langle w, r \rangle). \tag{7.110}$$

A closed-form expression for w is given by [520] as

$$w(t) = w_0 s(t) + \sum_{j=1}^{m} w_j\, p_j(t), \tag{7.111}$$

where $\boldsymbol{w}^T = [w_0,\ w_1,\ \ldots,\ w_m]$ is the first row of the matrix

$$\boldsymbol{C} = \left[\boldsymbol{R} + \sigma^2 \boldsymbol{A}^{-2}\right]^{-1} \tag{7.112}$$

and $\boldsymbol{A} = \mathrm{diag}\{A,\ A_I,\ \ldots,\ A_I\}$. Substituting (7.107) into (7.112), we have

$$\boldsymbol{C} = \begin{bmatrix} 1 + \dfrac{\sigma^2}{A^2} & \boldsymbol{\rho}^T \\ \boldsymbol{\rho} & \left(1 + \dfrac{\sigma^2}{A_I^2}\right) \boldsymbol{I}_m \end{bmatrix}^{-1}. \tag{7.113}$$

The following matrix identity can be easily verified,

$$\begin{bmatrix} \alpha & \boldsymbol{\rho}^T \\ \boldsymbol{\rho} & \beta \boldsymbol{I}_m \end{bmatrix}^{-1} = \begin{bmatrix} \dfrac{1}{\alpha\gamma} & -\dfrac{1}{\alpha\beta\gamma}\boldsymbol{\rho}^T \\ -\dfrac{1}{\alpha\beta\gamma}\boldsymbol{\rho} & \dfrac{1}{\beta}\boldsymbol{I}_m + \dfrac{1}{\alpha\beta^2\gamma}\boldsymbol{\rho}\boldsymbol{\rho}^T \end{bmatrix}, \tag{7.114}$$

where $\gamma = 1 - \boldsymbol{\rho}^T \boldsymbol{\rho} / \alpha\beta$.

Now on defining $\alpha = 1 + \sigma^2/A^2$ and $\beta = 1 + \sigma^2/A_I^2$, the first row of $\boldsymbol{C}$ in (7.113) is then given by

$$\boldsymbol{w}^T = \left[\left(1 + \frac{\sigma^2}{A^2}\right)\left(1 + \frac{\sigma^2}{A_I^2}\right) - \boldsymbol{\rho}^T \boldsymbol{\rho}\right]^{-1} \cdot \left[1 + \frac{\sigma^2}{A_I^2},\ -\rho_1,\ -\rho_2,\ \ldots,\ -\rho_m\right]. \tag{7.115}$$

Substituting (7.115) into (7.111) we get an expression for the linear MMSE detector for the SS user:

$$w(t) = \left[\left(1+\frac{\sigma^2}{A^2}\right)\left(1+\frac{\sigma^2}{A_I^2}\right) - \boldsymbol{\rho}^T\boldsymbol{\rho}\right]^{-1}\left[\left(1+\frac{\sigma^2}{A_I^2}\right)s(t) - \sum_{k=1}^{m}\rho_k p_k(t)\right]. \quad (7.116)$$

Using (7.74), the SINR at the output of the linear MMSE detector $w(t)$ becomes

$$\text{SINR} = \frac{A^2\langle w, s\rangle^2}{A_I^2\sum_{j=1}^{m}\langle w, p_j\rangle^2 + \sigma^2\langle w, w\rangle}. \quad (7.117)$$

That is, the SINR is the ratio of the desired SS signal power to the sum of the powers due to narrowband interference and noise at the output of the filter $w(t)$. Substituting (7.116) into (7.117), we obtain

$$\begin{aligned}\text{SINR} &= \frac{A^2\left(1+\frac{\sigma^2}{A_I^2} - \boldsymbol{\rho}^T\boldsymbol{\rho}\right)^2}{A_I^2\left(\frac{\sigma^2}{A_I^2}\right)^2\boldsymbol{\rho}^T\boldsymbol{\rho} + \sigma^2\left[\left(1+\frac{\sigma^2}{A_I^2}\right)^2 - 2\left(1+\frac{\sigma^2}{A_I^2}\right)\boldsymbol{\rho}^T\boldsymbol{\rho} + \boldsymbol{\rho}^T\boldsymbol{\rho}\right]} \\ &= \frac{A^2}{\sigma^2}\left(1 - \frac{\boldsymbol{\rho}^T\boldsymbol{\rho}}{1+\frac{\sigma^2}{A_I^2}}\right). \end{aligned} \quad (7.118)$$

Figure 7.12 illustrates the virtual multiuser system for the asynchronous case. Let t_0 be the fixed time lag between the spread-spectrum bit and the nearest previous start of a narrowband bit (i.e., $0 \le t_0 \le T/m$). We see that because of the time lag t_0, the virtual user 1 in Fig. 7.11 effectively contributes two interference signals during an SS bit interval: at the beginning and end of the SS bit interval, respectively. We can therefore treat the asynchronous system as a synchronous system with one additional virtual user (i.e., a synchronous system with one SS user) and $m+1$ virtual users. The preceding analysis therefore holds in the asynchronous case as well, with only minor modification.

7.5 Performance Comparisons of NBI Suppression Techniques

In Section 7.4, we have derived closed-form expressions for the performance measure (SINR) of the linear MMSE detector against three types of NBI. In this section we compare its performance against NBI with performance bounds for the linear and nonlinear NBI suppression methods discussed in Sections 7.2 and 7.3.

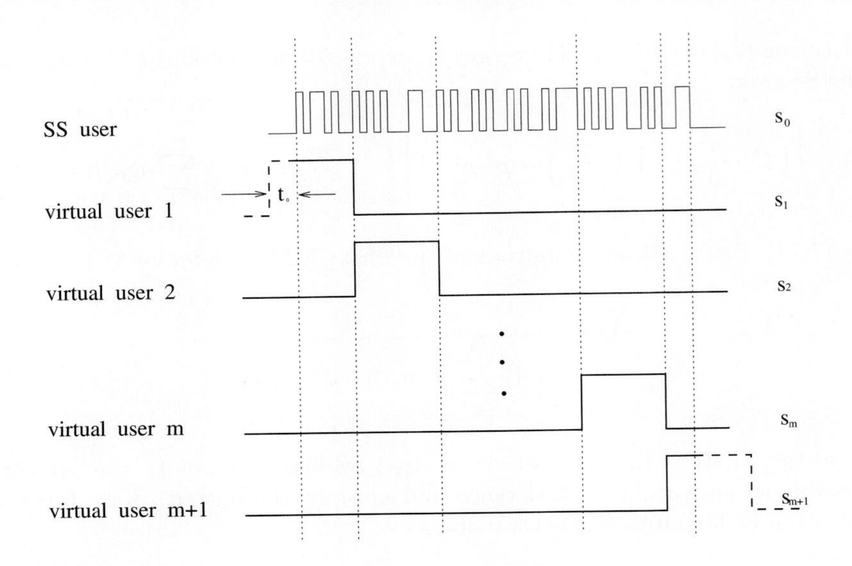

Figure 7.12. Virtual CDMA system: asynchronous case.

7.5.1 Matched Filter

For the conventional detector, the received signal $\boldsymbol{r}$ in (7.67) is sent directly to a single filter matched to the spreading sequence (i.e., $\boldsymbol{w} = \boldsymbol{s}$). The mean and variance at the output of the matched filter are

$$E\left\{\boldsymbol{s}^T\boldsymbol{r} \mid b\right\} = A\,b, \tag{7.119}$$

$$\mathrm{Var}\left\{\boldsymbol{s}^T\,\boldsymbol{r} \mid b\right\} = \boldsymbol{s}^T\left(\boldsymbol{R}_i + \sigma^2\boldsymbol{I}_N\right)\boldsymbol{s}. \tag{7.120}$$

The output SINR is then given by

$$\mathrm{SINR(matched\ filter)} = \frac{A^2}{\boldsymbol{s}^T\left(\boldsymbol{R}_i + \sigma^2\boldsymbol{I}_N\right)\boldsymbol{s}}. \tag{7.121}$$

7.5.2 Linear Predictor and Interpolator

As mentioned before, linear or nonlinear predictive NBI suppression methods are based on the following idea. Since the spread-spectrum signal has a nearly flat spectrum, it cannot be predicted accurately from its past values without explicit use of knowledge of the spreading code. On the other hand, the interfering signal, being narrowband, can be predicted accurately. These methods essentially form a replica of the NBI, which can be subtracted from the received signal to enhance the wideband components. The linear methods have involved primarily the use of linear transversal prediction or interpolation filters to create the NBI replica. Such a filter forms a linear prediction of the received signal based on a fixed number of previous samples, or a linear interpolation based on a fixed number of past and future samples. This estimate is subtracted from the appropriately timed received signal to obtain the error signal to be used as input to the SS user signature sequence correlator.

Let $S_i(\omega)$ denote the power spectral density of the NBI signal. The following output SINR *upper bounds* for the linear prediction/interpolation methods can be found in [310, 311]:

$$\text{SINR(linear predictor)} \leq \frac{A^2}{2\pi \exp\left[\frac{1}{2\pi}\int_{-\pi}^{\pi} \ln\left(\frac{A^2/N+\sigma^2}{2\pi} + S_i(\omega)\right) \mathrm{d}\omega\right] - A^2/N}, \tag{7.122}$$

$$\text{SINR(linear interpolator)} \leq \frac{A^2 \int_{-\pi}^{\pi} \left[\frac{A^2/N+\sigma^2}{2\pi} + S_i(\omega)\right]^{-1} \mathrm{d}\omega}{(2\pi)^2 - \frac{A^2}{N}\int_{-\pi}^{\pi} \left[\frac{A^2/N+\sigma^2}{2\pi} + S_i(\omega)\right]^{-1} \mathrm{d}\omega}. \tag{7.123}$$

7.5.3 Nonlinear Predictor and Interpolator

For narrowband interference added to a spread-spectrum signal in an AWGN environment, the prediction of the interferer takes place in the presence of both Gaussian and non-Gaussian noise. The non-Gaussian noise is the spread-spectrum signal itself. In such a non-Gaussian environment, linear methods are no longer optimal, and nonlinear techniques offer improved suppression capability over linear methods, as demonstrated in Section 7.3. Essentially, the nonlinear filters provide decision feedback that suppresses the spread-spectrum signal from the observations. When the decision feedback is accurate, the filter adaptation is done in essentially Gaussian noise (i.e., observations from which the spread-spectrum signal has been removed).

Based on the discussion above, we can obtain similar SINR upper bounds for the nonlinear predictive/interpolative methods. The idea is that we assume that the decision feedback part of the nonlinear filter accurately estimates the SS signal, and the SS signal is always subtracted from the observations, so that the NBI signal is estimated only in the presence of Gaussian noise. More specifically, consider the signal model (7.5). Assume that for the purpose of estimating the NBI signal, a genie provides an SS-signal-free observation

$$y_n = i_n + u_n.$$

A linear predictor or interpolator is then employed to obtain and estimate $\hat{i}_n$ of the NBI signal, which is then subtracted from the received signal to form the decision statistic for the SS data bit b,

$$\begin{aligned} U &= \sum_{n=0}^{N-1} \left(r_n - \hat{i}_n \right) c_{n,1} \\ &= A\, b + \sum_{n=0}^{N-1} \varepsilon_n \, c_{n,1}, \end{aligned} \tag{7.124}$$

where ε_n is the prediction error of the linear predictor (interpolator) (i.e., $\varepsilon_n = i_n - \hat{i}_n + u_n$). Modeling $\{c_{n,1}\}$ as a sequence of i.i.d. random variables such that $c_{n,1} = \pm 1/\sqrt{N}$ with probability $\frac{1}{2}$, the output SINR of this ideal system is then

$$\text{SINR} = \frac{E\{U\}^2}{\text{Var}\{U\}} = \frac{A^2}{E\{\varepsilon_n^2\}}. \tag{7.125}$$

Substituting the lower bounds for the mean-square prediction errors $E\{\varepsilon_n^2\}$ given by the linear predictor and linear interpolator [310, 311], we obtain the following very optimistic SINR upper bounds for the nonlinear estimator–subtracter methods:

$$\text{SINR(nonlinear predictor)} \le \frac{A^2}{2\pi \exp\left[\frac{1}{2\pi} \int_{-\pi}^{\pi} \ln\left(\frac{\sigma^2}{2\pi} + S_i(\omega) \right) d\omega \right]}, \tag{7.126}$$

$$\text{SINR(nonlinear interpolator)} \le \frac{A^2 \int_{-\pi}^{\pi} \left[\frac{\sigma^2}{2\pi} + S_i(\omega) \right]^{-1} d\omega}{(2\pi)^2}. \tag{7.127}$$

Now assume that the NBI is a pth-order AR signal, given by (7.94). Its power spectral density function is given by

$$S_i(\omega) = \frac{1}{2\pi} \frac{\nu^2}{\left| 1 + \sum_{k=1}^{p} \phi_k e^{-j\omega k} \right|^2}. \tag{7.128}$$

Substituting (7.128) into (7.127) and letting $\sigma \to 0$, we get the SINR upper bound for the nonlinear interpolator when the NBI is an AR signal in the absence of background noise:

$$\begin{aligned} \text{SINR(nonlinear interpolator)} &\le \frac{A^2}{\nu^2} \cdot \frac{1}{2\pi} \int_{-\pi}^{\pi} \left| 1 + \sum_{k=1}^{p} \phi_k e^{-j\omega k} \right|^2 d\omega \\ &= \left(1 + \phi_1^2 + \cdots + \phi_p^2 \right) \frac{A^2}{\nu^2}. \end{aligned} \tag{7.129}$$

Notice that this is the same output SINR value for the linear MMSE detector when the NBI is an AR signal in the absence of noise, given by (7.106).

7.5.4 Numerical Examples

In order to compare the NBI suppression capabilities of the various techniques described above, we consider two numerical examples. In the first example, the narrowband interferer is a second-order AR signal with both poles at 0.99 (i.e., $\phi_1 = -1.98$ and $\phi_2 = 0.9801$). The noise power is held constant at $\sigma^2 = -20$ dB (relative to the SS signal after despreading), while the interference power is varied from -20 dB to 40 dB (all relative to a unity SS power signal). The spreading signature sequence is a length-31 m-sequence. In Fig. 7.13 we plot the output SINR performance of various NBI suppression techniques for this example. We see that the linear MMSE detector significantly outperforms the linear predictor/interpolator, and it almost achieves the loose SINR upper bound for the nonlinear interpolator.

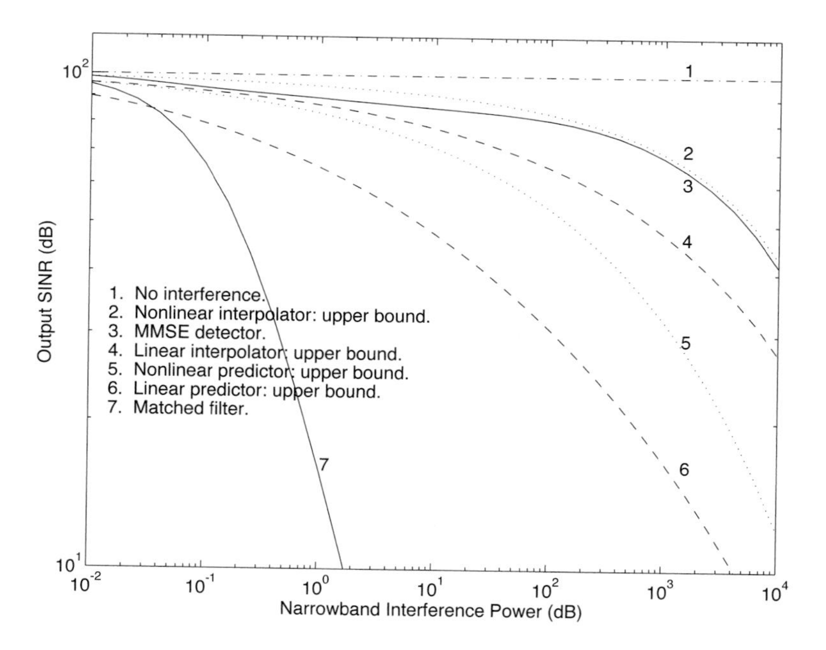

Figure 7.13. Comparison of the performance against the NBI of different NBI suppression techniques. The narrowband interferer is a second-order AR signal with both poles at 0.99. The noise power is held constant at $\sigma^2 = -20$ dB relative to the SS signal after despreading, while the interference power is varied from -20 dB to 40 dB relative to the SS signal. The spreading signature sequence is a length-31 m-sequence.

In the second example, the narrowband interferer is a digital signal with $m = 4$. Assuming that the digital NBI signal has a rectangular pulse waveform, the autocorrelation function of the chip-sampled NBI signal is then

$$R_i[k] = \begin{cases} \frac{A_i^2}{L}\left(1 - \frac{|k|}{L}\right), & |k| \leq L \\ 0, & |k| > L \end{cases}, \tag{7.130}$$

where $L = \lceil N/m \rceil$, and the power spectral density is

$$S_i(\omega) = \frac{1}{2\pi}\frac{A_i^2}{L^2}\left(\frac{\sin\frac{L\omega}{2}}{\sin\frac{\omega}{2}}\right)^2. \tag{7.131}$$

The performance of various methods against the digital NBI in this example is plotted in Fig. 7.14. The noise power is 20 dB below the SS signal power (after despreading), while the NBI signal power is varied from −20 dB to 40 dB relative to the SS signal power. It is seen that in this case the linear MMSE detector almost completely removes the NBI energy at the output, and the output SINR is held at 20 dB irrespective of the NBI power. This can be readily explained by (7.118). The other techniques are all clearly inferior to the linear MMSE detector in suppressing the digital NBI, and their performance degrades as the NBI power increases.

7.6 Near–Far Resistance to Both NBI and MAI by Linear MMSE Detector

We now consider the limiting behavior of the linear MMSE detector in the presence of NBI together with MAI, in which the energy of one or more of the interference signals (either MAI or NBI) can increase arbitrarily (i.e., the near–far situation). When the NBI is analog, the near–far resistance in the sense defined in [296, 297] is not easily determined, since in general the expression for the probability of error cannot be obtained. Another more intuitive view of a near–far resistant detector is that the output SINR is always greater than zero no matter how powerful the interference signal is [307]. In what follows we discuss the near–far resistance of the linear MMSE detector to both MAI and NBI in this sense.

7.6.1 Near–Far Resistance to NBI

We first consider the situation in which there is NBI but no MAI. Let the NBI signal $\boldsymbol{i}$ be an arbitrary discrete-time wide-sense stationary process, with autocorrelation matrix $\boldsymbol{R}_i$, which is nonnegative definite. Suppose that the spectral decomposition of $\boldsymbol{R}_i$ is given by

$$\boldsymbol{R}_i = \sum_{l=1}^{N} \lambda_l \boldsymbol{u}_l \boldsymbol{u}_l^T, \tag{7.132}$$

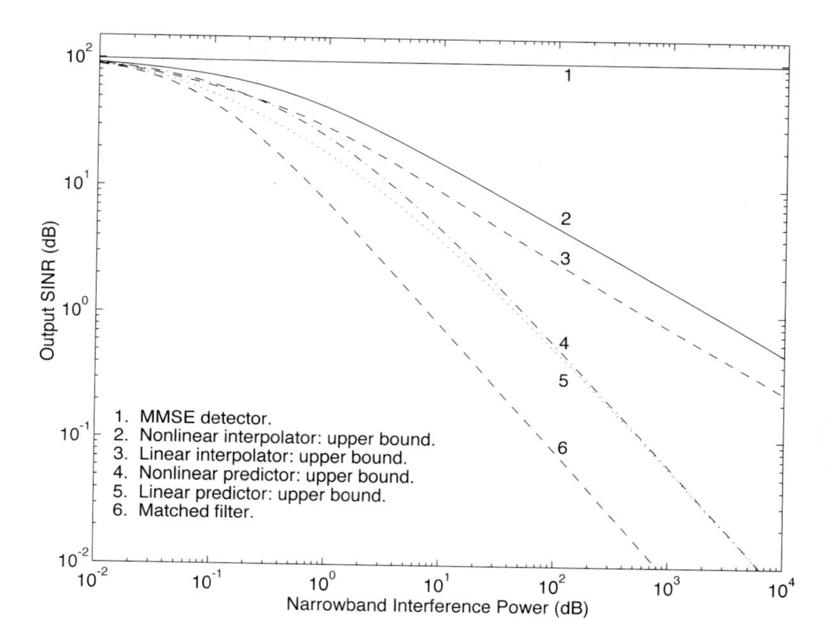

Figure 7.14. Comparison of the performance against the NBI of different NBI suppression techniques. The narrowband interferer is a digital signal with $m = 4$. The noise power is held constant at $\sigma^2 = -20$ dB relative to the SS signal after despreading, while the interference power is varied from -20 dB to 40 dB relative to the SS signal. The spreading signature sequence is a length-31 m-sequence.

where $\lambda_1, \ldots, \lambda_N$ and $\boldsymbol{u}_1, \ldots, \boldsymbol{u}_N$ are the nonnegative eigenvalues and the corresponding orthogonal eigenvectors of $\boldsymbol{R}_i$. Since

$$\boldsymbol{I}_N = \sum_{l=1}^{N} \boldsymbol{u}_l \boldsymbol{u}_l^T, \tag{7.133}$$

using (7.77) we obtain

$$\begin{aligned} \text{SINR} &= A^2 \boldsymbol{s}^T \left(\boldsymbol{R}_i + \sigma^2 \mathbf{I}_N \right)^{-1} \boldsymbol{s} \\ &= \sum_{l=1}^{N} \frac{A^2}{\sigma^2 + \lambda_l} \left| \boldsymbol{s}^T \boldsymbol{u}_l \right|^2 . \end{aligned} \tag{7.134}$$

When the NBI signal power is increased, the nonzero λ_l's increase proportionally. Therefore, it is seen from (7.134) that the near–far resistance to NBI is nonzero if and only if $\boldsymbol{R}_i$ has at least one zero eigenvalue and the corresponding eigenvector is not orthogonal to $\boldsymbol{s}$. On the other hand, if $\boldsymbol{R}_i$ has full rank, the near–far resistance to NBI is zero.

7.6.2 Near–Far Resistance to Both NBI and MAI

Now suppose that the interference in the system includes $(K-1)$ independent synchronous MAIs in addition to the NBI. Let the signature vector for the kth MAI be $\boldsymbol{s}_k$ and the power be P_k. It is straightforward to generalize (7.77) to include the effect of MAI, and we obtain the output SINR of the MMSE detector as

$$\begin{aligned}\text{SINR} &= A^2 \boldsymbol{s}^T \left(\sum_{k=1}^{K-1} P_k \, \boldsymbol{s}_k \, \boldsymbol{s}_k^T + \boldsymbol{R}_i + \sigma^2 \boldsymbol{I}_N \right)^{-1} \boldsymbol{s} \\ &= A^2 \boldsymbol{s}^T \left(\sum_{k=1}^{K-1} P_k \, \boldsymbol{s}_k \, \boldsymbol{s}_k^T + \sum_{l=1}^{N} \lambda_l \, \boldsymbol{u}_l \, \boldsymbol{u}_l^T + \sigma^2 \boldsymbol{I}_N \right)^{-1} \boldsymbol{s}. \end{aligned} \tag{7.135}$$

Equation (7.135) suggests that when we consider the output SINR for the linear MMSE detector, the NBI signal can be viewed as being equivalent to N *independent synchronous virtual MAIs*. The lth virtual MAI has signature vector $\boldsymbol{u}_l$ and power λ_l. Suppose that $r = \text{rank}(\boldsymbol{R}_i)$ and $\lambda_{r+1} = \cdots = \lambda_N = 0$. Then using the results from [307], the near–far resistance (to both MAI and NBI) is nonzero if and only if $\boldsymbol{s}$ is not contained in the subspace $\text{span}\{\boldsymbol{s}_1, \ldots, \boldsymbol{s}_K; \boldsymbol{u}_1, \ldots, \boldsymbol{u}_r\}$.

Next we consider the effect of NBI on the near–far resistance to MAI, by fixing the power of the NBI and increasing the power of the MAIs. It is shown in [307] that the linear MMSE solution $\boldsymbol{w}$ is asymptotically orthogonal to the subspace spanned by $\boldsymbol{s}_1, \ldots, \boldsymbol{s}_K$:

$$\boldsymbol{w} \perp \text{span}\{\boldsymbol{s}_1, \ldots, \boldsymbol{s}_K\}. \tag{7.136}$$

Such an asymptotic $\boldsymbol{w}$ can be found by solving the following constrained optimization problem:

$$\begin{aligned} \text{minimize} \quad & \text{MSE} = \boldsymbol{w}^T \boldsymbol{R}_r \boldsymbol{w} - 2\, A^2\, \boldsymbol{w}^T \boldsymbol{s} + A^2 \\ \text{s.t.} \quad & \boldsymbol{w}^T \boldsymbol{s}_l = 0, \quad l = 1, \ldots, K, \\ & \boldsymbol{w}^T \boldsymbol{s} = 1, \end{aligned}$$

where

$$\boldsymbol{R}_r \triangleq E\left\{\boldsymbol{r}\,\boldsymbol{r}^T\right\} = A^2 \boldsymbol{s}\boldsymbol{s}^T + \sum_{k=1}^{K-1} P_k \, \boldsymbol{s}_k \, \boldsymbol{s}_k^T + \boldsymbol{R}_i + \sigma^2 \boldsymbol{I}_N. \tag{7.137}$$

It then follows from the method of Lagrange multipliers that

$$\boldsymbol{w} \in \text{span}\left\{\boldsymbol{\Sigma}^{-1}\boldsymbol{s}, \boldsymbol{\Sigma}^{-1}\boldsymbol{s}_1, \ldots, \boldsymbol{\Sigma}^{-1}\boldsymbol{s}_K\right\}, \tag{7.138}$$

where $\boldsymbol{\Sigma} \triangleq \boldsymbol{R}_i + \sigma^2 \boldsymbol{I}_N$. Let $\boldsymbol{s} = \boldsymbol{s}^{\parallel} + \boldsymbol{s}^{\perp}$, where $\boldsymbol{s}^{\parallel} \in \text{span}\{\boldsymbol{s}_1, \ldots, \boldsymbol{s}_K\}$ and $\boldsymbol{s}^{\perp} \perp \text{span}\{\boldsymbol{s}_1, \ldots, \boldsymbol{s}_K\}$. The near–far resistance to MAI is nonzero if and only if $\boldsymbol{s}^T \boldsymbol{w} \neq 0$.

Therefore, from (7.136) and (7.138) it is easily seen that the near–far resistance to MAI is nonzero if and only if $\boldsymbol{s}^{\perp} \neq 0$ and $\boldsymbol{s}^{\perp} \in \text{span}\{\boldsymbol{\Sigma}^{-1}\boldsymbol{s}, \boldsymbol{\Sigma}^{-1}\boldsymbol{s}_1, \ldots, \boldsymbol{\Sigma}^{-1}\boldsymbol{s}_K\}$. Notice that if there is no NBI (i.e., $\boldsymbol{\Sigma} = \sigma^2 \boldsymbol{I}_N$), this condition for nonzero near–far resistance reduces to $\boldsymbol{s}^{\perp} \neq 0$ [307].

Simulation Examples

Figure 7.15 shows the output SINR of the linear MMSE detector in the presence of both MAI and NBI. The signal-to-noise ratio for the desired user in the absence of interference is fixed at 20 dB. The NBI is a second-order AR signal with both poles at 0.99. The MAIs are synchronous with the desired SS user, with random signature sequences and the same power. The processing gain is $N = 31$. Two cases are shown: three MAIs and six MAIs. For each case we vary the power of one type of interference (MAI or NBI) while keeping the power of the other fixed.

It is seen from Fig. 7.15 that the effects of the MAI and the NBI on the output SINR are different. The output SINR is insensitive to the power of the MAI while

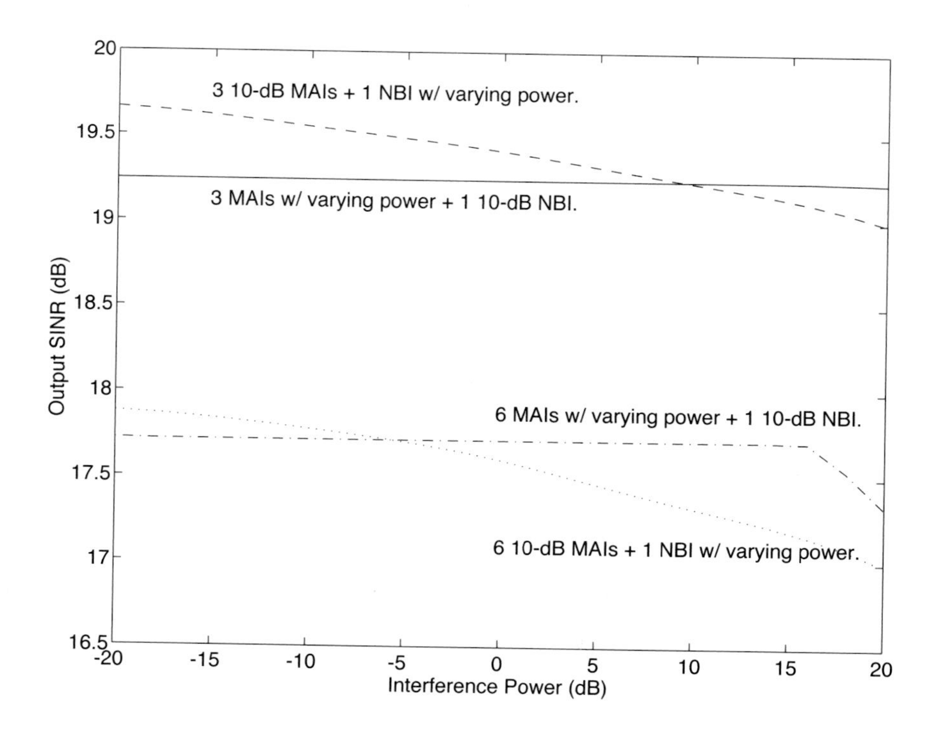

Figure 7.15. Output SINR of a linear MMSE detector in the presence of both MAI and NBI. The noise power is held constant at $\sigma^2 = -20$ dB relative to the SS signal after despreading. The NBI signal is a second-order AR signal with both poles at 0.99. The MAIs are synchronous with the SS user, with random signature sequences. The processing gain is $N = 31$.

it is more sensitive to the power of NBI. To see this, we consider a simple example where the CDMA system consists of the desired SS user signal, one MAI and one NBI, in the absence of background noise. Then by (7.135) the output SINR of the MMSE detector in this case is given by

$$\begin{aligned} \mathrm{SINR} &= A^2 \boldsymbol{s}^T \left(P_1 \boldsymbol{s}_1 \boldsymbol{s}_1^T + \boldsymbol{R}_i \right)^{-1} \boldsymbol{s} \\ &= A^2 \left(\boldsymbol{s}^T \boldsymbol{R}_i^{-1} \boldsymbol{s} \right) - \frac{P_1 \left(\boldsymbol{s}^T \boldsymbol{R}_i^{-1} \boldsymbol{s}_1 \right)^2}{1 + P_1 \left(\boldsymbol{s}^T \boldsymbol{R}_i^{-1} \boldsymbol{s} \right)}, \end{aligned} \tag{7.139}$$

where the second equality is obtained by using the matrix inversion lemma. Now because of the pseudorandomness of the signature vectors $\boldsymbol{s}$ and $\boldsymbol{s}_1$, $\boldsymbol{s}^T \boldsymbol{R}_i^{-1} \boldsymbol{s}_1 \ll \boldsymbol{s}^T \boldsymbol{R}_i^{-1} \boldsymbol{s}$. It is seen from (7.139) that the power of the MAI (P_1) affects the SINR only through the negligible second term in (7.139), while the power of the NBI affects the SINR through the dominant first term in (7.139).

Figure 7.16 is a plot of the probability of error of the MMSE detector, in the presence of strong MAI and NBI, in addition to ambient noise. The symbols o and + in this plot correspond to the data obtained from simulations, and the solid

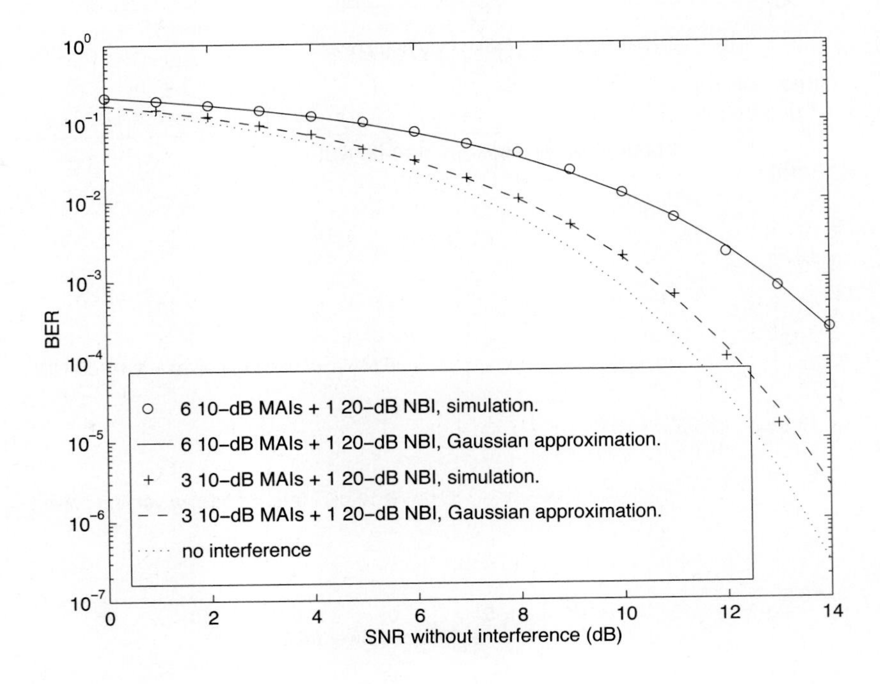

Figure 7.16. BER performance of a linear MMSE detector, in the presence of both MAI and NBI, in addition to ambient noise. The MAIs are synchronous with the SS user, with random signature sequence of length $N = 31$. The NBI signal is a second-order AR signal with both poles at 0.99.

and dashed lines correspond to Gaussian approximations of the probability of error (i.e., BER). It was shown in [386] that in an environment of MAI and AWGN, the error probability for the MMSE detector can be well approximated by assuming that the output MAI plus noise is Gaussian. This plot seems to suggest that even in the presence of NBI, the output NBI plus MAI plus noise is still approximately Gaussian, as one would expect, since the NBI here is Gaussian.

7.7 Adaptive Linear MMSE NBI Suppression

In Section 7.6, we saw that the linear MMSE detector is an excellent technique for suppressing NBI from spread-spectrum systems. A further advantage of the MMSE detector is that it is easily adapted to unknown NBI statistics. As discussed in Chapter 2, a number of adaptive algorithms for the linear MMSE detector as an MAI suppressor have been explored, including both those using a sequence of known training symbols and blind algorithms, which do not require such sequences [183, 237, 267, 325, 369, 403] (see [184] for a survey). Many of these studies have employed the LMS algorithm for adaptation because of its simplicity and overall good performance characteristics against wideband MAI. However, unlike the case of adaptive prediction-based NBI suppression discussed in Sections 7.2 and 7.3 (in which LMS features prominently), adaptation of the linear MMSE detector takes place at the symbol rate rather than at the chip rate. This does not cause difficulties with the LMS algorithm for wideband interference such as MAI. But for NBI, some problems may arise in using LMS at the symbol rate, due to resulting large eigenvalue spreads of the covariance matrix of the observations (cf. [603] for a review of the properties of LMS). These problems can be corrected by using instead the recursive-least-squares (RLS) algorithm, which may have better properties in such situations [389].

The use of the RLS algorithm for blind adaptation of the linear MMSE detector for MAI suppression is discussed in Section 2.3.2 of this book (cf. Algorithm 2.3). Exactly the same algorithm can be employed for adaptive suppression of both MAI and NBI. It is shown in the Appendix to this chapter (Section 7.9.5) that the steady-state SINR of the blind RLS linear MMSE detector is given by

$$\mathrm{SINR}^{\infty} = \frac{\mathrm{SINR}^*}{(1+d) + d \cdot \mathrm{SINR}^*}, \tag{7.140}$$

where SINR^* is the optimum SINR value given in (7.77) and where (recall that λ is the forgetting factor)

$$d \triangleq \frac{(1-\lambda)(N-1)}{2\lambda}. \tag{7.141}$$

Usually, the RLS algorithm operates in the range such that $d \ll 1$. From (7.140) it can be seen that the performance of the blind adaptive algorithm in terms of the steady-state SINR can be severely degraded from the optimum value SINR^*, especially when $\mathrm{SINR}^* \gg 1$. In fact, it is seen that the SINR in the steady state is upper bounded by $1/d$. This problem can be overcome by switching to the conventional

RLS algorithm that uses decision feedback, after the initial blind adaptation converges. The steady-state SINR of this scheme can be estimated via that of trained RLS, which is given in this case by [cf. the Appendix (Section 7.9.6)]

$$\text{SINR}^{\infty} = \frac{\text{SINR}^*}{(1+d) + d/\text{SINR}^*}. \tag{7.142}$$

It is seen from (7.142) that in contrast to the blind adaptive algorithm, when the adaptive algorithm has access to the transmitted symbols $b_1[i]$, the steady-state output SINR is close to its optimum possible value. Therefore, it is best to switch to a decision-directed adaptation mode as soon as the blind adaptation converges. However, decision-directed adaptation is subject to catastrophic error propagation in case of a sudden change in the environment. Whenever such a situation happens, the receiver should immediately switch back to the blind adaptation mode and stay in the blind mode until it converges, before it switches to the decision-directed mode again.

A difficulty with RLS relative to LMS is that RLS is more complex computationally. The complexity per update of RLS in this application is $\mathcal{O}(N^2)$ compared with $\mathcal{O}(N)$ for LMS, where we recall that N denotes the spreading gain. This complexity can be mitigated by using a parallel implementation on a systolic array first proposed in [316], as discussed in Section 2.3.3.

Simulation Examples

The first example illustrates the tracking capability of the RLS blind adaptive algorithm in a dynamic environment. Figure 7.17 shows a plot of time-averaged output SINR versus time of the RLS blind adaptive algorithm, in a synchronous CDMA system with processing gain $N = 31$ when the number and types of interferers in the system vary with time. The signal power to background noise power is 20 dB (after despreading). The simulation starts with one desired user's signal and six MAI signals, each at 10 dB above the desired user's signal. At time $n = 500$, a strong NBI signal of 20 dB above the desired user is added in the system. At time $n = 1000$, another strong MAI signal of 20 dB above the desired user is added. At time $n = 1500$, three of the original 10 dB MAI signals are removed from the system. The desired user's signature sequence is an m-sequence; and the signature sequences of the MAIs are generated randomly. The NBI signal is a second-order AR signal with both poles at 0.99. The forgetting factor is $\lambda = 0.995$. The data shown in the plot are values averaged over 100 simulations. It is seen that the RLS blind adaptive algorithm can adapt rapidly to the changing environment, which makes it suitable for practical use in a mobile environment.

The second example illustrates the difference between the steady-state SINRs of the blind adaptation rule and the decision-directed adaptation rule. Figure 7.18 shows a plot of time-averaged output SINR versus time for the RLS adaptive algorithms in a strong near–far environment. This example assumes a synchronous CDMA system with processing gain $N = 31$. There are three 10-dB MAIs, each with a randomly chosen signature sequence. In addition, there is a 20-dB NBI

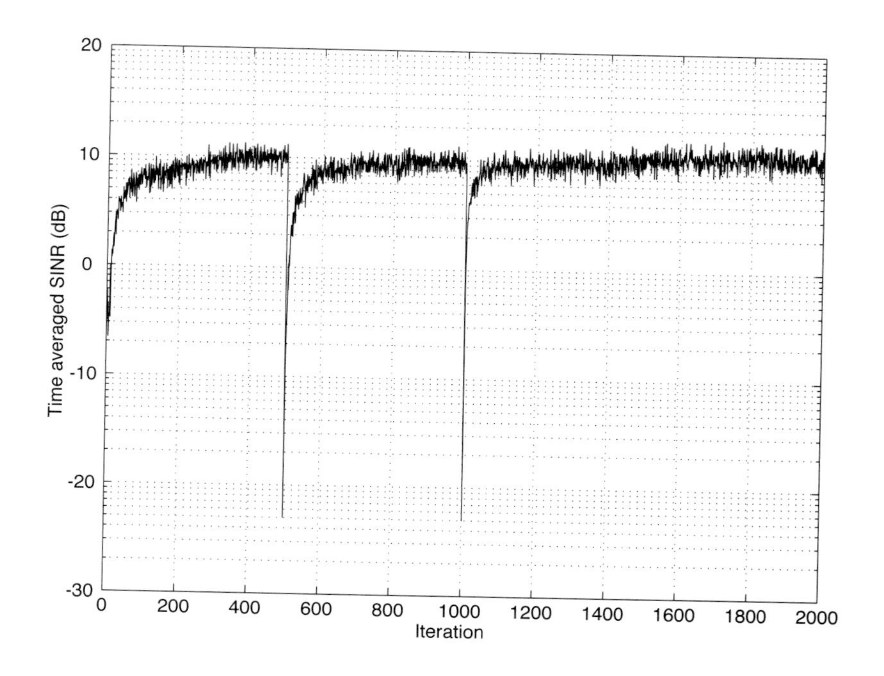

Figure 7.17. Tracking behavior of the RLS blind adaptive algorithm in a dynamic environment.

which is a second-order AR signal with both poles at 0.99. The signal power to background noise power is 20 dB. The blind adaptation rule is used for the first 500 iterations, and the conventional RLS algorithm using decision feedback is used thereafter. The forgetting factor is $\lambda = 0.995$. Again the data shown in the plot are values averaged over 100 simulations. It is seen from Fig. 7.18 that there is a significant gap between the steady-state SINR of the blind RLS algorithm and that of the conventional RLS algorithm, which can readily be explained by (7.140) and (7.142). Moreover, the steady-state SINR of the conventional RLS algorithm using decision feedback is very close to the optimal value of the MMSE detector, which is also plotted in Fig. 7.18 as the dashed line.

7.8 Maximum-Likelihood Code-Aided Method

In Section 7.7, we saw that the linear MMSE detector is a very useful tool for NBI suppression in DS-CDMA systems. A natural question to ask is whether its favorable performance properties can be improved upon. Within the context of linear code-aided methods, an optimal method was proposed and analyzed in [426], although without comparison to the linear MMSE technique (which had not yet

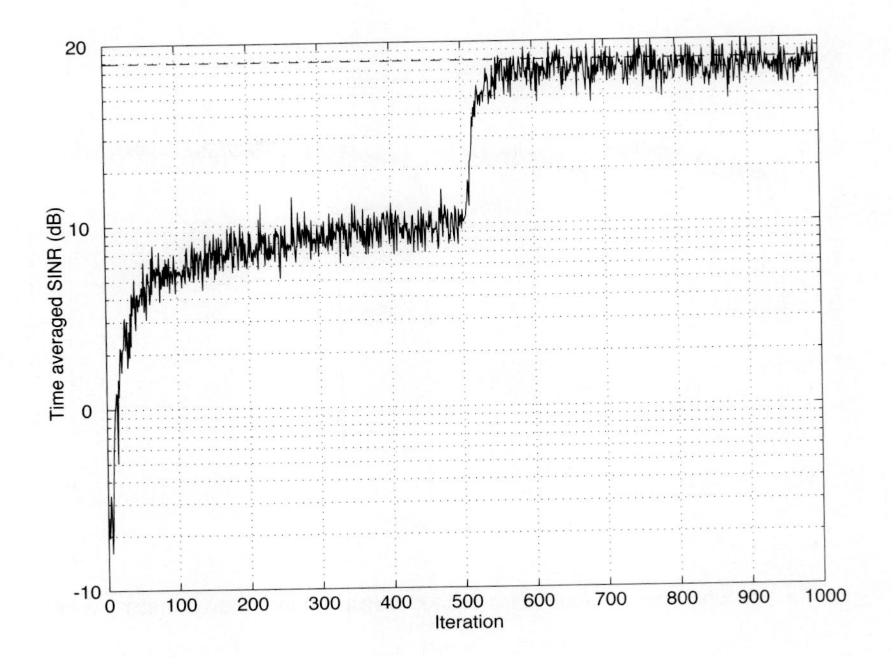

Figure 7.18. Time-averaged SINR for the blind adaptation and decision-directed adaptation rules.

been explored in this context at that time). In this section we look briefly at a more generally optimal, nonlinear, code-aided NBI suppression technique.

We saw in Sections 7.2 and 7.3 that in the context of predictive suppression, performance gains can be obtained by going from a linear to a nonlinear method to exploit signal structure. In the code-aided context, this suggests that it could be of use to progress from linear methods to optimal methods. One such method is maximum-likelihood detection, which is known to offer the ultimate performance improvement against MAI.

In the context of NBI suppression, we can examine a maximum-likelihood detector in the setting of digital NBI discussed in Section 7.4.4. To examine this situation, let us look at the signal model of (7.1)–(7.2) with a single spread user (i.e., $K = 1$) and with $\tau_1 = 0$, in which the NBI signal is given by

$$I(t) = A_I \sum_{j=0}^{m(M-1)} d_I[j]p(t - jT/m), \tag{7.143}$$

where $A_I > 0$ is the received amplitude of the NBI signal, m is the number of digital NBI symbols transmitted per spread-spectrum data symbol, $d_I[j]$ is the

jth (binary) symbol of the NBI, and $\{p(t)\}$ is the basic pulse shape (having unit energy and duration T/m) used by the digital NBI. To simplify the discussion, we assume that $\{p(t)\}$ is synchronous with $\{s_1(t)\}$ so that exactly m symbols of the NBI interfere with each symbol of the spread data signal. Similar to the situation in Fig. 7.11, we can think of the signal (7.143) as adding a set of m additional users to the channel, so that we have a multiple-access channel with $m+1$ total users.

We now consider the maximum-likelihood detection of the symbol stream $\{b_1[i]\}$ of the overlaid spread signal $\{S(t)\}$. Due to the synchrony and the assumption of white Gaussian noise, we can restrict attention to a single symbol interval. Examining the $i=0$ spread-data symbol interval, the log-likelihood function of the received waveform $\{r(t)\}$ can be shown straightforwardly to be proportional to (see, e.g., [381])

$$\mathcal{L}(\{r(t)\}) = A_I \sum_{j=0}^{m-1} d_I[j] x_I[j] + A_1 b_1[0] \Big[y_1[0] - A_I \sum_{j=0}^{m-1} d_I[j] \rho_j \Big], \tag{7.144}$$

where $y_1[0] = \int_0^T r(t) s_1(t)\, dt,$

$$x_I[j] = \int p(t - jT/m) r(t)\, dt, \quad j = 0, 1, \ldots, m-1, \tag{7.145}$$

$$\rho_j = \int p(t - jT/m) s_1(t)\, dt, \quad j = 0, 1, \ldots, m-1. \tag{7.146}$$

Let us examine the likelihood function (7.144) for a maximum over the unknown symbols $b_1[0], d_I[0], \ldots, d_I[m-1]$. Note that with the NBI symbols $d_I[0], \ldots, d_I[m-1]$ fixed, the maximum-likelihood choice of the spread-data symbol $b_1[0]$ is easily seen to be given by

$$\hat{b}_1[0] = \text{sign}\left\{ y_1[0] - A_I \sum_{j=0}^{m-1} d_I[j] \rho_j \right\}, \tag{7.147}$$

so that the maximum over $b_1[0]$ can we written as

$$\max_{b_1[0]} \mathcal{L}(\{r(t)\}) = A \sum_{j=0}^{m-1} d_I[j] x_I[j] + A_1 \left| y_1[0] - A_I \sum_{j=0}^{m-1} d_I[j] \rho_j \right|. \tag{7.148}$$

To find the global maximum-likelihood solution, we must maximize the quantity in (7.148) over the NBI data symbols, which generally requires direct search over the 2^m possible values for these m binary symbols. However, in a practical overlay system, the parameters A_I, A_1, and ρ_j should be such that the narrowband symbols can be detected by conventional methods with a relatively low probability of error. Thus, (7.148) is dominated by the first term on its right-hand side and so should be approximately maximized by the choice

$$\hat{d}_I[j] = \text{sign}\{\, x_I[j] \,\}, \qquad j = 0, 1, \ldots, m-1, \tag{7.149}$$

which maximizes this first term $A_I \sum_{j=0}^{m-1} d_I(j) x_I(j)$. So an approximate maximum-likelihood detector for the spread user's data symbol is

$$\hat{b}_1[0] = \text{sign}\left\{ y_1[0] - A_I \sum_{j=0}^{m-1} \text{sign}\left\{x_I[j]\right\} \rho_j \right\}. \tag{7.150}$$

This detector is essentially an "onion peeling" detector, in which the layer of NBI symbols is peeled off (i.e., detected and subtracted) using a conventional narrowband detector, and then the residual left after peeling is used for conventional detection of the spread user's symbol. Note that this detector fits the general mode of NBI suppression systems, in which a replica of the NBI is formed and then subtracted from the spread-spectrum signal before it is detected. A distinction is that here, this process takes place after despreading, so it fits within the code-aided framework. Note that multiple spread users can also be handled in this way, by first peeling off the NBI and then applying a standard multiuser detector on the residual. Similar ideas have been proposed in the context of multirate systems in [220, 299, 371, 509, 567].

Whether the detector of (7.150) offers general performance improvements over the linear code-aided methods of Section 7.7 is an interesting open question. Results from a simulation example comparing the maximum-likelihood and linear MMSE code-aided detectors for digital interferers with $N = 15$ and $m = 3$ are shown in Fig. 7.19. In this example it is seen that for a presuppression interference-to-signal ratio (ISR) of 0 dB the linear MMSE detector is better than the ML detector, but at ISR = 5 dB (and, of course, for larger values of ISR), the opposite behavior is observed. Also observe that for increasing ISR, the linear MMSE performance degrades (even though very slightly, in view of the near–far resistant feature of the linear MMSE receiver), whereas for increasing ISR, the performance of the ML detector improves. This matches with the intuition that for large ISR, the NBI can be better canceled with such a receiver.

There are many other techniques and aspects of the NBI suppression problem that we have not discussed in this chapter. Such contributions include a variety of other adaptive techniques [58, 72, 135, 153, 173, 285, 286, 352, 451, 480]; subspace-based methods [16, 118, 182]; Markov chain Monte Carlo (MCMC)–based Bayesian methods [594]; results for higher-order signaling [254, 539, 556]; other types of interference, such as chirp signals [151]; the effects of NBI suppression on tasks such as acquisition and tracking and on the correlation properties of spreading sequences (and vice versa) [155, 242, 328, 469]; and the explicit exploitation of cyclostationarity in this context [57]. The interested reader is referred to these sources for further details.

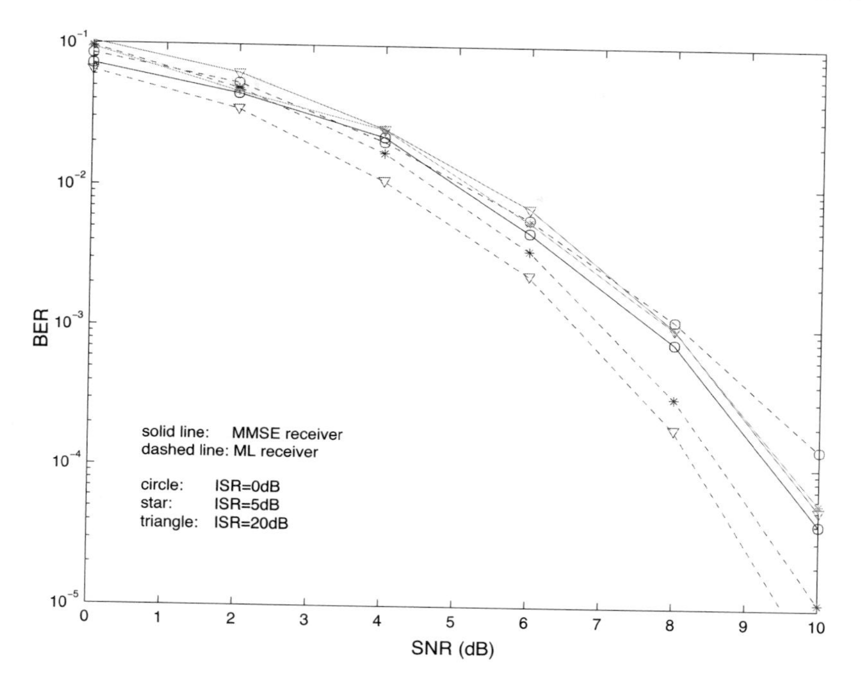

Figure 7.19. BER comparison of maximum-likelihood and MMSE code-aided suppression of digital NBI: $N = 15$, $m = 3$, and for several values of presuppression interference-to-signal ratio (ISR).

7.9 Appendix: Convergence of the RLS Linear MMSE Detector

7.9.1 Linear MMSE Detector and RLS Blind Adaptation Rule

Consider the following received signal model:

$$\boldsymbol{r} = \sum_{k=1}^{K} A_k b_k \boldsymbol{s}_k + \boldsymbol{i} + \boldsymbol{n}, \tag{7.151}$$

where A_k, b_k, and $\boldsymbol{s}_k$ denote, respectively, the received amplitude, data bit, and the spreading waveform of the kth user; $\boldsymbol{i}$ denotes the NBI signal; and $\boldsymbol{n} \sim \mathcal{N}(\boldsymbol{0}, \sigma^2 \boldsymbol{I}_N)$ is the Gaussian noise. Assume that user 1 is the user of interest, and for convenience we will use the following notations: $P \triangleq A_1^2$, $\boldsymbol{s} \triangleq \boldsymbol{s}_1$, $b \triangleq b_1$, and $P_k \triangleq A_k^2$. The weight vector of the linear MMSE detector is given by

$$\boldsymbol{w} = \frac{1}{\boldsymbol{s}^T \boldsymbol{R}_r^{-1} \boldsymbol{s}} \boldsymbol{R}_r^{-1} \boldsymbol{s}, \tag{7.152}$$

where $\boldsymbol{R}_r$ is the autocorrelation matrix of the received discrete signal $\boldsymbol{r}$:

$$\boldsymbol{R}_r \triangleq E\left\{\boldsymbol{r}\boldsymbol{r}^T\right\} = \sum_{k=1}^{K} P_k \boldsymbol{s}_k \boldsymbol{s}_k^T + \boldsymbol{R}_i + \sigma^2 \boldsymbol{I}_N. \tag{7.153}$$

The output SINR is given by

$$\mathrm{SINR}^* \triangleq \frac{E^2\left\{\boldsymbol{w}^T\boldsymbol{r}\right\}}{\mathrm{Var}\left\{\boldsymbol{w}^T\boldsymbol{r}\right\}} = P\boldsymbol{s}^T\boldsymbol{\Sigma}^{-1}\boldsymbol{s}, \tag{7.154}$$

where

$$\boldsymbol{\Sigma} \triangleq \boldsymbol{R}_r - P\boldsymbol{s}\boldsymbol{s}^T = \sum_{k=2}^{K} P_k \boldsymbol{s}_k \boldsymbol{s}_k^T + \boldsymbol{R}_i + \sigma^2 \boldsymbol{I}_N. \tag{7.155}$$

The mean output energy associated with $\boldsymbol{w}$, defined as the mean-square output value of $\boldsymbol{w}$ applied to $\boldsymbol{r}$, is

$$\bar{\xi} \triangleq E\left\{\left(\boldsymbol{w}^T\boldsymbol{r}\right)^2\right\} = \boldsymbol{w}^T\boldsymbol{R}_r\boldsymbol{w} = \frac{1}{\boldsymbol{s}^T\boldsymbol{R}_r^{-1}\boldsymbol{s}} = P + \frac{1}{\boldsymbol{s}^T\boldsymbol{\Sigma}^{-1}\boldsymbol{s}}, \tag{7.156}$$

where the last equality follows from (7.155) and the matrix inversion lemma. The mean-square error (MSE) at the output of $\boldsymbol{w}$ is

$$\bar{\epsilon} \triangleq E\left\{\left(\sqrt{P}b - \boldsymbol{w}^T\boldsymbol{r}\right)^2\right\} = P + \bar{\xi} - 2P\left(\boldsymbol{w}^T\boldsymbol{s}\right) = \bar{\xi} - P = \frac{1}{\boldsymbol{s}^T\boldsymbol{\Sigma}^{-1}\boldsymbol{s}}. \tag{7.157}$$

The exponentially windowed RLS algorithm selects the weight vector $\boldsymbol{w}[i]$ to minimize the sum of exponentially weighted output energies:

$$\text{minimize} \quad \sum_{m=1}^{i} \lambda^{i-m}\left(\boldsymbol{w}[i]^T\boldsymbol{r}[m]\right)^2 \qquad \text{s.t.} \quad \boldsymbol{s}^T\boldsymbol{w}[i] = 1,$$

where $0 < \lambda < 1$ is a forgetting factor $(1 - \lambda \ll 1)$. The purpose of λ is to ensure that the data in the distant past will be forgotten in order to provide tracking capability in nonstationary environments. The solution to this constrained optimization problem is given by

$$\boldsymbol{w}[i] = \frac{1}{\boldsymbol{s}^T\boldsymbol{R}_r^{-1}[i]\boldsymbol{s}}\boldsymbol{R}_r[n]^{-1}\boldsymbol{s}, \tag{7.158}$$

where

$$\boldsymbol{R}_r[i] \triangleq \sum_{m=1}^{i} \lambda^{i-m}\boldsymbol{r}[m]\,\boldsymbol{r}[m]^T. \tag{7.159}$$

A recursive procedure for updating $\boldsymbol{w}[i]$ is as follows:

$$\boldsymbol{k}[i] \triangleq \frac{\boldsymbol{R}_r^{-1}[i-1]\boldsymbol{r}[i]}{\lambda + \boldsymbol{r}[i]^T\boldsymbol{R}_r^{-1}[i-1]\boldsymbol{r}[i]}, \tag{7.160}$$

$$\boldsymbol{h}[i] \triangleq \boldsymbol{R}_r^{-1}[i]\boldsymbol{s} = \frac{1}{\lambda}\left(\boldsymbol{h}[i-1] - \boldsymbol{k}[i]\boldsymbol{r}[i]^T\boldsymbol{h}[i-1]\right), \tag{7.161}$$

$$\boldsymbol{w}[i] = \frac{1}{\boldsymbol{s}^T\boldsymbol{h}[i]}\boldsymbol{h}[i], \tag{7.162}$$

$$\boldsymbol{R}_r^{-1}[i] = \frac{1}{\lambda}\left(\boldsymbol{R}_r^{-1}[i-1] - \boldsymbol{k}[i]\boldsymbol{r}[i]^T\boldsymbol{R}_r^{-1}[i-1]\right). \tag{7.163}$$

In what follows we provide a convergence analysis for the algorithm above. In this analysis, we make use of three approximations/assumptions: (a) For large i, $\boldsymbol{R}_r[i]$ is approximated by its expected value [111, 301]; (b) the input data $\boldsymbol{r}[i]$ and the previous weight vector $\boldsymbol{w}[i-1]$ are assumed to be independent [175]; (c) some fourth-order statistic can be approximated in terms of a second-order statistic [175].

7.9.2 Convergence of the Mean Weight Vector

We start by deriving an explicit recursive relationship between $\boldsymbol{w}[i]$ and $\boldsymbol{w}[i-1]$. Denote

$$\alpha[i] \triangleq \frac{1}{\boldsymbol{s}^T\boldsymbol{R}_r^{-1}[i]\boldsymbol{s}} = \frac{1}{\boldsymbol{s}^T\boldsymbol{h}[i]}. \tag{7.164}$$

Premultiplying both sides of (7.161) by $\boldsymbol{s}^T$, we have

$$\alpha^{-1}[i] = \frac{1}{\lambda}\left(\alpha^{-1}[i-1] - \boldsymbol{s}^T\boldsymbol{k}[i]\boldsymbol{r}[i]^T\boldsymbol{h}[i-1]\right). \tag{7.165}$$

From (7.165) we obtain

$$\begin{aligned}\alpha[i] &= \lambda\left(\alpha[i-1] + \frac{\alpha^2[i-1]\boldsymbol{s}^T\boldsymbol{k}[i]\boldsymbol{r}[i]^T\boldsymbol{h}[i-1]}{1 - \boldsymbol{s}^T\boldsymbol{k}[i]\boldsymbol{r}[i]^T\boldsymbol{h}[i-1]\alpha[i-1]}\right)\\ &= \lambda\left(\alpha[i-1] + \alpha[i-1]\beta[i]\boldsymbol{r}[i]^T\boldsymbol{h}[i-1]\right),\end{aligned} \tag{7.166}$$

where

$$\beta[i] \triangleq \frac{\alpha[i-1]\boldsymbol{s}^T\boldsymbol{k}[i]}{1 - \boldsymbol{s}^T\boldsymbol{k}[i]\boldsymbol{r}[i]^T\boldsymbol{h}[i-1]\alpha[i-1]}. \tag{7.167}$$

Substituting (7.161) and (7.166) into (7.162), we can write

$$\begin{aligned}\boldsymbol{w}[i] &= \alpha[i]\boldsymbol{h}[i] = \lambda\alpha[i-1]\boldsymbol{h}[i] + \lambda\beta[i]\alpha[i-1]\boldsymbol{r}[i]^T\boldsymbol{h}[i-1]\boldsymbol{h}[i]\\ &= \alpha[i-1]\left(\boldsymbol{h}[i-1] - \boldsymbol{k}[i]\boldsymbol{r}[i]^T\boldsymbol{h}[i-1]\right) + \lambda\beta[i]e[i]\boldsymbol{h}[i]\\ &= \boldsymbol{w}[i-1] - e[i]\boldsymbol{k}[i] + \lambda\beta[i]e[i]\boldsymbol{h}[i],\end{aligned} \tag{7.168}$$

where

$$e[i] \triangleq \boldsymbol{r}[i]^T \boldsymbol{w}[i-1] = \alpha[i-1]\boldsymbol{r}[i]^T \boldsymbol{h}[i-1], \tag{7.169}$$

is the *a priori* least-squares estimate at time i. It is shown below that

$$\boldsymbol{k}[i] = \boldsymbol{R}_r^{-1}[i]\boldsymbol{r}[i], \tag{7.170}$$

$$\lambda\beta[i] = \alpha[i]\boldsymbol{s}^T\boldsymbol{k}[i]. \tag{7.171}$$

Substituting (7.161) and (7.170) into (7.168), we have

$$\boldsymbol{w}[i] = \boldsymbol{w}[i-1] - \boldsymbol{R}_r^{-1}[i]\boldsymbol{r}[i]e[i] + \boldsymbol{R}_r^{-1}[i]\boldsymbol{s}\lambda\beta[i]e[i]. \tag{7.172}$$

Premultiplying both sides of (7.172) by $\boldsymbol{R}_r[i]$, we get

$$\begin{aligned}\boldsymbol{R}_r[i]\boldsymbol{w}[i] &= \boldsymbol{R}_r[i]\boldsymbol{w}[i-1] - \boldsymbol{r}[i]e[i] + \boldsymbol{s}\lambda\beta[i]e[i] \\ &= \lambda\boldsymbol{R}_r[i-1]\boldsymbol{w}[i-1] + \boldsymbol{r}[i]\boldsymbol{r}[i]^T\boldsymbol{w}[i-1] - \boldsymbol{r}[i]e[i] + \boldsymbol{s}\lambda\beta[i]e[i] \\ &= \lambda\boldsymbol{R}_r[i-1]\boldsymbol{w}[i-1] + \boldsymbol{s}\lambda\beta[i]e[i],\end{aligned} \tag{7.173}$$

where we have used (7.159) and (7.169). Let $\boldsymbol{\theta}[i]$ be the weight error vector between the weight vector $\boldsymbol{w}[i]$ at time n and the optimal weight vector $\boldsymbol{w}$:

$$\boldsymbol{\theta}[i] \triangleq \boldsymbol{w}[i] - \boldsymbol{w}. \tag{7.174}$$

Then from (7.173) we can deduce that

$$\boldsymbol{R}_r[i]\boldsymbol{\theta}[i] = \lambda\boldsymbol{R}_r[i-1]\theta[i-1] + \left(\boldsymbol{s}\lambda\beta[i]e[i] - \boldsymbol{r}[i]\boldsymbol{r}[i]^T\boldsymbol{w}\right). \tag{7.175}$$

Therefore,

$$\boldsymbol{\theta}[i] = \lambda\boldsymbol{R}_r^{-1}[i]\boldsymbol{R}_r[i-1]\boldsymbol{\theta}[i-1] + \boldsymbol{R}_r^{-1}[i]\boldsymbol{y}[i], \tag{7.176}$$

where

$$\begin{aligned}\boldsymbol{y}[i] &\triangleq \boldsymbol{s}\lambda\beta[i]e[i] - \boldsymbol{r}[i]\boldsymbol{r}[i]^T\boldsymbol{w} \\ &= \alpha[i]\boldsymbol{s}\boldsymbol{s}^T\boldsymbol{k}[i]\boldsymbol{r}[i]^T\boldsymbol{w}[i-1] - \boldsymbol{r}[i]\boldsymbol{r}[i]^T\boldsymbol{w} \\ &= \alpha[i]\boldsymbol{s}\boldsymbol{s}^T\boldsymbol{k}[i]\boldsymbol{r}[i]^T\boldsymbol{\theta}[i-1] + \left(\alpha[i]\boldsymbol{s}\boldsymbol{s}^T\boldsymbol{k}[i]\boldsymbol{r}[i]^T\boldsymbol{w} - \boldsymbol{r}[i]\boldsymbol{r}[i]^T\boldsymbol{w}\right),\end{aligned} \tag{7.177}$$

in which we have used (7.171) and (7.169).

It has been shown [111,301] that for large i, the inverse autocorrelation estimate $\boldsymbol{R}_r^{-1}[i]$ behaves like a quasi-deterministic quantity when $N(1-\lambda) \ll 1$. Therefore, for large i, we can replace $\boldsymbol{R}_r^{-1}[i]$ by its expected value, which is given by [7,111,301]

$$\lim_{i\to\infty} \boldsymbol{R}_r^{-1}[i] \cong \lim_{i\to\infty} E\left\{\boldsymbol{R}_r^{-1}[i]\right\} = (1-\lambda)\boldsymbol{R}_r^{-1}. \tag{7.178}$$

Using this approximation, we have

$$\lim_{i\to\infty} \alpha[i]\boldsymbol{R}_r^{-1}[i]\boldsymbol{s} \cong \frac{(1-\lambda)\boldsymbol{R}_r^{-1}\boldsymbol{s}}{(1-\lambda)\boldsymbol{s}^T\boldsymbol{R}_r^{-1}\boldsymbol{s}} = \boldsymbol{w}. \tag{7.179}$$

Therefore, for large i,

$$\begin{aligned}
&\boldsymbol{R}_r^{-1}[i]\boldsymbol{y}[i] \\
&\quad \cong \alpha[i]\boldsymbol{R}_r^{-1}[i]\boldsymbol{s}\boldsymbol{s}^T\boldsymbol{k}[i]\boldsymbol{r}[i]^T\boldsymbol{\theta}[i-1] + \left(\alpha[i]\boldsymbol{R}_r^{-1}[i]\boldsymbol{s}\boldsymbol{s}^T\boldsymbol{k}[i]\boldsymbol{r}[i]^T - \boldsymbol{R}_r^{-1}[i]\boldsymbol{r}[i]\boldsymbol{r}[i]^T\right)\boldsymbol{w} \\
&\quad = \boldsymbol{w}\boldsymbol{s}^T\boldsymbol{k}[i]\boldsymbol{r}[i]^T\boldsymbol{\theta}[i-1] + \left(\boldsymbol{w}\boldsymbol{s}^T - \boldsymbol{I}_N\right)\boldsymbol{k}[i]\boldsymbol{r}[i]^T\boldsymbol{w}, \qquad (7.180)
\end{aligned}$$

where we have used (7.170) and (7.179). For large i, $\boldsymbol{R}_r[i]$ and $\boldsymbol{R}_r[i-1]$ can be assumed almost equal, and thus approximately [111, 301]

$$\lim_{i\to\infty} \boldsymbol{R}_r^{-1}[i]\boldsymbol{R}_r[i-1] \cong \boldsymbol{I}_N. \qquad (7.181)$$

Substituting (7.181) and (7.180) into (7.176), we then have

$$\boldsymbol{\theta}[i] \cong \left(\lambda\boldsymbol{I}_N + \boldsymbol{w}\boldsymbol{s}^T\boldsymbol{k}[i]\boldsymbol{r}[i]^T\right)\boldsymbol{\theta}[i-1] + \left(\boldsymbol{w}\boldsymbol{s}^T - \boldsymbol{I}_N\right)\boldsymbol{k}[i]\boldsymbol{r}[i]^T\boldsymbol{w}. \qquad (7.182)$$

Equation (7.182) is a recursive equation that the weight error vector $\boldsymbol{\theta}[i]$ satisfies for large i.

In what follows we assume that the present input $\boldsymbol{r}[i]$ and the previous weight error $\boldsymbol{\theta}[i-1]$ are independent. In this application of interference suppression, this assumption is satisfied when the interference signal consists of only MAI and white noise. If, in addition, there is NBI present, this assumption is not satisfied but is nevertheless assumed, as is the common practice in the analysis of adaptive algorithms [111, 175, 301]. Taking expectations on both sides of (7.182), we have

$$\begin{aligned}
&E\{\boldsymbol{\theta}[i]\} \\
&\quad \cong \lambda E\{\boldsymbol{\theta}[i-1]\} + \boldsymbol{w}\boldsymbol{s}^T E\left\{\boldsymbol{k}[i]\boldsymbol{r}[i]^T\right\} E\{\boldsymbol{\theta}[i-1]\} + \left(\boldsymbol{w}\boldsymbol{s}^T - \boldsymbol{I}_N\right) E\left\{\boldsymbol{k}[i]\boldsymbol{r}[i]^T\right\}\boldsymbol{w} \\
&\quad \cong \lambda E\{\boldsymbol{\theta}[i-1]\} + (1-\lambda)\left(\boldsymbol{w}\boldsymbol{s}^T E\{\boldsymbol{\theta}[i-1]\} + \boldsymbol{w}\boldsymbol{s}^T\boldsymbol{w} - \boldsymbol{w}\right) = \lambda E\{\boldsymbol{\theta}[i-1]\},
\end{aligned}$$

where we have used the facts that $\boldsymbol{s}^T\boldsymbol{w} = \boldsymbol{s}^T\boldsymbol{w}[i] = 1$, $\boldsymbol{s}^T\boldsymbol{\theta}[i] = \boldsymbol{s}^T\boldsymbol{w}[i] - \boldsymbol{s}^T\boldsymbol{w} = 0$ and

$$\begin{aligned}
&E\left\{\boldsymbol{k}[i]\boldsymbol{r}[i]^T\right\} \\
&\quad = E\left\{\boldsymbol{R}_r^{-1}[i]\boldsymbol{r}[i]\boldsymbol{r}[i]^T\right\} \cong (1-\lambda)\boldsymbol{R}_r^{-1}E\left\{\boldsymbol{r}[i]\boldsymbol{r}[i]^T\right\} = (1-\lambda)\boldsymbol{I}_N. \qquad (7.183)
\end{aligned}$$

Therefore, the expected weight error vector always converges to zero, and this convergence is independent of the eigenvalue distribution.

Finally, we verify (7.170) and (7.171). Postmultiplying both sides of (7.163) by $\boldsymbol{r}[i]$, we have

$$\boldsymbol{R}_r^{-1}[i]\boldsymbol{r}[i] = \frac{1}{\lambda}\left(\boldsymbol{R}_r^{-1}[i-1] - \boldsymbol{k}[i]\boldsymbol{r}[i]^T\boldsymbol{R}_r^{-1}[i-1]\boldsymbol{r}[i]\right). \qquad (7.184)$$

On the other hand, (7.160) can be rewritten as

$$\boldsymbol{k}[i] = \frac{1}{\lambda}\left(\boldsymbol{R}_r^{-1}[i-1] - \boldsymbol{k}[i]\boldsymbol{r}[i]^T\boldsymbol{R}_r^{-1}[i-1]\boldsymbol{r}[i]\right). \tag{7.185}$$

Equation (7.170) is obtained by comparing (7.184) and (7.185).

Multiplying both sides of (7.166) by $\boldsymbol{s}^T\boldsymbol{k}[i]$, we can write

$$\alpha[i]\boldsymbol{s}^T\boldsymbol{k}[i] = \lambda\left[\alpha[i-1]\boldsymbol{s}^T\boldsymbol{k}[i] + \alpha[i-1]\beta[i]\boldsymbol{r}[i]^T\boldsymbol{h}[i-1]\boldsymbol{s}^T\boldsymbol{k}[i]\right), \tag{7.186}$$

and (7.167) can be rewritten as

$$\beta[i] = \alpha[i-1]\boldsymbol{s}^T\boldsymbol{k}[i] + \alpha[i-1]\beta[i]\boldsymbol{r}[i]^T\boldsymbol{h}[i-1]\boldsymbol{s}^T\boldsymbol{k}[i]. \tag{7.187}$$

Equation (7.171) is obtained comparing (7.186) and (7.187).

7.9.3 Weight Error Correlation Matrix

We proceed to derive a recursive relationship for the time evolution of the correlation matrix of the weight error vector $\boldsymbol{\theta}[i]$, which is the key to analysis of the convergence of the MSE. Let $\boldsymbol{K}[i]$ be the weight error correlation matrix at time n. Taking the expectation of the outer product of the weight error vector $\boldsymbol{\theta}[i]$, we get

$$\begin{aligned}
\boldsymbol{K}[i] &\triangleq E\left\{\boldsymbol{\theta}[i]\,\boldsymbol{\theta}[i]^T\right\} \\
&= E\left\{\left(\lambda\boldsymbol{I}_N + \boldsymbol{w}\boldsymbol{s}^T\boldsymbol{k}[i]\boldsymbol{r}[i]^T\right)\boldsymbol{\theta}[i-1]\boldsymbol{\theta}[i-1]^T\left(\lambda\boldsymbol{I}_N + \boldsymbol{r}[i]\boldsymbol{k}[i]^T\boldsymbol{s}\boldsymbol{w}^T\right)\right\} \\
&\quad + E\left\{\left(\lambda\boldsymbol{I}_N + \boldsymbol{w}\boldsymbol{s}^T\boldsymbol{k}[i]\boldsymbol{r}[i]^T\right)\boldsymbol{\theta}[i-1]\boldsymbol{w}^T\boldsymbol{r}[i]\boldsymbol{k}[i]^T\left(\boldsymbol{s}\boldsymbol{w}^T - \boldsymbol{I}_N\right)\right\} \\
&\quad + E\left\{\left(\boldsymbol{w}\boldsymbol{s}^T - \boldsymbol{I}_N\right)\boldsymbol{k}[i]\boldsymbol{r}[i]^T\boldsymbol{w}\boldsymbol{\theta}[i-1]^T\left(\lambda\boldsymbol{I}_N + \boldsymbol{r}[i]\boldsymbol{k}[i]^T\boldsymbol{s}\boldsymbol{w}^T\right)\right\} \\
&\quad + E\left\{\left(\boldsymbol{w}\boldsymbol{s}^T - \boldsymbol{I}_N\right)\boldsymbol{k}[i]\boldsymbol{r}[i]^T\boldsymbol{w}\boldsymbol{w}^T\boldsymbol{r}[i]\boldsymbol{k}[i]^T\left(\boldsymbol{s}\boldsymbol{w}^T - \boldsymbol{I}_N\right)\right\}.
\end{aligned} \tag{7.188}$$

We next compute the four expectations appearing on the right-hand side of (7.188).

First term

$$\begin{aligned}
&= \lambda^2 E\left\{\boldsymbol{\theta}[i-1]\boldsymbol{\theta}[i-1]^T\right\} + \lambda\boldsymbol{w}\boldsymbol{s}^T E\left\{\boldsymbol{k}[i]\boldsymbol{r}[i]^T\right\}E\left\{\boldsymbol{\theta}[i-1]\boldsymbol{\theta}[i-1]^T\right\} \\
&\quad + \lambda E\left\{\boldsymbol{\theta}[i-1]\boldsymbol{\theta}[i-1]^T\right\}E\left\{\boldsymbol{r}[i]\boldsymbol{k}[i]^T\right\}\boldsymbol{s}\boldsymbol{w}^T \\
&\quad + \boldsymbol{w}\boldsymbol{s}^T E\left\{\boldsymbol{k}[i]\boldsymbol{r}[i]^T\boldsymbol{\theta}[i-1]\boldsymbol{\theta}[i-1]^T\boldsymbol{r}[i]\boldsymbol{k}[i]^T\right\}\boldsymbol{s}\boldsymbol{w}^T \\
&= \lambda^2\boldsymbol{K}[i-1] + \lambda(1-\lambda)\left(\boldsymbol{w}\boldsymbol{s}^T\boldsymbol{K}[i-1] + \boldsymbol{K}[i-1]\boldsymbol{s}\boldsymbol{w}^T\right) \\
&\quad + \boldsymbol{w}\boldsymbol{s}^T E\left\{\boldsymbol{k}[i]\boldsymbol{r}[i]^T\boldsymbol{\theta}[i-1]\boldsymbol{\theta}[i-1]^T\boldsymbol{r}[i]\boldsymbol{k}[i]^T\right\}\boldsymbol{s}\boldsymbol{w}^T && (7.189) \\
&= \lambda^2\boldsymbol{K}[i-1] + \boldsymbol{w}\boldsymbol{s}^T E\left\{\boldsymbol{k}[i]\boldsymbol{r}[i]^T\boldsymbol{\theta}[i-1]\boldsymbol{\theta}[i-1]^T\boldsymbol{r}[i]\boldsymbol{k}[i]^T\right\}\boldsymbol{s}\boldsymbol{w}^T && (7.190) \\
&= \lambda^2\boldsymbol{K}[i-1] + (1-\lambda)^2\boldsymbol{w}\boldsymbol{s}^T\left(2\boldsymbol{K}[i-1] + \mathrm{tr}\{\boldsymbol{R}_r\boldsymbol{K}[i-1]\}\boldsymbol{R}_r^{-1}\right)\boldsymbol{s}\boldsymbol{w}^T && (7.191) \\
&= \lambda^2\boldsymbol{K}[i-1] + (1-\lambda)^2\,\mathrm{tr}\{\boldsymbol{R}_r\boldsymbol{K}[i-1]\}\boldsymbol{w}\boldsymbol{s}^T\boldsymbol{R}_r^{-1}\boldsymbol{s}\boldsymbol{w}^T && (7.192) \\
&= \lambda^2\boldsymbol{K}[i-1] + (1-\lambda)^2\,\mathrm{tr}\{\boldsymbol{R}_r\boldsymbol{K}[i-1]\}\frac{\boldsymbol{R}_r^{-1}\boldsymbol{s}\boldsymbol{s}^T\boldsymbol{R}_r^{-1}}{\boldsymbol{s}^T\boldsymbol{R}_r^{-1}\boldsymbol{s}}, && (7.193)
\end{aligned}$$

where in (7.189) we have used (7.183); in (7.193) we have used (7.152); in (7.190) and (7.192) we have used the fact that $\boldsymbol{s}^T\boldsymbol{K}[i-1] = E\left\{\boldsymbol{s}^T\boldsymbol{\theta}[i-1]\boldsymbol{\theta}[i-1]^T\right\} = 0$; and in (7.191) we have used the following fact, which is derived below:

$$E\left\{\boldsymbol{k}[i]\boldsymbol{r}^T[i]\boldsymbol{\theta}[i-1]\boldsymbol{\theta}[i-1]^T\boldsymbol{r}[i]\boldsymbol{k}[i]^T\right\} \\ = (1-\lambda)^2\left(2\boldsymbol{K}[i-1] + \mathrm{tr}\{\boldsymbol{R}_r\boldsymbol{K}[i-1]\}\boldsymbol{R}_r^{-1}\right). \quad (7.194)$$

Second term

$$\begin{aligned} &= \lambda E\left\{\boldsymbol{\theta}[i-1]\right\}\boldsymbol{w}^T E\left\{\boldsymbol{r}[i]\boldsymbol{k}^T[i]\right\}\left(\boldsymbol{s}\boldsymbol{w}^T - \boldsymbol{I}_N\right) \\ &+ \boldsymbol{w}\boldsymbol{s}^T E\left\{\boldsymbol{k}[i]\boldsymbol{r}[i]^T\boldsymbol{\theta}[i-1]\boldsymbol{w}^T\boldsymbol{r}[i]\boldsymbol{k}[i]^T\right\}\left(\boldsymbol{s}\boldsymbol{w}^T - \boldsymbol{I}_N\right) \\ &\to \lambda(1-\lambda)E\left\{\boldsymbol{\theta}[i-1]\right\}\left(\boldsymbol{w}^T\boldsymbol{s}\boldsymbol{w}^T - \boldsymbol{w}^T\right) = 0, \quad \text{as } i \to \infty, \end{aligned} \quad (7.195)$$

where we have used (7.183) and the following fact, which is shown below:

$$E\left\{\boldsymbol{k}[i]\boldsymbol{r}[i]^T\boldsymbol{\theta}[i-1]\boldsymbol{w}^T\boldsymbol{r}[i]\boldsymbol{k}[i]^T\right\} \to 0, \quad \text{as } i \to \infty. \quad (7.196)$$

Therefore, the second term is a transient term.

Third term

The third term is the transpose of the second term, and therefore it is also a transient term.

Fourth term

$$\begin{aligned} &= \left(\boldsymbol{w}\boldsymbol{s}^T - \boldsymbol{I}_N\right)E\left\{\boldsymbol{k}[i]\boldsymbol{r}[i]^T\boldsymbol{w}\boldsymbol{w}^T\boldsymbol{r}[i]\boldsymbol{k}[i]^T\right\}\left(\boldsymbol{s}\boldsymbol{w}^T - \boldsymbol{I}_N\right) \\ &= 2(1-\lambda)^2\left(\boldsymbol{w}\boldsymbol{s}^T - \boldsymbol{I}_N\right)\boldsymbol{w}\boldsymbol{w}^T\left(\boldsymbol{s}\boldsymbol{w}^T - \boldsymbol{I}_N\right) \\ &\quad + (1-\lambda)^2\bar{\xi}\left(\boldsymbol{w}\boldsymbol{s}^T - \boldsymbol{I}_N\right)\boldsymbol{R}_r^{-1}\left(\boldsymbol{s}\boldsymbol{w}^T - \boldsymbol{I}_N\right) \qquad (7.197) \\ &= (1-\lambda)^2\bar{\xi}\left(\boldsymbol{R}_r^{-1} - \frac{\boldsymbol{R}_r^{-1}\boldsymbol{s}\boldsymbol{s}^T\boldsymbol{R}_r^{-1}}{\boldsymbol{s}^T\boldsymbol{R}_r^{-1}\boldsymbol{s}}\right), \qquad (7.198) \end{aligned}$$

where in (7.198) we have used (7.152), and in (7.197) we have used the following fact, which is derived below:

$$E\left\{\boldsymbol{k}[i]\boldsymbol{r}[i]^T\boldsymbol{w}\boldsymbol{w}^T\boldsymbol{r}[i]\boldsymbol{k}[i]^T\right\} = 2(1-\lambda)^2\boldsymbol{w}\boldsymbol{w}^T + (1-\lambda)^2\bar{\xi}\boldsymbol{R}_r^{-1}, \quad (7.199)$$

where $\bar{\xi}$ is the mean output energy defined in (7.156).

Now combining these four terms in (7.188), we obtain (for large i)

$$\boldsymbol{K}[i] \cong \lambda^2 \boldsymbol{K}[i-1] + (1-\lambda)^2 \operatorname{tr}\{\boldsymbol{R}_r \boldsymbol{K}[i-1]\} \frac{\boldsymbol{R}_r^{-1} \boldsymbol{s}\boldsymbol{s}^T \boldsymbol{R}_r^{-1}}{\boldsymbol{s}^T \boldsymbol{R}_r^{-1} \boldsymbol{s}} + (1-\lambda)^2 \bar{\xi} \left(\mathbf{R}^{-1} - \frac{\boldsymbol{R}_r^{-1} \boldsymbol{s}\boldsymbol{s}^T \boldsymbol{R}_r^{-1}}{\boldsymbol{s}^T \boldsymbol{R}_r^{-1} \boldsymbol{s}} \right). \quad (7.200)$$

Finally, we derive (7.194), (7.196), and (7.199).

Derivation of (7.194)

We use the notation $[\cdot]_{mn}$ to denote the (m,n)th entry of a matrix and $[\cdot]_k$ to denote the kth entry of a vector. Then

$$\begin{aligned} & E\left\{\boldsymbol{k}[i]\boldsymbol{r}[i]^T \boldsymbol{\theta}[i-1]\boldsymbol{\theta}[i-1]^T \boldsymbol{r}[i]\boldsymbol{k}[i]^T\right\}_{mn} \\ &= E\left\{\sum_{p=1}^{N}\sum_{q=1}^{N} \left[\boldsymbol{k}[i]\boldsymbol{r}[i]^T\right]_{mp} \left[\boldsymbol{\theta}[i-1]\boldsymbol{\theta}[i-1]^T\right]_{pq} \left[\boldsymbol{r}[i]\boldsymbol{k}[i]^T\right]_{qn}\right\} \\ &= \sum_{p=1}^{N}\sum_{q=1}^{N} E\left\{\Big[\boldsymbol{k}[i]\Big]_m \Big[\boldsymbol{r}[i]\Big]_p \Big[\boldsymbol{r}[i]\Big]_q \Big[\boldsymbol{k}[i]\Big]_n\right\} \Big[\boldsymbol{K}[i-1]\Big]_{pq}. \end{aligned} \quad (7.201)$$

Next we use the Gaussian moment factoring theorem to approximate the fourth-order moment introduced in (7.201). The Gaussian moment factoring theorem states that if z_1, z_2, z_3, and z_4 are four samples of a zero-mean, real Gaussian process, then [175]

$$E\{z_1 z_2 z_3 z_4\} = E\{z_1 z_2\} E\{z_3 z_4\} + E\{z_1 z_3\} E\{z_2 z_4\} + E\{z_1 z_4\} E\{z_2 z_3\}. \quad (7.202)$$

Using this approximation, we proceed with (7.201):

$$\begin{aligned} & E\left\{\boldsymbol{k}[i]\boldsymbol{r}[i]^T \boldsymbol{\theta}[i-1]\boldsymbol{\theta}[i-1]^T \boldsymbol{r}[i]\boldsymbol{k}[i]^T\right\}_{mn} \\ &= \sum_{p=1}^{N}\sum_{q=1}^{N} \Bigg[E\left\{\Big[\boldsymbol{k}[i]\Big]_n \Big[\boldsymbol{r}[i]\Big]_p\right\} E\left\{\Big[\boldsymbol{r}[i]\Big]_q \Big[\boldsymbol{k}[i]\Big]_n\right\} \\ &\quad + E\left\{\Big[\boldsymbol{k}[i]\Big]_m \Big[\boldsymbol{r}[i]\Big]_q\right\} E\left\{\Big[\boldsymbol{r}[i]\Big]_p \Big[\boldsymbol{k}[i]\Big]_n\right\} \\ &\quad + E\left\{\Big[\boldsymbol{k}[i]\Big]_m \Big[\boldsymbol{k}[i]\Big]_n\right\} E\left\{\Big[\boldsymbol{r}[i]\Big]_p \Big[\boldsymbol{r}[i]\Big]_q\right\}\Bigg] \Big[\boldsymbol{K}[i-1]\Big]_{pq} \\ &= \sum_{p=1}^{N}\sum_{q=1}^{N} \Big[E\left\{\boldsymbol{k}[i]\boldsymbol{r}[i]^T\right\}_{mp} E\left\{\boldsymbol{r}[i]\boldsymbol{k}[i]^T\right\}_{qn} + E\left\{\boldsymbol{k}[i]\boldsymbol{r}[i]^T\right\}_{mq} E\left\{\boldsymbol{r}[i]\boldsymbol{k}[i]^T\right\}_{pn} \\ &\quad + E\left\{\boldsymbol{k}[i]\boldsymbol{k}[i]^T\right\}_{mn} E\left\{\boldsymbol{r}[i]\boldsymbol{r}[i]^T\right\}_{pq}\Big] \Big[\boldsymbol{K}[i-1]\Big]_{pq} \\ &= 2\left[E\left\{\boldsymbol{k}[i]\boldsymbol{r}[i]^T\right\} \boldsymbol{K}[i-1] E\left\{\boldsymbol{r}[i]\boldsymbol{k}[i]^T\right\}\right]_{mn} \\ &\quad + \operatorname{tr}\left\{\boldsymbol{R}_r \boldsymbol{K}[i-1]\right\} E\left\{\boldsymbol{k}[i]\boldsymbol{k}[i]^T\right\}_{mn}. \end{aligned} \quad (7.203)$$

Therefore,

$$\begin{aligned}
&E\left\{\boldsymbol{k}[i]\boldsymbol{r}[i]^T\boldsymbol{\theta}[i-1]\boldsymbol{\theta}[i-1]^T\boldsymbol{r}[i]\boldsymbol{k}[i]^T\right\} \\
&\quad = 2E\left\{\boldsymbol{k}[i]\boldsymbol{r}[i]^T\right\}\boldsymbol{K}[i-1]E\left\{\boldsymbol{r}[i]\boldsymbol{k}[i]^T\right\} + \operatorname{tr}\left\{\boldsymbol{R}_r\boldsymbol{K}[i-1]\right\}E\left\{\boldsymbol{k}[i]\boldsymbol{k}[i]^T\right\} \\
&\quad = (1-\lambda)^2\left(2\boldsymbol{K}[i-1] + \operatorname{tr}\{\boldsymbol{R}_r\boldsymbol{K}[i-1]\}\boldsymbol{R}_r^{-1}\right),
\end{aligned}$$

where in the last equality we used (7.183) and the following fact:

$$\begin{aligned}
E\left\{\boldsymbol{k}[i]\boldsymbol{k}[i]^T\right\} &= E\left\{\boldsymbol{R}_r^{-1}[i]\boldsymbol{r}[i]\boldsymbol{r}[i]^T\boldsymbol{R}_r^{-1}[i]\right\} \\
&= (1-\lambda)^2\boldsymbol{R}_r^{-1}E\left\{\boldsymbol{r}[i]\boldsymbol{r}[i]^T\right\}\boldsymbol{R}_r^{-1} = (1-\lambda)^2\boldsymbol{R}_r^{-1}. \quad (7.204)
\end{aligned}$$

Derivation of (7.196)

Similarly, we use the approximation by the Gaussian moment factoring formula and obtain

$$\begin{aligned}
&E\left\{\boldsymbol{k}[i]\boldsymbol{r}[i]^T\boldsymbol{\theta}[i-1]\boldsymbol{w}^T\boldsymbol{r}[i]\boldsymbol{k}[i]^T\right\} \\
&\quad = E\left\{\boldsymbol{k}[i]\boldsymbol{r}[i]^T\right\}\left[E\left\{\boldsymbol{\theta}[i-1]\right\}\boldsymbol{w}^T + \boldsymbol{w}E\left\{\boldsymbol{\theta}[i-1]^T\right\}\right]E\left\{\boldsymbol{r}[i]\boldsymbol{k}[i]^T\right\} \\
&\qquad + \boldsymbol{w}^T\boldsymbol{R}_rE\left\{\boldsymbol{\theta}[i-1]\right\}E\left\{\boldsymbol{k}[i]\boldsymbol{k}[i]^T\right\} \\
&\quad = (1-\lambda^2)\left[E\left\{\boldsymbol{\theta}[i-1]\right\}\boldsymbol{w}^T + \boldsymbol{w}E\left\{\boldsymbol{\theta}[i-1]^T\right\} + \boldsymbol{w}^T\boldsymbol{R}_rE\left\{\boldsymbol{\theta}[i-1]\right\}\boldsymbol{R}_r^{-1}\right] \\
&\quad \to 0, \qquad \text{as } i \to \infty,
\end{aligned}$$

since $E\{\boldsymbol{\theta}[i]\} \to \mathbf{0}$.

Derivation of (7.199)

Using the Gaussian moment factoring formula, we obtain

$$\begin{aligned}
&E\left\{\boldsymbol{k}[i]\boldsymbol{r}[i]^T\boldsymbol{w}\boldsymbol{w}^T\boldsymbol{r}[i]\boldsymbol{k}[i]^T\right\} \\
&\quad = 2E\left\{\boldsymbol{k}[i]\boldsymbol{r}[i]^T\right\}\boldsymbol{w}\boldsymbol{w}^TE\left\{\boldsymbol{r}[i]\boldsymbol{k}[i]^T\right\} + \operatorname{tr}\left\{\boldsymbol{R}_r\boldsymbol{w}\boldsymbol{w}^T\right\}E\left\{\boldsymbol{k}[i]\boldsymbol{k}[i]^T\right\} \\
&\quad = 2(1-\lambda)^2\boldsymbol{w}\boldsymbol{w}^T + (1-\lambda)^2\bar{\xi}\boldsymbol{R}_r^{-1}.
\end{aligned}$$

7.9.4 Convergence of MSE

Next we consider the convergence of the output MSE. Let $\xi[i]$ denote the mean output energy at time i and $\epsilon[i]$ denote the MSE at time i:

$$\xi[i]) \triangleq E\left\{\left(\boldsymbol{w}[i-1]^T\boldsymbol{r}[i]\right)^2\right\}, \quad (7.205)$$

$$\epsilon[i] \triangleq E\left\{\left(\sqrt{P}\,b[i] - \boldsymbol{w}[i-1]^T\boldsymbol{r}[i]\right)^2\right\} = \xi[i] - P. \quad (7.206)$$

Since $\epsilon[i]$ and $\xi[i]$ differ only by a constant P, we can focus on the behavior of the mean output energy $\xi[i]$:

$$\begin{aligned}
\xi[i] &= E\left\{\left(\boldsymbol{w}^T + \boldsymbol{\theta}[i-1]^T\right)\boldsymbol{r}[i]\,\boldsymbol{r}[i]^T\left(\boldsymbol{w} + \boldsymbol{\theta}[i-1]\right)\right\} \\
&= \bar{\xi} + \operatorname{tr}\left\{\boldsymbol{R}_r\boldsymbol{K}[i-1]\right\} + 2\boldsymbol{w}^T\boldsymbol{R}_rE\left\{\boldsymbol{\theta}[i-1]\right\}. \quad (7.207)
\end{aligned}$$

Since $E\{\boldsymbol{\theta}[i]\} \to \mathbf{0}$, as $i \to \infty$, the last term in (7.207) is a transient term. Therefore, for large i, $\xi[i] \cong \bar{\xi} + \epsilon_{\text{ex}}[i]$, where $\epsilon_{\text{ex}}[i] \triangleq \text{tr}\{\boldsymbol{R}_r \boldsymbol{K}[i-1]\}$ is the average excess MSE at time i. We are interested in the asymptotic behavior of the excess MSE. Premultiplying both sides of (7.200) by $\boldsymbol{R}_r$ and then taking the trace on both sides, we obtain

$$\begin{aligned}\text{tr}\{\boldsymbol{R}_r\boldsymbol{K}[i]\} &\cong \lambda^2\,\text{tr}\{\boldsymbol{R}_r\boldsymbol{K}[i-1]\} + (1-\lambda)^2\text{tr}\{\boldsymbol{R}_r\boldsymbol{K}[i-1]\}\,\frac{\text{tr}\left\{\boldsymbol{s}\boldsymbol{s}^T\boldsymbol{R}_r^{-1}\right\}}{\boldsymbol{s}^T\boldsymbol{R}_r^{-1}\boldsymbol{s}} \\ &\quad + (1-\lambda)^2\bar{\xi}\left(\text{tr}\{\boldsymbol{I}_N\} - \frac{\text{tr}\left\{\boldsymbol{s}\boldsymbol{s}^T\boldsymbol{R}_r^{-1}\right\}}{\boldsymbol{s}^T\boldsymbol{R}_r^{-1}\boldsymbol{s}}\right) \\ &= \left[\lambda^2 + (1-\lambda)^2\right]\text{tr}\{\boldsymbol{R}_r\boldsymbol{K}[i-1]\} + (1-\lambda)^2(N-1)\bar{\xi}. \end{aligned} \tag{7.208}$$

Since $\lambda^2 + (1-\lambda)^2 < [\lambda + (1-\lambda)]^2 = 1$, the term $\text{tr}\{\boldsymbol{R}_r\boldsymbol{K}[i]\}$ converges. The steady-state excess mean-square error is then given by

$$\epsilon_{\text{ex}}(\infty) \triangleq \lim_{i\to\infty}\text{tr}\{\boldsymbol{R}_r\boldsymbol{K}[i]\} = \frac{1-\lambda}{2\lambda}(N-1)\bar{\xi}. \tag{7.209}$$

Again we see that the convergence of the MSE and the steady-state misadjustment are independent of the eigenvalue distribution of the data autocorrelation matrix, in contrast to the situation for the LMS version of the blind adaptive algorithm [183].

7.9.5 Steady-State SINR

We now consider the steady-state output SINR of the RLS blind adaptive algorithm. At time i the mean output value is

$$\begin{aligned}&E\left\{\boldsymbol{r}[i]^T\boldsymbol{w}[i-1]\right\} \\ &\quad = E\left\{\boldsymbol{r}[i]^T\right\}E\{\boldsymbol{w}[i-1]\} \to \sqrt{P}\,b[i]\,\boldsymbol{s}^T\boldsymbol{w} = \sqrt{P}\,b[i], \quad \text{as} \quad i\to\infty. \end{aligned} \tag{7.210}$$

The variance of the output at time i is

$$\begin{aligned}&\text{Var}\left\{\boldsymbol{r}[i]^T\boldsymbol{w}[i-1]\right\} \\ &\quad = E\left\{\left(\boldsymbol{w}[i-1]^T\boldsymbol{r}[i]\right)^2\right\} - E^2\left\{\boldsymbol{r}[i]^T\boldsymbol{w}[i-1]\right\} \to \xi(\infty) - P. \end{aligned} \tag{7.211}$$

Let $d \triangleq \dfrac{1-\lambda}{2\lambda}(N-1)$. Substituting (7.209) and (7.156) into (7.207), we get

$$\xi(\infty) = (1+d)\bar{\xi} = (1+d)\left(P + \frac{1}{\boldsymbol{s}^T\boldsymbol{\Sigma}^{-1}\boldsymbol{s}}\right). \tag{7.212}$$

Therefore the steady-state SINR is given by

$$\begin{aligned}\text{SINR}^{\infty} &\triangleq \lim_{i\to\infty}\frac{E^2\left\{\boldsymbol{r}[i]^T\boldsymbol{w}[i-1]\right\}}{\text{Var}\left\{\boldsymbol{r}[i]^T\boldsymbol{w}[i-1]\right\}} = \frac{P\boldsymbol{s}^T\boldsymbol{\Sigma}^{-1}\boldsymbol{s}}{(1+d) + dP\left(\boldsymbol{s}^T\boldsymbol{\Sigma}^{-1}\boldsymbol{s}\right)} \\ &= \frac{\text{SINR}^*}{(1+d) + d\cdot\text{SINR}^*}, \end{aligned} \tag{7.213}$$

where SINR^* is the optimum SINR value given in (7.154).

7.9.6 Comparison with Training-Based RLS Algorithm

We now compare the preceding results with the analogous results for the conventional RLS algorithms in which the data symbols $b[i]$ are assumed to be known to the receiver. This condition can be achieved by using either a training sequence or decision feedback. In this case, the exponentially windowed RLS algorithm chooses $\boldsymbol{w}[i]$ to minimize the cost function

$$\sum_{m=1}^{i} \lambda^{i-m} \left(\sqrt{P}\, b[m] - \boldsymbol{w}[i]^T \boldsymbol{r}[m] \right)^2 . \tag{7.214}$$

The RLS adaptation rule in this case is given by [175]

$$\varepsilon_p[i] = \sqrt{P}\, b[i] - \boldsymbol{w}[i-1]^T \boldsymbol{r}[i], \tag{7.215}$$

$$\boldsymbol{w}[i] = \boldsymbol{w}[i-1] + \varepsilon_p[i] \boldsymbol{k}[i], \tag{7.216}$$

where $\varepsilon_p[i]$ is the prediction error at time i and $\boldsymbol{k}[i]$ is the Kalman gain vector defined in (7.160). Using the results from [111], we conclude that the mean weight vector $\boldsymbol{w}[i]$ converges to $\boldsymbol{w}$ (i.e., $E\{\boldsymbol{w}[i]\} \to \boldsymbol{w}$, as $i \to \infty$), where $\boldsymbol{w}$ is the optimal linear MMSE solution:

$$\boldsymbol{w} = P \boldsymbol{R}_r^{-1} \boldsymbol{s}. \tag{7.217}$$

The MSE $\epsilon[i] = \varepsilon_p^2[i]$ also converges, $\epsilon[i] \to \epsilon^* + \epsilon_{\text{ex}}(\infty)$, as $i \to \infty$, where ϵ^* is the mean-square error of the optimum filter $\boldsymbol{w}$, given by

$$\begin{aligned}\epsilon^* &\triangleq E\left\{ \left(\boldsymbol{w}^T \boldsymbol{r} - \sqrt{P}\, b \right)^2 \right\} = \boldsymbol{w}^T \boldsymbol{R}_r \boldsymbol{w} - 2P \boldsymbol{w}^T \boldsymbol{s} + P = P\left[1 - P\left(\boldsymbol{s}^T \boldsymbol{R}_r^{-1} \boldsymbol{s}\right)\right] \\ &= \frac{P}{1 + P\left(\boldsymbol{s}^T \boldsymbol{\Sigma}^{-1} \boldsymbol{s}\right)} = \frac{P}{1 + \text{SINR}^*}. \end{aligned} \tag{7.218}$$

The steady-state excess mean-square error is given by [111]

$$\epsilon_{\text{ex}}(\infty) = \frac{1-\lambda}{1+\lambda} N \epsilon^* \cong d\epsilon^*, \tag{7.219}$$

where we have used the approximation that $\frac{1-\lambda}{1+\lambda} N \cong \frac{1-\lambda}{2\lambda}(N-1) \triangleq d$, since $1-\lambda \ll 1$ and $N \gg 1$. Next we consider the steady-state output SINR of this adaptation rule in which the data symbols $b[i]$ are known. At time i, the mean output value is

$$\begin{aligned} E\left\{\boldsymbol{r}[i]^T \boldsymbol{w}[i-1]\right\} &= E\left\{\boldsymbol{r}[i]^T\right\} E\left\{\boldsymbol{w}[i-1]\right\} \\ &\to \sqrt{P}\, b[i]\, \boldsymbol{s}^T \boldsymbol{w} = \sqrt{P}\, b[i]\, P\, \boldsymbol{s}^T \boldsymbol{R}_r^{-1} \boldsymbol{s} \\ &= \frac{\text{SINR}^*}{1 + \text{SINR}^*} \sqrt{P}\, b[i], \end{aligned} \tag{7.220}$$

where the last equality follows from (7.156). The output MSE at time i is

$$\begin{aligned}\epsilon[i] &= E\left\{\left(\sqrt{P}\,b[i] - \boldsymbol{r}[i]^T\boldsymbol{w}[i-1]\right)^2\right\}\\ &= P + E\left\{\left(\boldsymbol{r}[i]^T\boldsymbol{w}[i-1]\right)^2\right\} - 2\sqrt{P}\,b[i]E\left\{\boldsymbol{r}[i]^T\boldsymbol{w}[i-1]\right\}. \end{aligned} \tag{7.221}$$

Therefore,

$$\begin{aligned}E\left\{\left(\boldsymbol{r}[i]^T\boldsymbol{w}[i-1]\right)^2\right\} &= \epsilon[i] + 2\sqrt{P}\,b[i]E\left\{\boldsymbol{r}[i]^T\boldsymbol{w}[i-1]\right\} - P\\ &\to \epsilon(\infty) - P\left(1 - 2\boldsymbol{s}^T\boldsymbol{w}\right). \end{aligned} \tag{7.222}$$

Using (7.220) and (7.222), after some manipulation, we have

$$\begin{aligned}\mathrm{Var}\left\{\left(\boldsymbol{r}[i]^T\boldsymbol{w}[i-1]\right)^2\right\} &= E\left\{\left(\boldsymbol{r}[i]^T\boldsymbol{w}[i-1]\right)^2\right\} - E^2\left\{\left(\boldsymbol{r}[i]^T\boldsymbol{w}[i-1]\right)\right\}\\ &\to \epsilon(\infty) - P\left(1 - \boldsymbol{s}^T\boldsymbol{w}\right)^2\\ &= \frac{(1+d)\mathrm{SINR}^* + d}{(1+\mathrm{SINR}^*)^2}P. \end{aligned} \tag{7.223}$$

Therefore, the output SINR in the steady state is given by

$$\mathrm{SINR}^\infty \triangleq \lim_{i\to\infty} \frac{E^2\left\{\boldsymbol{r}[i]^T\boldsymbol{w}[i-1]\right\}}{\mathrm{Var}\left\{\boldsymbol{r}[i]^T\boldsymbol{w}[i-1]\right\}} = \frac{\mathrm{SINR}^*}{(1+d) + d/\mathrm{SINR}^*}. \tag{7.224}$$

Chapter 8

MONTE CARLO BAYESIAN SIGNAL PROCESSING

8.1 Introduction

Advanced statistical methods can result in substantial signal processing gain in wireless systems. Among the most powerful such techniques are the Monte Carlo Bayesian methodologies that have recently emerged in statistics. These methods provide a novel paradigm for the design of low-complexity signal processing techniques with performance approaching theoretical optima, for fast and reliable communication in the severe and highly dynamic wireless environments. Over the past decade or so in the field of statistics, a large body of methods has emerged based on iterative Monte Carlo techniques that is useful, especially in computing the Bayesian solutions to the optimal signal reception problems encountered in wireless communications. When employed in the signal processing engines of the digital receivers in wireless networks, these powerful statistical tools hold the potential of closing the substantial gap between the performance of current state-of-the-art wireless receivers and the ultimate optimal performance predicted by statistical communication theory.

In this chapter we provide an overview of the theories and applications in the emerging field of Monte Carlo signal processing [544]. These methods in general fall into two categories: Markov chain Monte Carlo (MCMC) methods for batch signal processing, and sequential Monte Carlo (SMC) methods for adaptive signal processing. For each category we outline the general theory and provide a signal processing example found in wireless communications to illustrate its application. Specifically, we apply the MCMC technique to the problem of Bayesian multiuser detection in unknown channels; and we apply the SMC technique to the problem of adaptive blind equalization in multiple-input/multiple-output (MIMO) intersymbol interference (ISI) channels. The remainder of this chapter is organized as follows. In Section 8.2 we describe the general Bayesian signal processing framework. In Section 8.3 we introduce the MCMC techniques for Bayesian computation. In Section 8.4 we illustrate the application of MCMC by treating the problem of Bayesian multiuser detection in unknown channels. In Section 8.5 we discuss the SMC paradigm for Bayesian computing. In Section 8.6 we illustrate the application of SMC by treating the problem of blind adaptive equalization in MIMO ISI channels. Finally, Section 8.7 contains some mathematical derivations and proofs.

The following is a list of the algorithms appearing in this chapter.

- *Algorithm 8.1:* Metropolis–Hastings algorithm—Form I
- *Algorithm 8.2:* Metropolis–Hastings algorithm—Form II
- *Algorithm 8.3:* Random-scan Gibbs sampler
- *Algorithm 8.4:* Systematic-scan Gibbs sampler
- *Algorithm 8.5:* Gibbs multiuser detector in Gaussian noise
- *Algorithm 8.6:* Gibbs multiuser detector in impulsive noise
- *Algorithm 8.7:* Sequential importance sampling (SIS)
- *Algorithm 8.8:* Sequential Monte Carlo filter for dynamical systems
- *Algorithm 8.9:* Residual resampling
- *Algorithm 8.10:* Mixture Kalman filter for conditional dynamical linear models
- *Algorithm 8.11:* SMC-based blind adaptive equalizer in MIMO channels

8.2 Bayesian Signal Processing

8.2.1 Bayesian Framework

A typical statistical signal processing problem can be stated as follows: Given a set of observations $\boldsymbol{Y} \triangleq [\boldsymbol{y}_1, \boldsymbol{y}_2, \ldots, \boldsymbol{y}_m]$, we would like to make statistical inferences about some unknown quantities $\boldsymbol{X} = [\boldsymbol{x}_1, \boldsymbol{x}_2, \ldots, \boldsymbol{x}_n]$. Typically, the observations $\boldsymbol{Y}$ are functions of the desired unknown quantities $\boldsymbol{X}$ and some unknown "nuisance" parameters $\boldsymbol{\Theta} = [\boldsymbol{\theta}_1, \boldsymbol{\theta}_2, \ldots, \boldsymbol{\theta}_l]$. To illustrate this, consider the following classical signal processing example of equalization.

Example 1: Equalization Suppose that we want to transmit binary symbols $b[1], b[2], \ldots, b[i] \in \{+1, -1\}$, through an intersymbol interference (ISI) channel whose input–output relationship is given by

$$y[i] = \sum_{\ell=0}^{L-1} g_\ell b[i-\ell] + n[i], \tag{8.1}$$

where $\{g_\ell\}_{\ell=0}^{L-1}$ represents the unknown complex channel response and $\{n[i]\}$ contains i.i.d. Gaussian noise samples, with $n[i] \sim \mathcal{N}_c(0, \sigma^2)$. The inference problem of interest is to estimate the transmitted symbols $\boldsymbol{X} = \{b[i]\}$, based on the received signal $\boldsymbol{Y} = \{y[i]\}$. The nuisance parameters are $\boldsymbol{\Theta} = \{g_0, \ldots, g_{L-1}, \sigma^2\}$.

In the Bayesian approach to statistical signal processing problems, all unknown quantities [i.e., $(\boldsymbol{X}, \boldsymbol{\Theta})$] are treated as random variables with some prior distribution described by a probability density $p(\boldsymbol{X}, \boldsymbol{\Theta})$. The Bayesian inference is made based on the joint posterior distribution of these unknowns, described by the density

$$p(\boldsymbol{X}, \boldsymbol{\Theta} \mid \boldsymbol{Y}) \propto p(\boldsymbol{Y} \mid \boldsymbol{X}, \boldsymbol{\Theta})\, p(\boldsymbol{X}, \boldsymbol{\Theta}). \tag{8.2}$$

Note that typically, the joint posterior distribution is completely known up to a normalizing constant. If we are interested in making an inference about the ith component $\boldsymbol{x}_i$ of $\boldsymbol{X}$ [i.e., we wish to compute $E\{h(\boldsymbol{x}_i) \mid \boldsymbol{Y}\}$ for some function $h(\cdot)$], this quantity is given by

$$E\{h(\boldsymbol{x}_i) \mid \boldsymbol{Y}\} = \int h(\boldsymbol{x}_i)\, p(\boldsymbol{x}_i \mid \boldsymbol{Y})\, \mathrm{d}\boldsymbol{x}_i \tag{8.3}$$

$$= \int h(\boldsymbol{x}_i) \int p(\boldsymbol{X}, \boldsymbol{\Theta} \mid \boldsymbol{Y})\, \mathrm{d}\boldsymbol{X}_{[-i]}\, \mathrm{d}\boldsymbol{\Theta}\, \mathrm{d}\boldsymbol{x}_i, \tag{8.4}$$

where $\boldsymbol{X}_{[-i]} \triangleq \boldsymbol{X} \backslash \boldsymbol{x}_i$. These computations are usually not easy in practice.

For an introductory treatment of the Bayesian philosophy, including the selection of prior distributions, see the textbooks [51, 136, 248]. An account of criticism of the Bayesian approach to data analysis can be found in [33, 422], and a defense of the Bayesian choice can be found in [419].

8.2.2 Batch Processing versus Adaptive Processing

Depending on how the data are processed and on how the inference is made, most signal processing methods fall into one of two categories: batch processing and adaptive (i.e., sequential) processing. In *batch signal processing*, the entire data block $\boldsymbol{Y}$ is received and stored before it is processed, and the inference about $\boldsymbol{X}$ is made based on the entire data block $\boldsymbol{Y}$. In *adaptive processing*, on the other hand, inference is made sequentially (i.e., online) as the data are being received. For example, at time t, after a new sample $\boldsymbol{y}_t$ is received, an update on the inference about some or all elements of $\boldsymbol{X}$ is made. In this chapter we focus on *optimal* signal processing under the Bayesian framework for both batch and adaptive processing. We next illustrate batch and adaptive Bayesian signal processing, respectively, using the equalization example above.

Example 1: Batch Equalization Consider the equalization problem mentioned above. Let $\boldsymbol{Y} \triangleq \begin{bmatrix} y[0] & y[1] & \cdots & y[M-1] \end{bmatrix}^T$ be the received signal and $\boldsymbol{X} \triangleq \begin{bmatrix} b[0]\; b[1] & \cdots & b[M-1] \end{bmatrix}^T$ be the transmitted symbols. Denote $\boldsymbol{g} \triangleq [g_0\; g_1 \;\cdots\; g_{L-1}]^T$. An optimal *batch* processing procedure for this problem is as follows. Assume that the unknown quantities $\boldsymbol{g}$, σ^2, and $\boldsymbol{X}$ are independent of each other and have prior densities $p(\boldsymbol{g})$, $p(\sigma^2)$, and $p(\boldsymbol{X})$, respectively. Since $\{n[i]\}$ is a sequence of i.i.d. Gaussian samples, the joint posterior density of these unknown quantities $(\boldsymbol{g}, \sigma^2, \boldsymbol{X})$ based on the received signal $\boldsymbol{Y}$ takes the form

$$p\left(\boldsymbol{g}, \sigma^2, \boldsymbol{X} \mid \boldsymbol{Y}\right) = p\left(\boldsymbol{Y} \mid \boldsymbol{g}, \sigma^2, \boldsymbol{X}\right) p\left(\boldsymbol{g}\right) p\left(\sigma^2\right) p\left(\boldsymbol{X}\right) / p(\boldsymbol{Y}) \tag{8.5}$$

$$\propto \left(\frac{1}{\sigma^2}\right)^{M/2} \exp\left(-\frac{1}{\sigma^2} \sum_{i=0}^{M-1} \left| y[i] - \sum_{\ell=0}^{L-1} h_\ell b[i-\ell] \right|^2\right) p\left(\boldsymbol{g}\right) p\left(\sigma^2\right) p\left(\boldsymbol{X}\right). \tag{8.6}$$

The *a posteriori* probabilities of the transmitted symbols can then be calculated from the joint posterior distribution (8.6) according to

$$P\left(b[i] = +1 \mid \boldsymbol{Y}\right) = \sum_{\boldsymbol{X}: b[i]=+1} P\left(\boldsymbol{X} \mid \boldsymbol{Y}\right) \tag{8.7}$$

$$= \sum_{\boldsymbol{X}: b[i]=+1} \int p\left(\boldsymbol{g}, \sigma^2, \boldsymbol{X} \mid \boldsymbol{Y}\right) \mathrm{d}\boldsymbol{g}\, \mathrm{d}\sigma^2. \tag{8.8}$$

Clearly, the direct computation in (8.8) involves 2^{M-1} multidimensional integrals, which is certainly infeasible for most practical implementations in which M might be on the order of hundreds.

Example 2: Adaptive Equalization Again consider the equalization problem above. Define $\boldsymbol{Y}[i] \triangleq \left[y[0]\ y[1]\ \cdots\ y[i]\right]^T$ and $\boldsymbol{X}[i] \triangleq \left[b[0]\ b[1]\ \cdots\ b[i]\right]^T$ for any i. We now look at the problem of online estimation of the symbol $b[i]$ based on the received signals up to time $i+\Delta$ for some fixed delay $\Delta > 0$. This problem is one of making Bayesian inference with respect to the posterior density

$$\begin{aligned} & p(\boldsymbol{g}, \sigma^2, \boldsymbol{X}[i+\Delta] \mid \boldsymbol{Y}[i+\Delta]) \\ & \propto \left(\frac{1}{\sigma^2}\right)^{(i+\Delta+1)/2} \exp\left(-\frac{1}{2\sigma^2} \sum_{j=0}^{i+\Delta} \left| y[j] - \sum_{\ell=0}^{L-1} g_\ell b[j-\ell] \right|^2\right) \\ & \quad \cdot p\left(\boldsymbol{g}\right) p\left(\sigma^2\right) p\left(\boldsymbol{X}[i+\Delta]\right), \qquad i = 1, 2, \ldots. \end{aligned} \tag{8.9}$$

An online symbol estimate can then be obtained from the marginal posterior distribution

$$\begin{aligned} & P(b[i] = +1 \mid \boldsymbol{Y}[i+\Delta]) \\ & = \sum_{\boldsymbol{X}[i+\Delta]:\, b[i]=+1} \int p(\boldsymbol{g}, \sigma^2, \boldsymbol{X}[i+\Delta] \mid \boldsymbol{Y}[i+\Delta])\, \mathrm{d}\boldsymbol{g}\, \mathrm{d}\sigma^2. \end{aligned} \tag{8.10}$$

Again we see that direct implementation of the optimal sequential Bayesian equalization above involves $2^{i+\Delta}$ multidimensional integrals at time i, which is increasing exponentially in time.

It is seen from the discussions above that although the optimal (i.e., Bayesian) signal processing procedures achieve the best performance (i.e., the Bayesian solutions achieve the minimum probability of error on symbol detection), they exhibit prohibitively high computational complexity and thus are not generally implementable in practice. The recently developed Monte Carlo methods for Bayesian computation have provided a viable approach to solving many such optimal signal processing problems with reasonable computational cost.

8.2.3 Monte Carlo Methods

In a typical Bayesian analysis, the computations involved in eliminating the missing parameters and other unknown quantities are so difficult that one has to resort to some numerical approaches to complete the required summations and integrations. Among all the numerical approaches, Monte Carlo methods are perhaps the most versatile, flexible, and powerful [275].

Suppose that we can generate random samples (either independent or dependent)

$$\left(\boldsymbol{X}^{(1)}, \boldsymbol{\Theta}^{(1)}\right), \left(\boldsymbol{X}^{(2)}, \boldsymbol{\Theta}^{(2)}\right), \ldots, \left(\boldsymbol{X}^{(n)}, \boldsymbol{\Theta}^{(n)}\right)$$

from the joint posterior distribution (8.2); or we can generate random samples

$$\boldsymbol{X}^{(1)}, \boldsymbol{X}^{(1)}, \ldots, \boldsymbol{X}^{(n)}$$

from the marginal posterior distribution $p(\boldsymbol{X}|\boldsymbol{Y})$. Then we can approximate the marginal posterior $p(\boldsymbol{x}_i|\boldsymbol{Y})$ by the empirical distribution (i.e., the histogram) based on the corresponding component in the Monte Carlo sample (i.e., $\boldsymbol{x}_i^{(1)}, \boldsymbol{x}_i^{(2)}, \ldots, \boldsymbol{x}_i^{(n)}$) and approximate the inference (8.4) by

$$E\{h(\boldsymbol{x}_i) \mid \boldsymbol{Y}\} \cong \frac{1}{n} \sum_{j=1}^{n} h\left(\boldsymbol{x}_i^{(j)}\right). \tag{8.11}$$

As noted in Section 8.1, most Monte Carlo techniques fall into one of the following two categories: Markov chain Monte Carlo methods, corresponding to batch processing, and sequential Monte Carlo methods, corresponding to adaptive processing. These are discussed in the remainder of this chapter.

8.3 Markov Chain Monte Carlo Signal Processing

Markov chain Monte Carlo refers to a class of algorithms that allow one to draw (pseudo-) random samples from an arbitrary target probability density, $p(\boldsymbol{x})$, known up to a normalizing constant. The basic idea behind these algorithms is that one can achieve the sampling from the target density $p(\boldsymbol{x})$ by running a Markov chain whose equilibrium density is exactly $p(\boldsymbol{x})$. Here we describe two basic MCMC algorithms, the Metropolis algorithm and the Gibbs sampler, which have been widely used in diverse fields. The validity of both algorithms can be proved using basic Markov chain theory [420].

The roots of MCMC methods can be traced back to the well-known Metropolis algorithm [318], which was used initially to investigate the equilibrium properties of molecules in a gas. The first use of the Metropolis algorithm in a statistical context is found in [174]. The Gibbs sampler, which is a special case of the Metropolis algorithm, was so termed in the seminal paper [137] on image processing. It was brought to statistical prominence by [134], where it was observed that many Bayesian computations could be carried out via the Gibbs sampler. For tutorials on the Gibbs sampler, see [21, 68].

8.3.1 Metropolis–Hastings Algorithm

Let $p(\boldsymbol{x}) = c\exp\{-f(\boldsymbol{x})\}$ be the target probability density from which we want to simulate random draws. The normalizing constant c may be unknown to us. Metropolis et al. introduced the fundamental idea of evolving a Markov process to achieve the sampling of $p(\boldsymbol{x})$ [318]. Hastings later provided a more general algorithm, which is described below [174].

Starting with any configuration $\boldsymbol{x}^{(0)}$, the algorithm evolves from the current state $\boldsymbol{x}^{(t)} = \boldsymbol{x}$ to the next state $\boldsymbol{x}^{(t+1)}$ as follows.

Algorithm 8.1: [Metropolis–Hastings algorithm—Form I]

- *Propose a random perturbation of the current state (i.e., $\boldsymbol{x} \to \boldsymbol{x}'$). More precisely, $\boldsymbol{x}'$ is generated from a proposal transition $T(\boldsymbol{x}^{(t)} \to \boldsymbol{x}')$ [i.e., $\sum_{\boldsymbol{x}'} T(\boldsymbol{x} \to \boldsymbol{x}') = 1$ for all $\boldsymbol{x}$], which is nearly arbitrary (of course, some are better than others in terms of efficiency) and is completely specified by the user.*

- *Compute the Metropolis ratio:*

$$r(\boldsymbol{x}, \boldsymbol{x}') = \frac{p(\boldsymbol{x}')T(\boldsymbol{x}' \to \boldsymbol{x})}{p(\boldsymbol{x})T(\boldsymbol{x} \to \boldsymbol{x}')}. \tag{8.12}$$

- *Generate a random number $u \sim$ uniform(0,1). Let $\boldsymbol{x}^{(t+1)} = \boldsymbol{x}'$ if $u \leq r(\boldsymbol{x}, \boldsymbol{x}')$, and let $\boldsymbol{x}^{(t+1)} = \boldsymbol{x}^{(t)}$ otherwise.*

A better-known form of the Metropolis algorithm is as follows. At each iteration:

Algorithm 8.2: [Metropolis–Hastings algorithm—Form II]

- *A small but random perturbation of the current configuration is made.*

- *The gain (or loss) of an objective function [corresponding to $\log p(\boldsymbol{x}) = f(\boldsymbol{x})$] resulting from this perturbation is computed.*

- *A random number $u \sim$ uniform(0,1) is generated independently.*

- *The new configuration is accepted if $\log(u)$ is smaller than or equal to the gain and rejected otherwise.*

Heuristically, this algorithm is constructed based on a trial-and-error strategy. Metropolis et al. restricted their choice of the perturbation function to be the symmetric ones [318]. Intuitively, this means that there is no trend bias at the proposal stage. That is, the chance of getting $\boldsymbol{x}'$ from perturbing $\boldsymbol{x}$ is the same as that of getting $\boldsymbol{x}$ from perturbing $\boldsymbol{x}'$. Since any proper random perturbation can be represented by a Markov transition function T, the symmetry condition can be written as $T(\boldsymbol{x} \to \boldsymbol{x}') = T(\boldsymbol{x}' \to \boldsymbol{x})$.

Hastings generalized the choice of T to all those that satisfy the property $T(\boldsymbol{x} \to \boldsymbol{x}') > 0$ if and only if $T(\boldsymbol{x}' \to \boldsymbol{x}) > 0$ [174]. It is easy to prove that the Metropolis–Hasting transition rule results in an "actual" transition function $A(\boldsymbol{x}, \boldsymbol{x}')$ (it is different from T because an acceptance/rejection step is involved) that satisfies the detailed balance condition

$$p(\boldsymbol{x})A(\boldsymbol{x}, \boldsymbol{x}') = p(\boldsymbol{x}')A(\boldsymbol{x}', \boldsymbol{x}), \tag{8.13}$$

which necessarily leads to a *reversible* Markov chain with $p(\boldsymbol{x})$ as its invariant distribution. Thus, the sequence $\boldsymbol{x}^{(0)}, \boldsymbol{x}^{(1)}, \ldots$ is drawn (asymptotically) from the desired distribution.

The Metropolis algorithm has been used extensively in statistical physics over the past four decades and is the cornerstone of all MCMC techniques recently adopted and generalized in the statistics community. Another class of MCMC algorithms, the Gibbs sampler [137], differs from the Metropolis algorithm in that it uses conditional distributions based on $p(\boldsymbol{x})$ to construct Markov chain moves.

8.3.2 Gibbs Sampler

Let $\boldsymbol{X} = (\boldsymbol{x}_1, \ldots, \boldsymbol{x}_d)$, where $\boldsymbol{x}_i$ is either a scalar or a vector. Suppose that we want to draw samples of $\boldsymbol{X}$ from an underlying density $p(\boldsymbol{X})$. In the Gibbs sampler, one randomly or systematically chooses a coordinate, say $\boldsymbol{x}_i$, and then updates its value with a new sample $\boldsymbol{x}_i'$ drawn from the conditional distribution $p(\boldsymbol{x}_i|\boldsymbol{X}_{[-i]})$, where we define $\boldsymbol{X}_{[-i]} \triangleq \boldsymbol{X}\backslash\boldsymbol{x}_i$. Algorithmically, the Gibbs sampler can be implemented as follows:

Algorithm 8.3: [Random-scan Gibbs sampler] *Suppose that the current sample is* $\boldsymbol{X}^{(t)} = \left(\boldsymbol{x}_1^{(t)}, \ldots, \boldsymbol{x}_d^{(t)}\right)$. *Then:*

- *Select i randomly from the index set $\{1, \ldots, d\}$ according to a given probability vector $(\pi_1, \ldots, \pi_d)$.*
- *Draw $\boldsymbol{x}_i^{(t+1)}$ from the conditional distribution $p\left(\boldsymbol{x}_i \mid \boldsymbol{X}_{[-i]}^{(t)}\right)$, and let $\boldsymbol{X}_{[-i]}^{(t+1)} = \boldsymbol{X}_{[-i]}^{(t)}$.*

Algorithm 8.4: [Systematic-scan Gibbs sampler] *Let the current sample be* $\boldsymbol{X}^{(t)} = \left(\boldsymbol{x}_1^{(t)}, \ldots, \boldsymbol{x}_d^{(t)}\right)$. *For $i = 1, \ldots, d$, draw $\boldsymbol{x}_i^{(t+1)}$ from the conditional distribution*

$$p\left(\boldsymbol{x}_i \mid \boldsymbol{x}_1^{(t+1)}, \ldots, \boldsymbol{x}_{i-1}^{(t+1)}, \boldsymbol{x}_{i+1}^{(t)}, \ldots, \boldsymbol{x}_d^{(t)}\right). \tag{8.14}$$

It is easy to check that *every* individual conditional update leaves $p(\cdot)$ invariant. Suppose that currently $\boldsymbol{X}^{(t)} \sim p(\cdot)$. Then $\boldsymbol{X}^{(t)}_{[-i]}$ follows its marginal distribution under $p(\cdot)$. Thus,

$$p\left(\boldsymbol{x}_i^{(t+1)} \mid \boldsymbol{X}^{(t)}_{[-i]}\right) \cdot p\left(\boldsymbol{X}^{(t)}_{[-i]}\right) = p\left(\boldsymbol{x}_i^{(t+1)}, \boldsymbol{X}^{(t)}_{[-i]}\right), \tag{8.15}$$

implying that the joint distribution of $\left(\boldsymbol{X}^{(t)}_{[-i]}, \boldsymbol{x}_i^{(t+1)}\right)$ is unchanged at $p(\cdot)$ after one update.

Under regularity conditions, in the steady state, the sequence of vectors $\ldots$, $\boldsymbol{X}^{(t-1)}, \boldsymbol{X}^{(t)}, \boldsymbol{X}^{(t+1)}, \ldots$ is a realization of a homogeneous Markov chain with the transition kernel from state $\boldsymbol{X}'$ to state $\boldsymbol{X}$ given by

$$\begin{aligned} &K\left(\boldsymbol{X}', \boldsymbol{X}\right) \\ &= p\left(\boldsymbol{x}_1 \mid \boldsymbol{x}_2', \ldots, \boldsymbol{x}_d'\right) p\left(\boldsymbol{x}_2 \mid \boldsymbol{x}_1, \boldsymbol{x}_3', \ldots, \boldsymbol{x}_d'\right) \cdots p\left(\boldsymbol{x}_d \mid \boldsymbol{x}_1, \ldots, \boldsymbol{x}_{d-1}\right). \end{aligned} \tag{8.16}$$

The convergence behavior of the Gibbs sampler is investigated in [71, 134, 137, 278, 436, 473] and general conditions are given for the following two results:

- The distribution of $\boldsymbol{X}^{(t)}$ converges geometrically to $p(\boldsymbol{X})$, as $t \to \infty$.
- $\frac{1}{T}\sum_{t=1}^{T} f\left(\boldsymbol{X}^{(t)}\right) \overset{\text{a.s.}}{\to} \int f(\boldsymbol{X})\, p(\boldsymbol{X})\, \mathrm{d}\boldsymbol{X}$, as $T \to \infty$, for any integrable function f.

The Gibbs sampler requires an initial transient period to converge to equilibrium. An initial period of length t_0 is known as the burning-in period, and the first t_0 samples should always be discarded. Convergence is usually detected in some ad hoc way; some methods for this are found in [472]. One such method is to monitor a sequence of weights that measure the discrepancy between the sampled and desired distributions [472]. The samples generated by the Gibbs sampler are not independent, hence care needs to be taken in assessing the accuracy of such estimators. By grouping variables together (i.e., drawing samples of several elements of $\boldsymbol{X}$ simultaneously), one can usually accelerate the convergence and generate less-correlated data [274, 278]. To reduce the dependence between samples, one can extract every rth sample to be used in the estimation procedure. When r is large, this approach generates *almost* independent samples.

Other Techniques

A main problem with all the MCMC algorithms is that they may, for various reasons, move very slowly in the configuration space or may become trapped in a local mode. This phenomenon is generally called *slow mixing* of the chain. When a chain is slow mixing, estimation based on the resulting Monte Carlo samples becomes very inaccurate. Some recent techniques suitable for designing more efficient MCMC samplers include parallel tempering [141], the multiple-try method [279], and evolutionary Monte Carlo [264].

8.4 Bayesian Multiuser Detection via MCMC

In this section we illustrate the application of MCMC signal processing (in particular, the Gibbs sampler) by treating three related problems in multiuser detection under a general Bayesian framework: (1) optimal multiuser detection in the presence of unknown channel parameters, (2) optimal multiuser detection in non-Gaussian ambient noise, and (3) multiuser detection in coded CDMA systems. The methods discussed in this section were first developed in [540]. We begin with a perspective on the related works in these three areas.

The optimal multiuser detection algorithms with known channel parameters, that is, the multiuser maximum-likelihood sequence detector (MLSD) and the multiuser maximum *a posteriori* probability (MAP) detector, were first investigated in [517, 518] (cf. [520]). When the channel parameters (e.g., the received signal amplitudes and the noise variance) are unknown, it is of interest to study the problem of joint multiuser channel parameter estimation and data detection from the received waveform. This problem was first treated in [375], where a solution based on the expectation-maximization (EM) algorithm is derived. In [460], the problem of sequential multiuser amplitude estimation in the presence of unknown data is studied, and an approach based on stochastic approximation is proposed. In [581], a tree-search algorithm is given for joint data detection and amplitude estimation. Other works concerning multiuser detection with unknown channel parameters include [119,206,208,326,339,464]. For systems employing channel coding, the optimal decoding scheme for convolutionally coded CDMA is studied in [143], which is shown to have a prohibitive computational complexity. In [144], some low-complexity receivers that perform multiuser symbol detection and decoding either separately or jointly are studied. The powerful turbo multiuser detection techniques for coded CDMA systems are discussed in Chapter 6. Finally, robust multiuser detection methods in non-Gaussian ambient noise CDMA systems are treated in Chapter 4.

In what follows we present Bayesian multiuser detection techniques with unknown channel parameters in both Gaussian and non-Gaussian ambient noise channels. The Gibbs sampler is employed to calculate the Bayesian estimates of the unknown multiuser symbols from the received waveforms. The Bayesian multiuser detector can naturally be used in conjunction with the MAP channel decoding algorithm to accomplish turbo multiuser detection in unknown channels. Note that although in this section we treat only the simple synchronous CDMA signal model, the techniques discussed here can be generalized to treat more complicated systems, such as intersymbol-interference (ISI) channels [541], asynchronous CDMA with multipath fading [594], nonlinearly modulated CDMA systems [372], multicarrier CDMA systems with space-time coding [591], and systems with Gaussian minimum-shift-keying (GMSK) modulation over multipath fading channels [592].

8.4.1 System Description

As in Chapter 6, we consider a coded discrete-time synchronous real-valued baseband CDMA system with K users, employing normalized modulation waveforms $\boldsymbol{s}_1, \boldsymbol{s}_2, \ldots, \boldsymbol{s}_K$, and signaling through a channel with additive white noise. The block diagram of the transmitter end of such a system is shown in Fig. 8.1. The binary information bits $\{d_k[n]\}_n$ for user k are encoded using a channel code (e.g., block code, convolutional code, or turbo code). A code-bit interleaver is used to reduce the influence of the error bursts at the input of the channel decoder. The interleaved code bits are then mapped to BPSK symbols, yielding a symbol stream $\{b_k[i]\}_i$. Each data symbol is then modulated by a spreading waveform $\boldsymbol{s}_k$ and transmitted through the channel. The received signal is the superposition of the K users' transmitted signals plus the ambient noise, given by

$$\boldsymbol{r}[i] = \sum_{k=1}^{K} A_k \, b_k[i] \, \boldsymbol{s}_k + \boldsymbol{n}[i], \qquad i = 0, \ldots, M-1. \tag{8.17}$$

In (8.17), M is the number of data symbols per user per frame; A_k, $b_k[i]$, and $\boldsymbol{s}_k$ denote, respectively, the amplitude, the ith symbol, and the normalized spreading waveform of the kth user; and $\boldsymbol{n}[i] = \begin{bmatrix} n_0[i] & n_1[i] & \cdots & n_{N-1}[i] \end{bmatrix}^T$ is a zero-mean white noise vector. The spreading waveform is of the form

$$\boldsymbol{s}_k = \frac{1}{\sqrt{N}} \left[c_{0,k} \; c_{1,k} \; \cdots \; c_{N-1,k}\right]^T, \qquad c_{j,k} \in \{+1, -1\}, \tag{8.18}$$

where N is the spreading factor. It is assumed that the receiver knows the spreading waveforms of all active users in the system. Define the following *a priori* symbol probabilities:

$$\rho_k[i] \triangleq P(b_k[i] = +1), \qquad i = 0, \ldots, M-1, \quad k = 1, \ldots, K. \tag{8.19}$$

Note that when no prior information is available, we choose $\rho_k[i] = \frac{1}{2}$ (i.e., all symbols are assumed to be equally likely).

It is further assumed that the additive ambient channel noise $\{\boldsymbol{n}[i]\}$ is a sequence of zero-mean i.i.d. random vectors independent of the symbol sequences $\{b_k[i]\}_{i;k}$. Moreover, each noise vector $\boldsymbol{n}[i]$ is assumed to consist of i.i.d. samples $\{n_j[i]\}_j$. Here we consider two types of noise distributions, corresponding to additive Gaussian noise and additive impulsive noise, respectively. For the former case, the noise $n_j[i]$ is of course assumed to have a Gaussian distribution:

$$n_j[i] \sim \mathcal{N}\left(0, \sigma^2\right), \tag{8.20}$$

where σ^2 is the variance of the noise. For the latter case, the noise $n_j[i]$ is assumed to have a two-term Gaussian mixture distribution (cf. Chapter 4):

$$n_j[i] \sim (1-\epsilon)\mathcal{N}\left(0, \sigma_1^2\right) + \epsilon \mathcal{N}\left(0, \sigma_2^2\right), \tag{8.21}$$

with $0 < \epsilon < 1$ and $\sigma_1^2 < \sigma_2^2$. Here the term $\mathcal{N}\left(0, \sigma_1^2\right)$ represents the nominal ambient noise, and the term $\mathcal{N}\left(0, \sigma_2^2\right)$ represents an impulsive component, with ϵ representing the probability that an impulse occurs. The total noise variance under distribution (8.21) is given by

$$\sigma^2 = (1-\epsilon)\sigma_1^2 + \epsilon\sigma_2^2. \tag{8.22}$$

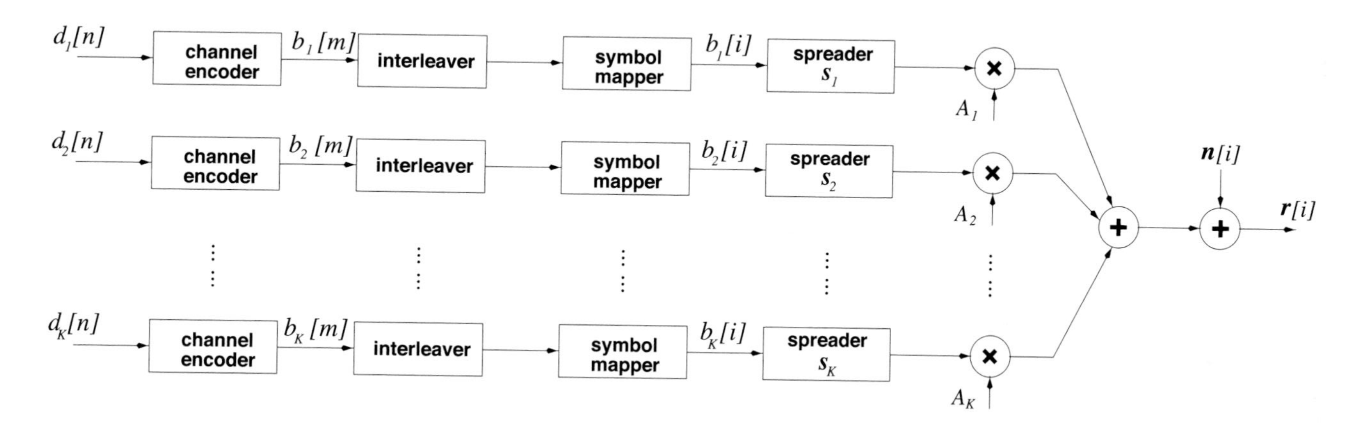

Figure 8.1. Coded synchronous CDMA communication system.

Denote $\boldsymbol{Y} \triangleq \left[\boldsymbol{r}[0]\ \boldsymbol{r}[1]\ \cdots\ \boldsymbol{r}[M-1]\right]$. We consider the problem of estimating the *a posteriori* probabilities of the transmitted symbols

$$P\left(b_k[i] = +1 \mid \boldsymbol{Y}\right), \qquad i = 0, \ldots, M-1, \quad k = 1, \ldots, K, \tag{8.23}$$

based on the received signals $\boldsymbol{Y}$ and the prior information $\{\rho_k[i]\}_{i;k}$, without knowing the channel amplitudes $\{A_k\}$ and the noise parameters (i.e., σ^2 for Gaussian noise; ϵ, σ_1^2, and σ_2^2 for non-Gaussian noise). These *a posteriori* probabilities are then used by the channel decoder to decode the information bits $\{d_k[n]\}_{n;k}$ shown in Fig. 8.1, which is discussed in Section 8.4.4.

8.4.2 Bayesian Multiuser Detection in Gaussian Noise

We now consider the problem of computing the *a posteriori* probabilities in (8.23) under the assumption that the ambient noise distribution is Gaussian; that is, the pdf of $\boldsymbol{n}[i]$ in (8.17) is given by

$$p(\boldsymbol{n}[i]) = \frac{1}{(2\pi\sigma^2)^{\frac{N}{2}}} \exp\left(-\frac{\|\boldsymbol{n}[i]\|^2}{2\sigma^2}\right). \tag{8.24}$$

Define the following notation:

$$\begin{aligned}
\boldsymbol{b}[i] &\triangleq \left[b_1[i]\ b_2[i]\ \cdots\ b_K[i]\right]^T, \qquad i = 0, 1, \ldots, M-1,\\
\boldsymbol{B}[i] &\triangleq \text{diag}\left\{b_1[i], b_2[i], \ldots, b_K[i]\right\}, \qquad i = 0, 1, \ldots, M-1,\\
\boldsymbol{X} &\triangleq \left[\boldsymbol{b}[0]\ \boldsymbol{b}[1]\ \cdots\ \boldsymbol{b}[M-1]\right],\\
\boldsymbol{Y} &\triangleq \left[\boldsymbol{r}[0]\ \boldsymbol{r}[1]\ \cdots\ \boldsymbol{r}[M-1]\right],\\
\boldsymbol{a} &\triangleq [A_1\ A_2\ \cdots\ A_K]^T,\\
\boldsymbol{A} &\triangleq \text{diag}\{A_1, A_2, \ldots, A_K\},\\
\boldsymbol{S} &\triangleq [\boldsymbol{s}_1\ \boldsymbol{s}_2\ \cdots\ \boldsymbol{s}_K].
\end{aligned}$$

Then (8.17) can be written as

$$\begin{aligned}
\boldsymbol{r}[i] &= \boldsymbol{S}\boldsymbol{A}\boldsymbol{b}[i] + \boldsymbol{n}[i] &(8.25)\\
&= \boldsymbol{S}\boldsymbol{B}[i]\boldsymbol{a} + \boldsymbol{n}[i], \qquad i = 0, 1, \ldots, M-1. &(8.26)
\end{aligned}$$

We approach this problem using a Bayesian framework: First, the unknown quantities $\boldsymbol{a}$, σ^2, and $\boldsymbol{X}$ are regarded as realizations of random variables with some prior distributions. The Gibbs sampler, a Monte Carlo method, is then employed to calculate the maximum *a posteriori* probability estimates of these unknowns.

Bayesian Inference

Assume that the unknown quantities $\boldsymbol{a}$, σ^2, and $\boldsymbol{X}$ are independent of each other and have prior densities $p(\boldsymbol{a})$, $p(\sigma^2)$, and $p(\boldsymbol{X})$, respectively. Since $\{\boldsymbol{n}[i]\}$ is a sequence of independent Gaussian vectors, using (8.24) and (8.25), the joint posterior density of these unknown quantities $(\boldsymbol{a}, \sigma^2, \boldsymbol{X})$ based on the received signal $\boldsymbol{Y}$ takes the form

$$\begin{aligned} p\left(\boldsymbol{a}, \sigma^2, \boldsymbol{X} \mid \boldsymbol{Y}\right) &= p\left(\boldsymbol{Y} \mid \boldsymbol{a}, \sigma^2, \boldsymbol{X}\right) p\left(\boldsymbol{a}\right) p\left(\sigma^2\right) p\left(\boldsymbol{X}\right) / p(\boldsymbol{Y}) \\ &\propto \left(\frac{1}{\sigma^2}\right)^{MN/2} \exp\left(-\frac{1}{2\sigma^2} \sum_{i=0}^{M-1} \|\boldsymbol{r}[i] - \boldsymbol{S}\boldsymbol{A}\boldsymbol{b}[i]\|^2\right) p\left(\boldsymbol{a}\right) p\left(\sigma^2\right) p\left(\boldsymbol{X}\right). \end{aligned} \quad (8.27)$$

The *a posteriori* probabilities (8.23) of the transmitted symbols can then be calculated from the joint posterior distribution (8.27) according to

$$\begin{aligned} P\left(b_k[i] = +1 \mid \boldsymbol{Y}\right) &= \sum_{\boldsymbol{X}:\, b_k[i]=+1} P\left(\boldsymbol{X} \mid \boldsymbol{Y}\right) \\ &= \sum_{\boldsymbol{X}:\, b_k[i]=+1} \int p\left(\boldsymbol{a}, \sigma^2, \boldsymbol{X} \mid \boldsymbol{Y}\right) \mathrm{d}\boldsymbol{a}\, \mathrm{d}\sigma^2. \end{aligned} \quad (8.28)$$

The computation in (8.28) involves 2^{KM-1} multidimensional integrals, which is clearly infeasible for any practical implementations with typical values of K and M. To avoid the direct evaluation of the Bayesian estimate (8.28), we resort to the Gibbs sampler discussed in Section 8.3. The basic idea is to generate ergodic random samples $\left\{\boldsymbol{a}^{(n)}, \sigma^{2(n)}, \boldsymbol{X}^{(n)} : n = n_0, n_0+1, \ldots\right\}$ from the posterior distribution (8.27), and then to average $\{b_k[i]^{(n)} : n = n_0, n_0+1, \ldots\}$ drawn from an appropriate conditional distribution to obtain an approximation of the *a posteriori* probabilities in (8.28).

Prior Distributions

In Bayesian analysis, prior distributions are used to incorporate the prior knowledge about the unknown parameters. When such prior knowledge is limited, the prior distributions should be chosen such that they have a minimal impact on the posterior distribution. Such priors are termed *noninformative*. The rationale for using noninformative prior distributions is to "let the data speak for themselves," so that inferences are unaffected by information external to current data [51, 136, 248].

Another consideration in the selection of the prior distributions is to simplify computations. To that end, *conjugate priors* are generally used to obtain simple analytical forms for the resulting posterior distributions. The property that the posterior distribution belongs to the same distribution family as the prior distribution is called *conjugacy*. Conjugate families of distributions are mathematically convenient in that the posterior distribution follows a known parametric form [51, 136, 248]. Finally, to make the Gibbs sampler more computationally efficient, the priors should also be chosen such that the conditional posterior distributions are easy to simulate.

Following these general guidelines in Bayesian analysis, we choose the conjugate prior distributions for the unknown parameters $p(\boldsymbol{a})$, $p\left(\sigma^2\right)$, and $p(\boldsymbol{X})$, as follows.

For the unknown amplitude vector $\boldsymbol{a}$, a truncated Gaussian prior distribution is assumed,

$$p(\boldsymbol{a}) \propto \mathcal{N}(\boldsymbol{a}_0, \boldsymbol{\Sigma}_0)\, I_{\{\boldsymbol{a}>\boldsymbol{0}\}}, \tag{8.29}$$

where $I_{\{\boldsymbol{a}>\boldsymbol{0}\}}$ is an indicator which is 1 if all elements of $\boldsymbol{a}$ are positive and is zero otherwise. Note that large values of $\boldsymbol{\Sigma}_0$ correspond to less informative priors. For the noise variance σ^2, an inverse chi-square prior distribution is assumed,

$$\begin{aligned} p\left(\sigma^2\right) &= \frac{(\nu_0\lambda_0/2)^{\nu_0/2}}{\Gamma(\nu_0/2)} \left(\frac{1}{\sigma^2}\right)^{(\nu_0/2)+1} \exp\left(-\frac{\nu_0\lambda_0}{2\sigma^2}\right) \\ &\sim \chi^{-2}(\nu_0,\, \lambda_0), \end{aligned} \tag{8.30}$$

or

$$\frac{\nu_0\lambda_0}{\sigma^2} \sim \chi^2(\nu_0). \tag{8.31}$$

Small values of ν_0 correspond to the less informative priors (roughly, the prior knowledge is worth ν_0 data points). The value of $\nu_0\lambda_0$ reflects the prior belief of the value of σ^2. Finally, since the symbols $\{b_k[i]\}_{i;k}$ are assumed to be independent, the prior distribution $p(\boldsymbol{X})$ can be expressed in terms of the prior symbol probabilities defined in (8.19) as

$$p(\boldsymbol{X}) = \prod_{i=0}^{M-1} \prod_{k=1}^{K} \rho_k[i]^{\delta_{k,i}} \left(1 - \rho_k[i]\right)^{1-\delta_{k,i}}, \tag{8.32}$$

where

$$\delta_{k,i} = \begin{cases} 1 & \text{if } b_k[i] = +1, \\ 0 & \text{if } b_k[i] = -1. \end{cases} \tag{8.33}$$

Conditional Posterior Distributions

The following conditional posterior distributions are required by the Gibbs multiuser detector in Gaussian noise. The derivations are given in the Appendix (Section 8.7.1).

1. The conditional distribution of the amplitude vector $\boldsymbol{a}$ given σ^2, $\boldsymbol{X}$, and $\boldsymbol{Y}$ is given by

$$p\left(\boldsymbol{a} \mid \sigma^2, \boldsymbol{X}, \boldsymbol{Y}\right) \propto \mathcal{N}(\boldsymbol{a}_*,\, \boldsymbol{\Sigma}_*)\, I_{\{\boldsymbol{a}>\boldsymbol{0}\}}, \tag{8.34}$$

with

$$\boldsymbol{\Sigma}_*^{-1} \triangleq \boldsymbol{\Sigma}_0^{-1} + \frac{1}{\sigma^2} \sum_{i=0}^{M-1} \boldsymbol{B}[i]\boldsymbol{R}\boldsymbol{B}[i], \tag{8.35}$$

$$\boldsymbol{a}_* \triangleq \boldsymbol{\Sigma}_* \left(\boldsymbol{\Sigma}_0^{-1}\boldsymbol{a}_0 + \frac{1}{\sigma^2} \sum_{i=0}^{M-1} \boldsymbol{B}[i]\boldsymbol{S}^T\boldsymbol{r}[i]\right), \tag{8.36}$$

where, in (8.35), we have used $\boldsymbol{R} \triangleq \boldsymbol{S}^T\boldsymbol{S}$ as usual to denote the cross-correlation matrix of the signaling set.

2. The conditional distribution of the noise variance σ^2 given $\boldsymbol{a}$, $\boldsymbol{X}$, and $\boldsymbol{Y}$ is given by

$$p\left(\sigma^2 \mid \boldsymbol{a}, \boldsymbol{X}, \boldsymbol{Y}\right) \sim \chi^{-2}\left(\nu_0 + MN, \frac{\nu_0\lambda_0 + s^2}{\nu_0 + MN}\right), \tag{8.37}$$

or

$$\frac{\nu_0\lambda_0 + s^2}{\sigma^2} \sim \chi^2\left(\nu_0 + MN\right), \tag{8.38}$$

with

$$s^2 \triangleq \sum_{i=0}^{M-1} \left\| \boldsymbol{r}[i] - \boldsymbol{S}\boldsymbol{A}\boldsymbol{b}[i] \right\|^2. \tag{8.39}$$

3. The conditional probabilities of $b_k[i] = \pm 1$, given $\boldsymbol{a}$, σ^2, $\boldsymbol{X}_{ki}$, and $\boldsymbol{Y}$ can be obtained from (where $\boldsymbol{X}_{ki} \triangleq \boldsymbol{X} \backslash b_k[i]$)

$$\frac{P\left(b_k[i] = +1 \mid \boldsymbol{a}, \sigma^2, \boldsymbol{X}_{ki}, \boldsymbol{Y}\right)}{P\left(b_k[i] = -1 \mid \boldsymbol{a}, \sigma^2, \boldsymbol{X}_{ki}, \boldsymbol{Y}\right)} = \frac{\rho_k[i]}{1 - \rho_k[i]} \exp\left\{\frac{2A_k}{\sigma^2} \boldsymbol{s}_k^T \left(\boldsymbol{r}[i] - \boldsymbol{S}\boldsymbol{A}\boldsymbol{b}_k^0[i]\right)\right\},$$
$$i = 0, \ldots, M-1, \quad k = 1, \ldots, K, \tag{8.40}$$

where $\boldsymbol{b}_k^0[i] \triangleq \left[b_1[i], \ldots, b_{k-1}[i], 0, b_{k+1}[i], \ldots, b_K[i]\right]^T$.

Gibbs Multiuser Detector in Gaussian Noise

Using the conditional posterior distributions above, the Gibbs sampling implementation of the Bayesian multiuser detector in Gaussian noise proceeds iteratively as follows.

Algorithm 8.5: [Gibbs multiuser detector in Gaussian noise] *Given initial values of the unknown quantities* $\left\{\boldsymbol{a}^{(0)}, {\sigma^2}^{(0)}, \boldsymbol{X}^{(0)}\right\}$ *drawn from their prior distributions, proceed as follows. For* $n = 1, 2, \ldots$*:*

- *Draw* $\boldsymbol{a}^{(n)}$ *from* $p\left(\boldsymbol{a} \mid {\sigma^2}^{(n-1)}, \boldsymbol{X}^{(n-1)}, \boldsymbol{Y}\right)$ *given by (8.34).*

- *Draw* ${\sigma^2}^{(n)}$ *from* $p\left(\sigma^2 \mid \boldsymbol{a}^{(n)}, \boldsymbol{X}^{(n-1)}, \boldsymbol{Y}\right)$ *given by (8.38).*

- *For* $i = 0, 1, \ldots, M-1$

  For $k = 1, 2, \ldots, K$

Draw $b_k[i]^{(n)}$ *from* $P\left(b_k[i] \mid \boldsymbol{a}^{(n)}, \sigma^{2(n)}, \boldsymbol{X}_{ki}^{(n)}, \boldsymbol{Y}\right)$ *given by (8.40), where*

$$\boldsymbol{X}_{ki}^{(n)} \triangleq \{\boldsymbol{b}[0]^{(n)}, \ldots, \boldsymbol{b}[i-1]^{(n)}, b_1[i]^{(n)}, \ldots b_{k-1}[i]^{(n)}, b_{k+1}[i]^{(n-1)}, \ldots, b_K[i]^{(n-1)}, \boldsymbol{b}[i+1]^{(n-1)}, \ldots, \boldsymbol{b}[M-1]^{(n-1)}\}.$$

Note that to draw samples of $\boldsymbol{a}$ from (8.29) or (8.34), the *rejection method* [527] can be used. For instance, after a sample is drawn from $\mathcal{N}(\boldsymbol{a}_0, \boldsymbol{\Sigma}_0)$ or $\mathcal{N}(\boldsymbol{a}_*, \boldsymbol{\Sigma}_*)$, check to see if the constraint $A_k > 0$, $k = 1, \ldots, K$, is satisfied; if not, the sample is rejected and a new sample is drawn from the same distribution. The procedure continues until a sample is obtained that satisfies the constraint.

To ensure convergence, the procedure above is usually carried out for $(n_0 + N_0)$ iterations for suitably chosen n_0 and N_0, and samples from the last N_0 iterations are used to calculate the Bayesian estimates of the unknown quantities. In particular, the *a posteriori* symbol probabilities in (8.28) are approximated as

$$P\left(b_k[i] = +1 \mid \boldsymbol{Y}\right) \cong \frac{1}{N_0} \sum_{n=n_0+1}^{n_0+N_0} \delta_{ki}^{(n)}, \tag{8.41}$$

where

$$\delta_{ki}^{(n)} \triangleq \begin{cases} 1 & \text{if } b_k^{(n)}[i] = +1, \\ 0 & \text{if } b_k^{(n)}[i] = -1. \end{cases} \tag{8.42}$$

A MAP decision on the symbol $b_k[i]$ is then given by

$$\widehat{b_k[i]} = \arg \max_{b \in \{+1,-1\}} P\left(b_k[i] = b \mid \boldsymbol{Y}\right). \tag{8.43}$$

Furthermore, if desired, estimates of the amplitude vector $\boldsymbol{a}$ and the noise variance σ^2 can also be obtained from the corresponding sample means:

$$E\left\{\boldsymbol{a} \mid \boldsymbol{Y}\right\} \cong \frac{1}{N_0} \sum_{n=n_0+1}^{n_0+N_0} \boldsymbol{a}^{(n)}, \tag{8.44}$$

$$E\left\{\sigma^2 \mid \boldsymbol{Y}\right\} \cong \frac{1}{N_0} \sum_{n=n_0+1}^{n_0+N_0} \sigma^{2(n)}. \tag{8.45}$$

The posterior variances of $\boldsymbol{a}$ and σ^2, which reflect the uncertainty in estimating these quantities on the basis of $\boldsymbol{Y}$, can also be approximated by the sample variances as

$$\begin{aligned} \operatorname{Cov}\left\{\boldsymbol{a} \mid \boldsymbol{Y}\right\} \cong & \frac{1}{N_0} \sum_{n=n_0+1}^{n_0+N_0} \left[\boldsymbol{a}^{(n)}\right]\left[\boldsymbol{a}^{(n)}\right]^T \\ & - \frac{1}{N_0^2}\left[\sum_{n=n_0+1}^{n_0+N_0} \boldsymbol{a}^{(n)}\right]\left[\sum_{n=n_0+1}^{n_0+N_0} \boldsymbol{a}^{(n)}\right]^T, \end{aligned} \tag{8.46}$$

$$\operatorname{Var}\left\{\sigma^2 \mid \boldsymbol{Y}\right\} \cong \frac{1}{N_0} \sum_{n=n_0+1}^{n_0+N_0} \left[\sigma^{2(n)}\right]^2 - \frac{1}{N_0^2}\left[\sum_{n=n_0+1}^{n_0+N_0} \sigma^{2(n)}\right]^2. \tag{8.47}$$

Note that the computations above are exact in the limit as $N_0 \to \infty$. However, since they involve only a finite number of samples, we think of them as approximations but realize that in theory any order of precision can be achieved given a sufficiently large sample size N_0.

The complexity of the Gibbs multiuser detector above per iteration is $\mathcal{O}\left(K^2 + KM\right)$; that is, it has a term that is quadratic with respect to the number of users K, due to the inversion of the positive-definite symmetric matrix in (8.35), and a term that is linear with respect to the symbol block size M. The total complexity is then $\mathcal{O}\left[\left(K^2 + KM\right)(n_0 + N_0)\right]$. For practical values of K and M, this is a substantial complexity reduction compared with the direct implementation of the Bayesian symbol estimate (8.28), whose complexity is $\mathcal{O}\left(2^{KM}\right)$.

Simulation Examples

We consider a five-user ($K = 5$) synchronous CDMA channel with processing gain $N = 10$. The user spreading waveform matrix $\boldsymbol{S}$ and the corresponding correlation matrix $\boldsymbol{R}$ are given, respectively, by

$$\boldsymbol{S}^T = \frac{1}{\sqrt{10}} \begin{bmatrix} -1 & -1 & 1 & 1 & -1 & 1 & -1 & 1 & -1 & 1 \\ 1 & 1 & -1 & -1 & -1 & -1 & -1 & 1 & 1 & 1 \\ 1 & -1 & -1 & 1 & -1 & -1 & -1 & -1 & 1 & -1 \\ -1 & -1 & 1 & -1 & -1 & -1 & 1 & 1 & -1 & 1 \\ 1 & 1 & -1 & -1 & -1 & 1 & -1 & -1 & -1 & -1 \end{bmatrix},$$

$$\boldsymbol{R} = \boldsymbol{S}^T\boldsymbol{S} = \frac{1}{10} \begin{bmatrix} 10 & -2 & -2 & 4 & -2 \\ -2 & 10 & 2 & 0 & 2 \\ -2 & 2 & 10 & -4 & 2 \\ 4 & 0 & -4 & 10 & -4 \\ -2 & 2 & 2 & -4 & 10 \end{bmatrix}.$$

The following noninformative conjugate prior distributions are used in the Gibbs sampler for the case of Gaussian noise:

$$p\left(\boldsymbol{a}^{(0)}\right) \sim \mathcal{N}(\boldsymbol{a}_0, \Sigma_0)\, I_{\{\boldsymbol{a}^{(0)} > \boldsymbol{0}\}} \longrightarrow \boldsymbol{a}_0 = [1\,1\,1\,1\,1]^T, \quad \boldsymbol{\Sigma}_0 = 1000\, \boldsymbol{I}_K;$$

$$p\left(\sigma^{2(0)}\right) \sim \chi^{-2}(\nu_0, \lambda_0) \longrightarrow \nu_0 = 1, \quad \lambda_0 = 0.1.$$

Note that the performance of the Gibbs sampler is insensitive to the values of the parameters in these priors as long as the priors are noninformative.

We illustrate the convergence behavior of the Bayesian multiuser detector in Gaussian noise. In this example, the user amplitudes and the noise variance are taken to be $A_1^2 = -4$ dB, $A_2^2 = -2$ dB, $A_3^2 = 0$ dB, $A_4^2 = 2$ dB, $A_5^2 = 4$ dB, and $\sigma^2 = -2$ dB. The data block size of each user is $M = 256$. In Fig. 8.2 we plot the first 100 samples drawn by the Gibbs sampler of the parameters $b_3[50]$, $b_4[100]$, A_1, A_5, and σ^2. The corresponding true values of these quantities are shown in the same figure as horizontal lines. Note that in this case, the number of unknown parameters is $K + KM + 1 = 1286$ (i.e., $\boldsymbol{a}$, $\boldsymbol{X}$, and σ^2). Remarkably, it is seen that

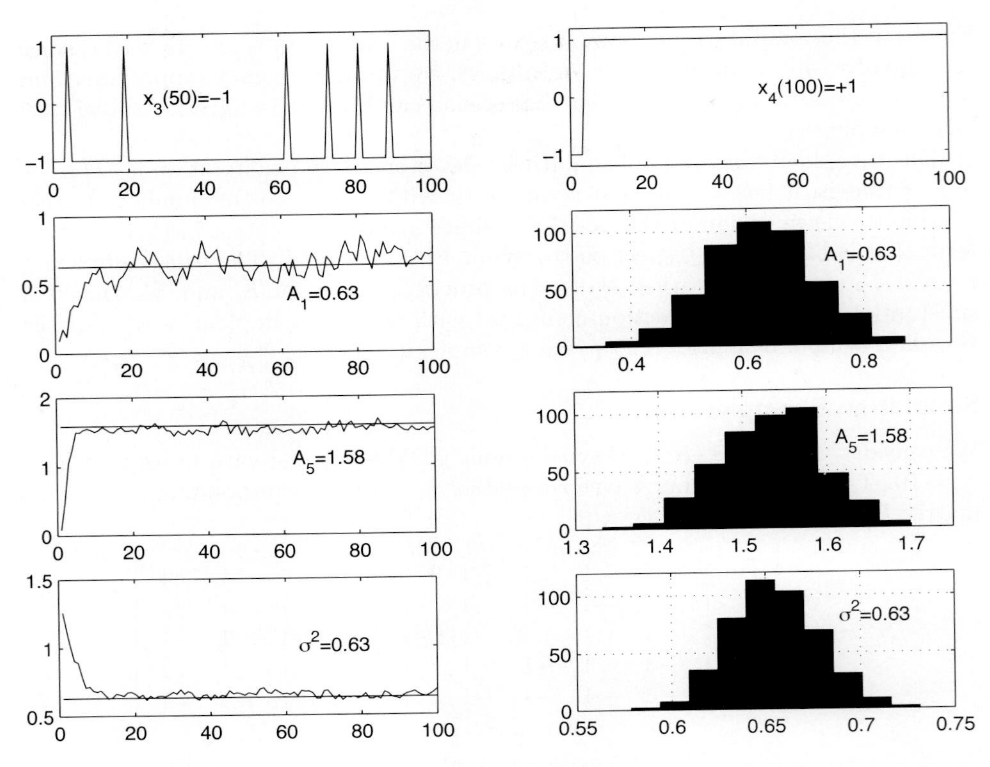

Figure 8.2. Samples and histograms: Gaussian noise. $A_1^2 = -4$ dB, $A_2^2 = -2$ dB, $A_3^2 = 0$ dB, $A_4^2 = 2$ dB, $A_5^2 = 4$ dB, and $\sigma^2 = -2$ dB. The histograms are based on 500 samples collected after the initial 50 iterations.

the Gibbs sampler reaches convergence within about 20 iterations. The marginal posterior distributions of the unknown parameters A_1, A_5, and σ^2 in the steady state can be illustrated by the corresponding histograms, which are also shown in Fig. 8.2. The histograms are based on 500 samples collected after the initial 50 iterations.

8.4.3 Bayesian Multiuser Detection in Impulsive Noise

We next discuss Bayesian multiuser detection via the Gibbs sampler in non-Gaussian impulsive noise. As discussed above, it is assumed that the noise samples $\{n_j[i]\}_j$ of $\boldsymbol{n}[i]$ in (8.17) are independent with a common two-term Gaussian mixture pdf given by

$$p(n_j[i]) = \frac{1-\epsilon}{\sqrt{2\pi\sigma_1^2}} \exp\left(-\frac{n_j[i]^2}{2\sigma_1^2}\right) + \frac{\epsilon}{\sqrt{2\pi\sigma_2^2}} \exp\left(-\frac{n_j[i]^2}{2\sigma_2^2}\right), \tag{8.48}$$

with $0 < \epsilon < 1$ and $\sigma_1^2 < \sigma_2^2$.

Prior Distributions

Define the following indicator random variable to indicate distribution of the noise sample $n_j[i]$:

$$I_j[i] = \begin{cases} 1 & \text{if } n_j[i] \sim \mathcal{N}\left(0, \sigma_1^2\right), \\ 2 & \text{if } n_j[i] \sim \mathcal{N}\left(0, \sigma_2^2\right), \end{cases} \qquad i = 0, \ldots, M-1, \quad j = 0, \ldots, N-1. \quad (8.49)$$

Denote $\boldsymbol{I} \triangleq \{I_j[i]\}_{j;i}$ and

$$\boldsymbol{\Lambda}[i] \triangleq \operatorname{diag}\left\{\sigma^2_{I_0[i]}, \sigma^2_{I_1[i]}, \ldots, \sigma^2_{I_{N-1}[i]}\right\}, \quad i = 0, \ldots, M-1. \quad (8.50)$$

The unknown quantities in this case are $\boldsymbol{a}, \sigma_1^2, \sigma_2^2, \epsilon, \boldsymbol{I}$, and $\boldsymbol{X}$. The joint posterior distribution of these unknown quantities based on the received signal $\boldsymbol{Y}$ takes the form

$$\begin{aligned} & p\left(\boldsymbol{a}, \sigma_1^2, \sigma_2^2, \epsilon, \boldsymbol{I}, \boldsymbol{X} \mid \boldsymbol{Y}\right) \\ & \propto p\left(\boldsymbol{Y} \mid \boldsymbol{a}, \sigma_1^2, \sigma_2^2, \epsilon, \boldsymbol{I}, \boldsymbol{X}\right) p\left(\boldsymbol{a}\right) p\left(\sigma_1^2\right) p\left(\sigma_2^2\right) p\left(\epsilon\right) p\left(\boldsymbol{I} \mid \epsilon\right) p(\boldsymbol{X}) \\ & \propto \exp\left\{-\frac{1}{2} \sum_{i=0}^{M-1} \left(\boldsymbol{r}[i] - \boldsymbol{S}\boldsymbol{A}\boldsymbol{b}[i]\right)^T \boldsymbol{\Lambda}[i]^{-1} \left(\boldsymbol{r}[i] - \boldsymbol{S}\boldsymbol{A}\boldsymbol{b}[i]\right)\right\} \\ & \cdot \left(\frac{1}{\sigma_1^2}\right)^{\frac{1}{2}\sum_{i=0}^{M-1} m_1[i]} \left(\frac{1}{\sigma_2^2}\right)^{\frac{1}{2}\sum_{i=0}^{M-1} m_2[i]} \\ & \cdot p\left(\boldsymbol{a}\right) p\left(\sigma_1^2\right) p\left(\sigma_2^2\right) p\left(\epsilon\right) p\left(\boldsymbol{I} \mid \epsilon\right) p(\boldsymbol{X}), \end{aligned} \quad (8.51)$$

where $m_l[i]$ is the number of l's in $\{I_0[i], I_1[i], \ldots, I_{N-1}[i]\}$, $l = 1, 2$. Note that $m_1[i] + m_2[i] = N$. We next specify the conjugate prior distributions of the unknown quantities in (8.51).

As in the case of Gaussian noise, the prior distributions $p(\boldsymbol{a})$ and $p(\boldsymbol{X})$ are given, respectively, by (8.29) and (8.32). For the noise variances σ_l^2, $l = 1, 2$, independent inverse chi-square distributions are assumed:

$$p(\sigma_l^2) \sim \chi^{-2}(\nu_l,\ \lambda_l), \quad l = 1, 2, \ \text{ with } \nu_1\lambda_1 < \nu_2\lambda_2. \quad (8.52)$$

For the impulse probability ϵ, a beta prior distribution is assumed:

$$p(\epsilon) = \frac{\Gamma(a_0 + b_0)}{\Gamma(a_0)\Gamma(b_0)} \epsilon^{a_0 - 1} (1 - \epsilon)^{b_0 - 1} \sim \operatorname{beta}(a_0, b_0). \quad (8.53)$$

Note that the value $a_0/(a_0 + b_0)$ reflects the prior knowledge of the value of ϵ. Moreover, $(a_0 + b_0)$ reflects the strength of the prior belief [i.e., roughly the prior knowledge is worth $(a_0 + b_0)$ data points]. Given ϵ, the conditional distribution of the indicator $I_j[i]$ is then

$$P(I_j[i] = 1 \mid \epsilon) = 1 - \epsilon \text{ and } \quad P(I_j[i] = 2 \mid \epsilon) = \epsilon, \quad (8.54)$$

$$\Rightarrow \ p(\boldsymbol{I} \mid \epsilon) \ = \ (1 - \epsilon)^{m_1} \epsilon^{m_2}, \quad (8.55)$$

with

$$m_1 \triangleq \sum_{i=0}^{M-1} m_1[i] \quad \text{and} \quad m_2 \triangleq \sum_{i=0}^{M-1} m_2[i] = MN - m_1.$$

Conditional Posterior Distributions

The following conditional posterior distributions are required by the Gibbs multiuser detector in non-Gaussian noise. The derivations are given in the Appendix (Section 8.7.2).

1. The conditional distribution of the amplitude vector $\boldsymbol{a}$ given σ_1^2, σ_2^2, ϵ, $\boldsymbol{I}$, $\boldsymbol{X}$, and $\boldsymbol{Y}$ is given by

$$p\left(\boldsymbol{a} \mid \sigma_1^2, \sigma_2^2, \epsilon, \boldsymbol{I}, \boldsymbol{X}, \boldsymbol{Y}\right) \sim \mathcal{N}\left(\boldsymbol{a}_*, \boldsymbol{\Sigma}_*\right) I_{\{\boldsymbol{a}>\boldsymbol{0}\}}, \tag{8.56}$$

with

$$\boldsymbol{\Sigma}_*^{-1} \triangleq \boldsymbol{\Sigma}_0^{-1} + \sum_{i=0}^{M-1} \boldsymbol{B}[i]\boldsymbol{S}^T\boldsymbol{\Lambda}[i]^{-1}\boldsymbol{S}\boldsymbol{B}[i], \tag{8.57}$$

$$\boldsymbol{a}_* \triangleq \boldsymbol{\Sigma}_* \left(\boldsymbol{\Sigma}_0^{-1}\boldsymbol{a}_0 + \sum_{i=0}^{M-1} \boldsymbol{B}[i]\boldsymbol{S}^T\boldsymbol{\Lambda}[i]^{-1}\boldsymbol{r}[i]\right). \tag{8.58}$$

2. The conditional distribution of the noise variance σ_l^2 given $\boldsymbol{a}$, $\sigma_{\bar{l}}^2$, ϵ, $\boldsymbol{I}$, $\boldsymbol{X}$, and $\boldsymbol{Y}$ is given by [here $\bar{l} = 2$ if $l = 1$, and $\bar{l} = 1$ if $l = 2$]

$$\begin{aligned} &p\left(\sigma_l^2 \mid \boldsymbol{a}, \sigma_{\bar{l}}^2, \epsilon, \boldsymbol{I}, \boldsymbol{X}, \boldsymbol{Y}\right) \\ &\quad \sim \chi^{-2}\left(\left[\nu_l + \textstyle\sum_{i=0}^{M-1} m_l[i]\right], \frac{\nu_l\lambda_l + s_l^2}{\nu_l + \sum_{i=0}^{M-1} m_l[i]}\right), \end{aligned} \tag{8.59}$$

with

$$s_l^2 \triangleq \sum_{i=0}^{M-1}\sum_{j=0}^{N-1} \left(r_j[i] - \boldsymbol{\xi}_j^T \boldsymbol{A}\boldsymbol{b}[i]\right)^2 \cdot 1_{\{I_j[i]=l\}}, \quad l = 1, 2. \tag{8.60}$$

In (8.60) $1_{\{I_j[i]=l\}}$ is 1 if $I_j[i] = l$ and is 0 if $I_j[i] \neq l$; and $\boldsymbol{\xi}_j^T$ is the jth row of the spreading waveform matrix $\boldsymbol{S}$, $j = 0, \ldots, N-1$.

3. The conditional probability of $b_k[i] = \pm 1$ given $\boldsymbol{a}$, σ_1^2, σ_2^2, ϵ, $\boldsymbol{I}$, $\boldsymbol{X}_{ki}$, and $\boldsymbol{Y}$ can be obtained from (where $\boldsymbol{X}_{ki} \triangleq \boldsymbol{X} \backslash b_k[i]$)

$$\frac{P\left(b_k[i] = +1 \mid \boldsymbol{a}, \sigma_1^2, \sigma_2^2, \epsilon, \boldsymbol{I}, \boldsymbol{X}_{ki}, \boldsymbol{Y}\right)}{P\left(b_k[i] = -1 \mid \boldsymbol{a}, \sigma_1^2, \sigma_2^2, \epsilon, \boldsymbol{I}, \boldsymbol{X}_{ki}, \boldsymbol{Y}\right)}$$
$$= \frac{\rho_k[i]}{1-\rho_k[i]} \exp\left\{2 A_k \boldsymbol{s}_k^T \boldsymbol{\Lambda}[i]^{-1} \left(\boldsymbol{r}[i] - \boldsymbol{S}\boldsymbol{A}\boldsymbol{b}^0[i]\right)\right\},$$
$$k = 1, \ldots, K, \quad i = 0, \ldots, M-1, \quad (8.61)$$

where $\boldsymbol{b}_k^0[i] \triangleq \left[b_1[i], \ldots, b_{k-1}[i], 0, b_{k+1}[i], \ldots, b_K[i]\right]^T$.

4. The conditional distribution of $I_j[i]$ given $\boldsymbol{a}$, σ_1^2, σ_2^2, ϵ, $\boldsymbol{I}_{ji}$, $\boldsymbol{X}$, and $\boldsymbol{Y}$ can be obtained from (where $\boldsymbol{I}_{ji} \triangleq \boldsymbol{I} \backslash I_j[i]$)

$$\frac{P\left(I_j[i] = 1 \mid \boldsymbol{a}, \sigma_1^2, \sigma_2^2, \epsilon, \boldsymbol{I}_{ji}, \boldsymbol{X}, \boldsymbol{Y}\right)}{P\left(I_j[i] = 2 \mid \boldsymbol{a}, \sigma_1^2, \sigma_2^2, \epsilon, \boldsymbol{I}_{ji}, \boldsymbol{X}, \boldsymbol{Y}\right)}$$
$$= \frac{1-\epsilon}{\epsilon}\left(\frac{\sigma_2^2}{\sigma_1^2}\right)^{1/2} \exp\left\{\frac{1}{2}\left(r_j[i] - \boldsymbol{\xi}_j^T \boldsymbol{A}\boldsymbol{b}[i]\right)^2 \left(\frac{1}{\sigma_2^2} - \frac{1}{\sigma_1^2}\right)\right\},$$
$$j = 0, \ldots, N-1, \; i = 0, \ldots, M-1. \quad (8.62)$$

5. The conditional distribution of ϵ given $\boldsymbol{a}$, σ_1^2, σ_2^2, $\boldsymbol{I}$, $\boldsymbol{X}$, and $\boldsymbol{Y}$ is given by

$$p\left(\epsilon \mid \boldsymbol{a}, \sigma_1^2, \sigma_2^2, \boldsymbol{I}, \boldsymbol{X}, \boldsymbol{Y}\right) = \text{beta}\left(a_0 + \sum_{i=0}^{M-1} m_2[i],\; b_0 + \sum_{i=0}^{M-1} m_1[i]\right). \quad (8.63)$$

Gibbs Multiuser Detector in Impulsive Noise

Using the conditional posterior distributions above, the Gibbs sampling implementation of the Bayesian multiuser detector in impulsive noise proceeds iteratively as follows.

Algorithm 8.6: [Gibbs multiuser detector in impulsive noise] *Given initial values of the unknown quantities* $\left\{\boldsymbol{a}^{(0)}, \sigma_1^{2(0)}, \sigma_2^{2(0)}, \epsilon^{(0)}, \boldsymbol{I}^{(0)}, \boldsymbol{X}^{(0)}\right\}$ *drawn from their prior distributions, proceed as follows. For* $n = 1, 2, \ldots$*:*

- *Draw* $\boldsymbol{a}^{(n)}$ *from* $p\left(\boldsymbol{a} \mid \sigma_1^{2(n-1)}, \sigma_2^{2(n-1)}, \epsilon^{(n-1)}, \boldsymbol{I}^{(n-1)}, \boldsymbol{X}^{(n-1)}, \boldsymbol{Y}\right)$ *given by (8.56).*

- *Draw* $\sigma_1^{2(n)}$ *from* $p\left(\sigma_1^2 \mid \boldsymbol{a}^{(n)}, \sigma_2^{2(n-1)}, \epsilon^{(n-1)}, \boldsymbol{I}^{(n-1)}, \boldsymbol{X}^{(n-1)}, \boldsymbol{Y}\right)$ *given by (8.59); draw* $\sigma_2^{2(n)}$ *from* $p\left(\sigma_2^2 \mid \boldsymbol{a}^{(n)}, \sigma_1^{2(n)}, \epsilon^{(n-1)}, \boldsymbol{I}^{(n-1)}, \boldsymbol{X}^{(n-1)}, \boldsymbol{Y}\right)$ *given by (8.59).*

- *For* $i = 0, 1, \ldots, M-1$

  For $k = 1, 2, \ldots, K$

Draw $b_k[i]^{(n)}$ *from* $P\left(b_k[i] \mid \boldsymbol{a}^{(n)}, \sigma_1^{2(n)}, \sigma_2^{2(n)}, \epsilon^{(n-1)}, \boldsymbol{I}^{(n-1)}, \boldsymbol{X}_{ki}^{(n)}, \boldsymbol{Y}\right)$ *given by (8.61), where*

$$\boldsymbol{X}_{ki}^{(n)} \triangleq \{\boldsymbol{b}[0]^{(n)}, \ldots, \boldsymbol{b}[i-1]^{(n)}, b_1[i]^{(n)}, \ldots, b_{k-1}[i]^{(n)}, b_{k+1}[i]^{(n-1)}, \ldots, b_K[i]^{(n-1)}, \boldsymbol{b}[i+1]^{(n-1)}, \ldots, \boldsymbol{b}[M-1]^{(n-1)}\}.$$

- *For* $i = 0, 1, \ldots, M-1$

  For $j = 0, 1, \ldots, N-1$

  Draw $I_j[i]^{(n)}$ *from* $P\left(I_j[i] \mid \boldsymbol{a}^{(n)}, \sigma_1^{2(n)}, \sigma_2^{2(n)}, \epsilon^{(n-1)}, \boldsymbol{I}_{ji}^{(n)}, \boldsymbol{X}^{(n)}, \boldsymbol{Y}\right)$ *given by (8.62), where*

$$\boldsymbol{I}_{ji}^{(n)} \triangleq \{I_0[0]^{(n)}, \ldots, I_{N-1}[0]^{(n)}, \ldots, I_0[i-1]^{(n)}, \ldots, I_{N-1}[i-1]^{(n)}, I_0[i]^{(n)}, \ldots, I_{j-1}[i]^{(n)}, I_{j+1}[i]^{(n-1)}, \ldots, I_{N-1}[i]^{(n-1)}, \ldots, I_{N-1}[M-1]^{(n-1)}\}.$$

- *Draw* $\epsilon^{(n)}$ *from* $p\left(\epsilon \mid \boldsymbol{a}^{(n)}, \sigma_1^{2(n)}, \sigma_2^{2(n)}, \boldsymbol{I}^{(n)}, \boldsymbol{X}^{(n)}, \boldsymbol{Y}\right)$ *given by (8.63).*

As in the case of Gaussian noise, the *a posteriori* symbol probabilities $P(b_k[i] = +1 \mid \boldsymbol{Y})$ are computed using (8.41). The *a posteriori* means and variances of the other unknown quantities can be computed similar to (8.44)–(8.47). The complexity of the Gibbs multiuser detector above is $\mathcal{O}\left(K^2 + KM + MN\right)$ per iteration. Note that the direct implementation of the Bayesian symbol estimate based on (8.51) has a computational complexity of $\mathcal{O}\left(2^{KM+MN}\right)$.

Simulation Examples

The simulated CDMA system is the same as that in Section 8.4.2, except that the noise samples are generated according to the two-term Gaussian model (8.21) with the parameters

$$\epsilon = 0.1, \quad \sigma_2^2/\sigma_1^2 = 100, \quad \sigma^2 \triangleq (1-\epsilon)\sigma_1^2 + \epsilon\sigma_2^2 = 7\,\text{dB}.$$

The data block size of each user is $M = 256$. The following noninformative conjugate prior distributions are used in the Gibbs sampler:

$$p\left(\boldsymbol{a}^{(0)}\right) \sim \mathcal{N}(\boldsymbol{a}_0, \Sigma_0)\, I_{\{\boldsymbol{a}^{(0)} > \boldsymbol{0}\}} \longrightarrow \boldsymbol{a}_0 = [1\,1\,1\,1\,1]^T, \quad \boldsymbol{\Sigma}_0 = 1000\,\boldsymbol{I}_K;$$

$$p\left(\sigma_1^{2(0)}\right) \sim \chi^{-2}(\nu_1, \lambda_1) \longrightarrow \nu_1 = 1, \quad \lambda_1 = 0.1;$$

$$p\left(\sigma_2^{2(0)}\right) \sim \chi^{-2}(\nu_1, \lambda_1) \longrightarrow \nu_1 = 1, \quad \lambda_1 = 1;$$

$$p\left(\epsilon^{(0)}\right) \sim \text{beta}(a_0, b_0) \longrightarrow a_0 = 1, \quad b_0 = 2.$$

The first 100 samples drawn by the Gibbs sampler of the parameters $b_3[100]$, $I_5[75]$, A_3, σ_1^2, σ_2^2, and ϵ are shown in Fig. 8.3. The corresponding true values of these quantities are shown in the same figure as horizontal lines. Note that in this case, the number of unknown parameters is $K + KM + NM + 3 = 3848$ (i.e., $\boldsymbol{a}$, $\boldsymbol{X}$,

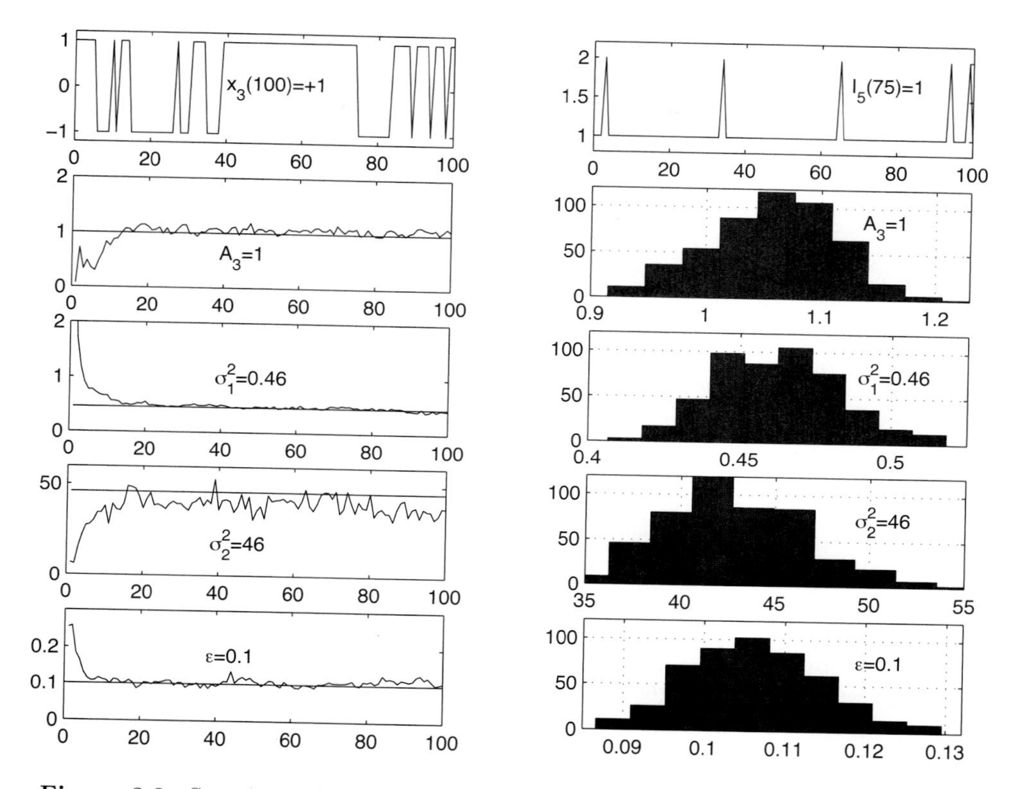

Figure 8.3. Samples and histograms: non-Gaussian noise. $A_1^2 = -4\,\text{dB}$, $A_2^2 = -2\,\text{dB}$, $A_3^2 = 0\,\text{dB}$, $A_4^2 = 2\,\text{dB}$, $A_5^2 = 4\,\text{dB}$, $\epsilon = 0.1$, $\sigma_2^2/\sigma_1^2 = 100$, $\sigma^2 \triangleq (1-\epsilon)\sigma_1^2 + \epsilon\sigma_2^2 = 7\,\text{dB}$. The histograms are based on 500 samples collected after the initial 50 iterations.

$\boldsymbol{I}$, σ_1^2, σ_2^2, and ϵ)! It is seen that as in the Gaussian noise case, the Gibbs sampler converges within about 20 samples. The histograms of the unknown parameters A_3, σ_1^2, σ_2^2, and ϵ are also shown in Fig. 8.3, which are based on 500 samples collected after the initial 50 iterations.

8.4.4 Bayesian Multiuser Detection in Coded Systems

Turbo Multiuser Detection in Unknown Channels

Because it utilizes *a priori* symbol probabilities and produces symbol (or bit) *a posteriori* probabilities, the Bayesian multiuser detector discussed in this section is well suited for iterative processing that allows the Bayesian multiuser detector to refine its processing based on the information from the decoding stage, and vice versa. In Chapter 6, turbo multiuser receivers are described for a number of systems under the assumption that the channels are known to the receiver. The Bayesian multiuser detectors discussed in Sections 8.4.2 and 8.4.3 make it possible to accomplish turbo multiuser detection in coded CDMA systems with unknown channels.

The turbo receiver structure discussed in Section 6.3.1 is shown in Fig. 8.4. It consists of two stages: a Bayesian multiuser detector followed by K MAP channel decoders. The two stages are separated by deinterleavers and interleavers. The Bayesian multiuser detector delivers the *a posteriori* symbol probabilities $\{P(b_k[i] = +1|\boldsymbol{Y})\}_{i;k}$. Based on these, we first compute the *a posteriori* log-likelihood ratios of a transmitted "+1" symbol and a transmitted "−1" symbol,

$$\Lambda_1(b_k[i]) \triangleq \log \frac{P(b_k[i] = +1 \mid \boldsymbol{Y})}{P(b_k[i] = -1 \mid \boldsymbol{Y})}, \quad k = 1, \ldots, K; \quad i = 0, \ldots, M-1. \tag{8.64}$$

Using Bayes' formula, (8.64) can be written as

$$\Lambda_1(b_k[i]) = \underbrace{\log \frac{p(\boldsymbol{Y} \mid b_k[i] = +1)}{p(\boldsymbol{Y} \mid b_k[i] = -1)}}_{\lambda_1(b_k[i])} + \underbrace{\log \frac{P(b_k[i] = +1)}{P(b_k[i] = -1)}}_{\lambda_2(b_k[i])}, \tag{8.65}$$

where the second term in (8.65), denoted by $\lambda_2(b_k[i])$, represents the *a priori* LLR of the code bit $b_k[i]$, which is computed by the channel decoder in the previous iteration, interleaved, and then fed back to the adaptive Bayesian multiuser detector. For the first iteration, assuming equally likely code bits (i.e., no prior information available), we then have $\lambda_2(b_k[i]) = 0$, $k = 1, \ldots, K$, $i = 0, \ldots, M-1$. The first term in (8.65), denoted by $\lambda_1(b_k[i])$, represents the extrinsic information delivered by the Bayesian multiuser detector, based on the received signals $\boldsymbol{Y}$, the structure of the multiuser signal given by (8.17), and prior information about all other code bits. The extrinsic information $\lambda_1(b_k[i])$, which is not influenced by the *a priori* information $\lambda_2(b_k[i])$ provided by the channel decoder, is then reverse interleaved (we denote the deinterleaved code bit sequence of the kth user as $\{b_k^\pi[i]\}_i$) and fed into the channel decoder as the *a priori* information in the next iteration.

Based on the extrinsic information of the code bits $\{\lambda_1(b_k^\pi[i])\}_i$ and the structure of the channel code, the kth user's MAP channel decoder computes the *a posteriori* LLR of each code bit (see Section 6.2 for the MAP decoding algorithm):

$$\begin{aligned}\Lambda_2(b_k^\pi[i]) &\triangleq \log \frac{P\left(b_k^\pi[i] = +1 \mid \{\lambda_1(b_k[i])\}_i \text{ , code structure }\right)}{P\left(b_k^\pi[i] = -1 \mid \{\lambda_1(b_k[i])\}_i \text{ , code structure }\right)} \\ &= \lambda_2(b_k^\pi[i]) + \lambda_1(b_k^\pi[i]).\end{aligned} \tag{8.66}$$

It is seen from (8.66) that the output of the MAP channel decoder is the sum of the prior information $\lambda_1(b_k^\pi[i])$ and the extrinsic information $\lambda_2(b_k^\pi[i])$ delivered by the channel decoder. This extrinsic information is the information about the code bit $b_k^\pi[i]$ gleaned from the prior information about the other code bits, $\{b_k^\pi[l]\}_{l \neq i}$, based on the constraint structure of the code. The MAP channel decoder also computes the *a posteriori* LLR of every information bit, which is used to make a decision on the decoded bit at the last iteration. After interleaving, the extrinsic information delivered by the channel decoder $\{\lambda_2(b_k[i])\}_{i;k}$ is then used to compute the *a priori*

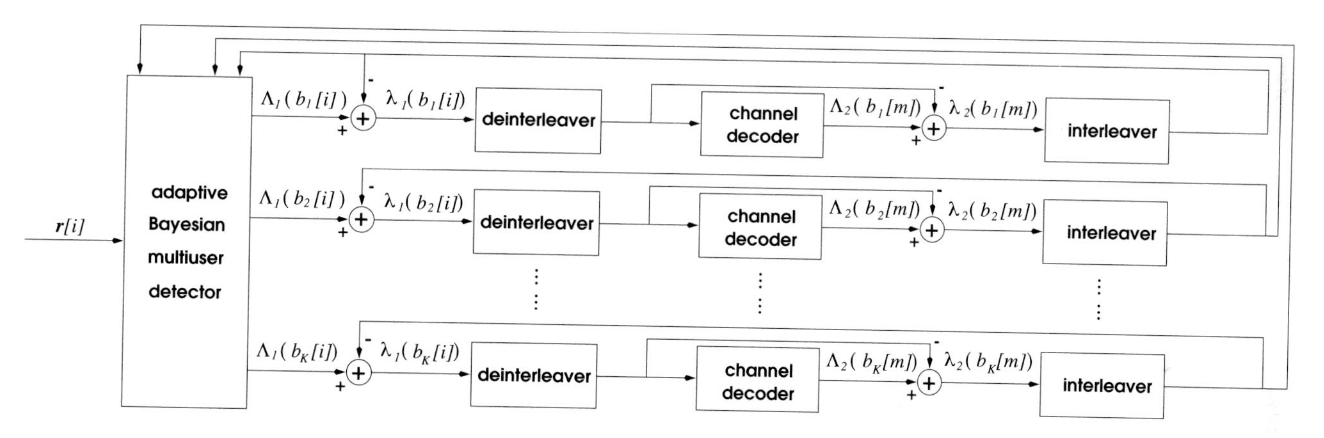

Figure 8.4. Turbo multiuser detection in unknown channels.

symbol distributions $\{\rho_k[i]\}_{i;k}$ defined in (8.23) from the corresponding LLRs as follows:

$$\begin{aligned}\rho_k[i] &\triangleq P(b_k[i] = +1) \\ &= \frac{1}{2}\left\{1 + \tanh\left[\frac{1}{2}\lambda_2(b_k[i])\right]\right\}, \end{aligned} \tag{8.67}$$

where the derivation of (8.67) is given in Section 6.3.2 [cf. (6.39)]. The symbol probabilities $\{\rho_k[i]\}_{i;k}$ are then fed back to the Bayesian multiuser detector as the prior information for the next iteration.

Decoder-Assisted Convergence Assessment

Detecting convergence in the Gibbs sampler is usually done in some ad hoc way. Some methods can be found in [472]. One of them is to monitor a sequence of weights that measure the discrepancy between the sampled and desired distributions. In the application considered here, since the adaptive multiuser detector is followed by a bank of channel decoders, we can assess convergence by monitoring the number of bit corrections made by the channel decoders. If this number exceeds some predetermined threshold, we decide that convergence is not achieved. In that case the Gibbs multiuser detector will be applied again to the same data block. The rationale is that if the Gibbs sampler has reached convergence, the symbol (and bit) error rate after multiuser detection should be relatively small. On the other hand, if convergence is not reached, the code bits generated by the multiuser detector are virtually random and do not satisfy the constraints imposed by the code trellis. Hence the channel decoders will make a large number of corrections. Note that there is no additional computational complexity for such a convergence detection: We need only compare the signs of the code bit log-likelihood ratios at the input and output of the soft channel decoder to determine the number of corrections made.

Code-Constrained Gibbs Multiuser Detector

Another approach to exploiting the coded signal structure in Bayesian multiuser detection is to make use of the code constraints in the Gibbs sampler. For instance, suppose that the user information bits are encoded by some block code of length L and the code bits are *not* interleaved. Then one signal frame of M symbols contains $J = M/L$ code words, with the jth code word given by

$$\underline{b}_k[j] = \Big[b_k[jL],\ b_k[jL+1], \ldots, b_k[jL+L-1]\Big]^T,$$
$$j = 0, 1, \ldots, M/L-1, \quad k = 1, \ldots, K.$$

Let $\boldsymbol{\mathcal{B}}_k$ be the set of all valid code words for user k. Now in the Gibbs sampler, instead of drawing the individual symbols $b_k[i]$ one at a time according to (8.40) or (8.61), we draw a code word $\underline{b}_k[j]$ of L symbols from $\boldsymbol{\mathcal{B}}_k$ each time. Specifically, let $-\underline{1}$ denote the code word with all entries being "-1". (This is the *all-zero code word* and is always a valid code word for any block code [565].) If the ambient channel noise is Gaussian, then for any code word $\underline{u} \in \mathcal{B}_k$, the conditional probability of $\underline{b}_k[j] = \underline{u}$, given the values of the rest of the unknowns, can be obtained from

$$\frac{P\left(\underline{b}_k[j] = \underline{u} \mid \boldsymbol{a}, \sigma^2, \boldsymbol{X}_{kj}, \boldsymbol{Y}\right)}{P\left(\underline{b}_k[j] = -\underline{1} \mid \boldsymbol{a}, \sigma^2, \boldsymbol{X}_{kj}, \boldsymbol{Y}\right)}$$

$$= \frac{\rho_{kj}(\underline{u})}{1-\rho_{kj}(\underline{u})} \cdot \exp\left\{\frac{2A_k}{\sigma^2}\boldsymbol{s}_k^T \sum_{\substack{l=0 \\ u[l]\neq -1}}^{L-1} \left(\boldsymbol{r}[jL+l] - \boldsymbol{S}\boldsymbol{A}\boldsymbol{b}_k^0[jL+l]\right)\right\},$$

$$k = 1, \ldots, K; \quad j = 0, 1, \ldots, M/L - 1, \qquad (8.68)$$

where $\boldsymbol{X}_{kj} \triangleq \boldsymbol{X} \backslash \underline{x}_k[j]$; $\rho_{kj}(\underline{u}) \triangleq P(\underline{b}_k[j] = \underline{u})$; and $\boldsymbol{b}_k^0[i] \triangleq \left[b_1[i], \ldots, b_{k-1}[i], 0, b_{k+1}[i], \ldots, b_K[i]\right]^T$. On the other hand, if the ambient channel noise is non-Gaussian, we have

$$\frac{P\left(\underline{b}_k[j] = \underline{u} \mid \boldsymbol{a}, \sigma_1^2, \sigma_2^2, \epsilon, \boldsymbol{I}, \boldsymbol{X}_{kj}, \boldsymbol{Y}\right)}{P\left(\underline{b}_k[j] = -\underline{1} \mid \boldsymbol{a}, \sigma_1^2, \sigma_2^2, \epsilon, \boldsymbol{I}, \boldsymbol{X}_{kj}, \boldsymbol{Y}\right)}$$

$$= \frac{\rho_{kj}(\underline{u})}{1-\rho_{kj}(\underline{u})} \cdot \exp\left\{2A_k\boldsymbol{s}_k^T \sum_{\substack{l=0 \\ u[l]\neq -1}}^{L-1} \boldsymbol{\Lambda}[jL+l]^{-1}\left(\boldsymbol{r}[jL+l] - \boldsymbol{S}\boldsymbol{A}\boldsymbol{b}^0[jL+l]\right)\right\},$$

$$k = 1, \ldots, K; \quad j = 0, \ldots, M/L - 1. \qquad (8.69)$$

The conditional distributions for sampling the other unknowns remain the same as before. The advantage of sampling a code word instead of sampling an individual symbol is that it can significantly improve the accuracy of samples drawn by the Gibbs sampler, since only valid code words are drawn. This will be demonstrated by simulation examples below.

Relationship between the Gibbs Sampler and the EM Algorithm

As noted previously, the expectation-maximization (EM) algorithm has also been applied to joint parameter estimation and multiuser detection [375]. The major advantage of the Gibbs sampling technique over the EM algorithm is that the Gibbs sampler is a *global* optimization technique. The EM algorithm is a *local* optimization method and it can become trapped by local extrema on the likelihood surface. The EM algorithm performs well if the initial estimates of the channel and symbols are close to their true values. On the other hand, the Gibbs sampler is guaranteed to converge to the global optimum with any random initialization. Of course, the convergence rate depends crucially on the shape of the joint posterior density surface. When the posterior distribution has several modes separated by very low density regions (energy gap), the Gibbs sampler, which generates "random walks" according to the distribution, may have difficulties crossing such gaps to visit all the modes. If such a gap is severe, the random walk may get stuck within one mode for a long time before it moves to another mode. Many modifications of the

Gibbs sampler have been developed to combat the "large energy gap" situation (see, e.g., [162, 575]).

Simulation Examples

We now illustrate the performance in coded systems of the turbo multiuser detectors above. The channel code for each user is a rate-$\frac{1}{2}$ constraint length-5 convolutional code (with generators 23 and 35 in octal notation). The interleaver of each user is independently and randomly generated and is fixed for all simulations. The block size of the information bits is 128 (i.e., the code bit block size is $M = 256$). The code bits are BPSK modulated (i.e., $b_k[i] \in \{+1, -1\}$). All users have the same signal-to-noise ratio (SNR). The symbol posterior probabilities are computed according to (8.41) with $n_0 = N_0 = 50$.

For the first iteration, the prior symbol probabilities $\rho_k[i] \triangleq P(b_k[i] = +1) = \frac{1}{2}$ for all symbols; in subsequent iterations, the prior symbol probabilities are provided by the channel decoder, as given by (6.39). The decoder-assisted convergence assessment is employed. Specifically, if the number of bit corrections made by the decoder exceeds one-third of the total number of bits (i.e., $M/3$), it is decided that convergence is not reached and the Gibbs sampler is applied to the same data block again.

Figure 8.5 illustrates the BER performance of the Gibbs turbo multiuser detector for users 1 and 3. The code bit-error rate at the output of the Bayesian multiuser detector is plotted for the first three iterations. The curve corresponding to the first iteration is the uncoded bit-error rate at the output of the Bayesian multiuser detector. The uncoded and coded bit-error-rate curves in a single-user additive white Gaussian noise (AWGN) channel are also shown in the figure (as, respectively, the dash-dotted and the dashed lines). It is seen that by incorporating the extrinsic information provided by the channel decoder as the prior symbol probabilities, the turbo multiuser detector approaches single-user performance in an AWGN channel within a few iterations. The BER performance of the turbo multiuser detector in impulsive noise is illustrated in Fig. 8.6, where the code bit-error rates at the output of the Bayesian multiuser detector for the first three iterations are shown. The uncoded and coded bit-error-rate curves in a single-user additive white impulsive noise (AWIN) channel are also shown in the figure (as the dash-dotted and dashed lines, respectively), where the conventional matched-filter receiver is employed for demodulation. Note that at high SNR, the performance of user 3 after the first iteration is actually better than the single-user performance. This is because the matched-filter receiver is not the optimal single-user receiver in non-Gaussian noise. Indeed, when $K = 1$, the maximum likelihood detector for the signal model (8.17) is given by

$$\hat{b}_1[i] = \text{sign}\left\{ \sum_{j=1}^{N} \frac{r_j[i] c_{j,k}}{\sigma_j^2[i]} \right\}.$$

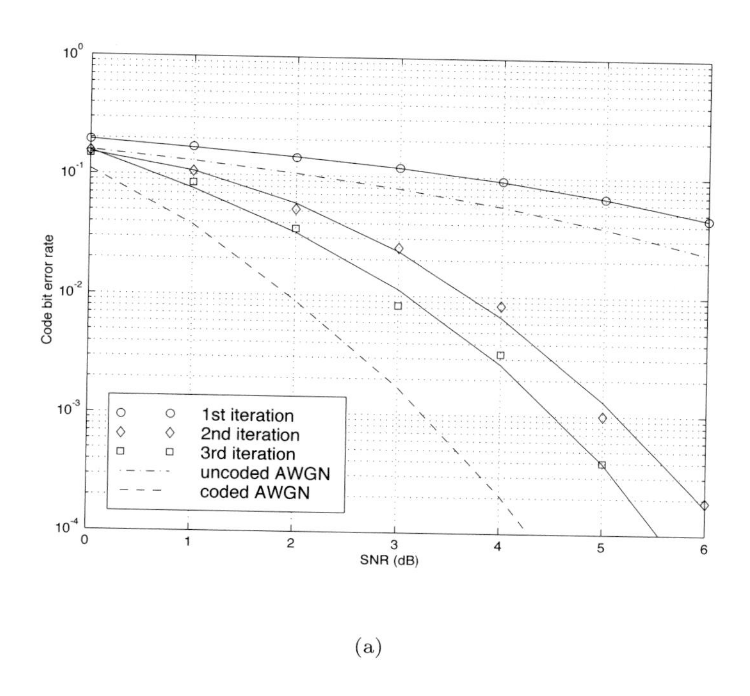

(a)

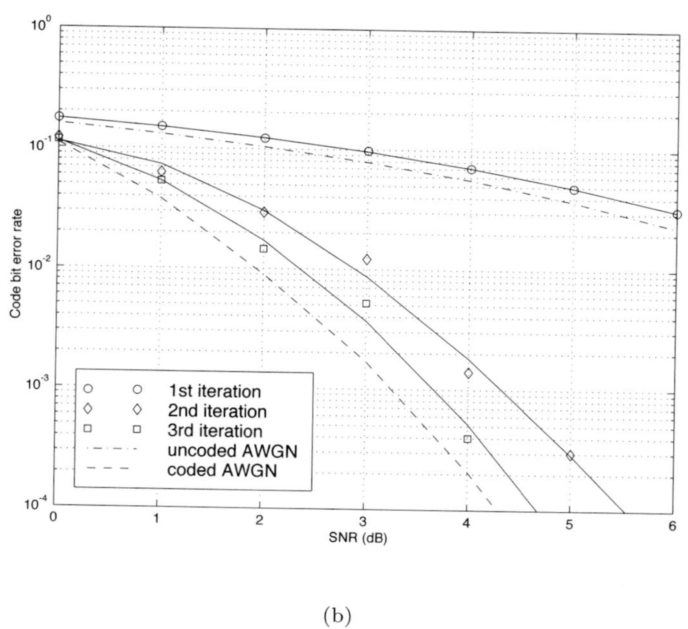

(b)

Figure 8.5. BER performance of a Gibbs turbo multiuser detector: convolutional code, Gaussian noise. (a) User 1; (b) user 3. Both users have the same amplitudes.

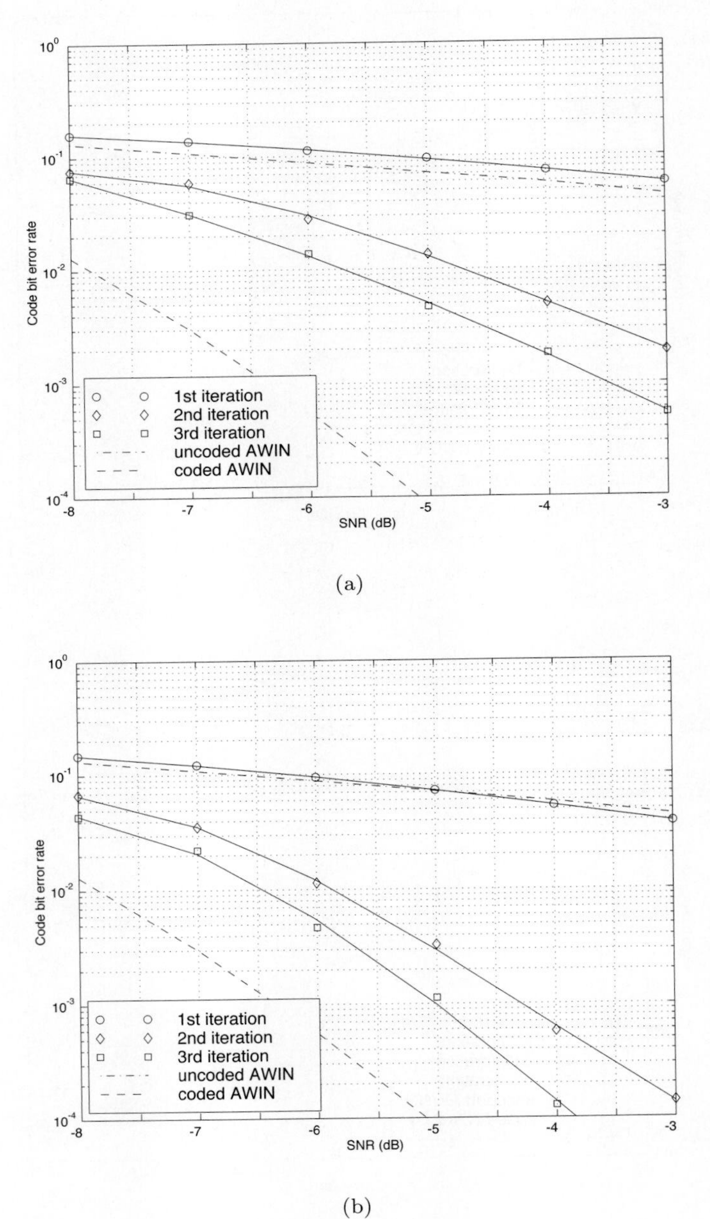

Figure 8.6. BER performance of a Gibbs turbo multiuser detector: convolutional code, impulsive noise. (a) User 1; (b) user 3. Both users have the same amplitudes. $\sigma_2^2/\sigma_1^2 = 100$ and $\epsilon = 0.1$.

Finally, we consider the performance of the code-constrained Gibbs multiuser detectors. We assume that each user employs the (7,4) cyclic block code with eight possible code words [565]:

$$\boldsymbol{\mathcal{B}} = \left\{ \begin{array}{c} (\ -1\ \ -1\ \ -1\ \ -1\ \ -1\ \ -1\ \ -1\ ) \\ (\ \ 1\ \ -1\ \ \ 1\ \ \ 1\ \ \ 1\ \ -1\ \ -1\ ) \\ (\ \ 1\ \ \ 1\ \ \ 1\ \ -1\ \ -1\ \ \ 1\ \ -1\ ) \\ (\ -1\ \ \ 1\ \ \ 1\ \ \ 1\ \ -1\ \ -1\ \ \ 1\ ) \\ (\ \ 1\ \ \ 1\ \ -1\ \ -1\ \ \ 1\ \ -1\ \ \ 1\ ) \\ (\ -1\ \ -1\ \ \ 1\ \ -1\ \ \ 1\ \ \ 1\ \ \ 1\ ) \\ (\ -1\ \ \ 1\ \ -1\ \ \ 1\ \ \ 1\ \ \ 1\ \ -1\ ) \\ (\ \ 1\ \ -1\ \ -1\ \ \ 1\ \ -1\ \ \ 1\ \ \ 1\ ) \end{array} \right\}.$$

The BER performance of the code-constrained Gibbs multiuser detector in Gaussian noise is shown in Fig. 8.7. In this case the Gibbs sampler draws a code word from $\boldsymbol{\mathcal{B}}$ at each time, according to (8.68). In the figure, the unconstrained Gibbs multiuser detector performance before and after decoding is also plotted. It is seen that by exploiting the code constraints in the Gibbs sampler, significant performance gain is achieved. The performance of the code-constrained Gibbs multiuser detector in non-Gaussian noise is shown in Fig. 8.8, and similar performance gain over the unconstrained Gibbs multiuser detector is evident.

8.5 Sequential Monte Carlo Signal Processing

8.5.1 Sequential Importance Sampling

Importance sampling is one of the most well-known and elementary Monte Carlo techniques. Suppose that we want to make an inference about some random quantity $\boldsymbol{X} \sim p(\boldsymbol{X})$ using the Monte Carlo method. Sometimes drawing samples directly from $p(\boldsymbol{X})$ is difficult, but it may be easier (or otherwise advantageous) to draw samples from a *trial* density, say, $q(\boldsymbol{X})$. Note that the desired inference can be written as

$$E\{h(\boldsymbol{X})\} = \int h(\boldsymbol{X})\, p(\boldsymbol{X})\, \mathrm{d}\boldsymbol{X} \tag{8.70}$$

$$= \int h(\boldsymbol{X})\, w(\boldsymbol{X})\, q(\boldsymbol{X})\, \mathrm{d}\boldsymbol{X}, \tag{8.71}$$

where

$$w(\boldsymbol{X}) \triangleq \frac{p(\boldsymbol{X})}{q(\boldsymbol{X})} \tag{8.72}$$

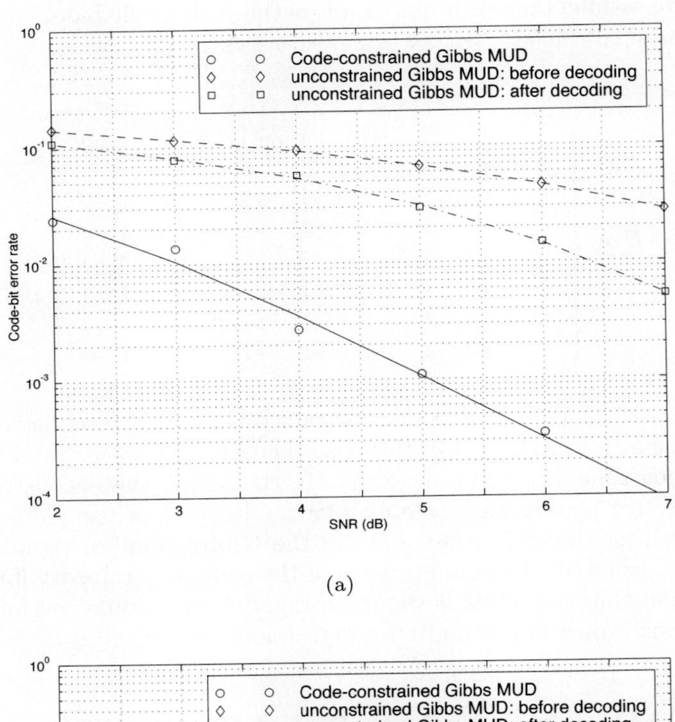

(a)

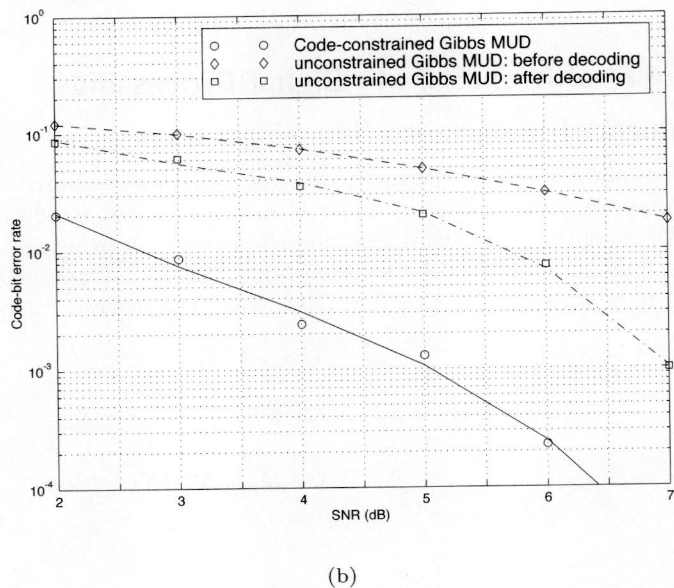

(b)

Figure 8.7. BER performance of a Gibbs turbo multiuser detector: block code, Gaussian noise. (a) User 1; (b) user 3. Both users have the same amplitudes.

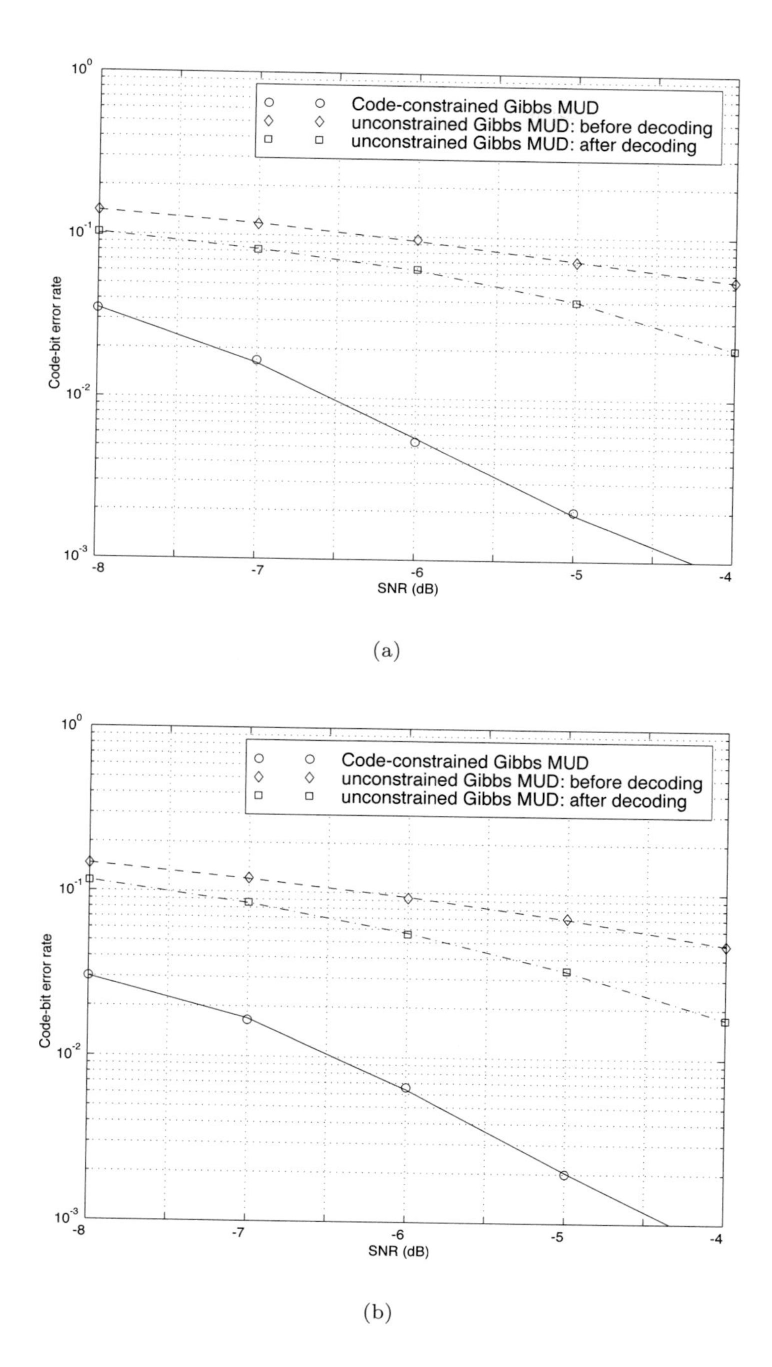

Figure 8.8. BER performance of a Gibbs turbo multiuser: block code, impulsive noise. (a) User 1; (b) user 3. Both users have the same amplitudes. $\sigma_2^2/\sigma_1^2 = 100$ and $\epsilon = 0.1$.

is called the *importance weight*. In importance sampling, we draw samples $\boldsymbol{X}^{(1)}, \boldsymbol{X}^{(2)}, \ldots, \boldsymbol{X}^{(n)}$ according to the trial distribution $q(\cdot)$. We then approximate the inference (8.71) by

$$E\{h(\boldsymbol{X})\} \cong \frac{1}{n} \sum_{j=1}^{n} h\left(\boldsymbol{X}^{(j)}\right) w\left(\boldsymbol{X}^{(j)}\right). \tag{8.73}$$

This technique is widely used, for example, for reducing the sample-size requirements in BER estimation.

However, it is usually difficult to design a good trial density function in high-dimensional problems. One of the most useful strategies in these problems is to build up the trial density sequentially. Suppose that we can decompose $\boldsymbol{X}$ as $\boldsymbol{X} = (\boldsymbol{x}_1, \ldots, \boldsymbol{x}_d)$, where each of the $\boldsymbol{x}_j$ may be either a scalar or a vector. Then our trial density can be constructed as

$$q(\boldsymbol{X}) = q_1(\boldsymbol{x}_1)\, q_2(\boldsymbol{x}_2 \mid \boldsymbol{x}_1) \cdots q_d(\boldsymbol{x}_d \mid \boldsymbol{x}_1, \ldots, \boldsymbol{x}_{d-1}), \tag{8.74}$$

by which we hope to obtain some guidance from the target density while building up the trial density. Corresponding to the decomposition of $\boldsymbol{X}$, we can rewrite the target density as

$$p(\boldsymbol{x}) = p_1(\boldsymbol{x}_1)\, p_2(\boldsymbol{x}_2 \mid \boldsymbol{x}_1) \cdots p_d(\boldsymbol{x}_d \mid \boldsymbol{x}_1, \ldots, \boldsymbol{x}_{d-1}), \tag{8.75}$$

and the importance weight as

$$w(\boldsymbol{X}) = \frac{p_1(\boldsymbol{x}_1)\, p_2(\boldsymbol{x}_2 \mid \boldsymbol{x}_1) \cdots p_d(\boldsymbol{x}_d \mid \boldsymbol{x}_1, \ldots, \boldsymbol{x}_{d-1})}{q_1(\boldsymbol{x}_1)\, q_2(\boldsymbol{x}_2 \mid \boldsymbol{x}_1) \cdots q_d(\boldsymbol{x}_d \mid \boldsymbol{x}_1, \ldots, \boldsymbol{x}_{d-1})}. \tag{8.76}$$

Equation (8.76) suggests a recursive way of computing and monitoring the importance weight. That is, by denoting $\boldsymbol{X}_t = (\boldsymbol{x}_1, \ldots, \boldsymbol{x}_t)$ (thus, $\boldsymbol{X}_d \equiv \boldsymbol{X}$), we have

$$w_t(\boldsymbol{X}_t) = w_{t-1}(\boldsymbol{X}_{t-1}) \cdot \frac{p_t(\boldsymbol{x}_t \mid \boldsymbol{X}_{t-1})}{q_t(\boldsymbol{x}_t \mid \boldsymbol{X}_{t-1})}. \tag{8.77}$$

Then $w_d(\boldsymbol{X}_d)$ is equal to $w(\boldsymbol{X})$ in (8.76). Potential advantages of this recursion and (8.75) are that we can stop generating further components of $\boldsymbol{X}$ if the partial weight derived from the sequentially generated partial sample is too small, and that we can take advantage of $p_t(\boldsymbol{x}_t|\boldsymbol{X}_{t-1})$ in designing $q_t(\boldsymbol{x}_t|\boldsymbol{X}_{t-1})$. In other words, the marginal distribution $p(\boldsymbol{X}_t)$ can be used to guide the generation of $\boldsymbol{X}$.

Although the idea above sounds interesting, the trouble is that the decomposition of $p(\boldsymbol{X})$ as in (8.75) and that of $w(\boldsymbol{X})$ as in (8.76) are not practical at all! The reason is that in order to get (8.75), one needs to have the marginal distribution

$$p(\boldsymbol{X}_t) = \int p(\boldsymbol{x}_1, \ldots, \boldsymbol{x}_d) \mathrm{d}\boldsymbol{x}_{t+1} \cdots \mathrm{d}\boldsymbol{x}_d, \tag{8.78}$$

whose computation involves integrating out components $\boldsymbol{x}_{t+1}, \ldots, \boldsymbol{x}_d$ in $p(\boldsymbol{X})$ and is as difficult as—or even more difficult than—the original problem.

To carry out the sequential sampling idea, we need to introduce another layer of complexity. Suppose that we can find a sequence of *auxiliary distributions*,

$$\pi_1(\boldsymbol{X}_1),\ \pi_2(\boldsymbol{X}_2),\ \ldots,\ \pi_d(\boldsymbol{X}),$$

so that $\pi_t(\boldsymbol{X}_t)$ is a reasonable approximation to the marginal distribution $p_t(\boldsymbol{X}_t)$ for $t = 1, \ldots, d-1$ and $\pi_d(\boldsymbol{X}) = p(\boldsymbol{X})$. We emphasize that $\{\pi_t(\boldsymbol{X}_t)\}$ are required to be known only up to a normalizing constant, and they serve only as "guides" to our construction of the entire sample $\boldsymbol{X} = (\boldsymbol{x}_1, \ldots, \boldsymbol{x}_d)$. The *sequential importance sampling* (SIS) *method* can then be defined as noted in the following recursive procedure.

Algorithm 8.7: [Sequential importance sampling (SIS)] *For* $t = 2, \ldots, d$*:*

- *Draw* $\boldsymbol{x}_t$ *from* $q_t(\boldsymbol{x}_t \mid \boldsymbol{X}_{t-1})$ *and let* $\boldsymbol{X}_t = (\boldsymbol{X}_{t-1}, \boldsymbol{x}_t)$.
- *Compute*

$$u_t \triangleq \frac{\pi_t(\boldsymbol{X}_t)}{\pi_{t-1}(\boldsymbol{X}_{t-1})\, q_t(\boldsymbol{x}_t \mid \boldsymbol{X}_{t-1})}, \tag{8.79}$$

and let $w_t = w_{t-1}\, u_t$.

In the SIS step, we call u_t an *incremental weight*. It is easy to show that $\boldsymbol{X}_t$ is properly weighted by w_t with respect to π_t provided that $\boldsymbol{X}_{t-1}$ is properly weighted by w_{t-1} with respect to π_{t-1}. Thus, the entire sample $\boldsymbol{X}$ obtained in this sequential fashion is properly weighted by the final importance weight, w_d, with respect to the target density $p(\boldsymbol{X})$. One reason for the sequential buildup of the trial density is that it breaks a difficult task into manageable pieces. The SIS framework is particularly attractive, as it can use the auxiliary distributions $\pi_1, \pi_2, \ldots, \pi_d$ to help construct more efficient trial distributions:

- We can build q_t in light of π_t. For example, one can choose (if possible)

$$q_t(\boldsymbol{x}_t \mid \boldsymbol{X}_{t-1}) = \pi_t(\boldsymbol{x}_t \mid \boldsymbol{X}_{t-1}). \tag{8.80}$$

  Then the incremental weight becomes

$$u_t = \frac{\pi_t(\boldsymbol{X}_t)}{\pi_{t-1}(\boldsymbol{X}_{t-1})}. \tag{8.81}$$

- When we observe that w_t is getting too small, we can choose to reject the sample halfway and restart. In this way we avoid wasting time on generating samples that are doomed to have little effect in the final estimation. However, as an outright rejection incurs bias, the rejection control techniques can be used to correct such bias [279].

- Another problem with SIS is that the resulting importance weights are still very skewed, especially when d is large. An important recent advance in sequential Monte Carlo to address this problem is the resampling technique [160, 276].

8.5.2 SMC for Dynamical Systems

Consider the following dynamical system modeled in state-space form as

$$\begin{array}{ll} \text{state equation} & \boldsymbol{z}_t = f_t(\boldsymbol{z}_{t-1}, \boldsymbol{u}_t) \\ \text{observation equation} & \boldsymbol{y}_t = g_t(\boldsymbol{z}_t, \boldsymbol{v}_t), \end{array} \tag{8.82}$$

where $\boldsymbol{z}_t$, $\boldsymbol{y}_t$, $\boldsymbol{u}_t$, and $\boldsymbol{v}_t$ are, respectively, the state variable, the observation, the state noise, and the observation noise at time t. These quantities can be either scalars or vectors.

Let $\boldsymbol{Z}_t = (\boldsymbol{z}_0, \boldsymbol{z}_1, \ldots, \boldsymbol{z}_t)$ and let $\boldsymbol{Y}_t = (\boldsymbol{y}_0, \boldsymbol{y}_1, \ldots, \boldsymbol{y}_t)$. Suppose an online inference of $\boldsymbol{Z}_t$ is of interest; that is, at current time t we wish to make a timely estimate of a function of the state variable $\boldsymbol{Z}_t$, say $h(\boldsymbol{Z}_t)$, based on the currently available observation, $\boldsymbol{Y}_t$. From Bayes' formula, we realize that the optimal solution to this problem is

$$E\{h(\boldsymbol{Z}_t) \mid \boldsymbol{Y}_t\} = \int h(\boldsymbol{Z}_t) p(\boldsymbol{Z}_t \mid \boldsymbol{Y}_t)\, \mathrm{d}\boldsymbol{Z}_t. \tag{8.83}$$

In most cases an exact evaluation of this expectation is analytically intractable because of the complexity of such a dynamical system. Monte Carlo methods provide us with a viable alternative to the required computation. Specifically, if we can draw m random samples $\left\{\boldsymbol{Z}_t^{(j)}\right\}_{j=1}^m$ from the distribution $p(\boldsymbol{Z}_t|\boldsymbol{Y}_t)$, we can approximate $E\{h(\boldsymbol{Z}_t)|\boldsymbol{Y}_t\}$ by

$$E\{h(\boldsymbol{Z}_t) \mid \boldsymbol{Y}_t\} \cong \frac{1}{m} \sum_{j=1}^m h\left(\boldsymbol{Z}_t^{(j)}\right). \tag{8.84}$$

Very often, direct simulation from $p(\boldsymbol{Z}_t|\boldsymbol{Y}_t)$ is not feasible, but drawing samples from some trial distribution is easy. In this case we can use the idea of importance sampling discussed above. Suppose that a set of random samples $\left\{\boldsymbol{Z}_t^{(j)}\right\}_{j=1}^m$ is generated from the trial distribution $q(\boldsymbol{Z}_t|\boldsymbol{Y}_t)$. By associating the weight

$$w_t^{(j)} = \frac{p\left(\boldsymbol{Z}_t^{(j)} \mid \boldsymbol{Y}_t\right)}{q\left(\boldsymbol{Z}_t^{(j)} \mid \boldsymbol{Y}_t\right)} \tag{8.85}$$

with the sample $\boldsymbol{Z}_t^{(j)}$, we can approximate the quantity of interest, $E\{h(\boldsymbol{Z}_t)|\boldsymbol{Y}_t\}$, as

$$E_p\{h(\boldsymbol{Z}_t) \mid \boldsymbol{Y}_t\} \cong \frac{1}{W_t} \sum_{j=1}^m h\left(\boldsymbol{Z}_t^{(j)}\right) w_t^{(j)}, \tag{8.86}$$

with

$$W_t \triangleq \sum_{j=1}^m w_t^{(j)}. \tag{8.87}$$

The pair $\left(\boldsymbol{Z}_t^{(j)}, w_t^{(j)}\right)$, $j = 1, \ldots, m$, is called a *properly weighted sample* with respect to distribution $p(\boldsymbol{Z}_t|\boldsymbol{Y}_t)$. A trivial but important observation is that the $\boldsymbol{z}_t^{(j)}$ (one of the components of $\boldsymbol{Z}_t^{(j)}$) is also properly weighted by the $w_t^{(j)}$ with respect to the marginal distribution $p(\boldsymbol{z}_t|\boldsymbol{Y}_t)$.

Another possible estimate of $E\{h(\boldsymbol{Z}_t)|\boldsymbol{Y}_t\}$ is

$$E_p\{h(\boldsymbol{Z}_t) \mid \boldsymbol{Y}_t\} \cong \frac{1}{m} \sum_{j=1}^{m} h\left(\boldsymbol{Z}_t^{(j)}\right) w_t^{(j)}. \tag{8.88}$$

The main reasons for preferring the ratio estimate (8.86) to the unbiased estimate (8.88) in an importance sampling framework are that the estimate (8.86) usually has a smaller mean-square error than that of (8.88), and that the normalizing constants of both the trial and target distributions are not required in using (8.86) (where these constants are canceled out); in such cases the weights $\{w_t^{(j)}\}$ are evaluated only up to a multiplicative constant. For example, the target distribution $p(\boldsymbol{Z}_t|\boldsymbol{Y}_t)$ in a typical dynamical system (and many Bayesian models) can be evaluated easily up to a normalizing constant (e.g., the likelihood multiplied by a prior distribution), whereas sampling from the distribution directly and evaluating the normalizing constant analytically are impossible.

To implement Monte Carlo techniques for a dynamical system, a set of random samples properly weighted with respect to $p(\boldsymbol{Z}_t|\boldsymbol{Y}_t)$ is needed for any time t. Because the state equation in system (8.82) possesses a Markovian structure, we can implement a recursive importance sampling strategy, which is the basis of all sequential Monte Carlo techniques [277]. Suppose that a set of properly weighted samples $\left\{(\boldsymbol{Z}_{t-1}^{(j)}, w_{t-1}^{(j)})\right\}_{j=1}^{m}$ with respect to $p(\boldsymbol{Z}_{t-1}|\boldsymbol{Y}_{t-1})$ is given at time $(t-1)$. A *sequential Monte Carlo filter* generates from the set a new one, $\left\{\boldsymbol{Z}_t^{(j)}, w_t^{(j)}\right\}_{j=1}^{m}$, which is properly weighted at time t with respect to $p(\boldsymbol{Z}_t|\boldsymbol{Y}_t)$. The algorithm is described as follows.

Algorithm 8.8: [Sequential Monte Carlo filter for dynamical systems] *For* $j = 1, \ldots, m$*:*

- *Draw a sample* $\boldsymbol{z}_t^{(j)}$ *from a trial distribution* $q\left(\boldsymbol{z}_t|\boldsymbol{Z}_{t-1}^{(j)}, \boldsymbol{Y}_t\right)$ *and let* $\boldsymbol{Z}_t^{(j)} = \left(\boldsymbol{Z}_{t-1}^{(j)}, \boldsymbol{z}_t^{(j)}\right)$.

- *Compute the importance weight:*

$$w_t^{(j)} = w_{t-1}^{(j)} \frac{p\left(\boldsymbol{Z}_t^{(j)} \mid \boldsymbol{Y}_t\right)}{p\left(\boldsymbol{Z}_{t-1}^{(j)} \mid \boldsymbol{Y}_{t-1}\right) q\left(\boldsymbol{z}_t^{(j)} \mid \boldsymbol{Z}_{t-1}^{(j)}, \boldsymbol{Y}_t\right)}. \tag{8.89}$$

The algorithm is initialized by drawing a set of i.i.d. samples $\boldsymbol{z}_0^{(1)}, \ldots, \boldsymbol{z}_0^{(m)}$ from $p(\boldsymbol{z}_0|\boldsymbol{y}_0)$. When $\boldsymbol{y}_0$ represents the "null" information, $p(\boldsymbol{z}_0|\boldsymbol{y}_0)$ corresponds to the prior distribution of z_0. The samples and weights drawn by the algorithm above are characterized by the following result, the proof of which is given in the Appendix (Section 8.7.3).

Proposition 8.1: *The weighted samples generated by Algorithm 8.8 satisfy*

$$E\left\{h\left(\boldsymbol{Z}_t^{(j)}\right) w_t^{(j)}\right\} = E\{h(\boldsymbol{Z}_t) \mid \boldsymbol{Y}_t\}, \tag{8.90}$$

$$E\left\{w_t^{(j)}\right\} = 1. \tag{8.91}$$

The result above, together with the law of large numbers, implies that

$$\begin{aligned}\frac{1}{W_t}\sum_{j=1}^{m} h\left(\boldsymbol{Z}_t^{(j)}\right) w_t^{(j)} &= \frac{\sum_{j=1}^{m} h\left(\boldsymbol{Z}_t^{(j)}\right)/m}{W_t/m} \\ &\xrightarrow{a.s.} E\{h(\boldsymbol{Z}_t) \mid \boldsymbol{Y}_t\}, \qquad \text{as } m \to \infty.\end{aligned} \tag{8.92}$$

There are a few important issues regarding the design and implementation of a sequential Monte Carlo filter, such as the choice of the trial distribution $q(\cdot)$ and the use of resampling (see Section 8.5.3). Specifically, a useful choice of the trial distribution $q\left(\boldsymbol{z}_t|\boldsymbol{Z}_{t-1}^{(j)}, \boldsymbol{Y}_t\right)$ for the state-space model (8.82) is of the form

$$\begin{aligned} q\left(\boldsymbol{z}_t \mid \boldsymbol{Z}_{t-1}^{(j)}, \boldsymbol{Y}_t\right) &= p\left(\boldsymbol{z}_t \mid \boldsymbol{Z}_{t-1}^{(j)}, \boldsymbol{Y}_t\right) \\ &= \frac{p\left(\boldsymbol{y}_t \mid \boldsymbol{z}_t, \boldsymbol{Z}_{t-1}^{(j)}, \boldsymbol{Y}_{t-1}\right) p\left(\boldsymbol{z}_t \mid \boldsymbol{Z}_{t-1}^{(j)}, \boldsymbol{Y}_{t-1}\right)}{p\left(\boldsymbol{y}_t \mid \boldsymbol{Z}_{t-1}^{(j)}, \boldsymbol{Y}_{t-1}\right)} \\ &= \frac{p(\boldsymbol{y}_t \mid \boldsymbol{z}_t)\, p\left(\boldsymbol{z}_t \mid \boldsymbol{z}_{t-1}^{(j)}\right)}{p\left(\boldsymbol{y}_t \mid \boldsymbol{z}_{t-1}^{(j)}\right)}, \end{aligned} \tag{8.93}$$

where in (8.93) we used the facts that

$$p\left(\boldsymbol{y}_t \mid \boldsymbol{z}_t, \boldsymbol{Z}_{t-1}^{(j)}, \boldsymbol{Y}_{t-1}\right) = p(\boldsymbol{y}_t \mid \boldsymbol{z}_t), \tag{8.94}$$

$$p\left(\boldsymbol{z}_t \mid \boldsymbol{Z}_{t-1}^{(j)}, \boldsymbol{Y}_{t-1}\right) = p\left(\boldsymbol{z}_t \mid \boldsymbol{z}_{t-1}^{(j)}\right), \tag{8.95}$$

both of which follow directly from the state-space model (8.82). For this trial distribution, the importance weight is updated according to

$$\begin{aligned} w_t^{(j)} &= w_{t-1}^{(j)} \frac{p\left(\boldsymbol{Z}_t^{(j)} \mid \boldsymbol{Y}_t\right)}{p\left(\boldsymbol{Z}_{t-1}^{(j)} \mid \boldsymbol{Y}_{t-1}\right) p\left(\boldsymbol{z}_t^{(j)} \mid \boldsymbol{Z}_{t-1}^{(j)}, \boldsymbol{Y}_t\right)} \\ &= w_{t-1}^{(j)} \frac{p\left(\boldsymbol{Z}_{t-1}^{(j)} \mid \boldsymbol{Y}_t\right)}{p\left(\boldsymbol{Z}_{t-1}^{(j)} \mid \boldsymbol{Y}_{t-1}\right)} \end{aligned} \tag{8.96}$$

$$
\begin{aligned}
&= w_{t-1}^{(j)} \frac{p\left(\boldsymbol{y}_t, \boldsymbol{Z}_{t-1}^{(j)}, \boldsymbol{Y}_{t-1}\right) p\left(\boldsymbol{Y}_{t-1}\right)}{p\left(\boldsymbol{Z}_{t-1}^{(j)}, \boldsymbol{Y}_{t-1}\right) p\left(\boldsymbol{Y}_t\right)} \\
&\propto w_{t-1}^{(j)} p\left(\boldsymbol{y}_t \mid \boldsymbol{Z}_{t-1}^{(j)}, \boldsymbol{Y}_{t-1}\right) \\
&= w_{t-1}^{(j)} p\left(\boldsymbol{y}_t \mid \boldsymbol{z}_{t-1}^{(j)}\right),
\end{aligned}
\tag{8.97}
$$

where (8.96) follows from the fact that

$$
p\left(\boldsymbol{Z}_t^{(j)} \mid \boldsymbol{Y}_t\right) = p\left(\boldsymbol{Z}_{t-1}^{(j)} \mid \boldsymbol{Y}_t\right) p\left(\boldsymbol{z}_t^{(j)} \mid \boldsymbol{Z}_{t-1}^{(j)}, \boldsymbol{Y}_t\right); \tag{8.98}
$$

and the last equality is due to the conditional independence property of the state-space model (8.82). See [277] for the general SMC framework and a detailed discussion on various implementation issues. SMC techniques have been employed to tackle a number of problems in wireless communications and networks, including blind detection in fading channels [76], blind detection in OFDM systems [593], and joint mobility tracking and hand-off detection in cellular networks [595].

8.5.3 Resampling Procedures

The importance sampling weight $w_t^{(j)}$ measures the quality of the corresponding imputed sequence $\boldsymbol{Z}_t^{(j)}$. A relatively small weight implies that the sample is drawn far from the main body of the posterior distribution and has a small contribution in the final estimation. Such a sample is said to be *ineffective*. If there are too many ineffective samples, the Monte Carlo procedure becomes inefficient. This can be detected by observing a large *coefficient of variation* in the importance weight. Suppose that $\left\{w_t^{(j)}\right\}_{j=1}^m$ is a sequence of importance weights. Then the coefficient of variation, v_t, is defined as

$$
\begin{aligned}
v_t^2 &= \frac{\frac{1}{m}\sum_{j=1}^m \left(w_t^{(j)} - \bar{w}_t\right)^2}{\bar{w}_t^2} \\
&= \frac{1}{m}\sum_{j=1}^m \left(\frac{w_t^{(j)}}{\bar{w}_t} - 1\right)^2,
\end{aligned}
\tag{8.99}
$$

with

$$
\bar{w}_t = \frac{1}{m}\sum_{j=1}^m w_t^{(j)}. \tag{8.100}
$$

Note that if the samples are drawn exactly from the target distribution, all the weights are equal, implying that $v_t = 0$. A large coefficient of variation in the importance weights indicates ineffective samples. It is shown in [232] that the importance weights resulting from an SMC filter form a martingale sequence. As more and more data are processed, the coefficient of variation of the weights increases—that is, the number of ineffective samples increases—rapidly.

A useful method for reducing ineffective samples and enhancing effective ones is *resampling*, which was suggested in [160, 276] under the SMC setting. Roughly speaking, resampling allows those "bad" samples (with small importance weights) to be discarded and those "good" ones (with large importance weights) to replicate so as to accommodate the dynamic change of the system. Specifically, let $\left\{(\boldsymbol{Z}_t^{(j)}, w_t^{(j)})\right\}_{j=1}^m$ be the original properly weighted samples at time t. A *residual resampling* strategy forms a new set of weighted samples $\left\{(\tilde{\boldsymbol{Z}}_t^{(j)}, \tilde{w}_t^{(j)})\right\}_{j=1}^m$ according to the following algorithm (assume that $\sum_{j=1}^m w_t^{(j)} = m$).

Algorithm 8.9: [Residual resampling]

- *For $j = 1, \ldots, m$, retain $k_j = \left\lfloor w_t^{(j)} \right\rfloor$ copies of the sample $\left(\boldsymbol{Z}_t^{(j)}, \kappa_t^{(j)}\right)$. Denote $K_r = m - \sum_{j=1}^m k_j$.*
- *Obtain K_r i.i.d. draws from the original sample set $\left\{\boldsymbol{Z}_t^{(j)}\right\}_{j=1}^m$, with probabilities proportional to $\left(w_t^{(j)} - k_j\right)$, $j = 1, \ldots, m$.*
- *Assign equal weight (i.e., set $\tilde{w}_t^{(j)} = 1$) to each new sample.*

The correctness of the residual resampling procedure above is stated by the following result, whose proof is given in the Appendix (Section 8.7.4).

Proposition 8.2: *The samples drawn by the residual resampling procedure (Algorithm 8.9) are properly weighted with respect to $p(\boldsymbol{Z}_t|\boldsymbol{Y}_t)$ for $m \to \infty$.*

Alternatively, we can use the following simple resampling procedure, which also produces properly weighted samples with respect to $p(\boldsymbol{Z}_t|\boldsymbol{Y}_t)$:

- For $j = 1, \ldots, m$, draw i.i.d. random numbers l_j from the set $\{1, 2, \ldots, m\}$, with probability distribution

$$P(l_j = k) \propto w_t^{(k)}, \qquad k = 1, \ldots, m. \tag{8.101}$$

- The m new samples and the corresponding weights $\{(\tilde{\boldsymbol{Z}}_t^{(j)}, \tilde{w}_t^{(j)})\}_{j=1}^m$ are given by

$$\tilde{\boldsymbol{Z}}_t^{(j)} = \boldsymbol{Z}_t^{(l_j)}, \quad \tilde{w}_t^{(j)} = \frac{1}{m}, \qquad j = 1, \ldots, m. \tag{8.102}$$

In practice, when small to modest m is used (we used $m = 50$ in our simulations), the resampling procedure can be seen as trading off between bias and variance. That is, the new samples with their weights resulting from the resampling procedure are only approximately proper, which introduces small bias in Monte Carlo estimation. On the other hand, however, resampling greatly reduces Monte Carlo variance for the future samples.

Resampling can be done at any time. However, resampling too often adds computational burden and decreases "diversities" of the Monte Carlo filter (i.e., it decreases the number of distinctive filters and loses information). On the other hand, resampling too rarely may result in a loss of efficiency. It is thus desirable to give guidance on when to do resampling. A measure of the efficiency of an importance sampling scheme is the *effective sample size* $\bar{m}_t$, defined as

$$\bar{m}_t \triangleq \frac{m}{1+v_t^2}. \tag{8.103}$$

Heuristically, $\bar{m}_t$ reflects the equivalent size of a set of i.i.d. samples for the set of m weighted ones. It is suggested in [277] that resampling should be performed when the effective sample size becomes small (e.g., $\bar{m}_t \leq m/10$). Alternatively, one can conduct resampling at every fixed-length time interval (say, every five steps).

8.5.4 Mixture Kalman Filter

Many dynamical system models belong to the class of conditional dynamical linear models (CDLMs) of the form

$$\begin{aligned} \boldsymbol{x}_t &= \boldsymbol{F}_{\lambda_t}\boldsymbol{x}_{t-1} + \boldsymbol{G}_{\lambda_t}\boldsymbol{u}_t, \\ \boldsymbol{y}_t &= \boldsymbol{H}_{\lambda_t}\boldsymbol{x}_t + \boldsymbol{K}_{\lambda_t}\boldsymbol{v}_t, \end{aligned} \tag{8.104}$$

where $\boldsymbol{u}_t \sim \mathcal{N}_c(\boldsymbol{0}, \boldsymbol{I})$, $\boldsymbol{v}_t \sim \mathcal{N}_c(\boldsymbol{0}, \boldsymbol{I})$, and λ_t is a random indicator variable. The matrices $\boldsymbol{F}_{\lambda_t}$, $\boldsymbol{G}_{\lambda_t}$, $\boldsymbol{H}_{\lambda_t}$, and $\boldsymbol{K}_{\lambda_t}$ are known, given λ_t. In this model, the state variable $\boldsymbol{z}_t$ corresponds to $(\boldsymbol{x}_t, \lambda_t)$.

We observe that for a given trajectory of the indicator λ_t in a CDLM, the system is both linear and Gaussian, for which the Kalman filter provides the complete statistical characterization of the system dynamics. Recently, a novel SMC method, the mixture Kalman filter (MKF), was developed in [75] for online filtering and prediction of CDLMs; it exploits the conditional Gaussian property and uses a marginalization operation to improve the algorithmic efficiency. Instead of dealing with both $\boldsymbol{x}_t$ and λ_t, the MKF draws Monte Carlo samples only in the indicator space and uses a mixture of Gaussian distributions to approximate the target distribution. Compared with the generic SMC method, the MKF is substantially more efficient (e.g., it gives more accurate results with the same computing resources). However, the MKF often needs more computational power for its proper implementation, as the formulas required are more complicated. Additionally, the MKF requires the CDLM structure, which is not applicable to all problems of interest.

Let $\boldsymbol{Y}_t = (\boldsymbol{y}_0, \boldsymbol{y}_1, \ldots, \boldsymbol{y}_t)$ and let $\boldsymbol{\Lambda}_t = (\lambda_0, \lambda_1, \ldots, \lambda_t)$. By recursively generating a set of properly weighted random samples $\left\{(\boldsymbol{\Lambda}_t^{(j)}, w_t^{(j)})\right\}_{j=1}^{m}$ to represent

$p(\boldsymbol{\Lambda}_t|\boldsymbol{Y}_t)$, the MKF approximates the target distribution $p(\boldsymbol{x}_t|\boldsymbol{Y}_t)$ by a random mixture of Gaussian distributions

$$\sum_{j=1}^{m} w_t^{(j)} \mathcal{N}_c\left(\boldsymbol{\mu}_t^{(j)}, \boldsymbol{\Sigma}_t^{(j)}\right), \tag{8.105}$$

where $\kappa_t^{(j)} \triangleq \left[\boldsymbol{\mu}_t^{(j)}, \boldsymbol{\Sigma}_t^{(j)}\right]$ is obtained by implementing a Kalman filter for the given indicator trajectory $\boldsymbol{\Lambda}_t^{(j)}$. Thus, a key step in the MKF is the production at time t of the weighted samples of indicators, $\left\{(\boldsymbol{\Lambda}_t^{(j)}, \kappa_t^{(j)}, w_t^{(j)})\right\}_{j=1}^{m}$, based on the set of samples, $\left\{(\boldsymbol{\Lambda}_{t-1}^{(j)}, \kappa_{t-1}^{(j)}, w_{t-1}^{(j)})\right\}_{j=1}^{m}$, at the previous time $(t-1)$ according to the following algorithm.

Algorithm 8.10: [Mixture Kalman filter for conditional dynamical linear models]
For $j = 1, \ldots, m$:

- *Draw a sample $\lambda_t^{(j)}$ from a trial distribution $q\left(\lambda_t \mid \boldsymbol{\Lambda}_{t-1}^{(j)}, \kappa_{t-1}^{(j)}, \boldsymbol{Y}_t\right)$.*
- *Run a one-step Kalman filter based on $\lambda_t^{(j)}$, $\kappa_{t-1}^{(j)}$, and $\boldsymbol{y}_t$ to obtain $\kappa_t^{(j)}$.*
- *Compute the weight:*

$$w_t^{(j)} \propto w_{t-1}^{(j)} \frac{p\left(\boldsymbol{\Lambda}_{t-1}^{(j)}, \lambda_t^{(j)} \mid \boldsymbol{Y}_t\right)}{p\left(\boldsymbol{\Lambda}_{t-1}^{(j)} \mid \boldsymbol{Y}_{t-1}\right) q\left(\lambda_t^{(j)} \mid \boldsymbol{\Lambda}_{t-1}^{(j)}, \kappa_{t-1}^{(j)}, \boldsymbol{Y}_t\right)}. \tag{8.106}$$

The MKF can be extended to handle the *partial CDLM*, where the state variable has a linear component and a nonlinear component. See [75] for a detailed treatment of the MKF and the extended MKF.

8.6 Blind Adaptive Equalization of MIMO Channels via SMC

Many systems can be modeled as multiple-input/multiple-output (MIMO) systems, where the signals observed are superpositions of several linearly distorted signals from different sources. Examples of MIMO systems include space-division multiple access (SDMA) in wireless communications, speech processing, seismic exploration, and some biological systems. The problem of blind source separation for MIMO systems with unknown parameters is of fundamental importance and its solutions find wide applications in many areas. Recently, there has been much interest in solving this problem, and there are primarily two approaches: an approach based on second-order statistics [5, 99, 470, 495], and an approach based on the constant-modulus algorithm [218, 261, 494]. In this section we treat the problem of blind adaptive signal separation in MIMO channels using the SMC method outlined in Section 8.5. The application of SMC technique to blind equalization of single-user ISI channels with single transmit and receive antennas was first treated in [276] and then generalized to multiuser MIMO channels in [543].

8.6.1 System Description

Consider an SDMA communications system with K users. The kth user transmits data symbols $\{b_k[i]\}_i$ in the same frequency band at the same time, where $b_k[i] \in \Omega$ and Ω is a signal constellation set. The receiver employs an antenna array consisting of P antenna elements. The signal received at the pth antenna element is the superposition of the convolutively distorted signals from all users plus the ambient noise, given by

$$\begin{aligned} y_p[i] &= \sum_{k=1}^{K} \sum_{\ell=0}^{L-1} g_{p,\ell,k} b_k[i-\ell] + n_p[i], \\ &= \boldsymbol{g}_p^H \boldsymbol{b}[i] + n_p[i], \qquad p = 1, \ldots, P, \end{aligned} \tag{8.107}$$

where $n_p[i] \overset{\text{i.i.d}}{\sim} \mathcal{N}_c(0, \sigma^2)$, L is the length of the channel dispersion in terms of number of symbols, and

$$\begin{aligned} \boldsymbol{g}_p &\triangleq [g_{p,0,1} \cdots g_{p,L-1,1} \cdots g_{p,0,K} \cdots g_{p,L-1,K}]^H, \\ \boldsymbol{b}[i] &\triangleq \Big[b_1[i] \cdots b_1[i-L+1] \cdots b_K[i] \cdots b_K[i-L+1]\Big]^T. \end{aligned}$$

Denote

$$\begin{aligned} \boldsymbol{y}[i] &\triangleq \Big[y_1[i] \cdots y_P[i]\Big]^T, \\ \boldsymbol{G} &\triangleq \Big[\boldsymbol{g}_1^H \cdots \boldsymbol{g}_P^H\Big], \\ \boldsymbol{n}[i] &\triangleq \Big[n_1[i] \ \cdots \ n_P[i]\Big]^T. \end{aligned}$$

Then (8.107) can be written as

$$\boldsymbol{y}[i] = \boldsymbol{G}\boldsymbol{b}[i] + \boldsymbol{n}[i]. \tag{8.108}$$

We now look at the problem of online estimation of the multiuser symbols

$$\underline{b}[i] \triangleq \Big[b_1[i] \cdots b_K[i]\Big]^T$$

and the channels $\boldsymbol{H}$ based on the received signals up to time i, $\{\boldsymbol{y}[j]\}_{j=1}^{i}$. Assume that the multiuser symbol streams are i.i.d. uniformly *a priori*, [i.e., $p(b_k[i] = a_l \in \Omega) = 1/|\Omega|$]. Denote

$$\begin{aligned} \boldsymbol{X}[i] &\triangleq \Big[\underline{b}[0] \cdots \underline{b}[i]\Big], \\ \boldsymbol{Y}[i] &\triangleq \Big[\boldsymbol{y}[0] \cdots \boldsymbol{y}[i]\Big]. \end{aligned}$$

Then the problem becomes one of making Bayesian inference with respect to the posterior density

$$p\Big(\boldsymbol{X}[i], \boldsymbol{H}, \sigma^2 \mid \boldsymbol{Y}[i]\Big) \propto \left(\pi\sigma^2\right)^{-(i+1)} \exp\Big\{ -\frac{1}{\sigma^2} \sum_{j=0}^{i} \Big\| \boldsymbol{y}[j] - \boldsymbol{H}\boldsymbol{b}[j] \Big\|^2 \Big\}. \tag{8.109}$$

For example, an online multiuser symbol estimation can be obtained from the marginal posterior distribution $p(\underline{b}[i]|\boldsymbol{Y}[i])$, and an online channel state estimation can be obtained from the marginal posterior distribution $p(\boldsymbol{H}|\boldsymbol{Y}[i])$. Although the joint distribution (8.109) can be written out explicitly up to a normalizing constant, the computation of the corresponding marginal distributions involves very high dimensional integration and is infeasible in practice. Our approach to this problem is the sequential Monte Carlo technique.

8.6.2 SMC Blind Adaptive Equalizer for MIMO Channels

For simplicity, assume that the noise variance σ^2 is known. The SMC principle suggests the following basic approach to the blind MIMO signal separation problem discussed above. At time i, draw m random samples,

$$\left\{\underline{b}^{(j)}[i]\right\}_{j=1}^{m} \sim q\Big(\underline{b}[i] \mid \boldsymbol{X}^{(j)}[i-1], \boldsymbol{Y}[i]\Big),$$

from some trial distribution $q(\cdot)$. Then update the important weights $\{w^{(j)}[i]\}_{j=1}^{m}$ according to (8.97). The *a posteriori* symbol probability of each user can then be estimated as

$$\begin{aligned} P\Big(b[i] = a_l \mid \boldsymbol{Y}[i]\Big) &= E\Big\{I(b[i] = a_l) \mid \boldsymbol{Y}[i]\Big\} \\ &= \frac{1}{W[i]} \sum_{j=1}^{m} I\left(b^{(j)}[i] = a_l\right) w^{(j)}[i], \end{aligned} \tag{8.110}$$

with

$$W[i] = \sum_{j=1}^{m} w^{(j)}[i]$$

for $a_l \in \Omega$, where $I(\cdot)$ is an indicator function such that $I(b[i] = a_l) = 1$ if $b[i] = a_l$ and $I(b[i] = a_l) = 0$ otherwise.

Following the discussions above, the trial distribution is chosen to be

$$q\Big(\underline{b}[i] \mid \boldsymbol{X}^{(j)}[i-1], \boldsymbol{Y}[i]\Big] = p\Big(\underline{b}[i] \mid \boldsymbol{X}^{(j)}[i-1], \boldsymbol{Y}[i]\Big); \tag{8.111}$$

and the importance weight is updated according to

$$w^{(j)}[i] \propto w^{(j)}[i-1] p\left(\boldsymbol{y}[i] \mid \boldsymbol{X}^{(j)}[i-1], \boldsymbol{Y}[i-1]\right). \tag{8.112}$$

We next specify the computation of the two predictive densities (8.111) and (8.112).

Assume that the channel $\boldsymbol{g}_p$ has an *a priori* Gaussian distribution:

$$\boldsymbol{g}_p \sim \mathcal{N}_c(\bar{\boldsymbol{g}}_p, \bar{\boldsymbol{\Sigma}}_p). \tag{8.113}$$

Then the conditional distribution of $\boldsymbol{g}_p$, conditioned on $\boldsymbol{X}[i]$ and $\boldsymbol{Y}[i]$ can be computed as

$$p\Big(\boldsymbol{g}_p \mid \boldsymbol{X}[i], \boldsymbol{Y}[i]\Big) \propto p\Big(\boldsymbol{X}[i], \boldsymbol{Y}[i] \mid \boldsymbol{g}_p\Big)\, p(\boldsymbol{g}_p)$$
$$\sim \mathcal{N}_c\Big(\boldsymbol{g}_p[i],\, \boldsymbol{\Sigma}_p[i]\Big), \tag{8.114}$$

where

$$\boldsymbol{g}_p[i] \triangleq \boldsymbol{\Sigma}_p[i]\Big(\bar{\boldsymbol{\Sigma}}_p^{-1}\bar{\boldsymbol{g}}_p + \frac{1}{\sigma^2}\sum_{j=0}^{i}\boldsymbol{b}[j]y_p[j]^*\Big), \tag{8.115}$$

$$\boldsymbol{\Sigma}_p[i] \triangleq \Big(\bar{\boldsymbol{\Sigma}}_p^{-1} + \frac{1}{\sigma^2}\sum_{j=0}^{n}\boldsymbol{b}[j]\boldsymbol{b}[j]^H\Big)^{-1}. \tag{8.116}$$

Hence the predictive density in (8.112) is given by

$$p\Big(\boldsymbol{y}[i] \mid \boldsymbol{X}[i-1], \boldsymbol{Y}[i-1]\Big)$$
$$\propto \sum_{\underline{a}_l \in \Omega^K} p\Big(\boldsymbol{y}[i] \mid \boldsymbol{X}[i-1], \boldsymbol{Y}[i-1], \underline{b}[i] = \underline{a}_l\Big)$$
$$= \sum_l \prod_{p=1}^{P} p\Big(y_p[i] \mid \boldsymbol{X}[i-1], \boldsymbol{Y}[i-1], \underline{b}[i] = \underline{a}_l\Big), \tag{8.117}$$

where

$$p\Big(y_p[i] \mid \boldsymbol{X}[i-1], \boldsymbol{Y}[i-1], \underline{b}[i] = \underline{a}_l\Big)$$
$$= \int p\Big(y_p[i] \mid \boldsymbol{X}[i-1], \boldsymbol{Y}[i-1], \underline{b}[i] = \underline{a}_l, \boldsymbol{g}_p\Big)$$
$$p\Big(\boldsymbol{g}_p \mid \boldsymbol{X}[i-1], \boldsymbol{Y}[i-1]\Big)\, \mathrm{d}\boldsymbol{g}_p. \tag{8.118}$$

Note that the above is an integral of a Gaussian pdf with respect to another Gaussian pdf. The resulting distribution is still Gaussian:

$$p\Big(y_p[i] \mid \boldsymbol{X}[i-1], \boldsymbol{Y}[i-1], \underline{b}[i] = \underline{a}_l\Big) \sim \mathcal{N}_c\Big(\mu_{p,l}[i], \sigma^2_{p,l}[i]\Big), \tag{8.119}$$

with mean and variance given, respectively, by

$$\mu_{p,l}[i] \triangleq E\Big\{y_p[i] \mid \boldsymbol{X}[i-1], \boldsymbol{Y}[i-1], \underline{b}[i] = \underline{a}_l\Big\}$$
$$= \boldsymbol{g}_p[i-1]^H \boldsymbol{b}[i] \mid_{\underline{b}[i]=\underline{a}_l}, \tag{8.120}$$
$$\sigma^2_{p,l}[i] \triangleq \mathrm{Var}\Big\{y_p[i] \mid \boldsymbol{X}[i-1], \boldsymbol{Y}[i-1], \underline{b}[i] = \underline{a}_l\Big\}$$
$$= \sigma^2 + \boldsymbol{b}[i]^H \boldsymbol{\Sigma}_p[i-1]\boldsymbol{b}[i] \mid_{\underline{b}[i]=\underline{a}_l}. \tag{8.121}$$

Therefore, (8.117) becomes

$$p\Big(\boldsymbol{y}[i] \mid \boldsymbol{X}[i-1], \boldsymbol{Y}[i-1]\Big) \propto \sum_{l} \prod_{p=1}^{P} \rho_{p,l}[i], \tag{8.122}$$

with

$$\rho_{p,l}[i] \triangleq \frac{1}{\sigma_{p,l}^2[i]} \exp\left\{ -\frac{|y_p[i] - \mu_{p,l}[i]|^2}{\sigma_{p,l}^2[i]} \right\}. \tag{8.123}$$

The filtering density in (8.111) can be computed as follows:

$$\begin{aligned} p\Big(\underline{b}[i] = \underline{a}_l \mid \boldsymbol{X}[i-1], \boldsymbol{Y}[i]\Big) &\propto p\Big(\boldsymbol{X}[i-1], \boldsymbol{Y}[i], \underline{b}[i] = \underline{a}_l\Big) \\ &\propto p\Big(\boldsymbol{y}[i] \mid \boldsymbol{X}[i], \boldsymbol{Y}[i-1], \underline{b}[i] = \underline{a}_l\Big) \\ &\propto \prod_{p=1}^{P} \rho_{p,l}[i]. \end{aligned} \tag{8.124}$$

Note that the *a posteriori* mean and covariance of the channel in (8.115) and (8.116) can be updated recursively as follows. At time i, after a new sample of $\underline{b}[i]$ is drawn, we combine it with the past samples $\boldsymbol{b}[i-1]$ to form $\boldsymbol{b}[i]$. Let $\mu_p[i]$ and $\sigma_p^2[i]$ be the quantities computed by (8.120) and (8.121) for the imputed $\underline{b}[i]$. It then follows from the matrix inversion lemma that (8.115) and (8.116) become

$$\boldsymbol{g}_p[i] = \boldsymbol{g}_p[i-1] + \left(\frac{y_p[i] - \mu_p[i]}{\sigma_p^2[i]} \right)^* \boldsymbol{\xi}[i], \tag{8.125}$$

$$\boldsymbol{\Sigma}_p[i] = \boldsymbol{\Sigma}_p[i-1] - \frac{1}{\sigma_p^2[i]} \boldsymbol{\xi}_p[i] \boldsymbol{\xi}_p[i]^H, \tag{8.126}$$

with

$$\boldsymbol{\xi}_p[i] \triangleq \boldsymbol{\Sigma}_p[i-1] \boldsymbol{b}[i]. \tag{8.127}$$

Finally, we summarize the SMC-based blind adaptive equalizer in MIMO channels as follows:

Algorithm 8.11: [SMC-based blind adaptive equalizer in MIMO channels]

- *Initialization: The initial samples of the channel vectors are drawn from the following* a priori *distribution:*

  $$\boldsymbol{g}_p^{(j)}[0] \sim \mathcal{N}_c(\boldsymbol{0}, 1000 \boldsymbol{I}_{KL}), \qquad j = 1, \ldots, m, \quad p = 1, \ldots, P.$$

  All importance weights are initialized as $w_0^{(j)}[0] = 1$, $j = 1, \ldots, m$. Since the data symbols are assumed to be independent, initial symbols are not needed.

  The following steps are implemented at time i to update each weighted sample. For $j = 1, \ldots, m$:

- *For each* $\underline{a}_l \in \Omega^K$ *and* $p = 1, \ldots, P$, *compute the following quantities:*

$$\mu_{p,l}^{(j)}[i] = \boldsymbol{g}_p^{(j)}[i-1]^H \boldsymbol{b}_l^{(j)}[i], \tag{8.128}$$

$$\sigma_{p,l}^2[i]^{(j)} = \sigma^2 + \boldsymbol{b}_l^{(j)}[i]^H \boldsymbol{\Sigma}_p^{(j)}[i-1] \boldsymbol{b}_l^{(j)}[i], \tag{8.129}$$

$$\rho_{p,l}^{(j)}[i] = \left[\sigma_{p,l}^2[i]^{(j)}\right]^{-1} \exp\left\{ - \frac{\left| y_p[i] - \mu_{p,l}^{(j)}[i] \right|^2}{\sigma_{p,l}^2[i]^{(j)}} \right\}, \tag{8.130}$$

with $\boldsymbol{b}_l^{(j)}[i] \triangleq \boldsymbol{b}^{(j)}[i] \mid_{\underline{b}^{(j)}[i] = \underline{a}_l}$.

- *Impute the multiuser symbol* $\underline{b}^{(j)}[i]$*: Draw* $\underline{b}^{(j)}[i]$ *from the set* Ω^K *with probability*

$$p\left(\underline{b}^{(j)}[i] = \underline{a}_l\right) \propto \prod_{p=1}^{P} \rho_{p,l}^{(j)}[i], \quad \underline{a}_l \in \Omega^K. \tag{8.131}$$

- *Compute the importance weight:*

$$w^{(j)}[i] \propto w^{(j)}[i-1] \cdot \sum_{\underline{a}_l \in \Omega^K} \prod_{p=1}^{P} \rho_{p,l}^{(j)}[i].$$

Let $\mu_p^{(j)}[i]$ *and* $\sigma_p^2[i]^{(j)}$ *be the quantities computed in the second step with* $\underline{a}_l$ *corresponding to the imputed symbol* $\underline{b}^{(j)}[i]$.

- *Update the* a posteriori *mean and covariance of channels:*

$$\boldsymbol{g}_p^{(j)}[i] = \boldsymbol{g}_p^{(j)}[i-1] + \left(\frac{y_p[i] - \mu_p^{(j)}[i]}{\sigma_p^2[i]^{(j)}} \right)^* \boldsymbol{\xi}_p^{(j)}[i],$$

$$\boldsymbol{\Sigma}_p^{(j)}[i] = \boldsymbol{\Sigma}_p^{(j)}[i-1] - \frac{1}{\sigma_p^2[i]^{(j)}} \boldsymbol{\xi}_p^{(j)}[i] \boldsymbol{\xi}_p^{(j)}[i]^H,$$

with

$$\boldsymbol{\xi}_p^{(j)}[i] \triangleq \boldsymbol{\Sigma}_p^{(j)}[i-1] \boldsymbol{b}^{(j)}[i].$$

- *Do resampling according to Algorithm 8.9 when the effective sample size* $\bar{m}_t$ *in (8.103) is below a threshold.*

As an example, we consider a single-user system with single transmit and single receive antenna and with channel length $L = 4$. In Fig. 8.9 we plot the channel estimates as a function of time by the SMC adaptive equalizer. It is seen that the channel can be tracked quickly. Note that, in general, when multiple users and/or multiple antennas are present, there is an ambiguity problem associated with any

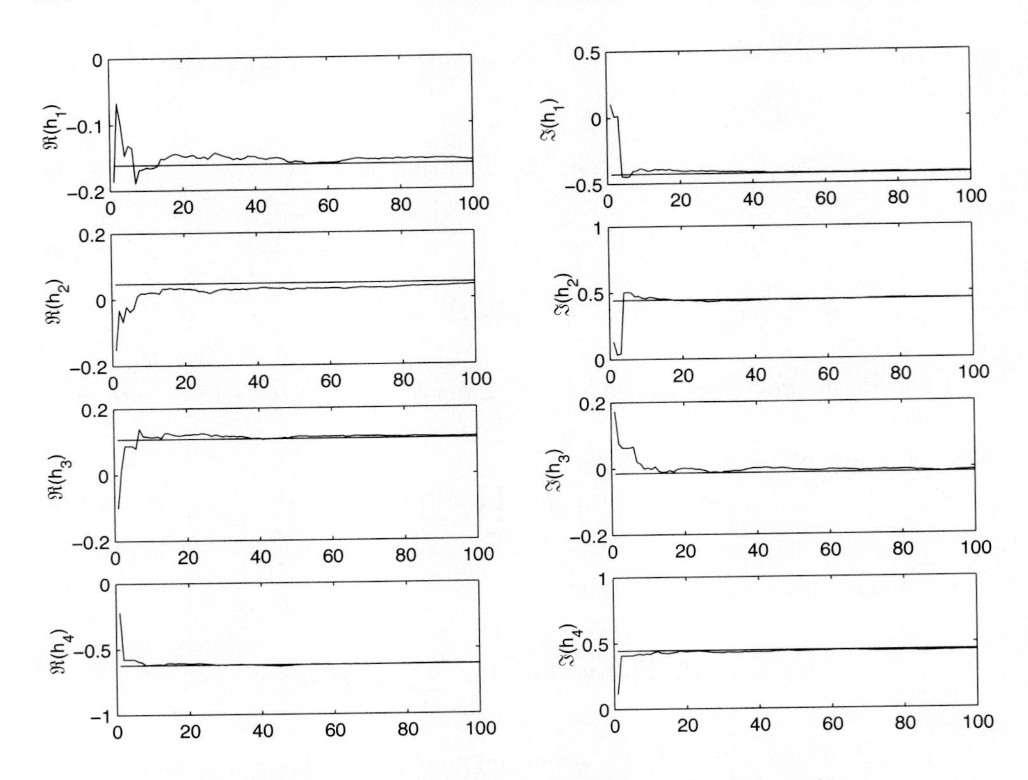

Figure 8.9. Convergence of the SMC blind adaptive equalizer.

blind methods, which can be resolved by periodically inserting a certain pattern of pilot symbols. For more discussions on the SMC blind adaptive equalizer, see [276, 277]. Note also that it is possible (and sometimes desirable) to make an inference of the current symbols $\underline{b}[i]$ based on both the current and future observations, $\boldsymbol{Y}[i+\Delta]$ for some $\Delta > 0$ [i.e., to make an inference with respect to $p(\underline{b}[i] | \boldsymbol{Y}[i+\Delta])$] [76, 542]. Called *delayed estimation*, such approaches are elaborated in Chapter 9. Moreover, when K is large, the choice of sampling density $p(\underline{b}[i] = \underline{a}_l \in \Omega^K)$ becomes computationally expensive. It is possible to devise more efficient trial sampling density.

8.7 Appendix

8.7.1 Derivations for Section 8.4.2

Derivation of (8.34)

$$
\begin{aligned}
p(\boldsymbol{a} \mid \sigma^2, \boldsymbol{X}, \boldsymbol{Y}) &= p(\boldsymbol{a}, \sigma^2, \boldsymbol{X} \mid \boldsymbol{Y}) / \underbrace{p(\sigma^2, \boldsymbol{X} \mid \boldsymbol{Y})}_{\text{not a function of } \boldsymbol{a}} \propto p(\boldsymbol{a}, \sigma^2, \boldsymbol{X} \mid \boldsymbol{Y}) \\
&\propto \exp\Big\{ -\frac{1}{2\sigma^2} \sum_{k=0}^{M-1} \|\boldsymbol{r}[i] - \boldsymbol{S}\boldsymbol{B}[i]\boldsymbol{a}\|^2 \Big\} \exp\Big\{ -\frac{1}{2}(\boldsymbol{a} - \boldsymbol{a}_0)^T \boldsymbol{\Sigma}_0^{-1} (\boldsymbol{a} - \boldsymbol{a}_0) \Big\} \\
&\propto \exp\Big\{ -\frac{1}{2}\boldsymbol{a}^T \underbrace{\Big(\boldsymbol{\Sigma}_0^{-1} + \frac{1}{\sigma^2} \sum_{k=0}^{M-1} \boldsymbol{B}[i]\boldsymbol{S}^T\boldsymbol{S}\boldsymbol{B}[i]\Big)}_{\boldsymbol{\Sigma}_*^{-1}} \boldsymbol{a} \\
&\quad + \boldsymbol{a}^T \underbrace{\Big(\boldsymbol{\Sigma}_0^{-1}\boldsymbol{a}_0 + \frac{1}{\sigma^2} \sum_{k=0}^{M-1} \boldsymbol{B}[i]\boldsymbol{S}^T\boldsymbol{r}[i]\Big)}_{\boldsymbol{\Sigma}_*^{-1}\boldsymbol{a}_*} \Big\} \\
&\propto \exp\Big\{ -\frac{1}{2}(\boldsymbol{a} - \boldsymbol{a}_*)^T \boldsymbol{\Sigma}_*^{-1} (\boldsymbol{a} - \boldsymbol{a}_*) \Big\} \sim \mathcal{N}(\boldsymbol{a}_*, \boldsymbol{\Sigma}_*).
\end{aligned}
\tag{8.132}
$$

Derivation of (8.38)

$$
\begin{aligned}
p(\sigma^2 \mid \boldsymbol{a}, \boldsymbol{X}, \boldsymbol{Y}) &= p(\boldsymbol{a}, \sigma^2, \boldsymbol{X} \mid \boldsymbol{Y}) / \underbrace{p(\boldsymbol{a}, \boldsymbol{X} \mid \boldsymbol{Y})}_{\text{not a function of } \sigma^2} \propto p(\boldsymbol{a}, \sigma^2, \boldsymbol{X} \mid \boldsymbol{Y}) \\
&\propto \Big(\frac{1}{\sigma^2}\Big)^{\frac{MN}{2}} \exp\Big\{ -\frac{1}{2\sigma^2} \underbrace{\sum_{k=0}^{M-1} \|\boldsymbol{r}[i] - \boldsymbol{S}\boldsymbol{A}\boldsymbol{b}[i]\|^2}_{s^2} \Big\} \cdot \Big(\frac{1}{\sigma^2}\Big)^{\frac{\nu_0}{2}+1} \exp\Big\{ -\frac{\nu_0\lambda_0}{2\sigma^2} \Big\} \\
&= \Big(\frac{1}{\sigma^2}\Big)^{\frac{\nu_0+MN}{2}+1} \exp\Big\{ -\frac{\nu_0\lambda_0 + s^2}{2\sigma^2} \Big\} \sim \chi^{-2}\Big(\nu_0 + MN,\ \frac{\nu_0\lambda_0 + s^2}{\nu_0 + MN}\Big).
\end{aligned}
\tag{8.133}
$$

Derivation of (8.40)

$$P(b_k[i] = +1 \mid \boldsymbol{a}, \sigma^2, \boldsymbol{X}_{ki}, \boldsymbol{Y}) = p(\boldsymbol{a}, \sigma^2, \boldsymbol{X} \mid \boldsymbol{Y}) / \underbrace{p(\boldsymbol{a}, \sigma^2, \boldsymbol{X}_{ki} \mid \boldsymbol{Y})}_{\text{not a function of } b_k[i]}$$

$$\propto p(\boldsymbol{a}, \sigma^2, \boldsymbol{X} \mid \boldsymbol{Y}) \;\propto\; \rho_k[i] \exp\left\{ -\frac{1}{2\sigma^2} \sum_{l=0}^{M-1} \|\boldsymbol{r}[l] - \boldsymbol{S}\boldsymbol{A}\boldsymbol{b}[l]\|^2 \right\}$$

$$\propto \rho_k[i] \exp\left\{ -\frac{1}{2\sigma^2} \|\boldsymbol{r}[i] - \boldsymbol{S}\boldsymbol{A}\boldsymbol{b}[i]\|^2 \right\} \tag{8.134}$$

$$\Rightarrow \frac{P(b_k[i] = +1 \mid \boldsymbol{a}, \sigma^2, \boldsymbol{X}_{ki}, \boldsymbol{Y})}{P(b_k[i] = -1 \mid \boldsymbol{a}, \sigma^2, \boldsymbol{X}_{ki}, \boldsymbol{Y})}$$

$$= \frac{\rho_k[i]}{1-\rho_k[i]} \exp\left\{ \frac{1}{2\sigma^2} \left(\|\boldsymbol{r}[i] - \boldsymbol{S}\boldsymbol{A}(\boldsymbol{b}_k^0[i] - \boldsymbol{e}_k)\|^2 - \|\boldsymbol{r}[i] - \boldsymbol{S}\boldsymbol{A}(\boldsymbol{b}_k^0[i] + \boldsymbol{e}_k)\|^2 \right) \right\}$$

$$= \frac{\rho_k[i]}{1-\rho_k[i]} \exp\left\{ \frac{2}{\sigma^2} (\boldsymbol{S}\boldsymbol{A}\boldsymbol{e}_k)^T (\boldsymbol{r}[i] - \boldsymbol{S}\boldsymbol{A}\boldsymbol{b}_k^0[i]) \right\}$$

$$= \frac{\rho_k[i]}{1-\rho_k[i]} \exp\left\{ \frac{2A_k}{\sigma^2} \boldsymbol{s}_k^T (\boldsymbol{r}[i] - \boldsymbol{S}\boldsymbol{A}\boldsymbol{b}_k^0[i]) \right\}. \tag{8.135}$$

($\boldsymbol{e}_k$ is a K-dimensional vector with all-zero entries except for the kth entry, which is 1.)

8.7.2 Derivations for Section 8.4.3

Derivation of (8.56)

$$p(\boldsymbol{a} \mid \sigma_1^2, \sigma_2^2, \epsilon, \boldsymbol{I}, \boldsymbol{X}, \boldsymbol{Y})$$

$$= p(\boldsymbol{a}, \sigma_1^2, \sigma_2^2, \epsilon, \boldsymbol{I}, \boldsymbol{X} \mid \boldsymbol{Y}) / \underbrace{p(\sigma_1^2, \sigma_2^2, \epsilon, \boldsymbol{I}, \boldsymbol{X} \mid \boldsymbol{Y})}_{\text{not a function of } \boldsymbol{a}} \;\propto\; p(\boldsymbol{a}, \sigma_1^2, \sigma_2^2, \epsilon, \boldsymbol{I}, \boldsymbol{X} \mid \boldsymbol{Y})$$

$$\propto \exp\left\{ -\frac{1}{2} \sum_{i=0}^{M-1} (\boldsymbol{r}[i] - \boldsymbol{S}\boldsymbol{B}[i]\boldsymbol{a})^T \boldsymbol{\Lambda}[i]^{-1} (\boldsymbol{r}[i] - \boldsymbol{S}\boldsymbol{B}[i]\boldsymbol{a}) \right\}$$

$$\exp\left\{ -\frac{1}{2} (\boldsymbol{a} - \boldsymbol{a}_0)^T \boldsymbol{\Sigma}_0^{-1} (\boldsymbol{a} - \boldsymbol{a}_0) \right\}$$

$$\propto \exp\left\{ -\frac{1}{2} \boldsymbol{a}^T \underbrace{\left(\boldsymbol{\Sigma}_0^{-1} + \sum_{i=0}^{M-1} \boldsymbol{B}[i]\boldsymbol{S}^T \boldsymbol{\Lambda}[i]^{-1} \boldsymbol{S}\boldsymbol{B}[i] \right) \boldsymbol{a}}_{\boldsymbol{\Sigma}_*^{-1}} \right.$$

$$\left. + \underbrace{\boldsymbol{a}^T \left[\boldsymbol{\Sigma}_0^{-1} \boldsymbol{a}_0 + \sum_{i=0}^{M-1} \boldsymbol{B}[i]\boldsymbol{S}^T \boldsymbol{\Lambda}[i]^{-1} \boldsymbol{r}[i] \right]}_{\boldsymbol{\Sigma}_*^{-1}\boldsymbol{a}_*} \right\}$$

$$\propto \exp\left\{ -\frac{1}{2} (\boldsymbol{a} - \boldsymbol{a}_*)^T \boldsymbol{\Sigma}_*^{-1} (\boldsymbol{a} - \boldsymbol{a}_*) \right\} \sim \mathcal{N}(\boldsymbol{a}_*, \boldsymbol{\Sigma}_*). \tag{8.136}$$

Derivation of (8.59)

$$
\begin{aligned}
& p(\sigma_{\bar{l}}^2 \mid \boldsymbol{a}, \sigma_{\bar{l}}^2, \epsilon, \boldsymbol{I}, \boldsymbol{X}, \boldsymbol{Y}) \\
& = p(\boldsymbol{a}, \sigma_1^2, \sigma_2^2, \epsilon, \boldsymbol{I}, \boldsymbol{X} \mid \boldsymbol{Y}) / \underbrace{p(\boldsymbol{a}, \sigma_{\bar{l}}^2, \epsilon, \boldsymbol{I}, \boldsymbol{X} \mid \boldsymbol{Y})}_{\text{not a function of of } \sigma_l^2} \propto p(\boldsymbol{a}, \sigma_1^2, \sigma_2^2, \epsilon, \boldsymbol{I}, \boldsymbol{X} \mid \boldsymbol{Y}) \\
& \propto \left(\frac{1}{\sigma_l^2}\right)^{\frac{1}{2}\sum_{i=0}^{M-1} m_l[i]} \exp\Big\{-\frac{1}{2\sigma_l^2} \underbrace{\sum_{i=0}^{M-1}\sum_{j=0}^{N-1}\Big(r_j[i] - \boldsymbol{\xi}_j^T \boldsymbol{A}\boldsymbol{b}[i]\Big)^2 1_{\{I_j[i]=l\}}}_{s_l^2}\Big\} \\
& \quad \cdot \left(\frac{1}{\sigma_l^2}\right)^{(\nu_l/2)+1} \exp\Big\{-\frac{\nu_l \lambda_l}{2\sigma_l^2}\Big\} \\
& = \left(\frac{1}{\sigma_l^2}\right)^{(\nu_l/2)+\frac{1}{2}\sum_{i=0}^{M-1} m_l[i]+1} \exp\Big\{-\frac{\nu_l\lambda_l + s_l^2}{2\sigma_l^2}\Big\} \\
& \sim \chi^{-2}\Big(\nu_l + \textstyle\sum_{i=0}^{M-1} m_l[i],\ \dfrac{\nu_l \lambda_l + s_l^2}{\nu_l + \sum_{i=0}^{M-1} m_l[i]}\Big).
\end{aligned} \tag{8.137}
$$

Derivation of (8.61)

$$
\begin{aligned}
& P(b_k[i] = +1 \mid \boldsymbol{a}, \sigma_1^2, \sigma_2^2, \epsilon, \boldsymbol{I}, \boldsymbol{X}_{ki}, \boldsymbol{Y}) \\
& \quad = p(\boldsymbol{a}, \sigma_1^2, \sigma_2^2, \epsilon, \boldsymbol{I}, \boldsymbol{X} \mid \boldsymbol{Y}) / \underbrace{p(\boldsymbol{a}, \sigma_1^2, \sigma_2^2, \epsilon, \boldsymbol{I}, \boldsymbol{X}_{ki} \mid \boldsymbol{Y})}_{\text{not a function of } b_k(i)} \\
& \propto p(\boldsymbol{a}, \sigma_1^2, \sigma_2^2, \epsilon, \boldsymbol{I}, \boldsymbol{X} \mid \boldsymbol{Y}) \\
& \quad \propto \rho_k[i] \exp\Big\{-\frac{1}{2}\sum_{l=0}^{M-1} (\boldsymbol{r}[l] - \boldsymbol{S}\boldsymbol{A}\boldsymbol{b}[l])^T \boldsymbol{\Lambda}[l]^{-1} (\boldsymbol{r}[l] - \boldsymbol{S}\boldsymbol{A}\boldsymbol{b}[l])\Big\}
\end{aligned}
$$

$$
\propto \rho_k[i] \exp\Big\{-\frac{1}{2}(\boldsymbol{r}[i] - \boldsymbol{S}\boldsymbol{A}\boldsymbol{b}[i])^T \boldsymbol{\Lambda}[i]^{-1} (\boldsymbol{r}[i] - \boldsymbol{S}\boldsymbol{A}\boldsymbol{b}[i])\Big\} \tag{8.138}
$$

$$
\begin{aligned}
& \Rightarrow \frac{P(b_k[i] = +1 \mid \boldsymbol{a}, \sigma_1^2, \sigma_2^2, \epsilon, \boldsymbol{X}_{ki}, \boldsymbol{Y})}{P(b_k[i] = -1 \mid \boldsymbol{a}, \sigma_1^2, \sigma_2^2, \epsilon, \boldsymbol{X}_{ki}, \boldsymbol{Y})} \\
& = \frac{\rho_k[i]}{1 - \rho_k[i]} \exp\Big\{\frac{1}{2}\Big[\boldsymbol{r}[i] - \boldsymbol{S}\boldsymbol{A}(\boldsymbol{b}[i] - \boldsymbol{e}_k)\Big]^T \boldsymbol{\Lambda}[i]^{-1} \Big[\boldsymbol{r}[i] - \boldsymbol{S}\boldsymbol{A}(\boldsymbol{b}[i] - \boldsymbol{e}_k)\Big] \\
& \qquad\qquad - \Big[\boldsymbol{r}[i] - \boldsymbol{S}\boldsymbol{A}(\boldsymbol{b}[i] + \boldsymbol{e}_k)\Big]^T \boldsymbol{\Lambda}[i]^{-1} \Big[\boldsymbol{r}[i] - \boldsymbol{S}\boldsymbol{A}(\boldsymbol{b}[i] + \boldsymbol{e}_k)\Big]\Big\} \\
& = \frac{\rho_k[i]}{1 - \rho_k[i]} \exp\Big\{2(\boldsymbol{S}\boldsymbol{A}\boldsymbol{e}_k)^T \boldsymbol{\Lambda}[i]^{-1} (\boldsymbol{r}[i] - \boldsymbol{S}\boldsymbol{A}\boldsymbol{b}_k^0[i])\Big\} \\
& = \frac{\rho_k[i]}{1 - \rho_k[i]} \exp\Big\{2A_k \boldsymbol{s}_k^T \boldsymbol{\Lambda}[i]^{-1} (\boldsymbol{r}[i] - \boldsymbol{S}\boldsymbol{A}\boldsymbol{b}_k^0[i])\Big\}.
\end{aligned} \tag{8.139}
$$

Derivation of (8.62)

$$
\begin{aligned}
&P[I_j(i) = l \mid \boldsymbol{a}, \sigma_1^2, \sigma_2^2, \epsilon, \boldsymbol{I}_{ji}, \boldsymbol{X}, \boldsymbol{Y}] \\
&\quad = p(\boldsymbol{a}, \sigma_1^2, \sigma_2^2, \epsilon, \boldsymbol{I}, \boldsymbol{X} \mid \boldsymbol{Y}) / \underbrace{p(\boldsymbol{a}, \sigma_1^2, \sigma_2^2, \epsilon, \boldsymbol{I}_{ji}, \boldsymbol{X} \mid \boldsymbol{Y})}_{\text{not a function of of } I_j(i)} \\
&\propto p(\boldsymbol{a}, \sigma_1^2, \sigma_2^2, \epsilon, \boldsymbol{I}, \boldsymbol{X} \mid \boldsymbol{Y}) \propto P[I_j(i) \\
&\quad = l \mid \epsilon] \frac{1}{\sqrt{\sigma_l^2}} \exp\left\{ -\frac{1}{2\sigma_l^2} \left(r_j[i] - \boldsymbol{\xi}_j^T \boldsymbol{A} \boldsymbol{b}[i] \right)^2 \right\} && (8.140) \\
&\Rightarrow \frac{P(I_j[i] = 1 \mid \boldsymbol{a}, \sigma_1^2, \sigma_2^2, \epsilon, \boldsymbol{I}_{ji}, \boldsymbol{X}, \boldsymbol{Y})}{P(I_j[i] = 2 \mid \boldsymbol{a}, \sigma_1^2, \sigma_2^2, \epsilon, \boldsymbol{I}_{ji}, \boldsymbol{X}, \boldsymbol{Y})} \\
&= \frac{1-\epsilon}{\epsilon} \sqrt{\frac{\sigma_2^2}{\sigma_1^2}} \exp\left\{ \frac{1}{2} \left(r_j[i] - \boldsymbol{\xi}_j^T \boldsymbol{A} \boldsymbol{b}[i] \right)^2 \left(\frac{1}{\sigma_2^2} - \frac{1}{\sigma_1^2} \right) \right\}. && (8.141)
\end{aligned}
$$

Derivation of (8.63)

$$
\begin{aligned}
&p(\epsilon \mid \boldsymbol{a}, \sigma_1^2, \sigma_2^2, \boldsymbol{I}, \boldsymbol{X}, \boldsymbol{Y}) \\
&\quad = p(\boldsymbol{a}, \sigma_1^2, \sigma_2^2, \epsilon, \boldsymbol{I}, \boldsymbol{X} \mid \boldsymbol{Y}) / \underbrace{p(\boldsymbol{a}, \sigma_1^2, \sigma_2^2, \boldsymbol{I}, \boldsymbol{X} \mid \boldsymbol{Y})}_{\text{not a function of } \epsilon} \\
&\propto p(\boldsymbol{a}, \sigma_1^2, \sigma_2^2, \epsilon, \boldsymbol{I}, \boldsymbol{X} \mid \boldsymbol{Y}) \;\propto\; p(\epsilon)\, p(\boldsymbol{I} \mid \epsilon) \\
&\propto \epsilon^{a_0 - 1} (1-\epsilon)^{b_0 - 1} \cdot \epsilon^{\sum_{i=0}^{M-1} m_2[i]} (1-\epsilon)^{\sum_{i=0}^{M-1} m_1[i]} \\
&\sim \text{beta}\left(a_0 + \textstyle\sum_{i=0}^{M-1} m_2[i],\, b_0 + \sum_{i=0}^{M-1} m_1[i] \right). && (8.142)
\end{aligned}
$$

8.7.3 Proof of Proposition 8.1 in Section 8.5.2

Note that

$$
\begin{aligned}
w_t &= w_{t-1} \frac{p(\boldsymbol{Z}_t \mid \boldsymbol{Y}_t)}{p(\boldsymbol{Z}_{t-1} \mid \boldsymbol{Y}_{t-1}) q(\boldsymbol{z}_t \mid \boldsymbol{Z}_{t-1}, \boldsymbol{Y}_t)} \qquad (\text{with } w_0 = 1) \\
&= \prod_{i=1}^{t} \frac{p(\boldsymbol{Z}_i \mid \boldsymbol{Y}_i)}{p(\boldsymbol{Z}_{i-1} \mid \boldsymbol{Y}_{i-1})\, q(\boldsymbol{z}_i \mid \boldsymbol{Z}_{i-1}, \boldsymbol{Y}_i)} \\
&= \frac{p(\boldsymbol{Z}_t \mid \boldsymbol{Y}_t)}{p(\boldsymbol{z}_0 \mid \boldsymbol{y}_0) \prod_{i=1}^{t} q(\boldsymbol{z}_i \mid \boldsymbol{Z}_{i-1}, \boldsymbol{Y}_i)}. && (8.143)
\end{aligned}
$$

The numerator in (8.143) is the target distribution, and the denominator is the sampling distribution from which $\boldsymbol{Z}_t$ was generated. Hence, for any measurable function $h(\cdot)$, we have

$$
\begin{aligned}
E\left\{h(\boldsymbol{Z}_t^{(j)})w_t^{(j)}\right\} &= \int h(\boldsymbol{Z}_t)\frac{p(\boldsymbol{Z}_t \mid \boldsymbol{Y}_t)}{p(\boldsymbol{z}_0 \mid \boldsymbol{y}_0)\prod_{i=1}^t q(\boldsymbol{z}_i \mid \boldsymbol{Z}_{i-1}, \boldsymbol{Y}_i)} \\
&\qquad \left[p(\boldsymbol{z}_0 \mid \boldsymbol{y}_0)\prod_{i=1}^t q(\boldsymbol{z}_i \mid \boldsymbol{Z}_{i-1}, \boldsymbol{Y}_i)\right] \mathrm{d}\boldsymbol{Z}_t \\
&= \int h(\boldsymbol{Z}_t)\, p(\boldsymbol{Z}_t \mid \boldsymbol{Y}_t)\mathrm{d}\boldsymbol{Z}_t = E\left\{h(\boldsymbol{Z}_t) \mid \boldsymbol{Y}_t\right\}. \qquad (8.144)
\end{aligned}
$$

Finally note that both (8.90) and (8.91) are special cases of (8.144).

8.7.4 Proof of Proposition 8.2 in Section 8.5.3

In this section we verify the correctness of the residual resampling under a general setting. Let $(\boldsymbol{x}_t^{(j)}, w_t^{(j)})$ be a properly weighted sample with respect to $p(\boldsymbol{x}_t|\boldsymbol{Y}_t)$—without loss of generality, we assume that $\sum_{j=1}^m w_t^{(j)} = m$—and let $\left\{\tilde{\boldsymbol{x}}_t^{(j)}\right\}_{j=1}^m$ be the set of samples generated from the residual resampling scheme. The new set consists of $k_j = \lfloor w_t^{(j)} \rfloor$ copies of the sample $\boldsymbol{x}_t^{(j)}$ for $j = 1, \ldots, m$, and $K_r = m - \sum_{j=1}^m k_j$ i.i.d. samples drawn from set $\left\{\boldsymbol{x}_t^{(j)}\right\}_{j=1}^m$ with probability proportional to $\frac{1}{K_r}\left(w_t^{(j)} - \lfloor w_t^{(j)} \rfloor\right)$. The weights for the new samples are set to 1. Hence,

$$
\begin{aligned}
E\left\{\frac{1}{m}\sum_{j=1}^m h(\tilde{\boldsymbol{x}}_t^{(j)})\right\} &= E\left\{E\left[\frac{1}{m}\sum_{j=1}^m h(\tilde{\boldsymbol{x}}_t^{(j)}) \;\middle|\; \left\{\boldsymbol{x}_t^{(j')}, w_t^{(j')}\right\}_{j'=1}^m\right]\right\} \\
&= \frac{1}{m}E\left\{\sum_{j=1}^m h(\boldsymbol{x}_t^{(j)})\lfloor w_t^{(j)} \rfloor\right. \\
&\qquad \left. + \sum_{j=m-K_r+1}^m E\left[h(\tilde{\boldsymbol{x}}_t^{(j)}) \;\middle|\; \left\{\boldsymbol{x}_t^{(j')}, w_t^{(j')}\right\}_{j'=1}^m\right]\right\} \\
&= \frac{1}{m}E\left\{\sum_{j=1}^m h(\boldsymbol{x}_t^{(j)})\lfloor w_t^{(j)} \rfloor + K_r E\left[h(\tilde{\boldsymbol{x}}_t) \;\middle|\; \left\{\boldsymbol{x}_t^{(j')}, w_t^{(j')}\right\}_{j'=1}^m\right]\right\} \\
&= \frac{1}{m}E\left\{\sum_{j=1}^m h(\boldsymbol{x}_t^{(j)})\lfloor w_t^{(j)} \rfloor + K_r \sum_{j=1}^m h(\boldsymbol{x}_t^{(j)})\frac{w_t^{(j)} - \lfloor w_t^{(j)} \rfloor}{K_r}\right\} \\
&= \frac{1}{m}E\left\{\sum_{j=1}^m h(\boldsymbol{x}_t^{(j)})w_t^{(j)}\right\} = E\{h(\boldsymbol{x}_t) \mid \boldsymbol{Y}_t\}. \qquad (8.145)
\end{aligned}
$$

Furthermore,

$$
\begin{aligned}
\mathrm{Var}\left\{\frac{1}{m}\sum_{j=1}^{m} h(\tilde{\boldsymbol{x}}_t^{(j)})\right\} &= \mathrm{Var}\left\{E\left[\frac{1}{m}\sum_{j=1}^{m} h(\tilde{\boldsymbol{x}}_t^{(j)}) \;\middle|\; \left\{\boldsymbol{x}_t^{(j')}, w_t^{(j')}\right\}_{j'=1}^{m}\right]\right\} \\
&\quad + E\left\{\mathrm{Var}\left[\frac{1}{m}\sum_{j=1}^{m} h(\tilde{\boldsymbol{x}}_t^{(j)}) \;\middle|\; \left\{\boldsymbol{x}_t^{(j')}, w_t^{(j')}\right\}_{j'=1}^{m}\right]\right\} \\
&= \mathrm{Var}\left\{\frac{1}{m}\sum_{j=1}^{m} h(\boldsymbol{x}_t^{(j)}) w_t^{(j)}\right\} \\
&\quad + E\left\{\frac{K_r}{m^2}\mathrm{Var}\left[h(\tilde{\boldsymbol{x}}_t) \;\middle|\; \left\{\boldsymbol{x}_t^{(j')}, w_t^{(j')}\right\}_{j'=1}^{m}\right]\right\} \\
&\leq \frac{1}{m}\mathrm{Var}\{h(\boldsymbol{x}_t)w_t\} + E\left\{\frac{K_r}{m^2}\sum_{j=1}^{m}(h(\boldsymbol{x}_t^{(j)}))^2 \frac{(w_t^{(j)} - \lfloor w_t^{(j)} \rfloor)}{K_r}\right\} \\
&\leq \frac{1}{m}\mathrm{Var}\{h(\boldsymbol{x}_t)w_t\} + \frac{1}{m^2}E\left\{\sum_{j=1}^{m}(h(\boldsymbol{x}_t^{(j)}))^2 \min\{1, w_t^{(j)}\}\right\} \\
&\leq \frac{1}{m}\mathrm{Var}\{h(\boldsymbol{x}_t)w_t\} + \frac{1}{m}E\{(h(\boldsymbol{x}_t))^2 w_t\} \to 0, \text{ as } m \to \infty.
\end{aligned}
\tag{8.146}
$$

Here we assume that $\mathrm{Var}\{h(\boldsymbol{x}_t)w_t\} < \infty$. Hence, $\frac{1}{m}\sum_{j=1}^{m} h(\tilde{\boldsymbol{x}}_t^{(j)}) \to E\{h(\boldsymbol{x}_t)|\boldsymbol{Y}_t\}$ in probability.

Chapter 9

SIGNAL PROCESSING FOR FADING CHANNELS

9.1 Introduction

As noted in Chapter 1, mobile wireless communication systems are affected by propagation anomalies due to terrain or buildings which cause multipath reception, producing extreme variations in both amplitude and apparent frequency in the received signals, a phenomenon known as *fading*. Signal reception in such channels presents new challenges, and dealing with these is the main theme of this chapter. Channel estimation and data detection in various fading channels have been the subjects of intensive research over the past two decades. In what follows we provide a brief overview of the literature in this area.

Single-User Receivers in Frequency-Flat Fading Channels

Narrowband mobile communications for voice and data can be modeled as signaling over flat, i.e., frequency-nonselective, Rayleigh fading channels. Depending on the fading rate relative to the data rate, the fading process can be categorized as either *slow* (*time-nonselective*) *fading*, where the fading process is assumed to remain constant over one symbol interval and to vary from symbol to symbol; or *fast* (*time-selective*) *fading*, where the fading process is assumed to vary within the symbol interval. A considerable amount of recent research has addressed the problem of data detection in flat fading channels. Specifically, various techniques for maximum-likelihood sequence detection (MLSD) in slow-fading channels have been proposed. The optimal solutions under several fading models are studied in [167, 284, 308], and the exact implementations of these solutions involve very high dimensional filtering. Most suboptimal schemes employ a two-stage receiver structure, with a channel estimation stage followed by a sequence detection stage. Channel estimation is typically implemented by a Kalman filter or a linear predictor, and is facilitated by per-survivor processing [399, 525], decision feedback [167, 223, 280], pilot symbols [70, 94, 233, 337], or a combination of the above [215]. Other alternative solutions to MLSD in slow-fading channels include a method based on a combination of hidden Markov model and Kalman filtering [85, 86], and an approach based on the expectation-maximization (EM) algorithm [138]. Moreover, in [139, 181], turbo receiver techniques for joint demodulation and decoding in flat-fading channels are developed.

Data detection over fast-fading channels has also been addressed in the recent literature. In [171, 524, 526], a linearly time-varying model is used to approximate the time variation within a symbol interval of a time-selective fading process, and several double-filtering receiver structures are developed. Another approach [360] that has been investigated is to sample the received signal at a multiple of the symbol rate, and to track the channel variation within a symbol interval using a nonlinear filter. Extensions of this method have been made to address the issue of tracking the random phase drift [249, 250] and the carrier frequency Doppler shift [359] in fast-fading channels.

Single-User Receivers in Frequency-Selective Fading Channels

Multipath effects over fading channels that cause time-varying intersymbol interference (ISI) constitutes a severe impediment to high-speed wireless communications. Although equalization of time-invariant channels has been an active research area for almost four decades [175], equalization of time-varying fading channels presents substantial new challenges and has received significant attention only recently, due to its potential for widespread application in high-speed wireless data/multimedia applications. Maximum-likelihood sequence estimation receivers for time-varying ISI channels with known channel state information are studied in [49, 171], which are generalizations of the Ungerboeck receiver for time-invariant channels [502]. In [93, 314, 489, 599], several MLSD receiver structures are developed that are based on the known second-order statistics of the fading process, instead of the actual channel state. When the fading statistics are unknown, they are usually estimated from the data in a training-assisted mode or decision-directed mode [96, 172, 240, 246, 268, 503]. Furthermore, symbol-by-symbol maximum *a posteriori* probability (MAP) schemes for equalizing time-varying fading channels have also been studied [22–24, 481], where channel estimation is facilitated by some *ad hoc* Kalman-type nonlinear estimators, which take as inputs the *a posteriori* probabilities of the ISI channel state and the received signal. Related to these methods are the Bayesian equalization techniques developed for time-invariant ISI channels [77, 149, 213, 245], which essentially model the channel coefficients as slowly time-varying processes. Moreover, orthogonal frequency-division modulation (OFDM) techniques convert a frequency-selective fading channel into a set of parallel frequency-flat fading channels. Channel estimation and data detection methods in OFDM systems are developed in [110, 259, 260, 262, 263].

Another approach to equalization of time-varying channels, found in the signal processing literature [271, 486, 488], is to model the time-varying channel impulse response function by a superposition of deterministic time-varying basis functions (e.g., complex exponentials) with time-invariant coefficients [147]. Such a model effectively converts the time-varying ISI channel into a time-invariant ISI channel. High-order statistic (HOS)–based and second-order statistic (SOS)–based equalization methods for time-invariant channels can then be employed to identify the channel coefficients and thus identify and equalize the time-varying channel.

Multiuser Receivers in Fading Channels

Data detection in multiuser fading channels has been addressed from a number of perspectives. Derivation and analysis of the optimum multiuser detection schemes under various fading channels are found in [65, 434, 513–516, 547, 615]. Suboptimal linear multiuser detection methods for fading channels are developed in [227, 434, 468, 547, 577, 616, 617]. Techniques for joint fading channel estimation and multiuser detection that are based on the EM algorithm are proposed in [91, 119]. Moreover, adaptive linear multiuser detection in fading channels has been studied in [26, 184, 188, 547, 613].

A few recent works have addressed the exploitation of coded signal structure in sequence estimation. In [597] the reduced-state sequence estimation [106, 113, 114, 560, 561] technique is integrated with an error-detection code for channel equalization. In this method, some subset of the set of all possible paths in the trellis is generated to satisfy the code constraints. Similar ideas have also been applied to joint channel and data estimation where the estimation procedure is forced to yield valid code-constrained path sequences [54, 55, 212].

In this chapter we present several advanced methods for signal reception in fast-fading channels. This presentation is organized as follows. In Section 9.2 we discuss statistical modeling of multipath fading channels. In Section 9.3 we present coherent receiver techniques for fading channels based on the expectation-maximization algorithm. In Section 9.4 we discuss decision-feedback-based low-complexity differential receiver techniques for fading channels. Finally, in Section 9.5 we present adaptive receiver techniques for fading channels that are based on the sequential Monte Carlo methodology described in Chapter 8.

The following is a list of the algorithms appearing in this chapter.

- *Algorithm 9.1:* EM algorithm for pilot-symbol-aided receiver in flat-fading channels
- *Algorithm 9.2:* Multiple-symbol decision-feedback differential detection
- *Algorithm 9.3:* Differential space-time decoding
- *Algorithm 9.4:* Multiple-symbol decision-feedback space-time differential decoding
- *Algorithm 9.5:* SMC for adaptive detection in flat-fading channels—Gaussian noise
- *Algorithm 9.6:* Delayed-sample SMC algorithm for adaptive detection in flat-fading channels—Gaussian noise

- *Algorithm 9.7:* SMC algorithm for adaptive decoding in flat-fading channels—Gaussian noise
- *Algorithm 9.8:* SMC algorithm for adaptive detection in flat-fading channels—impulsive noise.

9.2 Statistical Modeling of Multipath Fading Channels

We first describe the statistical modeling of mobile wireless channels. We follow [396] closely. For a typical terrestrial wireless channel, we can assume the existence of multiple propagation paths between the transmitter and the receiver. With each transmission path we can associate a propagation delay and an attenuation factor, which are usually time-varying due to changes in propagation conditions resulting primarily from transceiver mobility. In the absence of additive noise, the received complex baseband signal in such a channel is given by

$$y(t) = \sum_n \alpha_n(t)\, x(t - \tau_n(t))\, e^{-\jmath 2\pi f_c \tau_n(t)}, \tag{9.1}$$

where $x(t)$ is the transmitted baseband signal; $\alpha_n(t)$ and $\tau_n(t)$ are, respectively, the path attenuation and the propagation delay for the signal received on the nth path; and f_c is the carrier frequency. By inspecting (9.1), we can see that we can model the multipath fading channel by a time-varying linear filter with impulse response $g(\tau, t)$ given by

$$g(\tau, t) = \sum_n \alpha_n(t)\, e^{-\jmath 2\pi f_c \tau_n(t)}. \tag{9.2}$$

For some mobile channels, we can further assume that the received signal consists of a continuum of multipath components. Accordingly, for these channels, (9.1) is modified as follows:

$$y(t) = \int_{-\infty}^{\infty} \alpha(\tau, t)\, x(t - \tau)\, e^{-\jmath 2\pi f_c \tau}\, \mathrm{d}\tau, \tag{9.3}$$

where $\alpha(\tau, t)$ denotes the attenuation factor associated with a path delayed by τ at time instant t. The corresponding baseband time-varying impulse response of the channel is then

$$g(\tau, t) = \alpha(\tau, t)\, e^{-\jmath 2\pi f_c \tau}. \tag{9.4}$$

By the central limit theorem, assuming a large enough number of paths between the transmitter and the receiver, and by further assuming that the associated attenuations per path are independent and identically distributed, the impulse response $g(\tau, t)$ can be modeled by a complex-valued Gaussian random process. If the received signal $r(t)$ has only a diffuse multipath component, $g(\tau, t)$ is characterized by a zero-mean complex Gaussian random variable (i.e., $|g(\tau, t)|$ has a Rayleigh distribution). In this case the channel is called a *Rayleigh fading channel.* Alternatively, if there are fixed scatterers or signal reflections in the medium, $g(\tau, t)$ has a nonzero mean value and therefore $|g(\tau, t)|$ has a Rician distribution. In this case the channel is a *Rician fading channel.*

We will assume that the fading process $g(\tau, t)$ is wide-sense stationary in t, and define its corresponding autocorrelation function as

$$R_g(\tau_1, \tau_2; \Delta t) = \frac{1}{2} E\{g(\tau_1, t)\, g(\tau_2, t + \Delta t)^*\}. \tag{9.5}$$

A further reasonable assumption for most mobile communication channels is that the attenuation and phase shift associated with path delay τ_1 are uncorrelated with the corresponding attenuation and phase shift associated with a different path delay τ_2. This situation is known as *uncorrelated scattering.* Thus (9.5) can be expressed as

$$R_g(\tau_1, \tau_2; \Delta t) = R_g(\tau_1, \Delta t)\delta(\tau_1 - \tau_2), \tag{9.6}$$

where $R_g(\tau, \Delta t)$ represents the average channel power as a function of the time delay τ and the difference Δt in observation time. The *multipath spread* of the channel, T_m, is the range of values of the path delay τ for which $R_g(\tau, 0)$ is essentially constant. Let $S_g(f, \Delta t) = \mathcal{F}_\tau\{R_g(\tau, \Delta t)\}$ [i.e., the Fourier transform of $R_g(\tau, \Delta t)$ with respect to τ]. Then $S_g(f, \Delta t)$ is essentially the frequency response function of the linear time-varying channel. The *coherence bandwidth* of the channel, B_c, is the range of values of frequency f for which $S_g(f, 0)$ is essentially constant. Hence the multipath delay spread T_m and the coherence bandwidth B_c are related reciprocally (i.e., $B_c \approx 1/T_m$). Roughly speaking, the channel frequency response remains the same within the coherence bandwidth B_c. Let W be the bandwidth of the transmitted signal. When $W > B_c$, the channel is called *frequency-selective fading*; and when $W < B_c$, the channel is called *frequency nonselective fading* or *flat fading.*

We can also take the Fourier transform of $R_g(\tau, \Delta t)$ with respect to Δt to obtain the *scattering function* $S_g(\tau, \lambda) = \mathcal{F}_{\Delta t}\{R_g(\tau, \Delta t)\}$. The *Doppler spread* of the channel, B_d, is the range of values of frequency λ for which $S_g(0, \lambda)$ is essentially constant. The channel *coherence time* is given by $T_c \approx 1/B_d$. Roughly speaking, the channel time response remains the same within the coherence time T_c. Let T be the symbol interval of the transmitted signal. When $T < T_c$ (i.e., small Doppler), the channel is said to be *time-nonselective fading* or *slow fading*; and when $T > T_c$ (i.e., large Doppler), the channel is said to be *time-selective fading* or *fast fading.*

9.2.1 Frequency-Nonselective Fading Channels

Note from (9.3) and (9.4) that we have

$$y(t) = x(\tau) \star g(\tau, t) = \int X(f) G(f, t) e^{j2\pi f t} \mathrm{d}f, \tag{9.7}$$

where $X(f) = \mathcal{F}\{x(\tau)\}$ and $G(f, t) = \mathcal{F}_\tau\{g(\tau, t)\}$. Assume that the channel fading is frequency-nonselective (flat) (i.e., $W < B_c$); then the channel frequency response

$G(f,t)$ is approximately constant over the signal bandwidth [i.e., $G(f,t) = G(t)$]. In this case, (9.7) can be written as

$$y(t) = G(t) \int X(f) e^{\jmath 2\pi f t} \, \mathrm{d}f = G(t) x(t). \tag{9.8}$$

Hence the effect of a flat-fading channel can be modeled as a time-varying multiplicative distortion. Note that since $g(\tau, t)$ is assumed to be a complex Gaussian process, $G(t)$ is also a complex Gaussian process. When the fading is Rayleigh, we have $E\{G(t)\} = 0$. For mobile communications, the autocorrelation function of $G(t)$ is typically modeled by the *Jakes model* [216]:

$$R_G(\Delta t) \triangleq \frac{1}{2} E\left\{G(t)\, G(t+\Delta t)^*\right\} \;=\; P J_0(2\pi B_d \Delta t), \tag{9.9}$$

where P is the average power of the fading process (i.e., $P = E\left\{|G(t)|^2\right\}$) and $J_0(\cdot)$ is the Bessel function of the first kind and zeroth order. The corresponding Doppler power spectrum of the channel is then given by

$$S_G(\lambda) = \mathcal{F}\{R_G(\Delta t)\} = \frac{P}{\pi \sqrt{1 - (\lambda/B_d)^2}}. \tag{9.10}$$

9.2.2 Frequency-Selective Fading Channels

Now assume that the transmitted baseband signal has a bandwidth of W and that $W > B_c$ (i.e., the channel exhibits frequency-selective fading). By the sampling theorem, we have

$$x(t) = \sum_{n=-\infty}^{\infty} x\left(\frac{n}{W}\right) \operatorname{sinc}\left[\pi W \left(t - \frac{n}{W}\right)\right], \tag{9.11}$$

$$X(f) = \mathcal{F}\{x(t)\} = \frac{1}{W} \sum_{n=-\infty}^{\infty} x\left(\frac{n}{W}\right) e^{-\jmath(2\pi f n/W)}, \quad |f| \leq \frac{W}{2}. \tag{9.12}$$

Hence the noiseless received signal is given by

$$\begin{aligned} y(t) &= \int X(f) G(f;t) e^{\jmath 2\pi f t} \, \mathrm{d}f \\ &= \frac{1}{W} \sum_{n=-\infty}^{\infty} x\left(\frac{n}{W}\right) \int G(f,t) e^{\jmath 2\pi f (t - n/W)} \, \mathrm{d}f \\ &= \frac{1}{W} \sum_{n=-\infty}^{\infty} x\left(\frac{n}{W}\right) g\left(t - \frac{n}{W}, t\right) \\ &= \frac{1}{W} \sum_{n=-\infty}^{\infty} g\left(\frac{n}{W}, t\right) x\left(t - \frac{n}{W}\right). \end{aligned} \tag{9.13}$$

Let $L \triangleq \lceil WT_m + 1 \rceil$; then for practical purposes we can use the following truncated tapped-delay-line model to describe the frequency-selective fading channel [396]:

$$y(t) = \sum_{\ell=0}^{L-1} \alpha_\ell(t)\, x\left(t - \frac{\ell}{W}\right), \tag{9.14}$$

where $\alpha_\ell(t) \triangleq h(\ell/W, t)/W$, and $\{\alpha_\ell(t)\}_{\ell=0}^{L-1}$ comprises independent complex Gaussian processes.

9.3 Coherent Detection in Fading Channels Based on the EM Algorithm

As will be seen below, the maximum-likelihood sequence detector in fading channels typically has prohibitive computational complexity. The EM algorithm is an iterative technique for solving complex maximum-likelihood estimation problems. In this section we discuss sequence detection in fading channels based on the EM algorithm. Both a batch algorithm and a sequential algorithm are discussed.

9.3.1 Expectation-Maximization Algorithm

Suppose that $\boldsymbol{\theta}$ is a set of parameters to be estimated from some observed data $\boldsymbol{Y}$. The maximum-likelihood estimate $\hat{\boldsymbol{\theta}}$ of $\boldsymbol{\theta}$ is given by

$$\hat{\boldsymbol{\theta}} = \arg\max_{\boldsymbol{\theta}} p(\boldsymbol{Y} \mid \boldsymbol{\theta}), \tag{9.15}$$

where $p(\boldsymbol{Y}|\boldsymbol{\theta})$ denotes the probability density of $\boldsymbol{Y}$ with $\boldsymbol{\theta}$ fixed. In many cases, an explicit expression for the conditional density $p(\boldsymbol{Y}|\boldsymbol{\theta})$ does not exist. In other cases, the maximization problem above is very difficult to solve, even though the conditional density can be expressed explicitly. The EM algorithm [248, 315] is an iterative procedure for solving the ML estimation problem above in many such situations.

In the EM algorithm, the observation $\boldsymbol{Y}$ is termed *incomplete data*. The algorithm postulates that one has access to *complete data* $\boldsymbol{X}$, which is such that $\boldsymbol{Y}$ can be obtained through a many-to-one mapping. Typically, the complete data is chosen such that the conditional density $p(\boldsymbol{X}|\boldsymbol{\theta})$ is easy to obtain and maximize over $\boldsymbol{\theta}$. Starting from some initial estimate $\boldsymbol{\theta}^{(0)}$, the EM algorithm solves the ML estimation problem (9.15) by the following iterative procedure:

- *E-step:* Compute

$$Q\left(\boldsymbol{\theta} \mid \boldsymbol{\theta}^{(i)}\right) = E\left\{\log p(\boldsymbol{X}|\boldsymbol{\theta}) \mid \boldsymbol{Y}, \boldsymbol{\theta}^{(j)}\right\}. \tag{9.16}$$

- *M-step:* Solve

$$\boldsymbol{\theta}^{(j+1)} = \arg\max_{\boldsymbol{\theta}} Q\left(\boldsymbol{\theta} \mid \boldsymbol{\theta}^{(j)}\right). \tag{9.17}$$

It is known that the sequence $\left\{\boldsymbol{\theta}^{(j)}\right\}$ obtained in the EM algorithm above monotonically increases the incomplete-data likelihood function:

$$p\left(\boldsymbol{Y} \mid \boldsymbol{\theta}^{(j+1)}\right) \geq p\left(\boldsymbol{Y} \mid \boldsymbol{\theta}^{(j)}\right). \tag{9.18}$$

Moreover, if the function $Q(\boldsymbol{\theta};\boldsymbol{\theta}')$ is continuous in both $\boldsymbol{\theta}$ and $\boldsymbol{\theta}'$, all limit points of an EM sequence $\left\{\boldsymbol{\theta}^{(j)}\right\}$ are stationary points of $p(\boldsymbol{Y}|\boldsymbol{\theta})$ (i.e., local maxima or saddle points) and $p\left(\boldsymbol{Y}|\boldsymbol{\theta}^{(j)}\right)$ converges monotonically to $p\left(\boldsymbol{Y}|\hat{\boldsymbol{\theta}}\right)$ for some stationary point $\hat{\boldsymbol{\theta}}$ [248, 315].

9.3.2 EM-Based Receiver in Flat-Fading Channels

We consider the following discrete-time flat-fading channel

$$r[i] = \alpha[i]\, b[i] + n[i], \quad i = 0, 1, \ldots, M-1, \tag{9.19}$$

where $\{\alpha[i]\}$ is the complex Gaussian fading process, $\{b[i]\}$ is a sequence of transmitted phase-shift-keying (PSK) symbols ($|b[i]| = 1$), and $\{n[i]\}$ is a sequence of i.i.d. Gaussian noise samples. Define the following notation:

$$\begin{aligned}
\boldsymbol{r} &= \Big[r[0]\; r[1]\; \cdots\; r[M-1]\Big]^T, \\
\boldsymbol{b} &= \Big[b[0]\; b[1]\; \cdots\; b[M-1]\Big]^T, \\
\boldsymbol{B} &= \operatorname{diag}\Big\{b[0]\; b[1]\; \cdots\; b[M-1]\Big\}, \\
\boldsymbol{\alpha} &= \Big[\alpha[0]\; \alpha[1]\; \cdots\; \alpha[M-1]\Big]^T, \\
\boldsymbol{n} &= \Big[n[0]\; n[1]\; \cdots\; n[M-1]\Big]^T.
\end{aligned}$$

Then (9.19) can be written as

$$\boldsymbol{r} = \boldsymbol{B}\boldsymbol{\alpha} + \boldsymbol{n}. \tag{9.20}$$

Note that both $\boldsymbol{\alpha}$ and $\boldsymbol{n}$ are complex Gaussian vectors:

$$\boldsymbol{\alpha} \sim \mathcal{N}_c(\boldsymbol{0},\; E_s\, \boldsymbol{\Sigma}_M), \tag{9.21}$$

$$\boldsymbol{n} \sim \mathcal{N}_c(\boldsymbol{0},\; \sigma^2 \boldsymbol{I}_M), \tag{9.22}$$

where E_s is the average received signal energy. For mobile fading channels, the normalized $M \times M$ autocorrelation matrix has elements given by the Jakes model as

$$\boldsymbol{\Sigma}_M[i,j] \;=\; J_0\Big(2\pi B_d T(i-j)\Big), \tag{9.23}$$

where $B_d T$ is the symbol-rate-normalized Doppler shift and $J_0(\cdot)$ is the Bessel function of the first kind and zeroth order. Hence $\boldsymbol{r}$ in (9.20) has the following complex Gaussian distribution:

$$\boldsymbol{r} \sim \mathcal{N}_c\Big(\boldsymbol{0},\ \underbrace{E_s \boldsymbol{B}\boldsymbol{\Sigma}_M \boldsymbol{B}^H + \sigma^2 \boldsymbol{I}_M}_{\boldsymbol{Q}}\Big), \tag{9.24}$$

and the log-likelihood function of $\boldsymbol{r}$ given $\boldsymbol{B}$ is thus given by

$$\log p(\boldsymbol{r} \mid \boldsymbol{b}) = -\boldsymbol{r}^H \boldsymbol{Q}^{-1} \boldsymbol{r} - \log \det(\boldsymbol{Q}) - M \log \pi. \tag{9.25}$$

Note that

$$\begin{aligned} \boldsymbol{Q}^{-1} &= \left[\boldsymbol{B}\left(E_s \boldsymbol{\Sigma}_M + \sigma^2 \boldsymbol{I}_M\right)\boldsymbol{B}^H\right]^{-1} \\ &= E_s^{-1}\, \boldsymbol{B}\left(\boldsymbol{\Sigma}_M + \frac{\sigma^2}{E_s}\boldsymbol{I}_M\right)^{-1} \boldsymbol{B}^H, \end{aligned} \tag{9.26}$$

$$\begin{aligned} \det(\boldsymbol{Q}) &= \det\left[\boldsymbol{B}\left(E_s \boldsymbol{\Sigma}_M + \sigma^2 \boldsymbol{I}_M\right)\boldsymbol{B}^H\right] \\ &= \det(\boldsymbol{B}) \det\left(E_s \boldsymbol{\Sigma}_M + \sigma^2 \boldsymbol{I}_M\right) \det\left(\boldsymbol{B}^H\right) \\ &= \det\left(E_s \boldsymbol{\Sigma}_M + \sigma^2 \boldsymbol{I}_M\right), \end{aligned} \tag{9.27}$$

where we have used the facts that $\boldsymbol{B}\boldsymbol{B}^H = \boldsymbol{B}^H\boldsymbol{B} = \boldsymbol{I}_M$ and $\det(\boldsymbol{B})\det(\boldsymbol{B}^H) = \det(\boldsymbol{B}\boldsymbol{B}^H) = 1$, since $\boldsymbol{B}$ is a diagonal matrix containing PSK symbols. Hence the ML estimate of $\boldsymbol{b}$ becomes

$$\hat{\boldsymbol{b}} = \arg\min_{\boldsymbol{b}}\ \boldsymbol{r}^H \boldsymbol{B}\left(\boldsymbol{\Sigma}_M + \frac{\sigma^2}{E_s}\boldsymbol{I}_M\right)^{-1} \boldsymbol{B}^H \boldsymbol{r}. \tag{9.28}$$

The optimal solution involves an exhaustive enumeration of all possible PSK sequences of length M, which is certainly prohibitively complex even for moderate M.

The EM algorithm was applied to solve the fading channel detection problem above in [138]. To use the EM algorithm, we define the complete data as consisting of the incomplete data $\boldsymbol{r}$ together with the fading process $\boldsymbol{\alpha}$ [i.e., $\boldsymbol{x} = (\boldsymbol{r}, \boldsymbol{\alpha})$]. Then the log-likelihood function of the complete data is

$$\begin{aligned} \log p(\boldsymbol{x} \mid \boldsymbol{b}) &= -\left\|\boldsymbol{r} - \boldsymbol{B}\boldsymbol{\alpha}\right\|^2 + \text{terms not depending on } \boldsymbol{B} \\ &= 2\Re\left\{\boldsymbol{r}^H \boldsymbol{B}\boldsymbol{\alpha}\right\} + \text{terms not depending on } \boldsymbol{B}. \end{aligned} \tag{9.29}$$

Hence the E-step computes the following quantity:

$$\begin{aligned} Q\left(\boldsymbol{b} \mid \boldsymbol{b}^{(j)}\right) &= E\left\{\Re\left\{\boldsymbol{r}^H \boldsymbol{B}\boldsymbol{\alpha}\right\} \mid \boldsymbol{r}, \boldsymbol{b}^{(j)}\right\} \\ &= \Re\left\{\boldsymbol{r}^H \boldsymbol{B}\, E\left\{\boldsymbol{\alpha} \mid \boldsymbol{r}, \boldsymbol{b}^{(j)}\right\}\right\}. \end{aligned} \tag{9.30}$$

Since given $\boldsymbol{b} = \boldsymbol{b}^{(j)}$, $\boldsymbol{r}$ and $\boldsymbol{\alpha}$ are jointly Gaussian, we then have

$$\begin{aligned} \hat{\boldsymbol{\alpha}}^{(j)} \triangleq E\left\{\boldsymbol{\alpha} \mid \boldsymbol{r}, \boldsymbol{b}^{(j)}\right\} &= \operatorname{Cov}(\boldsymbol{\alpha}, \boldsymbol{r}) \operatorname{Cov}(\boldsymbol{r})^{-1} \boldsymbol{r} \\ &= E_s \boldsymbol{\Sigma}_M \boldsymbol{B}^{(j)H} \boldsymbol{Q}^{-1} \boldsymbol{r} \\ &= \boldsymbol{\Sigma}_M \left(\boldsymbol{\Sigma}_M + \frac{\sigma^2}{E_s} \boldsymbol{I}_M \right)^{-1} \boldsymbol{B}^{(j)H} \boldsymbol{r}. \end{aligned} \tag{9.31}$$

The maximization step becomes

$$\begin{aligned} \boldsymbol{b}^{(j+1)} &= \arg\max \Re\left\{\boldsymbol{r}^H \boldsymbol{B} \hat{\boldsymbol{\alpha}}^{(j)}\right\} \\ \Longrightarrow \quad b[i]^{(j+1)} &= \arg\max_{b[i]} \Re\left\{ r[i]^* \, b[i] \, \hat{\alpha}[i]^{(j)} \right\}, \quad i = 0, \ldots, M-1. \end{aligned} \tag{9.32}$$

An initial estimate of the data symbol sequence $\boldsymbol{b}^{(0)}$ can be obtained with the aid of pilot symbols as follows. Suppose we choose M such that $M = NL + 1$, where N and L are positive integers. Suppose further that the symbols in positions $0, N, 2N, \ldots, (L-1)N$ are known. Then the initial channel estimates at these positions are given by

$$\alpha[i]^{(-1)} = r[i] \, b[i]^*, \quad i = 0,\ N,\ \ldots, LN, \tag{9.33}$$

and the initial channel estimates at other positions are obtained by linear interpolation; that is,

$$\begin{aligned} \alpha[kN+m]^{(-1)} = \alpha[kN]^{(-1)} + \frac{m}{N} \left(\alpha[(k+1)N]^{(-1)} - \alpha[kN]^{(-1)} \right), \\ k = 0, \ldots, L-1, \quad m = 1, \ldots, N-1. \end{aligned} \tag{9.34}$$

Substituting the initial channel estimate $\boldsymbol{\alpha}^{(-1)}$ above into (9.32), we obtain the initial symbol estimate $\boldsymbol{b}^{(0)}$.

Finally, we summarize the EM-based pilot-symbol-aided receiver algorithm in flat-fading channels as follows.

Algorithm 9.1: [EM algorithm for pilot-symbol-aided receiver in flat-fading channels]

- *Initialization: Based on the pilot symbol information, obtain an initial channel estimate* $\boldsymbol{\alpha}^{(-1)}$ *using (9.33) and (9.34). Substitute* $\boldsymbol{\alpha}^{(-1)}$ *into (9.32) and compute the initial symbol estimate* $\boldsymbol{b}^{(0)}$.

- *For* $j = 1, \ldots, I$, *iterate the following E and M steps (where* I *is the number of EM iterations):*

  - *E-step: Compute* $\boldsymbol{\alpha}^{(j)}$ *according to (9.31).*
  - *M-step: Compute* $\boldsymbol{b}^{(j+1)}$ *according to (9.32).*

9.3.3 Linear Multiuser Detection in Flat-Fading Synchronous CDMA Channels

The EM-based receiver discussed above can easily be applied to synchronous CDMA systems in flat-fading channels [555]. The basic idea is to use a linear multiuser detector (e.g., the decorrelating detector) to separate the multiuser signals and then to employ an EM receiver for each user to demodulate its data. This procedure is discussed briefly next.

We consider the following simple K-user synchronous CDMA system signaling through flat-fading channels. The received signal during the ith symbol interval is given by

$$r(t) = \sum_{k=1}^{K} A_k \alpha_k[i] b_k[i] s_k(t - iT) + n(t), \qquad iT \leq t < (i+1)T, \tag{9.35}$$

where $\alpha_k[i]$ is the complex fading gain of the kth user's channel during the ith symbol interval; A_k is the amplitude of the kth user; $b_k[i] \in \{+1, -1\}$ is the ith bit of the kth user; $\{s_k(t),\ 0 \leq t \leq T\}$ is the unit-energy spreading waveform of the kth user; and $n(t)$ is white complex Gaussian noise with power spectral density σ^2. The received signal is correlated with the signature waveform of each user, to obtain the decision statistic:

$$\begin{aligned} y_k[i] &= \int_{iT}^{(i+1)T} r(t)\, s_k(t - iT)\, \mathrm{d}t \\ &= \sum_{j=1}^{K} A_j \rho_{jk} \alpha_j[i] b_j[i] + n_k[i], \end{aligned} \tag{9.36}$$

where $\rho_{jk} = \int_0^T s_j(t)\, s_k(t)\, \mathrm{d}t$ and $n_k[i] = \int_0^T n(t)\, s_k(t - iT)\, \mathrm{d}t$. On denoting $\boldsymbol{y}[i] = \left[y_1[i]\ y_2[i]\ \cdots\ y_K[i]\right]^T$, we can write

$$\boldsymbol{y}[i] = \boldsymbol{R}\boldsymbol{A}\boldsymbol{\Phi}[i]\boldsymbol{b}[i] + \boldsymbol{n}[i], \tag{9.37}$$

where $\boldsymbol{R} = [\rho_{jk}]$, $\boldsymbol{A} = \mathrm{diag}\{A_1, \ldots, A_K\}$, $\boldsymbol{\Phi}[i] = \mathrm{diag}\{\alpha_1[i], \ldots, \alpha_K[i]\}$, $\boldsymbol{b}[i] = \left[b_1[i]\ b_2[i]\ \cdots\ b_K[i]\right]^T$, and $\boldsymbol{n}[i] = \left[n_1[i]\ n_2[i]\ \cdots\ n_K[i]\right]^T$. Note that $\boldsymbol{n}[i] \sim \mathcal{N}_c\left(\boldsymbol{0}, \sigma^2\boldsymbol{R}\right)$.

The multiuser signals $\boldsymbol{y}[i]$ in (9.37) can be separated by a linear decorrelator, to obtain

$$\boldsymbol{z}[i] = \boldsymbol{R}^{-1}\boldsymbol{y}[i] \;=\; \boldsymbol{A}\boldsymbol{\Phi}[i]\boldsymbol{b}[i] + \boldsymbol{u}[i], \tag{9.38}$$

with $\boldsymbol{u}[i] \sim \mathcal{N}_c\left(\boldsymbol{0}, \sigma^2\boldsymbol{R}^{-1}\right)$. We can write (9.38) in scalar form as

$$z_k[i] = A_k\, \alpha_k[i]\, b_k[i] + u_k[i], \qquad k = 1, \ldots, K, \tag{9.39}$$

with $u_k[i] \sim \mathcal{N}_c\left(0,\, \sigma^2[\boldsymbol{R}^{-1}]_{kk}\right)$. We see that for each user, the output of the decorrelating detector (9.39) is of exactly the same form as (9.19). Hence with the aid of pilot symbols, the EM receiver discussed in Section 9.3.2 can be employed to demodulate the kth user's data $\{b_k[i]\}_i$, $k = 1, \ldots, K$.

An alternative suboptimal receiver structure for demodulating the kth user's data uses a Kalman filter to track the fading channel $\{\alpha_k[i]\}_i$, based on training symbols or decision feedback [547, 578, 579]. For example, in the simplest setting, the fading coefficients $\{\alpha_k[i]\}_i$ may be modeled by a second-order autoregressive (AR) process:

$$\alpha_k[i] = -a_1\alpha_k[i-1] - a_2\alpha_k[i-2] + w[i], \tag{9.40}$$

where $w[i]$ is a zero-mean white complex Gaussian process. The parameters a_1 and a_2 are chosen to fit the spectrum of the AR process to that of the underlying Rayleigh fading process. On the other hand, a statistically equivalent representation of the linear decorrelator output (9.39) is

$$z_k[i]b_k[i] = A_k\alpha_k[i] + u_k[i], \tag{9.41}$$

where we have invoked the symmetry of the distribution of $u_k[i]$. Based on the state equation (9.40) and the observation equation (9.41), we can use a Kalman filter to track the fading channel coefficients $\{\alpha_k[i]\}_i$ and subsequentially detect the data symbols. Note that in (9.41), the data symbols $\{b_k[i]\}_i$ are assumed known, corresponding to the case when they are training symbols. When these symbols are unknown, they are replaced by the detected symbols $\{\hat{b}_k[i]\}_i$. Such a decision-directed scheme is subject to error propagation, of course, and thus requires periodic insertion of training symbols.

9.3.4 Sequential EM Algorithm

The EM algorithm discussed above is a batch algorithm. We next briefly describe a sequential version of the EM algorithm [482, 563]. Suppose that $\boldsymbol{y}_1, \boldsymbol{y}_2, \ldots$ is a sequence of observations with marginal pdf $f(\boldsymbol{y}|\boldsymbol{\theta})$, where $\boldsymbol{\theta} \in \mathbb{C}^m$ is a static parameter vector. A class of sequential estimators derived from the maximum-likelihood principle is given by

$$\boldsymbol{\theta}^{(i+1)} = \boldsymbol{\theta}^{(i)} + \boldsymbol{\Pi}\left(\boldsymbol{y}_{i+1}, \boldsymbol{\theta}^{(i)}\right) \boldsymbol{s}\left(\boldsymbol{y}_{i+1}, \boldsymbol{\theta}^{(i)}\right), \tag{9.42}$$

where $\boldsymbol{\theta}^{(i)}$ is the estimate of $\boldsymbol{\theta}$ at the ith step, $\boldsymbol{\Pi}\left(\boldsymbol{y}_{i+1}, \boldsymbol{\theta}^{(i)}\right)$ is an $m \times m$ matrix defined later in this section, and

$$\boldsymbol{s}\left(\boldsymbol{y}_{i+1}, \boldsymbol{\theta}^{(i)}\right) \triangleq \left[\frac{\partial}{\partial\theta_1^*}\log f(\boldsymbol{y}_{i+1}|\boldsymbol{\theta}), \ldots, \frac{\partial}{\partial\theta_m^*}\log f(\boldsymbol{y}_{i+1}|\boldsymbol{\theta})\right]^T \Bigg|_{\boldsymbol{\theta}=\boldsymbol{\theta}^{(i)}} \tag{9.43}$$

is the update score (i.e., the gradient of the log-likelihood function). Let $\boldsymbol{H}\left(\boldsymbol{y}_i, \boldsymbol{\theta}^{(i)}\right)$ denote the Hessian matrix of $\log f(\boldsymbol{y}_i|\boldsymbol{\theta}^{(i)})$:

$$H_{j,k}\left(\boldsymbol{y}_i, \boldsymbol{\theta}^{(i)}\right) = \frac{\partial^2}{\partial\theta_j^* \partial\theta_k} \log f(\boldsymbol{y}_i|\boldsymbol{\theta})\bigg|_{\boldsymbol{\theta}=\boldsymbol{\theta}^{(i)}},$$

$$j = 1, \ldots, m, \quad k = 1, \ldots, m. \tag{9.44}$$

Let $\boldsymbol{x}_i$ denote a "complete" data set related to $\boldsymbol{y}_i$ for $i = 1, 2, \ldots$. The complete data set $\boldsymbol{x}_i$ is selected in the (sequential) EM algorithm such that $\boldsymbol{y}_i$ can be obtained through a many-to-one mapping $\boldsymbol{x}_i \to \boldsymbol{y}_i$, and so that its knowledge makes the estimation problem easy [e.g., the conditional density $f(\boldsymbol{x}_i|\boldsymbol{\theta})$ can easily be obtained]. Denote the Fisher information matrices of the data $\boldsymbol{y}_i$ and $\boldsymbol{x}_i$, respectively, as

$$\boldsymbol{I}\left(\boldsymbol{\theta}^{(i)}\right) = -E\left\{\boldsymbol{H}\left(\boldsymbol{y}_i, \boldsymbol{\theta}^{(i)}\right)\right\} \quad \text{and} \quad \boldsymbol{I}_c\left(\boldsymbol{\theta}^{(i)}\right) = -E\left\{\boldsymbol{H}\left(\boldsymbol{x}_i, \boldsymbol{\theta}^{(i)}\right)\right\}.$$

Different versions of sequential estimation algorithms are characterized by different choices of the function $\boldsymbol{\Pi}\left(\boldsymbol{y}_{i+1}, \boldsymbol{\theta}^{(i)}\right)$ in (9.42), as follows.

- *The sequential EM algorithm:*

$$\boldsymbol{\Pi}\left(\boldsymbol{y}_{i+1}, \boldsymbol{\theta}^{(i)}\right) = \frac{1}{i}\boldsymbol{I}_c^{-1}\left(\boldsymbol{\theta}^{(i)}\right). \tag{9.45}$$

  The consistency and asymptotic normality of this algorithm are considered in [482]. Applications of the sequential EM algorithm to communications and signal processing problems are reported in [123, 238, 239, 563, 601, 602].

- *The Newton–Raphson algorithm:*

$$\boldsymbol{\Pi}\left(\boldsymbol{y}_{i+1}, \boldsymbol{\theta}^{(i)}\right) = -\boldsymbol{H}^{-1}\left(\boldsymbol{y}_{i+1}, \boldsymbol{\theta}^{(i)}\right). \tag{9.46}$$

- *A stochastic approximation procedure:*

$$\boldsymbol{\Pi}\left(\boldsymbol{y}_{i+1}, \boldsymbol{\theta}^{(i)}\right) = \frac{1}{i}\boldsymbol{I}^{-1}\left(\boldsymbol{\theta}^{(i)}\right). \tag{9.47}$$

  Note that for i.i.d. observations $\{\boldsymbol{y}_i\}$, if i in (9.47) is replaced by $(i+1)$, we obtain the maximum-likelihood estimator of $\boldsymbol{\theta}$ for exponential families [482]. The asymptotic distribution of this procedure can be found in [115, 428].

- If $\boldsymbol{\Pi}\left(\boldsymbol{y}_{i+1}, \boldsymbol{\theta}^{(i)}\right)$ is a constant diagonal matrix with small elements, then (9.42) is the conventional steepest-descent algorithm. Some related choices of $\boldsymbol{\Pi}\left(\boldsymbol{y}_{i+1}, \boldsymbol{\theta}^{(i)}\right)$ are suggested in [482].

- For *time-varying* parameters $\{\boldsymbol{\theta}(i)\}$, a conventional approach suggested in [123, 283] is to substitute the converging series $1/i$ in (9.45) with a small positive constant λ_0. The new estimator is given by

$$\hat{\boldsymbol{\theta}}(i+1) = \hat{\boldsymbol{\theta}}(i) + \lambda_0 \, \boldsymbol{I}_c^{-1}\Big(\hat{\boldsymbol{\theta}}(i)\Big) \; \boldsymbol{s}\Big(\boldsymbol{y}_{i+1}, \hat{\boldsymbol{\theta}}(i)\Big), \tag{9.48}$$

  where $\hat{\boldsymbol{\theta}}(i)$ is the estimate of $\boldsymbol{\theta}(i)$.

9.4 Decision-Feedback Differential Detection in Fading Channels

9.4.1 Decision-Feedback Differential Detection in Flat-Fading Channels

The coherent detection methods discussed in Section 9.3 require explicit or implicit estimation of the fading channel, which in turn requires the transmission of pilot or training symbols. In this section we discuss decision-feedback differential detection in flat-fading channels, which does not require channel estimation. Consider again the signal model (9.19). Assume that the transmitted symbols $\{b[i]\}$ are the outputs of a differential encoder:

$$b[i] = d[i] \, b[i-1], \tag{9.49}$$

where $\{d[i]\}$ is a sequence of PSK information symbols. In simple differential detection, the complex plane is divided into M disjoint sectors, where M is the size of the PSK signaling alphabet. The detected information symbol $\hat{d}[i]$ is determined by the sector into which the complex number $(r[i]r[i-1]^*)$ falls. Such a simple differential detection rule incurs a 3 dB performance loss compared with coherent detection in AWGN channels [396]. In flat-fading channels, however, it exhibits an irreducible error floor in the high-SNR region [222]. For example, for binary DPSK we have

$$\lim_{\text{SNR}\to\infty} P_e\,(\text{SNR}) = \frac{1-\rho}{2}, \tag{9.50}$$

where ρ is the correlation coefficient between the fading gains at two consecutive symbol intervals.

Multiple-symbol decision-feedback differential detection was developed in [441]. This method makes use of the correlation function of the channel and can significantly reduce the error floor of simple differential detection. In multiple-symbol differential detection [101, 102, 179, 304], an observation interval of length N is introduced. Define the following quantities:

$$\boldsymbol{r}[i] = \Big[r[i] \; r[i-1] \; \cdots \; r[i-N+1]\Big]^T,$$
$$\boldsymbol{\alpha}[i] = \Big[\alpha[i] \; \alpha[i-1] \; \cdots \; \alpha[i-N+1]\Big]^T,$$

$$\boldsymbol{n}[i] = \Big[n[i]\; n[i-1]\; \cdots\; n[i-N+1]\Big]^T,$$
$$\boldsymbol{B}[i] = \text{diag}\Big\{b[i], b[i-1], \ldots, b[i-N+1]\Big\},$$
$$\boldsymbol{d}[i] = \Big[d[i]\; d[i-1]\; \cdots\; d[i-N+2]\Big]^T.$$

Similar to (9.24)–(9.25), we can write the log-likelihood function as

$$\log p\Big(\boldsymbol{r}[i] \mid \boldsymbol{d}[i]\Big) = -\boldsymbol{r}[i]^H \boldsymbol{Q}_d^{-1} \boldsymbol{r}[i] - \log\det(\boldsymbol{Q}_d) - N\log\pi, \tag{9.51}$$

where

$$\boldsymbol{Q}_d^{-1} = E_s^{-1} \boldsymbol{B}[i] \underbrace{\left(\boldsymbol{\Sigma}_N + \frac{\sigma^2}{E_s}\boldsymbol{I}_N\right)^{-1}}_{\boldsymbol{T}} \boldsymbol{B}[i]^H, \tag{9.52}$$

$$\det(\boldsymbol{Q}_d) = \det\left(E_s \boldsymbol{\Sigma}_N + \sigma^2 \boldsymbol{I}_N\right). \tag{9.53}$$

The maximum-likelihood decision metric thus becomes

$$\hat{\boldsymbol{d}}[i] = \arg\min_{\boldsymbol{d}[i]} \left\{\rho(\boldsymbol{d}[i]) \triangleq \boldsymbol{r}[i]^H \boldsymbol{B}[i] \boldsymbol{T} \boldsymbol{B}[i]^H \boldsymbol{r}[i]\right\}. \tag{9.54}$$

Since $\boldsymbol{T}^{-1}$ is symmetric, so is $\boldsymbol{T}$. That is, if we denote $\boldsymbol{T} = [t_{i,j}]$, then $t_{i,j} = t_{j,i}$. Hence we can write

$$\begin{aligned}
\rho(\boldsymbol{d}[i]) &= \sum_{l=0}^{N-1}\sum_{j=0}^{N-1} t_{l,j}\, b[i-l]\, b[i-j]^*\, r[i-l]^*\, r[i-j] \\
&= \sum_{l=0}^{N-1} t_{l,l}\, |b[i-l]|^2\, |r[i-l]|^2 \\
&\quad + 2\Re\left\{\sum_{l=0}^{N-1}\sum_{j=i+1}^{N-1} t_{l,j}\, r[i-l]\, r[i-j]^* \prod_{m=l}^{j-1} d[i-m]\right\},
\end{aligned} \tag{9.55}$$

where (9.55) follows from (9.49). In decision-feedback differential detection, the previous information symbols $d[i-1], d[i-2], \ldots, d[i-N+2]$ in (9.55) are assumed to take values given by the previous decisions (i.e., $\hat{d}[i-1], \hat{d}[i-2], \ldots, \hat{d}[i-N+2]$), and we will make a decision on the current information symbol $d[i]$ to minimize the cost function $\rho(\boldsymbol{d}[i])$ above. To that end, such a decision rule can be simplified to [441]

$$\hat{d}[i] = \arg\max_{d[i]} \Re\left\{ d[i]^* r[i] \left(\sum_{l=1}^{N-1} t_{0,l} r[i-l] \prod_{j=1}^{l-1} \hat{d}[i-j]\right)^*\right\}. \tag{9.56}$$

Based on (9.56), we arrive at the following decision rule: Divide the complex plane into M disjoint sectors and determine $\hat{d}[i]$ by the sector into which the complex number

$$g[i] \triangleq r[i] \left(\sum_{l=1}^{N-1} t_{0,l} r[i-l] \prod_{j=1}^{i-1} \hat{d}[i-j] \right)^* \tag{9.57}$$

falls. The multiple-symbol decision-feedback differential detection algorithm is summarized as follows.

Algorithm 9.2: [Multiple-symbol decision-feedback differential detection] *Given the decision memory order N, fading statistics $\mathbf{\Sigma}_N$, and signal-to-noise ratio E_s/σ^2:*

- *Compute the feedback filter coefficients from* $\boldsymbol{T} = \left(\mathbf{\Sigma}_N + \frac{\sigma^2}{E_s} \boldsymbol{I}_N \right)^{-1}$.

- *Estimate the initial information symbols $\hat{d}[1], \ldots, \hat{d}[N-1]$ by simple differential detection.*

- *For $i = N, N+1, \ldots,$ estimate $\hat{d}[i]$ according to (9.55).*

The corresponding multiple-symbol decision-feedback differential receiver structure is shown in Fig. 9.1, where the coefficients of the feedback filter are given by the metric coefficients $t_j = t_{0,j}, 1 \leq j \leq N-1$. Note that when $N = 2$, this receiver reduces to the simple differential detector.

Simulation Examples

For all simulation results presented below, a differential QPSK constellation is used. The feedback metric coefficients are obtained from the sample autocorrelation of the simulated fading process. Figures 9.2, 9.3, and 9.4 show the BER performance of the decision-feedback differential detector in flat-fading channels with normalized Doppler frequencies $B_d T$ equal to 0.003, 0.0075, and 0.01, respectively. It is seen that the error floors of the simple differential detector ($N = 2$) are reduced by the decision-feedback differential detector (DF-DD) with $N = 3$ and $N = 4$.

9.4.2 Decision-Feedback Space-Time Differential Decoding

In what follows we extend the decision-feedback differential detection method to systems employing multiple transmit antennas and space-time differential block coding, and develop a decision-feedback space-time differential decoding technique for flat-fading channels. The material presented here was developed in [282].

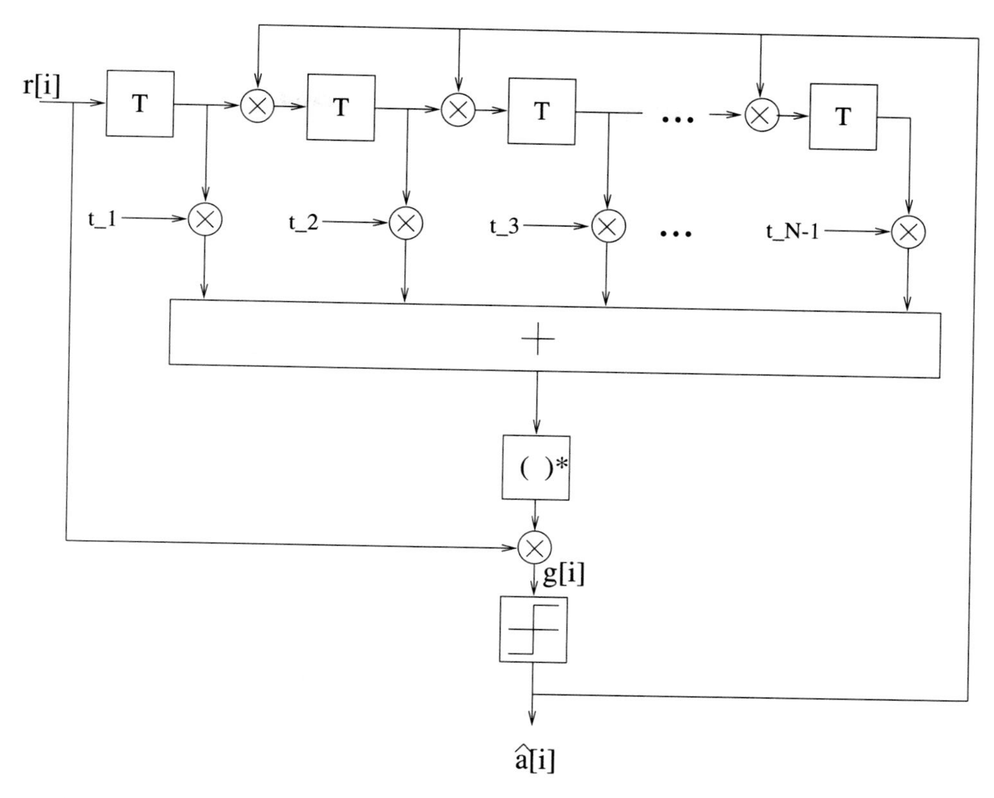

Figure 9.1. Structure of a multiple-symbol decision-feedback differential detector.

Space-Time Differential Block Coding

Space-time differential block coding was developed in [204, 474]. Consider a communication system with two transmit antennas and one receive antenna. Let the information PSK symbols at time i be

$$s[i] \in \mathcal{A} \triangleq \left\{ \frac{1}{\sqrt{2}} e^{\jmath(2\pi k/M)},\ k = 0, 1, \ldots, M-1 \right\}.$$

Define the following matrices:

$$\underline{G}[i] \triangleq \begin{bmatrix} s[2i] & s[2i+1] \\ -s[2i+1]^* & s[2i]^* \end{bmatrix}, \quad i = 1, 2, \ldots. \tag{9.58}$$

It is easy to see that $\underline{G}[i]$ is an orthogonal matrix (i.e., $\underline{G}[i]\underline{G}[i]^H = \underline{G}[i]^H\underline{G}[i] = \boldsymbol{I}_2$). The space-time differential block code is recursively defined as follows. Denote

$$\underline{X}[0] \triangleq \begin{bmatrix} x[0] & x[1] \\ -x[1]^* & x[0]^* \end{bmatrix}, \tag{9.59}$$

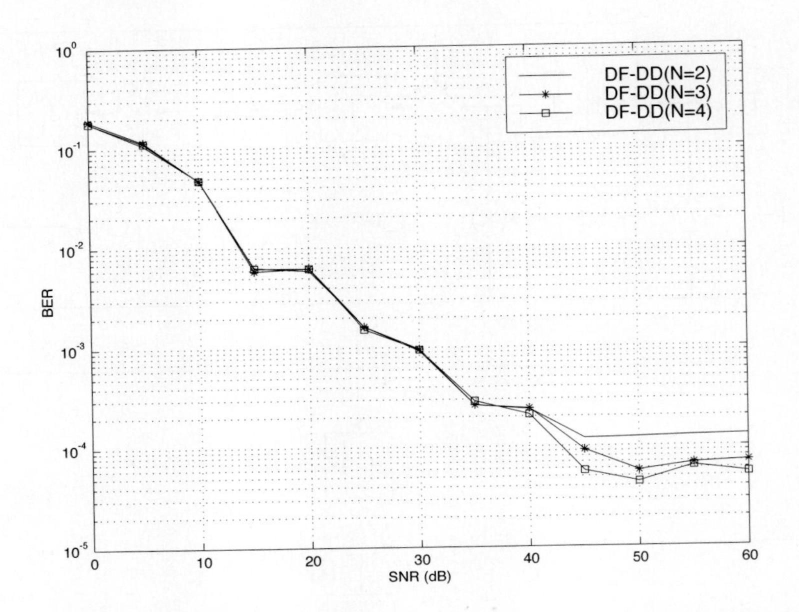

Figure 9.2. BER performance of a decision-feedback differential detector in a flat-fading channel with $B_dT = 0.003$.

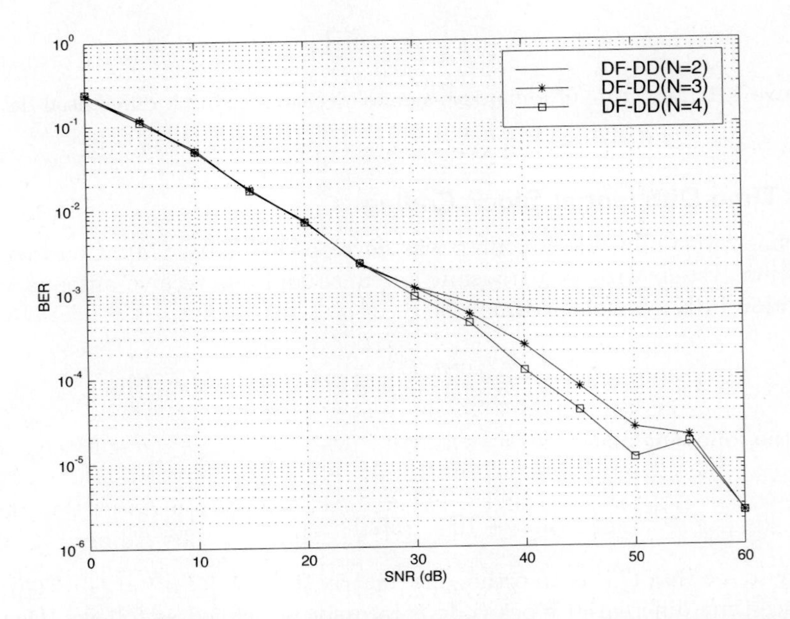

Figure 9.3. BER performance of a decision-feedback differential detector in a flat-fading channel with $B_dT = 0.0075$.

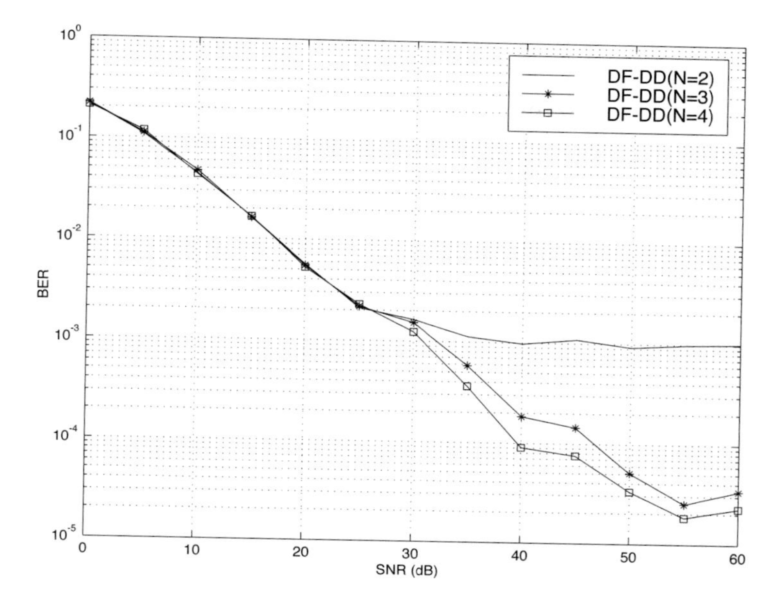

Figure 9.4. BER performance of a decision-feedback differential detector in a flat-fading channel with $B_dT = 0.01$.

where $x[0], x[1] \in \mathcal{A}$. Then define $X[i]$ as

$$\underline{X}[i] = \underline{G}[i]\underline{X}[i-1], \quad i = 1, 2, \ldots. \tag{9.60}$$

By simple induction, it is easy to show that the matrix $\underline{X}[i]$ has the form

$$\underline{X}[i] \triangleq \begin{bmatrix} x[2i] & x[2i+1] \\ -x[2i+1]^* & x[2i]^* \end{bmatrix}. \tag{9.61}$$

Hence $\underline{X}[i]$ is also an orthogonal matrix, and by (9.60), we have

$$\underline{X}[i]\underline{X}[i-1]^H = \underline{G}[i]. \tag{9.62}$$

At time slot $2i$, the symbols on the first row of $\underline{X}[i]$, $x[2i]$ and $x[2i+1]$, are transmitted simultaneously from antenna 1 and antenna 2, respectively. At time slot $2i+1$, the symbols on the second row of $\underline{X}[i]$, $-x[2i+1]^*$ and $x[2i]^*$, are transmitted simultaneously from the two antennas. We first consider the case where the channel is static. Let α_1 and α_2 be the respective complex fading gains between the two transmit antennas and the receive antenna. The received signals in time slots $2i$ and $2i+1$ are given, respectively, by

$$y[2i] = \alpha_1\, x[2i] + \alpha_2\, x[2i+1] + n[2i], \tag{9.63}$$

$$\text{and} \quad y[2i+1] = -\alpha_1\, x[2i+1]^* + \alpha_2\, x[2i]^* + n[2i+1], \tag{9.64}$$

where $n[2i]$ and $n[2i+1]$ are independent complex Gaussian noise samples. Note that from (9.63) and (9.64), in the absence of noise, we can write the following:

$$\underbrace{\begin{bmatrix} y[2i]^* & y[2i+1]^* \\ y[2i+1] & -y[2i] \end{bmatrix}}_{\underline{Y}[i]} = \underbrace{\begin{bmatrix} \alpha_1^* & \alpha_2^* \\ \alpha_2 & -\alpha_1 \end{bmatrix}}_{\underline{H}} \underbrace{\begin{bmatrix} x[2i]^* & -x[2i+1] \\ x[2i+1]^* & x[2i] \end{bmatrix}}_{\underline{X}[i]^H}. \tag{9.65}$$

Since

$$\underline{H}^H \underline{H} = \left(|\alpha_1|^2 + |\alpha_2|^2\right) \boldsymbol{I}_2, \tag{9.66}$$

using (9.65) and (9.62) we have

$$\begin{aligned} \underline{Y}[i]^H \underline{Y}[i-1] &= \left(|\alpha_1|^2 + |\alpha_2|^2\right) \underline{X}[i] \underline{X}[i-1]^H \\ &= \left(|\alpha_1|^2 + |\alpha_2|^2\right) \underline{G}[i]. \end{aligned} \tag{9.67}$$

Based on the discussion above, we arrive at the following differential space-time decoding algorithm.

Algorithm 9.3: [Differential space-time decoding] *Form* $\underline{Y}[0]$ *according to (9.65) using* $y[0]$ *and* $y[1]$. *For* $i = 1, 2, \ldots$:

- *Form the matrix* $\underline{Y}[i]$ *according to (9.65) using* $y[2i]$ *and* $y[2i+1]$.
- *Obtain an estimate* $\hat{\underline{G}}[i]$ *of* $\underline{G}[i]$ *that is closest to* $\underline{Y}[i]^H \underline{Y}[i-1]$.

Decision-Feedback Space-Time Differential Decoding in Flat-Fading Channels

We now consider decoding of the space-time differential block code in flat-fading channels. In such channels the received signals become

$$y[2i] = \alpha_1[2i]\, x[2i] + \alpha_2[2i]\, x[2i+1] + n[2i], \tag{9.68}$$

$$y[2i+1] = -\alpha_1[2i+1]\, x[2i+1]^* + \alpha_2[2i+1]\, x[2i]^* + n[2i+1], \tag{9.69}$$

where $\{\alpha_1[i]\}$ and $\{\alpha_2[i]\}$ are the fading processes associated with the channels between the two transmit antennas and the receive antenna, which as before, are modeled as mutually independent complex Gaussian processes with Jakes' correlation structure. To simplify the receiver structure, we make the assumption that the channels remain constant over two consecutive symbol intervals (i.e., $\alpha_1[2i] = \alpha_1[2i+1]$ and $\alpha_2[2i] = \alpha_2[2i+1]$). Then (9.68) and (9.69) can be written as

$$\underbrace{\begin{bmatrix} y[2i] \\ y[2i+1] \end{bmatrix}}_{\underline{Y}[i]} = \underbrace{\begin{bmatrix} x[2i] & x[2i+1] \\ -x[2i+1]^* & x[2i]^* \end{bmatrix}}_{\underline{X}[i]} \underbrace{\begin{bmatrix} \alpha_1[2i] \\ \alpha_1[2i] \end{bmatrix}}_{\underline{\alpha}[i]} + \underbrace{\begin{bmatrix} n[2i] \\ n[2i] \end{bmatrix}}_{\underline{n}[i]}. \tag{9.70}$$

As before, denote

$$\begin{aligned}
\boldsymbol{y}[i] &= \Big[\underline{y}[i]^T \ \underline{y}[i-1]^T \ \cdots \ \underline{y}[i-N+1]^T\Big]^T, \\
\boldsymbol{\alpha}[i] &= \Big[\underline{\alpha}[i]^T \ \underline{\alpha}[i-1]^T \ \cdots \ \underline{\alpha}[i-N+1]^T\Big]^T, \\
\boldsymbol{n}[i] &= \Big[\underline{n}[i]^T \ \underline{n}[i-1]^T \ \cdots \ \underline{n}[i-N+1]^T\Big]^T, \\
\boldsymbol{X}[i] &= \operatorname{diag}\Big\{\underline{X}[i], \underline{X}[i-11], \ldots, \underline{X}[i-N+1]\Big\}, \\
\text{and} \quad \boldsymbol{G}[i] &= \Big[\underline{G}[i] \ \underline{G}[i-1] \ \cdots \ \underline{G}[i-N+2]\Big].
\end{aligned}$$

Then we have

$$\boldsymbol{y}[i] = \boldsymbol{X}[i]\boldsymbol{\alpha}[i] + \boldsymbol{n}[i]. \tag{9.71}$$

The conditional log-likelihood function is given by

$$\log p\Big(\boldsymbol{y}[i] \mid \boldsymbol{G}[i]\Big) = -\boldsymbol{y}[i]^H \boldsymbol{Q}_G^{-1} \boldsymbol{y}[i] - \log\det(\boldsymbol{Q}_G) - 2N \log \pi, \tag{9.72}$$

where

$$\begin{aligned}
\boldsymbol{Q}_G \triangleq E\left\{\boldsymbol{y}[i]\boldsymbol{y}[i]^H\right\} &= \boldsymbol{X}[i]\, E\left\{\boldsymbol{\alpha}[i]\boldsymbol{\alpha}[i]^H\right\} \boldsymbol{X}[i]^H + \sigma^2 \boldsymbol{I}_{2N} \\
&= E_s\, \boldsymbol{X}[i]\Big(\boldsymbol{\Sigma}_N \otimes \boldsymbol{I}_2\Big)\boldsymbol{X}[i]^H + \sigma^2 \boldsymbol{I}_{2N},
\end{aligned} \tag{9.73}$$

where $\otimes$ denotes the Kronecker matrix product. Note that $\boldsymbol{X}[i]\boldsymbol{X}[i]^H = \boldsymbol{X}[i]^H\boldsymbol{X}[i] = \boldsymbol{I}_{2N}$, and hence we have

$$\boldsymbol{Q}_G^{-1} = E_s^{-1}\, \boldsymbol{X}[i] \left[\left(\boldsymbol{\Sigma}_N + \frac{\sigma^2}{E_s}\boldsymbol{I}_N\right) \otimes \boldsymbol{I}_2\right]^{-1} \boldsymbol{X}[i]^H, \tag{9.74}$$

$$\det(\boldsymbol{Q}_G) = \det(E_s \boldsymbol{\Sigma}_N \otimes \boldsymbol{I}_2 + \sigma^2 \boldsymbol{I}_{2N}). \tag{9.75}$$

As before, denote $\boldsymbol{T} \triangleq \left(\boldsymbol{\Sigma}_N + \frac{\sigma^2}{E_s}\boldsymbol{I}_N\right)^{-1}$; then we have

$$\left[\left(\boldsymbol{\Sigma}_N + \frac{\sigma^2}{E_s}\boldsymbol{I}_N\right) \otimes \boldsymbol{I}_2\right]^{-1} = \boldsymbol{T} \otimes \boldsymbol{I}_2. \tag{9.76}$$

The maximum-likelihood decoding metric becomes

$$\hat{\boldsymbol{G}}[i] = \arg\min_{\boldsymbol{G}[i]} \ \rho(\boldsymbol{G}[i]) \triangleq \boldsymbol{y}[i]^H \boldsymbol{X}[i]\Big(\boldsymbol{T} \otimes \boldsymbol{I}_2\Big)\boldsymbol{X}[i]^H \boldsymbol{y}[i]. \tag{9.77}$$

Note that since $\boldsymbol{T} = [t_{ij}]$ is symmetric, the cost function $\rho(\boldsymbol{G}_n)$ above can be written as

$$\begin{aligned}
\rho(\boldsymbol{G}[i]) &= \sum_{l=0}^{N-1}\sum_{j=0}^{N-1} t_{l,j}\, \underline{y}[i-l]^H\, \underline{X}[i-l]\, \underline{X}[i-j]^H\, \underline{y}[i-j] \\
&= \sum_{l=0}^{N-1} t_{l,l} \underline{y}[i-l]^H\, \underline{X}[i-l]\, \underline{X}[i-l]^H\, \underline{y}[i-l] \\
&\quad + \sum_{l=0}^{N-1}\sum_{j\neq l} t_{l,j}\, \underline{y}[i-l]^H\, \underline{X}[i-l]\, \underline{X}[i-j]^H\, \underline{y}[i-j] \\
&= \sum_{l=0}^{N-1} t_{l,l}\, \|\underline{y}[i-l]\|^2 \\
&\quad + 2\Re\left\{\sum_{l=0}^{N-1}\sum_{j=l+1}^{N-1} t_{l,j}\underline{y}[i-l]^H \left(\prod_{m=l}^{j-1} \underline{G}[i-m]\right) \underline{y}[i-j]\right\},
\end{aligned} \tag{9.78}$$

where (9.78) follows from (9.60). Since the first term in (9.78) is independent of $\boldsymbol{G}[i]$, the decision rule (9.77) becomes

$$\hat{\boldsymbol{G}}[i] = \arg\max_{\boldsymbol{G}[i]} \Re\left\{\sum_{l=0}^{N-1}\sum_{j=l+1}^{N-1} t_{l,j}\underline{y}[i-l]^H \left(\prod_{m=l}^{j-1} \underline{G}_{i-m}\right) \underline{y}[i-j]\right\}. \tag{9.79}$$

Finally, by replacing the previous symbol matrices $\underline{G}[i-1], \ldots, \underline{G}[i-N+2]$ by their estimates $\hat{\underline{G}}[i-1], \ldots, \hat{\underline{G}}[i-N+2]$, we obtain the following decision-feedback decoding rule for $\underline{G}[i]$:

$$\hat{\underline{G}}[i] = \arg\max_{\underline{G}[i]} \Re\left\{\underline{y}[i]^H \underline{G}[i] \sum_{j=1}^{N-1} t_{0,j} \left(\prod_{l=1}^{j-1} \hat{\underline{G}}[i-l]\right) \underline{y}[i-j]\right\}. \tag{9.80}$$

The decision-feedback space-time differential decoding algorithm is summarized as follows.

Algorithm 9.4: [Multiple-symbol decision-feedback space-time differential decoding]

- *Based on the decision memory order N, the fading statistics $\boldsymbol{\Sigma}_N$, and the signal-to-noise ratio E_s/σ^2, compute the feedback metric coefficients from $\boldsymbol{T} = \left(\boldsymbol{\Sigma}_N + (\sigma^2 \boldsymbol{I}_N/E_s)\right)^{-1}$.*

- *Estimate the initial symbol matrices: for $i = 1, 2, \ldots, N-1$, estimate $\hat{\underline{G}}[i]$ simply by quantizing $\underline{Y}[i]^H \underline{Y}[i-1]$.*

- *For $i = N, N+1, \ldots,$ estimate $\hat{\underline{G}}[i]$ according to (9.80).*

The structure of a decision-feedback space-time differential decoder is shown in Fig. 9.5.

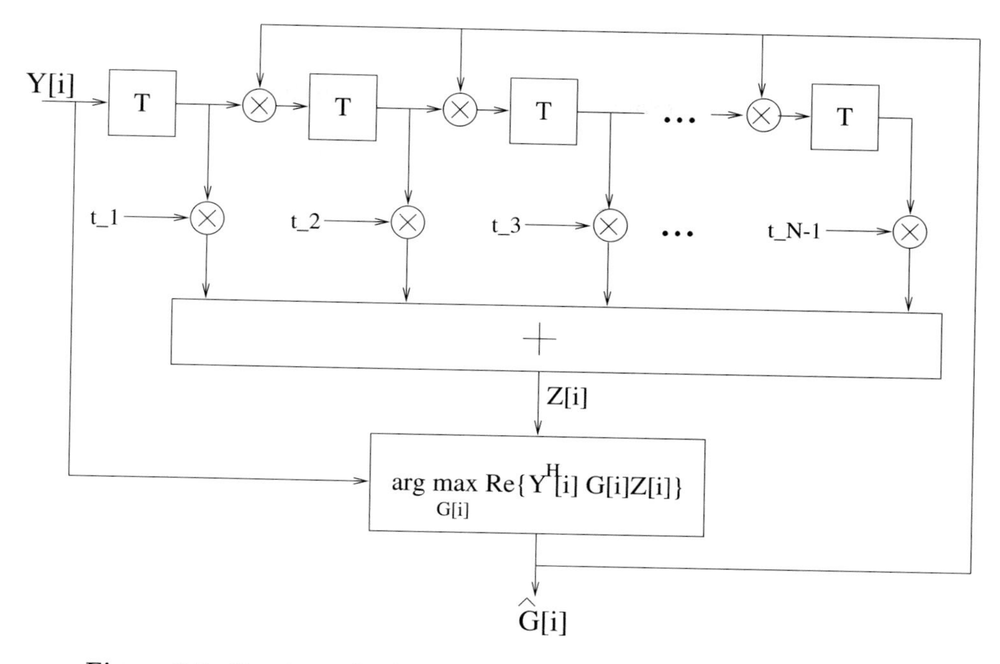

Figure 9.5. Structure of a decision-feedback space-time differential decoder.

Simulation Examples

Assume two transmit antennas and one receive antenna. By assuming that the fading process remains constant over the duration of two symbol intervals, Figs. 9.6, 9.7, and 9.8 show the BER performance of the decision-feedback space-time differential decoder in flat-fading channels with normalized Doppler $B_d T = 0.003$, 0.0075, and 0.01, respectively. The performance of a single-antenna system is also shown. It is seen that space-time coding provides diversity gains over single-antenna systems. Moreover, the multiple-symbol decision-feedback decoding scheme reduces the error floor exhibited by the simple space-time differential decoding method in fading channels. Although the multiple-symbol decoding scheme above is derived based on the assumption that the fading remains constant over two consecutive symbols, little performance degradation is incurred when the channels actually vary from symbol to symbol. This is illustrated in Figs. 9.9, 9.10, and 9.11, where the simulation conditions are the same as before except that the fading processes now vary from symbol to symbol. It is seen that the performance degradation due to such a modeling mismatch is negligible for practical Doppler frequencies.

9.5 Adaptive SMC Receivers for Flat-Fading Channels

In this section we describe an adaptive receiver technique for signal reception and decoding in flat-fading channels based on a Bayesian formulation of the problem

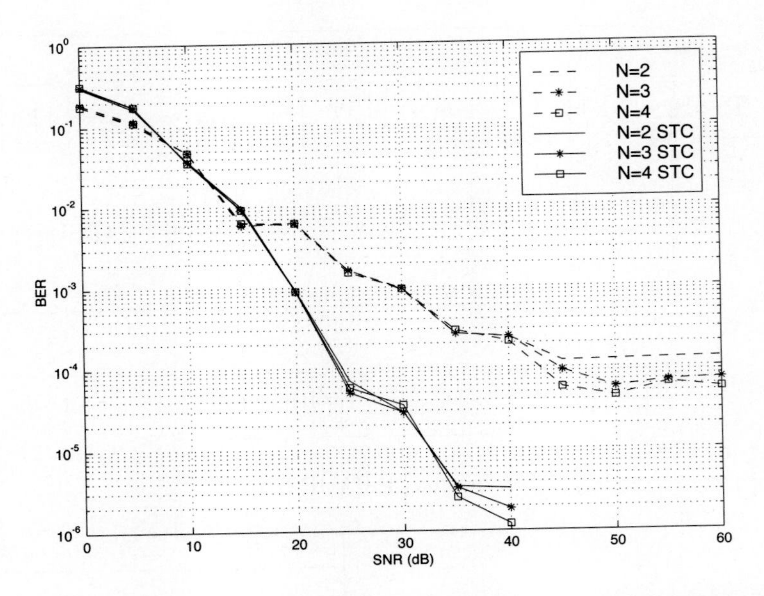

Figure 9.6. BER performance of decision-feedback space-time differential decoding in flat-fading channels with normalized Doppler $B_dT = 0.003$. Channels vary every two symbols.

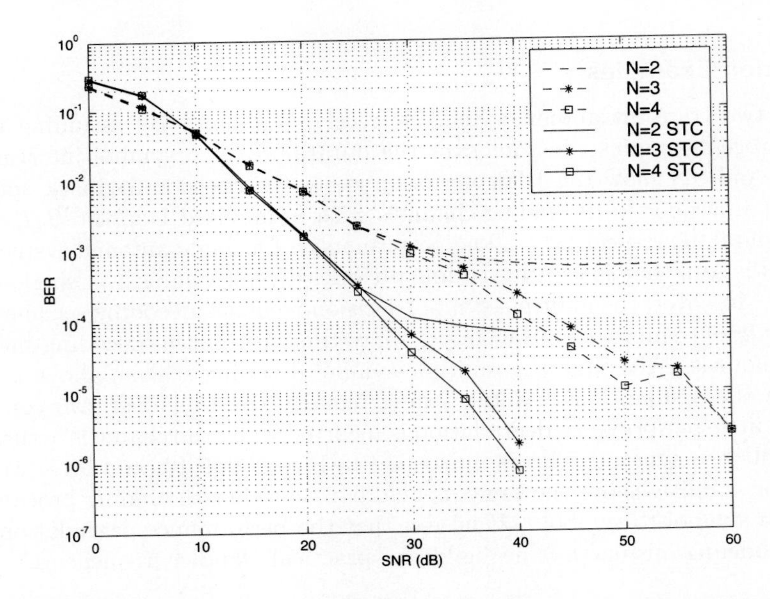

Figure 9.7. BER performance of decision-feedback space-time differential decoding in flat-fading channels with normalized Doppler $B_dT = 0.0075$. Channels vary every two symbols.

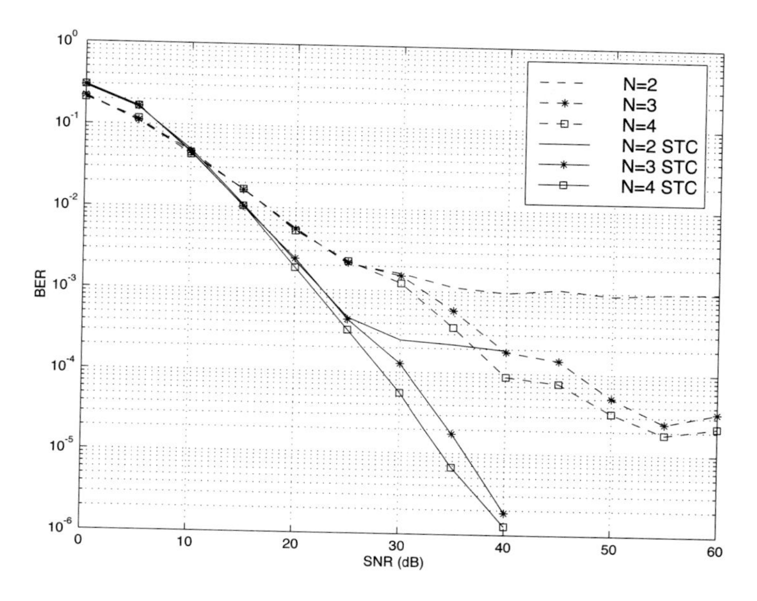

Figure 9.8. BER performance of decision-feedback space-time differential decoding in flat-fading channels with normalized Doppler $B_d T = 0.01$. Channels vary every two symbols.

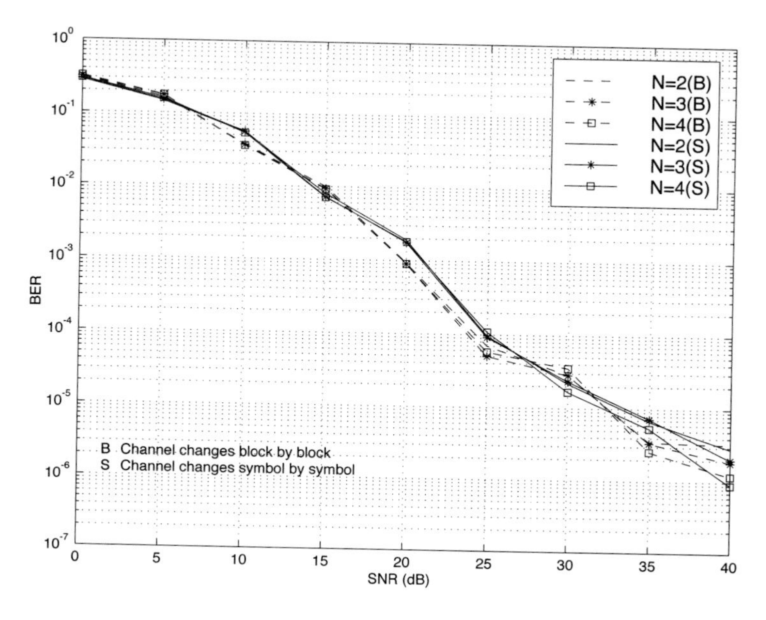

Figure 9.9. BER performance of decision-feedback space-time differential decoding in flat-fading channels with normalized Doppler $B_d T = 0.003$. Channels vary every symbol.

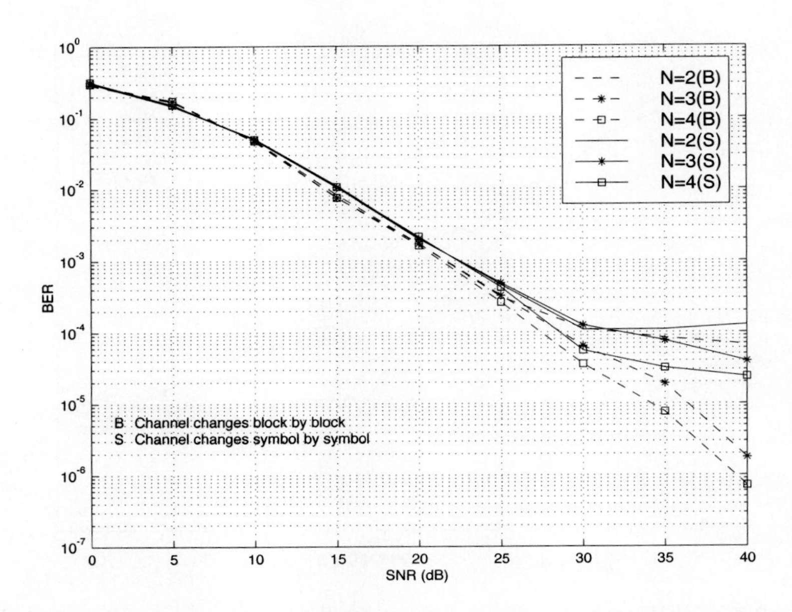

Figure 9.10. BER performance of decision-feedback space-time differential decoding in flat-fading channels with normalized Doppler $B_dT = 0.0075$. Channels vary every symbol.

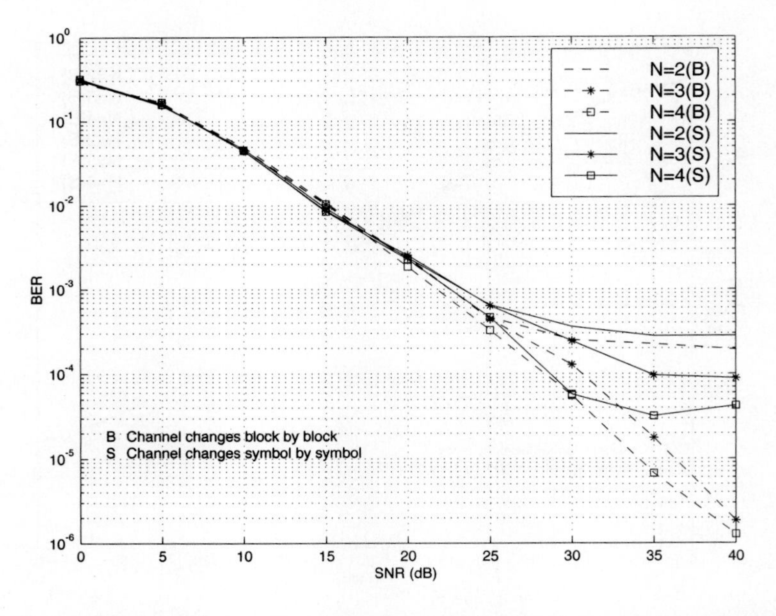

Figure 9.11. BER performance of decision-feedback space-time differential decoding in flat-fading channels with normalized Doppler $B_dT = 0.01$. Channels vary every symbol.

and the sequential Monte Carlo filtering technique outlined in Chapter 8. The techniques presented in this section were first developed in [76]. The basic idea is to treat the transmitted signals as missing data and to impute multiple samples of them sequentially based on the current observation. The importance weight for each of the imputed signal sequences is computed according to its relative ability in predicting the future observation. Then the imputed signal sequences, together with their importance weights, can be used to approximate the Bayesian estimates of the transmitted signals and the fading coefficients of the channel. The novel features of such an approach include the following:

- The algorithm is self-adaptive and no training/pilot symbols or decision feedback is needed.
- The tracking of fading channels and the estimation of data symbols are naturally integrated.
- The ambient channel noise can be either Gaussian or impulsive.
- If the system employs channel coding, the coded signal structure can easily be exploited to substantially improve the accuracy of both channel and data estimation.
- The resulting receiver structure exhibits massive parallelism and is ideally suited for high-speed parallel implementation using VLSI systolic array technology.

9.5.1 System Description

We consider a channel-coded communication system signaling through a flat-fading channel with additive ambient noise. The block diagram of such a system is shown in Fig. 9.12. The input binary information bits $\{d_t\}$ are encoded using a channel code, resulting in a code-bit stream $\{b_t\}$. The code bits are passed to a symbol mapper, yielding complex data symbols $\{s_t\}$ which take values from a finite alphabet $\mathcal{A} = \{a_1, \dots, a_{|\mathcal{A}|}\}$. Each symbol is then transmitted through a flat-fading channel whose input–output relationship is given by

$$y_t = \alpha_t s_t + n_t, \qquad t = 0, 1, \dots, \tag{9.81}$$

where y_t, α_t, s_t, and n_t are the received signal, fading channel coefficient, transmitted symbol, and ambient additive noise at time t, respectively. The processes $\{\alpha_t\}$, $\{s_t\}$, and $\{n_t\}$ are assumed to be mutually independent. Note that, although (9.81) is exactly the same flat-fading channel model as (9.19), here we use a different notation for the sake of notational clarity in describing the sequential Monte Carlo algorithms later.

It is assumed that the additive noise $\{n_t\}$ is a sequence of i.i.d. zero-mean complex random variables. In this section we consider two types of noise distributions. In the first type, n_t assumes a complex Gaussian distribution,

$$n_t \sim \mathcal{N}_c(0, \sigma^2); \tag{9.82}$$

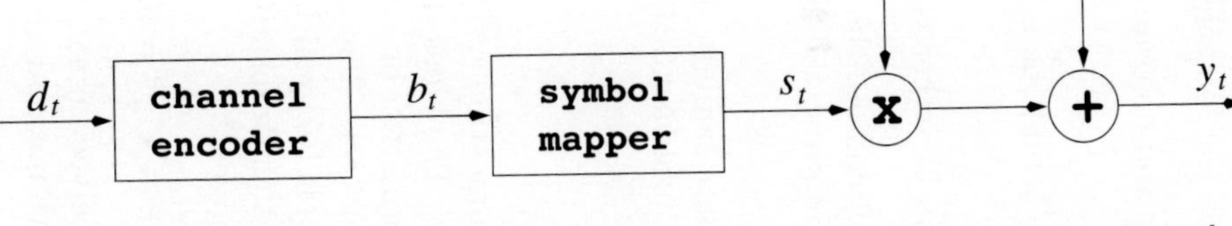

Figure 9.12. Coded communication system signaling through a flat-fading channel.

whereas in the second type, n_t follows a two-term mixture Gaussian distribution,

$$n_t \sim (1-\epsilon)\mathcal{N}_c(0,\sigma_1^2) + \epsilon\mathcal{N}_c(0,\sigma_2^2), \tag{9.83}$$

where $0 < \epsilon < 1$ and $\sigma_2^2 > \sigma_1^2$. Here, as in Chapters 4 and 8, the term $\mathcal{N}_c(0,\sigma_1^2)$ represents the nominal ambient noise and the term $\mathcal{N}_c(0,\sigma_2^2)$ represents an impulsive component. The probability that impulses occur is ϵ. Note that the overall variance of the noise is $(1-\epsilon)\sigma_1^2 + \epsilon\sigma_2^2$.

It is further assumed that the channel-fading process is Rayleigh; that is, the fading coefficients $\{\alpha_t\}$ form a complex zero-mean Gaussian process. Moreover, to model the temporal behavior of the fading, we assume that it can be represented by the output of a lowpass Butterworth filter of order r driven by white Gaussian noise,

$$\{\alpha_t\} = \frac{\Theta(D)}{\Phi(D)}\{u_t\}, \tag{9.84}$$

where D is the back-shift operator $D^k\, u_t \triangleq u_{t-k}$;

$$\Phi(z) \triangleq \phi_r z^r + \cdots + \phi_1 z + 1, \tag{9.85}$$

$$\Theta(z) \triangleq \theta_r z^r + \cdots + \theta_1 z + \theta_0, \tag{9.86}$$

and $\{u_t\}$ is a white complex Gaussian noise sequence with unit variance and independent real and complex components. The coefficients $\{\phi_i\}$ and $\{\theta_i\}$, as well as the order r of the Butterworth filter, are chosen so that the transfer function of the filter matches the power spectral density of the fading process, which in turn is determined by the channel Doppler frequency. In this section we assume that the statistical properties of the fading process are known *a priori*. Consequently, the order and coefficients of the Butterworth filter in (9.84) are known.

We next write system (9.81) and (9.84) in state-space form, which is instrumental in developing the adaptive Bayesian receiver. Define

$$\{x_t\} \triangleq \Theta^{-1}(D)\{\alpha_t\} \quad \Longrightarrow \quad \Phi(D)\{x_t\} = \{u_t\}. \tag{9.87}$$

Denote $\boldsymbol{x}_t \triangleq [x_t, \ldots, x_{t-r+1}]^T$. By (9.84) we then have

$$\boldsymbol{x}_t = \boldsymbol{F}\boldsymbol{x}_{t-1} + \boldsymbol{g}u_t, \qquad u_t \overset{\text{i.i.d.}}{\sim} \mathcal{N}_c(0,1), \tag{9.88}$$

where

$$\boldsymbol{F} \triangleq \begin{pmatrix} -\phi_1 & -\phi_2 & \cdots & -\phi_r & 0 \\ 1 & 0 & \cdots & 0 & 0 \\ 0 & 1 & \cdots & 0 & 0 \\ \vdots & \vdots & \ddots & \vdots & \vdots \\ 0 & 0 & \cdots & 1 & 0 \end{pmatrix} \quad \text{and} \quad \boldsymbol{g} \triangleq \begin{pmatrix} 1 \\ 0 \\ \vdots \\ 0 \end{pmatrix}.$$

Because of (9.87), the fading coefficient sequence $\{\alpha_t\}$ can be written as

$$\alpha_t = \boldsymbol{h}^H \boldsymbol{x}_t, \qquad \text{where} \qquad \boldsymbol{h} \triangleq [\theta_0\ \theta_1\ \cdots\ \theta_r]^H. \tag{9.89}$$

If the additive noise in (9.81) is Gaussian [i.e., $n_t \sim \mathcal{N}_c(0, \sigma^2)$], we have the following state-space model for the system defined by (9.81) and (9.84):

$$\boldsymbol{x}_t = \boldsymbol{F}\boldsymbol{x}_{t-1} + \boldsymbol{g}u_t, \tag{9.90}$$

$$y_t = s_t \boldsymbol{h}^H \boldsymbol{x}_t + \sigma v_t, \tag{9.91}$$

where $\{v_t\}$ in (9.91) is a white complex Gaussian noise sequence with unit variance and independent real and imaginary components.

On the other hand, if the additive noise in (9.81) is impulsive and is modeled by (9.83) as in Chapter 8, we introduce an indicator random variable I_t, $t = 0, 1, \ldots$:

$$I_t \triangleq \begin{cases} 1 & \text{if } n_t \sim \mathcal{N}_c(0, \sigma_1^2), \\ 2 & \text{if } n_t \sim \mathcal{N}_c(0, \sigma_2^2), \end{cases} \tag{9.92}$$

with $P(I_t = 1) = \epsilon$ and $P(I_t = 2) = 1 - \epsilon$. Because n_t is an i.i.d. sequence, so is I_t. We then have the state-space signal model for this case given by

$$\boldsymbol{x}_t = \boldsymbol{F}\boldsymbol{x}_{t-1} + \boldsymbol{g}u_t, \tag{9.93}$$

$$y_t = s_t \boldsymbol{h}^H \boldsymbol{x}_t + \sigma_{I_t} v_t. \tag{9.94}$$

We now look at the problem of online estimation of the symbol s_t and the channel coefficient α_t based on the received signals up to time t, $\{y_i\}_{i=0}^t$. Consider the simple case when the ambient channel noise is Gaussian and the symbols are independent and identically distributed uniformly *a priori* [i.e., $p(s_i) = 1/|\mathcal{A}|$]. Then the problem becomes one of making Bayesian inferences with respect to the posterior distribution

$$p(x_0, \ldots, x_t, s_0, \ldots, s_t \mid y_0, \ldots, y_t) \propto \prod_{j=1}^{t} p(x_j \mid \boldsymbol{x}_{j-1}) p(s_j) p(y_j \mid \boldsymbol{x}_j, s_j)$$

$$\propto \prod_{j=1}^{t} \exp\left(-\|x_j + \sum_{i=1}^{r} \phi_i x_{j-i}\|^2 - \frac{1}{\sigma^2}\|y_j - s_j \boldsymbol{h}^T \boldsymbol{x}_j\|^2 \right), \quad t = 0, 1, \ldots. \tag{9.95}$$

For example, an online symbol estimation can be obtained from the marginal posterior distribution $p(s_t|y_0, \ldots, y_t)$, and an online channel state estimation can be obtained from the marginal posterior distribution $p(\boldsymbol{x}_t|y_0, \ldots, y_t)$. Although the joint distribution (9.95) can be written out explicitly up to a normalizing constant, computation of the corresponding marginal distributions involves very high dimensional integration and is infeasible in practice. An effective approach to this problem is the sequential Monte Carlo filtering technique discussed in Chapter 8.

9.5.2 Adaptive Receiver in Fading Gaussian Noise Channels: Uncoded Case

MKF-Based Sequential Monte Carlo Receiver

Consider the flat-fading channel with additive Gaussian noise given by (9.90) and (9.91). Denote $\boldsymbol{Y}_t \triangleq (y_0, \ldots, y_t)$ and $\boldsymbol{S}_t \triangleq (s_0, \ldots, s_t)$. We first consider the case of an uncoded system, where the transmitted symbols are assumed to be independent:

$$P(s_t = a_i \mid \boldsymbol{S}_{t-1}) = P(s_t = a_i), \quad a_i \in \mathcal{A}. \tag{9.96}$$

When no prior information about the symbols is available, the symbols are assumed to take each possible value in $\mathcal{A}$ with equal probability [i.e., $P(s_t = a_i) = 1/|\mathcal{A}|$ for $i = 1, \ldots, |\mathcal{A}|$]. We are interested in estimating the symbol s_t and the channel coefficient $\alpha_t = \boldsymbol{h}^H \boldsymbol{x}_t$ at time t based on the observation $\boldsymbol{Y}_t$. The Bayesian solution to this problem requires the posterior distribution

$$p(\boldsymbol{x}_t, s_t \mid \boldsymbol{Y}_t) = \int p(\boldsymbol{x}_t \mid \boldsymbol{S}_t, \boldsymbol{Y}_t)\, p(\boldsymbol{S}_t \mid \boldsymbol{Y}_t)\, \mathrm{d}\boldsymbol{S}_{t-1}. \tag{9.97}$$

Note that with a given $\boldsymbol{S}_t$, the state-space model (9.90)–(9.91) becomes a linear Gaussian system. Hence,

$$p(\boldsymbol{x}_t \mid \boldsymbol{S}_t, \boldsymbol{Y}_t) \sim \mathcal{N}_c\Big(\boldsymbol{\mu}_t(\boldsymbol{S}_t), \boldsymbol{\Sigma}_t(\boldsymbol{S}_t)\Big), \tag{9.98}$$

where the mean $\boldsymbol{\mu}_t(\boldsymbol{S}_t)$ and covariance matrix $\boldsymbol{\Sigma}_t(\boldsymbol{S}_t)$ can be obtained by a Kalman filter with the given $\boldsymbol{S}_t$.

To implement the MKF, we need to obtain a set of Monte Carlo samples of the transmitted symbols, $\left\{(\boldsymbol{S}_t^{(j)}, w_t^{(j)})\right\}_{j=1}^{m}$, properly weighted with respect to the distribution $p(\boldsymbol{S}_t|\boldsymbol{Y}_t)$. Then for any integrable function $h(\boldsymbol{x}_t, s_t)$, we can approximate the quantity of interest $E\{h(\boldsymbol{x}_t, s_t)|\boldsymbol{Y}_t\}$ as follows:

$$\begin{aligned} E\left\{h(\boldsymbol{x}_t, s_t) \mid \boldsymbol{Y}_t\right\} &= \int\int h(\boldsymbol{x}_t, s_t)\, p(\boldsymbol{x}_t, s_t \mid \boldsymbol{Y}_t)\, \mathrm{d}\boldsymbol{x}_t\, \mathrm{d}s_t \\ &= \int\int h(\boldsymbol{x}_t, s_t)\, p(\boldsymbol{x}_t \mid \boldsymbol{S}_t, \boldsymbol{Y}_t)\, p(\boldsymbol{S}_t \mid \boldsymbol{Y}_t)\, \mathrm{d}\boldsymbol{x}_t\, \mathrm{d}\boldsymbol{S}_t && (9.99) \\ &= \int \underbrace{\left[\int h(\boldsymbol{x}, s_t)\, \phi\left(\boldsymbol{x}; \boldsymbol{\mu}_t(\boldsymbol{S}_t), \boldsymbol{\Sigma}_t(\boldsymbol{S}_t)\right) \mathrm{d}\boldsymbol{x}\right]}_{\xi(\boldsymbol{S}_t)} p(\boldsymbol{S}_t \mid \boldsymbol{Y}_t)\, \mathrm{d}\boldsymbol{S}_t && (9.100) \\ &\cong \frac{1}{W_t} \sum_{j=1}^{m} \xi\left(\boldsymbol{S}_t^{(j)}\right) w_t^{(j)}, && (9.101) \end{aligned}$$

with

$$W_t = \sum_{j=1}^{m} w_t^{(j)}, \tag{9.102}$$

where (9.99) follows from (9.97); (9.100) follows from (9.98); and in (9.100), $\phi(\cdot; \boldsymbol{\mu}, \boldsymbol{\Sigma})$ denotes a complex Gaussian density function with mean $\boldsymbol{\mu}$ and covariance matrix $\boldsymbol{\Sigma}$. In particular, the MMSE channel estimate is given by

$$\begin{aligned} E\{\alpha_t \mid \boldsymbol{Y}_t\} &= \boldsymbol{h}^H E\{\boldsymbol{x}_t \mid \boldsymbol{Y}_t\} \\ &\cong \frac{1}{W_t} \boldsymbol{h}^H \left[\sum_{j=1}^{m} \boldsymbol{\mu}_t \left(\boldsymbol{S}_t^{(j)} \right) w_t^{(j)} \right]. \end{aligned} \tag{9.103}$$

In other words, we can let $h(\boldsymbol{x}_t, s_t) \triangleq \boldsymbol{h}^H \boldsymbol{x}_t$, implying that $\xi(\boldsymbol{S}_t) \triangleq \boldsymbol{h}^H \boldsymbol{\mu}_t(\boldsymbol{S}_t)$ in (9.100). Moreover, the *a posteriori* symbol probability can be estimated as

$$\begin{aligned} P(s_t = a_i \mid \boldsymbol{Y}_t) &= E\{\delta(s_t = a_i) \mid \boldsymbol{Y}_t\} \\ &\cong \frac{1}{W_t} \sum_{j=1}^{m} \delta(s_t^{(j)} = a_i)\, w_t^{(j)}, \qquad i = 1, \ldots, |\mathcal{A}|, \end{aligned} \tag{9.104}$$

where $\delta(\cdot)$ is an indicator function such that $\delta(s_t = a_i) = 1$ if $s_t = a_i$, and $\delta(s_t = a_i) = 0$ otherwise. This corresponds to having $h(\boldsymbol{x}_t, s_t) \triangleq \delta(s_t = a_i)$ and $\xi(\boldsymbol{S}_t) \triangleq \delta(s_t = a_i)$.

Note that a hard decision on the symbol s_t is obtained by

$$\begin{aligned} \hat{s}_t &= \arg\max_{a_i \in \mathcal{A}} P(s_t = a_i \mid \boldsymbol{Y}_t) \\ &\cong \arg\max_{a_i \in \mathcal{A}} \sum_{j=1}^{m} \delta(s_t^{(j)} = a_i) w_t^{(j)}. \end{aligned} \tag{9.105}$$

When MPSK signals are transmitted [i.e., $a_i = \exp\left(\jmath(2\pi i/|\mathcal{A}|)\right)$ for $i = 0, \ldots, |\mathcal{A}| - 1$, where $\jmath = \sqrt{-1}$], the estimated symbol $\hat{s}_t$ may have a phase ambiguity. For instance, for BPSK signals, $s_t \in \{+1, -1\}$. It is easily seen from (9.81) that if both the symbol sequence $\{s_t\}$ and the channel value sequence $\{\alpha_t\}$ are phase-shifted by π (resulting in $\{-s_t\}$ and $\{-\alpha_t\}$, respectively), no change is incurred in the observed signal $\{y_t\}$. Alternatively, in the state-space model (9.90)–(9.91), a phase shift of π on both the symbol sequence $\{s_t\}$ and the state sequence $\{\boldsymbol{x}_t\}$ yields the same model for the observations. Hence such a phase ambiguity necessitates the use of differential encoding and decoding.

Hereafter, we let $\boldsymbol{\mu}_t^{(j)} \triangleq \boldsymbol{\mu}_t\left(\boldsymbol{S}_t^{(j)}\right)$, $\boldsymbol{\Sigma}_t^{(j)} \triangleq \boldsymbol{\Sigma}_t\left(\boldsymbol{S}_t^{(j)}\right)$, and $\kappa_t^{(j)} \triangleq \left[\boldsymbol{\mu}_t^{(j)}, \boldsymbol{\Sigma}_t^{(j)}\right]$. By applying the MKF techniques outlined in Section 8.3 to the flat-fading channel system, we describe the following algorithm for generating properly weighted Monte Carlo samples $\left\{\left(\boldsymbol{S}_t^{(j)}, \kappa_t^{(j)}, w_t^{(j)}\right)\right\}_{j=1}^{m}$.

Algorithm 9.5: [SMC for adaptive detection in flat-fading channels—Gaussian noise]

- *Initialization: Each Kalman filter is initialized as* $\kappa_0^{(j)} = \left(\boldsymbol{\mu}_0^{(j)}, \boldsymbol{\Sigma}_0^{(j)}\right)$, *with* $\boldsymbol{\mu}_0^{(j)} = \boldsymbol{0}$, $\boldsymbol{\Sigma}_0^{(j)} = 2\boldsymbol{\Sigma}$, $j = 1, \ldots, m$, *where* $\boldsymbol{\Sigma}$ *is the stationary covariance of*

$\boldsymbol{x}_t$ *and is computed analytically from (9.87). (The factor 2 is to accommodate the initial uncertainty.) All importance weights are initialized as* $w_0^{(j)} = 1$, $j = 1, \ldots, m$. *Since the data symbols are assumed to be independent, initial symbols are not needed.*

Based on the state-space model (9.90)–(9.91), the following steps are implemented at time t *to update each weighted sample. For* $j = 1, \ldots, m$:

- *Compute the one-step predictive update of each Kalman filter* $\kappa_{t-1}^{(j)}$:

$$\boldsymbol{K}_t^{(j)} = \boldsymbol{F}\boldsymbol{\Sigma}_{t-1}^{(j)}\boldsymbol{F}^H + \boldsymbol{g}\boldsymbol{g}^H, \tag{9.106}$$

$$\gamma_t^{(j)} = \boldsymbol{h}^H \boldsymbol{K}_t^{(j)} \boldsymbol{h} + \sigma^2, \tag{9.107}$$

$$\eta_t^{(j)} = \boldsymbol{h}^H \boldsymbol{F} \boldsymbol{\mu}_{t-1}^{(j)}. \tag{9.108}$$

- *Compute the trial sampling density: For each* $a_i \in \mathcal{A}$, *compute*

$$\begin{aligned}
\rho_{t,i}^{(j)} &\triangleq P\left(s_t = a_i \mid \boldsymbol{S}_{t-1}^{(j)}, \boldsymbol{Y}_t\right) \\
&\propto p\left(y_t, \boldsymbol{Y}_{t-1}, s_t = a_i, \boldsymbol{S}_{t-1}^{(j)}\right) \\
&= p\left(y_t \mid s_t = a_i, \boldsymbol{S}_{t-1}^{(j)}, \boldsymbol{Y}_{t-1}\right) P\left(s_t = a_i \mid \boldsymbol{S}_{t-1}^{(j)}, \boldsymbol{Y}_{t-1}\right) \\
&= p\left(y_t \mid s_t = a_i, \boldsymbol{S}_{t-1}^{(j)}, \boldsymbol{Y}_{t-1}\right) P(s_t = a_i),
\end{aligned} \tag{9.109}$$

where (9.109) holds because s_t *is independent of* $\boldsymbol{S}_{t-1}$ *and* $\boldsymbol{Y}_{t-1}$. *Furthermore, we observe that*

$$p\left(y_t \mid s_t = a_i, \boldsymbol{S}_{t-1}^{(j)}, \boldsymbol{Y}_{t-1}\right) \sim \mathcal{N}_c\left(a_i \eta_t^{(j)}, \gamma_t^{(j)}\right). \tag{9.110}$$

- *Impute the symbol* s_t: *Draw* $s_t^{(j)}$ *from the set* $\mathcal{A}$ *with probability*

$$P\left(s_t^{(j)} = a_i\right) \propto \rho_{t,i}^{(j)}, \qquad a_i \in \mathcal{A}. \tag{9.111}$$

Append $s_t^{(j)}$ *to* $\boldsymbol{S}_{t-1}^{(j)}$ *and obtain* $\boldsymbol{S}_t^{(j)}$.

- *Compute the importance weight:*

$$\begin{aligned}
w_t^{(j)} &= w_{t-1}^{(j)} p\left(y_t \mid \boldsymbol{S}_{t-1}^{(j)}, \boldsymbol{Y}_{t-1}\right) \\
&= w_{t-1}^{(j)} \sum_{a_i \in \mathcal{A}} p\left(y_t \mid s_t = a_i, \boldsymbol{S}_{t-1}^{(j)}, \boldsymbol{Y}_{t-1}\right) P(s_t = a_i) \\
&\propto w_{t-1}^{(j)} \sum_{a_i \in \mathcal{A}} \rho_{t,i}^{(j)},
\end{aligned} \tag{9.112}$$

where (9.112) follows from (9.109).

- *Compute the one-step filtering update of the Kalman filter $\kappa_{t-1}^{(j)}$: Based on the imputed symbol $s_t^{(j)}$ and the observation y_t, complete the Kalman filter update to obtain $\kappa_t^{(j)} = \left[\boldsymbol{\mu}_t^{(j)}, \boldsymbol{\Sigma}_t^{(j)}\right]$, as follows:*

$$\boldsymbol{\mu}_t^{(j)} = \boldsymbol{F}\boldsymbol{\mu}_{t-1}^{(j)} + \frac{1}{\gamma_t^{(j)}}\left(y_t - s_t^{(j)}\eta_t^{(j)}\right)\boldsymbol{K}_t^{(j)}\boldsymbol{h}\, s_t^{(j)}, \tag{9.113}$$

$$\boldsymbol{\Sigma}_t^{(j)} = \boldsymbol{K}_t^{(j)} - \frac{1}{\gamma_t^{(j)}}\boldsymbol{K}_t^{(j)}\boldsymbol{h}\boldsymbol{h}^H\boldsymbol{K}_t^{(j)}. \tag{9.114}$$

- *Perform resampling according to Algorithm 8.9 when $\bar{m}_t$ in (8.103) is below a threshold.*

The correctness of Algorithm 9.5 is stated by the following result, whose proof is given in the Appendix (Section 9.6.1).

Proposition 9.1: *The samples $\left\{(\boldsymbol{S}_t^{(j)}, \kappa_t^{(j)}, w_t^{(j)})\right\}_{j=1}^m$ drawn by Algorithm 9.5 are properly weighted with respect to $p(\boldsymbol{S}_t|\boldsymbol{Y}_t)$, provided that $\left\{(\boldsymbol{S}_{t-1}^{(j)}, \kappa_{t-1}^{(j)}, w_{t-1}^{(j)})\right\}_{j=1}^m$ are proper at time $(t-1)$.*

Algorithm 9.5 is depicted in Fig. 9.13. It is seen that at any time t, the only quantities that need to be stored are $\left\{\kappa_t^{(j)}, w_t^{(j)}\right\}_{j=1}^m$. At each time t, the dominant computation in this receiver involves the m one-step Kalman filter updates. Since the m samplers operate independently and in parallel, such a sequential Monte Carlo receiver is well suited for massively parallel implementation using VLSI systolic array technology [241].

9.5.3 Delayed Estimation

Since the fading process is highly correlated, the future received signals contain information about current data and channel state. A delayed estimate is usually more accurate than the concurrent estimate. This is true for any channel with memory and is especially prominent when the transmitted symbols are coded, in which case not only the channel states but also the data symbols are highly correlated. In delayed estimation, instead of making an inference on $(\boldsymbol{x}_t, s_t)$ instantaneously with the posterior distribution $p(\boldsymbol{x}_t, s_t|\boldsymbol{Y}_t)$, we delay this inference to a later time $(t+\Delta)$, $\Delta \geq 0$, with the distribution $p(\boldsymbol{x}_t, s_t|\boldsymbol{Y}_{t+\Delta})$. Here we discuss two methods for delayed estimation: the delayed-weight method and the delayed-sample method.

Delayed-Weight Method

From Algorithm 9.5 we note by induction that if the set $\left\{(\boldsymbol{S}_t^{(j)}, w_t^{(j)})\right\}_{j=1}^m$ is properly weighted with respect to $p(\boldsymbol{S}_t|\boldsymbol{Y}_t)$, the set $\left\{(\boldsymbol{S}_{t+\delta}^{(j)}, w_{t+\delta}^{(j)})\right\}_{j=1}^m$ is properly weighted

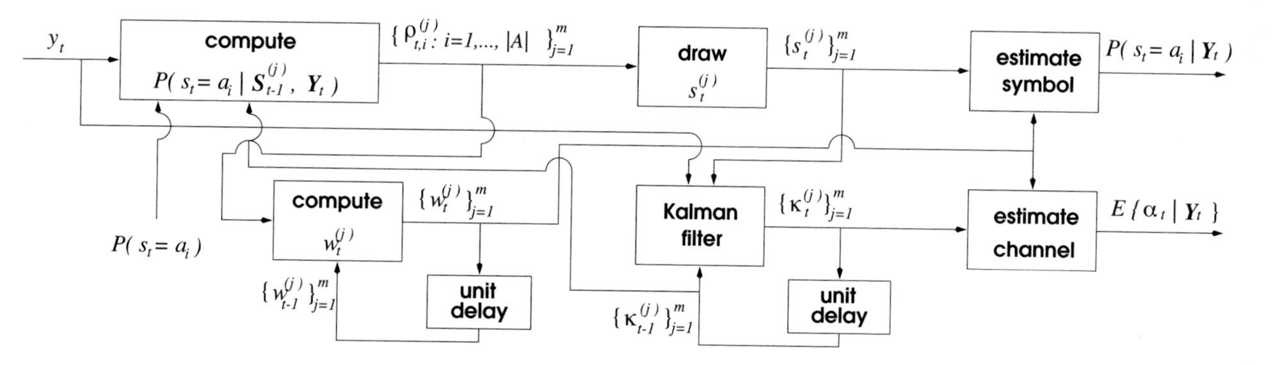

Figure 9.13. Adaptive Bayesian receiver for flat-fading Gaussian noise channels based on mixture Kalman filtering.

with respect to $p(\boldsymbol{S}_{t+\delta}|\boldsymbol{Y}_{t+\delta})$, $\delta > 0$. Hence, if we focus our attention on $\boldsymbol{S}_t$ at time $(t+\delta)$ and let $h(\boldsymbol{x}_t, s_t) = \delta(s_t = a_i)$ as in (9.104), we obtain a delayed estimate of the symbol:

$$P(s_t = a_i \mid \boldsymbol{Y}_{t+\delta}) \cong \frac{1}{W_{t+\delta}} \sum_{j=1}^{m} \delta\left(s_t^{(j)} = a_i\right) w_{t+\delta}^{(j)}, \quad i = 1, \ldots, |\mathcal{A}|. \tag{9.115}$$

Since the weights $\left\{w_{t+\delta}^{(j)}\right\}_{j=1}^{m}$ contain information about the signals $(y_{t+1}, \ldots, y_{t+\delta})$, the estimate in (9.115) is usually more accurate. Note that such a delayed estimation method incurs no additional computation, but it requires extra memory for storing $\left\{(s_{t+1}^{(j)}, \ldots, s_{t+\delta}^{(j)})\right\}_{j=1}^{m}$. As will be seen in the simulation examples, for uncoded systems this simple delayed-weight method is quite effective for improving the detection performance over the concurrent method. However, for coded systems, this method is not sufficient for exploiting the constraint structures of both the channel and the symbols, and we must resort to the delayed-sample method, which is described next.

Delayed-Sample Method

An alternative method is to generate both the delayed samples and the weights $\left\{(s_t^{(j)}, w_t^{(j)})\right\}_{j=1}^{m}$ based on the signals $\boldsymbol{Y}_{t+\Delta}$, hence making $p(\boldsymbol{S}_t|\boldsymbol{Y}_{t+\Delta})$ the target distribution at time $(t+\Delta)$. This procedure will provide better Monte Carlo samples since it uses future information $(y_{t+1}, \ldots, y_{t+\Delta})$ in generating the current sample of s_t. But the algorithm is also more demanding both analytically and computationally because of the need to marginalize out s_{t+d} for $d = 1, \ldots, \Delta$.

For each possible "future" symbol sequence at time $t + \Delta - 1$ [i.e., $(s_t, s_{t+1}, \ldots, s_{t+\Delta-1}) \in \mathcal{A}^{\Delta}$ (a total of $|\mathcal{A}|^{\Delta}$ possibilities)], we keep the value of a Δ-step Kalman filter $\left\{\kappa_{t+\tau}^{(j)}(\underline{s}_t^{t+\tau})\right\}_{\tau=0}^{\Delta-1}$, where

$$\kappa_{t+\tau}^{(j)}(\underline{s}_t^{t+\tau}) \triangleq \left[\boldsymbol{\mu}_{t+\tau}\left(\boldsymbol{S}_{t-1}^{(j)}, \underline{s}_t^{t+\tau}\right), \; \boldsymbol{\Sigma}_{t+\tau}\left(\boldsymbol{S}_{t-1}^{(j)}, \underline{s}_t^{t+\tau}\right)\right], \quad \tau = 0, 1, \ldots, \Delta - 1,$$

with $\underline{s}_a^b \triangleq (s_a, s_{a+1}, \ldots, s_b)$. Denote

$$\boldsymbol{\kappa}_{t-1}^{(j)} \triangleq \left\{\kappa_{t-1}^{(j)}; \; \left\{\kappa_{t+\tau}^{(j)}(\underline{s}_t^{t+\tau})\right\}_{\tau=0}^{\Delta-1} : \underline{s}_t^{t+\tau} \in \mathcal{A}^{\tau+1}\right\}.$$

The following is the delayed-sample algorithm for adaptive detection in flat-fading channels with Gaussian noise.

Algorithm 9.6: [Delayed-sample SMC algorithm for adaptive detection in flat-fading channels—Gaussian noise]

- *Initialization: Each Kalman filter is initialized as* $\kappa_0^{(j)} = \left(\boldsymbol{\mu}_0^{(j)}, \boldsymbol{\Sigma}_0^{(j)}\right)$, *with* $\boldsymbol{\mu}_0^{(j)} = \boldsymbol{0}$ *and* $\boldsymbol{\Sigma}_0^{(j)} = 2\boldsymbol{\Sigma}$, $j = 1, \ldots, m$, *where* $\boldsymbol{\Sigma}$ *is the stationary covariance of* $\boldsymbol{x}_t$. *All importance weights are initialized as* $w_0^{(j)} = 1$, $j = 1, \ldots, m$. *Since the data symbols are assumed to be independent, initial symbols are not needed.*

At time $(t+\Delta)$, *we perform the following updates for* $j = 1, \ldots, m$ *to propagate from the sample* $\left\{\left(\boldsymbol{S}_{t-1}^{(j)}, \kappa_{t-1}^{(j)}, w_{t-1}^{(j)}\right)\right\}_{j=1}^{m}$, *properly weighted for* $p(\boldsymbol{S}_{t-1}|\boldsymbol{Y}_{t+\Delta-1})$, *to that for* $p(\boldsymbol{S}_{t-1}|\boldsymbol{Y}_{t+\Delta})$.

- *Compute the one-step predictive update for each of the* $|\mathcal{A}|^{\Delta}$ *Kalman filters: For each* $\underline{s}_t^{t+\Delta-1} \in \mathcal{A}^{\Delta}$, *perform the update on the Kalman filter* $\kappa_{t+\Delta-1}^{(j)}\left(\underline{s}_t^{t+\Delta-1}\right)$, *according to equations (9.106)–(9.108) to obtain* $\boldsymbol{K}_{t+\Delta}^{(j)}\left(\underline{s}_t^{t+\Delta-1}\right)$, $\gamma_{t+\Delta}^{(j)}\left(\underline{s}_t^{t+\Delta-1}\right)$, *and* $\eta_{t+\Delta}^{(j)}\left(\underline{s}_t^{t+\Delta-1}\right)$. *(Here we make it explicit that these quantities are functions of* $\underline{s}_t^{t+\Delta-1}$.*)*

- *Compute the trial sampling density: For each* $a_i \in \mathcal{A}$, *compute*

$$
\begin{aligned}
\rho_{t,i}^{(j)} &\triangleq P\left(s_t = a_i \mid \boldsymbol{S}_{t-1}^{(j)}, \boldsymbol{Y}_{t+\Delta}\right) \\
&\propto p\left(\boldsymbol{Y}_{t+\Delta}, \boldsymbol{S}_{t-1}^{(j)}, s_t = a_i\right) \\
&= \sum_{\underline{s}_{t+1}^{t+\Delta} \in \mathcal{A}^{\Delta}} p\left(\boldsymbol{Y}_{t+\Delta}, \boldsymbol{S}_{t-1}^{(j)}, s_t = a_i, \underline{s}_{t+1}^{t+\Delta}\right) \\
&\propto \sum_{\underline{s}_{t+1}^{t+\Delta} \in \mathcal{A}^{\Delta}} \prod_{\tau=0}^{\Delta} \underbrace{p\left(y_{t+\tau} \mid \boldsymbol{Y}_{t+\tau-1}, \boldsymbol{S}_{t-1}^{(j)}, s_t = a_i, \underline{s}_{t+1}^{t+\tau}\right)}_{\mathcal{N}_c\left(s_{t+\tau}\, \gamma_{t+\tau}^{(j)}(\underline{s}_t^{t+\tau-1}),\; \eta_{t+\tau}^{(j)}(\underline{s}_t^{t+\tau-1})\right)} p(s_t = a_i) \\
&\quad \cdot \prod_{\tau=0}^{\Delta} p(s_{t+\tau}).
\end{aligned}
\tag{9.116}
$$

- *Impute the symbol* s_t: *Draw* $s_t^{(j)}$ *with probability*

$$
P\left(s_t^{(j)} = a_i\right) \propto \rho_{t,i}^{(j)}, \quad a_i \in \mathcal{A}. \tag{9.117}
$$

  Append $s_t^{(j)}$ *to* $\boldsymbol{S}_{t-1}^{(j)}$ *and obtain* $\boldsymbol{S}_t^{(j)}$.

- *Compute the importance weight:*

$$
\begin{aligned}
w_t^{(j)} &= w_{t-1}^{(j)} \frac{p\left(\boldsymbol{S}_t^{(j)} \mid \boldsymbol{Y}_{t+\Delta}\right)}{p\left(\boldsymbol{S}_{t-1}^{(j)} \mid \boldsymbol{Y}_{t+\Delta-1}\right) p\left(s_t^{(j)} \mid \boldsymbol{S}_{t-1}^{(j)}, \boldsymbol{Y}_{t+\Delta}\right)} \\
&= w_{t-1}^{(j)} \frac{p\left(\boldsymbol{S}_{t-1}^{(j)} \mid \boldsymbol{Y}_{t+\Delta}\right)}{p\left(\boldsymbol{S}_{t-1}^{(j)} \mid \boldsymbol{Y}_{t+\Delta-1}\right)}
\end{aligned}
$$

$$\propto w_{t-1}^{(j)} \frac{p\left(\boldsymbol{Y}_{t+\Delta}, \boldsymbol{S}_{t-1}^{(j)}\right)}{p\left(\boldsymbol{Y}_{t+\Delta-1}, \boldsymbol{S}_{t-1}^{(j)}\right)} \tag{9.118}$$

$$= w_{t-1}^{(j)} \frac{\sum_{\underline{s}_t^{t+\Delta} \in \mathcal{A}^{\Delta+1}} p\left(\underline{s}_t^{t+\Delta}, \boldsymbol{S}_{t-1}^{(j)}, \boldsymbol{Y}_{t+\Delta}\right)}{\sum_{\underline{s}_t^{t+\Delta-1} \in \mathcal{A}^{\Delta}} p\left(\underline{s}_t^{t+\Delta-1}, \boldsymbol{S}_{t-1}^{(j)}, \boldsymbol{Y}_{t+\Delta-1}\right)}$$

$$\propto w_{t-1}^{(j)} \frac{\sum_{\underline{s}_t^{t+\Delta} \in \mathcal{A}^{\Delta+1}} \left[\prod_{\tau=0}^{\Delta} p\left(y_{t+\tau} \mid \boldsymbol{Y}_{t+\tau-1}, \boldsymbol{S}_{t-1}^{(j)}, \underline{s}_t^{t+\tau}\right) \cdot \prod_{\tau=0}^{\Delta} p(s_{t+\tau})\right]}{\sum_{\underline{s}_t^{t+\Delta-1} \in \mathcal{A}^{\Delta}} \left[\prod_{\tau=0}^{\Delta-1} p\left(y_{t+\tau} \mid \boldsymbol{Y}_{t+\tau-1}, \boldsymbol{S}_{t-1}^{(j)}, \underline{s}_t^{t+\tau}\right) \cdot \prod_{\tau=0}^{\Delta-1} p(s_{t+\tau})\right]}, \tag{9.119}$$

where

$$p\left(y_{t+\tau} \mid \boldsymbol{Y}_{t+\tau-1}, \boldsymbol{S}_{t-1}^{(j)}, \underline{s}_t^{t+\tau}\right) \sim \mathcal{N}_c\left(s_{t+\tau}\, \gamma_{t+\tau}^{(j)}(\underline{s}_t^{t+\tau-1}),\ \eta_{t+\tau}^{(j)}(\underline{s}_t^{t+\tau-1})\right). \tag{9.120}$$

- *Compute the one-step filtering update for each of the* $|A|^{\Delta}$ *Kalman filters: Using the values of* $s_t^{(j)}$ *and* $y_{t+\Delta}$, *for each* $\underline{s}_{t+1}^{t+\Delta} \in \mathcal{A}^{\Delta}$ *perform a one-step filtering update on the Kalman filter* $\kappa_{t+\Delta-1}^{(j)}(\underline{s}_t^{t+\Delta-1})$ *according to (9.113)–(9.114) to obtain*

  $$\kappa_{t+\Delta}^{(j)}\left(\underline{s}_{t+1}^{t+\Delta}\right) \triangleq \left[\boldsymbol{\mu}_{t+\Delta}\left(\boldsymbol{S}_t^{(j)}, \underline{s}_{t+1}^{t+\Delta}\right),\ \boldsymbol{\Sigma}_{t+\Delta}\left(\boldsymbol{S}_t^{(j)}, \underline{s}_{t+1}^{t+\Delta}\right)\right].$$

  With this and the subset of $\left\{\kappa_{t+\tau}^{(j)}(\underline{s}_{t+1}^{t+\tau})\right\}_{\tau=0}^{\Delta-1}$ *corresponding to the sample* $s_t^{(j)}$, *which has been obtained in the previous iteration, form the new filter class* $\boldsymbol{\kappa}_t^{(j)}$.

- *Perform resampling according to Algorithm 8.9 when* $\bar{m}_t$ *in (8.103) is below a threshold.*

The dominant computation of the delayed-sample method above at each time t involves the $\left(m\,|\mathcal{A}|^{\Delta}\right)$ one-step Kalman filter updates, which—as before—can be carried out in parallel. Finally, we note that we can use the delayed-sample method in conjunction with the delayed-weight method. For example, using the delayed-sample method, we generate delayed samples and weights $\left\{(s_t^{(j)}, w_t^{(j)})\right\}_{j=1}^{m}$ based on the signals $\boldsymbol{Y}_{t+\Delta}$. Then with an additional delay δ, we can use the following delayed-weight method to estimate the symbol *a posteriori* probability:

$$P(s_t = a_i \mid \boldsymbol{Y}_{t+\Delta+\delta}) \cong \frac{1}{W_{t+\delta}} \sum_{j=1}^{m} \delta\left(s_t^{(j)} = a_i\right) w_{t+\delta}^{(j)}, \quad i = 1, \ldots, |\mathcal{A}|. \tag{9.121}$$

Simulation Examples

The fading process is modeled as the output of a Butterworth filter of order $r = 3$ driven by a complex white Gaussian noise process. The cutoff frequency of this filter is 0.05, corresponding to a normalized Doppler frequency (with respect to the symbol rate $1/T$) $f_d T = 0.05$, which is a fast-fading scenario. Specifically, the sequence of fading coefficients $\{\alpha_t\}$ is modeled by the following ARMA(3,3) process:

$$\begin{aligned}\alpha_t - 2.37409\alpha_{t-1} + 1.92936\alpha_{t-2} - 0.53208\alpha_{t-3}\\ = 10^{-2}(0.89409u_t + 2.68227u_{t-1} + 2.68227u_{t-2} + 0.89409u_{t-3}),\end{aligned} \qquad (9.122)$$

where $u_t \sim \mathcal{N}_c(0, 1)$. The filter coefficients in (9.122) are chosen such that $\text{Var}\{\alpha_t\} = 1$. It is assumed that BPSK modulation is employed (i.e., the transmitted symbols $s_t \in \{+1, -1\}$).

To demonstrate the high performance of the Monte Carlo adaptive receiver, in the following simulation examples we compare the performance (in terms of bit-error rate) of the Monte Carlo receivers with that of the following three receiver schemes:

- *Known channel lower bound:* In this case we assume that the fading coefficients $\{\alpha_t\}$ are known to the receiver. Then by (9.81), the optimal coherent detection rule is given by $\hat{s}_t = \text{sign}\left(\Re\left\{\alpha_t^* y_t\right\}\right)$ for both the Gaussian noise case (9.82) and the impulsive noise case (9.83).

- *Genie-aided lower bound:* In this case we assume that a genie provides the receiver with an observation of the modulation-free channel coefficient corrupted by additive noise with the same variance [i.e., $\tilde{y}_t = \alpha_t + \tilde{n}_t$, where $\tilde{n}_t \sim \mathcal{N}_c(0, \sigma^2)$ for the Gaussian noise case and $\tilde{n}_t \sim \mathcal{N}_c(0, \sigma_{I_t}^2)$ for the impulsive noise case]. In the case of impulsive noise, the genie also provides the receiver with the noise indicator I_t. The receiver then uses a Kalman filter to track the fading process based on the information provided by the genie (i.e., it computes $\hat{\alpha}_t = E\left\{\alpha_t \mid \tilde{\boldsymbol{Y}}_t, \boldsymbol{I}_t\right\}$). The transmitted symbols are then demodulated according to $\hat{s}_t = \text{sign}\left(\Re\left\{\hat{\alpha}_t^* y_t\right\}\right)$. It is clear that such a genie-aided bound is lower bounded by the known channel bound. It should also be noted that the genie is used only for calculating the lower bound. The SMC algorithms estimate the channel and the symbols simultaneously with no help from the genie.

- *Differential detector:* In this case no attempt is made to estimate the fading channel. Instead, the receiver detects the phase difference in two consecutively transmitted bits by using the simple rule of differential detection: $\widehat{b_t b_{t-1}} = \text{sign}\left(\Re\left\{y_t^* y_{t-1}\right\}\right)$.

We consider the performance of the sequential Monte Carlo receiver in a fading Gaussian noise channel without coding. In this case differential encoding and decoding are employed to resolve the phase ambiguity. The adaptive receiver implements Algorithm 9.5, described in Section 9.5.2. The number of Monte Carlo

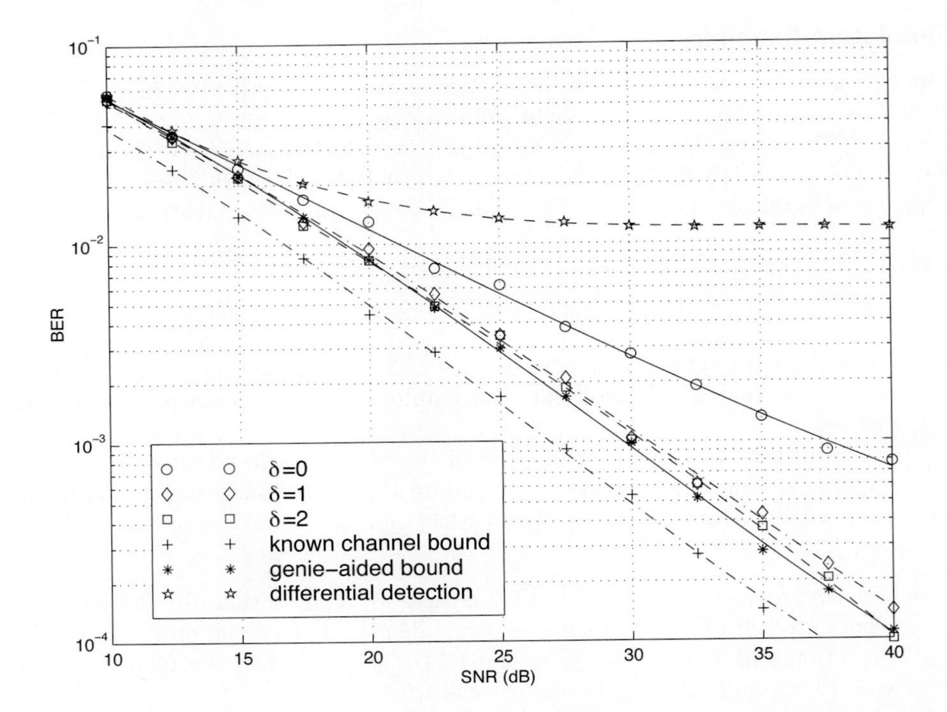

Figure 9.14. BER performance of a sequential Monte Carlo receiver in a fading channel with Gaussian noise and without coding. The delayed-weight method is used. The BER curves corresponding to delays $\delta = 0$, $\delta = 1$, and $\delta = 2$ are shown. Also shown are BER curves for the known channel lower bound, genie-aided lower bound, and differential detector.

samples drawn at each time was empirically set as $m = 50$. Simulation results have shown that the performance did not improve much when m was increased to 100, whereas it degraded notably when m was reduced to 20. Algorithm 8.9 for resampling was employed to maintain the efficiency of the algorithm, in which the effective sample size threshold is $\bar{m}_t = m/10$. The *delayed-weight method* discussed in Section 9.5.3 was used to extract further information from future received signals, which resulted in improved performance compared with concurrent estimation. In each simulation, the sequential Monte Carlo algorithm was run on 10,000 symbols (i.e., $t = 1, \ldots, 10,000$). In counting the symbol detection errors, the first 50 symbols were discarded to allow the algorithm to reach steady state. In Fig. 9.14, the bit-error rate performance versus the signal-to-noise ratio (defined as $\mathrm{Var}\{\alpha_t\}/\mathrm{Var}\{n_t\}$) corresponding to delay values $\delta = 0$ (concurrent estimate), $\delta = 1$, and $\delta = 2$ is plotted. In the figure we also plot the known channel lower bound, the genie-aided lower bound, and the BER curve of the differential detector. From this figure it is seen that for the uncoded case—with only a small amount of delay—the performance of the sequential Monte Carlo receiver can be improved significantly by the delayed-weight method compared with the concurrent estimate. Even with the concurrent estimate, the proposed adaptive receiver does not exhibit an error floor as does the differential detector. Moreover, with a delay $\delta = 2$, the proposed adaptive receiver essentially achieves the genie-aided lower bound. We have also implemented the delayed-sample method for this case and found that it offers little improvement over the delayed-weight method.

9.5.4 Adaptive Receiver in Fading Gaussian Noise Channels: Coded Case

So far we have considered the problem of detecting uncoded independent symbols in flat-fading channels. In what follows we extend the adaptive receiver technique presented in Section 9.5.2 and address the problem of sequential decoding of information bits in a convolutionally coded system signaling through a flat-fading channel.

Consider a binary rate k_0/n_0 convolutional encoder of overall constraint length $k_0\nu_0$. Suppose that the encoder starts with an all-zero state at time $t = 0$. The input to the encoder at time t is a block of information bits $\boldsymbol{d}_t = (d_{t,1}, \ldots, d_{t,k_0})$; the encoder output at time t is a block of code bits $\boldsymbol{b}_t = (b_{t,1}, \ldots, b_{t,n_0})$. For simplicity here we assume that BPSK modulation is employed. Then the transmitted symbols at time t are $\boldsymbol{s}_t = (s_{t,1}, \ldots, s_{t,n_0})$, where $s_{t,l} = 2b_{t,l} - 1$, $l = 1, \ldots, n_0$. (That is, $s_{t,l} = 1$ if $b_{t,l} = 1$, and $s_{t,l} = -1$ if $b_{t,l} = 0$.) Since $\boldsymbol{b}_t$ is determined by $(\boldsymbol{d}_t, \boldsymbol{d}_{t-1}, \ldots, \boldsymbol{d}_{t-\nu_0})$, so is $\boldsymbol{s}_t$. Hence we can write

$$\boldsymbol{s}_t = \psi(\boldsymbol{d}_t, \boldsymbol{d}_{t-1}, \ldots, \boldsymbol{d}_{t-\nu_0}) \tag{9.123}$$

for some function $\psi(\cdot)$ which is determined by the structure of the encoder.

Let $\boldsymbol{y}_t = (y_{t,1}, \ldots, y_{t,n_0})$ be the received signals at time t and let $\boldsymbol{\alpha}_t = (\alpha_{t,1}, \ldots, \alpha_{t,n_0})$ be the channel states corresponding to $\boldsymbol{b}_t$ and $\boldsymbol{d}_t$. Recall that $\alpha_{t-1,n_0} = \boldsymbol{h}^H \boldsymbol{x}_{t-1,n_0}$. Denote also $\boldsymbol{D}_t \triangleq (\boldsymbol{d}_0, \ldots, \boldsymbol{d}_t)$, $\boldsymbol{S}_t \triangleq (\boldsymbol{s}_0, \ldots, \boldsymbol{s}_t)$, and $\boldsymbol{Y}_t \triangleq (\boldsymbol{y}_0, \ldots, \boldsymbol{y}_t)$. The Monte Carlo samples recorded at time $(t-1)$ are $\left\{\left(\boldsymbol{D}_{t-1}^{(j)}, \kappa_{t-1,n_0}^{(j)}, w_{t-1}^{(j)}\right)\right\}_{j=1}^{m}$, where $\kappa_{t-1,n_0}^{(j)} \triangleq \left[\boldsymbol{\mu}_{t-1,n_0}^{(j)}, \boldsymbol{\Sigma}_{t-1,n_0}^{(j)}\right]$ contains the mean and covariance matrix of the state vector channel $\boldsymbol{x}_{t-1,n_0}$ conditioned on $\boldsymbol{D}_{t-1}^{(j)}$ and $\boldsymbol{Y}_{t-1}$. That is,

$$p\left(\boldsymbol{x}_{t-1,n_0} \mid \boldsymbol{D}_{t-1}^{(j)}, \boldsymbol{Y}_{t-1}\right) \sim \mathcal{N}_c\left(\boldsymbol{\mu}_{t-1,n_0}^{(j)}, \boldsymbol{\Sigma}_{t-1,n_0}^{(j)}\right). \tag{9.124}$$

As before, given the information bit sequence $\boldsymbol{D}_{t-1}^{(j)}$, the corresponding $\kappa_{t-1,n_0}^{(j)}$ is obtained by a Kalman filter. Our algorithm is as follows.

Algorithm 9.7: [SMC algorithm for adaptive decoding in flat-fading channels—Gaussian noise]

- *Initialization: Each Kalman filter is initialized as* $\kappa_{0,n_0}^{(j)} = \left(\boldsymbol{\mu}_{0,n_0}^{(j)}, \boldsymbol{\Sigma}_{0,n_0}^{(j)}\right)$, *with* $\boldsymbol{\mu}_{0,n_0}^{(j)} = \boldsymbol{0}$ *and* $\boldsymbol{\Sigma}_{0,n_0}^{(j)} = 2\boldsymbol{\Sigma}$, $j = 1, \ldots, m$, *where* $\boldsymbol{\Sigma}$ *is the stationary covariance of* $\boldsymbol{x}_t$*. All importance weights are initialized as* $w_0^{(j)} = 1$*,* $j = 1, \ldots, m$*. The initial* $\boldsymbol{D}_0^{(j)}$ *are randomly generated from the set* $\{0,1\}^{k_0}$*,* $j = 1, \ldots, m$*.*

At time t *we implement the following steps to update each sample* j*,* $j = 1, \ldots, m$*.*

- *Compute the* n_0*-step update of the Kalman filter:* *For each possible code vector* $\boldsymbol{d}_t = \boldsymbol{a}_i \in \{0,1\}^{k_0}$*, compute the corresponding symbol vector* $\boldsymbol{s}_t$ *using (9.123) to obtain*

$$\boldsymbol{s}_t^{(j)}(\boldsymbol{a}_i) = \psi\left(\boldsymbol{d}_t = \boldsymbol{a}_i, \boldsymbol{d}_{t-1}^{(j)}, \ldots, \boldsymbol{d}_{t-\nu_0}^{(j)}\right). \tag{9.125}$$

Let $\boldsymbol{\Sigma}_{t,0}^{(j)}(\boldsymbol{a}_i) \triangleq \boldsymbol{\Sigma}_{t-1,n_0}^{(j)}$ *and* $\boldsymbol{\mu}_{t,0}^{(j)}(\boldsymbol{a}_i) \triangleq \boldsymbol{\mu}_{t-1,n_0}^{(j)}$*. Perform* n_0 *steps of the Kalman filter update, using* $\boldsymbol{s}_t^{(j)}(\boldsymbol{a}_i)$ *and* $\boldsymbol{y}_t$*, as follows: For* $l = 1, \ldots, n_0$*, compute*

$$\boldsymbol{K}_{t,l}^{(j)}(\boldsymbol{a}_i) = \boldsymbol{F}\boldsymbol{\Sigma}_{t,l-1}^{(j)}(\boldsymbol{a}_i)\boldsymbol{F}^H + \boldsymbol{g}\boldsymbol{g}^H, \tag{9.126}$$

$$\gamma_{t,l}^{(j)}(\boldsymbol{a}_i) = \boldsymbol{h}^H\boldsymbol{K}_{t,l}^{(j)}(\boldsymbol{a}_i)\boldsymbol{h} + \sigma^2, \tag{9.127}$$

$$\eta_{t,l}^{(j)}(\boldsymbol{a}_i) = \boldsymbol{h}^H\boldsymbol{F}\boldsymbol{\mu}_{t,l-1}^{(j)}(\boldsymbol{a}_i), \tag{9.128}$$

$$\boldsymbol{\mu}_{t,l}^{(j)}(\boldsymbol{a}_i) = \boldsymbol{F}\boldsymbol{\mu}_{t,l-1}^{(j)}(\boldsymbol{a}_i) + \frac{1}{\gamma_{t,l}^{(j)}(\boldsymbol{a}_i)}\left[y_{t,l} - s_{t,l}^{(j)}(\boldsymbol{a}_i)\eta_{t,l}^{(j)}(\boldsymbol{a}_i)\right]\boldsymbol{K}_{t,l}^{(j)}(\boldsymbol{a}_i)\boldsymbol{h}, \tag{9.129}$$

$$\boldsymbol{\Sigma}_{t,l}^{(j)}(\boldsymbol{a}_i) = \boldsymbol{K}_{t,l}^{(j)}(\boldsymbol{a}_i) - \frac{1}{\gamma_{t,l}^{(j)}(\boldsymbol{a}_i)}\boldsymbol{K}_{t,l}^{(j)}(\boldsymbol{a}_i)\boldsymbol{h}\boldsymbol{h}^H\boldsymbol{K}_{t,l}^{(j)}(\boldsymbol{a}_i). \tag{9.130}$$

In (9.125)–(9.130) it is made explicit that the quantity on the left side of each equation is a function of the code-bit vector $\boldsymbol{a}_i$*. This yields* $\left\{\gamma_{t,l}^{(j)}(\boldsymbol{a}_i), \eta_{t,l}^{(j)}(\boldsymbol{a}_i)\right\}_{l=1}^{n_0}$ *and* $\left[\boldsymbol{\mu}_{t,n_0}^{(j)}(\boldsymbol{a}_i), \boldsymbol{\Sigma}_{t,n_0}^{(j)}(\boldsymbol{a}_i)\right]$ *for each* $\boldsymbol{a}_i \in \{0,1\}^{k_0}$*.*

- *Compute the trial sampling density: For each* $\boldsymbol{a}_i \in \{0,1\}^{k_0}$*, compute*

$$\begin{aligned}
\rho_{t,i}^{(j)} &\triangleq P\left(\boldsymbol{d}_t = \boldsymbol{a}_i \mid \boldsymbol{D}_{t-1}^{(j)}, \boldsymbol{Y}_t\right) \\
&\propto p\left(\boldsymbol{y}_t, \boldsymbol{Y}_{t-1}, \boldsymbol{d}_t = \boldsymbol{a}_i, \boldsymbol{D}_{t-1}^{(j)}\right) \\
&\propto p\left(\boldsymbol{y}_t, \boldsymbol{d}_t = \boldsymbol{a}_i \mid \boldsymbol{D}_{t-1}^{(j)}, \boldsymbol{Y}_{t-1}\right) && (9.131) \\
&= P\left(\boldsymbol{d}_t = \boldsymbol{a}_i \mid \boldsymbol{D}_{t-1}^{(j)}, \boldsymbol{Y}_{t-1}\right) p\left(\boldsymbol{y}_t \mid \boldsymbol{d}_t = \boldsymbol{a}_i, \boldsymbol{D}_{t-1}^{(j)}, \boldsymbol{Y}_{t-1}\right) \\
&= P(\boldsymbol{d}_t = \boldsymbol{a}_i)\, p\left[\boldsymbol{y}_t \mid \boldsymbol{s}_t^{(j)}(\boldsymbol{a}_i) = \psi\left(\boldsymbol{d}_t = \boldsymbol{a}_i, \boldsymbol{d}_{t-1}^{(j)}, \ldots, \boldsymbol{d}_{t-\nu_0}^{(j)}\right),\right. \\
&\qquad \left.\boldsymbol{S}_{t-1}^{(j)}, \boldsymbol{Y}_{t-1}\right] && (9.132)
\end{aligned}$$

$$\propto P(\boldsymbol{d}_t = \boldsymbol{a}_i)$$
$$\prod_{l=1}^{n_0} \underbrace{p\left[y_{t,l} \mid \boldsymbol{S}_{t-1}^{(j)}, s_{t,1}^{(j)}(\boldsymbol{a}_i), \ldots, s_{t,l}^{(j)}(\boldsymbol{a}_i), \boldsymbol{Y}_{t-1}, y_{t,1}, \ldots, y_{t,l-1}\right]}_{\mathcal{N}_c\left(s_{t,l}^{(j)}(\boldsymbol{a}_i)\,\eta_{t,l}^{(j)}(\boldsymbol{a}_i),\ \gamma_{t,l}^{(j)}(\boldsymbol{a}_i)\right)}, \quad (9.133)$$

where (9.132) follows from the fact that $\boldsymbol{d}_t$ is independent of $\boldsymbol{D}_{t-1}$ and $\boldsymbol{Y}_{t-1}$.

- *Impute the code bit vector $\boldsymbol{d}_t$: Draw $\boldsymbol{d}_t^{(j)}$ from the set $\{0,1\}^{k_0}$ with probability*

$$P\left(\boldsymbol{d}_t^{(j)} = \boldsymbol{a}_i\right) \propto \rho_{t,i}^{(j)}, \qquad \boldsymbol{a}_i \in \{0,1\}^{k_0}. \quad (9.134)$$

Append $\boldsymbol{d}_t^{(j)}$ to $\boldsymbol{D}_{t-1}^{(j)}$ and obtain $\boldsymbol{D}_t^{(j)}$. Pick the updated Kalman filter values $\boldsymbol{\mu}_{t,n_0}^{(j)} \triangleq \boldsymbol{\mu}_{t,n_0}^{(j)}\left(\boldsymbol{d}_t^{(j)}\right)$ and $\boldsymbol{\Sigma}_{t,n_0}^{(j)} = \boldsymbol{\Sigma}_{t,n_0}^{(j)}\left(\boldsymbol{d}_t^{(j)}\right)$ from the results in the first step, according to the value of the sample $\boldsymbol{d}_t^{(j)}$. This yields $\kappa_{t,n_0}^{(j)} = \left[\boldsymbol{\mu}_{t,n_0}^{(j)}, \boldsymbol{\Sigma}_{t,n_0}^{(j)}\right]$.

- *Compute the importance weight:*

$$\begin{aligned} w_t^{(j)} &= w_{t-1}^{(j)} \cdot p\left(\boldsymbol{y}_t \mid \boldsymbol{D}_{t-1}^{(j)}, \boldsymbol{Y}_{t-1}\right) \\ &= w_{t-1}^{(j)} \cdot \sum_{\boldsymbol{a}_i \in \{0,1\}^{k_0}} p\left(\boldsymbol{y}_t, \boldsymbol{d}_t = \boldsymbol{a}_i \mid \boldsymbol{D}_{t-1}^{(j)}, \boldsymbol{Y}_{t-1}\right) \\ &\propto w_{t-1}^{(j)} \sum_{\boldsymbol{a}_i \in \{0,1\}^{k_0}} \rho_{t,i}^{(j)}, \end{aligned} \quad (9.135)$$

where (9.135) follows from (9.131).

- *Perform resampling according to Algorithm 8.9 when $\bar{m}_t$ in (8.103) is below a threshold.*

Following the same line of proof as in Section 9.6.1, it can be shown that $\left\{\left(\boldsymbol{D}_t^{(j)}, \kappa_{t,n_0}^{(j)}, w_t^{(j)}\right)\right\}_{j=1}^m$ drawn by the procedure above are properly weighted with respect to $p(\boldsymbol{D}_t|\boldsymbol{Y}_t)$ provided that the samples $\left\{\left(\boldsymbol{D}_{t-1}^{(j)}, \kappa_{t-1,n_0}^{(j)}, w_{t-1}^{(j)}\right)\right\}_{j=1}^m$ are properly weighted with respect to $p(\boldsymbol{D}_{t-1}|\boldsymbol{Y}_{t-1})$. Note that in the coded case, phase ambiguity is prevented by the code constraint (9.123), and differential encoding is not needed.

At each time t, the major computation involved in the adaptive decoding algorithm above is the $\left(m\, n_0\, 2^{k_0}\right)$ one-step Kalman filter updates, which can be carried out by $\left(m\, 2^{k_0}\right)$ processing units, each computing an n_0-step update. (Note that $\boldsymbol{d}_t$ contains k_0 bits of information.) Furthermore, if the delayed-sample method outlined in Section 9.5.3 is employed for delayed estimation, then for a delay of Δ time units, a total of $\left(m\, n_0\, 2^{k_0(\Delta+1)}\right)$ one-step Kalman filter updates are needed at each time t, which can be distributed among $\left(m\, 2^{k_0(\Delta+1)}\right)$ processing units, each computing an n_0-step update.

Simulation Examples

We next show the performance of the proposed sequential Monte Carlo receiver in a coded system. The information bits are encoded using a rate-$\frac{1}{2}$ constraint-length-5 convolutional code (with generators 23 and 25 in octal notation). The receiver implements Algorithm 9.7 with a combination of delayed-sample and delayed-weight methods. That is, the information bit samples $\left\{ \boldsymbol{d}_t^{(j)} \right\}_{j=1}^m$ are drawn by using the delayed-sample method with delay Δ, whereas the importance weights $\left\{ w_{t+\delta}^{(j)} \right\}_{j=1}^m$ are obtained after a further delay of δ. The coded BER performance of this adaptive receiver with different delays—together with that of the known channel lower bound, the genie-aided lower bound, and the differential detector—is plotted in Fig. 9.15. It is seen that unlike the uncoded case, for coded systems the delayed-sample method is very effective in improving receiver performance. With a sample delay of $\Delta = 5$ and weight delay $\delta = 10$, the receiver performance is close to the genie-aided lower bound.

9.5.5 Adaptive Receivers in Fading Impulsive Noise Channels

As noted in Chapter 4, the ambient noise in many mobile communication channels is impulsive, due to various natural and human-made impulsive sources. In [546], a technique is developed for signal detection in fading channels with impulsive noise based on the Masreliez nonlinear filter [309] (cf. Section 7.3.1), making use of pilot symbols and decision feedback. In this section we discuss an adaptive receiver for flat-fading channels with impulsive ambient noise using the sequential Monte Carlo technique.

As in the case of Gaussian noise fading channels, we first develop adaptive receivers for uncoded systems. Consider the state-space system given by (9.93)–(9.94). Note that given both the symbol sequence $\boldsymbol{S}_t \triangleq (s_0, \ldots, s_t)$ and the noise indicator sequence $\boldsymbol{I}_t \triangleq (I_0, \ldots, I_t)$, this system is linear and Gaussian. Hence,

$$p(\boldsymbol{x}_t \mid \boldsymbol{S}_t, \boldsymbol{I}_t, \boldsymbol{Y}_t) \sim \mathcal{N}_c\Big(\boldsymbol{\mu}_t(\boldsymbol{S}_t, \boldsymbol{I}_t), \boldsymbol{\Sigma}_t(\boldsymbol{S}_t, \boldsymbol{I}_t)\Big), \tag{9.136}$$

where the mean $\boldsymbol{\mu}_t(\boldsymbol{S}_t, \boldsymbol{I}_t)$ and the covariance matrix $\boldsymbol{\Sigma}_t(\boldsymbol{S}_t, \boldsymbol{I}_t)$ can be obtained by a Kalman filter with given $\boldsymbol{S}_t$ and $\boldsymbol{I}_t$. As before, we seek to obtain properly weighted samples $\left\{ \boldsymbol{S}_t^{(j)}, \boldsymbol{I}_t^{(j)}, \kappa_t^{(j)}, w_t^{(j)} \right\}_{j=1}^m$ with respect to the distribution $p(\boldsymbol{S}_t, \boldsymbol{I}_t | \boldsymbol{Y}_t)$. These samples are then used to estimate the transmitted symbols and channel parameters.

Algorithm 9.8: [SMC algorithm for adaptive detection in flat-fading channels—impulsive noise]

- *Initialization: This step is the same as that in the Gaussian case. Note that no initial values for $I_0^{(j)}$ are needed due to independence.*

  At time t, the following updates are implemented for each sample j, $j = 1, \ldots, m$.

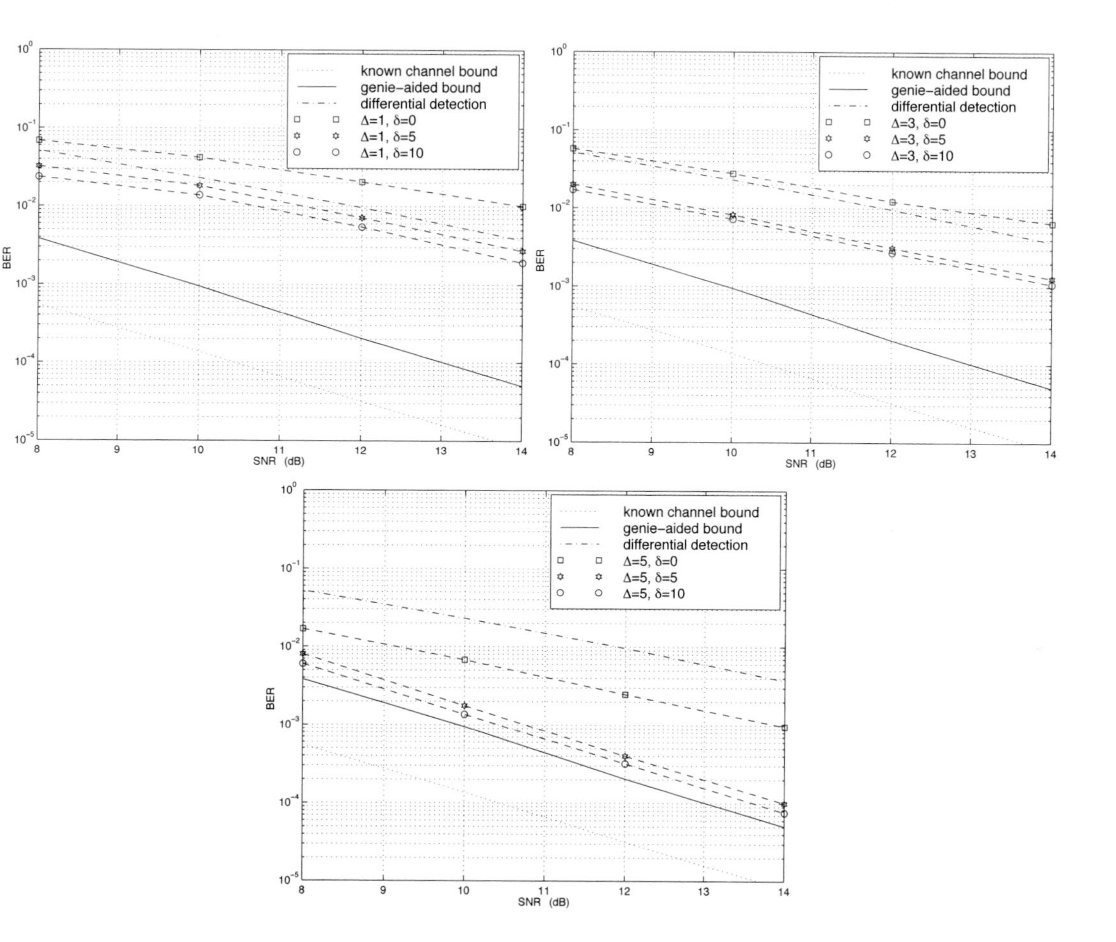

Figure 9.15. BER performance of a sequential Monte Carlo receiver in a fading channel with Gaussian noise for a convolutionally coded system. The convolutional code has rate-$\frac{1}{2}$ and constraint length 5. A combination delayed-sample (with delay Δ), delayed-weight (with delay δ) method is used. The BER curves corresponding to delays $\Delta = 1$, $\Delta = 3$, and $\Delta = 5$ are shown. Also shown are BER curves for the known channel lower bound, genie-aided lower bound, and differential detector.

- *Compute the one-step predictive update of the Kalman filter* $\kappa_{t-1}^{(j)}$:

$$\boldsymbol{K}_t^{(j)} = \boldsymbol{F}\boldsymbol{\Sigma}_{t-1}^{(j)}\boldsymbol{F}^H + \boldsymbol{g}\boldsymbol{g}^H, \tag{9.137}$$

$$\tilde{\gamma}_t^{(j)} = \boldsymbol{h}^H \boldsymbol{K}_t^{(j)} \boldsymbol{h}, \tag{9.138}$$

$$\eta_t^{(j)} = \boldsymbol{h}^H \boldsymbol{F} \boldsymbol{\mu}_{t-1}^{(j)}. \tag{9.139}$$

Conditioned on $\boldsymbol{S}_t^{(j)}$ *and* $\boldsymbol{I}_t^{(j)}$, *the predictive distribution is then given by*

$$p\left(y_t \mid \boldsymbol{S}_t^{(j)}, \boldsymbol{I}_t^{(j)}, \boldsymbol{Y}_{t-1}\right) \sim \mathcal{N}_c\left(s_t^{(j)}\eta_t^{(j)}, \tilde{\gamma}_t^{(j)} + \sigma^2_{I_t^{(j)}}\right). \tag{9.140}$$

- *Compute the trial sampling density: For each* $(a,\delta)_i \in \mathcal{A} \times \{1,2\}$, *compute*

$$\begin{aligned} \rho_{t,i}^{(j)} &\triangleq P\left[(s_t, I_t) = (a,\delta)_i \mid \boldsymbol{S}_{t-1}^{(j)}, \boldsymbol{I}_{t-1}^{(j)}, \boldsymbol{Y}_t\right] \\ &\propto p\left[y_t, \boldsymbol{Y}_{t-1}, (s_t, I_t) = (a,\delta)_i, \boldsymbol{S}_{t-1}^{(j)}, \boldsymbol{I}_{t-1}^{(j)}\right] \\ &= \underbrace{p\left[y_t \mid (s_t, I_t) = (a,\delta)_i, \boldsymbol{S}_{t-1}^{(j)}, \boldsymbol{I}_{t-1}^{(j)}, \boldsymbol{Y}_{t-1}\right]}_{\mathcal{N}_c\left(s_t\,\eta_t^{(j)},\ \tilde{\gamma}_t^{(j)} + \sigma^2_{I_t^{(j)}}\right)} P\left[(s_t, I_t) = (a,\delta)_i\right]. \end{aligned} \tag{9.141}$$

- *Impute the symbol and the noise indicator* (s_t, I_t): *Draw* $\left[s_t^{(j)}, I_t^{(j)}\right]$ *from the set* $\mathcal{A} \times \{1,2\}$ *with probability*

$$P\left[(s_t^{(j)}, I_t^{(j)}) = (a,\delta)_i\right] \propto \rho_{t,i}^{(j)}, \quad (a,\delta)_i \in \mathcal{A} \times \{1,2\}. \tag{9.142}$$

Append $\left(s_t^{(j)}, I_t^{(j)}\right)$ *to* $\left(\boldsymbol{S}_{t-1}^{(j)}, \boldsymbol{I}_{t-1}^{(j)}\right)$ *and obtain* $\left(\boldsymbol{S}_t^{(j)}, \boldsymbol{I}_t^{(j)}\right)$.

- *Compute the importance weight:*

$$\begin{aligned} w_t^{(j)} &= w_{t-1}^{(j)} p\left(y_t \mid \boldsymbol{S}_{t-1}^{(j)}, \boldsymbol{I}_{t-1}^{(j)}, \boldsymbol{Y}_{t-1}\right) \\ &= w_{t-1}^{(j)} \sum_{(a,\delta)_i \in \mathcal{A}\times\{1,2\}} p\left[y_t \mid (s_t, I_t) = (a,\delta)_i, \boldsymbol{S}_{t-1}^{(j)}, \boldsymbol{I}_{t-1}^{(j)}, \boldsymbol{Y}_{t-1}\right] \\ &\qquad P\left[(s_t, I_t) = (a,\delta)_i\right] \\ &= w_{t-1}^{(j)} \sum_{(a,\delta)_i \in \mathcal{A}\times\{1,2\}} \rho_{t,i}^{(j)}, \end{aligned} \tag{9.143}$$

where (9.143) follows from (9.141).

- *Compute the one-step filtering update of the Kalman filter: Based on the imputed symbol and indicator* $\left(s_t^{(j)}, I_t^{(j)}\right)$ *and the observation* y_t, *complete the Kalman filter update to obtain* $\kappa_t^{(j)} = \left[\boldsymbol{\mu}_t^{(j)}, \boldsymbol{\Sigma}_t^{(j)}\right]$ *according to (9.113) and (9.114) with* $\gamma_t^{(j)} = \tilde{\gamma}_t^{(j)} + \sigma^2_{I_t^{(j)}}$.

- *Perform resampling according to Algorithm 8.9 when the effective sample size $\bar{m}_t$ in (8.103) is below a threshold.*

The proof that Algorithm 9.8 gives properly weighted samples is similar to that for the Gaussian fading channels in Section 9.6.1. The dominant computation involved in the algorithm at each time t includes m one-step Kalman filter updates. If the delayed-sample method is employed for delayed estimation with a delay of Δ time units, then at each time t, $\left(m\,(2|\mathcal{A}|)^{\Delta}\right)$ one-step Kalman filter updates are needed because $|\mathcal{A} \times \{1,2\}| = 2|\mathcal{A}|$, which can be implemented in parallel.

Moreover, we can also develop the adaptive receiver algorithm for coded systems in impulsive noise flat-fading channels, similar to the one discussed in Section 9.5.4. For a rate-k_0/n_0 convolutional code, if the delayed-sample method is used with a delay of Δ time units, then at each time t a total of $\left(m\,n_0\,2^{(k_0+n_0)(\Delta+1)}\right)$ one-step Kalman filter updates are needed, which can be distributed among $\left(m\,2^{(k_0+n_0)(\Delta+1)}\right)$ processors, each computing one n_0-step update. (With a delay of Δ units, there are $2^{k_0(\Delta+1)}$ possible code vectors, and there are $2^{n_0(\Delta+1)}$ possible noise indicator vectors.)

Simulation Examples

The uncoded BER performance of the proposed adaptive receiver, together with that of the other three receiver schemes, in a fading channel with impulsive ambient noise is shown in Fig. 9.16. The noise distribution is given by the two-term Gaussian mixture model (9.83) with $\kappa = 100$ and $\epsilon = 0.1$. As mentioned earlier, in this case, for the genie-aided bound the genie not only provides the observation of the noise-corrupted modulation-free channel coefficients, but also the true noise indicator $\{I_t\}$, to the channel estimator. It is seen from this figure that, again, the delayed-weight method offers significant improvement over the concurrent estimate, although in this case the BER curve for $\delta = 2$ is slightly off the genie-aided lower bound. Furthermore, the proposed adaptive receiver does not have the error floor exhibited by the simple differential detector.

In summary, in this section we have discussed adaptive receiver algorithms for both uncoded and coded systems, where the delayed-weight method, the delayed-sample method, and a combination of both are employed to improve estimation accuracy. The Monte Carlo receiver techniques can also handle impulsive ambient noise. The computational complexities of the various algorithms discussed in this chapter are summarized in Table 9.1. Note that although the delayed-sample SMC estimation method offers a significant performance gain over the simple SMC method, it has a higher computational complexity. In [542], a number of alternative delayed estimation methods based on SMC are developed which trade off performance and complexity. Finally, note that the adaptive SMC receivers developed in this section require knowledge of the second-order statistics of the fading process. In [165], a nonparametric SMC receiver was developed that is based on wavelet modeling of the fading process and does not require knowledge of channel statistics.

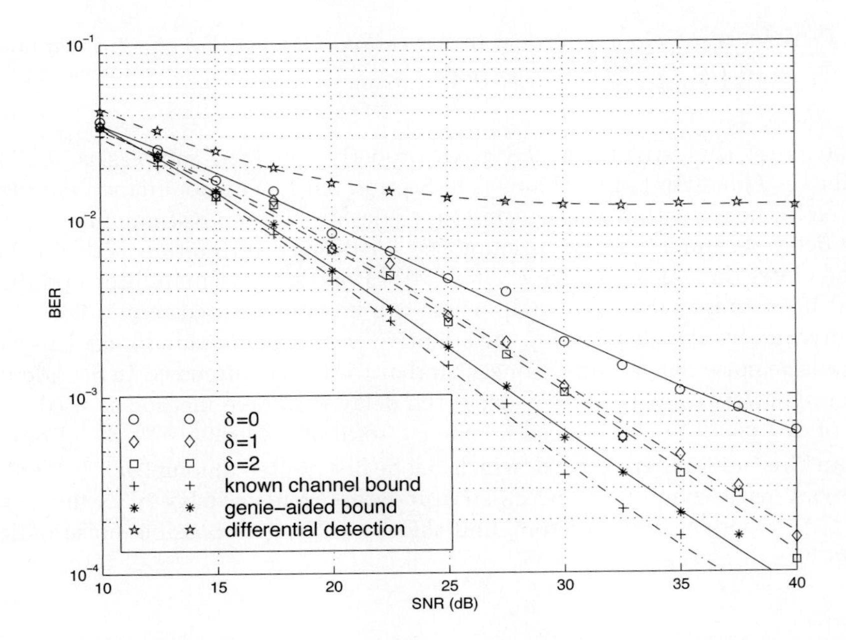

Figure 9.16. BER performance of a sequential Monte Carlo receiver in a fading channel with impulsive noise and without coding. $\epsilon = 0.1$, $\kappa = 100$. The delayed-weight method is used. BER curves corresponding to delays $\delta = 0$, $\delta = 1$, and $\delta = 2$ are shown. Also shown are BER curves for the known channel lower bound, genie-aided lower bound, and differential detector.

Table 9.1. Computational Complexities of the Proposed Sequential Monte Carlo Receiver Algorithms under Different Conditions in Terms of the Number of One-Step Kalman Filter Updates Needed at Each Time t.[a]

	Uncoded System		Coded System	
	Complexity	Degree of Parallelism	Complexity	Degree of Parallelism
Gaussian	$m\,\lvert\mathcal{A}\rvert^{\Delta}$	$m\,\lvert\mathcal{A}\rvert^{\Delta}$	$m\,n_0\,2^{k_0(\Delta+1)}$	$m\,2^{k_0(\Delta+1)}$
Impulsive	$m\,(2\lvert\mathcal{A}\rvert)^{\Delta}$	$m\,(2\lvert\mathcal{A}\rvert)^{\Delta}$	$m\,n_0\,2^{(k_0+n_0)(\Delta+1)}$	$m\,2^{(k_0+n_0)(\Delta+1)}$

[a] The degree of parallelism refers to the maximum number of computing units that can be employed to implement the algorithm in parallel. It is assumed that the delayed-sample method is used with a delay of Δ time units. The number of samples drawn at each time is m. For uncoded system, the cardinality of the symbol alphabet is $|\mathcal{A}|$. For coded system, a k_0/n_0-convolutional code is used. The impulsive noise is modeled by a two-term Gaussian mixture.

9.6 Appendix

9.6.1 Proof of Proposition 9.1 in Section 9.5.2

To show that the sample $(s_t^{(j)}, w_t^{(j)})$ given by (9.111) and (9.112) is a properly weighted sample with respect to $p(\boldsymbol{S}_t|\boldsymbol{Y}_t)$, we need to verify that (9.112) gives the correct weight. Assume that at time $(t-1)$, we have a properly weighted sample $(s_{t-1}^{(j)}, w_{t-1}^{(j)})$ with respect to $p(\boldsymbol{S}_{t-1}|\boldsymbol{Y}_{t-1})$. That is, assume that $s_{t-1}^{(j)}$ is drawn from some trial distribution $q(\boldsymbol{S}_{t-1} \mid \boldsymbol{Y}_{t-1})$ and that the importance weight is given by $w_{t-1}^{(j)} = \omega_{t-1}(\boldsymbol{S}_{t-1}^{(j)}|\boldsymbol{Y}_{t-1})$, with

$$\omega_{t-1}(\boldsymbol{S}_{t-1} \mid \boldsymbol{Y}_{t-1}) \triangleq \frac{p(\boldsymbol{S}_{t-1} \mid \boldsymbol{Y}_{t-1})}{q(\boldsymbol{S}_{t-1} \mid \boldsymbol{Y}_{t-1})}. \tag{9.144}$$

By (9.109) and (9.111), $s_t^{(j)}$ is drawn from the distribution $p(s_t|\boldsymbol{S}_{t-1}^{(j)}, \boldsymbol{Y}_t)$. Hence, the sampling distribution for $\boldsymbol{S}_t^{(j)}$ is given by $q(\boldsymbol{S}_{t-1}|\boldsymbol{Y}_{t-1})\, p(s_t|\boldsymbol{S}_{t-1}, \boldsymbol{Y}_t)$. Since the target distribution is $p(\boldsymbol{S}_t|\boldsymbol{Y}_t)$, the weight function at time t is then given by

$$\begin{aligned}
\omega_t(\boldsymbol{S}_t \mid \boldsymbol{Y}_t) &= \frac{p(\boldsymbol{S}_t \mid \boldsymbol{Y}_t)}{q(\boldsymbol{S}_{t-1} \mid \boldsymbol{Y}_{t-1})\, p(s_t \mid \boldsymbol{S}_{t-1}, \boldsymbol{Y}_t)} \\
&= \frac{p(\boldsymbol{S}_{t-1} \mid \boldsymbol{Y}_{t-1})}{q(\boldsymbol{S}_{t-1} \mid \boldsymbol{Y}_{t-1})} \frac{p(\boldsymbol{S}_{t-1} \mid \boldsymbol{Y}_t)}{p(\boldsymbol{S}_{t-1} \mid \boldsymbol{Y}_{t-1})} \\
&= \omega_{t-1}(\boldsymbol{S}_{t-1} \mid \boldsymbol{Y}_{t-1}) \frac{p(y_t \mid \boldsymbol{Y}_{t-1}, \boldsymbol{S}_{t-1})\, p(\boldsymbol{Y}_{t-1} \mid \boldsymbol{S}_{t-1})\, p(\boldsymbol{S}_{t-1})/p(\boldsymbol{Y}_t)}{p(\boldsymbol{Y}_{t-1}|\boldsymbol{S}_{t-1})\, p(\boldsymbol{S}_{t-1})/p(\boldsymbol{Y}_{t-1})} \\
&\propto \omega_{t-1}(\boldsymbol{S}_{t-1} \mid \boldsymbol{Y}_{t-1}) p(y_t \mid \boldsymbol{Y}_{t-1}, \boldsymbol{S}_{t-1}) \\
&= \omega_{t-1}(\boldsymbol{S}_{t-1} \mid \boldsymbol{Y}_{t-1}) \\
&\qquad \sum_{a_i \in \mathcal{A}} p(y_t \mid s_t = a_i, \boldsymbol{S}_{t-1}, \boldsymbol{Y}_{t-1}) p(s_t = a_i \mid \boldsymbol{S}_{t-1}, \boldsymbol{Y}_{t-1}) \\
&= \omega_{t-1}(\boldsymbol{S}_{t-1} \mid \boldsymbol{Y}_{t-1}) \sum_{a_i \in \mathcal{A}} \rho_{t,i}.
\end{aligned} \tag{9.145}$$

Hence, $w_t^{(j)} = \omega_t(\boldsymbol{S}_t^{(j)} \mid \boldsymbol{Y}_t) = w_{t-1}^{(j)} \sum_{a_i \in \mathcal{A}} \rho_{t,i}^{(j)}$. This verifies that (9.112) gives the correct importance weight at time t.

Chapter 10

ADVANCED SIGNAL PROCESSING FOR CODED OFDM SYSTEMS

10.1 Introduction

Orthogonal frequency-division multiplexing (OFDM) is a bandwidth-efficient signaling scheme for wideband digital communications. The main difference between traditional frequency-division multiplexing (FDM) and OFDM is that in OFDM, the spectra of the individual carriers overlap. Nevertheless, the OFDM carriers exhibit orthogonality on a symbol interval if they are spaced in frequency exactly at the reciprocal of the symbol interval, which can be accomplished by using the discrete Fourier transform (DFT). With the development of modern digital signal processing technology, OFDM has become practical to implement and has been proposed as an efficient modulation scheme for applications ranging from modems, digital audio broadcast, to next-generation high-speed wireless data communications. For example, as discussed in Chapter 1, the high-speed wireless LAN standard IEEE 802.11a is based on OFDM.

One of the principal advantages of OFDM is that it effectively converts a frequency-selective fading channel into a set of parallel flat-fading channels. Both the intersymbol interference (ISI) and intercarrier interference (ICI) can be eliminated completely by inserting between symbols a small time interval known as a *guard interval.* The length of the guard interval is made equal to or greater than the delay spread of the channel. If the symbol signal waveform is extended periodically into the guard interval (cyclic prefix), orthogonality of the carrier is maintained over the symbol period, thus eliminating ICI. Also, ISI is eliminated because successive symbols do not overlap due to the guard interval. Hence at the receiver there is no need to perform channel equalization, and the complexity of the receiver is quite low.

In this chapter we discuss receiver design for OFDM systems signaling through unknown frequency-selective fading channels. In particular, we focus on the design of turbo receivers in a number of OFDM systems, including an OFDM system with frequency offset, a space-time block coded OFDM system, and a space-time coded

OFDM system based on low-density parity-check (LDPC) codes. This chapter is organized as follows. In Section 10.2 we introduce the OFDM communication system. In Section 10.3 we present an MCMC-based blind turbo receiver for OFDM systems with unknown frequency offset and frequency-selective fading. In Section 10.4 we discuss a pilot-symbol-aided turbo receiver for space-time block coded OFDM systems. In Section 10.5 we present an LDPC-based space-time coded OFDM system and the corresponding turbo receiver structure.

The following is a list of the algorithms appearing in this chapter.

- *Algorithm 10.1:* MCMC-based OFDM blind demodulator in the presence of frequency offset and frequency-selective fading
- *Algorithm 10.2:* Metropolis–Hasting sampling of frequency offset
- *Algorithm 10.3:* Gibbs sampling of frequency offset
- *Algorithm 10.4:* LDPC decoding algorithm

10.2 OFDM Communication System

Figure 10.1 is a block diagram of an OFDM communication system. A serial-to-parallel buffer segments the information sequence into frames of Q symbols. An OFDM word at time i consists of Q data symbols $X_0[i], X_1[i], \ldots, X_{Q-1}[i]$. An inverse discrete Fourier transform (IDFT) is first applied to the OFDM word, to obtain

$$x_m[i] = \frac{1}{Q}\sum_{k=0}^{Q-1} X_k[i] \exp\left(\jmath\frac{2\pi km}{Q}\right), \qquad m = 0, 1, \ldots, Q-1. \tag{10.1}$$

A guard interval with cyclic prefix is then inserted to prevent possible intersymbol interference between OFDM words. After pulse shaping and parallel-to-serial conversion, the signals are then transmitted through a frequency-selective fading channel. The time-domain channel impulse response can be modeled as a tapped-delay line, given by

$$g(\tau; t) = \sum_{\ell=0}^{L-1} \alpha_\ell(t)\delta\left(\tau - \frac{\ell}{Q\Delta_f}\right), \tag{10.2}$$

where $L \triangleq \lceil \tau_m \Delta_f + 1 \rceil$ denotes the maximum number of resolvable channel taps, with τ_m being the maximum multipath spread and Δ_f being the carrier spacing. Assume that the channel taps remain constant over the interval of one OFDM word [i.e., $\alpha_\ell(t) \equiv \alpha_\ell[(i-1)T]$ for $(i-1)T \le t < iT$, where T is the duration of one OFDM word]. At the receiver end, after matched filtering and removing the cyclic prefix, the sampled received signal corresponding to the nth OFDM word becomes

$$y_m[i] = x_m[i] \star h_m[i] + n_m[i] \tag{10.3}$$

$$= \sum_{\ell=0}^{L-1} h_\ell[i]\, x_{m-\ell}[i] + n_m[i], \qquad m = 0, 1, \ldots, Q-1, \tag{10.4}$$

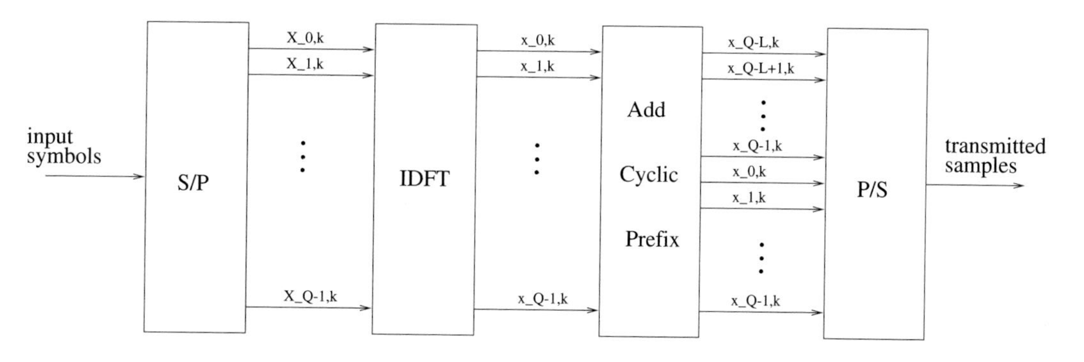

Figure 10.1. Block diagram of a simple OFDM transmitter.

where $\star$ denotes the convolution, $h_\ell[i] \triangleq \alpha_\ell[(n-1)T]$, and $\{n_m[i]\}_m$ are i.i.d. complex white Gaussian noise samples. A DFT is then applied to the received signals $\{y_m[i]\}_m$ to demultiplex the multicarrier signals:

$$Y_k[i] = \sum_{m=0}^{Q-1} y_m[i] \exp\left(-\jmath \frac{2\pi km}{Q}\right), \qquad k = 0, 1, \ldots, Q-1. \tag{10.5}$$

For OFDM systems with proper cyclic extensions and proper sample timing, with tolerable leakage, the received signal after demultiplexing at the kth subcarrier can be expressed as

$$Y_k[i] = X_k[i]H_k[i] + N_k[i], \qquad k = 0, 1, \ldots, Q-1, \tag{10.6}$$

where $\{N_k[i]\}_k$ contains the DFT of the noise samples $\{n_m[i]\}_m$, and $N_k \overset{\text{i.i.d.}}{\sim} \mathcal{N}_c(0, \sigma^2)$; and $\{H_k[i]\}_k$ contains the DFT of channel impulse response $\{h_m[i]\}_m$:

$$H_k[i] = \sum_{\ell=0}^{L-1} h_\ell[i] \exp\left(-\jmath \frac{2\pi k\ell}{Q}\right), \qquad k = 0, 1, \ldots, Q-1. \tag{10.7}$$

Assume that for each ℓ, $0 \le \ell < L$, $\{h_\ell[i]\}_i$ is a complex Gaussian process with an autocorrelation following the Jakes model:

$$E\left\{h_\ell[i]h_\ell[i+m]^*\right\} = P_\ell\, J_0(2\pi f_d T m), \tag{10.8}$$

where P_ℓ is the average power of the ℓth tap and f_d is the Doppler spread. Assume further that the L fading processes are mutually independent. Since $\{H_k[i]\}_k$ are linear transformations of $\{h_\ell[i]\}_\ell$, then for each k, $0 \le k < Q$, $\{H_k[i]\}_i$ is also a complex Gaussian process with autocorrelation

$$\begin{aligned}
E\left\{H_k[i]H_k[i+m]^*\right\} &= E\left\{\sum_{\ell=0}^{L-1} h_\ell[i] \exp\left(-\jmath \frac{2\pi k\ell}{Q}\right) \sum_{\ell=0}^{L-1} h_\ell[i+m]^* \exp\left(\jmath \frac{2\pi k\ell}{Q}\right)\right\} \\
&= E\left\{\sum_{\ell=0}^{L-1} h_\ell[i]h_\ell[i+m]^*\right\} \\
&= \sum_{\ell=0}^{L-1} E\left\{h_\ell[i]h_\ell[i+m]^*\right\} \\
&= J_0(2\pi f_d T m) \sum_{\ell=0}^{L-1} P_\ell.
\end{aligned} \tag{10.9}$$

Hence from (10.6) and (10.9) it is seen that the received frequency-domain signal at each subcarrier k follows a flat-fading model with the same fading autocorrelation function as that in the time domain. Hence the OFDM system effectively transforms a frequency-selective fading channel into a set of parallel flat-fading channels.

However, note that the frequency-domain channel responses of different carriers are correlated. In fact, we have

$$\begin{aligned}E\left\{H_{k_1}[i]H_{k_2}[i+m]^*\right\} &= E\left\{\sum_{\ell=0}^{L-1} h_\ell[i]\exp\left(-\jmath\frac{2\pi k_1 \ell}{Q}\right)\sum_{\ell=0}^{L-1} h_\ell[i+m]^*\exp\left(\jmath\frac{2\pi k_2 \ell}{Q}\right)\right\}\\ &= E\left\{\sum_{\ell=0}^{L-1} h_\ell[i]h_\ell[i+m]^*\exp\left(\jmath\frac{2\pi(k_2-k_1)\ell}{Q}\right)\right\}\\ &= J_0(2\pi f_d T m)\sum_{\ell=1}^{L-1} P_\ell \exp\left(\jmath\frac{2\pi(k_2-k_1)\ell}{Q}\right). \end{aligned} \tag{10.10}$$

10.3 Blind MCMC Receiver for Coded OFDM with Frequency-Selective Fading and Frequency Offset

In practical OFDM systems, the existence of frequency offset, which is caused by the mismatch between the oscillator in the transmitter and that in the receiver, destroys the orthogonality among OFDM subcarriers and leads to a performance degradation [374]. Several schemes of frequency offset estimation in OFDM systems have been investigated in [84, 89, 243, 295, 341, 440, 500, 507]. For OFDM applications over additive Gaussian white noise (AWGN) channels, the maximum-likelihood (ML) frequency offset estimates are derived in [89, 243, 295, 507]. Given that wireless channels typically exhibit frequency-selective fading, these methods designed for AWGN channels are not applicable in wireless OFDM systems. On the other hand, frequency offset estimators in frequency-selective fading channels are developed in [84, 341, 440], which require some particular form of data redundancy (e.g., data repetition [341] or pilot insertion [84, 440]). In [500], a blind subspace method for frequency offset estimation is proposed.

In wireless OFDM systems, in addition to the frequency offset, the frequency-selective fading channel states are also unknown to the receiver. The problem of channel estimation in OFDM systems has been studied in many previous works. The methods proposed in [260, 506] estimate the fading channel based on the pilot symbols, while blind estimation schemes based on second- or high-order statistics are proposed in [109, 345, 614]. Moreover, in [205, 367], subcarrier phase estimators are proposed by employing the expectation-maximization (EM) algorithm.

As an important remark, we note that the ultimate objective of the receiver is to recover the information-bearing data symbols from the received signals. Although the prevailing receiver-design paradigm is to estimate the unknown parameters first and then to use these estimated parameters in the detector, such an "estimate-then-plug-in" approach is ad hoc and bears no theoretical optimality. In this section we treat the problem of blind receiver design for coded OFDM systems in the presence of unknown frequency offset and frequency-selective fading, under the Markov chain Monte Carlo (MCMC) framework for Bayesian computation (cf. Chapter 8) and the principle of turbo processing (cf. Chapter 6). The techniques in this section were developed in [290].

10.3.1 System Description

Channel Model with Frequency Offset

When there is a carrier frequency offset in the OFDM channel, the received time-domain signal in (10.4) becomes [341]

$$y_m[i] = \frac{1}{Q}\sum_{k=0}^{Q-1} X_k H_k \exp\left(-\jmath\frac{2\pi m(k+\epsilon)}{Q}\right) + n_m[i], \qquad m = 0, 1, \ldots, Q-1, \quad (10.11)$$

where ϵ is the relative frequency offset of the channel (the ratio of the actual frequency offset to the intercarrier spacing). Note that for practical purposes, we assume that the absolute value of the frequency offset is no larger than half of the OFDM subcarrier spacing (i.e., $|\epsilon| < 0.5$). That is, any large frequency offset has already been compensated (e.g., by an automatic frequency control circuit [120]) and what remains is the residual frequency offset. We next write the signal model (10.11) in matrix form. Denote

$$\boldsymbol{h}[i] \triangleq \Big[h_0[i]\ h_1[i]\ \cdots\ h_{L-1}[i] \quad \underbrace{0\cdots 0}_{(Q-L)\ 0\text{'s}}\Big]^T,$$

$$\boldsymbol{H}[i] \triangleq \Big[H_0[i]\ H_1[i]\ \cdots\ H_{Q-1}[i]\Big]^T,$$

$$\boldsymbol{y}[i] \triangleq \Big[y_0[i]\ y_1[i]\ \cdots\ y_{Q-1}[i]\Big]^T,$$

$$\boldsymbol{Y}[i] \triangleq \Big[Y_0[i]\ Y_1[i]\ \cdots\ Y_{Q-1}[i]\Big]^T,$$

$$\boldsymbol{n}[i] \triangleq \Big[n_0[i]\ n_1[i]\ \cdots\ n_{Q-1}[i]\Big]^T,$$

$$\boldsymbol{N}[i] \triangleq \Big[N_0[i]\ N_1[i]\ \cdots\ N_{Q-1}[i]\Big]^T,$$

$$\boldsymbol{X}[i] \triangleq \mathrm{diag}\Big\{X_0[i]\ X_1[i]\ \cdots\ X_{Q-1}[i]\Big\},$$

$$\boldsymbol{W} = \begin{bmatrix} 1 & 1 & \cdots & 1 \\ 1 & \exp[-\jmath 2\pi/Q] & \cdots & \exp[-\jmath 2\pi(Q-1)/Q] \\ \vdots & \vdots & \ddots & \vdots \\ 1 & \exp[-\jmath 2\pi(Q-1)/Q] & \cdots & \exp[-\jmath 2\pi(Q-1)(Q-1)/Q] \end{bmatrix},$$

$$\boldsymbol{F}_\epsilon \triangleq \mathrm{diag}\Big\{1, \exp[\jmath 2\pi\epsilon/Q], \ldots, \exp[\jmath 2\pi\epsilon(Q-1)/Q]\Big\}.$$

Note that $\boldsymbol{W}$ is the DFT matrix and $(1/Q)\boldsymbol{W}^H$ is the inverse DFT matrix [i.e., $\boldsymbol{W}\boldsymbol{W}^H/Q = \boldsymbol{W}^H\boldsymbol{W}/Q = \boldsymbol{I}_Q$]. Hence $\boldsymbol{H}[i] = \boldsymbol{W}\boldsymbol{h}[i]$ and $\boldsymbol{N}[i] = \boldsymbol{W}\boldsymbol{n}[i]$.

Then upon applying a DFT to $\{y_m[i]\}_m$ in (10.4), we obtain the signal model

$$\boldsymbol{Y}[i] = \frac{1}{Q}\boldsymbol{W}\boldsymbol{F}_\epsilon\boldsymbol{W}^H\boldsymbol{X}[i]\boldsymbol{W}\boldsymbol{h}[i] + \boldsymbol{N}[i]. \quad (10.12)$$

For a better understanding of the effect of the frequency offset, we now take a closer look at the matrix $\boldsymbol{\Psi} \triangleq \boldsymbol{W}\boldsymbol{F}_\epsilon \boldsymbol{W}^H/Q$ in (10.12). Since $|\epsilon| < 0.5$, after some simple algebra, the (i,j)th element of the matrix $\boldsymbol{\Psi}$ can be expressed as

$$\psi(i,j) \triangleq \begin{cases} \dfrac{1-\exp\left[\jmath 2\pi(i-j+\epsilon)\right]}{Q\left\{1-\exp\left[\jmath 2\pi(i-j+\epsilon)/Q\right]\right\}}, & \epsilon \neq 0,\ \forall i,j, \\ \delta(i-j), & \epsilon = 0,\ \forall i,j; \end{cases}$$

with

$$|\psi(i,j)| \leq 1, \forall i,j \quad \text{and} \quad |\psi(i,j)| \geq |\psi(i',j')| \quad \text{if} \quad |i-j| \leq |i'-j'|.$$

Hence $\boldsymbol{\Psi} \neq \boldsymbol{I}$ when $\epsilon \neq 0$; and the spillover to off-diagonal elements of $\boldsymbol{\Psi}$, which corresponds to the ICI [341], increases as ϵ increases.

Bayesian Formulation of Optimal Demodulation

We consider a coded OFDM system with Q subcarriers, signaling through a frequency-selective fading channel in the presence of frequency offset. The system model, which has taken into account the frequency offset, is illustrated in Fig. 10.2. Each signal frame contains the information to be transmitted in one OFDM slot. The information bits of each signal frame are first encoded by a channel encoder; the encoded bits are then interleaved. After interleaving, the code bits $\{b_m\}$ are mapped into MPSK symbols $\{c_k\}$. Finally, the differentially encoded MPSK symbols $\{X_k\}$ are transmitted at the Q OFDM subcarriers. Note that the receiver processes each OFDM word independently, and hence in the remainder of this section, we drop the word index i in the signal model (10.12).

Since the receiver design problem addressed here is blind in nature, differential encoding is employed to resolve the phase ambiguity. For each signal frame, a block of MPSK symbols $\{c_1, \ldots, c_{Q-1}\}$ is input to the differential encoder and the output MPSK symbols $\{X_0, \ldots, X_{Q-1}\}$, with

$$X_k = \begin{cases} 1, & k = 0, \\ c_k X_{k-1}, & k = 1, \ldots, Q-1, \end{cases} \tag{10.13}$$

are then transmitted from the Q OFDM subcarriers. As will be seen in the following section, the decoding of the differentially encoded MPSK symbols is carried out implicitly in the Bayesian demodulator.

As the first step in the receiver, the code bits $\boldsymbol{b} \triangleq \{b_m\}$ are demodulated (cf. Fig. 10.2). The optimal demodulator computes the *a posteriori* probabilities of the code bits as

$$P\Big(b_m = +1 \mid \boldsymbol{Y}\Big) \propto \sum_{\boldsymbol{b}: b_m = +1} \int p(\boldsymbol{Y} \mid \boldsymbol{b}, \boldsymbol{h}, \epsilon) P(\boldsymbol{b}) p(\boldsymbol{h}) p(\epsilon)\, d\boldsymbol{h}\, d\epsilon, \quad \forall m, \tag{10.14}$$

where $p(\boldsymbol{Y}|\boldsymbol{b},\boldsymbol{h},\epsilon)$ is a complex Gaussian density function [cf. (10.12)]. The computation above is prohibitive and we therefore resort to the Markov chain Monte Carlo (MCMC) techniques introduced in Chapter 8 to calculate $P(b_m = +1|\boldsymbol{Y})$ in (10.14).

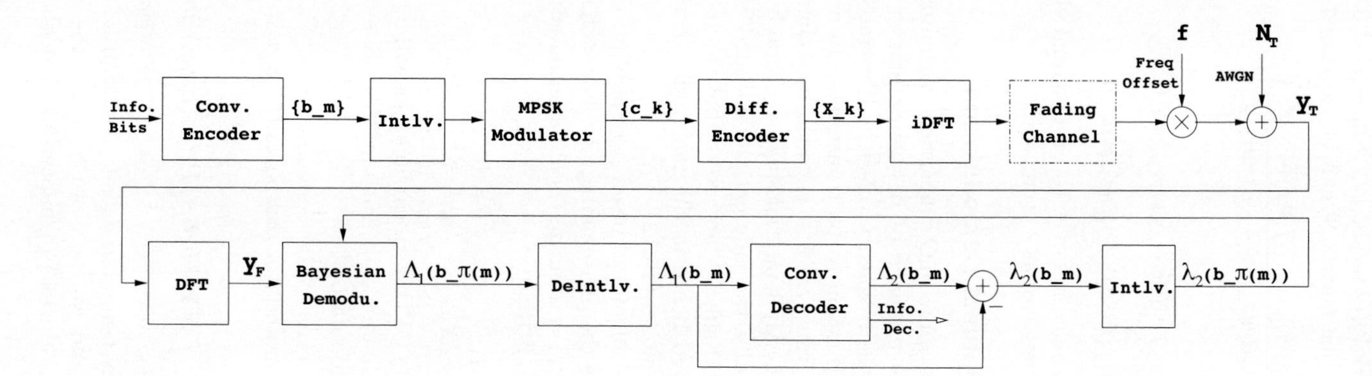

Figure 10.2. Block diagram of a coded OFDM system, including the transmitter, effect of frequency offset, frequency-selective fading channel, and receiver.

10.3.2 Bayesian MCMC Demodulator

In this section we focus on the design of the MCMC demodulator for OFDM systems in the presence of frequency offset and frequency-selective fading. The receiver algorithm is summarized as follows.

Algorithm 10.1: [MCMC-based OFDM blind demodulator in the presence of frequency offset and frequency-selective fading] *Given the initial samples* $\left\{\boldsymbol{X}^{(0)}, \boldsymbol{h}^{(0)}, \epsilon^{(0)}\right\}$ *drawn from their prior distributions, proceed as follows. For* $n = 1, 2, \ldots$:

- *Draw a sample of* $\boldsymbol{h}^{(n)}$ *from* $p\left(\boldsymbol{h} | \boldsymbol{Y}, \boldsymbol{X}^{(n-1)}, \epsilon^{(n-1)}\right)$.
- *For* $k = 0, 1, \ldots, Q-1$, *draw a sample of* $X_k^{(n)}$ *from* $P\left(X_k \,|\, \boldsymbol{Y}, \boldsymbol{h}^{(n)}, \epsilon^{(n-1)}, X_0^{(n)}, \ldots, X_{k-1}^{(n)}, X_{k+1}^{(n-1)}, \ldots, X_{Q-1}^{(n-1)}\right)$.
- *Draw a sample of* $\epsilon^{(n)}$ *from* $p\left(\epsilon \,|\, \boldsymbol{Y}, \boldsymbol{h}^{(n)}, \boldsymbol{X}^{(n)}\right)$.

We next elaborate on each step of the MCMC blind demodulator above.

Prior Distributions

The prior distributions of $\{\boldsymbol{X}, \boldsymbol{h}, \epsilon\}$ are assigned as follows.

1. The data sequence $\boldsymbol{X} = \{X_k\}$, which is differentially encoded from $\boldsymbol{c} = \{c_k\}$, forms a Markov chain. Its prior distribution can be expressed as

$$P(\boldsymbol{X}) = P(X_0) \prod_{k=1}^{Q-1} P(X_k \,|\, X_{k-1}) = P(X_0) \prod_{k=1}^{Q-1} P(c_k = X_k X_{k-1}^*). \tag{10.15}$$

In (10.15), $P(c_k = X_k X_{k-1}^*)$ can be computed from the extrinsic information fed from the channel decoder; and we set $P(X_0) = 1/|\Omega|$ to count for the phase ambiguity in X_0, where Ω represents the constellation of MPSK symbols.

2. For the unknown frequency-selective fading channel response $\boldsymbol{h}$, a complex Gaussian prior distribution is assumed:

$$p(\boldsymbol{h}) \sim \mathcal{N}_c(\boldsymbol{h}_0, \boldsymbol{\Sigma}_0). \tag{10.16}$$

We set $\boldsymbol{h}_0 = \boldsymbol{0}$ and $\boldsymbol{\Sigma}_0 = \alpha \boldsymbol{I}_L$, where α usually takes a large value corresponding to a noninformative prior of $\boldsymbol{h}$.

3. For the unknown frequency offset ϵ, a uniform prior distribution is assumed:

$$p(\epsilon) \sim \text{uniform}[-\varpi, \varpi], \tag{10.17}$$

where ϖ denotes the prior range of ϵ, which is a real number that satisfies $\varpi > \epsilon$ and $\varpi \in (0, 0.5)$.

Conditional Posterior Distributions

The following conditional posterior distributions are required in the MCMC blind demodulation algorithm.

1. The conditional posterior distribution of the channel response $\boldsymbol{h}$ given $\boldsymbol{Y}$, $\boldsymbol{X}$, and ϵ, as derived in the Appendix, is given by

$$p(\boldsymbol{h} \mid \boldsymbol{Y}, \boldsymbol{X}, \epsilon) \sim \mathcal{N}_c(\boldsymbol{h}_*, \boldsymbol{\Sigma}_*), \tag{10.18}$$

with

$$\boldsymbol{\Sigma}_*^{-1} \triangleq \frac{Q}{\sigma^2}\boldsymbol{I}_L + \boldsymbol{\Sigma}_0^{-1} \approx \frac{Q}{\sigma^2}\boldsymbol{I}_L, \tag{10.19}$$

$$\begin{aligned} \boldsymbol{h}_* &\triangleq \boldsymbol{\Sigma}_* \left[\frac{1}{Q\sigma^2}\boldsymbol{W}^H\boldsymbol{X}^H\boldsymbol{W}\boldsymbol{F}_\epsilon^H\boldsymbol{W}^H\boldsymbol{Y} + \boldsymbol{\Sigma}_0^{-1}\boldsymbol{h}_0\right] \\ &\approx \frac{1}{Q^2}\boldsymbol{W}^H\boldsymbol{X}^H\boldsymbol{W}\boldsymbol{F}_\epsilon^H\boldsymbol{W}^H\boldsymbol{Y}, \end{aligned} \tag{10.20}$$

where the approximations in (10.19) and (10.20) follow from the fact that α in (10.16) is large and hence $\boldsymbol{\Sigma}_0^{-1}$ can be neglected. Moreover, it is seen that due to the orthogonality property of OFDM modulation, no matrix inversion is involved in generating the Monte Carlo samples of $\boldsymbol{h}$; therefore, the computational complexity is low.

2. The conditional posterior distribution of the data symbol X_k is obtained by conditioning on $\boldsymbol{Y}$, $\boldsymbol{h}$, ϵ, and the samples of other data symbols $\boldsymbol{X}_{[-k]} \triangleq \left\{X_0, \ldots, X_{k-1}, X_{k+1}, \ldots, X_{Q-1}\right\}$. As shown in the Appendix, the conditional posterior distribution of X_k is given by

$$\begin{aligned} &P\Big(X_k = a_j \mid \boldsymbol{Y}, \boldsymbol{h}, \epsilon, \boldsymbol{X}_{[-k]}\Big) \\ &\quad\propto \exp\left[-\frac{2}{\sigma^2}\Re\left(\tilde{Y}_k^* a_j H_k\right)\right] P\left(c_k = a_j X_{k-1}^*\right) P\left(c_{k+1} = a_j^* X_{k+1}\right) \\ &\quad\approx \exp\left[-\frac{2}{\sigma^2}\Re\left(\tilde{Y}_k^* a_j H_k\right)\right] P\left(c_k = a_j X_{k-1}^*\right), \qquad j = 1, \ldots, |\Omega|, \end{aligned} \tag{10.21}$$

where $a_j \in \Omega$ and $\tilde{Y}_k$ is the kth element of $\tilde{\boldsymbol{Y}} \triangleq \boldsymbol{W}\boldsymbol{F}_\epsilon\boldsymbol{Y}$. The term $P(c_k = a_j X_{k-1}^*)$ in (10.21) is the *a priori* probability of the MPSK symbol c_k, which is delivered by the channel decoder. Through this term, the channel coding constraint embedded in $\{c_k\}$ is exploited in the demodulator. Note that in the final step of (10.21), the term $P\left(c_{k+1} = a_j^* X_{k+1}\right)$ is dropped. This is because any random samples of the data sequence $\boldsymbol{X}$ must satisfy the differential coding constraint [cf. Eq. (10.15)]; since the conditioned data sequence $\left\{X_0^{(n)}, \ldots, X_{k-1}^{(n)}, X_k, X_{k+1}^{(n-1)}, \ldots, X_{Q-1}^{(n-1)}\right\}$ in (10.21) may not satisfy this constraint, it is not a valid sample of the data sequence $\boldsymbol{X}$. To avoid this problem, we propose to compute the conditional posterior probability of X_k by conditioning only on those data samples drawn in the current Gibbs iteration as $\left\{X_0^{(n)}, \dots, X_{k-1}^{(n)}\right\}$, and correspondingly to drop the term related to the sample of the previous Gibbs iteration [i.e., $P\left(c_{k+1} = a_j^* X_{k+1}\right)$]. Our simulation results confirm that by neglecting this term, the Bayesian blind turbo receiver can yield much better performance through the turbo iterations.

3. The conditional posterior distribution of ϵ can be expressed as

$$p\Big(\epsilon \,|\, \boldsymbol{Y}, \boldsymbol{h}, \boldsymbol{X}\Big) \propto p\Big(\boldsymbol{Y} \,|\, \boldsymbol{h}, \boldsymbol{X}, \boldsymbol{F}_\epsilon\Big) \propto \exp\left\{-\frac{1}{\sigma^2}\left\|\boldsymbol{Y} - \frac{1}{Q}\boldsymbol{W}\boldsymbol{F}_\epsilon \boldsymbol{W}^H \boldsymbol{X}\boldsymbol{W}\boldsymbol{h}\right\|^2\right\}. \tag{10.22}$$

Due to the nonlinear signal model of ϵ in (10.12), the conditional posterior distribution of ϵ in (10.22) is not a commonly used distribution (e.g., Gaussian, chi-square, etc.); hence generally there does not exist an efficient way to draw the random samples of ϵ directly from such a distribution function as we did above for $\boldsymbol{h}$ and $\boldsymbol{X}$. As an important component in the Bayesian demodulator, we next discuss three methods for drawing samples of the frequency offset ϵ.

Sampling the Frequency Offset

We consider three methods for drawing samples of the frequency offset ϵ. The first two methods are within the Bayesian MCMC framework: the Metropolis–Hastings algorithm and the Gibbs sampler with local linearization. The third method simply ignores the frequency offset (i.e., it sets $\epsilon^{(n)} = 0$, $\forall n$).

Method I: Metropolis–Hastings Algorithm The Metropolis–Hastings algorithm was introduced briefly in Section 9.3.1. For the problem considered here, the target distribution is $p(\epsilon \,|\, \boldsymbol{Y}, \boldsymbol{h}, \boldsymbol{X})$; and the transition function is chosen as $T(\epsilon, \epsilon') = 1$. Note that such a transition function is by no means the optimal choice, but it has been widely adopted in practice due to its simplicity [424]. Following the Metropolis–Hastings algorithm, and by using the prior distribution (10.17) as well as the posterior distribution (10.22), the procedure for drawing random samples of the frequency offset is as follows:

Algorithm 10.2: [Metropolis–Hasting sampling of frequency offset]

- *Draw a sample* $\epsilon' \sim \text{uniform}[-\varpi, \varpi]$.
- *Compute the Metropolis ratio:*

$$r(\epsilon, \epsilon') = \frac{p(\epsilon' \,|\, \boldsymbol{Y}, \boldsymbol{h}, \boldsymbol{X})}{p(\epsilon \,|\, \boldsymbol{Y}, \boldsymbol{h}, \boldsymbol{X})} = \exp\left\{-\frac{2}{Q\sigma^2}\Re\left[\boldsymbol{Y}^H \boldsymbol{W}(\boldsymbol{F}_\epsilon - \boldsymbol{F}_{\epsilon'})\boldsymbol{W}^H \boldsymbol{X}\boldsymbol{W}\boldsymbol{h}\right]\right\}. \tag{10.23}$$

- *Generate $u \sim$ uniform$(0,1)$, and let*

$$\epsilon^{(n+1)} = \begin{cases} \epsilon' & \text{if } u \le r(\epsilon, \epsilon'), \\ \epsilon^{(n)} & \text{otherwise.} \end{cases} \tag{10.24}$$

Method II: Gibbs Sampler with Local Linearization As seen in (10.22), due to the nonlinear signal model of ϵ, we cannot directly apply the Gibbs sampler to draw samples of ϵ. However, following an idea that appeared in [104], we can first linearize the received signal model at its mode (the maximum value) with respect to ϵ; then based on the linearized signal model, we obtain a locally linear conditional posteriori distribution of ϵ. Specifically, applying an inverse DFT on both sides of (10.12), we obtain

$$\boldsymbol{y} = \frac{1}{Q}\boldsymbol{F}_\epsilon \boldsymbol{W}^H \boldsymbol{X} \boldsymbol{W} \boldsymbol{h} + \boldsymbol{v}. \tag{10.25}$$

A Taylor series expansion is applied around the mode of the signal model in (10.25) and the linearized signal model is given by

$$\begin{aligned} \boldsymbol{y} &\cong \frac{1}{Q}\boldsymbol{F}_{\hat{\epsilon}} \boldsymbol{W}^H \boldsymbol{X} \boldsymbol{W} \boldsymbol{h} + \left.\frac{\partial \boldsymbol{y}}{\partial \epsilon}\right|_{\epsilon=\hat{\epsilon}} (\epsilon - \hat{\epsilon}) + \boldsymbol{v} \\ &= \boldsymbol{F}_{\hat{\epsilon}} \overline{\boldsymbol{h}} + \overline{\boldsymbol{F}}_{\hat{\epsilon}} \overline{\boldsymbol{h}} (\epsilon - \hat{\epsilon}) + \boldsymbol{v}, \end{aligned} \tag{10.26}$$

with

$$\begin{aligned} \overline{\boldsymbol{F}}_{\hat{\epsilon}} &\triangleq \operatorname{diag}\left\{0, \frac{\jmath 2\pi}{Q} \exp[\jmath 2\pi \hat{\epsilon}/Q], \ldots, \frac{\jmath 2\pi (Q-1)}{Q} \exp[\jmath 2\pi \hat{\epsilon}(Q-1)/Q]\right\}, \\ \overline{\boldsymbol{h}} &\triangleq \frac{1}{Q} \boldsymbol{W}^H \boldsymbol{X} \boldsymbol{W} \boldsymbol{h}. \end{aligned}$$

Based on the locally linearized signal model (10.26), as derived in the Appendix, the conditional posterior distribution of ϵ is Gaussian, given by

$$p\Big(\epsilon \,|\, \boldsymbol{Y}, \boldsymbol{h}, \boldsymbol{X}\Big) \sim \mathcal{N}\Big(\mu_\epsilon, \sigma_\epsilon^2\Big), \tag{10.27}$$

with

$$\begin{aligned} \mu_\epsilon &= \hat{\epsilon} + \Re\Big[\big(\boldsymbol{y} - \boldsymbol{F}_{\hat{\epsilon}} \overline{\boldsymbol{h}}\big)^H \overline{\boldsymbol{F}}_{\hat{\epsilon}} \overline{\boldsymbol{h}}\Big] \Big(\overline{\boldsymbol{h}}^H \overline{\boldsymbol{F}}_{\hat{\epsilon}}^H \overline{\boldsymbol{F}}_{\hat{\epsilon}} \overline{\boldsymbol{h}}\Big)^{-1}, \\ \sigma_\epsilon^2 &= \frac{\sigma^2}{2Q} \Big(\overline{\boldsymbol{h}}^H \overline{\boldsymbol{F}}_{\hat{\epsilon}}^H \overline{\boldsymbol{F}}_{\hat{\epsilon}} \overline{\boldsymbol{h}}\Big)^{-1}. \end{aligned}$$

Note that μ_ϵ and σ_ϵ^2 are real. Using (10.22) and (10.27), the procedure for frequency offset sampling is as follows:

Algorithm 10.3: [Gibbs sampling of frequency offset]

- *Search for the mode $\hat{\epsilon}$ of $p(\epsilon \,|\, \boldsymbol{Y}, \boldsymbol{h}, \boldsymbol{X})$ from $\epsilon \in [-\varpi, \varpi]$.*
- *Draw a sample of ϵ from its linearized conditional posterior distribution: $p(\epsilon \,|\, \boldsymbol{Y}, \boldsymbol{h}, \boldsymbol{X}) \sim \mathcal{N}(\mu_\epsilon, \sigma_\epsilon^2)$.*

Method III: Null Sampling In this method, we simply ignore the frequency offset by setting $\epsilon^{(n)} = 0$, $\forall n$. Although bearing no theoretical optimality, this method can be used to test the robustness of the Bayesian demodulator against a modeling mismatch. That is, when $\epsilon = 0$ is assumed, we essentially ask the Gibbs sampler to fit an OFDM model with no frequency offset into an actual OFDM system with a certain frequency offset.

We consider a special case and see how the blind receiver behaves in the presence of a modeling mismatch. Let us revisit the system model in (10.12) and define

$$\widetilde{\boldsymbol{h}} \triangleq \frac{1}{Q} \boldsymbol{F} \boldsymbol{W}^H \boldsymbol{X} \boldsymbol{W} \boldsymbol{h}. \tag{10.28}$$

The vector $\widetilde{\boldsymbol{h}}$ can be seen as the time response of a "composite" channel with zero frequency offset, which incorporates the effect of the frequency offset, the data symbols, and the original frequency-selective fading channel. It is easy to see that when $\boldsymbol{X} = \boldsymbol{I}$, $\widetilde{\boldsymbol{h}} = \boldsymbol{F}\boldsymbol{h}$ preserves the same statistics as $\boldsymbol{h}$. In other words, no matter how large the frequency offset is, the blind receiver derived based on the statistics of $\boldsymbol{h}$ can also adapt to that zero-frequency-offset composite channel with response $\widetilde{\boldsymbol{h}}$, by setting $\epsilon^{(n)} = 0$, $\forall n$, in the Gibbs sampler.

When $\boldsymbol{X} \neq \boldsymbol{I}$, the statistics of $\widetilde{\boldsymbol{h}}$ are usually different from those of $\boldsymbol{h}$ and the receiver will suffer from a performance loss. A quantitative evaluation of such a performance loss due to the modeling mismatch is given by computer simulations later in this section.

Computing Data Posterior Probabilities

After collecting J random samples of the differentially encoded MPSK symbols $\{\boldsymbol{X}^{(n)}\}$, the *a posteriori* probabilities of the MPSK symbols $\boldsymbol{c}$ and the code bits $\boldsymbol{b}$ are computed as follows. First, the *a posteriori* probabilities of $\boldsymbol{c}$ are computed from J random samples of $\{\boldsymbol{X}^{(n)}\}$ as

$$P\Big(c_k = a_j \mid \boldsymbol{Y}\Big) \cong \frac{1}{J} \sum_{n=J_0+1}^{J_0+J} \delta_k^{(n)}, \qquad k = 1, \ldots, Q-1, \quad j = 1, \ldots, |\Omega|, \tag{10.29}$$

where $\delta_k^{(n)}$ is an indicator such that

$$\delta_k^{(n)} = \begin{cases} 1 & \text{if } X_k^{(n)} X_{k-1}^{(n)*} = a_j, \\ 0 & \text{otherwise.} \end{cases} \tag{10.30}$$

Furthermore, the *a posteriori* probabilities of the code bits $\boldsymbol{b}$ can easily be obtained from the *a posteriori* probabilities of the MPSK symbols $\boldsymbol{c}$ in (10.29). Assume that the symbol c_k is modulated from the code bits $\{b_k^1, \ldots, b_k^S\}$, with $S = \log_2 |\Omega|$. Then the *a posteriori* probability of code bit b_k^i is given by

$$P\Big(b_k^i = +1 \,|\, \boldsymbol{Y}\Big) \propto \sum_{a \in \mathcal{A}_+^i} P(c_k = a \,|\, \boldsymbol{Y}), \quad i = 1, \ldots, S, \tag{10.31}$$

where $\mathcal{A}_+^i \subset \Omega$ denotes all symbols in Ω that are modulated from the code bits with the ith bit as "+1".

So far, the Bayesian demodulator performs the soft demodulation of the code bits $\boldsymbol{b}$. To further exploit the channel coding constraints embedded in the code bits $\boldsymbol{b}$, we adopt a turbo receiver structure discussed in Chapter 6, which iteratively exchanges information about $\boldsymbol{b}$ between the Bayesian demodulator developed above and the channel decoder, to achieve successively improved receiver performance.

Bayesian Blind Turbo Receiver

The Bayesian blind turbo receiver consists of two stages—the Bayesian demodulator as developed in Section 10.2, followed by a MAP channel decoder—and these two stages are separated by an interleaver and a deinterleaver. The *a posteriori* log-likelihood ratios (LLRs) of the channel code bits $\boldsymbol{b}$ are iteratively exchanged between these two stages, to successively refine the receiver performance.

The Bayesian demodulator takes as input the interleaved *a priori* LLRs of code bits $\{\boldsymbol{\lambda}_2[b_{\pi(m)}]\}$ from the MAP channel decoder in the previous turbo iteration as well as the received signals $\boldsymbol{Y}$, where $\pi(\cdot)$ denotes the interleaving function. It computes as output the *a posteriori* LLRs of the code bits:

$$\boldsymbol{\Lambda}_1[b_{\pi(m)}] \triangleq \log \frac{P(b_{\pi(m)} = +1 \mid \boldsymbol{Y})}{P(b_{\pi(m)} = -1 \mid \boldsymbol{Y})}\,, \tag{10.32}$$

where $P(b_{\pi(m)} = +1 \mid \boldsymbol{Y})$ is the *a posteriori* probability of code bit $b_{\pi(m)}$ as computed in (10.31). Note that according to the original form of the turbo principle, the *a priori* LLRs $\{\boldsymbol{\lambda}_2[b_{\pi(m)}]\}$ are supposed to be deducted from the *a priori* LLRs $\{\boldsymbol{\Lambda}_1[b_{\pi(m)}]\}$ to yield extrinsic information. However, the posterior distribution delivered by the Gibbs-sampler-based Bayesian demodulator takes on a quantized value as $P(c_k = a_j \mid \boldsymbol{Y}) \in \{0, 1/J, 2/J, \ldots, 1\}$, due to the finite samples of $\boldsymbol{X}$ [cf. Eq. (10.29)]. Hence, to enhance numerical stability and the iterative receiver performance, for this particular receiver structure, we feed the complete posterior information $\{\boldsymbol{\Lambda}_1[b_{\pi(m)}]\}$ to the MAP channel decoder.

The MAP channel decoder employs the standard MAP decoding algorithm to compute the *a posteriori* LLRs of code bits:

$$\boldsymbol{\Lambda}_2[b_m] \triangleq \log \frac{P(b_m = +1 \mid \{\boldsymbol{\Lambda}_1(b_m)\})}{P(b_m = -1 \mid \{\boldsymbol{\Lambda}_1(b_m)\})} = \boldsymbol{\lambda}_2(b_m) + \boldsymbol{\Lambda}_1(b_m)\,. \tag{10.33}$$

In (10.33), the extrinsic information $\{\boldsymbol{\lambda}_2(b_m)\}$ is obtained by subtracting the prior information $\{\boldsymbol{\Lambda}_1(b_m)\}$ from the posterior information $\{\boldsymbol{\Lambda}_2(b_m)\}$. After being interleaved, this extrinsic information is fed back to the Bayesian demodulator as *a priori* information for the next iteration; and thus we complete one turbo iteration. At the last turbo iteration, the LLRs and hard decisions of information bits are computed.

In addition to exchanging extrinsic information with the Bayesian demodulator, the channel decoder can also help the Gibbs-sampler-based Bayesian demodulator to assess its convergence, as discussed in Section 8.4.4. The number of bit corrections made by the MAP channel decoder is monitored, where the number of corrections is counted by comparing the signs of the code-bit LLRs at the input and output of the MAP channel decoder. If this number exceeds some predetermined threshold, we decide that the convergence of the Gibbs-sampler-based Bayesian demodulator is not achieved. In this case, the Bayesian demodulator will again be applied to the same received data.

Simulation Examples

In this section we provide computer simulation results to illustrate the performance of the MCMC blind turbo receiver for coded OFDM systems with frequency offset and frequency-selective fading. In simulations the available bandwidth is 1 MHz and $Q = 256$ subcarriers are used for OFDM modulation. These correspond to a subcarrier symbol rate of 3.9 KHz and an OFDM word duration of $1/\Delta_f = 256\,\mu$s. For each OFDM word, a guard interval of $40\,\mu$s is added to combat the effect of intersymbol interference. Simulations are carried out through an equal-power four-tap frequency-selective fading channel, where the delays of these taps are $\tau_l = l/Q\Delta_f, l = 0, \ldots, 3$. The modulator employs a QPSK constellation. A four-state, rate-$\frac{1}{2}$ convolutional code with generator (5,7) in octal notation is chosen as the channel code. For each OFDM slot, $J_0 + J = 100$ samples are drawn by the MCMC demodulator, with the first $J_0 = 50$ samples discarded. After completing 100 MCMC iterations, the convergence is tested by counting the number of corrections made by the decoders. In a few cases, when convergence is not reached, it gets restarted for another round of 100 MCMC iterations. In the following, the performance is demonstrated in terms of the bit-error rate (BER) and OFDM word-error rate (WER) versus the signal-to-noise ratio (SNR), defined as $\text{SNR} \triangleq \|\boldsymbol{h}\|^2/\sigma^2$.

Performance Degradation Due to Frequency Offset First, we demonstrate the performance degradation due to the frequency offset in the coded OFDM system simulated here. The channel state information (CSI) (i.e., the channel response $\boldsymbol{h}$) is assumed to be known at the receiver. In Fig. 10.3 the performance of the turbo receiver under perfect CSI is shown for the coded OFDM system with different frequency offsets, $\epsilon = \{0.00, 0.09, 0.18\}$. These results confirm the analysis in previous works (i.e., [374]) that the receiver performance degrades rapidly as the frequency offset increases. Hence, appropriate measures should be taken to combat the frequency offset.

Performance of Various Frequency Offset Sampling Methods In Figs. 10.4–10.11 the impact of different methods for drawing samples of the frequency offset on the overall Bayesian blind turbo receiver performance is examined. For methods I and II, the impact of the prior range ($\varpi = \{0.1, 0.5\}$) on the receiver performance is examined as well. In particular, in method II the mode of the conditional posterior distribution of the frequency offset is found by a global search with a step size of $\delta\epsilon = 0.05$. Hence, the computational complexity of method II is still acceptable; and simulations show that only marginal performance improvement is obtained by using smaller step sizes.

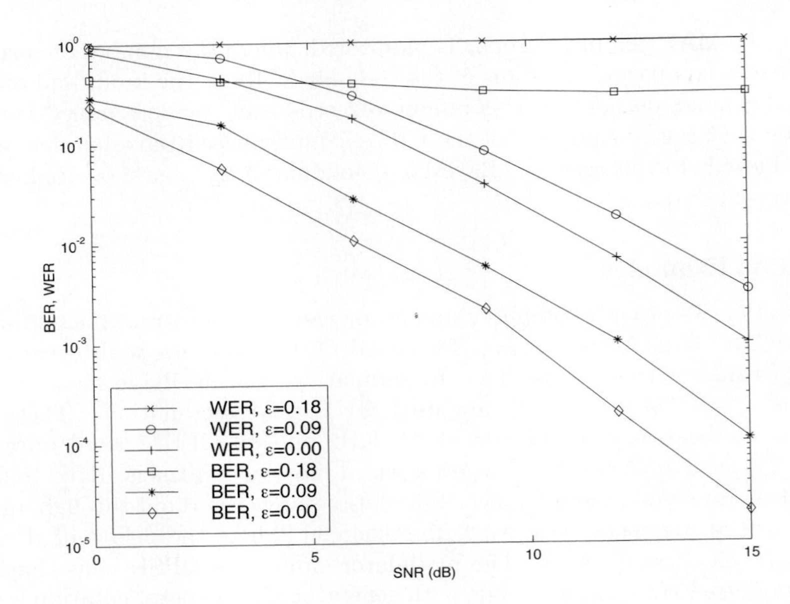

Figure 10.3. BER and WER in a coded OFDM system with frequency offset $\epsilon = \{0.00, 0.09, 0.18\}$. Perfect CSI (i.e., the channel response $\boldsymbol{h}$) is assumed at the receiver.

The performance of the receiver, when it employs method I or method II, is demonstrated through the BER and WER after the first, third, and fifth turbo iterations; whereas when method III is employed, for clarity, only the performance after the fifth turbo iteration is shown, denoted "WER,Iter#5,3rd" and "BER,Iter#5,3rd" in the figures. Moreover, for comparison, the ideal CSI performance of the system with zero frequency offset, which is approximately the best performance we can achieve in this system, is shown again in these figures and denoted "WER,CSI,ϵ=0" and "BER,CSI,ϵ=0."

Example 1: Small Frequency Offset In Figs. 10.4–10.7 we present the performance of the MCMC blind turbo receiver in a coded OFDM system with frequency offset $\epsilon = 0.09$. From the simulation results, several conclusions can be drawn. First, the receiver performance is improved significantly through turbo iterations and can approach the performance under perfect CSI at the BER of 10^{-3} and WER of 2×10^{-2}, as can be seen from the performance curves when the receiver employs either method I or method II for frequency offset sampling. Second, the receiver performance is not sensitive to the prior range ϖ of method I or method II, which can be seen by comparing Figs. 10.4 and 10.5 or Figs. 10.6 and 10.7. This is a favorable fact for receiver design, as we can always set the largest prior range of ϵ as $[-0.5, 0.5]$. Third, the robustness of the receiver is tested by employing method III in the Bayesian demodulator. Compared with the performance of the receiver that explicitly samples the frequency offset (i.e., by method I or method II), we do not

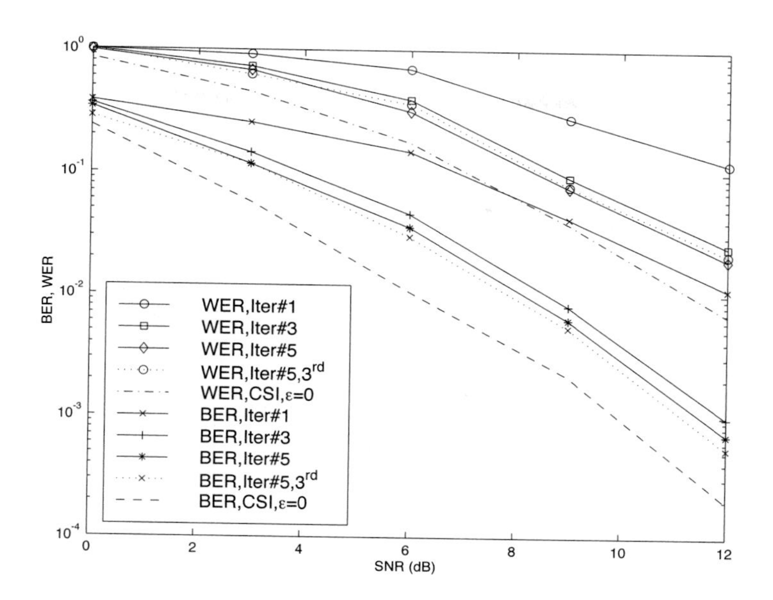

Figure 10.4. BER and WER in a coded OFDM system with frequency offset $\epsilon = 0.09$. The Metropolis–Hastings algorithm is employed to generate Monte Carlo sampling of the frequency offset, where the prior range $\varpi = 0.5$.

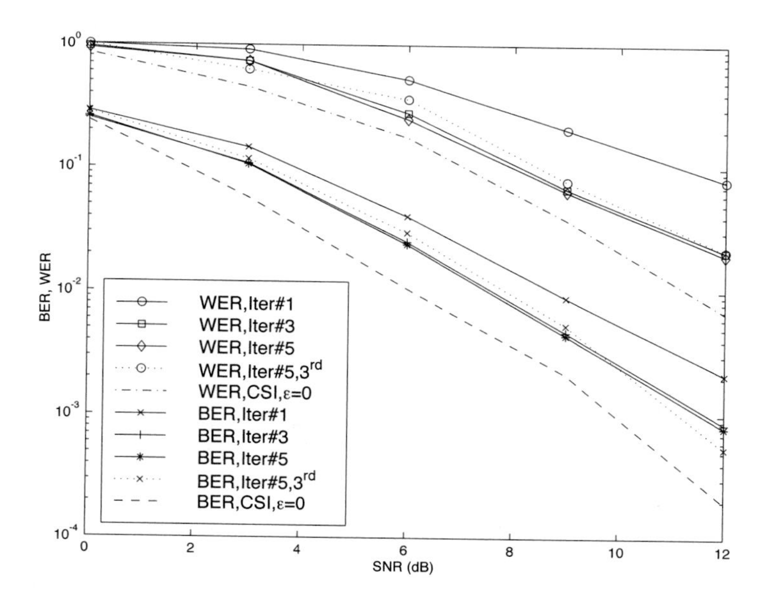

Figure 10.5. BER and WER in a coded OFDM system with frequency offset $\epsilon = 0.09$. The Metropolis–Hastings algorithm is employed to generate Monte Carlo sampling of the frequency offset, where the prior range $\varpi = 0.1$.

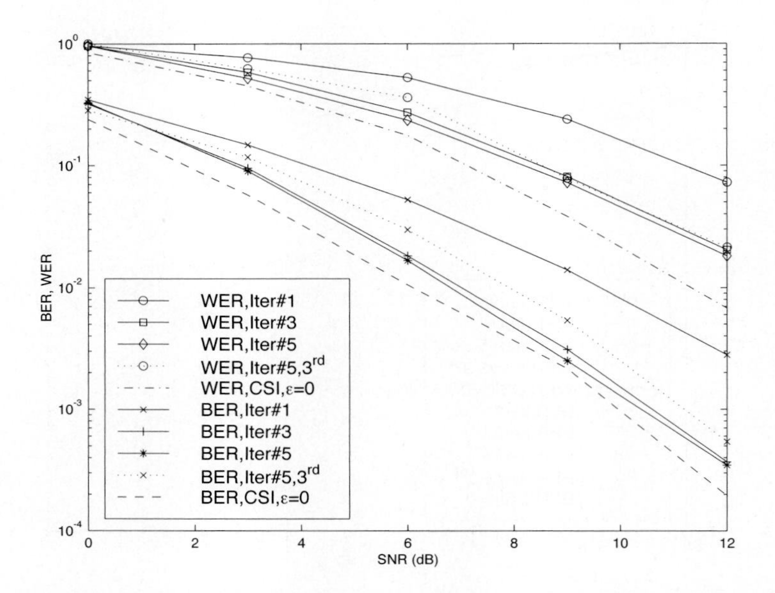

Figure 10.6. BER and WER in a coded OFDM system with frequency offset $\epsilon = 0.09$. The Gibbs sampler with local linearization is employed to generate Monte Carlo sampling of the frequency offset, where the prior range $\varpi = 0.5$ and the search step size $\delta\epsilon = 0.05$.

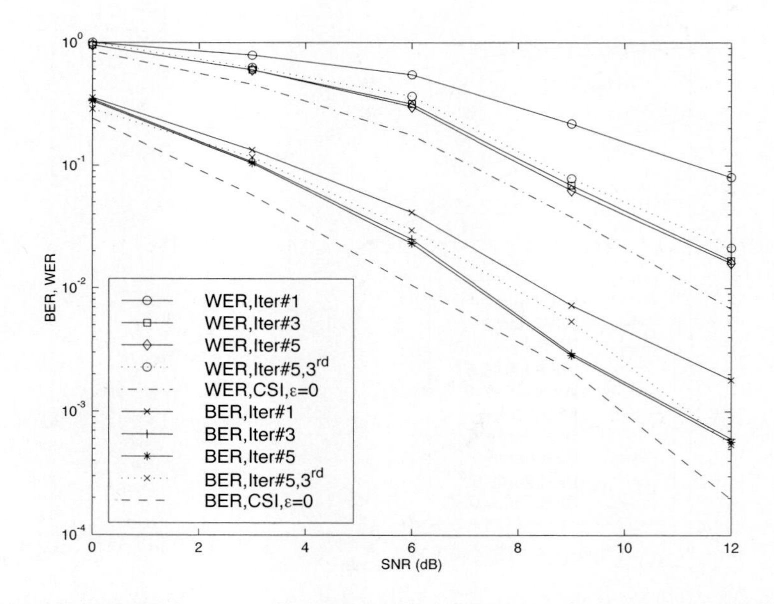

Figure 10.7. BER and WER in a coded OFDM system with frequency offset $\epsilon = 0.09$. The Gibbs sampler with local linearization is employed to generate Monte Carlo sampling of the frequency offset, where the prior range $\varpi = 0.1$ and the search step size $\delta\epsilon = 0.05$.

see any performance loss when the receiver samples null frequency offset (i.e., by method III). In other words, the MCMC blind turbo receiver is robust against a modeling mismatch.

Example 2: Large Frequency Offset In Figs. 10.8–10.11, in the same form as in Example 1, we present the receiver performance in a coded OFDM system with a larger frequency offset, $\epsilon = 0.18$. Recall that from Fig. 10.3, when no proper methods are employed to combat the frequency offset, the receiver assuming perfect CSI completely fails in the presence of such a large frequency offset.

From Figs. 10.8–10.11, in addition to the conclusions drawn in Example 1, some new observations are made. When method I or method II is employed, it is seen from the figures that the receiver still performs very well and can approach the performance under perfect CSI after three to five turbo iterations, in the presence of such a large frequency offset. However, when method III is employed, the receiver performance mildly degrades by about 1.5 dB, due to the modeling mismatch, compared to the performance in a smaller frequency offset system (i.e., the performance shown in Figs. 10.4–10.7).

Finally, based on all the simulation results shown above, we compare the efficiency of all three methods for frequency offset sampling in terms of both the performance and complexity. Method III has the lowest complexity by ignoring the frequency offset, but it leads to a noticeable receiver performance degradation as the frequency offset becomes large. Method I has lower complexity than method II, and it can yield almost the same receiver performance as method II. Moreover, since no approximation has been made in deriving method I, its convergence is guaranteed by the theory of MCMC. Therefore, we advocate the use of method I, the Metropolis–Hastings algorithm, to draw samples of the frequency offset in the MCMC blind turbo receiver.

10.4 Pilot-Symbol-Aided Turbo Receiver for Space-Time Block-Coded OFDM Systems

In Section 10.3 we have treated the problem of blind receiver design based on MCMC methods for OFDM systems. In this section we discuss the design of a pilot-symbol-aided receiver for OFDM communication systems operating over frequency-selective fading channels. Here we treat a general scenario where multiple transmit and receive antennas are employed. It is assumed that space-time block coding (STBC) (cf. Sections 5.5.2 and 6.7) is adopted at the transmitter end. The techniques in this section were developed in [292].

10.4.1 System Descriptions

We consider an STBC-OFDM system with Q subcarriers, N transmitter antennas, and M receiver antennas, signaling through frequency- and time-selective fading channels. As illustrated in Fig. 10.12, the information bits are first modulated by an MPSK modulator; then the modulated MPSK symbols are encoded by an

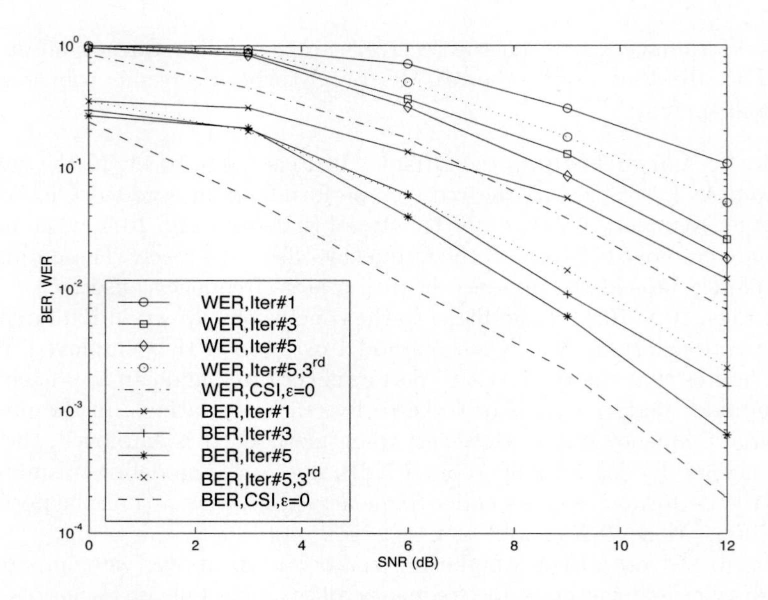

Figure 10.8. BER and WER in a coded OFDM system with frequency offset $\epsilon = 0.18$. The Metropolis–Hastings algorithm is employed to generate Monte Carlo sampling of the frequency offset, where the prior range $\varpi = 0.5$.

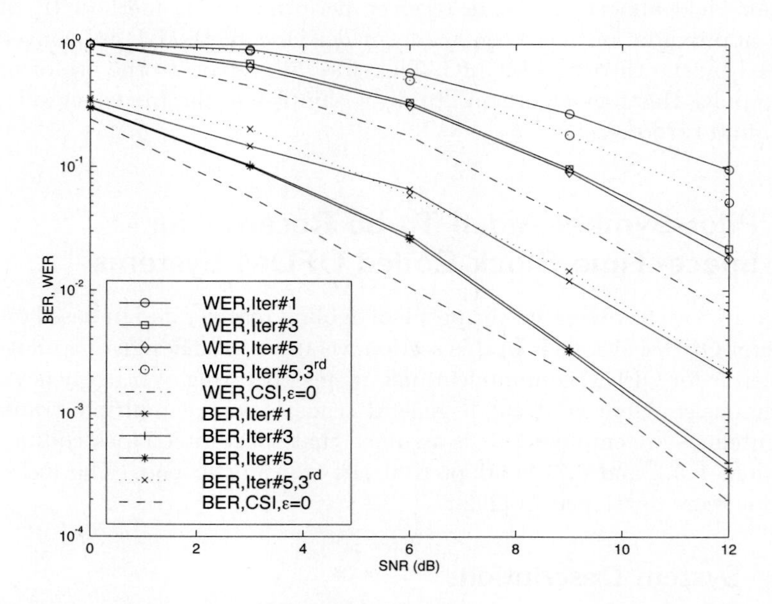

Figure 10.9. BER and WER in a coded OFDM system with frequency offset $\epsilon = 0.18$. The Metropolis–Hastings algorithm is employed to generate Monte Carlo sampling of the frequency offset, where the prior range $\varpi = 0.2$.

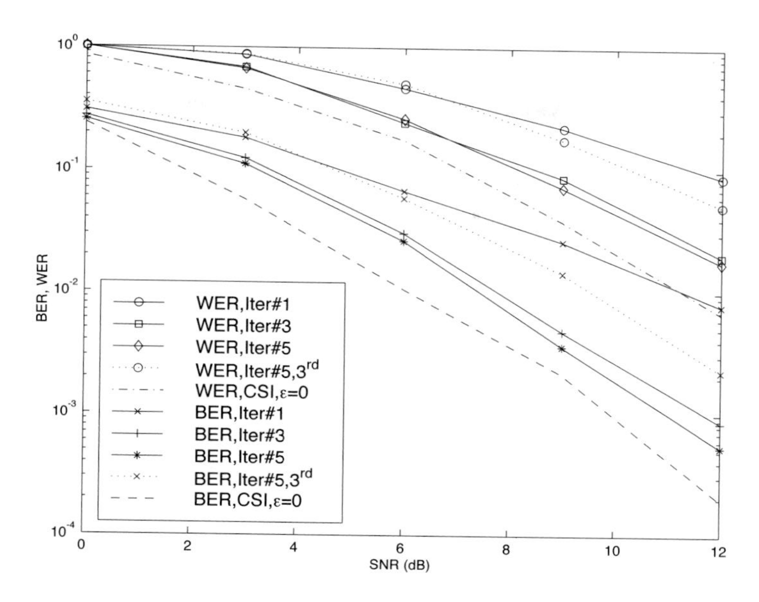

Figure 10.10. BER and WER in a coded OFDM system with frequency offset $\epsilon = 0.18$. The Gibbs sampler with local linearization is employed to generate Monte Carlo sampling of the frequency offset, where the prior range $\varpi = 0.5$ and the search step size $\delta\epsilon = 0.05$.

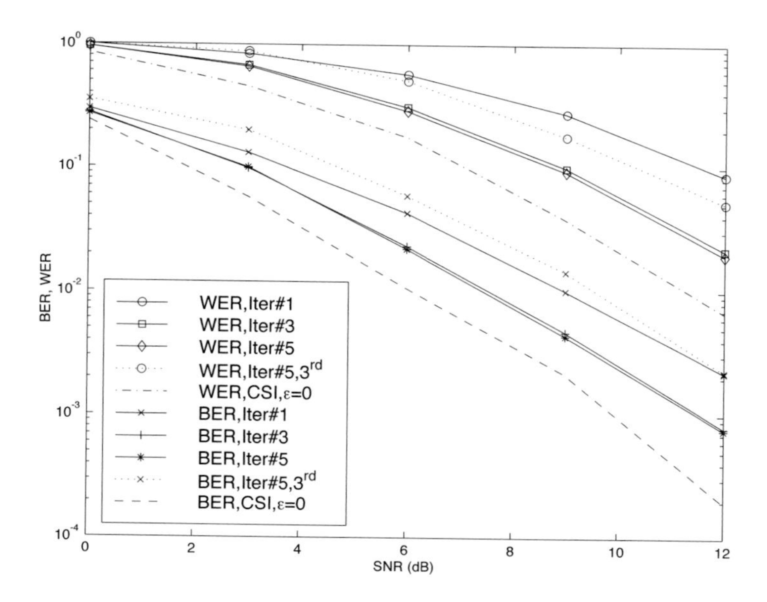

Figure 10.11. BER and WER in a coded OFDM system with frequency offset $\epsilon = 0.18$. The Gibbs sampler with local linearization is employed to generate Monte Carlo sampling of the frequency offset, where the prior range $\varpi = 0.2$ and the search step size $\delta\epsilon = 0.05$.

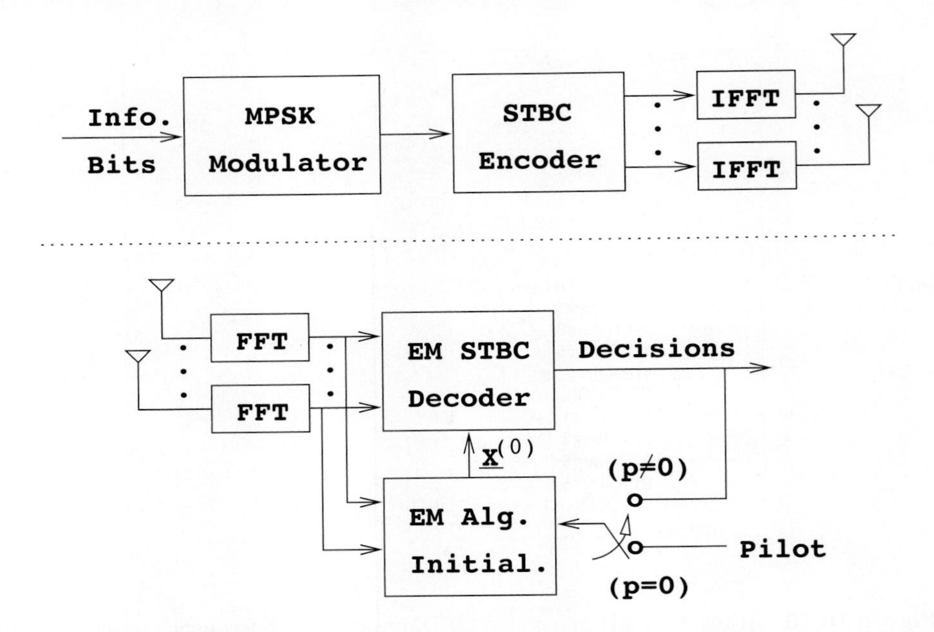

Figure 10.12. Transmitter and receiver structure for an STBC-OFDM system.

STBC encoder. Each STBC code word consists of (PN) STBC symbols, which are transmitted from N transmitter antennas and across P consecutive OFDM slots at a particular OFDM subcarrier. The STBC code words at different OFDM subcarriers are independently encoded, therefore, during P OFDM slots, altogether Q STBC code words [or (QPN) STBC code symbols] are transmitted. It is assumed that the fading processes remain static during each OFDM word (one time slot), but it varies from one OFDM word to another, and that the fading processes associated with different transmitter-receiver antenna pairs are uncorrelated.

At the receiver, the signals are received from M receiver antennas. After matched filtering and symbol-rate sampling, the discrete Fourier transform (DFT) is then applied to the received discrete-time signals, to obtain

$$\begin{aligned} \boldsymbol{y}_i[p] &= \sum_{j=1}^{N} \boldsymbol{X}_j[p]\boldsymbol{H}_{i,j}[p] + \boldsymbol{z}_i[p] \\ &= \boldsymbol{X}[p]\boldsymbol{H}_i[p] + \boldsymbol{z}_i[p] \,, \qquad i = 1, \ldots, M, \;\; p = 1, \ldots, P, \end{aligned} \tag{10.34}$$

with

$$\boldsymbol{X}[p] \triangleq \Big[\boldsymbol{X}_1[p], \ldots, \boldsymbol{X}_N[p]\Big]_{Q\times(NQ)},$$

$$\boldsymbol{X}_j[p] \triangleq \mathrm{diag}\Big\{x_j[p,0], \ldots, x_j[p,Q-1]\Big\}_{Q\times Q},$$

$$\boldsymbol{H}_i[p] \triangleq \left[\boldsymbol{H}_{i,1}^H[p,0], \ldots, \boldsymbol{H}_{i,N}^H[p,Q-1]\right]_{(NQ)\times 1}^H,$$

$$\boldsymbol{H}_{i,j}[p] \triangleq \left[H_{i,j}[p,0], \ldots, H_{i,j}[p,Q-1]\right]_{Q\times 1}^T,$$

where $\boldsymbol{H}_i[p]$ is the NQ-vector containing the complex channel frequency responses between the ith receiver antenna and all N transmitter antennas at the pth OFDM slot, which is explained below; $x_j[p,k]$ is the STBC symbol transmitted from the jth transmitter antenna at the kth subcarrier and at the pth OFDM slot; $\boldsymbol{y}_i[p]$ is the Q-vector of received signals from the ith receiver antenna and at the pth time slot; $\boldsymbol{z}_i[p]$ is the ambient noise, which is circularly symmetric complex Gaussian with covariance matrix $\sigma_z^2 \boldsymbol{I}$. Here we restrict our attention to MPSK signal constellations (i.e., $x_j[p,k] \in \Omega \triangleq \{e^{\jmath 0}, e^{\jmath(2\pi/|\Omega|)}, \ldots, e^{\jmath(2\pi/|\Omega|)(|\Omega|-1)}\}$).

Consider the channel response between the jth transmitter antenna and the ith receiver antenna. Following [396], the time-domain channel impulse response can be modeled as a tapped delay line, similar to (10.2), given by

$$g_{i,j}(\tau;t) = \sum_{\ell=0}^{L-1} \alpha_{i,j}(\ell;t)\delta\left(\tau - \frac{\ell}{Q\Delta_f}\right), \tag{10.35}$$

where $\delta(\cdot)$ is the Kronecker delta function; $L \triangleq \lceil \tau_m \Delta_f + 1 \rceil$ denotes the maximum number of resolvable taps, with τ_m being the maximum multipath spread and Δ_f being the tone spacing of the OFDM system; $\alpha_{i,j}(\ell;t)$ is the complex amplitude of the ℓth tap, whose delay is ℓ/Δ_f. For OFDM systems with proper cyclic extension and sample timing, with tolerable leakage, the channel frequency response between the jth transmitter antenna and the ith receiver antenna at the pth time slot and at the kth subcarrier can be expressed as [506]

$$H_{i,j}[p,k] \triangleq H_{i,j}(pT, k\Delta_f) = \sum_{\ell=0}^{L-1} h_{i,j}[\ell;p]e^{-\jmath 2\pi k l/Q} = \boldsymbol{w}_f^H(k)\boldsymbol{h}_{i,j}(p), \tag{10.36}$$

where $h_{i,j}[\ell;p] \triangleq \alpha_{i,j}(\ell;pT)$, T is the duration of one OFDM slot; $\boldsymbol{h}_{i,j}(p) \triangleq [\alpha_{i,j}(0;pT), \ \ldots, \ \alpha_{i,j}(L-1;pT)]^T$ is the L-vector containing the time responses of all the taps; and $\boldsymbol{w}_f(k) \triangleq \left[e^{-\jmath 0}, e^{-\jmath 2\pi k/Q}, \ldots, \ e^{-\jmath 2\pi k(L-1)/Q}\right]^H$ contains the corresponding DFT coefficients.

Using (10.36), the signal model in (10.34) can be further expressed as

$$\boldsymbol{y}_i[p] = \boldsymbol{X}[p]\boldsymbol{W}\boldsymbol{h}_i[p] + \boldsymbol{z}_i[p], \qquad i = 1,\ldots,M, \ \ p = 1,\ldots,P, \tag{10.37}$$

with

$$\boldsymbol{W} \triangleq \text{diag}\left\{\boldsymbol{W}_f, \ldots, \boldsymbol{W}_f\right\}_{(NQ)\times(NL)},$$

$$\boldsymbol{W}_f \triangleq [\boldsymbol{w}_f(0), \boldsymbol{w}_f(1), \ldots, \boldsymbol{w}_f(Q-1)]_{Q\times L}^H,$$

$$\boldsymbol{h}_i[p] \triangleq \left[\boldsymbol{h}_{i,1}^H(p), \ldots, \boldsymbol{h}_{i,N}^H(p)\right]_{(NL)\times 1}^H.$$

As discussed in Section 6.7, an STBC is defined by a $P \times N$ code matrix $\mathcal{G}$, where N denotes the number of transmitter antennas or the spatial transmitter diversity order, and P denotes the number of time slots for transmitting an STBC code word, i.e., the *temporal* transmitter diversity order. Each row of $\mathcal{G}$ is a permuted and transformed (i.e., negated and/or conjugated) form of the N-dimensional vector of complex data symbols $\boldsymbol{x}$. As a simple example, we consider a 2×2 STBC (i.e., $P = 2, N = 2$). Its code matrix $\mathcal{G}_1$ is defined by

$$\mathcal{G}_1 = \begin{bmatrix} x_1 & x_2 \\ -x_2^* & x_1^* \end{bmatrix}. \tag{10.38}$$

The input to this STBC is the data vector $\boldsymbol{x} = [x_1, x_2]^T$. During the first time slot, the two symbols in the first row $[x_1, x_2]$ of $\mathcal{G}_1$ are transmitted simultaneously from the two transmitter antennas; during the second time slot, the symbols in the second row $[-x_2^*, x_1^*]$ of $\mathcal{G}_1$ are transmitted.

In an STBC-OFDM system, we apply the STBC encoder above to data symbols transmitted at different subcarriers independently. For example, by using the STBC defined by $\mathcal{G}_1$, at the kth subcarrier, during the first OFDM slot, two data symbols $\Big[x_1[1,k], x_2[1,k]\Big]$ are transmitted simultaneously from two transmitter antennas; during the next OFDM slot, symbols $\Big[x_1[2,k], x_2[2,k]\Big] \equiv \Big[-x_2^*[1,k], x_1^*[1,k]\Big]$ are transmitted.

Simplified System Model

From the description above, it is seen that decoding in an STBC-OFDM system involves the received signals over P consecutive OFDM slots. To simplify the problem, we assume that the channel time responses $\boldsymbol{h}_i[p], p = 1, \ldots, P$, remain constant over the duration of one STBC code word (i.e., P consecutive OFDM slots). As will be seen, such an assumption simplifies the receiver design significantly. Using the channel model in (10.37) and considering the coding constraints of the STBC, the received signals over the duration of each STBC code word is obtained as

$$\underline{\boldsymbol{y}}_i = \underline{\boldsymbol{X}}\,\boldsymbol{W}\boldsymbol{h}_i + \underline{\boldsymbol{z}}_i, \qquad i = 1, \ldots, M, \tag{10.39}$$

with

$$\begin{aligned}
\underline{\boldsymbol{y}}_i &= \Big[\boldsymbol{y}_i^H[1], \ldots, \boldsymbol{y}_i^H[P]\Big]^H_{(PQ)\times 1}, \\
\underline{\boldsymbol{X}} &\triangleq \Big[\boldsymbol{X}^H[1], \ldots, \boldsymbol{X}^H[P]\Big]^H_{(PQ)\times(NQ)}, \\
\underline{\boldsymbol{z}}_i &\triangleq \Big[\boldsymbol{z}_i^H[1], \ldots, \boldsymbol{z}_i^H[P]\Big]^H_{(PQ)\times 1}, \\
\boldsymbol{h}_i &\triangleq \boldsymbol{h}_i[1] = \boldsymbol{h}_i[2] = \cdots = \boldsymbol{h}_i[P].
\end{aligned}$$

According to the definitions of $\boldsymbol{W}$ in (10.37) and $\underline{\boldsymbol{X}}$ in (10.39), we have

$$\boldsymbol{W}^H \underline{\boldsymbol{X}}^H \underline{\boldsymbol{X}}\, \boldsymbol{W} = \boldsymbol{W}^H \left(\sum_{p=1}^{P} \boldsymbol{X}^H[p] \boldsymbol{X}[p] \right) \boldsymbol{W}, \tag{10.40}$$

where $\sum_{p=1}^{P} \boldsymbol{X}^H[p]\boldsymbol{X}[p]$ is an $NQ \times NQ$ matrix, which is composed of N^2 submatrices of dimension $Q \times Q$ of the form

$$\begin{aligned} \sum_{p=1}^{P} \boldsymbol{X}_j^H[p]\boldsymbol{X}_{j'}[p] &= \text{diag}\left\{ \left[\sum_{p=1}^{P} x_j^*[p,1] x_{j'}[p,1] \right], \ldots, \left[\sum_{p=1}^{P} x_j^*[p,Q] x_{j'}[p,Q] \right] \right\} \\ &= \begin{cases} P \cdot \boldsymbol{I}, & j = j', \\ \underline{0}, & j \neq j', \end{cases} \qquad j = 1, \ldots, N, \;\; j' = 1, \ldots, N, \end{aligned} \tag{10.41}$$

where the last equality follows from the constant modulus property of the symbols $\{x_j[p,k]\}_{j,p,k}$, and the orthogonality property of the STBC [475] as well as that of the OFDM modulation. Hence, (10.40) reduces to

$$\boldsymbol{W}^H \underline{\boldsymbol{X}}^H \underline{\boldsymbol{X}}\, \boldsymbol{W} \;=\; (PQ)\, \boldsymbol{I}. \tag{10.42}$$

As will be seen in the following sections, (10.42) is the key equation in designing low-complexity iterative receivers for STBC-OFDM systems.

10.4.2 ML Receiver Based on the EM Algorithm

We next consider ML receiver design for STBC-OFDM systems. With ideal channel state information (CSI), the optimal decoder has been derived in [476]. However, in practice, CSI must be estimated by the receiver. We develop an EM-based ML receiver for STBC-OFDM systems operating in unknown fast-fading channels. As in a typical data communication scenario, communication is carried out in a bursty manner. A data burst is illustrated in Fig. 10.13. It consists of $(Pq+1)$ OFDM words, with the first OFDM word $(p=0)$ containing known pilot symbols and the remaining (Pq) OFDM words spanning the duration of q STBC code words.

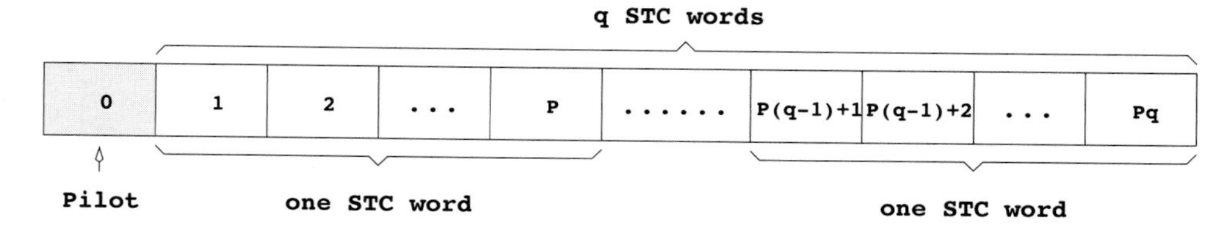

Figure 10.13. OFDM time slots allocation in data burst transmission. A data burst consists of $Pq+1$ OFDM words, with the first OFDM word containing known pilot symbols and the remaining Pq OFDM words spanning the duration of q STBC code words.

EM-Based STBC-OFDM Receiver

Without CSI, the maximum likelihood detection problem is written as

$$\begin{aligned}\hat{\underline{\boldsymbol{X}}} &= \arg\max_{\underline{\boldsymbol{X}}} \sum_{i=1}^{M} \log p(\underline{\boldsymbol{y}}_i|\underline{\boldsymbol{X}}) \\ &= \arg\max_{\underline{\boldsymbol{X}}} \sum_{i=1}^{M} \log \int p(\underline{\boldsymbol{y}}_i|\underline{\boldsymbol{X}}, \boldsymbol{h}_i)p(\boldsymbol{h}_i)\,\mathrm{d}\boldsymbol{h}_i,\end{aligned} \tag{10.43}$$

where the summation of log probabilities from all M receiver antennas follows from the assumption that the ambient noise processes at different receiver antennas are independent. It is seen in (10.43) that the direct computation of optimal ML decisions involves multidimensional integration over the unknown random vector $\boldsymbol{h}_i$, and hence is of prohibitive complexity. Next, we turn to the EM algorithm to solve (10.43).

As discussed in Section 9.3.1, the basic idea of the EM algorithm is to solve problem (10.43) iteratively according to the following two steps:

1. *E-step:* Compute $Q(\underline{\boldsymbol{X}}|\underline{\boldsymbol{X}}^{(\kappa)}) = E\left\{\left[\sum_{i=1}^{M} \log p(\underline{\boldsymbol{y}}_i|\underline{\boldsymbol{X}}, \boldsymbol{h}_i)\right] \middle| \underline{\boldsymbol{y}}_i, \underline{\boldsymbol{X}}^{(\kappa)}\right\}$, (10.44)
2. *M-step:* Solve $\underline{\boldsymbol{X}}^{(\kappa+1)} = \arg\max_{\underline{\boldsymbol{X}}} Q(\underline{\boldsymbol{X}}|\underline{\boldsymbol{X}}^{(\kappa)})$, (10.45)

where $\underline{\boldsymbol{X}}^{(\kappa)}$ contains hard decisions on the data symbols at the κth EM iteration and $\underline{\boldsymbol{X}}^{(\kappa)}$ satisfies the STBC coding constraints. It is known that the likelihood function $\sum_{i=1}^{M} \log p(\underline{\boldsymbol{y}}_i|\underline{\boldsymbol{X}}^{(\kappa)})$ is nondecreasing as a function of κ, and under regularity conditions the EM algorithm converges to a local stationary point [315].

In the E-step, the expectation is taken with respect to the "hidden" channel response $\boldsymbol{h}_i$ conditioned on $\underline{\boldsymbol{y}}_i$ and $\underline{\boldsymbol{X}}^{(\kappa)}$. It is easily seen that conditioned on $\underline{\boldsymbol{y}}_i$ and $\underline{\boldsymbol{X}}^{(\kappa)}$, $\boldsymbol{h}_i$ has a complex Gaussian distribution. Using (10.39) and (10.42), this distribution can be expressed as

$$\boldsymbol{h}_i|(\underline{\boldsymbol{y}}_i, \underline{\boldsymbol{X}}^{(\kappa)}) \sim \mathcal{N}_c(\hat{\boldsymbol{h}}_i, \hat{\boldsymbol{\Sigma}}_{h_i}), \qquad i = 1, \ldots, M, \tag{10.46}$$

with

$$\begin{aligned}\hat{\boldsymbol{h}}_i &\triangleq (\boldsymbol{W}^H \underline{\boldsymbol{X}}^{(\kappa)H} \boldsymbol{\Sigma}_z^{-1} \underline{\boldsymbol{X}}^{(\kappa)} \boldsymbol{W} + \boldsymbol{\Sigma}_{h_i}^{\dagger})^{-1} \boldsymbol{W}^H \underline{\boldsymbol{X}}^{(\kappa)H} \boldsymbol{\Sigma}_z^{-1} \underline{\boldsymbol{y}}_i \\ &= (\boldsymbol{W}^H \underline{\boldsymbol{X}}^{(\kappa)H} \underline{\boldsymbol{X}}^{(\kappa)} \boldsymbol{W} + \sigma_z^2 \boldsymbol{\Sigma}_{h_i}^{\dagger})^{-1} \boldsymbol{W}^H \underline{\boldsymbol{X}}^{(\kappa)H} \underline{\boldsymbol{y}}_i \\ &= [(PQ)\boldsymbol{I} + \sigma_z^2 \boldsymbol{\Sigma}_{h_i}^{\dagger}]^{-1} \boldsymbol{W}^H \underline{\boldsymbol{X}}^{(\kappa)H} \underline{\boldsymbol{y}}_i,\end{aligned} \tag{10.47}$$

$$\begin{aligned}\hat{\boldsymbol{\Sigma}}_{h_i} &\triangleq \boldsymbol{\Sigma}_{h_i} - (\boldsymbol{W}^H \underline{\boldsymbol{X}}^{(\kappa)H} \boldsymbol{\Sigma}_z^{-1} \underline{\boldsymbol{X}}^{(\kappa)} \boldsymbol{W} + \boldsymbol{\Sigma}_{h_i}^{\dagger})^{-1} \boldsymbol{W}^H \underline{\boldsymbol{X}}^{(\kappa)H} \boldsymbol{\Sigma}_z^{-1} \underline{\boldsymbol{X}}^{(\kappa)} \boldsymbol{W} \boldsymbol{\Sigma}_{h_i} \\ &= \boldsymbol{\Sigma}_{h_i} - (\boldsymbol{W}^H \underline{\boldsymbol{X}}^{(\kappa)H} \underline{\boldsymbol{X}}^{(\kappa)} \boldsymbol{W} + \sigma_z^2 \boldsymbol{\Sigma}_{h_i}^{\dagger})^{-1} \boldsymbol{W}^H \underline{\boldsymbol{X}}^{(\kappa)H} \underline{\boldsymbol{X}}^{(\kappa)} \boldsymbol{W} \boldsymbol{\Sigma}_{h_i} \\ &= \boldsymbol{\Sigma}_{h_i} - (PQ)[(PQ)\boldsymbol{I} + \sigma_z^2 \boldsymbol{\Sigma}_{h_i}^{\dagger}]^{-1} \boldsymbol{\Sigma}_{h_i},\end{aligned} \tag{10.48}$$

where $\boldsymbol{\Sigma}_z$ and $\boldsymbol{\Sigma}_{h_i}$ denote, respectively, the covariance matrix of the ambient white Gaussian noise $\underline{\boldsymbol{z}}_i$ and channel responses $\boldsymbol{h}_i$. According to the assumptions made above, both of these are diagonal matrices, given as

$$\boldsymbol{\Sigma}_z \triangleq E\{\underline{\boldsymbol{z}}_i \underline{\boldsymbol{z}}_i^H\} = \sigma_z^2 \boldsymbol{I}, \tag{10.49}$$

$$\boldsymbol{\Sigma}_{h_i} \triangleq E\{\boldsymbol{h}_i \boldsymbol{h}_i^H\} = \text{diag}\left\{\beta_{1,0}^2, \ldots, \beta_{1,L-1}^2, \ldots\ldots, \beta_{N,0}^2, \ldots, \beta_{N,L-1}^2\right\}, \tag{10.50}$$

where $\beta_{j,l}^2$ is the average power of the lth tap associated with the jth transmitter antenna; $\beta_{j,l}^2 = 0$ if the channel response at this tap is zero. Assuming that $\boldsymbol{\Sigma}_{h_i}$ is known (e.g., measured with the aid of pilot symbols), then

$$\boldsymbol{\Sigma}_{h_i}^{\dagger} \triangleq \text{diag}\left\{\gamma_{1,0}, \ldots, \gamma_{1,L-1}, \ldots\ldots, \gamma_{N,0}, \ldots, \gamma_{N,L-1}\right\}, \tag{10.51}$$

with

$$\gamma_{j,l} \triangleq \begin{cases} 1/\beta_{j,l}^2, & \beta_{j,l}^2 \neq 0, \\ 0, & \beta_{j,l}^2 = 0, \end{cases} \quad l = 0, \ldots, L-1 \;\; j = 1, \ldots, N. \tag{10.52}$$

It is seen that in the E-step, due to the orthogonality properties of the STBC and the OFDM modulation (10.42), no matrix inversion is involved. Therefore, the computational complexity of the E-step is reduced from $\mathcal{O}(MN^3L^3)$ to $\mathcal{O}(MNL)$ and the computation is also numerically more stable. Using (10.39) and (10.46), $Q(\underline{\boldsymbol{X}}|\underline{\boldsymbol{X}}^{(\kappa)})$ can be computed via

$$\begin{aligned}
&Q(\underline{\boldsymbol{X}}|\underline{\boldsymbol{X}}^{(\kappa)}) \\
&= -\frac{1}{\sigma_z^2}\sum_{i=1}^{M}\left\{E_{\boldsymbol{h}_i|(\underline{\boldsymbol{y}}_i, \underline{\boldsymbol{X}}^{(\kappa)})}\left[\|\underline{\boldsymbol{y}}_i - \underline{\boldsymbol{X}}\,\boldsymbol{W}\,\boldsymbol{h}_i\|^2\right]\right\} + \text{const.} \\
&= -\frac{1}{\sigma_z^2}\sum_{i=1}^{M}\left\{E_{\boldsymbol{h}_i|(\underline{\boldsymbol{y}}_i, \underline{\boldsymbol{X}}^{(\kappa)})}\left[\|(\underline{\boldsymbol{y}}_i - \underline{\boldsymbol{X}}\,\boldsymbol{W}\,\hat{\boldsymbol{h}}_i) + (\underline{\boldsymbol{X}}\,\boldsymbol{W}\,\hat{\boldsymbol{h}}_i - \underline{\boldsymbol{X}}\,\boldsymbol{W}\,\boldsymbol{h}_i)\|^2\right]\right\} + \text{const.} \\
&= -\frac{1}{\sigma_z^2}\sum_{i=1}^{M}\left\{\|\underline{\boldsymbol{y}}_i - \underline{\boldsymbol{X}}\,\boldsymbol{W}\,\hat{\boldsymbol{h}}_i\|^2 + \text{tr}\{\underline{\boldsymbol{X}}\,\boldsymbol{W}\,\hat{\boldsymbol{\Sigma}}_{h_i}\,\boldsymbol{W}^H\underline{\boldsymbol{X}}^H\}\right\} + \text{const.} \\
&= -\frac{1}{\sigma_z^2}\sum_{i=1}^{M}\sum_{p=1}^{P}\left\{\|\boldsymbol{y}_i[p] - \boldsymbol{X}[p]\,\boldsymbol{W}\,\hat{\boldsymbol{h}}_i\|^2 + \text{tr}\{\boldsymbol{X}[p]\,\boldsymbol{W}\,\hat{\boldsymbol{\Sigma}}_{h_i}\,\boldsymbol{W}^H\boldsymbol{X}^H[p]\}\right\} + \text{const.} \\
&= -\frac{1}{\sigma_z^2}\sum_{i=1}^{M}\sum_{p=1}^{P}\sum_{k=0}^{Q-1} \\
&\quad \underbrace{\left\{\left[y_i[p,k] - \boldsymbol{x}^H[p,k]\boldsymbol{W}_f'(k)\hat{\boldsymbol{h}}_i\right]^2 + \left[\boldsymbol{x}^H[p,k]\hat{\boldsymbol{\Sigma}}_{h_i}(k)\boldsymbol{x}[p,k]\right]\right\}}_{q_i^{(\kappa)}(\boldsymbol{x}[p,k])} + \text{const.},
\end{aligned} \tag{10.53}$$

with

$$\boldsymbol{x}[p,k] \triangleq [x_1[p,k],\ldots,x_N[p,k]]^H_{N\times 1},$$
$$\boldsymbol{W}'_f(k) \triangleq \mathrm{diag}\left[\boldsymbol{w}_f^H(k),\ldots,\boldsymbol{w}_f^H(k)\right]_{N\times(NL)},$$
$$\left[\hat{\boldsymbol{\Sigma}}_{h_i}(k)\right]_{(i',j')} \triangleq \left[\boldsymbol{W}\hat{\boldsymbol{\Sigma}}_{h_i}\boldsymbol{W}^H\right]_{\left((i'-1)Q+k+1,(j'-1)Q+k+1\right)}$$
$$i' = 1,\ldots,N, \quad j' = 1,\ldots,N,$$

where $\mathrm{tr}(\boldsymbol{A})$ denotes the trace of the matrix $\boldsymbol{A}$, and $[\boldsymbol{A}]_{(i',j')}$ denotes the (i',j')th element of the matrix $\boldsymbol{A}$.

Next, based on (10.53), the M-step in (10.45) proceeds as follows:

$$\begin{aligned}\underline{\boldsymbol{X}}^{(\kappa+1)} &= \arg\max_{\underline{\boldsymbol{X}}} Q(\underline{\boldsymbol{X}}|\underline{\boldsymbol{X}}^{(\kappa)})\\ &= \sum_{k=0}^{Q-1} \arg \min_{\{\boldsymbol{x}[p,k]\}_p} \left[\sum_{i=1}^{M}\sum_{p=1}^{P} q_i^{(\kappa)}(\boldsymbol{x}[p,k])\right].\end{aligned} \tag{10.54}$$

It is seen from (10.54) that the M-step can be decoupled into Q independent minimization problems, each of which can be solved by enumerating over all possible $\boldsymbol{x}[p,k] \in \Omega^N$, $\forall p$; and the coding constraints of STBC are taken into account when solving the M-step [i.e., $\boldsymbol{x}[p,k]$, $\forall p$, are different permutations and/or transformations of $\boldsymbol{x}[1,k]$ as defined in (10.38)]. Hence the complexity of the M-step is $\mathcal{O}(Q|\Omega|^N)$; and the total complexity of the EM algorithm is $[\mathcal{O}(MNL)+\mathcal{O}(Q|\Omega|^N)]$ per EM iteration.

Initialization of the EM Algorithm

The performance of the EM algorithm (and hence the overall receiver) is closely related to the quality of the initial value of $\underline{\boldsymbol{X}}^{(0)}$ [cf. Eq. (10.44)]. The initial estimate of $\underline{\boldsymbol{X}}^{(0)}$ is computed based on the method proposed in [260, 263] by the following steps. First, a linear estimator is used to estimate the channel with the aid of pilot symbols or decision feedback of the data symbols. Second, the resulting channel estimate is refined by a temporal filter to further exploit the time-domain correlation of the channel. Finally, conditioned on the temporally filtered channel estimate, $\underline{\boldsymbol{X}}^{(0)}$ is obtained through ML detection. We next elaborate on the linear channel estimator as well as the temporal filtering.

Least-Squares Channel Estimator In (10.47), by assuming perfect knowledge of $\boldsymbol{\Sigma}_{h_i}$, $\hat{\boldsymbol{h}}_i$ is simply the minimum mean-square-error estimate of the channel response $\boldsymbol{h}_i$. When $\boldsymbol{\Sigma}_{h_i}$ is not known to the receiver, a least-squares estimator can be applied to estimate the channel and to measure $\boldsymbol{\Sigma}_{h_i}$ as well. We next derive the least-squares channel estimator for STBC-OFDM systems. By treating $\boldsymbol{h}_i$ as an unknown vector without any prior information and using (10.39) and (10.42), the least-squares estimate $\hat{\boldsymbol{h}}_i$ can be expressed as

$$\hat{\boldsymbol{h}}_i = \left(\boldsymbol{W}^H \underline{\boldsymbol{X}}^H \underline{\boldsymbol{X}} \boldsymbol{W}\right)^{-1} \boldsymbol{W}^H \underline{\boldsymbol{X}}^H \underline{\boldsymbol{y}}_i = \boldsymbol{Q}^{-1} \boldsymbol{W}^H \left(\sum_{p=1}^{P} \boldsymbol{X}^H[p] \boldsymbol{y}_i[p]\right)$$

$$= \frac{1}{PQ} \boldsymbol{W}^H \left(\sum_{p=1}^{P} \boldsymbol{X}^H[p] \boldsymbol{y}_i[p]\right), \tag{10.55}$$

with

$$\boldsymbol{Q} \triangleq \boldsymbol{W}^H \underline{\boldsymbol{X}}^H \underline{\boldsymbol{X}} \boldsymbol{W} = (PQ)\boldsymbol{I}.$$

It is seen that in (10.55), unlike a typical least-squares estimator, no matrix inversion is involved here. Hence, its complexity is reduced from $\mathcal{O}(N^3 L^3)$ to only $\mathcal{O}(NL)$ and the computation is numerically more stable, which is very attractive in systems using many transmitter antennas (large N) and/or communicating in highly dispersive fading channels (large L).

Finally, the procedure for initializing the EM algorithm is listed in Table 10.1. Here, the ML detection in ($\star$) takes into account the STBC coding constraints of $\underline{\boldsymbol{X}}$. Freq-filter denotes the least-squares estimator, where $\boldsymbol{X}[0]$ represents the pilot symbols and $\underline{\boldsymbol{X}}^{(I)}[m]$, $m = 0, \ldots, q-1$, represent hard decisions of the data symbols $\underline{\boldsymbol{X}}[m]$ which are provided by the EM algorithm after a total of I EM iterations. Temp-filter denotes the temporal filter [260, 263], which is used to further exploit the time-domain correlation of the channel within one OFDM data burst [i.e., $(Pq+1)$ OFDM slots]:

$$\text{Temp-filter}\left\{\hat{\boldsymbol{h}}_i[p-1], \hat{\boldsymbol{h}}_i[p-2], \ldots, \hat{\boldsymbol{h}}_i[p-\iota]\right\} \triangleq \sum_{j=1}^{\iota} a_j \hat{\boldsymbol{h}}_i[p-j],$$

$$i = 1, \ldots, M, \tag{10.56}$$

where $\hat{\boldsymbol{h}}_i[p-j], j = 1, \ldots, \iota$, is computed from ($\star\star$); $\{a_j\}_{j=1}^{\iota}$ denotes the coefficients of an ι-length ($\iota \leq Pq$) temporal filter, which can be precomputed by solving the Wiener equation or from robust design as in [260, 263]. Furthermore, as suggested in [263], after receiving all $(Pq+1)$ OFDM words in a burst, an enhanced temporal filter can be applied as

$$\text{Temp-filter}_p\left\{\hat{\boldsymbol{h}}_i[Pq], \hat{\boldsymbol{h}}_i[Pq-1], \ldots, \hat{\boldsymbol{h}}_i[0]\right\} \triangleq \sum_{j=0}^{Pq} a_{p,j} \hat{\boldsymbol{h}}_i[Pq-j],$$

$$i = 1, \ldots, M, \tag{10.57}$$

where Temp-filter$_p$ computes $\tilde{\boldsymbol{h}}_i[p]$ by temporally filtering the "past" channel estimate $\hat{\boldsymbol{h}}_i[p-\iota], \iota = 1, 2, \ldots$, the "current" channel estimate $\hat{\boldsymbol{h}}_i[p]$, and the "future"

Table 10.1. Procedure for Computing $\underline{\boldsymbol{X}}^{(0)}$ for the EM Algorithm

$$\underline{\boldsymbol{y}}_i[m] \triangleq \left[\boldsymbol{y}_i^H[mP+1], \ldots, \boldsymbol{y}_i^H[mP+P]\right]^H$$

$$\left\{\underline{\boldsymbol{y}}_i[m]\right\}_i \triangleq \left\{\underline{\boldsymbol{y}}_1[m], \ldots, \underline{\boldsymbol{y}}_M[m]\right\}$$

$$\underline{\boldsymbol{X}}[m] \triangleq \left[\boldsymbol{X}^H[mP+1], \ldots, \boldsymbol{X}^H[mP+P]\right]^H$$

---

pilot slot: $\hat{\boldsymbol{h}}_i[0] = \text{Freq-filter}\left\{\boldsymbol{y}_i[0], \boldsymbol{X}[0]\right\}, \qquad i = 1, \ldots, M,$

data slots: for $m = 0, 1, \ldots, q-1$

  for $n = 1, 2, \ldots, P$

$$\tilde{\boldsymbol{h}}_i[mP+n] = \text{Temp-filter}\left\{\hat{\boldsymbol{h}}_i[mP+n-1], \hat{\boldsymbol{h}}_i[mP+n-2], \ldots, \hat{\boldsymbol{h}}[mP+n-\iota]\right\}, \qquad i = 1, \ldots, M,$$

  end

$$\underline{\boldsymbol{X}}^{(0)}[m] = \text{arg max}_{\underline{\boldsymbol{X}}}\left\{\sum_{i=1}^{M}\sum_{n'=1}^{P} \log p\left[\boldsymbol{y}_i[mP+n'] \middle| \underline{\boldsymbol{X}}, \tilde{\boldsymbol{h}}_i[mP+n']\right]\right\}, \qquad (\star)$$

$$\underline{\boldsymbol{X}}^{(I)}[m] = \text{EM}\left\{\left\{\underline{\boldsymbol{y}}_i[m]\right\}_i, \underline{\boldsymbol{X}}^{(0)}[m]\right\}, \qquad \text{[cf. Eq. (10.44)–(10.45)]}$$

  for $n = 1, 2, \ldots, P$

$$\hat{\boldsymbol{h}}_i[mP+n] = \text{Freq-filter}\left\{\underline{\boldsymbol{y}}_i[m], \underline{\boldsymbol{X}}^{(I)}[m]\right\}, \qquad i = 1, \ldots, M, \qquad (\star\star)$$

  end

end

channel estimate $\hat{\boldsymbol{h}}_i[p+\iota], \iota = 1, 2, \ldots$. From the discussions above, it is seen that the computation involved in initializing $\underline{\boldsymbol{X}}^{(0)}$ consists mainly of the ML detection of $\underline{\boldsymbol{X}}^{(0)}$ in ($\star$) and the estimation of $\hat{\boldsymbol{h}}_i$ in ($\star\star$), with a total complexity $[\mathcal{O}(Q|\Omega|^N) + \mathcal{O}(MNL)]$.

10.4.3 Pilot-Symbol-Aided Turbo Receiver

In practice, in order to impose the coding constraints across the various OFDM subcarriers and further improve receiver performance, an outer channel code (e.g., a convolutional code or turbo code) is usually applied in addition to the STBC. As illustrated in Fig. 10.14, the information bits are encoded by an outer-channel-code encoder and then interleaved. The interleaved code bits are modulated by an MPSK modulator. Finally, the modulated MPSK symbols are encoded by an STBC encoder and transmitted from N transmitter antennas across the P consecutive OFDM slots at a particular OFDM subcarrier. During P OFDM slots, altogether Q STBC code words [or (QPN) STBC symbols] are transmitted.

In what follows we discuss a turbo receiver employing the maximum *a posteriori* probability (MAP)-EM STBC decoding algorithm and the MAP outer-channel-code decoding algorithm for this concatenated STBC-OFDM system, as depicted in Fig. 10.14. It consists of a soft MAP-EM STBC decoder and a soft MAP outer-channel-code decoder. The MAP-EM STBC decoder takes as input the fast Fourier transform (FFT) of the received signals from M receiver antennas, and

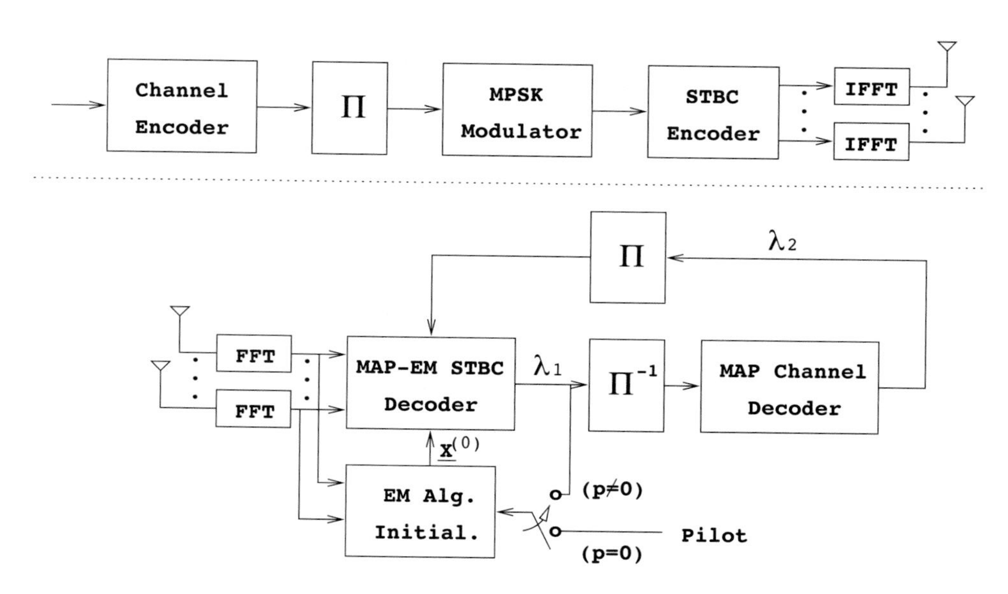

Figure 10.14. Transmitter and receiver structure for an STBC-OFDM system with outer channel coding. Π denotes the interleaver and Π^{-1} denotes the corresponding deinterleaver.

the interleaved extrinsic log-likelihood ratios of the outer-channel-code bits $\{\lambda_2^e\}$ [cf. Eq. (10.62)] (which is fed back by the outer-channel-code decoder). It computes as output the extrinsic *a posteriori* LLRs of the outer-channel-code bits $\{\lambda_1^e\}$ [cf. Eq. (10.62)]. The MAP outer-channel-code decoder takes as input the deinterleaved LLRs of the outer-channel-code bits from the MAP-EM STBC decoder and computes as output the extrinsic LLRs of the outer-channel-code bits as well as the hard decisions of the information bits at the last turbo iteration.

STBC-OFDM Receiver Based on the MAP-EM Algorithm

Without CSI, the MAP detection problem can be written as

$$\hat{\underline{\boldsymbol{X}}} = \arg\max_{\underline{\boldsymbol{X}}} \sum_{i=1}^{M} \log p(\underline{\boldsymbol{X}} | \underline{\boldsymbol{y}}_i). \tag{10.58}$$

The MAP-EM algorithm solves (10.58) iteratively according to the following two steps:

1. *E-step:* Compute $Q(\underline{\boldsymbol{X}}|\underline{\boldsymbol{X}}^{(\kappa)}) = E\left\{\left[\sum_{i=1}^{M} \log p(\underline{\boldsymbol{y}}_i|\underline{\boldsymbol{X}}, \boldsymbol{h}_i)\right] \middle| \underline{\boldsymbol{y}}_i, \underline{\boldsymbol{X}}^{(\kappa)}\right\}$. (10.59)
2. *M-step:* Solve $\underline{\boldsymbol{X}}^{(\kappa+1)} = \arg\max_{\underline{\boldsymbol{X}}} \left[Q(\underline{\boldsymbol{X}}|\underline{\boldsymbol{X}}^{(\kappa)}) + \log P(\underline{\boldsymbol{X}})\right]$. (10.60)

Comparing the MAP-EM algorithm in (10.59)–(10.60) with the maximum likelihood EM algorithm in (10.44)–(10.45), we see that the E-step is exactly the same, but the M-step of the MAP-EM algorithm includes an extra term $P(\underline{\boldsymbol{X}})$, which represents the *a priori* probability of $\underline{\boldsymbol{X}}$, which is fed back by the outer-channel-code decoder from the previous turbo iteration.

Similar to (10.54), the M-step for the MAP-EM can be written as

$$\begin{aligned}
\underline{\boldsymbol{X}}^{(\kappa+1)} &= \arg\max_{\underline{\boldsymbol{X}}} \left[Q(\underline{\boldsymbol{X}}|\underline{\boldsymbol{X}}^{(\kappa)}) + \log P(\underline{\boldsymbol{X}})\right] \\
&= \sum_{k=0}^{K-1} \arg \min_{\{\boldsymbol{x}[p,k]\}_p} \sum_{i=1}^{M} \sum_{p=1}^{P} \left[\frac{1}{\sigma_z^2} q_i^{(\kappa)}(\boldsymbol{x}[p,k]) - \log P(\boldsymbol{x}[p,k])\right] \\
&= \sum_{k=0}^{K-1} \arg \min_{\{\boldsymbol{x}[p,k]\}_p} \left\{\sum_{i=1}^{M} \sum_{p=1}^{P} \left[\frac{1}{\sigma_z^2} q_i^{(\kappa)}(\boldsymbol{x}[p,k])\right] - \log P(\boldsymbol{x}[1,k])\right\},
\end{aligned} \tag{10.61}$$

where the second equality in (10.61) holds by assuming that the outer-channel-code bits are ideally interleaved and hence $\boldsymbol{x}[p,k]$ at different OFDM subcarriers are independent. The last equality in (10.61) follows from the fact that $\boldsymbol{x}[p,k]$, $p = 2, \ldots, P$, are uniquely determined by $\boldsymbol{x}[1,k]$ according to the STBC coding constraints. Note that when computing the M-step in (10.61), we consider only the coding constraints of the STBC; the coding constraints induced by the outer channel code are exploited by the MAP outer-channel-code decoder and the turbo processing.

Within each turbo iteration, the E-step and M-step above are iterated I times. At the end of the Ith EM iteration, the extrinsic *a posteriori* LLRs of the outer-channel-code bits are computed, and then fed to the MAP outer-channel-code decoder. Recall that only the STBC symbols in the first OFDM slot are obtained from the MPSK modulation of outer-channel-code bits; the STBC symbols transmitted during the remaining $(P-1)$ OFDM slots are simply permutations and/or transformations of the STBC symbols in the first OFDM slot, as defined in (10.38). At each OFDM subcarrier, N transmitter antennas transmit N STC symbols, which correspond to $(N \log_2 |\Omega|)$ outer-channel-code bits. Based on (10.61), after I EM iterations, the extrinsic *a posteriori* LLR of the jth $(j = 1, \ldots, N \log_2 |\Omega|)$ outer-channel-code bit at the kth subcarrier $d^j(k)$ is computed at the output of the MAP-EM STBC decoder as follows:

$$
\begin{aligned}
\lambda_1(d^j[k]) &= \log \frac{\prod_{i=1}^{M} P\left[d^j(k) = +1 | \underline{\boldsymbol{y}}_i\right]}{\prod_{i=1}^{M} P(d^j[k] = -1 | \underline{\boldsymbol{y}}_i)} - \log \frac{P\left(d^j[k] = +1\right)}{P\left(d^j[k] = -1\right)} \\
&= \sum_{i=1}^{M} \log \frac{\sum_{\{\bar{\boldsymbol{x}}[p]\}_p \in \mathcal{C}_{j,p}^{+}} P\Big(\boldsymbol{x}[p,k] = \bar{\boldsymbol{x}}[p] | \underline{\boldsymbol{y}}_i\Big)}{\sum_{\{\bar{\boldsymbol{x}}[p]\}_p \in \mathcal{C}_{j,p}^{-}} P\Big(\boldsymbol{x}[p,k] = \bar{\boldsymbol{x}}[p] | \underline{\boldsymbol{y}}_i\Big)} - \lambda_2(d^j[k]) \\
&= \underbrace{\sum_{i=1}^{M} \log \frac{\sum_{\{\bar{\boldsymbol{x}}[p]\}_p \in \mathcal{C}_{j,p}^{+}} \left\{ \exp\left[-\frac{1}{\sigma_z^2} \sum_{p=1}^{P} q_i^{(I)}(\bar{\boldsymbol{x}}[p]) \right] \cdot P(\bar{\boldsymbol{x}}[1]) \right\}}{\sum_{\{\bar{\boldsymbol{x}}[p]\}_p \in \mathcal{C}_{j,p}^{-}} \left\{ \exp\left[-\frac{1}{\sigma_z^2} \sum_{p=1}^{P} q_i^{(I)}(\bar{\boldsymbol{x}}[p]) \right] \cdot P(\bar{\boldsymbol{x}}[1]) \right\}}}_{\Lambda_1(d^j[k])} \\
&\quad - \lambda_2(d^j[k]),
\end{aligned}
\tag{10.62}
$$

where $\mathcal{C}_{j,p}^{+}$ is the set of $\bar{\boldsymbol{x}}[p]$ for which the jth outer-channel-code bit is "+1," and $\mathcal{C}_{j,p}^{-}$ is similarly defined; $\{\bar{\boldsymbol{x}}[p]\}_p$ satisfy the STBC coding constraints. The extrinsic *a priori* LLRs $\{\lambda_2(d^j[k])\}_{j,k}$ are provided by the MAP outer-channel-code decoder at the previous turbo iteration. Finally, the extrinsic *a posteriori* LLRs $\{\lambda_1(d^j[k])\}_{j,k}$ are sent to the MAP outer-channel-code decoder, which in turn computes the extrinsic LLRs $\{\lambda_2(d^j[k])\}_{j,k}$ and then feeds them back to the MAP-EM STBC decoder, and thus completes one turbo iteration. At the end of the last turbo iteration, hard decisions of the information bits are output by the MAP outer-channel-code decoder.

The MAP-EM algorithm needs to be initialized at each turbo iteration. Except for the first turbo iteration, $\underline{\boldsymbol{X}}^{(0)}$ is simply taken as $\underline{\boldsymbol{X}}^{(I)}$ given by (10.61) from the previous turbo iteration. The procedure for computing $\underline{\boldsymbol{X}}^{(0)}$ at the first turbo iteration is similar to that described in Table 10.1.

Simulation Examples

In this section we provide computer simulation results to illustrate the performance of the proposed iterative receivers for STBC-OFDM systems, with or without outer channel coding. The receiver performance is simulated in three typical channel models with different delay profiles: the two-ray, typical urban (TU), and hilly terrain (HT) model with 50-Hz and 200-Hz Doppler frequencies [263]. In the following simulations the available bandwidth is 800 kHz and is divided into 128 subcarriers. These correspond to a subcarrier symbol rate of 5 kHz and OFDM word duration of 160 μs. In each OFDM word, a cyclic prefix interval of 40 μs is added to combat the effect of intersymbol interference, hence the duration of one OFDM word $T = 200\,\mu$s. For all simulations, two transmitter antennas and two receiver antennas are used; and the $\mathcal{G}_1$ STBC is adopted [see (10.38)]. The modulator uses a QPSK constellation. The OFDM system transmits in a burst manner as illustrated in Fig. 10.13. Each data burst includes 11 OFDM words ($q = 5, P = 2$), the first OFDM word contains the pilot symbols and the remaining 10 OFDM words span the duration of five STBC code words. Simulation results are shown in terms of the OFDM word-error rate versus the SNR.

Performance of the EM-Based ML Receiver In an STBC-OFDM system without outer channel coding, 512 information bits are transmitted from 128 subcarriers during two ($P = 2$) OFDM slots; therefore, the information rate is $2 \times 160/200 = 1.6$ bits/s per hertz, with 160/200 being the factor induced by the cyclic prefix interval. In Figs. 10.15–10.17, when ideal CSI is assumed to be available at the receiver side, the ML performance is shown in dashed lines, denoted by Ideal CSI. (Note that the ML performance difference between the 50-Hz and the 200-Hz Doppler fading channels is unnoticeable; hence, we only present the ML performance when $f_d = 50$ Hz.) Without the CSI, the EM-based ML receiver is employed. The performance after each EM iteration is demonstrated in curves denoted by EM Iter#1, EM Iter#2, and EM Iter#3. From the figures it is seen that the receiver performance is significantly improved through EM iteration. Furthermore, although the receiver is designed under the assumption that the fading channels remain static over one STBC code word (whereas the actual channels vary within one STBC code word), it can perform close to the ML performance with ideal CSI after two or three EM iterations for all three types of channels with a Doppler frequency as high as 200 Hz.

In Fig. 10.18, the performance of an EM-based ML receiver employing the causal temporal filtering scheme (denoted by C-T) is compared with that employing the noncausal temporal filtering scheme (denoted by N-T) in two-ray fading channels. It is seen that applying a second-round noncausal temporal filtering in addition to the first-round causal temporal filtering [263] does not bring much performance improvement to the EM-based ML receiver considered here, which is also true for the TU and HT fading channels. Because in the proposed EM-based ML receiver the performance improvement is achieved mainly by the EM iterations, we conclude that only causal temporal filtering is needed for initializing the EM algorithm.

Performance of the MAP-EM-Based Turbo Receiver A four-state rate-$\frac{1}{2}$ convolutional code with generator (5,7) in octal notation is adopted as the outer channel code, as depicted in Fig. 10.14. The overall information rate for this system is 0.8 bits/s per hertz. Figures 10.19–10.21 show the performance of the turbo receiver employing the MAP-EM algorithm for this concatenated STBC-OFDM system. The

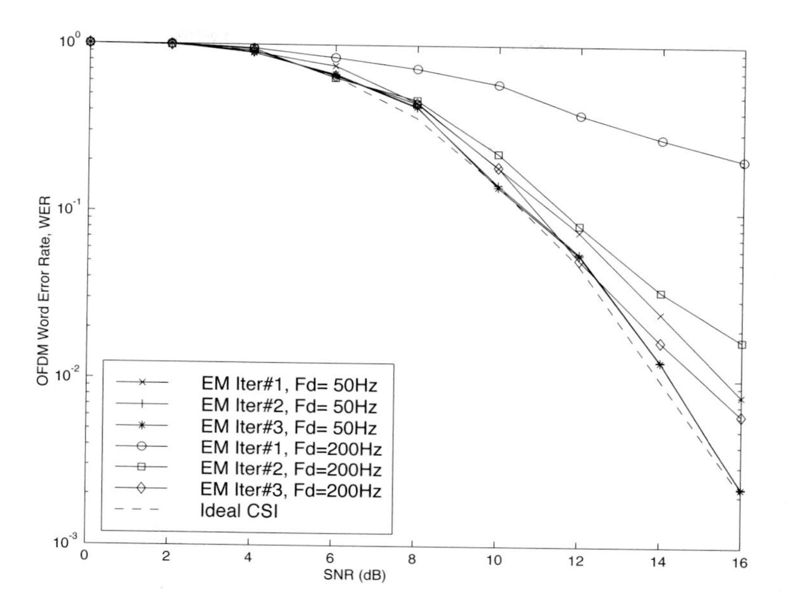

Figure 10.15. Word-error rate of a multiple-antenna ($N = 2, M = 2$) STBC-OFDM system in two-ray fading channels with Doppler frequencies $f_d = 50$ and 200 Hz.

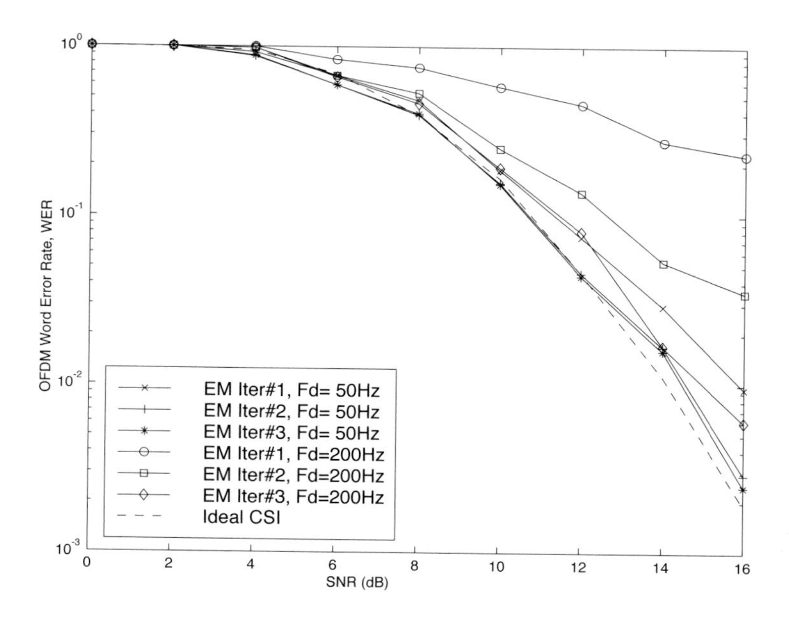

Figure 10.16. Word-error rate of a multiple-antenna ($N = 2, M = 2$) STBC-OFDM system in typical urban (TU) fading channels with Doppler frequencies $f_d = 50$ and 200 Hz.

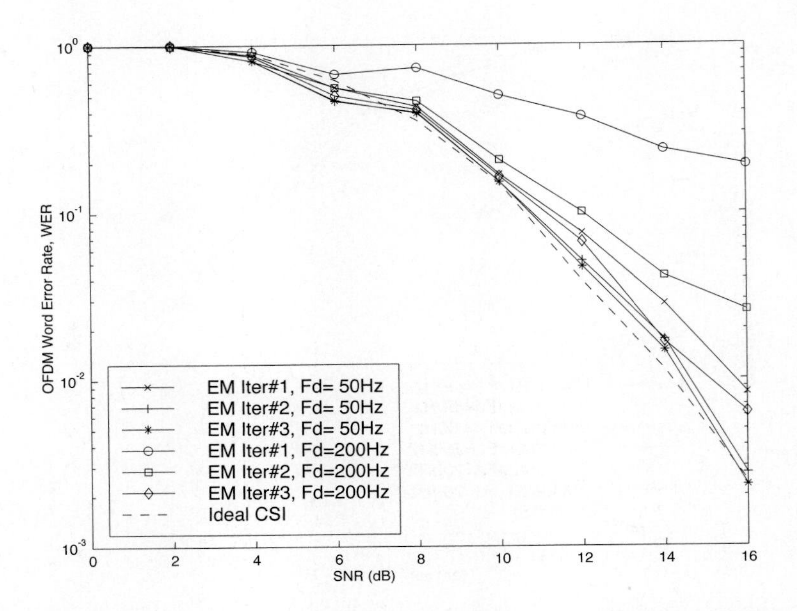

Figure 10.17. Word-error rate of a multiple-antenna ($N = 2, M = 2$) STBC-OFDM system in hilly terrain (HT) fading channels with Doppler frequencies $f_d = 50$ and 200 Hz.

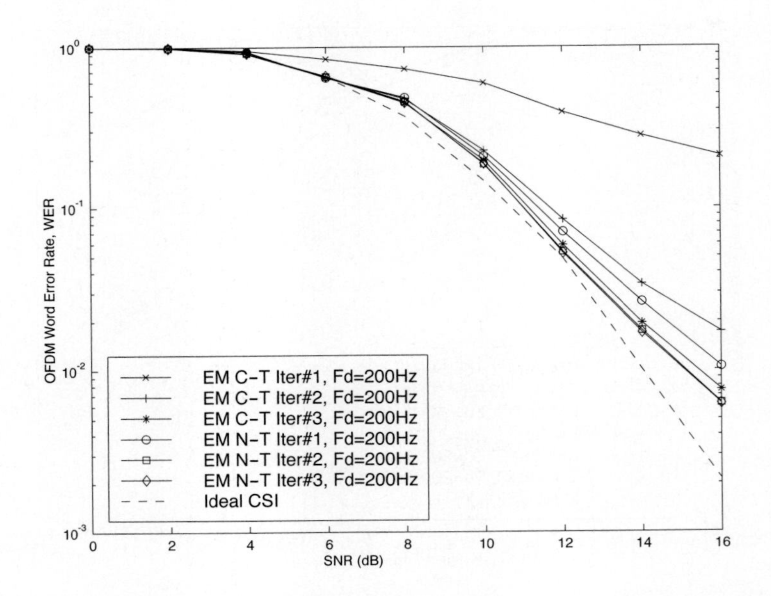

Figure 10.18. Word-error rate of a multiple-antenna ($N = 2, M = 2$) STBC-OFDM system in two-ray fading channels with Doppler frequency $f_d = 200$ Hz. Comparison of different temporal filtering schemes in initializing the EM algorithm.

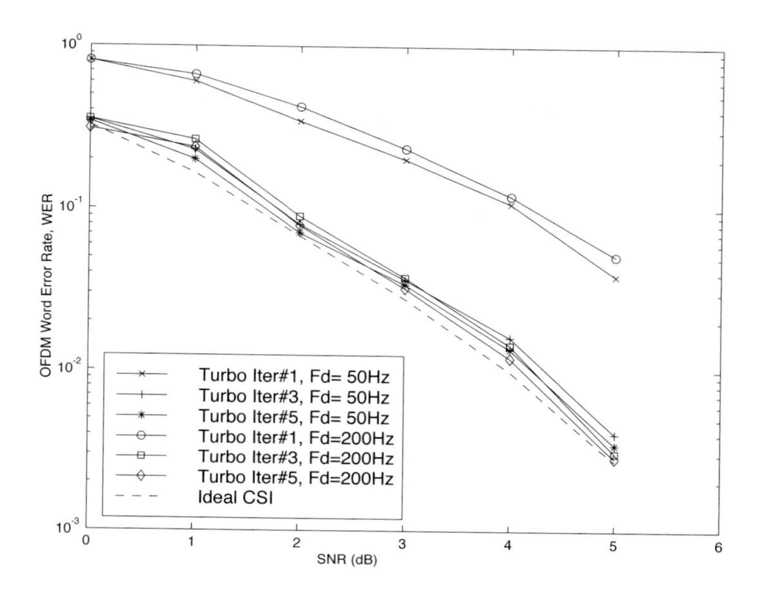

Figure 10.19. Word-error rate of a multiple-antenna ($N = 2, M = 2$) STBC-OFDM system with outer convolutional coding in two-ray fading channels with Doppler frequencies $f_d = 50$ and 200 Hz.

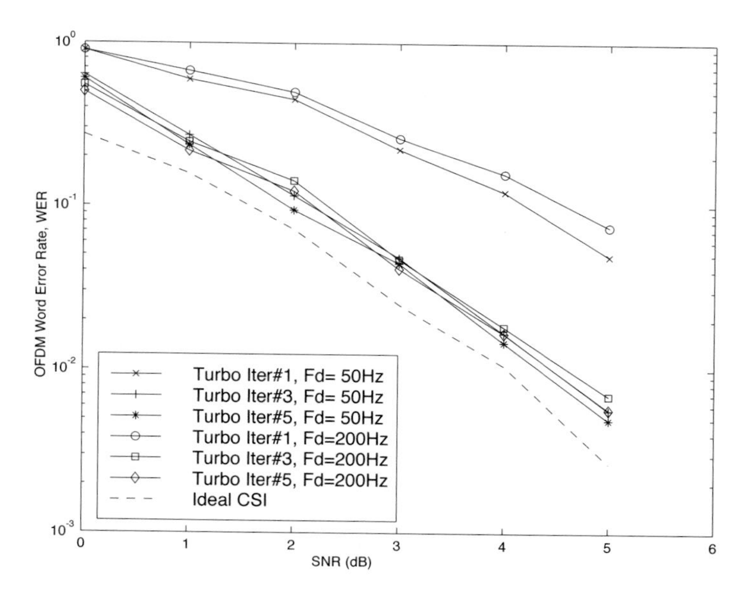

Figure 10.20. Word-error rate of a multiple-antenna ($N = 2, M = 2$) STBC-OFDM system with outer convolutional coding in typical urban (TU) fading channels with Doppler frequencies $f_d = 50$ and 200 Hz.

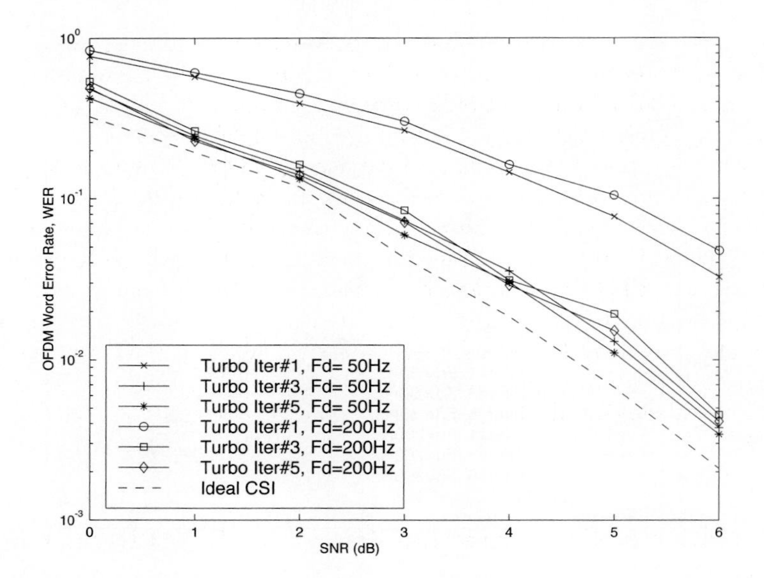

Figure 10.21. Word-error rate of a multiple-antenna ($N = 2, M = 2$) STBC-OFDM system with outer convolutional coding in hilly terrain (HT) fading channels with Doppler frequencies $f_d = 50$ and 200 Hz.

performance of the turbo receiver after the first, third, and fifth turbo iteration is demonstrated, respectively, in curves denoted by Turbo Iter#1, Turbo Iter#3, and Turbo Iter#5. During each turbo iteration, three EM iterations are carried out in the MAP-EM STBC decoder. Ideal CSI denotes the approximated ML lower bound, which is obtained by performing the MAP STBC decoder with ideal CSI and iterating a sufficient number of turbo iterations (three to four iterations are shown to be enough for the systems simulated here) between the MAP STBC decoder and the MAP convolutional decoder. From the simulation results, it is seen that by employing outer channel coding, the receiver performance is significantly improved (at the expense of lowering spectral efficiency). Moreover, without CSI, after three to five turbo iterations, the turbo receiver performs close to the approximated ML lower bound in all three types of channels with a Doppler frequency as high as 200 Hz.

10.5 LDPC-Based Space-Time Coded OFDM Systems

In this section we first examine the STC-OFDM system performance in correlated fading channels in terms of channel capacity and pairwise error probability (PEP). In [363], information-theoretic aspects of a two-ray propagation fading channel are studied. In [122, 478], the channel capacity of a multiple-antenna system in fading channels is investigated; and in [39], the limiting performance of a multiple-antenna system in block-fading channels is studied, under the assumption that the fading channels are uncorrelated and the channel state information is known to both the transmitter and the receiver. Here, we analyze the channel capacity of a multiple-antenna OFDM system over correlated frequency- and time-selective fading channels, assuming that the CSI is known only to the receiver. As a promising coding scheme to approach the channel capacity, STC is employed as the channel code in this system. The pairwise error probability analysis of the STC-OFDM system is also given. Moreover, based on the analysis of the channel capacity and the PEP, some STC design principles for the system under consideration are suggested. Since the STC based on state-of-the-art low-density parity-check (LDPC) codes [127, 302, 303, 416, 417] turns out to be a good candidate to meet these design principles, we then discuss an LDPC-based STC-OFDM system and a turbo receiver for this system. The materials in this section first appeared in [293]. Note that a simple space-time trellis code design method for OFDM systems is given in [289, 291].

10.5.1 Capacity Considerations for STC-OFDM Systems

System Model

We consider an STC-OFDM system with Q subcarriers, N transmitter antennas, and M receiver antennas, signaling through frequency- and time-selective fading channels, as illustrated in Fig. 10.22. Each STC code word spans P adjacent OFDM words, and each OFDM word consists of (NQ) STC symbols, transmitted simultaneously during one time slot. Each STC symbol is transmitted on a particular OFDM subcarrier and a particular transmitter antenna.

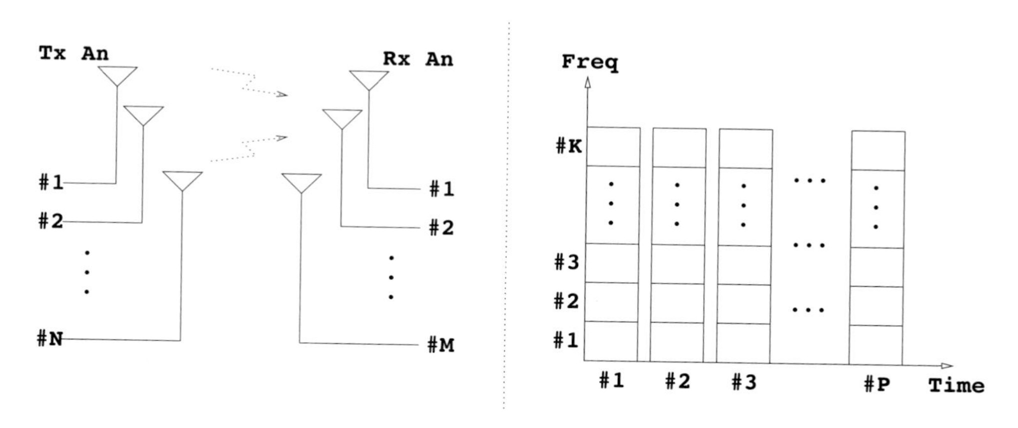

Figure 10.22. System description of a multiple-antenna STC-OFDM system over correlated fading channels. Each STC code word spans K subcarriers and P time slots in the system; on a particular subcarrier and in a particular time slot, STC symbols are transmitted from N transmitter antennas and received by M receiver antennas.

As in Section 10.4, it is assumed that the fading process remains static during each OFDM word (one time slot) but varies from one OFDM word to another; and the fading processes associated with different transmitter–receiver antenna pairs are uncorrelated. (However, as will be shown below, in a typical OFDM system, for a particular transmitter–receiver antenna pair, the fading processes are correlated in both frequency and time.)

At the receiver, the signals are received from M receiver antennas. After matched filtering and sampling, the DFT is applied to the received discrete-time signal to obtain

$$\boldsymbol{y}[p,k] = \boldsymbol{H}[p,k]\boldsymbol{x}[p,k] + \boldsymbol{z}[p,k], \qquad k = 0,\ldots,Q-1, \;\; p = 1,\ldots,P, \tag{10.63}$$

where $\boldsymbol{H}[p,k] \in \mathbb{C}^{M\times N}$ is the matrix of complex channel frequency responses at the kth subcarrier and at the pth time slot, which is explained below; $\boldsymbol{x}[p,k] \in \mathbb{C}^N$ and $\boldsymbol{y}[p,k] \in \mathbb{C}^M$ are, respectively, the transmitted and received signals at the kth subcarrier and at the pth time slot; $\boldsymbol{z}[p,k] \in \mathbb{C}^M$ is the ambient noise, which is circularly symmetric complex Gaussian with unit variance.

Consider the channel response between the jth transmitter antenna and the ith receiver antenna. As before, the time-domain channel impulse response can be modeled as a tapped-delay line. With only the nonzero taps considered, it can be expressed as

$$g_{i,j}(\tau;t) = \sum_{\ell=1}^{L_f} \alpha_{i,j}(\ell;t)\delta\left(\tau - \frac{n_\ell}{Q\Delta_f}\right), \tag{10.64}$$

where $\delta(\cdot)$ is the Dirac delta function; L_f denotes the number of nonzero taps; $\alpha_{i,j}(\ell;t)$ is the complex amplitude of the ℓth nonzero tap, whose delay is $n_\ell/(Q\Delta_f)$, where n_ℓ is an integer and Δ_f is the tone spacing of the OFDM system. In mobile channels, for the particular (i,j)th antenna pair, the time-varying tap coefficients $\alpha_{i,j}(\ell;t)$ can be modeled as wide-sense stationary random processes with uncorrelated scattering (WSSUS) and with bandlimited Doppler power spectrum [396]. For the signal model in (10.63), we need only consider the time responses of $\alpha_{i,j}(\ell;t)$ within the time interval $t \in [0, PT]$, where T is the total time duration of one OFDM word plus its cyclic extension and PT is the total time involved in transmitting P adjacent OFDM words. Following [568], for the particular ℓth tap of the (i,j)th antenna pair, the dimension of the band- and time-limited random process $\alpha_{i,j}(\ell;t)$, $t \in [0, PT]$ (defined as the number of significant eigenvalues in the Karhunen–Loeve expansion of this random process) is approximately equal to $L_t \triangleq \lceil 2f_d PT + 1 \rceil$, where f_d is the maximum Doppler frequency. Hence, ignoring edge effects, the time response of $\alpha_{i,j}(\ell;t)$ can be expressed in terms of the Fourier expansion as

$$\alpha_{i,j}(\ell;t) \simeq \sum_{n=-f_d PT}^{f_d PT} \beta_{i,j}(\ell,n) e^{\jmath 2\pi n t/(PT)}, \tag{10.65}$$

where $\{\beta_{i,j}(\ell, n)\}_n$ is a set of independent circularly symmetric complex Gaussian random variables, indexed by n.

For OFDM systems with proper cyclic extension and sample timing, with tolerable leakage, the channel frequency response between the jth transmitter antenna and the ith receiver antenna at the pth time slot and the kth subcarrier, which is exactly the (i, j)th element of $\boldsymbol{H}[p, k]$ in (10.63), can be expressed as [506]

$$H_{i,j}[p,k] \triangleq H_{i,j}(pT, k\Delta_f) = \sum_{\ell=1}^{L_f} \alpha_{i,j}(\ell; pT) e^{-\jmath 2\pi k n_\ell / Q} = \boldsymbol{h}_{i,j}^H(p) \boldsymbol{w}_f(k), \quad (10.66)$$

where $\boldsymbol{h}_{i,j}(p) \triangleq [\alpha_{i,j}(1; pT), \ldots, \alpha_{i,j}(L_f; pT)]^H$ is the L_f-sized vector containing the time responses of all the nonzero taps, and $\boldsymbol{w}_f(k) \triangleq \left[e^{-\jmath 2\pi k n_1/Q}, \ldots, e^{-\jmath 2\pi k n_{L_f}/Q}\right]^T$ contains the corresponding DFT coefficients.

Using (10.65), $\alpha_{i,j}(\ell; pT)$ can be simplified to

$$\alpha_{i,j}(\ell; pT) = \sum_{n=-f_d PT}^{f_d PT} \beta_{i,j}(\ell, n) e^{\jmath 2\pi n p / P} = \boldsymbol{\beta}_{i,j}^H(\ell) \boldsymbol{w}_t(p), \quad (10.67)$$

where $\boldsymbol{\beta}_{i,j}(\ell) \triangleq [\beta_{i,j}(\ell, -f_d PT), \ldots, \beta_{i,j}(\ell, 0), \ldots, \beta_{i,j}(\ell, f_d PT)]^H$ is an L_t-length vector and $\boldsymbol{w}_t(p) \triangleq \left[e^{-\jmath 2\pi p f_d T}, \ldots, e^{\jmath 0}, \ \ldots, e^{\jmath 2\pi p f_d T}\right]^T$ contains the corresponding inverse DFT coefficients. Substituting (10.67) into (10.66), we have

$$H_{i,j}[p,k] = \boldsymbol{g}_{i,j}^H \boldsymbol{W}_t'(p) \boldsymbol{w}_f(k), \quad (10.68)$$

with

$$\boldsymbol{g}_{i,j} \triangleq \left[\boldsymbol{\beta}_{i,j}^H(1), \ldots, \boldsymbol{\beta}_{i,j}^H(L_f)\right]_{L \times 1}^H,$$
$$\boldsymbol{W}_t'(p) \triangleq \operatorname{diag}\{\boldsymbol{w}_t(p), \ldots, \boldsymbol{w}_t(p)\}_{L \times L_f}.$$

From (10.68) it is seen that due to the close spacing of OFDM subcarriers and the limited Doppler frequency, for a specific antenna pair (i, j) the channel responses $\{H_{i,j}[p, k]\}_{p,k}$ are different transformations [specified by $\boldsymbol{w}_t(p)$ and $\boldsymbol{w}_f(k)$] of the same random vector $\boldsymbol{g}_{i,j}$, and hence they are correlated in both frequency and time.

Channel Capacity

In this section we consider the channel capacity of the system described above. Assuming that the channel state information is known only at the receiver and the transmitter power is constrained as $\mathsf{trace}\left\{E\left[\boldsymbol{x}[p,k]\boldsymbol{x}^H[p,k]\right]\right\} \le \gamma$, the instantaneous channel capacity of this system, which is defined as the mutual information conditioned on the correlated fading channel values $\mathcal{H} \triangleq \{\boldsymbol{H}[p,k]\}_{p=1,k=0}^{P\ \ Q-1}$, can be computed as [39, 363]

$$\begin{aligned} I_{\mathcal{H}}(\gamma) &\triangleq I(\{\boldsymbol{y}[p,k]\}_{p,k}; \{\boldsymbol{x}[p,k]\}_{p,k}|\mathcal{H},\gamma) \\ &= \frac{1}{PQ}\sum_{p=1}^{P}\sum_{k=0}^{Q-1}\sum_{i=1}^{m}\log_2\left(1+\lambda_i(p,k)\gamma/N\right) \qquad \text{bits/s/Hz}, \end{aligned} \tag{10.69}$$

where $m \triangleq \min(N, M)$, and $\lambda_i(p,k)$ is the ith nonzero eigenvalue of the nonnegative definite Hermitian matrix $\boldsymbol{H}[p,k]\boldsymbol{H}^H[p,k]$. The maximization of $I_{\mathcal{H}}(\gamma)$ is achieved when $\{\boldsymbol{x}[p,k]\}_{p,k}$ consists of independent circularly symmetric complex Gaussian random variables with identical variances [39, 363]. (When the CSI is known to both the transmitter and the receiver, the instantaneous channel capacity is maximized by "water filling" [40].) The *ergodic channel capacity* is defined as $I(\gamma) \triangleq E_{\mathcal{H}}\{I_{\mathcal{H}}(\gamma)\}$. In the system considered, the concept of ergodic channel capacity $I(\gamma)$ is of less interest, because the fading processes are not ergodic, due to the limited number of antennas and the limited L_f and L_t.

Since $I_{\mathcal{H}}(\gamma)$ is a random variable, whose statistics are determined jointly by (γ, N, M) and the characteristics of correlated fading channels, we turn to another important concept—*outage capacity*, which is closely related to the code word error probability, as averaged over the random coding ensemble and over all channel realizations [39]. The outage probability is defined as the probability that the channel cannot support a given information rate R,

$$P_{\text{out}}(R,\gamma) = P(I_{\mathcal{H}}(\gamma) < R). \tag{10.70}$$

Since it is difficult to get an analytical expression for (10.70), we resort to Monte Carlo integration for its numerical evaluation.

In the following, we give some numerical results of the outage probability in (10.70) obtained by Monte Carlo integration. For simplicity, we assume that all elements in $\{\boldsymbol{g}_{i,j}\}_{i,j}$ have the same variance. Define the *selective-fading diversity order* L as the product of the number of nonzero delay taps L_f and the dimension of Doppler fading process L_t (i.e., $L \triangleq L_f L_t$). The following observations can be made from the numerical evaluations of (10.70).

1. From Figs. 10.23 and 10.24, it is seen that at a practical outage probability (e.g., $P_{\text{out}} = 1\%$), for fixed (N, M, γ) the highest achievable information rate increases as the selective-fading diversity order L increases, but the increase diminishes as L becomes larger. Eventually, as $L \to \infty$, the highest achievable information rate converges to the ergodic capacity. [Note that the ergodic capacity is the area above each curve in the figure: $I(\gamma) = \int_0^\infty P(I_{\mathcal{H}}(\gamma) > R)\,\mathrm{d}R$.]

2. Figure 10.24 compares the effects of the frequency-selectivity order L_f and the time-selectivity order L_t on the outage capacity. It shows that the frequency and time selectivity are essentially equivalent in terms of their effects on the

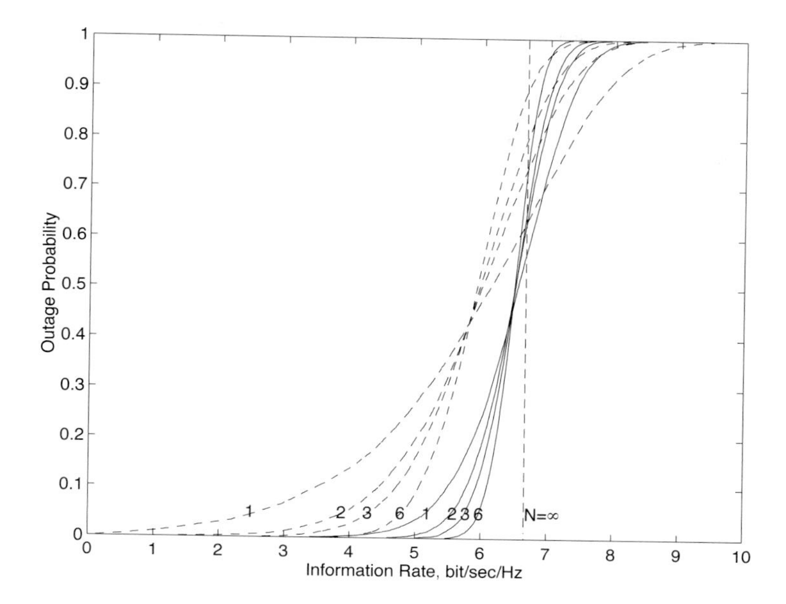

Figure 10.23. Outage probability versus information rate in a correlated fading OFDM system with $M = 1, Q = 256, P = 1$, SNR $= 20$ dB, where dashed lines represent a system with one transmitter antenna ($N = 1$) and solid lines represent a system with four transmitter antennas ($N = 4$). The vertical dash-dotted line represents the AWGN channel capacity (when SNR $= 20$ dB). The fading channels are frequency- and time-nonselective with $L_t = 1, L = L_f = \{1, 2, 3, 6\}$.

outage capacity. In other words, the selective-fading diversity order $L = L_f L_t$ ultimately affects the outage capacity.

3. From Fig. 10.23 it is seen that as the area above each curve, the ergodic channel capacity is invariant to the selective-fading diversity order L (which is the key parameter in determining the correlation characteristics of the fading channels) and it is determined only by the spatial diversity order (N, M) and the transmitted signal power γ [122, 478]. Moreover, it is seen that both the outage and ergodic capacities can be increased by fixing the number of receiver antennas and increasing the number of transmitter antennas (or vice versa) (e.g., by fixing $M = 1$ and letting $N \to \infty$, the ergodic capacity converges to the capacity of AWGN channels [353]).

In summary, we have seen the different effects of two diversity resources, spatial diversity and selective-fading diversity, on the channel capacity of a multiple-antenna correlated fading OFDM system. Increasing the spatial diversity order (i.e., N, M) can always bring capacity (outage capacity and/or ergodic capacity) increase at the expense of extra physical costs. By contrast, the selective-fading diversity is a free resource, but its effect on improving the channel capacity becomes

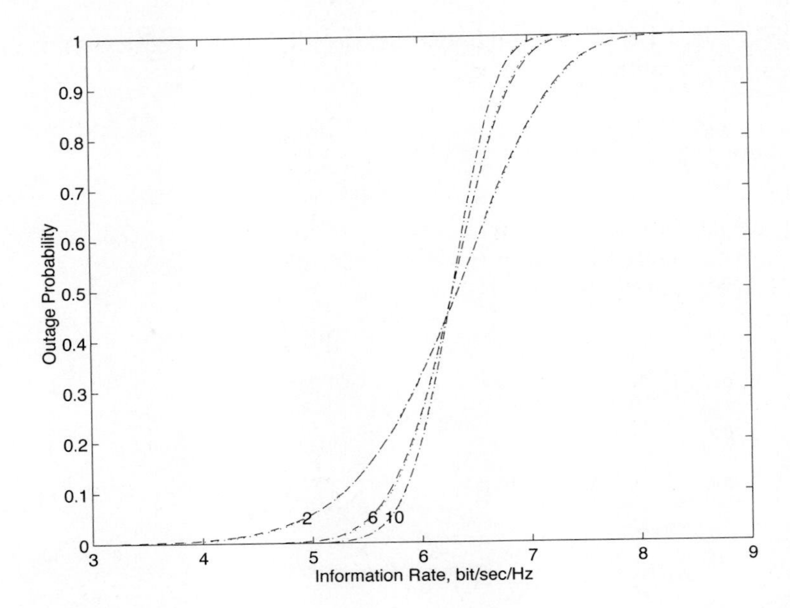

Figure 10.24. Outage probability versus information rate in a correlated fading OFDM system with $N = 2, M = 1, Q = 256, P = 10$, SNR $= 20$ dB. Dashed lines represent frequency-selective and time-nonselective channels with $L_t = 1, L = L_f = \{2, 6, 10\}$. Dotted lines represent frequency- and time-selective channels with $L_f = 2, L = 2L_t = \{2, 6, 10\}$. Note that for the same L, the dashed and dotted lines overlap, which shows the equivalent impacts of frequency- and time-selective fading on the outage probability.

less as L becomes larger. Since both diversity resources can improve the capacity of a multiple-antenna OFDM system, it is crucial to have an efficient channel coding scheme, which can take advantage of all available diversity resources of the system.

Pairwise Error Probability

To obtain further insight into coding design, it is of interest to analyze the pairwise error probability of this system with coded modulation.

With perfect CSI at the receiver, the maximum likelihood decision rule for the signal model (10.63) is given by

$$\hat{\boldsymbol{x}} = \arg\min_{\boldsymbol{x}} \sum_{i=1}^{M} \sum_{p=1}^{P} \sum_{k=0}^{Q-1} \left| y_i[p,k] - \sum_{j=1}^{N} H_{i,j}[p,k] x_j[p,k] \right|^2, \tag{10.71}$$

where the minimization is over all possible STC code words $\boldsymbol{x} = \{x_j[p,k]\}_{j,p,k}$. Assuming equal transmitted power at all transmitter antennas, using the Chernoff bound [396], the PEP of transmitting $\boldsymbol{x}$ and deciding in favor of another code word $\bar{\boldsymbol{x}}$ at the decoder is upper bounded by

$$P(\boldsymbol{x} \to \bar{\boldsymbol{x}}|\mathcal{H}) \leq \exp\left(-\frac{d^2(\boldsymbol{x},\bar{\boldsymbol{x}})\gamma}{8N}\right), \tag{10.72}$$

where γ is the total signal power transmitted from all N transmitted antennas. (Recall that the noise at each receiver antenna is assumed to have unit variance.) Using (10.66)–(10.68), $d^2(\boldsymbol{x},\bar{\boldsymbol{x}})$ is given by

$$\begin{aligned} d^2(\boldsymbol{x},\bar{\boldsymbol{x}}) &= \sum_{i=1}^{M}\sum_{p=1}^{P}\sum_{k=0}^{Q-1}\left|\sum_{j=1}^{N} H_{i,j}[p,k]e_j[p,k]\right|^2 \\ &= \sum_{i=1}^{M}\sum_{p=1}^{P}\sum_{k=0}^{Q-1}\left[\begin{array}{ccc} \boldsymbol{g}_{i,1}^H & \cdots & \boldsymbol{g}_{i,N}^H \end{array}\right]_{1\times(NL)} \Big\{\boldsymbol{W}_t(p)\Big[\boldsymbol{W}_f(k)\boldsymbol{e}[p,k] \\ &\quad \cdot \boldsymbol{e}^H[p,k]\boldsymbol{W}_f^H(k)\Big]\boldsymbol{W}_t^H(p)\Big\}_{(NL)\times(NL)} \left[\begin{array}{ccc} \boldsymbol{g}_{i,1}^H & \cdots & \boldsymbol{g}_{i,N}^H \end{array}\right]^H_{(NL)\times 1} \\ &= \sum_{i=1}^{M} \bar{\boldsymbol{g}}_i^H \boldsymbol{D}\bar{\boldsymbol{g}}_i, \end{aligned} \tag{10.73}$$

with

$$\begin{aligned} e_j[p,k] &\triangleq x_j[p,k] - \bar{x}_j[p,k], \\ \boldsymbol{e}[p,k] &\triangleq [e_1[p,k],\ldots,e_N[p,k]]^T_{N\times 1}, \\ \boldsymbol{W}_f(k) &\triangleq \operatorname{diag}\{\boldsymbol{w}_f(k),\ldots,\boldsymbol{w}_f(k)\}_{(NL_f)\times N}, \\ \boldsymbol{W}_t(p) &\triangleq \operatorname{diag}\{\boldsymbol{W}_t'(p),\ldots,\boldsymbol{W}_t'(p)\}_{(NL)\times(NL_f)}, \\ \boldsymbol{D} &\triangleq \left\{\sum_{p=1}^{P}\sum_{k=0}^{Q-1}\boldsymbol{W}_t(p)\boldsymbol{W}_f(k)\boldsymbol{e}[p,k]\boldsymbol{e}^H[p,k]\boldsymbol{W}_f^H(k)\boldsymbol{W}_t^H(p)\right\}_{(NL)\times(NL)}, \end{aligned} \tag{10.74}$$

$$\bar{\boldsymbol{g}}_i \triangleq [\boldsymbol{g}_{i,1}^H,\ldots,\boldsymbol{g}_{i,N}^H]^H_{(NL)\times 1}. \tag{10.75}$$

In (10.74), $(\boldsymbol{e}[p,k]\boldsymbol{e}^H[p,k])$ is a rank-one matrix, which is equal to a zero matrix if the entries of code words $\boldsymbol{x}$ and $\bar{\boldsymbol{x}}$ corresponding to the kth subcarrier and pth time slot are the same. Let D denote the number of instances when $\boldsymbol{e}[p,k]\boldsymbol{e}^H[p,k] \neq \underline{0}$; similar to [438], D_{eff}, which is the minimum D over every two possible code word pair, is called the *effective length* of the code. Denote $r \triangleq \mathsf{rank}(\boldsymbol{D})$; it is easily seen that $\min_{\boldsymbol{x},\bar{\boldsymbol{x}}} r \leq \min(D_{\text{eff}}, NL)$. Since $\boldsymbol{w}_f(k)$ and $\boldsymbol{w}_t(p)$ vary with different multipath delay profiles and Doppler power spectrum shapes, the matrix $\boldsymbol{D}$ also varies with different channel environments. However, since $\boldsymbol{D}$ is a nonnegative definite Hermitian matrix, by an eigendecomposition, it can be written as

$$\boldsymbol{D} = \boldsymbol{V}\boldsymbol{\Lambda}\boldsymbol{V}^H, \tag{10.76}$$

where $\boldsymbol{V}$ is a unitary matrix and $\boldsymbol{\Lambda} \triangleq \text{diag}\{\lambda_1, \ldots, \lambda_r, 0, \ldots, 0\}$, with $\{\lambda_j\}_{j=1}^r$ being the positive eigenvalues of $\boldsymbol{D}$. Moreover, by assumption, all the (NML) elements of $\{\boldsymbol{g}_{i,j}\}_{i,j}$ are i.i.d. circularly symmetric complex Gaussian with zero means. So (10.72) can be rewritten as

$$P(\boldsymbol{x} \rightarrow \bar{\boldsymbol{x}}|\mathcal{H}) \leq \prod_{i=1}^{M} \exp\left(-\frac{\gamma}{8N}\sum_{j=1}^{r}\lambda_j|\tilde{\beta}_i(j)|^2\right), \tag{10.77}$$

where $\tilde{\beta}_i(j) \triangleq \left[\boldsymbol{V}^H\bar{\boldsymbol{g}}_i\right]_j$ is the jth element of $\boldsymbol{V}^H\bar{\boldsymbol{g}}_i$. Since $\boldsymbol{V}$ is unitary, $\{\tilde{\beta}_i(j)\}_{i,j}$ are also i.i.d. circularly symmetric complex Gaussian with zero means, and their magnitudes $\{|\tilde{\beta}_i(j)|\}_{i,j}$ are i.i.d. Rayleigh distributed. By averaging the conditional PEP in (10.77) over the Rayleigh pdf (probability density function), the PEP bound for a multiple-antenna STC-OFDM system over correlated fading channels can finally be written as

$$P(\boldsymbol{x} \rightarrow \bar{\boldsymbol{x}}) \leq \left(\frac{1}{\prod_{j=1}^{r}\left(1+\frac{\lambda_j\gamma}{8N}\right)}\right)^M \leq \left(\prod_{j=1}^{r}\lambda_j\right)^{-M}\left(\frac{\gamma}{8N}\right)^{-rM}. \tag{10.78}$$

It is seen from (10.78) that the highest possible diversity order the STC-OFDM system can provide is NML: the product of the number of transmitter antennas, the number of receiver antennas, and the selective-fading diversity order of the channels. In other words, the attractiveness of the STC-OFDM system lies in its ability to exploit all the available diversity resources. However, note that although in the analysis of PEP the three parameters (N, M, L) appear equivalent in improving the system performance, they actually play different roles from the capacity viewpoint, as indicated above.

10.5.2 Low-Density Parity-Check Codes

First proposed by Gallager in 1962 [127] and reexamined in [302, 303, 416, 417], low-density parity-check codes have been shown to be a very promising coding technique for approaching the channel capacity in AWGN channels. For example, a carefully constructed rate-$\frac{1}{2}$ irregular LDPC code with long block length has a bit-error probability of 10^{-6} just 0.04 dB away from the Shannon capacity of AWGN channels [82].

As the name suggests, an LDPC code is a linear block code specified by a very sparse parity-check matrix as illustrated in Fig. 10.25. The parity-check matrix $\boldsymbol{H}$

$$
\begin{array}{cccccccccccccccccccc}
\hline
1 & 1 & 1 & 1 & 0 & 0 & 0 & 0 & 0 & 0 & 0 & 0 & 0 & 0 & 0 & 0 & 0 & 0 & 0 & 0 \\
0 & 0 & 0 & 0 & 1 & 1 & 1 & 1 & 0 & 0 & 0 & 0 & 0 & 0 & 0 & 0 & 0 & 0 & 0 & 0 \\
0 & 0 & 0 & 0 & 0 & 0 & 0 & 0 & 1 & 1 & 1 & 1 & 0 & 0 & 0 & 0 & 0 & 0 & 0 & 0 \\
0 & 0 & 0 & 0 & 0 & 0 & 0 & 0 & 0 & 0 & 0 & 0 & 1 & 1 & 1 & 1 & 0 & 0 & 0 & 0 \\
0 & 0 & 0 & 0 & 0 & 0 & 0 & 0 & 0 & 0 & 0 & 0 & 0 & 0 & 0 & 0 & 1 & 1 & 1 & 1 \\
\hline
1 & 0 & 0 & 0 & 1 & 0 & 0 & 0 & 1 & 0 & 0 & 0 & 1 & 0 & 0 & 0 & 0 & 0 & 0 & 0 \\
0 & 1 & 0 & 0 & 0 & 1 & 0 & 0 & 0 & 1 & 0 & 0 & 0 & 0 & 0 & 0 & 1 & 0 & 0 & 0 \\
0 & 0 & 1 & 0 & 0 & 0 & 1 & 0 & 0 & 0 & 0 & 0 & 0 & 1 & 0 & 0 & 0 & 1 & 0 & 0 \\
0 & 0 & 0 & 1 & 0 & 0 & 0 & 0 & 0 & 0 & 1 & 0 & 0 & 0 & 1 & 0 & 0 & 0 & 1 & 0 \\
0 & 0 & 0 & 0 & 0 & 0 & 0 & 1 & 0 & 0 & 0 & 1 & 0 & 0 & 0 & 1 & 0 & 0 & 0 & 1 \\
\hline
1 & 0 & 0 & 0 & 0 & 1 & 0 & 0 & 0 & 0 & 0 & 1 & 0 & 0 & 0 & 0 & 0 & 1 & 0 & 0 \\
0 & 1 & 0 & 0 & 0 & 0 & 1 & 0 & 0 & 0 & 1 & 0 & 0 & 0 & 0 & 1 & 0 & 0 & 0 & 0 \\
0 & 0 & 1 & 0 & 0 & 0 & 0 & 1 & 0 & 0 & 0 & 0 & 1 & 0 & 0 & 0 & 0 & 0 & 1 & 0 \\
0 & 0 & 0 & 1 & 0 & 0 & 0 & 0 & 1 & 0 & 0 & 0 & 0 & 1 & 0 & 0 & 1 & 0 & 0 & 0 \\
0 & 0 & 0 & 0 & 1 & 0 & 0 & 0 & 0 & 1 & 0 & 0 & 0 & 0 & 1 & 0 & 0 & 0 & 0 & 1 \\
\hline
\end{array}
$$

Figure 10.25. Example of a parity-check matrix $\boldsymbol{P}$ for an $(n, k, t, j) = (20, 5, 3, 4)$ regular LDPC code with code rate-$\frac{1}{4}$, block length $n = 20$, column weight $t = 3$, and row weight $j = 4$.

of a *regular* (N, K, s, t) LDPC code of rate $R = K/N$ is an $(N - K) \times N$ matrix, which has s 1s in each column and $t > s$ 1s in each row where $s \ll N$. Apart from these constraints, the 1s are typically placed at random in the parity-check matrix. When the number of 1s in every column is not the same, the code is known as an *irregular* LDPC code. It should be noted that the parity-check matrix is not constructed in systematic form. Consequently, to obtain the generator matrix $\boldsymbol{G}$, we first apply Gaussian elimination to reduce the parity-check matrix to a form $\boldsymbol{H} = [\boldsymbol{I}_{N-K}|\boldsymbol{P}^T]$, where $\boldsymbol{I}_{N-K}$ is an $(N-K) \times (N-K)$ identity matrix. Then the generator matrix is given by $\boldsymbol{G} = [\boldsymbol{P}|\boldsymbol{I}_K]$. In contrast to $\boldsymbol{P}$, the generator matrix $\boldsymbol{G}$ is dense. Consequently, the number of bit operations required to encode is $\mathcal{O}(n^2)$, which is larger than that for other linear codes. Similar to turbo codes, LDPC codes can be efficiently decoded by a suboptimal iterative belief propagation algorithm, which is explained in detail in [127]. At the end of each iteration, the parity check is performed. If the parity check is correct, the decoding is terminated; otherwise, the decoding continues until it reaches the maximum number of iterations (e.g., 30).

The code with parity-check matrix $\boldsymbol{H}$ can be represented by a bipartite graph which consists of two types of nodes, variable nodes and check nodes. Each code bit is a variable node, while each parity check or each row of the parity-check matrix represents a check node. An edge in the graph is placed between variable node i and check node j if $H_{j,i} = 1$. That is, each check node is connected to code bits whose sum modulo-2 should be zero. Irregular LDPC codes are specified by two polynomials $\lambda(x) = \sum_{i=1}^{d_{l\max}} \lambda_i \, x^{i-1}$ and $\rho(x) = \sum_{i=1}^{d_{r\max}} \rho_i \, x^{i-1}$, where λ_i is the fraction of edges in the bipartite graph that are connected to variable nodes of degree i, and ρ_i is the fraction of edges that are connected to check nodes of degree i. Equivalently, the degree profiles can also be specified from the node perspective:

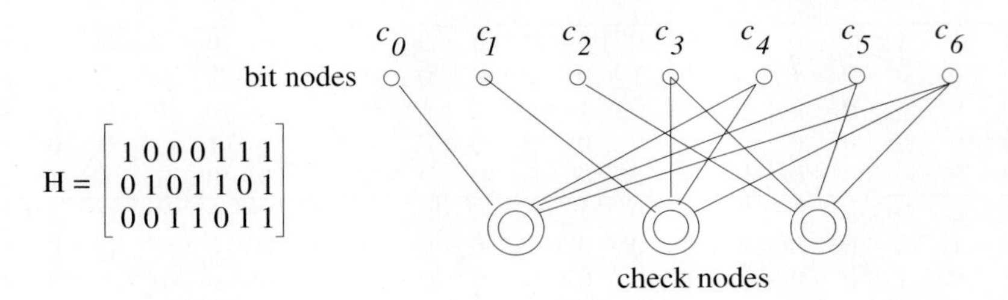

Figure 10.26. Bipartite graph representing parity-check and bit nodes of an irregular LDPC code.

two polynomials $\tilde{\lambda}(x) = \sum_{i=1}^{d_{l\max}} \tilde{\lambda}_i \, x^{i-1}$ and $\tilde{\rho}(x) = \sum_{i=1}^{d_{r\max}} \tilde{\rho}_i \, x^{i-1}$, where $\tilde{\lambda}_i$ is the fraction of variable nodes of degree i and $\tilde{\rho}_i$ is the fraction of check nodes of degree i. The parity-check matrix for an irregular $(7,4)$ code and its associated bipartite graph is shown in Fig. 10.26 as an example. The degree profiles for this code from the edge perspective are $\lambda(x) = \frac{1}{4} + \frac{1}{2}x + \frac{1}{2}x^2$ and $\rho(x) = x^3$. The degree profiles from the node perspective are $\tilde{\lambda}(x) = \frac{3}{7} + \frac{3}{7}x + \frac{1}{7}x^2$ and $\tilde{\rho}(x) = x^3$.

Before we discuss the LDPC decoding algorithm, we first establish the following notation. All extrinsic messages (information) are in log-likelihood form, and the variable L is used to refer to extrinsic messages. Superscript p is used to denote quantities during the pth iteration of LDPC decoding. A subscript $b \rightarrow c$ or $b \leftarrow c$ denotes quantities passed between the bit nodes and the check nodes of the LDPC code. The variable (bit) nodes in the bipartite graph of the LDPC code are numbered from 1 to N and the check nodes from 1 to $N-K$ (in any order). The degree of the ith bit node is denoted by ν_i and the degree of the ith check node is denoted by Δ_i. Denote by $\{e^b_{i,1}, e^b_{i,2}, \ldots, e^b_{i,\nu_i}\}$ the set of edges connected to the ith bit node and by $\{e^c_{i,1}, e^c_{i,2}, \ldots, e^c_{i,\Delta_i}\}$ the set of edges connected to the ith check node. That is, $e^b_{i,k}$ denotes the kth edge connected to the ith bit node, and $e^c_{i,k}$ denotes the kth edge connected to the ith check node. The particular edge or bit associated with a piece of extrinsic information is denoted as the argument of L. For example, $L^p_{b\rightarrow c}(e^b_{i,j})$ denotes the extrinsic LLR passed from a bit node to a check node along the jth edge connected to the ith bit node, during the pth iteration within the LDPC decoder.

The LDPC decoding algorithm is summarized as follows.

Algorithm 10.4: [LDPC decoding algorithm] *Initially, all extrinsic messages are assumed to be zeros [i.e.,* $L^{0,0}_{b\leftarrow c}(e^b_{i,k}) = 0, \forall(i,k)$*].*

- *Iterate between bit node update and check node update: For* $p = 1, 2, \ldots, P$*:*
  - *Bit node update: For each of the bit nodes* $i = 1, 2, \ldots, N$*, for every edge connected to the bit node, compute the extrinsic message passed from the bit node to the check node along the edge, given by*

$$L_{b\rightarrow c}^{p,q}(e_{i,j}^{b}) = L_{eq\rightarrow L}^{q}(b_i) + \sum_{k=1,k\neq j}^{\nu_i} L_{b\leftarrow c}^{p-1,q}(e_{i,k}^{b}). \tag{10.79}$$

– *Check node update: For each of the check nodes $i = 1, 2, \ldots, N-K$, for all edges that are connected to the check node, compute the extrinsic message passed from the check node to the bit node, given by*

$$L_{b\leftarrow c}^{p,q}(e_{i,j}^{c}) = 2\tanh^{-1}\left[\prod_{k=1,k\neq j}^{\Delta_i} \tanh\left(\frac{L_{b\rightarrow c}^{p,q}(e_{i,k}^{c})}{2}\right)\right]. \tag{10.80}$$

- *Make final hard decisions on information and parity bits:*

$$\hat{b}_i = \text{sign}\left[\sum_{k=1}^{\nu_i} L_{b\leftarrow c}^{P,q}(e_{i,k}^{b})\right]. \tag{10.81}$$

10.5.3 LDPC-Based STC-OFDM System

In this section, we consider coding design for STC-OFDM systems. We assume that the CSI is known only at the receiver.

Coding Design Principles

The capacity and PEP analyses of a general STC-OFDM system shed some light on the STC coding design problem:

1. The dominant exponent in the PEP (10.78) that is related to the structure of the code is r, the rank of the matrix $\boldsymbol{D}$. Recall that $\min_{\boldsymbol{x},\tilde{\boldsymbol{x}}} r \leq \min(D_{\text{eff}}, NL)$; to achieve the maximum diversity (NML), it is necessary that $D_{\text{eff}} \geq NL$ [i.e., the effective length of the code must be larger than the dimension of matrix $\boldsymbol{D}$ in (10.74)]. Since L is associated with the channel characteristic, which is not known to the transmitter (or the STC encoder) in advance, it is preferable to have an STC code with large effective length.

2. Another factor in the PEP is $\prod_{j=1}^{r} \lambda_j$, the product of eigenvalues of matrix $\boldsymbol{D}$. Since $\boldsymbol{D}$ changes with different channel setup, the optimal design of $\prod_{j=1}^{r} \lambda_j$ is not feasible. However, as observed in [477], the space-time trellis codes (STTCs) with higher state numbers (and essentially larger effective length) have better performance, which suggests that increasing the effective length of the STC beyond the minimum requirement (e.g., NL in our system) may help to improve the factor $\prod_{j=1}^{r} \lambda_j$.

3. Also, as seen from (10.69), to achieve the channel capacity, all the NKP transmitted STC symbols are required to be independent. Therefore, after introducing the coding constraints to the coded symbols, an interleaver is needed to scramble the coded symbols in order to satisfy the independence condition. From the standpoint of PEP analysis, such an interleaver helps to improve the factor $\prod_{j=1}^{r} \lambda_j$ as well.

In summary, in the system considered here, because of the diverse fading profiles of the wireless channels and the assumption that the CSI is known only at the receiver, the systematic coding design (e.g., by computer search) is less helpful; instead, two general principles should be met in choosing STC codes in order to robustly exploit the rich diversity resources in this system, namely, *large effective length* and *ideal interleaving*.

Space-time trellis codes have been proposed for multiple-antenna systems over flat-fading channels [477]. However, the complexity of the STTC increases dramatically as the effective length increases, and therefore it may not be a good candidate for the OFDM system considered here. Another family of STCs comprises the turbo-code-based STCs [281, 459], but their decoding complexity is high and they are not flexible in terms of scalability (e.g., when employed in systems with different requirements of the information rate). Here, we consider an alternative STC scheme—LDPC-based STC [293].

LDPC-Based STC

The LDPC codes have the following advantages for the STC-OFDM system considered here: (1) The LDPC decoder usually has a lower computational complexity than the turbo-code decoder. In addition to this, since the decoding complexity of each iteration in an LDPC decoder is much less than a turbo-code decoder, a finer resolution in the performance–complexity trade-off can be obtained by varying the maximum number of iterations. Moreover, the decoding of LDPC is highly parallelizable. (2) The minimum distance of binary LDPC codes increases linearly with the block length with probability close to 1 [127]. (3) It is easier to design a competitive LDPC code with any block length and any code rate, which makes it easier for the LDPC-based STC to scale according to different system requirements (e.g., different numbers of antennas or different information rates). (4) LDPC codes do not typically show an error floor, which is suitable for short-frame applications. (5) Due to the random generation of the parity-check matrix (or equivalently, the encoder matrix), the coded bits have been effectively interleaved; therefore, no extra interleaver is needed.

The transmitter structure of an LDPC-based STC-OFDM system is illustrated in Fig. 10.27. Denote by Ω the set of all possible STC symbols, which is up to a constant $\sqrt{\gamma}$ of the traditional constellation (e.g., MPSK or QAM). (Recall that the additive noise is assumed to have unit variance.) The $PK \log_2 |\Omega|$ information bits are first encoded by a rate $R = 1/N$ LDPC encoder into $NPK \log_2 |\Omega|$ coded bits, and then the binary LDPC coded bits are modulated into NPK STC symbols by an MPSK (or QAM) modulator. These NPK STC symbols, which correspond to an STC code word, are split into N streams; the PK STC symbols of each stream are transmitted from one particular transmitter antenna at K subcarriers and over P adjacent OFDM slots. Note that in the bit-interleaved coded-modulation system proposed above, the built-in random interleaver of the LDPC codes is also helpful to minimize loss in the effective length between the binary LDPC code bits and the modulated STC code symbols, which is caused by the MPSK (or QAM) modulation.

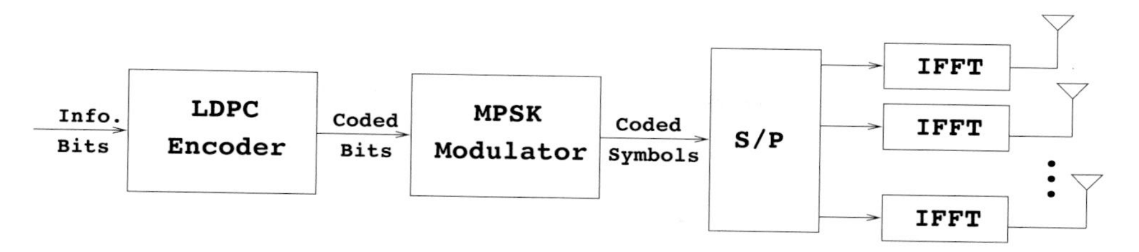

Figure 10.27. Transmitter structure of an LDPC-based STC-OFDM system with multiple antennas.

As an example, consider a regular binary LDPC code with column weight $t = 3$, rate $R = \frac{1}{2}$, and block length $n = 1024$, in which case the minimum distance is around 100 [127]. The STC based on this LDPC code is configured with a QPSK modulator and two transmitter antennas; therefore, the effective length of this LDPC-based STC is at least 25, which is more than enough to satisfy the minimum effective length requirement for a two-transmitter antenna ($N = 2$) OFDM system in a six-tap ($L = 6$) frequency-selective fading channel. Together with its built-in random interleaver, this LDPC code can well satisfy the two coding design principles mentioned earlier and therefore is an empirically good STC for the OFDM system considered in this chapter. Since the minimum distance of binary LDPC codes increase linearly with the block length, further performance improvement is possible by increasing the block length. Note that we do not claim the optimality of the proposed LDPC-based STC; but rather, we argue that with its low decoding complexity, flexible scalability, and high performance, the LDPC-based STC is a promising coding technique for reliable high-speed data communication in multiple-antenna OFDM systems with frequency- and time-selective fading.

As in a typical data communication scenario, communication is carried out in a bursty manner. A data burst is illustrated in Fig. 10.13. It spans $(Pq + 1)$ OFDM words, with the first OFDM word containing known pilot symbols. The remaining Pq OFDM words contain q STC code words.

10.5.4 Turbo Receiver

We next consider receiver design for the proposed LDPC-based STC-OFDM system. Even with ideal CSI, the optimal decoding algorithm for this system has exponential complexity. Hence we apply the turbo receiver structure. As a standard procedure, to demodulate each STC code word, the turbo receiver consists of two stages, the *soft demodulator* and the *soft LDPC decoder*, and the extrinsic information is iteratively exchanged between these two stages to successively improve the receiver performance.

However, in practice, the channel state information must be estimated by the receiver. In the following we discuss a turbo receiver for unknown fast-fading channels based on the MAP-EM algorithm. The turbo receiver for the LDPC-based STC-OFDM system is illustrated in Fig. 10.28. It consists of a soft maximum *a posteriori*

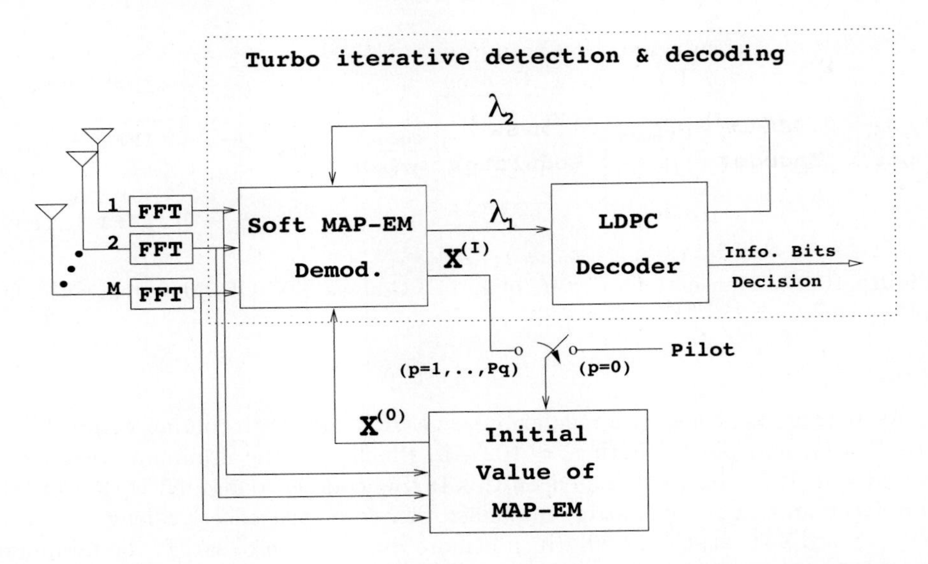

Figure 10.28. Turbo receiver structure which employs a MAP-EM demodulator and a soft LDPC decoder for multiple-antenna LDPC-based STC-OFDM systems in unknown fading channels.

expectation-maximization (MAP-EM) demodulator and a soft LDPC decoder, both of which are iterative devices themselves. The soft MAP-EM demodulator takes as input the FFT of the received signals from M receiver antennas, and the extrinsic log-likelihood ratios of the LDPC coded bits $\{\lambda_2\}$ [cf. (10.62)] (which are fed back by the soft LDPC decoder). It computes as output the extrinsic *a posteriori* LLRs of the LDPC coded bits $\{\lambda_1^e\}$ [cf. (10.62)]. (As an important issue in the EM algorithm, the initialization of the MAP-EM demodulator will be discussed later in this section.) The soft LDPC decoder takes as input the LLRs of the LDPC coded bits from the MAP-EM demodulator and computes as output the extrinsic LLRs of the LDPC coded bits as well as the hard decisions of the information bits at the last turbo iteration. It is assumed that the q STC words in a data burst are encoded independently. Therefore, each STC word (consisting of P OFDM words) is decoded independently by turbo processing. We next describe each component of the receiver shown in Fig. 10.28.

MAP-EM Demodulator

Here we apply the MAP-EM algorithm discussed in Section 10.4.3. For notational simplicity, here we consider an LDPC-based STC-OFDM system with two transmitter antennas and one receiver antenna. The results can easily be extended to a system with N transmitter antennas and M receiver antennas. Note that for the purpose of performance analysis, the $\boldsymbol{h}_{i,j}(p)$ defined in (10.66) contains only the time responses of L_f nonzero taps; whereas for the purpose of receiver design, especially when channel state information (CSI) is not available, the $\boldsymbol{h}_{i,j}(p)$ needs to be redefined to contain the time responses of all the taps within the maximum multipath spread. That is, $\boldsymbol{h}_{i,j}(p) \triangleq \left[h_{i,j}[1;p], \ldots, h_{i,j}[L'_f;p]\right]^T$, with $L'_f \triangleq \lceil \tau_m Q \Delta_f + 1 \rceil \geq L_f$ and τ_m being the maximum multipath spread; and $\boldsymbol{w}_f(k)$ is correspondingly redefined as $\boldsymbol{w}_f(k) \triangleq \left[e^{\jmath 0}, \ldots, e^{-\jmath 2\pi k(L'_f-1)/Q}\right]^H$. The received signal during one data burst can be written as

$$\boldsymbol{y}[p] = \boldsymbol{X}[p]\boldsymbol{W}\boldsymbol{h}[p] + \boldsymbol{z}[p], \qquad p = 0, 1, \ldots, Pq, \tag{10.82}$$

with

$$\begin{aligned}
\boldsymbol{X}[p] &\triangleq [\boldsymbol{X}_1[p], \boldsymbol{X}_2[p]]_{K\times(2K)}, \\
\boldsymbol{X}_j[p] &\triangleq \operatorname{diag}\{x_j[p,0], x_j[p,1], \ldots, x_j[p,Q-1]\}_{K\times K}, \qquad j = 1, 2, \\
\boldsymbol{W} &\triangleq \operatorname{diag}[\boldsymbol{W}_f, \boldsymbol{W}_f]_{(2Q)\times(2L'_f)}, \\
\boldsymbol{W}_f &\triangleq [\boldsymbol{w}_f(0), \boldsymbol{w}_f(1), \ldots, \boldsymbol{w}_f(Q-1)]^H_{Q\times L'_f}, \\
\boldsymbol{h}[p] &\triangleq \left[\boldsymbol{h}^H_{1,1}(p), \boldsymbol{h}^H_{1,2}(p)\right]^H_{(2L'_f)\times 1},
\end{aligned}$$

where $\boldsymbol{y}[p]$ and $\boldsymbol{z}[p]$ are Q-sized vectors which contain, respectively, the received signals and the ambient Gaussian noise at all Q subcarriers and at the pth time slot; the diagonal elements of $\boldsymbol{X}_j[p]$ are the Q STC symbols transmitted from the jth transmitter antenna and at the pth time slot.

Without CSI, the MAP detection problem is written as,

$$\hat{\boldsymbol{X}}[p] = \arg\max_{\boldsymbol{X}[p]} \log p(\boldsymbol{X}[p]|\boldsymbol{y}[p]), \quad p = 1, 2, \ldots, Pq. \tag{10.83}$$

(Recall that $\boldsymbol{X}[0]$ contains pilot symbols.) As in Section 10.4, we use the EM algorithm to solve (10.83).

In the E-step, the expectation is taken with respect to the "hidden" channel response $\boldsymbol{h}$ conditioned on $\boldsymbol{y}$ and $\boldsymbol{X}^{(i)}$. It is easily seen that conditioned on $\boldsymbol{y}$ and $\boldsymbol{X}^{(i)}$, $\boldsymbol{h}$ has a complex Gaussian distribution given by

$$\boldsymbol{h}|(\boldsymbol{y}, \boldsymbol{X}^{(i)}) \sim \mathcal{N}_c(\hat{\boldsymbol{h}}, \hat{\boldsymbol{\Sigma}}_h), \tag{10.84}$$

with

$$\begin{aligned}
\hat{\boldsymbol{h}} &\triangleq (\boldsymbol{W}^H\boldsymbol{X}^{(i)H}\boldsymbol{\Sigma}_z^{-1}\boldsymbol{X}^{(i)}\boldsymbol{W} + \boldsymbol{\Sigma}_h^{\dagger})^{-1}\boldsymbol{W}^H\boldsymbol{X}^{(i)H}\boldsymbol{\Sigma}_z^{-1}\boldsymbol{y} \\
&= (\boldsymbol{W}^H\boldsymbol{X}^{(i)H}\boldsymbol{X}^{(i)}\boldsymbol{W} + \boldsymbol{\Sigma}_h^{\dagger})^{-1}\boldsymbol{W}^H\boldsymbol{X}^{(i)H}\boldsymbol{y}, \\
\hat{\boldsymbol{\Sigma}}_h &\triangleq \boldsymbol{\Sigma}_h - (\boldsymbol{W}^H\boldsymbol{X}^{(i)H}\boldsymbol{\Sigma}_z^{-1}\boldsymbol{X}^{(i)}\boldsymbol{W} + \boldsymbol{\Sigma}_h^{\dagger})^{-1}\boldsymbol{W}^H\boldsymbol{X}^{(i)H}\boldsymbol{\Sigma}_z^{-1}\boldsymbol{X}^{(i)}\boldsymbol{W}\boldsymbol{\Sigma}_h \\
&= \boldsymbol{\Sigma}_h - (\boldsymbol{W}^H\boldsymbol{X}^{(i)H}\boldsymbol{X}^{(i)}\boldsymbol{W} + \boldsymbol{\Sigma}_h^{\dagger})^{-1}\boldsymbol{W}^H\boldsymbol{X}^{(i)H}\boldsymbol{X}^{(i)}\boldsymbol{W}\boldsymbol{\Sigma}_h,
\end{aligned}$$

where $\boldsymbol{\Sigma}_z$ and $\boldsymbol{\Sigma}_h$ denote, respectively, the covariance matrix of the ambient white Gaussian noise $\boldsymbol{z}$ and channel response $\boldsymbol{h}$. As before, by assumption, both of them are diagonal matrices as $\boldsymbol{\Sigma}_z \triangleq E\{\boldsymbol{z}\boldsymbol{z}^H\} = \boldsymbol{I}$ and $\boldsymbol{\Sigma}_h \triangleq E\left\{\boldsymbol{h}\boldsymbol{h}^H\right\} = \text{diag}\left\{\sigma_{1,1}^2, \ldots, \sigma_{1,L'_f}^2, \sigma_{2,1}^2, \ldots, \sigma_{2,L'_f}^2\right\}$, where $\sigma_{j,l}^2$ is the average power of the lth tap related with the jth transmitter antenna; $\sigma_{j,l}^2 = 0$ if the channel response at this tap is zero. Assuming that $\boldsymbol{\Sigma}_h$ is known (e.g., measured with the aid of pilot symbols), $\boldsymbol{\Sigma}_h^\dagger \triangleq \text{diag}\left\{\gamma_{1,1}, \ldots, \gamma_{1,L'_f}, \gamma_{2,1}, \ldots, \gamma_{2,L'_f}\right\}$ is defined as the pseudo-inverse of $\boldsymbol{\Sigma}_h$ as

$$\gamma_{j,l} = \begin{cases} 1/\sigma_{j,l}^2 & \sigma_{j,l}^2 \neq 0, \\ 0 & \sigma_{j,l}^2 = 0, \end{cases} \qquad l = 1, \ldots, L'_f, \quad j = 1, 2\,. \tag{10.85}$$

Using (10.82) and (10.84), $Q(\boldsymbol{X}|\boldsymbol{X}^{(i)})$ is computed as

$$\begin{aligned} &Q(\boldsymbol{X}|\boldsymbol{X}^{(i)}) \\ &= -E_{\boldsymbol{h}|(\boldsymbol{y},\boldsymbol{X}^{(i)})}\left\{\|\boldsymbol{y} - \boldsymbol{X}\boldsymbol{W}\boldsymbol{h}\|^2\right\} + \text{const.} \\ &= -E_{\boldsymbol{h}|(\boldsymbol{y},\boldsymbol{X}^{(i)})}\left\{\|(\boldsymbol{y} - \boldsymbol{X}\boldsymbol{W}\hat{\boldsymbol{h}}) + (\boldsymbol{X}\boldsymbol{W}\hat{\boldsymbol{h}} - \boldsymbol{X}\boldsymbol{W}\boldsymbol{h})\|^2\right\} + \text{const.} \\ &= -\|\boldsymbol{y} - \boldsymbol{X}\boldsymbol{W}\hat{\boldsymbol{h}}\|^2 - E_{\boldsymbol{h}|(\boldsymbol{y},\boldsymbol{X}^{(i)})}\left\{(\boldsymbol{h} - \hat{\boldsymbol{h}})^H\boldsymbol{W}^H\boldsymbol{X}^H\boldsymbol{X}\boldsymbol{W}(\boldsymbol{h} - \hat{\boldsymbol{h}})\right\} + \text{const.} \\ &= -\|\boldsymbol{y} - \boldsymbol{X}\boldsymbol{W}\hat{\boldsymbol{h}}\|^2 - \text{trace}\{\boldsymbol{X}\boldsymbol{W}\hat{\boldsymbol{\Sigma}}_h\boldsymbol{W}^H\boldsymbol{X}^H\} + \text{const.} \\ &= -\sum_{k=0}^{K-1} \underbrace{\left\{\left[y[k] - \boldsymbol{x}^H[k]\boldsymbol{W}'_f(k)\hat{\boldsymbol{h}}\right]^2 + \left[\boldsymbol{x}^H[k]\hat{\boldsymbol{\Sigma}}_h(k)\boldsymbol{x}[k]\right]\right\}}_{q(\boldsymbol{x}[k])} + \text{const.}, \end{aligned} \tag{10.86}$$

with

$$\begin{aligned} \boldsymbol{x}[k] &\triangleq [x_1[k], x_2[k]]_{2\times 1}^H, \\ \boldsymbol{W}'_f(k) &\triangleq \begin{bmatrix} \boldsymbol{w}_f^H(k) & \boldsymbol{0} \\ \boldsymbol{0} & \boldsymbol{w}_f^H(k) \end{bmatrix}_{2\times(2L'_f)}, \\ \hat{\boldsymbol{\Sigma}}_h(k) &\triangleq \begin{bmatrix} [\boldsymbol{W}\hat{\boldsymbol{\Sigma}}_h\boldsymbol{W}^H]_{(k+1,k+1)} & [\boldsymbol{W}\hat{\boldsymbol{\Sigma}}_h\boldsymbol{W}^H]_{(Q+k+1,k+1)} \\ [\boldsymbol{W}\hat{\boldsymbol{\Sigma}}_h\boldsymbol{W}^H]_{(k+1,Q+k+1)} & [\boldsymbol{W}\hat{\boldsymbol{\Sigma}}_h\boldsymbol{W}^H]_{(Q+k+1,Q+k+1)} \end{bmatrix}_{2\times 2}, \end{aligned}$$

where $[\boldsymbol{W}\hat{\boldsymbol{\Sigma}}_h\boldsymbol{W}^H]_{(i,j)}$ denotes the (i,j)th element of the matrix $[\boldsymbol{W}\hat{\boldsymbol{\Sigma}}_h\boldsymbol{W}^H]$.

Next, based on (10.86), the M-step proceeds as follows:

$$\begin{aligned} \boldsymbol{X}^{(i+1)} &= \arg\max_{\boldsymbol{X}}\left[Q(\boldsymbol{X}|\boldsymbol{X}^{(i)}) + \log P(\boldsymbol{X})\right] \\ &= \arg\max_{\boldsymbol{X}}\left[-\sum_{k=0}^{Q-1} q(\boldsymbol{x}[k]) + \sum_{k=0}^{Q-1}\log P(\boldsymbol{x}[k])\right] \end{aligned} \tag{10.87}$$

$$= \sum_{k=0}^{Q-1} \arg\min_{\boldsymbol{x}[k]} \left[q(\boldsymbol{x}[k]) - \log P(\boldsymbol{x}[k]) \right],$$

or

$$\boldsymbol{x}^{(i+1)}[k] = \arg\min_{\boldsymbol{x}[k]} \left[q(\boldsymbol{x}[k]) - \log P(\boldsymbol{x}[k]) \right], \qquad k = 0, 1, \ldots, Q-1, \tag{10.88}$$

where (10.87) follows from the assumption that $\boldsymbol{X}$ contains independent symbols. It is seen from (10.88) that the M-step can be decoupled into Q independent minimization problems, each of which can be solved by enumeration over all possible $\boldsymbol{x} \in \Omega \times \Omega$. (Recall that Ω denotes the set of all STC symbols.) Hence the total complexity of the maximization step is $\mathcal{O}(Q|\Omega|^2)$.

Within each turbo iteration, the E-step and M-step above are iterated I times. At the end of the Ith EM iteration, the extrinsic *a posteriori* LLRs of the LDPC code bits are computed and then fed to the soft LDPC decoder. At each OFDM subcarrier, two transmitter antennas transmit two STC symbols, which correspond to $2\log_2|\Omega|$ LDPC code bits. Based on (10.88), after I EM iterations, the extrinsic *a posteriori* LLR of the jth ($j = 1, \ldots, 2\log_2|\Omega|$) LDPC code bit at the kth subcarrier $d^j[k]$ is computed at the output of the MAP-EM demodulator as follows:

$$\begin{aligned}
\lambda_1(d^j[k]) &= \log \frac{P\left(d^j[k] = +1 | \boldsymbol{y}\right)}{P\left(d^j[k] = -1 | \boldsymbol{y}\right)} - \log \frac{P\left(d^j[k] = +1\right)}{P\left(d^j[k] = -1\right)} \\
&= \log \frac{\sum_{\boldsymbol{x} \in \mathcal{C}_j^+} P\Big(\boldsymbol{x}[k] = \boldsymbol{x} | \boldsymbol{y}\Big)}{\sum_{\boldsymbol{x} \in \mathcal{C}_j^-} P\Big(\boldsymbol{x}[k] = \boldsymbol{x} | \boldsymbol{y}\Big)} - \lambda_2(d^j[k]) \\
&= \underbrace{\log \frac{\sum_{\boldsymbol{x} \in \mathcal{C}_j^+} \exp\Big[- q(\boldsymbol{x}) + \log P(\boldsymbol{x})\Big]}{\sum_{\boldsymbol{x} \in \mathcal{C}_j^-} \exp\Big[- q(\boldsymbol{x}) + \log P(\boldsymbol{x})\Big]}}_{\Lambda_1(d^j[k])} - \lambda_2(d^j[k]),
\end{aligned} \tag{10.89}$$

where $\mathcal{C}_j^+$ is the set of $\boldsymbol{x}$ for which the jth LDPC coded bit is "+1" and $\mathcal{C}_j^-$ is defined similarly. The extrinsic *a priori* LLRs $\{\lambda_2(d^j[k])\}_{j,k}$ are provided by the soft LDPC decoder at the previous turbo iteration. Finally, the extrinsic *a posteriori* LLRs $\{\lambda_1(d^j[k])\}_{j,k}$ are sent to the soft LDPC decoder, which in turn iteratively computes the extrinsic LLRs $\{\lambda_2(d^j[k])\}_{j,k}$ and then feeds them back to the MAP-EM demodulator and thus completes one turbo iteration. At the end of the last turbo iteration, hard decisions of the information bits are output by the LDPC decoder.

Initialization of MAP-EM Demodulator

The performance of the MAP-EM demodulator (and hence the overall receiver) is closely related to the quality of the initial value of $\boldsymbol{X}^{(0)}[p]$ [cf. (10.44)]. At each turbo iteration, $\boldsymbol{X}^{(0)}[p]$ needs to be specified to initialize the MAP-EM demodulator. Except for the first turbo iteration, $\boldsymbol{X}^{(0)}[p]$ is simply taken as $\boldsymbol{X}^{(I)}[p]$ given by (10.87) from the previous turbo iteration. We next discuss the procedure for computing $\boldsymbol{X}^{(0)}[p]$ at the first turbo iteration.

The initial estimate of $\boldsymbol{X}^{(0)}[p]$ is based on the method proposed in [260, 263], which makes use of pilot symbols and decision feedback as well as spatial and temporal filtering for channel estimates. The procedure is listed in Table 10.2, where Freq-filter denotes either the least-squares estimator (LSE) or the MMSE estimator as

$$\begin{aligned}\text{LSE:} \quad & \textsf{Freq-filter}\Big\{\boldsymbol{y}, \boldsymbol{X}\Big\} \triangleq (\boldsymbol{W}^H\boldsymbol{X}^H\boldsymbol{X}\boldsymbol{W})^{-1}\boldsymbol{W}^H\boldsymbol{X}^H\boldsymbol{y}, \\ \text{MMSE:} \quad & \textsf{Freq-filter}\Big\{\boldsymbol{y}, \boldsymbol{X}\Big\} \triangleq (\boldsymbol{W}^H\boldsymbol{X}^H\boldsymbol{X}\boldsymbol{W} + \boldsymbol{\Sigma}_h^{\dagger})^{-1}\boldsymbol{W}^H\boldsymbol{X}^H\boldsymbol{y},\end{aligned} \tag{10.90}$$

where $\boldsymbol{X}$ represents either the pilot symbols or $\boldsymbol{X}^{(I)}$ provided by the MAP-EM demodulator. Comparing these two estimators, the LSE does not need any statistical information of $\boldsymbol{h}$, but the MMSE offers better performance in terms of mean-square error (MSE). Hence, in the pilot slot, the LSE is used to estimate channels and to measure $\boldsymbol{\Sigma}_h$; and in the remaining data slots, the MMSE is used. In Table 10.2, Temp-filter denotes the temporal filter, which is used to further exploit the time-domain correlation of the channel:

$$\textsf{Temp-filter}\Big\{\hat{\boldsymbol{h}}[p-1], \hat{\boldsymbol{h}}[p-2], \ldots, \hat{\boldsymbol{h}}[p-\iota]\Big\} \triangleq \sum_{j=1}^{\iota} a_j \hat{\boldsymbol{h}}[p-j], \tag{10.91}$$

where $\hat{\boldsymbol{h}}[p-j]$, $j = 1, \ldots, \iota$, is computed from ($\star\star$) (cf. Table 10.1); $\{a_j\}_{j=1}^{\iota}$ denotes the coefficients of an ι-length ($\iota \leq Pq$) temporal filter, which can be obtained by solving the Wiener equation or from robust design as in [260, 263]. From the discussion above, it is seen that the computation involved in initializing $\boldsymbol{X}^{(0)}[p]$ consists mainly of the ML detection of $\boldsymbol{X}^{(0)}[p]$ in ($\star$) and the estimation of $\hat{\boldsymbol{h}}[p]$ in ($\star\star$). In general, for an STC-OFDM system with parameters (N, M, Q, L'_f), the total complexity in initializing $\boldsymbol{X}^{(0)}[p]$ is $\mathcal{O}[(Q|\Omega|^N) + M(NL'_f)^3]$.

Simulation Examples

In this section we provide simulation results to illustrate the performance of the proposed LDPC-based STC-OFDM system in frequency- and time-selective fading channels. The correlated fading processes are generated by using the methods in [180]. In the following simulations the available bandwidth is 1 MHz and is divided into 256 subcarriers. These correspond to a subcarrier symbol rate of 3.9 kHz and OFDM word duration of 256 μs. In each OFDM word, a guard interval of 40 μs is added to combat the effect of intersymbol interference; hence $T = 296\,\mu$s. For all simulations, 512 information bits are transmitted from 256 subcarriers at each OFDM slot, and therefore the information rate is $2 \times 256/296 = 1.73$ bits/s per hertz. Unless otherwise specified, all the LDPC codes used in simulations are regular LDPC codes with column weight $t = 3$ in the parity-check matrices and

Table 10.2. Procedure for Computing $\boldsymbol{X}^{(0)}[p]$ for a MAP-EM Demodulator (at the First Turbo Iteration)

pilot slot:		$\hat{\boldsymbol{h}}[0]$	$=$	$\text{Freq-filter}\left\{\boldsymbol{y}[0], \boldsymbol{X}[0]\right\}$	
data slots:	for	$p = 1, 2, \ldots, Pq$			
		$\tilde{\boldsymbol{h}}[p]$	$=$	$\text{Temp-filter}\left\{\hat{\boldsymbol{h}}[p-1], \hat{\boldsymbol{h}}[p-2], \ldots, \hat{\boldsymbol{h}}[p-\iota]\right\}$	
		$\boldsymbol{X}^{(0)}[p]$	$=$	$\text{arg max}_{\boldsymbol{X}}\left\{\log p\left[\boldsymbol{y}[p] \middle\vert \boldsymbol{X}, \tilde{\boldsymbol{h}}[p]\right]\right\}$	$(\star)$
		$\boldsymbol{X}^{(I)}[p]$	$=$	$\text{MAP-EM}\left\{\boldsymbol{y}[p], \boldsymbol{X}^{(0)}[p]\right\}$ [cf. Eq. (10.44)–(10.45)]	
		$\hat{\boldsymbol{h}}[p]$	$=$	$\text{Freq-filter}\left\{\boldsymbol{y}[p], \boldsymbol{X}^{(I)}[p]\right\}$	$(\star\star)$
	end				

with appropriate block lengths and code rates. The modulator uses the QPSK constellation. Simulation results are shown in terms of the OFDM word-error rate (WER) versus the SNR γ.

Performance with Ideal CSI Figures 10.29 and 10.30 show the performance of multiple-antenna (N transmitter antennas and one receiver antenna) LDPC-based STC-OFDM systems by using turbo detection and decoding with ideal CSI. Performance is compared for systems with different fading profiles and different numbers of transmitter antennas. Namely, Ch1 denotes a channel with a single tap at 0 μs, Ch2a denotes a channel with two equal-power taps at 0 and 5 μs, Ch2b denotes a channel with two equal-power taps at 0 and 40 μs, and Ch6a denotes a channel with six equal-power taps equally spaced from 0 to 40 μs. Suffix N2 denotes a system with two transmitter antennas ($N = 2$), and similarly for N3; suffix P1 denotes that each STC code word spans one OFDM slot ($P = 1$), and similarly for P5 and P10. Unless otherwise specified, all the STC-OFDM systems are assumed to use two transmitter antennas ($N = 2$) and each STC code word spans one OFDM slot ($P = 1$).

First, Fig. 10.29 shows the performance of the LDPC-based STC-OFDM system in frequency- and time-nonselective channels. The dash-dotted curves represent the performance after the first turbo iteration; and the solid curves represent the performance after the fifth iteration. It is seen that the receiver performance is improved significantly through turbo iteration. During each turbo iteration, in the LDPC decoder, the maximum number of iterations is 30; and as observed in simulations, the average number of iterations needed in LDPC decoding is less than 10 when WER is less than 10^{-2}. Compared with the conventional trellis-based STC-OFDM system (see figures in [8]), the LDPC-based STC-OFDM system improves performance significantly (e.g., there is around a 5-dB performance improvement in Ch2a/Ch2b channels and even more improvement in Ch6a channels). Moreover, due to the inherent interleaving in the LDPC encoder, the proposed LDPC-based STC narrows the performance difference between Ch2a and Ch2b channels (essentially the outage capacity of these two channels are the same). As the selective-fading diversity order L increases from Ch1 to Ch6a, LDPC-based STC can efficiently take advantage of the available diversity resources and hence can significantly improve the system performance. Moreover, in a highly frequency-selective channel, Ch6a, the LDPC-based STC performs only 3.0 dB away from the outage capacity of this channel (at a high information rate, 1.73 bits/s per hertz) at WER of 2×10^{-4}.

Next, Fig. 10.30 shows the performance of the LDPC-based STC-OFDM system in frequency- and time-selective fading channels. The maximum Doppler frequency is 200 Hz (i.e., the normalized Doppler frequency is $f_d T = 0.059$). Again, it is seen that the performance of the system improves as the selective-fading diversity order L (including both frequency and time selectivities) increases.

Finally, Fig. 10.29 also compares the performance of LDPC-based STC-OFDM systems with the same multipath delay profiles (Ch2a) but with different numbers of transmitter antennas ($N = 2$ or $N = 3$). Since Ch2bN3 has a larger outage capacity than Ch2bN2, it is seen that at medium to high SNRs, Ch2bN3 starts to perform better than Ch2bN2 with a steeper slope, which shows that the LDPC-based

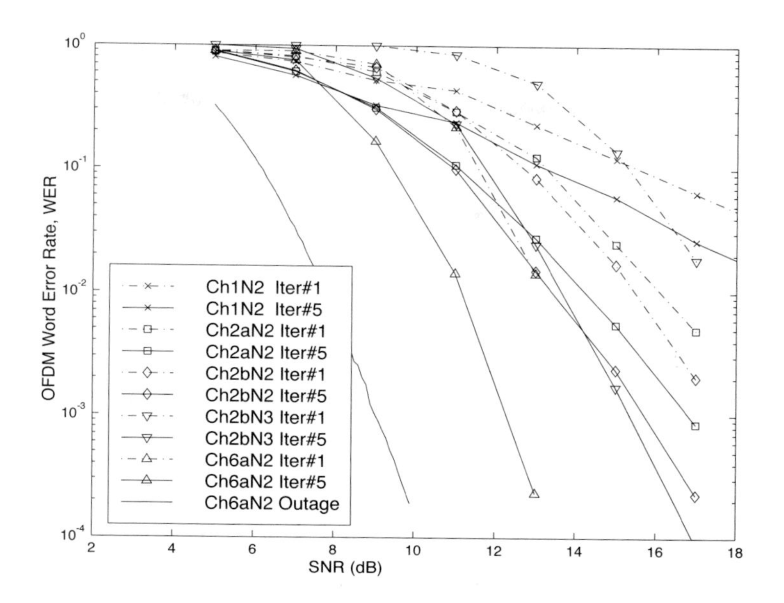

Figure 10.29. Word-error rate of an LDPC-based STC-OFDM system with multiple antennas ($N = \{2, 3\}, M = 1$) in frequency- and time-nonselective fading channels, with ideal CSI.

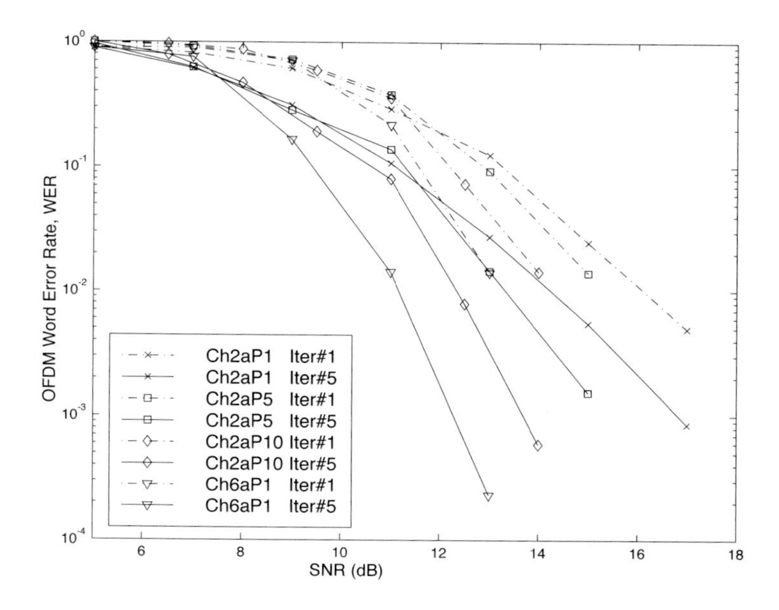

Figure 10.30. Word-error rate of an LDPC-based STC-OFDM system with multiple antennas ($N = 2, M = 1$) in frequency- and time-selective fading channels, with ideal CSI.

STC can be flexibly scaled according to a different number of transmitter antennas and can still improve the performance by exploiting the increased spatial diversity, especially at low WER (which is attractive in data communication applications).

Performance with Unknown CSI In the following simulations, the receiver performance with unknown CSI is shown. The system transmits in a burst manner, as illustrated in Fig. 10.13. Each data burst includes 10 OFDM words ($q = 9, P = 1$), the first OFDM word contains the pilot symbols, and the remaining nine OFDM words contain the information data symbols. Simulations are carried out in two-tap (two equal-power taps at 0 and 1 μs) frequency- and time-selective fading channels. The maximum Doppler frequency of the fading channels is 50 or 150 Hz (with normalized Doppler frequencies 0.015 and 0.044, respectively). Note that in Figs. 10.31 and 10.32, the energy consumption of transmitting pilot symbols is not taken into account in computing SNRs.

The turbo receiver performance of a regular LDPC-based STC-OFDM system is shown in Fig. 10.31, whereas that of an irregular LDPC-based STC-OFDM system is shown in Fig. 10.32. (The average column weight in the parity-check matrix of the irregular LDPC code is 2.30.) TurboDD denotes the turbo receiver as before, except that the perfect CSI is replaced by the pilot/decision-directed channel estimates as proposed in [262], and TurboEM denotes the turbo receiver with the MAP-EM demodulator as proposed in Section 10.5.4. The temporal filter parameters are taken from [260]. The performance of these two receiver structures is compared when using either the regular LDPC codes or the irregular LDPC codes. From the simulations it is seen that with ideal CSI the receiver performance of regular and irregular LDPC-based STC-OFDM systems is quite similar. When CSI is not available, the proposed TurboEM receiver reduces the error floor significantly. Moreover, it is observed that by using the irregular LDPC codes, both the TurboDD and TurboEM receivers improve their performance, and the TurboEM receiver can even approach the receiver performance with ideal CSI in low to medium SNRs. A possible reason for the better performance of irregular LDPC-based STC than that of regular LDPC-based STC in the presence of nonideal CSI is the better performance of the irregular LDPC codes at low SNRs. In simulations, the turbo receiver takes three turbo iterations, and at each turbo iteration, the MAP-EM demodulator takes three EM iterations. At the cost of 10% pilot insertion and a modest complexity, the turbo receiver with the MAP-EM demodulator is a promising receiver technique, especially for application in fast-fading channels.

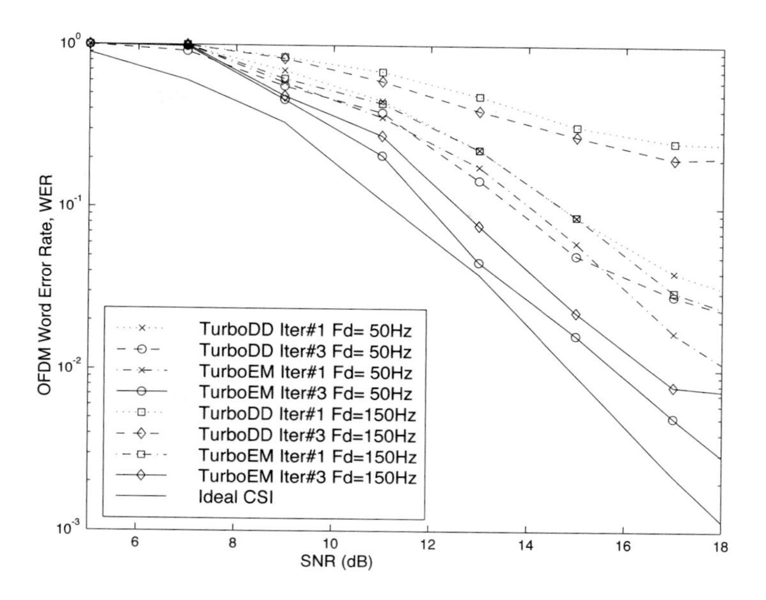

Figure 10.31. Word-error rate of a regular LDPC-based STC-OFDM system with multiple antennas ($N = 2, M = 1$) in two-tap ($L = 2$) frequency-selective fading channels, without CSI.

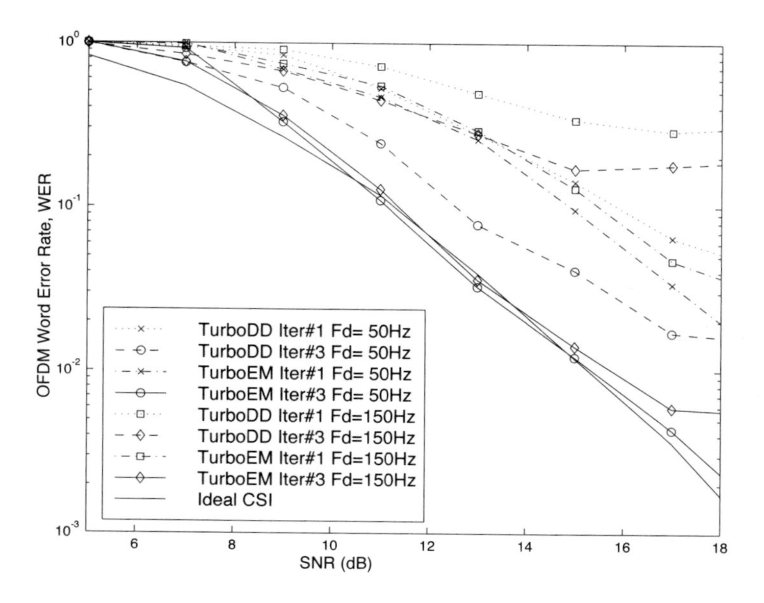

Figure 10.32. Word-error rate of an irregular LDPC-based STC-OFDM system with multiple antennas ($N = 2, M = 1$) in two-tap ($L = 2$) frequency-selective fading channels, without CSI.

10.6 Appendix

10.6.1 Derivations for Section 10.3

Note that the parameters with one-to-one mapping relationships, such as $\boldsymbol{b} \Leftrightarrow \boldsymbol{c} \Leftrightarrow \boldsymbol{X}$, $\epsilon \Leftrightarrow \boldsymbol{F}_\epsilon$, or $\boldsymbol{Y} \Leftrightarrow \boldsymbol{y}$, are equivalent to be conditioned upon [e.g., $p(\cdot|\boldsymbol{Y}) \equiv p(\cdot|\boldsymbol{y})$].

Derivation of (10.18)–(10.20)

$$
\begin{aligned}
p(\boldsymbol{h} \,|\, \boldsymbol{Y}, \boldsymbol{X}, \epsilon) &\propto p(\boldsymbol{Y} \,|\, \boldsymbol{h}, \boldsymbol{X}, \epsilon)\, p(\boldsymbol{h}) \\
&\propto \exp\Big\{ -\frac{1}{\sigma^2}\|\boldsymbol{Y} - \frac{1}{Q}\boldsymbol{W}\boldsymbol{F}_\epsilon \boldsymbol{W}^H \boldsymbol{X}\boldsymbol{W}\boldsymbol{h}\|^2 \Big\} \exp\Big\{ -(\boldsymbol{h}-\boldsymbol{h}_0)^H \boldsymbol{\Sigma}_0^{-1}(\boldsymbol{h}-\boldsymbol{h}_0) \Big\} \\
&\propto \exp\Big\{ -\boldsymbol{h}^H \underbrace{(\frac{Q}{\sigma^2}\boldsymbol{I}_L + \boldsymbol{\Sigma}_0^{-1})}_{\boldsymbol{\Sigma}_*^{-1}} \boldsymbol{h} \Big\} \\
&\qquad \cdot \exp\Big\{ 2\Re\Big[\boldsymbol{h}^H \underbrace{(\frac{1}{Q\sigma^2}\boldsymbol{W}^H \boldsymbol{X}^H \boldsymbol{W}\boldsymbol{F}_\epsilon^H \boldsymbol{W}^H \boldsymbol{Y} + \boldsymbol{\Sigma}_0^{-1}\boldsymbol{h}_0)}_{\boldsymbol{\Sigma}_*^{-1}\boldsymbol{h}_*} \Big] \Big\} \\
&\propto \exp\left\{ -(\boldsymbol{h}-\boldsymbol{h}_*)^H \boldsymbol{\Sigma}_*^{-1} (\boldsymbol{h}-\boldsymbol{h}_*) \right\} \sim \mathcal{N}_c\Big(\boldsymbol{h}_*, \boldsymbol{\Sigma}_*\Big).
\end{aligned}
\tag{10.92}
$$

Derivation of (10.21)

$$
\begin{aligned}
P(X_k = a_j \,|\, \boldsymbol{Y}, \boldsymbol{h}, \epsilon, \boldsymbol{X}_{[-k]}) &\propto p(\boldsymbol{Y} \,|\, \boldsymbol{h}, \epsilon, X_k = a_j, \boldsymbol{X}_{[-k]}) P(X_k = a_j, \boldsymbol{X}_{[-k]}) \\
&\propto \exp\Big\{ -\frac{1}{\sigma^2}\|\boldsymbol{Y} - \frac{1}{Q}\boldsymbol{W}\boldsymbol{F}_\epsilon \boldsymbol{W}^H \widetilde{\boldsymbol{X}}_{k,j}\boldsymbol{W}\boldsymbol{h}\|^2 \Big\} \\
&\qquad \cdot P(X_k = a_j \,|\, X_{k-1}) P(X_{k+1} \,|\, X_k = a_j) \\
&\propto \exp\Big\{ -\frac{Q}{\sigma^2}\|\boldsymbol{y} - \frac{1}{Q}\boldsymbol{F}_\epsilon \boldsymbol{W}^H \widetilde{\boldsymbol{X}}_{k,j}\boldsymbol{W}\boldsymbol{h}\|^2 \Big\} \\
&\qquad \cdot P(c_k = a_j X_{k-1}^*) P(c_{k+1} = a_j^* X_{k+1}) \\
&\propto \exp\Big\{ -\frac{1}{\sigma^2}\|\widetilde{\boldsymbol{Y}} - \widetilde{\boldsymbol{X}}_{k,j}\boldsymbol{H}\|^2 \Big\} P(c_k = a_j X_{k-1}^*) P(c_{k+1} = a_j^* X_{k+1}) \\
&\propto \exp\Big(-\frac{2}{\sigma^2}\Re\Big\{ \widetilde{Y}_k^* a_j H_k \Big\} \Big) P(c_k = a_j X_{k-1}^*) P(c_{k+1} = a_j^* X_{k+1}),
\end{aligned}
\tag{10.93}
$$

with

$$
\begin{aligned}
\widetilde{\boldsymbol{X}}_{k,j} &\triangleq \operatorname{diag}\Big\{ X_0^{(n)}, \ldots, X_{k-1}^{(n)}, a_j, X_{k+1}^{(n-1)}, \ldots, X_{Q-1}^{(n-1)} \Big\}, \\
\widetilde{\boldsymbol{Y}} &\triangleq \boldsymbol{W}\boldsymbol{F}_\epsilon^H \boldsymbol{y}, \qquad\qquad j = 1, \ldots, |\Omega|.
\end{aligned}
$$

Derivation of (10.27)

$$
\begin{aligned}
p(\epsilon \,|\, \boldsymbol{Y}, \boldsymbol{h}, \boldsymbol{X}) &\propto p(\boldsymbol{Y} \mid \boldsymbol{h}, \epsilon, \boldsymbol{X}) \propto \exp\left\{ -\frac{Q}{\sigma^2} \|\boldsymbol{Y} - \boldsymbol{F}_{\hat{\epsilon}}\overline{\boldsymbol{h}} - \overline{\boldsymbol{F}}_{\hat{\epsilon}}\overline{\boldsymbol{h}}(\epsilon - \hat{\epsilon})\|^2 \right\} \\
&\propto \exp\left\{ -\epsilon^2 \underbrace{\left(\frac{Q}{\sigma^2} \overline{\boldsymbol{h}}^H \overline{\boldsymbol{F}}_{\hat{\epsilon}}^H \overline{\boldsymbol{F}}_{\hat{\epsilon}} \overline{\boldsymbol{h}} \right)}_{(2\sigma_\epsilon^2)^{-1}} \right\} \exp\left\{ 2\epsilon \underbrace{\frac{Q}{\sigma^2} \Re\left[\left(\boldsymbol{y} - \boldsymbol{F}_{\hat{\epsilon}}\overline{\boldsymbol{h}} + \overline{\boldsymbol{F}}_{\hat{\epsilon}}\overline{\boldsymbol{h}}\hat{\epsilon} \right)^H \overline{\boldsymbol{F}}_{\hat{\epsilon}}\overline{\boldsymbol{h}} \right]}_{\mu_\epsilon (2\sigma_\epsilon^2)^{-1}} \right\} \\
&\propto \exp\left\{ -\frac{(\epsilon - \mu_\epsilon)^2}{2\sigma_\epsilon^2} \right\} \sim \mathcal{N}\left(\mu_\epsilon, \sigma_\epsilon^2 \right).
\end{aligned}
\qquad (10.94)
$$

ACRONYMS

3G	third generation
ACM	approximate conditional mean
AIC	Akaike information criterion
AME	asymptotic multiuser efficiency
AMPS	advanced mobile phone system
AR	autoregressive
ARMA	autoregressive moving average
AWGN	additive white Gaussian noise
BCJR	Bahl, Cocke, Jelinek, and Raviv
BER	bit-error rate
BPSK	binary phase-shift keying
CCD	canonical correlation decomposition
CCI	co-channel interference
CDLM	conditional dynamic linear model
CDMA	code-division multiple access
CSI	channel state information
DFE	decision feedback equalizer
DFT	discrete Fourier transform
DMI	direct matrix inversion
DS	direct sequence
DS-SS	direct-sequence spread spectrum
ED	eigendecomposition
EM	expectation-maximization
FDMA	frequency-division multiple access
FER	frame-error rate
FH	frequency hopping
FIR	finite impulse response
GSM	global system for mobile
HMM	hidden Markov model
IEEE	Institute of Electrical and Electronics Engineers
IF	influence function
IIR	infinite impulse response
IS-95	Interim Standard 95

ISI	inter-symbol interference
ISM	industrial, scientific, and medical
LAN	local area network
LCMV	linearly constrained minimum variance
LDPC	low-density parity check
LLR	log-likelihood ratio
LMS	least mean squares
MAI	multiple-access interference
MAN	metropolitan area network
MAP	maximum *a posteriori* probability
MCMC	Markov chain Monte Carlo
MDL	minimum description length
MF	matched filter
MIMO	multiple-input/multiple-output
MKF	mixture Kalman filter
ML	maximum likelihood
MLSD	maximum-likelihood sequence detector
MMSE	minimum mean-square error
MOE	minimum output energy
MPSK	M-ary phase-shift keying
MSE	mean-square error
MSIR	maximum signal-to-interference ratio
MUD	multiuser detection
NAHJ	noise-averaged Hermitian Jacobi
NBI	narrowband interference
NMT	Nordic mobile telephone
OFDM	orthogonal frequency-division multiplexing
PAN	personal area network
PASTd	projection approximation subspace tracking with deflation
PEP	pairwise error probability
QAM	quadrature amplitude modulation
QPSK	quadrature phase-shift keying
RLS	recursive least squares
SDMA	space-division multiple access
SINR	signal-to-interference-plus-noise ratio
SIR	signal-to-interference ratio
SIS	sequential importance sampling
SISO	soft-input/soft-output
SMC	sequential Monte Carlo
SNR	signal-to-noise ratio
STC	space-time code
STBC	space-time block code
STTC	space-time trellis code
SVD	singular value decomposition
TCM	trellis-coded modulation

TDL	tapped delay line
TDMA	time-division multiple access
ULA	uniform linear array
UWB	ultrawideband
VHF	very high frequency
VLSI	very large scale integrated
WCDMA	wideband CDMA
WER	word-error rate
WLL	wireless local loop
ZF	zero forcing

BIBLIOGRAPHY

[1] B. Aazhang and H. V. Poor. Performance of DS/SSMA communications in impulsive channels: I. Linear correlation receivers. *IEEE Trans. Commun.*, COM-35(11):1179–1187, Nov. 1987.

[2] B. Aazhang and H. V. Poor. Performance of DS/SSMA communications in impulsive channels: II. Hard-limiting correlation receivers. *IEEE Trans. Commun.*, 36(1):88–96, Jan. 1988.

[3] B. Aazhang and H. V. Poor. An analysis of nonlinear direct-sequence correlators. *IEEE Trans. Commun.*, 37(7):723–731, July 1989.

[4] A. Abdulrahman, D. D. Falconer, and A. U. Sheikh. Decision feedback equalization for CDMA in indoor wireless communications. *IEEE J. Select. Areas Commun.*, 12(4):698–706, May 1994.

[5] K. Abed-Meriam, P. Loubaton, and E. Moulines. A subspace algorithm for certain blind identification problems. *IEEE Trans. Inform. Theory*, 43(2):499–511, Mar. 1997.

[6] D. Achilles and D. N. Li. Narrow-band interference suppression in spread spectrum communication systems. *Signal Process.*, 24(1):61–76, 1991.

[7] T. Adali and S. H. Ardalan. On the effect of input signal correlation on weight misadjustment in the RLS algorithm. *IEEE Trans. Signal Process.*, 43(4):988–991, Apr. 1995.

[8] D. Agrawal, V. Tarokh, A. Naguib, and N. Seshadri. Space-time coded OFDM for high data-rate wireless communication over wideband channels. In *Proc. 1998 IEEE Veh. Technol. Conf.*, pages 2232–2236, 1998.

[9] Y. Akaiwa. *Introduction to Digital Mobile Communications.* Wiley, New York, 1997.

[10] E. Aktas and U. Mitra. Complexity reduction in subspace-based blind channel identification for DS/CDMA systems. *IEEE Trans. Commun.*, 48(8):1392–1404, Aug. 2000.

[11] A. S. Al-Ruwais, A. Al-Obaid, and S. A. Alshebeili. Suppression of narrowband interference in CDMA cellular radio telephone using higher order statistics. *Arab. J. Sci. Eng.*, 24(2C):51–68, Dec. 1999.

[12] S. M. Alamouti. A simple transmit diversity technique for wireless communications. *IEEE J. Select. Areas Commun.*, 16(8):1451–1458, Oct. 1998.

[13] P. D. Alexander, A. J. Grant, and M. C. Reed. Performance analysis of an iterative decoder for code-division multiple-access. *Eur. Trans. Telecommun.*, 9(5):419–426, Sept./Oct. 1998.

[14] P. D. Alexander, M. C. Reed, J. A. Asenstorfer, and C. B. Schlegel. Iterative multiuser interference reduction: Turbo CDMA. *IEEE Trans. Commun.*, 47(7):1008–1014, July 1999.

[15] S. A. Alshebeili, A. M. Al-Qurainy, and A. S. Al-Ruwais. Nonlinear approaches for narrowband interference suppression in DS spread spectrum systems. *Signal Process.*, 77(1):11–20, Aug. 1999.

[16] S. A. Alshebeili et al. Subspace-based approaches for narrowband interference suppression in DS spread spectrum systems. *J. Franklin Inst.*, 336(8):1199–1207, Nov. 1999.

[17] M. G. Amin, C. S. Wang, and A. R. Lindsey. Optimum interference excision in spread spectrum communications using open-loop adaptive filters. *IEEE Trans. Signal Process.*, 47(7):1966–1976, July 1999.

[18] T. W. Anderson. *An Introduction to Multivariate Statistical Analysis.* Wiley, New York, 2nd edition, 1984.

[19] A. Ansari and R. Viswanathan. Performance study of maximum-likelihood receivers and transversal filters for the detection of direct-sequence spread-spectrum signals in narrow-band interference. *IEEE Trans. Commun.*, 42(2/3/4):1939–1946, Feb./Mar./Apr. 1994.

[20] A. Ansari and R. Viswanathan. On SNR as a measure of performance for narrowband interference rejection in direct sequence spread spectrum. *IEEE Trans. Commun.*, 43(2/3/4):1318–1322, Feb./Mar./Apr. 1995.

[21] S. F. Arnold. Gibbs sampling. In C. R. Rao, editor, *Handbook of Statistics*, volume 9, pages 599–625. Elsevier, New York, 1993.

[22] E. Baccarelli and R. Cusani. Combined channel estimation and data detection using soft statistics for frequency-selective fast fading digital links. *IEEE Trans. Commun.*, 46(4):424–427, Apr. 1998.

[23] E. Baccarelli, R. Cusani, and S. Galli. An application of the HMM theory to optimal nonlinear equalization of quantised-output digital ISI channels. *Signal Process.*, 58(1):95–196, 1997.

[24] E. Baccarelli, R. Cusani, and S. Galli. A novel adaptive receiver with enhanced channel tracking capability for TDMA-based mobile radio communications. *IEEE Trans. Commun.*, 16(9):1630–1639, Dec. 1998.

[25] L. R. Bahl, J. Cocke, F. Jelinek, and J. Raviv. Optimal decoding of linear codes for minimizing symbol error rate. *IEEE Trans. Inform. Theory*, 20(3):284–287, Mar. 1974.

[26] A. N. Barbosa and S. L. Miller. Adaptive detection of DS/CDMA signals in fading channels. *IEEE Trans. Commun.*, 46(1):115–124, Jan. 1998.

[27] S. N. Batalama, M. J. Medley, and D. A. Pados. Robust adaptive recovery of spread-spectrum signals with short data records. *IEEE Trans. Commun.*, 48(10):1725–1731, Oct. 2000.

[28] S. N. Batalama, M. J. Medley, and I. N. Psaromiligkos. Adaptive robust spread-spectrum receivers. *IEEE Trans. Commun.*, 47(6):905–917, June 1999.

[29] S. Benedetto et al. Soft-output decoding algorithms in iterative decoding of turbo codes. *TDA Prog. Rep.*, 42(124):63–87, Aug. 1995.

[30] S. Benedetto and G. Monotorsi. Design of parallel concatenated convolutional codes. *IEEE Trans. Commun.*, 44(5):591–600, July 1999.

[31] S. E. Bensley and B. Aazhang. Subspace-based channel estimation for code-division multiple-access communication systems. *IEEE Trans. Commun.*, 44(8):1009–1020, Aug. 1996.

[32] S. E. Bensley and B. Aazhang. Maximum likelihood synchronization of a single user for CDMA communication systems. *IEEE Trans. Commun.*, 46(3):392–399, Mar. 1998.

[33] J. O. Berger. *Statistical Decision Theory and Bayesian Analysis*. Springer-Verlag, New York, 2nd edition, 1985.

[34] X. Bernstein and A. M. Haimovic. Space-time processing for increased capacity of wireless CDMA. In *Proc. 1996 IEEE Int. Conf. Commun.*, pages 597–601, 1996.

[35] C. Berrou and A. Glavieux. Near optimum error-correcting coding and decoding: Turbo codes. *IEEE Trans. Commun.*, 44(10):1261–1271, Oct. 1996.

[36] C. Berrou, A. Glavieux, and P. Thitimajshima. Near Shannon limit error-correction coding and decoding: Turbo codes. In *Proc. 1993 IEEE Int. Conf. Commun.*, pages 1064–1070, 1993.

[37] N. Bershad. Error probabilities of DS spread-spectrum using ALE for narrow-band interference rejection. *IEEE Trans. Commun.*, 36(5):587–595, May 1988.

[38] H. Bertoni. *Radio Propagation for Modern Wireless Systems*. Prentice Hall, Upper Saddle River, NJ, 2000.

[39] E. Biglieri, G. Caire, and G. Taricco. Limiting performance of block-fading channels with multiple antennas. *IEEE Trans. Inform. Theory*, 47(4):1273–1289, May 2001.

[40] E. Biglieri, J. Proakis, and S. Shamai. Fading channels: Information-theoretic and communication aspects. *IEEE Trans. Inform. Theory*, 44(6):2619–2692, Oct. 1998.

[41] B. Bing. *High-Speed Wireless ATM and LANs*. Artech House, Norwood, MA, 2000.

[42] B. Bing, R. Van Nee, and V. Hayes, editors. Wireless local area and home networks. *IEEE Commun. Mag.*, 39(11):63–157, Nov. 2001.

[43] C. H. Bischof and G. M. Shroff. On updating signal subspaces. *IEEE Trans. Signal Process.*, 40(1):96–105, Jan. 1992.

[44] K. L. Blackard, T. S. Rappaport, and C. W. Bostian. Measurements and models of radio frequency impulsive noise for indoor wireless communications. *IEEE J. Select. Areas Commun.*, 11(7):991–1001, Sept. 1993.

[45] T. K. Blankenship, D. M. Krizman, and T. S. Rappaport. Measurements and simulation of radio frequency impulsive noise in hospitals and clinics. In *Proc. 1997 IEEE Veh. Technol. Conf.*, pages 1942–1946, 1997.

[46] N. Blaustein. *Radio Propagation in Cellular Networks*. Artech House, Norwood, MA, 2000.

[47] R. S. Blum, R. J. Kozick, and B. M. Sadler. An adaptive spatial diversity receiver for non-Gaussian interference and noise. *IEEE Trans. Signal Process.*, 47(8):2100–2111, Aug. 1999.

[48] G. E. Bottomley. Adaptive arrays and MLSE equalization. In *Proc. 1995 IEEE Veh. Technol. Conf.*, pages 50–54, 1995.

[49] G. E. Bottomley and S. Chennakeshu. Unification of MLSE receivers and extension to time-varying channels. *IEEE Trans. Commun.*, 46(4):464–472, Apr. 1998.

[50] J. Boutros and G. Caire. Iterative multiuser joint decoding: Unified framework and asymptotic analysis. *IEEE Trans. Inform. Theory*, 48(4):1772–1773, July 2002.

[51] G. E. Box and G. C. Tiao. *Bayesian Inference in Statistical Analysis*. Addison-Wesley, Reading, MA, 1973.

[52] P. L. Brockett, M. Hinich, and G. R. Wilson. Nonlinear and non-Gaussian ocean noise. *J. Acoust. Soc. Am.*, 82:1286–1399, 1987.

[53] D. R. Brown III, M. Motani, V. V. Veeravalli, H. V. Poor, and C. R. Johnson. On the performance of linear parallel interference cancellation. *IEEE Trans. Inform. Theory*, 45(5):1957–1970, July 2001.

[54] G. D. Brushe, V. Krishnamurthy, and L. B. White. A reduced complexity online state sequence and parameter estimator for superimposed convolutional coded signals. *IEEE Trans. Commun.*, 45(12):1565–1574, Dec. 1997.

[55] G. D. Brushe and L. B. White. Spatial filtering of superimposed convolutional coded signals. *IEEE Trans. Commun.*, 45(9):1144–1153, Sept. 1997.

[56] M. F. Bugallo, J. Miguez, and L. Castedo. A maximum likelihood approach to blind multiuser interference cancellation. *IEEE Trans. Signal Process.*, 49(6):1228–1239, June 2001.

[57] S. Buzzi, M. Lops, and H. V. Poor. Code-aided interference suppression for DS/CDMA overlay systems. *Proc. IEEE*, 90(3):394–435, Mar. 2002.

[58] S. Buzzi, M. Lops, and A. M. Tulino. MMSE RAKE reception for asynchronous DS/CDMA overlay systems and frequency-selective fading channels. *Wireless Personal Commun.*, 13(3):295–318, June 2000.

[59] S. Buzzi, M. Lops, and A. M. Tulino. A new family of MMSE multiuser receivers for interference suppression in DS/CDMA systems employing BPSK modulation. *IEEE Trans. Commun.*, 49(1):154–167, Jan. 2001.

[60] S. Buzzi, M. Lops, and A. M. Tulino. Partially blind adaptive MMSE interference rejection in asynchronous DS/CDMA networks over frequency-selective fading. *IEEE Trans. Commun.*, 49(1):94–108, Jan. 2001.

[61] S. Buzzi and H. V. Poor. Channel estimation and multiuser detection in long-code DS/CDMA systems. *IEEE J. Select. Areas Commun.*, 19(8):1476–1487, Aug. 2001.

[62] S. Buzzi and H. V. Poor. Timing-free blind multiuser detection in differentially encoded DS/CDMA systems. *IEEE Trans. Commun.*, 49(12):2077–2082, Dec. 2001.

[63] G. Caire. Two-stage non-data-aided adaptive linear receivers for DS/CDMA. *IEEE Trans. Commun.*, 48(10):1712–1724, Oct. 2000.

[64] G. Caire and U. Mitra. Structured multiuser channel esitmation for block-synchronous DS/CDMA. *IEEE Trans. Commun.*, 49(9):1605–1617, Sept. 2001.

[65] G. Caire, G. Taricco, J. Ventura-Traveset, and E. Biglieri. A multiuser approach to narrowband cellular communications. *IEEE Trans. Inform. Theory*, 43(5):1503–1517, Sept. 1997.

[66] G. Carayannis, D. Manolakis, and N. Kalouptsidis. A fast sequential algorithm for least squares filtering and prediction. *IEEE Trans. Acoust. Speech Signal Process.*, 31(6):1392–1402, June 1983.

[67] C. Carlemalm, H. V. Poor, and A. Logothetis. Suppression of multiple narrowband interferers in a spread-spectrum communication system. *IEEE J. Select. Areas Commun.*, 18(8):1365–1374, Aug. 2000.

[68] G. Casella and E. I. George. Explaining the Gibbs sampler. *Am. Stat.*, 46:167–174, 1992.

[69] R. T. Causey and J. R. Barry. Blind multiuser detection using linear prediction. *IEEE J. Select. Areas Commun.*, 16(9):1702–1710, Dec. 1998.

[70] J. K. Cavers. An analysis of pilot symbol assisted modulation for Rayleigh fading channels. *IEEE Trans. Veh. Technol.*, 40(4):686–693, Nov. 1991.

[71] K. S. Chan. Asymptotic behavior of the Gibbs sampler. *J. Am. Stat. Assoc.*, 88:320–326, 1993.

[72] P. R. Chang and J. T. Hu. Narrow-band interference suppression in spread-spectrum CDMA communications using pipelined recurrent neural networks. *IEEE Trans. Veh. Technol.*, 48(2):467–477, Mar. 1999.

[73] D.-S. Chen and S. Roy. An adaptive multiuser receiver for CDMA systems. *IEEE J. Select. Areas Commun.*, 12(5):808–816, June 1994.

[74] J. Chen and U. Mitra. Analysis of decorrelator-based receivers for multi-rate CDMA communications. *IEEE Trans. Veh. Technol.*, 48(6):1966–1983, Nov. 1999.

[75] R. Chen and J. S. Liu. Mixture Kalman filters. *J. R. Stat. Soc. B*, 62(3):493–509, 2000.

[76] R. Chen, X. Wang, and J. S. Liu. Adaptive joint detection and decoding in flat-fading channels via mixture Kalman filtering. *IEEE Trans. Inform. Theory*, 46(6):2079–2094, Sept. 2000.

[77] S. Chen, G. J. Gibson, C. F. Cowan, and P. M. Grant. Adaptive equalization of finite nonlinear channels using multilayer perceptrons. *Signal Process.*, 20:107–119, Sept. 1990.

[78] A. Chevreuil and P. Loubaton. MIMO blind second-order equalization method and conjugate cyclostationarity. *IEEE Trans. Signal Process.*, 47(2):572–578, Feb. 1999.

[79] C.-Y. Chi and C.-H. Chen. Cumulant-based inverse filter criteria for MIMO blind deconvolution: Properties, algorithms and applications to DS/CDMA systems in multipath. *IEEE Trans. Signal Process.*, 49(7):1282–1299, July 2001.

[80] J. H. Cho and J. S. Lehnert. Blind adaptive multiuser detection for DS/SSMA communications with generalized random spreading. *IEEE Trans. Commun.*, 49(6):1082–1091, June 2001.

[81] T.-C. Chuah, B. S. Sharif, and O. R. Hinton. Robust adaptive spread-spectrum receiver with neural net preprocessing in non-Gaussian noise. *IEEE Trans. Neural Networks*, 12(3):546–558, May 2001.

[82] S.-Y. Chung, G. D. Forney, T. J. Richardson, and R. Urbanke. On the design of low-density parity-check codes within 0.0045 dB of the Shannon limit. *IEEE Commun. Lett.*, 5(2):58–60, Feb. 2001.

[83] J. M. Cioffi and T. Kailath. Fast recursive least squares transversal filters for adaptive filtering. *IEEE Trans. Acoust. Speech Signal Process.*, 32(2):304–337, Feb. 1984.

[84] F. Classen and H. Meyr. Frequency synchronization algorithms for OFDM systems suitable for communication over frequency selective fading channels. In *Proc. 1994 IEEE Veh. Technol. Conf.*, pages 1655–1659, 1994.

[85] I. B. Collings and J. B. Moore. An HMM approach to adaptive demodulation of QAM signals in fading channels. *Int. J. Adapt. Control Signal Process.*, 8:457–474, 1994.

[86] I. B. Collings and J. B. Moore. An adaptive hidden Markov model approach to FM and M-ary DPSK demodulation in noisy fading channels. *Signal Process.*, 47:71–84, 1995.

[87] P. Comon and G. H. Golub. Tracking a few extreme singular values and vectors in signal processing. *Proc. IEEE*, 78(8):1327–1343, Aug. 1990.

[88] R. Cusani and D. Crea. Soft Bayesian recursive interference cancellation and turbo decoding for TDD-CDMA receivers. *IEEE J. Select. Areas Commun.*, 19(9):1804–1809, Sept. 2001.

[89] F. Daffara and A. Chouly. Maximum likelihood frequency detectors for orthogonal multicarrier systems. In *Proc. 1993 IEEE Int. Conf. Commun.*, pages 766–771, 1993.

[90] D. Dahlhaus et al. Joint demodulation in DS/CDMA systems exploiting the space and time diversity of the mobile radio channel. In *Proc. 1997 Personal, Indoor and Mobile Radio Commun. Conf.*, pages 47–52, 1997.

[91] D. Dahlhaus, B. H. Fleury, and A. Radovic. A sequential algorithm for joint parameter estimation and multiuser detection in DS/CDMA systems with multipath propagation. *Wireless Personal Commun.*, 6(1/2):169–196, Jan. 1998.

[92] H. Dai and H. V. Poor. Iterative space-time processing for multiuser detection in multipath CDMA channels. *IEEE Trans. Signal Process.*, 50(9):2116–2127, Sept. 2002.

[93] Q. Dai and E. Shwedyk. Detection of bandlimited signals over frequency-selective Rayleigh fading channels. *IEEE Trans. Commun.*, 42(2/3/4):941–950, Feb./Mar./Apr. 1994.

[94] A. N. D'Andrea, A. Diglio, and U. Mengali. Symbol-aided channel estimation with nonselective Rayleigh fading channels. *IEEE Trans. Commun.*, 44(1):41–48, Feb. 1995.

[95] S. Davidovici and E. G. Kanterakis. Narrow-band interference rejection using real-time Fourier transforms. *IEEE Trans. Commun.*, 37(7):713–722, July 1989.

[96] L. M. Davis, I. B. Collings, and R. J. Evans. Coupled estimators for equalization of fast-fading mobile channels. *IEEE Trans. Commun.*, 46(10):1262–1265, Oct. 1998.

[97] R. D. DeGroat. Noniterative subspace tracking. *IEEE Trans. Signal Process.*, 40(3):571–577, Mar. 1992.

[98] R. D. DeGroat, E. M. Dowling, and D. A. Linebarger. Subspace tracking. In V. K. Madisetti and D. B. Williams, editors, *The Digital Signal Processing Handbook*. IEEE Press, Piscataway, NJ, 1998.

[99] Z. Ding. Matrix outer-product decomposition method for blind multiple channel identification. *IEEE Trans. Signal Process.*, 45(12):3053–3061, Dec. 1997.

[100] D. Divsalar and F. Pollara. Multiple turbo codes. In *Proc. 1995 IEEE Military Commun. Conf.*, pages 279–284, 1995.

[101] D. Divsalar and M. K. Simon. Multiple-symbol differential detection of MPSK. *IEEE Trans. Commun.*, 38(3):300–308, Mar. 1990.

[102] D. Divsalar and M. K. Simon. Maximum-likelihood differential detection of uncoded and trellis coded amplitude phase modulation over AWGN and fading channels: Metrics and performance. *IEEE Trans. Commun.*, 43(7):76–89, Jan. 1994.

[103] S. Dolinar and D. Divsalar. Weight distributions for turbo codes using random and nonrandom permutations. *TDA Prog. Rep.*, 42(122):56–65, Aug. 1995.

[104] L. Dou and R. J. W. Hodgson. Bayesian inference and Gibbs sampling in spectral analysis and parameter estimation: I. *Inverse Probl.*, 11:1069–1085, 1995.

[105] C. Douillard, M. Jezequel, C. Berrou, A. Picart, P. Didier, and A. Glevieux. Iterative correction of intersymbol interference: Turbo equalization. *Eur. Trans. Telecommun.*, 6(5):507–511, Sept.–Oct. 1995.

[106] A. Duel-Hallen and C. Heegard. Delayed decision-feedback sequence estimation. *IEEE Trans. Commun.*, 37:428–436, 1989.

[107] M. L. Dukic and Z. S. Dobrosavljevic. The performance analysis of the adaptive two-stage DFB narrowband interference suppression filter in DSSS receiver. *Kybernetika*, 33(1):51–60, 1997.

[108] M. L. Dukic, I. S. Stojanovic, and Z. D. Stojanovic. Performance of direct-sequence spread-spectrum receiver using decision feedback and transversal filters for combating narrow-band interference. *IEEE J. Select. Areas Commun.*, 8(5):907–914, July 1990.

[109] O. Edfors et al. OFDM channel estimation by singular value decomposition. *IEEE Trans. Commun.*, 46(7):931–939, July 1998.

[110] O. Edfors, M. Sandell, J. J. van de Beek, S. K. Wilson, and P. O. Borjesson. OFDM channel estimation by singular value decomposition. *IEEE Trans. Commun.*, 47(7):931–939, July 1998.

[111] E. Eleftheriou and D. D. Falconer. Tracking properties and steady-state performance of RLS adaptive filter algorithms. *IEEE Trans. Acoust. Speech Signal Process.*, ASSP-34(5):1097–1109, Oct. 1986.

[112] E. Ertin, U. Mitra, and S. Siwamogsatham. Maximum-likelihood-based multipath channel estimation for code-division multiple-access systems. *IEEE Trans. Commun.*, 49(2):290–302, Feb. 2001.

[113] M. V. Eyuboglu and S. U. Qureshi. Reduced-state sequence estimation with set partitioning and decision feedback. *IEEE Trans. Commun.*, 36(1):13–20, Jan. 1988.

[114] M. V. Eyuboglu and S. U. Qureshi. Reduced-state sequence estimation for coded modulation on intersymbol interference channels. *IEEE J. Select. Areas Commun.*, 7(6):989–995, Aug. 1989.

[115] V. Fabian. On asymptotic normality in stochastic approximation. *Ann. Math. Stat.*, 39(4):1327–1332, 1968.

[116] D. D. Falconer and L. Ljung. Applications of fast Kalman estimation to adaptive equalization. *IEEE Trans. Commun.*, COM-26(10):1439–1446, Oct. 1978.

[117] H. H. Fan and X. Li. Linear prediction approach for joint blind equalization and blind multiuser detection in CDMA systems. *IEEE Trans. Signal Process.*, 48(11):3134–3145, Nov. 2000.

[118] H. Fathallah and L. A. Rusch. A subspace approach to adaptive narrowband interference suppression in DSSS. *IEEE Trans. Commun.*, 45(12):1575–1585, Dec. 1997.

[119] U. Fawer and B. Aazhang. A multiuser receiver for code-division multiple-access communications over multipath channels. *IEEE Trans. Commun.*, 43(2/3/4):1556–1565, Feb./Mar./Apr. 1995.

[120] S. A. Fechtel. OFDM carrier and sampling frequency synchronization and its performance on stationary and mobile channels. *IEEE Trans. Consumer Electron.*, 46(8):438–441, Aug. 2000.

[121] J. R. Fonollosa et al. Blind multiuser detection with array observations. *Wireless Personal Commun.*, 6:179–196, 1998.

[122] G. J. Foschini and M. J. Gans. On limits of wireless communications in a fading environment when using multiple antennas. *Wireless Personal Commun.*, 6(3):311–335, 1998.

[123] L. Frenkel and M. Feder. Recursive expectation-maximization (EM) algorithms for time-varying parameters with applications to multiple target tracking. *IEEE Trans. Signal Process.*, 47(2):306–320, Feb. 1999.

[124] B. Friedlander. Lattice filters for adaptive filtering. *Proc. IEEE*, 70(8):829–867, Aug. 1982.

[125] O. L. Frost. An algorithm for linearly constrained adaptive array processing. *Proc. IEEE*, 60(8):926–935, Aug. 1972.

[126] K. Fujimoto and J. R. James. *Mobile Antenna Systems Handbook*. Artech House, Norwood, MA, 2nd edition, 2001.

[127] R. G. Gallager. *Low-Density Parity-Check Codes*. MIT Press, Cambridge, MA, 1963.

[128] R. G. Gallager. *Information Theory and Reliable Communication*. Wiley, New York, 1968.

[129] H. El Gamal and E. Geraniotis. Iterative multiuser detection for coded CDMA signals in AWGN and fading channels. *IEEE J. Select. Areas Commun.*, 18(1):30–41, Jan. 2000.

[130] V. K. Garg. *IS-95 and cdma2000*. Prentice Hall, Upper Saddle River, NJ, 2000.

[131] V. K. Garg, K. Smolik, and J. E. Wilkes. *Applications of CDMA in Wireless/Personal Communications*. Prentice-Hall, Upper Saddle River, NJ, 1997.

[132] V. K. Garg and J. E. Wilkes. *Wireless and Personal Communications Systems*. Prentice Hall, Upper Saddle River, NJ, 1996.

[133] L. M. Garth and H. V. Poor. Narrowband interference suppression techniques in impulsive channels. *IEEE Trans. Aerosp. Electron. Syst.*, 28(1):15–34, Jan. 1992.

[134] A. E. Gelfand and A. F. W. Smith. Sampling-based approaches to calculating marginal densities. *J. Am. Stat. Assoc.*, 85:398–409, 1990.

[135] G. Gelli, L. Paura, and A. M. Tulino. Cyclostationarity-based filtering for narrowband interference suppression in direct-sequence spread-spectrum systems. *IEEE J. Select. Areas Commun.*, 16(9):1747–1755, Dec. 1998.

[136] A. Gelman, J. B. Carlin, H. S. Stern, and D. B. Rubin. *Bayesian Data Analysis.* Chapman & Hall, London, 1995.

[137] S. Geman and D. Geman. Stochastic relaxation, Gibbs distribution, and the Bayesian restoration of images. *IEEE Trans. Pattern Anal. Machine Intell.*, PAMI-6(11):721–741, Nov. 1984.

[138] C. N. Georghiades and J. C. Han. Sequence estimation in the presence of random parameters via the EM algorithm. *IEEE Trans. Commun.*, 45(3):300–308, Mar. 1997.

[139] M. J. Gertsman and J. H. Lodge. Symbol-by-symbol MAP demodulation of CPM and PSK signals on Rayleigh flat fading channels. *IEEE Trans. Commun.*, 45(7):788–799, July 1997.

[140] J. Gevargiz, P. K. Das, and L. B. Milstein. Adaptive narrow-band interference rejection in a DS spread-spectrum intercept receiver using transform domain signal-processing techniques. *IEEE Trans. Commun.*, 37(12):1359–1366, Dec. 1989.

[141] C. J. Geyer. Markov chain Monte Carlo maximum likelihood. In E. M. Keramigas, editor, *Computing Science and Statistics: Proceedings of the 23rd Symposium on the Interface*, pages 156–163, Interface Foundation, Fairfax, VA, 1991.

[142] V. Ghazi-Moghadam and M. Kaveh. CDMA interference cancelling receiver with an adaptive blind array. In *Proc. 1997 IEEE Int. Conf. Acoust. Speech Signal Process.*, pages 4037–4040, 1997.

[143] T. R. Giallorenzi and S. G. Wilson. Multiuser ML sequence estimator for convolutional coded asynchronous DS-CDMA systems. *IEEE Trans. Commun.*, 44(8):997–1008, Aug. 1996.

[144] T. R. Giallorenzi and S. G. Wilson. Suboptimum multiuser receivers for convolutionally coded asynchronous DS-CDMA systems. *IEEE Trans. Commun.*, 44(9):1183–1196, Sept. 1996.

[145] G. B. Giannakis et al., editors. *Signal Processing Advances in Wireless and Mobile Communications,* Vol. 1, *Trends in Channel Estimation and Equalization.* Prentice Hall, Upper Saddle River, NJ, 2000.

[146] G. B. Giannakis et al., editors. *Signal Processing Advances in Wireless and Mobile Communications,* Vol. 2, *Trends in Single- and Multi-User Systems.* Prentice Hall, Upper Saddle River, NJ, 2000.

[147] G. B. Giannakis and C. Tepedelenliglu. Basis expansion models and diversity techniques for blind identification and equalization of time-varying channels. *Proc. IEEE,* 86(10):1969–1986, Oct. 1998.

[148] J. D. Gibson, editor. *The Mobile Communications Handbook.* CRC Press, Boca Raton, FL, 1996.

[149] K. Giridhar, J. J. Shynk, and R. A. Iltis. Bayesian/decision-feedback algorithm for blind equalization. *Opt. Eng.*, 31:1211–1223, June 1992.

[150] S. Glisic and B. Vucetic. *Spread Spectrum CDMA Systems for Wireless Communications.* Artech House, Norwood, MA, 1997.

[151] S. G. Glisic et al. Rejection of frequency sweeping signal in DS spread-spectrum systems using complex adaptive filters. *IEEE Trans. Commun.*, 43(1):136–145, Jan. 1995.

[152] S. G. Glisic et al. Performance enhancement of DSSS systems: Two-dimensional interference suppression. *IEEE Trans. Commun.*, 47(10):1549–1560, Oct. 1999.

[153] S. G. Glisic, Z. Nikolic, and B. Dimitrijevic. Adaptive self-reconfigurable interference suppression schemes for CDMA wireless networks. *IEEE Trans. Commun.*, 47(4):598–607, Apr. 1999.

[154] J. Goldberg and J. R. Fonollosa. Downlink beamforming for cellular mobile communications. In *Proc. 1997 IEEE Veh. Technol. Conf.*, pages 632–636, 1997.

[155] S. H. Goldberg and R. A. Iltis. PN code synchronization effects on narrow-band interference rejection in a direct-sequence spread-spectrum receiver. *IEEE Trans. Commun.*, 36(4):420–429, Apr. 1988.

[156] J. S. Goldstein, I. S. Reed, and L. L. Scharf. A multistage representation of the Wiener filter based on orthogonal projections. *IEEE Trans. Inform. Theory,* 44(6):2943–2959, Nov. 1998.

[157] S. Gollamudi and Y.-F. Huang. Iterative nonlinear MMSE multiuser detection. In *Proc. 1999 IEEE Int. Conf. Acoust. Speech Signal Process.*, pages 2595–2598, 1999.

[158] G. H. Golub and C. F. Van Loan. *Matrix Computations.* Johns Hopkins University Press, Baltimore, 2nd edition, 1989.

[159] D. J. Goodman. *Wireless Personal Communications Systems.* Addison-Wesley, Reading, MA, 1997.

[160] N. J. Gordon, D. J. Salmon, and A. F. M. Smith. A novel approach to nonlinear/non-Gaussian Bayesian state estimation. *IEE Proc. Radar Signal Process.*, 140:107–113, 1993.

[161] S. S. Gorshe and Z. Papir, editors. Topics in broadband access. *IEEE Commun. Mag.*, 40(4):84–133, Apr. 2002.

[162] P. J. Green. Revisible jump Markov chain Monte Carlo computation and Bayesian model determination. *Biometrika*, 82:711–732, 1985.

[163] J. B. Groe and L. E. Larson. *CDMA Mobile Radio Design.* Artech House, Norwood, MA, 2000.

[164] M. Guiziani et al., editors. Wideband wireless access technologies to broadband internet. *IEEE Commun. Mag.*, 40(4):34–83, Apr. 2002.

[165] D. Guo, X. Wang, and R. Chen. Wavelet-based sequential Monte Carlo blind receivers in fading channels. In *Proc. 2002 IEEE Int. Conf. Commun.*, pages 821–825, 2002.

[166] A. Hać. *Multimedia Applications Support for Wireless ATM Networks.* Prentice Hall, Upper Saddle River, NJ, 2000.

[167] R. Haeb and H. Meyr. A systematic approach to carrier recovery and detection of digitally phase modulated signals on fading channels. *IEEE Trans. Commun.*, 37(7):748–754, July 1989.

[168] J. Hagenauer. Foward error correcting for CDMA systems. In *Proc. 1996 Int. Symp. Spread Spectrum Tech. Appl.*, pages 566–569, Mainz, Germany, Sept. 1996.

[169] J. Hagenauer. The Turbo principle: Tutorial introduction and state of the art. In *Proc. Int. Symp. Turbo Codes Relat. Top.*, pages 1–11, Brest, France, Sept. 1997.

[170] F. R. Hampel et al. *Robust Statistics: The Approach Based on Influence Functions.* Wiley, New York, 1986.

[171] B. D. Hart and D. P. Taylor. Extended MLSE receiver for the frequency-flat fast-fading channel. *IEEE Trans. Veh. Technol.*, 46(2):381–389, May 1997.

[172] B. D. Hart and D. P. Taylor. Maximum-likelihood synchronization, equalization and sequence estimation for unknown time-varying frequency-selective Rician channels. *IEEE Trans. Commun.*, 46(2):211–221, Feb. 1998.

[173] M. A. Hasan, J. C. Lee, and V. K. Bhargava. A narrow-band interference canceler with an adjustable center weight. *IEEE Trans. Commun.*, 42(2/3/4):877–880, Feb./Mar./Apr. 1994.

[174] W. K. Hastings. Monte Carlo sampling methods using Markov chains and their applications. *Biometrika*, 57:97–109, 1970.

[175] S. Haykin, editor. *Blind Deconvolution.* Prentice Hall, Upper Saddle River, NJ, 1994.

[176] S. Haykin. *Adaptive Filter Theory.* Prentice Hall, Upper Saddle River, NJ, 4th edition, 2002.

[177] C. Heegard and S. B. Wicker. *Turbo Coding.* Kluwer Academic Publishers, Norwell, MA, 1995.

[178] G. Heine. *GSM Networks: Protocols, Terminology, and Implementation.* Artech House, Norwood, MA, 1998.

[179] P. Ho and D. Fung. Error performance of multiple-symbol differential detection of PSK signals transmitted over correlated Rayleigh fading channels. *IEEE Trans. Commun.*, 40(7):25–29, Oct. 1992.

[180] P. Hoeher. A statistical discrete-time model for the WSSUS multipath channel. *IEEE Trans. Veh. Technol.*, 41(6):461–468, Nov. 1992.

[181] P. Hoeher and J. H. Lodge. Turbo DPSK: Iterative differential PSK demodulation and channel decodings. *IEEE Trans. Commun.*, 47(6):837–843, June 1999.

[182] S. H. Hong and B. G. Lee. Time-frequency localised subspace tracking for narrowband interference suppression of DS/CDMA systems. *Electron. Lett.*, 35(13):1063–1064, 1999.

[183] M. L. Honig, U. Madhow, and S. Verdú. Blind adaptive multiuser detection. *IEEE Trans. Inform. Theory*, 41(4):944–960, July 1995.

[184] M. L. Honig and H. V. Poor. Adaptive interference suppression. In H. V. Poor and G. W. Wornell, editors, *Wireless Communications: A Signal Processing Perspective*, pages 64–128. Prentice Hall, Upper Saddle River, NJ, 1998.

[185] M. L. Honig and M. K. Tsatsanis. Adaptive techniques for multiuser CDMA receivers. *IEEE Trans. Signal Process.*, 17(3):49–61, May 2000.

[186] M. L. Honig and J. S. Goldstein. Adaptive reduced-rank residual correlation algorithms for DS-CDMA interference suppression. In *Proc. 1998 Asilomar Conf. Signals Syst. Comput.*, pages 1106–1110, 1998.

[187] M. L. Honig, S. L. Miller, M. J. Shensa, and L. B. Milstein. Performance of adaptive linear interference suppression in the presence of dynamic fading. *IEEE Trans. Commun.*, 49(4):635–645, Apr. 2001.

[188] M. L. Honig, M. Sensha, S. Miller, and L. Milstein. Performance of adaptive linear interference suppression for DS-CDMA in the presence of flat Rayleigh fading. In *Proc. 1997 IEEE Veh. Technol. Conf.*, pages 2191–2195, 1997.

[189] R. A. Horn and C. R. Johnson. *Matrix Analysis.* Cambridge University Press, Cambridge, 1985.

[190] A. Høst-Madsen. Semi-blind multiuser detectors for CDMA: Subspace methods. Technical Report. TR Labs, University of Calgary, Calgary, Alberta, Canada, 1997.

[191] A. Høst-Madsen and K.-S. Cho. MMSE/PIC multi-user detection for DS/CDMA systems with inter- and intra- interference. *IEEE Trans. Commun.*, 47(2):291–299, Feb. 1999.

[192] A. Høst-Madsen and X. Wang. Performance bounds for linear blind and group-blind multiuser detectors. To appear in *IEEE Trans. Signal Process.*

[193] A. Høst-Madsen and X. Wang. Performance of blind and group-blind multiuser detectors. In *Proc. 38th Annual Allerton Conf. Commun. Comput. Control*, Monticello, IL, Oct. 2000.

[194] A. Høst-Madsen and X. Wang. Performance of blind multiuser detectors. In *Proc. 10th Int. Symp. Inform. Theory Its Appl.*, Honolulu, HI, Nov. 2000.

[195] A. Høst-Madsen and X. Wang. Cramer-Rao bounds on blind multiuser detectors. In *Proc. 39th Annual Allerton Conf. Commun. Control Comput.*, Monticello, IL, Oct. 2001.

[196] A. Høst-Madsen and X. Wang. Performance of subspace-based multiuser detection. In *Proc. 2001 IEEE Int. Symp. Inform. Theory*, page 10, 2001.

[197] A. Høst-Madsen and X. Wang. Performance of blind and group-blind multiuser detection. *IEEE Trans. Inform. Theory*, 48(7):1849–1872, July 2002.

[198] A. Høst-Madsen and J. Yu. Hybrid semi-blind multiuser detection: Subspace tracking method. In *Proc. 1999 IEEE Int. Conf. Acoust. Speech Signal Process.*, pages III.352–III.355, 1999.

[199] F. M. Hsu and A. A. Giordano. Digital whitening techniques for improving spread-spectrum communications performance in the presence of narrow-band jamming and interference. *IEEE Trans. Commun.*, 26(2):587–595, Feb. 1978.

[200] J.-M. Hsu and C.-L. Wang. A low-complexity iterative multiuser receiver for turbo-coded DS-CDMA systems. *IEEE J. Select. Areas Commun.*, 19(9):1775–1783, Sept. 2001.

[201] C. Huang. An analysis of CDMA 3G wireless communications standards. In *Proc. 1999 IEEE Veh. Technol. Conf.*, pages 342–345, Houston, TX, May 1999.

[202] H. C. Huang, S. C. Schwartz, and S. Verdú. Combined multipath and spatial resolution for multiuser detection: Potentials and problems. In *Proc. 1995 IEEE Int. Symp. Inform. Theory*, page 205, Whistler, BC, Canada, 1995.

[203] P. J. Huber. *Robust Statistics*. Wiley, New York, 1981.

[204] B. L. Hughes. Differential space-time modulation. *IEEE Trans. Inform. Theory*, 46(7):2567–2578, Nov. 2000.

[205] A. T. Huq, E. Panayirci, and C. N. Georghiades. ML NDA carrier phase recovery for OFDM systems. In *Proc. 1999 IEEE Int. Conf. Commun.*, pages 786–790, 1999.

[206] R. A. Iltis. An EKF-based joint estimator for interference, multipath and code delay in a DS spread-spectrum receiver. *IEEE Trans. Commun.*, 42(2/3/4):1288–1299, Feb./Mar./Apr. 1994.

[207] R. A. Iltis. A DS-CDMA tracking mode receiver with joint channel/delay estimation and MMSE detection. *IEEE Trans. Commun.*, 49(10):1770–1779, Oct. 2001.

[208] R. A. Iltis and L. Mailaender. An adaptive multiuser detector with joint amplitude and delay estimation. *IEEE J. Select. Areas Commun.*, 12(5):774–785, June 1994.

[209] R. A. Iltis and L. B. Milstein. Performance analysis of narrow-band interference rejection techniques in DS spread-spectrum systems. *IEEE Trans. Commun.*, 32(11):1169–1177, Nov. 1984.

[210] R. A. Iltis and L. B. Milstein. An approximate statistical analysis of the Widrow LMS algorithm with application to narrow-band interference rejection. *IEEE Trans. Commun.*, 33(2):121–130, Feb. 1985.

[211] R. A. Iltis, J. A. Ritcey, and L. B. Milstein. Interference rejection in FFH systems using least-squares estimation techniques. *IEEE Trans. Commun.*, 38(12):2174–2183, Dec. 1990.

[212] R. A. Iltis, J. J. Shynk, and K. Giridhar. Bayesian blind equalization for coded waveforms. In *Proc. 1992 IEEE Military Commun. Conf.*, pages 211–215, 1992.

[213] R. A. Iltis, J. J. Shynk, and K. Giridhar. Bayesian algorithms for blind equalization using parallel adaptive filters. *IEEE Trans. Commun.*, 42(3):1017–1032, Mar. 1994.

[214] Y. Inouye and T. Sato. Iterative algorithms based on multistage criteria for multichannel blind deconvolution. *IEEE Trans. Signal Process.*, 47(6):1759–1764, June 1999.

[215] G. T. Irvine and P. J. McLane. Symbol-aided plus decision-directed reception for PSK/TCM modulation on shadowed mobile satellite fading channels. *IEEE J. Select. Areas Commun.*, 10(8):1289–1299, Oct. 1992.

[216] W. C. Jakes. *Microwave Mobile Communications.* Wiley, New York, 1974.

[217] S. K. Jayaweera and H. V. Poor. Low-complexity receiver structures for space-time coded multiple-access systems. *EURASIP J. Appl. Signal Process.*, 2002(3):275–288, Mar. 2002.

[218] C. R. Johnson, Jr. et al. Blind equalization using the constant modulus criterion: A review. *Proc. IEEE*, 86(10):1927–1950, Oct. 1998.

[219] L. Johnston and V. Krishnamurthy. Hidden Markov model algorithms for narrowband interference suppression in CDMA spread spectrum systems. *Signal Process.*, 79(3):315–324, Dec. 1999.

[220] M. J. Juntti. Multiuser detector performance comparison in multirate CDMA systems. In *Proc. 1998 IEEE Veh. Technol. Conf.*, pages 31–35, 1998.

[221] P. Kalidas and K. M. M. Prabhu. Improved LMS adaptive algorithm for narrow-band interference suppression in direct-sequence spread-spectrum. *IEEE Trans. Aerosp. Electron. Syst.*, 31(2):1198–1201, July 1995.

[222] P. Y. Kam. Bit error probabilities of MDPSK over the nonselective Rayleigh fading channel with diversity reception. *IEEE Trans. Commun.*, 39(2):220–224, Feb. 1991.

[223] P. Y. Kam and H. M. Ching. Sequence estimation over the slow nonselective Rayleigh fading channel with diversity reception and its application to Viterbi decoding. *IEEE J. Select. Areas Commun.*, 10(3):562–570, 1992.

[224] I. Karasalo. Estimating the covariance matrix by signal subspace averaging. *IEEE Trans. Acoust. Speech Signal Process.*, ASSP-34(1):8–12, Jan. 1986.

[225] T. Kasparis, M. Georgiopoulos, and Q. Memon. Direct-sequence spread-spectrum with transform domain interference suppression. *J. Circuit Syst. Comput.*, 5(2):167–179, June 1995.

[226] S. A. Kassam and H. V. Poor. Robust techniques for signal processing: A survey. *Proc. IEEE*, 73(3):433–481, Mar. 1985.

[227] T. Kawahara and T. Matsumoto. Joint decorrelating multiuser detection and channel estimation in asychronous CDMA mobile communication channels. *IEEE Trans. Veh. Technol.*, 44(3):506–515, Aug. 1995.

[228] J. W. Ketchum and J. G. Proakis. Adaptive algorithms for estimating and suppressing narrow-band interference in PN spread-spectrum systems. *IEEE Trans. Commun.*, 30(5):913–924, May 1982.

[229] I. G. Kim and D. Kim. Spectral overlay of multicarrier CDMA on existing CDMA mobile systems. *IEEE Commun. Lett.*, 3(1):15–17, Jan. 1999.

[230] J. Y. Kim. Narrowband interference suppression for PN code acquisition in a DS/CDMA overlay environment. *IEICE Trans. Commun.*, E83B(8):1631–1639, Aug. 2000.

[231] R. Kohno et al. Combination of an adaptive array antenna and a canceller of interference for direct-sequence spread-spectrum multiple-access system. *IEEE J. Select. Areas Commun.*, 8(4):641–649, May 1990.

[232] A. Kong, J. S. Liu, and W. H. Wong. Sequential imputations and Bayesian missing data problems. *J. Am. Stat. Assoc.*, 89:278–288, 1994.

[233] H. Kong and E. Shwedyk. On channel estimation and sequence detection of interleaved coded signals over frequency nonselective Rayleigh fading channels. *IEEE Trans. Veh. Technol.*, 47(2):558–565, May 1998.

[234] F. Koorevaar and J. Ruprecht. Frequency overlay of GSM and cellular B-CDMA. *IEEE Trans. Veh. Technol.*, 48(3):696–707, May 1999.

[235] B. W. Kosminchuk and A. U. H. Sheikh. A Kalman filter-based architecture for interference excision. *IEEE Trans. Commun.*, 43(2/3/4):574–580, Feb./Mar./Apr. 1995.

[236] R. J. Kozick and B. M. Sadler. Maximum-likelihood array processing in non-Gaussian noise with Gaussian mixtures. *IEEE Trans. Signal Process.*, 48(12):3520–3535, Dec. 2000.

[237] V. Krishnamurthy. Averaged stochastic gradient algorithms for adaptive blind multiuser detection in DS/CDMA systems. *IEEE Trans. Commun.*, 48(1):125–134, Jan. 2000.

[238] V. Krishnamurthy and A. Logothetis. Adaptive nonlinear filters for narrow-band interference suppression in spread-spectrum CDMA systems. *IEEE Trans. Commun.*, 47(5):742–753, May 1999.

[239] V. Krishnamurthy and J. B. Moore. On-line estimation of hidden Markov model parameters based on the Kullback-Leibler information measure. *IEEE Trans. Signal Process.*, 41(8):2557–2573, Aug. 1993.

[240] H. Kubo, K. Murakami, and T. Fujino. An adaptive maximum likelihood sequence estimator for fast time-varying intersymbol interference channels. *IEEE Trans. Commun.*, 43(2/3/4):1872–1880, Feb./Mar./Apr. 1994.

[241] S. Y. Kung. *VLSI Array Processing.* Prentice Hall, Englewood Cliffs, NJ, 1988.

[242] H. Kusaka et al. Rejection of narrow-band interference in a delay-lock loop using prediction error filters. *IEICE Trans. Commun.*, E76B(8):37–62, May 1993.

[243] N. Lashkarian and S. Kiaei. Class of cyclic-based estimators for frequency-offset estimation of OFDM systems. *IEEE Trans. Commun.*, 48(12):2139–2149, Dec. 2000.

[244] J. D. Laster and J. H. Reed. Interference rejection in digital wireless communications. *IEEE Signal Process. Mag.*, 14(3):37–62, May 1997.

[245] G.-K. Lee, S. B. Gelfand, and M. P. Fitz. Bayesian techniques for blind deconvolution. *IEEE Trans. Commun.*, 44(7):826–835, July 1996.

[246] H.-N. Lee and G. J. Pottie. Fast adaptive equalization/diversity combining for time-varying dispersive channels. *IEEE Trans. Commun.*, 46(9):1146–1162, Sept. 1998.

[247] J. S. Lee and L. E. Miller. *CDMA Systems Engineering Handbook*. Artech House, Norwood, MA, 1998.

[248] E. L. Lehmann and G. Casella. *Theory of Point Estimation*. Springer-Verlag, New York, 2nd edition, 1998.

[249] J. Leitao and F. D. Dunes. A nonlinear filtering approach to carrier tracking and symbol detection in digital phase modulation. In *Proc. 1994 IEEE Int. Conf. Commun.*, pages 386–390, New Orleans, LA, May 1994.

[250] J. Leitao and J. Moura. Nonlinear phase estimation based on the Kullback distance. In *Proc. 1994 IEEE Int. Conf. Acoust. Speech Signal Process.*, pages 521–524, Adelaide, Australia, Apr. 1994.

[251] C.-N. Li, G.-R. Hu, and M.-J. Liu. Narrow-band interference excision in spread-spectrum systems using self-orthogonalizing transform-domain adaptive filters. *IEEE J. Select. Areas Commun.*, 18(3):403–406, Mar. 2000.

[252] K. Li and H. Liu. Joint channel and carrier offset estimation in CDMA communications. *IEEE Trans. Signal Process.*, 47(7):1811–1822, July 1999.

[253] L.-M. Li and L. B. Milstein. Rejection of narrowband interference in PN spread spectrum signals using transversal filters. *IEEE Trans. Commun.*, 30(5):925–928, May 1982.

[254] L.-M. Li and L. B. Milstein. Rejection of CW interference in QPSK systems using decision-feedback filters. *IEEE Trans. Commun.*, 31(4):473–483, Apr. 1983.

[255] L.-M. Li and L. B. Milstein. Rejection of pulsed CW interference in PN spread-spectrum systems using complex adaptive filters. *IEEE Trans. Commun.*, 31(1):10–20, Jan. 1983.

[256] Q. Li, X. Wang, and C. Georghiades. Iterative multiuser detection for turbo coded CDMA in multipath fading channels. In *Proc. 2000 IEEE Int. Conf. Commun.*, pages 904–908, New Orleans, LA, June 2000.

[257] Q. Li, X. Wang, and C. Georghiades. Turbo multiuser detection for turbo coded CDMA in multipath fading channels. *IEEE Trans. Veh. Technol.*, 51(5):1096–1108, Sept. 2002.

[258] X. Li and H. H. Fan. Direct blind multiuser detection for CDMA in multipath without channel estimation. *IEEE Trans. Signal Process.*, 49(1):63–73, Jan. 2001.

[259] Y. Li. Simplified channel estimation for OFDM systems with multiple transmit antennas. *IEEE Trans. Wireless Commun.*, 1(1):67–75, Jan. 2002.

[260] Y. Li, L. J. Cimini, and N. R. Sollenberger. Robust channel estimation for OFDM systems with rapid dispersive fading channels. *IEEE Trans. Commun.*, 46(7):902–915, July 1998.

[261] Y. Li and K. J. R. Liu. Adaptive blind source separation and equalization for multiple-input/multiple-output systems. *IEEE Trans. Inform. Theory*, 44(7):2864–2876, Nov. 1998.

[262] Y. Li, N. Seshadri, and S. Ariyavisitakul. Channel estimation for OFDM systems with transmitter diversity in mobile wireless channels. *IEEE J. Select. Areas Commun.*, 17(3):461–471, Mar. 1999.

[263] Y. Li and N. R. Sollenberger. Adaptive antenna arrays for OFDM systems with cochannel interference. *IEEE Trans. Commun.*, 47(2):217–229, Feb. 1998.

[264] F. Liang and W.H. Wong. Evolutionary Monte Carlo: Applications to c_p model sampling and change point problem. *Stat. Sini.*, 10:317–342, 2000.

[265] P. C. Liang and W. E. Stark. Iterative multiuser detection for turbo-coded FHMA communications. *IEEE J. Select. Areas Commun.*, 19(9):1804–1810, Sept. 2001.

[266] J. C. Liberti and T. S. Rappaport. *Smart Antennas for Wireless Communications: IS-95 and Third Generation CDMA Applications.* Prentice Hall, Upper Saddle River, NJ, 1999.

[267] T. J. Lim, Y. Gong, and B. Farhang-Boroujeny. Convergence analysis of chip- and fractionally spaced LMS adaptive multiuser CDMA detectors. *IEEE Trans. Signal Process.*, 48(8):2219–2228, Aug. 2000.

[268] J. Lin, J. G. Proakis, F. Ling, and H. Lev-Ari. Optimum tracking of time-varying channels: A frequency domain approach for known and new algorithms. *IEEE J. Select. Areas Commun.*, 13(1):141–153, Jan. 1995.

[269] J. Litva and T. K.-Y. Lo. *Digital Beamforming in Wireless Communications.* Artech House, Norwood, MA, 1996.

[270] H. Liu. *Signal Processing Applications in CDMA Communications.* Artech House, Norwood, MA, 2000.

[271] H. Liu, G. B. Giannakis, and M. K. Tsatsanis. Time-varying system identification: A deterministic blind approach using antenna arrays. In *Proc. 1996 Conf. Inform. Sci. Syst.*, pages 880–884, Princeton University, Princeton, NJ, Mar. 1996.

[272] H. Liu and G. Xu. A subspace method for signal waveform estimation in synchronous CDMA systems. *IEEE Trans. Commun.*, 44(10):1346–1354, Oct. 1996.

[273] H. Liu and M. D. Zoltowski. Blind equalization in antenna array CDMA systems. *IEEE Trans. Signal Process.*, 45(1):161–172, Jan. 1997.

[274] J. S. Liu. The collapsed Gibbs sampler with applications to a gene regulation problem. *J. Am. Stat. Assoc.*, 89:958–966, 1994.

[275] J. S. Liu. *Monte Carlo Methods for Scientific Computing.* Springer-Verlag, New York, 2001.

[276] J. S. Liu and R. Chen. Blind deconvolution via sequential imputations. *J. Am. Stat. Assoc.*, 90:567–576, 1995.

[277] J. S. Liu and R. Chen. Sequential Monte Carlo methods for dynamic systems. *J. Am. Stat. Assoc.*, 93:1032–1044, 1998.

[278] J. S. Liu, A. Kong, and W. H. Wong. Covariance structure of the Gibbs sampler with applications to the comparisons of estimators and augmentation schemes. *Biometrika*, 81:27–40, 1994.

[279] J. S. Liu, F. Ling, and W. H. Wong. The use of multiple-try method and local optimization in Metropolis sampling. *J. Am. Stat. Assoc.*, 95:121–134, 2000.

[280] Y. Liu and S. D. Bolstein. Identification of frequency non-selective fading channels using decision feedback and adaptive linear prediction. *IEEE Trans. Commun.*, 43(2/3/4):1484–1492, Feb./Mar./Apr. 1995.

[281] Y. Liu and M. P. Fitz. Space-time turbo codes. In *Proc. 37th Annual Allerton Conf. Commun. Comput. Control.* Monticello, IL, 1999.

[282] Y. Liu and X. Wang. Multiple-symbol decision-feedback differential space-time decoding in fading channels. *EURASIP J. Appl. Signal Process.*, 2002(3):297–304, Mar. 2002.

[283] L. Ljung and T. Söderström. *Theory and Practice of Recursive Identification.* MIT Press, Cambridge, MA, 1987.

[284] J. H. Lodge and M. L. Moher. Maximum likelihood sequence estimation of CPM signals transmitted over Rayleigh flat-fading channels. *IEEE Trans. Commun.*, 38(6):787–794, June 1990.

[285] M. Lops, G. Ricci, and A. M. Tulino. Narrow-band-interference suppression in multiuser CDMA systems. *IEEE Trans. Commun.*, 46(9):1163–1175, Sept. 1998.

[286] M. Lops and A. M. Tulino. Automatic suppression of narrow-band interference in direct-sequence spread-spectrum systems. *IEEE Trans. Commun.*, 47(8):1133–1136, Aug. 1999.

[287] P. Loubaton and E. Moulines. On blind multiuser forward link channel estimation by the subspace method: Identifiability results. *IEEE Trans. Signal Process.*, 48(8):2366–2376, Aug. 2000.

[288] B. Lu and X. Wang. Iterative receivers for multiuser space-time coding systems. *IEEE J. Select. Areas Commun.*, 18(11):2322–2335, Nov. 2000.

[289] B. Lu and X. Wang. Space-time code design in OFDM systems. In *Proc. 2000 IEEE Global Telecommun. Conf.*, pages 1000–1004, San Francisco, CA, Nov. 2000.

[290] B. Lu and X. Wang. Bayesian blind turbo receiver for coded OFDM systems with frequency offset and frequency-selective fading. *IEEE J. Select. Areas Commun.*, 19(12):2516–2527, Dec. 2001.

[291] B. Lu and X. Wang. A space-time trellis code design method for OFDM systems. *Wireless Personal Commun.*, 24(3):403–418, Feb. 2003.

[292] B. Lu, X. Wang, and Y. Li. Iterative receivers for space-time block coded OFDM systems in dispersive fading channels. *IEEE Trans. Wireless Commun.*, 1(2):213–225, Apr. 2001.

[293] B. Lu, X. Wang, and K. R. Narayanan. LDPC-based space-time coded OFDM systems over correlated fading channels: Performance analysis and receiver design. *IEEE Trans. Commun.*, 50(1):74–88, Jan. 2002.

[294] Lucent Technologies. Downlink diversity improvements through space-time spreading. Proposed 3GPP2-C30-19990817-014 to the CDMA-2000 standard, Aug. 1999.

[295] M. Luise and R. Reggiannini. Carrier frequency acquisition and tracking for OFDM systems. *IEEE Trans. Commun.*, 44(11):1590–1598, Nov. 1996.

[296] R. Lupas and S. Verdú. Linear multi-user detectors for synchronous code-division multiple-access channels. *IEEE Trans. Inform. Theory*, 35(1):123–136, Jan. 1989.

[297] R. Lupas and S. Verdú. Near-far resistance of multi-user detectors in asynchronous channels. *IEEE Trans. Commun.*, 38:496–508, Apr. 1990.

[298] C. T. Ma, Z. Ding, and S. F. Yau. A two-stage algorithm for MIMO blind deconvolution of nonstationary colored signals. *IEEE Trans. Signal Process.*, 48(4):1187–1192, Apr. 2000.

[299] J. Ma and H. Ge. Groupwise successive interference cancellation for multirate CDMA based on MMSE criterion. In *Proc. 2000 IEEE Int. Conf. Commun.*, pages 1174–1178, 2000.

[300] Y. Ma and T. J. Lim. Linear and nonlinear chip-rate minimum mean-squared-error multiuser CDMA detection. *IEEE Trans. Commun.*, 49(3):530–542, Mar. 2001.

[301] O. D. Macchi and N. J. Bershad. Adaptive recovery of a chirped sinusoid in noise: I. Performance of the RLS algorithm. *IEEE Trans. Signal Process.*, 39(3):583–594, Mar. 1991.

[302] D. J. C. MacKay. Good error correcting codes based on very sparce matrices. *IEEE Trans. Inform. Theory*, 45(2):399–431, Mar. 1999.

[303] D. J. C. MacKay, S. T. Wilson, and M. C. Davey. Comparison of constructions of irregular Gallager codes. *IEEE Trans. Commun.*, 47(10):1449–1454, Oct. 1999.

[304] K. M. Mackenthun. A fast algorithm for multiple-symbol differential detection of MPSK. *IEEE Trans. Commun.*, 42(2/3/4):1471–1474, Feb./Mar./Apr. 1994.

[305] U. Madhow. Blind adaptive interference suppression for the near-far resistant acquisition and demodulation of direct-sequence CDMA signals. *IEEE Trans. Signal Process.*, 45(1):124–136, Jan. 1997.

[306] U. Madhow. Blind adaptive interference suppression for CDMA. *Proc. IEEE*, 86(10):2049–2069, Oct. 1998.

[307] U. Madhow and M. Honig. MMSE interference suppression for direct-sequence spread-spectrum CDMA. *IEEE Trans. Commun.*, 42(12):3178–3188, Dec. 1994.

[308] D. Makrakis, P. T. Mathiopoulos, and D. P. Bouras. Optimal decoding of coded PSK and QAM signals in correlated fast fading channels and AWGN: A combined envelop, multiple differential and coherent detection approach. *IEEE Trans. Commun.*, 42(1):63–75, Jan. 1994.

[309] C. J. Masreliez. Approximate non-Gaussian filtering with linear state and observation relations. *IEEE Trans. Automat. Contr.*, AC-20(1):107–110, Jan. 1975.

[310] E. Masry. Closed-form analytical results for the rejection of narrow-band interference in PN spread-spectrum systems: I. Linear prediction filters. *IEEE Trans. Commun.*, COM-32:888–896, Aug. 1984.

[311] E. Masry. Closed-form analytical results for the rejection of narrow-band interference in PN spread-spectrum systems: II. Linear interpolation filters. *IEEE Trans. Commun.*, COM-33:10–19, Jan. 1985.

[312] E. Masry and L. B. Milstein. Performance of DS spread-spectrum receiver employing interference-suppression filters under a worst-case jamming condition. *IEEE Trans. Commun.*, COM-34(1):13–21, Jan. 1986.

[313] R. Mathias. Accurate eigensystem computations by Jacobi methods. *SIAM J. Matrix Anal. Appl.*, 1(3):977–1003, July 1995.

[314] D. W. Matolak and S. G. Wilson. Detection for a statistically known time-varying dispersive channel. *IEEE Trans. Commun.*, 44(12):1673–1682, Dec. 1996.

[315] G. J. McLachlan and T. Krishnan. *The EM Algorithm and Extensions.* Wiley, New York, 1997.

[316] J. G. McWhirter. Recursive least-square minimization using a systolic array. In K. Bromley, editor, *Proc. SPIE Real Time Signal Process. VI*, pages 105–112, Aug. 1983.

[317] M. J. Medley, G. J. Saulnier, and P. K. Das. Narrow-band interference excision in spread spectrum systems using lapped transforms. *IEEE Trans. Commun.*, 45(11):1444–1455, Nov. 1997.

[318] N. Metropolis, A. W. Rosenbluth, A. H. Teller, and E. Teller. Equations of state calculations by fast computing machines. *J. Chem. Phys.*, 21:1087–1091, 1953.

[319] D. Middleton. Man-made noise in urban environments and transportation systems: Models and measurement. *IEEE Trans. Commun.*, COM-21(11):1232–1241, Nov. 1973.

[320] D. Middleton. Statistical–physical models of electromagnetic interference. *IEEE Trans. Electromagn. Compat.*, EMC-19:106–127, 1977.

[321] D. Middleton. Channel modeling and threshold signal processing in underwater acoustics: An analytical overview. *IEEE J. Ocean. Eng.*, 12:4–28, 1987.

[322] D. Middleton. Non-Gaussian noise models in signal processing for telecommunications: New methods and results for class A and class B noise models. *IEEE Trans. Inform. Theory*, 45(4):1122–1129, May 1999.

[323] D. Middleton and A. D. Spaulding. Elements of weak signal detection in non-Gaussian noise. In H. V. Poor and J. B. Thomas, editors, *Advances in Statistical Signal Processing*, Vol. 2, *Signal Detection.* JAI Press, Greenwich, CT, 1993.

[324] S. Miller and S. C. Schwartz. Integrated spatial-temporal detectors for asynchronous Gaussian multiple access channels. *IEEE Trans. Commun.*, 43(2/3/4):396–411, Feb./Mar./Apr. 1995.

[325] S. L. Miller. An adaptive direct-sequence code-division multiple-access receiver for multiuser interference rejection. *IEEE Trans. Commun.*, 43(2/3/4):1556–1565, Feb./Mar./Apr. 1995.

[326] S. Y. Miller. *Detection and Estimation in Multiple-Access Channels.* PhD thesis, Department of Electrical Engineering, Princeton University, 1989.

[327] L. B. Milstein. Interference rejection techniques in spread spectrum communications. *Proc. IEEE*, 76(6):657–671, June 1988.

[328] L. B. Milstein. Interference suppression to aid acquisition in direct-sequence spread-spectrum communications. *IEEE Trans. Commun.*, 36(11):1200–1202, Nov. 1988.

[329] L. B. Milstein. Wideband code division multiple access. *IEEE J. Select. Areas Commun.*, 18(8):1344–1354, Aug. 2000.

[330] L. B. Milstein and P. K. Das. An analysis of a real-time transform domain filtering digital communication system: I. Narrowband interference rejection. *IEEE Trans. Commun.*, COM-28(6):816–824, June 1980.

[331] L. B. Milstein et al. On the feasibility of a CDMA overlay for personal communications networks. *IEEE J. Select. Areas Commun.*, 10(3):655–668, May 1992.

[332] L. B. Milstein and J. Wang. Interference suppression for CDMA overlays of narrowband waveforms. In S. Glisic and P. Leppanen, editors, *Code Division Multiple Access Communications.* Kluwer Academic, Norwell, MA, 1995.

[333] U. Mitra. Comparison of ML-based detection for two multi-rate access schemes for CDMA signals. *IEEE Trans. Commun.*, 47(1):64–77, Jan. 1999.

[334] U. Mitra and H. V. Poor. Adaptive receiver algorithms for near-far resistant CDMA. *IEEE Trans. Commun.*, 43(2/3/4):1713–1724, Feb./Mar./Apr. 1995.

[335] U. Mitra and H. V. Poor. Analysis of an adaptive decorrelating detector for synchronous CDMA channels. *IEEE Trans. Commun.*, 44(2):257–268, Feb. 1996.

[336] M. Moher. An iterative multiuser decoder for near-capacity communications. *IEEE Trans. Commun.*, 46(7):870–880, July 1998.

[337] M. L. Moher and J. H. Lodge. TCMP: A modulation and coding strategy for Rician fading channels. *IEEE J. Select. Areas Commun.*, 7(9):1347–1355, Dec. 1989.

[338] A. F. Molisch, editor. *Wideband Wireless Digital Communications.* Prentice Hall, Upper Saddle River, NJ, 2001.

[339] T. K. Moon, Z. Xie, C. K. Rushforth, and R. T. Short. Parameter estimation in a multiuser communication system. *IEEE Trans. Commun.*, 42(8):2553–2560, Aug. 1994.

[340] M. Moonen, V. Dooren, and J. Vandewalle. A singular value decomposition algorithm for subspace tracking. *SIAM J. Matrix Anal. Appl.*, 13(4):1015–1038, Oct. 1992.

[341] P. H. Moose. A technique for orthogonal frequency division multiplexing frequency offset correction. *IEEE Trans. Commun.*, 42(10):2908–2914, Oct. 1994.

[342] E. Moulines, P. Duhamel, J. Cardoso, and S. Mayrargue. Subspace methods for blind identification of multichannel FIR channels. *IEEE Trans. Signal Process.*, 43(2):526–525, Feb. 1995.

[343] G. V. Moustakides and H. V. Poor. On the relative error probabilities of linear multiuser detectors. *IEEE Trans. Inform. Theory*, 47(1):450–456, Jan. 2001.

[344] N. J. Muller. *Wireless Data Networking*. Artech House, Norwood, MA, 1995.

[345] B. Muquet and M. de Courville. Blind and semi-blind channel identification methods using second order statistics for OFDM systems. In *Proc. 1999 IEEE Int. Conf. Acoust. Speech Signal Process.*, pages 2745–2748, 1999.

[346] S. Nagaraj et al. Adaptive interference suppression for CDMA systems with a worst-case error criterion. *IEEE Trans. Signal Process.*, 48(1):284–289, Jan. 2000.

[347] M. Nagatsuka and R. Kohno. A spatially temporally optimal multiuser receiver using an array antenna for DS/CDMA. *IEICE Trans. Commun.*, E78-b(11):1489–1497, Nov. 1995.

[348] A. Naguib and A. Paulraj. Performance of CDMA cellular networks with base-station antenna arrays. In *Proc. Int. Zurich Semin. Digital Commun.*, pages 87–100, Zurich, Switzerland, 1994.

[349] A. Naguib, V. Tarokh, N. Seshadri, and A. R. Calderbank. A space-time coding modem for high-data-rate wireless communications. *IEEE J. Select. Areas Commun.*, 16(8):1459–1478, Oct. 1998.

[350] A. F. Naguib, N. Seshadri, and A. R. Calderbank. Increasing data rate over wireless channels: Space-time coding and signal processing for high data rate wireless communications. *IEEE Signal Process. Mag.*, 17(3):76–92, May 2000.

[351] A. Napolitano and M. Tanda. Blind parameter estimation in multiple-access systems. *IEEE Trans. Commun.*, 49(4):688–698, Apr. 2001.

[352] K. R. Narayanan and J. F. Doherty. A convex projections method for improved narrowband interference rejection in direct-sequence spread-spectrum systems. *IEEE Trans. Commun.*, 45(7):772–774, July 1997.

[353] A. Narula, M. D. Trott, and G. W. Wornell. Performance limits of coded diversity methods for transmitter antenna arrays. *IEEE Trans. Inform. Theory*, 45(6):2418–2433, Nov. 1999.

[354] C. J. Nasser et al. *Multi-carrier Technologies for Wireless Communication*. Kluwer Academic, Norwood, MA, 2002.

[355] L. B. Nelson and H. V. Poor. Iterative multiuser receivers for the synchronous CDMA channel: An EM approach. *IEEE Trans. Commun.*, 44(12):1700–1710, Dec. 1996.

[356] B. C. Ng, J. T. Chen, and A. Paulraj. Space-time processing for fast fading channels with co-channel interferers. In *Proc. 1996 IEEE Int. Conf. Commun.*, pages 1491–1495, 1996.

[357] C. L. Nikias and M. Shao. *Signal Processing with α-Stable Distributions and Applications.* Wiley, New York, 1995.

[358] L. Noel and T. Widdowson. Experimental CDMA data overlay of GSM network. *Electron. Lett.*, 35(8):614–615, 1999.

[359] F. D. Nunes and J. Leitao. Doppler shift robust receiver for mobile satellite communications. In *Proc. 1996 IEEE Global Telecommun. Conf.*, pages 405–409, London, UK, Nov. 1996.

[360] F. D. Nunes and J. M. N. Leitao. A nonlinear filtering approach to estimation and detection in mobile communications. *IEEE J. Select. Areas Commun.*, 16(9):1649–1659, Dec. 1998.

[361] T. Ojanperä and R. Prasad, editors. *Wideband CDMA for Third Generation Mobile Communications.* Artech House, Norwood, MA, 1998.

[362] T. Ojanperä and R. Prasad, editors. *WCDMA: Towards IP Mobility and Mobile Internet.* Artech House, Norwood, MA, 2001.

[363] L. H. Ozarow, S. Shamai, and A. D. Wyner. Information theoretic considerations for cellular mobile radio. *IEEE Trans. Veh. Technol.*, 43(3):359–378, May 1994.

[364] D. A. Pados and S. N. Batalama. Low-complexity blind detection of DS/CDMA signals: Auxiliary-vector receivers. *IEEE Trans. Commun.*, 45(12):1586–1594, Dec. 1997.

[365] K. Pahlavan and A. H. Levesque. *Wireless Information Networks.* Wiley, New York, 1995.

[366] E. Panayirci and Y. Barness. Performance of direct-sequence spread-spectrum systems employing minimum redundant transversal filters for narrow-band interference cancellation. *AEU Arch. Elektron.*, 49(4):183–191, July 1995.

[367] E. Panayirci and C. N. Georghiades. Carrier phase synchronization of OFDM systems over frequency-selective channels via the EM algorithm. In *Proc. 1999 IEEE Veh. Technol. Conf.*, pages 675–679, 1999.

[368] E. Papproth and G. K. Kaleh. A CDMA overlay system using frequency-diversity spread spectrum. *IEEE Trans. Veh. Technol.*, 48(2):397–404, Mar. 1999.

[369] C. N. Pateros and G. J. Saulnier. An adaptive correlator receiver for direct-sequence spread-spectrum communications. *IEEE Trans. Commun.*, 44(11):1543–1552, Nov. 1996.

[370] A. J. Paulraj and C. B. Papadias. Space-time processing for wireless communications. *IEEE Signal Process.*, 14(6):49–83, Nov. 1997.

[371] K. I. Pedersen and P.E. Moigensen. Vector groupwise interference cancellation in multirate DS-CDMA systems. In *Proc. 2000 IEEE Veh. Technol. Conf.—Spring*, pages 750–754, Tokyo, Japan, 2000.

[372] V. D. Phan and X. Wang. Bayesian turbo multiuser detection for nonlinearly modulated CDMA. *Signal Process.*, 82(1):42–68, Jan. 2002.

[373] R. L. Pickholz, L. B. Milstein, and D. L. Schilling. Spread spectrum for mobile communications. *IEEE Trans. Veh. Technol.*, 40(2):313–322, May 1991.

[374] T. Pollet, M. V. Bladel, and M. Moeneclaey. BER sensitivity of OFDM systems to carrier frequency offset and Wiener phase noise. *IEEE Trans. Commun.*, 43(2/3/4):191–193, Feb./Mar./Apr. 1995.

[375] H. V. Poor. On parameter estimation in DS/CDMA formats. In W. A. Porter and S. C. Kak, editors, *Advances in Communications and Signal Processing.* Springer-Verlag, New York, 1989.

[376] H. V. Poor. Signal processing for wideband communications. *IEEE Inform. Theory Soc. Newsl.*, 42(2):1–10, June 1992.

[377] H. V. Poor. *An Introduction to Signal Detection and Estimation.* Springer-Verlag, New York, 2nd edition, 1994.

[378] H. V. Poor. Non-Gaussian signal processing problems in multiple-access communications. In *Proc. 1996 USC/CRASP Workshop Non-Gaussian Signal Process.*, Ft. George Meade, MD, May 1996.

[379] H. V. Poor. Turbo multiuser detection: A primer. *J. Commun. Networks*, 3(3):196–201, Sept. 2001.

[380] H. V. Poor. Dynamic programming in digital communications: Viterbi decoding to turbo multiuser detection. *J. Optim. Theory Appl.*, 115(3):629–657, Dec. 2002.

[381] H. V. Poor and L. A. Rusch. Narrowband interference suppression in spread spectrum CDMA. *IEEE Personal Commun.*, 1(3):14–27, Aug. 1994.

[382] H. V. Poor and M. Tanda. Robust multiuser detection in frequency-selective non-Gaussian channels. *Eur. Trans. Telecommun.*, 14(3):255–263, May/Jun. 2003.

[383] H. V. Poor and M. Tanda. Multiuser detection in impulsive channels. *Ann. Telecommun.*, 54(7–8):392–400, Jul.–Aug. 1999.

[384] H. V. Poor and M. Tanda. Multiuser detection in flat-fading non-Gaussian channels. *IEEE Trans. Commun.*, 50(11):1769–1777, Nov. 2002.

[385] H. V. Poor and L. Tong, editors. *Signal Processing for Wireless Communication Systems.* Kluwer Academic, Norwell, MA, 2002.

[386] H. V. Poor and S. Verdú. Probability of error in MMSE multiuser detection. *IEEE Trans. Inform. Theory*, 43(3):858–871, May 1997.

[387] H. V. Poor and R. Vijayan. Analysis of a class of adaptive nonlinear predictors. In W. A. Porter and S. C. Kak, editors, *Advances in Communications and Signal Processing*, pages 231–241. Springer-Verlag, New York, 1989.

[388] H. V. Poor and X. Wang. Code-aided interference suppression in DS/CDMA communications: I. Interference suppression capability. *IEEE Trans. Commun.*, 45(9):1101–1111, Sept. 1997.

[389] H. V. Poor and X. Wang. Code-aided interference suppression in DS/CDMA communications: II. Parallel blind adaptive implementations. *IEEE Trans. Commun.*, 45(9):1112–1122, Sept. 1997.

[390] H. V. Poor and X. Wang. Blind adaptive suppression of narrowband digital interferers from spread-spectrum signals. *Wireless Personal Commun.*, 6(1/2):69–96, Jan. 1998.

[391] H. V. Poor and G. W. Wornell, editors. *Wireless Communications: A Signal Processing Perspective.* Prentice Hall, Upper Saddle River, NJ, 1998.

[392] R. Prasad, editor. *CDMA for Wireless Personal Communications.* Artech House, Norwood, MA, 1996.

[393] R. Prasad. *Universal Wireless Personal Communications.* Artech House, Norwood, MA, 1998.

[394] R. Prasad, editor. *Towards a Global 3G System: Advanced Mobile Communications in Europe,* Vol. 1. Artech House, Norwood, MA, 2001.

[395] R. Prasad, W. Mohr, and W. Konhäuser, editors. *Third Generation Mobile Communication Systems.* Artech House, Norwood, MA, 2000.

[396] J. G. Proakis. *Digital Communications.* McGraw-Hill, New York, 4th edition, 2001.

[397] Z. Qin, K.C. Teh, and E. Gunawan. Iterative multiuser detection for asynchronous CDMA with concatenated convolutional coding. *IEEE J. Select. Areas Commun.*, 19(9):1784–1792, Sept. 2001.

[398] D. J. Rabideau. Fast, rank-adaptive subspace tracking. *IEEE Trans. Signal Process.*, 44(9):1525–1541, June 1998.

[399] R. Raheli, A. Polydoros, and C. Tzou. Per-survivor processing: A general approach to MLSE in uncertain environments. *IEEE Trans. Commun.*, 43(2/3/4):354–364, Feb./Mar./Apr. 1995.

[400] B. J. Rainbolt and S. L. Miller. Multicarrier CDMA for cellular overlay systems. *IEEE J. Select. Areas Commun.*, 17(10):1807–1814, Oct. 1999.

[401] B. J. Rainbolt and S. L. Miller. CDMA transmitter filtering for cellular overlay systems. *IEEE Trans. Commun.*, 48(2):290–297, Feb. 2000.

[402] A. Ranheim. Narrowband interference rejection in direct-sequence spread-spectrum system using time-frequency decomposition. *IEE Proc. Commun.*, 142(6):393–400, Dec. 1995.

[403] P. B. Rapajić and B. S. Vučetić. Adaptive receiver structures for asynchronous CDMA systems. *IEEE J. Select. Areas Commun.*, 12(4):685–697, May 1994.

[404] T. S. Rappaport, editor. *Smart Antennas: Adaptive Arrays, Algorithms and Wireless Position Location.* IEEE Press, Piscataway, NJ, 1998.

[405] T. S. Rappaport. *Wireless Communications: Principles and Practice.* Prentice Hall, Upper Saddle River, NJ, 2nd edition, 2002.

[406] E. C. Real, D. W. Tufts, and J. W. Cooley. Two algorithms for fast approximate subspace tracking. *IEEE Trans. Signal Process.*, 47(7):1936–1945, July 1999.

[407] S. M. Redl, M. K. Weber, and M. W. Oliphant. *An Introduction to GSM.* Artech House, Norwood, MA, 1995.

[408] S. M. Redl, M. K. Weber, and M. W. Oliphant. *GSM and Personal Communications Handbook.* Artech House, Norwood, MA, 1998.

[409] M. C. Reed and P. D. Alexander. Iterative detection using antenna arrays and FEC on multipath channels. *IEEE J. Select. Areas Commun.*, 17(12):2082–2089, Dec. 1999.

[410] M. C. Reed, C. B. Schlegel, P. D. Alexander, and J. A. Asenstorfer. Iterative multiuser detection for CDMA with FEC: Near single user performance. *IEEE Trans. Commun.*, 46(12):1693–1699, Dec. 1998.

[411] D. Reynolds and X. Wang. Group-blind multiuser detection based on subspace tracking. In *Proc. 2000 Conf. Inform. Sci. Syst.*, Princeton, NJ, Mar. 2000.

[412] D. Reynolds and X. Wang. Adaptive group-blind multiuser detection based on a new subspace tracking algorithm. *IEEE Trans. Commun.*, 49(7):1135–1141, July 2001.

[413] D. Reynolds and X. Wang. Low-complexity turbo equalization for diversity channels. *Signal Process.*, 81(5):989–995, May 2001.

[414] D. Reynolds and X. Wang. Turbo multiuser detection with unknown interferers. *IEEE Trans. Commun.*, 50(4):616–622, Apr. 2002.

[415] D. Reynolds, X. Wang, and H. V. Poor. Blind adaptive space-time multiuser detection with multiple transmitter and receiver antennas. *IEEE Trans. Signal Process.*, 50(6):1261–1276, June 2002.

[416] T. Richardson, A. Shokrollahi, and R. Urbanke. Design of capacity-approaching irregular low-density parity-check codes. *IEEE Trans. Inform. Theory*, 47(2):619–637, Feb. 2001.

[417] T. Richardson and R. Urbanke. The capacity of low density parity check codes under message passing decoding. *IEEE Trans. Inform. Theory*, 47(2):599–618, Feb. 2001.

[418] A. Rifkin. Narrow-band interference rejection using real-time Fourier transforms: Comment. *IEEE Trans. Commun.*, 39(9):1292–1294, Sept. 1991.

[419] C. P. Robert. *The Bayesian Choice: A Decision-Theoretic Motivation.* Springer-Verlag, New York, 1994.

[420] C. P. Robert and G. Casella. *Monte Carlo Statistical Methods.* Springer-Verlag, New York, 1999.

[421] G. S. Rogers. *Matrix Derivatives: Lecture Notes in Statistics,* Vol. 2. Marcel Dekker, New York, 1980.

[422] T. J. Rothenberg. The Bayesian approach and alternatives in econometrics. In S. Fienberg and A. Zellners, editors, *Studies in Bayesian Econometrics and Statistics,* Vol. 1, pages 55–75. North-Holland, Amsterdam, 1977.

[423] D. N. Rowitch and L. B. Milstein. Convolutionally coded multicarrier DS-CDMA systems in a multipath fading channel: II. Narrow-band interference suppression. *IEEE Trans. Commun.*, 47(11):1729–1736, Nov. 1999.

[424] J. K. Ruanaidh and J. J. O'Ruanaidh. *Numerical Bayesian Methods Applied to Signal Processing.* Springer-Verlag, New York, 1996.

[425] L. A. Rusch and H. V. Poor. Narrowband interference suppression in CDMA spread spectrum communications. *IEEE Trans. Commun.*, 42(2/3/4):1969–1979, Feb./Mar./Apr. 1994.

[426] L. A. Rusch and H. V. Poor. Multiuser detection techniques for narrowband interference suppression in spread spectrum communications. *IEEE Trans. Commun.*, 43(2/3/4):1725–1737, Feb./Mar./Apr. 1995.

[427] A. Sabharwal, U. Mitra, and R. Moses. MMSE receivers for multirate DS-CDMA systems. *IEEE Trans. Commun.*, 49(12):2184–2197, Dec. 2001.

[428] J. Sacks. Asymptotic distribution of stochastic approximation procedures. *Ann. Math. Stat.*, 29(2):373–405, 1958.

[429] S. Sampei. *Applications of Digital Wireless Technologies to Global Wireless Communications.* Prentice Hall, Upper Saddle River, NJ, 1997.

[430] S. D. Sandberg et al. Some alternatives in transform-domain suppression of narrowband interference for signal detection and demodulation. *IEEE Trans. Commun.*, 43(12):3025–3036, Dec. 1995.

[431] A. Sathyendram et al. Capacity estimation of 3rd generation CDMA cellular systems. In *Proc. 1999 IEEE Veh. Technol. Conf.*, Houston, Texas, 1999.

[432] G. J. Saulnier. Suppression of narrowband jammers in a spread-spectrum receiver using transform-domain adaptive filtering. *IEEE J. Select. Areas Commun.*, 10(4):742–749, June 1992.

[433] G. J. Saulnier, P. Das, and L. B. Milstein. Suppression of narrow-band interference in a PN spread-spectrum receiver using a CTD-based adaptive filter. *IEEE Trans. Commun.*, 32(11):1227–1232, Nov. 1984.

[434] A. M. Sayeed, A. Sendonaris, and B. Aazhang. Multiuser detection in fast-fading multipath environments. *IEEE J. Select. Areas Commun.*, 16(9):1691–1701, Dec. 1998.

[435] A. Scaglione, G. B. Giannakis, and S. Barbarossa. Lagrange/Vandermonde MUI eliminating user codes for quasi-synchronous CDMA in unknown multipath. *IEEE Trans. Signal Process.*, 48(7):2057–2073, July 2000.

[436] M. J. Schervish and B. P. Carlin. On the convergence of successive substitution sampling. *J. Cmp. Gr. St.*, 1:111–127, 1992.

[437] J. Schiller. *Mobile Communications.* Addison-Wesley, Reading, MA, 2000.

[438] C. Schlegel and D. J. Costello. Bandwidth efficient coding for fading channels: Code construction and performance analysis. *IEEE J. Select. Areas Commun.*, 7(9):1356–1368, Dec. 1989.

[439] C. B. Schlegel, S. Roy, P. D. Alexander, and Z. Xiang. Multiuser projection receivers. *IEEE J. Select. Areas Commun.*, 14(8):1610–1618, Oct. 1996.

[440] T. M. Schmidl and D. C. Cox. Robust frequency and timing synchronization for OFDM. *IEEE Trans. Commun.*, 45(12):1613–1621, Dec. 1997.

[441] R. Schober, W. H. Gerstacker, and J. B. Huber. Decision-feedback differential detection of MDPSK for flat Rayleigh fading channels. *IEEE Trans. Commun.*, 47(7):1025–1035, July 1999.

[442] C. Sengupta, J. R. Cavallaro, and B. Aazhang. On multipath channel estimation for CDMA systems using multiple sensors. *IEEE Trans. Commun.*, 49(3):543–554, Mar. 2001.

[443] R. J. Serfling. *Approximation Theorems of Mathematical Statistics.* Wiley, New York, 1980.

[444] N. Seshadri and J. H. Winters. Two signaling schemes for improving the error performance of frequency-division duplex (FDD) transmission systems using transmitter antenna diversity. *Int. J. Wireless Inform. Networks*, 1(1), 1994.

[445] L. Setian. *Practical Communication Antennas with Wireless Applications.* Prentice Hall, Upper Saddle River, NJ, 1998.

[446] J. Shen and Z. Ding. Zero-forcing blind equalization based on subspace estimation for multiuser systems. *IEEE Trans. Commun.*, 49(2):262–271, Feb. 2001.

[447] N. D. Sidiropoulos, G. B. Giannakis, and R. Bro. Blind PARAFAC receivers for DS-CDMA systems. *IEEE Trans. Signal Process.*, 48(3):810–813, Mar. 2000.

[448] P. H. Siegel, D. Dvisalar, E. Eleftheriou, J. Hagenauer, and D. Eowitch, editors. Special issue on the turbo principle: From theory to practice I and II. *IEEE J. Select. Areas Commun.*, 19(5/9), May/Sept. 2001.

[449] C. A. Siller, A. D. Gelman, and G.-S. Kuo. In-home networking. *IEEE Commun. Mag.*, 39(12):76–115, Dec. 2001.

[450] K. Siwiak. *Radiowave Propagation and Antennas for Personal Communications.* Artech House, Norwood, MA, 2nd edition, 1998.

[451] M. A. Soderstrand et al. Suppression of multiple narrow-band interference using real-time adaptive notch filters. *IEEE Trans. Circuits Syst. II*, 44(3):217–225, Mar. 1997.

[452] H. Sorenson and D. Alspach. Recursive Bayesian estimation using Gaussian sums. *Automatica*, 7:465–479, 1971.

[453] P. Spasojević. *Sequence and Channel Estimation for Channels with Memory.* PhD thesis, Department of Electrical Engineering, Texas A&M University, 1999.

[454] P. Spasojević and C. N. Georghiades. The slowest descent method and its application to sequence estimation. *IEEE Trans. Commun.*, 49(9):1592–1604, Sept. 2001.

[455] P. Spasojević and X. Wang. Improved robust multiuser detection in non-Gaussian channels. *IEEE Signal Process Lett.*, 8(3):83–86, Mar. 2001.

[456] P. Spasojević, X. Wang, and A. Høst-Madsen. Nonlinear group-blind multiuser detection. *IEEE Trans. Commun.*, 49(9), Sept. 2001.

[457] W. Stallings. *Wireless Communications and Networks.* Prentice Hall, Upper Saddle River, NJ, 2002.

[458] R. Steele. *Mobile Radio Communications.* Pentech Press, London, 1992.

[459] A. Stefanov and T. M. Duman. Turbo coded modulation for wireless communications with antenna diversity. In *Proc. 1999 IEEE Veh. Technol. Conf.*, pages 1565–1569, 1999.

[460] Y. Steinberg and H. V. Poor. Sequential amplitude estimation in multiuser communications. *IEEE Trans. Inform. Theory*, 40(1):11–20, Jan. 1994.

[461] G. W. Stewart. An updating algorithm for subspace tracking. *IEEE Trans. Signal Process.*, 40(6):1535–1541, June 1992.

[462] P. Stoica and M. Jansson. MIMO system indentification: State-space and subspace approximations versus transfer function and instrumental variables. *IEEE Trans. Signal Process.*, 48(11):3087–3099, Nov. 2000.

[463] P. Strobach. *Linear Prediction Theory: A Mathematical Basis for Adaptive Systems.* Springer-Verlag, Heidelberg, 1990.

[464] E. Strom and S. Miller. *Asynchronous DS-CDMA Systems: Low Complexity Near–far Resistant Receiver and Parameter Estimation.* Technical Report. Department of Electrical Engineering, University of Florida, Gainesville, FL, Jan. 1994.

[465] G. Stüber. *Principles of Mobile Communications.* Kluwer Academic, Norwell, MA, 1996.

[466] Y. T. Su, L. D. Jeng, and F. B. Ueng. Cochannel interference rejection in multipath channels. *IEICE Trans. Commun.*, E80B(12):1797–1804, Dec. 1997.

[467] B. Suard et al. Performance of CDMA mobile communications systems using antenna arrays. In *Proc. 1993 IEEE Int. Conf. Acoust. Speech Signal Process.*, Vol. 4, pages 153–156, New York, 1993.

[468] P. Sung and K. Chen. A linear minimum mean square error multiuser receiver in Rayleigh-fading channels. *IEEE J. Select. Areas Commun.*, 14(8):1583–1594, Oct. 1996.

[469] S. M. Sussman and E. J. Ferrari. The effects of notch filters on the correlation properties of PN signals. *IEEE Trans. Aerosp. Electron. Syst.*, 10:385–390, May 1974.

[470] S. Talwar, M. Viberg, and A. Paulraj. Blind separation of synchronous co-channel digital signals using an antenna array: I. Algorithms. *IEEE Trans. Signal Process.*, 44(5):1184–1197, May 1996.

[471] K. Tang, L. B. Milstein, and P. H. Siegel. Combined MMSE interference suppression and turbo coding for a coherent DS-CDMA system. *IEEE J. Select. Areas Commun.*, 19(9):1793–1803, Sept. 2001.

[472] M. A. Tanner. *Tools for Statistics Inference.* Springer-Verlag, New York, 1991.

[473] M. A. Tanner and W. H. Wong. The calculation of posterior distribution by data augmentation (with discussion). *J. Am. Stat. Assoc.*, 82:528–550, 1987.

[474] V. Tarokh and H. Jafarkhani. A differential detection scheme for transmit diversity. *IEEE J. Select. Areas Commun.*, 18(7):1169–1174, July 2000.

[475] V. Tarokh, H. Jafarkhani, and A. R. Calderbank. Space-time block codes from orthogonal designs. *IEEE Trans. Inform. Theory*, 45(4):1456–1467, July 1999.

[476] V. Tarokh, H. Jafarkhani, and A. R. Calderbank. Space-time block coding for wireless communications: Performance results. *IEEE J. Select. Areas Commun.*, 17(3):451–460, Mar. 1999.

[477] V. Tarokh, N. Seshadri, and A. R. Calderbank. Space-time codes for high data rate wireless communications: Performance criterion and code construction. *IEEE Trans. Inform. Theory*, 44(2):744–765, Mar. 1998.

[478] E. Telatar. *Capacity of Multi-antenna Gaussian Channels.* AT&T-Bell Labs Internal Technical Memorandum, June 1995.

[479] Texas Instruments. Space-time block coded transmit antenna diversity for WCDMA. Proposed TDOC 662/98 to ETSI SMG2 UMTS standards, Dec. 1998.

[480] S. Theodosidis et al. Interference rejection in PN spread-spectrum systems with LS linear-phase FIR filters. *IEEE Trans. Commun.*, 37(9):991–994, Sept. 1989.

[481] J. Thielecke. A soft-decision state-space equalizer for FIR channels. *IEEE Trans. Commun.*, 45(10):1208–1217, Oct. 1997.

[482] D. M. Titterington. Recursive parameter estimation using incomplete date. *J. R. Stat. Soc.*, 46(2):257–267, 1984.

[483] L. Tong and Q. Zhao. Joint order detection and blind channel estimation by least squares smoothing. *IEEE Trans. Signal Process.*, 47(9):2345–2355, Sept. 1999.

[484] M. Torlak and G. Xu. Blind multiuser channel estimation in asynchronous CDMA systems. *IEEE Trans. Signal Process.*, 45(1):137–147, Jan. 1997.

[485] M. K. Tsatsanis. Inverse filtering criteria for CDMA systems. *IEEE Trans. Signal Process.*, 45(1):102–112, Jan. 1997.

[486] M. K. Tsatsanis and G. B. Giannakis. Equalization of rapidly fading channels: Self recovering methods. *IEEE Trans. Commun.*, 44(5):619–630, May 1996.

[487] M. K. Tsatsanis and G. B. Giannakis. Blind estimation of direct sequence spread spectrum signals in multipath. *IEEE Trans. Signal Process.*, 45(5):1241–1252, May 1997.

[488] M. K. Tsatsanis and G. B. Giannakis. Subspace methods for blind estimation of time-varying channels. *IEEE Trans. Signal Process.*, 45(12):3084–3093, Dec. 1997.

[489] M. K. Tsatsanis, G. B. Giannakis, and G. Zhou. Estimation and equalization of fading channels with random coefficients. *Signal Process.*, 53(2/3):211–229, Sept. 1996.

[490] M. K. Tsatsanis and Z. Xu. Performance analysis of minimum variance CDMA receivers. *IEEE Trans. Signal Process.*, 46(11):3014–3022, Nov. 1998.

[491] G. A. Tsihrintzis. Statistical modeling and receiver design for multi-user communication networks. In R. J. Adler, R. E. Feldman, and M. S. Taqqu, editors, *A Practical Guide to Heavy Tails*, pages 406–431. Birkhauser, Boston, 1998.

[492] G. Tsoulos, M. Beach, and J. McGeehan. Wireless personal communications for the 21st century: European technological advances in adaptive antennas. *IEEE Commun. Mag.*, 35(9):102–109, Sept. 1997.

[493] D. W. Tufts and C. D. Melissinos. Simple, effective computation of principal eigenvectors and their eigenvalues and application to high resolution estimation of frequencies. *IEEE Trans. Acoust. Speech Signal Process.*, ASSP-34(5):1046–1053, Oct. 1986.

[494] J. K. Tugnait. Blind spatio-temporal equalization and impulse response estimation for MIMO channels using Goddard cost function. *IEEE Trans. Signal Process.*, 45(1):268–271, Jan. 1997.

[495] J. K. Tugnait. Multistep linear prediction-based blind identification and equalization of multiple-input multiple-output channels. *IEEE Trans. Signal Process.*, 48(1):26–38, Jan. 2000.

[496] J. K. Tugnait and B. Huang. Multistep linear predictor-based blind identification and equalization of multiple-input multiple-output channels. *IEEE Trans. Signal Process.*, 48(1):26–38, Jan. 2000.

[497] J. K. Tugnait and B. Huang. On a whitening approach to partial channel estimation and blind equalization in FIR/IIR multiple-input multiple-output channels. *IEEE Trans. Signal Process.*, 48(3):832–845, Mar. 2000.

[498] J. K. Tugnait and T. Li. Blind detection of asynchronous CDMA signals in multipath channels using code-constrained inverse filter criterion. *IEEE Trans. Signal Process.*, 49(7):1300–1309, July 2001.

[499] J. W. Tukey. A survey of sampling from contaminated distribution. In I. Olkin, editor, *Contributions to Probability and Statistics*, pages 448–485. Stanford University Press, Stanford, CA, 1960.

[500] U. Tureli, H. Liu, and M. D. Zoltowski. OFDM blind carrier offset estimation: ESPRIT. *IEEE Trans. Commun.*, 48(9):1459–1461, Sept. 2000.

[501] S. Ulukus and R. D. Yates. A blind adaptive decorrelating detector for CDMA systems. *IEEE J. Select. Areas Commun.*, 16(8):1530–1541, Oct. 1998.

[502] G. Ungerboeck. Adaptive maximum likelihood receiver for carrier modulated data transmission systems. *IEEE Trans. Commun.*, COM-22(5):624–635, May 1974.

[503] T. Vaidis and C. L. Weber. Block adaptive techniques for channel identification and data demodulation over band-limited channels. *IEEE Trans. Commun.*, 46(2):232–243, Feb. 1998.

[504] C. Vaidyanathan and K. Buckley. An adaptive decision feedback equalizer antenna array for multiuser CDMA wireless communications. In *Proc. 30th Asilomar Conf. Circuits Syst. Comput.*, pages 340–344, Pacific Grove, CA, Nov. 1996.

[505] C. Vaidyanathan, K. Buckley, and S. Hosur. A blind adaptive antenna array for CDMA systems. In *Proc. 29th Asilomar Conf. Circuits Syst. Comput.*, pages 1373–1377, Pacific Grove, CA, Nov. 1995.

[506] J.-J. van de Beek et al. On channel estimation in OFDM systems. In *Proc. 1995 IEEE Veh. Technol. Conf.*, pages 815–819, 1995.

[507] J.-J. van de Beek, M. Sandell, and P. O. Börjesson. ML estimation of time and frequency offset in OFDM systems. *IEEE Trans. Signal Process.*, 45(4):1800–1805, July 1997.

[508] M. van der Heijden and M. Taylor, editors. *Understanding WAP: Wireless Applications, Devices and Services.* Artech House, Norwood, MA, 2000.

[509] G. van der Wijk, G. M. Janssen, and R. Prasad. Groupwise successive interference cancellation in a DS/CDMA system. In *Proc. 6th IEEE Int. Symp. Personal Indoor Mobile Radio Commun.*, pages 742–745, Tokyo, Japan, 1995.

[510] P. van Rooyen, M. Lötter, and D. van Wyk. *Space-Time Processing for CDMA Mobile Communications.* Kluwer Academic, Norwell, MA, 2000.

[511] B. Van Veen and K. Buckley. Beamforming: A versatile approach to spatial filtering. *IEEE Acoust. Speech Signal Process.*, 5(2):4–24, Apr. 1988.

[512] M. K. Varanasi and B. Aazhang. Near-optimum detection in synchronous code-division multiple-access systems. *IEEE Trans. Commun.*, 39(5):725–736, May 1991.

[513] M. K. Varanasi and S. Vasudevan. Multiuser detectors for synchronous CDMA communication over non-selective Rician fading channels. *IEEE Trans. Commun.*, 42(2/3/4):711–722, Feb./Mar./Apr. 1994.

[514] S. Vasudevan and M. K. Varanasi. Optimum diversity combiner based multiuser detection for time-dispersive Rician fading channels. *IEEE J. Select. Areas Commun.*, 12(4):580–592, May 1994.

[515] S. Vasudevan and M. K. Varanasi. Achieving near-optimum asymptotic efficiency and fading resistance over the time-varying Rayleigh-faded CDMA channel. *IEEE Trans. Commun.*, 44(9):1130–1143, Sept. 1996.

[516] S. Vasudevan and M. K. Varanasi. A near-optimum receiver for CDMA communications over time-varying Rayleigh fading channel. *IEEE Trans. Commun.*, 44(9):1130–1143, Sept. 1996.

[517] S. Verdú. *Optimum Multi-user Signal Detection.* PhD thesis, Department of Electrical and Computer Engineering, University of Illinois at Urbana-Champaign, 1984.

[518] S. Verdú. Minimum probability of error for asynchronous Gaussian multiple-access channels. *IEEE Trans. Inform. Theory*, 32(1):85–96, Jan. 1986.

[519] S. Verdú. Computational complexity of optimum multiuser detection. *Algorithmica*, 4:303–312, 1989.

[520] S. Verdú. *Multiuser Detection.* Cambridge University Press, Cambridge, UK, 1998.

[521] M. Viberg and B. Ottersten. Sensor array processing based on subspace fitting. *IEEE Trans. Signal Process.*, 39(5):1110–1120, May 1991.

[522] R. Vijayan and H. V. Poor. Nonlinear techniques for interference suppression in spread spectrum systems. *IEEE Trans. Commun.*, 38(7):1060–1065, July 1990.

[523] A. J. Viterbi. *CDMA: Principles of Spread Spectrum Communications.* Addison-Wesley, Reading, MA, 1995.

[524] G. M. Vitetta, U. Mengali, and D. P. Taylor. Double-filtering differential detection of PSK signals transmitted over linearly time-selective Rayleigh fading channels. *IEEE Trans. Commun.*, 47(2):239–247, Feb. 1999.

[525] G. M. Vitetta and D. P. Taylor. Maximum likelihood decoding of uncoded and coded PSK signal sequences transmitted over Rayleigh flat-fading channels. *IEEE Trans. Commun.*, 43(11):2750–2758, Nov. 1995.

[526] G. M. Vitetta, D. P. Taylor, and U. Mengali. Double-filtering receivers for PSK signals transmitted over Rayleigh frequency-flat fading channels. *IEEE Trans. Commun.*, 44(6):686–695, June 1996.

[527] J. von Neumann. Various techniques used in connection with random digits. *Natl. Bureau of Standards Appl. Math. Ser.*, 12:36–38, 1951.

[528] J. Wang. Cellular CDMA overlay systems. *IEE Proc. Commun.*, 143(6):389–395, Dec. 1996.

[529] J. Wang. Cellular CDMA overlay situations. *IEE Proc. Commun.*, 143(6):402–404, Dec. 1997.

[530] J. Wang. On the use of a suppression filter for CDMA overlay situations. *IEEE Trans. Veh. Technol.*, 48(2):405–414, Mar. 1999.

[531] J. Wang and L. B. Milstein. CDMA overlay situations for microcellular mobile communications. *IEEE Trans. Commun.*, 43(2):603–614, Feb. 1995.

[532] J. Wang and L. B. Milstein. Interference suppression for CDMA overlay of a narrowband waveform. *Electron. Lett.*, 31(24):2074–2075, 1995.

[533] J. Wang and L. B. Milstein. Adaptive LMS filters for cellular CDMA overlay situations. *IEEE J. Select. Areas Commun.*, 14(8):1548–1559, Oct. 1996.

[534] J. Wang and V. Prahatheesan. A lattice filter for CDMA overlay. *IEEE Trans. Commun.*, 48(5):820–828, May 2000.

[535] K. J. Wang and Y. Yao. Nonlinear narrowband interference canceller in DS spread spectrum systems. *Electron. Lett.*, 34(24):2311–2312, 1998.

[536] K. J. Wang and Y. Yao. New nonlinear algorithms for narrow-band interference suppression in CDMA spread-spectrum systems. *IEEE J. Select. Areas Commun.*, 17(12):2148–2153, Dec. 1999.

[537] K. J. Wang, Z. C. Zhou, and Y. Yao. Closed-form analytical results for interference rejection of nonlinear prediction filters in DS-SS systems. *Electron. Lett.*, 33(16):1354–1355, 1997.

[538] K. J. Wang, Z. C. Zhou, and Y. Yao. Performance analysis of nonlinear interpolation filters in DS spread spectrum systems under narrowband interference condition. *Electron. Lett.*, 34(15):1464–1465, 1998.

[539] Q. Wang et al. Multiple-symbol detection of MPSK in narrow-band interference and AWGN. *IEEE Trans. Commun.*, 46(4):460–463, Apr. 1998.

[540] X. Wang and R. Chen. Adaptive Bayesian multiuser detection for synchronous CDMA in Gaussian and impulsive noise. *IEEE Trans. Signal Process.*, 48(7):2013–2028, July 2000.

[541] X. Wang and R. Chen. Blind turbo equalization in Gaussian and impulsive noise. *IEEE Trans. Veh. Technol.*, 50(4):1092–1105, July 2001.

[542] X. Wang, R. Chen, and D. Guo. Delayed-pilot sampling for mixture Kalman filter with application in fading channels. *IEEE Trans. Signal Process.*, 50(2):241–254, Feb. 2002.

[543] X. Wang, R. Chen, and J. S. Liu. Blind adaptive multiuser detection in MIMO systems via sequential Monte Carlo. In *Proc. 2000 Conf. Inform. Sci. Syst.*, Princeton University, Princeton, NJ, Mar. 2000.

[544] X. Wang, R. Chen, and J. S. Liu. Monte Carlo signal processing for wireless communications. *J. VLSI Signal Process.*, 30(1/3):89–105, Jan./Mar. 2002.

[545] X. Wang and A. Høst-Madsen. Group-blind multiuser detection for uplink CDMA. *IEEE J. Select. Areas Commun.*, 17(11):1971–1984, Nov. 1999.

[546] X. Wang and H. V. Poor. Joint channel estimation and symbol detection in Rayleigh flat fading channels with impulsive noise. *IEEE Commun. Lett.*, 1(1):19–21, Jan. 1997.

[547] X. Wang and H. V. Poor. Adaptive joint multiuser detection and channel estimation in multipath fading CDMA channels. *ACM/Baltzer Wireless Networks*, 4(1):453–470, 1998.

[548] X. Wang and H. V. Poor. Blind equalization and multiuser detection for CDMA communications in dispersive channels. *IEEE Trans. Commun.*, 46(1):91–103, Jan. 1998.

[549] X. Wang and H. V. Poor. Blind multiuser detection: A subspace approach. *IEEE Trans. Inform. Theory*, 44(2):677–691, Mar. 1998.

[550] X. Wang and H. V. Poor. Robust adaptive array for wireless communications. *IEEE J. Select. Areas Commun.*, 16(8):1352–1367, Oct. 1998.

[551] X. Wang and H. V. Poor. Blind joint equalization and multiuser detection for DS-CDMA in unknown correlated noise. *IEEE Trans. Circuits Syst. II: Analog Digital Signal Process.*, 46(7):886–895, July 1999.

[552] X. Wang and H. V. Poor. Iterative (turbo) soft interference cancellation and decoding for coded CDMA. *IEEE Trans. Commun.*, 47(7):1046–1061, July 1999.

[553] X. Wang and H. V. Poor. Robust multiuser detection in non-Gaussian channels. *IEEE Trans. Signal Process.*, 47(2):289–305, Feb. 1999.

[554] X. Wang and H. V. Poor. Space-time multiuser detection in multipath CDMA channels. *IEEE Trans. Signal Process.*, 47(9):2356–2374, Sept. 1999.

[555] X. Wang and H. V. Poor. Subspace methods for blind adaptive multiuser detection. *ACM/Baltzer Wireless Networks*, 6(1):59–71, 2000.

[556] Y.-C. Wang and L. B. Milstein. Rejection of multiple narrowband interference in both BPSK and QPSK DS spread-spectrum systems. *IEEE Trans. Commun.*, 36(2):195–204, Feb. 1988.

[557] M. Wax and T. Kailath. Detection of signals by information theoretic criteria. *IEEE Trans. Acoust. Speech Signal Process.*, ASSP-33(2):387–392, Apr. 1985.

[558] W. Webb. *Understanding Cellular Radio.* Artech House, Norwood, MA, 1998.

[559] W. Webb. *Introduction to Wireless Local Loop: Broadband and Narrowband Systems.* Artech House, Norwood, MA, 2nd edition, 2000.

[560] L. Wei and L. K. Rasmussen. Near optimal tree-search detection schemes for bit synchronous multiuser CDMA systems over Gaussian and two-path Rayleigh fading channels. *IEEE Trans. Commun.*, 45(6):691–700, June 1997.

[561] L. Wei and C. Schlegel. Synchronous DS-SSMA with improved decorrelating decision-feedback multiuser detection. *IEEE Trans. Veh. Technol.*, 43:767–772, Aug. 1994.

[562] P. Wei, J. R. Zeidler, and W. H. Ku. Adaptive interference suppression for CDMA overlay systems. *IEEE J. Select. Areas Commun.*, 12(9):1510–1523, Dec. 1994.

[563] E. Weinstein, A. V. Oppenheim, M. Feder, and J. R. Buck. Iterative and sequential algorithms for multisensor signal enhancement. *IEEE Trans. Signal Process.*, 42(4):846–859, Apr. 1994.

[564] A. J. Weiss and B. Friedlander. Synchronous DS-CDMA downlink with frequency selective fading. *IEEE Trans. Signal Process.*, 47(1):158–167, Jan. 1999.

[565] S. B. Wicker. *Error Control Systems for Digital Communication and Storage.* Prentice Hall, Upper Saddle River, NJ, 1995.

[566] T. Widdowson and L. Noel. Uplink and downlink experimental CDMA overlay of GSM network in fading environment. *Electron. Lett.*, 35(17):1440–1441, 1999.

[567] C.S. Wijting et al. Groupwise serial multiuser detectors for multirate DS-CDMA. In *Proc. 1999 IEEE Veh. Technol. Conf.*, pages 836–840, 1999.

[568] S. G. Wilson. *Digital Modulation and Coding.* Prentice Hall, Upper Saddle River, NJ, 1996.

[569] M. Z. Win and R. A. Scholtz. Ultra-wide bandwidth time-hopping spread-spectrum impulse radio for wireless multiple-access communications. *IEEE Trans. Commun.*, 48(4):679–691, Apr. 2000.

[570] J. H. Winters. Signal acquisition and tracking with adaptive arrays in the digital mobile radio system IS-54 with flat fading. *IEEE Trans. Veh. Technol.*, 42(5):377–384, May 1993.

[571] J. H. Winters, J. Salz, and R. D. Gitlin. The impact of antenna diversity on the capacity of wireless communication systems. *IEEE Trans. Commun.*, 42(2/3/4):1740–1750, Feb./Mar./Apr. 1994.

[572] A. Wittneben. Base station modulation diversity for digital SIMULCAST. In *Proc. 1993 IEEE Veh. Technol. Conf.*, pages 505–511, 1993.

[573] A. Wittneben. A new bandwidth efficient transmit antenna modulation diversity scheme for linear digital modulation. In *Proc. 1993 IEEE Int. Conf. Commun.*, pages 1630–1634, 1993.

[574] T. F. Wong, T. M. Lok, J. S. Lehnert, and M. D. Zoltowski. A linear receiver for direct-sequence spread-spectrum multiple-access with antenna array and blind adaptation. *IEEE Trans. Inform. Theory*, 44(2):659–677, Mar. 1998.

[575] W. H. Wong and F. Liang. Dynamic importance weighting in Monte Carlo and optimization. *Proc. Natl. Acad. Sci.*, 94:14220–14224, 1997.

[576] G. Woodward, M. Honig, and P. Alexander. Adaptive multiuser parallel decision-feedback with iterative decoding. In *Proc. 2000 IEEE Int. Symp. Inform. Theory*, page 335, Sorrento, Italy, June 2000.

[577] H.-Y. Wu and A. Duel-Hallen. Performance comparison of multiuser detectors with channel estimation for flat Rayleigh fading CDMA channels. *Wireless Personal Commun.*, 6(1/2):137–160, Jan. 1998.

[578] H.-Y. Wu and A. Duel-Hallen. Performance comparison of multiuser detectors with channel estimation for flat Rayleigh fading CDMA channels. *Wireless Personal Commun.*, 6(1/2):137–160, Jan. 1998.

[579] H.-Y. Wu and A. Duel-Hallen. Multiuser detectors with disjoint Kalman channel estimators for synchronous CDMA mobile radio channels. *IEEE Trans. Commun.*, 48(5):752–756, May 2000.

[580] Q. Wu and K. M. Wong. UN-MUSIC and UN-CLE: An application of generalized correlation analysis to the estimation of the direction of arrival of signals in unknown correlated noise. *IEEE Trans. Signal Process.*, 42(9):2331–2343, Sept. 1994.

[581] Z. Xie, C. K. Rushforth, R. T. Short, and T. K. Moon. Joint signal detection and parameter estimation in multiuser communications. *IEEE Trans. Commun.*, 41(8):1208–1216, Aug. 1993.

[582] C. Xu, G. Feng, and K. S. Kwak. A modified constrained constant modulus approach to blind adaptive multiuser detection. *IEEE Trans. Commun.*, 49(9):1642–1648, Sept. 2001.

[583] Z. Xu and M. K. Tsatsanis. Blind adaptive algorithms for minimum variance CDMA receivers. *IEEE Trans. Commun.*, 49(1):180–194, Jan. 2001.

[584] Z. D. Xu and M. K. Tsatsanis. Blind channel estimation for long code multiuser CDMA systems. *IEEE Trans. Signal Process.*, 48(4):988–1001, Apr. 2000.

[585] B. Yang. An extension of the PASTd algorithm to both rank and subspace tracking. *IEEE Signal Process. Lett.*, 2(9):179–182, Sept. 1995.

[586] B. Yang. Projection approximation subspace tracking. *IEEE Trans. Signal Process.*, 44(1):95–107, Jan. 1995.

[587] B. Yang. Asymptotic convergence analysis of the projection approximation subspace tracking algorithms. *Signal Process.*, 50:123–136, 1996.

[588] B. Yang and J. F. Böhme. Rotation-based RLS algorithms: Unified derivations, numerical properties and parallel implementations. *IEEE Trans. Signal Process.*, 40(5):1151–1166, May 1992.

[589] S. C. Yang. *CDMA RF System Engineering.* Artech House, Norwood, MA, 1998.

[590] W. M. Yang and G. G. Bi. Adaptive wavelet packet transform-based narrow-band interference canceller in DSSS systems. *Electron. Lett.*, 33(14):1189–1190, 1997.

[591] Z. Yang, B. Lu, and X. Wang. Bayesian Monte Carlo multiuser receiver for space-time coded multi-carrier CDMA systems. *IEEE J. Select. Areas Commun.*, 19(8):1625–1637, Aug. 2001.

[592] Z. Yang and X. Wang. Turbo equalization for GMSK signaling over multipath channels based on the Gibbs sampler. *IEEE J. Select. Areas Commun.*, 19(9):1753–1763, Sept. 2001.

[593] Z. Yang and X. Wang. Blind detection of OFDM signals in multipath fading channels via sequential Monte Carlo. *IEEE Trans. Signal Process.*, 50(2):271–280, Feb. 2002.

[594] Z. Yang and X. Wang. Blind turbo multiuser detection for long-code multipath CDMA. *IEEE Trans. Commun.*, 50(1):112–125, Jan. 2002.

[595] Z. Yang and X. Wang. Joint mobility tracking and handoff in cellular networks via sequential monte carlo filtering. *IEEE Trans. Signal Process.*, 51(1):269–281, Jan. 2003.

[596] Y. Yao and H. V. Poor. Eavesdropping in the synchronous CDMA channel: An EM-based approach. *IEEE Trans. Signal Process.*, 49(8):1748–1756, Aug. 2001.

[597] D. Yellin, A. Vardy, and O. Amrani. Joint equalization and coding for intersymbol interference channels. *IEEE Trans. Inform. Theory*, 43(2):409–425, Mar. 1997.

[598] D. Young. *Iterative Solution of Large Linear Systems.* Academic Press, New York, 1971.

[599] X. Yu and S. Pasupathy. Innovation-based MLSE for Rayleigh fading channels. *IEEE Trans. Commun.*, 43(2/3/4):1534–1544, Feb./Mar./Apr. 1995.

[600] S. M. Zabin and H. V. Poor. Efficient estimation of the class A parameters via the EM algorithm. *IEEE Trans. Inform. Theory*, 37(1):60–72, Jan 1991.

[601] H. Zamiri-Jafarian and S. Pasupathy. Adaptive MLSDE using the EM algorithm. *IEEE Trans. Commun.*, 47(8):1181–1193, Aug. 1999.

[602] H. Zamiri-Jafarian and S. Pasupathy. EM-based recursive estimation of channel parameters. *IEEE Trans. Commun.*, 47(9):1297–1302, Sept. 1999.

[603] J. R. Zeidler. Performance analysis of LMS adaptive prediction filters. *Proc. IEEE*, 78(12):1781–1806, Dec. 1990.

[604] J. Zhang and X. Wang. Large-system asymptotic performance of blind and group-blind multiuser detection. *IEEE Trans. Inform. Theory*, 48(9):2507–2523, Sept. 2002.

[605] R. Zhang and M. K. Tsatsanis. Blind startup of MMSE receivers for CDMA systems. *IEEE Trans. Signal Process.*, 49(7):1492–1500, July 2001.

[606] X.-D. Zhang and W. Wei. Blind adaptive multiuser detection based on Kalman filtering. *IEEE Trans. Signal Process.*, 50(1):87–95, Jan. 2002.

[607] Y. Zhang and R. S. Blum. An adaptive receiver with an antenna array for channels with correlated non-Gaussian interference and noise using the SAGE algorithm. *IEEE Trans. Signal Process.*, 48(7):2172–2175, July 2000.

[608] Y. Zhang and R. S. Blum. Multistage multiuser detection for CDMA with space-time coding. In *Proc. 2000 IEEE Workshop Stat. Array Process.*, pages 1–5, Aug. 2000.

[609] L. C. Zhao, P. R. Krishnaiah, and Z. D. Bai. On detection of the number of signals in the presence of white noise. *J. Multivar. Anal.*, 20(1):1–25, 1986.

[610] Q. Zhao and L. Tong. Adaptive blind channel estimation by least squares smoothing. *IEEE Trans. Signal Process.*, 47(11):3000–3012, Nov. 1999.

[611] S. N. Zhdanov and I. A. Tiskin. Noise-immunity of a spread-spectrum noise-type signal receiver with the time-frequency rejection of narrow-band interference. *Isv. Radioelektron.*, 33(4):73–75, Apr. 1990.

[612] D. Zheng, J. Li, S. L. Miller, and E. G. Strom. An efficient code-timing estimator for DS-CDMA signals. *IEEE Trans. Signal Process.*, 45(1):82–89, Jan. 1997.

[613] L. J. Zhu and U. Madhow. Adaptive interference suppression for direct sequence CDMA over severely time-varying channels. In *Proc. 1997 IEEE Global Telecommun. Conf.*, pages 917–922, Nov. 1997.

[614] X. Zhuang, Z. Ding, and A. Swindlehurtst. A statistical subspace method for blind channel identification in OFDM communications. In *Proc. 2000 IEEE Int. Conf. Acoust. Speech Signal Process.*, pages 2493–2496, 2000.

[615] Z. Zvonar and D. Brady. Multiuser detection in single-path fading channels. *IEEE Trans. Commun.*, 42(2/3/4):1729–1739, Feb./Mar./Apr. 1994.

[616] Z. Zvonar and D. Brady. Suboptimum multiuser detector for frequency-selective Rayleigh fading synchronous CDMA channels. *IEEE Trans. Commun.*, 43(2/3/4):154–157, Feb./Mar./Apr. 1995.

[617] Z. Zvonar and D. Brady. Linear multipath-decorrelating receivers for CDMA frequency-selective fading channels. *IEEE Trans. Commun.*, 44(6):650–653, Apr. 1996.

INDEX

B

C

D

E

F

G

H

I

O

P

Q

R

S

U

V

W

Z

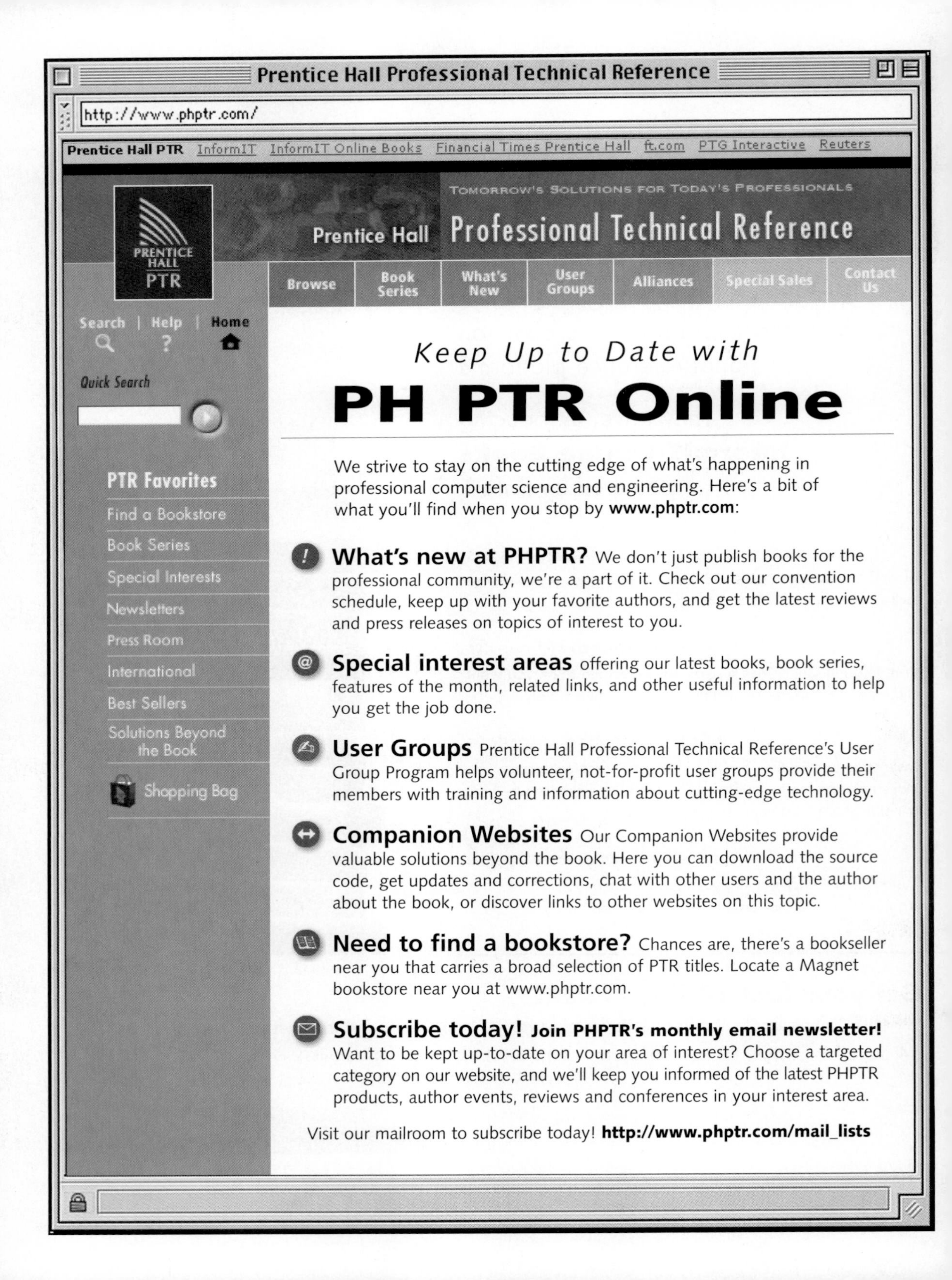
Prentice Hall Professional Technical Reference
http://www.phptr.com/
Prentice Hall PTR
InformIT
InformIT Online Books
Financial Times Prentice Hall
ft.com
PTG Interactive
Reuters
PRENTICE HALL PTR
TOMORROW'S SOLUTIONS FOR TODAY'S PROFESSIONALS
Prentice Hall Professional Technical Reference
Browse
Book Series
What's New
User Groups
Alliances
Special Sales
Contact Us
Search
Help
Home
Quick Search
PTR Favorites
Find a Bookstore
Book Series
Special Interests
Newsletters
Press Room
International
Best Sellers
Solutions Beyond the Book
Shopping Bag
Keep Up to Date with
PH PTR Online
We strive to stay on the cutting edge of what's happening in professional computer science and engineering. Here's a bit of what you'll find when you stop by www.phptr.com:
What's new at PHPTR? We don't just publish books for the professional community, we're a part of it. Check out our convention schedule, keep up with your favorite authors, and get the latest reviews and press releases on topics of interest to you.
Special interest areas offering our latest books, book series, features of the month, related links, and other useful information to help you get the job done.
User Groups Prentice Hall Professional Technical Reference's User Group Program helps volunteer, not-for-profit user groups provide their members with training and information about cutting-edge technology.
Companion Websites Our Companion Websites provide valuable solutions beyond the book. Here you can download the source code, get updates and corrections, chat with other users and the author about the book, or discover links to other websites on this topic.
Need to find a bookstore? Chances are, there's a bookseller near you that carries a broad selection of PTR titles. Locate a Magnet bookstore near you at www.phptr.com.
Subscribe today! Join PHPTR's monthly email newsletter!
Want to be kept up-to-date on your area of interest? Choose a targeted category on our website, and we'll keep you informed of the latest PHPTR products, author events, reviews and conferences in your interest area.
Visit our mailroom to subscribe today! http://www.phptr.com/mail_lists